Design and Analysis of Experiments

International Student Version

Eighth Edition

DOUGLAS C. MONTGOMERY
Arizona State University

WILEY

Design and Analysis of Experiments
Eighth Edition

Authorised reprint by Wiley India Pvt. Ltd., 4435-36/7, Ansari Road, Daryaganj, New Delhi-110002.

Reprint: 2017

Printed at: Durga Printographics.

ISBN: 978-81-265-4050-1

Design and Analysis of Experiments

International Student Version

Preface

Audience

This is an introductory textbook dealing with the design and analysis of experiments. It is based on college-level courses in design of experiments that I have taught over nearly 40 years at Arizona State University, the University of Washington, and the Georgia Institute of Technology. It also reflects the methods that I have found useful in my own professional practice as an engineering and statistical consultant in many areas of science and engineering, including the research and development activities required for successful technology commercialization and product realization.

The book is intended for students who have completed a first course in statistical methods. This background course should include at least some techniques of descriptive statistics, the standard sampling distributions, and an introduction to basic concepts of confidence intervals and hypothesis testing for means and variances. Chapters 10, 11, and 12 require some familiarity with matrix algebra.

Because the prerequisites are relatively modest, this book can be used in a second course on statistics focusing on statistical design of experiments for undergraduate students in engineering, the physical and chemical sciences, statistics, mathematics, and other fields of science. For many years I have taught a course from the book at the first-year graduate level in engineering. Students in this course come from all of the fields of engineering, materials science, physics, chemistry, mathematics, operations research life sciences, and statistics. I have also used this book as the basis of an industrial short course on design of experiments for practicing technical professionals with a wide variety of backgrounds. There are numerous examples illustrating all of the design and analysis techniques. These examples are based on real-world applications of experimental design and are drawn from many different fields of engineering and the sciences. This adds a strong applications flavor to an academic course for engineers and scientists and makes the book useful as a reference tool for experimenters in a variety of disciplines.

About the Book

The eighth edition is a major revision of the book. I have tried to maintain the balance between design and analysis topics of previous editions; however, there are many new topics and examples, and I have reorganized much of the material. There is much more emphasis on the computer in this edition.

Design-Expert, JMP, and Minitab Software

During the last few years a number of excellent software products to assist experimenters in both the design and analysis phases of this subject have appeared. I have included output from three of these products, Design-Expert, JMP, and Minitab at many points in the text. Minitab and JMP are widely available general-purpose statistical software packages that have good data analysis capabilities and that handles the analysis of experiments with both fixed and random factors (including the mixed model). Design-Expert is a package focused exclusively on experimental design. All three of these packages have many capabilities for construction and evaluation of designs and extensive analysis features. I urge all instructors who use this book to incorporate computer software into your course. (In my course, I bring a laptop computer and use a computer projector in every lecture, and every design or analysis topic discussed in class is illustrated with the computer.)

Empirical Model

I have continued to focus on the connection between the experiment and the model that the experimenter can develop from the results of the experiment. Engineers (and physical, chemical and life scientists to a large extent) learn about physical mechanisms and their underlying mechanistic models early in their academic training, and throughout much of their professional careers they are involved with manipulation of these models. Statistically designed experiments offer the engineer a valid basis for developing an *empirical* model of the system being investigated. This empirical model can then be manipulated (perhaps through a response surface or contour plot, or perhaps mathematically) just as any other engineering model. I have discovered through many years of teaching that this viewpoint is very effective in creating enthusiasm in the engineering community for statistically designed experiments. Therefore, the notion of an underlying empirical model for the experiment and response surfaces appears early in the book and receives much more emphasis.

Factorial Designs

I have expanded the material on factorial and fractional factorial designs (Chapters 5–9) in an effort to make the material flow more effectively from both the reader's and the instructor's viewpoint and to place more emphasis on the empirical model. There is new material on a number of important topics, including follow-up experimentation following a fractional factorial, nonregular and nonorthogonal designs, and small, efficient resolution IV and V designs. Nonregular fractions as alternatives to traditional minimum aberration fractions in 16 runs and analysis methods for these design are discussed and illustrated.

Additional Important Changes

I have added a lot of material on optimal designs and their application. The chapter on response surfaces (Chapter 11) has several new topics and problems. I have expanded Chapter 12 on robust parameter design and process robustness experiments. Chapters 13 and 14 discuss experiments involving random effects and some applications of these concepts to nested and split-plot designs. The residual maximum likelihood method is now widely available in software and I have emphasized this technique throughout the book. Because there is expanding industrial interest in nested and split-plot designs, Chapters 13 and 14 have several new topics. Chapter 15 is an overview of important design and analysis topics: nonnormality of the response, the Box–Cox method for selecting the form of a transformation, and other alternatives; unbalanced factorial experiments; the analysis of covariance, including covariates in a factorial design, and repeated measures. I have also added new examples and problems from various fields, including biochemistry and biotechnology.

Experimental Design

Throughout the book I have stressed the importance of experimental design as a tool for engineers and scientists to use for product design and development as well as process development and improvement. The use of experimental design in developing products that are robust to environmental factors and other sources of variability is illustrated. I believe that the use of experimental design early in the product cycle can substantially reduce development lead time and cost, leading to processes and products that perform better in the field and have higher reliability than those developed using other approaches.

The book contains more material than can be covered comfortably in one course, and I hope that instructors will be able to either vary the content of each course offering or discuss some topics in greater depth, depending on class interest. There are problem sets at the end of each chapter. These problems vary in scope from computational exercises, designed to reinforce the fundamentals, to extensions or elaboration of basic principles.

Course Suggestions

My own course focuses extensively on factorial and fractional factorial designs. Consequently, I usually cover Chapter 1, Chapter 2 (very quickly), most of Chapter 3, Chapter 4 (excluding the material on incomplete blocks and only mentioning Latin squares briefly), and I discuss Chapters 5 through 8 on factorials and two-level factorial and fractional factorial designs in detail. To conclude the course, I introduce response surface methodology (Chapter 11) and give an overview of random effects models (Chapter 13) and nested and split-plot designs (Chapter 14). I always require the students to complete a term project that involves designing, conducting, and presenting the results of a statistically designed experiment. I require them to do this in teams because this is the way that much industrial experimentation is conducted. They must present the results of this project, both orally and in written form.

The Supplemental Text Material

For the eighth edition I have prepared supplemental text material for each chapter of the book. Often, this supplemental material elaborates on topics that could not be discussed in greater detail in the book. I have also presented some subjects that do not appear directly in the book, but an introduction to them could prove useful to some students and professional practitioners. Some of this material is at a higher mathematical level than the text. I realize that instructors use this book with a wide array of audiences, and some more advanced design courses could possibly benefit from including several of the supplemental text material topics. This material is in electronic form on the World Wide Website for this book, located at www.wiley.com/go/global/montgomery.

Website

Current supporting material for instructors and students is available at the website www.wiley.com/go/global/montgomery. This site will be used to communicate information about innovations and recommendations for effectively using this text. The supplemental text material described above is available at the site, along with electronic versions of data sets used for examples and homework problems, a course syllabus, and some representative student term projects from the course at Arizona State University.

Student Companion Site (www.wiley.com/go/global/montogomery)

The student's section of the textbook website contains the following:

1. The supplemental text material described above
2. Data sets from the book examples and homework problems, in electronic form
3. Sample Student Projects

Instructor Resources

The instructor's section of the textbook website contains the following:

4. Solutions to the text problems
5. Data sets from the book examples and homework problems, in electronic form
6. Sample Student Projects
7. The supplemental text material described above
8. PowerPoint lecture slides
9. Figures from the text in electronic format, for easy inclusion in lecture slides

To request, please send a mail to acadmktg@wiley.com.

Acknowledgments

I express my appreciation to the many students, instructors, and colleagues who have used the six earlier editions of this book and who have made helpful suggestions for its revision. The contributions of Dr. Raymond H. Myers, Dr. G. Geoffrey Vining, Dr. Brad Jones, Dr. Christine Anderson-Cook, Dr. Connie M. Borror, Dr. Scott Kowalski, Dr. Dennis Lin, Dr. John Ramberg, Dr. Joseph Pignatiello, Dr. Lloyd S. Nelson, Dr. Andre Khuri, Dr. Peter Nelson, Dr. John A. Cornell, Dr. Saeed Maghsoodlo, Dr. Don Holcomb, Dr. George C. Runger, Dr. Bert Keats, Dr. Dwayne Rollier, Dr. Norma Hubele, Dr. Murat Kulahci, Dr. Cynthia Lowry, Dr. Russell G. Heikes, Dr. Harrison M. Wadsworth, Dr. William W. Hines, Dr. Arvind Shah, Dr. Jane Ammons, Dr. Diane Schaub, Mr. Mark Anderson, Mr. Pat Whitcomb, Dr. Pat Spagon, and Dr. William DuMouche were particularly valuable. My current and former Department Chairs, Dr. Ron Askin and Dr. Gary Hogg, have provided an intellectually stimulating environment in which to work.

The contributions of the professional practitioners with whom I have worked have been invaluable. It is impossible to mention everyone, but some of the major contributors include Dr. Dan McCarville of Mindspeed Corporation, Dr. Lisa Custer of the George Group; Dr. Richard Post of Intel; Mr. Tom Bingham, Mr. Dick Vaughn, Dr. Julian Anderson, Mr. Richard Alkire, and Mr. Chase Neilson of the Boeing Company; Mr. Mike Goza, Mr. Don Walton, Ms. Karen Madison, Mr. Jeff Stevens, and Mr. Bob Kohm of Alcoa; Dr. Jay Gardiner, Mr. John Butora, Mr. Dana Lesher, Mr. Lolly Marwah, Mr. Leon Mason of IBM; Dr. Paul Tobias of IBM and Sematech; Ms. Elizabeth A. Peck of The Coca-Cola Company; Dr. Sadri Khalessi and Mr. Franz Wagner of Signetics; Mr. Robert V. Baxley of Monsanto Chemicals; Mr. Harry Peterson-Nedry and Dr. Russell Boyles of Precision Castparts Corporation; Mr. Bill New and Mr. Randy Schmid of Allied-Signal Aerospace; Mr. John M. Fluke, Jr. of the John Fluke Manufacturing Company; Mr. Larry Newton and Mr. Kip Howlett of Georgia-Pacific; and Dr. Ernesto Ramos of BBN Software Products Corporation.

I am indebted to Professor E. S. Pearson and the *Biometrika* Trustees, John Wiley & Sons, Prentice Hall, The American Statistical Association, The Institute of Mathematical Statistics, and the editors of *Biometrics* for permission to use copyrighted material. Dr. Lisa Custer and Dr. Dan McCorville did an excellent job of preparing the solutions that appear in the Instructor's Solutions Manual, and Dr. Cheryl Jennings and Dr. Sarah Street provided effective and very helpful proofreading assistance. I am grateful to NASA, the Office of Naval Research, the National Science Foundation, the member companies of the NSF/Industry/University Cooperative Research Center in Quality and Reliability Engineering at Arizona State University, and the IBM Corporation for supporting much of my research in engineering statistics and experimental design.

DOUGLAS C. MONTGOMERY
TEMPE, ARIZONA

Contents

3
Analysis of Variance 65

4
Experiments with Blocking Factors 139

8

Two-Level Fractional Factorial Designs 320

9

Other Topics on Factorial and Fractional Factorial Designs 394

CHAPTER 1

Introduction to Designed Experiments

CHAPTER OUTLINE

The supplemental material is on the textbook website www.wiley.com/go/global/montgomery.

1.1 Strategy of Experimentation

Observing a system or process while it is in operation is an important part of the learning process, and is an integral part of understanding and learning about how systems and processes work. The great New York Yankees catcher Yogi Berra said that ". . . you can observe a lot just by watching." However, to understand what happens to a process when you change certain input factors, you have to do more than just watch—you actually have to change the factors. This means that to really understand cause-and-effect relationships in a system you must deliberately change the input variables to the system and observe the changes in the system output that these changes to the inputs produce. In other words, you need to conduct **experiments** on the system. Observations on a system or process can lead to theories or hypotheses about what makes the system work, but experiments of the type described above are required to demonstrate that these theories are correct.

Investigators perform experiments in virtually all fields of inquiry, usually to discover something about a particular process or system. Each experimental **run** is a **test**. More formally, we can define an **experiment** as a test or series of runs in which purposeful changes are made to the input variables of a process or system so that we may observe and identify the reasons for changes that may be observed in the output response. We may want to determine which input variables are responsible for the observed changes in the response, develop a model relating the response to the important input variables and to use this model for process or system improvement or other decision-making.

This book is about planning and conducting experiments and about analyzing the resulting data so that valid and objective conclusions are obtained. Our focus is on experiments in engineering and science. Experimentation plays an important role in **technology**

commercialization and **product realization** activities, which consist of new product design and formulation, manufacturing process development, and process improvement. The objective in many cases may be to develop a **robust** process, that is, a process affected minimally by external sources of variability. There are also many applications of designed experiments in a non-manufacturing or non-product-development setting, such as marketing, service operations, and general business operations.

As an example of an experiment, suppose that a metallurgical engineer is interested in studying the effect of two different hardening processes, oil quenching and saltwater quenching, on an aluminum alloy. Here the objective of the **experimenter** (the engineer) is to determine which quenching solution produces the maximum hardness for this particular alloy. The engineer decides to subject a number of alloy specimens or test coupons to each quenching medium and measure the hardness of the specimens after quenching. The average hardness of the specimens treated in each quenching solution will be used to determine which solution is best.

As we consider this simple experiment, a number of important questions come to mind:

1. Are these two solutions the only quenching media of potential interest?
2. Are there any other factors that might affect hardness that should be investigated or controlled in this experiment (such as, the temperature of the quenching media)?
3. How many coupons of alloy should be tested in each quenching solution?
4. How should the test coupons be assigned to the quenching solutions, and in what order should the data be collected?
5. What method of data analysis should be used?
6. What difference in average observed hardness between the two quenching media will be considered important?

All of these questions, and perhaps many others, will have to be answered satisfactorily before the experiment is performed.

Experimentation is a vital part of the **scientific** (or **engineering**) **method**. Now there are certainly situations where the scientific phenomena are so well understood that useful results including mathematical models can be developed directly by applying these well-understood principles. The models of such phenomena that follow directly from the physical mechanism are usually called **mechanistic models**. A simple example is the familiar equation for current flow in an electrical circuit, Ohm's law, $E = IR$. However, most problems in science and engineering require **observation** of the system at work and **experimentation** to elucidate information about why and how it works. Well-designed experiments can often lead to a model of system performance; such experimentally determined models are called **empirical models**. Throughout this book, we will present techniques for turning the results of a designed experiment into an empirical model of the system under study. These empirical models can be manipulated by a scientist or an engineer just as a mechanistic model can.

A well-designed experiment is important because the results and conclusions that can be drawn from the experiment depend to a large extent on the manner in which the data were collected. To illustrate this point, suppose that the metallurgical engineer in the above experiment used specimens from one heat in the oil quench and specimens from a second heat in the saltwater quench. Now, when the mean hardness is compared, the engineer is unable to say how much of the observed difference is the result of the quenching media and how much is the result of inherent differences between the heats.[1] Thus, the method of data collection has adversely affected the conclusions that can be drawn from the experiment.

[1] A specialist in experimental design would say that the effect of quenching media and heat were *confounded*; that is, the effects of these two factors cannot be separated.

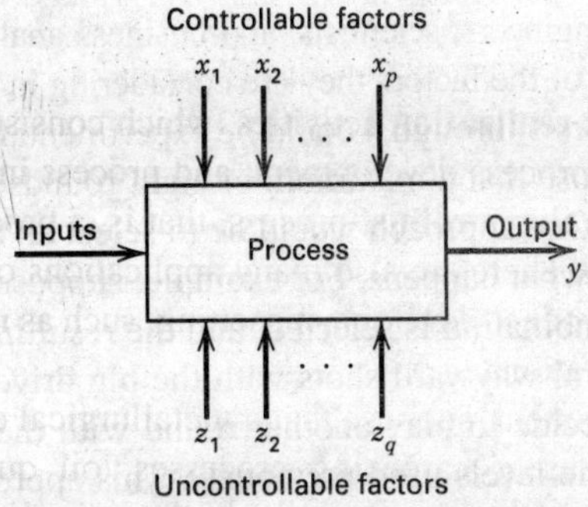

■ **FIGURE 1.1 General model of a process or system**

In general, experiments are used to study the performance of processes and systems. The process or system can be represented by the model shown in Figure 1.1. We can usually visualize the process as a combination of operations, machines, methods, people, and other resources that transforms some input (often a material) into an output that has one or more observable **response** variables. Some of the process variables and material properties x_1, x_2, . . . , x_p are **controllable**, whereas other variables z_1, z_2, . . . , z_q are **uncontrollable** (although they may be controllable for purposes of a test). The objectives of the experiment may include the following:

1. Determining which variables are most influential on the response y
2. Determining where to set the influential x's so that y is almost always near the desired nominal value
3. Determining where to set the influential x's so that variability in y is small
4. Determining where to set the influential x's so that the effects of the uncontrollable variables z_1, z_2, . . . , z_q are minimized.

As you can see from the foregoing discussion, experiments often involve several factors. Usually, an objective of the **experimenter** is to determine the influence that these factors have on the output response of the system. The general approach to planning and conducting the experiment is called the **strategy of experimentation**. An experimenter can use several strategies. We will illustrate some of these with a very simple example.

I really like to play golf. Unfortunately, I do not enjoy practicing, so I am always looking for a simpler solution to lowering my score. Some of the factors that I think may be important, or that may influence my golf score, are as follows:

1. The type of driver used (oversized or regular sized)
2. The type of ball used (balata or three piece)
3. Walking and carrying the golf clubs or riding in a golf cart
4. Drinking water or drinking "something else" while playing
5. Playing in the morning or playing in the afternoon
6. Playing when it is cool or playing when it is hot
7. The type of golf shoe spike worn (metal or soft)
8. Playing on a windy day or playing on a calm day.

Obviously, many other factors could be considered, but let's assume that these are the ones of primary interest. Furthermore, based on long experience with the game, I decide that factors 5 through 8 can be ignored; that is, these factors are not important because their effects are so small that they have no practical value. Engineers, scientists, and business analysts, often must make these types of decisions about some of the factors they are considering in real experiments.

Now, let's consider how factors 1 through 4 could be experimentally tested to determine their effect on my golf score. Suppose that a maximum of eight rounds of golf can be played over the course of the experiment. One approach would be to select an arbitrary combination of these factors, test them, and see what happens. For example, suppose the oversized driver, balata ball, golf cart, and water combination is selected, and the resulting score is 87. During the round, however, I noticed several wayward shots with the big driver (long is not always good in golf), and, as a result, I decide to play another round with the regular-sized driver, holding the other factors at the same levels used previously. This approach could be continued almost indefinitely, switching the levels of one or two (or perhaps several) factors for the next test, based on the outcome of the current test. This strategy of experimentation, which we call the **best-guess approach**, is frequently used in practice by engineers and scientists. It often works reasonably well, too, because the experimenters often have a great deal of technical or theoretical knowledge of the system they are studying, as well as considerable practical experience. The best-guess approach has at least two disadvantages. First, suppose the initial best-guess does not produce the desired results. Now the experimenter has to take another guess at the correct combination of factor levels. This could continue for a long time, without any guarantee of success. Second, suppose the initial best-guess produces an acceptable result. Now the experimenter is tempted to stop testing, although there is no guarantee that the *best* solution has been found.

Another strategy of experimentation that is used extensively in practice is the **one-factor-at-a-time (OFAT)** approach. The OFAT method consists of selecting a starting point, or **baseline** set of levels, for each factor, and then successively varying each factor over its range with the other factors held constant at the baseline level. After all tests are performed, a series of graphs are usually constructed showing how the response variable is affected by varying each factor with all other factors held constant. Figure 1.2 shows a set of these graphs for the golf experiment, using the oversized driver, balata ball, walking, and drinking water levels of the four factors as the baseline. The interpretation of this graph is straightforward; for example, because the slope of the mode of travel curve is negative, we would conclude that riding improves the score. Using these one-factor-at-a-time graphs, we would select the optimal combination to be the regular-sized driver, riding, and drinking water. The type of golf ball seems unimportant.

The major disadvantage of the OFAT strategy is that it fails to consider any possible **interaction** between the factors. An interaction is the failure of one factor to produce the same effect on the response at different levels of another factor. Figure 1.3 shows an interaction between the type of driver and the beverage factors for the golf experiment. Notice that if I use the regular-sized driver, the type of beverage consumed has virtually no effect on the score, but if I use the oversized driver, much better results are obtained by drinking water instead of beer. Interactions between factors are very common, and if they occur, the one-factor-at-a-time strategy will usually produce poor results. Many people do not recognize this, and, consequently,

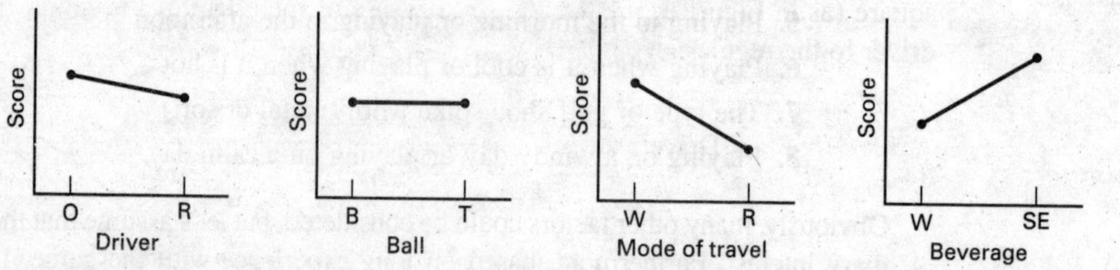

■ **FIGURE 1.2** **Results of the one-factor-at-a-time strategy for the golf experiment**

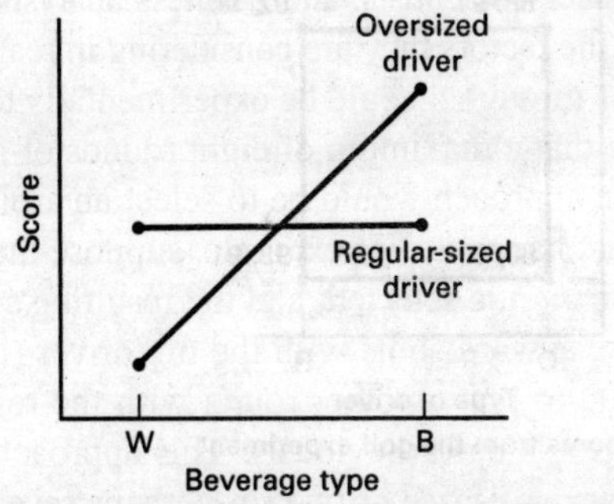

■ **FIGURE 1.3 Interaction between type of driver and type of beverage for the golf experiment**

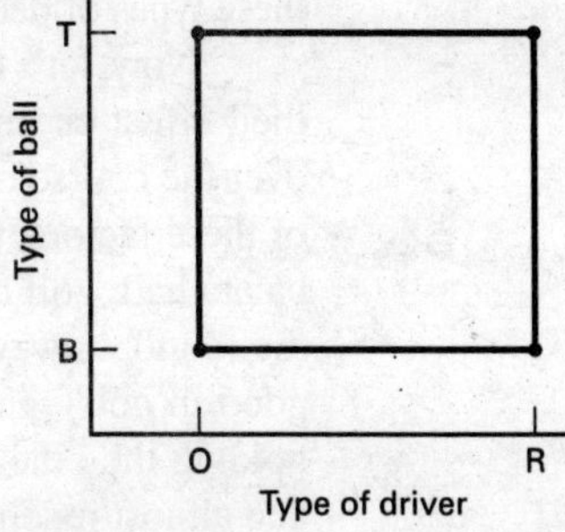

■ **FIGURE 1.4 A two-factor factorial experiment involving type of driver and type of ball**

OFAT experiments are run frequently in practice. (Some individuals actually think that this strategy is related to the scientific method or that it is a "sound" engineering principle.) One-factor-at-a-time experiments are always less efficient than other methods based on a statistical approach to design. We will discuss this in more detail in Chapter 5.

The correct approach to dealing with several factors is to conduct a **factorial** experiment. This is an experimental strategy in which factors are varied *together*, instead of one at a time. The factorial experimental design concept is extremely important, and several chapters in this book are devoted to presenting basic factorial experiments and a number of useful variations and special cases.

To illustrate how a factorial experiment is conducted, consider the golf experiment and suppose that only two factors, type of driver and type of ball, are of interest. Figure 1.4 shows a two-factor factorial experiment for studying the joint effects of these two factors on my golf score. Notice that this factorial experiment has both factors at two levels and that all possible combinations of the two factors across their levels are used in the design. Geometrically, the four runs form the corners of a square. This particular type of factorial experiment is called a **2^2 factorial design** (two factors, each at two levels). Because I can reasonably expect to play eight rounds of golf to investigate these factors, a reasonable plan would be to play two rounds of golf at each combination of factor levels shown in Figure 1.4. An experimental designer would say that we have **replicated** the design twice. This experimental design would enable the experimenter to investigate the individual effects of each factor (or the **main** effects) and to determine whether the factors interact.

Figure 1.5*a* shows the results of performing the factorial experiment in Figure 1.4. The scores from each round of golf played at the four test combinations are shown at the corners of the square. Notice that there are four rounds of golf that provide information about using the regular-sized driver and four rounds that provide information about using the oversized driver. By finding the average difference in the scores on the right- and left-hand sides of the square (as in Figure 1.5*b*), we have a measure of the effect of switching from the oversized driver to the regular-sized driver, or

$$\text{Driver effect} = \frac{92 + 94 + 93 + 91}{4} - \frac{88 + 91 + 88 + 90}{4}$$
$$= 3.25$$

That is, on average, switching from the oversized to the regular-sized driver increases the score by 3.25 strokes per round. Similarly, the average difference in the four scores at the top

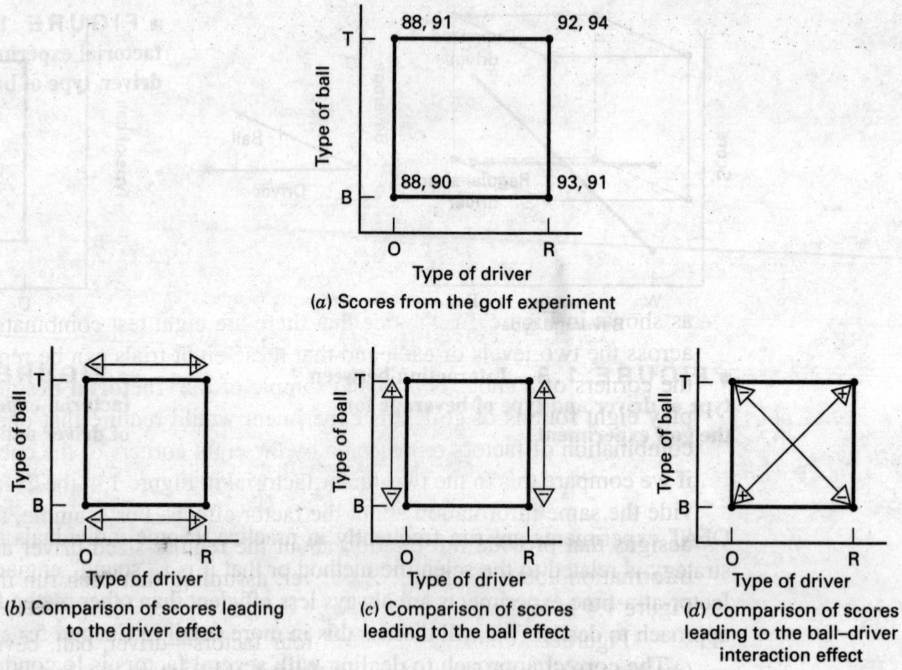

■ FIGURE 1.5 Scores from the golf experiment in Figure 1.4 and calculation of the factor effects

of the square and the four scores at the bottom measures the effect of the type of ball used (see Figure 1.5*c*):

$$\text{Ball effect} = \frac{88 + 91 + 92 + 94}{4} - \frac{88 + 90 + 93 + 91}{4}$$
$$= 0.75$$

Finally, a measure of the interaction effect between the type of ball and the type of driver can be obtained by subtracting the average scores on the left-to-right diagonal in the square from the average scores on the right-to-left diagonal (see Figure 1.5*d*), resulting in

$$\text{Ball–driver interaction effect} = \frac{92 + 94 + 88 + 90}{4} - \frac{88 + 91 + 93 + 91}{4}$$
$$= 0.25$$

The results of this factorial experiment indicate that driver effect is larger than either the ball effect or the interaction. Statistical testing could be used to determine whether any of these effects differ from zero. In fact, it turns out that there is reasonably strong statistical evidence that the driver effect differs from zero and the other two effects do not. Therefore, this experiment indicates that I should always play with the oversized driver.

One very important feature of the factorial experiment is evident from this simple example; namely, factorials make the most efficient use of the experimental data. Notice that this experiment included eight observations, and all eight observations are used to calculate the driver, ball, and interaction effects. No other strategy of experimentation makes such an efficient use of the data. This is an important and useful feature of factorials.

We can extend the factorial experiment concept to three factors. Suppose that I wish to study the effects of type of driver, type of ball, and the type of beverage consumed on my golf score. Assuming that all three factors have two levels, a factorial design can be set up

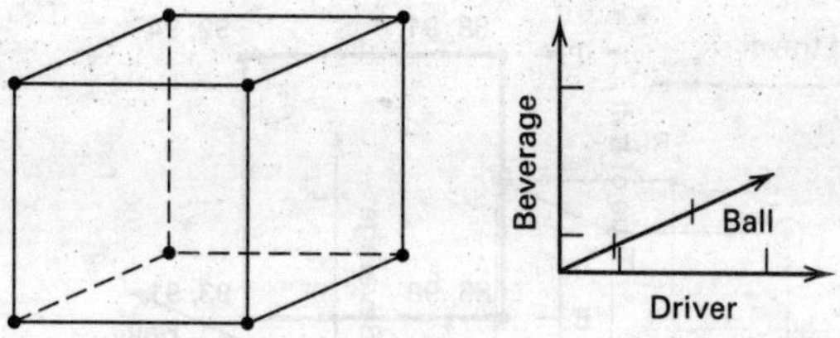

■ **FIGURE 1.6** **A three-factor factorial experiment involving type of driver, type of ball, and type of beverage**

as shown in Figure 1.6. Notice that there are eight test combinations of these three factors across the two levels of each and that these eight trials can be represented geometrically as the corners of a cube. This is an example of a **2^3 factorial design**. Because I only want to play eight rounds of golf, this experiment would require that one round be played at each combination of factors represented by the eight corners of the cube in Figure 1.6. However, if we compare this to the two-factor factorial in Figure 1.4, the 2^3 factorial design would provide the same information about the factor effects. For example, there are four tests in both designs that provide information about the regular-sized driver and four tests that provide information about the oversized driver, assuming that each run in the two-factor design in Figure 1.4 is replicated twice.

Figure 1.7 illustrates how all four factors—driver, ball, beverage, and mode of travel (walking or riding)—could be investigated in a **2^4 factorial design**. As in any factorial design, all possible combinations of the levels of the factors are used. Because all four factors are at two levels, this experimental design can still be represented geometrically as a cube (actually a hypercube).

Generally, if there are k factors, each at two levels, the factorial design would require 2^k runs. For example, the experiment in Figure 1.7 requires 16 runs. Clearly, as the number of factors of interest increases, the number of runs required increases rapidly; for instance, a 10-factor experiment with all factors at two levels would require 1024 runs. This quickly becomes infeasible from a time and resource viewpoint. In the golf experiment, I can only play eight rounds of golf, so even the experiment in Figure 1.7 is too large.

Fortunately, if there are four to five or more factors, it is usually unnecessary to run all possible combinations of factor levels. A **fractional factorial experiment** is a variation of the basic factorial design in which only a subset of the runs is used. Figure 1.8 shows a fractional factorial design for the four-factor version of the golf experiment. This design requires only 8 runs instead of the original 16 and would be called a **one-half fraction**. If I can play only eight rounds of golf, this is an excellent design in which to study all four factors. It will provide good information about the main effects of the four factors as well as some information about how these factors interact.

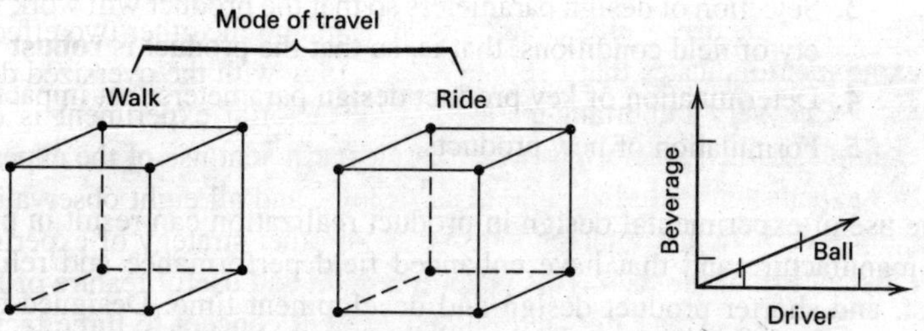

■ **FIGURE 1.7** **A four-factor factorial experiment involving type of driver, type of ball, type of beverage, and mode of travel**

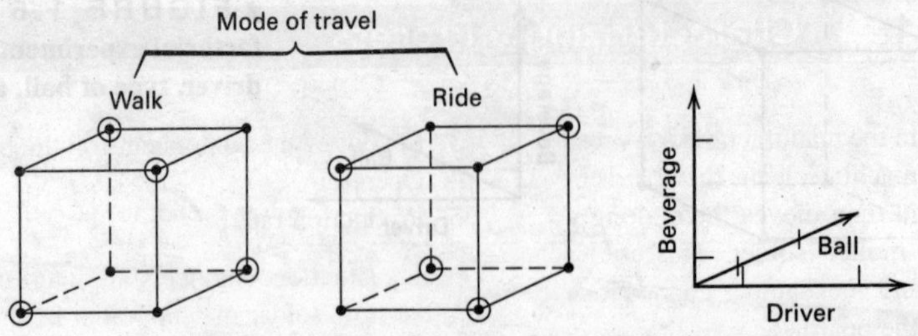

■ **FIGURE 1.8 A four-factor fractional factorial experiment involving type of driver, type of ball, type of beverage, and mode of travel**

Fractional factorial designs are used extensively in industrial research and development, and for process improvement. These designs will be discussed in Chapters 8 and 9.

1.2 Some Typical Applications of Experimental Design

Experimental design methods have found broad application in many disciplines. As noted previously, we may view experimentation as part of the scientific process and as one of the ways by which we learn about how systems or processes work. Generally, we learn through a series of activities in which we make conjectures about a process, perform experiments to generate data from the process, and then use the information from the experiment to establish new conjectures, which lead to new experiments, and so on.

Experimental design is a critically important tool in the scientific and engineering world for improving the product realization process. Critical components of these activities are in new manufacturing process design and development, and process management. The application of experimental design techniques early in process development can result in

1. Improved process yields
2. Reduced variability and closer conformance to nominal or target requirements
3. Reduced development time
4. Reduced overall costs.

Experimental design methods are also of fundamental importance in **engineering design** activities, where new products are developed and existing ones improved. Some applications of experimental design in engineering design include

1. Evaluation and comparison of basic design configurations
2. Evaluation of material alternatives
3. Selection of design parameters so that the product will work well under a wide variety of field conditions, that is, so that the product is **robust**
4. Determination of key product design parameters that impact product performance
5. Formulation of new products.

The use of experimental design in product realization can result in products that are easier to manufacture and that have enhanced field performance and reliability, lower product cost, and shorter product design and development time. Designed experiments also have extensive applications in marketing, market research, transactional and service operations, and general business operations. We now present several examples that illustrate some of these ideas.

EXAMPLE 1.1 Characterizing a Process

A flow solder machine is used in the manufacturing process for printed circuit boards. The machine cleans the boards in a flux, preheats the boards, and then moves them along a conveyor through a wave of molten solder. This solder process makes the electrical and mechanical connections for the leaded components on the board.

The process currently operates around the 1 percent defective level. That is, about 1 percent of the solder joints on a board are defective and require manual retouching. However, because the average printed circuit board contains over 2000 solder joints, even a 1 percent defective level results in far too many solder joints requiring rework. The process engineer responsible for this area would like to use a designed experiment to determine which machine parameters are influential in the occurrence of solder defects and which adjustments should be made to those variables to reduce solder defects.

The flow solder machine has several variables that can be controlled. They include

1. Solder temperature
2. Preheat temperature
3. Conveyor speed
4. Flux type
5. Flux specific gravity
6. Solder wave depth
7. Conveyor angle.

In addition to these controllable factors, several other factors cannot be easily controlled during routine manufacturing, although they could be controlled for the purposes of a test. They are

1. Thickness of the printed circuit board
2. Types of components used on the board
3. Layout of the components on the board
4. Operator
5. Production rate.

In this situation, engineers are interested in **characterizing** the flow solder machine; that is, they want to determine which factors (both controllable and uncontrollable) affect the occurrence of defects on the printed circuit boards. To accomplish this, they can design an experiment that will enable them to estimate the magnitude and direction of the factor effects; that is, how much does the response variable (defects per unit) change when each factor is changed, and does changing the factors *together* produce different results than are obtained from individual factor adjustments—that is, do the factors interact? Sometimes we call an experiment such as this a **screening experiment**. Typically, screening or characterization experiments involve using fractional factorial designs, such as in the golf example in Figure 1.8.

The information from this screening or characterization experiment will be used to identify the critical process factors and to determine the direction of adjustment for these factors to reduce further the number of defects per unit. The experiment may also provide information about which factors should be more carefully controlled during routine manufacturing to prevent high defect levels and erratic process performance. Thus, one result of the experiment could be the application of techniques such as control charts to one or more **process variables** (such as solder temperature), in addition to control charts on process output. Over time, if the process is improved enough, it may be possible to base most of the process control plan on controlling process input variables instead of control charting the output.

EXAMPLE 1.2 Optimizing a Process

In a characterization experiment, we are usually interested in determining which process variables affect the response. A logical next step is to optimize, that is, to determine the region in the important factors that leads to the best possible response. For example, if the response is yield, we would look for a region of maximum yield, whereas if the response is variability in a critical product dimension, we would seek a region of minimum variability.

Suppose that we are interested in improving the yield of a chemical process. We know from the results of a characterization experiment that the two most important process variables that influence the yield are operating temperature and reaction time. The process currently runs at 145°F and 2.1 hours of reaction time, producing yields of around 80 percent. Figure 1.9 shows a view of the time–temperature region from above. In this graph, the lines of constant yield are connected to form response **contours**, and we have shown the contour lines for yields of 60, 70, 80, 90, and 95 percent. These contours are projections on the time–temperature region of cross sections of the yield surface corresponding to the aforementioned percent yields. This surface is sometimes called a **response surface**. The true response surface in Figure 1.9 is unknown to the process personnel, so experimental methods will be required to optimize the yield with respect to time and temperature.

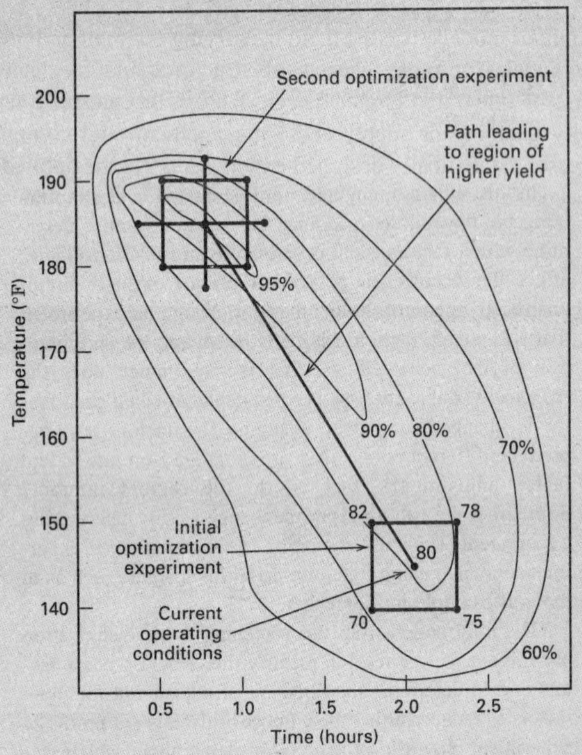

■ **FIGURE 1.9 Contour plot of yield as a function of reaction time and reaction temperature, illustrating experimentation to optimize a process**

To locate the optimum, it is necessary to perform an experiment that varies both time and temperature together, that is, a factorial experiment. The results of an initial factorial experiment with both time and temperature run at two levels is shown in Figure 1.9. The responses observed at the four corners of the square indicate that we should move in the general direction of increased temperature and decreased reaction time to increase yield. A few additional runs would be performed in this direction, and this additional experimentation would lead us to the region of maximum yield.

Once we have found the region of the optimum, a second experiment would typically be performed. The objective of this second experiment is to develop an empirical model of the process and to obtain a more precise estimate of the optimum operating conditions for time and temperature. This approach to process optimization is called **response surface methodology**, and it is explored in detail in Chapter 11. The second design illustrated in Figure 1.9 is a **central composite design**, one of the most important experimental designs used in process optimization studies.

EXAMPLE 1.3 Designing a Product—I

A biomedical engineer is designing a new pump for the intravenous delivery of a drug. The pump should deliver a constant quantity or dose of the drug over a specified period of time. She must specify a number of variables or design parameters. Among these are the diameter and length of the cylinder, the fit between the cylinder and the plunger, the plunger length, the diameter and wall thickness of the tube connecting the pump and the needle inserted into the patient's vein, the material to use for fabricating both the cylinder and the tube, and the nominal pressure at which the system must operate. The impact of some of these parameters on the design can be evaluated by building prototypes in which these factors can be varied over appropriate ranges. Experiments can then be designed and the prototypes tested to investigate which design parameters are most influential on pump performance. Analysis of this information will assist the engineer in arriving at a design that provides reliable and consistent drug delivery.

EXAMPLE 1.4 Designing a Product—II

An engineer is designing an aircraft engine. The engine is a commercial turbofan, intended to operate in the cruise configuration at 40,000 ft and 0.8 Mach. The design parameters include inlet flow, fan pressure ratio, overall pressure, stator outlet temperature, and many other factors. The output response variables in this system are specific fuel consumption and engine thrust. In designing this system, it would be prohibitive to build prototypes or actual test articles early in the design process, so the engineers use a **computer model** of the system that allows them to focus on the key design parameters of the engine and to vary them in an effort to optimize the performance of the engine. Designed experiments can be employed with the computer model of the engine to determine the most important design parameters and their optimal settings.

Designers frequently use computer models to assist them in carrying out their activities. Examples include finite element models for many aspects of structural and mechanical design, electrical circuit simulators for integrated circuit design, factory or enterprise-level models for scheduling and capacity planning or supply chain management, and computer models of complex chemical processes. Statistically designed experiments can be applied to these models just as easily and successfully as they can to actual physical systems and will result in reduced development lead time and better designs.

EXAMPLE 1.5 Formulating a Product

A biochemist is formulating a diagnostic product to detect the presence of a certain disease. The product is a mixture of biological materials, chemical reagents, and other materials that when combined with human blood react to provide a diagnostic indication. The type of experiment used here is a **mixture experiment**, because various ingredients that are combined to form the diagnostic make up 100 percent of the mixture composition (on a volume, weight, or mole ratio basis), and the response is a function of the mixture proportions that are present in the product. Mixture experiments are a special type of response surface experiment that we will study in Chapter 11. They are very useful in designing biotechnology products, pharmaceuticals, foods and beverages, paints and coatings, consumer products such as detergents, soaps, and other personal care products, and a wide variety of other products.

EXAMPLE 1.6 Designing a Web Page

A lot of business today is conducted via the World Wide Web. Consequently, the design of a business' web page has potentially important economic impact. Suppose that the website has the following components: (1) a photoflash image, (2) a main headline, (3) a subheadline, (4) a main text copy, (5) a main image on the right side, (6) a background design, and (7) a footer. We are interested in finding the factors that influence the click-through rate; that is, the number of visitors who click through into the site divided by the total number of visitors to the site. Proper selection of the important factors can lead to an optimal web page design. Suppose that there are four choices for the photoflash image, eight choices for the main headline, six choices for the subheadline, five choices for the main text copy, four choices for the main image, three choices for the background design, and seven choices for the footer. If we use a factorial design, web pages for all possible combinations of these factor levels must be constructed and tested. This is a total of $4 \times 8 \times 6 \times 5 \times 4 \times 3 \times 7 = 80{,}640$ web pages. Obviously, it is not feasible to design and test this many combinations of web pages, so a complete factorial experiment cannot be considered. However, a fractional factorial experiment that uses a small number of the possible web page designs would likely be successful. This experiment would require a fractional factorial where the factors have different numbers of levels. We will discuss how to construct these designs in Chapter 9.

1.3 Basic Principles

If an experiment such as the ones described in Examples 1.1 through 1.6 is to be performed most efficiently, a scientific approach to planning the experiment must be employed. **Statistical design of experiments** refers to the process of planning the experiment so that appropriate data will be collected and analyzed by statistical methods, resulting in valid and objective conclusions. The statistical approach to experimental design is necessary if we wish to draw meaningful conclusions from the data. When the problem involves data that are subject to experimental errors, statistical methods are the only **objective** approach to analysis. Thus, there are two aspects to any experimental problem: the design of the experiment and the statistical analysis of the data. These two subjects are closely related because the method of analysis depends directly on the design employed. Both topics will be addressed in this book.

The three basic principles of experimental design are **randomization**, **replication**, and **blocking**. Sometimes we add the **factorial principle** to these three. Randomization is the cornerstone underlying the use of statistical methods in experimental design. By randomization we mean that both the allocation of the experimental material and the order in which the individual runs of the experiment are to be performed are randomly determined. Statistical methods require that the observations (or errors) be independently distributed random variables. Randomization usually makes this assumption valid. By properly randomizing the experiment, we also assist in "averaging out" the effects of extraneous factors that may be present. For example, suppose that the specimens in the hardness experiment are of slightly different thicknesses and that the effectiveness of the quenching medium may be affected by specimen thickness. If all the specimens subjected to the oil quench are thicker than those subjected to the saltwater quench, we may be introducing systematic bias into the experimental results. This bias handicaps one of the quenching media and consequently invalidates our results. Randomly assigning the specimens to the quenching media alleviates this problem.

Computer software programs are widely used to assist experimenters in selecting and constructing experimental designs. These programs often present the runs in the experimental design in random order. This random order is created by using a random number generator. Even with such a computer program, it is still often necessary to assign units of experimental material (such as the specimens in the hardness example mentioned above), operators, gauges or measurement devices, and so forth for use in the experiment.

Sometimes experimenters encounter situations where randomization of some aspect of the experiment is difficult. For example, in a chemical process, temperature may be a very hard-to-change variable as we may want to change it less often than we change the levels of other factors. In an experiment of this type, **complete randomization** would be difficult because it would add time and cost. There are statistical design methods for dealing with restrictions on randomization. Some of these approaches will be discussed in subsequent chapters (see in particular Chapter 14).

By **replication** we mean an independent **repeat run** of each factor combination. In the metallurgical experiment discussed in Section 1.1, replication would consist of treating a specimen by oil quenching and treating a specimen by saltwater quenching. Thus, if five specimens are treated in each quenching medium, we say that five **replicates** have been obtained. Each of the 10 observations should be run in random order. Replication has two important properties. First, it allows the experimenter to obtain an estimate of the experimental error. This estimate of error becomes a basic unit of measurement for determining whether observed differences in the data are really *statistically* different. Second, if the sample mean ($\bar{y}$) is used to estimate the true mean response for one of the factor levels in the experiment, replication permits the experimenter to obtain a more precise estimate of this parameter. For example; if σ^2 is the variance of an individual observation and there are n replicates, the variance of the sample mean is

$$\sigma_{\bar{y}}^2 = \frac{\sigma^2}{n}$$

The practical implication of this is that if we had $n = 1$ replicates and observed $y_1 = 145$ (oil quench) and $y_2 = 147$ (saltwater quench), we would probably be unable to make satisfactory inferences about the effect of the quenching medium—that is, the observed difference could be the result of experimental error. The point is that without replication we have no way of knowing why the two observations are different. On the other hand, if n was reasonably large and the experimental error was sufficiently small and if we observed sample averages $\bar{y}_1 < \bar{y}_2$, we would be reasonably safe in concluding that saltwater quenching produces a higher hardness in this particular aluminum alloy than does oil quenching.

Often when the runs in an experiment are randomized, two (or more) consecutive runs will have exactly the same levels for some of the factors. For example, suppose we have three factors in an experiment: pressure, temperature, and time. When the experimental runs are randomized, we find the following:

Run number	Pressure (psi)	Temperature (°C)	Time (min)
i	30	100	30
$i + 1$	30	125	45
$i + 2$	40	125	45

Notice that between runs i and $i + 1$, the levels of pressure are identical and between runs $i + 1$ and $i + 2$, the levels of both temperature and time are identical. To obtain a true replicate, the experimenter needs to "twist the pressure knob" to an intermediate setting between runs i and $i + 1$, and reset pressure to 30 psi for run $i + 1$. Similarly, temperature and time should be reset to intermediate levels between runs $i + 1$ and $i + 2$ before being set to their design levels for run $i + 2$. Part of the experimental error is the variability associated with hitting and holding factor levels.

There is an important distinction between **replication** and **repeated measurements**. For example, suppose that a silicon wafer is etched in a single-wafer plasma etching process, and a critical dimension (CD) on this wafer is measured three times. These measurements are not replicates; they are a form of repeated measurements, and in this case the observed variability in the three repeated measurements is a direct reflection of the inherent variability in the measurement system or gauge and possibly the variability in this CD at different locations on the wafer where the measurement were taken. As another illustration, suppose that as part of an experiment in semiconductor manufacturing four wafers are processed simultaneously in an oxidation furnace at a particular gas flow rate and time and then a measurement is taken on the oxide thickness of each wafer. Once again, the measurements on the four wafers are not replicates but repeated measurements. In this case, they reflect differences among the wafers and other sources of variability within that particular furnace run. Replication reflects sources of variability both **between** runs and (potentially) **within** runs.

Blocking is a design technique used to improve the precision with which comparisons among the factors of interest are made. Often blocking is used to reduce or eliminate the variability transmitted from **nuisance factors**—that is, factors that may influence the experimental response but in which we are not directly interested. For example, an experiment in a chemical process may require two batches of raw material to make all the required runs. However, there could be differences between the batches due to supplier-to-supplier variability, and if we are not specifically interested in this effect, we would think of the batches of raw material as a nuisance factor. Generally, a block is a set of relatively homogeneous experimental conditions. In the chemical process example, each batch of raw material would form a block, because the variability within a batch would be expected to be smaller than the variability between batches. Typically, as in this example, each level of the nuisance factor becomes a block. Then the experimenter divides the observations from the statistical design into groups that are run in each block. We study blocking in detail in several places in the text, including Chapters 4, 5, 7, 8, 9, 11, and 13. A simple example illustrating the blocking principal is given in Section 2.5.1.

The three basic principles of experimental design, randomization, replication, and blocking are part of every experiment. We will illustrate and emphasize them repeatedly throughout this book.

1.4 Guidelines for Designing Experiments

To use the statistical approach in designing and analyzing an experiment, it is necessary for everyone involved in the experiment to have a clear idea in advance of exactly what is to be studied, how the data are to be collected, and at least a qualitative understanding of how these data are to be analyzed. An outline of the recommended procedure is shown in Table 1.1. We now give a brief discussion of this outline and elaborate on some of the key points. For more details, see Coleman and Montgomery (1993), and the references therein. The **supplemental text material** for this chapter is also useful.

1. Recognition of and statement of the problem. This may seem to be a rather obvious point, but in practice often neither it is simple to realize that a problem requiring experimentation exists, nor is it simple to develop a clear and generally accepted statement of this problem. It is necessary to develop all ideas about the objectives of the experiment. Usually, it is important to solicit input from all concerned parties: engineering, quality assurance, manufacturing, marketing, management, customer, and operating personnel (who usually have much insight and who are too often ignored). For this reason, a **team approach** to designing experiments is recommended.

It is usually helpful to prepare a list of specific problems or questions that are to be addressed by the experiment. A clear statement of the problem often contributes substantially to better understanding of the phenomenon being studied and the final solution of the problem.

It is also important to keep the overall objectives of the experiment in mind. There are several broad reasons for running experiments and each type of experiment will generate its own list of specific questions that need to be addressed. Some (but by no means all) of the reasons for running experiments include:

a. ***Factor screening or characterization.*** When a system or process is new, it is usually important to learn which factors have the most influence on the response(s) of interest. Often there are a lot of factors. This usually indicates that the experimenters do not know much about the system so screening is essential if we are to efficiently get the desired performance from the system. Screening experiments are extremely important when working with new systems or technologies so that valuable resources will not be wasted using best guess and OFAT approaches.

b. ***Optimization.*** After the system has been characterized and we are reasonably certain that the important factors have been identified, the next objective is usually optimization, that is, find the settings or levels of

■ **TABLE 1.1**
Guidelines for Designing an Experiment

1. Recognition of and statement of the problem	Pre-experimental planning
2. Selection of the response variable[a]	
3. Choice of factors, levels, and ranges[a]	
4. Choice of experimental design	
5. Performing the experiment	
6. Statistical analysis of the data	
7. Conclusions and recommendations	

[a]In practice, steps 2 and 3 are often done simultaneously or in reverse order.

the important factors that result in desirable values of the response. For example, if a screening experiment on a chemical process results in the identification of time and temperature as the two most important factors, the optimization experiment may have as its objective finding the levels of time and temperature that maximize yield, or perhaps maximize yield while keeping some product property that is critical to the customer within specifications. An optimization experiment is usually a follow-up to a screening experiment. It would be very unusual for a screening experiment to produce the optimal settings of the important factors.

c. ***Confirmation.*** In a confirmation experiment, the experimenter is usually trying to verify that the system operates or behaves in a manner that is consistent with some theory or past experience. For example, if theory or experience indicates that a particular new material is equivalent to the one currently in use and the new material is desirable (perhaps less expensive, or easier to work with in some way), then a confirmation experiment would be conducted to verify that substituting the new material results in no change in product characteristics that impact its use. Moving a new manufacturing process to full-scale production based on results found during experimentation at a pilot plant or development site is another situation that often results in confirmation experiments—that is, are the same factors and settings that were determined during development work appropriate for the full-scale process?

d. ***Discovery.*** In discovery experiments, the experimenters are usually trying to determine what happens when we explore new materials, or new factors, or new ranges for factors. In the pharmaceutical industry, scientists are constantly conducting discovery experiments to find new materials or combinations of materials that will be effective in treating disease.

e. ***Robustness.*** These experiments often address questions such as under what conditions do the response variables of interest seriously degrade? Or what conditions would lead to unacceptable variability in the response variables? A variation of this is determining how we can set the factors in the system that we can control to minimize the variability transmitted into the response from factors that we cannot control very well. We will discuss some experiments of this type in Chapter 12.

Obviously, the specific questions to be addressed in the experiment relate directly to the overall objectives. An important aspect of problem formulation is the recognition that one large comprehensive experiment is unlikely to answer the key questions satisfactorily. A single comprehensive experiment requires the experimenters to know the answers to a lot of questions, and if they are wrong, the results will be disappointing. This leads to wasting time, materials, and other resources and may result in never answering the original research questions satisfactorily. A **sequential** approach employing a series of smaller experiments, each with a specific objective, such as factor screening, is a better strategy.

2. ***Selection of the response variable.*** In selecting the response variable, the experimenter should be certain that this variable really provides useful information about the process under study. Most often, the average or standard deviation (or both) of the measured characteristic will be the response variable. Multiple responses are not unusual. The experimenters must decide how each response will be measured, and address issues such as how will any measurement system be calibrated and

how this calibration will be maintained during the experiment. The gauge or measurement system capability (or measurement error) is also an important factor. If gauge capability is inadequate, only relatively large factor effects will be detected by the experiment or perhaps additional replication will be required. In some situations where gauge capability is poor, the experimenter may decide to measure each experimental unit several times and use the average of the repeated measurements as the observed response. It is usually critically important to identify issues related to defining the responses of interest and how they are to be measured *before* conducting the experiment. Sometimes designed experiments are employed to study and improve the performance of measurement systems. For an example, see Chapter 13.

3. *Choice of factors, levels, and range.* (As noted in Table 1.1, steps 2 and 3 are often done simultaneously or in the reverse order.) When considering the factors that may influence the performance of a process or system, the experimenter usually discovers that these factors can be classified as either **potential design factors** or nuisance factors. The potential design factors are those factors that the experimenter may wish to vary in the experiment. Often we find that there are a lot of potential design factors, and some further classification of them is helpful. Some useful classifications are **design factors, held-constant factors**, and **allowed-to-vary** factors. The design factors are the factors actually selected for study in the experiment. Held-constant factors are variables that may exert some effect on the response, but for purposes of the present experiment these factors are not of interest, so they will be held at a specific level. For example, in an etching experiment in the semiconductor industry, there may be an effect that is unique to the specific plasma etch tool used in the experiment. However, this factor would be very difficult to vary in an experiment, so the experimenter may decide to perform all experimental runs on one particular (ideally "typical") etcher. Thus, this factor has been held constant. As an example of allowed-to-vary factors, the experimental units or the "materials" to which the design factors are applied are usually nonhomogeneous, yet we often ignore this unit-to-unit variability and rely on randomization to balance out any material or experimental unit effect. We often assume that the effects of held-constant factors and allowed-to-vary factors are relatively small.

Nuisance factors, on the other hand, may have large effects that must be accounted for, yet we may not be interested in them in the context of the present experiment. Nuisance factors are often classified as **controllable, uncontrollable**, or **noise factors**. A controllable nuisance factor is one whose levels may be set by the experimenter. For example, the experimenter can select different batches of raw material or different days of the week when conducting the experiment. The blocking principle, discussed in the previous section, is often useful in dealing with controllable nuisance factors. If a nuisance factor is uncontrollable in the experiment, but it can be measured, an analysis procedure called the **analysis of covariance** can often be used to compensate for its effect. For example, the relative humidity in the process environment may affect process performance, and if the humidity cannot be controlled, it probably can be measured and treated as a covariate. When a factor that varies naturally and uncontrollably in the process can be controlled for purposes of an experiment, we often call it a noise factor. In such situations, our objective is usually to find the settings of the controllable design factors that minimize the variability transmitted from the noise factors. This is sometimes called a process robustness study or a robust design problem. Blocking, analysis of covariance, and process robustness studies are discussed later in the text.

Once the experimenter has selected the design factors, he or she must choose the ranges over which these factors will be varied and the specific levels at which runs will be made. Thought must also be given to how these factors are to be controlled at the desired values and how they are to be measured. For instance, in the flow solder experiment, the engineer has defined 12 variables that may affect the occurrence of solder defects. The experimenter will also have to decide on a region of interest for each variable (that is, the range over which each factor will be varied) and on how many levels of each variable to use. **Process knowledge** is required to do this. This process knowledge is usually a combination of practical experience and theoretical understanding. It is important to investigate all factors that may be of importance and to be not overly influenced by past experience, particularly when we are in the early stages of experimentation or when the process is not very mature.

When the objective of the experiment is **factor screening** or **process characterization**, it is usually best to keep the number of factor levels low. Generally, two levels work very well in factor screening studies. Choosing the region of interest is also important. In factor screening, the region of interest should be relatively large—that is, the range over which the factors are varied should be broad. As we learn more about which variables are important and which levels produce the best results, the region of interest in subsequent experiments will usually become narrower.

The **cause-and-effect diagram** can be a useful technique for organizing some of the information generated in pre-experimental planning. Figure 1.10 is the cause-and-effect diagram constructed while planning an experiment to resolve problems with wafer charging (a charge accumulation on the wafers) encountered in an etching tool used in semiconductor manufacturing. The cause-and-effect diagram is also known as a **fishbone diagram** because the "effect" of interest or the response variable is drawn along the spine of the diagram and the potential causes or design factors are organized in a series of ribs. The cause-and-effect diagram uses the traditional causes of measurement, materials, people, environment, methods, and machines to organize the information and potential design factors. Notice that some of the individual causes will probably lead directly to a design factor that

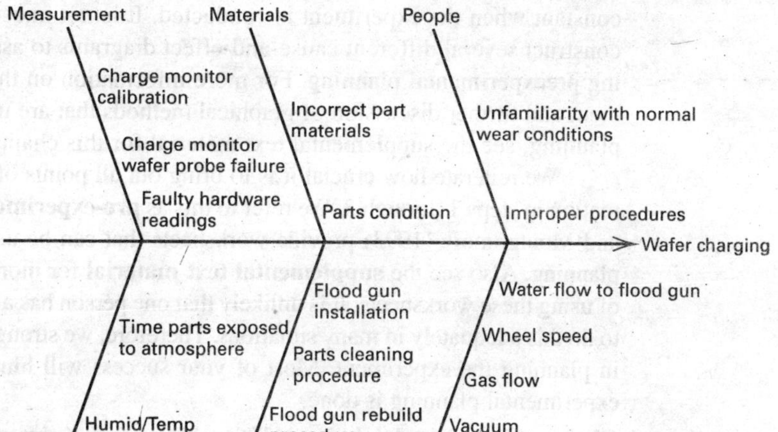

■ **FIGURE 1.10 A cause-and-effect diagram for the etching process experiment**

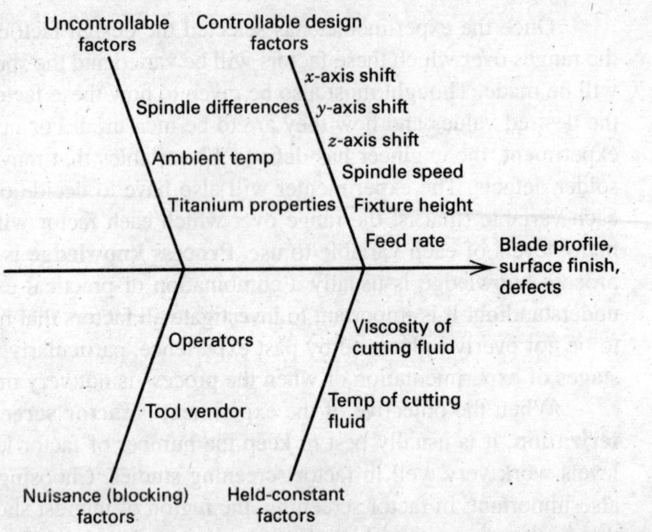

■ **FIGURE 1.11 A cause-and-effect diagram for the CNC machine experiment**

will be included in the experiment (such as wheel speed, gas flow, and vacuum), while others represent potential areas that will need further study to turn them into design factors (such as operators following improper procedures), and still others will probably lead to either factors that will be held constant during the experiment or blocked (such as temperature and relative humidity). Figure 1.11 is a cause-and-effect diagram for an experiment to study the effect of several factors on the turbine blades produced on a computer-numerical-controlled (CNC) machine. This experiment has three response variables: blade profile, blade surface finish, and surface finish defects in the finished blade. The causes are organized into groups of controllable factors from which the design factors for the experiment may be selected, uncontrollable factors whose effects will probably be balanced out by randomization, nuisance factors that may be blocked, and factors that may be held constant when the experiment is conducted. It is not unusual for experimenters to construct several different cause-and-effect diagrams to assist and guide them during preexperimental planning. For more information on the CNC machine experiment and further discussion of graphical methods that are useful in preexperimental planning, see the supplemental text material for this chapter.

We reiterate how crucial it is to bring out all points of view and process information in steps 1 through 3. We refer to this as **pre-experimental planning**. Coleman and Montgomery (1993) provide worksheets that can be useful in pre-experimental planning. Also see the **supplemental text material** for more details and an example of using these worksheets. It is unlikely that one person has all the knowledge required to do this adequately in many situations. Therefore, we strongly argue for a team effort in planning the experiment. Most of your success will hinge on how well the pre-experimental planning is done.

4. ***Choice of experimental design.*** If the above pre-experimental planning activities are done correctly, this step is relatively easy. Choice of design involves consideration of sample size (number of replicates), selection of a suitable run order for the experimental trials, and determination of whether or not blocking or other randomization restrictions are involved. This book discusses some of the more important types of

experimental designs, and it can ultimately be used as a guide for selecting an appropriate experimental design for a wide variety of problems.

There are also several interactive statistical software packages that support this phase of experimental design. The experimenter can enter information about the number of factors, levels, and ranges, and these programs will either present a selection of designs for consideration or recommend a particular design. (We usually prefer to see several alternatives instead of relying entirely on a computer recommendation in most cases.) Most software packages also provide some diagnostic information about how each design will perform. This is useful in evaluation of different design alternatives for the experiment. These programs will usually also provide a worksheet (with the order of the runs randomized) for use in conducting the experiment.

Design selection also involves thinking about and selecting a tentative **empirical model** to describe the results. The model is just a quantitative relationship (equation) between the response and the important design factors. In many cases, a low-order polynomial model will be appropriate. A **first-order** model in two variables is

$$y = \beta_0 + \beta_1 x_1 + \beta_2 x_2 + \varepsilon$$

where y is the response, the x's are the design factors, the β's are unknown parameters that will be estimated from the data in the experiment, and ε is a random error term that accounts for the experimental error in the system that is being studied. The first-order model is also sometimes called a **main effects** model. First-order models are used extensively in screening or characterization experiments. A common extension of the first-order model is to add an **interaction** term, say

$$y = \beta_0 + \beta_1 x_1 + \beta_2 x_2 + \beta_{12} x_1 x_2 + \varepsilon$$

where the cross-product term $x_1 x_2$ represents the two-factor interaction between the design factors. Because interactions between factors is relatively common, the first-order model with interaction is widely used. Higher-order interactions can also be included in experiments with more than two factors if necessary. Another widely used model is the **second-order** model

$$y = \beta_0 + \beta_1 x_1 + \beta_2 x_2 + \beta_{12} x_1 x_2 + \beta_{11} x_{11}^2 + \beta_{22} x_2^2 + \varepsilon$$

Second-order models are often used in optimization experiments.

In selecting the design, it is important to keep the experimental objectives in mind. In many engineering experiments, we already know at the outset that some of the factor levels will result in different values for the response. Consequently, we are interested in identifying *which* factors cause this difference and in estimating the *magnitude* of the response change. In other situations, we may be more interested in verifying uniformity. For example, two production conditions A and B may be compared, A being the standard and B being a more cost-effective alternative. The experimenter will then be interested in demonstrating that, say, there is no difference in yield between the two conditions.

5. ***Performing the experiment.*** When running the experiment, it is vital to monitor the process carefully to ensure that everything is being done according to plan. Errors in experimental procedure at this stage will usually destroy experimental validity. One of the most common mistakes that I have encountered is that the people conducting the experiment failed to set the variables to the proper levels on some runs. Someone should be assigned to check factor settings before each run. Up-front planning to prevent mistakes like this is crucial to success. It is easy to

underestimate the logistical and planning aspects of running a designed experiment in a complex manufacturing or research and development environment.

Coleman and Montgomery (1993) suggest that prior to conducting the experiment a few trial runs or pilot runs are often helpful. These runs provide information about consistency of experimental material, a check on the measurement system, a rough idea of experimental error, and a chance to practice the overall experimental technique. This also provides an opportunity to revisit the decisions made in steps 1–4, if necessary.

6. ***Statistical analysis of the data.*** Statistical methods should be used to analyze the data so that results and conclusions are **objective** rather than judgmental in nature. If the experiment has been designed correctly and performed according to the design, the statistical methods required are not elaborate. There are many excellent software packages designed to assist in data analysis, and many of the programs used in step 4 to select the design provide a seamless, direct interface to the statistical analysis. Often we find that simple **graphical methods** play an important role in data analysis and interpretation. Because many of the questions that the experimenter wants to answer can be cast into an hypothesis-testing framework, hypothesis testing and confidence interval estimation procedures are very useful in analyzing data from a designed experiment. It is also usually very helpful to present the results of many experiments in terms of an **empirical model**, that is, an equation derived from the data that express the relationship between the response and the important design factors. Residual analysis and model adequacy checking are also important analysis techniques. We will discuss these issues in detail later.

   Remember that statistical methods cannot prove that a factor (or factors) has a particular effect. They only provide guidelines as to the reliability and validity of results. When properly applied, statistical methods do not allow anything to be proved experimentally, but they do allow us to measure the likely error in a conclusion or to attach a level of confidence to a statement. The primary advantage of statistical methods is that they add objectivity to the decision-making process. Statistical techniques coupled with good engineering or process knowledge and common sense will usually lead to sound conclusions.

7. ***Conclusions and recommendations.*** Once the data have been analyzed, the experimenter must draw *practical* conclusions about the results and recommend a course of action. Graphical methods are often useful in this stage, particularly in presenting the results to others. **Follow-up runs** and **confirmation testing** should also be performed to validate the conclusions from the experiment.

Throughout this entire process, it is important to keep in mind that experimentation is an important part of the learning process, where we tentatively formulate hypotheses about a system, perform experiments to investigate these hypotheses, and on the basis of the results formulate new hypotheses, and so on. This suggests that experimentation is **iterative**. It is usually a major mistake to design a single, large, comprehensive experiment at the start of a study. A successful experiment requires knowledge of the important factors, the ranges over which these factors should be varied, the appropriate number of levels to use, and the proper units of measurement for these variables. Generally, we do not perfectly know the answers to these questions, but we learn about them as we go along. As an experimental program progresses, we often drop some input variables, add others, change the region of exploration for some factors, or add new response variables. Consequently, we usually experiment **sequentially**, and as a general rule, no more than about 25 percent of the available resources should be invested in the first experiment. This will ensure that sufficient resources are available to perform confirmation runs and ultimately accomplish the final objective of the experiment.

Finally, it is important to recognize that **all** experiments are designed experiments. The important issue is whether they are well designed or not. Good pre-experimental planning will usually lead to a good, successful experiment. Failure to do such planning usually leads to wasted time, money, and other resources and often poor or disappointing results.

1.5 A Brief History of Statistical Design

There have been four eras in the modern development of statistical experimental design. The agricultural era was led by the pioneering work of Sir Ronald A. Fisher in the 1920s and early 1930s. During that time, Fisher was responsible for statistics and data analysis at the Rothamsted Agricultural Experimental Station near London, England. Fisher recognized that flaws in the way the experiment that generated the data had been performed often hampered the analysis of data from systems (in this case, agricultural systems). By interacting with scientists and researchers in many fields, he developed the insights that led to the three basic principles of experimental design that we discussed in Section 1.3: randomization, replication, and blocking. Fisher systematically introduced statistical thinking and principles into designing experimental investigations, including the factorial design concept and the analysis of variance. His two books [the most recent editions are Fisher (1958, 1966)] had profound influence on the use of statistics, particularly in agricultural and related life sciences. For an excellent biography of Fisher, see Box (1978).

Although applications of statistical design in industrial settings certainly began in the 1930s, the second, or industrial, era was catalyzed by the development of response surface methodology (RSM) by Box and Wilson (1951). They recognized and exploited the fact that many industrial experiments are fundamentally different from their agricultural counterparts in two ways: (1) the response variable can usually be observed (nearly) immediately, and (2) the experimenter can quickly learn crucial information from a small group of runs that can be used to plan the next experiment. Box (1999) calls these two features of industrial experiments **immediacy** and **sequentiality**. Over the next 30 years, RSM and other design techniques spread throughout the chemical and the process industries, mostly in research and development work. George Box was the intellectual leader of this movement. However, the application of statistical design at the plant or manufacturing process level was still not extremely widespread. Some of the reasons for this include an inadequate training in basic statistical concepts and methods for engineers and other process specialists and the lack of computing resources and user-friendly statistical software to support the application of statistically designed experiments.

It was during this second or industrial era that work on **optimal** design of experiments began. Kiefer (1959, 1961) and Kiefer and Wolfowitz (1959) proposed a formal approach to selecting a design based on specific objective optimality criteria. Their initial approach was to select a design that would result in the model parameters being estimated with the best possible precision. This approach did not find much application because of the lack of computer tools for its implementation. However, there have been great advances in both algorithms for generating optimal designs and computing capability over the last 25 years. Optimal designs have great application and are discussed at several places in the book.

The increasing interest of Western industry in quality improvement that began in the late 1970s ushered in the third era of statistical design. The work of Genichi Taguchi [Taguchi and Wu (1980), Kackar (1985), and Taguchi (1987, 1991)] had a significant impact on expanding the interest in and use of designed experiments. Taguchi advocated using designed experiments for what he termed **robust parameter design**, or

1. Making processes insensitive to environmental factors or other factors that are difficult to control
2. Making products insensitive to variation transmitted from components
3. Finding levels of the process variables that force the mean to a desired value while simultaneously reducing variability around this value.

Taguchi suggested highly fractionated factorial designs and other orthogonal arrays along with some novel statistical methods to solve these problems. The resulting methodology generated much discussion and controversy. Part of the controversy arose because Taguchi's methodology was advocated in the West initially (and primarily) by entrepreneurs, and the underlying statistical science had not been adequately peer reviewed. By the late 1980s, the results of peer review indicated that although Taguchi's engineering concepts and objectives were well founded, there were substantial problems with his experimental strategy and methods of data analysis. For specific details of these issues, see Box (1988), Box, Bisgaard, and Fung (1988), Hunter (1985, 1989), Myers, Montgomery and Anderson-Cook (2009), and Pignatiello and Ramberg (1992). Many of these concerns are also summarized in the extensive panel discussion in the May 1992 issue of *Technometrics* [see Nair et al. (1992)].

There were several positive outcomes of the Taguchi controversy. First, designed experiments became more widely used in the discrete parts industries, including automotive and aerospace manufacturing, electronics and semiconductors, and many other industries that had previously made little use of the technique. Second, the fourth era of statistical design began. This era has included a renewed general interest in statistical design by both researchers and practitioners and the development of many new and useful approaches to experimental problems in the industrial world, including alternatives to Taguchi's technical methods that allow his engineering concepts to be carried into practice efficiently and effectively. Some of these alternatives will be discussed and illustrated in subsequent chapters, particularly in Chapter 12. Third, computer software for construction and evaluation of designs has improved greatly with many new features and capability. Forth, formal education in statistical experimental design is becoming part of many engineering programs in universities, at both undergraduate and graduate levels. The successful integration of good experimental design practice into engineering and science is a key factor in future industrial competitiveness.

Applications of designed experiments have grown far beyond the agricultural origins. There is not a single area of science and engineering that has not successfully employed statistically designed experiments. In recent years, there has been a considerable utilization of designed experiments in many other areas, including the service sector of business, financial services, government operations, and many nonprofit business sectors. An article appeared in *Forbes* magazine on March 11, 1996, entitled "The New Mantra: MVT," where MVT stands for "multivariable testing," a term authors use to describe factorial designs. The article notes the many successes that a diverse group of companies have had through their use of statistically designed experiments.

1.6 Summary: Using Statistical Techniques in Experimentation

Much of the research in engineering, science, and industry is empirical and makes extensive use of experimentation. Statistical methods can greatly increase the efficiency of these experiments and often strengthen the conclusions so obtained. The proper use of statistical techniques in experimentation requires that the experimenter keep the following points in mind:

1. ***Use your nonstatistical knowledge of the problem.*** Experimenters are usually highly knowledgeable in their fields. For example, a civil engineer working on a problem in hydrology typically has considerable practical experience and formal academic training in this area. In some fields, there is a large body of physical theory on which to draw in explaining relationships between factors and responses. This type of nonstatistical knowledge is invaluable in choosing factors, determining factor levels, deciding how many replicates to run, interpreting the results of the analysis, and so forth. Using a designed experiment is no substitute for thinking about the problem.
2. ***Keep the design and analysis as simple as possible.*** Don't be overzealous in the use of complex, sophisticated statistical techniques. Relatively simple design and analysis methods are almost always best. This is a good place to reemphasize steps 1–3 of the procedure recommended in Section 1.4. If you do the pre-experiment planning carefully and select a reasonable design, the analysis will almost always be relatively straightforward. In fact, a well-designed experiment will sometimes almost analyze itself! However, if you botch the pre-experimental planning and execute the experimental design badly, it is unlikely that even the most complex and elegant statistics can save the situation.
3. ***Recognize the difference between practical and statistical significance.*** Just because two experimental conditions produce mean responses that are statistically different, there is no assurance that this difference is large enough to have any practical value. For example, an engineer may determine that a modification to an automobile fuel injection system may produce a true mean improvement in gasoline mileage of 0.1 mi/gal and be able to determine that this is a statistically significant result. However, if the cost of the modification is \$1000, the 0.1 mi/gal difference is probably too small to be of any practical value.
4. ***Experiments are usually iterative.*** Remember that in most situations it is unwise to design too comprehensive an experiment at the start of a study. Successful design requires the knowledge of important factors, the ranges over which these factors are varied, the appropriate number of levels for each factor, and the proper methods and units of measurement for each factor and response. Generally, we are not well equipped to answer these questions at the beginning of the experiment, but we learn the answers as we go along. This argues in favor of the **iterative**, or **sequential**, approach discussed previously. Of course, there are situations where comprehensive experiments are entirely appropriate, but as a general rule most experiments should be iterative. Consequently, we usually should not invest more than about 25 percent of the resources of experimentation (runs, budget, time, etc.) in the initial experiment. Often these first efforts are just learning experiences, and some resources must be available to accomplish the final objectives of the experiment.

1.7 Problems

1.1. Suppose that you want to design an experiment to study the proportion of unpopped kernels of popcorn. Complete steps 1–3 of the guidelines for designing experiments in Section 1.4. Are there any major sources of variation that would be difficult to control?

1.2. Suppose that you want to investigate the factors that potentially affect cooking rice.

(a) What would you use as a response variable in this experiment? How would you measure the response?

(b) List all of the potential sources of variability that could impact the response.

(c) Complete the first three steps of the guidelines for designing experiments in Section 1.4.

1.3. Suppose that you want to compare the growth of garden flowers with different conditions of sunlight, water, fertilizer, and soil conditions. Complete steps 1–3 of the guidelines for designing experiments in Section 1.4.

1.4. Select an experiment of interest to you. Complete steps 1–3 of the guidelines for designing experiments in Section 1.4.

1.5. Search the World Wide Web for information about Sir Ronald A. Fisher and his work on experimental design in agricultural science at the Rothamsted Experimental Station.

1.6. Find a website for a business that you are interested in. Develop a list of factors that you would use in an experiment to improve the effectiveness of this website.

1.7. Almost everyone is concerned about the rising price of gasoline. Construct a cause and effect diagram identifying the factors that potentially influence the gasoline mileage that you get in your car. How would you go about conducting an experiment to determine any of these factors actually affect your gasoline mileage?

1.8. What is replication? Why do we need replication in an experiment? Present an example that illustrates the difference between replication and repeated measurements.

1.9. Why is randomization important in an experiment?

1.10. What are the potential risks of a single large, comprehensive experiment in contrast to a sequential approach?

CHAPTER 2

Basic Statistical Methods

CHAPTER OUTLINE

The supplemental material is on the textbook website www.wiley.com/go/global/montgomery.

In this chapter, we consider experiments to compare two **conditions** (sometimes called **treatments**). These are often called **simple comparative experiments**. We begin with an example of an experiment performed to determine whether two different formulations of a product give equivalent results. The discussion leads to a review of several basic statistical concepts, such as random variables, probability distributions, random samples, sampling distributions, and tests of hypotheses.

2.1 Introduction

An engineer is studying the formulation of a Portland cement mortar. He has added a polymer latex emulsion during mixing to determine if this impacts the curing time and tension bond strength of the mortar. The experimenter prepared 10 samples of the original formulation and 10 samples of the modified formulation. We will refer to the two different formulations as two **treatments** or as two **levels** of the **factor** formulations. When the cure process

■ **TABLE 2.1**
Tension Bond Strength Data for the Portland Cement Formulation Experiment

j	Modified Mortar y_{1j}	Unmodified Mortar y_{2j}
1	16.85	16.62
2	16.40	16.75
3	17.21	17.37
4	16.35	17.12
5	16.52	16.98
6	17.04	16.87
7	16.96	17.34
8	17.15	17.02
9	16.59	17.08
10	16.57	17.27

was completed, the experimenter did find a very large reduction in the cure time for the modified mortar formulation. Then he began to address the tension bond strength of the mortar. If the new mortar formulation has an adverse effect on bond strength, this could impact its usefulness.

The tension bond strength data from this experiment are shown in Table 2.1 and plotted in Figure 2.1. The graph is called a **dot diagram**. Visual examination of these data gives the impression that the strength of the unmodified mortar may be greater than the strength of the modified mortar. This impression is supported by comparing the *average* tension bond strengths, $\bar{y}_1 = 16.76$ kgf/cm^2 for the modified mortar and $\bar{y}_2 = 17.04$ kgf/cm^2 for the unmodified mortar. The average tension bond strengths in these two samples differ by what seems to be a modest amount. However, it is not obvious that this difference is large enough to imply that the two formulations really *are* different. Perhaps this observed difference in average strengths is the result of sampling fluctuation and the two formulations are really identical. Possibly another two samples would give opposite results, with the strength of the modified mortar exceeding that of the unmodified formulation.

A technique of statistical inference called **hypothesis testing** can be used to assist the experimenter in comparing these two formulations. Hypothesis testing allows the comparison of the two formulations to be made on **objective** terms, with knowledge of the risks associated with reaching the wrong conclusion. Before presenting procedures for hypothesis testing in simple comparative experiments, we will briefly summarize some elementary statistical concepts.

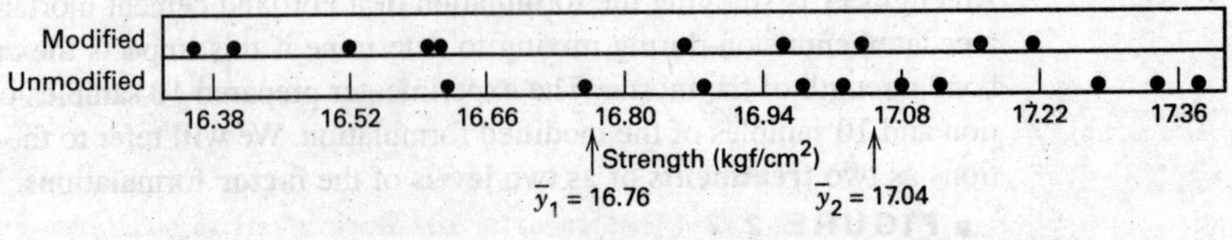

■ **FIGURE 2.1** **Dot diagram for the tension bond strength data in Table 2.1**

2.2 Basic Statistical Concepts

Each of the observations in the Portland cement experiment described above would be called a **run**. Notice that the individual runs differ, so there is fluctuation, or **noise**, in the observed bond strengths. This noise is usually called **experimental error** or simply **error**. It is a **statistical error**, meaning that it arises from variation that is uncontrolled and generally unavoidable. The presence of error or noise implies that the response variable, tension bond strength, is a **random variable**. A random variable may be either **discrete** or **continuous**. If the set of all possible values of the random variable is either finite or countably infinite, then the random variable is discrete, whereas if the set of all possible values of the random variable is an interval, then the random variable is continuous.

Graphical Description of Variability. We often use simple graphical methods to assist in analyzing the data from an experiment. The **dot diagram**, illustrated in Figure 2.1, is a very useful device for displaying a small body of data (say up to about 20 observations). The dot diagram enables the experimenter to see quickly the general **location** or **central tendency** of the observations and their **spread** or **variability**. For example, in the Portland cement tension bond experiment, the dot diagram reveals that the two formulations may differ in mean strength but that both formulations produce about the same variability in strength.

If the data are fairly numerous, the dots in a dot diagram become difficult to distinguish and a **histogram** may be preferable. Figure 2.2 presents a histogram for 200 observations on the metal recovery, or yield, from a smelting process. The histogram shows the central tendency, spread, and general shape of the distribution of the data. Recall that a histogram is constructed by dividing the horizontal axis into bins (usually of equal length) and drawing a rectangle over the jth bin with the area of the rectangle proportional to n_j, the number of observations that fall in that bin. The histogram is a large-sample tool. When the sample size is small the shape of the histogram can be very sensitive to the number of bins, the width of the bins, and the starting value for the first bin. Histograms should not be used with fewer than 75–100 observations.

The **box plot** (or **box-and-whisker plot**) is a very useful way to display data. A box plot displays the minimum, the maximum, the lower and upper quartiles (the 25th percentile and the 75th percentile, respectively), and the median (the 50th percentile) on a rectangular box aligned either horizontally or vertically. The box extends from the lower quartile to the

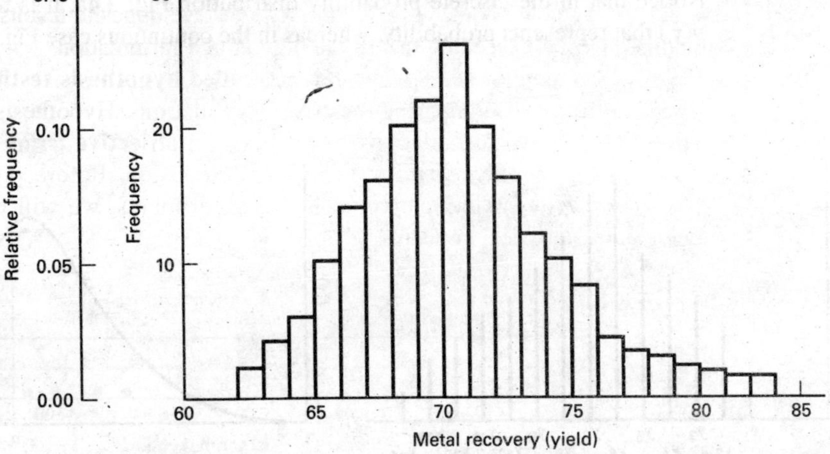

■ **FIGURE 2.2** **Histogram for 200 observations on metal recovery (yield) from a smelting process**

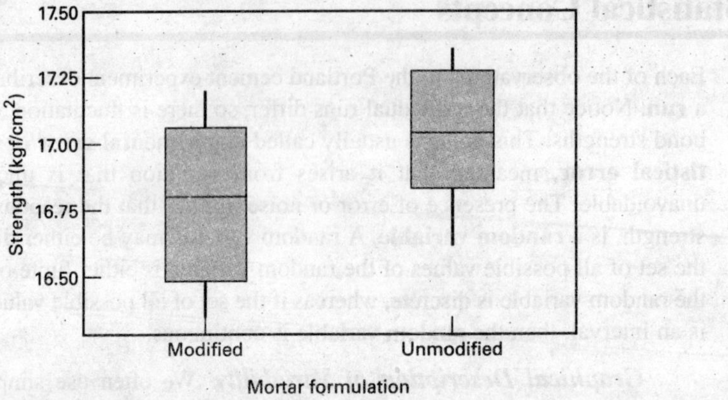

■ **FIGURE 2.3 Box plots for the Portland cement tension bond strength experiment**

upper quartile, and a line is drawn through the box at the median. Lines (or whiskers) extend from the ends of the box to (typically) the minimum and maximum values. [There are several variations of box plots that have different rules for denoting the extreme sample points. See Montgomery and Runger (2011) for more details.]

Figure 2.3 presents the box plots for the two samples of tension bond strength in the Portland cement mortar experiment. This display indicates some difference in mean strength between the two formulations. It also indicates that both formulations produce reasonably symmetric distributions of strength with similar variability or spread.

Dot diagrams, histograms, and box plots are useful for summarizing the information in a **sample** of data. To describe the observations that might occur in a sample more completely, we use the concept of the probability distribution.

Probability Distributions. The probability structure of a random variable, say y, is described by its **probability distribution**. If y is discrete, we often call the probability distribution of y, say $p(y)$, the probability mass function of y. If y is continuous, the probability distribution of y, say $f(y)$, is often called the probability density function for y.

Figure 2.4 illustrates hypothetical discrete and continuous probability distributions. Notice that in the discrete probability distribution Fig. 2.4a, it is the height of the function $p(y_j)$ that represents probability, whereas in the continuous case Fig. 2.4b, it is the area under

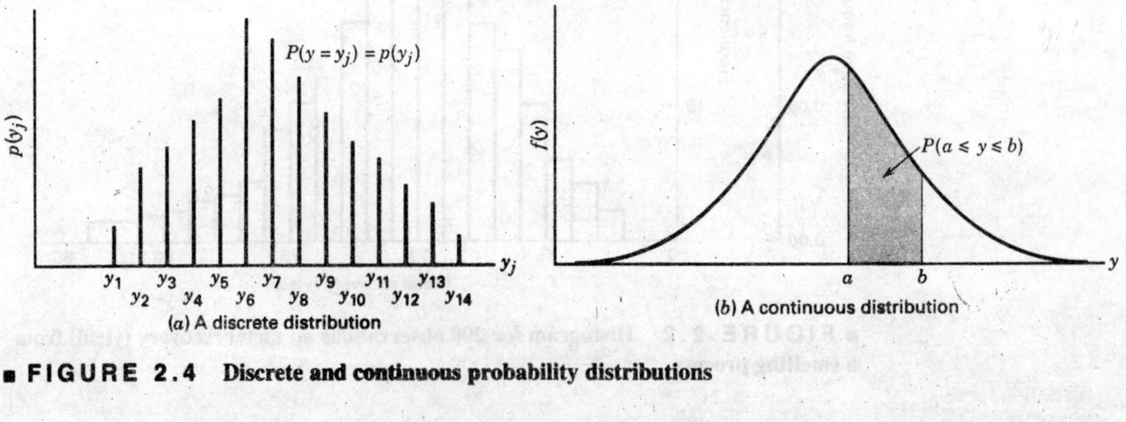

■ **FIGURE 2.4 Discrete and continuous probability distributions**

the curve $f(y)$ associated with a given interval that represents probability. The properties of probability distributions may be summarized quantitatively as follows:

$$y \text{ discrete:} \quad 0 \le p(y_j) \le 1 \quad \text{all values of } y_j$$

$$P(y = y_j) = p(y_j) \quad \text{all values of } y_j$$

$$\sum_{\substack{\text{all values}\\ \text{of } y_j}} p(y_j) = 1$$

$$y \text{ continuous:} \quad 0 \le f(y)$$

$$P(a \le y \le b) = \int_a^b f(y)\,dy$$

$$\int_{-\infty}^{\infty} f(y)\,dy = 1$$

Mean, Variance, and Expected Values. The **mean**, μ, of a probability distribution is a measure of its central tendency or location. Mathematically, we define the mean as

$$\mu = \begin{cases} \int_{-\infty}^{\infty} yf(y)\,dy & y \text{ continuous} \\ \sum_{\text{all } y} yp(y) & y \text{ discrete} \end{cases} \tag{2.1}$$

We may also express the mean in terms of the **expected value** or the long-run average value of the random variable y as

$$\mu = E(y) = \begin{cases} \int_{-\infty}^{\infty} yf(y)\,dy & y \text{ continuous} \\ \sum_{\text{all } y} yp(y) & y \text{ discrete} \end{cases} \tag{2.2}$$

where E denotes the **expected value operator**.

The variability or dispersion of a probability distribution can be measured by the **variance**, defined as

$$\sigma^2 = \begin{cases} \int_{-\infty}^{\infty} (y - \mu)^2 f(y)\,dy & y \text{ continuous} \\ \sum_{\text{all } y} (y - \mu)^2 p(y) & y \text{ discrete} \end{cases} \tag{2.3}$$

Note that the variance can be expressed entirely in terms of expectation because

$$\sigma^2 = E[(y - \mu)^2] \tag{2.4}$$

Finally, the variance is used so extensively that it is convenient to define a **variance operator** V such that

$$V(y) = E[(y - \mu)^2] = \sigma^2 \tag{2.5}$$

The concepts of expected value and variance are used extensively throughout this book, and it may be helpful to review several elementary results concerning these operators. If y is a random variable with mean μ and variance σ^2 and c is a constant, then

1. $E(c) = c$
2. $E(y) = \mu$

3. $E(cy) = cE(y) = c\mu$
4. $V(c) = 0$
5. $V(y) = \sigma^2$
6. $V(cy) = c^2V(y) = c^2\sigma^2$

If there are two random variables, say, y_1 with $E(y_1) = \mu_1$ and $V(y_1) = \sigma_1^2$ and y_2 with $E(y_2) = \mu_2$ and $V(y_2) = \sigma_2^2$, we have

7. $E(y_1 + y_2) = E(y_1) + E(y_2) = \mu_1 + \mu_2$

It is possible to show that

8. $V(y_1 + y_2) = V(y_1) + V(y_2) + 2\,\mathrm{Cov}(y_1, y_2)$

where

$$\mathrm{Cov}(y_1, y_2) = E\,[(y_1 - \mu_1)(y_2 - \mu_2)] \tag{2.6}$$

is the **covariance** of the random variables y_1 and y_2. The covariance is a measure of the linear association between y_1 and y_2. More specifically, we may show that if y_1 and y_2 are independent,[1] then $\mathrm{Cov}(y_1, y_2) = 0$. We may also show that

9. $V(y_1 - y_2) = V(y_1) + V(y_2) - 2\,\mathrm{Cov}(y_1, y_2)$

If y_1 and y_2 are **independent**, we have

10. $V(y_1 \pm y_2) = V(y_1) + V(y_2) = \sigma_1^2 + \sigma_2^2$

and

11. $E(y_1 \cdot y_2) = E(y_1) \cdot E(y_2) = \mu_1 \cdot \mu_2$

However, note that, in general

12. $E\left(\dfrac{y_1}{y_2}\right) \neq \dfrac{E(y_1)}{E(y_2)}$

regardless of whether or not y_1 and y_2 are independent.

2.3 Sampling and Sampling Distributions

Random Samples, Sample Mean, and Sample Variance. The objective of statistical inference is to draw conclusions about a population using a sample from that population. Most of the methods that we will study assume that **random samples** are used. A **random sample** is a sample that has been selected from the population in such a way that every possible sample has an equal probability of being selected. In practice, it is sometimes difficult to obtain random samples, and random numbers generated by a computer program may be helpful.

Statistical inference makes considerable use of quantities computed from the observations in the sample. We define a **statistic** as any function of the observations in a sample that

[1] Note that the converse of this is not necessarily so; that is, we may have $\mathrm{Cov}(y_1, y_2) = 0$ and yet this does not imply independence. For an example, see Hines et al. (2003).

does not contain unknown parameters. For example, suppose that $y_1, y_2, \ldots, y_n$ represents a sample. Then the **sample mean**

$$\bar{y} = \frac{\sum_{i=1}^{n} y_i}{n} \tag{2.7}$$

and the **sample variance**

$$S^2 = \frac{\sum_{i=1}^{n} (y_i - \bar{y})^2}{n - 1} \tag{2.8}$$

are both statistics. These quantities are measures of the central tendency and dispersion of the sample, respectively. Sometimes $S = \sqrt{S^2}$, called the **sample standard deviation**, is used as a measure of dispersion. Experimenters often prefer to use the standard deviation to measure dispersion because its units are the same as those for the variable of interest y.

Properties of the Sample Mean and Variance. The sample mean $\bar{y}$ is a point estimator of the population mean μ, and the sample variance S^2 is a point estimator of the population variance σ^2. In general, an **estimator** of an unknown parameter is a statistic that corresponds to that parameter. Note that a point estimator is a random variable. A particular numerical value of an estimator, computed from sample data, is called an **estimate**. For example, suppose we wish to estimate the mean and variance of the suspended solid material in the water of a lake. A random sample of $n = 25$ observation is tested, and the mg/l of suspended solid material is recorded for each. The sample mean and variance are computed according to Equations 2.7 and 2.8, respectively, and are $\bar{y} = 18.6$ and $S^2 = 1.20$. Therefore, the estimate of μ is $\bar{y} = 18.6$, and the estimate of σ^2 is $S^2 = 1.20$.

Several properties are required of good point estimators. Two of the most important are the following:

1. The point estimator should be **unbiased**. That is, the long-run average or expected value of the point estimator should be equal to the parameter that is being estimated. Although unbiasedness is desirable, this property alone does not always make an estimator a good one.
2. An unbiased estimator should have **minimum variance**. This property states that the minimum variance point estimator has a variance that is smaller than the variance of any other estimator of that parameter.

We may easily show that $\bar{y}$ and S^2 are unbiased estimators of μ and σ^2, respectively. First consider $\bar{y}$. Using the properties of expectation, we have

$$\begin{aligned} E(\bar{y}) &= E\left(\frac{\sum_{i=1}^{n} y_i}{n}\right) \\ &= \frac{1}{n}\sum_{i=1}^{n} E(y_i) \\ &= \frac{1}{n}\sum_{i=1}^{n} \mu \\ &= \mu \end{aligned}$$

because the expected value of each observation y_i is μ. Thus, $\bar{y}$ is an unbiased estimator of μ.

Now consider the sample variance S^2. We have

$$E(S^2) = E\left[\frac{\sum_{i=1}^{n} (y_i - \bar{y})^2}{n - 1}\right]$$

$$= \frac{1}{n-1} E\left[\sum_{i=1}^{n} (y_i - \bar{y})^2\right]$$

$$= \frac{1}{n-1} E(SS)$$

where $SS = \sum_{i=1}^{n}(y_i - \bar{y})^2$ is the **corrected sum of squares** of the observations y_i. Now

$$E(SS) = E\left[\sum_{i=1}^{n} (y_i - \bar{y})^2\right] \tag{2.9}$$

$$= E\left[\sum_{i=1}^{n} y_i^2 - n\bar{y}^2\right]$$

$$= \sum_{i=1}^{n} (\mu^2 + \sigma^2) - n(\mu^2 + \sigma^2/n)$$

$$= (n - 1)\sigma^2 \tag{2.10}$$

Therefore,

$$E(S^2) = \frac{1}{n-1} E(SS) = \sigma^2$$

and we see that S^2 is an unbiased estimator of σ^2.

Degrees of Freedom. The quantity $n - 1$ in Equation 2.10 is called the **number of degrees of freedom** of the sum of squares SS. This is a very general result; that is, if y is a random variable with variance σ^2 and $SS = \sum(y_i - \bar{y})^2$ has v degrees of freedom, then

$$E\left(\frac{SS}{v}\right) = \sigma^2 \tag{2.11}$$

The number of degrees of freedom of a sum of squares is equal to the number of independent elements in that sum of squares. For example, $SS = \sum_{i=1}^{n}(y_i - \bar{y})^2$ in Equation 2.9 consists of the sum of squares of the n elements $y_1 - \bar{y}, y_2 - \bar{y}, \ldots, y_n - \bar{y}$. These elements are not all independent because $\sum_{i=1}^{n}(y_i - \bar{y}) = 0$; in fact, only $n - 1$ of them are independent, implying that SS has $n - 1$ degrees of freedom.

The Normal and Other Sampling Distributions. Often we are able to determine the probability distribution of a particular statistic if we know the probability distribution of the population from which the sample was drawn. The probability distribution of a statistic is called a **sampling distribution**. We will now briefly discuss several useful sampling distributions.

One of the most important sampling distributions is the **normal distribution**. If y is a normal random variable, the probability distribution of y is

$$f(y) = \frac{1}{\sigma\sqrt{2\pi}} e^{-(1/2)[(y-\mu)/\sigma]^2} \qquad -\infty < y < \infty \tag{2.12}$$

where $-\infty < \mu < \infty$ is the mean of the distribution and $\sigma^2 > 0$ is the variance. The normal distribution is shown in Figure 2.5.

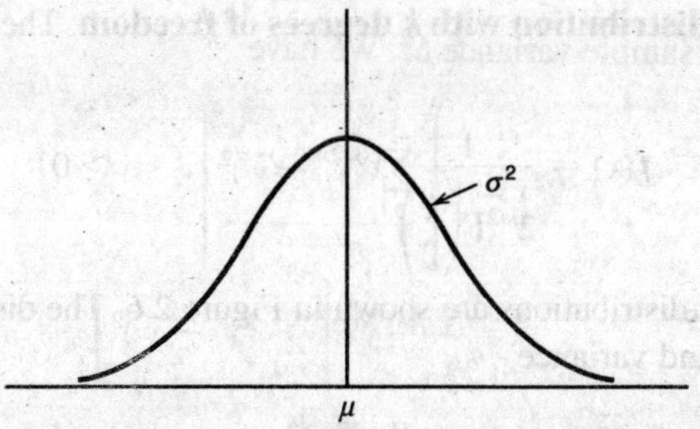

■ **FIGURE 2.5** **The normal distribution**

Because sample runs that differ as a result of experimental error often are well described by the normal distribution, the normal plays a central role in the analysis of data from designed experiments. Many important sampling distributions may also be defined in terms of normal random variables. We often use the notation $y \sim N(\mu, \sigma^2)$ to denote that y is distributed normally with mean μ and variance σ^2.

An important special case of the normal distribution is the **standard normal distribution**; that is, $\mu = 0$ and $\sigma^2 = 1$. We see that if $y \sim N(\mu, \sigma^2)$, the random variable

$$z = \frac{y - \mu}{\sigma} \tag{2.13}$$

follows the standard normal distribution, denoted $z \sim N(0, 1)$. The operation demonstrated in Equation 2.13 is often called **standardizing** the normal random variable y. The cumulative standard normal distribution is given in Table I of the Appendix.

Many statistical techniques assume that the random variable is normally distributed. The central limit theorem is often a justification of approximate normality.

THEOREM 2-1
The Central Limit Theorem

If $y_1, y_2, \ldots, y_n$ is a sequence of n independent and identically distributed random variables with $E(y_i) = \mu$ and $V(y_i) = \sigma^2$ (both finite) and $x = y_1 + y_2 + \cdots + y_n$, then the limiting form of the distribution of

$$z_n = \frac{x - n\mu}{\sqrt{n\sigma^2}}$$

as $n \to \infty$, is the standard normal distribution.

This result states essentially that the sum of n independent and identically distributed random variables is approximately normally distributed. In many cases, this approximation is good for very small n, say $n < 10$, whereas in other cases large n is required, say $n > 100$. Frequently, we think of the error in an experiment as arising in an additive manner from several independent sources; consequently, the normal distribution becomes a plausible model for the combined experimental error.

An important sampling distribution that can be defined in terms of normal random variables is the **chi-square** or **χ^2 distribution**. If $z_1, z_2, \ldots, z_k$ are normally and independently distributed random variables with mean 0 and variance 1, abbreviated NID(0, 1), then the random variable

$$x = z_1^2 + z_2^2 + \cdots + z_k^2$$

follows the **chi-square distribution with *k* degrees of freedom**. The density function of chi-square is

$$f(x) = \frac{1}{2^{k/2}\Gamma\left(\frac{k}{2}\right)} x^{(k/2)-1} e^{-x/2} \qquad x > 0 \tag{2.14}$$

Several chi-square distributions are shown in Figure 2.6. The distribution is asymmetric, or **skewed**, with mean and variance

$$\mu = k$$
$$\sigma^2 = 2k$$

respectively. Percentage points of the chi-square distribution are given in Table III of the Appendix.

As an example of a random variable that follows the chi-square distribution, suppose that $y_1, y_2, \ldots, y_n$ is a random sample from an $N(\mu, \sigma^2)$ distribution. Then

$$\frac{SS}{\sigma^2} = \frac{\sum_{i=1}^{n} (y_i - \bar{y})^2}{\sigma^2} \sim \chi^2_{n-1} \tag{2.15}$$

That is, SS/σ^2 is distributed as chi-square with $n - 1$ degrees of freedom.

Many of the techniques used in this book involve the computation and manipulation of sums of squares. The result given in Equation 2.15 is extremely important and occurs repeatedly; a sum of squares in normal random variables when divided by σ^2 follows the chi-square distribution.

Examining Equation 2.8, we see that the sample variance can be written as

$$S^2 = \frac{SS}{n-1} \tag{2.16}$$

If the observations in the sample are NID(μ, σ^2), then the distribution of S^2 is $[\sigma^2/(n-1)]\chi^2_{n-1}$. Thus, the sampling distribution of the sample variance is a constant times the chi-square distribution if the population is normally distributed.

If z and χ^2_k are independent standard normal and chi-square random variables, respectively, the random variable

$$t_k = \frac{z}{\sqrt{\chi^2_k/k}} \tag{2.17}$$

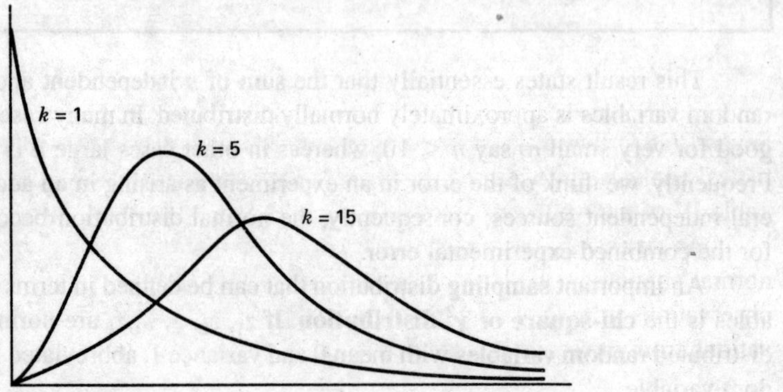

■ **FIGURE 2.6** Several Chi-square distributions

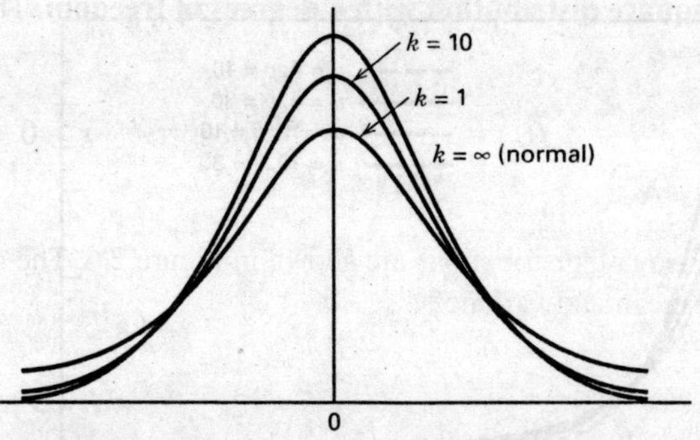

■ **FIGURE 2.7** Several t distributions

follows the ***t* distribution with *k* degrees of freedom**, denoted t_k. The density function of t is

$$f(t) = \frac{\Gamma[(k+1)/2]}{\sqrt{k\pi}\,\Gamma(k/2)} \frac{1}{[(t^2/k)+1]^{(k+1)/2}} \qquad -\infty < t < \infty \tag{2.18}$$

and the mean and variance of t are $\mu = 0$ and $\sigma^2 = k/(k-2)$ for $k > 2$, respectively. Several t distributions are shown in Figure 2.7. Note that if $k = \infty$, the t distribution becomes the standard normal distribution. The percentage points of the t distribution are given in Table II of the Appendix. If $y_1, y_2, \ldots, y_n$ is a random sample from the $N(\mu, \sigma^2)$ distribution, then the quantity

$$t = \frac{\bar{y} - \mu}{S/\sqrt{n}} \tag{2.19}$$

is distributed as t with $n - 1$ degrees of freedom.

The final sampling distribution that we will consider is the ***F* distribution.** If χ_u^2 and χ_v^2 are two independent chi-square random variables with u and v degrees of freedom, respectively, then the ratio

$$F_{u,v} = \frac{\chi_u^2/u}{\chi_v^2/v} \tag{2.20}$$

follows the ***F* distribution with *u numerator* degrees of freedom and *v denominator* degrees of freedom**. If x is an F random variable with u numerator and v denominator degrees of freedom, then the probability distribution of x is

$$h(x) = \frac{\Gamma\left(\frac{u+v}{2}\right)\left(\frac{u}{v}\right)^{u/2} x^{(u/2)-1}}{\Gamma\left(\frac{u}{x}\right)\Gamma\left(\frac{v}{2}\right)\left[\left(\frac{u}{v}\right)x + 1\right]^{(u+v)/2}} \qquad 0 < x < \infty \tag{2.21}$$

Several F distributions are shown in Figure 2.8. This distribution is very important in the statistical analysis of designed experiments. Percentage points of the F distribution are given in Table IV of the Appendix.

As an example of a statistic that is distributed as F, suppose we have two independent normal populations with common variance σ^2. If $y_{11}, y_{12}, \ldots, y_{1n_1}$ is a random sample of n_1 observations from the first population, and if $y_{21}, y_{22}, \ldots, y_{2n_2}$ is a random sample of n_2 observations from the second, then

$$\frac{S_1^2}{S_2^2} \sim F_{n_1-1,\, n_2-1} \tag{2.22}$$

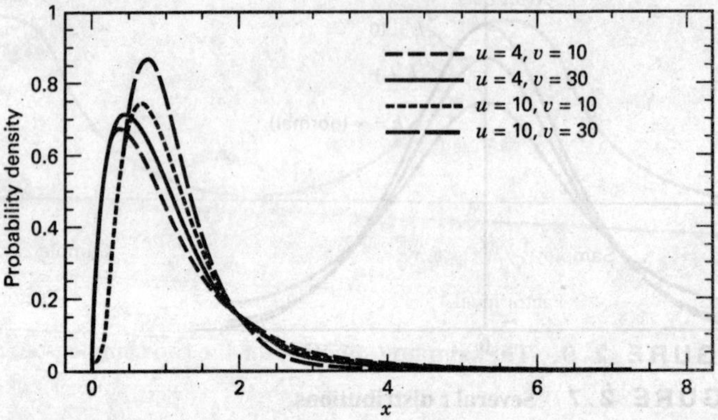

■ **FIGURE 2.8** **Several *F* distributions**

where S_1^2 and S_2^2 are the two sample variances. This result follows directly from Equations 2.15 and 2.20.

2.4 Inferences About the Differences in Means, Randomized Designs

We are now ready to return to the Portland cement mortar problem posed in Section 2.1. Recall that two different formulations of mortar were being investigated to determine if they differ in tension bond strength. In this section we discuss how the data from this simple comparative experiment can be analyzed using **hypothesis testing** and **confidence interval** procedures for comparing two treatment means.

Throughout this section we assume that a **completely randomized experimental design** is used. In such a design, the data are usually viewed as if they were a random sample from a normal distribution.

2.4.1 Hypothesis Testing

We now reconsider the Portland cement experiment introduced in Section 2.1. Recall that we are interested in comparing the strength of two different formulations: an unmodified mortar and a modified mortar. In general, we can think of these two formulations as two **levels of the factor** "formulations." Let $y_{11}, y_{12}, \ldots, y_{1n_1}$ represent the n_1 observations from the first factor level and $y_{21}, y_{22}, \ldots, y_{2n_2}$ represent the n_2 observations from the second factor level. We assume that the samples are drawn at random from two independent normal populations. Figure 2.9 illustrates the situation.

A Model for the Data. We often describe the results of an experiment with a **model**. A simple statistical model that describes the data from an experiment such as we have just described is

$$y_{ij} = \mu_i + \epsilon_{ij}\begin{cases} i = 1, 2 \\ j = 1, 2, \ldots, n_i \end{cases} \tag{2.23}$$

where y_{ij} is the jth observation from factor level i, μ_i is the mean of the response at the ith factor level, and ϵ_{ij} is a normal random variable associated with the ijth observation. We assume

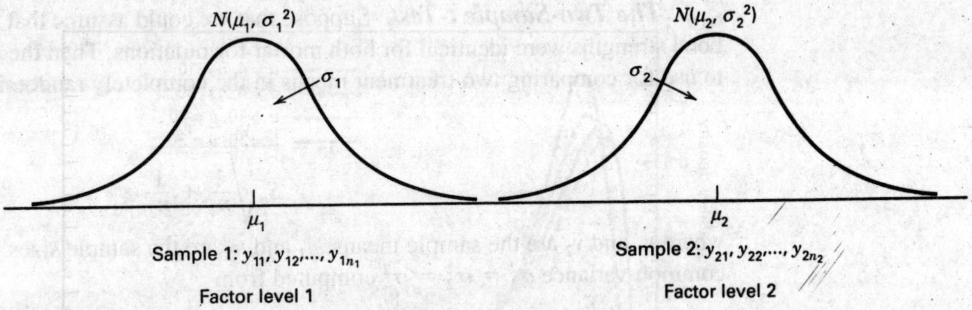

■ **FIGURE 2.9 The sampling situation for the two-sample *t*-test**

that ϵ_{ij} are NID$(0, \sigma_i^2)$, $i = 1, 2$. It is customary to refer to ϵ_{ij} as the **random error** component of the model. Because the means μ_1 and μ_2 are constants, we see directly from the model that y_{ij} are NID(μ_i, σ_i^2), $i = 1, 2$, just as we previously assumed. For more information about models for the data, refer to the supplemental text material.

Statistical Hypotheses. A **statistical hypothesis** is a statement either about the parameters of a probability distribution or the parameters of a model. The hypothesis reflects some **conjecture** about the problem situation. For example, in the Portland cement experiment, we may think that the mean tension bond strengths of the two mortar formulations are equal. This may be stated formally as

$$H_0: \mu_1 = \mu_2$$
$$H_1: \mu_1 \neq \mu_2$$

where μ_1 is the mean tension bond strength of the modified mortar and μ_2 is the mean tension bond strength of the unmodified mortar. The statement $H_0: \mu_1 = \mu_2$ is called the **null hypothesis** and $H_1: \mu_1 \neq \mu_2$ is called the **alternative hypothesis**. The alternative hypothesis specified here is called a **two-sided alternative hypothesis** because it would be true if $\mu_1 < \mu_2$ or if $\mu_1 > \mu_2$.

To test a hypothesis, we devise a procedure for taking a random sample, computing an appropriate **test statistic**, and then rejecting or failing to reject the null hypothesis H_0 based on the computed value of the test statistic. Part of this procedure is specifying the set of values for the test statistic that leads to rejection of H_0. This set of values is called the **critical region** or **rejection region** for the test.

Two kinds of errors may be committed when testing hypotheses. If the null hypothesis is rejected when it is true, a **type I** error has occurred. If the null hypothesis is *not* rejected when it is false, a **type II error** has been made. The probabilities of these two errors are given special symbols

$$\alpha = P(\text{type I error}) = P(\text{reject } H_0 | H_0 \text{ is true})$$
$$\beta = P(\text{type II error}) = P(\text{fail to reject } H_0 | H_0 \text{ is false})$$

Sometimes it is more convenient to work with the **power** of the test, where

$$\text{Power} = 1 - \beta = P(\text{reject } H_0 | H_0 \text{ is false})$$

The general procedure in hypothesis testing is to specify a value of the probability of type I error α, often called the **significance level** of the test, and then design the test procedure so that the probability of type II error β has a suitably small value.

The Two-Sample t-Test. Suppose that we could assume that the variances of tension bond strengths were identical for both mortar formulations. Then the appropriate test statistic to use for comparing two treatment means in the completely randomized design is

$$t_0 = \frac{\bar{y}_1 - \bar{y}_2}{S_p\sqrt{\frac{1}{n_1} + \frac{1}{n_2}}} \tag{2.24}$$

where $\bar{y}_1$ and $\bar{y}_2$ are the sample means, n_1 and n_2 are the sample sizes, S_p^2 is an estimate of the common variance $\sigma_1^2 = \sigma_2^2 = \sigma^2$ computed from

$$S_p^2 = \frac{(n_1 - 1)S_1^2 + (n_2 - 1)S_2^2}{n_1 + n_2 - 2} \tag{2.25}$$

and S_1^2 and S_2^2 are the two individual sample variances. The quality $S_p\sqrt{\frac{1}{n_1} + \frac{1}{n_2}}$ in the denominator of Equation 2.24 is often called the **standard error** of the difference in means in the numerator, abbreviated $se\,(\bar{y}_1 - \bar{y}_2)$. To determine whether to reject $H_0 : \mu_1 = \mu_2$, we would compare t_0 to the t distribution with $n_1 + n_2 - 2$ degrees of freedom. If $|t_0| > t_{\alpha/2, n_1+n_2-2}$, where $t_{\alpha/2, n_1+n_2-2}$ is the upper $\alpha/2$ percentage point of the t distribution with $n_1 + n_2 - 2$ degrees of freedom, we would *reject* H_0 and conclude that the mean strengths of the two formulations of Portland cement mortar differ. This test procedure is usually called the **two-sample *t*-test.**

This procedure may be justified as follows. If we are sampling from independent normal distributions, then the distribution of $\bar{y}_1 - \bar{y}_2$ is $N[\mu_1 - \mu_2, \sigma^2(1/n_1 + 1/n_2)]$. Thus, if σ^2 were known, and if $H_0 : \mu_1 = \mu_2$ were true, the distribution of

$$Z_0 = \frac{\bar{y}_1 - \bar{y}_2}{\sigma\sqrt{\frac{1}{n_1} + \frac{1}{n_2}}} \tag{2.26}$$

would be $N(0, 1)$. However, in replacing σ in Equation 2.26 by S_p, the distribution of Z_0 changes from standard normal to t with $n_1 + n_2 - 2$ degrees of freedom. Now if H_0 is true, t_0 in Equation 2.24 is distributed as $t_{n_1+n_2-2}$ and, consequently, we would expect $100(1 - \alpha)$ percent of the values of t_0 to fall between $-t_{\alpha/2, n_1+n_2-2}$ and $t_{\alpha/2, n_1+n_2-2}$. A sample producing a value of t_0 outside these limits would be unusual if the null hypothesis were true and is evidence that H_0 should be rejected. Thus the t distribution with $n_1 + n_2 - 2$ degrees of freedom is the appropriate **reference distribution** for the test statistic t_0. That is, it describes the behavior of t_0 when the null hypothesis is true. Note that α is the probability of type I error for the test. Sometimes α is called the **significance level** of the test.

In some problems, one may wish to reject H_0 only if one mean is larger than the other. Thus, one would specify a **one-sided alternative hypothesis** $H_1 : \mu_1 > \mu_2$ and would reject H_0 only if $t_0 > t_{\alpha, n_1+n_2-2}$. If one wants to reject H_0 only if μ_1 is less than μ_2, then the alternative hypothesis is $H_1 : \mu_1 < \mu_2$, and one would reject H_0 if $t_0 < -t_{\alpha, n_1+n_2-2}$.

To illustrate the procedure, consider the Portland cement data in Table 2.1. For these data, we find that

Modified Mortar	**Unmodified Mortar**
$\bar{y}_1 = 16.76$ kgf/cm^2	$\bar{y}_2 = 17.04$ kgf/cm^2
$S_1^2 = 0.100$	$S_2^2 = 0.061$
$S_1 = 0.316$	$S_2 = 0.248$
$n_1 = 10$	$n_2 = 10$

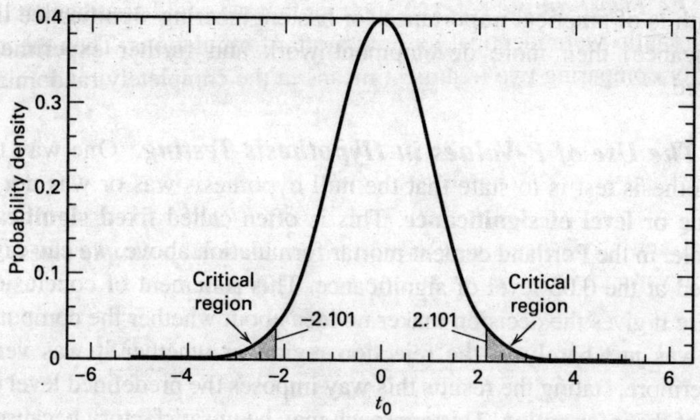

■ **FIGURE 2.10** **The t distribution with 18 degrees of freedom with the critical region $\pm t_{0.025,18} = \pm 2.101$**

Because the sample standard deviations are reasonably similar, it is not unreasonable to conclude that the population standard deviations (or variances) are equal. Therefore, we can use Equation 2.24 to test the hypotheses

$$H_0: \mu_1 = \mu_2$$
$$H_1: \mu_1 \neq \mu_2$$

Furthermore, $n_1 + n_2 - 2 = 10 + 10 - 2 = 18$, and if we choose $\alpha = 0.05$, then we would reject $H_0: \mu_1 = \mu_2$ if the numerical value of the test statistic $t_0 > t_{0.025,18} = 2.101$, or if $t_0 < -t_{0.025,18} = -2.101$. These boundaries of the critical region are shown on the reference distribution (t with 18 degrees of freedom) in Figure 2.10.

Using Equation 2.25 we find that

$$S_p^2 = \frac{(n_1 - 1)S_1^2 + (n_2 - 1)S_2^2}{n_1 + n_2 - 2} = \frac{9(0.100) + 9(0.061)}{10 + 10 - 2} = 0.081$$
$$S_p = 0.284$$

and the test statistic is

$$t_0 = \frac{\bar{y}_1 - \bar{y}_2}{S_p\sqrt{\frac{1}{n_1} + \frac{1}{n_2}}} = \frac{16.76 - 17.04}{0.284\sqrt{\frac{1}{10} + \frac{1}{10}}} = \frac{-0.28}{0.127} = -2.20$$

Because $t_0 = -2.20 < -t_{0.025,18} = -2.101$, we would reject H_0 and conclude that the mean tension bond strengths of the two formulations of Portland cement mortar are different. This is a potentially important engineering finding. The change in mortar formulation had the desired effect of reducing the cure time, but there is evidence that the change also affected the tension bond strength. One can conclude that the modified formulation reduces the bond strength (just because we conducted a two-sided test, this does not preclude drawing a one-sided conclusion when the null hypothesis is rejected). If the reduction in mean bond strength is of practical importance (or has engineering significance in addition to statistical significance) then more development work and further experimentation will likely be required.

The Use of P-Values in Hypothesis Testing. One way to report the results of a hypothesis test is to state that the null hypothesis was or was not rejected at a specified α-value or **level of significance**. This is often called **fixed significance level testing.** For example, in the Portland cement mortar formulation above, we can say that $H_0 : \mu_1 = \mu_2$ was rejected at the 0.05 level of significance. This statement of conclusions is often inadequate because it gives the decision maker no idea about whether the computed value of the test statistic was just barely in the rejection region or whether it was very far into this region. Furthermore, stating the results this way imposes the predefined level of significance on other users of the information. This approach may be unsatisfactory because some decision makers might be uncomfortable with the risks implied by $\alpha = 0.05$.

To avoid these difficulties, the ***P*-value approach** has been adopted widely in practice. The *P*-value is the probability that the test statistic will take on a value that is at least as extreme as the observed value of the statistic when the null hypothesis H_0 is true. Thus, a *P*-value conveys much information about the weight of evidence against H_0, and so a decision maker can draw a conclusion at *any* specified level of significance. More formally, we define the ***P*-value** as the smallest level of significance that would lead to rejection of the null hypothesis H_0.

It is customary to call the test statistic (and the data) significant when the null hypothesis H_0 is rejected; therefore, we may think of the *P*-value as the smallest level α at which the data are significant. Once the *P*-value is known, the decision maker can determine how significant the data are without the data analyst formally imposing a preselected level of significance.

It is not always easy to compute the exact *P*-value for a test. However, most modern computer programs for statistical analysis report *P*-values, and they can be obtained on some handheld calculators. We will show how to approximate the *P*-value for the Portland cement mortar experiment. Because $|t_0| = 2.20 > t_{0.025,18} = 2.101$, we know that the *P*-value is less than 0.05. From Appendix Table II, for a *t* distribution with 18 degrees of freedom, and tail area probability 0.01 we find $t_{0.01,18} = 2.552$. Now $|t_0| = 2.20 < 2.552$, so because the alternative hypothesis is two sided, we know that the *P*-value must be between 0.05 and 2(0.01) = 0.02. Some handheld calculators have the capability to calculate *P*-values. One such calculator is the HP-48. From this calculator, we obtain the *P*-value for the value $t_0 = -2.20$ in the Portland cement mortar formulation experiment as $P = 0.0411$. Thus the null hypothesis $H_0 : \mu_1 = \mu_2$ would be rejected at any level of significance $\alpha >$ 0.0411.

Computer Solution. Many statistical software packages have capability for statistical hypothesis testing. The output from both the Minitab and the JMP two-sample *t*-test procedure applied to the Portland cement mortar formulation experiment is shown in Table 2.2. Notice that the output includes some summary statistics about the two samples (the abbreviation "SE mean" in the Minitab section of the table refers to the standard error of the mean, $s/\sqrt{n}$) as well as some information about confidence intervals on the difference in the two means (which we will discuss in the next section). The programs also test the hypothesis of interest, allowing the analyst to specify the nature of the alternative hypothesis ("not =" in the Minitab output implies $H_1 : \mu_1 \neq \mu_2$).

The output includes the computed value of t_0, the value of the test statistic t_0 (JMP reports a positive value of t_0 because of how the sample means are subtracted in the numerator

■ **TABLE 2.2**
Computer Output for the Two-Sample *t*-Test

```
Minitab
Two-sample T for Modified vs Unmodified

                 N        Mean       Std. Dev.        SE Mean
Modified        10      16.764          0.316           0.10
Unmodified      10      17.042          0.248           0.078

Difference = mu (Modified) – mu (Unmodified)
Estimate for difference: -0.278000
95% CI for difference: (-0.545073, -0.010927)
T-Test of difference = 0 (vs not = ): T-Value = -2.19
P-Value = 0.042  DF = 18
Both use Pooled Std. Dev. = 0.2843

JMP t-test
Unmodified-Modified
Assuming equal variances

Difference     0.278000  t Ratio   2.186876
Std Err Dif    0.127122  DF              18
Upper CL Dif   0.545073  Prob> |t|   0.0422
Lower CL Dif   0.010927  Prob> t     0.0211
Confidence         0.95  Prob< t     0.9789
```

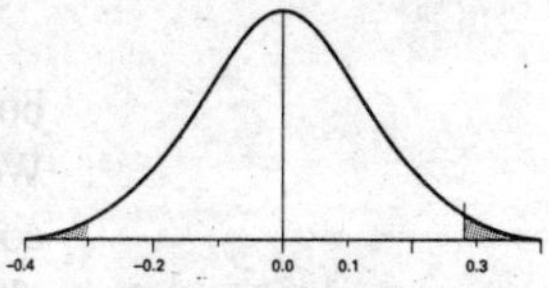

of the test statistic), and the *P*-value. Notice that the computed value of the *t* statistic differs slightly from our manually calculated value and that the *P*-value is reported to be $P = 0.042$. JMP also reports the *P*-values for the one-sided alternative hypothesis. Many software packages will not report an actual *P*-value less than some predetermined value such as 0.0001 and instead will return a "default" value such as "<0.001" or in some cases, zero.

Checking Assumptions in the t-Test. In using the *t*-test procedure we make the assumptions that both samples are random samples that are drawn from independent populations that can be described by a normal distribution, and that the standard deviation or variances of both populations are equal. The assumption of independence is critical, and if the run order is randomized (and, if appropriate, other experimental units and materials are selected at random), this assumption will usually be satisfied. The equal variance and normality assumptions are easy to check using a **normal probability plot.**

Generally, probability plotting is a graphical technique for determining whether sample data conform to a hypothesized distribution based on a subjective visual examination of the data. The general procedure is very simple and can be performed quickly with most statistics software packages. The **supplemental text material** discusses manual construction of normal probability plots.

To construct a probability plot, the observations in the sample are first ranked from smallest to largest. That is, the sample $y_1, y_2, \ldots, y_n$ is arranged as $y_{(1)}, y_{(2)}, \ldots, y_{(n)}$ where $y_{(1)}$ is the smallest observation, $y_{(2)}$ is the second smallest observation, and so forth, with $y_{(n)}$ the largest. The ordered observations $y_{(j)}$ are then plotted against their observed cumulative frequency $(j - 0.5)/n$.

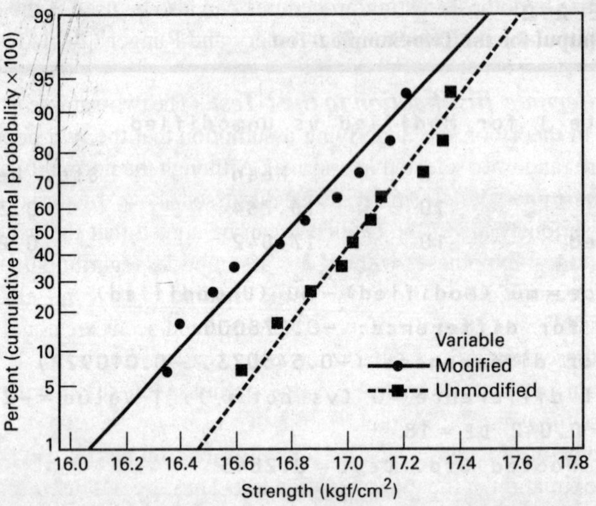

■ **FIGURE 2.11** **Normal probability plots of tension bond strength in the Portland cement experiment**

The cumulative frequency scale has been arranged so that if the hypothesized distribution adequately describes the data, the plotted points will fall approximately along a straight line; if the plotted points deviate significantly from a straight line, the hypothesized model is not appropriate. Usually, the determination of whether or not the data plot as a straight line is subjective.

To illustrate the procedure, suppose that we wish to check the assumption that tension bond strength in the Portland cement mortar formulation experiment is normally distributed. We initially consider only the observations from the unmodified mortar formulation. A computer-generated normal probability plot is shown in Figure 2.11. Most normal probability plots present $100(j - 0.5)/n$ on the left vertical scale (and occasionally $100[1 - (j - 0.5)/n]$ is plotted on the right vertical scale), with the variable value plotted on the horizontal scale. Some computer-generated normal probability plots convert the cumulative frequency to a standard normal z score. A straight line, chosen subjectively, has been drawn through the plotted points. In drawing the straight line, you should be influenced more by the points near the middle of the plot than by the extreme points. A good rule of thumb is to draw the line approximately between the 25th and 75th percentile points. This is how the lines in Figure 2.11 for each sample were determined. In assessing the "closeness" of the points to the straight line, imagine a fat pencil lying along the line. If all the points are covered by this imaginary pencil, a normal distribution adequately describes the data. Because the points for each sample in Figure 2.11 would pass the fat pencil test, we conclude that the normal distribution is an appropriate model for tension bond strength for both the modified and the unmodified mortar.

We can obtain an estimate of the mean and standard deviation directly from the normal probability plot. The mean is estimated as the 50th percentile on the probability plot, and the standard deviation is estimated as the difference between the 84th and 50th percentiles. This means that we can verify the assumption of equal population variances in the Portland cement experiment by simply comparing the slopes of the two straight lines in Figure 2.11. Both lines have very similar slopes, and so the assumption of equal variances is a reasonable one. If this assumption is violated, you should use the version of the t-test described in Section 2.4.4. The supplemental text material has more information about checking assumptions on the t-test.

When assumptions are badly violated, the performance of the t-test will be affected. Generally, small to moderate violations of assumptions are not a major concern, but *any* failure of the independence assumption and strong indications of nonnormality should not be ignored. Both the significance level of the test and the ability to detect differences between the means will be adversely affected by departures from assumptions. **Transformations** are one approach to dealing with this problem. We will discuss this in more detail in Chapter 3.

Nonparametric hypothesis testing procedures can also be used if the observations come from nonnormal populations. Refer to Montgomery and Runger (2011) for more details.

An Alternate Justification to the t-Test. The two-sample t-test we have just presented depends in theory on the underlying assumption that the two populations from which the samples were randomly selected are normal. Although the normality assumption is required to develop the test procedure formally, as we discussed above, moderate departures from normality will not seriously affect the results. It can be argued that the use of a randomized design enables one to test hypotheses without *any* assumptions regarding the form of the distribution. Briefly, the reasoning is as follows. If the treatments have no effect, all [20!/(10!10!)] = 184,756 possible ways that the 20 observations could occur are equally likely. Corresponding to each of these 184,756 possible arrangements is a value of t_0. If the value of t_0 actually obtained from the data is unusually large or unusually small with reference to the set of 184,756 possible values, it is an indication that $\mu_1 \neq \mu_2$.

This type of procedure is called a **randomization test**. It can be shown that the t-test is a good approximation of the randomization test. Thus, we will use t-tests (and other procedures that can be regarded as approximations of randomization tests) without extensive concern about the assumption of normality. This is one reason a simple procedure such as normal probability plotting is adequate to check the assumption of normality.

2.4.2 Confidence Intervals

Although hypothesis testing is a useful procedure, it sometimes does not tell the entire story. It is often preferable to provide an interval within which the value of the parameter or parameters in question would be expected to lie. These interval statements are called **confidence intervals**. In many engineering and industrial experiments, the experimenter already knows that the means μ_1 and μ_2 differ; consequently, hypothesis testing on $\mu_1 = \mu_2$ is of little interest. The experimenter would usually be more interested in knowing how much the means differ. A confidence interval on the difference in means $\mu_1 - \mu_2$ is used in answering this question.

To define a confidence interval, suppose that θ is an unknown parameter. To obtain an interval estimate of θ, we need to find two statistics L and U such that the probability statement

$$P(L \leqslant \theta \leqslant U) = 1 - \alpha \tag{2.27}$$

is true. The interval

$$L \leqslant \theta \leqslant U \tag{2.28}$$

is called a **100(1 − α) percent confidence interval** for the parameter θ. The interpretation of this interval is that if, in repeated random samplings, a large number of such intervals are constructed, 100(1 − α) percent of them will contain the true value of θ. The statistics L and U are called the **lower** and **upper confidence limits**, respectively, and $1 - \alpha$ is called the **confidence coefficient**. If $\alpha = 0.05$, Equation 2.28 is called a 95 percent confidence interval for θ. Note that confidence intervals have a frequency interpretation; that is, we do not know if the statement is true for this specific sample, but we do know that the *method* used to produce the confidence interval yields correct statements 100(1 − α) percent of the time.

Suppose that we wish to find a 100(1 − α) percent confidence interval on the true difference in means $\mu_1 - \mu_2$ for the Portland cement problem. The interval can be derived in the following way. The statistic

$$\frac{\bar{y}_1 - \bar{y}_2 - (\mu_1 - \mu_2)}{S_p\sqrt{\dfrac{1}{n_1} + \dfrac{1}{n_2}}}$$

is distributed as $t_{n_1+n_2-2}$. Thus,

$$P\left(-t_{\alpha/2,n_1+n_2-2} \le \frac{\bar{y}_1 - \bar{y}_2 - (\mu_1 - \mu_2)}{S_p\sqrt{\frac{1}{n_1} + \frac{1}{n_2}}} \le t_{\alpha/2,n_1+n_2-2}\right) = 1 - \alpha$$

or

$$P\left(\bar{y}_1 - \bar{y}_2 - t_{\alpha/2,n_1+n_2-2}\, S_p\sqrt{\frac{1}{n_1} + \frac{1}{n_2}} \le \mu_1 - \mu_2 \le \bar{y}_1 - \bar{y}_2 + t_{\alpha/2,n_1+n_2-2}\, S_p\sqrt{\frac{1}{n_1} + \frac{1}{n_2}}\right) = 1 - \alpha \quad (2.29)$$

Comparing Equations 2.29 and 2.27, we see that

$$\bar{y}_1 - \bar{y}_2 - t_{\alpha/2,n_1+n_2-2}\, S_p\sqrt{\frac{1}{n_1} + \frac{1}{n_2}} \le \mu_1 - \mu_2 \le \bar{y}_1 - \bar{y}_2 + t_{\alpha/2,n_1+n_2-2}\, S_p\sqrt{\frac{1}{n_1} + \frac{1}{n_2}} \quad (2.30)$$

is a $100(1 - \alpha)$ percent confidence interval for $\mu_1 - \mu_2$.

The actual 95 percent confidence interval estimate for the difference in mean tension bond strength for the formulations of Portland cement mortar is found by substituting in Equation 2.30 as follows:

$$16.76 - 17.04 - (2.101)0.284\sqrt{\tfrac{1}{10} + \tfrac{1}{10}} \le \mu_1 - \mu_2 \le 16.76 - 17.04 + (2.101)0.284\sqrt{\tfrac{1}{10} + \tfrac{1}{10}}$$

$$-0.28 - 0.27 \le \mu_1 - \mu_2 \le -0.28 + 0.27$$

$$-0.55 \le \mu_1 - \mu_2 \le -0.01$$

Thus, the 95 percent confidence interval estimate on the difference in means extends from -0.55 to -0.01 kgf/cm^2. Put another way, the confidence interval is $\mu_1 - \mu_2 = -0.28 \pm 0.27$ kgf/cm^2, or the difference in mean strengths is -0.28 kgf/cm^2, and the accuracy of this estimate is ± 0.27 kgf/cm^2. Note that because $\mu_1 - \mu_2 = 0$ is *not* included in this interval, the data do not support the hypothesis that $\mu_1 = \mu_2$ at the 5 percent level of significance (recall that the *P*-value for the two-sample *t*-test was 0.042, just slightly less than 0.05). It is likely that the mean strength of the unmodified formulation exceeds the mean strength of the modified formulation. Notice from Table 2.2 that both Minitab and JMP reported this confidence interval when the hypothesis testing procedure was conducted.

2.4.3 Choice of Sample Size

Selection of an appropriate sample size is one of the most important parts of any experimental design problem. One way to do this is to consider the impact of sample size on the estimate of the difference in two means. From Equation 2.30 we know that the $100(1 - \alpha)\%$ confidence interval on the difference in two means is a measure of the precision of estimation of the difference in the two means. The length of this interval is determined by

$$t_{\alpha/2,\, n_1+n_2-2}\, S_p\sqrt{\frac{1}{n_1} + \frac{1}{n_2}}$$

We consider the case where the sample sizes from the two populations are equal, so that $n_1 = n_2 = n$. Then the length of the CI is determined by

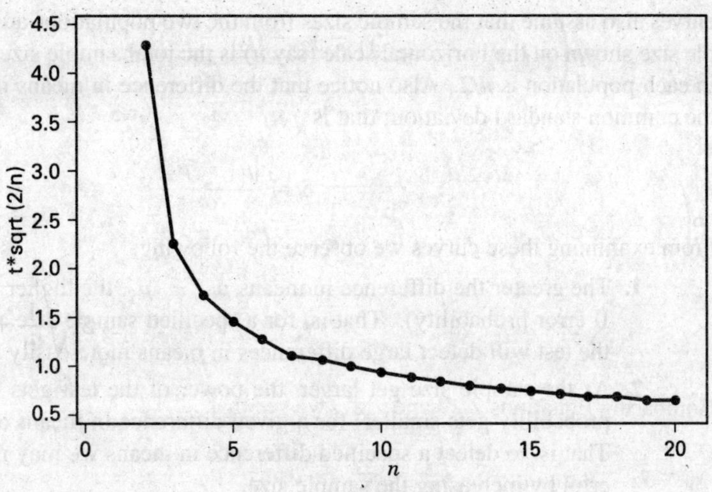

■ **FIGURE 2.12** **Plot of $t_{\alpha/2,\,2n-2}\sqrt{2/n}$ versus sample size in each population n for $\alpha = 0.05$**

$$t_{\alpha/2,\,2n-2}\,S_p\sqrt{\frac{2}{n}}$$

Consequently the precision with which the difference in the two means is estimated depends on two quantities—S_p, over which we have no control, and $t_{\alpha/2,\,2n-2}\sqrt{2/n}$, which we can control by choosing the sample size n. Figure 2.12 is a plot of $t_{\alpha/2,\,2n-2}\sqrt{2/n}$ versus n for $\alpha = 0.05$. Notice that the curve descends rapidly as n increases up to about $n = 10$ and less rapidly beyond that. Since S_p is relatively constant and $t_{\alpha/2,\,2n-2}\sqrt{2/n}$ isn't going to change much for sample sizes beyond $n = 10$ or 12, we can conclude that choosing a sample size of $n = 10$ or 12 from each population in a two-sample 95% CI will result in a CI that results in about the best precision of estimation for the difference in the two means that is possible given the amount of inherent variability that is present in the two populations.

We can also use a hypothesis testing framework to determine sample size. The choice of sample size and the probability of type II error β are closely connected. Suppose that we are testing the hypotheses

$$H_0\colon \mu_1 = \mu_2$$
$$H_1\colon \mu_1 \neq \mu_2$$

and that the means are *not* equal so that $\delta = \mu_1 - \mu_2$. Because $H_0\colon \mu_1 = \mu_2$ is not true, we are concerned about wrongly failing to reject H_0. The probability of type II error depends on the true difference in means δ. A graph of β versus δ for a particular sample size is called the **operating characteristic curve**, or **O.C. curve** for the test. The β error is also a function of sample size. Generally, for a given value of δ, the β error decreases as the sample size increases. That is, a specified difference in means is easier to detect for larger sample sizes than for smaller ones.

An alternative to the OC curve is a **power curve**, which typically plots power or $1 - \beta$, versus sample size for a specified difference in the means. Some software packages perform power analysis and will plot power curves. A set of power curves constructed using JMP for the hypotheses

$$H_0\colon \mu_1 = \mu_2$$
$$H_1\colon \mu_1 \neq \mu_2$$

is shown in **Figure 2.13** for the case where the two population variances σ_1^2 and σ_2^2 are unknown but equal ($\sigma_1^2 = \sigma_2^2 = \sigma^2$) and for a level of significance of $\alpha = 0.05$. These power curves also assume that the sample sizes from the two populations are equal and that the sample size shown on the horizontal scale (say n) is the total sample size, so that the sample size in each population is $n/2$. Also notice that the difference in means is expressed as a ratio to the common standard deviation; that is

$$\delta = \frac{|\mu_1 - \mu_2|}{\sigma}$$

From examining these curves we observe the following:

1. The greater the difference in means $\mu_1 - \mu_2$, the higher the power (smaller type II error probability). That is, for a specified sample size and significance level α, the test will detect large differences in means more easily than small ones.
2. As the sample size get larger, the power of the test gets larger (the type II error probability gets smaller) for a given difference in means and significance level α. That is, to detect a specified difference in means we may make the test more powerful by increasing the sample size.

Operating curves and power curves are often helpful in selecting a sample size to use in an experiment. For example, consider the Portland cement mortar problem discussed previously. Suppose that a difference in mean strength of 0.5 kgf/cm^2 has practical impact on the use of the mortar, so if the difference in means is at least this large, we would like to detect it with a high probability. Thus, because $\mu_1 - \mu_2 = 0.5$ kgf/cm^2 is the "critical" difference in means that we wish to detect, we find that the power curve parameter would be $\delta = 0.5/\sigma$. Unfortunately, δ involves the unknown standard deviation σ. However, suppose on the basis of past experience we think that it is very unlikely that the standard deviation will exceed 0.25 kgf/cm^2. Then substituting $\sigma = 0.25$ kgf/cm^2 into the expression for δ results in $\delta = 2$. If we wish to reject the null hypothesis when the difference in means $\mu_1 - \mu_2 = 0.5$ with probability at least 0.95 (power = 0.95) with $\alpha = 0.05$, then referring to Figure 2.13 we find that the required sample size on the horizontal axis is 16, approximately. This is the total sample size, so the sample size in each population should be

$$n = 16/2 = 8.$$

In our example, the experimenter actually used a sample size of 10. The experimenter could have elected to increase the sample size slightly to guard against the possibility that the prior estimate of the common standard deviation σ was too conservative and was likely to be somewhat larger than 0.25.

Operating characteristic curves often play an important role in the choice of sample size in experimental design problems. Their use in this respect is discussed in subsequent chapters. For a discussion of the uses of operating characteristic curves for other simple comparative experiments similar to the two-sample t-test, see Montgomery and Runger (2011).

Many statistics software packages can also assist the experimenter in performing power and sample size calculations. The following boxed display illustrates several computations for the Portland cement mortar problem from the power and sample size routine for the two-sample t test in Minitab. The first section of output repeats the analysis performed with the OC curves; find the sample size necessary for detecting the critical difference in means of 0.5 kgf/cm^2, assuming that the standard deviation of strength is 0.25 kgf/cm^2. Notice that the answer obtained from Minitab, $n_1 = n_2 = 8$, is identical to the value obtained from the OC curve analysis. The second section of the output computes the power for the case where the critical difference in means is much smaller; only 0.25 kgf/cm^2. The power has dropped considerably, from over 0.95 to 0.562. The final section determines the sample sizes that would be necessary to detect an actual difference in means of 0.25 kgf/cm^2 with a power of at least 0.9. The required sample size turns out to be considerably larger, $n_1 = n_2 = 23$.

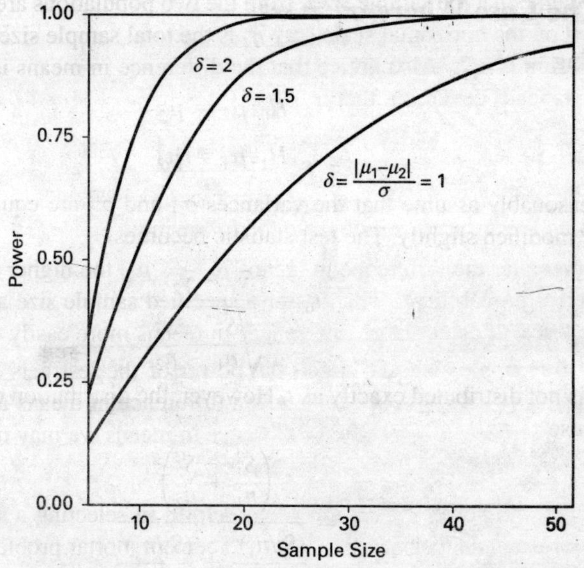

■ **FIGURE 2.13 Power Curves (from JMP) for the Two-Sample *t*-Test Assuming Equal Varianes and $\alpha = 0.05$. The Sample Size on the Horizontal Axis is the Total sample Size, so the sample Size in Each population is n = sample size from graph/2.**

```
Power and Sample Size

2-Sample t-Test
Testing mean 1 = mean 2 (versus not =)
Calculating power for mean 1 = mean 2 + difference
Alpha = 0.05   Sigma = 0.25

                 Sample      Target      Actual
Difference        Size        Power       Power
0.5                8         0.9500      0.9602

Power and Sample Size
2-Sample t-Test

Testing mean 1 = mean 2 (versus not =)
Calculating power for mean 1 = mean 2 + difference
Alpha = 0.05 Sigma = 0.25

                 Sample
Difference        Size        Power
0.25               10        0.5620

Power and Sample Size
2-Sample t-Test

Testing mean 1 = mean 2 (versus not =)
Calculating power for mean 1 = mean 2 + difference
Alpha = 0.05   Sigma = 0.25

                 Sample      Target      Actual
Difference        Size        Power       Power
0.25               23        0.9000      0.9125
```

2.4.4 The Case Where $\sigma_1^2 \neq \sigma_2^2$

If we are testing

$$H_0: \mu_1 = \mu_2$$
$$H_1: \mu_1 \neq \mu_2$$

and cannot reasonably assume that the variances σ_1^2 and σ_2^2 are equal, then the two-sample t-test must be modified slightly. The test statistic becomes

$$t_0 = \frac{\bar{y}_1 - \bar{y}_2}{\sqrt{\dfrac{S_1^2}{n_1} + \dfrac{S_2^2}{n_2}}} \tag{2.31}$$

This statistic is not distributed exactly as t. However, the distribution of t_0 is well approximated by t if we use

$$v = \frac{\left(\dfrac{S_1^2}{n_1} + \dfrac{S_2^2}{n_2}\right)^2}{\dfrac{(S_1^2/n_1)^2}{n_1 - 1} + \dfrac{(S_2^2/n_2)^2}{n_2 - 1}} \tag{2.32}$$

as the number of degrees of freedom. A strong indication of unequal variances on a normal probability plot would be a situation calling for this version of the t-test. You should be able to develop an equation for finding that confidence interval on the difference in mean for the unequal variances case easily.

EXAMPLE 2.1

Nerve preservation is important in surgery because accidental injury to the nerve can lead to post-surgical problems such as numbness, pain, or paralysis. Nerves are usually identified by their appearance and relationship to nearby structures or detected by local electrical stimulation (electromyography), but it is relatively easy to overlook them. An article in *Nature Biotechnology* ("Fluorescent Peptides Highlight Peripheral Nerves During Surgery in Mice," Vol. 29, 2011) describes the use of a fluorescently labeled peptide that binds to nerves to assist in identification. Table 2.3 shows the normalized fluorescence after two hours for nerve and muscle tissue for 12 mice (the data were read from a graph in the paper).

We would like to test the hypothesis that the mean normalized fluorescence after two hours is greater for nerve tissue then for muscle tissue. That is, if μ_δ is the mean normalized fluorescence for nerve tissue and is the mean normalized fluorescence for muscle tissue, we want to test

$$H_0: \mu_1 = \mu_2$$
$$H_1: \mu_1 > \mu_2$$

The descriptive statistics output from Minitab is shown below:

Variable	N	Mean	StDev	Minimum	Median	Maximum
Nerve	12	4228	1918	450	4825	6625
Non-nerve	12	2534	961	1130	2650	3900

TABLE 2.3
Normalized Fluorescence After Two Hours

Observation	Nerve	Muscle
1	6625	3900
2	6000	3500
3	5450	3450
4	5200	3200
5	5175	2980
6	4900	2800
7	4750	2500
8	4500	2400
9	3985	2200
10	900	1200
11	450	1150
12	2800	1130

Notice that the two sample standard deviations are quite different, so the assumption of equal variances in the pooled *t*-test may not be appropriate. Figure 2.14 is the normal probability plot from Minitab for the two samples. This plot also indicates that the two population variances are probably not the same.

Because the equal variance assumption is not appropriate here, we will use the two-sample *t*-test described in this section to test the hypothesis of equal means. The test statistic, Equation 2.31, is

$$t_0 = \frac{\bar{y}_1 - \bar{y}_2}{\sqrt{\frac{S_1^2}{n_1} + \frac{S_2^2}{n_2}}} = \frac{4228 - 2534}{\sqrt{\frac{(1918)^2}{12} + \frac{(961)^2}{12}}} = 2.7354$$

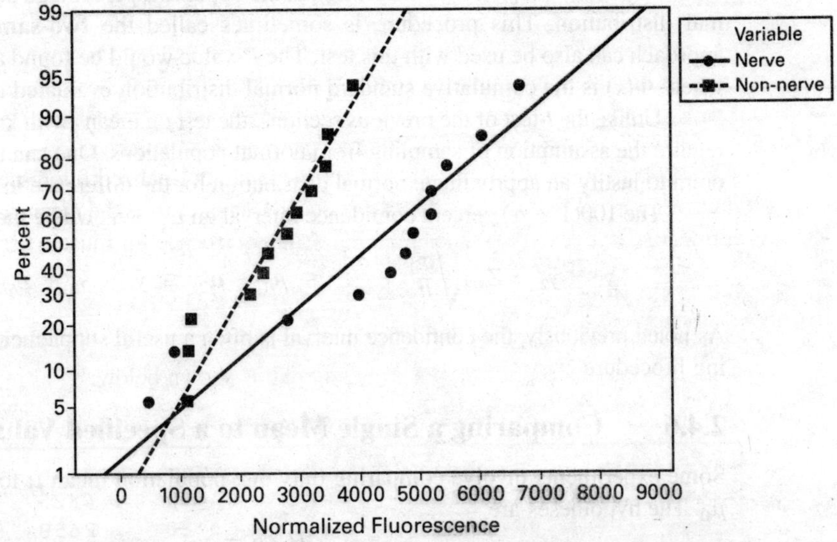

■ **FIGURE 2.14** Normalized Fluorescence Data from Table 2.3

The number of degrees of freedom are calculated from Equation 2.32:

$$v = \frac{\left(\frac{S_1^2}{n_1} + \frac{S_2^2}{n_2}\right)^2}{\frac{(S_1^2 / n_1)^2}{n_1 - 1} + \frac{(S_2^2 / n_2)^2}{n_2 - 1}} = \frac{\left(\frac{(1918)^2}{12} + \frac{(961)^2}{12}\right)^2}{\frac{[(1918)^2 / 12]^2}{11} + \frac{[(961)^2 / 12]^2}{11}} = 16.1955$$

If we are going to find a P-value from a table of the t-distribution, we should round the degrees of freedom down to 16. Most computer programs interpolate to determine the P-value. The Minitab output for the two-sample t-test is shown below. Since the P-value reported is small (0.015), we would reject the null hypothesis and conclude that the mean normalized fluorescence for nerve tissue is greater than the mean normalized fluorescence for muscle tissue.

```
Difference = mu (Nerve) - mu (Non-nerve)
Estimate for difference:   1694
95% lower bound for difference:   613
T-Test of difference = 0 (vs >): T-Value = 2.74   P-Value = 0.007   DF = 16
```

2.4.5 The Case Where σ_1^2 and σ_2^2 Are Known

If the variances of both populations are **known**, then the hypotheses

$$H_0: \mu_1 = \mu_2$$
$$H_1: \mu_1 \neq \mu_2$$

may be tested using the statistic

$$Z_0 = \frac{\bar{y}_1 - \bar{y}_2}{\sqrt{\frac{\sigma_1^2}{n_1} + \frac{\sigma_2^2}{n_2}}} \tag{2.33}$$

If both populations are normal, or if the sample sizes are large enough so that the central limit theorem applies, the distribution of Z_0 is $N(0, 1)$ if the null hypothesis is true. Thus, the critical region would be found using the normal distribution rather than the t. Specifically, we would reject H_0 if $|Z_0| > Z_{\alpha/2}$, where $Z_{\alpha/2}$ is the upper $\alpha/2$ percentage point of the standard normal distribution. This procedure is sometimes called the **two-sample Z-test**. A P-value approach can also be used with this test. The P-value would be found as $P = 2\,[1 - \Phi(|Z_0|)]$, where $\Phi(x)$ is the cumulative standard normal distribution evaluated at the point x.

Unlike the t-test of the previous sections, the test on means with known variances does not require the assumption of sampling from normal populations. One can use the central limit theorem to justify an approximate normal distribution for the difference in sample means $\bar{y}_1 - \bar{y}_2$

The $100(1 - \alpha)$ percent confidence interval on $\mu_1 - \mu_2$ where the variances are known is

$$\bar{y}_1 - \bar{y}_2 - Z_{\alpha/2}\sqrt{\frac{\sigma_1^2}{n_1} + \frac{\sigma_2^2}{n_2}} \le \mu_1 - \mu_2 \le \bar{y}_1 - \bar{y}_2 + Z_{\alpha/2}\sqrt{\frac{\sigma_1^2}{n_1} + \frac{\sigma_2^2}{n_2}} \tag{2.34}$$

As noted previously, the confidence interval is often a useful supplement to the hypothesis testing procedure.

2.4.6 Comparing a Single Mean to a Specified Value

Some experiments involve comparing only one population mean μ to a specified value, say. μ_0. The hypotheses are

$$H_0: \mu = \mu_0$$
$$H_1: \mu \neq \mu_0$$

If the population is normal with known variance, or if the population is nonnormal but the sample size is large enough so that the central limit theorem applies, then the hypothesis may be tested using a direct application of the normal distribution. The **one-sample Z-test** statistic is

$$Z_0 = \frac{\bar{y} - \mu_0}{\sigma/\sqrt{n}} \tag{2.35}$$

If $H_0: \mu = \mu_0$ is true, then the distribution of Z_0 is $N(0, 1)$. Therefore, the decision rule for $H_0: \mu = \mu_0$ is to reject the null hypothesis if $|Z_0| > Z_{\alpha/2}$. A P-value approach could also be used.

The value of the mean μ_0 specified in the null hypothesis is usually determined in one of three ways. It may result from past evidence, knowledge, or experimentation. It may be the result of some theory or model describing the situation under study. Finally, it may be the result of contractual specifications.

The $100(1 - \alpha)$ percent confidence interval on the true population mean is

$$\bar{y} - Z_{\alpha/2}\sigma/\sqrt{n} \le \mu \le \bar{y} + Z_{\alpha/2}\sigma/\sqrt{n} \tag{2.36}$$

EXAMPLE 2.2

A supplier submits lots of fabric to a textile manufacturer. The customer wants to know if the lot average breaking strength exceeds 200 psi. If so, she wants to accept the lot. Past experience indicates that a reasonable value for the variance of breaking strength is $100(\text{psi})^2$. The hypotheses to be tested are

$$H_0: \mu = 200$$
$$H_1: \mu > 200$$

Note that this is a one-sided alternative hypothesis. Thus, we would accept the lot only if the null hypothesis $H_0: \mu = 200$ could be rejected (i.e., if $Z_0 > Z_\alpha$).

Four specimens are randomly selected, and the average breaking strength observed is $\bar{y} = 214$ psi. The value of the test statistic is

$$Z_0 = \frac{\bar{y} - \mu_0}{\sigma/\sqrt{n}} = \frac{214 - 200}{10/\sqrt{4}} = 2.80$$

If a type I error of $\alpha = 0.05$ is specified, we find $Z_\alpha = Z_{0.05} = 1.645$ from Appendix Table I. The P-value would be computed using only the area in the upper tail of the standard normal distribution, because the alternative hypothesis is one-sided. The P-value is $P = 1 - \Phi(2.80) = 1 - 0.99744 = 0.00256$. Thus H_0 is rejected, and we conclude that the lot average breaking strength exceeds 200 psi.

If the variance of the population is unknown, we must make the additional assumption that the population is normally distributed, although moderate departures from normality will not seriously affect the results.

To test $H_0: \mu = \mu_0$ in the variance unknown case, the sample variance S^2 is used to estimate σ^2. Replacing σ with S in Equation 2.35, we have the **one-sample *t*-test** statistic

$$t_0 = \frac{\bar{y} - \mu_0}{S/\sqrt{n}} \tag{2.37}$$

The null hypothesis $H_0: \mu = \mu_0$ would be rejected if $|t_0| > t_{\alpha/2,n-1}$, where $t_{\alpha/2,n-1}$ denotes the upper $\alpha/2$ percentage point of the t distribution with $n - 1$ degrees of freedom. A P-value approach could also be used. The $100(1 - \alpha)$ percent confidence interval in this case is

$$\bar{y} - t_{\alpha/2,n-1}S/\sqrt{n} \le \mu \le \bar{y} + t_{\alpha/2,n-1}S/\sqrt{n} \tag{2.38}$$

2.4.7 Summary

Tables 2.4 and 2.5 summarize the t-test and z-test procedures discussed above for sample means. Critical regions are shown for both two-sided and one-sided alternative hypotheses.

■ **TABLE 2.4**
Tests on Means with Variance Known

Hypothesis	Test Statistic	Fixed Significance Level Criteria for Rejection	*P*-Value
$H_0: \mu = \mu_0$ $H_1: \mu \neq \mu_0$		$\lvert Z_0 \rvert > Z_{\alpha/2}$	$P = 2[1 - \Phi(\lvert Z_0 \rvert)]$
$H_0: \mu = \mu_0$ $H_1: \mu < \mu_0$	$Z_0 = \dfrac{\bar{y} - \mu_0}{\sigma/\sqrt{n}}$	$Z_0 < -Z_\alpha$	$P = \Phi(Z_0)$
$H_0: \mu = \mu_0$ $H_1: \mu > \mu_0$		$Z_0 > Z_\alpha$	$P = 1 - \Phi(Z_0)$
$H_0: \mu_1 = \mu_2$ $H_1: \mu_1 \neq \mu_2$		$\lvert Z_0 \rvert > Z_{\alpha/2}$	$P = 2[1 - \Phi(\lvert Z_0 \rvert)]$
$H_0: \mu_1 = \mu_2$ $H_1: \mu_1 < \mu_2$	$Z_0 = \dfrac{\bar{y}_1 - \bar{y}_2}{\sqrt{\dfrac{\sigma_1^2}{n_1} + \dfrac{\sigma_2^2}{n_2}}}$	$Z_0 < -Z_\alpha$	$P = \Phi(Z_0)$
$H_0: \mu_1 = \mu_2$ $H_1: \mu_1 > \mu_2$		$Z_0 > Z_\alpha$	$P = 1 - \Phi(Z_0)$

■ **TABLE 2.5**
Tests on Means of Normal Distributions, Variance Unknown

Hypothesis	Test Statistic	Fixed Significance Level Criteria for Rejection	*P*-Value
$H_0: \mu = \mu_0$ $H_1: \mu \neq \mu_0$		$\lvert t_0 \rvert > t_{\alpha/2,n-1}$	sum of the probability above t_0 and below $-t_0$
$H_0: \mu = \mu_0$ $H_1: \mu < \mu_0$	$t_0 = \dfrac{\bar{y} - \mu_0}{S/\sqrt{n}}$	$t_0 < -t_{\alpha,n-1}$	probability below t_0
$H_0: \mu = \mu_0$ $H_1: \mu > \mu_0$		$t_0 > t_{\alpha,n-1}$	probability above t_0
$H_0: \mu_1 = \mu_2$ $H_1: \mu_1 \neq \mu_2$	if $\sigma_1^2 = \sigma_2^2$ $t_0 = \dfrac{\bar{y}_1 - \bar{y}_2}{S_p\sqrt{\dfrac{1}{n_1} + \dfrac{1}{n_2}}}$ $v = n_1 + n_2 - 2$	$\lvert t_0 \rvert > t_{\alpha/2,v}$	sum of the probability above t_0 and below $-t_0$
$H_0: \mu_1 = \mu_2$ $H_1: \mu_1 < \mu_2$	if $\sigma_1^2 \neq \sigma_2^2$ $t_0 = \dfrac{\bar{y}_1 - \bar{y}_2}{\sqrt{\dfrac{S_1^2}{n_1} + \dfrac{S_2^2}{n_2}}}$	$t_0 < -t_{\alpha,v}$	probability below t_0
$H_0: \mu_1 = \mu_2$ $H_1: \mu_1 > \mu_2$	$v = \dfrac{\left(\dfrac{S_1^2}{n_1} + \dfrac{S_2^2}{n_2}\right)^2}{\dfrac{(S_1^2/n_1)^2}{n_1 - 1} + \dfrac{(S_2^2/n_2)^2}{n_2 - 1}}$	$t_0 > t_{\alpha,v}$	probability above t_0

2.5 Inferences About the Differences in Means, Paired Comparison Designs

2.5.1 The Paired Comparison Problem

In some simple comparative experiments, we can greatly improve the precision by making comparisons within matched pairs of experimental material. For example, consider a hardness testing machine that presses a rod with a pointed tip into a metal specimen with a known force. By measuring the depth of the depression caused by the tip, the hardness of the specimen is determined. Two different tips are available for this machine, and although the precision (variability) of the measurements made by the two tips seems to be the same, it is suspected that one tip produces different mean hardness readings than the other.

An experiment could be performed as follows. A number of metal specimens (e.g., 20) could be randomly selected. Half of these specimens could be tested by tip 1 and the other half by tip 2. The exact assignment of specimens to tips would be randomly determined. Because this is a completely randomized design, the average hardness of the two samples could be compared using the *t*-test described in Section 2.4.

A little reflection will reveal a serious disadvantage in the completely randomized design for this problem. Suppose the metal specimens were cut from different bar stock that were produced in different heats or that were not exactly homogeneous in some other way that might affect the hardness. This lack of homogeneity between specimens will contribute to the variability of the hardness measurements and will tend to inflate the experimental error, thus making a true difference between tips harder to detect.

To protect against this possibility, consider an alternative experimental design. Assume that each specimen is large enough so that *two* hardness determinations may be made on it. This alternative design would consist of dividing each specimen into two parts, then randomly assigning one tip to one-half of each specimen and the other tip to the remaining half. The order in which the tips are tested for a particular specimen would also be randomly selected. The experiment, when performed according to this design with 10 specimens, produced the (coded) data shown in Table 2.6.

We may write a **statistical model** that describes the data from this experiment as

$$y_{ij} = \mu_i + \beta_j + \epsilon_{ij} \begin{cases} i = 1, 2 \\ j = 1, 2, \ldots, 10 \end{cases} \tag{2.39}$$

■ **TABLE 2.6**
Data for the Hardness Testing Experiment

Specimen	Tip 1	Tip 2
1	7	6
2	3	3
3	3	5
4	4	3
5	8	8
6	3	2
7	2	4
8	9	9
9	5	4
10	4	5

where y_{ij} is the observation on hardness for tip i on specimen j, μ_i is the true mean hardness of the ith tip, β_j is an effect on hardness due to the jth specimen, and ϵ_{ij} is a random experimental error with mean zero and variance σ_i^2. That is, σ_1^2 is the variance of the hardness measurements from tip 1, and σ_2^2 is the variance of the hardness measurements from tip 2.

Note that if we compute the jth paired difference

$$d_j = y_{1j} - y_{2j} \qquad j = 1, 2, \ldots, 10 \tag{2.40}$$

the expected value of this difference is

$$\begin{aligned}
\mu_d &= E(d_j) \\
&= E(y_{1j} - y_{2j}) \\
&= E(y_{1j}) - E(y_{2j}) \\
&= \mu_1 + \beta_j - (\mu_2 + \beta_j) \\
&= \mu_1 - \mu_2
\end{aligned}$$

That is, we may make inferences about the difference in the mean hardness readings of the two tips $\mu_1 - \mu_2$ by making inferences about the mean of the differences μ_d. Notice that the additive effect of the specimens β_j cancels out when the observations are paired in this manner.

Testing $H_0: \mu_1 = \mu_2$ is equivalent to testing

$$H_0: \mu_d = 0$$
$$H_1: \mu_d \neq 0$$

This is a single-sample t-test. The test statistic for this hypothesis is

$$t_0 = \frac{\bar{d}}{S_d/\sqrt{n}} \tag{2.41}$$

where

$$\bar{d} = \frac{1}{n}\sum_{j=1}^{n} d_j \tag{2.42}$$

is the sample mean of the differences and

$$S_d = \left[\frac{\sum_{j=1}^{n}(d_j - \bar{d})^2}{n-1}\right]^{1/2} = \left[\frac{\sum_{j=1}^{n} d_j^2 - \frac{1}{n}\left(\sum_{j=1}^{n} d_j\right)^2}{n-1}\right]^{1/2} \tag{2.43}$$

is the sample standard deviation of the differences. $H_0: \mu_d = 0$ would be rejected if $|t_0| > t_{\alpha/2,n-1}$. A P-value approach could also be used. Because the observations from the factor levels are "paired" on each experimental unit, this procedure is usually called the **paired *t*-test**.

For the data in Table 2.6, we find

$$\begin{aligned}
d_1 &= 7 - 6 = 1 & d_6 &= 3 - 2 = 1 \\
d_2 &= 3 - 3 = 0 & d_7 &= 2 - 4 = -2 \\
d_3 &= 3 - 5 = -2 & d_8 &= 9 - 9 = 0 \\
d_4 &= 4 - 3 = 1 & d_9 &= 5 - 4 = 1 \\
d_5 &= 8 - 8 = 0 & d_{10} &= 4 - 5 = -1
\end{aligned}$$

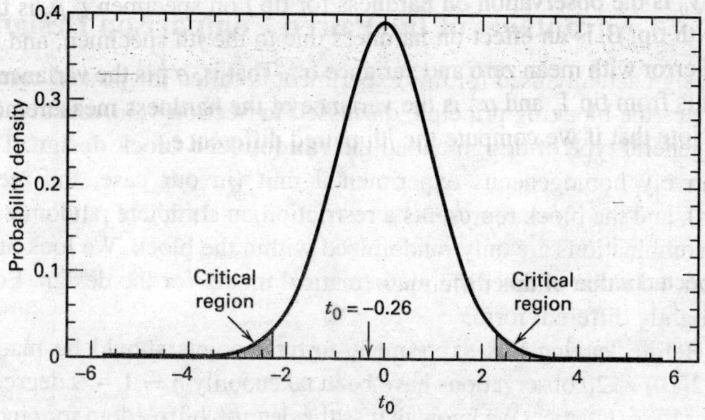

■ **FIGURE 2.15 The reference distribution (t with 9 degrees of freedom) for the hardness testing problem**

Thus,

$$\bar{d} = \frac{1}{n}\sum_{j=1}^{n} d_j = \frac{1}{10}(-1) = -0.10$$

$$S_d = \left[\frac{\sum_{j=1}^{n} d_j^2 - \frac{1}{n}\left(\sum_{j=1}^{n} d_j\right)^2}{n-1}\right]^{1/2} = \left[\frac{13 - \frac{1}{10}(-1)^2}{10-1}\right]^{1/2} = 1.20$$

Suppose we choose $\alpha = 0.05$. Now to make a decision, we would compute t_0 and reject H_0 if $|t_0| > t_{0.025,9} = 2.262$.

The computed value of the paired t-test statistic is

$$t_0 = \frac{\bar{d}}{S_d/\sqrt{n}} = \frac{-0.10}{1.20/\sqrt{10}} = -0.26$$

and because $|t_0| = 0.26 \ngtr t_{0.025,9} = 2.262$, we cannot reject the hypothesis $H_0: \mu_d = 0$. That is, there is no evidence to indicate that the two tips produce different hardness readings. Figure 2.15 shows the t_0 distribution with 9 degrees of freedom, the reference distribution for this test, with the value of t_0 shown relative to the critical region.

Table 2.7 shows the computer output from the Minitab paired t-test procedure for this problem. Notice that the P-value for this test is $P \simeq 0.80$, implying that we cannot reject the null hypothesis at *any* reasonable level of significance.

■ **TABLE 2.7**
Minitab Paired t-Test Results for the Hardness Testing Example

Paired T for Tip 1−Tip 2

	N	Mean	Std. Dev.	SE Mean
Tip 1	10	4.800	2.394	0.757
Tip 2	10	4.900	2.234	0.706
Difference	10	−0.100	1.197	0.379

95% CI for mean difference: (−0.956, 0.756)

t-Test of mean difference = 0 (vs not = 0):

T-Value = −0.26 P-Value = 0.798

2.5.2 Advantages of the Paired Comparison Design

The design actually used for this experiment is called the **paired comparison design**, and it illustrates the blocking principle discussed in Section 1.3. Actually, it is a special case of a more general type of design called the **randomized block design**. The term *block* refers to a relatively homogeneous experimental unit (in our case, the metal specimens are the blocks), and the block represents a restriction on complete randomization because the treatment combinations are only randomized within the block. We look at designs of this type in Chapter 4. In that chapter the mathematical model for the design, Equation 2.39, is written in a slightly different form.

Before leaving this experiment, several points should be made. Note that, although $2n = 2(10) = 20$ observations have been taken, only $n - 1 = 9$ degrees of freedom are available for the t statistic. (We know that as the degrees of freedom for t increase, the test becomes more sensitive.) By blocking or pairing we have effectively "lost" $n - 1$ degrees of freedom, but we hope we have gained a better knowledge of the situation by eliminating an additional source of variability (the difference between specimens).

We may obtain an indication of the quality of information produced from the paired design by comparing the standard deviation of the differences S_d with the pooled standard deviation S_p that would have resulted had the experiment been conducted in a completely randomized manner and the data of Table 2.5 been obtained. Using the data in Table 2.5 as two independent samples, we compute the pooled standard deviation from Equation 2.25 to be $S_p = 2.32$. Comparing this value to $S_d = 1.20$, we see that blocking or pairing has reduced the estimate of variability by nearly 50 percent.

Generally, when we don't block (or pair the observations) when we really should have, S_p will always be larger than S_d. It is easy to show this formally. If we pair the observations, it is easy to show that S_d^2 is an unbiased estimator of the variance of the differences d_j under the model in Equation 2.39 because the block effects (the β_j) cancel out when the differences are computed. However, if we don't block (or pair) and treat the observations as two independent samples, then S_p^2 is not an unbiased estimator of σ^2 under the model in Equation 2.39. In fact, assuming that both population variances are equal,

$$E(S_p^2) = \sigma^2 + \sum_{j=1}^{n} \beta_j^2$$

That is, the block effects β_j inflate the variance estimate. This is why blocking serves as a **noise reduction** design technique.

We may also express the results of this experiment in terms of a confidence interval on $\mu_1 - \mu_2$. Using the paired data, a 95 percent confidence interval on $\mu_1 - \mu_2$ is

$$\bar{d} \pm t_{0.025,9} S_d/\sqrt{n}$$

$$-0.10 \pm (2.262)(1.20)/\sqrt{10}$$

$$-0.10 \pm 0.86$$

Conversely, using the pooled or independent analysis, a 95 percent confidence interval on $\mu_1 - \mu_2$ is

$$\bar{y}_1 - \bar{y}_2 \pm t_{0.025,18} S_p \sqrt{\frac{1}{n_1} + \frac{1}{n_2}}$$

$$4.80 - 4.90 \pm (2.101)(2.32)\sqrt{\tfrac{1}{10} + \tfrac{1}{10}}$$

$$-0.10 \pm 2.18$$

The confidence interval based on the paired analysis is much narrower than the confidence interval from the independent analysis. This again illustrates the **noise reduction** property of blocking.

Blocking is not always the best design strategy. If the within-block variability is the same as the between-block variability, the variance of $\bar{y}_1 - \bar{y}_2$ will be the same regardless of which design is used. Actually, blocking in this situation would be a poor choice of design because blocking results in the loss of $n - 1$ degrees of freedom and will actually lead to a wider confidence interval on $\mu_1 - \mu_2$. A further discussion of blocking is given in Chapter 4.

2.6 Inferences About the Variances of Normal Distributions

In many experiments, we are interested in possible differences in the mean response for two treatments. However, in some experiments it is the comparison of variability in the data that is important. In the food and beverage industry, for example, it is important that the variability of filling equipment be small so that all packages have close to the nominal net weight or volume of content. In chemical laboratories, we may wish to compare the variability of two analytical methods. We now briefly examine tests of hypotheses and confidence intervals for variances of normal distributions. Unlike the tests on means, the procedures for tests on variances are rather sensitive to the normality assumption. A good discussion of the normality assumption is in Appendix 2A of Davies (1956).

Suppose we wish to test the hypothesis that the variance of a normal population equals a constant, for example, σ_0^2. Stated formally, we wish to test

$$H_0{:}\sigma^2 = \sigma_0^2$$
$$H_1{:}\sigma^2 \neq \sigma_0^2 \tag{2.44}$$

The test statistic for Equation 2.44 is

$$\chi_0^2 = \frac{SS}{\sigma_0^2} = \frac{(n-1)S^2}{\sigma_0^2} \tag{2.45}$$

where $SS = \Sigma_{i=1}^n (y_i - \bar{y})^2$ is the corrected sum of squares of the sample observations. The appropriate reference distribution for χ_0^2 is the chi-square distribution with $n - 1$ degrees of freedom. The null hypothesis is rejected if $\chi_0^2 > \chi_{\alpha/2,n-1}^2$ or if $\chi_0^2 < \chi_{1-(\alpha/2),n-1}^2$, where $\chi_{\alpha/2,n-1}^2$ and $\chi_{1-(\alpha/2),n-1}^2$ are the upper $\alpha/2$ and lower $1 - (\alpha/2)$ percentage points of the chi-square distribution with $n - 1$ degrees of freedom, respectively. Table 2.8 gives the critical regions for the one-sided alternative hypotheses. The $100(1 - \alpha)$ percent confidence interval on σ^2 is

$$\frac{(n-1)S^2}{\chi_{\alpha/2,n-1}^2} \leq \sigma^2 \leq \frac{(n-1)S^2}{\chi_{1-(\alpha/2),n-1}^2} \tag{2.46}$$

Now consider testing the equality of the variances of two normal populations. If independent random samples of size n_1 and n_2 are taken from populations 1 and 2, respectively, the test statistic for

$$H_0{:}\sigma_1^2 = \sigma_2^2$$
$$H_1{:}\sigma_1^2 \neq \sigma_2^2 \tag{2.47}$$

is the ratio of the sample variances

$$F_0 = \frac{S_1^2}{S_2^2} \tag{2.48}$$

■ **TABLE 2.8**
Tests on Variances of Normal Distributions

Hypothesis	Test Statistic	Fixed Significance Level Criteria for Rejection
$H_0: \sigma^2 = \sigma_0^2$ $H_1: \sigma^2 \neq \sigma_0^2$		$\chi_0^2 > \chi^2_{\alpha/2,n-1}$ or $\chi_0^2 < \chi^2_{1-\alpha/2,n-1}$
$H_0: \sigma^2 = \sigma_0^2$ $H_1: \sigma^2 < \sigma_0^2$	$\chi_0^2 = \dfrac{(n-1)S^2}{\sigma_0^2}$	$\chi_0^2 < \chi^2_{1-\alpha,n-1}$
$H_0: \sigma^2 = \sigma_0^2$ $H_1: \sigma^2 > \sigma_0^2$		$\chi_0^2 > \chi^2_{\alpha,n-1}$
$H_0: \sigma_1^2 = \sigma_2^2$ $H_1: \sigma_1^2 \neq \sigma_2^2$	$F_0 = \dfrac{S_1^2}{S_2^2}$	$F_0 > F_{\alpha/2,n_1-1,n_2-1}$ or $F_0 < F_{1-\alpha/2,n_1-1,n_2-1}$
$H_0: \sigma_1^2 = \sigma_2^2$ $H_1: \sigma_1^2 < \sigma_2^2$	$F_0 = \dfrac{S_2^2}{S_1^2}$	$F_0 > F_{\alpha,n_2-1,n_1-1}$
$H_0: \sigma_1^2 = \sigma_2^2$ $H_1: \sigma_1^2 > \sigma_2^2$	$F_0 = \dfrac{S_1^2}{S_2^2}$	$F_0 > F_{\alpha,n_1-1,n_2-1}$

The appropriate reference distribution for F_0 is the F distribution with $n_1 - 1$ numerator degrees of freedom and $n_2 - 1$ denominator degrees of freedom. The null hypothesis would be rejected if $F_0 > F_{\alpha/2,n_1-1,n_2-1}$ or if $F_0 < F_{1-(\alpha/2),n_1-1,n_2-1}$, where $F_{\alpha/2,n_1-1,n_2-1}$ and $F_{1-(\alpha/2),n_1-1,n_2-1}$ denote the upper $\alpha/2$ and lower $1 - (\alpha/2)$ percentage points of the F distribution with $n_1 - 1$ and $n_2 - 1$ degrees of freedom. Table IV of the Appendix gives only upper-tail percentage points of F; however, the upper- and lower-tail points are related by

$$F_{1-\alpha,v_1,v_2} = \frac{1}{F_{\alpha,v_2,v_1}} \tag{2.49}$$

Critical values for the one-sided alternative hypothesis are given in Table 2.8. Test procedures for more than two variances are discussed in Section 3.4.3. We will also discuss the use of the variance or standard deviation as a response variable in more general experimental settings.

EXAMPLE 2.3

A chemical engineer is investigating the inherent variability of two types of test equipment that can be used to monitor the output of a production process. He suspects that the old equipment, type 1, has a larger variance than the new one. Thus, he wishes to test the hypothesis

$$H_0: \sigma_1^2 = \sigma_2^2$$
$$H_1: \sigma_1^2 > \sigma_2^2$$

Two random samples of $n_1 = 12$ and $n_2 = 10$ observations are taken, and the sample variances are $S_1^2 = 14.5$ and $S_2^2 = 10.8$. The test statistic is

$$F_0 = \frac{S_1^2}{S_2^2} = \frac{14.5}{10.8} = 1.34$$

From Appendix Table IV we find that $F_{0.05,11,9} = 3.10$, so the null hypothesis cannot be rejected. That is, we have found insufficient statistical evidence to conclude that the variance of the old equipment is greater than the variance of the new equipment.

The $100(1-\alpha)$ confidence interval for the ratio of the population variances σ_1^2/σ_2^2 is

$$\frac{S_1^2}{S_2^2}F_{1-\alpha/2,n_2-1,n_1-1} \leq \frac{\sigma_1^2}{\sigma_2^2} \leq \frac{S_1^2}{S_2^2}F_{\alpha/2,n_2-1,n_1-1} \quad \textbf{(2.50)}$$

To illustrate the use of Equation 2.50, the 95 percent confidence interval for the ratio of variances σ_1^2/σ_2^2 in Example 2.2 is, using $F_{0.025,9,11} = 3.59$ and $F_{0.975,9,11} = 1/F_{0.025,11,9} = 1/3.92 = 0.255$,

$$\frac{14.5}{10.8}(0.255) \leq \frac{\sigma_1^2}{\sigma_2^2} \leq \frac{14.5}{10.8}(3.59)$$

$$0.34 \leq \frac{\sigma_1^2}{\sigma_2^2} \leq 4.82$$

2.7 Problems

2.1. Computer output for a random sample of data is shown below. Some of the quantities are missing. Compute the values of the missing quantities.

Variable	N	Mean	SE Mean	Std. Dev.	Variance	Minimum	Maximum
Y	8	19.96	?	3.12	?	15.94	27.16

2.2. Computer output for a random sample of data is shown below. Some of the quantities are missing. Compute the values of the missing quantities.

Variable	N	Mean	SE Mean	Std. Dev.	Sum
Y	20	?	0.159	?	399.851

2.3. Consider the computer output shown below.

```
One-Sample Z
Test of mu = 30 vs not = 30
The assumed standard deviation = 1.2

N    Mean      SE Mean       95% CI               Z    P
16   31.2000   0.3000    (30.6120, 31.7880)      ?    ?
```

(a) Fill in the missing values in the output. What conclusion would you draw?

(b) Is this a one-sided or two-sided test?

(c) Use the output and the normal table to find a 99 percent CI on the mean.

(d) What is the *P*-value if the alternative hypothesis is $H_1: \mu > 30$?

2.4. Suppose that we are testing $H_0: \mu = \mu_0$ versus $H_1: \mu \neq \mu_0$. Calculate the *P*-value for the following observed values of the test statistic:

(a) $Z_0 = 2.35$ **(b)** $Z_0 = 1.60$ **(c)** $Z_0 = 2.15$
(d) $Z_0 = 1.89$ **(e)** $Z_0 = -0.90$

2.5. Suppose that we are testing $H_0: \mu = \mu_0$ versus $H_1: \mu > \mu_0$. Calculate the *P*-value for the following observed values of the test statistic:

(a) $Z_0 = 2.25$ **(b)** $Z_0 = -1.65$ **(c)** $Z_0 = 2.05$
(d) $Z_0 = 1.90$ **(e)** $Z_0 = -2.25$

2.6. Suppose that we are testing $H_0: \mu_1 = \mu_2$ versus $H_0: \mu_1 \neq \mu_2$ where the two sample sizes are $n_1 = n_2 = 10$. Both sample variances are unknown but assumed equal. Find bounds on the *P*-value for the following observed values of the test statistic.

(a) $t_0 = 2.45$ **(b)** $t_0 = 3.25$ **(c)** $t_0 = 1.90$ **(d)** $t_0 = -2.50$

2.7. Suppose that we are testing $H_0: \mu_1 = \mu_2$ versus $H_0: \mu_1 > \mu_2$ where the two sample sizes are $n_1 = n_2 = 12$. Both sample variances are unknown but assumed equal. Find bounds on the *P*-value for the following observed values of the test statistic.

(a) $t_0 = 2.31$ **(b)** $t_0 = 3.60$ **(c)** $t_0 = 1.95$ **(d)** $t_0 = 2.19$

2.8. Consider the following sample data: 9.37, 13.04, 11.69, 8.21, 11.18, 10.41, 13.15, 11.51, 13.21, and 7.75. Is it reasonable to assume that this data is a sample from a normal distribution? Is there evidence to support a claim that the mean of the population is 10?

2.9. A computer program has produced the following output for a hypothesis-testing problem:

```
Difference in sample means:   2.35
Degrees of freedom: 18
Standard error of the difference in sample means: ?
Test statistic:    t0 = 2.01
P-value: 0.0298
```

(a) What is the missing value for the standard error?

(b) Is this a two-sided or a one-sided test?

(c) If $\alpha = 0.05$, what are your conclusions?

(d) Find a 90% two-sided CI on the difference in means.

2.10. A computer program has produced the following output for a hypothesis-testing problem:

```
Difference in sample means:  11.5
Degrees of freedom: 24
Standard error of the difference in sample means: ?
Test statistic:   t0 = -1.88
P-value: 0.0723
```

(a) What is the missing value for the standard error?

(b) Is this a two-sided or a one-sided test?

(c) If $\alpha = 0.05$, what are your conclusions?

(d) Find a 95% two-sided CI on the difference in means.

2.11. Suppose that we are testing $H_0: \mu = \mu_0$ versus $H_1: \mu > \mu_0$ with a sample size of $n = 20$. Calculate bounds on the P-value for the following observed values of the test statistic:

(a) $t_0 = 2.35$ (b) $t_0 = 3.55$ (c) $t_0 = 2.00$ (d) $t_0 = 1.55$

2.12. Suppose that we are testing $H_0: \mu = \mu_0$ versus $H_1: \mu \neq \mu_0$ with a sample size of $n = 16$. Calculate bounds on the P-value for the following observed values of the test statistic:

(a) $t_0 = 2.48$ (b) $t_0 = -3.95$ (c) $t_0 = 2.69$

(d) $t_0 = 1.88$ (e) $t_0 = -1.25$

2.13. Consider the computer output shown below.

```
One-Sample T: Y

Test of mu = 91 vs. not = 91

Variable N   Mean     Std. Dev. SE Mean    95% CI          T   P
   Y    25 92.5805     ?       0.4673  (91.6160, ?)   ?  0.002
```

(a) Fill in the missing values in the output. Can the null hypothesis be rejected at the 0.05 level? Why?

(b) Is this a one-sided or a two-sided test?

(c) If the hypotheses had been $H_0: \mu = 90$ versus $H_1: \mu \neq 90$ would you reject the null hypothesis at the 0.05 level?

(d) Use the output and the t table to find a 99 percent two-sided CI on the mean.

(e) What is the P-value if the alternative hypothesis is $H_1: \mu > 91$?

2.14. Consider the computer output shown below.

```
One-Sample T: Y

Test of mu = 25 vs > 25

                                             95% Lower
Variable   N   Mean    Std. Dev. SE Mean   Bound     T    P
Y         12  25.6818     ?      0.3360      ?       ?  0.034
```

(a) How many degrees of freedom are there on the t-test statistic?

(b) Fill in the missing information.

2.15. Consider the computer output shown below.

```
Two-Sample T-Test and CI: Y1, Y2

Two-sample T for Y1 vs Y2

      N    Mean    Std. Dev.   SE Mean
Y1   20   50.19     1.71       0.38
Y2   20   52.52     2.48       0.55

Difference = mu (X1) - mu (X2)
Estimate for difference: -2.33341
95% CI for difference: (-3.69547, -0.97135)
T-Test of difference = 0 (vs not =) : T-Value = -3.47
P-Value = 0.001 DF = 38
Both use Pooled Std. Dev. = 2.1277
```

(a) Can the null hypothesis be rejected at the 0.025 level? Why?

(b) Is this a one-sided or a two-sided test?

(c) If the hypotheses had been $H_0: \mu_1 - \mu_2 = 2$ versus $H_1: \mu_1 - \mu_2 \neq 2$ would you reject the null hypothesis at the 0.05 level?

(d) If the hypotheses had been $H_0: \mu_1 - \mu_2 = 2$ versus $H_1: \mu_1 - \mu_2 < 2$ would you reject the null hypothesis at the 0.05 level? Can you answer this question without doing any additional calculations? Why?

(e) Use the output and the t table to find a 95 percent upper confidence bound on the difference in means.

(f) What is the P-value if the hypotheses are $H_0: \mu_1 - \mu_2 = 2$ versus $H_1: \mu_1 - \mu_2 \neq 2$?

2.16. The breaking strength of a fiber is required to be at least 160 psi. Past experience has indicated that the standard deviation of breaking strength is $\sigma = 3$ psi. A random sample of four specimens is tested, and the results are $y_1 = 145$, $y_2 = 153$, $y_3 = 150$, and $y_4 = 149$.

(a) State the hypotheses that you think should be tested in this experiment.

(b) Test these hypotheses using $\alpha = 0.05$. What are your conclusions?

(c) Find the P-value for the test in part (b).

(d) Construct a 95 percent confidence interval on the mean breaking strength.

2.17. The viscosity of a liquid detergent is supposed to average 800 centistokes at 25°C. A random sample of 10 batches of detergent is collected, and the average viscosity is 812. Suppose we know that the standard deviation of viscosity is $\sigma = 20$ centistokes.

(a) State the hypotheses that should be tested.

(b) Test these hypotheses using $\alpha = 0.05$. What are your conclusions?

(c) What is the P-value for the test?

(d) Find a 95 percent confidence interval on the mean.

2.18. A normally distributed random variable has an unknown mean μ and a known variance $\sigma^2 = 9$. Find the sample size required to construct a 95 percent confidence interval on the mean that has total length of 1.0.

2.19. The diameters of steel shafts produced by a certain manufacturing process should have a mean diameter of 0.255 inches. The diameter is known to have a standard deviation of $\sigma = 0.0001$ inch. A random sample of 12 shafts has an average diameter of 0.2545 inch.

(a) Set up appropriate hypotheses on the mean μ.

(b) Test these hypotheses using $\alpha = 0.05$. What are your conclusions?

(c) Find the P-value for this test.

(d) Construct a 95 percent confidence interval on the mean shaft diameter.

2.20. The shelf life of a carbonated beverage is of interest. Ten bottles are randomly selected and tested, and the following results are obtained:

Days	
108	138
124	163
124	159
106	134
115	139

(a) We would like to demonstrate that the mean shelf life exceeds 120 days. Set up appropriate hypotheses for investigating this claim.

(b) Test these hypotheses using $\alpha = 0.01$. What are your conclusions?

(c) Find the P-value for the test in part (b).

(d) Construct a 99 percent confidence interval on the mean shelf life.

2.21. Consider the shelf life data in Problem 2.20. Can shelf life be described or modeled adequately by a normal distribution? What effect would the violation of this assumption have on the test procedure you used in solving Problem 2.20?

2.22. The time to repair an electronic instrument is a normally distributed random variable measured in hours. The repair times for 16 such instruments chosen at random are as follows:

Hours			
189	280	201	215
224	379	179	264
222	362	168	250
149	260	485	170

(a) You wish to know if the mean repair time exceeds 225 hours. Set up appropriate hypotheses for investigating this issue.

(b) Test the hypotheses you formulated in part (a). What are your conclusions? Use $\alpha = 0.05$.

(c) Find the P-value for the test.

(d) Construct a 95 percent confidence interval on mean repair time.

2.23. Reconsider the repair time data in Problem 2.22. Can repair time, in your opinion, be adequately modeled by a normal distribution?

2.24. Two machines are used for filling plastic bottles with a net volume of 16.0 ounces. The filling processes can be assumed to be normal, with standard deviations of $\sigma_1 = 0.012$ and $\sigma_2 = 0.020$. The quality engineering department suspects that both machines fill to the same net volume, whether or not this volume is 16.0 ounces. An experiment is performed by taking a random sample from the output of each machine.

Machine 1		Machine 2	
16.03	16.01	16.02	16.03
16.04	15.96	15.97	16.04
16.05	15.98	15.96	16.02
16.05	16.02	16.01	16.01
16.02	15.99	15.99	16.00

(a) State the hypotheses that should be tested in this experiment.

(b) Test these hypotheses using $\alpha = 0.05$. What are your conclusions?

(c) Find the P-value for this test.

(d) Find a 95 percent confidence interval on the difference in mean fill volume for the two machines.

2.25. The following are the burning times (in minutes) of chemical flares of two different formulations. The design engineers are interested in both the mean and variance of the burning times.

Type 1		Type 2	
65	82	64	56
81	67	71	69
57	59	83	74
66	75	59	82
82	70	65	79

(a) Test the hypothesis that the two variances are equal. Use $\alpha = 0.05$.

(b) Using the results of (a), test the hypothesis that the mean burning times are equal. Use $\alpha = 0.05$. What is the P-value for this test?

(c) Discuss the role of the normality assumption in this problem. Check the assumption of normality for both types of flares.

2.26. Two types of plastic are suitable for use by an electronic calculator manufacturer. The breaking strength of this plastic is important. It is known that $\sigma_1 = \sigma_2 = 1.0$ psi. From random samples of $n_1 = 10$ and $n_2 = 12$ we obtain $\bar{y}_1 = 162.5$ and $\bar{y}_2 = 155.0$. The company will not adopt plastic 1 unless its breaking strength exceeds that of plastic 2 by at least 10 psi. Based on the sample information, should they use plastic

1? In answering this question, set up and test appropriate hypotheses using $\alpha = 0.01$. Construct a 99 percent confidence interval on the true mean difference in breaking strength.

2.27. An article in *Solid State Technology*, "Orthogonal Design for Process Optimization and Its Application to Plasma Etching" by G. Z. Yin and D. W. Jillie (May 1987) describes an experiment to determine the effect of the C_2F_6 flow rate on the uniformity of the etch on a silicon wafer used in integrated circuit manufacturing. All of the runs were made in random order. Data for two flow rates are as follows:

C_2F_6 Flow (SCCM)	Uniformity Observation 1	2	3	4	5	6
125	2.7	4.6	2.6	3.0	3.2	3.8
200	4.6	3.4	2.9	3.5	4.1	5.1

(a) Does the C_2F_6 flow rate affect average etch uniformity? Use $\alpha = 0.05$.

(b) What is the *P*-value for the test in part (a)?

(c) Does the C_2F_6 flow rate affect the wafer-to-wafer variability in etch uniformity? Use $\alpha = 0.05$.

(d) Draw box plots to assist in the interpretation of the data from this experiment.

2.28. Photoresist is a light-sensitive material applied to semiconductor wafers so that the circuit pattern can be imaged on to the wafer. After application, the coated wafers are baked to remove the solvent in the photoresist mixture and to harden the resist. Here are measurements of photoresist thickness (in kA) for eight wafers baked at two different temperatures. Assume that all of the runs were made in random order.

95 °C	100 °C
11.176	5.265
7.089	6.751
8.097	7.463
11.739	6.995
11.291	8.103
10.759	7.525
6.467	3.789
8.315	8.895

(a) Is there evidence to support the claim that the higher baking temperature results in wafers with a lower mean photoresist thickness? Use $\alpha = 0.05$.

(b) What is the *P*-value for the test conducted in part (a)?

(c) Find a 95 percent confidence interval on the difference in means. Provide a practical interpretation of this interval.

(d) Draw dot diagrams to assist in interpreting the results from this experiment.

(e) Check the assumption of normality of the photoresist thickness.

(f) Find the power of this test for detecting an actual difference in means of 2.5 kA.

(g) What sample size would be necessary to detect an actual difference in means of 1.5 kA with a power of at least 0.9?

2.29. A new filtering device is installed in a chemical unit. Before its installation, a random sample yielded the following information about the percentage of impurity: $\bar{y}_1 = 12.5$, $S_1^2 = 101.17$, and $n_1 = 8$. After installation, a random sample yielded $\bar{y}_2 = 10.2$, $S_2^2 = 94.73$, $n_2 = 9$.

(a) Can you conclude that the two variances are equal? Use $\alpha = 0.05$.

(b) Has the filtering device reduced the percentage of impurity significantly? Use $\alpha = 0.05$.

2.30. Twenty observations on etch uniformity on silicon wafers are taken during a qualification experiment for a plasma etcher. The data are as follows:

5.34	6.65	4.76	5.98	7.25
6.00	7.55	5.54	5.62	6.21
5.97	7.35	5.44	4.39	4.98
5.25	6.35	4.61	6.00	5.32

(a) Construct a 95 percent confidence interval estimate of σ^2.

(b) Test the hypothesis that $\sigma^2 = 1.0$. Use $\alpha = 0.05$. What are your conclusions?

(c) Discuss the normality assumption and its role in this problem.

(d) Check normality by constructing a normal probability plot. What are your conclusions?

2.31. Front housings for cell phones are manufactured in an injection molding process. The time the part is allowed to cool in the mold before removal is thought to influence the occurrence of a particularly troublesome cosmetic defect, flow lines, in the finished housing. After manufacturing, the housings are inspected visually and assigned a score between 1 and 10 based on their appearance, with 10 corresponding to a perfect part and 1 corresponding to a completely defective part. An experiment was conducted using two cool-down times, 10 and 20 seconds, and 20 housings were evaluated at each level of cool-down time. All 40 observations in this experiment were run in random order. The data are as follows.

10 seconds		20 seconds	
1	3	7	6
2	6	8	9
1	5	5	5
3	3	9	7
5	2	5	4

1	1	8	6
5	6	6	8
2	8	4	5
3	2	6	8
5	3	7	7

(a) Is there evidence to support the claim that the longer cool-down time results in fewer appearance defects? Use $\alpha = 0.05$.

(b) What is the *P*-value for the test conducted in part (a)?

(c) Find a 95 percent confidence interval on the difference in means. Provide a practical interpretation of this interval.

(d) Draw dot diagrams to assist in interpreting the results from this experiment.

(e) Check the assumption of normality for the data from this experiment.

2.32. The diameter of a ball bearing was measured by 10 inspectors, each using two different kinds of calipers. The results were

Inspector	Caliper 1	Caliper 2
1	0.265	0.264
2	0.265	0.265
3	0.266	0.264
4	0.267	0.266
5	0.267	0.267
6	0.265	0.268
7	0.267	0.264
8	0.267	0.265
9	0.265	0.265
10	0.268	0.267

(a) Is there a significant difference between the means of the population of measurements from which the two samples were selected? Use $\alpha = 0.05$.

(b) Find the *P*-value for the test in part (a).

(c) Construct a 95 percent confidence interval on the difference in mean diameter measurements for the two types of calipers.

2.33. An article in the journal *Neurology* (1998, Vol. 50, pp. 1246–1252) observed that monozygotic twins share numerous physical, psychological, and pathological traits. The investigators measured an intelligence score of 10 pairs of twins. The data obtained are as follows:

Pair	Birth Order: 1	Birth Order: 2
1	6.08	5.73
2	6.22	5.80
3	7.99	8.42
4	7.44	6.84
5	6.48	6.43
6	7.99	8.76
7	6.32	6.32
8	7.60	7.62
9	6.03	6.59
10	7.52	7.67

(a) Is the assumption that the difference in score is normally distributed reasonable?

(b) Find a 95% confidence interval on the difference in mean score. Is there any evidence that mean score depends on birth order?

(c) Test an appropriate set of hypotheses indicating that the mean score does not depend on birth order.

2.34. An article in the *Journal of Strain Analysis* (Vol. 18, No. 2, 1983) compares several procedures for predicting the shear strength for steel plate girders. Data for nine girders in the form of the ratio of predicted to observed load for two of these procedures, the Karlsruhe and Lehigh methods, are as follows:

Girder	Karlsruhe Method	Lehigh Method
S1/1	1.186	1.061
S2/1	1.151	0.992
S3/1	1.322	1.063
S4/1	1.339	1.062
S5/1	1.200	1.065
S2/1	1.402	1.178
S2/2	1.365	1.037
S2/3	1.537	1.086
S2/4	1.559	1.052

(a) Is there any evidence to support a claim that there is a difference in mean performance between the two methods? Use $\alpha = 0.05$.

(b) What is the *P*-value for the test in part (a)?

(c) Construct a 95 percent confidence interval for the difference in mean predicted to observed load.

(d) Investigate the normality assumption for both samples.

(e) Investigate the normality assumption for the difference in ratios for the two methods.

(f) Discuss the role of the normality assumption in the paired *t*-test.

2.35. The deflection temperature under load for two different formulations of ABS plastic pipe is being studied. Two samples of 9 observations each are prepared using each formulation and the deflection temperatures (in °F) are reported below:

Formulation 1			Formulation 2		
216	193	192	187	176	198
198	207	210	197	185	188
205	185	194	206	200	189

(a) Construct normal probability plots for both samples. Do these plots support assumptions of normality and equal variance for both samples?

(b) Do the data support the claim that the mean deflection temperature under load for formulation 1 exceeds that of formulation 2? Use $\alpha = 0.05$.

(c) What is the P-value for the test in part (a)?

2.36. Refer to the data in Problem 2.35. Do the data support a claim that the mean deflection temperature under load for formulation 1 exceeds that of formulation 2 by at least 3°F?

2.37. In semiconductor manufacturing wet chemical etching is often used to remove silicon from the backs of wafers prior to metalization. The etch rate is an important characteristic of this process. Two different etching solutions are being evaluated. Six randomly selected wafers have been etched in each solution, and the observed etch rates (in mils/min) are as follows.

Solution 1		Solution 2	
9.9	10.6	10.2	10.6
9.4	10.3	10.0	10.2
10.0	9.3	10.7	10.4

(a) Do the data indicate that the claim that both solutions have the same mean etch rate is valid? Use $\alpha = 0.05$ and assume equal variances.

(b) Find a 95 percent confidence interval on the difference in mean etch rates.

(c) Use normal probability plots to investigate the adequacy of the assumptions of normality and equal variances.

2.38. Two popular pain medications are being compared on the basis of the speed of absorption by the body. Specifically, tablet 1 is claimed to be absorbed twice as fast as tablet 2. Assume that σ_1^2 and σ_2^2 are known. Develop a test statistic for

$$H_0: 2\mu_1 = \mu_2$$
$$H_1: 2\mu_1 \neq \mu_2$$

2.39. *Continuation of Problem 2.38*. An article in *Nature* (1972, pp. 225–226) reported on the levels of monoamine oxidase in blood platelets for a sample of 43 schizophrenic patients resulting in $\bar{y}_1 = 2.69$ and $s_1 = 2.30$ while for a sample of 45 normal patients the results were $\bar{y}_2 = 6.35$ and $s_2 = 4.03$. The units are nm/mg protein/h. Use the results of the previous problem to test the claim that the mean monoamine oxidase level for normal patients is at last twice the mean level for schizophrenic patients. Assume that the sample sizes are large enough to use the sample standard deviations as the true parameter values.

2.40. Suppose we are testing

$$H_0: \mu_1 = \mu_2$$
$$H_1: \mu_1 \neq \mu_2$$

where $\sigma_1^2 > \sigma_2^2$ are known. Our sampling resources are constrained such that $n_1 + n_2 = N$. Show that an allocation of the observation n_1 and n_2 to the two sample that leads to the most powerful test is in the ratio $n_1/n_2 = \sigma_1/\sigma_2$.

2.41. ***Continuation of Problem 2.40***. Suppose that we want to construct a 95% two-sided confidence interval on the difference in two means where the two sample standard deviations are known to be $\sigma_1 = 4$ and $\sigma_2 = 8$. The total sample size is restricted to $N = 30$. What is the length of the 95% CI if the sample sizes used by the experimenter are $n_1 = n_2 = 15$? How much shorter would the 95% CI have been if the experimenter had used an optimal sample size allocation?

2.42. Develop Equation 2.46 for a $100(1 - \alpha)$ percent confidence interval for the variance of a normal distribution.

2.43. Develop Equation 2.50 for a $100(1 - \alpha)$ percent confidence interval for the ratio σ_1^2/σ_2^2, where σ_1^2 and σ_2^2 are the variances of two normal distributions.

2.44. Develop an equation for finding a $100\,(1 - \alpha)$ percent confidence interval on the difference in the means of two normal distributions where $\sigma_1^2 \neq \sigma_2^2$. Apply your equation to the Portland cement experiment data, and find a 95 percent confidence interval.

2.45. Construct a data set for which the paired t-test statistic is very large, but for which the usual two-sample or pooled t-test statistic is small. In general, describe how you created the data. Does this give you any insight regarding how the paired t-test works?

2.46. Consider the experiment described in Problem 2.25. If the mean burning times of the two flares differ by as much as 2 minutes, find the power of the test. What sample size would be required to detect an actual difference in mean burning time of 1 minute with a power of at least 0.90?

2.47. Reconsider the bottle filling experiment described in Problem 2.24. Rework this problem assuming that the two population variances are unknown but equal.

2.48. Consider the data from Problem 2.24. If the mean fill volume of the two machines differ by as much as 0.25 ounces, what is the power of the test used in Problem 2.24? What sample size would result in a power of at least 0.9 if the actual difference in mean fill volume is 0.25 ounces?

CHAPTER 3

Analysis of Variance

CHAPTER OUTLINE

he supplemental material is on the textbook website www.wiley.com/go/global/montgomery.

In Chapter 2, we discussed methods for comparing two conditions or treatments. For example, the Portland cement tension bond experiment involved two different mortar formulations. Another way to describe this experiment is as a single-factor experiment with two levels of the factor, where the factor is mortar formulation and the two levels are the two different formulation methods. Many experiments of this type involve more than two levels of the factor. This chapter focuses on methods for the design and analysis of single-factor experiments with an arbitrary number *a* levels of the factor (or *a* treatments). We will assume that the experiment has been completely randomized.

3.1 An Example

In many integrated circuit manufacturing steps, wafers are completely coated with a layer of material such as silicon dioxide or a metal. The unwanted material is then selectively removed by etching through a mask, thereby creating circuit patterns, electrical interconnects, and areas in which diffusions or metal depositions are to be made. A plasma etching process is widely used for this operation, particularly in small geometry applications. Figure 3.1 shows the important features of a typical single-wafer etching tool. Energy is supplied by a radio-frequency (RF) generator causing plasma to be generated in the gap between the electrodes. The chemical species in the plasma are determined by the particular gases used. Fluorocarbons, such as CF_4 (tetrafluoromethane) or C_2F_6 (hexafluoroethane), are often used in plasma etching, but other gases and mixtures of gases are relatively common, depending on the application.

An engineer is interested in investigating the relationship between the RF power setting and the etch rate for this tool. The objective of an experiment like this is to model the relationship between etch rate and RF power, and to specify the power setting that will give a desired target etch rate. She is interested in a particular gas (C_2F_6) and gap (0.80 cm) and wants to test four levels of RF power: 160, 180, 200, and 220 W. She decided to test five wafers at each level of RF power.

This is an example of a single-factor experiment with $a = 4$ **levels** of the factor and $n = 5$ **replicates**. The 20 runs should be made in random order. A very efficient way to generate the run order is to enter the 20 runs in a spreadsheet (Excel), generate a column of random numbers using the RAND ( ) function, and then sort by that column.

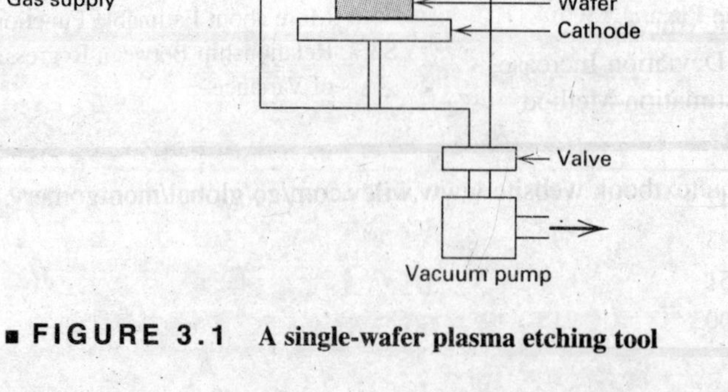

■ **FIGURE 3.1** **A single-wafer plasma etching tool**

Suppose that the test sequence obtained from this process is given as below:

Test Sequence	Excel Random Number (Sorted)	Power
1	12417	200
2	18369	220
3	21238	220
4	24621	160
5	29337	160
6	32318	180
7	36481	200
8	40062	160
9	43289	180
10	49271	200
11	49813	220
12	52286	220
13	57102	160
14	63548	160
15	67710	220
16	71834	180
17	77216	180
18	84675	180
19	89323	200
20	94037	200

This randomized test sequence is necessary to prevent the effects of unknown nuisance variables, perhaps varying out of control during the experiment, from contaminating the results. To illustrate this, suppose that we were to run the 20 test wafers in the original nonrandomized order (that is, all five 160 W power runs are made first, all five 180 W power runs are made next, and so on). If the etching tool exhibits a warm-up effect such that the longer it is on, the lower the observed etch rate readings will be, the warm-up effect will potentially contaminate the data and destroy the validity of the experiment.

Suppose that the engineer runs the experiment that we have designed in the random order. The observations that she obtains on etch rate are shown in Table 3.1.

It is always a good idea to examine experimental data **graphically**. Figure 3.2*a* presents **box plots** for etch rate at each level of RF power, and Figure 3.2*b* a **scatter diagram** of etch rate versus RF power. Both graphs indicate that etch rate increases as the power setting increases. There

■ **TABLE 3.1**
Etch Rate Data (in Å/min) from the Plasma Etching Experiment

Power (W)	Observations 1	2	3	4	5	Totals	Averages
160	575	542	530	539	570	2756	551.2
180	565	593	590	579	610	2937	587.4
200	600	651	610	637	629	3127	625.4
220	725	700	715	685	710	3535	707.0

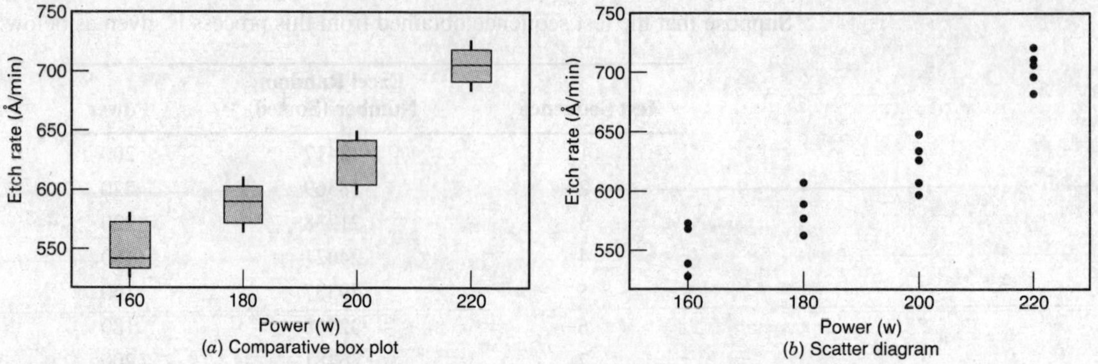

■ **FIGURE 3.2** **Box plots and scatter diagram of the etch rate data**

is no strong evidence to suggest that the variability in etch rate around the average depends on the power setting. On the basis of this simple graphical analysis, we strongly suspect that (1) RF power setting affects the etch rate and (2) higher power settings result in increased etch rate.

Suppose that we wish to be more **objective** in our analysis of the data. Specifically, suppose that we wish to test for differences between the mean etch rates at all $a = 4$ levels of RF power. Thus, we are interested in testing the equality of all four means. It might seem that this problem could be solved by performing a t-test for all six possible pairs of means. However, this is not the best solution to this problem. First of all, performing all six pairwise t-tests is inefficient. It takes a lot of effort. Second, conducting all these pairwise comparisons inflates the type I error. Suppose that all four means are equal, so if we select $\alpha = 0.05$, the probability of reaching the correct decision on any single comparison is 0.95. However, the probability of reaching the correct conclusion on all six comparisons is considerably less than 0.95, so the type I error is inflated.

The appropriate procedure for testing the equality of several means is the **analysis of variance**. However, the analysis of variance has a much wider application than the problem above. It is probably the most useful technique in the field of statistical inference.

3.2 The Analysis of Variance

Suppose we have a **treatments** or different **levels** of a **single factor** that we wish to compare. The observed response from each of the a treatments is a random variable. The data would appear as in Table 3.2. An entry in Table 3.2 (e.g., y_{ij}) represents the jth observation taken under factor level or treatment i. There will be, in general, n observations under the ith treatment. Notice that Table 3.2 is the general case of the data from the plasma etching experiment in Table 3.1.

■ **TABLE 3.2**
Typical Data for a Single-Factor Experiment

Treatment (Level)	Observations				Totals	Averages
1	y_{11}	y_{12}	$\cdots$	y_{1n}	$y_{1.}$	$\bar{y}_{1.}$
2	y_{21}	y_{22}	$\cdots$	y_{2n}	$y_{2.}$	$\bar{y}_{2.}$
$\vdots$	$\vdots$	$\vdots$	$\vdots\vdots\vdots$	$\vdots$	$\vdots$	$\vdots$
a	y_{a1}	y_{a2}	$\cdots$	y_{an}	$\underline{y_{a.}}$	$\underline{\bar{y}_{a.}}$
					$y_{..}$	$\bar{y}_{..}$

Models for the Data. We will find it useful to describe the observations from an experiment with a **model**. One way to write this model is

$$y_{ij} = \mu_i + \epsilon_{ij} \begin{cases} i = 1, 2, \ldots, a \\ j = 1, 2, \ldots, n \end{cases} \tag{3.1}$$

where y_{ij} is the ijth observation, μ_i is the mean of the ith factor level or treatment, and ϵ_{ij} is a **random error** component that incorporates all other sources of variability in the experiment including measurement, variability arising from uncontrolled factors, differences between the experimental units (such as test material, etc.) to which the treatments are applied, and the general background noise in the process (such as variability over time, effects of environmental variables, and so forth). It is convenient to think of the errors as having mean zero, so that $E(y_{ij}) = \mu_i$.

Equation 3.1 is called the **means model**. An alternative way to write a model for the data is to define

$$\mu_i = \mu + \tau_i, \qquad i = 1, 2, \ldots, a$$

so that Equation 3.1 becomes

$$y_{ij} = \mu + \tau_i + \epsilon_{ij} \begin{cases} i = 1, 2, \ldots, a \\ j = 1, 2, \ldots, n \end{cases} \tag{3.2}$$

In this form of the model, μ is a parameter common to all treatments called the **overall mean**, and τ_i is a parameter unique to the ith treatment called the ***i*th treatment effect.** Equation 3.2 is usually called the **effects model**.

Both the means model and the effects model are **linear statistical models**; that is, the response variable y_{ij} is a linear function of the model parameters. Although both forms of the model are useful, the effects model is more widely encountered in the experimental design literature. It has some intuitive appeal in that μ is a constant and the treatment effects τ_i represent deviations from this constant when the specific treatments are applied.

Equation 3.2 (or 3.1) is also called the **one-way** or **single-factor analysis of variance (ANOVA)** model because only one factor is investigated. Furthermore, we will require that the experiment be performed in random order so that the environment in which the treatments are applied (often called the **experimental units**) is as uniform as possible. Thus, the experimental design is a **completely randomized design**. Our objectives will be to test appropriate hypotheses about the treatment means and to estimate them. For hypothesis testing, the model errors are assumed to be normally and independently distributed random variables with mean zero and variance σ^2. The variance σ^2 is assumed to be constant for all levels of the factor. This implies that the observations

$$y_{ij} \sim N(\mu + \tau_i, \sigma^2)$$

and that the observations are mutually independent.

Fixed or Random Factor? The statistical model, Equation 3.2, describes two different situations with respect to the treatment effects. First, the a treatments could have been specifically chosen by the experimenter. In this situation, we wish to test hypotheses about the treatment means, and our conclusions will apply only to the factor levels considered in the analysis. The conclusions cannot be extended to similar treatments that were not explicitly considered. We may also wish to estimate the model parameters (μ, τ_i, σ^2). This is called the **fixed effects model**. Alternatively, the a treatments could be a **random sample** from a larger population of treatments. In this situation, we should like to be able to extend the conclusions (which are based on the sample of treatments) to all treatments in the population, whether or not they were explicitly considered in the analysis. Here, the τ_i are random variables, and knowledge about the particular ones investigated is relatively useless. Instead, we test hypotheses about the variability of the τ_i and try to estimate this variability. This is called the **random effects model** or **components of variance model**. We discuss the single-factor random effects model in Section 3.9. However, we will defer a more complete discussion of experiments with random factors to Chapter 13.

3.3 Analysis of the Fixed Effects Model

In this section, we develop the single-factor analysis of variance for the fixed effects model. Recall that $y_{i.}$ represents the total of the observations under the ith treatment. Let $\bar{y}_{i.}$ represent the average of the observations under the ith treatment. Similarly, let $y_{..}$ represent the grand total of all the observations and $\bar{y}_{..}$ represent the grand average of all the observations. Expressed symbolically,

$$y_{i.} = \sum_{j=1}^{n} y_{ij} \qquad \bar{y}_{i.} = y_{i.}/n \qquad i = 1, 2, \ldots, a$$

$$y_{..} = \sum_{i=1}^{a}\sum_{j=1}^{n} y_{ij} \qquad \bar{y}_{..} = y_{..}/N \tag{3.3}$$

where $N = an$ is the total number of observations. We see that the "dot" subscript notation implies summation over the subscript that it replaces.

We are interested in testing the equality of the a treatment means; that is, $E(y_{ij}) = \mu + \tau_i = \mu_i$, $i = 1, 2, \ldots, a$. The appropriate hypotheses are

$$\begin{aligned} H_0&: \mu_1 = \mu_2 = \cdots = \mu_a \\ H_1&: \mu_i \neq \mu_j \quad \text{for at least one pair } (i, j) \end{aligned} \tag{3.4}$$

In the effects model, we break the ith treatment mean μ_i into two components such that $\mu_i = \mu + \tau_i$. We usually think of μ as an overall mean so that

$$\frac{\sum_{i=1}^{a} \mu_i}{a} = \mu$$

This definition implies that

$$\sum_{i=1}^{a} \tau_i = 0$$

That is, the treatment or factor effects can be thought of as deviations from the overall mean.[1] Consequently, an equivalent way to write the above hypotheses is in terms of the treatment effects τ_i, say

$$\begin{aligned} H_0&: \tau_1 = \tau_2 = \cdots \tau_a = 0 \\ H_1&: \tau_i \neq 0 \quad \text{for at least one } i \end{aligned}$$

Thus, we speak of testing the equality of treatment means or testing that the treatment effects (the τ_i) are zero. The appropriate procedure for testing the equality of a treatment means is the analysis of variance.

[1] For more information on this subject, refer to the supplemental text material for Chapter 3.

3.3.1 Decomposition of the Total Sum of Squares

The name **analysis of variance** is derived from a partitioning of total variability into its component parts. The total corrected sum of squares

$$SS_T = \sum_{i=1}^{a} \sum_{j=1}^{n} (y_{ij} - \bar{y}_{..})^2$$

is used as a measure of overall variability in the data. Intuitively, this is reasonable because if we were to divide SS_T by the appropriate number of degrees of freedom (in this case, $an - 1 = N - 1$), we would have the **sample variance** of the y's. The sample variance is, of course, a standard measure of variability.

Note that the total corrected sum of squares SS_T may be written as

$$\sum_{i=1}^{a} \sum_{j=1}^{n} (y_{ij} - \bar{y}_{..})^2 = \sum_{i=1}^{a} \sum_{j=1}^{n} [(\bar{y}_{i.} - \bar{y}_{..}) + (y_{ij} - \bar{y}_{i.})]^2 \tag{3.5}$$

or

$$\sum_{i=1}^{a} \sum_{j=1}^{n} (y_{ij} - \bar{y}_{..})^2 = n \sum_{i=1}^{a} (\bar{y}_{i.} - \bar{y}_{..})^2 + \sum_{i=1}^{a} \sum_{j=1}^{n} (y_{ij} - \bar{y}_{i.})^2 + 2 \sum_{i=1}^{a} \sum_{j=1}^{n} (\bar{y}_{i.} - \bar{y}_{..})(y_{ij} - \bar{y}_{i.})$$

However, the cross-product term in this last equation is zero, because

$$\sum_{j=1}^{n} (y_{ij} - \bar{y}_{i.}) = y_{i.} - n\bar{y}_{i.} = y_{i.} - n(y_{i.}/n) = 0$$

Therefore, we have

$$\sum_{i=1}^{a} \sum_{j=1}^{n} (y_{ij} - \bar{y}_{..})^2 = n \sum_{i=1}^{a} (\bar{y}_{i.} - \bar{y}_{..})^2 + \sum_{i=1}^{a} \sum_{j=1}^{n} (y_{ij} - \bar{y}_{i.})^2 \tag{3.6}$$

Equation 3.6 is the fundamental ANOVA identity. It states that the total variability in the data, as measured by the total corrected sum of squares, can be partitioned into a sum of squares of the differences **between** the treatment averages and the grand average plus a sum of squares of the differences of observations **within** treatments from the treatment average. Now, the difference between the observed treatment averages and the grand average is a measure of the differences between treatment means, whereas the differences of observations within a treatment from the treatment average can be due to only random error. Thus, we may write Equation 3.6 symbolically as

$$SS_T = SS_{\text{Treatments}} + SS_E$$

where $SS_{\text{Treatments}}$ is called the sum of squares due to treatments (i.e., between treatments), and SS_E is called the sum of squares due to error (i.e., within treatments). There are $an = N$ total observations; thus, SS_T has $N - 1$ degrees of freedom. There are a levels of the factor (and a treatment means), so $SS_{\text{Treatments}}$ has $a - 1$ degrees of freedom. Finally, there are n replicates within any treatment providing $n - 1$ degrees of freedom with which to estimate the experimental error. Because there are a treatments, we have $a(n - 1) = an - a = N - a$ degrees of freedom for error.

It is instructive to examine explicitly the two terms on the right-hand side of the fundamental ANOVA identity. Consider the error sum of squares

$$SS_E = \sum_{i=1}^{a} \sum_{j=1}^{n} (y_{ij} - \bar{y}_{i.})^2 = \sum_{i=1}^{a} \left[\sum_{j=1}^{n} (y_{ij} - \bar{y}_{i.})^2 \right]$$

In this form, it is easy to see that the term within square brackets, if divided by $n - 1$, is the sample variance in the ith treatment, or

$$S_i^2 = \frac{\sum_{j=1}^{n} (y_{ij} - \bar{y}_{i.})^2}{n - 1} \qquad i = 1, 2, \ldots, a$$

Now a sample variances may be combined to give a single estimate of the common population variance as follows:

$$\frac{(n-1)S_1^2 + (n-1)S_2^2 + \cdots + (n-1)S_a^2}{(n-1) + (n-1) + \cdots + (n-1)} = \frac{\sum_{i=1}^{a}\left[\sum_{j=1}^{n} (y_{ij} - \bar{y}_{i.})^2\right]}{\sum_{i=1}^{a} (n-1)}$$

$$= \frac{SS_E}{(N - a)}$$

Thus, $SS_E/(N - a)$ is a **pooled estimate** of the common variance within each of the a treatments.

Similarly, if there were no differences between the a treatment means, we could use the variation of the treatment averages from the grand average to estimate σ^2. Specifically,

$$\frac{SS_{\text{Treatments}}}{a - 1} = \frac{n \sum_{i=1}^{a} (\bar{y}_{i.} - \bar{y}_{..})^2}{a - 1}$$

is an estimate of σ^2 if the treatment means are equal. The reason for this may be intuitively seen as follows: The quantity $\Sigma_{i=1}^{a}(\bar{y}_{i.} - \bar{y}_{..})^2/(a - 1)$ estimates σ^2/n, the variance of the treatment averages, so $n\Sigma_{i=1}^{a}(\bar{y}_{i.} - \bar{y}_{..})^2/(a - 1)$ must estimate σ^2 if there are no differences in treatment means.

We see that the ANOVA identity (Equation 3.6) provides us with two estimates of σ^2—one based on the inherent variability within treatments and the other based on the variability between treatments. If there are no differences in the treatment means, these two estimates should be very similar, and if they are not, we suspect that the observed difference must be caused by differences in the treatment means. Although we have used an intuitive argument to develop this result, a somewhat more formal approach can be taken.

The quantities

$$MS_{\text{Treatments}} = \frac{SS_{\text{Treatments}}}{a - 1}$$

and

$$MS_E = \frac{SS_E}{N - a}$$

are called **mean squares**. We now examine the **expected values** of these mean squares. Consider

$$E(MS_E) = E\left(\frac{SS_E}{N - a}\right) = \frac{1}{N - a} E\left[\sum_{i=1}^{a}\sum_{j=1}^{n} (y_{ij} - \bar{y}_{i.})^2\right]$$

$$= \frac{1}{N - a} E\left[\sum_{i=1}^{a}\sum_{j=1}^{n} (y_{ij}^2 - 2y_{ij}\bar{y}_{i.} + \bar{y}_{i.}^2)\right]$$

$$= \frac{1}{N-a} E\left[\sum_{i=1}^{a}\sum_{j=1}^{n} y_{ij}^2 - 2n\sum_{i=1}^{a} \bar{y}_{i.}^2 + n\sum_{i=1}^{a} \bar{y}_{i.}^2\right]$$

$$= \frac{1}{N-a} E\left[\sum_{i=1}^{a}\sum_{j=1}^{n} y_{ij}^2 - \frac{1}{n}\sum_{i=1}^{a} y_{i.}^2\right]$$

Substituting the model (Equation 3.1) into this equation, we obtain

$$E(MS_E) = \frac{1}{N-a} E\left[\sum_{i=1}^{a}\sum_{j=1}^{n} (\mu + \tau_i + \epsilon_{ij})^2 - \frac{1}{n}\sum_{i=1}^{a}\left(\sum_{j=1}^{n} \mu + \tau_i + \epsilon_{ij}\right)^2\right]$$

Now when squaring and taking expectation of the quantity within the brackets, we see that terms involving ϵ_{ij}^2 and $\epsilon_{i.}^2$ are replaced by σ^2 and $n\sigma^2$, respectively, because $E(\epsilon_{ij}) = 0$. Furthermore, all cross products involving ϵ_{ij} have zero expectation. Therefore, after squaring and taking expectation, the last equation becomes

$$E(MS_E) = \frac{1}{N-a}\left[N\mu^2 + n\sum_{i=1}^{a} \tau_i^2 + N\sigma^2 - N\mu^2 - n\sum_{i=1}^{a} \tau_i^2 - a\sigma^2\right]$$

or

$$E(MS_E) = \sigma^2$$

By a similar approach, we may also show that[2]

$$E(MS_{\text{Treatments}}) = \sigma^2 + \frac{n\sum_{i=1}^{a} \tau_i^2}{a-1}$$

Thus, as we argued heuristically, $MS_E = SS_E/(N-a)$ estimates σ^2, and, if there are no differences in treatment means (which implies that $\tau_i = 0$), $MS_{\text{Treatments}} = SS_{\text{Treatments}}/(a-1)$ also estimates σ^2. However, note that if treatment means do differ, the expected value of the treatment mean square is greater than σ^2.

It seems clear that a test of the hypothesis of no difference in treatment means can be performed by comparing $MS_{\text{Treatments}}$ and MS_E. We now consider how this comparison may be made.

3.3.2 Statistical Analysis

We now investigate how a formal test of the hypothesis of no differences in treatment means ($H_0: \mu_1 = \mu_2 = \cdots = \mu_a$, or equivalently, $H_0: \tau_1 = \tau_2 = \cdots = \tau_a = 0$) can be performed. Because we have assumed that the errors ϵ_{ij} are normally and independently distributed with mean zero and variance σ^2, the observations y_{ij} are normally and independently distributed with mean $\mu + \tau_i$ and variance σ^2. Thus, SS_T is a sum of squares in normally distributed random variables; consequently, it can be shown that SS_T/σ^2 is distributed as chi-square with $N-1$ degrees of freedom. Furthermore, we can show that SS_E/σ^2 is chi-square with $N-a$ degrees of freedom and that $SS_{\text{Treatments}}/\sigma^2$ is chi-square with $a-1$ degrees of freedom if the null hypothesis $H_0: \tau_i = 0$ is true. However, all three sums of squares are not necessarily independent because $SS_{\text{Treatments}}$ and SS_E add to SS_T. The following theorem, which is a special form of one attributed to William G. Cochran, is useful in establishing the independence of SS_E and $SS_{\text{Treatments}}$.

[2] Refer to the supplemental text material for Chapter 3.

THEOREM 3-1
Cochran's Theorem

Let Z_i be NID(0, 1) for $i = 1, 2, \ldots, \nu$ and

$$\sum_{i=1}^{\nu} Z_i^2 = Q_1 + Q_2 + \cdots + Q_s$$

where $s \leq \nu$, and Q_i has ν_i degrees of freedom ($i = 1, 2, \ldots, s$). Then $Q_1, Q_2, \ldots, Q_s$ are independent chi-square random variables with $\nu_1, \nu_2, \ldots, \nu_s$ degrees of freedom, respectively, if and only if

$$\nu = \nu_1 + \nu_2 + \cdots + \nu_s$$

Because the degrees of freedom for $SS_{\text{Treatments}}$ and SS_E add to $N - 1$, the total number of degrees of freedom, Cochran's theorem implies that $SS_{\text{Treatments}}/\sigma^2$ and SS_E/σ^2 are independently distributed chi-square random variables. Therefore, if the null hypothesis of no difference in treatment means is true, the ratio

$$F_0 = \frac{SS_{\text{Treatments}}/(a-1)}{SS_E/(N-a)} = \frac{MS_{\text{Treatments}}}{MS_E} \tag{3.7}$$

is distributed as F with $a - 1$ and $N - a$ degrees of freedom. Equation 3.7 is the **test statistic** for the hypothesis of no differences in treatment means.

From the expected mean squares we see that, in general, MS_E is an unbiased estimator of σ^2. Also, under the null hypothesis, $MS_{\text{Treatments}}$ is an unbiased estimator of σ^2. However, if the null hypothesis is false, the expected value of $MS_{\text{Treatments}}$ is greater than σ^2. Therefore, under the alternative hypothesis, the expected value of the numerator of the test statistic (Equation 3.7) is greater than the expected value of the denominator, and we should reject H_0 on values of the test statistic that are too large. This implies an upper-tail, one-tail critical region. Therefore, we should reject H_0 and conclude that there are differences in the treatment means if

$$F_0 > F_{\alpha,a-1,N-a}$$

where F_0 is computed from Equation 3.7. Alternatively, we could use the P-value approach for decision making. The table of F percentages in the Appendix (Table IV) can be used to find bounds on the P-value.

The sums of squares may be computed in several ways. One direct approach is to make use of the definition

$$y_{ij} - \bar{y}_{..} = (\bar{y}_{i.} - \bar{y}_{..}) + (y_{ij} - \bar{y}_{i.})$$

Use a spreadsheet to compute these three terms for each observation. Then, sum up the squares to obtain SS_T, $SS_{\text{Treatments}}$, and SS_E. Another approach is to rewrite and simplify the definitions of $SS_{\text{Treatments}}$ and SS_T in Equation 3.6, which results in

$$SS_T = \sum_{i=1}^{a}\sum_{j=1}^{n} y_{ij}^2 - \frac{y_{..}^2}{N} \tag{3.8}$$

$$SS_{\text{Treatments}} = \frac{1}{n}\sum_{i=1}^{a} y_{i.}^2 - \frac{y_{..}^2}{N} \tag{3.9}$$

and

$$SS_E = SS_T - SS_{\text{Treatments}} \tag{3.10}$$

■ **TABLE 3.3**
The Analysis of Variance Table for the Single-Factor, Fixed Effects Model

Source of Variation	Sum of Squares	Degrees of Freedom	Mean Square	F_0
Between treatments	$SS_{\text{Treatments}} = n\sum_{i=1}^{a}(\bar{y}_{i.} - \bar{y}_{..})^2$	$a - 1$	$MS_{\text{Treatments}}$	$F_0 = \dfrac{MS_{\text{Treatments}}}{MS_E}$
Error (within treatments)	$SS_E = SS_T - SS_{\text{Treatments}}$	$N - a$	MS_E	
Total	$SS_T = \sum_{i=1}^{a}\sum_{j=1}^{n}(y_{ij} - \bar{y}_{..})^2$	$N-1$		

This approach is nice because some calculators are designed to accumulate the sum of entered numbers in one register and the sum of the squares of those numbers in another, so each number only has to be entered once. In practice, we use computer software to do this.

The test procedure is summarized in Table 3.3. This is called an **analysis of variance** (or **ANOVA**) table.

EXAMPLE 3.1 The Plasma Etching Experiment

To illustrate the analysis of variance, return to the first example discussed in Section 3.1. Recall that the engineer is interested in determining if the RF power setting affects the etch rate, and she has run a completely randomized experiment with four levels of RF power and five replicates. For convenience, we repeat here the data from Table 3.1:

We will use the analysis of variance to test $H_0: \mu_1 = \mu_2 = \mu_3 = \mu_4$ against the alternative H_1: some means are different. The sums of squares required are computed using Equations 3.8, 3.9, and 3.10 as follows:

RF Power (W)	Observed Etch Rate (Å/min) 1	2	3	4	5	Totals $y_{i.}$	Averages $\bar{y}_{i.}$
160	575	542	530	539	570	2756	551.2
180	565	593	590	579	610	2937	587.4
200	600	651	610	637	629	3127	625.4
220	725	700	715	685	710	3535	707.0
						$y_{..} = 12{,}355$	$\bar{y}_{..} = 617.75$

$$SS_T = \sum_{i=1}^{4}\sum_{j=1}^{5} y_{ij}^2 - \frac{y_{..}^2}{N}$$

$$= (575)^2 + (542)^2 + \cdots + (710)^2 - \frac{(12{,}355)^2}{20}$$

$$= 72{,}209.75$$

$$SS_{\text{Treatments}} = \frac{1}{n}\sum_{i=1}^{4} y_{i.}^2 - \frac{y_{..}^2}{N}$$

$$= \frac{1}{5}[(2756)^2 + \cdots + (3535)^2] - \frac{(12{,}355)^2}{20}$$

$$= 66{,}870.55$$

$$SS_E = SS_T - SS_{\text{Treatments}}$$

$$= 72{,}209.75 - 66{,}870.55 = 5339.20$$

Usually, these calculations would be performed on a computer, using a software package with the capability to analyze data from designed experiments.

The ANOVA is summarized in Table 3.4. Note that the RF power or between-treatment mean square (22,290.18) is many times larger than the within-treatment or error mean square (333.70). This indicates that it is unlikely that the treatment means are equal. More formally, we

■ **TABLE 3.4**
ANOVA for the Plasma Etching Experiment

Source of Variation	Sum of Squares	Degrees of Freedom	Mean Square	F_0	P-Value
RF Power	66,870.55	3	22,290.18	$F_0 = 66.80$	<0.01
Error	5339.20	16	333.70		
Total	72,209.75	19			

can compute the F ratio $F_0 = 22{,}290.18/333.70 = 66.80$ and compare this to an appropriate upper-tail percentage point of the $F_{3,16}$ distribution. To use a fixed significance level approach, suppose that the experimenter has selected $\alpha = 0.05$. From Appendix Table IV we find that $F_{0.05,3,16} = 3.24$. Because $F_0 = 66.80 > 3.24$, we reject H_0 and conclude that the treatment means differ; that is, the RF power setting significantly affects the mean etch rate. We could also compute a P-value for this test statistic. Figure 3.3 shows the reference distribution ($F_{3,16}$) for the test statistic F_0. Clearly, the P-value is very small in this case. From Appendix Table A-4, we find that $F_{0.01,3,16} = 5.29$ and because $F_0 > 5.29$, we can conclude that an upper bound for the P-value is 0.01; that is, $P < 0.01$ (the exact P-value is $P = 2.88 \times 10^{-9}$).

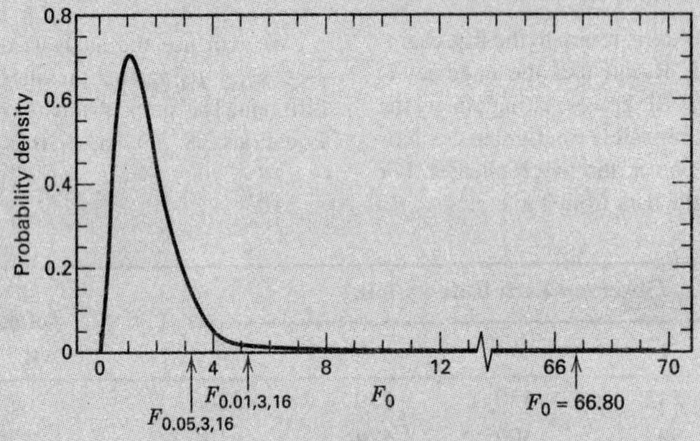

■ **FIGURE 3.3 The reference distribution ($F_{3,16}$) for the test statistic F_0 in Example 3.1**

Coding the Data. Generally, we need not be too concerned with computing because there are many widely available computer programs for performing the calculations. These computer programs are also helpful in performing many other analyses associated with experimental design (such as residual analysis and model adequacy checking). In many cases, these programs will also assist the experimenter in setting up the design.

However, when hand calculations are necessary, it is sometimes helpful to code the observations. This is illustrated in the next example.

EXAMPLE 3.2 Coding the Observations

The ANOVA calculations may often be made more easily or accurately by **coding** the observations. For example, consider the plasma etching data in Example 3.1. Suppose we subtract 600 from each observation. The coded data are shown in Table 3.5. It is easy to verify that

$$SS_T = (-25)^2 + (-58)^2 + \cdots + (110)^2 - \frac{(355)^2}{20} = 72{,}209.75$$

$$SS_{\text{Treatments}} = \frac{(-244)^2 + (-63)^2 + (127)^2 + (535)^2}{5} - \frac{(355)^2}{20} = 66{,}870.55$$

and

$$SS_E = 5339.20$$

Comparing these sums of squares to those obtained in Example 3.1, we see that subtracting a constant from the original data does not change the sums of squares.

Now suppose that we multiply each observation in Example 3.1 by 2. It is easy to verify that the sums of squares for the transformed data are $SS_T = 288{,}839.00$, $SS_{\text{Treatments}} = 267{,}482.20$, and $SS_E = 21{,}356.80$. These sums of squares appear to differ considerably from those obtained in Example 3.1. However, if they are divided by 4 (i.e., 2^2), the results are identical. For example, for the treatment sum of squares $267{,}482.20/4 = 66{,}870.55$. Also, for the coded data, the F ratio is $F = (267{,}482.20/3)/(21{,}356.80/16) = 66.80$, which is identical to the F ratio for the original data. Thus, the ANOVAs are equivalent.

■ **TABLE 3.5**
Coded Etch Rate Data for Example 3.2

RF Power (W)	Observations 1	2	3	4	5	Totals $y_{i.}$
160	−25	−58	−70	−61	−30	−244
180	−35	−7	−10	−21	10	−63
200	0	51	10	37	29	127
220	125	100	115	85	110	535

Randomization Tests and Analysis of Variance. In our development of the ANOVA F test, we have used the assumption that the random errors ϵ_{ij} are normally and independently distributed random variables. The F test can also be justified as an approximation to a **randomization test.** To illustrate this, suppose that we have five observations on each of two treatments and that we wish to test the equality of treatment means. The data would look like this:

Treatment 1	Treatment 2
y_{11}	y_{21}
y_{12}	y_{22}
y_{13}	y_{23}
y_{14}	y_{24}
y_{15}	y_{25}

We could use the ANOVA F test to test $H_0: \mu_1 = \mu_2$. Alternatively, we could use a somewhat different approach. Suppose we consider all the possible ways of allocating the 10 numbers in the above sample to the two treatments. There are 10!/5!5! = 252 possible arrangements of the 10 observations. If there is no difference in treatment means, all 252 arrangements are equally likely. For each of the 252 arrangements, we calculate the value of the F statistic using Equation 3.7. The distribution of these F values is called a **randomization distribution**, and a large value of F indicates that the data are not consistent with the hypothesis $H_0: \mu_1 = \mu_2$. For example, if the value of F actually observed was exceeded by only five of the values of the randomization distribution, this would correspond to rejection of $H_0: \mu_1 = \mu_2$ at a significance level of $\alpha = 5/252 = 0.0198$ (or 1.98 percent). Notice that no normality assumption is required in this approach.

The difficulty with this approach is that, even for relatively small problems, it is computationally prohibitive to enumerate the exact randomization distribution. However, numerous studies have shown that the exact randomization distribution is well approximated by the usual normal-theory F distribution. Thus, even without the normality assumption, the ANOVA F test can be viewed as an approximation to the randomization test. For further reading on randomization tests in the analysis of variance, see Box, Hunter, and Hunter (2005).

3.3.3 Estimation of the Model Parameters

We now present estimators for the parameters in the single-factor model

$$y_{ij} = \mu + \tau_i + \epsilon_{ij}$$

and confidence intervals on the treatment means. We will prove later that reasonable estimates of the overall mean and the treatment effects are given by

$$\hat{\mu} = \bar{y}_{..}$$

$$\hat{\tau}_i = \bar{y}_{i.} - \bar{y}_{..}, \qquad i = 1, 2, \ldots, a \tag{3.11}$$

These estimators have considerable intuitive appeal; note that the overall mean is estimated by the grand average of the observations and that any treatment effect is just the difference between the treatment average and the grand average.

A **confidence interval** estimate of the ith treatment mean may be easily determined. The mean of the ith treatment is

$$\mu_i = \mu + \tau_i$$

A point estimator of μ_i would be $\hat{\mu}_i = \hat{\mu} + \hat{\tau}_i = \bar{y}_{i.}$. Now, if we assume that the errors are normally distributed, each treatment average $\bar{y}_{i.}$ is distributed NID($\mu_i, \sigma^2/n$). Thus, if σ^2 were known, we could use the normal distribution to define the confidence interval. Using the MS_E as an estimator of σ^2, we would base the confidence interval on the t distribution. Therefore, a 100(1 − α) percent confidence interval on the ith treatment mean μ_i is

$$\bar{y}_{i.} - t_{\alpha/2,N-a}\sqrt{\frac{MS_E}{n}} \le \mu_i \le \bar{y}_{i.} + t_{\alpha/2,N-a}\sqrt{\frac{MS_E}{n}} \tag{3.12}$$

Differences in treatments are frequently of great practical interest. A 100(1 − α) percent confidence interval on the difference in any two treatments means, say $\mu_i - \mu_j$, would be

$$\bar{y}_{i.} - \bar{y}_{j.} - t_{\alpha/2,N-a}\sqrt{\frac{2MS_E}{n}} \le \mu_i - \mu_j \le \bar{y}_{i.} - \bar{y}_{j.} + t_{\alpha/2,N-a}\sqrt{\frac{2MS_E}{n}} \tag{3.13}$$

EXAMPLE 3.3

Using the data in Example 3.1, we may find the estimates of the overall mean and the treatment effects as $\hat{\mu} = 12{,}355/20 = 617.75$ and

$$\hat{\tau}_1 = \bar{y}_{1.} - \bar{y}_{..} = 551.20 - 617.75 = -66.55$$
$$\hat{\tau}_2 = \bar{y}_{2.} - \bar{y}_{..} = 587.40 - 617.75 = -30.35$$
$$\hat{\tau}_3 = \bar{y}_{3.} - \bar{y}_{..} = 625.40 - 617.75 = 7.65$$
$$\hat{\tau}_4 = \bar{y}_{4.} - \bar{y}_{..} = 707.00 - 617.75 = 89.25$$

A 95 percent confidence interval on the mean of treatment 4 (220W of RF power) is computed from Equation 3.12 as

$$707.00 - 2.120\sqrt{\frac{333.70}{5}} \le \mu_4 \le 707.00 + 2.120\sqrt{\frac{333.70}{5}}$$

or

$$707.00 - 17.32 \le \mu_4 \le 707.00 + 17.32$$

Thus, the desired 95 percent confidence interval is $689.68 \le \mu_4 \le 724.32$.

Simultaneous Confidence Intervals. The confidence interval expressions given in Equations 3.12 and 3.13 are **one-at-a-time** confidence intervals. That is, the confidence level $1 - \alpha$ applies to only one particular estimate. However, in many problems, the experimenter may wish to calculate several confidence intervals, one for each of a number of means or differences between means. If there are r such $100(1 - \alpha)$ percent confidence intervals of interest, the probability that the r intervals will **simultaneously** be correct is at least $1 - r\alpha$. The probability $r\alpha$ is often called the **experimentwise error rate** or overall confidence coefficient. The number of intervals r does not have to be large before the set of confidence intervals becomes relatively uninformative. For example, if there are $r = 5$ intervals and $\alpha = 0.05$ (a typical choice), the simultaneous confidence level for the set of five confidence intervals is at least 0.75, and if $r = 10$ and $\alpha = 0.05$, the simultaneous confidence level is at least 0.50.

One approach to ensuring that the simultaneous confidence level is not too small is to replace $\alpha/2$ in the one-at-a-time confidence interval Equations 3.12 and 3.13 with $\alpha/(2r)$. This is called the **Bonferroni method**, and it allows the experimenter to construct a set of r simultaneous confidence intervals on treatment means or differences in treatment means for which the overall confidence level is at least $100(1 - \alpha)$ percent. When r is not too large, this is a very nice method that leads to reasonably short confidence intervals. For more information, refer to the **supplemental text material** for Chapter 3.

3.3.4 Unbalanced Data

In some single-factor experiments, the number of observations taken within each treatment may be different. We then say that the design is **unbalanced**. The analysis of variance described above may still be used, but slight modifications must be made in the sum of squares formulas. Let n_i observations be taken under treatment i ($i = 1, 2, \ldots, a$) and $N = \sum_{i=1}^{a} n_i$. The manual computational formulas for SS_T and $SS_{\text{Treatments}}$ become

$$SS_T = \sum_{i=1}^{a}\sum_{j=1}^{n_i} y_{ij}^2 - \frac{y_{..}^2}{N} \tag{3.14}$$

and

$$SS_{\text{Treatments}} = \sum_{i=1}^{a} \frac{y_{i.}^2}{n_i} - \frac{y_{..}^2}{N} \tag{3.15}$$

No other changes are required in the analysis of variance.

There are two advantages in choosing a balanced design. First, the test statistic is relatively insensitive to small departures from the assumption of equal variances for the a treatments if the sample sizes are equal. This is not the case for unequal sample sizes. Second, the power of the test is maximized if the samples are of equal size.

3.4 Model Adequacy Checking

The decomposition of the variability in the observations through an analysis of variance identity (Equation 3.6) is a purely algebraic relationship. However, the use of the partitioning to test formally for no differences in treatment means requires that certain assumptions be satisfied. Specifically, these assumptions are that the observations are adequately described by the model

$$y_{ij} = \mu + \tau_i + \epsilon_{ij}$$

and that the errors are normally and independently distributed with mean zero and constant but unknown variance σ^2. If these assumptions are valid, the analysis of variance procedure is an exact test of the hypothesis of no difference in treatment means.

In practice, however, these assumptions will usually not hold exactly. Consequently, it is usually unwise to rely on the analysis of variance until the validity of these assumptions has been checked. Violations of the basic assumptions and model adequacy can be easily investigated by the examination of **residuals**. We define the residual for observation j in treatment i as

$$e_{ij} = y_{ij} - \hat{y}_{ij} \tag{3.16}$$

where $\hat{y}_{ij}$ is an estimate of the corresponding observation y_{ij} obtained as follows:

$$\begin{aligned} \hat{y}_{ij} &= \hat{\mu} + \hat{\tau}_i \\ &= \bar{y}_{..} + (\bar{y}_{i.} - \bar{y}_{..}) \\ &= \bar{y}_{i.} \end{aligned} \tag{3.17}$$

Equation 3.17 gives the intuitively appealing result that the estimate of any observation in the ith treatment is just the corresponding treatment average.

Examination of the residuals should be an automatic part of any analysis of variance. If the model is adequate, the residuals should be **structureless**; that is, they should contain no obvious patterns. Through analysis of residuals, many types of model inadequacies and violations of the underlying assumptions can be discovered. In this section, we show how model diagnostic checking can be done easily by graphical analysis of residuals and how to deal with several commonly occurring abnormalities.

3.4.1 The Normality Assumption

A check of the normality assumption could be made by plotting a histogram of the residuals. If the NID(0, σ^2) assumption on the errors is satisfied, this plot should look like a sample from a normal distribution centered at zero. Unfortunately, with small samples, considerable fluctuation in the shape of a histogram often occurs, so the appearance of a moderate departure from normality does not necessarily imply a serious violation of the assumptions. Gross deviations from normality are potentially serious and require further analysis.

An extremely useful procedure is to construct a **normal probability plot** of the residuals. Recall from Chapter 2 that we used a normal probability plot of the raw data to check the assumption of normality when using the t-test. In the analysis of variance, it is usually more effective (and straightforward) to do this with the **residuals**. If the underlying error distribution is normal, this plot will resemble a straight line. In visualizing the straight line, place more emphasis on the central values of the plot than on the extremes.

■ **TABLE 3.6**
Etch Rate Data and Residuals from Example 3.1[a]

Power (w)	Observations (j) 1	2	3	4	5	$\hat{y}_{ij} = \bar{y}_{i.}$
	23.8	−9.2	−21.2	−12.2	18.8	
160	575 (13)	542 (14)	530 (8)	539 (5)	570 (4)	551.2
	−22.4	5.6	2.6	−8.4	22.6	
180	565 (18)	593 (9)	590 (6)	579 (16)	610 (17)	587.4
	−25.4	25.6	−15.4	11.6	3.6	
200	600 (7)	651 (19)	610 (10)	637 (20)	629 (1)	625.4
	18.0	−7.0	8.0	−22.0	3.0	
220	725 (2)	700 (3)	715 (15)	685 (11)	710 (12)	707.0

[a]The residuals are shown in the box in each cell. The numbers in parentheses indicate the order in which each experimental run was made.

Table 3.6 shows the original data and the residuals for the etch rate data in Example 3.1. The normal probability plot is shown in Figure 3.4. The general impression from examining this display is that the error distribution is approximately normal. The tendency of the normal probability plot to bend down slightly on the left side and upward slightly on the right side implies that the tails of the error distribution are somewhat *thinner* than would be anticipated in a normal distribution; that is, the largest residuals are not quite as large (in absolute value) as expected. This plot is not grossly nonnormal, however.

In general, moderate departures from normality are of little concern in the fixed effects analysis of variance (recall our discussion of randomization tests in Section 3.3.2). An error distribution that has considerably thicker or thinner tails than the normal is of more concern than a skewed distribution. Because the F test is only slightly affected, we say that the analysis of

■ **FIGURE 3.4**
Normal probability plot of residuals for Example 3.1

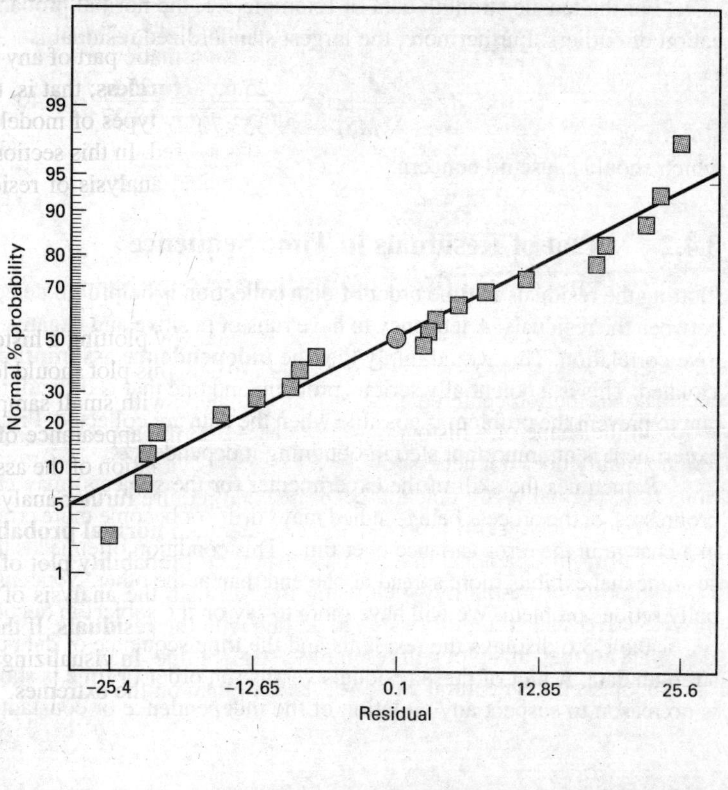

variance (and related procedures such as multiple comparisons) is **robust** to the normality assumption. Departures from normality usually cause both the true significance level and the power to differ slightly from the advertised values, with the power generally being lower. The random effects model that we will discuss in Section 3.9 and Chapter 13 is more severely affected by nonnormality.

A very common defect that often shows up on normal probability plots is one residual that is very much larger than any of the others. Such a residual is often called an **outlier**. The presence of one or more outliers can seriously distort the analysis of variance, so when a potential outlier is located, careful investigation is called for. Frequently, the cause of the outlier is a mistake in calculations or a data coding or copying error. If this is not the cause, the experimental circumstances surrounding this run must be carefully studied. If the outlying response is a particularly desirable value (high strength, low cost, etc.), the outlier may be more informative than the rest of the data. We should be careful not to reject or discard an outlying observation unless we have reasonably nonstatistical grounds for doing so. At worst, you may end up with two analyses; one with the outlier and one without.

Several formal statistical procedures may be used for detecting outliers [e.g., see Stefansky (1972), John and Prescott (1975), and Barnett and Lewis (1994)]. Some statistical software packages report the results of a statistical test for normality (such as the Anderson-Darling test) on the normal probability plot of residuals. This should be viewed with caution as those tests usually assume that the data to which they are applied are independent and residuals are not independent.

A rough check for outliers may be made by examining the **standardized residuals**

$$d_{ij} = \frac{e_{ij}}{\sqrt{MS_E}} \tag{3.18}$$

If the errors ϵ_{ij} are $N(0, \sigma^2)$, the standardized residuals should be approximately normal with mean zero and unit variance. Thus, about 68 percent of the standardized residuals should fall within the limits ± 1, about 95 percent of them should fall within ± 2, and virtually all of them should fall within ± 3. A residual bigger than 3 or 4 standard deviations from zero is a potential outlier.

For the tensile strength data of Example 3.1, the normal probability plot gives no indication of outliers. Furthermore, the largest standardized residual is

$$d_1 = \frac{e_1}{\sqrt{MS_E}} = \frac{25.6}{\sqrt{333.70}} = \frac{25.6}{18.27} = 1.40$$

which should cause no concern.

3.4.2 Plot of Residuals in Time Sequence

Plotting the residuals in time order of data collection is helpful in detecting strong **correlation** between the residuals. A tendency to have runs of positive and negative residuals indicates positive correlation. This would imply that the **independence assumption** on the errors has been violated. This is a potentially serious problem and one that is difficult to correct, so it is important to prevent the problem if possible when the data are collected. Proper randomization of the experiment is an important step in obtaining independence.

Sometimes the skill of the experimenter (or the subjects) may change as the experiment progresses, or the process being studied may "drift" or become more erratic. This will often result in a change in the error variance over time. This condition often leads to a plot of residuals versus time that exhibits more spread at one end than at the other. Nonconstant variance is a potentially serious problem. We will have more to say on the subject in Sections 3.4.3 and 3.4.4.

Table 3.6 displays the residuals and the time sequence of data collection for the tensile strength data. A plot of these residuals versus run order or time is shown in Figure 3.5. There is no reason to suspect any violation of the independence or constant variance assumptions.

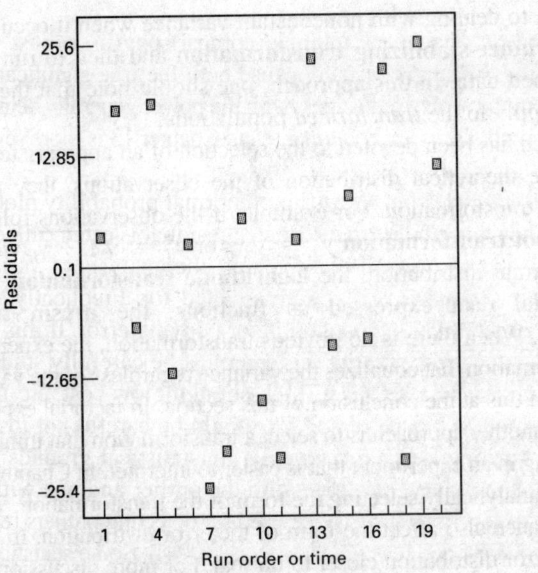

■ **FIGURE 3.5 Plot of residuals versus run order or time**

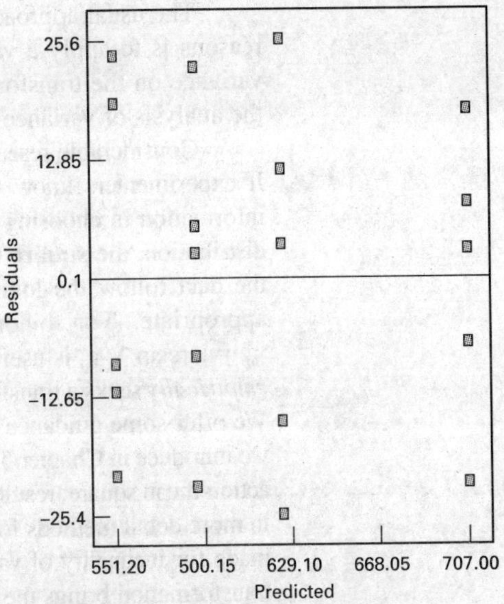

■ **FIGURE 3.6 Plot of residuals versus fitted values**

3.4.3 Plot of Residuals Versus Fitted Values

If the model is correct and the assumptions are satisfied, the residuals should be structureless; in particular, they should be unrelated to any other variable including the predicted response. A simple check is to plot the residuals versus the fitted values $\hat{y}_{ij}$. (For the single-factor experiment model, remember that $\hat{y}_{ij} = \bar{y}_{i.}$, the ith treatment average.) This plot should not reveal any obvious pattern. Figure 3.6 plots the residuals versus the fitted values for the tensile strength data of Example 3.1. No unusual structure is apparent.

A defect that occasionally shows up on this plot is **nonconstant variance**. Sometimes the variance of the observations increases as the magnitude of the observation increases. This would be the case if the error or background noise in the experiment was a constant percentage of the size of the observation. (This commonly happens with many measuring instruments—error is a percentage of the scale reading.) If this were the case, the residuals would get larger as y_{ij} gets larger, and the plot of residuals versus $\hat{y}_{ij}$ would look like an outward-opening funnel or megaphone. Nonconstant variance also arises in cases where the data follow a nonnormal, skewed distribution because in skewed distributions the variance tends to be a function of the mean.

If the assumption of homogeneity of variances is violated, the F test is only slightly affected in the balanced (equal sample sizes in all treatments) fixed effects model. However, in unbalanced designs or in cases where one variance is very much larger than the others, the problem is more serious. Specifically, if the factor levels having the larger variances also have the smaller sample sizes, the actual type I error rate is larger than anticipated (or confidence intervals have lower actual confidence levels than were specified). Conversely, if the factor levels with larger variances also have the larger sample sizes, the significance levels are smaller than anticipated (confidence levels are higher). This is a good reason for choosing **equal sample sizes** whenever possible. For random effects models, unequal error variances can significantly disturb inferences on variance components even if balanced designs are used.

Inequality of variance also shows up occasionally on the plot of residuals versus run order. An outward-opening funnel pattern indicates that variability is increasing over time. This could result from operator/subject fatigue, accumulated stress on equipment, changes in material properties such as catalyst degradation, or tool wear, or any of a number of causes.

The usual approach to dealing with nonconstant variance when it occurs for the above reasons is to apply a **variance-stabilizing transformation** and then to run the analysis of variance on the transformed data. In this approach, one should note that the conclusions of the analysis of variance apply to the *transformed* populations.

Considerable research has been devoted to the selection of an appropriate transformation. If experimenters know the theoretical distribution of the observations, they may utilize this information in choosing a transformation. For example, if the observations follow the Poisson distribution, the **square root transformation** $y_{ij}^* = \sqrt{y_{ij}}$ or $y_{ij}^* = \sqrt{1 + y_{ij}}$ would be used. If the data follow the lognormal distribution, the **logarithmic transformation** $y_{ij}^* = \log y_{ij}$ is appropriate. For binomial data expressed as fractions, the **arcsin transformation** $y_{ij}^* = \arcsin \sqrt{y_{ij}}$ is useful. When there is no obvious transformation, the experimenter usually *empirically* seeks a transformation that equalizes the variance regardless of the value of the mean. We offer some guidance on this at the conclusion of this section. In factorial experiments, which we introduce in Chapter 5, another approach is to select a transformation that minimizes the interaction mean square, resulting in an experiment that is easier to interpret. In Chapter 15, we discuss in more detail methods for analytically selecting the form of the transformation. Transformations made for inequality of variance also affect the form of the error distribution. In most cases, the transformation brings the error distribution closer to normal. For more discussion of transformations, refer to Bartlett (1947), Dolby (1963), Box and Cox (1964), and Draper and Hunter (1969).

Statistical Tests for Equality of Variance. Although residual plots are frequently used to diagnose inequality of variance, several statistical tests have also been proposed. These tests may be viewed as formal tests of the hypotheses

$$H_0: \sigma_1^2 = \sigma_2^2 = \cdots = \sigma_a^2$$
$$H_1: \text{above not true for at least one } \sigma_i^2$$

A widely used procedure is **Bartlett's test**. The procedure involves computing a statistic whose sampling distribution is closely approximated by the chi-square distribution with $a - 1$ degrees of freedom when the a random samples are from independent normal populations. The test statistic is

$$\chi_0^2 = 2.3026 \frac{q}{c} \tag{3.19}$$

where

$$q = (N - a)\log_{10} S_p^2 - \sum_{i=1}^{a} (n_i - 1)\log_{10} S_i^2$$

$$c = 1 + \frac{1}{3(a-1)}\left(\sum_{i=1}^{a} (n_i - 1)^{-1} - (N - a)^{-1}\right)$$

$$S_p^2 = \frac{\sum_{i=1}^{a} (n_i - 1)S_i^2}{N - a}$$

and S_i^2 is the sample variance of the ith population.

The quantity q is large when the sample variances S_i^2 differ greatly and is equal to zero when all S_i^2 are equal. Therefore, we should reject H_0 on values of χ_0^2 that are too large; that is, we reject H_0 only when

$$\chi_0^2 > \chi_{\alpha,a-1}^2$$

where $\chi_{\alpha,a-1}^2$ is the upper α percentage point of the chi-square distribution with $a - 1$ degrees of freedom. The *P*-value approach to decision making could also be used.

Bartlett's test is very sensitive to the normality assumption. Consequently, when the validity of this assumption is doubtful, Bartlett's test should not be used.

EXAMPLE 3.4

In the plasma etch experiment, the normality assumption is not in question, so we can apply Bartlett's test to the etch rate data. We first compute the sample variances in each treatment and find that $S_1^2 = 400.7$, $S_2^2 = 280.3$, $S_3^2 = 421.3$, and $S_4^2 = 232.5$. Then

$$S_p^2 = \frac{4(400.7) + 4(280.3) + 4(421.3) + 4(232.5)}{16} = 333.7$$

$$q = 16 \log_{10}(333.7) - 4[\log_{10}400.7 + \log_{10}280.3 + \log_{10}421.3 + \log_{10}232.5] = 0.21$$

$$c = 1 + \frac{1}{3(3)}\left(\frac{4}{4} - \frac{1}{16}\right) = 1.10$$

and the test statistic is

$$\chi_0^2 = 2.3026 \frac{(0.21)}{(1.10)} = 0.43$$

From Appendix Table III, we find that $\chi^2_{0.05,3} = 7.81$ (the P-value is $P = 0.934$), so we cannot reject the null hypothesis. There is no evidence to counter the claim that all five variances are the same. This is the same conclusion reached by analyzing the plot of residuals versus fitted values.

Because Bartlett's test is sensitive to the normality assumption, there may be situations where an alternative procedure would be useful. Anderson and McLean (1974) present a useful discussion of statistical tests for equality of variance. The **modified Levene test** [see Levene (1960) and Conover, Johnson, and Johnson (1981)] is a very nice procedure that is robust to departures from normality. To test the hypothesis of equal variances in all treatments, the modified Levene test uses the absolute deviation of the observations y_{ij} in each treatment from the treatment median, say, $\tilde{y}_i$. Denote these deviations by

$$d_{ij} = |y_{ij} - \tilde{y}_i| \begin{cases} i = 1, 2, \ldots, a \\ j = 1, 2, \ldots, n_i \end{cases}$$

The modified Levene test then evaluates whether or not the means of these deviations are equal for all treatments. It turns out that if the mean deviations are equal, the variances of the observations in all treatments will be the same. The test statistic for Levene's test is simply the usual ANOVA F statistic for testing equality of means applied to the absolute deviations.

EXAMPLE 3.5

A civil engineer is interested in determining whether four different methods of estimating flood flow frequency produce equivalent estimates of peak discharge when applied to the same watershed. Each procedure is used six times on the watershed, and the resulting discharge data (in cubic feet per second) are shown in the upper panel of Table 3.7. The analysis of variance for the data, summarized in Table 3.8, implies that there is a difference in mean peak discharge estimates given by the four procedures. The plot of residuals versus fitted values, shown in Figure 3.7, is disturbing because the outward-opening funnel shape indicates that the constant variance assumption is not satisfied.

We will apply the modified Levene test to the peak discharge data. The upper panel of Table 3.7 contains the treatment medians $\tilde{y}_i$ and the lower panel contains the deviations d_{ij} around the medians. Levene's test consists of conducting a standard analysis of variance on the d_{ij}. The F test statistic that results from this is $F_0 = 4.55$, for which the P-value is $P = 0.0137$. Therefore, Levene's test rejects the null hypothesis of equal variances, essentially confirming the diagnosis we made from visual examination of Figure 3.7. The peak discharge data are a good candidate for data transformation.

■ **TABLE 3.7**
Peak Discharge Data

Estimation Method	Observations						$\bar{y}_{i.}$	$\tilde{y}_i$	S_i
1	0.34	0.12	1.23	0.70	1.75	0.12	0.71	0.520	0.66
2	0.91	2.94	2.14	2.36	2.86	4.55	2.63	2.610	1.09
3	6.31	8.37	9.75	6.09	9.82	7.24	7.93	7.805	1.66
4	17.15	11.82	10.95	17.20	14.35	16.82	14.72	15.59	2.77

Estimation Method	Deviations d_{ij} for the Modified Levene Test					
1	0.18	0.40	0.71	0.18	1.23	0.40
2	1.70	0.33	0.47	0.25	0.25	1.94
3	1.495	0.565	1.945	1.715	2.015	0.565
4	1.56	3.77	4.64	1.61	1.24	1.23

■ **TABLE 3.8**
Analysis of Variance for Peak Discharge Data

Source of Variation	Sum of Squares	Degrees of Freedom	Mean Square	F_0	*P*-Value
Methods	708.3471	3	236.1157	76.07	<0.001
Error	62.0811	20	3.1041		
Total	770.4282	23			

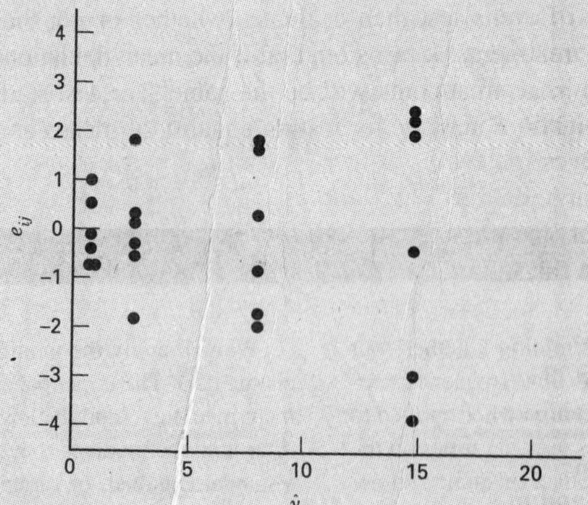

■ **FIGURE 3.7** Plot of residuals versus $\hat{y}_{ij}$ for Example 3.5

Empirical Selection of a Transformation. We observed above that if experimenters knew the relationship between the variance of the observations and the mean, they could use this information to guide them in selecting the form of the transformation. We now elaborate on this point and show one method for empirically selecting the form of the required transformation from the data.

Let $E(y) = \mu$ be the mean of y, and suppose that the standard deviation of y is proportional to a power of the mean of y such that

$$\sigma_y \propto \mu^\alpha$$

We want to find a transformation on y that yields a constant variance. Suppose that the transformation is a power of the original data, say

$$y^* = y^\lambda \tag{3.20}$$

Then it can be shown that

$$\sigma_{y^*} \propto \mu^{\lambda+\alpha-1} \tag{3.21}$$

Clearly, if we set $\lambda = 1 - \alpha$, the variance of the transformed data y^* is constant.

Several of the common transformations discussed previously are summarized in Table 3.9. Note that $\lambda = 0$ implies the log transformation. These transformations are arranged in order of increasing **strength**. By the strength of a transformation, we mean the amount of curvature it induces. A mild transformation applied to data spanning a narrow range has little effect on the analysis, whereas a strong transformation applied over a large range may have dramatic results. Transformations often have little effect unless the ratio $y_{\max}/y_{\min}$ is larger than 2 or 3.

In many experimental design situations where there is replication, we can empirically estimate α from the data. Because in the ith treatment combination $\sigma_{y_i} \propto \mu_i^\alpha = \theta\mu_i^\alpha$, where θ is a constant of proportionality, we may take logs to obtain

$$\log \sigma_{y_i} = \log \theta + \alpha \log \mu_i \tag{3.22}$$

Therefore, a plot of $\log \sigma_{y_i}$ versus $\log \mu_i$ would be a straight line with slope α. Because we don't know σ_{y_i} and μ_i, we may substitute reasonable estimates of them in Equation 3.22 and use the slope of the resulting straight line fit as an estimate of α. Typically, we would use the standard deviation S_i and the average $\bar{y}_{i.}$ of the ith treatment (or, more generally, the ith treatment combination or set of experimental conditions) to estimate σ_{y_i} and μ_i.

To investigate the possibility of using a variance-stabilizing transformation on the peak discharge data from Example 3.5, we plot $\log S_i$ versus $\log \bar{y}_{i.}$ in Figure 3.8. The slope of a straight line passing through these four points is close to 1/2 and from Table 3.9 this implies that the square root transformation may be appropriate. The analysis of variance for

■ TABLE 3.9
Variance-Stabilizing Transformations

Relationship Between σ_y and μ	α	$\lambda = 1 - \alpha$	Transformation	Comment
$\sigma_y \propto$ constant	0	1	No transformation	
$\sigma_y \propto \mu^{1/2}$	1/2	1/2	Square root	Poisson (count) data
$\sigma_y \propto \mu$	1	0	Log	
$\sigma_y \propto \mu^{3/2}$	3/2	−1/2	Reciprocal square root	
$\sigma_y \propto \mu^2$	2	−1	Reciprocal	

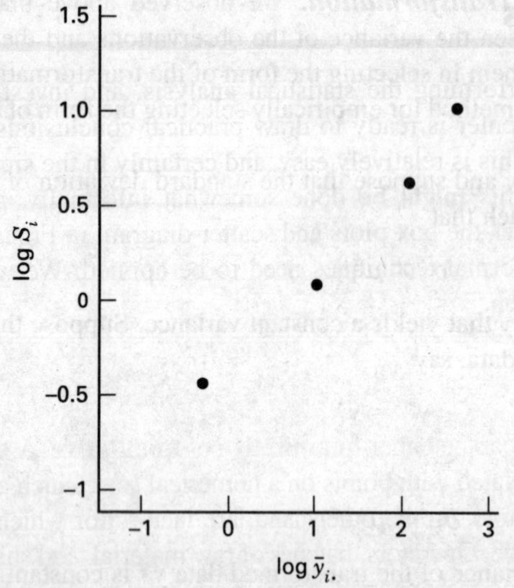

■ **FIGURE 3.8** **Plot of log S_i versus log $\bar{y}_{i.}$ for the peak discharge data from Example 3.5**

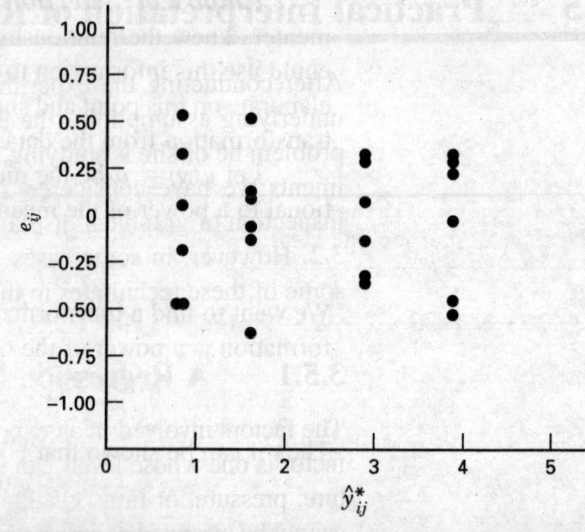

■ **FIGURE 3.9** **Plot of residuals from transformed data versus $\hat{y}^*_{ij}$ for the peak discharge data in Example 3.5**

the transformed data $y^* = \sqrt{y}$ is presented in Table 3.10, and a plot of residuals versus the predicted response is shown in Figure 3.9. This residual plot is much improved in comparison to Figure 3.7, so we conclude that the square root transformation has been helpful. Note that in Table 3.10 we have reduced the degrees of freedom for error and total by 1 to account for the use of the data to estimate the transformation parameter α.

In practice, many experimenters select the form of the transformation by simply trying several alternatives and observing the effect of each transformation on the plot of residuals versus the predicted response. The transformation that produced the most satisfactory residual plot is then selected. Alternatively, there is a formal method called the **Box-Cox Method** for selecting a variance-stability transformation. In Chapter 15 we discuss and illustrate this procedure. It is widely used and implemented in many software packages.

3.4.4 Plots of Residuals Versus Other Variables

If data have been collected on any other variables that might possibly affect the response, the residuals should be plotted against these variables. For example, in the tensile strength experiment of Example 3.1, strength may be significantly affected by the thickness of the fiber, so the residuals should be plotted versus fiber thickness. If different testing machines were used to collect the data, the residuals should be plotted against machines. Patterns in such residual plots imply that the variable affects the response. This suggests that the variable should be either controlled more carefully in future experiments or included in the analysis.

■ **TABLE 3.10**
Analysis of Variance for Transformed Peak Discharge Data, $y^* = \sqrt{y}$

Source of Variation	Sum of Squares	Degrees of Freedom	Mean Square	F_0	*P*-Value
Methods	32.6842	3	10.8947	76.99	<0.001
Error	2.6884	19	0.1415		
Total	35.3726	22			

3.5 Practical Interpretation of Results

After conducting the experiment, performing the statistical analysis, and investigating the underlying assumptions, the experimenter is ready to draw practical conclusions about the problem he or she is studying. Often this is relatively easy, and certainly in the simple experiments we have considered so far, this might be done somewhat informally, perhaps by inspection of graphical displays such as the box plots and scatter diagram in Figures 3.1 and 3.2. However, in some cases, more formal techniques need to be applied. We will present some of these techniques in this section.

3.5.1 A Regression Model

The factors involved in an experiment can be either **quantitative** or **qualitative**. A quantitative factor is one whose levels can be associated with points on a numerical scale, such as temperature, pressure, or time. Qualitative factors, on the other hand, are factors for which the levels cannot be arranged in order of magnitude. Operators, batches of raw material, and shifts are typical qualitative factors because there is no reason to rank them in any particular numerical order.

Insofar as the initial design and analysis of the experiment are concerned, both types of factors are treated identically. The experimenter is interested in determining the differences, if any, between the levels of the factors. In fact, the analysis of variance treat the design factor as if it were qualitative or categorical. If the factor is really qualitative, such as operators, it is meaningless to consider the response for a subsequent run at an intermediate level of the factor. However, with a quantitative factor such as time, the experimenter is usually interested in the entire range of values used, particularly the response from a subsequent run at an intermediate factor level. That is, if the levels 1.0, 2.0, and 3.0 hours are used in the experiment, we may wish to predict the response at 2.5 hours. Thus, the experimenter is frequently interested in developing an interpolation equation for the response variable in the experiment. This equation is an **empirical model** of the process that has been studied.

The general approach to fitting empirical models is called **regression analysis**, which is discussed extensively in Chapter 10. See also the **supplemental text material** for this chapter. This section briefly illustrates the technique using the etch rate data of Example 3.1.

Figure 3.10 presents scatter diagrams of etch rate y versus the power x for the experiment in Example 3.1. From examining the scatter diagram, it is clear that there is a strong relationship between etch rate and power. As a first approximation, we could try fitting a **linear model** to the data, say

$$y = \beta_0 + \beta_1 x + \epsilon$$

where β_0 and β_1 are unknown parameters to be estimated and ϵ is a random error term. The method often used to estimate the parameters in a model such as this is the **method of least squares**. This consists of choosing estimates of the β's such that the sum of the squares of the errors (the ϵ's) is minimized. The least squares fit in our example is

$$\hat{y} = 137.62 + 2.527x$$

(If you are unfamiliar with regression methods, see Chapter 10 and the supplemental text material for this chapter.)

This linear model is shown in Figure 3.10*a*. It does not appear to be very satisfactory at the higher power settings. Perhaps an improvement can be obtained by adding a quadratic term in x. The resulting **quadratic model** fit is

$$\hat{y} = 1147.77 - 8.2555\,x + 0.028375\,x^2$$

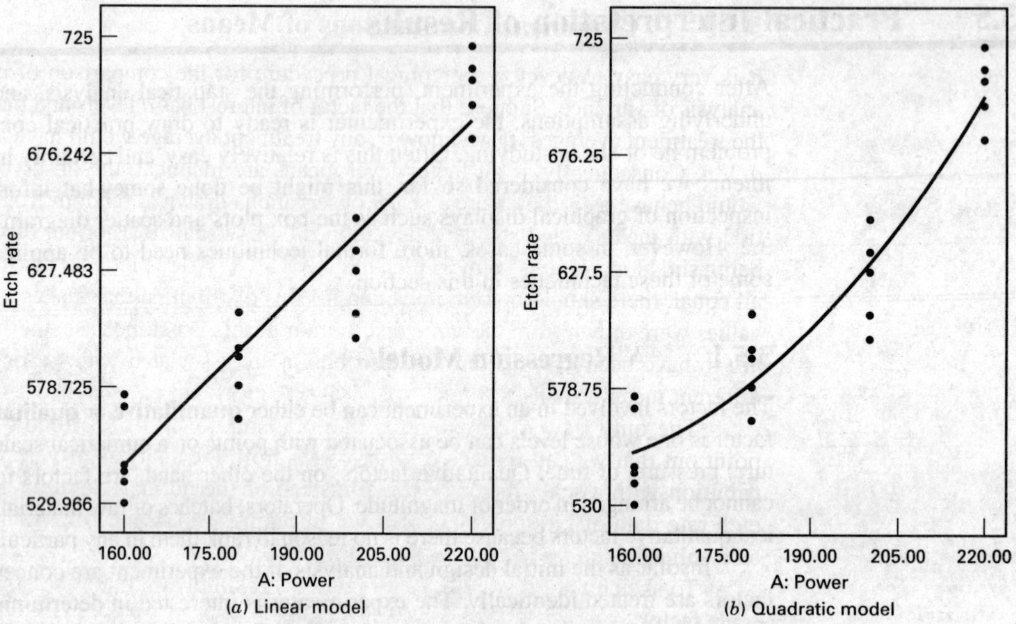

■ **FIGURE 3.10** **Scatter diagrams and regression models for the etch rate data of Example 3.1**

This quadratic fit is shown in Figure 3.10*b*. The quadratic model appears to be superior to the linear model because it provides a better fit at the higher power settings.

In general, we would like to fit the lowest order polynomial that adequately describes the system or process. In this example, the quadratic polynomial seems to fit better than the linear model, so the extra complexity of the quadratic model is justified. Selecting the order of the approximating polynomial is not always easy, however, and it is relatively easy to overfit, that is, to add high-order polynomial terms that do not really improve the fit but increase the complexity of the model and often damage its usefulness as a predictor or interpolation equation.

In this example, the empirical model could be used to predict etch rate at power settings within the region of experimentation. In other cases, the empirical model could be used for **process optimization**, that is, finding the levels of the design variables that result in the best values of the response. We will discuss and illustrate these problems extensively later in the book.

3.5.2 Comparisons Among Treatment Means

Suppose that in conducting an analysis of variance for the fixed effects model the null hypothesis is rejected. Thus, there are differences between the treatment means but exactly *which* means differ is not specified. Sometimes in this situation, further comparisons and analysis among **groups** of treatment means may be useful. The ith treatment mean is defined as $\mu_i = \mu + \tau_i$, and μ_i is estimated by $\bar{y}_{i.}$. Comparisons between treatment means are made in terms of either the treatment totals $\{y_{i.}\}$ or the treatment averages $\{\bar{y}_{i.}\}$. The procedures for making these comparisons are usually called **multiple comparison methods**. In the next several sections, we discuss methods for making comparisons among individual treatment means or groups of these means.

3.5.3 Graphical Comparisons of Means

It is very easy to develop a graphical procedure for the comparison of means following an analysis of variance. Suppose that the factor of interest has a levels and that $\bar{y}_{1.}, \bar{y}_{2.}, \ldots, \bar{y}_{a.}$ are the treatment averages. If we know σ, any treatment average would have a standard deviation $\sigma/\sqrt{n}$. Consequently, if all factor level means are identical, the observed sample means $\bar{y}_{i.}$ would behave as if they were a set of observations drawn at random from a normal distribution with mean $\bar{y}_{..}$ and standard deviation $\sigma/\sqrt{n}$. Visualize a normal distribution capable of being slid along an axis below which the $\bar{y}_{1.}, \bar{y}_{2.}, \ldots, \bar{y}_{a.}$ are plotted. If the treatment means are all equal, there should be some position for this distribution that makes it obvious that the $\bar{y}_{i.}$ values were drawn from the same distribution. If this is not the case, the $\bar{y}_{i.}$ values that appear *not* to have been drawn from this distribution are associated with factor levels that produce different mean responses.

The only flaw in this logic is that σ is unknown. Box, Hunter, and Hunter (2005) point out that we can replace σ with $\sqrt{MS_E}$ from the analysis of variance and use a t distribution with a scale factor $\sqrt{MS_E/n}$ instead of the normal. Such an arrangement for the etch rate data of Example 3.1 is shown in Figure 3.11. Focus on the t distribution shown as a solid line curve in the middle of the display.

To sketch the t distribution in Figure 3.11, simply multiply the abscissa t value by the scale factor

$$\sqrt{MS_E/n} = \sqrt{330.70/5} = 8.13$$

and plot this against the ordinate of t at that point. Because the t distribution looks much like the normal, except that it is a little flatter near the center and has longer tails, this sketch is usually easily constructed by eye. If you wish to be more precise, there is a table of abscissa t values and the corresponding ordinates in Box, Hunter, and Hunter (2005). The distribution can have an arbitrary origin, although it is usually best to choose one in the region of the $\bar{y}_{i.}$ values to be compared. In Figure 3.11, the origin is 615 Å/min.

Now visualize sliding the t distribution in Figure 3.11 along the horizontal axis as indicated by the dashed lines and examine the four means plotted in the figure. Notice that there is no location for the distribution such that all four averages could be thought of as typical, randomly selected observations from the distribution. This implies that all four means are not equal; thus, the figure is a graphical display of the ANOVA results. Furthermore, the figure indicates that all four levels of power (160, 180, 200, 220 W) produce mean etch rates that differ from each other. In other words, $\mu_1 \neq \mu_2 \neq \mu_3 \neq \mu_4$.

This simple procedure is a rough but effective technique for many multiple comparison problems. However, there are more formal methods. We now give a brief discussion of some of these procedures.

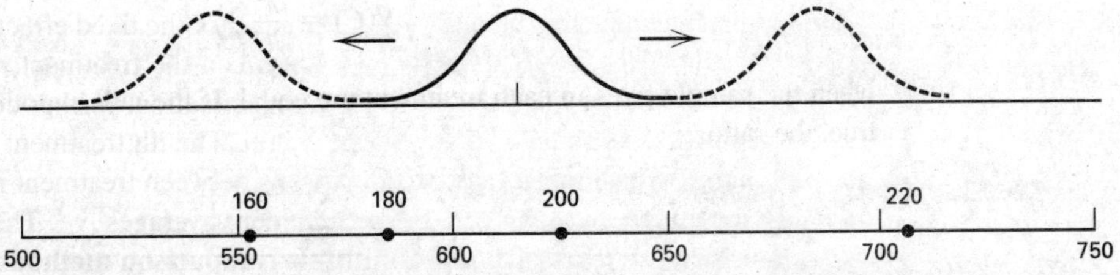

■ **FIGURE 3.11** **Etch rate averages from Example 3.1 in relation to a t distribution with scale factor $\sqrt{MS_E/n} = \sqrt{330.70/5} = 8.13$**

3.5.4 Contrasts

Many multiple comparison methods use the idea of a **contrast**. Consider the plasma etching experiment of Example 3.1. Because the null hypothesis was rejected, we know that some power settings produce different etch rates than others, but which ones actually cause this difference? We might suspect at the outset of the experiment that 200 W and 220 W produce the same etch rate, implying that we would like to test the hypothesis

$$H_0: \mu_3 = \mu_4$$
$$H_1: \mu_3 \neq \mu_4$$

or equivalently

$$H_0: \mu_3 - \mu_4 = 0$$
$$H_1: \mu_3 - \mu_4 \neq 0 \tag{3.23}$$

If we had suspected at the start of the experiment that the *average* of the lowest levels of power did not differ from the *average* of the highest levels of power, then the hypothesis would have been

$$H_0: \mu_1 + \mu_2 = \mu_3 + \mu_4$$
$$H_1: \mu_1 + \mu_2 \neq \mu_3 + \mu_4$$

or

$$H_0: \mu_1 + \mu_2 - \mu_3 - \mu_4 = 0$$
$$H_1: \mu_1 + \mu_2 - \mu_3 - \mu_4 \neq 0 \tag{3.24}$$

In general, a **contrast** is a linear combination of parameters of the form

$$\Gamma = \sum_{i=1}^{a} c_i \mu_i$$

where the **contrast constants** $c_1, c_2, \ldots, c_a$ sum to zero; that is, $\Sigma_{i=1}^{a} c_i = 0$. Both of the above hypotheses can be expressed in terms of contrasts:

$$H_0: \sum_{i=1}^{a} c_i \mu_i = 0$$
$$H_1: \sum_{i=1}^{a} c_i \mu_i \neq 0 \tag{3.25}$$

The contrast constants for the hypotheses in Equation 3.23 are $c_1 = c_2 = 0$, $c_3 = +1$, and $c_4 = -1$, whereas for the hypotheses in Equation 3.24, they are $c_1 = c_2 = +1$ and $c_3 = c_4 = -1$.

Testing hypotheses involving contrasts can be done in two basic ways. The first method uses a *t*-test. Write the contrast of interest in terms of the **treatment averages**, giving

$$C = \sum_{i=1}^{a} c_i \bar{y}_{i.}$$

The variance of *C* is

$$V(C) = \frac{\sigma^2}{n} \sum_{i=1}^{a} c_i^2 \tag{3.26}$$

when the sample sizes in each treatment are equal. If the null hypothesis in Equation 3.25 is true, the ratio

$$\frac{\sum_{i=1}^{a} c_i \bar{y}_{i.}}{\sqrt{\frac{\sigma^2}{n} \sum_{i=1}^{a} c_i^2}}$$

has the $N(0, 1)$ distribution. Now we would replace the unknown variance σ^2 by its estimate, the mean square error MS_E and use the statistic

$$t_0 = \frac{\sum_{i=1}^{a} c_i \bar{y}_{i.}}{\sqrt{\frac{MS_E}{n}\sum_{i=1}^{a} c_i^2}} \tag{3.27}$$

to test the hypotheses in Equation 3.25. The null hypothesis would be rejected if $|t_0|$ in Equation 3.27 exceeds $t_{\alpha/2,N-a}$.

The second approach uses an F test. Now the square of a t random variable with ν degrees of freedom is an F random variable with 1 numerator and ν denominator degrees of freedom. Therefore, we can obtain

$$F_0 = t_0^2 = \frac{\left(\sum_{i=1}^{a} c_i \bar{y}_{i.}\right)^2}{\frac{MS_E}{n}\sum_{i=1}^{a} c_i^2} \tag{3.28}$$

as an F statistic for testing Equation 3.25. The null hypothesis would be rejected if $F_0 > F_{\alpha,1,N-a}$. We can write the test statistic of Equation 3.28 as

$$F_0 = \frac{MS_C}{MS_E} = \frac{SS_C/1}{MS_E}$$

where the single degree of freedom contrast sum of squares is

$$SS_C = \frac{\left(\sum_{i=1}^{a} c_i \bar{y}_{i.}\right)^2}{\frac{1}{n}\sum_{i=1}^{a} c_i^2} \tag{3.29}$$

Confidence Interval for a Contrast. Instead of testing hypotheses about a contrast, it may be more useful to construct a confidence interval. Suppose that the contrast of interest is

$$\Gamma = \sum_{i=1}^{a} c_i \mu_i$$

Replacing the treatment means with the treatment averages yields

$$C = \sum_{i=1}^{a} c_i \bar{y}_{i.}$$

Because

$$E\left(\sum_{i=1}^{a} c_i \bar{y}_{i.}\right) = \sum_{i=1}^{a} c_i \mu_i \quad \text{and} \quad V(C) = \sigma^2/n \sum_{i=1}^{a} c_i^2$$

the $100(1-\alpha)$ percent confidence interval on the contrast $\Sigma_{i=1}^{a} c_i \mu_i$ is

$$\sum_{i=1}^{a} c_i \bar{y}_{i.} - t_{\alpha/2,N-a}\sqrt{\frac{MS_E}{n}\sum_{i=1}^{a} c_i^2} \le \sum_{i=1}^{a} c_i \mu_i \le \sum_{i=1}^{a} c_i \bar{y}_{i.} + t_{\alpha/2,N-a}\sqrt{\frac{MS_E}{n}\sum_{i=1}^{a} c_i^2} \tag{3.30}$$

Note that we have used MS_E to estimate σ^2. Clearly, if the confidence interval in Equation 3.30 includes zero, we would be unable to reject the null hypothesis in Equation 3.25.

Standardized Contrast. When more than one contrast is of interest, it is often useful to evaluate them on the same scale. One way to do this is to standardize the contrast so that it has variance σ^2. If the contrast $\Sigma_{i=1}^a c_i\mu_i$ is written in terms of treatment averages as $\Sigma_{i=1}^a c_i\bar{y}_{i.}$, dividing it by $\sqrt{(1/n)\Sigma_{i=1}^a c_i^2}$ will produce a standardized contrast with variance σ^2. Effectively, then, the **standardized contrast** is

$$\sum_{i=1}^{a} c_i^* \bar{y}_{i.}$$

where

$$c_i^* = \frac{c_i}{\sqrt{\frac{1}{n}\sum_{i=1}^{a} c_i^2}}$$

Unequal Sample Sizes. When the sample sizes in each treatment are different, minor modifications are made in the above results. First, note that the definition of a contrast now requires that

$$\sum_{i=1}^{a} n_i c_i = 0$$

Other required changes are straightforward. For example, the t statistic in Equation 3.27 becomes

$$t_0 = \frac{\sum_{i=1}^{a} c_i \bar{y}_{i.}}{\sqrt{MS_E \sum_{i=1}^{a} \frac{c_i^2}{n_i}}}$$

and the contrast sum of squares from Equation 3.29 becomes

$$SS_C = \frac{\left(\sum_{i=1}^{a} c_i \bar{y}_{i.}\right)^2}{\sum_{i=1}^{a} \frac{c_i^2}{n_i}}$$

3.5.5 Orthogonal Contrasts

A useful special case of the procedure in Section 3.5.4 is that of **orthogonal contrasts.** Two contrasts with coefficients $\{c_i\}$ and $\{d_i\}$ are orthogonal if

$$\sum_{i=1}^{a} c_i d_i = 0$$

or, for an unbalanced design, if

$$\sum_{i=1}^{a} n_i c_i d_i = 0$$

For a treatments, the set of $a - 1$ orthogonal contrasts partition the sum of squares due to treatments into $a - 1$ independent single-degree-of-freedom components. Thus, tests performed on orthogonal contrasts are independent.

There are many ways to choose the orthogonal contrast coefficients for a set of treatments. Usually, something in the nature of the experiment should suggest which comparisons will be of interest. For example, if there are $a = 3$ treatments, with treatment 1 a control and treatments 2 and 3 actual levels of the factor of interest to the experimenter, appropriate orthogonal contrasts might be as follows:

Treatment	Coefficients for Orthogonal Contrasts	
1 (control)	−2	0
2 (level 1)	1	−1
3 (level 2)	1	1

Note that contrast 1 with $c_i = -2, 1, 1$ compares the average effect of the factor with the control, whereas contrast 2 with $d_i = 0, \ -1, 1$ compares the two levels of the factor of interest.

Generally, the method of contrasts (or orthogonal contrasts) is useful for what are called **preplanned comparisons**. That is, the contrasts are specified prior to running the experiment and examining the data. The reason for this is that if comparisons are selected after examining the data, most experimenters would construct tests that correspond to large observed differences in means. These large differences could be the result of the presence of real effects, or they could be the result of random error. If experimenters consistently pick the largest differences to compare, they will inflate the type I error of the test because it is likely that, in an unusually high percentage of the comparisons selected, the observed differences will be the result of error. Examining the data to select comparisons of potential interest is often called **data snooping**. The Scheffé method for all comparisons, discussed in the next section, permits data snooping.

EXAMPLE 3.6

Consider the plasma etching experiment in Example 3.1. There are four treatment means and three degrees of freedom between these treatments. Suppose that prior to running the experiment the following set of comparisons among the treatment means (and their associated contrasts) were specified:

Hypothesis	Contrast
$H_0: \mu_1 = \mu_2$	$C_1 = \bar{y}_{1.} - \bar{y}_{2.}$
$H_0: \mu_1 + \mu_2 = \mu_3 + \mu_4$	$C_2 = \bar{y}_{1.} + \bar{y}_{2.} - \bar{y}_{3.} - \bar{y}_{4.}$
$H_0: \mu_3 = \mu_4$	$C_3 = \bar{y}_{3.} - \bar{y}_{4.}$

Notice that the contrast coefficients are orthogonal. Using the data in Table 3.4, we find the numerical values of the contrasts and the sums of squares to be as follows:

$$C_1 = +1(551.2) - 1(587.4) = -36.2$$

$$SS_{C_1} = \frac{(-36.2)^2}{\frac{1}{5}(2)} = 3276.10$$

$$C_2 = \begin{matrix} +1(551.2) + 1(587.4) \\ -1(625.4) - 1(707.0) \end{matrix} = -193.8$$

$$SS_{C_2} = \frac{(-193.8)^2}{\frac{1}{5}(4)} = 46{,}948.05$$

$$C_3 = +1(625.4) - 1(707.6) = -81.6$$

$$SS_{C_3} = \frac{(-81.6)^2}{\frac{1}{5}(2)} = 16{,}646.40$$

These contrast sums of squares completely partition the treatment sum of squares. The tests on such orthogonal contrasts are usually incorporated in the ANOVA, as shown in Table 3.11. We conclude from the P-values that there are significant differences in mean etch rates between levels 1 and 2 and between levels 3 and 4 of the power settings, and that the *average* of levels 1 and 2 does differ significantly from the average of levels 3 and 4 at the $\alpha = 0.05$ level.

■ **TABLE 3.11**

Analysis of Variance for the Plasma Etching Experiment

Source of Variation	Sum of Squares	Degrees of Freedom	Mean Square	F_0	*P*-Value
Power setting	66,870.55	3	22,290.18	66.80	<0.001
Orthogonal contrasts					
$C_1: \mu_1 = \mu_2$	(3276.10)	1	3276.10	9.82	<0.01
$C_2: \mu_1 + \mu_3 = \mu_3 + \mu_4$	(46,948.05)	1	46,948.05	140.69	<0.001
$C_3: \mu_3 = \mu_4$	(16,646.40)	1	16,646.40	49.88	<0.001
Error	5,339.20	16	333.70		
Total	72,209.75	19			

3.5.6 Scheffé's Method for Comparing All Contrasts

In many situations, experimenters may not know in advance which contrasts they wish to compare, or they may be interested in more than $a - 1$ possible comparisons. In many exploratory experiments, the comparisons of interest are discovered only after preliminary examination of the data. Scheffé (1953) has proposed a method for comparing any and all possible contrasts between treatment means. In the Scheffé method, the type I error is at most α for any of the possible comparisons.

Suppose that a set of m contrasts in the treatment means

$$\Gamma_u = c_{1u}\mu_1 + c_{2u}\mu_2 + \cdots + c_{au}\mu_a \quad u = 1, 2, \ldots, m \tag{3.31}$$

of interest have been determined. The corresponding contrast in the treatment averages $\bar{y}_{i.}$ is

$$C_u = c_{1u}\bar{y}_{1.} + c_{2u}\bar{y}_{2.} + \cdots + c_{au}\bar{y}_{a.} \quad u = 1, 2, \ldots, m \tag{3.32}$$

and the **standard error** of this contrast is

$$S_{C_u} = \sqrt{MS_E \sum_{i=1}^{a} (c_{iu}^2/n_i)} \tag{3.33}$$

where n_i is the number of observations in the ith treatment. It can be shown that the critical value against which C_u should be compared is

$$S_{\alpha,u} = S_{C_u}\sqrt{(a-1)F_{\alpha,a-1,N-a}} \tag{3.34}$$

To test the hypothesis that the contrast Γ_u differs significantly from zero, refer C_u to the critical value. If $|C_u| > S_{\alpha,u}$, the hypothesis that the contrast Γ_u equals zero is rejected.

The Scheffé procedure can also be used to form confidence intervals for all possible contrasts among treatment means. The resulting intervals, say $C_u - S_{\alpha,u} \leq \Gamma_u \leq C_u + S_{\alpha,u}$, are **simultaneous confidence intervals** in that the probability that all of them are simultaneously true is at least $1 - \alpha$.

To illustrate the procedure, consider the data in Example 3.1 and suppose that the contrasts of interests are

$$\Gamma_1 = \mu_1 + \mu_2 - \mu_3 - \mu_4$$

and

$$\Gamma_2 = \mu_1 - \mu_4$$

The numerical values of these contrasts are

$$\begin{aligned} C_1 &= \bar{y}_{1.} + \bar{y}_{2.} - \bar{y}_{3.} - \bar{y}_{4.} \\ &= 551.2 + 587.4 - 625.4 - 707.0 = -193.80 \end{aligned}$$

and

$$\begin{aligned} C_2 &= \bar{y}_{1.} - \bar{y}_{4.} \\ &= 551.2 - 707.0 = -155.8 \end{aligned}$$

The standard errors are found from Equation 3.33 as

$$S_{C_1} = \sqrt{MS_E \sum_{i=1}^{5} (c_{i1}^2/n_i)} = \sqrt{333.70(1+1+1+1)/5} = 16.34$$

and

$$S_{C_2} = \sqrt{MS_E \sum_{i=1}^{5} (c_{i2}^2/n_i)} = \sqrt{333.70(1+1)/5} = 11.55$$

From Equation 3.34, the 1 percent critical values are

$$S_{0.01,1} = S_{C_1}\sqrt{(a-1)F_{0.01,a-1,N-a}} = 16.34\sqrt{3(5.29)} = 65.09$$

and

$$S_{0.01,2} = S_{C_2}\sqrt{(a-1)F_{0.01,a-1,N-a}} = 11.55\sqrt{3(5.29)} = 45.97$$

Because $|C_1| > S_{0.01,1}$, we conclude that the contrast $\Gamma_1 = \mu_1 + \mu_2 - \mu_3 - \mu_4$ does not equal zero; that is, we conclude that the mean etch rates of power settings 1 and 2 as a group differ from the means of power settings 3 and 4 as a group. Furthermore, because $|C_2| > S_{0.01,2}$, we conclude that the contrast $\Gamma_2 = \mu_1 - \mu_4$ does not equal zero; that is, the mean etch rates of treatments 1 and 4 differ significantly.

3.5.7 Comparing Pairs of Treatment Means

In many practical situations, we will wish to compare only **pairs of means**. Frequently, we can determine which means differ by testing the differences between *all* pairs of treatment means. Thus, we are interested in contrasts of the form $\Gamma = \mu_i - \mu_j$ for all $i \neq j$. Although the Scheffé method described in the previous section could be easily applied to this problem, it is not the most sensitive procedure for such comparisons. We now turn to a consideration of methods specifically designed for pairwise comparisons between all a population means.

Suppose that we are interested in comparing all pairs of a treatment means and that the null hypotheses that we wish to test are $H_0: \mu_i = \mu_j$ for all $i \neq j$. There are numerous procedures available for this problem. We now present two popular methods for making such comparisons.

Tukey's Test. Suppose that, following an ANOVA in which we have rejected the null hypothesis of equal treatment means, we wish to test all pairwise mean comparisons:

$$H_0: \mu_i = \mu_j$$
$$H_1: \mu_i \neq \mu_j$$

for all $i \neq j$. Tukey (1953) proposed a procedure for testing hypotheses for which the overall significance level is exactly α when the sample sizes are equal and at most α when the sample sizes are unequal. His procedure can also be used to construct confidence intervals on the differences in all pairs of means. For these intervals, the simultaneous confidence level is $100(1 - \alpha)$ percent when the sample sizes are equal and at least $100(1 - \alpha)$ percent when sample sizes are unequal. In other words, the Tukey procedure controls the **experimentwise** or "family" error rate at the selected level α. This is an excellent data snooping procedure when interest focuses on pairs of means.

Tukey's procedure makes use of the distribution of the **studentized range statistic**

$$q = \frac{\bar{y}_{max} - \bar{y}_{min}}{\sqrt{MS_E/n}}$$

where $\bar{y}_{max}$ and $\bar{y}_{min}$ are the largest and smallest sample means, respectively, out of a group of p sample means. The values of $q_\alpha(p, f)$, the upper α percentage points of q, where f is the number of degrees of freedom associated with the MS_E, can be seen from footnote *. For equal sample sizes, Tukey's test declares two means significantly different if the absolute value of their sample differences exceeds

$$T_\alpha = q_\alpha(a, f)\sqrt{\frac{MS_E}{n}} \tag{3.35}$$

Equivalently, we could construct a set of $100(1 - \alpha)$ percent confidence intervals for all pairs of means as follows:

$$\bar{y}_{i.} - \bar{y}_{j.} - q_\alpha(a, f)\sqrt{\frac{MS_E}{n}} \leq \mu_i - \mu_j$$
$$\leq \bar{y}_{i.} - \bar{y}_{j.} + q_\alpha(a, f)\sqrt{\frac{MS_E}{n}}, \quad i \neq j. \tag{3.36}$$

When sample sizes are not equal, Equations 3.35 and 3.36 become

$$T_\alpha = \frac{q_\alpha(a, f)}{\sqrt{2}}\sqrt{MS_E\left(\frac{1}{n_i} + \frac{1}{n_j}\right)} \tag{3.37}$$

and

$$\bar{y}_{i.} - \bar{y}_{j.} - \frac{q_\alpha(a, f)}{\sqrt{2}}\sqrt{MS_E\left(\frac{1}{n_i} + \frac{1}{n_j}\right)} \leq \mu_i - \mu_j$$
$$\leq \bar{y}_{i.} - \bar{y}_{j.} + \frac{q_\alpha(a, f)}{\sqrt{2}}\sqrt{MS_E\left(\frac{1}{n_i} + \frac{1}{n_j}\right)}, \; i \neq j \tag{3.38}$$

respectively. The unequal sample size version is sometimes called the **Tukey–Kramer procedure.**

*From J. M. May, "Extended and Corrected Tables of the Upper Percentage Points of the Studentized Range," *Biometrika*, Vol. 39, pp. 192–193, 1952.

EXAMPLE 3.7

To illustrate Tukey's test, we use the data from the plasma etching experiment in Example 3.1. With $\alpha = 0.05$ and $f = 16$ degrees of freedom for error, we get $q_{0.05}(4, 16) = 4.05$.* Therefore, from Equation 3.35,

$$T_{0.05} = q_{0.05}(4, 16)\sqrt{\frac{MS_E}{n}} = 4.05\sqrt{\frac{333.70}{5}} = 33.09$$

Thus, any pairs of treatment averages that differ in absolute value by more than 33.09 would imply that the corresponding pair of population means are significantly different. The four treatment averages are

$$\bar{y}_{1.} = 551.2 \quad \bar{y}_{2.} = 587.4$$
$$\bar{y}_{3.} = 625.4 \quad \bar{y}_{4.} = 707.0$$

and the differences in averages are

$$\bar{y}_{1.} - \bar{y}_{2.} = 551.2 - 587.4 = -36.20^*$$
$$\bar{y}_{1.} - \bar{y}_{3.} = 551.2 - 625.4 = -74.20^*$$
$$\bar{y}_{1.} - \bar{y}_{4.} = 551.2 - 707.0 = -155.8^*$$
$$\bar{y}_{2.} - \bar{y}_{3.} = 587.4 - 625.4 = -38.0^*$$
$$\bar{y}_{2.} - \bar{y}_{4.} = 587.4 - 707.0 = -119.6^*$$
$$\bar{y}_{3.} - \bar{y}_{4.} = 625.4 - 707.0 = -81.60^*$$

The starred values indicate the pairs of means that are significantly different. Note that the Tukey procedure indicates that all pairs of means differ. Therefore, each power setting results in a mean etch rate that differs from the mean etch rate at any other power setting.

When using any procedure for pairwise testing of means, we occasionally find that the overall F test from the ANOVA is significant, but the pairwise comparison of means fails to reveal any significant differences. This situation occurs because the F test is simultaneously considering all possible contrasts involving the treatment means, not just pairwise comparisons. That is, in the data at hand, the significant contrasts may not be of the form $\mu_i - \mu_j$.

The derivation of the Tukey confidence interval of Equation 3.36 for equal sample sizes is straightforward. For the studentized range statistic q, we have

$$P\left(\frac{\max(\bar{y}_{i.} - \mu_i) \quad \min(\bar{y}_{i.} - \mu_i)}{\sqrt{MS_E/n}} \le q_\alpha(a, f)\right) = 1 - \alpha$$

If $\max(\bar{y}_{i.} - \mu_i) - \min(\bar{y}_{i.} - \mu_i)$ is less than or equal to $q_\alpha(a, f)\sqrt{MS_E/n}$, it must be true that $|(\bar{y}_{i.} - \mu_i) - (\bar{y}_{j.} - \mu_j)| \le q_\alpha(a, f)\sqrt{MS_E/n}$ for every pair of means. Therefore

$$P\left(-q_\alpha(a, f)\sqrt{\frac{MS_E}{n}} \le \bar{y}_{i.} - \bar{y}_{j.} - (\mu_i - \mu_j) \le q_\alpha(a, f)\sqrt{\frac{MS_E}{n}}\right) = 1 - \alpha$$

Rearranging this expression to isolate $\mu_i - \mu_j$ between the inequalities will lead to the set of $100(1 - \alpha)$ percent simultaneous confidence intervals given in Equation 3.38.

The Fisher Least Significant Difference (LSD) Method. The Fisher method for comparing all pairs of means controls the error rate α for each individual pairwise comparison but does not control the experimentwise or family error rate. This procedure uses the t statistic for testing $H_0: \mu_i = \mu_j$

$$t_0 = \frac{\bar{y}_{i.} - \bar{y}_{j.}}{\sqrt{MS_E\left(\frac{1}{n_i} + \frac{1}{n_j}\right)}} \tag{3.39}$$

*From J. M. May, "Extended and Corrected Tables of the Upper Percentage Points of the Studentized Range," *Biometrika*, Vol. 39, pp. 192–193, 1952.

Assuming a two-sided alternative, the pair of means μ_i and μ_j would be declared significantly different if $|\bar{y}_{i.} - \bar{y}_{j.}| > t_{\alpha/2,N-a}\sqrt{MS_E(1/n_i + 1/n_j)}$. The quantity

$$\text{LSD} = t_{\alpha/2,N-a}\sqrt{MS_E\left(\frac{1}{n_i} + \frac{1}{n_j}\right)} \tag{3.40}$$

is called the **least significant difference**. If the design is balanced, $n_1 = n_2 = \cdots = n_a = n$, and

$$\text{LSD} = t_{\alpha/2,N-a}\sqrt{\frac{2MS_E}{n}} \tag{3.41}$$

To use the Fisher LSD procedure, we simply compare the observed difference between each pair of averages to the corresponding LSD. If $|\bar{y}_{i.} - \bar{y}_{j.}| > \text{LSD}$, we conclude that the population means μ_i and μ_j differ. The t statistic in Equation 3.39 could also be used.

EXAMPLE 3.8

To illustrate the procedure, if we use the data from the experiment in Example 3.1, the LSD at $\alpha = 0.05$ is

$$\text{LSD} = t_{.025,16}\sqrt{\frac{2MS_E}{n}} = 2.120\sqrt{\frac{2(333.70)}{5}} = 24.49$$

Thus, any pair of treatment averages that differ in absolute value by more than 24.49 would imply that the corresponding pair of population means are significantly different. The differences in averages are

$$\bar{y}_{1.} - \bar{y}_{2.} = 551.2 - 587.4 = -36.2^*$$

$$\bar{y}_{1.} - \bar{y}_{3.} = 551.2 - 625.4 = -74.2^*$$

$$\bar{y}_{1.} - \bar{y}_{4.} = 551.2 - 707.0 = -155.8^*$$

$$\bar{y}_{2.} - \bar{y}_{3.} = 587.4 - 625.4 = -38.0^*$$

$$\bar{y}_{2.} - \bar{y}_{4.} = 587.4 - 707.0 = -119.6^*$$

$$\bar{y}_{3.} - \bar{y}_{4.} = 625.4 - 707.0 = -81.6^*$$

The starred values indicate pairs of means that are significantly different. Clearly, all pairs of means differ significantly.

Note that the overall α risk may be considerably inflated using this method. Specifically, as the number of treatments a gets larger, the experimentwise or family type I error rate (the ratio of the number of experiments in which at least one type I error is made to the total number of experiments) becomes large.

Which Pairwise Comparison Method Do I Use? Certainly, a logical question at this point is, Which one of these procedures should I use? Unfortunately, there is no clear-cut answer to this question, and professional statisticians often disagree over the utility of the various procedures. Carmer and Swanson (1973) have conducted Monte Carlo simulation studies of a number of multiple comparison procedures, including others not discussed here. They report that the least significant difference method is a very effective test for detecting true differences in means if it is applied *only after* the *F* test in the ANOVA is significant at 5 percent. However, this method does not contain the experimentwise error rate. Because the Tukey method does control the overall error rate, many statisticians prefer to use it.

As indicated above, there are several other multiple comparison procedures. For articles describing these methods, see O'Neill and Wetherill (1971), Miller (1977), and Nelson (1989). The books by Miller (1991) and Hsu (1996) are also recommended.

3.5.8 Comparing Treatment Means with a Control

In many experiments, one of the treatments is a **control**, and the analyst is interested in comparing each of the other $a - 1$ treatment means with the control. Thus, only $a - 1$ comparisons are to be made. A procedure for making these comparisons has been developed by Dunnett (1964). Suppose that treatment a is the control and we wish to test the hypotheses

$$H_0: \mu_i = \mu_a$$
$$H_1: \mu_i \neq \mu_a$$

for $i = 1, 2, \ldots, a - 1$. Dunnett's procedure is a modification of the usual t-test. For each hypothesis, we compute the observed differences in the sample means

$$|\bar{y}_{i.} - \bar{y}_{a.}| \qquad i = 1, 2, \ldots, a - 1$$

The null hypothesis $H_0: \mu_i = \mu_a$ is rejected using a type I error rate α if

$$|\bar{y}_{i.} - \bar{y}_{a.}| > d_\alpha(a - 1, f)\sqrt{MS_E\left(\frac{1}{n_i} + \frac{1}{n_a}\right)} \tag{3.42}$$

Note that α is the **joint significance level** associated with all $a - 1$ tests.

EXAMPLE 3.9

To illustrate Dunnett's test, consider the experiment from Example 3.1 with treatment 4 considered as the control. In this example, $a = 4$, $a - 1 = 3$, $f = 16$, and $n_i = n = 5$. At the 5 percent level, we find $d_{0.05}(3, 16) = 2.59$.* Thus, the critical difference becomes

$$d_{0.05}(3, 16)\sqrt{\frac{2MS_E}{n}} = 2.59\sqrt{\frac{2(333.70)}{5}} = 29.92$$

(Note that this is a simplification of Equation 3.42 resulting from a balanced design.) Thus, any treatment mean that differs in absolute value from the control by more than 29.92 would be declared significantly different. The observed differences are

$$1 \text{ vs. } 4: \bar{y}_{1.} - \bar{y}_{4.} = 551.2 - 707.0 = -155.8$$
$$2 \text{ vs. } 4: \bar{y}_{2.} - \bar{y}_{4.} = 587.4 - 707.0 = -119.6$$
$$3 \text{ vs. } 4: \bar{y}_{3.} - \bar{y}_{4.} = 625.4 - 707.0 = -81.6$$

Note that all differences are significant. Thus, we would conclude that all power settings are different from the control.

When comparing treatments with a control, it is a good idea to use more observations for the control treatment (say n_a) than for the other treatments (say n), assuming equal numbers of observations for the remaining $a - 1$ treatments. The ratio n_a/n should be chosen to be approximately equal to the square root of the total number of treatments. That is, choose $n_a/n = \sqrt{a}$.

* From C. W. Dunnett, "New Tables for Multiple Comparison with a Control," *Biometrics*, Vol. 20, No. 3, 1964, and from C. W. Dunnett, "A Multiple Comparison Procedure for Comparing Several Treatments with a Control," *Journal of the American Statistical Association*, Vol. 50, 1955.

3.6 Sample Computer Output

Computer programs for supporting experimental design and performing the analysis of variance are widely available. The output from one such program, Design-Expert, is shown in Figure 3.12, using the data from the plasma etching experiment in Example 3.1. The sum of squares corresponding to the "Model" is the usual $SS_{\text{Treatments}}$ for a single-factor design. That source is further identified as "A." When there is more than one factor in the experiment, the model sum of squares will be decomposed into several sources (A, B, etc.). Notice that the analysis of variance summary at the top of the computer output contains the usual sums of squares, degrees of freedom, mean squares, and test statistic F_0. The column "Prob > F" is the P-value (actually, the upper bound on the P-value because probabilities less than 0.0001 are defaulted to 0.0001).

In addition to the basic analysis of variance, the program displays some other useful information. The quantity "R-squared" is defined as

$$R^2 = \frac{SS_{\text{Model}}}{SS_{\text{Total}}} = \frac{66{,}870.55}{72{,}209.75} = 0.9261$$

and is loosely interpreted as the proportion of the variability in the data "explained" by the ANOVA model. Thus, in the plasma etching experiment, the factor "power" explains about 92.61 percent of the variability in etch rate. Clearly, we must have $0 \le R^2 \le 1$, with larger values being more desirable. There are also some other R^2-like statistics displayed in the output. The "adjusted" R^2 is a variation of the ordinary R^2 statistic that reflects the number of factors in the model. It can be a useful statistic for more complex experiments with several design factors when we wish to evaluate the impact of increasing or decreasing the number of model terms. "Std. Dev." is the square root of the error mean square, $\sqrt{333.70} = 18.27$, and "C.V." is the coefficient of variation, defined as $(\sqrt{MS_E}/\bar{y})100$. The coefficient of variation measures the unexplained or residual variability in the data as a percentage of the mean of the response variable. "PRESS" stands for "prediction error sum of squares," and it is a measure of how well the model for the experiment is likely to predict the responses in a *new experiment*. Small values of PRESS are desirable. Alternatively, one can calculate an R^2 for prediction based on PRESS (we will show how to do this later). This R^2_{Pred} in our problem is 0.8845, which is not unreasonable, considering that the model accounts for about 93 percent of the variability in the current experiment. The "adequate precision" statistic is computed by dividing the difference between the maximum predicted response and the minimum predicted response by the average standard deviation of all predicted responses. Large values of this quantity are desirable, and values that exceed four usually indicate that the model will give reasonable performance in prediction.

Treatment means are estimated, and the standard error (or sample standard deviation of each treatment mean, $\sqrt{MS_E/n}$) is displayed. Differences between pairs of treatment means are investigated by using a hypothesis testing version of the Fisher LSD method described in Section 3.5.7.

The computer program also calculates and displays the residuals, as defined in Equation 3.16. The program will also produce all of the residual plots that we discussed in Section 3.4. There are also several other residual diagnostics displayed in the output. Some of these will be discussed later. Design-Expert also displays the studentized residual (called "Student Residual" in the output), calculate as

$$r_{ij} = \frac{e_{ij}}{\sqrt{MS_E(1 - \text{Leverage}_{ij})}}$$

where Leverage_{ij} is a measure of the influence of the ij^{th} observation on the model. We will discuss leverage in more detail and show how it is calculated in chapter 10. Studentized residuals

Response: Etch Rate

ANOVA for Selected Factorial Model
Analysis of variance table [Partial sum of squares]

Source	Sum of Squares	DF	Mean Square	F Value	Prob > F
Model	66870.55	3	22290.18	66.80	<0.0001 significant
A	*66870.55*	*3*	*22290.18*	*66.80*	*<0.0001*
Pure Error	5338.20	16	333.70		
Cor Total	72209.75	19			

The Model F-value of 66.80 implies that the model is significant. There is only a 0.01% chance that a "Model F-Value" this large could occur due to noise.

Values of "Prob > F" less than 0.0500 indicate that model terms are significant.
In this case, A are significant model terms.
Values greater than 0.1000 indicate that the model terms are not significant.
If there are many insignificant model terms (not counting those required to support hierarchy), model reduction may improve your model.

Std. Dev.	18.27	R-Squared	0.9261
Mean	617.75	Adj R-Squared	0.9122
C.V.	2.96	Pred R-Squared	0.8846
PRESS	8342.50	Adeq Precision	19.071

The "Pred R-Squared" of 0.8845 is in reasonable agreement with the "Adj R-Squared" of 0.9122.

"Adeq Precision" measures the signal-to-noise ratio. A ratio greater than four is disirable. Your ratio of 19.071 indicates an adequate signal. This model can be used to navigate the design space.

Treatment Means (Adjusted, If Necessary)

	Estimated Mean	Standard Error
1-160	551.20	8.17
2-180	587.40	8.17
3-200	625.40	8.17
4-220	707.00	8.17

Treatment	Mean Difference	DF	Standard Error	t for H_0 Coeff=0	Prob > |t|
1 vs 2	−36.20	1	11.55	−3.13	0.0064
1 vs 3	−74.20	1	11.55	−6.42	<0.0001
1 vs 4	−155.80	1	11.55	−13.49	<0.0001
2 vs 3	−38.00	1	11.55	−3.29	0.0046
2 vs 4	−119.60	1	11.55	−10.35	<0.0001
3 vs 4	−81.60	1	11.55	−7.06	<0.0001

Values of "Prob > |t|" less than 0.0500 indicate that the difference in the treatment means is significant.
Values of "Prob > |t|" greater than 0.1000 indicate that the difference in the two treatment means is not significant.

Diagnostics Case Statistics

Standard Order	Actual Value	Predicted Value	Residual	Leverage	Student Residual	Cook's Distance	Outlier *t*	Run Order
1	575.00	551.20	23.80	0.200	1.457	0.133	1.514	13
2	542.00	551.20	−9.20	0.200	−0.563	0.020	−0.551	14
3	530.00	551.20	−21.20	0.200	−1.298	0.105	−1.328	8
4	539.00	551.20	−12.20	0.200	−0.747	0.035	−0.736	5
5	570.00	551.20	18.80	0.200	1.151	0.083	1.163	4
6	565.00	587.40	−22.40	0.200	−1.371	0.117	−1.413	18
7	593.00	587.40	5.60	0.200	0.343	0.007	0.333	9
8	590.00	587.40	2.60	0.200	0.159	0.002	0.154	6
9	579.00	587.40	−8.40	0.200	−0.514	0.017	−0.502	16
10	610.00	587.40	22.60	0.200	1.383	0.120	1.427	17
11	600.00	625.40	−25.40	0.200	−1.555	0.151	−1.634	7
12	651.00	625.40	25.60	0.200	1.567	0.153	1.649	19
13	610.00	625.40	−15.40	0.200	−0.943	0.056	−0.939	10
14	637.00	625.40	11.60	0.200	0.710	0.032	0.699	20
15	629.00	625.40	3.60	0.200	0.220	0.003	0.214	1
16	725.00	707.00	18.00	0.200	1.102	0.076	1.110	2
17	700.00	707.00	−7.00	0.200	−0.428	0.011	−0.417	3
18	715.00	707.00	8.00	0.200	0.490	0.015	0.478	15
19	685.00	707.00	−22.00	0.200	−1.346	0.113	−1.385	11
20	710.00	707.00	3.00	0.200	0.184	0.002	0.178	12

Proceed to Diagnostic Plots (the next icon in progression). Be sure to look at the
(1) Normal probability plot of the studentized residuals to check for normality of residuals.
(2) Studentized residuals versus predicted values to check for constant error.
(3) Outlier *t* versus run order to look for outliers, i.e., influential values.
(4) Box-Cox plot for power transformations.

If all the model statistics and diagnostic plots are OK, finish up with the Model Graphs icon.

■ **FIGURE 3.12** **Design-Expert computer output for Example 3.1**

```
One-way ANOVA: Etch Rate versus Power

Source   DF     SS      MS      F      P
Power     3  66871   22290  66.80  0.000
Error    16   5339     334
Total    19  72210

S = 18.27   R-Sq = 92.61%   R-Sq (adj) = 91.22%

                                 Individual 95% CIs For Mean Based on
                                 Pooled StDev
Level  N    Mean  Std.Dev.     ---+---------+---------+---------+------
160    5  551.20     20.02     (---*---)
180    5  587.40     16.74           (---*---)
200    5  625.40     20.53                   (---*---)
220    5  707.00     15.25                                 (---*---)
                               ---+---------+---------+---------+------
                                550       600       650       700

Pooled Std. Dev. = 18.27

Turkey 95% Simultaneous Confidence Intervals
All Pairwise Comparisons among Levels of Power

Individual confidence level = 98.87%

Power = 160 subtracted from

Power   Lower  Center   Upper  ------+---------+---------+---------+---
180      3.11   36.20   69.29                  (---*---)
200     41.11   74.20  107.29                     (---*---)
220    122.71  155.80  188.89                            (---*---)
                               ------+---------+---------+---------+---
                                  -100         0       100       200

Power = 180 subtracted from

Power   Lower  Center   Upper  ------+---------+---------+---------+---
200      4.91   38.00   71.09                  (---*---)
220     86.51  119.60  152.69                        (---*---)
                               ------+---------+---------+---------+---
                                  -100         0       100       200

Power = 200 subtracted from

Power   Lower  Center   Upper  ------+---------+---------+---------+---
220     48.51   81.60  114.69                      (---*---)
                               ------+---------+---------+---------+---
                                  -100         0       100       200
```

■ **FIGURE 3.13 Minitab computer output for Example 3.1**

are considered to be more effective in identifying potential rather than either the ordinary residuals or standardized residuals.

Finally, notice that the computer program also has some interpretative guidance embedded in the output. This "advisory" information is fairly standard in many PC-based statistics packages. Remember in reading such guidance that it is written in very general terms and may not exactly suit the report writing requirements of any specific experimenter. This advisory output may be hidden upon request by the user.

Figure 3.13 presents the output from Minitab for the plasma etching experiment. The output is very similar to the Design-Expert output in Figure 3.12. Note that confidence intervals on each individual treatment mean are provided and that the pairs of means are compared using Tukey's method. However, the Tukey method is presented using the confidence interval format instead of the hypothesis-testing format that we used in Section 3.5.7. None of the Tukey confidence intervals includes zero, so we would conclude that all of the means are different.

Figure 3.14 is the output from JMP for the plasma etch experiment in Example 3.1. The output information is very similar to that from Design-Expert and Minitab. The plots of actual observations versus the predicted values and residuals versus the predicted values are default output. There is an option in JMP to provide the Fisher LSD procedure or Tukey's method to compare all pairs of means.

3.7 Determining Sample Size

In any experimental design problem, a critical decision is the choice of sample size—that is, determining the number of replicates to run. Generally, if the experimenter is interested in detecting small effects, more replicates are required than if the experimenter is interested in detecting large effects. In this section, we discuss several approaches to determining sample size. Although our discussion focuses on a single-factor design, most of the methods can be used in more complex experimental situations.

3.7.1 Operating Characteristic Curves

Recall that an **operating characteristic (OC) curve** is a plot of the type II error probability of a statistical test for a particular sample size versus a parameter that reflects the extent to which the null hypothesis is false. These curves can be used to guide the experimenter in selecting the number of replicates so that the design will be sensitive to important potential differences in the treatments.

We consider the probability of type II error of the fixed effects model for the case of equal sample sizes per treatment, say

$$\begin{aligned}\beta &= 1 - P\{\text{Reject } H_0 | H_0 \text{ is false}\} \\ &= 1 - P\{F_0 > F_{\alpha,a-1,N-a} | H_0 \text{ is false}\}\end{aligned} \quad (3.43)$$

To evaluate the probability statement in Equation 3.43, we need to know the distribution of the test statistic F_0 if the null hypothesis is false. It can be shown that, if H_0 is false, the statistic $F_0 = MS_{\text{Treatments}}/MS_E$ is distributed as a **noncentral *F*** random variable with $a - 1$ and $N - a$ degrees of freedom and the noncentrality parameter δ. If $\delta = 0$, the noncentral F distribution becomes the usual (central) F distribution.

Operating characteristic curves (for Fixed Effects Model Analysis of Variance)* are used to evaluate the probability statement in Equation 3.43. These curves plot the probability of type II error (β) against a parameter Φ, where

$$\Phi^2 = \frac{n \sum_{i=1}^{a} \tau_i^2}{a\sigma^2} \quad (3.44)$$

The quantity Φ^2 is related to the noncentrality parameter δ. Curves are available for $\alpha = 0.05$ and $\alpha = 0.01$ and a range of degrees of freedom for numerator and denominator.

In using the OC curves, the experimenter must specify the parameter Φ and the value of σ^2. This is often difficult to do in practice. One way to determine Φ is to choose the actual values of the treatment means for which we would like to reject the null hypothesis with high probability. Thus, if $\mu_1, \mu_2, \ldots, \mu_a$ are the specified treatment means, we find the τ_i in Equation 3.48 as $\tau_i = \mu_i - \bar{\mu}$, where $\bar{\mu} = (1/a)\Sigma_{i=1}^{a}\mu_i$ is the average of the individual treatment means. The estimate of σ^2 may be available from prior experience, a previous experiment or a preliminary test (as suggested in Chapter 1), or a judgment estimate. When we are uncertain about the value of σ^2, sample sizes could be determined for a range of likely values of σ^2 to study the effect of this parameter on the required sample size before a final choice is made.

---

* From *Biometrika Tables for Statisticians*, Vol. 2, by E. S. Pearson and H. O. Hartley, (1972).

Response Etch rate
Whole Model

Actual by Predicted Plot

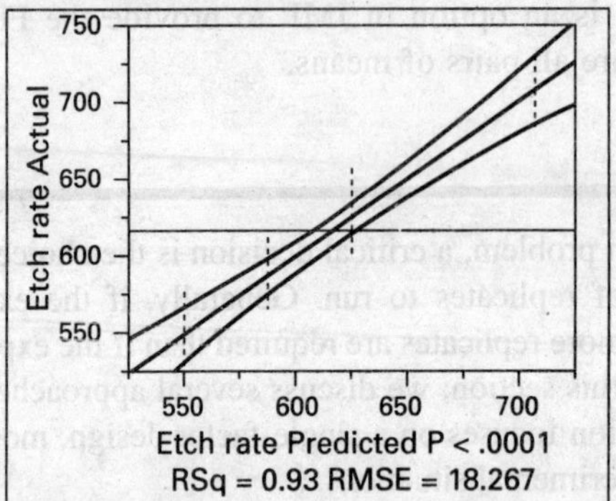

Summary of Fit

RSquare	0.92606
RSquare Adj	0.912196
Root Mean Square Error	18.26746
Mean of Response	617.75
Observations (or Sum Wgts)	20

Analysis of Variance

Source	DF	Sum of Squares	Mean Square	*F* Ratio
Model	3	66870.550	22290.2	66.7971
Error	16	5339.200	333.7	Prob> F
C. Total	19	72209.750		<.0001

Effect Tests

Source	Nparm	DF	Sum of Squares	*F* Ratio	Prob > *F*
RF power	3	3	66870.550	66.7971	<.0001

Residual by Predicted Plot

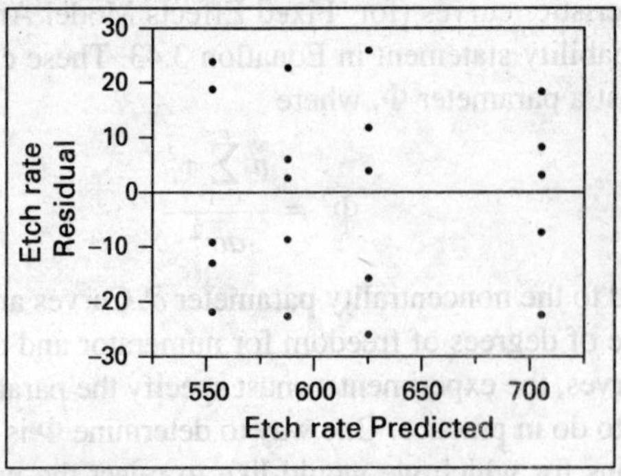

RF power

Least Squares Means Table

Level	Least Sq Mean	Std Error	Mean
160	551.20000	8.1694553	551.200
180	587.40000	8.1694553	587.400
200	625.40000	8.1694553	625.400
220	707.00000	8.1694553	707.000

■ **FIGURE 3.14** **JMP output from Example 3.1**

EXAMPLE 3.10

Consider the plasma etching experiment described in Example 3.1. Suppose that the experimenter is interested in rejecting the null hypothesis with a probability of at least 0.90 if the four treatment means are

$$\mu_1 = 575 \quad \mu_2 = 600 \quad \mu_3 = 650 \quad \text{and} \quad \mu_4 = 675$$

She plans to use $\alpha = 0.01$. In this case, because $\sum_{i=1}^{4}\mu_i = 2500$, we have $\overline{\mu} = (1/4)2500 = 625$ and

$$\tau_1 = \mu_1 - \overline{\mu} = 575 - 625 = -50$$
$$\tau_2 = \mu_2 - \overline{\mu} = 600 - 625 = -25$$
$$\tau_3 = \mu_3 - \overline{\mu} = 650 - 625 = 25$$
$$\tau_4 = \mu_4 - \overline{\mu} = 675 - 625 = 50$$

Thus, $\sum_{i=1}^{4}\tau_i^2 = 6250$. Suppose the experimenter feels that the standard deviation of etch rate at any particular level of power will be no larger than $\sigma = 25$ Å/min. Then, by using Equation 3.44, we have

$$\Phi^2 = \frac{n\sum_{i=1}^{4}\tau_i^2}{a\sigma^2} = \frac{n(6{,}250)}{4(25)^2} = 2.5n$$

We use the OC curve for $a - 1 = 4 - 1 = 3$ with $N - a = a(n-1) = 4(n-1)$ error degrees of freedom and $\alpha = 0.01$. As a first guess at the required sample size, try $n = 3$ replicates. This yields $\Phi^2 = 2.5n = 2.5(3) = 7.5$, $\Phi = 2.74$, and $4(2) = 8$ error degrees of freedom. Consequently, from Chart V, we find that $\beta \simeq 0.25$. Therefore, the power of the test is approximately $1 - \beta = 1 - 0.25 = 0.75$, which is less than the required 0.90, and so we conclude that $n = 3$ replicates are not sufficient. Proceeding in a similar manner, we can construct the following display:

n	Φ^2	Φ	$a(n-1)$	β	Power $(1-\beta)$
3	7.5	2.74	8	0.25	0.75
4	10.0	3.16	12	0.04	0.96
5	12.5	3.54	16	<0.01	>0.99

Thus, 4 or 5 replicates are sufficient to obtain a test with the required power.

A significant problem with this approach to using OC curves is that it is usually difficult to select a set of treatment means on which the sample size decision should be based. An alternate approach is to select a sample size such that if the difference between any two treatment means exceeds a specified value, the null hypothesis should be rejected. If the difference between any two treatment means is as large as D, it can be shown that the minimum value of Φ^2 is

$$\Phi^2 = \frac{nD^2}{2a\sigma^2} \tag{3.45}$$

Because this is a minimum value of Φ^2, the corresponding sample size obtained from the operating characteristic curve is a conservative value; that is, it provides a power at least as great as that specified by the experimenter.

To illustrate this approach, suppose that in the plasma etching experiment from Example 3.1, the experimenter wished to reject the null hypothesis with probability at least 0.90 if any two treatment means differed by as much as 75 Å/min and $\alpha = 0.01$. Then, assuming that $\sigma = 25$ psi, we find the minimum value of Φ^2 to be

$$\Phi^2 = \frac{n(75)^2}{2(4)(25^2)} = 1.125n$$

Now we can use the OC curves exactly as in Example 3.10. Suppose we try $n = 4$ replicates. This results in $\Phi^2 = 1.125(4) = 4.5$, $\Phi = 2.12$, and $4(3) = 12$ degrees of freedom for error. From the OC curve, we find that the power is approximately 0.65. For $n = 5$ replicates, we have $\Phi^2 = 5.625$, $\Phi = 2.37$, and $4(4) = 16$ degrees of freedom for error. From the OC curve, the power is approximately 0.8. For $n = 6$ replicates, we have $\Phi^2 = 6.75$, $\Phi = 2.60$, and $4(5) = 20$ degrees of freedom for error. From the OC curve, the power exceeds 0.90, so $n = 6$ replicates are required.

Minitab uses this approach to perform power calculations and find sample sizes for single-factor ANOVAs. Consider the following display:

```
Power and Sample Size

One-way ANOVA

Alpha = 0.01 Assumed standard deviation = 25
Number of Levels = 4

              Sample                          Maximum
SS Means        Size          Power        Difference
  2812.5           5       0.804838                75

The sample size is for each level.

Power and Sample Size

One-way ANOVA

Alpha = 0.01 Assumed standard deviation = 25
Number of Levels 5 4

              Sample    Target                            Maximum
SS Means        Size     Power     Actual Power        Difference
  2812.5           6       0.9         0.915384                75

The sample size is for each level.
```

In the upper portion of the display, we asked Minitab to calculate the power for $n = 5$ replicates when the maximum difference in treatment means is 75. Notice that the results closely match those obtained from the OC curves. The bottom portion of the display the output when the experimenter requests the sample size to obtain a target power of at least 0.90. Once again, the results agree with those obtained from the OC curve.

3.7.2 Specifying a Standard Deviation Increase

This approach is occasionally helpful in choosing the sample size. If the treatment means do not differ, the standard deviation of an observation chosen at random is σ. If the treatment means are different, however, the standard deviation of a randomly chosen observation is

$$\sqrt{\sigma^2 + \left(\sum_{i=1}^{a} \tau_i^2/a\right)}$$

If we choose a percentage P for the increase in the standard deviation of an observation beyond which we wish to reject the hypothesis that all treatment means are equal, this is equivalent to choosing

$$\frac{\sqrt{\sigma^2 + \left(\sum_{i=1}^{a} \tau_i^2/a\right)}}{\sigma} = 1 + 0.01P \quad (P = \text{percent})$$

or

$$\frac{\sqrt{\sum_{i=1}^{a} \tau_i^2/a}}{\sigma} = \sqrt{(1 + 0.01P)^2 - 1}$$

so that

$$\Phi = \frac{\sqrt{\sum_{i=1}^{a} \tau_i^2/a}}{\sigma/\sqrt{n}} = \sqrt{(1 + 0.01P)^2 - 1}(\sqrt{n}) \tag{3.46}$$

Thus, for a specified value of P, we may compute Φ from Equation 3.46 and then use the operating characteristic curves to determine the required sample size.

For example, in the plasma etching experiment from Example 3.1, suppose that we wish to detect a standard deviation increase of 20 percent with a probability of at least 0.90 and $\alpha = 0.05$. Then

$$\Phi = \sqrt{(1.2)^2 - 1}(\sqrt{n}) = 0.66\sqrt{n}$$

Reference to the operating characteristic curves shows that $n = 10$ replicates would be required to give the desired sensitivity.

3.7.3 Confidence Interval Estimation Method

This approach assumes that the experimenter wishes to express the final results in terms of confidence intervals and is willing to specify in advance how wide he or she wants these confidence intervals to be. For example, suppose that in the plasma etching experiment from Example 3.1, we wanted a 95 percent confidence interval on the difference in mean etch rate for any two power settings to be ±30 Å/min and a prior estimate of σ is 25. Then, using Equation 3.13, we find that the accuracy of the confidence interval is

$$\pm t_{\alpha/2,N-a}\sqrt{\frac{2MS_E}{n}}$$

Suppose that we try $n = 5$ replicates. Then, using $\sigma^2 = (25)^2 = 625$ as an estimate of MS_E, the accuracy of the confidence interval becomes

$$\pm 2.120\sqrt{\frac{2(625)}{5}} = \pm 33.52$$

which does not meet the requirement. Trying $n = 6$ gives

$$\pm 2.086\sqrt{\frac{2(625)}{6}} = \pm 30.11$$

Trying $n = 7$ gives

$$\pm 2.064\sqrt{\frac{2(625)}{7}} = \pm 27.58$$

Clearly, $n = 7$ is the smallest sample size that will lead to the desired accuracy.

The quoted level of significance in the above illustration applies only to one confidence interval. However, the same general approach can be used if the experimenter wishes to prespecify a *set* of confidence intervals about which a **joint** or **simultaneous confidence statement** is made (see the comments about simultaneous confidence intervals in Section 3.3.3). Furthermore, the confidence intervals could be constructed about more general contrasts in the treatment means than the pairwise comparison illustrated above.

3.8 Other Examples of Single-Factor Experiments

3.8.1 Chocolate and Cardiovascular Health

An article in *Nature* describes an experiment to investigate the effect of consuming chocolate on cardiovascular health ("Plasma Antioxidants from Chocolate," *Nature*, Vol. 424, 2003, pp. 1013). The experiment consisted of using three different types of chocolates: 100 g of dark chocolate, 100 g of dark chocolate with 200 mL of full-fat milk, and 200 g of milk chocolate. Twelve subjects were used, 7 women and 5 men, with an average age range of 32.2 ± 1 years, an average weight of 65.8 ± 3.1 kg, and body-mass index of 21.9 ± 0.4 kg m^{-2}. On different days a subject consumed one of the chocolate-factor levels and one hour later the total antioxidant capacity of their blood plasma was measured in an assay. Data similar to that summarized in the article are shown in Table 3.12.

Figure 3.15 presents box plots for the data from this experiment. The result is an indication that the blood antioxidant capacity one hour after eating the dark chocolate is higher than for the other two treatments. The variability in the sample data from all three treatments seems very similar. Table 3.13 is the Minitab ANOVA output. The test statistic is highly significant (Minitab reports a *P*-value of 0.000, which is clearly wrong because *P*-values cannot be zero; this means that the *P*-value is less than 0.001), indicating that some of the treatment means are different. The output also contains the Fisher LSD analysis for this experiment. This indicates that the mean antioxidant capacity after consuming dark chocolate is higher than after consuming dark chocolate plus milk or milk chocolate alone, and the mean antioxidant capacity after consuming dark chocolate plus milk or milk chocolate alone are equal. Figure 3.16 is the normal probability plot of the residual and Figure 3.17 is the plot of residuals versus predicted values. These plots do not suggest any problems with model assumptions. We conclude that consuming dark chocolate results in higher mean blood antioxidant capacity after one hour than consuming either dark chocolate plus milk or milk chocolate alone.

3.8.2 A Real Economy Application of a Designed Experiment

Designed experiments have had tremendous impact on manufacturing industries, including the design of new products and the improvement of existing ones, development of new

■ **TABLE 3.12**
Blood Plasma Levels One Hour Following Chocolate Consumption

	Subjects (Observations)											
Factor	**1**	**2**	**3**	**4**	**5**	**6**	**7**	**8**	**9**	**10**	**11**	**12**
DC	118.8	122.6	115.6	113.6	119.5	115.9	115.8	115.1	116.9	115.4	115.6	107.9
DC+MK	105.4	101.1	102.7	97.1	101.9	98.9	100.0	99.8	102.6	100.9	104.5	93.5
MC	102.1	105.8	99.6	102.7	98.8	100.9	102.8	98.7	94.7	97.8	99.7	98.6

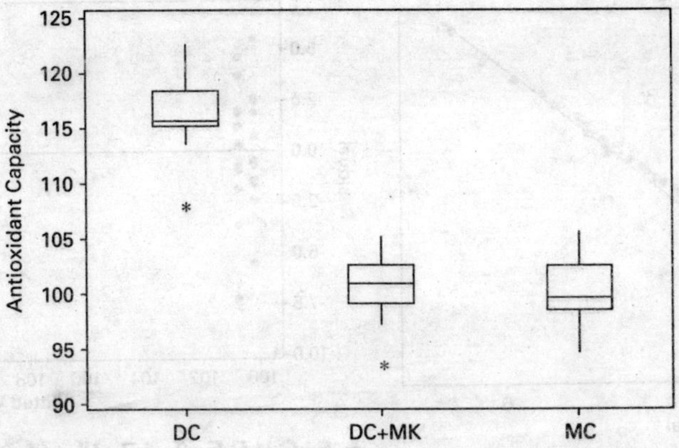

■ **FIGURE 3.15** **Box plots of the blood antioxidant capacity data from the chocolate consumption experiment**

■ **TABLE 3.13**
Minitab ANOVA Output, Chocolate Consumption Experiment

One-way ANOVA: DC, DC+MK, MC

```
Source  DF      SS     MS      F      P
Factor   2  1952.6  976.3  93.58  0.000
Error   33   344.3   10.4
Total   35  2296.9

S = 3.230   R-Sq = 85.01%   R-Sq(adj) = 84.10%

                           Individual 95% CIs For Mean Based on
                           Pooled StDev
Level   N    Mean  StDev   ---+---------+---------+---------+------
DC     12  116.06   3.53                                  (---*---)
DC+MK  12  100.70   3.24    (--*---)
MC     12  100.18   2.89   (--*--)
                           ---+---------+---------+---------+------
                           100.0     105.0     110.0     115.0

Pooled StDev = 3.23

Fisher 95% Individual Confidence Intervals
All Pairwise Comparisons
Simultaneous confidence level = 88.02
DC subtracted from:

         Lower   Center    Upper   -+---------+---------+---------+---
DC+MK  -18.041  -15.358  -12.675   (---*----)
MC     -18.558  -15.875  -13.192  (----*---)
                                   -+---------+---------+---------+---
                                 -18.0     -12.0      -6.0       0.0

DC+MK subtracted from:

      Lower  Center  Upper   -+---------+---------+---------+--------
MC   -3.200  -0.517  2.166                                 (---*----)
                             -+---------+---------+---------+--------
                          -18.0     -12.0      -6.0       0.0
```

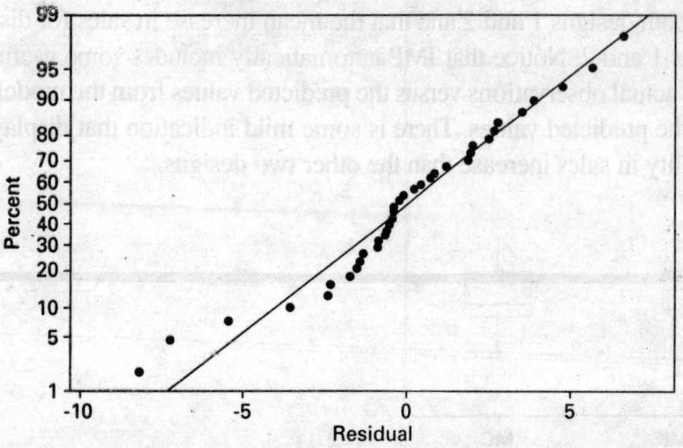

■ **FIGURE 3.16 Normal probability plot of the residuals from the chocolate consumption experiment**

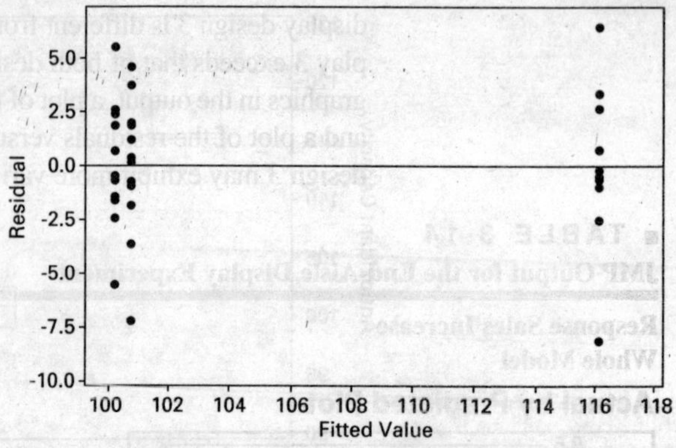

■ **FIGURE 3.17 Plot of residuals versus the predicted values from the chocolate consumption experiment**

manufacturing processes, and process improvement. In the last 15 years, designed experiments have begun to be widely used outside of this traditional environment. These applications are in financial services, telecommunications, health care, e-commerce, legal services, marketing, logistics and transporation, and many of the nonmanufacturing components of manufacturing businesses. These types of businesses are sometimes referred to as the real economy. It has been estimated that manufacturing accounts for only about 20 percent of the total US economy, so applications of experimental design in the real economy are of growing importance. In this section, we present an example of a designed experiment in marketing.

A soft drink distributor knows that end-aisle displays are an effective way to increase sales of the product. However, there are several ways to design these displays: by varying the text displayed, the colors used, and the visual images. The marketing group has designed three new end-aisle displays and wants to test their effectiveness. They have identified 15 stores of similar size and type to participate in the study. Each store will test one of the displays for a period of one month. The displays are assigned at random to the stores, and each display is tested in five stores. The response variable is the percentage increase in sales activity over the typical sales for that store when the end-aisle display is not in use. The data from this experiment are shown in Table 3.13.

Table 3.14 shows the analysis of the end-asile display experiment. This analysis was conducted using JMP. The *P*-value for the model *F* statistic in the ANOVA indicates that there is a difference in the mean percentage increase in sales between the three display types. In this application, we had JMP use the Fisher LSD procedure to compare the pairs of treatment means (JMP labels these as the least squares means). The results of this comparison are presented as confidence intervals on the difference in pairs of means. For pairs of means where the confidence interval includes zero, we would not declare that pair of means are different. The JMP output indicates that display designs 1 and 2 are similar in that they result in the same mean increase in sales, but that

■ **TABLE 3.13**
The End-Aisle Display Experimental Design

Display Design	Sample Observations, Percent Increase in Sales				
1	5.43	5.71	6.22	6.01	5.29
2	6.24	6.71	5.98	5.66	6.60
3	8.79	9.20	7.90	8.15	7.55

display design 3 is different from both designs 1 and 2 and that the mean increase in sales for display 3 exceeds that of both designs 1 and 2. Notice that JMP automatically includes some useful graphics in the output, a plot of the actual observations versus the predicted values from the model, and a plot of the residuals versus the predicted values. There is some mild indication that display design 3 may exhibit more variability in sales increase than the other two designs.

■ **TABLE 3.14**
JMP Output for the End-Aisle Display Experiment

Response Sales Increase
Whole Model

Actual by Predicted Plot

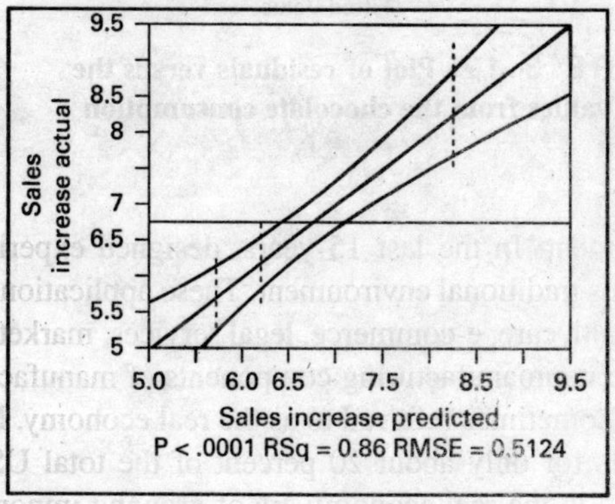

Summary of Fit

RSquare	0.856364
RSquare Adj	0.832425
Root Mean Square Error	0.512383
Mean of Response	6.762667
Observations (or Sum Wgts)	15

Analysis of Variance

Source	DF	Sum of Squares	Mean Square	F Ratio
Model	2	18.783053	9.39153	35.7722
Error	12	3.150440	0.26254	Prob>*F*
C.Total	14	21.933493		<.0001

Effect Tests

Source	Nparm	DF	Sum of Squares	F Ratio	Prob > *F*
Display	2	2	18.783053	35.7722	<.001

Residual by Predicted Plot

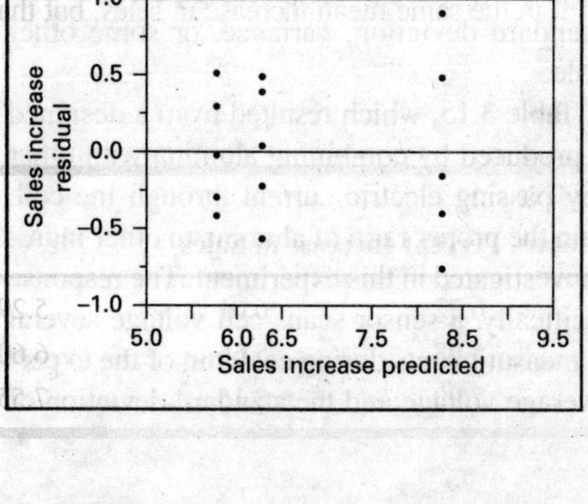

■ **TABLE 3.14** *(Continued)*

Least Squares Means Table

Level	Least Sq Mean	Std Error	Mean
1	5.7320000	0.22914479	5.73200
2	6.2380000	0.22914479	6.23800
3	8.3180000	0.22914479	8.31800

LSMeans Differences Student's t

a = 0.050 *t* = 2.17881

LSMean[i] By LSMean [i]

Mean[i]-Mean [i] Std Err Dif Lower CL Dif Upper CL Dif	1	2	3
1	0 0 0 0	−0.506 0.32406 −1.2121 0.20007	−2.586 0.32406 −3.2921 −1.8799
2	0.506 0.32406 −0.2001 1.21207	0 0 0 0	−2.08 0.32406 −2.7861 −1.3739
3	2.586 0.32406 1.87993 3.29207	2.08 0.32406 1.37393 2.78607	0 0 0 0

Level		Least Sq Mean
3	A	8.3180000
2	B	6.2380000
1	B	5.7320000

Levels not connected by same letter are significantly different.

3.8.3 Discovering Dispersion Effects

We have focused on using the analysis of variance and related methods to determine which factor levels result in differences among treatment or factor level means. It is customary to refer to these effects as **location effects**. If there was inequality of variance at the different factor levels, we used transformations to stabilize the variance to improve our inference on the location effects. In some problems, however, we are interested in discovering whether the different factor levels affect **variability**; that is, we are interested in discovering potential **dispersion effects**. This will occur whenever the standard deviation, variance, or some other measure of variability is used as a response variable.

To illustrate these ideas, consider the data in Table 3.15, which resulted from a designed experiment in an aluminum smelter. Aluminum is produced by combining alumina with other ingredients in a reaction cell and applying heat by passing electric current through the cell. Alumina is added continuously to the cell to maintain the proper ratio of alumina to other ingredients. Four different ratio control algorithms were investigated in this experiment. The response variables studied were related to cell voltage. Specifically, a sensor scans cell voltage several times each second, producing thousands of voltage measurements during each run of the experiment. The process engineers decided to use the average voltage and the standard deviation of

■ **TABLE 3.15**
Data for the Smelting Experiment

Ratio Control Algorithm	Observations 1	2	3	4	5	6
1	4.93(0.05)	4.86(0.04)	4.75(0.05)	4.95(0.06)	4.79(0.03)	4.88(0.05)
2	4.85(0.04)	4.91(0.02)	4.79(0.03)	4.85(0.05)	4.75(0.03)	4.85(0.02)
3	4.83(0.09)	4.88(0.13)	4.90(0.11)	4.75(0.15)	4.82(0.08)	4.90(0.12)
4	4.89(0.03)	4.77(0.04)	4.94(0.05)	4.86(0.05)	4.79(0.03)	4.76(0.02)

■ **TABLE 3.16**
Analysis of Variance for the Natural Logarithm of Pot Noise

Source of Variation	Sum of Squares	Degrees of Freedom	Mean Square	F_0	*P*-Value
Ratio control algorithm	6.166	3	2.055	21.96	<0.001
Error	1.872	20	0.094		
Total	8.038	23			

cell voltage (shown in parentheses) over the run as the response variables. The average voltage is important because it affects cell temperature, and the standard deviation of voltage (called "pot noise" by the process engineers) is important because it affects the overall cell efficiency.

An analysis of variance was performed to determine whether the different ratio control algorithms affect average cell voltage. This revealed that the ratio control algorithm had no **location effect**; that is, changing the ratio control algorithms does not change the average cell voltage. (Refer to Problem 3.38.)

To investigate dispersion effects, it is usually best to use

$$\log(s) \quad \text{or} \quad \log(s^2)$$

as a response variable since the log transformation is effective in stabilizing variability in the distribution of the sample standard deviation. Because all sample standard deviations of pot voltage are less than unity, we will use

$$y = -\ln(s)$$

as the response variable. Table 3.16 presents the analysis of variance for this response, the natural logarithm of "pot noise." Notice that the choice of a ratio control algorithm affects pot noise; that is, the ratio control algorithm has a **dispersion effect**. Standard tests of model adequacy, including normal probability plots of the residuals, indicate that there are no problems with experimental validity. (Refer to Problem 3.39.)

Figure 3.18 plots the average log pot noise for each ratio control algorithm and also presents a scaled *t* distribution for use as a **reference distribution** in discriminating between ratio control algorithms. This plot clearly reveals that ratio control algorithm 3 produces

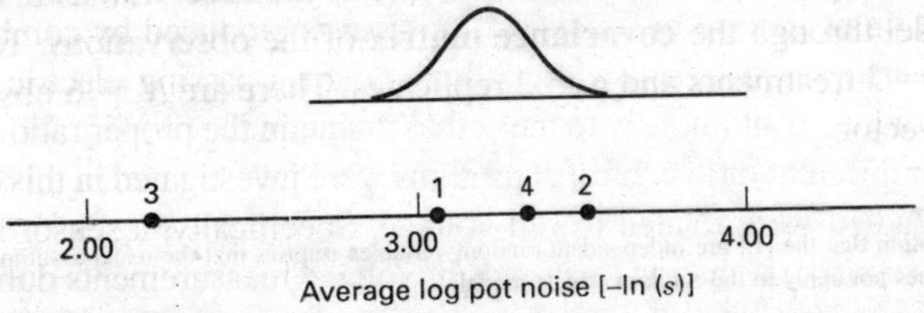

■ **FIGURE 3.18** **Average log pot noise [$-\ln(s)$] for four ratio control algorithms relative to a scaled *t* distribution with scale factor $\sqrt{MS_E/n} = \sqrt{0.094/6} = 0.125$**

greater pot noise or greater cell voltage standard deviation than the other algorithms. There does not seem to be much difference between algorithms 1, 2, and 4.

3.9 The Random Effects Model

3.9.1 A Single Random Factor

An experimenter is frequently interested in a factor that has a large number of possible levels. If the experimenter randomly selects a of these levels from the population of factor levels, then we say that the factor is **random**. Because the levels of the factor actually used in the experiment were chosen randomly, inferences are made about the entire population of factor levels. We assume that the population of factor levels is either of infinite size or is large enough to be considered infinite. Situations in which the population of factor levels is small enough to employ a finite population approach are not encountered frequently. Refer to Bennett and Franklin (1954) and Searle and Fawcett (1970) for a discussion of the finite population case.

The linear statistical model is

$$y_{ij} = \mu + \tau_i + \epsilon_{ij} \quad \begin{cases} i = 1, 2, \ldots, a \\ j = 1, 2, \ldots, n \end{cases} \tag{3.47}$$

where both the treatment effects τ_i and ϵ_{ij} are random variables. We will assume that the treatment effects τ_i are NID $(0, \sigma_\tau^2)$ random variables[1] and that the errors are NID $(0, \sigma^2)$, random variables, and that the τ_i and ϵ_{ij} are independent. Because τ_i is independent of ϵ_{ij}, the variance of any observation is

$$V(y_{ij}) = \sigma_\tau^2 + \sigma^2$$

The variances σ_τ^2 and σ^2 are called **variance components**, and the model (Equation 3.47) is called the **components of variance** or **random effects model**. The observations in the random effects model are normally distributed because they are linear combinations of the two normally and independently distributed random variables τ_i and ε_{ij}. However, unlike the fixed effects case in which all of the observations y_{ij} are independent, in the random model the observations y_{ij} are only independent if they come from different factor levels. Specifically, we can show that the covariance of any two observations is

$$Cov\left(y_{ij}, y_{ij'}\right) = \sigma_\tau^2 \qquad j \neq j'$$

$$Cov\left(y_{ij}, y_{i'j'}\right) = 0 \qquad i \neq i'$$

Note that the observations within a specific factor level all have the same covariance, because before the experiment is conducted, we expect the observations at that factor level to be similar because they all have the same random component. Once the experiment has been conducted, we can assume that all observations can be assumed to be independent, because the parameter τ_i has been determined and the observations in that treatment differ only because of random error.

We can express the covariance structure of the observations in the single-factor random effects model through the **covariance matrix** of the observations. To illustrate, suppose that we have $a = 3$ treatments and $n = 2$ replicates. There are $N = 6$ observations, which we can write as a vector

[1] The as assumption that the $[\tau_i]$ are independent random variables implies that the usual assumption of $\Sigma_{i=1}^{a} \tau_i = 0$ from the fixed effects model does not apply to the random effects model.

$$\mathbf{y} = \begin{bmatrix} y_{11} \\ y_{12} \\ y_{21} \\ y_{22} \\ y_{31} \\ y_{32} \end{bmatrix}$$

and the 6 × 6 covariance matrix of these observations is

$$Cov(\mathbf{y}) = \begin{bmatrix} \sigma_\tau^2 + \sigma^2 & \sigma_\tau^2 & 0 & 0 & 0 & 0 \\ \sigma_\tau^2 & \sigma_\tau^2 + \sigma^2 & 0 & 0 & 0 & 0 \\ 0 & 0 & \sigma_\tau^2 + \sigma^2 & \sigma_\tau^2 & 0 & 0 \\ 0 & 0 & \sigma_\tau^2 & \sigma_\tau^2 + \sigma^2 & 0 & 0 \\ 0 & 0 & 0 & 0 & \sigma_\tau^2 + \sigma^2 & \sigma^2 \\ 0 & 0 & 0 & 0 & \sigma_\tau^2 & \sigma_\tau^2 + \sigma^2 \end{bmatrix}$$

The main diagonals of this matrix are the variances of each individual observation and every off-diagonal element is the covariance of a pair of observations.

3.9.2 Analysis of Variance for the Random Model

The basic ANOVA sum of squares identity

$$SS_T = SS_{\text{Treatments}} + SS_E \tag{3.48}$$

is still valid. That is, we partition the total variability in the observations into a component that measures the variation between treatments ($SS_{\text{Treatments}}$) and a component that measures the variation within treatments (SS_E). Testing hypotheses about individual treatment effects is not very meaningful because they were selected randomly, we are more interested in the **population** of treatments, so we test hypotheses about the variance component σ_τ^2.

$$\begin{aligned} H_0&: \sigma_\tau^2 = 0 \\ H_1&: \sigma_\tau^2 > 0 \end{aligned} \tag{3.49}$$

If $\sigma_\tau^2 = 0$, all treatments are identical; but if $\sigma_\tau^2 = 0$, variability exists between treatments. As before, SS_E/σ^2 is distributed as chi-square with $N - a$ degrees of freedom and, under the null hypothesis, $SS_{\text{Treatments}}/\sigma^2$ is distributed as chi-square with $a - 1$ degrees of freedom. Both random variables are independent. Thus, under the null hypothesis $\sigma_\tau^2 = 0$, the ratio

$$F_0 = \frac{\dfrac{SS_{\text{Treatments}}}{a-1}}{\dfrac{SS_E}{N-a}} = \frac{MS_{\text{Treatments}}}{MS_E} \tag{3.50}$$

is distributed as F with $a - 1$ and $N - a$ degrees of freedom. However, we need to examine the expected mean squares to fully describe the test procedure.

Consider

$$E(MS_{\text{Treatments}}) = \frac{1}{a-1} E(SS_{\text{Treatments}}) = \frac{1}{a-1} E\left[\sum_{i=1}^{a} \frac{y_{i.}^2}{n} - \frac{y_{..}^2}{N}\right]$$

$$= \frac{1}{a-1} E\left[\frac{1}{n}\sum_{i=1}^{a}\left(\sum_{j=1}^{n} \mu + \tau_i + \epsilon_{ij}\right)^2 - \frac{1}{N}\left(\sum_{i=1}^{a}\sum_{j=1}^{n} \mu + \tau_i + \epsilon_{ij}\right)^2\right]$$

When squaring and taking expectation of the quantities in brackets, we see that terms involving τ_i^2 are replaced by σ_τ^2 as $E(\tau_i) = 0$. Also, terms involving $\epsilon_{i.}^2$, $\epsilon_{..}^2$, and $\sum_{i=1}^{a}\sum_{j=1}^{n}\tau_i^2$ are replaced by $n\sigma^2$, $an\sigma^2$, and an^2, respectively. Furthermore, all cross-product terms involving τ_i and ϵ_{ij} have zero expectation. This leads to

$$E(MS_{\text{Treatments}}) = \frac{1}{a-1}\left[N\mu^2 + N\sigma_\tau^2 + a\sigma^2 - N\mu^2 - n\sigma_\tau^2 - \sigma^2\right]$$

or

$$E(MS_{\text{Treatments}}) = \sigma^2 + n\sigma_\tau^2 \tag{3.51}$$

Similarly, we may show that

$$E(MS_E) = \sigma^2 \tag{3.52}$$

From the expected mean squares, we see that under H_0 both the numerator and denominator of the test statistic (Equation 3.50) are unbiased estimators of σ^2, whereas under H_1 the expected value of the numerator is greater than the expected value of the denominator. Therefore, we should reject H_0 for values of F_0 that are too large. This implies an upper-tail, one-tail critical region, so we reject H_0 if $F_0 > F_{\alpha,a-1,N-a}$.

The computational procedure and ANOVA for the random effects model are identical to those for the fixed effects case. The conclusions, however, are quite different because they apply to the entire population of treatments.

3.9.3 Estimating the Model Parameters

We are usually interested in estimating the variance components (σ^2 and σ_τ^2) in the model. One very simple procedure that we can use to estimate σ^2 and σ_τ^2 is called the **analysis of variance method** because it makes use of the lines in the analysis of variance table. The procedure consists of equating the expected mean squares to their observed values in the ANOVA table and solving for the variance components. In equating observed and expected mean squares in the single-factor random effects model, we obtain

$$MS_{\text{Treatments}} = \sigma^2 + n\sigma_\tau^2$$

and

$$MS_E = \sigma^2$$

Therefore, the estimators of the variance components are

$$\hat{\sigma}^2 = MS_E \tag{3.53}$$

and

$$\hat{\sigma}_\tau^2 = \frac{MS_{\text{Treatments}} - MS_E}{n} \tag{3.54}$$

For unequal sample sizes, replace n in Equation 13.8 by

$$n_0 = \frac{1}{a-1}\left[\sum_{i=1}^{a} n_i - \frac{\sum_{i=1}^{a} n_i^2}{\sum_{i=1}^{a} n_i}\right] \tag{3.55}$$

The analysis of variance method of variance component estimation is a **method of moments procedure**. It does not require the normality assumption. It does yield estimators of σ^2 and σ_τ^2 that are best quadratic unbiased (i.e., of all unbiased quadratic functions of the observations, these estimators have minimum variance). There is a different method based on maximum likelihood that can be used to estimate the variance components that will be introduced later.

Occasionally, the analysis of variance method produces a negative estimate of a variance component. Clearly, variance components are by definition nonnegative, so a negative estimate of a variance component is viewed with some concern. One course of action is to accept the estimate and use it as evidence that the true value of the variance component is zero, assuming that sampling variation led to the negative estimate. This has intuitive appeal, but it suffers from some theoretical difficulties. For instance, using zero in place of the negative estimate can disturb the statistical properties of other estimates. Another alternative is to reestimate the negative variance component using a method that always yields nonnegative estimates. Still another alternative is to consider the negative estimate as evidence that the assumed linear model is incorrect and reexamine the problem. Comprehensive treatment of variance component estimation is given by Searle (1971a, 1971b), Searle, Casella, and McCullogh (1992), and Burdick and Graybill (1992).

EXAMPLE 3.11

A textile company weaves a fabric on a large number of looms. It would like the looms to be homogeneous so that it obtains a fabric of uniform strength. The process engineer suspects that, in addition to the usual variation in strength within samples of fabric from the same loom, there may also be significant variations in strength between looms. To investigate this, she selects four looms at random and makes four strength determinations on the fabric manufactured on each loom. This experiment is run in random order, and the data obtained are shown in Table 3.17. The ANOVA is conducted and is shown in Table 3.18. From the ANOVA, we conclude that the looms in the plant differ significantly.

TABLE 3.17
Strength Data for Example 3.11

	Observations				
Looms	**1**	**2**	**3**	**4**	$y_{i.}$
1	98	97	99	96	390
2	91	90	93	92	366
3	96	95	97	95	383
4	95	96	99	98	388
					$1527 = y_{..}$

The variance components are estimated by $\hat{\sigma}^2 = 1.90$ and

$$\hat{\sigma}_\tau^2 = \frac{29.73 - 1.90}{4} = 6.96$$

Therefore, the variance of any observation on strength is estimated by

$$\hat{\sigma}_y = \hat{\sigma}^2 + \hat{\sigma}_\tau^2 = 1.90 + 6.96 = 8.86.$$

Most of this variability is attributable to differences *between* looms.

TABLE 3.18
Analysis of Variance for the Strength Data

Source of Variation	Sum of Squares	Degrees of Freedom	Mean Square	F_0	*P*-Value
Looms	89.19	3	29.73	15.68	<0.001
Error	22.75	12	1.90		
Total	111.94	15			

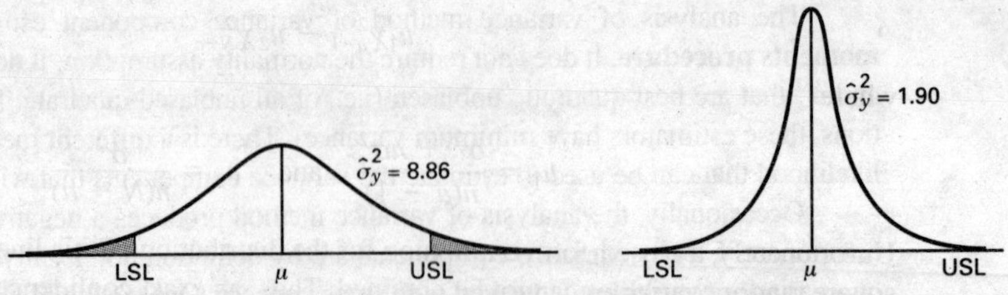

(*a*) Variability of process output. (*b*) Variability of process output if $\sigma_\tau^2 = 0$.

■ **FIGURE 3.19** **Process output in the fiber strength problem**

This example illustrates an important use of variance components—isolating different sources of variability that affect a product or system. The problem of product variability frequently arises in quality assurance, and it is often difficult to isolate the sources of variability. For example, this study may have been motivated by an observation that there is too much variability in the strength of the fabric, as illustrated in Figure 3.19*a*. This graph displays the process output (fiber strength) modeled as a normal distribution with variance $\hat{\sigma}_y^2 = 8.86$. (This is the estimate of the variance of any observation on strength from Example 3.11.) Upper and lower specifications on strength are also shown in Figure 3.19*a*, and it is relatively easy to see that a fairly large proportion of the process output is outside the specifications (the shaded tail areas in Figure 3.19*a*). The process engineer has asked why so much fabric is defective and must be scrapped, reworked, or downgraded to a lower quality product. The answer is that most of the product strength variability is the result of differences between looms. Different loom performance could be the result of faulty setup, poor maintenance, ineffective supervision, poorly trained operators, defective input fiber, and so forth.

The process engineer must now try to isolate the specific causes of the differences in loom performance. If she could identify and eliminate these sources of between-loom variability, the variance of the process output could be reduced considerably, perhaps to as low as $\hat{\sigma}_y^2 = 1.90$, the estimate of the within-loom (error) variance component in Example 3.11. Figure 3.19*b* shows a normal distribution of fiber strength with $\hat{\sigma}_y^2 = 1.90$. Note that the proportion of defective product in the output has been dramatically reduced. Although it is unlikely that *all* of the between-loom variability can be eliminated, it is clear that a significant reduction in this variance component would greatly increase the quality of the fiber produced.

We may easily find a confidence interval for the variance component σ^2. If the observations are normally and independently distributed, then $(N-a)MS_E/\sigma^2$ is distributed as χ^2_{N-a}. Thus,

$$P\left[\chi^2_{1-(\alpha/2),N-a} \le \frac{(N-a)MS_E}{\sigma^2} \le \chi^2_{\alpha/2,N-a}\right] = 1-\alpha$$

and a $100(1-\alpha)$ percent confidence interval for σ^2 is

$$\frac{(N-a)MS_E}{\chi^2_{\alpha/2,N-a}} \le \sigma^2 \le \frac{(N-a)MS_E}{\chi^2_{1-(\alpha/2),N-a}} \tag{3.56}$$

Since $MS_E = 190$, $N = 16$, $a = 4$, $\chi^2_{0.025,12} = 23{,}3367$ and $\chi^2_{0.975,\,12} = 4.4038$, the 95% CI on σ^2 is $0.9770 \le \sigma^2 \le 5.1775$.

Now consider the variance component σ_τ^2. The point estimator of σ_τ^2 is

$$\hat{\sigma}_\tau^2 = \frac{MS_{\text{Treatments}} - MS_E}{n}$$

The random variable $(a-1)MS_{\text{Treatments}}/(\sigma^2 + n\sigma_\tau^2)$ is distributed as χ^2_{a-1}, and $(N-a)MS_E/\sigma^2$ is distributed as χ^2_{N-a}. Thus, the probability distribution of $\hat{\sigma}_\tau^2$ is a linear combination of two chi-square random variables, say

$$u_1\chi^2_{a-1} - u_2\chi^2_{N-a}$$

where

$$u_1 = \frac{\sigma^2 + n\sigma^2_\tau}{n(a-1)} \quad \text{and} \quad u_2 = \frac{\sigma^2}{n(N-a)}$$

Unfortunately, a closed-form expression for the distribution of this linear combination of chi-square random variables cannot be obtained. Thus, an exact confidence interval for σ^2_τ cannot be constructed. Approximate procedures are given in Graybill (1961) and Searle (1971a). Also see Section 13.6 of Chapter 13.

It is easy to find an exact expression for a confidence interval on the ratio $\sigma^2_\tau/(\sigma^2_\tau + \sigma^2)$. This ratio is called the **intraclass correlation coefficient**, and it reflects the *proportion* of the variance of an observation [recall that $V(y_{ij}) = \sigma^2_\tau + \sigma^2$] that is the result of differences between treatments. To develop this confidence interval for the case of a balanced design, note that $MS_{\text{Treatments}}$ and MS_E are independent random variables and, furthermore, it can be shown that

$$\frac{MS_{\text{Treatments}}/(n\sigma^2_\tau + \sigma^2)}{MS_E/\sigma^2} \sim F_{a-1,N-a}$$

Thus,

$$\left(F_{1-\alpha/2,a-1,N-a} \le \frac{MS_{\text{Treatments}}}{MS_E}\,\frac{\sigma^2}{n\sigma^2_\tau + \sigma^2} \le F_{\alpha/2,a-1,N-a}\right) = 1 - \alpha \tag{3.57}$$

By rearranging Equation 13.11, we may obtain the following:

$$P\left(L \le \frac{\sigma^2_\tau}{\sigma^2} \le U\right) = 1 - \alpha \tag{3.58}$$

where

$$L = \frac{1}{n}\left(\frac{MS_{\text{Treatments}}}{MS_E}\,\frac{1}{F_{\alpha/2,a-1,N-a}} - 1\right) \tag{3.59a}$$

and

$$U = \frac{1}{n}\left(\frac{MS_{\text{Treatments}}}{MS_E}\,\frac{1}{F_{1-\alpha/2,a-1,N-a}} - 1\right) \tag{3.59b}$$

Note that L and U are $100(1-\alpha)$ percent lower and upper confidence limits, respectively, for the ratio σ^2_τ/σ^2. Therefore, a $100(1-\alpha)$ percent confidence interval for $\sigma^2_\tau/(\sigma^2_\tau + \sigma^2)$ is

$$\frac{L}{1+L} \le \frac{\sigma^2_\tau}{\sigma^2_\tau + \sigma^2} \le \frac{U}{1+U} \tag{3.60}$$

To illustrate this procedure, we find a 95 percent confidence interval on $\sigma^2_\tau/(\sigma^2_\tau + \sigma^2)$ for the strength data in Example 3.11. Recall that $MS_{\text{Treatments}} = 29.73$, $MS_E = 1.90$, $a = 4$, $n = 4$, $F_{0.025,3,12} = 4.47$, and $F_{0.975,3,12} = 1/F_{0.025,12,3} = 1/14.34 = 0.070$. Therefore, from Equation 3.59*a* and *b*,

$$L = \frac{1}{4}\left[\left(\frac{29.73}{1.90}\right)\left(\frac{1}{4.47}\right) - 1\right] = 0.625$$

$$U = \frac{1}{4}\left[\left(\frac{29.73}{1.90}\right)\left(\frac{1}{0.070}\right) - 1\right] = 55.633$$

and from Equation 3.60, the 95 percent confidence interval on $\sigma_\tau^2/(\sigma_\tau^2 + \sigma^2)$ is

$$\frac{0.625}{1.625} \leq \frac{\sigma_\tau^2}{\sigma_\tau^2 + \sigma^2} \leq \frac{55.633}{56.633}$$

or

$$0.38 \leq \frac{\sigma_\tau^2}{\sigma_\tau^2 + \sigma^2} \leq 0.98$$

We conclude that variability between looms accounts for between 38 and 98 percent of the variability in the observed strength of the fabric produced. This confidence interval is relatively wide because of the small number of looms used in the experiment. Clearly, however, the variability between looms (σ_τ^2) is not negligible.

Estimation of the Overall Mean μ. In many random effects experiments the experimenter is interested in estimating the overall mean μ. From the basic model assumptions it is easy to see that the expected value of any observation is just the overall mean. Consequently, an unbiased estimator of the overall mean is

$$\hat{\mu} = \bar{y}_{..}$$

So for Example 3.11 the estimate of the overall mean strength is

$$\hat{\mu} = \bar{y}_{..} = \frac{y_{..}}{N} = \frac{1527}{16} = 95.44$$

It is also possible to find a $100(1 - \alpha)\%$ confidence interval on the overall mean. The variance of $\bar{y}$ is

$$V(\bar{y}_{..}) = V\left(\frac{\sum_{i=1}^{a}\sum_{j=1}^{n} y_{ij}}{an}\right) = \frac{n\sigma_\tau^2 + \sigma^2}{an}$$

The numerator of this ratio is estimated by the treatment mean square, so an unbiased estimator of $V(\bar{y})$ is

$$\hat{V}(\bar{y}_{..}) = \frac{MS_{\text{Treatments}}}{an}$$

Therefore, the $100(1 - \alpha)\%$ CI on the overall mean is

$$\bar{y}_{..} - t_{\alpha/2,\, a(n-1)}\sqrt{\frac{MS_{\text{Treatments}}}{an}} \leq \mu \leq \bar{y}_{..} + t_{\alpha/2,\, a(n-1)}\sqrt{\frac{MS_{\text{Treatments}}}{an}} \tag{3.61}$$

To find a 95% CI on the overall mean in the fabric strength experiment from Example 3.11, we need $MS_{Treatments} = 29.73$ and $t_{0.025,12} = 2.18$. The CI is computed from Equation 3.61 as follows:

$$\bar{y}_{..} - t_{\alpha/2,a(n-1)}\sqrt{\frac{MS_{\text{Treatments}}}{an}} \leq \mu \leq \bar{y}_{..} + t_{\alpha/2,a(n-1)}\sqrt{\frac{MS_{\text{Treatments}}}{an}}$$

$$95.44 - 2.18\sqrt{\frac{29.73}{20}} \leq \mu \leq 95.44 + 2.18\sqrt{\frac{29.73}{20}}$$

$$92.78 \leq \mu \leq 98.10$$

So, at 95 percent confidence the mean strength of the fabric produced by the looms in this facility is between 92.78 and 98.10. This is a relatively wide confidence interval because a small number of looms were sampled and there is a large difference between looms as reflected by the large portion of total variability that is accounted for by the differences between looms.

Maximum Likelihood Estimation of the Variance Components. Earlier in this section we presented the analysis of variance method of variance component estimation. This method is relatively straightforward to apply and makes use of familiar quantities—the mean squares in the analysis of variance table. However, the method has some disadvantages. As we pointed out previously, it is a **method of moments estimator**, a technique that mathematical statisticians generally do not prefer to use for parameter estimation because it often results in parameter estimates that do not have good statistical properties. One obvious problem is that it does not always lead to an easy way to construct confidence intervals on the variance components of interest. For example, in the single-factor random model there is not a simple way to construct confidence intervals on σ_τ^2, which is certainly a parameter of primary interest to the experimenter. The preferred parameter estimation technique is called the **method of maximum likelihood**. The implementation of this method can be somewhat involved, particularly for an experimental design model, but it has been incorporated in some modern computer software packages that support designed experiments, including JMP.

A complete presentation of the method of maximum likelihood is beyond the scope of this book, but the general idea can be illustrated very easily. Suppose that x is a random variable with probability distribution $f(x, \theta)$, where θ is an unknown parameter. Let $x_1, x_2, \ldots, x_n$ be a random sample of n observations. The joint probability distribution of the sample is $\prod_{i=1}^{n} f(x_i, \theta)$. The **likelihood function** is just this joint probability distribution with the sample observations consider fixed and the parameter θ unknown. Note that the likelihood function, say

$$L(x_1, x_2, \ldots, x_n; \theta) = \prod_{i=1}^{n} f(x_i, \theta)$$

is now a function of only the unknown parameter θ. The **maximum likelihood estimator** of θ is the value of θ that maximizes the likelihood function $L(x_1, x_2, \ldots, x_n; \theta)$. To illustrate how this applies to an experimental design model with random effects, let $\mathbf{y}$ be the $an \times 1$ vector of observations for a single-factor random effects model with a treatments and n replicates and let Σ be the $an \times an$ covariance matrix of the observations. Refer to Section 3.9.1 where we developed this covariance matrix for the special case where $a = 3$ and $n = 2$. The likelihood function is

$$L(x_{11}, x_{12}, \ldots, x_{a,n};\ \mu, \sigma_\tau^2, \sigma^2) = \frac{1}{(2\pi)^{N/2}|\Sigma|^{1/2}} \exp\left[-\frac{1}{2}(\mathbf{y} - \mathbf{j}_N\mu)'\Sigma^{-1}(\mathbf{y} - \mathbf{j}_N\mu)\right]$$

where $N = an$ is the total number of observations, $\mathbf{j}_N$ is an $N \times 1$ vector of 1s, and μ is the overall mean in the model. The maximum likelihood estimates of the parameters μ, σ_τ^2, and σ^2 are the values of these quantities that maximize the likelihood function.

Maximum likelihood estimators (MLEs) have some very useful properties. For large samples, they are unbiased, and they have a normal distribution. Furthermore, the inverse of the matrix of second derivatives of the likelihood function (multiplied by -1) is the covariance matrix of the MLEs. This makes it relatively easy to obtain approximate confidence intervals on the MLEs.

The standard variant of maximum likelihood estimation that is used for estimating variance components is known as the **residual maximum likelihood (REML) method**. It is popular because it produces unbiased estimators and like all MLEs, it is easy to find CIs. The basic characteristic of REML is that it takes the location parameters in the model into account when estimating the random effects. As a simple example, suppose that we want to estimate the mean and variance of a normal distribution using the method of maximum likelihood. It is easy to show that the MLEs are

$$\hat{\mu} = \frac{\sum_{i=1}^{n} y_i}{n} = \bar{y}$$

$$\hat{\sigma}^2 = \frac{\sum_{i=1}^{n} (y_i - \bar{y})^2}{n}$$

Notice that the MLE $\hat{\sigma}^2$ is not the familiar sample standard deviation. It does not take the estimation of the location parameter μ into account. The REML estimator would be

$$S^2 = \frac{\sum_{i=1}^{n} (y_i - \bar{y})^2}{n - 1}$$

The REML estimator is unbiased.

To illustrate the REML method, Table 3.19 presents the JMP output for the loom experiment in Example 3.11. The REML estimates of the model parameters μ, σ_τ^2, and σ^2 are shown in the output. Note that the REML estimates of the variance components are identical to those found earlier by the ANOVA method. These two methods will agree for balanced designs. However, the REML output also contains the covariance matrix of the variance components. The square roots of the main diagonal elements of this matrix are the standard

■ **TABLE 3.19**
JMP Output for the Loom Experiment in Example 3.11

Response Y

Summary of Fit

RSquare	0.793521
RSquare Adj	0.793521
Root Mean Square Error	1.376893
Mean of Response	95.4375
Observations (or Sum Wgts)	16

Parameter Estimates

Term	Estimate	Std Error	DFDen	t Ratio	Prob>|t|
Intercept	95.4375	1.363111	3	70.01	<.0001*

REML Variance Component Estimates

Random Effect	Var Ratio	Var Component	Std Error	95% Lower	95% Upper	Pct of Total
X1	3.6703297	6.9583333	6.0715247	−4.941636	18.858303	78.588
Residual		1.8958333	0.7739707	0.9748608	5.1660065	21.412
Total		8.8541667				100.000

Covariance Matrix of Variance Component Estimates

Random Effect	X1	Residual
X1	36.863412	−0.149758
Residual	−0.149758	0.5990307

errors of the variance components. If $\hat{\theta}$ is the MLE of θ and $\hat{\sigma}(\hat{\theta})$ is its estimated standard error, then the approximate $100(1-\alpha)\%$ CI on θ is

$$\hat{\theta} - Z_{\alpha/2}\hat{\sigma}(\hat{\theta}) \leq \theta \leq \hat{\theta} + Z_{\alpha/2}\hat{\sigma}(\hat{\theta})$$

JMP uses this approach to find the approximate CIs σ_τ^2 and σ^2 shown in the output. The 95 percent CI from REML for σ^2 is very similar to the chi-square based interval computed earlier in Section 3.9.

3.10 The Regression Approach to the Analysis of Variance

We have given an intuitive or heuristic development of the analysis of variance. However, it is possible to give a more formal development. The method will be useful later in understanding the basis for the statistical analysis of more complex designs. Called the **general regression significance test**, the procedure essentially consists of finding the reduction in the total sum of squares for fitting the model with all parameters included and the reduction in sum of squares when the model is restricted to the null hypotheses. The difference between these two sums of squares is the treatment sum of squares with which a test of the null hypothesis can be conducted. The procedure requires the least squares estimators of the parameters in the analysis of variance model. We have given these parameter estimates previously (in Section 3.3.3); however, we now give a formal development.

3.10.1 Least Squares Estimation of the Model Parameters

We now develop estimators for the parameter in the single-factor ANOVA fixed-effects model

$$y_{ij} = \mu + \tau_i + \epsilon_{ij}$$

using the method of least squares. To find the least squares estimators of μ and τ_i, we first form the sum of squares of the errors

$$L = \sum_{i=1}^{a}\sum_{j=1}^{n} \epsilon_{ij}^2 = \sum_{i=1}^{a}\sum_{j=1}^{n} (y_{ij} - \mu - \tau_i)^2 \tag{3.61}$$

and then choose values of μ and τ_i, say $\hat{\mu}$ and $\hat{\tau}_i$, that minimize L. The appropriate values would be the solutions to the $a + 1$ simultaneous equations

$$\left.\frac{\partial L}{\partial \mu}\right|_{\hat{\mu},\hat{\tau}_i} = 0$$

$$\left.\frac{\partial L}{\partial \tau_i}\right|_{\hat{\mu},\hat{\tau}_i} = 0 \quad i = 1, 2, \ldots, a$$

Differentiating Equation 3.61 with respect to μ and τ_i and equating to zero, we obtain

$$-2\sum_{i=1}^{a}\sum_{j=1}^{n} (y_{ij} - \hat{\mu} - \hat{\tau}_i) = 0$$

and

$$-2\sum_{j=1}^{n} (y_{ij} + \hat{\mu} - \hat{\tau}_i) = 0 \qquad i = 1, 2, \ldots, a$$

which, after simplification, yield

$$\begin{aligned} N\hat{\mu} + n\hat{\tau}_1 + n\hat{\tau}_2 + \cdots + n\hat{\tau}_a &= y_{..} \\ n\hat{\mu} + n\hat{\tau}_1 \qquad\qquad\qquad &= y_{1.} \end{aligned}$$

$$n\hat{\mu} \qquad + n\hat{\tau}_2 \qquad = y_{2.}$$
$$\vdots$$
$$n\hat{\mu} \qquad\qquad + n\hat{\tau}_a = y_{a.} \tag{3.62}$$

The $a + 1$ equations (Equation 3.62) in $a + 1$ unknowns are called the **least squares normal equations**. Notice that if we add the last a normal equations, we obtain the first normal equation. Therefore, the normal equations are not linearly independent, and no unique solution for $\mu, \tau_1, \ldots, \tau_a$ exists. This has happened because the effects model is **overparameterized.** This difficulty can be overcome by several methods. Because we have defined the treatment effects as deviations from the overall mean, it seems reasonable to apply the **constraint**

$$\sum_{i=1}^{a} \hat{\tau}_i = 0 \tag{3.63}$$

Using this constraint, we obtain as the solution to the normal equations

$$\hat{\mu} = \bar{y}_{..}$$
$$\hat{\tau}_i = \bar{y}_{i.} - \bar{y}_{..} \qquad i = 1, 2, \ldots, a \tag{3.64}$$

This solution is obviously not unique and depends on the constraint (Equation 3.63) that we have chosen. At first this may seem unfortunate because two different experimenters could analyze the same data and obtain different results if they apply different constraints. However, certain **functions** of the model parameters *are* uniquely estimated, regardless of the constraint. Some examples are $\tau_i - \tau_j$, which would be estimated by $\hat{\tau}_i - \hat{\tau}_j = \bar{y}_{i.} - \bar{y}_{j.}$, and the ith treatment mean $\mu_i = \mu + \tau_i$, which would be estimated by $\hat{\mu}_i = \hat{\mu} + \hat{\tau}_i = \bar{y}_{i.}$.

Because we are usually interested in differences among the treatment effects rather than their actual values, it causes no concern that the τ_i cannot be uniquely estimated. In general, any function of the model parameters that is a linear combination of the left-hand side of the normal equations (Equations 3.48) can be uniquely estimated. Functions that are uniquely estimated regardless of which constraint is used are called **estimable functions**. For more information, see the **supplemental material** for this chapter. We are now ready to use these parameter estimates in a general development of the analysis of variance.

3.10.2 The General Regression Significance Test

A fundamental part of this procedure is writing the normal equations for the model. These equations may always be obtained by forming the least squares function and differentiating it with respect to each unknown parameter, as we did in Section 3.9.1. However, an easier method is available. The following rules allow the normal equations for *any* experimental design model to be written directly:

RULE 1. There is one normal equation for each parameter in the model to be estimated.

RULE 2. The right-hand side of any normal equation is just the sum of all observations that contain the parameter associated with that particular normal equation.

To illustrate this rule, consider the single-factor model. The first normal equation is for the parameter μ; therefore, the right-hand side is $y_{..}$ because *all* observations contain μ.

RULE 3. The left-hand side of any normal equation is the sum of all model parameters, where each parameter is multiplied by the number of times it appears in the total on the right-hand side. The parameters are written with a circumflex (ˆ) to indicate that they are **estimators** and not the true parameter values.

For example, consider the first normal equation in a single-factor experiment. According to the above rules, it would be

$$N\hat{\mu} + n\hat{\tau}_1 + n\hat{\tau}_2 + \cdots + n\hat{\tau}_a = y_{..}$$

because μ appears in all N observations, τ_1 appears only in the n observations taken under the first treatment, τ_2 appears only in the n observations taken under the second treatment, and so on. From Equation 3.62, we verify that the equation shown above is correct. The second normal equation would correspond to τ_1 and is

$$n\hat{\mu} + n\hat{\tau}_1 = y_{1.}$$

because only the observations in the first treatment contain τ_1 (this gives $y_{1.}$ as the right-hand side), μ and τ_1 appear exactly n times in $y_{1.}$, and all other τ_i appear zero times. In general, the left-hand side of any normal equation is the expected value of the right-hand side.

Now, consider finding the reduction in the sum of squares by fitting a particular model to the data. By fitting a model to the data, we "explain" some of the variability; that is, we reduce the unexplained variability by some amount. The reduction in the unexplained variability is always the sum of the parameter estimates, each multiplied by the right-hand side of the normal equation that corresponds to that parameter. For example, in a single-factor experiment, the reduction due to fitting the **full model** $y_{ij} = \mu + \tau_i + \epsilon_{ij}$ is

$$\begin{aligned} R(\mu, \tau) &= \hat{\mu}y_{..} + \hat{\tau}_1 y_{1.} + \hat{\tau}_2 y_{2.} + \cdots + \hat{\tau}_a y_{a.} \\ &= \hat{\mu}y_{..} + \sum_{i=1}^{a} \hat{\tau}_i y_{i.} \end{aligned} \tag{3.65}$$

The notation $R(\mu, \tau)$ means that reduction in the sum of squares from fitting the model containing μ and $\{\tau_i\}$. $R(\mu, \tau)$ is also sometimes called the "regression" sum of squares for the full model $y_{ij} = \mu + \tau_i + \epsilon_{ij}$. The number of degrees of freedom associated with a reduction in the sum of squares, such as $R(\mu, \tau)$, is always equal to the number of linearly independent normal equations. The remaining variability unaccounted for by the model is found from

$$SS_E = \sum_{i=1}^{a}\sum_{j=1}^{n} y_{ij}^2 - R(\mu, \tau) \tag{3.66}$$

This quantity is used in the denominator of the test statistic for $H_0\!:\tau_1 = \tau_2 = ... = \tau_a = 0$.

We now illustrate the general regression significance test for a single-factor experiment and show that it yields the usual one-way analysis of variance. The model is $y_{ij} = \mu + \tau_i + \epsilon_{ij}$, and the normal equations are found from the above rules as

$$\begin{aligned} N\hat{\mu} + n\hat{\tau}_1 + n\hat{\tau}_2 + \cdots + n\hat{\tau}_a &= y_{..} \\ n\hat{\mu} + n\hat{\tau}_1 \qquad\qquad\qquad &= y_{1.} \\ n\hat{\mu} \qquad + n\hat{\tau}_2 \qquad\qquad &= y_{2.} \\ \vdots \qquad\qquad\qquad\qquad &\;\;\vdots \\ n\hat{\mu} \qquad\qquad\qquad + n\hat{\tau}_a &= y_{a.} \end{aligned}$$

Compare these normal equations with those obtained in Equation 3.62.

Applying the constraint $\Sigma_{i=1}^{a}\hat{\tau}_i = 0$, we find that the estimators for μ and τ_i are

$$\hat{\mu} = \bar{y}_{..} \qquad \hat{\tau}_i = \bar{y}_{i.} - \bar{y}_{..} \qquad i = 1, 2, \ldots, a$$

The reduction in the sum of squares due to fitting this full model is found from Equation 3.51 as

$$\begin{aligned} R(\mu, \tau) &= \hat{\mu}y_{..} + \sum_{i=1}^{a} \hat{\tau}_i y_{i.} \\ &= (\bar{y}_{..})y_{..} + \sum_{i=1}^{a} (\bar{y}_{i.} - \bar{y}_{..})y_{i.} \\ &= \frac{y_{..}^2}{N} + \sum_{i=1}^{a} \bar{y}_{i.}y_{i.} - \bar{y}_{..}\sum_{i=1}^{a} y_{i.} \\ &= \sum_{i=1}^{a} \frac{y_{i.}^2}{n} \end{aligned}$$

which has a degrees of freedom because there are a linearly independent normal equations. The error sum of squares is, from Equation 3.66,

$$SS_E = \sum_{i=1}^{a}\sum_{j=1}^{n} y_{ij}^2 - R(\mu, \tau)$$

$$= \sum_{i=1}^{a}\sum_{j=1}^{n} y_{ij}^2 - \sum_{i=1}^{a} \frac{y_{i.}^2}{n}$$

and has $N - a$ degrees of freedom.

To find the sum of squares resulting from the treatment effects (the $\{\tau_i\}$), we consider a **reduced model**; that is, the model to be restricted to the null hypothesis ($\tau_i = 0$ for all i). The reduced model is $y_{ij} = \mu + \epsilon_{ij}$. There is only one normal equation for this model:

$$N\hat{\mu} = y_{..}$$

and the estimator of μ is $\hat{\mu} = \bar{y}_{..}$ Thus, the reduction in the sum of squares that results from fitting the reduced model containing only μ is

$$R(\mu) = (\bar{y}_{..})(y_{..}) = \frac{y_{..}^2}{N}$$

Because there is only one normal equation for this reduced model, $R(\mu)$ has one degree of freedom. The sum of squares due to the $\{\tau_i\}$, given that μ is already in the model, is the difference between $R(\mu, \tau)$ and $R(\mu)$, which is

$$R(\tau|\mu) = R(\mu, \tau) - R(\mu)$$

$$= R(\text{Full Model}) - R(\text{Reduced Model})$$

$$= \frac{1}{n}\sum_{i=1}^{a} y_{i.}^2 - \frac{y_{..}^2}{N}$$

with $a - 1$ degrees of freedom, which we recognize from Equation 3.9 as $SS_{\text{Treatments}}$. Making the usual normality assumption, we obtain appropriate statistic for testing $H_0{:}\tau_1 = \tau_2 = \cdots = \tau_a = 0$

$$F_0 = \frac{R(\tau|\mu)(/(a-1)}{\left[\sum_{i=1}^{a}\sum_{j=1}^{n} y_{ij}^2 - R(\mu, \tau)\right] \Big/ (N-a)}$$

which is distributed as $F_{a-1, N-a}$ under the null hypothesis. This is, of course, the test statistic for the single-factor analysis of variance.

3.11 Nonparametric Methods in the Analysis of Variance

3.11.1 The Kruskal–Wallis Test

In situations where the normality assumption is unjustified, the experimenter may wish to use an alternative procedure to the F test analysis of variance that does not depend on this assumption. Such a procedure has been developed by Kruskal and Wallis (1952). The Kruskal–Wallis test is used to test the null hypothesis that the a treatments are identical against the alternative hypothesis that some of the treatments generate observations that are larger than others. Because the procedure is designed to be sensitive for testing differences in means, it is sometimes convenient to think of the Kruskal–Wallis test as a test for equality of treatment means. The Kruskal–Wallis test is a **nonparametric alternative** to the usual analysis of variance.

To perform a Kruskal–Wallis test, first rank the observations y_{ij} in ascending order and replace each observation by its rank, say R_{ij}, with the smallest observation having rank 1. In the case of ties (observations having the same value), assign the average rank to each of the tied observations. Let $R_{i.}$ be the sum of the ranks in the ith treatment. The test statistic is

$$H = \frac{1}{S^2}\left[\sum_{i=1}^{a} \frac{R_{i.}^2}{n_i} - \frac{N(N+1)^2}{4}\right] \tag{3.67}$$

where n_i is the number of observations in the ith treatment, N is the total number of observations, and

$$S^2 = \frac{1}{N-1}\left[\sum_{i=1}^{a}\sum_{j=1}^{n_i} R_{ij}^2 - \frac{N(N+1)^2}{4}\right] \tag{3.68}$$

Note that S^2 is just the variance of the ranks. If there are no ties, $S^2 = N(N+1)/12$ and the test statistic simplifies to

$$H = \frac{12}{N(N+1)} \sum_{i=1}^{a} \frac{R_{i.}^2}{n_i} - 3(N+1) \tag{3.69}$$

When the number of ties is moderate, there will be little difference between Equations 3.68 and 3.69, and the simpler form (Equation 3.69) may be used. If the n_i are reasonably large, say $n_i \geq 5$, H is distributed approximately as χ^2_{a-1} under the null hypothesis. Therefore, if

$$H > \chi^2_{\alpha,a-1}$$

the null hypothesis is rejected. The P-value approach could also be used.

EXAMPLE 3.12

The data from Example 3.1 and their corresponding ranks are shown in Table 3.20. There are ties, so we use Equation 3.67 as the test statistic. From Equation 3.67

$$S^2 = \frac{1}{19}\left[2869.50 - \frac{20(21)^2}{4}\right] = 34.97$$

and the test statistic is

$$H = \frac{1}{S^2}\left[\sum_{i=1}^{a} \frac{R_{i.}^2}{n_i} - \frac{N(N+1)^2}{4}\right]$$

$$= \frac{1}{34.97}[2796.30 - 2205]$$

$$= 16.91$$

■ **TABLE 3.20**
Data and Ranks for the Plasma Etching Experiment in Example 3.1

Power							
160		180		200		220	
y_{1j}	R_{1j}	y_{2j}	R_{2j}	y_{3j}	R_{3j}	y_{4j}	R_{4j}
575	6	565	4	600	10	725	20
542	3	593	9	651	15	700	17
530	1	590	8	610	11.5	715	19
539	2	579	7	637	14	685	16
570	5	610	11.5	629	13	710	18
$R_{i.}$	17		39.5		63.5		90

Because $H > \chi^2_{0.01,3} = 11.34$, we would reject the null hypothesis and conclude that the treatments differ. (The P-value for $H = 16.91$ is $P = 7.38 \times 10^{-4}$.) This is the same conclusion as given by the usual analysis of variance F test.

3.11.2 General Comments on the Rank Transformation

The procedure used in the previous section of replacing the observations by their ranks is called the **rank transformation.** It is a very powerful and widely useful technique. If we were to apply the ordinary F test to the ranks rather than to the original data, we would obtain

$$F_0 = \frac{H/(a-1)}{(N-1-H)/(N-a)} \tag{3.70}$$

as the test statistic [see Conover (1980), p. 337]. Note that as the Kruskal–Wallis statistic H increases or decreases, F_0 also increases or decreases, so the Kruskal–Wallis test is equivalent to applying the usual analysis of variance to the ranks.

The rank transformation has wide applicability in experimental design problems for which no nonparametric alternative to the analysis of variance exists. This includes many of the designs in subsequent chapters of this book. If the data are ranked and the ordinary F test is applied, an approximate procedure that has good statistical properties results [see Conover and Iman (1976, 1981)]. When we are concerned about the normality assumption or the effect of outliers or "wild" values, we recommend that the usual analysis of variance be performed on both the original data and the ranks. When both procedures give similar results, the analysis of variance assumptions are probably satisfied reasonably well, and the standard analysis is satisfactory. When the two procedures differ, the rank transformation should be preferred because it is less likely to be distorted by nonnormality and unusual observations. In such cases, the experimenter may want to investigate the use of transformations for nonnormality and examine the data and the experimental procedure to determine whether outliers are present and why they have occurred.

3.12 Problems

3.1. An experimenter has conducted a single-factor experiment with five levels of the factor, and each factor level has been replicated six times. The computed value of the F-statistic is $F_0 = 3.26$. Find bounds on the P-value.

3.2. An experimenter has conducted a single-factor experiment with six levels of the factor, and each factor level has been replicated three times. The computed value of the F-statistic is $F_0 = 5.81$. Find bounds on the P-value.

3.3. A computer ANOVA output is shown below. Fill in the blanks. You may give bounds on the P-value.

One-way ANOVA

Source	DF	SS	MS	F	P
Factor	3	38.25	?	?	?
Error	?	?	?		
Total	19	196.04			

3.4. A computer ANOVA output is shown below. Fill in the blanks. You may give bounds on the P-value.

One-way ANOVA

Source	DF	SS	MS	F	P
Factor	?	?	246.93	?	?
Error	25	186.53	?		
Total	29	1174.24			

3.5. An article in *Bioelectromagnetics* ("Electromagnetic Effects on Forearm Disuse Osteopenia: A Randomized, double-blind, sham-controlled study," Vol. 32, 2011, pp. 273–282) described a randomized, double-blind, sham-controlled, feasibility and dosing study to determine if a common pulsing electromagnetic field (PEMF) treatment could moderate the substantial osteopenia that occurs after forearm disuse. Subjects were randomized into four groups after a distal radius fracture, or carpal surgery requiring immobilization in a cast. Active or identical sham PEMF transducers were worn on the distal forearm for 1, 2, or 4h/day for 8 weeks starting after cast removal ("baseline") when bone density continues to decline. Bone mineral density (BMD) and bone geometry were measured in the distal forearm by dual energy X-ray absorptiometry (DXA) and peripheral quantitative computed tomography (pQCT). The data below are the percent losses in BMD measurements on the radius after 16 weeks for patients wearing the active or sham PEMF transducers for 1, 2, or 4h/day (data were constructed to match the means and standard deviations read from a graph in the paper).

(a) Is there evidence to support a claim that PEMF usage affects BMD loss? If so, analyze the data to determine which specific treatments produce the differences.

(b) Analyze the residuals from this experiment and comment on the underlying assumptions and model adequacy.

Sham	PEMF 1 h/day	PEMF 2 h/day	PEMF 4 h/day
4.51	5.32	4.73	7.03
7.95	6.00	5.81	4.65
4.97	5.12	5.69	6.65
3.00	7.08	3.86	5.49
7.97	5.48	4.06	6.98
2.23	6.52	6.56	4.85
3.95	4.09	8.34	7.26
5.64	6.28	3.01	5.92
9.35	7.77	6.71	5.58
6.52	5.68	6.51	7.91
4.96	8.47	1.70	4.90
6.10	4.58	5.89	4.54
7.19	4.11	6.55	8.18
4.03	5.72	5.34	5.42
2.72	5.91	5.88	6.03
9.19	6.89	7.50	7.04
5.17	6.99	3.28	5.17
5.70	4.98	5.38	7.60
5.85	9.94	7.30	7.90
6.45	6.38	5.46	7.91

3.6. An article appeared in *The Wall Street Journal* on Tuesday, April 27, 2010, with the title "Eating Chocolate Is Linked to Depression." The article reported on a study funded by the National Heart, Lung and Blood Institute (part of the National Institutes of Health) and conducted by faculty at the University of California, San Diego, and the University of California, Davis. The research was also published in the *Archives of Internal Medicine* (2010, pp. 699–703). The study examined 931 adults who were not taking antidepressants and didn't have known cardiovascular disease or diabetes. The group was about 70 percent men and the average age of the group was reported to be about 58. The participants were asked about chocolate consumption and then screened for depression using a questionnaire. People who score less than 16 on the questionnaire are not considered depressed, while those with scores above 16 and less than or equal to 22 are considered possibly depressed, while those with scores above 22 are considered likely to be depressed. The survey found that people who weren't depressed ate an average 5.4 servings of chocolate per month, possibly depressed individuals ate an average of 8.4 servings of chocolate per month, while those individuals who scored above 22 and were likely to be depressed ate the most chocolate, an average of 11.8 servings per month. No differentiation was made between dark and milk chocolate. Other foods were also examined, but no pattern emerged between other foods and depression. Is this study really a designed experiment? Does it establish a cause-and-effect link between chocolate consumption and depression? How would the study have to be conducted to establish such a cause-and-effect link?

3.7. The tensile strength of Portland cement is being studied. Four different mixing techniques can be used economically. A completely randomized experiment was conducted and the following data were collected:

Mixing Technique	Tensile Strength (lb/in^2)			
1	3135	3001	2865	2890
2	3210	3311	2975	3150
3	2815	2910	2985	3050
4	2619	2706	2612	2765

(a) Test the hypothesis that mixing techniques affect the strength of the cement. Use $\alpha = 0.05$.

(b) Construct a graphical display as described in Section 3.5.3 to compare the mean tensile strengths for the four mixing techniques. What are your conclusions?

(c) Use the Fisher LSD method with $\alpha = 0.05$ to make comparisons between pairs of means.

(d) Construct a normal probability plot of the residuals. What conclusion would you draw about the validity of the normality assumption?

(e) Plot the residuals versus the predicted tensile strength. Comment on the plot.

(f) Prepare a scatter plot of the results to aid the interpretation of the results of this experiment.

3.8(a) Rework part (c) of Problem 3.7 using Tukey's test with $\alpha = 0.05$. Do you get the same conclusions from Tukey's test that you did from the graphical procedure and/or the Fisher LSD method?

(b) Explain the difference between the Tukey and Fisher procedures.

3.9. Reconsider the experiment in Problem 3.7. Find a 95 percent confidence interval on the mean tensile strength of the Portland cement produced by each of the four mixing techniques. Also find a 95 percent confidence interval on the difference in means for techniques 1 and 3. Does this aid you in interpreting the results of the experiment?

3.10. A product developer is investigating the tensile strength of a new synthetic fiber that will be used to make cloth for men's shirts. Strength is usually affected by the percentage of cotton used in the blend of materials for the fiber. The engineer conducts a completely randomized experiment with five levels of cotton content and replicates the experiment five times. The data are shown in the following table.

Cotton Weight Percent	Observations				
15	7	7	15	11	9
20	12	17	12	18	18
25	14	19	19	18	18
30	19	25	22	19	23
35	7	10	11	15	11

(a) Is there evidence to support the claim that cotton content affects the mean tensile strength? Use $\alpha = 0.05$.

(b) Use the Fisher LSD method to make comparisons between the pairs of means. What conclusions can you draw?

(c) Analyze the residuals from this experiment and comment on model adequacy.

3.11. A pharmaceutical manufacturer wants to investigate the bioactivity of a new drug. A completely randomized single-factor experiment was conducted with three dosage levels, and the following results were obtained.

Dosage	Observations			
20 g	24	28	37	30
30 g	37	44	31	35
40 g	42	47	52	38

(a) Is there evidence to indicate that dosage level affects bioactivity? Use $\alpha = 0.05$.

(b) If it is appropriate to do so, make comparisons between the pairs of means. What conclusions can you draw?

(c) Analyze the residuals from this experiment and comment on model adequacy.

3.12. Reconsider the experiment described in Problem 3.10. Suppose that 30 percent cotton content is a control. Use Dunnett's test with $\alpha = 0.05$ to compare all of the other means with the control.

3.13. A rental car company wants to investigate whether the type of car rented affects the length of the rental period. An experiment is run for one week at a particular location, and 12 rental contracts are selected at random for each car type. The results are shown in the following table.

Type of Car	Observations											
Subcompact	3	5	3	7	6	5	3	2	1	6	5	3
Compact	1	3	4	7	5	6	3	2	1	7	6	1
Midsize	4	1	3	5	7	1	2	4	2	7	2	4
Full size	3	5	7	5	10	3	4	7	2	7	9	7

(a) Is there evidence to support a claim that the type of car rented affects the length of the rental contract? Use $\alpha = 0.05$. If so, which types of cars are responsible for the difference?

(b) Analyze the residuals from this experiment and comment on model adequacy.

(c) Notice that the response variable in this experiment is a count. Should this cause any potential concerns about the validity of the analysis of variance?

3.14. I belong to a golf club in my neighborhood. I divide the year into three golf seasons: summer (June–September), winter (November–March), and shoulder (October, April, and May). I believe that I play my best golf during the summer (because I have more time and the course isn't crowded) and shoulder (because the course isn't crowded) seasons, and my worst golf is during the winter (because when all of the part-year residents show up, the course is crowded, play is slow, and I get frustrated). Data from the last year are shown in the following table.

Season	Observations									
Summer	83	85	85	87	90	88	88	84	91	90
Shoulder	91	87	84	87	85	86	83			
Winter	94	91	87	85	87	91	92	86		

(a) Do the data indicate that my opinion is correct? Use $\alpha = 0.05$.

(b) Analyze the residuals from this experiment and comment on model adequacy.

3.15. An experiment was run to determine whether four specific firing temperatures affect the density of a certain type of brick. A completely randomized experiment led to the following data:

Temperature	Density				
100	21.8	21.9	21.7	21.6	21.7
125	21.7	21.4	21.5		
150	21.9	21.8	21.8	21.6	21.5
175	21.9	21.7	21.8	21.4	21.3

(a) Does the firing temperature affect the density of the bricks? Use $\alpha = 0.05$.

(b) Is it appropriate to compare the means using the Fisher LSD method (for example) in this experiment?

(c) Analyze the residuals from this experiment. Are the analysis of variance assumptions satisfied?

(d) Construct a graphical display of the treatment as described in Section 3.5.3. Does this graph adequately summarize the results of the analysis of variance in part (a)?

3.16. A regional opera company has tried three approaches to solicit donations from 24 potential sponsors. The 24 potential sponsors were randomly divided into three groups of eight, and one approach was used for each group. The dollar amounts of the resulting contributions are shown in the following table.

Approach	Contributions (in $)							
1	1000	1500	1200	1800	1600	1100	1000	1250
2	1500	1800	2000	1200	2000	1700	1800	1900
3	900	1000	1200	1500	1200	1550	1000	1100

(a) Do the data indicate that there is a difference in results obtained from the three different approaches? Use $\alpha = 0.05$.

(b) Analyze the residuals from this experiment and comment on model adequacy.

3.17. Rework part (d) of Problem 3.15 using the Tukey method. What conclusions can you draw? Explain carefully how you modified the technique to account for unequal sample sizes.

3.18. A manufacturer of television sets is interested in the effect on tube conductivity of four different types of coating for color picture tubes. A completely randomized experiment is conducted and the following conductivity data are obtained:

Coating Type	Conductivity			
1	143	141	150	146
2	152	149	137	143
3	134	136	132	127
4	129	127	132	129

(a) Is there a difference in conductivity due to coating type? Use $\alpha = 0.05$.

(b) Estimate the overall mean and the treatment effects.

(c) Compute a 95 percent confidence interval estimate of the mean of coating type 4. Compute a 99 percent confidence interval estimate of the mean difference between coating types 1 and 4.

(d) Test all pairs of means using the Fisher LSD method with $\alpha = 0.05$.

(e) Use the graphical method discussed in Section 3.5.3 to compare the means. Which coating type produces the highest conductivity?

(f) Assuming that coating type 4 is currently in use, what are your recommendations to the manufacturer? We wish to minimize conductivity.

3.19. Reconsider the experiment from Problem 3.18. Analyze the residuals and draw conclusions about model adequacy.

3.20. An article in the *ACI Materials Journal* (Vol. 84, 1987, pp. 213–216) describes several experiments investigating the rodding of concrete to remove entrapped air. A 3-inch × 6-inch cylinder was used, and the number of times this rod was used is the design variable. The resulting compressive strength of the concrete specimen is the response. The data are shown in the following table:

Rodding Level	Compressive Strength		
10	1530	1530	1440
15	1610	1650	1500
20	1560	1730	1530
25	1500	1490	1510

(a) Is there any difference in compressive strength due to the rodding level? Use $\alpha = 0.05$.

(b) Find the P-value for the F statistic in part (a).

(c) Analyze the residuals from this experiment. What conclusions can you draw about the underlying model assumptions?

(d) Construct a graphical display to compare the treatment means as described in Section 3.5.3.

3.21. An article in *Environment International* (Vol. 18, No. 4, 1992) describes an experiment in which the amount of radon released in showers was investigated. Radon-enriched water was used in the experiment, and six different orifice diameters were tested in shower heads. The data from the experiment are shown in the following table:

Orifice Diameter	Radon Released (%)			
0.37	80	83	83	85
0.51	75	75	79	79
0.71	74	73	76	77
1.02	67	72	74	74
1.40	62	62	67	69
1.99	60	61	64	66

(a) Does the size of the orifice affect the mean percentage of radon released? Use $\alpha = 0.05$.

(b) Find the P-value for the F statistic in part (a).

(c) Analyze the residuals from this experiment.

(d) Find a 95 percent confidence interval on the mean percent of radon released when the orifice diameter is 1.40.

(e) Construct a graphical display to compare the treatment means as described in Section 3.5.3. What conclusions can you draw?

3.22. The response time in milliseconds was determined for three different types of circuits that could be used in an automatic valve shutoff mechanism. The results from a completely randomized experiment are shown in the following table:

Circuit Type	Response Time				
1	9	12	10	8	15
2	20	21	23	17	30
3	6	5	8	16	7

(a) Test the hypothesis that the three circuit types have the same response time. Use $\alpha = 0.01$.

(b) Use Tukey's test to compare pairs of treatment means. Use $\alpha = 0.01$.

(c) Use the graphical procedure in Section 3.5.3 to compare the treatment means. What conclusions can you draw? How do they compare with the conclusions from part (b)?

(d) Construct a set of orthogonal contrasts, assuming that at the outset of the experiment you suspected the response time of circuit type 2 to be different from the other two.

(e) If you were the design engineer and you wished to minimize the response time, which circuit type would you select?

(f) Analyze the residuals from this experiment. Are the basic analysis of variance assumptions satisfied?

3.23. Four different designs for a digital computer circuit are being studied to compare the amount of noise present. The following data have been obtained:

Circuit Design	Noise Observed					
1	19	20	19	30	8	16
2	80	61	73	56	80	73
3	47	26	25	35	50	32
4	95	46	83	78	97	87

(a) Is the same amount of noise present for all four designs? Use $\alpha = 0.05$.

(b) Analyze the residuals from this experiment. Are the analysis of variance assumptions satisfied?

(c) Which circuit design would you select for use? Low noise is best.

3.24. The effective life of insulating fluids at an accelerated load of 35 kV is being studied. Test data have been obtained for four types of fluids. The results from a completely randomized experiment were as follows:

Fluid Type	Life (in h) at 35 kV Load					
1	17.6	18.9	16.3	17.4	20.1	21.6
2	16.9	15.3	18.6	17.1	19.5	20.3
3	21.4	23.6	19.4	18.5	20.5	22.3
4	19.3	21.1	16.9	17.5	18.3	19.8

(a) Is there any indication that the fluids differ? Use $\alpha = 0.05$.

(b) Which fluid would you select, given that the objective is long life?

(c) Analyze the residuals from this experiment. Are the basic analysis of variance assumptions satisfied?

3.25. Three brands of batteries are under study. It is suspected that the lives (in weeks) of the three brands are different. Five randomly selected batteries of each brand are tested with the following results:

Weeks of Life		
Brand 1	Brand 2	Brand 3
100	76	108
96	80	100
92	75	96
96	84	98
92	82	100

(a) Are the lives of these brands of batteries different?

(b) Analyze the residuals from this experiment.

(c) Construct a 95 percent confidence interval estimate on the mean life of battery brand 2. Construct a 99 percent confidence interval estimate on the mean difference between the lives of battery brands 2 and 3.

(d) Which brand would you select for use? If the manufacturer will replace without charge any battery that fails in less than 85 weeks, what percentage would the company expect to replace?

3.26. Four catalysts that may affect the concentration of one component in a three-component liquid mixture are being investigated. The following concentrations are obtained from a completely randomized experiment:

Catalyst			
1	2	3	4
58.2	56.3	50.1	52.9
57.2	54.5	54.2	49.9
58.4	57.0	55.4	50.0
55.8	55.3		51.7
54.9			

(a) Do the four catalysts have the same effect on the concentration?

(b) Analyze the residuals from this experiment.

(c) Construct a 99 percent confidence interval estimate of the mean response for catalyst 1.

3.27. Four chemists are asked to determine the percentage of methyl alcohol in a certain chemical compound. Each chemist makes three determinations, and the results are the following:

Chemist	Percentage of Methyl Alcohol		
1	84.99	84.04	84.38
2	85.15	85.13	84.88
3	84.72	84.48	85.16
4	84.20	84.10	84.55

(a) Do chemists differ significantly? Use $\alpha = 0.05$.

(b) Analyze the residuals from this experiment.

(c) If chemist 2 is a new employee, construct a meaningful set of orthogonal contrasts that might have been useful at the start of the experiment.

3.28. A semiconductor manufacturer has developed three different methods for reducing particle counts on wafers. All three methods are tested on six different wafers and the after treatment particle count obtained. The data are shown below:

Method	Count					
1	31	10	21	4	1	11
2	62	40	24	30	35	43
3	53	27	120	97	68	80

(a) Do all methods have the same effect on mean particle count?

(b) Plot the residuals versus the predicted response. Construct a normal probability plot of the residuals. Are there potential concerns about the validity of the assumptions?

(c) Based on your answer to part (b) conduct another analysis of the particle count data and draw appropriate conclusions.

3.29. An experiment was performed to investigate the effectiveness of five insulating materials. Four samples of each material were tested at an elevated voltage level to accelerate the time to failure. The failure times (in minutes) are shown below:

Material	Failure Time (minutes)			
1	110	157	194	178
2	1	2	4	18
3	880	1256	5276	4355
4	495	7040	5307	10,050
5	7	5	29	2

(a) Do all five materials have the same effect on mean failure time?

(b) Plot the residuals versus the predicted response. Construct a normal probability plot of the residuals. What information is conveyed by these plots?

(c) Based on your answer to part (b) conduct another analysis of the failure time data and draw appropriate conclusions.

3.30. Several ovens in a metal working shop are used to heat metal specimens. All the ovens are supposed to operate at the same temperature, although it is suspected that this may not be true. Three ovens are selected at random, and their temperatures on successive heats are noted. The data collected are as follows:

Oven	Temperature					
1	491.50	498.30	498.10	493.50	493.60	
2	488.50	484.65	479.90	477.35		
3	490.10	484.80	488.25	473.00	471.85	478.65

(a) Is there significant variation in temperature between ovens? Use $\alpha = 0.05$.

(b) Estimate the components of variance for this model.

(c) Analyze the residuals from this experiment and draw conclusions about model adequacy.

3.31. A manufacturer suspects that the batches of raw material furnished by his supplier differ significantly in calcium content. There are a large number of batches currently in the warehouse. Five of these are randomly selected for study. A chemist makes five determinations on each batch and obtains the following data:

Batch 1	Batch 2	Batch 3	Batch 4	Batch 5
23.46	23.59	23.51	23.28	23.29
23.48	23.46	23.64	23.40	23.46
23.56	23.42	23.46	23.37	23.37
23.39	23.49	23.52	23.46	23.32
23.40	23.50	23.49	23.39	23.38

(a) Is there significant variation in calcium content from batch to batch? Use $\alpha = 0.05$.

(b) Estimate the components of variance.

(c) Find a 95 percent confidence interval for $\sigma_\tau^2 / (\sigma_\tau^2 + \sigma^2)$.

(d) Analyze the residuals from this experiment. Are the analysis of variance assumptions satisfied?

3.32. An article in the *Journal of the Electrochemical Society* (Vol. 139, No. 2, 1992, pp. 524–532) describes an experiment to investigate the low-pressure vapor deposition of polysilicon. The experiment was carried out in a large-capacity reactor at Sematech in Austin, Texas. The reactor has several wafer positions, and four of these positions are selected at random. The response variable is film thickness uniformity. Three replicates of the experiment were run, and the data are as follows:

Wafer Position	Uniformity		
1	2.76	5.67	4.49
2	1.43	1.70	2.19
3	2.34	1.97	1.47
4	0.94	1.36	1.65

(a) Is there a difference in the wafer positions? Use $\alpha = 0.05$.

(b) Estimate the variability due to wafer positions.

(c) Estimate the rendom error component.

(d) Analyze the residuals from this experiment and comment on model adequacy.

3.33. Consider the vapor-deposition experiment described in Problem 3.32.

(a) Estimate the total variability in the uniformity response.

(b) How much of the total variability in the uniformity response is due to the difference between positions in the reactor?

(c) To what level could the variability in the uniformity response be reduced if the position-to-position variability in the reactor could be eliminated? Do you believe this is a significant reduction?

3.34. An article in the *Journal of Quality Technology* (Vol. 13, No. 2, 1981, pp. 111–114) describes an experiment that investigates the effects of four bleaching chemicals on pulp brightness. These four chemicals were selected at random from a large population of potential bleaching agents. The data are as follows:

Chemical	Pulp Brightness				
1	77.199	74.466	92.746	76.208	82.876
2	80.522	79.306	81.914	80.346	73.385
3	79.417	78.017	91.596	80.802	80.626
4	78.001	78.358	77.544	77.364	77.386

(a) Is there a difference in the chemical types? Use $\alpha = 0.05$.

(b) Estimate the variability due to chemical types.

(c) Estimate the variability due to random error.

(d) Analyze the residuals from this experimental and comment on model adequacy.

3.35. Consider the single-factor random effects model discussed in this chapter. Develop a procedure for finding a $100(1 - \alpha)\%$ confidence interval on the ratio $\sigma^2/(\sigma_\tau^2 + \sigma^2)$. Assume that the experiment is balanced.

3.36. Consider testing the equality of the means of two normal populations, where the variances are unknown but are assumed to be equal. The appropriate test procedure is the pooled *t*-test. Show that the pooled *t*-test is equivalent to the single-factor analysis of variance.

3.37. Show that the variance of the linear combination $\Sigma_{i=1}^a c_i y_{i.}$ is $\sigma^2 \Sigma_{i=1}^a n_i c_i^2$.

3.38. In a fixed effects experiment, suppose that there are n observations for each of the four treatments. Let Q_1^2, Q_2^2, Q_3^2 be single-degree-of-freedom components for the orthogonal contrasts. Prove that $SS_{\text{Treatments}} = Q_1^2 + Q_2^2 + Q_3^2$.

3.39. Use Bartlett's test to determine if the assumption of equal variances is satisfied in Problem 3.24. Use $\alpha = 0.05$. Did you reach the same conclusion regarding equality of variances by examining residual plots?

3.40. Use the modified Levene test to determine if the assumption of equal variances is satisfied in Problem 3.25. Use $\alpha = 0.05$. Did you reach the same conclusion regarding the equality of variances by examining residual plots?

3.41. Refer to Problem 3.22. If we wish to detect a maximum difference in mean response times of 10 milliseconds with a probability of at least 0.90, what sample size should be used? How would you obtain a preliminary estimate of σ^2?

3.42. Refer to Problem 3.25.

(a) If we wish to detect a maximum difference in battery life of 10 hours with a probability of at least 0.90, what sample size should be used? Discuss how you would obtain a preliminary estimate of σ^2 for answering this question.

(b) If the difference between brands is great enough so that the standard deviation of an observation is increased by 25 percent, what sample size should be used if we wish to detect this with a probability of at least 0.90?

3.43. Consider the experiment in Problem 3.25. If we wish to construct a 95 percent confidence interval on the difference in two mean battery lives that has an accuracy of ±2 weeks, how many batteries of each brand must be tested?

3.44. Suppose that four normal populations have means of $\mu_1 = 50$, $\mu_2 = 60$, $\mu_3 = 50$, and $\mu_4 = 60$. How many observations should be taken from each population so that the probability of rejecting the null hypothesis of equal population means is at least 0.90? Assume that $\alpha = 0.05$ and that a reasonable estimate of the error variance is $\sigma^2 = 25$.

3.45. Refer to Problem 3.44.

(a) How would your answer change if a reasonable estimate of the experimental error variance were $\sigma^2 = 36$?

(b) How would your answer change if a reasonable estimate of the experimental error variance were $\sigma^2 = 49$?

(c) Can you draw any conclusions about the sensitivity of your answer in this particular situation about how your estimate of σ affects the decision about sample size?

(d) Can you make any recommendations about how we should use this general approach to choosing n in practice?

3.46. Refer to the aluminum smelting experiment described in Section 3.8.3. Verify that ratio control methods do not affect average cell voltage. Construct a normal probability plot of the residuals. Plot the residuals versus the predicted values. Is there an indication that any underlying assumptions are violated?

3.47. Refer to the aluminum smelting experiment in Section 3.8.3. Verify the ANOVA for pot noise summarized in Table 3.16. Examine the usual residual plots and comment on the experimental validity.

3.48. Four different feed rates were investigated in an experiment on a CNC machine producing a component part used in an aircraft auxiliary power unit. The manufacturing engineer in charge of the experiment knows that a critical part dimension of interest may be affected by the feed rate. However, prior experience has indicated that only dispersion effects are likely to be present. That is, changing the feed rate does not affect the *average* dimension, but it could affect dimensional variability. The engineer makes five production runs at each feed rate and obtains the standard deviation of the critical dimension (in 10^{-3} mm). The data are shown below. Assume that all runs were made in random order.

	Production Run				
Feed Rate (in/min)	**1**	**2**	**3**	**4**	**5**
10	0.09	0.10	0.13	0.08	0.07
12	0.06	0.09	0.12	0.07	0.12
14	0.11	0.08	0.08	0.05	0.06
16	0.19	0.13	0.15	0.20	0.11

(a) Does feed rate have any effect on the standard deviation of this critical dimension?

(b) Use the residuals from this experiment to investigate model adequacy. Are there any problems with experimental validity?

3.49. Consider the data shown in Problem 3.22.

(a) Write out the least squares normal equations for this problem and solve them for $\hat{\mu}$ and $\hat{\tau}_i$, using the usual constraint $(\sum_{i=1}^{3} \hat{\tau}_i = 0)$. Estimate $\tau_1 - \tau_2$.

(b) Solve the equations in (a) using the constraint $\hat{\tau}_3 = 0$. Are the estimators $\hat{\tau}_i$ and $\hat{\mu}$ the same as you found in (a)? Why? Now estimate $\tau_1 - \tau_2$ and compare your answer with that for (a). What statement can you make about estimating contrasts in the τ_i?

(c) Estimate $\mu + \tau_1$, $2\tau_1 - \tau_2 - \tau_3$, and $\mu + \tau_1 + \tau_2$ using the two solutions to the normal equations. Compare the results obtained in each case.

3.50. Apply the general regression significance test to the experiment in Example 3.5. Show that the procedure yields the same results as the usual analysis of variance.

3.51. Use the Kruskal–Wallis test for the experiment in Problem 3.24. Compare the conclusions obtained with those from the usual analysis of variance.

3.52. Use the Kruskal–Wallis test for the experiment in Problem 3.24. Are the results comparable to those found by the usual analysis of variance?

3.53. Consider the experiment in Example 3.5. Suppose that the largest observation on etch rate is incorrectly recorded as 250 Å/min. What effect does this have on the usual analysis of variance? What effect does it have on the Kruskal–Wallis test?

3.54. A textile mill has a large number of looms. Each loom is supposed to provide the same output of cloth per minute. To investigate this assumption, five looms are chosen at random, and their output is noted at different times. The following data are obtained:

Loom	Output (lb/min)				
1	14.0	14.1	14.2	14.0	14.1
2	13.9	13.8	13.9	14.0	14.0
3	14.1	14.2	14.1	14.0	13.9
4	13.6	13.8	14.0	13.9	13.7
5	13.8	13.6	13.9	13.8	14.0

(a) Explain why this is a random effects experiment. Are the looms equal in output? Use $\alpha = 0.05$.

(b) Estimate the variability between looms.

(c) Estimate the experimental error variance.

(d) Find a 95 percent confidence interval for $\sigma_\tau^2 / (\sigma_\tau^2 + \sigma^2)$.

(e) Analyze the residuals from this experiment. Do you think that the analysis of variance assumptions are satisfied?

(f) Use the REML method to analyze this data. Compare the 95% CI on the error variance from REML with the exact chi-square confidence interval.

3.55. A manufacturer suspects that the batches of raw material furnished by his supplier differ significantly in calcium content. There are a large number of batches currently in the warehouse. Five of these are randomly selected for study.

A chemist makes five determinations on each batch and obtains the following data:

Batch 1	Batch 2	Batch 3	Batch 4	Batch 5
23.46	23.59	23.51	23.28	23.29
23.48	23.46	23.64	23.40	23.46
23.56	23.42	23.46	23.37	23.37
23.39	23.49	23.52	23.46	23.32
23.40	23.50	23.49	23.39	23.38

(a) Is there significant variation in calcium content from batch to batch? Use $\alpha = 0.05$.

(b) Estimate the components of variance.

(c) Find a 95 percent confidence interval for $\sigma_\tau^2 / (\sigma_\tau^2 + \sigma^2)$.

(d) Analyze the residuals from this experiment. Are the analysis of variance assumptions satisfied?

(e) Use the REML method to analyze this data. Compare the 95 percent confidence interval on the error variance from REML with the exact chi-square confidence interval.

CHAPTER 4

Experiments with Blocking Factors

CHAPTER OUTLINE

The supplemental material is on the textbook website www.wiley.com/go/global/montgomery.

4.1 The Randomized Complete Block Design

In any experiment, variability arising from a nuisance factor can affect the results. Generally, we define a **nuisance factor** as a design factor that probably has an effect on the response, but we are not interested in that effect. Sometimes a nuisance factor is **unknown and uncontrolled**; that is, we don't know that the factor exists, and it may even be changing levels while we are conducting the experiment. **Randomization** is the design technique used to guard against such a "lurking" nuisance factor. In other cases, the nuisance factor is **known but uncontrollable**. If we can at least observe the value that the nuisance factor takes on at each run of the experiment, we can compensate for it in the statistical analysis by using the **analysis of covariance**, a technique we will discuss in Chapter 14. When the nuisance source of variability is **known and controllable**, a design technique called **blocking** can be used to systematically eliminate its effect on the statistical comparisons among treatments. Blocking is an extremely important design technique used extensively in industrial experimentation and is the subject of this chapter.

To illustrate the general idea, reconsider the hardness testing experiment first described in Section 2.5.1. Suppose now that we wish to determine whether or not four different tips produce different readings on a hardness testing machine. An experiment such as this might be part of a

■ TABLE 4.1
Randomized Complete Block Design for the Hardness Testing Experiment

Test Coupon (Block)			
1	2	3	4
Tip 3	Tip 3	Tip 2	Tip 1
Tip 1	Tip 4	Tip 1	Tip 4
Tip 4	Tip 2	Tip 3	Tip 2
Tip 2	Tip 1	Tip 4	Tip 3

gauge capability study. The machine operates by pressing the tip into a metal test coupon, and from the depth of the resulting depression, the hardness of the coupon can be determined. The experimenter has decided to obtain four observations on Rockwell C-scale hardness for each tip. There is only one factor—tip type—and a completely randomized single-factor design would consist of randomly assigning each one of the $4 \times 4 = 16$ runs to an **experimental unit**, that is, a metal coupon, and observing the hardness reading that results. Thus, 16 different metal test coupons would be required in this experiment, one for each run in the design.

There is a potentially serious problem with a completely randomized experiment in this design situation. If the metal coupons differ slightly in their hardness, as might happen if they are taken from ingots that are produced in different heats, the experimental units (the coupons) will contribute to the variability observed in the hardness data. As a result, the experimental error will reflect *both* random error *and* variability between coupons.

We would like to make the experimental error as small as possible; that is, we would like to remove the variability between coupons from the experimental error. A design that would accomplish this requires the experimenter to test each tip once on each of four coupons. This design, shown in Table 4.1, is called a **randomized complete block design (RCBD)**. The word "complete" indicates that each block (coupon) contains all the treatments (tips). By using this design, the blocks, or coupons, form a more homogeneous experimental unit on which to compare the tips. Effectively, this design strategy improves the accuracy of the comparisons among tips by eliminating the variability among the coupons. Within a block, the order in which the four tips are tested is randomly determined. Notice the similarity of this design problem to the paired t-test of Section 2.5.1. The randomized complete block design is a generalization of that concept.

The RCBD is one of the most widely used experimental designs. Situations for which the RCBD is appropriate are numerous. Units of test equipment or machinery are often different in their operating characteristics and would be a typical blocking factor. Batches of raw material, people, and time are also common nuisance sources of variability in an experiment that can be systematically controlled through blocking.[1]

Blocking may also be useful in situations that do not necessarily involve nuisance factors. For example, suppose that a chemical engineer is interested in the effect of catalyst feed rate on the viscosity of a polymer. She knows that there are several factors, such as raw material source, temperature, operator, and raw material purity that are very difficult to control in the full-scale process. Therefore she decides to test the catalyst feed rate factor in blocks, where each block consists of some combination of these uncontrollable factors. In effect, she is using the blocks to test the **robustness** of her process variable (feed rate) to conditions she cannot easily control. For more discussion of this, see Coleman and Montgomery (1993).

[1] A special case of blocking occurs where the blocks are experimental units such as people, and each block receives the treatments own time or the treatment effects are measured at different times. These are called **repeated measures** designs. They are discussed in chapter 15.

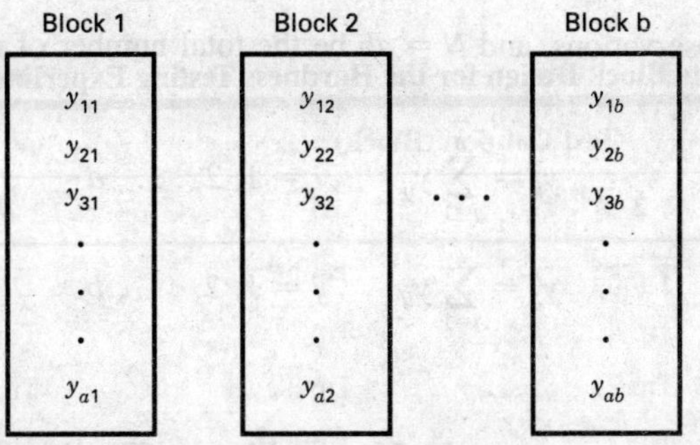

■ **FIGURE 4.1** The randomized complete block design

4.1.1 Statistical Analysis of the RCBD

Suppose we have, in general, *a* treatments that are to be compared and *b* blocks. The randomized complete block design is shown in Figure 4.1. There is one observation per treatment in each block, and the order in which the treatments are run within each block is determined randomly. Because the only randomization of treatments is within the blocks, we often say that the blocks represent a **restriction on randomization**.

The **statistical model** for the RCBD can be written in several ways. The traditional model is an **effects model**:

$$y_{ij} = \mu + \tau_i + \beta_j + \epsilon_{ij} \qquad \begin{cases} i = 1, 2, \ldots, a \\ j = 1, 2, \ldots, b \end{cases} \tag{4.1}$$

where μ is an overall mean, τ_i is the effect of the *i*th treatment, β_j is the effect of the *j*th block, and ϵ_{ij} is the usual NID $(0, \sigma^2)$ random error term. We will initially consider treatments and blocks to be fixed factors. The case of random blocks, which is very important, is considerd in Section 4.1.3. Just as in the single-factor experimental design model in Chapter 3, the effects model for the RCBD is an overspecified model. Consequently, we usually think of the treatment and block effects as deviations from the overall mean so that

$$\sum_{i=1}^{a} \tau_i = 0 \quad \text{and} \quad \sum_{j=1}^{b} \beta_j = 0$$

It is also possible to use a **means model** for the RCBD, say

$$y_{ij} = \mu_{ij} + \epsilon_{ij} \qquad \begin{cases} i = 1, 2, \ldots, a \\ j = 1, 2, \ldots, b \end{cases}$$

where $\mu_{ij} = \mu + \tau_i + \beta_j$. However, we will use the effects model in Equation 4.1 throughout this chapter.

In an experiment involving the RCBD, we are interested in testing the equality of the treatment means. Thus, the hypotheses of interest are

$$H_0\colon \mu_1 = \mu_2 = \cdots = \mu_a$$
$$H_1\colon \text{at least one } \mu_i \neq \mu_j$$

Because the *i*th treatment mean $\mu_i = (1/b)\Sigma_{j=1}^{b}(\mu + \tau_i + \beta_j) = \mu + \tau_i$, an equivalent way to write the above hypotheses is in terms of the treatment effects, say

$$H_0\colon \tau_1 = \tau_2 = \cdots = \tau_a = 0$$
$$H_1\colon \tau_i \neq 0 \text{ at least one } i$$

The analysis of variance can be easily extended to the RCBD. Let $y_{i.}$ be the total of all observations taken under treatment *i*, $y_{.j}$ be the total of all observations in block *j*, $y_{..}$ be the grand total of all observations, and $N = ab$ be the total number of observations. Expressed mathematically,

$$y_{i.} = \sum_{j=1}^{b} y_{ij} \qquad i = 1, 2, \ldots, a \tag{4.2}$$

$$y_{.j} = \sum_{i=1}^{a} y_{ij} \qquad j = 1, 2, \ldots, b \tag{4.3}$$

and

$$y_{..} = \sum_{i=1}^{a}\sum_{j=1}^{b} y_{ij} = \sum_{i=1}^{a} y_{i.} = \sum_{j=1}^{b} y_{.j} \tag{4.4}$$

Similarly, $\bar{y}_{i.}$ is the average of the observations taken under treatment i, $\bar{y}_{.j}$ is the average of the observations in block j, and $\bar{y}_{..}$ is the grand average of all observations. That is,

$$\bar{y}_{i.} = y_{i.}/b \quad \bar{y}_{.j} = y_{.j}/a \quad \bar{y}_{..} = y_{..}/N \tag{4.5}$$

We may express the total corrected sum of squares as

$$\sum_{i=1}^{a}\sum_{j=1}^{b}(y_{ij} - \bar{y}_{..})^2 = \sum_{i=1}^{a}\sum_{j=1}^{b}[(\bar{y}_{i.} - \bar{y}_{..}) + (\bar{y}_{.j} - \bar{y}_{..}) + (y_{ij} - \bar{y}_{i.} - \bar{y}_{.j} + \bar{y}_{..})]^2 \tag{4.6}$$

By expanding the right-hand side of Equation 4.6, we obtain

$$\begin{aligned}\sum_{i=1}^{a}\sum_{j=1}^{b}(y_{ij} - \bar{y}_{..})^2 &= b\sum_{i=1}^{a}(\bar{y}_{i.} - \bar{y}_{..})^2 + a\sum_{j=1}^{b}(\bar{y}_{.j} - \bar{y}_{..})^2 \\ &+ \sum_{i=1}^{a}\sum_{j=1}^{b}(y_{ij} - \bar{y}_{i.} - \bar{y}_{.j} + \bar{y}_{..})^2 + 2\sum_{i=1}^{a}\sum_{j=1}^{b}(\bar{y}_{i.} - \bar{y}_{..})(\bar{y}_{.j} - \bar{y}_{..}) \\ &+ 2\sum_{i=1}^{a}\sum_{j=1}^{b}(\bar{y}_{.j} - \bar{y}_{..})(y_{ij} - \bar{y}_{i.} - \bar{y}_{.j} + \bar{y}_{..}) \\ &+ 2\sum_{i=1}^{a}\sum_{j=1}^{b}(\bar{y}_{i.} - \bar{y}_{..})(y_{ij} - \bar{y}_{i.} - \bar{y}_{.j} + \bar{y}_{..})\end{aligned}$$

Simple but tedious algebra proves that the three cross products are zero. Therefore,

$$\sum_{i=1}^{a}\sum_{j=1}^{b}(y_{ij} - \bar{y}_{..})^2 = b\sum_{i=1}^{a}(\bar{y}_{i.} - \bar{y}_{..})^2 + a\sum_{j=1}^{b}(\bar{y}_{.j} - \bar{y}_{..})^2 + \sum_{i=1}^{a}\sum_{j=1}^{b}(y_{ij} - \bar{y}_{.j} - \bar{y}_{i.} + \bar{y}_{..})^2 \tag{4.7}$$

represents a partition of the total sum of squares. This is the fundamental ANOVA equation for the RCBD. Expressing the sums of squares in Equation 4.7 symbolically, we have

$$SS_T = SS_{\text{Treatments}} + SS_{\text{Blocks}} + SS_E \tag{4.8}$$

Because there are N observations, SS_T has $N - 1$ degrees of freedom. There are a treatments and b blocks, so $SS_{\text{Treatments}}$ and SS_{Blocks} have $a - 1$ and $b - 1$ degrees of freedom, respectively. The error sum of squares is just a sum of squares between cells minus the sum of squares for treatments and blocks. There are ab cells with $ab - 1$ degrees of freedom between them, so SS_E has $ab - 1 - (a - 1) - (b - 1) = (a - 1)(b - 1)$ degrees of freedom. Furthermore, the degrees of freedom on the right-hand side of Equation 4.8 add to the total on the left; therefore, making the usual normality assumptions on the errors, one may use Theorem 3-1 to show that $SS_{\text{Treatments}}/\sigma^2$, $SS_{\text{Blocks}}/\sigma^2$, and SS_E/σ^2 are independently distributed chi-square random variables. Each sum of squares divided by its degrees of freedom is a mean square. The expected value of the mean squares, if treatments and blocks are fixed, can be shown to be

$$E(MS_{\text{Treatments}}) = \sigma^2 + \frac{b \sum_{i=1}^{a} \tau_i^2}{a - 1}$$

$$E(MS_{\text{Blocks}}) = \sigma^2 + \frac{a \sum_{j=1}^{b} \beta_j^2}{b - 1}$$

$$E(MS_E) = \sigma^2$$

Therefore, to test the equality of treatment means, we would use the test statistic

$$F_0 = \frac{MS_{\text{Treatments}}}{MS_E}$$

which is distributed as $F_{a-1,(a-1)(b-1)}$ if the null hypothesis is true. The critical region is the upper tail of the F distribution, and we would reject H_0 if $F_0 > F_{\alpha,a-1,(a-1)(b-1)}$. A P-value approach can also be used.

We may also be interested in comparing block means because, if these means do not differ greatly, blocking may not be necessary in future experiments. From the expected mean squares, it seems that the hypothesis $H_0{:}\beta_j = 0$ may be tested by comparing the statistic $F_0 = MS_{\text{Blocks}}/MS_E$ to $F_{\alpha,b-1,(a-1)(b-1)}$. However, recall that randomization has been applied only to treatments *within* blocks; that is, the blocks represent a **restriction on randomization**. What effect does this have on the statistic $F_0 = MS_{\text{Blocks}}/MS_E$? Some differences in treatment of this question exist. For example, Box, Hunter, and Hunter (2005) point out that the usual analysis of variance F test can be justified on the basis of randomization only,[2] without direct use of the normality assumption. They further observe that the test to compare block means cannot appeal to such a justification because of the randomization restriction; but if the errors are NID(0, σ^2), the statistic $F_0 = MS_{\text{Blocks}}/MS_E$ can be used to compare block means. On the other hand, Anderson and McLean (1974) argue that the randomization restriction prevents this statistic from being a meaningful test for comparing block means and that this F ratio really is a test for the equality of the block means plus the randomization restriction [which they call a restriction error; see Anderson and McLean (1974) for further details].

In practice, then, what do we do? Because the normality assumption is often questionable, to view $F_0 = MS_{\text{Blocks}}/MS_E$ as an exact F test on the equality of block means is not a good general practice. For that reason, we exclude this F test from the analysis of variance table. However, as an approximate procedure to investigate the effect of the blocking variable, examining the ratio of MS_{Blocks} to MS_E is certainly reasonable. If this ratio is large, it implies that the blocking factor has a large effect and that the noise reduction obtained by blocking was probably helpful in improving the precision of the comparison of treatment means.

The procedure is usually summarized in an ANOVA table, such as the one shown in Table 4.2. The computing would usually be done with a statistical software package. However, computing formulas for the sums of squares may be obtained for the elements in Equation 4.7 by working directly with the identity

$$y_{ij} - \bar{y}_{..} = (\bar{y}_{i.} - \bar{y}_{..}) + (\bar{y}_{.j} - \bar{y}_{..}) + (y_{ij} - \bar{y}_{i.} - \bar{y}_{.j} + \bar{y}_{..})$$

[2] Actually, the normal-theory F distribution is an approximation to the randomization distribution generated by calculating F_0 from every possible assignment of the responses to the treatments.

■ **TABLE 4.2**
Analysis of Variance for a Randomized Complete Block Design

Source of Variation	Sum of Squares	Degrees of Freedom	Mean Square	F_0
Treatments	$SS_{\text{Treatments}}$	$a - 1$	$\frac{SS_{\text{Treatments}}}{a-1}$	$\frac{MS_{\text{Treatments}}}{MS_E}$
Blocks	SS_{Blocks}	$b - 1$	$\frac{SS_{\text{Blocks}}}{b-1}$	
Error	SS_E	$(a-1)(b-1)$	$\frac{SS_E}{(a-1)(b-1)}$	
Total	SS_T	$N - 1$		

These quantities can be computed in the columns of a spreadsheet (Excel). Then each column can be squared and summed to produce the sum of squares. Alternatively, computing formulas can be expressed in terms of treatment and block totals. These formulas are

$$SS_T = \sum_{i=1}^{a}\sum_{j=1}^{b} y_{ij}^2 - \frac{y_{..}^2}{N} \tag{4.9}$$

$$SS_{\text{Treatments}} = \frac{1}{b}\sum_{i=1}^{a} y_{i.}^2 - \frac{y_{..}^2}{N} \tag{4.10}$$

$$SS_{\text{Blocks}} = \frac{1}{a}\sum_{j=1}^{b} y_{.j}^2 - \frac{y_{..}^2}{N} \tag{4.11}$$

and the error sum of squares is obtained by subtraction as

$$SS_E = SS_T - SS_{\text{Treatments}} - SS_{\text{Blocks}} \tag{4.12}$$

EXAMPLE 4.1

A medical device manufacturer produces vascular grafts (artificial veins). These grafts are produced by extruding billets of polytetrafluoroethylene (PTFE) resin combined with a lubricant into tubes. Frequently, some of the tubes in a production run contain small, hard protrusions on the external surface. These defects are known as "flicks." The defect is cause for rejection of the unit.

The product developer responsible for the vascular grafts suspects that the extrusion pressure affects the occurrence of flicks and therefore intends to conduct an experiment to investigate this hypothesis. However, the resin is manufactured by an external supplier and is delivered to the medical device manufacturer in batches. The engineer also suspects that there may be significant batch-to-batch variation, because while the material should be consistent with respect to parameters such as molecular weight, mean particle size, retention, and peak height ratio, it probably isn't due to manufacturing variation at the resin supplier and natural variation in the material. Therefore, the product developer decides to investigate the effect of four different levels of extrusion pressure on flicks using a randomized complete block design considering batches of resin as blocks. The RCBD is shown in Table 4.3. Note that there are four levels of extrusion pressure (treatments) and six batches of resin (blocks). Remember that the order in which the extrusion pressures are tested within each block is random. The response variable is yield, or the percentage of tubes in the production run that did not contain any flicks.

■ **TABLE 4.3**
Randomized Complete Block Design for the Vascular Graft Experiment

Extrusion Pressure (PSI)	Batch of Resin (Block) 1	2	3	4	5	6	Treatment Total
8500	90.3	89.2	98.2	93.9	87.4	97.9	556.9
8700	92.5	89.5	90.6	94.7	87.0	95.8	550.1
8900	85.5	90.8	89.6	86.2	88.0	93.4	533.5
9100	82.5	89.5	85.6	87.4	78.9	90.7	514.6
Block Totals	350.8	359.0	364.0	362.2	341.3	377.8	$y_{..} = 2155.1$

To perform the analysis of variance, we need the following sums of squares:

$$SS_T = \sum_{i=1}^{4}\sum_{j=1}^{6} y_{ij}^2 - \frac{y_{..}^2}{N}$$

$$= 193{,}999.31 - \frac{(2155.1)^2}{24} = 480.31$$

$$SS_{\text{Treatments}} = \frac{1}{b}\sum_{i=1}^{4} y_{i.}^2 - \frac{y_{..}^2}{N}$$

$$= \frac{1}{6}[(556.9)^2 + (550.1)^2 + (533.5)^2 + (514.6)^2] - \frac{(2155.1)^2}{24} = 178.17$$

$$SS_{\text{Blocks}} = \frac{1}{a}\sum_{j=1}^{6} y_{.j}^2 - \frac{y_{..}^2}{N}$$

$$= \frac{1}{4}[(350.8)^2 + (359.0)^2 + \cdots + (377.8)^2] - \frac{(2155.1)^2}{24} = 192.25$$

$$SS_E = SS_T - SS_{\text{Treatments}} - SS_{\text{Blocks}}$$

$$= 480.31 - 178.17 - 192.25 = 109.89$$

The ANOVA is shown in Table 4.4. Using $\alpha = 0.05$, the critical value of F is $F_{0.05,3,15} = 3.29$. Because $8.11 > 3.29$, we conclude that extrusion pressure affects the mean yield. The P-value for the test is also quite small. Also, the resin batches (blocks) seem to differ significantly, because the mean square for blocks is large relative to error.

■ **TABLE 4.4**
Analysis of Variance for the Vascular Graft Experiment

Source of Variation	Sum of Squares	Degrees of Freedom	Mean Square	F_0	P-Value
Treatments (extrusion pressure)	178.17	3	59.39	8.11	0.0019
Blocks (batches)	192.25	5	38.45		
Error	109.89	15	7.33		
Total	480.31	23			

It is interesting to observe the results we would have obtained from this experiment had we not been aware of randomized block designs. Suppose that this experiment had been run as a completely randomized design, and (by chance) the same design resulted as in Table 4.3. The incorrect analysis of these data as a completely randomized single-factor design is shown in Table 4.5.

Because the P-value is less than 0.05, we would still reject the null hypothesis and conclude that extrusion pressure significantly affects the mean yield. However, note that the mean

■ **TABLE 4.5**

Incorrect Analysis of the Vascular Graft Experiment as a Completely Randomized Design

Source of Variation	Sum of Squares	Degrees of Freedom	Mean Square	F_0	*P*-Value
Extrusion pressure	178.17	3	59.39	3.95	0.0235
Error	302.14	20	15.11		
Total	480.31	23			

square for error has more than doubled, increasing from 7.33 in the RCBD to 15.11. All of the variability due to blocks is now in the error term. This makes it easy to see why we sometimes call the RCBD a noise-reducing design technique; it effectively increases the signal-to-noise ratio in the data, or it improves the precision with which treatment means are compared. This example also illustrates an important point. If an experimenter fails to block when he or she should have, the effect may be to inflate the experimental error, and it would be possible to inflate the error so much that important differences among the treatment means could not be identified.

Sample Computer Output. Condensed computer output for the vascular graft experiment in Example 4.1, obtained from Design-Expert and JMP is shown in Figure 4.2. The Design-Expert output is in Figure 4.2*a* and the JMP output is in Figure 4.2*b*. Both outputs are very similar, and match the manual computation given earlier. Note that JMP computes an *F*-statistic for blocks (the batches). The sample means for each treatment are shown in the output. At 8500 psi, the mean yield is $\bar{y}_{1.} = 92.82$, at 8700 psi the mean yield is $\bar{y}_{2.} = 91.68$, at 8900 psi the mean yield is $\bar{y}_{3.} = 88.92$, and at 9100 psi the mean yield is $\bar{y}_{4.} = 85.77$. Remember that these sample mean yields estimate the treatment means μ_1, μ_2, μ_3, and μ_4. The model residuals are shown at the bottom of the Design-Expert output. The residuals are calculated from

$$e_{ij} = y_{ij} - \hat{y}_{ij}$$

and, as we will later show, the fitted values are $\hat{y}_{ij} = \bar{y}_{i.} + \bar{y}_{.j} - \bar{y}_{..}$, so

$$e_{ij} = y_{ij} - \bar{y}_{i.} - \bar{y}_{.j} + \bar{y}_{..} \quad \textbf{(4.13)}$$

In the next section, we will show how the residuals are used in **model adequacy checking.**

Multiple Comparisons. If the treatments in an RCBD are fixed, and the analysis indicates a significant difference in treatment means, the experimenter is usually interested in multiple comparisons to discover *which* treatment means differ. Any of the multiple comparison procedures discussed in Section 3.5 may be used for this purpose. In the formulas of Section 3.5, simply replace the number of replicates in the single-factor completely randomized design (n) by the number of blocks (b). Also, remember to use the number of error degrees of freedom for the randomized block $[(a-1)(b-1)]$ instead of those for the completely randomized design $[a(n-1)]$.

The Design-Expert output in Figure 4.2 illustrates the Fisher LSD procedure. Notice that we would conclude that $\mu_1 = \mu_2$, because the *P*-value is very large. Furthermore, μ_1 differs from all other means. Now the *P*-value for $H_0{:}\mu_2 = \mu_3$ is 0.097, so there is some evidence to conclude that $\mu_2 \neq \mu_3$, and $\mu_2 \neq \mu_4$ because the *P*-value is 0.0018. Overall, we would conclude that lower extrusion pressures (8500 psi and 8700 psi) lead to fewer defects.

Response: Yield
ANOVA for Selected Factorial Model
Analysis of Variance Table [Partial Sum of Squares]

Source	Sum of Squares	DF	Mean Square	F Value	Prob > F
Block	192.25	5	38.45		
Model	178.17	3	59.39	8.11	0.0019
A	*178.17*	*3*	*59.39*	*8.11*	*0.0019*
Residual	109.89	15	7.33		
Cor Total	480.31	23			

Std. Dev.	2.71	R-Squared	0.6185
Mean	89.80	Adj R-Squared	0.5422
C.V.	3.01	Pred R-Squared	0.0234
PRESS	281.31	Adeq Precision	9.759

Treatment Means (Adjusted, If Necessary)

	Estimated Mean	Standard Error
1–8500	92.82	1.10
2-8700	91.68	1.10
3-8900	88.92	1.10
4-9100	85.77	1.10

Treatment	Mean Difference	DF	Standard Error	t for H_0 Coeff=0	Prob > |t|
1 vs.2	1.13	1	1.56	0.73	0.4795
1 vs.3	3.90	1	1.56	2.50	0.0247
1 vs.4	7.05	1	1.56	4.51	0.0004
2 vs.3	2.77	1	1.56	1.77	0.0970
2 vs.4	5.92	1	1.56	3.79	0.0018
3 vs.4	3.15	1	1.56	2.02	0.0621

Diagnostics Case Statistics

Standard Order	Actual Value	Predicted Value	Residual	Leverage	Student Residual	Cook's Distance	Outlier t	Run Order
1	90.30	90.72	−0.42	0.375	−0.197	0.003	−0.190	1
2	89.20	92.77	−3.57	0.375	−1.669	0.186	−1.787	6
3	98.20	94.02	4.18	0.375	1.953	0.254	2.185	9
4	93.90	93.57	0.33	0.375	0.154	0.002	0.149	13
5	87.40	88.35	−0.95	0.375	−0.442	0.013	−0.430	19
6	97.90	97.47	0.43	0.375	0.201	0.003	0.194	23
7	92.50	89.59	2.91	0.375	1.361	0.124	1.405	4
8	89.50	91.64	−2.14	0.375	−0.999	0.067	−0.999	5
9	90.60	92.89	−2.29	0.375	−1.069	0.076	−1.075	10
10	94.70	92.44	2.26	0.375	1.057	0.075	1.062	16
11	87.00	87.21	−0.21	0.375	−0.099	0.001	−0.096	20
12	95.80	96.34	−0.54	0.375	−0.251	0.004	−0.243	21
13	85.50	86.82	−1.32	0.375	−0.617	0.025	−0.604	3
14	90.80	88.87	1.93	0.375	0.902	0.054	0.896	8
15	89.60	90.12	−0.52	0.375	−0.243	0.004	−0.236	12
16	86.20	89.67	−3.47	0.375	−1.622	0.175	−1.726	15
17	88.00	84.45	3.55	0.375	1.661	0.184	1.776	17
18	93.40	93.57	−0.17	0.375	−0.080	0.000	−0.077	22
19	82.50	83.67	−1.17	0.375	−0.547	0.020	−0.534	2
20	89.50	85.72	3.78	0.375	1.766	0.208	1.917	7
21	85.60	86.97	−1.37	0.375	−0.641	0.027	−0.628	11
22	87.40	86.52	0.88	0.375	0.411	0.011	0.399	14
23	78.90	81.30	−2.40	0.375	−1.120	0.084	−1.130	18
24	90.70	90.42	0.28	0.375	0.130	0.001	0.126	24

Note: Predicted values include block corrections.

(a)

■ **FIGURE 4.2** Computer output for Example 4.1. (*a*) Design-Expert; (*b*) JMP

Oneway Analysis of Yield By Pressure
Block
Batch

Oneway Anova
Summary of Fit

Rsquare	0.771218
Adj Rsquare	0.649201
Root Mean Square Error	2.706612
Mean of Response	89.79583
Observations (or Sum Wgts)	24

Analysis of Variance

Source	DF	Sum of Squares	Mean Square	*F* Ratio	Prob > F
Pressure	3	178.17125	59.3904	8.1071	0.0019
Batch	5	192.25208	38.4504	5.2487	0.0055
Error	15	109.88625	7.3257		
C.Total	23	480.30958			

Means for Oneway Anova

Level	Number	Mean	Std. Error	Lower 95%	Upper 95%
8500	6	92.8167	1.1050	90.461	95.172
8700	6	91.6833	1.1050	89.328	94.039
8900	6	88.9167	1.1050	86.561	91.272
9100	6	85.7667	1.1050	83.411	88.122

Std. Error uses a pooled estimate of error variance

Block Means

Batch	Mean	Number
1	87.7000	4
2	89.7500	4
3	91.0000	4
4	90.5500	4
5	85.3250	4
6	94.4500	4

(b)

■ **FIGURE 4.2** *(Continued)*

We can also use the graphical procedure of Section 3.5.1 to compare mean yield at the four extrusion pressures. Figure 4.3 plots the four means from Example 4.1 relative to a scaled *t* distribution with a scale factor $\sqrt{MS_E/b} = \sqrt{7.33/6} = 1.10$. This plot indicates that the two lowest pressures result in the same mean yield, but that the mean yields for 8700 psi and

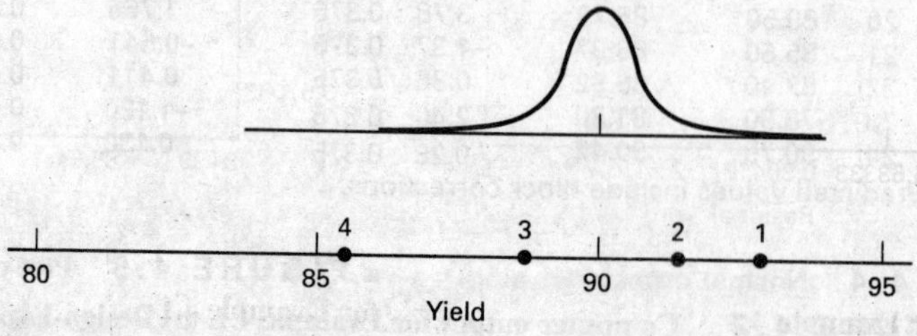

■ **FIGURE 4.3** **Mean yields for the four extrusion pressures relative to a scaled *t* distribution with a scale factor** $\sqrt{MS_E/b} = \sqrt{7.33/6} = 1.10$

8900 psi (μ_2 and μ_3) are also similar. The highest pressure (9100 psi) results in a mean yield that is much lower than all other means. This figure is a useful aid in interpreting the results of the experiment and the Fisher LSD calculations in the Design-Expert output in Figure 4.2.

4.1.2 Model Adequacy Checking

We have previously discussed the importance of checking the adequacy of the assumed model. Generally, we should be alert for potential problems with the normality assumption, unequal error variance by treatment or block, and block–treatment interaction. As in the completely randomized design, residual analysis is the major tool used in this diagnostic checking. The residuals for the randomized block design in Example 4.1 are listed at the bottom of the Design-Expert output in Figure 4.2.

A normal probability plot of these residuals is shown in Figure 4.4. There is no severe indication of nonnormality, nor is there any evidence pointing to possible outliers. Figure 4.5 plots the residuals versus the fitted values $\hat{y}_{ij}$. There should be no relationship between the size of the residuals and the fitted values $\hat{y}_{ij}$. This plot reveals nothing of unusual interest. Figure 4.6 shows plots of the residuals by treatment (extrusion pressure) and by batch of resin or block. These plots are potentially very informative. If there is more scatter in the residuals for a particular treatment, that could indicate that this treatment produces more erratic response readings than the others. More scatter in the residuals for a particular block could indicate that the block is not homogeneous. However, in our example, Figure 4.6 gives no indication of inequality of variance by treatment but there is an indication that there is less variability in the yield for batch 6. However, since all of the other residual plots are satisfactory, we will ignore this.

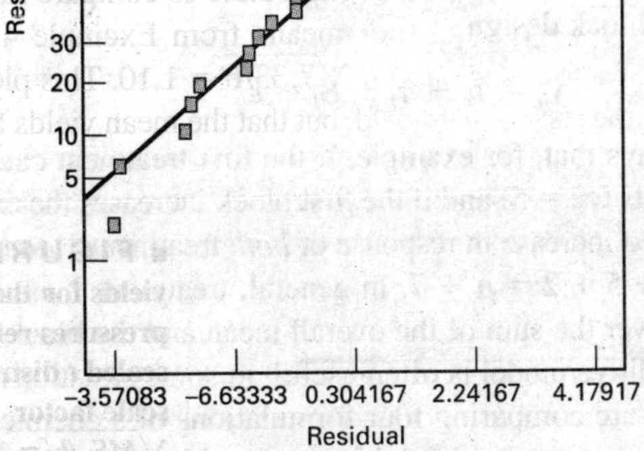

■ **FIGURE 4.4 Normal probability plot of residuals for Example 4.1**

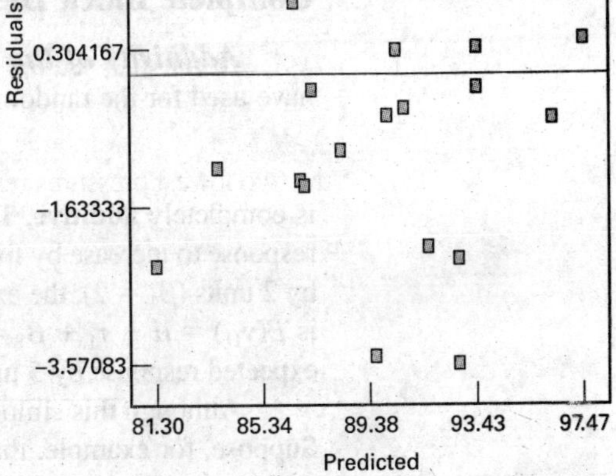

■ **FIGURE 4.5 Plot of residuals versus $\hat{y}_{ij}$ for Example 4.1**

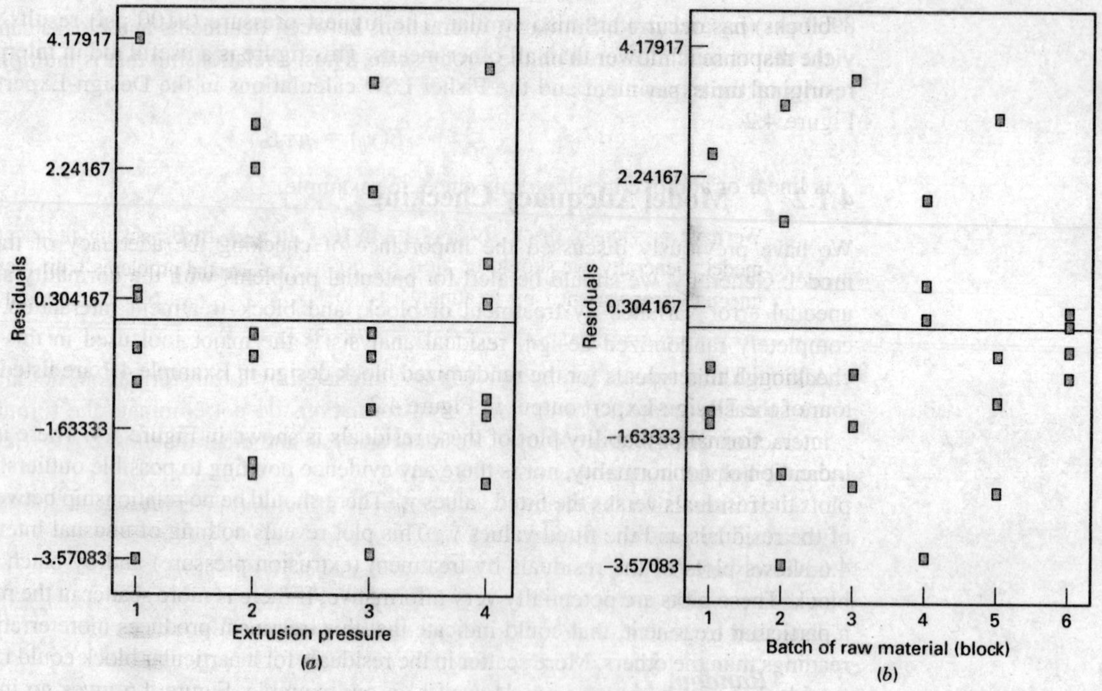

■ **FIGURE 4.6 Plot of residuals by extrusion pressure (treatment) and by batches of resin (block) for Example 4.1**

Sometimes the plot of residuals versus $\hat{y}_{ij}$ has a curvilinear shape; for example, there may be a tendency for negative residuals to occur with low $\hat{y}_{ij}$ values, positive residuals with intermediate $\hat{y}_{ij}$ values, and negative residuals with high $\hat{y}_{ij}$ values. This type of pattern is suggestive of **interaction** between blocks and treatments. If this pattern occurs, a transformation should be used in an effort to eliminate or minimize the interaction. In Section 5.3.7, we describe a statistical test that can be used to detect the presence of interaction in a randomized block design.

4.1.3 Some Other Aspects of the Randomized Complete Block Design

Additivity of the Randomized Block Model. The linear statistical model that we have used for the randomized block design

$$y_{ij} = \mu + \tau_i + \beta_j + \epsilon_{ij}$$

is completely **additive**. This says that, for example, if the first treatment causes the expected response to increase by five units ($\tau_1 = 5$) and if the first block increases the expected response by 2 units ($\beta_1 = 2$), the expected increase in response of *both* treatment 1 *and* block 1 together is $E(y_{11}) = \mu + \tau_1 + \beta_1 = \mu + 5 + 2 = \mu + 7$. In general, treatment 1 *always* increases the expected response by 5 units over the sum of the overall mean and the block effect.

Although this simple additive model is often useful, in some situations it is inadequate. Suppose, for example, that we are comparing four formulations of a chemical product using six batches of raw material; the raw material batches are considered blocks. If an impurity in batch 2 affects formulation 2 adversely, resulting in an unusually low yield, but does not affect the other formulations, an **interaction** between formulations (or treatments) and batches (or blocks) has occurred. Similarly, interactions between treatments and blocks can occur when the response is measured on the wrong scale. Thus, a relationship that is multiplicative in the original units, say

$$E(y_{ij}) = \mu\tau_i\beta_j$$

is linear or additive in a log scale since, for example,

$$\ln E(y_{ij}) = \ln \mu + \ln \tau_i + \ln \beta_j$$

or

$$E(y_{ij}^*) = \mu^* + \tau_i^* + \beta_j^*$$

Although this type of interaction can be eliminated by a transformation, not all interactions are so easily treated. For example, transformations do not eliminate the formulation–batch interaction discussed previously. Residual analysis and other diagnostic checking procedures can be helpful in detecting nonadditivity.

If interaction is present, it can seriously affect and possibly invalidate the analysis of variance. In general, the presence of interaction inflates the error mean square and may adversely affect the comparison of treatment means. In situations where both factors, as well as their possible interaction, are of interest, **factorial designs** must be used. These designs are discussed extensively in Chapters 5 through 9.

Random Treatments and Blocks. Our presentation of the randomized complete block design thus far has focused on the case when both the treatments and blocks were considered as fixed factors. There are many situations where either treatments or blocks (or both) are random factors. It is very common to find that the blocks are random. This is usually what the experimenter would like to do, because we would like for the conclusions from the experiment to be valid across the population of blocks that the ones selected for the experiments were sampled from. First, we consider the case where the treatments are fixed and the blocks are random. Equation 4.1 is still the appropriate statistical model, but now the block effects are random, that is, we assume that the β_j, $j = 1, 2, \ldots, b$ are $NID(0, \sigma^2{}_\beta)$ random variables. This is a special case of a mixed model (because it contains both fixed and random factors). In Chapters 13 and 14 we will discuss mixed models in more detail and provide several examples of situations where they occur. Our discussion here is limited to the RCBD.

Assuming that the RCBD model Equation 4.1 is appropriate, if the blocks are random and the treatments are fixed we can show that:

$$\begin{aligned} E(y_{ij}) &= \mu + \tau_i, \qquad i = 1, 2, \ldots, a \\ V(y_{ij}) &= \sigma_\beta^2 + \sigma^2 \\ Cov(y_{ij}, y_{i'j'}) &= 0, \quad j \neq j' \\ Cov(y_{ij}, y_{i'j}) &= \sigma_\beta^2 \; i \neq i' \end{aligned} \tag{4.14}$$

Thus, the variance of the observations is constant, the covariance between any two observations in different blocks is zero, but the covariance between two observations from the same block is σ_β^2. The expected mean squares from the usual ANOVA partitioning of the total sum of squares are

$$\begin{aligned} E(MS_{\text{Treatments}}) &= \sigma^2 + \frac{b\sum_{i=1}^{a}\tau_i^2}{a-1} \\ E(MS_{\text{Blocks}}) &= \sigma^2 + a\sigma_\beta^2 \\ E(MS_E) &= \sigma^2 \end{aligned} \tag{4.15}$$

The appropriate statistic for testing the null hypothesis of no treatment effects (all $\tau_i = 0$) is

$$F_0 = \frac{MS_{\text{Treatment}}}{MS_E}$$

which is exactly the same test statistic we used in the case where the blocks were fixed. Based on the expected mean squares, we can obtain an ANOVA-type estimator of the variance component for blocks as

$$\hat{\sigma}_\beta^2 = \frac{MS_{\text{Blocks}} - MS_E}{a} \tag{4.16}$$

For example, for the vascular graft experiment in Example 4.1 the estimate of σ_β^2 is

$$\hat{\sigma}_\beta^2 = \frac{MS_{\text{Blocks}} - MS_E}{a} = \frac{38.45 - 7.33}{4} = 7.78$$

This is a method-of-moments estimate and there is no simple way to find a confidence interval on the block variance component σ_β^2. The REML method would be preferred here. Table 4.6 is the JMP output for Example 4.1 assuming that blocks are random. The REML estimate of σ_β^2 is exactly the same as the ANOVA estimate, but REML automatically produces the standard error of the estimate (6.116215) and the approximate 95 percent confidence interval. JMP gives the test for the fixed effect (pressure), and the results are in agreement with those originally reported in Example 4.1. REML also produces the point estimate and CI for the error variance σ^2. The ease with which confidence intervals can be constructed is a major reason why REML has been so widely adopted.

Now consider a situation where there is an interaction between treatments and blocks. This could be accounted for by adding an interaction term to the original statistical model Equation 4.1. Let $(\tau\beta)_{ij}$ be the interaction effect of treatment I in block j. Then the model is

$$y_{ij} = \mu + \tau_i + \beta_j + (\tau\beta)_{ij} + \varepsilon_{ij} \begin{cases} i = 1, 2, \ldots, a \\ j = 1, 2, \ldots, b \end{cases} \tag{4.17}$$

The interaction effect is assumed to be random because it involves the random block effects. If $\sigma_{\tau\beta}^2$ is the variance component for the block treatment interaction, then we can show that the expected mean squares are

$$\begin{aligned} E(MS_{\text{Treatments}}) &= \sigma^2 + \sigma_{\tau\beta}^2 + \frac{b\sum_{i=1}^{a}\tau_i^2}{a-1} \\ E(MS_{\text{Blocks}}) &= \sigma^2 + a\sigma_\beta^2 \\ E(MS_E) &= \sigma^2 + \sigma_{\tau\beta}^2 \end{aligned} \tag{4.18}$$

From the expected mean squares, we see that the usual F-statistic $F = MS_{\text{Treatments}}/MS_E$ would be used to test for no treatment effects. So another advantage of the random block model is that the assumption of no interaction in the RCBD is not important. However, if blocks are fixed and there is interaction, then the interaction effect is not in the expected mean square for treatments but it is in the error expected mean square, so there would not be a statistical test for the treatment effects.

TABLE 4.6
JMP Output for Example 4.1 with Blocks Assumed Random

Response Y

Summary of Fit

RSquare	0.756688
RSquare Adj	0.720192
Root Mean Square Error	2.706612
Mean of Response	89.79583
Observations (or Sum Wgts)	24

REML Variance Component Estimates

Random Effect	Var Ratio	Var Component	Std Error	95% Lower	95% Upper	Pct of Total
Block	1.0621666	7.7811667	6.116215	−4.206394	19.768728	51.507
Residual		7.32575	2.6749857	3.9975509	17.547721	48.493
Total		15.106917				100.000

Covariance Matrix of Variance Component Estimates

Random Effect	Block	Residual
Block	37.408085	−1.788887
Residual	−1.788887	7.1555484

Fixed Effect Tests

Source	Nparm	DF	DFDen	F Ratio	Prob > F
Pressure	3	3	15	8.1071	0.0019*

Choice of Sample Size. Choosing the **sample size**, or the **number of blocks** to run, is an important decision when using an RCBD. Increasing the number of blocks increases the number of replicates and the number of error degrees of freedom, making design more sensitive. Any of the techniques discussed in Section 3.7 for selecting the number of replicates to run in a completely randomized single-factor experiment may be applied directly to the RCBD. For the case of a fixed factor, the operating characteristic curves may be used with

$$\Phi^2 = \frac{b\sum_{i=1}^{a}\tau_i^2}{a\sigma^2} \tag{4.19}$$

where there are $a - 1$ numerator degrees of freedom and $(a - 1)(b - 1)$ denominator degrees of freedom.

EXAMPLE 4.2

Consider the RCBD for the vascular grafts described in Example 4.1. Suppose that we wish to determine the appropriate number of blocks to run if we are interested in detecting a true maximum difference in yield of 6 with a reasonably high probability and an estimate of the standard deviation of the errors is $\sigma = 3$. From Equation 3.45, the minimum value of Φ^2 is (writing b, the number of blocks, for n)

$$\Phi^2 = \frac{bD^2}{2a\sigma^2}$$

where D is the maximum difference we wish to detect. Thus,

$$\Phi^2 = \frac{b(6)^2}{2(4)(3)^2} = 0.5b$$

If we use $b = 5$ blocks, $\Phi = \sqrt{0.5b} = \sqrt{0.5(5)} = 1.58$, and there are $(a - 1)(b - 1) = 3(4) = 12$ error degrees of freedom. Using the Operating Characteristics Curves for the Fixed Effects Model Analysis of Variance* with $\nu_1 = a - 1 = 3$ and $\alpha = 0.05$ indicates that the β risk for this design is approximately 0.55 (power $= 1 - \beta = 0.45$). If we use $b = 6$ blocks, $\Phi = \sqrt{0.5b} = \sqrt{0.5(6)} = 1.73$, with $(a - 1)(b - 1) = 3(5) = 15$ error degrees of freedom, and the corresponding β risk is approximately 0.4 (power $= 1 - \beta = 0.6$). Because the batches of resin are expensive and the cost of experimentation is high, the experimenter decides to use six blocks, even though the power is only about 0.6 (actually many experiments work very well with power values of only 0.5 or higher).

Estimating Missing Values. When using the RCBD, sometimes an observation in one of the blocks is missing. This may happen because of carelessness or error or for reasons beyond our control, such as unavoidable damage to an experimental unit. A missing observation introduces a new problem into the analysis because treatments are no longer **orthogonal to blocks**; that is, every treatment does not occur in every block. There are two general approaches to the missing value problem. The first is an **approximate analysis** in which the missing observation is estimated and the usual analysis of variance is performed just as if the estimated observation were real data, with the error degrees of freedom reduced by 1. This approximate analysis is the subject of this section. The second is an **exact analysis**, which is discussed in Section 4.1.4.

Suppose the observation y_{ij} for treatment i in block j is missing. Denote the missing observation by x. As an illustration, suppose that in the vascular graft experiment of Example 4.1 there was a problem with the extrusion machine when the 8700 psi run was conducted in the fourth batch of material, and the observation y_{24} could not be obtained. The data might appear as in Table 4.7.

In general, we will let y'_{ij} represent the grand total with one missing observation, $y'_{i.}$ represent the total for the treatment with one missing observation, and $y'_{.j}$ be the total for the block with one missing observation. Suppose we wish to estimate the missing observation x

■ **TABLE 4.7**
Randomized Complete Block Design for the Vascular Graft Experiment with One Missing Value

Extrusion Pressures (PSI)	Batch of Resin (Block) 1	2	3	4	5	6	
8500	90.3	89.2	98.2	93.9	87.4	97.9	556.9
8700	92.5	89.5	90.6	x	87.0	95.8	455.4
8900	85.5	90.8	89.6	86.2	88.0	93.4	533.5
9100	82.5	89.5	85.6	87.4	78.9	90.7	514.6
Block totals	350.8	359.0	364.0	267.5	341.3	377.8	$y'_{..} = 2060.4$

* From *Biometrika Tables for Statisticians*, Vol. 2, by E. S. Pearson and H. O. Hartley, (1972).

■ **TABLE 4.8**
Approximate Analysis of Variance for Example 4.1 with One Missing Value

Source of Variation	Sum of Squares	Degrees of Freedom	Mean Square	F_0	*P*-Value
Extrusion pressure	166.14	3	55.38	7.63	0.0029
Batches of raw material	189.52	5	37.90		
Error	101.70	14	7.26		
Total	457.36	23			

so that x will have a minimum contribution to the error sum of squares. Because $SS_E = \sum_{i=1}^{a}\sum_{j=1}^{b}(y_{ij} - \bar{y}_{i.} - \bar{y}_{.j} + \bar{y}_{..})^2$, this is equivalent to choosing x to minimize

$$SS_E = \sum_{i=1}^{a}\sum_{j=1}^{b} y_{ij}^2 - \frac{1}{b}\sum_{i=1}^{a}\left(\sum_{j=1}^{b} y_{ij}\right)^2 - \frac{1}{a}\sum_{j=1}^{b}\left(\sum_{i=1}^{a} y_{ij}\right)^2 + \frac{1}{ab}\left(\sum_{i=1}^{a}\sum_{j=1}^{b} y_{ij}\right)^2$$

or

$$SS_E = x^2 - \frac{1}{b}(y'_{i.} + x)^2 - \frac{1}{a}(y'_{.j} + x)^2 + \frac{1}{ab}(y'_{..} + x)^2 + R \tag{4.20}$$

where R includes all terms not involving x. From $dSS_E/dx = 0$, we obtain

$$x = \frac{ay'_{i.} + by'_{.j} - y'_{..}}{(a-1)(b-1)} \tag{4.21}$$

as the estimate of the missing observation.

For the data in Table 4.7, we find that $y'_{2.} = 455.4$, $y'_{.4} = 267.5$, and $y'_{..} = 2060.4$. Therefore, from Equation 4.16,

$$x \equiv y_{24} = \frac{4(455.4) + 6(267.5) - 2060.4}{(3)(5)} = 91.08$$

The usual analysis of variance may now be performed using $y_{24} = 91.08$ and reducing the error degrees of freedom by 1. The analysis of variance is shown in Table 4.8. Compare the results of this approximate analysis with the results obtained for the full data set (Table 4.4).

If several observations are missing, they may be estimated by writing the error sum of squares as a function of the missing values, differentiating with respect to each missing value, equating the results to zero, and solving the resulting equations. Alternatively, we may use Equation 4.21 iteratively to estimate the missing values. To illustrate the iterative approach, suppose that two values are missing. Arbitrarily estimate the first missing value, and then use this value along with the real data and Equation 4.21 to estimate the second. Now Equation 4.21 can be used to reestimate the first missing value, and following this, the second can be reestimated. This process is continued until convergence is obtained. In any missing value problem, the error degrees of freedom are reduced by one for each missing observation.

4.1.4 Estimating Model Parameters and the General Regression Significance Test

If both treatments and blocks are fixed, we may estimate the parameters in the RCBD model by least squares. Recall that the linear statistical model is

$$y_{ij} = \mu + \tau_i + \beta_j + \epsilon_{ij} \quad \begin{cases} i = 1, 2, \ldots, a \\ j = 1, 2, \ldots, b \end{cases} \tag{4.22}$$

Applying the rules in Section 3.9.2 for finding the normal equations for an experimental design model, we obtain

$$\begin{aligned}
\mu&: ab\hat{\mu} + b\hat{\tau}_1 + b\hat{\tau}_2 + \cdots + b\hat{\tau}_a + a\hat{\beta}_1 + a\hat{\beta}_2 + \cdots + a\hat{\beta}_b = y_{..}\\
\tau_1&: b\hat{\mu} + b\hat{\tau}_1 + \hat{\beta}_1 + \hat{\beta}_2 + \cdots + \hat{\beta}_b = y_{1.}\\
\tau_2&: b\hat{\mu} + b\hat{\tau}_2 + \hat{\beta}_1 + \hat{\beta}_2 + \cdots + \hat{\beta}_b = y_{2.}\\
&\vdots\\
\tau_a&: b\hat{\mu} + b\hat{\tau}_a + \hat{\beta}_1 + \hat{\beta}_2 + \cdots + \hat{\beta}_b = y_{a.}\\
\beta_1&: a\hat{\mu} + \hat{\tau}_1 + \hat{\tau}_2 + \cdots + \hat{\tau}_a + a\hat{\beta}_1 = y_{.1}\\
\beta_2&: a\hat{\mu} + \hat{\tau}_1 + \hat{\tau}_2 + \cdots + \hat{\tau}_a + a\hat{\beta}_2 = y_{.2}\\
&\vdots\\
\beta_b&: a\hat{\mu} + \hat{\tau}_1 + \hat{\tau}_2 + \cdots + \hat{\tau}_a + a\hat{\beta}_b = y_{.b}
\end{aligned} \tag{4.23}$$

Notice that the second through the $(a + 1)$st equations in Equation 4.23 sum to the first normal equation, as do the last b equations. Thus, there are two linear dependencies in the normal equations, implying that two constraints must be imposed to solve Equation 4.23. The usual constraints are

$$\sum_{i=1}^{a} \hat{\tau}_i = 0 \qquad \sum_{j=1}^{b} \hat{\beta}_j = 0 \tag{4.24}$$

Using these constraints helps simplify the normal equations considerably. In fact, they become

$$\begin{aligned}
ab\,\hat{\mu} &= y_{..}\\
b\hat{\mu} + b\hat{\tau}_i &= y_{i.} \quad i = 1, 2, \ldots, a\\
a\hat{\mu} + a\hat{\beta}_j &= y_{.j} \quad j = 1, 2, \ldots, b
\end{aligned} \tag{4.25}$$

whose solution is

$$\begin{aligned}
\hat{\mu} &= \bar{y}_{..}\\
\hat{\tau}_i &= \bar{y}_{i.} - \bar{y}_{..} \quad i = 1, 2, \ldots, a\\
\hat{\beta}_j &= \bar{y}_{.j} - \bar{y}_{..} \quad j = 1, 2, \ldots, b
\end{aligned} \tag{4.26}$$

Using the solution to the normal equation in Equation 4.26, we may find the estimated or fitted values of y_{ij} as

$$\begin{aligned}
\hat{y}_{ij} &= \hat{\mu} + \hat{\tau}_i + \hat{\beta}_j\\
&= \bar{y}_{..} + (\bar{y}_{i.} - \bar{y}_{..}) + (\bar{y}_{.j} - \bar{y}_{..})\\
&= \bar{y}_{i.} + \bar{y}_{.j} - \bar{y}_{..}
\end{aligned}$$

This result was used previously in Equation 4.13 for computing the residuals from a randomized block design.

The general regression significance test can be used to develop the analysis of variance for the randomized complete block design. Using the solution to the normal equations given by Equation 4.26, the reduction in the sum of squares for fitting the **full model** is

$$\begin{aligned} R(\mu, \tau, \beta) &= \hat{\mu} y_{..} + \sum_{i=1}^{a} \hat{\tau}_i y_{i.} + \sum_{j=1}^{b} \hat{\beta}_j y_{.j} \\ &= \bar{y}_{..} y_{..} + \sum_{i=1}^{a} (\bar{y}_{i.} - \bar{y}_{..}) y_{i.} + \sum_{j=1}^{b} (\bar{y}_{.j} - \bar{y}_{..}) y_{.j} \\ &= \frac{y_{..}^2}{ab} + \sum_{i=1}^{a} \bar{y}_{i.} y_{i.} - \frac{y_{..}^2}{ab} + \sum_{j=1}^{b} \bar{y}_{.j} y_{.j} - \frac{y_{..}^2}{ab} \\ &= \sum_{i=1}^{a} \frac{y_{i.}^2}{b} + \sum_{j=1}^{b} \frac{y_{.j}^2}{a} - \frac{y_{..}^2}{ab} \end{aligned}$$

with $a + b - 1$ degrees of freedom, and the error sum of squares is

$$\begin{aligned} SS_E &= \sum_{i=1}^{a} \sum_{j=1}^{b} y_{ij}^2 - R(\mu, \tau, \beta) \\ &= \sum_{i=1}^{a} \sum_{j=1}^{b} y_{ij}^2 - \sum_{i=1}^{a} \frac{y_{i.}^2}{b} - \sum_{j=1}^{b} \frac{y_{.j}^2}{a} + \frac{y_{..}^2}{ab} \\ &= \sum_{i=1}^{a} \sum_{j=1}^{b} (y_{ij} - \bar{y}_{i.} - \bar{y}_{.j} + \bar{y}_{..})^2 \end{aligned}$$

with $(a - 1)(b - 1)$ degrees of freedom. Compare this last equation with SS_E in Equation 4.7.

To test the hypothesis H_0: $\tau_i = 0$, the **reduced model** is

$$y_{ij} = \mu + \beta_j + \epsilon_{ij}$$

which is just a single-factor analysis of variance. By analogy with Equation 3.5, the reduction in the sum of squares for fitting the reduced model is

$$R(\mu, \beta) = \sum_{j=1}^{b} \frac{y_{.j}^2}{a}$$

which has b degrees of freedom. Therefore, the sum of squares due to $\{\tau_i\}$ after fitting μ and $\{\beta_j\}$ is

$$\begin{aligned} R(\tau | \mu, \beta) &= R(\mu, \tau, \beta) - R(\mu, \beta) \\ &= R(\text{full model}) - R(\text{reduced model}) \\ &= \sum_{i=1}^{a} \frac{y_{i.}^2}{b} + \sum_{j=1}^{b} \frac{y_{.j}^2}{a} - \frac{y_{..}^2}{ab} - \sum_{j=1}^{b} \frac{y_{.j}^2}{a} \\ &= \sum_{i=1}^{a} \frac{y_{i.}^2}{b} - \frac{y_{..}^2}{ab} \end{aligned}$$

which we recognize as the treatment sum of squares with $a - 1$ degrees of freedom (Equation 4.10).

The block sum of squares is obtained by fitting the **reduced model**

$$y_{ij} = \mu + \tau_i + \epsilon_{ij}$$

which is also a single-factor analysis. Again, by analogy with Equation 3.5, the reduction in the sum of squares for fitting this model is

$$R(\mu, \tau) = \sum_{i=1}^{a} \frac{y_{i.}^2}{b}$$

with a degrees of freedom. The sum of squares for blocks $\{\beta_j\}$ after fitting μ and $\{\tau_i\}$ is

$$\begin{aligned} R(\beta|\mu, \tau) &= R(\mu, \tau, \beta) - R(\mu, \tau) \\ &= \sum_{i=1}^{a} \frac{y_{i.}^2}{b} + \sum_{j=1}^{b} \frac{y_{.j}^2}{a} - \frac{y_{..}^2}{ab} - \sum_{i=1}^{a} \frac{y_{i.}^2}{b} \\ &= \sum_{j=1}^{b} \frac{y_{.j}^2}{a} - \frac{y_{..}^2}{ab} \end{aligned}$$

with $b - 1$ degrees of freedom, which we have given previously as Equation 4.11.

We have developed the sums of squares for treatments, blocks, and error in the randomized complete block design using the general regression significance test. Although we would not ordinarily use the general regression significance test to actually analyze data in a randomized complete block, the procedure occasionally proves useful in more general randomized block designs, such as those discussed in Section 4.4.

Exact Analysis of the Missing Value Problem. In Section 4.1.3 an approximate procedure for dealing with missing observations in the RCBD was presented. This approximate analysis consists of estimating the missing value so that the error mean square is minimized. It can be shown that the approximate analysis produces a biased mean square for treatments in the sense that $E(MS_{\text{Treatments}})$ is larger than $E(MS_E)$ if the null hypothesis is true. Consequently, too many significant results are reported.

The missing value problem may be analyzed exactly by using the general regression significance test. The missing value causes the design to be **unbalanced**, and because all the treatments do not occur in all blocks, we say that the treatments and blocks are not **orthogonal**. This method of analysis is also used in more general types of randomized block designs; it is discussed further in Section 4.4. Many computer packages will perform this analysis.

4.2 The Latin Square Design

In Section 4.1 we introduced the randomized complete block design as a design to reduce the residual error in an experiment by removing variability due to a known and controllable nuisance variable. There are several other types of designs that utilize the blocking principle. For example, suppose that an experimenter is studying the effects of five different formulations of a rocket propellant used in aircrew escape systems on the observed burning rate. Each formulation is mixed from a batch of raw material that is only large enough for five formulations to be tested. Furthermore, the formulations are prepared by several operators, and there may be substantial differences in the skills and experience of the operators. Thus, it would seem that there are two nuisance factors to be "averaged out" in the design: batches of raw material and operators. The appropriate design for this problem consists of testing each formulation exactly once in each batch of raw material and for each formulation to be prepared exactly once by each of five operators. The resulting design, shown in Table 4.9, is called a **Latin square design**. Notice that the design is a square arrangement and that the five formulations (or treatments) are denoted by the Latin letters A, B, C, D, and E; hence the name Latin square.

■ **TABLE 4.9**
Latin Square Design for the Rocket Propellant Problem

Batches of Raw Material	Operators 1	2	3	4	5
1	$A = 24$	$B = 20$	$C = 19$	$D = 24$	$E = 24$
2	$B = 17$	$C = 24$	$D = 30$	$E = 27$	$A = 36$
3	$C = 18$	$D = 38$	$E = 26$	$A = 27$	$B = 21$
4	$D = 26$	$E = 31$	$A = 26$	$B = 23$	$C = 22$
5	$E = 22$	$A = 30$	$B = 20$	$C = 29$	$D = 31$

We see that both batches of raw material (rows) and operators (columns) are orthogonal to treatments.

The Latin square design is used to eliminate two nuisance sources of variability; that is, it systematically allows blocking in two directions. Thus, the rows and columns actually represent **two restrictions on randomization**. In general, a Latin square for p factors, or a $p \times p$ Latin square, is a square containing p rows and p columns. Each of the resulting p^2 cells contains one of the p letters that corresponds to the treatments, and each letter occurs once and only once in each row and column. Some examples of Latin squares are

4 × 4	5 × 5	6 × 6
A B D C	*A D B E C*	*A D C E B F*
B C A D	*D A C B E*	*B A E C F D*
C D B A	*C B E D A*	*C E D F A B*
D A C B	*B E A C D*	*D C F B E A*
	E C D A B	*F B A D C E*
		E F B A D C

Latin squares are closely related to a popular puzzle called a sudoku puzzle that originated in Japan (sudoku means "single number" in Japanese). The puzzle typically consists of a 9×9 grid, with nine additional 3×3 blocks contained within. A few of the spaces contain numbers and the others are blank. The goal is to fill the blanks with the integers from 1 to 9 so that each row, each column, and each of the nine 3×3 blocks making up the grid contains just one of each of the nine integers. The additional constraint that a standard 9×9 sudoku puzzle have 3×3 blocks that also contain each of the nine integers reduces the large number of possible 9×9 Latin squares to a smaller but still quite large number, approximately 6×10^{21}.

Depending on the number of clues and the size of the grid, sudoku puzzles can be extremely difficult to solve. Solving an $n \times n$ sudoku puzzle belongs to a class of computational problems called *NP*-complete (the *NP* refers to non-polynomial computing time). An *NP*-complete problem is one for which it's relatively easy to check whether a particular answer is correct but may require an impossibly long time to solve by any simple algorithm as n gets larger.

Solving a sudoku puzzle is also equivalent to "coloring" a graph—an array of points (vertices) and lines (edges) in a particular way. In this case, the graph has 81 vertices, one for each cell of the grid. Depending on the puzzle, only certain pairs of vertices are joined by an edge. Given that some vertices have already been assigned a "color" (chosen from the nine number possibilities), the problem is to "color" the remaining vertices so that any two vertices joined by an edge don't have the same "color."

The **statistical model** for a Latin square is

$$y_{ijk} = \mu + \alpha_i + \tau_j + \beta_k + \epsilon_{ijk} \begin{cases} i = 1, 2, \ldots, p \\ j = 1, 2, \ldots, p \\ k = 1, 2, \ldots, p \end{cases} \quad \textbf{(4.27)}$$

where y_{ijk} is the observation in the ith row and kth column for the jth treatment, μ is the overall mean, α_i is the ith row effect, τ_j is the jth treatment effect, β_k is the kth column effect, and ϵ_{ijk} is the random error. Note that this is an **effects model.** The model is completely **additive**; that is, there is no interaction between rows, columns, and treatments. Because there is only one observation in each cell, only two of the three subscripts i, j, and k are needed to denote a particular observation. For example, referring to the rocket propellant problem in Table 4.8, if $i = 2$ and $k = 3$, we automatically find $j = 4$ (formulation D), and if $i = 1$ and $j = 3$ (formulation C), we find $k = 3$. This is a consequence of each treatment appearing exactly once in each row and column.

The analysis of variance consists of partitioning the total sum of squares of the $N = p^2$ observations into components for rows, columns, treatments, and error, for example,

$$SS_T = SS_{\text{Rows}} + SS_{\text{Columns}} + SS_{\text{Treatments}} + SS_E \quad \textbf{(4.28)}$$

with respective degrees of freedom

$$p^2 - 1 = p - 1 + p - 1 + p - 1 + (p - 2)(p - 1)$$

Under the usual assumption that ϵ_{ijk} is NID $(0, \sigma^2)$, each sum of squares on the right-hand side of Equation 4.28 is, upon division by σ^2, an independently distributed chi-square random variable. The appropriate statistic for testing for no differences in treatment means is

$$F_0 = \frac{MS_{\text{Treatments}}}{MS_E}$$

which is distributed as $F_{p-1,(p-2)(p-1)}$ under the null hypothesis. We may also test for no row effect and no column effect by forming the ratio of MS_{Rows} or MS_{Columns} to MS_E. However, because the rows and columns represent restrictions on randomization, these tests may not be appropriate.

The computational procedure for the ANOVA in terms of treatment, row, and column totals is shown in Table 4.10. From the computational formulas for the sums of squares, we see that the analysis is a simple extension of the RCBD, with the sum of squares resulting from rows obtained from the row totals.

■ **TABLE 4.10**
Analysis of Variance for the Latin Square Design

Source of Variation	Sum of Squares	Degrees of Freedom	Mean Square	F_0
Treatments	$SS_{\text{Treatments}} = \frac{1}{p}\sum_{j=1}^{p} y_{.j.}^2 - \frac{y_{...}^2}{N}$	$p - 1$	$\frac{SS_{\text{Treatments}}}{p-1}$	$F_0 = \frac{MS_{\text{Treatments}}}{MS_E}$
Rows	$SS_{\text{Rows}} = \frac{1}{p}\sum_{i=1}^{p} y_{i..}^2 - \frac{y_{...}^2}{N}$	$p - 1$	$\frac{SS_{\text{Rows}}}{p-1}$	
Columns	$SS_{\text{Columns}} = \frac{1}{p}\sum_{k=1}^{p} y_{..k}^2 - \frac{y_{...}^2}{N}$	$p - 1$	$\frac{SS_{\text{Columns}}}{p-1}$	
Error	SS_E (by subtraction)	$(p-2)(p-1)$	$\frac{SS_E}{(p-2)(p-1)}$	
Total	$SS_T = \sum_i\sum_j\sum_k y_{ijk}^2 - \frac{y_{...}^2}{N}$	$p^2 - 1$		

EXAMPLE 4.3

Consider the rocket propellant problem previously described, where both batches of raw material and operators represent randomization restrictions. The design for this experiment, shown in Table 4.8, is a 5 × 5 Latin square. After coding by subtracting 25 from each observation, we have the data in Table 4.11. The sums of squares for the total, batches (rows), and operators (columns) are computed as follows:

$$SS_T = \sum_i \sum_j \sum_k y_{ijk}^2 - \frac{y_{...}^2}{N}$$

$$= 680 - \frac{(10)^2}{25} = 676.00$$

$$SS_{\text{Batches}} = \frac{1}{p}\sum_{i=1}^{p} y_{i..}^2 - \frac{y_{...}^2}{N}$$

$$= \frac{1}{5}[(-14)^2 + 9^2 + 5^2 + 3^2 + 7^2] - \frac{(10)^2}{25} = 68.00$$

$$SS_{\text{Operators}} = \frac{1}{p}\sum_{k=1}^{p} y_{..k}^2 - \frac{y_{...}^2}{N}$$

$$= \frac{1}{5}[(-18)^2 + 18^2 + (-4)^2 + 5^2 + 9^2] - \frac{(10)^2}{25} = 150.00$$

The totals for the treatments (Latin letters) are

Latin Letter	Treatment Total
A	$y_{.1.} = 18$
B	$y_{.2.} = -24$
C	$y_{.3.} = -13$
D	$y_{.4.} = 24$
E	$y_{.5.} = 5$

The sum of squares resulting from the formulations is computed from these totals as

$$SS_{\text{Formulations}} = \frac{1}{p}\sum_{j=1}^{p} y_{.j.}^2 - \frac{y_{...}^2}{N}$$

$$= \frac{18^2 + (-24)^2 + (-13)^2 + 24^2 + 5^2}{5} - \frac{(10)^2}{25} = 330.00$$

The error sum of squares is found by subtraction

$$SS_E = SS_T - SS_{\text{Batches}} - SS_{\text{Operators}} - SS_{\text{Formulations}}$$
$$= 676.00 - 68.00 - 150.00 - 330.00 = 128.00$$

The analysis of variance is summarized in Table 4.12. We conclude that there is a significant difference in the mean burning rate generated by the different rocket propellant formulations. There is also an indication that differences between operators exist, so blocking on this factor was a good precaution. There is no strong evidence of a difference between batches of raw material, so it seems that in this particular experiment we were unnecessarily concerned about this source of variability. However, blocking on batches of raw material is usually a good idea.

■ **TABLE 4.11**
Coded Data for the Rocket Propellant Problem

Batches of Raw Material	Operators 1	2	3	4	5	$y_{i..}$
1	$A = -1$	$B = -5$	$C = -6$	$D = -1$	$E = -1$	-14
2	$B = -8$	$C = -1$	$D = 5$	$E = 2$	$A = 11$	9
3	$C = -7$	$D = 13$	$E = 1$	$A = 2$	$B = -4$	5
4	$D = 1$	$E = 6$	$A = 1$	$B = -2$	$C = -3$	3
5	$E = -3$	$A = 5$	$B = -5$	$C = 4$	$D = 6$	7
$y_{..k}$	-18	18	-4	5	9	$10 = y_{...}$

■ **TABLE 4.12**
Analysis of Variance for the Rocket Propellant Experiment

Source of Variation	Sum of Squares	Degrees of Freedom	Mean Square	F_0	*P*-Value
Formulations	330.00	4	82.50	7.73	0.0025
Batches of raw material	68.00	4	17.00		
Operators	150.00	4	37.50		
Error	128.00	12	10.67		
Total	676.00	24			

As in any design problem, the experimenter should investigate the adequacy of the model by inspecting and plotting the residuals. For a Latin square, the residuals are given by

$$\begin{aligned} e_{ijk} &= y_{ijk} - \hat{y}_{ijk} \\ &= y_{ijk} - \bar{y}_{i..} - \bar{y}_{.j.} - \bar{y}_{..k} + 2\bar{y}_{...} \end{aligned}$$

The reader should find the residuals for Example 4.3 and construct appropriate plots.

A Latin square in which the first row and column consists of the letters written in alphabetical order is called a **standard Latin square**, which is the design shown in Example 4.4. A standard Latin square can always be obtained by writing the first row in alphabetical order and then writing each successive row as the row of letters just above shifted one place to the left. Table 4.13 summarizes several important facts about Latin squares and standard Latin squares.

As with any experimental design, the observations in the Latin square should be taken in random order. The proper randomization procedure is to select the particular square employed at random. As we see in Table 4.13, there are a large number of Latin squares of a particular size, so it is impossible to enumerate all the squares and select one randomly. The usual procedure is

■ **TABLE 4.13**
Standard Latin Squares and Number of Latin Squares of Various Sizes[a]

Size	3 × 3	4 × 4	5 × 5	6 × 6	7 × 7	*p* × *p*
Examples of standard squares	*A B C*	*A B C D*	*A B C D E*	*A B C D E F*	*A B C D E F G*	*ABC . . . P*
	B C A	*B C D A*	*B A E C D*	*B C F A D E*	*B C D E F G A*	*BCD . . . A*
	C A B	*C D A B*	*C D A E B*	*C F B E A D*	*C D E F G A B*	*CDE . . . B*
		D A B C	*D E B A C*	*D E A B F C*	*D E F G A B C*	⋮
			E C D B A	*E A D F C B*	*E F G A B C D*	
				F D E C B A	*F G A B C D E*	*PAB . . .* $(P-1)$
					G A B C D E F	
Number of standard squares	1	4	56	9408	16,942,080	—
Total number of Latin squares	12	576	161,280	818,851,200	61,479,419,904,000	$p!(p-1)! \times$ (number of standard squares)

[a]Some of the information in this table is found in Fisher and Yates (1953). Little is known about the properties of Latin squares larger than 7 × 7.

to select an arbitrary Latin square from a table of such designs, as in Fisher and Yates (1953), or start with a standard square, and then arrange the order of the rows, columns, and letters at random. This is discussed more completely in Fisher and Yates (1953).

Occasionally, one observation in a Latin square is missing. For a $p \times p$ Latin square, the missing value may be estimated by

$$y_{ijk} = \frac{p(y'_{i..} + y'_{.j.} + y'_{..k}) - 2y'_{...}}{(p-2)(p-1)} \tag{4.29}$$

where the primes indicate totals for the row, column, and treatment with the missing value, and $y'_{...}$ is the grand total with the missing value.

Latin squares can be useful in situations where the rows and columns represent factors the experimenter actually wishes to study and where there are no randomization restrictions. Thus, three factors (rows, columns, and letters), each at p levels, can be investigated in only p^2 runs. This design assumes that there is no interaction between the factors. More will be said later on the subject of interaction.

Replication of Latin Squares. A disadvantage of small Latin squares is that they provide a relatively small number of error degrees of freedom. For example, a 3×3 Latin square has only two error degrees of freedom, a 4×4 Latin square has only six error degrees of freedom, and so forth. When small Latin squares are used, it is frequently desirable to replicate them to increase the error degrees of freedom.

A Latin square may be replicated in several ways. To illustrate, suppose that the 5×5 Latin square used in Example 4.4 is replicated n times. This could have been done as follows:

1. Use the same batches and operators in each replicate.
2. Use the same batches but different operators in each replicate (or, equivalently, use the same operators but different batches).
3. Use different batches and different operators.

The analysis of variance depends on the method of replication.

Consider case 1, where the same levels of the row and column blocking factors are used in each replicate. Let y_{ijkl} be the observation in row i, treatment j, column k, and replicate l. There are $N = np^2$ total observations. The ANOVA is summarized in Table 4.14.

■ TABLE 4.14
Analysis of Variance for a Replicated Latin Square, Case 1

Source of Variation	Sum of Squares	Degrees of Freedom	Mean Square	F_0
Treatments	$\frac{1}{np}\sum_{j=1}^{p} y_{.j..}^2 - \frac{y_{....}^2}{N}$	$p-1$	$\frac{SS_{\text{Treatments}}}{p-1}$	$\frac{MS_{\text{Treatments}}}{MS_E}$
Rows	$\frac{1}{np}\sum_{i=1}^{p} y_{i...}^2 - \frac{y_{....}^2}{N}$	$p-1$	$\frac{SS_{\text{Rows}}}{p-1}$	
Columns	$\frac{1}{np}\sum_{k=1}^{p} y_{..k.}^2 - \frac{y_{....}^2}{N}$	$p-1$	$\frac{SS_{\text{Columns}}}{p-1}$	
Replicates	$\frac{1}{p^2}\sum_{l=1}^{n} y_{...l}^2 - \frac{y_{....}^2}{N}$	$n-1$	$\frac{SS_{\text{Replicates}}}{n-1}$	
Error	Subtraction	$(p-1)[n(p+1)-3]$	$\frac{SS_E}{(p-1)[n(p+1)-3]}$	
Total	$\sum\sum\sum\sum y_{ijkl}^2 - \frac{y_{....}^2}{N}$	np^2-1		

■ **TABLE 4.15**
Analysis of Variance for a Replicated Latin Square, Case 2

Source of Variation	Sum of Squares	Degrees of Freedom	Mean Square	F_0
Treatments	$\frac{1}{np}\sum_{j=1}^{p} y_{.j..}^2 - \frac{y_{....}^2}{N}$	$p-1$	$\frac{SS_{\text{Treatments}}}{p-1}$	$\frac{MS_{\text{Treatments}}}{MS_E}$
Rows	$\frac{1}{p}\sum_{l=1}^{n}\sum_{i=1}^{p} y_{i..l}^2 - \sum_{l=1}^{n}\frac{y_{...l}^2}{p^2}$	$n(p-1)$	$\frac{SS_{\text{Rows}}}{n(p-1)}$	
Columns	$\frac{1}{np}\sum_{k=1}^{p} y_{..k.}^2 - \frac{y_{....}^2}{N}$	$p-1$	$\frac{SS_{\text{Columns}}}{p-1}$	
Replicates	$\frac{1}{p^2}\sum_{l=1}^{n} y_{...l}^2 - \frac{y_{....}^2}{N}$	$n-1$	$\frac{SS_{\text{Replicates}}}{n-1}$	
Error	Subtraction	$(p-1)(np-1)$	$\frac{SS_E}{(p-1)(np-1)}$	
Total	$\sum_i\sum_j\sum_k\sum_l y_{ijkl}^2 - \frac{y_{....}^2}{N}$	np^2-1		

Now consider case 2 and assume that new batches of raw material but the same operators are used in each replicate. Thus, there are now five new rows (in general, p new rows) within each replicate. The ANOVA is summarized in Table 4.15. Note that the source of variation for the rows really measures the variation between rows within the n replicates.

Finally, consider case 3, where new batches of raw material and new operators are used in each replicate. Now the variation that results from both the rows and columns measures the variation resulting from these factors within the replicates. The ANOVA is summarized in Table 4.16.

There are other approaches to analyzing replicated Latin squares that allow some interactions between treatments and squares (refer to Problem 4.30).

Crossover Designs and Designs Balanced for Residual Effects. Occasionally, one encounters a problem in which time periods are a factor in the experiment. In general, there are p treatments to be tested in p time periods using np experimental units. For example, a human performance analyst is studying the effect of two replacement fluids on dehydration

■ **TABLE 4.16**
Analysis of Variance for a Replicated Latin Square, Case 3

Source of Variation	Sum of Squares	Degrees of Freedom	Mean Square	F_0
Treatments	$\frac{1}{np}\sum_{j=1}^{p} y_{.j..}^2 - \frac{y_{....}^2}{N}$	$p-1$	$\frac{SS_{\text{Treatments}}}{p-1}$	$\frac{MS_{\text{Treatments}}}{MS_E}$
Rows	$\frac{1}{p}\sum_{l=1}^{n}\sum_{i=1}^{p} y_{i..l}^2 - \sum_{l=1}^{n}\frac{y_{...l}^2}{p^2}$	$n(p-1)$	$\frac{SS_{\text{Rows}}}{n(p-1)}$	
Columns	$\frac{1}{p}\sum_{l=1}^{n}\sum_{k=1}^{p} y_{..kl}^2 - \sum_{l=1}^{n}\frac{y_{...l}^2}{p^2}$	$n(p-1)$	$\frac{SS_{\text{Columns}}}{n(p-1)}$	
Replicates	$\frac{1}{p^2}\sum_{l=1}^{n} y_{...l}^2 - \frac{y_{....}^2}{N}$	$n-1$	$\frac{SS_{\text{Replicates}}}{n-1}$	
Error	Subtraction	$(p-1)[n(p-1)-1]$	$\frac{SS_E}{(p-1)[n(p-1)-1]}$	
Total	$\sum_i\sum_j\sum_k\sum_l y_{ijkl}^2 - \frac{y_{....}^2}{N}$	np^2-1		

	Latin Squares																			
	I		II		III		IV		V		VI		VII		VIII		IX		X	
Subject	1	2	3	4	5	6	7	8	9	10	11	12	13	14	15	16	17	18	19	20
Period 1	*A*	*B*	*B*	*A*	*B*	*A*	*A*	*B*	*A*	*B*	*B*	*A*	*A*	*B*	*A*	*B*	*A*	*B*	*A*	*B*
Period 2	*B*	*A*	*A*	*B*	*A*	*B*	*B*	*A*	*B*	*A*	*A*	*B*	*B*	*A*	*B*	*A*	*B*	*A*	*B*	*A*

FIGURE 4.7 A crossover design

TABLE 4.17
Analysis of Variance for the Crossover Design in Figure 4.7

Source of Variation	Degrees of Freedom
Subjects (columns)	19
Periods (rows)	1
Fluids (letters)	1
Error	18
Total	39

in 20 subjects. In the first period, half of the subjects (chosen at random) are given fluid *A* and the other half fluid *B*. At the end of the period, the response is measured and a period of time is allowed to pass in which any physiological effect of the fluids is eliminated. Then the experimenter has the subjects who took fluid *A* take fluid *B* and those who took fluid *B* take fluid *A*. This design is called a **crossover design**. It is analyzed as a set of 10 Latin squares with two rows (time periods) and two treatments (fluid types). The two columns in each of the 10 squares correspond to subjects.

The layout of this design is shown in Figure 4.7. Notice that the rows in the Latin square represent the time periods and the columns represent the subjects. The 10 subjects who received fluid *A* first (1, 4, 6, 7, 9, 12, 13, 15, 17, and 19) are randomly determined.

An abbreviated analysis of variance is summarized in Table 4.17. The subject sum of squares is computed as the corrected sum of squares among the 20 subject totals, the period sum of squares is the corrected sum of squares among the rows, and the fluid sum of squares is computed as the corrected sum of squares among the letter totals. For further details of the statistical analysis of these designs see Cochran and Cox (1957), John (1971), and Anderson and McLean (1974).

It is also possible to employ Latin square type designs for experiments in which the treatments have a **residual effect**—that is, for example, if the data for fluid *B* in period 2 still reflected some effect of fluid *A* taken in period 1. Designs balanced for residual effects are discussed in detail by Cochran and Cox (1957) and John (1971).

4.3 The Graeco-Latin Square Design

Consider a $p \times p$ Latin square, and superimpose on it a second $p \times p$ Latin square in which the treatments are denoted by Greek letters. If the two squares when superimposed have the property that each Greek letter appears once and only once with each Latin letter, the two Latin squares are said to be **orthogonal**, and the design obtained is called a **Graeco-Latin square**. An example of a 4×4 Graeco-Latin square is shown in Table 4.18.

■ **TABLE 4.18**
4 × 4 Graeco-Latin Square Design

	Column			
Row	**1**	**2**	**3**	**4**
1	$A\alpha$	$B\beta$	$C\gamma$	$D\delta$
2	$B\delta$	$A\gamma$	$D\beta$	$C\alpha$
3	$C\beta$	$D\alpha$	$A\delta$	$B\gamma$
4	$D\gamma$	$C\delta$	$B\alpha$	$A\beta$

The Graeco-Latin square design can be used to control systematically three sources of extraneous variability, that is, to block in *three* directions. The design allows investigation of four factors (rows, columns, Latin letters, and Greek letters), each at p levels in only p^2 runs. Graeco-Latin squares exist for all $p \geq 3$ except $p = 6$.

The statistical model for the Graeco-Latin square design is

$$y_{ijkl} = \mu + \theta_i + \tau_j + \omega_k + \Psi_l + \epsilon_{ijkl} \begin{cases} i = 1, 2, \ldots, p \\ j = 1, 2, \ldots, p \\ k = 1, 2, \ldots, p \\ l = 1, 2, \ldots, p \end{cases} \tag{4.30}$$

where y_{ijkl} is the observation in row i and column l for Latin letter j and Greek letter k, θ_i is the effect of the ith row, τ_j is the effect of Latin letter treatment j, ω_k is the effect of Greek letter treatment k, Ψ_l is the effect of column l, and ϵ_{ijkl} is an NID $(0, \sigma^2)$ random error component. Only two of the four subscripts are necessary to completely identify an observation.

The analysis of variance is very similar to that of a Latin square. Because the Greek letters appear exactly once in each row and column and exactly once with each Latin letter, the factor represented by the Greek letters is orthogonal to rows, columns, and Latin letter treatments. Therefore, a sum of squares due to the Greek letter factor may be computed from the Greek letter totals, and the experimental error is further reduced by this amount. The computational details are illustrated in Table 4.19. The null hypotheses of equal row, column, Latin letter, and Greek letter treatments would be tested by dividing the corresponding mean square by mean square error. The rejection region is the upper tail point of the $F_{p-1,(p-3)(p-1)}$ distribution.

■ **TABLE 4.19**
Analysis of Variance for a Graeco-Latin Square Design

Source of Variation	Sum of Squares	Degrees of Freedom
Latin letter treatments	$SS_L = \frac{1}{p}\sum_{j=1}^{p} y_{.j..}^2 - \frac{y_{....}^2}{N}$	$p - 1$
Greek letter treatments	$SS_G = \frac{1}{p}\sum_{k=1}^{p} y_{..k.}^2 - \frac{y_{....}^2}{N}$	$p - 1$
Rows	$SS_{\text{Rows}} = \frac{1}{p}\sum_{i=1}^{p} y_{i...}^2 - \frac{y_{....}^2}{N}$	$p - 1$
Columns	$SS_{\text{Columns}} = \frac{1}{p}\sum_{l=1}^{p} y_{...l}^2 - \frac{y_{....}^2}{N}$	$p - 1$
Error	SS_E (by subtraction)	$(p - 3)(p - 1)$
Total	$SS_T = \sum_i\sum_j\sum_k\sum_l y_{ijkl}^2 - \frac{y_{....}^2}{N}$	$p^2 - 1$

EXAMPLE 4.4

Suppose that in the rocket propellant experiment of Example 4.3 an additional factor, test assemblies, could be of importance. Let there be five test assemblies denoted by the Greek letters α, β, γ, δ, and ϵ. The resulting 5×5 Graeco-Latin square design is shown in Table 4.20.

Notice that, because the totals for batches of raw material (rows), operators (columns), and formulations (Latin letters) are identical to those in Example 4.3, we have

$$SS_{\text{Batches}} = 68.00, \quad SS_{\text{Operators}} = 150.00, \quad \text{and} \quad SS_{\text{Formulations}} = 330.00$$

The totals for the test assemblies (Greek letters) are

Greek Letter	Test Assembly Total
α	$y_{..1.} = 10$
β	$y_{..2.} = -6$
γ	$y_{..3.} = -3$
δ	$y_{..4.} = -4$
ϵ	$y_{..5.} = 13$

Thus, the sum of squares due to the test assemblies is

$$\begin{aligned} SS_{\text{Assemblies}} &= \frac{1}{p}\sum_{k=1}^{p} y_{..k.}^2 - \frac{y_{....}^2}{N} \\ &= \frac{1}{5}[10^2 + (-6)^2 + (-3)^2 \\ &\quad + (-4)^2 + 13^2] - \frac{(10)^2}{25} = 62.00 \end{aligned}$$

The complete ANOVA is summarized in Table 4.21. Formulations are significantly different at 1 percent. In comparing Tables 4.21 and 4.12, we observe that removing the variability due to test assemblies has decreased the experimental error. However, in decreasing the experimental error, we have also reduced the error degrees of freedom from 12 (in the Latin square design of Example 4.3) to 8. Thus, our estimate of error has fewer degrees of freedom, and the test may be less sensitive.

TABLE 4.20
Graeco-Latin Square Design for the Rocket Propellant Problem

Batches of Raw Material	Operators 1	2	3	4	5	$y_{i...}$
1	$A\alpha = -1$	$B\gamma = -5$	$C\epsilon = -6$	$D\beta = -1$	$E\delta = -1$	−14
2	$B\beta = -8$	$C\delta = -1$	$D\alpha = 5$	$E\gamma = 2$	$A\epsilon = 11$	9
3	$C\gamma = -7$	$D\epsilon = 13$	$E\beta = 1$	$A\delta = 2$	$B\alpha = -4$	5
4	$D\delta = 1$	$E\alpha = 6$	$A\gamma = 1$	$B\epsilon = -2$	$C\beta = -3$	3
5	$E\epsilon = -3$	$A\beta = 5$	$B\delta = -5$	$C\alpha = 4$	$D\gamma = 6$	7
$y_{...l}$	−18	18	−4	5	9	$10 = y_{....}$

TABLE 4.21
Analysis of Variance for the Rocket Propellant Problem

Source of Variation	Sum of Squares	Degrees of Freedom	Mean Square	F_0	P-Value
Formulations	330.00	4	82.50	10.00	0.0033
Batches of raw material	68.00	4	17.00		
Operators	150.00	4	37.50		
Test assemblies	62.00	4	15.50		
Error	66.00	8	8.25		
Total	676.00	24			

The concept of orthogonal pairs of Latin squares forming a Graeco-Latin square can be extended somewhat. A $p \times p$ **hypersquare** is a design in which three or more orthogonal $p \times p$ Latin squares are superimposed. In general, up to $p + 1$ factors could be studied if a complete set of $p - 1$ orthogonal Latin squares is available. Such a design would utilize all $(p + 1)(p - 1) = p^2 - 1$ degrees of freedom, so an independent estimate of the error variance is necessary. Of course, there must be no interactions between the factors when using hypersquares.

4.4 Balanced Incomplete Block Designs

In certain experiments using randomized block designs, we may not be able to run all the treatment combinations in each block. Situations like this usually occur because of shortages of experimental apparatus or facilities or the physical size of the block. For example, in the vascular graft experiment (Example 4.1), suppose that each batch of material is only large enough to accommodate testing three extrusion pressures. Therefore, each pressure cannot be tested in each batch. For this type of problem it is possible to use randomized block designs in which every treatment is not present in every block. These designs are known as **randomized incomplete block designs**.

When all treatment comparisons are equally important, the treatment combinations used in each block should be selected in a balanced manner, so that any pair of treatments occur together the same number of times as any other pair. Thus, a **balanced incomplete block design (BIBD)** is an incomplete block design in which any two treatments appear together an equal number of times. Suppose that there are a treatments and that each block can hold exactly k $(k < a)$ treatments. A balanced incomplete block design may be constructed by taking $\binom{a}{k}$ blocks and assigning a different combination of treatments to each block. Frequently, however, balance can be obtained with fewer than $\binom{a}{k}$ blocks. Tables of BIBDs are given in Fisher and Yates (1953), Davies (1956), and Cochran and Cox (1957).

As an example, suppose that a chemical engineer thinks that the time of reaction for a chemical process is a function of the type of catalyst employed. Four catalysts are currently being investigated. The experimental procedure consists of selecting a batch of raw material, loading the pilot plant, applying each catalyst in a separate run of the pilot plant, and observing the reaction time. Because variations in the batches of raw material may affect the performance of the catalysts, the engineer decides to use batches of raw material as blocks. However, each batch is only large enough to permit three catalysts to be run. Therefore, a randomized incomplete block design must be used. The balanced incomplete block design for this experiment, along with the observations recorded, is shown in Table 4.22. The order in which the catalysts are run in each block is randomized.

4.4.1 Statistical Analysis of the BIBD

As usual, we assume that there are a treatments and b blocks. In addition, we assume that each block contains k treatments, that each treatment occurs r times in the design (or is replicated

■ **TABLE 4.22**
Balanced Incomplete Block Design for Catalyst Experiment

Treatment	Block (Batch of Raw Material)				
(Catalyst)	1	2	3	4	$y_{i.}$
1	73	74	—	71	218
2	—	75	67	72	214
3	73	75	68	—	216
4	75	—	72	75	222
$y_{.j}$	221	224	207	218	$870 = y_{..}$

r times), and that there are $N = ar = bk$ total observations. Furthermore, the number of times each pair of treatments appears in the same block is

$$\lambda = \frac{r(k-1)}{a-1}$$

If $a = b$, the design is said to be **symmetric**.

The parameter λ must be an integer. To derive the relationship for λ, consider any treatment, say treatment 1. Because treatment 1 appears in r blocks and there are $k - 1$ other treatments in each of those blocks, there are $r(k - 1)$ observations in a block containing treatment 1. These $r(k - 1)$ observations also have to represent the remaining $a - 1$ treatments λ times. Therefore, $\lambda(a - 1) = r(k - 1)$.

The **statistical model** for the BIBD is

$$y_{ij} = \mu + \tau_i + \beta_j + \epsilon_{ij} \tag{4.31}$$

where y_{ij} is the ith observation in the jth block, μ is the overall mean, τ_i is the effect of the ith treatment, β_j is the effect of the jth block, and ϵ_{ij} is the NID $(0, \sigma^2)$ random error component. The total variability in the data is expressed by the total corrected sum of squares:

$$SS_T = \sum_i \sum_j y_{ij}^2 - \frac{y_{..}^2}{N} \tag{4.32}$$

Total variability may be partitioned into

$$SS_T = SS_{\text{Treatments(adjusted)}} + SS_{\text{Blocks}} + SS_E$$

where the sum of squares for treatments is **adjusted** to separate the treatment and the block effects. This adjustment is necessary because each treatment is represented in a different set of r blocks. Thus, differences between unadjusted treatment totals $y_{1.}, y_{2.}, \ldots, y_{a.}$ are also affected by differences between blocks.

The block sum of squares is

$$SS_{\text{Blocks}} = \frac{1}{k}\sum_{j=1}^{b} y_{.j}^2 - \frac{y_{..}^2}{N} \tag{4.33}$$

where $y_{.j}$ is the total in the jth block. SS_{Blocks} has $b - 1$ degrees of freedom. The adjusted treatment sum of squares is

$$SS_{\text{Treatments(adjusted)}} = \frac{k\sum_{i=1}^{a} Q_i^2}{\lambda a} \tag{4.34}$$

where Q_i is the adjusted total for the ith treatment, which is computed as

$$Q_i = y_{i.} - \frac{1}{k}\sum_{j=1}^{b} n_{ij} y_{.j} \qquad i = 1, 2, \ldots, a \tag{4.35}$$

with $n_{ij} = 1$ if treatment i appears in block j and $n_{ij} = 0$ otherwise. The adjusted treatment totals will always sum to zero. $SS_{\text{Treatments(adjusted)}}$ has $a - 1$ degrees of freedom. The error sum of squares is computed by subtraction as

$$SS_E = SS_T - SS_{\text{Treatments(adjusted)}} - SS_{\text{Blocks}} \tag{4.36}$$

and has $N - a - b + 1$ degrees of freedom.

The appropriate statistic for testing the equality of the treatment effects is

$$F_0 = \frac{MS_{\text{Treatments(adjusted)}}}{MS_E}$$

The ANOVA is summarized in Table 4.23.

■ **TABLE 4.23**
Analysis of Variance for the Balanced Incomplete Block Design

Source of Variation	Sum of Squares	Degrees of Freedom	Mean Square	F_0
Treatments (adjusted)	$\dfrac{k\sum Q_i^2}{\lambda a}$	$a-1$	$\dfrac{SS_{\text{Treatments(adjusted)}}}{a-1}$	$F_0 = \dfrac{MS_{\text{Treatments(adjusted)}}}{MS_E}$
Blocks	$\dfrac{1}{k}\sum y_{.j}^2 - \dfrac{y_{..}^2}{N}$	$b-1$	$\dfrac{SS_{\text{Blocks}}}{b-1}$	
Error	SS_E (by subtraction)	$N-a-b+1$	$\dfrac{SS_E}{N-a-b+1}$	
Total	$\sum\sum y_{ij}^2 - \dfrac{y_{..}^2}{N}$	$N-1$		

EXAMPLE 4.5

Consider the data in Table 4.22 for the catalyst experiment. This is a BIBD with $a = 4$, $b = 4$, $k = 3$, $r = 3$, $\lambda = 2$, and $N = 12$. The analysis of this data is as follows. The total sum of squares is

$$SS_T = \sum_i\sum_j y_{ij}^2 - \frac{y_{..}^2}{12}$$

$$= 63{,}156 - \frac{(870)^2}{12} = 81.00$$

The block sum of squares is found from Equation 4.33 as

$$SS_{\text{Blocks}} = \frac{1}{3}\sum_{j=1}^{4} y_{.j}^2 - \frac{y_{..}^2}{12}$$

$$= \frac{1}{3}[(221)^2 + (207)^2 + (224)^2 + (218)^2] - \frac{(870)^2}{12}$$
$$= 55.00$$

To compute the treatment sum of squares adjusted for blocks, we first determine the adjusted treatment totals using Equation 4.35 as

$$Q_1 = (218) - \tfrac{1}{3}(221 + 224 + 218) = -9/3$$
$$Q_2 = (214) - \tfrac{1}{3}(207 + 224 + 218) = -7/3$$
$$Q_3 = (216) - \tfrac{1}{3}(221 + 207 + 224) = -4/3$$
$$Q_4 = (222) - \tfrac{1}{3}(221 + 207 + 218) = 20/3$$

The adjusted sum of squares for treatments is computed from Equation 4.34 as

$$SS_{\text{Treatments(adjusted)}} = \frac{k\sum_{i=1}^{4} Q_i^2}{\lambda a}$$

$$= \frac{3[(-9/3)^2 + (-7/3)^2 + (-4/3)^2 + (20/3)^2]}{(2)(4)}$$

$$= 22.75$$

The error sum of squares is obtained by subtraction as

$$SS_E = SS_T - SS_{\text{Treatments(adjusted)}} - SS_{\text{Blocks}}$$
$$= 81.00 - 22.75 - 55.00 = 3.25$$

The analysis of variance is shown in Table 4.24. Because the P-value is small, we conclude that the catalyst employed has a significant effect on the time of reaction.

■ **TABLE 4.24**
Analysis of Variance for Example 4.5

Source of Variation	Sum of Squares	Degrees of Freedom	Mean Square	F_0	P-Value
Treatments (adjusted for blocks)	22.75	3	7.58	11.66	0.0107
Blocks	55.00	3	—		
Error	3.25	5	0.65		
Total	81.00	11			

If the factor under study is fixed, tests on individual treatment means may be of interest. If orthogonal contrasts are employed, the contrasts must be made on the **adjusted treatment totals**, the $\{Q_i\}$ rather than the $\{y_{i.}\}$. The contrast sum of squares is

$$SS_C = \frac{k\left(\sum_{i=1}^{a} c_i Q_i\right)^2}{\lambda a \sum_{i=1}^{a} c_i^2}$$

where $\{c_i\}$ are the contrast coefficients. Other multiple comparison methods may be used to compare all the pairs of adjusted treatment effects, which we will find in Section 4.4.2, are estimated by $\hat{\tau}_i = kQ_i/(\lambda a)$. The standard error of an adjusted treatment effect is

$$S = \sqrt{\frac{kMS_E}{\lambda a}} \tag{4.37}$$

In the analysis that we have described, the total sum of squares has been partitioned into an adjusted sum of squares for treatments, an unadjusted sum of squares for blocks, and an error sum of squares. Sometimes we would like to assess the block effects. To do this, we require an alternate partitioning of SS_T, that is,

$$SS_T = SS_{\text{Treatments}} + SS_{\text{Blocks(adjusted)}} + SS_E$$

Here $SS_{\text{Treatments}}$ is unadjusted. If the design is symmetric, that is, if $a = b$, a simple formula may be obtained for $SS_{\text{Blocks(adjusted)}}$. The adjusted block totals are

$$Q'_j = y_{.j} - \frac{1}{4}\sum_{i=1}^{a} n_{ij} y_{i.} \quad j = 1, 2, \ldots, b \tag{4.38}$$

and

$$SS_{\text{Blocks(adjusted)}} = \frac{r\sum_{j=1}^{b}(Q'_j)^2}{\lambda b} \tag{4.39}$$

The BIBD in Example 4.5 is symmetric because $a = b = 4$. Therefore,

$$\begin{aligned}
Q'_1 &= (221) - \tfrac{1}{3}(218 + 216 + 222) = 7/3 \\
Q'_2 &= (224) - \tfrac{1}{3}(218 + 214 + 216) = 24/3 \\
Q'_3 &= (207) - \tfrac{1}{3}(214 + 216 + 222) = -31/3 \\
Q'_4 &= (218) - \tfrac{1}{3}(218 + 214 + 222) = 0
\end{aligned}$$

and

$$SS_{\text{Blocks(adjusted)}} = \frac{3[(7/3)^2 + (24/3)^2 + (-31/3)^2 + (0)^2]}{(2)(4)} = 66.08$$

Also,

$$SS_{\text{Treatments}} = \frac{(218)^2 + (214)^2 + (216)^2 + (222)^2}{3} - \frac{(870)^2}{12} = 11.67$$

A summary of the analysis of variance for the symmetric BIBD is given in Table 4.25. Notice that the sums of squares associated with the mean squares in Table 4.25 do not add to the total sum of squares, that is,

$$SS_T \neq SS_{\text{Treatments(adjusted)}} + SS_{\text{Blocks(adjusted)}} + SS_E$$

This is a consequence of the nonorthogonality of treatments and blocks.

■ **TABLE 4.25**
Analysis of Variance for Example 4.5, Including Both Treatments and Blocks

Source of Variation	Sum of Squares	Degrees of Freedom	Mean Square	F_0	*P*-Value
Treatments (adjusted)	22.75	3	7.58	11.66	0.0107
Treatments (unadjusted)	11.67	3			
Blocks (unadjusted)	55.00	3			
Blocks (adjusted)	66.08	3	22.03	33.90	0.0010
Error	3.25	5	0.65		
Total	81.00	11			

Computer Output. There are several computer packages that will perform the analysis for a balanced incomplete block design. The SAS General Linear Models procedure is one of these and Minitab and JMP are others. The upper portion of Table 4.26 is the Minitab General Linear Model output for Example 4.5. Comparing Tables 4.26 and 4.25, we see that Minitab has computed the adjusted treatment sum of squares and the adjusted block sum of squares (they are called "Adj *SS*" in the Minitab output).

The lower portion of Table 4.26 is a multiple comparison analysis, using the Tukey method. Confidence intervals on the differences in all pairs of means and the Tukey test are displayed. Notice that the Tukey method would lead us to conclude that catalyst 4 is different from the other three.

4.4.2 Least Squares Estimation of the Parameters

Consider estimating the treatment effects for the BIBD model. The least squares normal equations are

$$
\begin{aligned}
\mu &: N\hat{\mu} + r\sum_{i=1}^{a}\hat{\tau}_i + k\sum_{j=1}^{b}\hat{\beta}_j = y_{..} \\
\tau_i &: r\hat{\mu} + r\hat{\tau}_i + \sum_{j=1}^{b} n_{ij}\hat{\beta}_j = y_{i.} \quad i = 1, 2, \ldots, a \\
\beta_j &: k\hat{\mu} + \sum_{i=1}^{a} n_{ij}\hat{\tau}_i + k\hat{\beta}_j = y_{.j} \quad j = 1, 2, \ldots, b
\end{aligned}
\tag{4.40}
$$

Imposing $\Sigma\hat{\tau}_i = \Sigma\hat{\beta}_j = 0$, we find that $\hat{\mu} = \bar{y}_{..}$. Furthermore, using the equations for $\{\beta_j\}$ to eliminate the block effects from the equations for $\{\tau_i\}$, we obtain

$$
rk\hat{\tau}_i - r\hat{\tau}_i - \sum_{j=1}^{b}\sum_{\substack{p=1 \\ p\neq 1}}^{a} n_{ij}n_{pj}\,\hat{\tau}_p = ky_{i.} - \sum_{j=1}^{b} n_{ij}y_{.j} \tag{4.41}
$$

Note that the right-hand side of Equation 4.36 is kQ_i, where Q_i is the ith adjusted treatment total (see Equation 4.29). Now, because $\Sigma_{j=1}^{b} n_{ij}n_{pj} = \lambda$ if $p \neq i$ and $n_{pj}^2 = n_{pj}$ (because $n_{pj} = 0$ or 1), we may rewrite Equation 4.41 as

$$
r(k-1)\hat{\tau}_i - \lambda\sum_{\substack{p=1 \\ p\neq 1}}^{a}\hat{\tau}_p = kQ_i \quad i = 1, 2, \ldots, a \tag{4.42}
$$

■ **TABLE 4.26**
Minitab (General Linear Model) Analysis for Example 4.5

```
General Linear Model
Factor       Type     Levels      Values
Catalyst     fixed         4     1 2 3 4
Block        fixed         4     1 2 3 4

Analysis of Variance for Time, using Adjusted SS for Tests
Source       DF       Seq SS      Adj SS      Adj MS          F        P
Catalyst      3       11.667      22.750       7.583      11.67    0.011
Block         3       66.083      66.083      22.028      33.89    0.001
Error         5        3.250       3.250       0.650
Total        11       81.000

Tukey 95.0% Simultaneous Confidence Intervals
Response Variable Time
All Pairwise Comparisons among Levels of Catalyst

Catalyst = 1 subtracted from:

Catalyst    Lower    Center    Upper   --------+---------+---------+-----
2          -2.327    0.2500    2.827   (--------*--------)
3          -1.952    0.6250    3.202    (--------*--------)
4           1.048    3.6250    6.202           (--------*--------)
                                       --------+---------+---------+-----
                                              0.0       2.5       5.0

Catalyst = 2 subtracted from:

Catalyst    Lower    Center    Upper   --------+---------+---------+-----
3          -2.202    0.3750    2.952   (--------*--------)
4           0.798    3.3750    5.952          (--------*--------)
                                       --------+---------+---------+-----
                                              0.0       2.5       5.0

Catalyst = 3 subtracted from:

Catalyst    Lower    Center    Upper   --------+---------+---------+-----
4          0.4228     3.000    5.577          (--------*--------)
                                       --------+---------+---------+-----
                                              0.0       2.5       5.0

Tukey Simultaneous Tests
Response Variable Time
All Pairwise Comparisons among Levels of Catalyst

Catalyst = 1 subtracted from:

Level         Difference          SE of                      Adjusted
Catalyst        of Means     Difference      T-Value          P-Value
2                 0.2500         0.6982       0.3581           0.9825
3                 0.6250         0.6982       0.8951           0.8085
4                 3.6250         0.6982       5.1918           0.0130

Catalyst = 2 subtracted from:

Level         Difference          SE of                      Adjusted
Catalyst        of Means     Difference      T-Value          P-Value
3                 0.3750         0.6982       0.5371           0.9462
4                 3.3750         0.6982       4.8338           0.0175

Catalyst = 3 subtracted from:

Level         Difference          SE of                      Adjusted
Catalyst        of Means     Difference      T-Value          P-Value
4                  3.000         0.6982        4.297           0.0281
```

Finally, note that the constraint $\Sigma_{i=1}^{a}\hat{\tau}_i = 0$ implies that $\Sigma_{\substack{p=1\\p\neq i}}^{a}\hat{\tau}_p = -\hat{\tau}_i$ and recall that $r(k-1) = \lambda(a-1)$ to obtain

$$\lambda a\hat{\tau}_i = kQ_i \qquad i = 1, 2, \ldots, a \tag{4.43}$$

Therefore, the least squares estimators of the treatment effects in the balanced incomplete block model are

$$\hat{\tau}_i = \frac{kQ_i}{\lambda a} \qquad i = 1, 2, \ldots, a \tag{4.44}$$

As an illustration, consider the BIBD in Example 4.5. Because $Q_1 = -9/3$, $Q_2 = -7/3$, $Q_3 = -4/3$, and $Q_4 = 20/3$, we obtain

$$\hat{\tau}_1 = \frac{3(-9/3)}{(2)(4)} = -9/8 \qquad \hat{\tau}_2 = \frac{3(-7/3)}{(2)(4)} = -7/8$$

$$\hat{\tau}_3 = \frac{3(-4/3)}{(2)(4)} = -4/8 \qquad \hat{\tau}_4 = \frac{3(20/3)}{(2)(4)} = 20/8$$

as we found in Section 4.4.1.

4.4.3 Recovery of Interblock Information in the BIBD

The analysis of the BIBD given in Section 4.4.1 is usually called the **intrablock analysis** because block differences are eliminated and all contrasts in the treatment effects can be expressed as comparisons between observations in the same block. This analysis is appropriate regardless of whether the blocks are fixed or random. Yates (1940) noted that, if the block effects are uncorrelated random variables with zero means and variance σ_β^2, one may obtain additional information about the treatment effects τ_i. Yates called the method of obtaining this additional information the **interblock analysis**.

Consider the block totals $y_{.j}$ as a collection of b observations. The model for these observations [following John (1971)] is

$$y_{.j} = k\mu + \sum_{i=1}^{a} n_{ij}\tau_i + \left(k\beta_j + \sum_{i=1}^{a}\epsilon_{ij}\right) \tag{4.45}$$

where the term in parentheses may be regarded as error. The interblock estimators of μ and τ_i are found by minimizing the least squares function

$$L = \sum_{j=1}^{b}\left(y_{.j} - k\mu - \sum_{i=1}^{a} n_{ij}\tau_i\right)^2$$

This yields the following least squares normal equations:

$$\mu : N\tilde{\mu} + r\sum_{i=1}^{a}\tilde{\tau}_i = y_{..}$$

$$\tau_i : kr\tilde{\mu} + r\tilde{\tau}_i + \lambda\sum_{\substack{p=1\\p\neq 1}}^{a}\tilde{\tau}_p = \sum_{j=1}^{b} n_{ij}y_{.j} \qquad i = 1, 2, \ldots, a \tag{4.46}$$

where $\tilde{\mu}$ and $\tilde{\tau}_i$ denote the **interblock estimators**. Imposing the constraint $(_{i=1}^{a}\hat{\tau}_i = 0$, we obtain the solutions to Equations 4.46 as

$$\tilde{\mu} = \bar{y}_{..} \tag{4.47}$$

$$\tilde{\tau}_i = \frac{\sum_{j=1}^{b} n_{ij}y_{.j} - kr\bar{y}_{..}}{r - \lambda} \qquad i = 1, 2, \ldots, a \tag{4.48}$$

It is possible to show that the interblock estimators $\{\tilde{\tau}_i\}$ and the intrablock estimators $\{\hat{\tau}_i\}$ are uncorrelated.

The interblock estimators $\{\tilde{\tau}_i\}$ can differ from the intrablock estimators $\{\hat{\tau}_i\}$. For example, the interblock estimators for the BIBD in Example 4.5 are computed as follows:

$$\tilde{\tau}_1 = \frac{663 - (3)(3)(72.50)}{3 - 2} = 10.50$$

$$\tilde{\tau}_2 = \frac{649 - (3)(3)(72.50)}{3 - 2} = -3.50$$

$$\tilde{\tau}_3 = \frac{652 - (3)(3)(72.50)}{3 - 2} = -0.50$$

$$\tilde{\tau}_4 = \frac{646 - (3)(3)(72.50)}{3 - 2} = -6.50$$

Note that the values of $\Sigma_{j=1}^{b} n_{ij} y_{.j}$ were used previously on page 169 in computing the adjusted treatment totals in the intrablock analysis.

Now suppose we wish to combine the interblock and intrablock estimators to obtain a single, unbiased, minimum variance estimate of each τ_i. It is possible to show that both $\hat{\tau}_i$ and $\tilde{\tau}_i$ are unbiased and also that

$$V(\hat{\tau}_i) = \frac{k(a-1)}{\lambda a^2} \sigma^2 \qquad \text{(intrablock)}$$

and

$$V(\tilde{\tau}_i) = \frac{k(a-1)}{a(r-\lambda)} (\sigma^2 + k\sigma_\beta^2) \qquad \text{(interblock)}$$

We use a linear combination of the two estimators, say

$$\tau_i^* = \alpha_1 \hat{\tau}_i + \alpha_2 \tilde{\tau}_i \tag{4.49}$$

to estimate τ_i. For this estimation method, the minimum variance unbiased combined estimator τ_i^* should have weights $\alpha_1 = u_1/(u_1 + u_2)$ and $\alpha_2 = u_2/(u_1 + u_2)$, where $u_1 = 1/V(\hat{\tau}_i)$ and $u_2 = 1/V(\tilde{\tau}_i)$. Thus, the optimal weights are inversely proportional to the variances of $\hat{\tau}_i$ and $\tilde{\tau}_i$. This implies that the best combined estimator is

$$\tau_i^* = \frac{\hat{\tau}_i \dfrac{k(a-1)}{a(r-\lambda)} (\sigma^2 + k\sigma_\beta^2) + \tilde{\tau}_i \dfrac{k(a-1)}{\lambda a^2} \sigma^2}{\dfrac{k(a-1)}{\lambda a^2} \sigma^2 + \dfrac{k(a-1)}{a(r-\lambda)} (\sigma^2 + k\sigma_\beta^2)} \qquad i = 1, 2, \ldots, a$$

which can be simplified to

$$\tau_i^* = \frac{kQ_i(\sigma^2 + k\sigma_\beta^2) + \left(\sum_{j=1}^{b} n_{ij} y_{.j} - kr\bar{y}_{..}\right)\sigma^2}{(r-\lambda)\sigma^2 + \lambda a(\sigma^2 + k\sigma_\beta^2)} \qquad i = 1, 2, \ldots, a \tag{4.50}$$

Unfortunately, Equation 4.50 cannot be used to estimate the τ_i because the variances σ^2 and σ_β^2 are unknown. The usual approach is to estimate σ^2 and σ_β^2 from the data and replace these parameters in Equation 4.50 by the estimates. The estimate usually taken for σ^2 is the error mean square from the intrablock analysis of variance, or the **intrablock error**. Thus,

$$\hat{\sigma}^2 = MS_E$$

The estimate of σ_β^2 is found from the mean square for blocks adjusted for treatments. In general, for a balanced incomplete block design, this mean square is

$$MS_{\text{Blocks(adjusted)}} = \frac{\left(\dfrac{k\sum_{i=1}^{a} Q_i^2}{\lambda a} + \sum_{j=1}^{b} \dfrac{y_{.j}^2}{k} - \sum_{i=1}^{a} \dfrac{y_{i.}^2}{r}\right)}{(b-1)} \tag{4.51}$$

and its expected value [which is derived in Graybill (1961)] is

$$E[MS_{\text{Blocks(adjusted)}}] = \sigma^2 + \frac{a(r-1)}{b-1}\sigma_\beta^2$$

Thus, if $MS_{\text{Blocks(adjusted)}} > MS_E$, the estimate of $\hat{\sigma}_\beta^2$ is

$$\hat{\sigma}_\beta^2 = \frac{[MS_{\text{Blocks(adjusted)}} - MS_E](b-1)}{a(r-1)} \tag{4.52}$$

and if $MS_{\text{Blocks(adjusted)}} \leq MS_E$, we set $\hat{\sigma}_\beta^2 = 0$. This results in the combined estimator

$$\tau_i^* = \begin{cases} \dfrac{kQ_i(\hat{\sigma}^2 + k\hat{\sigma}_\beta^2) + \left(\sum_{j=1}^{b} n_{ij}y_{.j} - kr\bar{y}_{..}\right)\hat{\sigma}^2}{(r-\lambda)\hat{\sigma}^2 + \lambda a(\hat{\sigma}^2 + k\hat{\sigma}_\beta^2)}, & \hat{\sigma}_\beta^2 > 0 \quad (4.53a) \\ \dfrac{y_{i.} - (1/a)y_{..}}{r}, & \hat{\sigma}_\beta^2 = 0 \quad (4.53b) \end{cases}$$

We now compute the combined estimates for the data in Example 4.5. From Table 4.25 we obtain $\hat{\sigma}^2 = MS_E = 0.65$ and $MS_{\text{Blocks(adjusted)}} = 22.03$. (Note that in computing $MS_{\text{Blocks(adjusted)}}$ we make use of the fact that this is a symmetric design. In general, we must use Equation 4.51. Because $MS_{\text{Blocks(adjusted)}} > MS_E$, we use Equation 4.52 to estimate σ_β^2 as

$$\hat{\sigma}_\beta^2 = \frac{(22.03 - 0.65)(3)}{4(3-1)} = 8.02$$

Therefore, we may substitute $\hat{\sigma}^2 = 0.65$ and $\hat{\sigma}_\beta^2 = 8.02$ into Equation 4.53a to obtain the combined estimates listed below. For convenience, the intrablock and interblock estimates are also given. In this example, the combined estimates are close to the intrablock estimates because the variance of the interblock estimates is relatively large.

Parameter	Intrablock Estimate	Interblock Estimate	Combined Estimate
τ_1	−1.12	10.50	−1.09
τ_2	−0.88	−3.50	−0.88
τ_3	−0.50	−0.50	−0.50
τ_4	2.50	−6.50	2.47

4.5 Problems

4.1. The ANOVA from a randomized complete block experiment output is shown below.

```
Source      DF       SS        MS       F  P
Treatment    4  1010.56         ? 29.84   ?
Block        ?        ?    64.765       ?  ?
Error       20   169.33         ?
Total       29  1503.71
```

(a) Fill in the blanks. You may give bounds on the *P*-value.

(b) How many blocks were used in this experiment?

(c) What conclusions can you draw?

4.2. Consider the single-factor completely randomized single factor experiment shown in Problem 3.4. Suppose that this experiment had been conducted in a randomized complete block design, and that the sum of squares for blocks was 80.00. Modify the ANOVA for this experiment to show the correct analysis for the randomized complete block experiment.

4.3. Three different washing solutions are being compared to study their effectiveness in retarding bacteria growth in 5-gallon milk containers. The analysis is done in a laboratory, and only three trials can be run on any day. Because days could represent a potential source of variability, the experimenter decides to use a randomized block design. Observations are taken for four days, and the data are shown here. Analyze the data from this experiment (use $\alpha = 0.05$) and draw conclusions.

	Days			
Solution	**1**	**2**	**3**	**4**
1	13	22	18	39
2	16	25	16	44
3	5	4	2	22

4.4. A chemist wishes to test the effect of four chemical agents on the strength of a particular type of cloth. Because there might be variability from one bolt to another, the chemist decides to use a randomized block design, with the bolts of cloth considered as blocks. She selects five bolts and applies all four chemicals in random order to each bolt. The resulting tensile strengths follow. Analyze the data from this experiment (use $\alpha = 0.05$) and draw appropriate conclusions.

	Bolt				
Chemical	**1**	**2**	**3**	**4**	**5**
1	73	68	74	71	67
2	73	67	75	72	70
3	75	68	78	73	68
4	73	71	75	75	69

4.5. Plot the mean tensile strengths observed for each chemical type in Problem 4.4 and compare them to an appropriately scaled t distribution. What conclusions would you draw from this display?

4.6. Plot the average bacteria counts for each solution in Problem 4.3 and compare them to a scaled t distribution. What conclusions can you draw?

4.7. Consider the hardness testing experiment described in Section 4.1. Suppose that the experiment was conducted as described and that the following Rockwell C-scale data (coded by subtracting 40 units) obtained:

	Coupon			
Tip	**1**	**2**	**3**	**4**
1	9.3	9.4	9.6	10.0
2	9.4	9.3	9.8	9.9
3	9.2	9.4	9.5	9.7
4	9.7	9.6	10.0	10.2

(a) Analyze the data from this experiment.

(b) Use the Fisher LSD method to make comparisons among the four tips to determine specifically which tips differ in mean hardness readings.

(c) Analyze the residuals from this experiment.

4.8. A consumer products company relies on direct mail marketing pieces as a major component of its advertising campaigns. The company has three different designs for a new brochure and wants to evaluate their effectiveness, as there are substantial differences in costs between the three designs. The company decides to test the three designs by mailing 5000 samples of each to potential customers in four different regions of the country. Since there are known regional differences in the customer base, regions are considered as blocks. The number of responses to each mailing is as follows:

	Region			
Design	**NE**	**NW**	**SE**	**SW**
1	250	350	219	375
2	400	525	390	580
3	275	340	200	310

(a) Analyze the data from this experiment.

(b) Use the Fisher LSD method to make comparisons among the three designs to determine specifically which designs differ in the mean response rate.

(c) Analyze the residuals from this experiment.

4.9. An article in the *Fire Safety Journal* ("The Effect of Nozzle Design on the Stability and Performance of Turbulent Water Jets," Vol. 4, August 1981) describes an experiment in which a shape factor was determined for several different nozzle designs at six levels of jet efflux velocity. Interest focused on potential differences between nozzle designs, with velocity considered as a nuisance variable. The data are shown below:

Nozzle	Jet Efflux Velocity (m/s)					
Design	11.73	14.37	16.59	20.43	23.46	28.74
1	0.78	0.80	0.81	0.75	0.77	0.78
2	0.85	0.85	0.92	0.86	0.81	0.83
3	0.93	0.92	0.95	0.89	0.89	0.83
4	1.14	0.97	0.98	0.88	0.86	0.83
5	0.97	0.86	0.78	0.76	0.76	0.75

(a) Does nozzle design affect the shape factor? Compare the nozzles with a scatter plot and with an analysis of variance, using $\alpha = 0.05$.

(b) Analyze the residuals from this experiment.

(c) Which nozzle designs are different with respect to shape factor? Draw a graph of the average shape factor for each nozzle type and compare this to a scaled t distribution. Compare the conclusions that you draw from this plot to those from Duncan's multiple range test.

4.10. The effect of three different lubricating oils on fuel economy in diesel truck engines is being studied. Fuel economy is measured using brake-specific fuel consumption after the engine has been running for 15 minutes. Five different truck engines are available for the study, and the experimenters conduct the following randomized complete block design.

	Truck				
Oil	1	2	3	4	5
1	0.500	0.634	0.487	0.329	0.512
2	0.535	0.675	0.520	0.435	0.540
3	0.513	0.595	0.488	0.400	0.510

(a) Analyze the data from this experiment.

(b) Use the Fisher LSD method to make comparisons among the three lubricating oils to determine specifically which oils differ in brake-specific fuel consumption.

(c) Analyze the residuals from this experiment.

4.11. An article in *Nature Genetics* (2003, Vol. 34, pp. 85–90) "Treatment-specific changes in gene expression discriminate in vivo drug response in human leukemia cells" studied gene expression as a function of different treatments for leukemia. Three treatment groups are: mercaptopurine (MP) only; low-dose methotrexate (LDMTX) and MP; and high-dose methotrexate (HDMTX) and MP. Each group contained ten subjects. The responses from a specific gene are shown in the table below.

(a) Is there evidence to support the claim that the treatment means differ?

(b) Check the normality assumption. Can we assume these samples are from normal populations?

(c) Take the logarithm of the raw data. Is there evidence to support the claim that the treatment means differ for the transformed data?

(d) Analyze the residuals from the transformed data and comment on model adequacy.

Treatments	Observations									
MP ONLY	334.5	31.6	701	41.2	61.2	69.6	67.5	66.6	120.7	881.9
MP + HDMTX	919.4	404.2	1024.8	54.1	62.8	671.6	882.1	354.2	321.9	91.1
MP + LDMTX	108.4	26.1	240.8	191.1	69.7	242.8	62.7	396.9	23.6	290.4

4.12. An article in *Communications of the ACM* (Vol. 30, No. 5, 1987) studied different algorithms for estimating software development costs. Six algorithms were applied to several different software development projects and the percent error in estimating the development cost was observed. Some of the data from this experiment is shown in the table below.

(a) Do the algorithms differ in their mean cost estimation accuracy?

(b) Analyze the residuals from this experiment.

(c) Which algorithm would you recommend for use in practice?

	Project					
Algorithm	1	2	3	4	5	6
1(SLIM)	1244	21	82	2221	905	839
2(COCOMO-A)	281	129	396	1306	336	910
3(COCOMO-R)	220	84	458	543	300	794
4(COCONO-C)	225	83	425	552	291	826
5(FUNCTION POINTS)	19	11	−34	121	15	103
6(ESTIMALS)	−20	35	−53	170	104	199

4.13. Consider the ratio control algorithm experiment described in Section 3.8. The experiment was actually conducted as a randomized block design, where six time periods were selected as the blocks, and all four ratio control algorithms were tested in each time period. The average cell voltage and the standard deviation of voltage (shown in parentheses) for each cell are as follows:

Ratio Control Algorithm	Time Period 1	2	3
1	4.93 (0.05)	4.86 (0.04)	4.75 (0.05)
2	4.85 (0.04)	4.91 (0.02)	4.79 (0.03)
3	4.83 (0.09)	4.88 (0.13)	4.90 (0.11)
4	4.89 (0.03)	4.77 (0.04)	4.94 (0.05)

Ratio Control Algorithm	Time Period 4	5	6
1	4.95 (0.06)	4.79 (0.03)	4.88 (0.05)
2	4.85 (0.05)	4.75 (0.03)	4.85 (0,02)
3	4.75 (0.15)	4.82 (0.08)	4.90 (0.12)
4	4.86 (0.05)	4.79 (0.03)	4.76 (0.02)

(a) Analyze the average cell voltage data. (Use $\alpha = 0.05$.) Does the choice of ratio control algorithm affect the average cell voltage?

(b) Perform an appropriate analysis on the standard deviation of voltage. (Recall that this is called "pot noise.") Does the choice of ratio control algorithm affect the pot noise?

(c) Conduct any residual analyses that seem appropriate.

(d) Which ratio control algorithm would you select if your objective is to reduce both the average cell voltage *and* the pot noise?

4.14. An aluminum master alloy manufacturer produces grain refiners in ingot form. The company produces the product in four furnaces. Each furnace is known to have its own unique operating characteristics, so any experiment run in the foundry that involves more than one furnace will consider furnaces as a nuisance variable. The process engineers suspect that stirring rate affects the grain size of the product. Each furnace can be run at four different stirring rates. A randomized block design is run for a particular refiner, and the resulting grain size data is as follows:

Stirring Rate (rpm)	Furnace 1	2	3	4
5	8	3	5	6
10	14	5	7	9
15	14	6	9	3
20	17	9	2	6

(a) Is there any evidence that stirring rate affects grain size?

(b) Graph the residuals from this experiment on a normal probability plot. Interpret this plot.

(c) Plot the residuals versus furnace and stirring rate. Does this plot convey any useful information?

(d) What should the process engineers recommend concerning the choice of stirring rate and furnace for this particular grain refiner if small grain size is desirable?

4.15. Analyze the data in Problem 4.3 using the general regression significance test.

4.16. Assuming that chemical types and bolts are fixed, estimate the model parameters τ_i and β_j in Problem 4.4.

4.17. Draw an operating characteristic curve for the design in Problem 4.3. Does the test seem to be sensitive to small differences in the treatment effects?

4.18. Suppose that the observation for chemical type 2 and bolt 3 is missing in Problem 4.4. Analyze the problem by estimating the missing value. Perform the exact analysis and compare the results.

4.19. Consider the hardness testing experiment in Problem 4.7. Suppose that the observation for tip 2 in coupon 3 is missing. Analyze the problem by estimating the missing value.

4.20. ***Two missing values in a randomized block.*** Suppose that in Problem 4.4 the observations for chemical type 2 and bolt 3 and chemical type 4 and bolt 4 are missing.

(a) Analyze the design by iteratively estimating the missing values, as described in Section 4.1.3.

(b) Differentiate SS_E with respect to the two missing values, equate the results to zero, and solve for estimates of the missing values. Analyze the design using these two estimates of the missing values.

(c) Derive general formulas for estimating two missing values when the observations are in *different* blocks.

(d) Derive general formulas for estimating two missing values when the observations are in the *same* block.

4.21. An industrial engineer is conducting an experiment on eye focus time. He is interested in the effect of the distance of the object from the eye on the focus time. Four different distances are of interest. He has five subjects available for the experiment. Because there may be differences among individuals, he decides to conduct the experiment in a randomized block design. The data obtained follow. Analyze the data from this experiment (use $\alpha = 0.05$) and draw appropriate conclusions.

Distance (ft)	Subject 1	2	3	4	5
4	10	6	6	6	6
6	7	6	6	1	6
8	5	3	3	2	5
10	6	4	4	2	3

4.22. An industrial engineer is investigating the effect of four assembly methods (*A, B, C, D*) on the assembly time for a color television component. Four operators are selected for the study. Furthermore, the engineer knows that each assembly method produces such fatigue that the time required for the last assembly may be greater than the time required for the first, regardless of the method. That is, a trend develops in the required assembly time. To account for this source of variability, the engineer uses the Latin square design shown below. Analyze the data from this experiment ($\alpha = 0.05$) and draw appropriate conclusions.

Order of Assembly	Operator 1	Operator 2	Operator 3	Operator 4
1	$C = 10$	$D = 14$	$A = 7$	$B = 8$
2	$B = 7$	$C = 18$	$D = 11$	$A = 8$
3	$A = 5$	$B = 10$	$C = 11$	$D = 9$
4	$D = 10$	$A = 10$	$B = 12$	$C = 14$

4.23. The effect of five different ingredients (*A, B, C, D, E*) on the reaction time of a chemical process is being studied. Each batch of new material is only large enough to permit five runs to be made. Furthermore, each run requires approximately $1\frac{1}{2}$ hours, so only five runs can be made in one day. The experimenter decides to run the experiment as a Latin square so that day and batch effects may be systematically controlled. She obtains the data that follow. Analyze the data from this experiment (use $\alpha = 0.05$) and draw conclusions.

Batch	Day 1	Day 2	Day 3	Day 4	Day 5
1	$A = 8$	$B = 7$	$D = 1$	$C = 7$	$E = 3$
2	$C = 11$	$E = 2$	$A = 7$	$D = 3$	$B = 8$
3	$B = 4$	$A = 9$	$C = 10$	$E = 1$	$D = 5$
4	$D = 6$	$C = 8$	$E = 6$	$B = 6$	$A = 10$
5	$E = 4$	$D = 2$	$B = 3$	$A = 8$	$C = 8$

4.24. Consider the randomized complete block design in Problem 4.3. Assume that the days are random. Estimate the block variance component.

4.25. Consider the randomized complete block design in Problem 4.7. Assume that the coupons are random. Estimate the block variance component.

4.26. Consider the randomized complete block design in Problem 4.10. Assume that the trucks are random. Estimate the block variance component.

4.27. Consider the randomized complete block design in Problem 4.12. Assume that the software projects that were used as blocks are random. Estimate the block variance component.

4.28. Consider the gene expression experiment in Problem 4.11. Assume that the subjects used in this experiment are random. Estimate the block variance component.

4.29. Suppose that in Problem 4.23 the observation from batch 3 on day 4 is missing. Estimate the missing value and perform the analysis using the value.

4.30. Consider a $p \times p$ Latin square with rows (α_i), columns (β_k), and treatments (τ_j) fixed. Obtain least squares estimates of the model parameters α_i, β_k, and τ_j.

4.31. Derive the missing value formula (Equation 4.27) for the Latin square design.

4.32. ***Designs involving several Latin squares.*** [See Cochran and Cox (1957), John (1971).] The $p \times p$ Latin square contains only p observations for each treatment. To obtain more replications the experimenter may use several squares, say n. It is immaterial whether the squares used are the same or different. The appropriate model is

$$y_{ijkh} = \begin{matrix} \mu + \rho_h + \alpha_{i(h)} \\ + \tau_j + \beta_{k(h)} \\ + (\tau\rho)_{jh} + \epsilon_{ijkh} \end{matrix} \quad \begin{cases} i = 1,2,\ldots,p \\ j = 1,2,\ldots,p \\ k = 1,2,\ldots,p \\ h = 1,2,\ldots,n \end{cases}$$

where y_{ijkh} is the observation on treatment j in row i and column k of the hth square. Note that $\alpha_{i(h)}$ and $\beta_{k(h)}$ are the row and column effects in the hth square, ρ_h is the effect of the hth square, and $(\tau\rho)_{jh}$ is the interaction between treatments and squares.

(a) Set up the normal equations for this model, and solve for estimates of the model parameters. Assume that appropriate side conditions on the parameters are $\sum_h \hat{\rho}_h = 0$, $\sum_i \hat{\alpha}_{i(h)} = 0$, and $\sum_k \hat{\beta}_{k(h)} = 0$ for each h, $\sum_j \hat{\tau}_j = 0$, $\sum_j (\widehat{\tau\rho})_{jh} = 0$ for each h, and $\sum_h (\widehat{\tau\rho})_{jh} = 0$ for each j.

(b) Write down the analysis of variance table for this design.

4.33. Discuss how the operating characteristics curves may be used with the Latin square design.

4.34. Suppose that in Problem 4.23 the data taken on day 5 were incorrectly analyzed and had to be discarded. Develop an appropriate analysis for the remaining data.

4.35. The yield of a chemical process was measured using five batches of raw material, five acid concentrations, five standing times (*A, B, C, D, E*), and five catalyst concentrations ($\alpha, \beta, \gamma, \delta, \epsilon$). The Graeco-Latin square that follows was used. Analyze the data from this experiment (use $\alpha = 0.05$) and draw conclusions.

Batch	Acid Concentration 1	2	3
1	$A\alpha = 26$	$B\beta = 16$	$C\gamma = 19$
2	$B\gamma = 18$	$C\delta = 21$	$D\epsilon = 18$
3	$C\epsilon = 20$	$D\alpha = 12$	$E\beta = 16$
4	$D\beta = 15$	$E\gamma = 15$	$A\delta = 22$
5	$E\delta = 10$	$A\epsilon = 24$	$B\alpha = 17$

Batch	Acid Concentration 4	5
1	$D\delta = 16$	$E\epsilon = 13$
2	$E\alpha = 11$	$A\beta = 21$
3	$A\gamma = 25$	$B\delta = 13$
4	$B\epsilon = 14$	$C\alpha = 17$
5	$C\beta = 17$	$D\gamma = 14$

4.36. Suppose that in Problem 4.22 the engineer suspects that the workplaces used by the four operators may represent an additional source of variation. A fourth factor, workplace (α, β, γ, δ) may be introduced and another experiment conducted, yielding the Graeco-Latin square that follows. Analyze the data from this experiment (use $\alpha = 0.05$) and draw conclusions.

Order of Assembly	Operator 1	2	3	4
1	$C\beta = 11$	$B\gamma = 10$	$D\delta = 14$	$A\alpha = 8$
2	$B\alpha = 8$	$C\delta = 12$	$A\gamma = 10$	$D\beta = 12$
3	$A\delta = 9$	$D\alpha = 11$	$B\beta = 7$	$C\gamma = 15$
4	$D\gamma = 9$	$A\beta = 8$	$C\alpha = 18$	$B\delta = 6$

4.37. Construct a 5×5 hypersquare for studying the effects of five factors. Exhibit the analysis of variance table for this design.

4.38. Consider the data in Problems 4.22 and 4.36. Suppressing the Greek letters in Problem 4.36, analyze the data using the method developed in Problem 4.32.

4.39. Consider the randomized block design with one missing value in Problem 4.19. Analyze this data by using the exact analysis of the missing value problem discussed in Section 4.1.4. Compare your results to the approximate analysis of these data given from Problem 4.19.

4.40. An engineer is studying the mileage performance characteristics of five types of gasoline additives. In the road test he wishes to use cars as blocks; however, because of a time constraint, he must use an incomplete block design. He runs the balanced design with the five blocks that follow. Analyze the data from this experiment (use $\alpha = 0.05$) and draw conclusions.

Additive	Car 1	2	3	4	5
1		17	14	13	12
2	14	14		13	10
3	12		13	12	9
4	13	11	11	12	
5	11	12	10		8

4.41. Construct a set of orthogonal contrasts for the data in Problem 4.40. Compute the sum of squares for each contrast.

4.42. Seven different hardwood concentrations are being studied to determine their effect on the strength of the paper produced. However, the pilot plant can only produce three runs each day. As days may differ, the analyst uses the balanced incomplete block design that follows. Analyze the data from this experiment (use $\alpha = 0.05$) and draw conclusions.

Hardwood Concentration (%)	Days 1	2	3	4
2	114			
4	126	120		
6		137	117	
8	141		129	149
10		145		150
12			120	
14				136

Hardwood Concentration (%)	Days 5	6	7
2	120		117
4		119	
6			134
8			
10	143		
12	118	123	
14		130	127

4.43. Analyze the data in Example 4.5 using the general regression significance test.

4.44. Prove that $k\sum_{i=1}^{a} Q_i^2/(\lambda a)$ is the adjusted sum of squares for treatments in a BIBD.

4.45. An experimenter wishes to compare four treatments in blocks of two runs. Find a BIBD for this experiment with six blocks.

4.46. An experimenter wishes to compare eight treatments in blocks of four runs. Find a BIBD with 14 blocks and $\lambda = 3$.

4.47. Perform the interblock analysis for the design in Problem 4.40.

4.48. Perform the interblock analysis for the design in Problem 4.42.

4.49. Verify that a BIBD with the parameters $a = 8$, $r = 8$, $k = 4$, and $b = 16$ does not exist.

4.50. Show that the variance of the intrablock estimators $\{\hat{\tau}_i\}$ is $k(a-1)\sigma^2/(\lambda a^2)$.

4.51. ***Extended incomplete block designs.*** Occasionally, the block size obeys the relationship $a < k < 2a$. An extended incomplete block design consists of a single replicate of each treatment in each block along with an incomplete block design with $k^* = k - a$. In the balanced case, the incomplete block design will have parameters $k^* = k - a$, $r^* = r - b$, and λ^*. Write out the statistical analysis. (*Hint:* In the extended incomplete block design, we have $\lambda = 2r - b + \lambda^*$.)

4.52. Suppose that a single-factor experiment with five levels of the factor has been conducted. There are three replicates and the experiment has been conducted as a complete randomized design. If the experiment had been conducted in blocks, the pure error degrees of freedom would be reduced by (choose the correct answer):

(a) 3

(b) 5

(c) 2

(d) 4

(e) None of the above

4.53. Physics graduate student Laura Van Ertia has conducted a CRD with a single factor, hoping to solve the mystery of the unified theory and complete her dissertation. The results of this experiment are summarized in the following ANOVA display:

Source	DF	SS	MS	F
Factor	–	–	14.18	–
Error	–	37.75	–	
Total	23	108.63		

Answer the following questions about this experiment.

(a) The sum of squares for the factor is ______.

(b) The number of degrees of freedom for the single factor in the experiment is______.

(c) The number of degrees of freedom for error is ______.

(d) The mean square for error is______.

(e) The value of the test statistic is______.

(f) If the significance level is 0.05, your conclusions are not to reject the null hypothesis. (Yes or No)

(g) An upper bound on the *P*-value for the test statistic is______.

(h) A lower bound on the *P*-value for the test statistic is______.

(i) Laura used ______ levels of the factor in this experiment.

(j) Laura replicated this experiment ______ times.

(k) Suppose that Laura had actually conducted this experiment as an RCBD and the sum of squares for blocks was 12. Reconstruct the ANOVA display above to reflect this new situation. How much has blocking reduced the estimate of experimental error?

4.54. Consider the direct mail marketing experiment in Problem 4.8. Suppose that this experiment had been run as a complete randomized design, ignoring potential regional differences, but that exactly the same data was obtained. Reanalyze the experiment under this new assumption. What difference would ignoring blocking have on the results and conclusions?

CHAPTER 5

Factorial Experiments

CHAPTER OUTLINE

The supplemental material is on the textbook website www.wiley.com/go/global/montgomery.

5.1 Basic Definitions and Principles

Many experiments involve the study of the effects of two or more factors. In general, **factorial designs** are most efficient for this type of experiment. By a factorial design, we mean that in each complete trial or replicate of the experiment all possible combinations of the levels of the factors are investigated. For example, if there are a levels of factor A and b levels of factor B, each replicate contains all ab treatment combinations. When factors are arranged in a factorial design, they are often said to be **crossed**.

The effect of a factor is defined to be the change in response produced by a change in the level of the factor. This is frequently called a **main effect** because it refers to the primary factors of interest in the experiment. For example, consider the simple experiment in Figure 5.1. This is a two-factor factorial experiment with both design factors at two levels. We have called these levels "low" and "high" and denoted them "−" and "+," respectively. The main effect of factor A in this two-level design can be thought of as the difference between the average response at the low level of A and the average response at the high level of A. Numerically, this is

$$A = \frac{40 + 52}{2} - \frac{20 + 30}{2} = 21$$

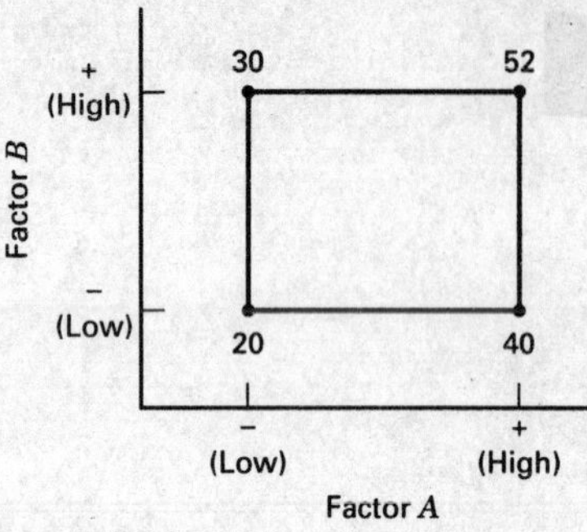

■ **FIGURE 5.1** **A two-factor factorial experiment, with the response (*y*) shown at the corners**

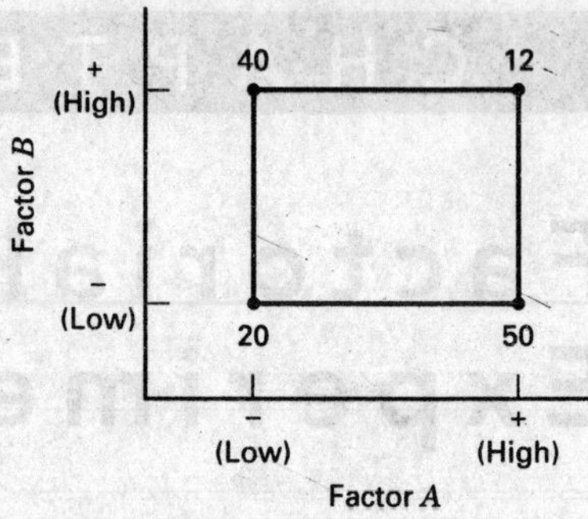

■ **FIGURE 5.2** **A two-factor factorial experiment with interaction**

That is, increasing factor A from the low level to the high level causes an **average response increase** of 21 units. Similarly, the main effect of B is

$$B = \frac{30 + 52}{2} - \frac{20 + 40}{2} = 11$$

If the factors appear at more than two levels, the above procedure must be modified because there are other ways to define the effect of a factor. This point is discussed more completely later.

In some experiments, we may find that the difference in response between the levels of one factor is not the same at all levels of the other factors. When this occurs, there is an **interaction** between the factors. For example, consider the two-factor factorial experiment shown in Figure 5.2. At the low level of factor B (or B^-), the A effect is

$$A = 50 - 20 = 30$$

and at the high level of factor B (or B^+), the A effect is

$$A = 12 - 40 = -28$$

Because the effect of A depends on the level chosen for factor B, we see that there is interaction between A and B. The magnitude of the interaction effect is the average *difference* in these two A effects, or $AB = (-28 - 30)/2 = -29$. Clearly, the interaction is large in this experiment.

These ideas may be illustrated graphically. Figure 5.3 plots the response data in Figure 5.1 against factor A for both levels of factor B. Note that the B^- and B^+ lines are approximately parallel, indicating a lack of interaction between factors A and B. Similarly, Figure 5.4

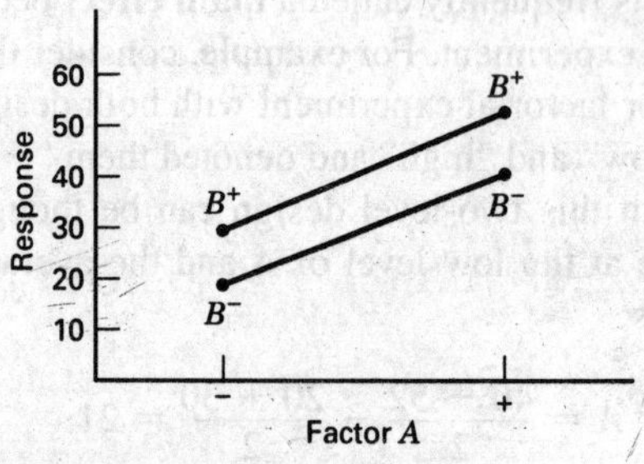

■ **FIGURE 5.3** **A factorial experiment without interaction**

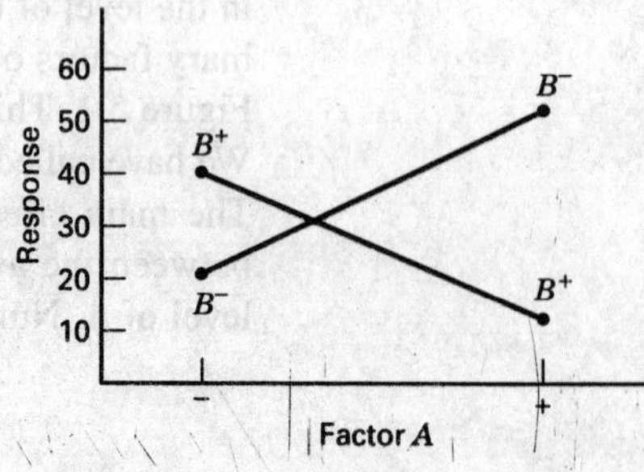

■ **FIGURE 5.4** **A factorial experiment with interaction**

plots the response data in Figure 5.2. Here we see that the B^- and B^+ lines are not parallel. This indicates an interaction between factors A and B. Two-factor interaction graphs such as these are frequently very useful in interpreting significant interactions and in reporting results to nonstatistically trained personnel. However, they should not be utilized as the sole technique of data analysis because their interpretation is subjective and their appearance is often misleading.

There is another way to illustrate the concept of interaction. Suppose that both of our design factors are **quantitative** (such as temperature, pressure, time, etc.). Then a **regression model representation** of the two-factor factorial experiment could be written as

$$y = \beta_0 + \beta_1 x_1 + \beta_2 x_2 + \beta_{12} x_1 x_2 + \epsilon$$

where y is the response, the β's are parameters whose values are to be determined, x_1 is a variable that represents factor A, x_2 is a variable that represents factor B, and ϵ is a random error term. The variables x_1 and x_2 are defined on a **coded scale** from -1 to $+1$ (the low and high levels of A and B), and $x_1 x_2$ represents the interaction between x_1 and x_2.

The parameter estimates in this regression model turn out to be related to the effect estimates. For the experiment shown in Figure 5.1 we found the main effects of A and B to be $A = 21$ and $B = 11$. The estimates of β_1 and β_2 are one-half the value of the corresponding main effect; therefore, $\hat{\beta}_1 = 21/2 = 10.5$ and $\hat{\beta}_2 = 11/2 = 5.5$. The interaction effect in Figure 5.1 is $AB = 1$, so the value of interaction coefficient in the regression model is $\hat{\beta}_{12} = 1/2 = 0.5$. The parameter β_0 is estimated by the average of all four responses, or $\hat{\beta}_0 = (20 + 40 + 30 + 52)/4 = 35.5$. Therefore, the fitted regression model is

$$\hat{y} = 35.5 + 10.5x_1 + 5.5x_2 + 0.5x_1 x_2$$

The parameter estimates obtained in the manner for the factorial design with all factors at two levels ($-$ and $+$) turn out to be **least squares estimates** (more on this later).

The interaction coefficient ($\hat{\beta}_{12} = 0.5$) is small relative to the main effect coefficients $\hat{\beta}_1$ and $\hat{\beta}_2$. We will take this to mean that interaction is small and can be ignored. Therefore, dropping the term $0.5x_1 x_2$ gives us the model

$$\hat{y} = 35.5 + 10.5x_1 + 5.5x_2$$

Figure 5.5 presents graphical representations of this model. In Figure 5.5*a* we have a plot of the plane of y-values generated by the various combinations of x_1 and x_2. This three-dimensional graph is called a **response surface plot**. Figure 5.5*b* shows the contour lines of constant response y in the x_1, x_2 plane. Notice that because the response surface is a plane, the contour plot contains parallel straight lines.

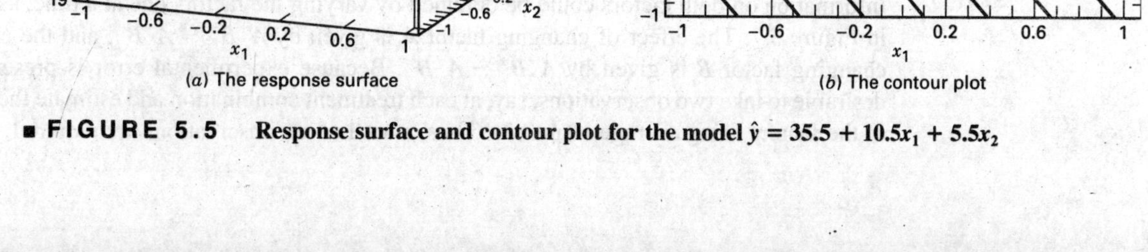

■ **FIGURE 5.5** **Response surface and contour plot for the model $\hat{y} = 35.5 + 10.5x_1 + 5.5x_2$**

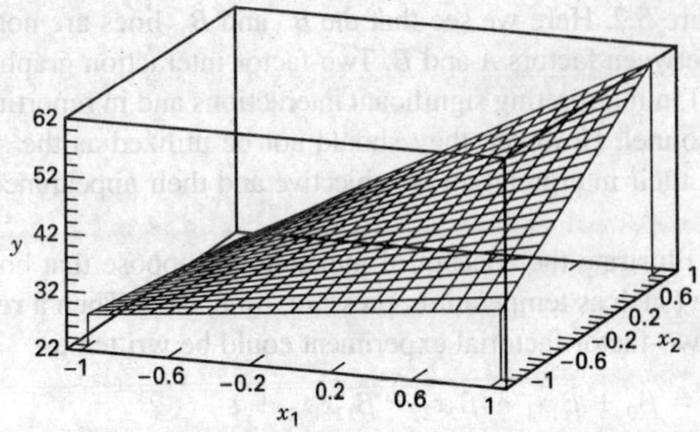

(a) The response surface

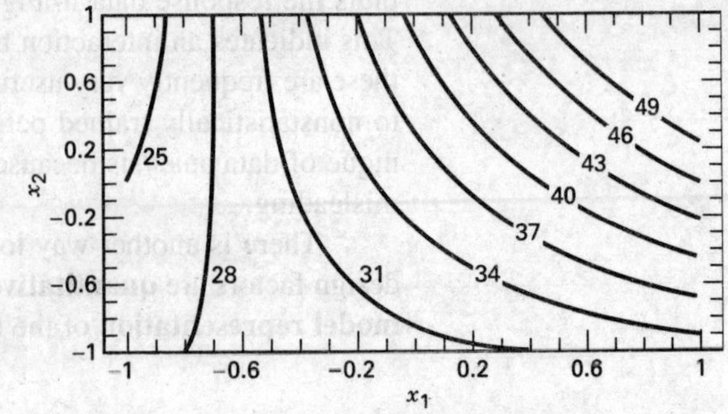

(b) The contour plot

■ **FIGURE 5.6 Response surface and contour plot for the model $\hat{y} = 35.5 + 10.5x_1 + 5.5x_2 + 8x_1x_2$**

Now suppose that the interaction contribution to this experiment was not negligible; that is, the coefficient β_{12} was not small. Figure 5.6 presents the response surface and contour plot for the model

$$\hat{y} = 35.5 + 10.5x_1 + 5.5x_2 + 8x_1x_2$$

(We have let the interaction effect be the average of the two main effects.) Notice that the significant interaction effect "twists" the plane in Figure 5.6*a*. This twisting of the response surface results in curved contour lines of constant response in the x_1, x_2 plane, as shown in Figure 5.6*b*. Thus, **interaction is a form of curvature** in the underlying response surface model for the experiment.

The response surface model for an experiment is extremely important and useful. We will say more about it in Section 5.5 and in subsequent chapters.

Generally, when an interaction is large, the corresponding main effects have little practical meaning. For the experiment in Figure 5.2, we would estimate the main effect of *A* to be

$$A = \frac{50 + 12}{2} - \frac{20 + 40}{2} = 1$$

which is very small, and we are tempted to conclude that there is no effect due to *A*. However, when we examine the effects of *A* at *different levels of factor B*, we see that this is not the case. Factor *A* has an effect, but it *depends on the level of factor B*. That is, knowledge of the *AB* interaction is more useful than knowledge of the main effect. A significant interaction will often **mask** the significance of main effects. These points are clearly indicated by the interaction plot in Figure 5.4. In the presence of significant interaction, the experimenter must usually examine the levels of one factor, say *A*, with levels of the other factors fixed to draw conclusions about the main effect of *A*.

5.2 The Advantage of Factorials

The advantage of factorial designs can be easily illustrated. Suppose we have two factors *A* and *B*, each at two levels. We denote the levels of the factors by A^-, A^+, B^-, and B^+. Information on both factors could be obtained by varying the factors one at a time, as shown in Figure 5.7. The effect of changing factor *A* is given by $A^+B^- - A^-B^-$, and the effect of changing factor *B* is given by $A^-B^+ - A^-B^-$. Because experimental error is present, it is desirable to take two observations, say, at each treatment combination and estimate the effects of the factors using average responses. Thus, a total of six observations are required.

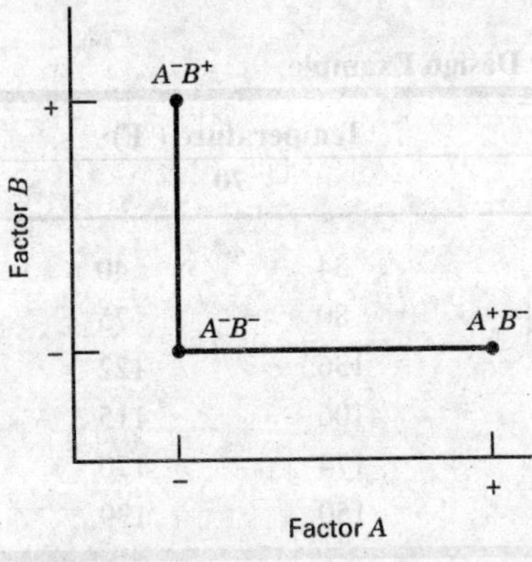

■ **FIGURE 5.7 A one-factor-at-a-time experiment**

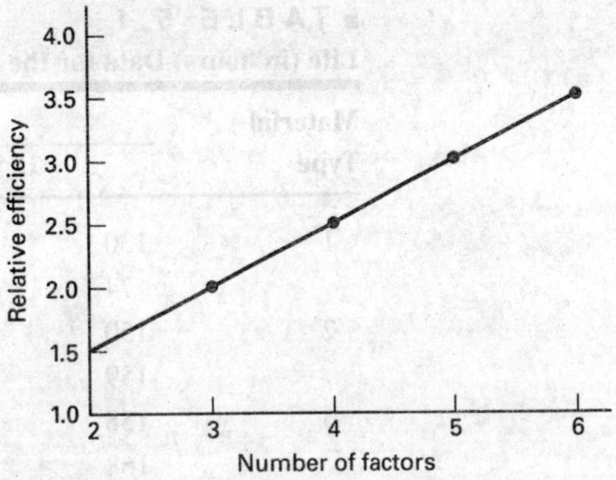

■ **FIGURE 5.8 Relative efficiency of a factorial design to a one-factor-at-a-time experiment (two factor levels)**

If a factorial experiment had been performed, an additional treatment combination, A^+B^+, would have been taken. Now, using just *four* observations, two estimates of the A effect can be made: $A^+B^- - A^-B^-$ and $A^+B^+ - A^-B^+$. Similarly, two estimates of the B effect can be made. These two estimates of each main effect could be averaged to produce average main effects that are *just as precise* as those from the single-factor experiment, but only four total observations are required and we would say that the relative efficiency of the factorial design to the one-factor-at-a-time experiment is $(6/4) = 1.5$. Generally, this relative efficiency will increase as the number of factors increases, as shown in Figure 5.8.

Now suppose interaction is present. If the one-factor-at-a-time design indicated that A^-B^+ and A^+B^- gave better responses than A^-B^-, a logical conclusion would be that A^+B^+ would be even better. However, if interaction is present, this conclusion may be *seriously in error*. For an example, refer to the experiment in Figure 5.2.

In summary, note that factorial designs have several advantages. They are more efficient than one-factor-at-a-time experiments. Furthermore, a factorial design is necessary when interactions may be present to avoid misleading conclusions. Finally, factorial designs allow the effects of a factor to be estimated at several levels of the other factors, yielding conclusions that are valid over a range of experimental conditions.

5.3 The Two-Factor Factorial Design

5.3.1 An Example

The simplest types of factorial designs involve only two factors or sets of treatments. There are a levels of factor A and b levels of factor B, and these are arranged in a factorial design; that is, each replicate of the experiment contains all ab treatment combinations. In general, there are n replicates.

As an example of a factorial design involving two factors, an engineer is designing a battery for use in a device that will be subjected to some extreme variations in temperature. The only design parameter that he can select at this point is the plate material for the battery, and he has three possible choices. When the device is manufactured and is shipped to the field, the engineer has no control over the temperature extremes that the device will encounter, and he knows from experience that temperature will probably affect the effective battery life. However, temperature can be controlled in the product development laboratory for the purposes of a test.

■ **TABLE 5.1**
Life (in hours) Data for the Battery Design Example

Material Type	Temperature (°F) 15		70		125	
1	130	155	34	40	20	70
	74	180	80	75	82	58
2	150	188	136	122	25	70
	159	126	106	115	58	45
3	138	110	174	120	96	104
	168	160	150	139	82	60

The engineer decides to test all three plate materials at three temperature levels—15, 70, and 125°F—because these temperature levels are consistent with the product end-use environment. Because there are two factors at three levels, this design is sometimes called a **3^2 factorial design**. Four batteries are tested at each combination of plate material and temperature, and all 36 tests are run in random order. The experiment and the resulting observed battery life data are given in Table 5.1.

In this problem the engineer wants to answer the following questions:

1. What effects do material type and temperature have on the life of the battery?
2. Is there a choice of material that would give *uniformly long life regardless of temperature*?

This last question is particularly important. It may be possible to find a material alternative that is not greatly affected by temperature. If this is so, the engineer can make the battery **robust** to temperature variation in the field. This is an example of using statistical experimental design for **robust product design**, a very important engineering problem.

This design is a specific example of the general case of a two-factor factorial. To pass to the general case, let y_{ijk} be the observed response when factor A is at the ith level ($i = 1, 2, \ldots, a$) and factor B is at the jth level ($j = 1, 2, \ldots, b$) for the kth replicate ($k = 1, 2, \ldots, n$). In general, a two-factor factorial experiment will appear as in Table 5.2. The order in which the abn observations are taken is selected at random so that this design is a **completely randomized design**.

The observations in a factorial experiment can be described by a model. There are several ways to write the model for a factorial experiment. The **effects model** is

$$y_{ijk} = \mu + \tau_i + \beta_j + (\tau\beta)_{ij} + \epsilon_{ijk} \quad \begin{cases} i = 1, 2, \ldots, a \\ j = 1, 2, \ldots, b \\ k = 1, 2, \ldots, n \end{cases} \tag{5.1}$$

where μ is the overall mean effect, τ_i is the effect of the ith level of the row factor A, β_j is the effect of the jth level of column factor B, $(\tau\beta)_{ij}$ is the effect of the interaction between τ_i and β_j, and ϵ_{ijk} is a random error component. Both factors are assumed to be **fixed**, and the treatment effects are defined as deviations from the overall mean, so $\sum_{i=1}^{a}\tau_i = 0$ and $\sum_{j=1}^{b}\beta_j = 0$. Similarly, the interaction effects are fixed and are defined such that $\sum_{i=1}^{a}(\tau\beta)_{ij} = \sum_{j=1}^{b}(\tau\beta)_{ij} = 0$. Because there are n replicates of the experiment, there are abn total observations.

■ **TABLE 5.2**
General Arrangement for a Two-Factor Factorial Design

		Factor B			
		1	2	...	b
Factor A	1	$y_{111}, y_{112}, \ldots, y_{11n}$	$y_{121}, y_{122}, \ldots, y_{12n}$		$y_{1b1}, y_{1b2}, \ldots, y_{1bn}$
	2	$y_{211}, y_{212}, \ldots, y_{21n}$	$y_{221}, y_{222}, \ldots, y_{22n}$		$y_{2b1}, y_{2b2}, \ldots, y_{2bn}$
	$\vdots$				
	a	$y_{a11}, y_{a12}, \ldots, y_{a1n}$	$y_{a21}, y_{a22}, \ldots, y_{a2n}$		$y_{ab1}, y_{ab2}, \ldots, y_{abn}$

Another possible model for a factorial experiment is the **means model**

$$y_{ijk} = \mu_{ij} + \epsilon_{ijk} \quad \begin{cases} i = 1, 2, \ldots, a \\ j = 1, 2, \ldots, b \\ k = 1, 2, \ldots, n \end{cases}$$

where the mean of the *ij*th cell is

$$\mu_{ij} = \mu + \tau_i + \beta_j + (\tau\beta)_{ij}$$

We could also use a **regression model** as in Section 5.1. Regression models are particularly useful when one or more of the factors in the experiment are quantitative. Throughout most of this chapter we will use the effects model (Equation 5.1) with an illustration of the regression model in Section 5.5.

In the two-factor factorial, both row and column factors (or treatments), *A* and *B*, are of equal interest. Specifically, we are interested in **testing hypotheses** about the equality of row treatment effects, say

$$\begin{aligned} &H_0{:}\tau_1 = \tau_2 = \cdots = \tau_a = 0 \\ &H_1{:}\text{at least one } \tau_i \neq 0 \end{aligned} \tag{5.2a}$$

and the equality of column treatment effects, say

$$\begin{aligned} &H_0{:}\beta_1 = \beta_2 = \cdots = \beta_b = 0 \\ &H_1{:}\text{at least one } \beta_j \neq 0 \end{aligned} \tag{5.2b}$$

We are also interested in determining whether row and column treatments *interact*. Thus, we also wish to test

$$\begin{aligned} &H_0{:}(\tau\beta)_{ij} = 0 \qquad \text{for all } i, j \\ &H_1{:}\text{at least one } (\tau\beta)_{ij} \neq 0 \end{aligned} \tag{5.2c}$$

We now discuss how these hypotheses are tested using a **two-factor analysis of variance**.

5.3.2 Statistical Analysis of the Fixed Effects Model

Let $y_{i..}$ denote the total of all observations under the *i*th level of factor *A*, $y_{.j.}$ denote the total of all observations under the *j*th level of factor *B*, $y_{ij.}$ denote the total of all observations in the

ijth cell, and $y_{...}$ denote the grand total of all the observations. Define $\bar{y}_{i..}$, $\bar{y}_{.j.}$, $\bar{y}_{ij.}$, and $\bar{y}_{...}$ as the corresponding row, column, cell, and grand averages. Expressed mathematically,

$$
\begin{aligned}
y_{i..} &= \sum_{j=1}^{b}\sum_{k=1}^{n} y_{ijk} & \bar{y}_{i..} &= \frac{y_{i..}}{bn} & i &= 1, 2, \ldots, a \\
y_{.j.} &= \sum_{i=1}^{a}\sum_{k=1}^{n} y_{ijk} & \bar{y}_{.j.} &= \frac{y_{.j.}}{an} & j &= 1, 2, \ldots, b \\
y_{ij.} &= \sum_{k=1}^{n} y_{ijk} & \bar{y}_{ij.} &= \frac{y_{ij.}}{n} & i &= 1, 2, \ldots, a \\
& & & & j &= 1, 2, \ldots, b \\
y_{...} &= \sum_{i=1}^{a}\sum_{j=1}^{b}\sum_{k=1}^{n} y_{ijk} & \bar{y}_{...} &= \frac{y_{...}}{abn}
\end{aligned}
\tag{5.3}
$$

The **total corrected sum of squares** may be written as

$$
\begin{aligned}
\sum_{i=1}^{a}\sum_{j=1}^{b}\sum_{k=1}^{n}(y_{ijk}-\bar{y}_{...})^2 &= \sum_{i=1}^{a}\sum_{j=1}^{b}\sum_{k=1}^{n}[(\bar{y}_{i..}-\bar{y}_{...})+(\bar{y}_{.j.}-\bar{y}_{...}) \\
&\quad + (\bar{y}_{ij.}-\bar{y}_{i..}-\bar{y}_{.j.}+\bar{y}_{...})+(y_{ijk}-\bar{y}_{ij.})]^2 \\
&= bn\sum_{i=1}^{a}(\bar{y}_{i..}-\bar{y}_{...})^2 + an\sum_{j=1}^{b}(\bar{y}_{.j.}-\bar{y}_{...})^2 \\
&\quad + n\sum_{i=1}^{a}\sum_{j=1}^{b}(\bar{y}_{ij.}-\bar{y}_{i..}-\bar{y}_{.j.}+\bar{y}_{...})^2 \\
&\quad + \sum_{i=1}^{a}\sum_{j=1}^{b}\sum_{k=1}^{n}(y_{ijk}-\bar{y}_{ij.})^2
\end{aligned}
\tag{5.4}
$$

because the six cross products on the right-hand side are zero. Notice that the total sum of squares has been partitioned into a sum of squares due to "rows," or factor A, (SS_A); a sum of squares due to "columns," or factor B, (SS_B); a sum of squares due to the interaction between A and B, (SS_{AB}); and a sum of squares due to error, (SS_E). This is the fundamental ANOVA equation for the two-factor factorial. From the last component on the right-hand side of Equation 5.4, we see that there must be at least two replicates ($n \geq 2$) to obtain an error sum of squares.

We may write Equation 5.4 symbolically as

$$
SS_T = SS_A + SS_B + SS_{AB} + SS_E \tag{5.5}
$$

The number of degrees of freedom associated with each sum of squares is

Effect	Degrees of Freedom
A	$a - 1$
B	$b - 1$
AB interaction	$(a - 1)(b - 1)$
Error	$ab(n - 1)$
Total	$abn - 1$

We may justify this allocation of the $abn - 1$ total degrees of freedom to the sums of squares as follows: The main effects A and B have a and b levels, respectively; therefore they have $a - 1$ and $b - 1$ degrees of freedom as shown. The interaction degrees of freedom are simply the number of degrees of freedom for cells (which is $ab - 1$) minus the number of degrees of freedom for the two main effects A and B; that is, $ab - 1 - (a - 1) - (b - 1) =$

$(a-1)(b-1)$. Within each of the ab cells, there are $n-1$ degrees of freedom between the n replicates; thus there are $ab(n-1)$ degrees of freedom for error. Note that the number of degrees of freedom on the right-hand side of Equation 5.5 adds to the total number of degrees of freedom.

Each sum of squares divided by its degrees of freedom is a **mean square**. The expected values of the mean squares are

$$E(MS_A) = E\left(\frac{SS_A}{a-1}\right) = \sigma^2 + \frac{bn\sum_{i=1}^{a}\tau_i^2}{a-1}$$

$$E(MS_B) = E\left(\frac{SS_B}{b-1}\right) = \sigma^2 + \frac{an\sum_{j=1}^{b}\beta_j^2}{b-1}$$

$$E(MS_{AB}) = E\left(\frac{SS_{AB}}{(a-1)(b-1)}\right) = \sigma^2 + \frac{n\sum_{i=1}^{a}\sum_{j=1}^{b}(\tau\beta)_{ij}^2}{(a-1)(b-1)}$$

and

$$E(MS_E) = E\left(\frac{SS_E}{ab(n-1)}\right) = \sigma^2$$

Notice that if the null hypotheses of no row treatment effects, no column treatment effects, and no interaction are true, then MS_A, MS_B, MS_{AB}, and MS_E all estimate σ^2. However, if there are differences between row treatment effects, say, then MS_A will be larger than MS_E. Similarly, if there are column treatment effects or interaction present, then the corresponding mean squares will be larger than MS_E. Therefore, to test the significance of both main effects and their interaction, simply divide the corresponding mean square by the error mean square. Large values of this ratio imply that the data do not support the null hypothesis.

If we assume that the model (Equation 5.1) is adequate and that the error terms ϵ_{ijk} are normally and independently distributed with constant variance σ^2, then each of the ratios of mean squares MS_A/MS_E, MS_B/MS_E, and MS_{AB}/MS_E is distributed as F with $a-1$, $b-1$, and $(a-1)(b-1)$ numerator degrees of freedom, respectively, and $ab(n-1)$ denominator degrees of freedom,[1] and the critical region would be the upper tail of the F distribution. The test procedure is usually summarized in an **analysis of variance table**, as shown in Table 5.3.

Computationally, we almost always employ a statistical software package to conduct an ANOVA. However, manual computing of the sums of squares in Equation 5.5 is straightforward. One could write out the individual elements of the ANOVA identity

$$y_{ijk} - \bar{y}_{...} = (\bar{y}_{i..} - \bar{y}_{...}) + (\bar{y}_{.j.} - \bar{y}_{...}) + (\bar{y}_{ij.} - \bar{y}_{i..} - \bar{y}_{.j.} + \bar{y}_{...}) + (y_{ijk} - \bar{y}_{ij.})$$

and calculate them in the columns of a spreadsheet. Then each column could be squared and summed to produce the ANOVA sums of squares. Computing formulas in terms of row, column, and cell totals can also be used. The total sum of squares is computed as usual by

$$SS_T = \sum_{i=1}^{a}\sum_{j=1}^{b}\sum_{k=1}^{n} y_{ijk}^2 - \frac{y_{...}^2}{abn} \tag{5.6}$$

[1]The F test may be viewed as an approximation to a randomization test, as noted previously.

■ **TABLE 5.3**
The Analysis of Variance Table for the Two-Factor Factorial, Fixed Effects Model

Source of Variation	Sum of Squares	Degrees of Freedom	Mean Square	F_0
A treatments	SS_A	$a - 1$	$MS_A = \frac{SS_A}{a-1}$	$F_0 = \frac{MS_A}{MS_E}$
B treatments	SS_B	$b - 1$	$MS_B = \frac{SS_B}{b-1}$	$F_0 = \frac{MS_B}{MS_E}$
Interaction	SS_{AB}	$(a-1)(b-1)$	$MS_{AB} = \frac{SS_{AB}}{(a-1)(b-1)}$	$F_0 = \frac{MS_{AB}}{MS_E}$
Error	SS_E	$ab(n-1)$	$MS_E = \frac{SS_E}{ab(n-1)}$	
Total	SS_T	$abn - 1$		

The sums of squares for the main effects are

$$SS_A = \frac{1}{bn}\sum_{i=1}^{a} y_{i..}^2 - \frac{y_{...}^2}{abn} \tag{5.7}$$

and

$$SS_B = \frac{1}{an}\sum_{j=1}^{b} y_{.j.}^2 - \frac{y_{...}^2}{abn} \tag{5.8}$$

It is convenient to obtain the SS_{AB} in two stages. First we compute the sum of squares between the ab cell totals, which is called the sum of squares due to "subtotals":

$$SS_{\text{Subtotals}} = \frac{1}{n}\sum_{i=1}^{a}\sum_{j=1}^{b} y_{ij.}^2 - \frac{y_{...}^2}{abn}$$

This sum of squares also contains SS_A and SS_B. Therefore, the second step is to compute SS_{AB} as

$$SS_{AB} = SS_{\text{Subtotals}} - SS_A - SS_B \tag{5.9}$$

We may compute SS_E by subtraction as

$$SS_E = SS_T - SS_{AB} - SS_A - SS_B \tag{5.10}$$

or

$$SS_E = SS_T - SS_{\text{Subtotals}}$$

EXAMPLE 5.1 The Battery Design Experiment

Table 5.4 presents the effective life (in hours) observed in the battery design example described in Section 5.3.1. The row and column totals are shown in the margins of the table, and the circled numbers are the cell totals.

■ **TABLE 5.4**
Life Data (in hours) for the Battery Design Experiment

Material Type	Temperature (°F) 15			70			125			$y_{i..}$
1	130	155		34	40		20	70		
	74	180	(539)	80	75	(229)	82	58	(230)	998
2	150	188		136	122		25	70		
	159	126	(623)	106	115	(479)	58	45	(198)	1300
3	138	110		174	120		96	104		
	168	160	(576)	150	139	(583)	82	60	(342)	1501
$y_{.j.}$	1738			1291			770			$3799 = y_{...}$

Using Equations 5.6 through 5.10, the sums of squares are computed as follows:

$$SS_T = \sum_{i=1}^{a}\sum_{j=1}^{b}\sum_{k=1}^{n} y_{ijk}^2 - \frac{y_{...}^2}{abn}$$

$$= (130)^2 + (155)^2 + (74)^2 + \cdots$$

$$+ (60)^2 - \frac{(3799)^2}{36} = 77{,}646.97$$

$$SS_{\text{Material}} = \frac{1}{bn}\sum_{i=1}^{a} y_{i..}^2 - \frac{y_{...}^2}{abn}$$

$$= \frac{1}{(3)(4)}[(998)^2 + (1300)^2 + (1501)^2]$$

$$- \frac{(3799)^2}{36} = 10{,}683.72$$

$$SS_{\text{Temperature}} = \frac{1}{an}\sum_{j=1}^{b} y_{.j.}^2 - \frac{y_{...}^2}{abn}$$

$$= \frac{1}{(3)(4)}[(1738)^2 + (1291)^2 + (770)^2]$$

$$- \frac{(3799)^2}{36} = 39{,}118.72$$

$$SS_{\text{Interaction}} = \frac{1}{n}\sum_{i=1}^{a}\sum_{j=1}^{b} y_{ij.}^2 - \frac{y_{...}^2}{abn} - SS_{\text{Material}}$$

$$- SS_{\text{Temperature}}$$

$$= \frac{1}{4}[(539)^2 + (229)^2 + \cdots + (342)^2]$$

$$- \frac{(3799)^2}{36} - 10{,}683.72$$

$$- 39{,}118.72 = 9613.78$$

and

$$SS_E = SS_T - SS_{\text{Material}} - SS_{\text{Temperature}} - SS_{\text{Interaction}}$$

$$= 77{,}646.97 - 10{,}683.72 - 39{,}118.72$$

$$- 9613.78 = 18{,}230.75$$

The ANOVA is shown in Table 5.5. Because $F_{0.05,4,27} = 2.73$, we conclude that there is a significant interaction between material types and temperature. Furthermore, $F_{0.05,2,27} = 3.35$, so the main effects of material type and temperature are also significant. Table 5.5 also shows the *P*-values for the test statistics.

To assist in interpreting the results of this experiment, it is helpful to construct a graph of the average responses at

■ **TABLE 5.5**
Analysis of Variance for Battery Life Data

Source of Variation	Sum of Squares	Degrees of Freedom	Mean Square	F_0	*P*-Value
Material types	10,683.72	2	5,341.86	7.91	0.0020
Temperature	39,118.72	2	19,559.36	28.97	< 0.0001
Interaction	9,613.78	4	2,403.44	3.56	0.0186
Error	18,230.75	27	675.21		
Total	77,646.97	35			

each treatment combination. This graph is shown in Figure 5.9. The significant interaction is indicated by the lack of parallelism of the lines. In general, longer life is attained at low temperature, regardless of material type. Changing from low to intermediate temperature, battery life with material type 3 may actually increase, whereas it decreases for types 1 and 2. From intermediate to high temperature, battery life decreases for material types 2 and 3 and is essentially unchanged for type 1. Material type 3 seems to give the best results if we want less loss of effective life as the temperature changes.

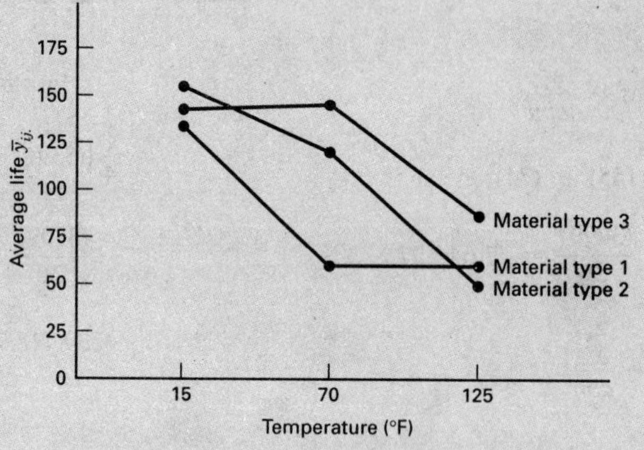

■ **FIGURE 5.9 Material type–temperature plot for Example 5.1**

Multiple Comparisons. When the ANOVA indicates that row or column means differ, it is usually of interest to make comparisons between the individual row or column means to discover the specific differences. The multiple comparison methods discussed in Chapter 3 are useful in this regard.

We now illustrate the use of Tukey's test on the battery life data in Example 5.1. Note that in this experiment, interaction is significant. When interaction is significant, comparisons between the means of one factor (e.g., A) may be obscured by the AB interaction. One approach to this situation is to fix factor B at a specific level and apply Tukey's test to the means of factor A at that level. To illustrate, suppose that in Example 5.1 we are interested in detecting differences among the means of the three material types. Because interaction is significant, we make this comparison at just one level of temperature, say level 2 (70°F). We assume that the best estimate of the error variance is the MS_E from the ANOVA table, utilizing the assumption that the experimental error variance is the same over all treatment combinations.

The three material type averages at 70°F arranged in ascending order are

$$\bar{y}_{12.} = 57.25 \quad \text{(material type 1)}$$
$$\bar{y}_{22.} = 119.75 \quad \text{(material type 2)}$$
$$\bar{y}_{32.} = 145.75 \quad \text{(material type 3)}$$

and

$$T_{0.05} = q_{0.05}(3, 27)\sqrt{\frac{MS_E}{n}} = 3.50\sqrt{\frac{675.21}{4}} = 45.47$$

where we obtained $q_{0.05}(3, 27) \simeq 3.50$ by interpolation. The pairwise comparisons yield

$$\text{3 vs. 1:} \quad 145.75 - 57.25 = 88.50 > T_{0.05} = 45.47$$
$$\text{3 vs. 2:} \quad 145.75 - 119.75 = 26.00 < T_{0.05} = 45.47$$
$$\text{2 vs. 1:} \quad 119.75 - 57.25 = 62.50 > T_{0.05} = 45.47$$

This analysis indicates that at the temperature level 70°F, the mean battery life is the same for material types 2 and 3, and that the mean battery life for material type 1 is significantly lower in comparison to both types 2 and 3.

If interaction is significant, the experimenter could compare *all ab* cell means to determine which ones differ significantly. In this analysis, differences between cell means include interaction effects as well as both main effects. In Example 5.1, this would give 36 comparisons between all possible pairs of the nine cell means.

Computer Output. Figure 5.10 presents condensed computer output for the battery life data in Example 5.1. Figure 5.10a contains Design-Expert output and Figure 5.10b contains JMP output. Note that

$$SS_{\text{Model}} = SS_{\text{Material}} + SS_{\text{Temperature}} + SS_{\text{Interaction}} = 10{,}683.72 + 39{,}118.72 + 9613.78 = 59{,}416.22$$

with 8 degrees of freedom. An F test is displayed for the model source of variation. The P-value is small (< 0.0001), so the interpretation of this test is that at least one of the three terms in the model is significant. The tests on the individual model terms (A, B, AB) follow. Also,

$$R^2 = \frac{SS_{\text{Model}}}{SS_{\text{Total}}} = \frac{59{,}416.22}{77{,}646.97} = 0.7652$$

That is, about 77 percent of the variability in the battery life is explained by the plate material in the battery, the temperature, and the material type–temperature interaction. The residuals from the fitted model are displayed on the Design-Expert computer output and the JMP output contains a plot of the residuals versus the predicted response. We now discuss the use of these residuals and residual plots in model adequacy checking.

Response: Life In Hours
ANOVA for Selected Factorial Model
Analysis of Variance Table [Partial Sum of Squares]

Source	Sum of Squares	DF	Mean Square	F Value	Prob > F	
Model	59416.22	8	7427.03	11.00	<0.0001	significant
A	*10683.72*	*2*	*5341.86*	*7.91*	*0.0020*	
B	*39118.72*	*2*	*19559.36*	*28.97*	*<0.0001*	
AB	*9613.78*	*4*	*2403.44*	*3.56*	*0.0186*	
Residual	18230.75	27	675.21			
Lack of Fit	*0.000*	*0*				
Pure Error	*18230.75*	*27*	*675.21*			
Cor Total	77646.97	35				

Std. Dev.	25.98	R-Squared	0.7652
Mean	105.53	Adj R-Squared	0.6956
C.V.	24.62	Pred R-Squared	0.5826
PRESS	32410.22	Adeq Precision	8.178

Diagnostics Case Statistics

Standard Order	Actual Value	Predicted Value	Residual	Leverage	Student Residual	Cook's Distance	Outlier *t*
1	130.00	134.75	−4.75	0.250	−0.211	0.002	−0.207
2	74.00	134.75	−60.75	0.250	−2.700	0.270	−3.100
3	155.00	134.75	20.25	0.250	0.900	0.030	0.897
4	180.00	134.75	45.25	0.250	2.011	0.150	2.140
5	150.00	155.75	−5.75	0.250	−0.256	0.002	−0.251
6	159.00	155.75	3.25	0.250	0.144	0.001	0.142
7	188.00	155.75	32.25	0.250	1.433	0.076	1.463
8	126.00	155.75	−29.75	0.250	−1.322	0.065	−1.341
9	138.00	144.00	−6.00	0.250	−0.267	0.003	−0.262
10	168.00	144.00	24.00	0.250	1.066	0.042	1.069
11	110.00	144.00	−34.00	0.250	−1.511	0.085	−1.550
12	160.00	144.00	16.00	0.250	0.711	0.019	0.704
13	34.00	57.25	−23.25	0.250	−1.033	0.040	−1.035
14	80.00	57.25	22.75	0.250	1.011	0.038	1.011
15	40.00	57.25	−17.25	0.250	−0.767	0.022	−0.761
16	75.00	57.25	17.75	0.250	0.789	0.023	0.783
17	136.00	119.75	16.25	0.250	0.722	0.019	0.716
18	106.00	119.75	−13.75	0.250	−0.611	0.014	−0.604
19	122.00	119.75	2.25	0.250	0.100	0.000	0.098
20	115.00	119.75	−4.75	0.250	−0.211	0.002	−0.207
21	174.00	145.75	28.25	0.250	1.255	0.058	1.269
22	150.00	145.75	4.25	0.250	0.189	0.001	0.185
23	120.00	145.75	−25.75	0.250	−1.144	0.048	−1.151
24	139.00	145.75	−6.75	0.250	−0.300	0.003	−0.295
25	20.00	57.50	−37.50	0.250	−1.666	0.103	−1.726
26	82.00	57.50	24.50	0.250	1.089	0.044	1.093
27	70.00	57.50	12.50	0.250	0.555	0.011	0.548
28	58.00	57.50	0.50	0.250	0.022	0.000	0.022
29	25.00	49.50	−24.50	0.250	−1.089	0.044	−1.093
30	58.00	49.50	8.50	0.250	0.378	0.005	0.372
31	70.00	49.50	20.50	0.250	0.911	0.031	0.908
32	45.00	49.50	−4.50	0.250	−0.200	0.001	−0.196
33	96.00	85.50	10.50	0.250	0.467	0.008	0.460
34	82.00	85.50	−3.50	0.250	−0.156	0.001	−0.153
35	104.00	85.50	18.50	0.250	0.822	0.025	0.817
36	60.00	85.50	−25.50	0.250	−1.133	0.048	−1.139

(a)

■ **FIGURE 5.10** **Computer output for Example 5.1. (a) Design-Expert output; (b) JMP output**

Response Life
Whole Model
Actual by Predicted Plot

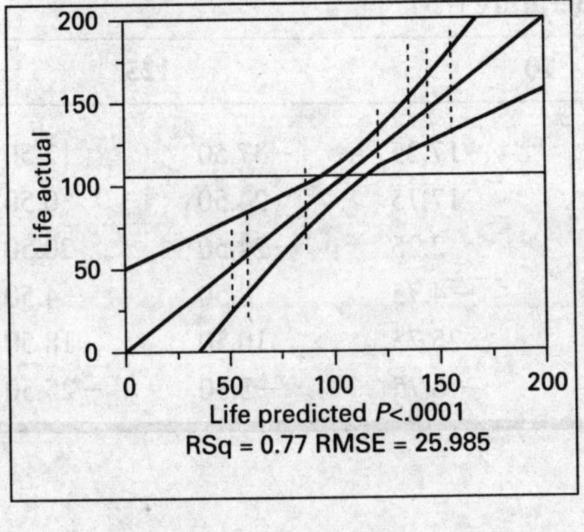

Summary of Fit

RSquare	0.76521
RSquare Adj	0.695642
Root Mean Square Error	25.98486
Mean of Response	105.5278
Observations (or Sum Wgts)	36

Analysis of Variance

Source	DF	Sum of Squares	Mean Square	*F* Ratio
Model	8	59416.222	7427.03	10.9995
Error	27	18230.750	675.21	Prob > F
C.Total	35	77646.972		<.001

Effect Tests

Source	Nparm	DF	Sum of Squares	*F* Ratio	Prob > *F*
Material Type	2	2	10683.722	7.9114	0.0020
Temperature	2	2	39118.722	28.9677	<.0001
Material Type Temperature	4	4	9613.778	3.5595	0.0186

Residual by Predicted Plot

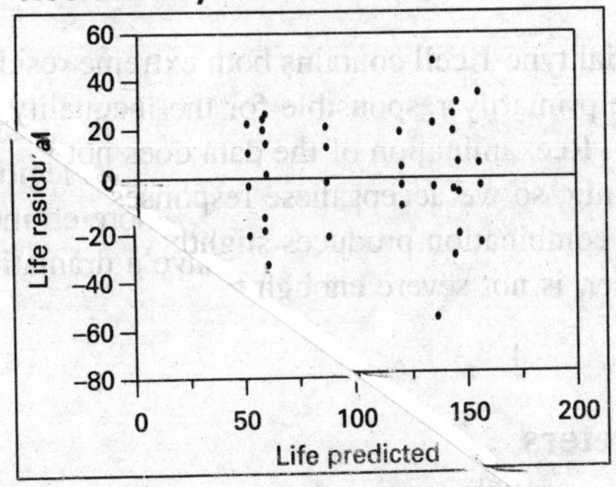

(b)

■ **FIGURE 5.10** ***(Continued)***

■ **TABLE 5.6**
Residuals for Example 5.1

Material Type	Temperature (°F) 15		70		125	
1	−4.75	20.25	−23.25	−17.25	−37.50	12.50
	−60.75	45.25	22.75	17.75	24.50	0.50
2	−5.75	32.25	16.25	2.25	−24.50	20.50
	3.25	−29.75	−13.75	−4.75	8.50	−4.50
3	−6.00	−34.00	28.25	−25.75	10.50	18.50
	24.00	16.00	4.25	−6.75	−3.50	−25.50

5.3.3 Model Adequacy Checking

Before the conclusions from the ANOVA are adopted, the adequacy of the underlying model should be checked. As before, the primary diagnostic tool is **residual analysis**. The residuals for the two-factor factorial model with interaction are

$$e_{ijk} = y_{ijk} - \hat{y}_{ijk} \tag{5.11}$$

and because the fitted value $\hat{y}_{ijk} = \bar{y}_{ij.}$ (the average of the observations in the ijth cell), Equation 5.11 becomes

$$e_{ijk} = y_{ijk} - \bar{y}_{ij.} \tag{5.12}$$

The residuals from the battery life data in Example 5.1 are shown in the Design-Expert computer output (Figure 5.10a) and in Table 5.6. The normal probability plot of these residuals (Figure 5.11) does not reveal anything particularly troublesome, although the largest negative residual (−60.75 at 15°F for material type 1) does stand out somewhat from the others. The standardized value of this residual is $-60.75/\sqrt{675.21} = -2.34$, and this is the only residual whose absolute value is larger than 2.

Figure 5.12 plots the residuals versus the fitted values $\hat{y}_{ijk}$. This plot was also shown in the JMP computer output in Figure 5.10*b*. There is some mild tendency for the variance of the residuals to increase as the battery life increases. Figures 5.13 and 5.14 plot the residuals versus material types and temperature, respectively. Both plots indicate mild inequality of variance, with the treatment combination of 15°F and material type 1 possibly having larger variance than the others.

From Table 5.6 we see that the 15°F-material type 1 cell contains both extreme residuals (−60.75 and 45.25). These two residuals are primarily responsible for the inequality ... variance detected in Figures 5.12, 5.13, and 5.14. Reexamination of the data does no... ...veal any obvious problem, such as an error in recording, so we accept these respons... ...as legitimate. It is possible that this particular treatment combination produces sligh... more erratic battery life than the others. The problem, however, is not severe enough ... have a dramatic impact on the analysis and conclusions.

5.3.4 Estimating the Model Parameters

The parameters in the effects model for two-factor f... ...orial

$$y_{ijk} = \mu + \tau_i + \ldots + (\tau\beta)_{ij} + \epsilon_{ijk} \tag{5.13}$$

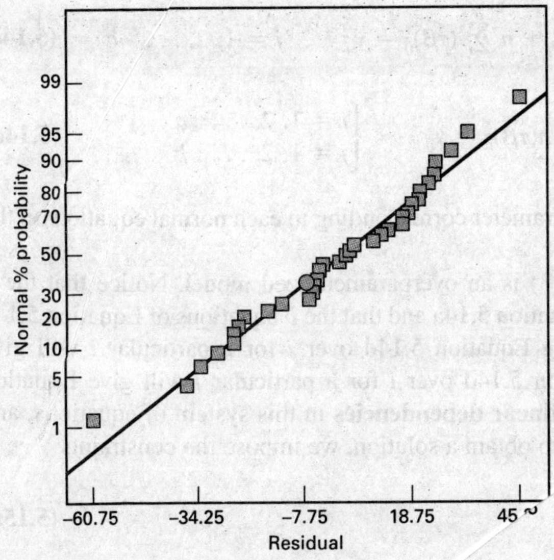

■ **FIGURE 5.11**
Normal probability plot of residuals for Example 5.1

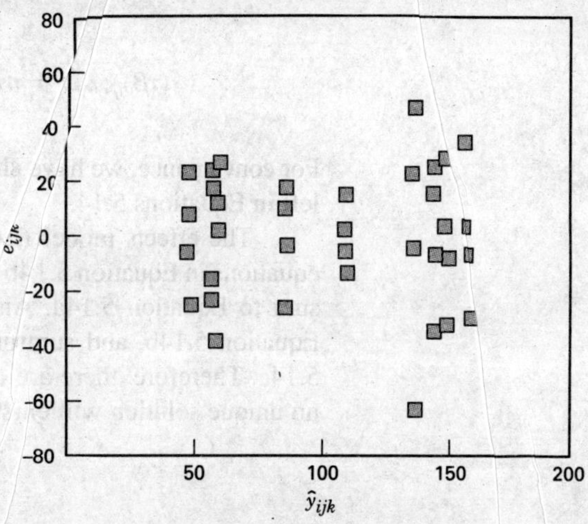

■ **FIGURE 5.12**
Plot of residuals versus $\hat{y}_{ijk}$ for Example 5.1

may be estimated by least squares. Because the model has $1 + a + b + ab$ parameters to be estimated, there are $1 + a + b + ab$ normal equations. Using the method of Section 3.9, we find that it is not difficult to show that the normal equations are

$$\mu: abn\hat{\mu} + bn\sum_{i=1}^{a}\hat{\tau}_i + an\sum_{j=1}^{b}\hat{\beta}_j + n\sum_{i=1}^{a}\sum_{j=1}^{b}(\widehat{\tau\beta})_{ij} = y_{...} \tag{5.14a}$$

$$\tau_i: bn\hat{\mu} + bn\hat{\tau}_i + n\sum_{j=1}^{b}\hat{\beta}_j + n\sum_{j=1}^{b}(\widehat{\tau\beta})_{ij} = y_{i..} \qquad i = 1, 2, \ldots, a \tag{5.14b}$$

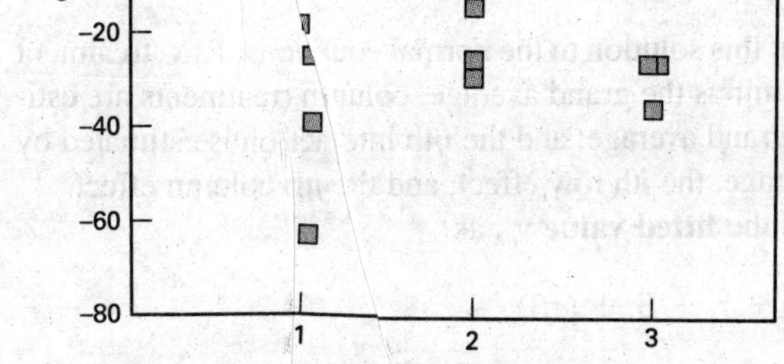

■ **FIGURE 5.13**
Plot of residuals versus material type for Example 5.1

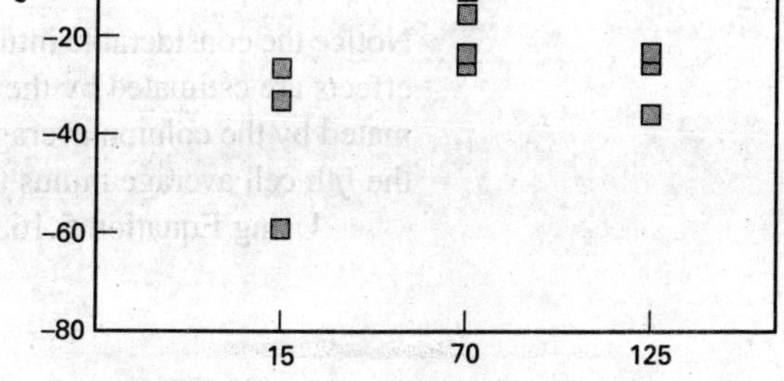

■ **FIGURE 5.14**
Plot of residuals versus temperature for Example 5.1

$$\beta_j : an\hat{\mu} + n\sum_{i=1}^{a}\hat{\tau}_i + an\hat{\beta}_j + n\sum_{i=1}^{a}(\widehat{\tau\beta})_{ij} = y_{.j.} \qquad j = 1, 2, \ldots, b \tag{5.14c}$$

$$(\tau\beta)_{ij} : n\hat{\mu} + n\hat{\tau}_i + n\hat{\beta}_j + n(\widehat{\tau\beta})_{ij} = y_{ij.} \qquad \begin{cases} i = 1, 2, \ldots, a \\ j = 1, 2, \ldots, b \end{cases} \tag{5.14d}$$

For convenience, we have shown the parameter corresponding to each normal equation on the left in Equations 5.14.

The effects model (Equation 5.13) is an overparameterized model. Notice that the a equations in Equation 5.14b sum to Equation 5.14a and that the b equations of Equation 5.14c sum to Equation 5.14a. Also summing Equation 5.14d over j for a particular i will give Equation 5.14b, and summing Equation 5.14d over i for a particular j will give Equation 5.14c. Therefore, there are $a + b + 1$ **linear dependencies** in this system of equations, and no unique solution will exist. In order to obtain a solution, we impose the constraints

$$\sum_{i=1}^{a}\hat{\tau}_i = 0 \tag{5.15a}$$

$$\sum_{j=1}^{b}\hat{\beta}_j = 0 \tag{5.15b}$$

$$\sum_{i=1}^{a}(\widehat{\tau\beta})_{ij} = 0 \qquad j = 1, 2, \ldots, b \tag{5.15c}$$

and

$$\sum_{j=1}^{b}(\widehat{\tau\beta})_{ij} = 0 \qquad i = 1, 2, \ldots, a \tag{5.15d}$$

Equations 5.15a and 5.15b constitute two constraints, whereas Equations 5.15c and 5.15d form $a + b - 1$ independent constraints. Therefore, we have $a + b + 1$ total constraints, the number needed.

Applying these constraints, the normal equations (Equations 5.14) simplify considerably, and we obtain the solution

$$\begin{aligned} \hat{\mu} &= \bar{y}_{...} \\ \hat{\tau}_i &= \bar{y}_{i..} - \bar{y}_{...} \qquad i = 1, 2, \ldots, a \\ \hat{\beta}_j &= \bar{y}_{.j.} - \bar{y}_{...} \qquad j = 1, 2, \ldots, b \\ (\widehat{\tau\beta})_{ij} &= \bar{y}_{ij.} - \bar{y}_{i..} - \bar{y}_{.j.} + \bar{y}_{...} \qquad \begin{cases} i = 1, 2, \ldots, a \\ j = 1, 2, \ldots, b \end{cases} \end{aligned} \tag{5.16}$$

Notice the considerable intuitive appeal of this solution to the normal equations. Row treatment effects are estimated by the row average minus the grand average; column treatments are estimated by the column average minus the grand average; and the ijth interaction is estimated by the ijth cell average minus the grand average, the ith row effect, and the jth column effect.

Using Equation 5.16, we may find the **fitted value** y_{ijk} as

$$\begin{aligned} \hat{y}_{ijk} &= \hat{\mu} + \hat{\tau}_i + \hat{\beta}_j + (\widehat{\tau\beta})_{ij} \\ &= \bar{y}_{...} + (\bar{y}_{i..} - \bar{y}_{...}) + (\bar{y}_{.j.} - \bar{y}_{...}) \\ &\quad + (\bar{y}_{ij.} - \bar{y}_{i..} - \bar{y}_{.j.} + \bar{y}_{...}) \\ &= \bar{y}_{ij.} \end{aligned}$$

That is, the kth observation in the ijth cell is estimated by the average of the n observations in that cell. This result was used in Equation 5.12 to obtain the residuals for the two-factor factorial model.

Because constraints (Equations 5.15) have been used to solve the normal equations, the model parameters are not uniquely estimated. However, certain important **functions** of the model parameters *are* estimable, that is, uniquely estimated regardless of the constraint chosen. An example is $\tau_i - \tau_u + \overline{(\tau\beta)}_{i.} - \overline{(\tau\beta)}_{u.}$, which might be thought of as the "true" difference between the ith and the uth levels of factor A. Notice that the true difference between the levels of any main effect includes an "average" interaction effect. It is this result that disturbs the tests on main effects in the presence of interaction, as noted earlier. In general, any function of the model parameters that is a linear combination of the left-hand side of the normal equations is **estimable**. This property was also noted in Chapter 3 when we were discussing the single-factor model. For more information, see the supplemental text material for this chapter.

5.3.5 Choice of Sample Size

The operating characteristic curves in Appendix Chart V can be used to assist the experimenter in determining an appropriate sample size (number of replicates, n) for a two-factor factorial design. The appropriate value of the parameter Φ^2 and the numerator and denominator degrees of freedom are shown in Table 5.7.

A very effective way to use these curves is to find the smallest value of Φ^2 corresponding to a specified difference between any two treatment means. For example, if the difference in any two row means is D, then the minimum value of Φ^2 is

$$\Phi^2 = \frac{nbD^2}{2a\sigma^2} \tag{5.17}$$

whereas if the difference in any two column means is D, then the minimum value of Φ^2 is

$$\Phi^2 = \frac{naD^2}{2b\sigma^2} \tag{5.18}$$

■ TABLE 5.7
Operating Characteristic Curve Parameters for the Two-Factor Factorial, Fixed Effects Model

Factor	Φ^2	Numerator Degrees of Freedom	Denominator Degrees of Freedom
A	$\dfrac{bn \sum_{i=1}^{a} \tau_i^2}{a\sigma^2}$	$a - 1$	$ab(n - 1)$
B	$\dfrac{an \sum_{j=1}^{b} \beta_j^2}{b\sigma^2}$	$b - 1$	$ab(n - 1)$
AB	$\dfrac{n \sum_{i=1}^{a} \sum_{j=1}^{b} (\tau\beta)_{ij}^2}{\sigma^2[(a-1)(b-1)+1]}$	$(a - 1)(b - 1)$	$ab(n - 1)$

Finally, the minimum value of Φ^2 corresponding to a difference of D between any two interaction effects is

$$\Phi^2 = \frac{nD^2}{2\sigma^2[(a-1)(b-1)+1]} \tag{5.19}$$

To illustrate the use of these equations, consider the battery life data in Example 5.1. Suppose that before running the experiment we decide that the null hypothesis should be rejected with a high probability if the difference in mean battery life between any two temperatures is as great as 40 hours. Thus a difference of $D = 40$ has engineering significance, and if we assume that the standard deviation of battery life is approximately 25, then Equation 5.18 gives

$$\begin{aligned}\Phi^2 &= \frac{naD^2}{2b\sigma^2} \\ &= \frac{n(3)(40)^2}{2(3)(25)^2} \\ &= 1.28n\end{aligned}$$

as the minimum value of Φ^2. Assuming that $\alpha = 0.05$, we can now use Characteristic Curves for the Fixed Effects Model Analysis of Variance* to construct the following display:

n	Φ^2	Φ	ν_1 = Numerator Degrees of Freedom	ν_2 = Error Degrees of Freedom	β
2	2.56	1.60	2	9	0.45
3	3.84	1.96	2	18	0.18
4	5.12	2.26	2	27	0.06

Note that $n = 4$ replicates give a β risk of about 0.06, or approximately a 94 percent chance of rejecting the null hypothesis if the difference in mean battery life at any two temperature levels is as large as 40 hours. Thus, we conclude that four replicates are enough to provide the desired sensitivity as long as our estimate of the standard deviation of battery life is not seriously in error. If in doubt, the experimenter could repeat the above procedure with other values of σ to determine the effect of misestimating this parameter on the sensitivity of the design.

5.3.6 The Assumption of No Interaction in a Two-Factor Model

Occasionally, an experimenter feels that a **two-factor model without interaction** is appropriate, say

$$y_{ijk} = \mu + \tau_i + \beta_j + \epsilon_{ijk} \quad \begin{cases} i = 1, 2, \ldots, a \\ j = 1, 2, \ldots, b \\ k = 1, 2, \ldots, n \end{cases} \tag{5.20}$$

We should be very careful in dispensing with the interaction terms, however, because the presence of significant interaction can have a dramatic impact on the interpretation of the data.

The statistical analysis of a two-factor factorial model without interaction is straightforward. Table 5.8 presents the analysis of the battery life data from Example 5.1, assuming that

* From *Biometrika Tables for Statisticians*, Vol. 2, by E. S. Pearson and H. O. Hartley, (1972).

■ **TABLE 5.8**

Analysis of Variance for Battery Life Data Assuming No Interaction

Source of Variation	Sum of Squares	Degrees of Freedom	Mean Square	F_0
Material types	10,683.72	2	5,341.86	5.95
Temperature	39,118.72	2	19,559.36	21.78
Error	27,844.52	31	898.21	
Total	77,646.96	35		

the no-interaction model (Equation 5.20) applies. As noted previously, both main effects are significant. However, as soon as a residual analysis is performed for these data, it becomes clear that the no-interaction model is inadequate. For the two-factor model without interaction, the fitted values are $\hat{y}_{ijk} = \bar{y}_{i..} + \bar{y}_{.j.} - \bar{y}_{...}$. A plot of $\bar{y}_{ij.} - \hat{y}_{ijk}$ (the cell averages minus the fitted value for that cell) versus the fitted value $\hat{y}_{ijk}$ is shown in Figure 5.15. Now the quantities $\bar{y}_{ij.} - \hat{y}_{ijk}$ may be viewed as the differences between the observed cell means and the estimated cell means assuming no interaction. Any pattern in these quantities is suggestive of the presence of interaction. Figure 5.15 shows a distinct pattern as the quantities $\bar{y}_{ij.} - \hat{y}_{ijk}$ move from positive to negative to positive to negative again. This structure is the result of interaction between material types and temperature.

5.3.7 One Observation per Cell

Occasionally, one encounters a two-factor experiment with only a **single replicate**, that is, only one observation per cell. If there are two factors and only one observation per cell, the effects model is

$$y_{ij} = \mu + \tau_i + \beta_j + (\tau\beta)_{ij} + \epsilon_{ij} \qquad \begin{cases} i = 1, 2, \ldots, a \\ j = 1, 2, \ldots, b \end{cases} \tag{5.21}$$

The analysis of variance for this situation is shown in Table 5.9, assuming that both factors are fixed.

From examining the expected mean squares, we see that the error variance σ^2 is *not estimable*; that is, the two-factor interaction effect $(\tau\beta)_{ij}$ and the experimental error cannot be separated in any obvious manner. Consequently, there are no tests on main effects unless the

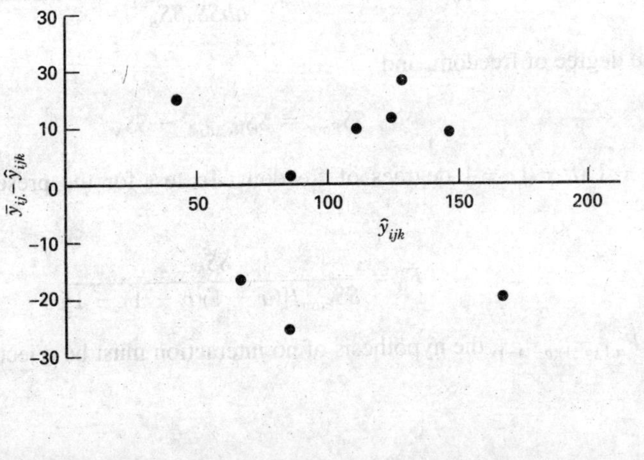

■ **FIGURE 5.15** **Plot of $\bar{y}_{ij.} - \hat{y}_{ijk}$ versus $\hat{y}_{ijk}$, battery life data**

■ **TABLE 5.9**
Analysis of Variance for a Two-Factor Model, One Observation per Cell

Source of Variation	Sum of Squares	Degrees of Freedom	Mean Square	Expected Mean Square
Rows (*A*)	$\sum_{i=1}^{a} \frac{y_{i.}^2}{b} - \frac{y_{..}^2}{ab}$	$a-1$	MS_A	$\sigma^2 + \frac{b\sum \tau_i^2}{a-1}$
Columns (*B*)	$\sum_{j=1}^{b} \frac{y_{.j}^2}{a} - \frac{y_{..}^2}{ab}$	$b-1$	MS_B	$\sigma^2 + \frac{a\sum \beta_j^2}{b-1}$
Residual or *AB*	Subtraction	$(a-1)(b-1)$	MS_{Residual}	$\sigma^2 + \frac{\sum\sum (\tau\beta)_{ij}^2}{(a-1)(b-1)}$
Total	$\sum_{i=1}^{a}\sum_{j=1}^{b} y_{ij}^2 - \frac{y_{..}^2}{ab}$	$ab-1$		

interaction effect is zero. If there is no interaction present, then $(\tau\beta)_{ij} = 0$ for all i and j, and a plausible model is

$$y_{ij} = \mu + \tau_i + \beta_j + \epsilon_{ij} \quad \begin{cases} i = 1, 2, \ldots, a \\ j = 1, 2, \ldots, b \end{cases} \tag{5.22}$$

If the model (Equation 5.22) is appropriate, then the residual mean square in Table 5.9 is an unbiased estimator of σ^2, and the main effects may be tested by comparing MS_A and MS_B to MS_{Residual}.

A test developed by Tukey (1949a) is helpful in determining whether interaction is present. The procedure assumes that the interaction term is of a particularly simple form, namely,

$$(\tau\beta)_{ij} = \gamma\tau_i\beta_j$$

where γ is an unknown constant. By defining the interaction term this way, we may use a regression approach to test the significance of the interaction term. The test partitions the residual sum of squares into a single-degree-of-freedom component due to nonadditivity (interaction) and a component for error with $(a-1)(b-1)-1$ degrees of freedom. Computationally, we have

$$SS_N = \frac{\left[\sum_{i=1}^{a}\sum_{j=1}^{b} y_{ij}y_{i.}y_{.j} - y_{..}\left(SS_A + SS_B + \frac{y_{..}^2}{ab}\right)\right]^2}{abSS_ASS_B} \tag{5.23}$$

with one degree of freedom, and

$$SS_{\text{Error}} = SS_{\text{Residual}} - SS_N \tag{5.24}$$

with $(a-1)(b-1)-1$ degrees of freedom. To test for the presence of interaction, we compute

$$F_0 = \frac{SS_N}{SS_{\text{Error}}/[(a-1)(b-1)-1]} \tag{5.25}$$

If $F_0 > F_{\alpha,1,(a-1)(b-1)-1}$, the hypothesis of no interaction must be rejected.

EXAMPLE 5.2

The impurity present in a chemical product is affected by two factors—pressure and temperature. The data from a single replicate of a factorial experiment are shown in Table 5.10. The sums of squares are

$$SS_A = \frac{1}{b}\sum_{i=1}^{a} y_{i.}^2 - \frac{y_{..}^2}{ab}$$

$$= \frac{1}{5}[23^2 + 13^2 + 8^2] - \frac{44^2}{(3)(5)} = 23.33$$

$$SS_B = \frac{1}{a}\sum_{j=1}^{b} y_{.j}^2 - \frac{y_{..}^2}{ab}$$

$$= \frac{1}{3}[9^2 + 6^2 + 13^2 + 6^2 + 10^2] - \frac{44^2}{(3)(5)} = 11.60$$

$$SS_T = \sum_{i=1}^{a}\sum_{j=1}^{b} y_{ij}^2 - \frac{y_{..}^2}{ab}$$

$$= 166 - 129.07 = 36.93$$

and

$$SS_{\text{Residual}} = SS_T - SS_A - SS_B$$

$$= 36.93 - 23.33 - 11.60 = 2.00$$

The sum of squares for nonadditivity is computed from Equation 5.23 as follows:

$$\sum_{i=1}^{a}\sum_{j=1}^{b} y_{ij}y_{i.}y_{.j} = (5)(23)(9) + (4)(23)(6) + \cdots + (2)(8)(10) = 7236$$

$$SS_N = \frac{\left[\sum_{i=1}^{a}\sum_{j=1}^{b} y_{ij}y_{i.}y_{.j} - y_{..}\left(SS_A + SS_B + \frac{y_{..}^2}{ab}\right)\right]^2}{abSS_ASS_B}$$

$$= \frac{[7236 - (44)(23.33 + 11.60 + 129.07)]^2}{(3)(5)(23.33)(11.60)}$$

$$= \frac{[20.00]^2}{4059.42} = 0.0985$$

and the error sum of squares is, from Equation 5.24,

$$SS_{\text{Error}} = SS_{\text{Residual}} - SS_N$$

$$= 2.00 - 0.0985 = 1.9015$$

The complete ANOVA is summarized in Table 5.11. The test statistic for nonadditivity is $F_0 = 0.0985/0.2716 = 0.36$, so we conclude that there is no evidence of interaction in these data. The main effects of temperature and pressure are significant.

TABLE 5.10
Impurity Data for Example 5.2

Temperature	Pressure					
(°F)	25	30	35	40	45	$y_{i.}$
100	5	4	6	3	5	23
125	3	1	4	2	3	13
150	1	1	3	1	2	8
$y_{.j}$	9	6	13	6	10	$44 = y_{..}$

TABLE 5.11
Analysis of Variance for Example 5.2

Source of Variation	Sum of Squares	Degrees of Freedom	Mean Square	F_0	*P*-Value
Temperature	23.33	2	11.67	42.97	0.0001
Pressure	11.60	4	2.90	10.68	0.0042
Nonadditivity	0.0985	1	0.0985	0.36	0.5674
Error	1.9015	7	0.2716		
Total	36.93	14			

In concluding this section, we note that the two-factor factorial model with one observation per cell (Equation 5.22) looks exactly like the randomized complete block model (Equation 4.1). In fact, the Tukey single-degree-of-freedom test for nonadditivity can be directly applied to test for interaction in the randomized block model. However, remember that the **experimental situations** that lead to the randomized block and factorial models are very different. In the factorial model, *all ab* runs have been made in random order, whereas in the randomized block model, randomization occurs only *within the block*. The blocks are a randomization restriction. Hence, the manner in which the experiments are run and the interpretation of the two models are quite different.

5.4 The General Factorial Design

The results for the two-factor factorial design may be extended to the general case where there are a levels of factor A, b levels of factor B, c levels of factor C, and so on, arranged in a factorial experiment. In general, there will be $abc \ldots n$ total observations if there are n replicates of the complete experiment. Once again, note that we must have at least two replicates ($n \geq 2$) to determine a sum of squares due to error if all possible interactions are included in the model.

If all factors in the experiment are fixed, we may easily formulate and test hypotheses about the main effects and interactions using the ANOVA. For a fixed effects model, test statistics for each main effect and interaction may be constructed by dividing the corresponding mean square for the effect or interaction by the mean square error. All of these F tests will be upper-tail, one-tail tests. The number of degrees of freedom for any main effect is the number of levels of the factor minus one, and the number of degrees of freedom for an interaction is the product of the number of degrees of freedom associated with the individual components of the interaction.

For example, consider the **three-factor analysis of variance model**:

$$y_{ijkl} = \mu + \tau_i + \beta_j + \gamma_k + (\tau\beta)_{ij} + (\tau\gamma)_{ik} + (\beta\gamma)_{jk} + (\tau\beta\gamma)_{ijk} + \epsilon_{ijkl} \quad \begin{cases} i = 1, 2, \ldots, a \\ j = 1, 2, \ldots, b \\ k = 1, 2, \ldots, c \\ l = 1, 2, \ldots, n \end{cases} \tag{5.26}$$

Assuming that A, B, and C are fixed, the **analysis of variance table** is shown in Table 5.12. The F tests on main effects and interactions follow directly from the expected mean squares.

Usually, the analysis of variance computations would be done using a statistics software package. However, manual computing formulas for the sums of squares in Table 5.12 are occasionally useful. The total sum of squares is found in the usual way as

$$SS_T = \sum_{i=1}^{a}\sum_{j=1}^{b}\sum_{k=1}^{c}\sum_{l=1}^{n} y_{ijkl}^2 - \frac{y_{....}^2}{abcn} \tag{5.27}$$

The sums of squares for the main effects are found from the totals for factors $A(y_{i...})$, $B(y_{.j..})$, and $C(y_{..k.})$ as follows:

$$SS_A = \frac{1}{bcn}\sum_{i=1}^{a} y_{i...}^2 - \frac{y_{....}^2}{abcn} \tag{5.28}$$

$$SS_B = \frac{1}{acn}\sum_{j=1}^{b} y_{.j..}^2 - \frac{y_{....}^2}{abcn} \tag{5.29}$$

$$SS_C = \frac{1}{abn}\sum_{k=1}^{c} y_{..k.}^2 - \frac{y_{....}^2}{abcn} \tag{5.30}$$

■ **TABLE 5.12**
The Analysis of Variance Table for the Three-Factor Fixed Effects Model

Source of Variation	Sum of Squares	Degrees of Freedom	Mean Square	Expected Mean Square	F_0
A	SS_A	$a-1$	MS_A	$\sigma^2 + \frac{bcn\sum \tau_i^2}{a-1}$	$F_0 = \frac{MS_A}{MS_E}$
B	SS_B	$b-1$	MS_B	$\sigma^2 + \frac{acn\sum \beta_j^2}{b-1}$	$F_0 = \frac{MS_B}{MS_E}$
C	SS_C	$c-1$	MS_C	$\sigma^2 + \frac{abn\sum \gamma_k^2}{c-1}$	$F_0 = \frac{MS_C}{MS_E}$
AB	SS_{AB}	$(a-1)(b-1)$	MS_{AB}	$\sigma^2 + \frac{cn\sum\sum (\tau\beta)_{ij}^2}{(a-1)(b-1)}$	$F_0 = \frac{MS_{AB}}{MS_E}$
AC	SS_{AC}	$(a-1)(c-1)$	MS_{AC}	$\sigma^2 + \frac{bn\sum\sum (\tau\gamma)_{ik}^2}{(a-1)(c-1)}$	$F_0 = \frac{MS_{AC}}{MS_E}$
BC	SS_{BC}	$(b-1)(c-1)$	MS_{BC}	$\sigma^2 + \frac{an\sum\sum (\beta\gamma)_{jk}^2}{(b-1)(c-1)}$	$F_0 = \frac{MS_{BC}}{MS_E}$
ABC	SS_{ABC}	$(a-1)(b-1)(c-1)$	MS_{ABC}	$\sigma^2 + \frac{n\sum\sum\sum (\tau\beta\gamma)_{ijk}^2}{(a-1)(b-1)(c-1)}$	$F_0 = \frac{MS_{ABC}}{MS_E}$
Error	SS_E	$abc(n-1)$	MS_E	σ^2	
Total	SS_T	$abcn-1$			

To compute the two-factor interaction sums of squares, the totals for the $A \times B$, $A \times C$, and $B \times C$ cells are needed. It is frequently helpful to collapse the original data table into three two-way tables to compute these quantities. The sums of squares are found from

$$SS_{AB} = \frac{1}{cn}\sum_{i=1}^{a}\sum_{j=1}^{b} y_{ij..}^2 - \frac{y_{....}^2}{abcn} - SS_A - SS_B$$

$$= SS_{\text{Subtotals}(AB)} - SS_A - SS_B \tag{5.31}$$

$$SS_{AC} = \frac{1}{bn}\sum_{i=1}^{a}\sum_{k=1}^{c} y_{i.k.}^2 - \frac{y_{....}^2}{abcn} - SS_A - SS_C$$

$$= SS_{\text{Subtotals}(AC)} - SS_A - SS_C \tag{5.32}$$

and

$$SS_{BC} = \frac{1}{an}\sum_{j=1}^{b}\sum_{k=1}^{c} y_{.jk.}^2 - \frac{y_{....}^2}{abcn} - SS_B - SS_C$$

$$= SS_{\text{Subtotals}(BC)} - SS_B - SS_C \tag{5.33}$$

Note that the sums of squares for the two-factor subtotals are found from the totals in each two-way table. The three-factor interaction sum of squares is computed from the three-way cell totals $\{y_{ijk.}\}$ as

$$SS_{ABC} = \frac{1}{n}\sum_{i=1}^{a}\sum_{j=1}^{b}\sum_{k=1}^{c} y_{ijk.}^2 - \frac{y_{....}^2}{abcn} - SS_A - SS_B - SS_C - SS_{AB} - SS_{AC} - SS_{BC} \tag{5.34a}$$

$$= SS_{\text{Subtotals}(ABC)} - SS_A - SS_B - SS_C - SS_{AB} - SS_{AC} - SS_{BC} \tag{5.34b}$$

The error sum of squares may be found by subtracting the sum of squares for each main effect and interaction from the total sum of squares or by

$$SS_E = SS_T - SS_{\text{Subtotals}(ABC)} \tag{5.35}$$

EXAMPLE 5.3 The Soft Drink Bottling Problem

A soft drink bottler is interested in obtaining more uniform fill heights in the bottles produced by his manufacturing process. The filling machine theoretically fills each bottle to the correct target height, but in practice, there is variation around this target, and the bottler would like to understand the sources of this variability better and eventually reduce it.

The process engineer can control three variables during the filling process: the percent carbonation (A), the operating pressure in the filler (B), and the bottles produced per minute or the line speed (C). The pressure and speed are easy to control, but the percent carbonation is more difficult to control during actual manufacturing because it varies with product temperature. However, for purposes of an experiment, the engineer can control carbonation at three levels: 10, 12, and 14 percent. She chooses two levels for pressure (25 and 30 psi) and two levels for line speed (200 and 250 bpm). She decides to run two replicates of a factorial design in these three factors, with all 24 runs taken in random order. The response variable observed is the average deviation from the target fill height observed in a production run of bottles at each set of conditions. The data that resulted from this experiment are shown in Table 5.13. Positive deviations are fill heights above the target, whereas negative deviations are fill heights below the target. The circled numbers in Table 5.13 are the three-way cell totals $y_{ijk.}$.

The total corrected sum of squares is found from Equation 5.27 as

$$SS_T = \sum_{i=1}^{a}\sum_{j=1}^{b}\sum_{k=1}^{c}\sum_{l=1}^{n} y_{ijkl}^2 - \frac{y_{....}^2}{abcn}$$

$$= 571 - \frac{(75)^2}{24} = 336.625$$

■ **TABLE 5.13**
Fill Height Deviation Data for Example 5.3

	Operating Pressure (B)				
	25 psi		30 psi		
	Line Speed (C)		Line Speed (C)		
Percent Carbonation (A)	**200**	**250**	**200**	**250**	$y_{i...}$
10	−3, −1 (−4)	−1, 0 (−1)	−1, 0 (−1)	1, 1 (2)	−4
12	0, 1 (1)	2, 1 (3)	2, 3 (5)	6, 5 (11)	20
14	5, 4 (9)	7, 6 (13)	7, 9 (16)	10, 11 (21)	59
$B \times C$ Totals $y_{.jk.}$	6	15	20	34	$75 = y_{....}$
$y_{.j..}$	21		54		

$A \times B$ Totals $y_{ij..}$

A	B = 25	30
10	−5	1
12	4	16
14	22	37

$A \times C$ Totals $y_{i.k.}$

A	C = 200	250
10	−5	1
12	6	14
14	25	34

and the sums of squares for the main effects are calculated from Equations 5.28, 5.29, and 5.30 as

$$SS_{\text{Carbonation}} = \frac{1}{bcn}\sum_{i=1}^{a} y_{i...}^2 - \frac{y_{....}^2}{abcn}$$
$$= \frac{1}{8}[(-4)^2 + (20)^2 + (59)^2] - \frac{(75)^2}{24} = 252.750$$

$$SS_{\text{Pressure}} = \frac{1}{acn}\sum_{j=1}^{b} y_{.j..}^2 - \frac{y_{....}^2}{abcn}$$
$$= \frac{1}{12}[(21)^2 + (54)^2] - \frac{(75)^2}{24} = 45.375$$

and

$$SS_{\text{Speed}} = \frac{1}{abn}\sum_{k=1}^{c} y_{..k.}^2 - \frac{y_{....}^2}{abcn}$$
$$= \frac{1}{12}[(26)^2 + (49)^2] - \frac{(75)^2}{24} = 22.042$$

To calculate the sums of squares for the two-factor interactions, we must find the two-way cell totals. For example, to find the carbonation–pressure or AB interaction, we need the totals for the $A \times B$ cells $\{y_{ij..}\}$ shown in Table 5.13. Using Equation 5.31, we find the sums of squares as

$$SS_{AB} = \frac{1}{cn}\sum_{i=1}^{a}\sum_{j=1}^{b} y_{ij..}^2 - \frac{y_{....}^2}{abcn} - SS_A - SS_B$$
$$= \frac{1}{4}[(-5)^2 + (1)^2 + (4)^2 + (16)^2 + (22)^2 + (37)^2] - \frac{(75)^2}{24} - 252.750 - 45.375$$
$$= 5.250$$

The carbonation–speed or AC interaction uses the $A \times C$ cell totals $\{y_{i.k.}\}$ shown in Table 5.13 and Equation 5.32:

$$SS_{AC} = \frac{1}{bn}\sum_{i=1}^{a}\sum_{k=1}^{c} y_{i.k.}^2 - \frac{y_{....}^2}{abcn} - SS_A - SS_C$$
$$= \frac{1}{4}[(-5)^2 + (1)^2 + (6)^2 + (14)^2 + (25)^2 + (34)^2] - \frac{(75)^2}{24} - 252.750 - 22.042$$
$$= 0.583$$

The pressure–speed or BC interaction is found from the $B \times C$ cell totals $\{y_{.jk.}\}$ shown in Table 5.13 and Equation 5.33:

$$SS_{BC} = \frac{1}{an}\sum_{j=1}^{b}\sum_{k=1}^{c} y_{.jk.}^2 - \frac{y_{....}^2}{abcn} - SS_B - SS_C$$
$$= \frac{1}{6}[(6)^2 + (15)^2 + (20)^2 + (34)^2] - \frac{(75)^2}{24} - 45.375 - 22.042$$
$$= 1.042$$

The three-factor interaction sum of squares is found from the $A \times B \times C$ cell totals $\{y_{ijk.}\}$, which are circled in Table 5.13. From Equation 5.34a, we find

$$SS_{ABC} = \frac{1}{n}\sum_{i=1}^{a}\sum_{j=1}^{b}\sum_{k=1}^{c} y_{ijk.}^2 - \frac{y_{....}^2}{abcn} - SS_A - SS_B - SS_C - SS_{AB} - SS_{AC} - SS_{BC}$$
$$= \frac{1}{2}[(-4)^2 + (-1)^2 + (-1)^2 + \cdots + (16)^2 + (21)^2] - \frac{(75)^2}{24} - 252.750 - 45.375 - 22.042 - 5.250 - 0.583 - 1.042$$
$$= 1.083$$

Finally, noting that

$$SS_{\text{Subtotals}(ABC)} = \frac{1}{n}\sum_{i=1}^{a}\sum_{j=1}^{b}\sum_{k=1}^{c} y_{ijk.}^2 - \frac{y_{....}^2}{abcn} = 328.125$$

we have

$$SS_E = SS_T - SS_{\text{Subtotals}(ABC)}$$
$$= 336.625 - 328.125$$
$$= 8.500$$

The ANOVA is summarized in Table 5.14. We see that the percentage of carbonation, operating pressure, and line speed significantly affect the fill volume. The carbonation-pressure interaction F ratio has a P-value of 0.0558, indicating some interaction between these factors.

The next step should be an analysis of the residuals from this experiment. We leave this as an exercise for the reader but point out that a normal probability plot of the residuals and the other usual diagnostics do not indicate any major concerns.

To assist in the practical interpretation of this experiment, Figure 5.16 presents plots of the three main effects and the AB (carbonation–pressure) interaction. The main effect plots are just graphs of the marginal response averages at the levels of the three factors. Notice that all three variables have *positive* main effects; that is, increasing the variable moves the average deviation from the fill target upward. The interaction between carbonation and pressure is fairly small, as shown by the similar shape of the two curves in Figure 5.16*d*.

Because the company wants the average deviation from the fill target to be close to zero, the engineer decides to recommend the low level of operating pressure (25 psi) and the high level of line speed (250 bpm, which will maximize the production rate). Figure 5.17 plots the average observed deviation from the target fill height at the three different carbonation levels for this set of operating conditions. Now the carbonation level cannot presently be perfectly controlled in the manufacturing process, and the normal distribution shown with the solid curve in Figure 5.17 approximates

■ **TABLE 5.14**
Analysis of Variance for Example 5.3

Source of Variation	Sum of Squares	Degrees of Freedom	Mean Square	F_0	*P*-Value
Percentage of carbonation (*A*)	252.750	2	126.375	178.412	<0.0001
Operating pressure (*B*)	45.375	1	45.375	64.059	<0.0001
Line speed (*C*)	22.042	1	22.042	31.118	0.0001
AB	5.250	2	2.625	3.706	0.0558
AC	0.583	2	0.292	0.412	0.6713
BC	1.042	1	1.042	1.471	0.2485
ABC	1.083	2	0.542	0.765	0.4867
Error	8.500	12	0.708		
Total	336.625	23			

the variability in the carbonation levels presently experienced. As the process is impacted by the values of the carbonation level drawn from this distribution, the fill heights will fluctuate considerably. This variability in the fill heights could be reduced if the distribution of the carbonation level values followed the normal distribution shown with the dashed line in Figure 5.17. Reducing the standard deviation of the carbonation level distribution was ultimately achieved by improving temperature control during manufacturing.

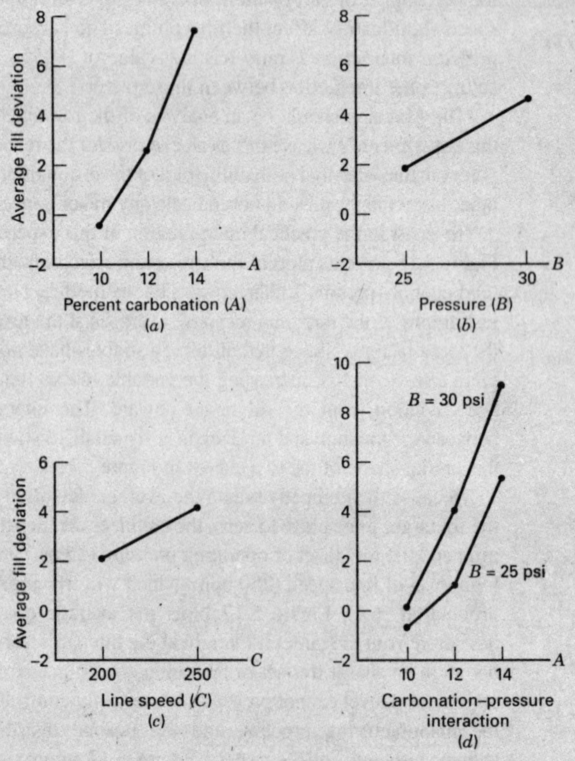

■ **FIGURE 5.16**
Main effects and interaction plots for Example 5.3. (*a*) Percentage of carbonation (*A*). (*b*) Pressure (*B*). (*c*) Line speed (*C*). (*d*) Carbonation–pressure interaction

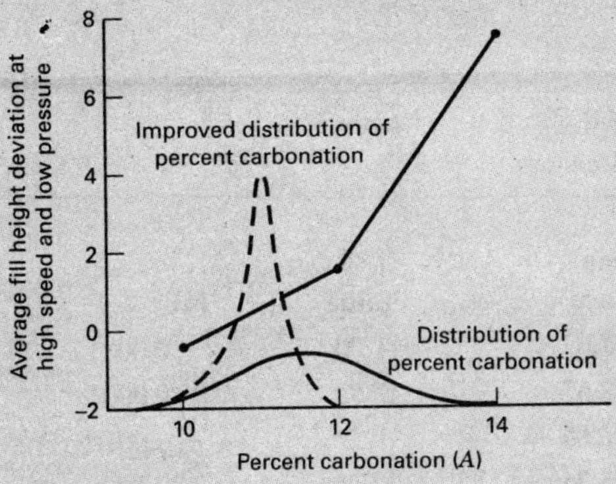

■ **FIGURE 5.17** **Average fill height deviation at high speed and low pressure for different carbonation levels**

We have indicated that if all the factors in a factorial experiment are fixed, test statistic construction is straightforward. The statistic for testing any main effect or interaction is always formed by dividing the mean square for the main effect or interaction by the mean square error. However, if the factorial experiment involves one or more **random factors**, the test statistic construction is not always done this way. We must examine the expected mean squares to determine the correct tests. We defer a complete discussion of experiments with random factors until Chapter 13.

5.5 Fitting Response Curves and Surfaces

The ANOVA always treats all of the factors in the experiment as if they were qualitative or categorical. However, many experiments involve at least one quantitative factor. It can be useful to fit a **response curve** to the levels of a quantitative factor so that the experimenter has an equation that relates the response to the factor. This equation might be used for interpolation, that is, for predicting the response at factor levels between those actually used in the experiment. When at least two factors are quantitative, we can fit a **response surface** for predicting y at various combinations of the design factors. In general, **linear regression methods** are used to fit these models to the experimental data. We illustrated this procedure in Section 3.5.1 for an experiment with a single factor. We now present two examples involving factorial experiments. In both examples, we will use a computer software package to generate the regression models. For more information about regression analysis, refer to Chapter 10 and the supplemental text material for this chapter.

EXAMPLE 5.4

Consider the battery life experiment described in Example 5.1. The factor temperature is quantitative, and the material type is qualitative. Furthermore, there are three levels of temperature. Consequently, we can compute a linear and a quadratic temperature effect to study how temperature affects the battery life. Table 5.15 presents condensed output from Design-Expert for this experiment and assumes that temperature is quantitative and material type is qualitative.

The ANOVA in Table 5.15 shows that the "model" source of variability has been subdivided into several components. The components "A" and "A^2" represent the

■ **TABLE 5.15**
Design-Expert Output for Example 5.4

Response: Life In Hours
ANOVA for Response Surface Reduced Cubic Model
Analysis of Variance Table [Partial Sum of Squares]

Source	Sum of Squares	DF	Mean Square	F Value	Prov > F	
Model	59416.22	8	7427.03	11.00	<0.0001	significant
A	*39042.67*	*1*	*39042.67*	*57.82*	<0.0001	
B	*10683.72*	2	*5341.86*	*7.91*	*0.0020*	
A^2	*76.06*	*1*	*76.06*	*0.11*	*0.7398*	
AB	*2315.08*	2	*1157.54*	*1.71*	*0.1991*	
A^2B	*7298.69*	2	*3649.35*	*5.40*	*0.0106*	
Residual	18230.75	27	675.21			
Lack of Fit	*0.000*	*0*				
Pure Error	*18230.75*	*27*	*675.21*			
Cor Total	77646.97	35				

Std. Dev.	25.98	R-Squared	0.7652
Mean	105.53	Adj R-Squared	0.6956
C.V.	24.62	Pred R-Squared	0.5826
PRESS	32410.22	Adeq Precision	8.178

Term	Coefficient Estimate	DF	Standard Error	95% CI Low	95% CI High	VIF
Intercept	107.58	1	7.50	92.19	122.97	
A-Temp	−40.33	1	5.30	−51.22	−29.45	1.00
$B[1]$	−50.33	1	10.61	−72.10	−28.57	
$B[2]$	12.17	1	10.61	−9.60	33.93	
A^2	−3.08	1	9.19	−21.93	15.77	1.00
$AB[1]$	1.71	1	7.50	−13.68	17.10	
$AB[2]$	−12.79	1	7.50	−28.18	2.60	
$A^2B[1]$	41.96	1	12.99	15.30	68.62	
$A^2B[2]$	−14.04	1	12.99	−40.70	12.62	

Final Equation in Terms of Coded Factors:

Life =
+107.58
−40.33 $*A$
−50.33 $*B[1]$
+12.17 $*B[2]$
−3.08 $*A^2$
+1.71 $*AB[1]$
−12.79 $*AB[2]$
+41.96 $*A^2B[1]$
−14.04 $*A^2[2]$

■ **TABLE 5.15** *(Continued)*

Final Equation in Terms of Actual Factors:

Material Type		1
Life =		
+169.38017		
−2.48860	*Temp	
+0.012851	*Temp^2	
Material Type		2
Life =		
+159.62397		
−0.17901	*Temp	
+0.41627	*Temp^2	
Material Type		3
Life =		
+132.76240		
+0.89264	*Temp	
−0.43218	*Temp^2	

linear and quadratic effects of temperature, and "B" represents the main effect of the material type factor. Recall that material type is a qualitative factor with three levels. The terms "AB" and "A^2B" are the interactions of the linear and quadratic temperature factor with material type.

The P-values indicate that A^2 and AB are not significant, whereas the A^2B term is significant. Often we think about removing nonsignificant terms or factors from a model, but in this case, removing A^2 and AB and retaining A^2B will result in a model that is not **hierarchical**. The **hierarchy principle** indicates that if a model contains a high-order term (such as A^2B), it should also contain all of the lower order terms that compose it (in this case A^2 and AB). Hierarchy promotes a type of internal

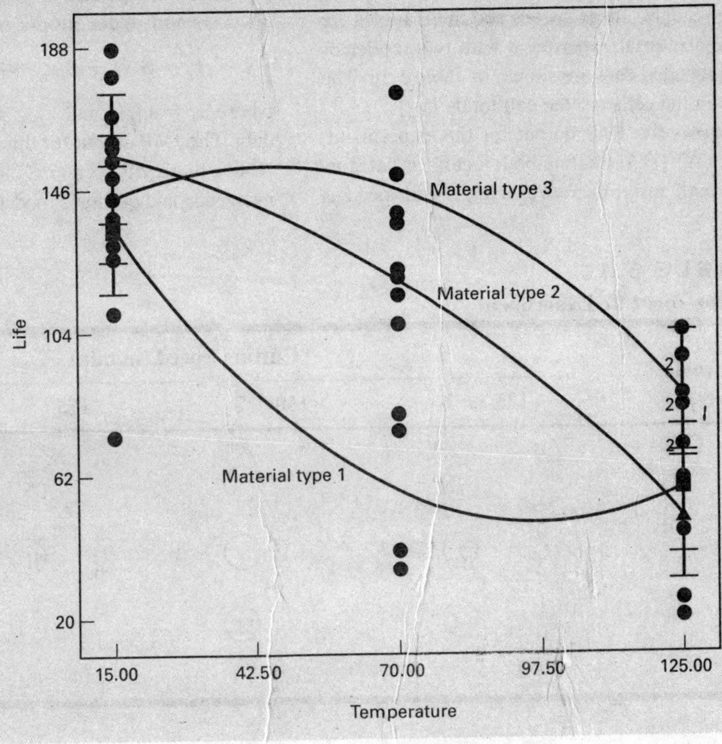

■ **FIGURE 5.18** **Predicted life as a function of temperature for the three material types, Example 5.4**

consistency in a model, and many statistical model builders rigorously follow the principle. However, hierarchy is not always a good idea, and many models actually work better as prediction equations without including the nonsignificant terms that promote hierarchy. For more information, see the supplemental text material for this chapter.

The computer output also gives model coefficient estimates and a final prediction equation for battery life in coded factors. In this equation, the levels of temperature are $A = -1, 0, +1$, respectively, when temperature is at the low, middle, and high levels (15, 70, and 125°C). The variables $B[1]$ and $B[2]$ are coded **indicator variables** that are defined as follows:

	Material Type		
	1	2	3
$B[1]$	1	0	−1
$B[2]$	0	1	−1

There are also prediction equations for battery life in terms of the actual factor levels. Notice that because material type is a qualitative factor there is an equation for predicted life as a function of temperature for each material type. Figure 5.18 shows the response curves generated by these three prediction equations. Compare them to the two-factor interaction graph for this experiment in Figure 5.9.

If several factors in a factorial experiment are quantitative a **response surface** may be used to model the relationship between y and the design factors. Furthermore, the quantitative factor effects may be represented by single-degree-of-freedom polynomial effects. Similarly, the interactions of quantitative factors can be partitioned into single-degree-of-freedom components of interaction. This is illustrated in the following example.

EXAMPLE 5.5

The effective life of a cutting tool installed in a numerically controlled machine is thought to be affected by the cutting speed and the tool angle. Three speeds and three angles are selected, and a 3^2 factorial experiment with two replicates is performed. The coded data are shown in Table 5.16. The circled numbers in the cells are the cell totals $\{y_{ij.}\}$.

Table 5.17 shows the JMP output for this experiment. This is a classical ANOVA, treating both factors as categorical. Notice that both design factors tool angle and speed as well as the angle–speed interaction are significant. Since the factors are quantitative, and both factors have three levels, a **second-order model** such as

$$y = \beta_0 + \beta_1 x_1 + \beta_2 x_2 + \beta_{12} x_1 x_2 + \beta_{11} x_1^2 + \beta_{22} x_2^2 + \varepsilon$$

where x_1 = angle and x_2 = speed could also be fit to the data. The JMP output for this model is shown in Table 5.18. Notice that JMP "centers" the predictors when forming the interaction and quadratic model terms. The second-order model

■ **TABLE 5.16**
Data for Tool Life Experiment

Total Angle (degrees)	Cutting Speed (in/min) 125		150		175		$y_{i..}$
15	−2, −1	(−3)	−3, 0	(−3)	2, 3	(5)	−1
20	0, 2	(2)	1, 3	(4)	4, 6	(10)	16
25	−1, 0	(−1)	5, 6	(11)	0, −1	(−1)	9
$y_{.j.}$	−2		12		14		$24 = y_{...}$

■ **TABLE 5.17**

JMP ANOVA for the Tool Life Experiment in Example 5.5

Response Tool Life
Whole Model
Actual by Predicted Plot

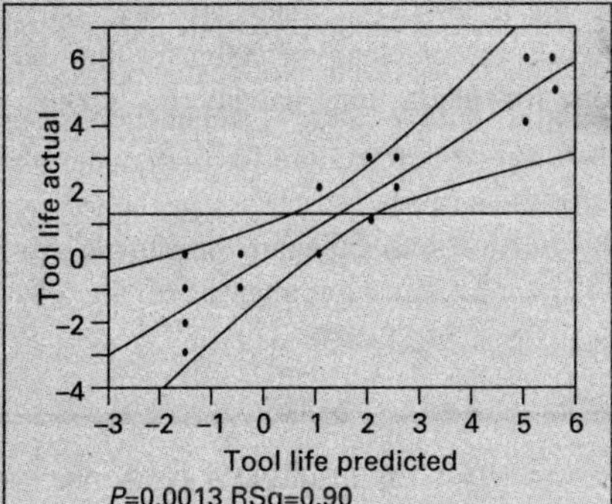

Summary of Fit

RSquare	0.895161
RSquare Adj	0.801971
Root Mean Square Error	1.20185
Mean of Response	1.333333
Observations (or Sum Wgts)	18

Analysis of Variance

Source	DF	Sum of Squares	Mean Square	*F* Ratio
Model	8	111.00000	13.8750	9.6058
Error	9	13.00000	1.4444	Prob > *F*
C. Total	17	124.00000		0.0013

Effect Tests

Source	Nparm	DF	Sum of Squares	*F* Ratio	Prob > *F*
Angle	2	2	24.333333	8.4231	0.0087
Speed	2	2	25.333333	8.7692	0.0077
Angle*Speed	4	4	61.333333	10.6154	0.0018

Residual by Predicted Plot

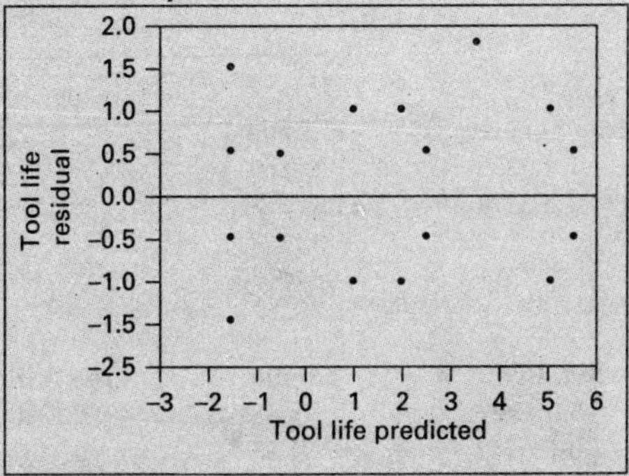

doesn't look like a very good fit to the data; the value of R^2 is only 0.465 (compared to $R^2 = 0.895$ in the categorical variable ANOVA) and the only significant factor is the linear term in speed for which the P-value is 0.0731. Notice that the mean square for error in the second-order model fit is 5.5278, considerably larger than it was in the classical categorical variable ANOVA of Table 5.17. The JMP output in Table 5.18 shows the **prediction profiler**, a graphical display showing the response variable life as a function of each design factor, angle and speed. The prediction profiler is very useful for optimization. Here it has been set to the levels of angle and speed that result in maximum predicted life.

Part of the reason for the relatively poor fit of the second-order model is that only one of the four degrees of freedom for interaction are accounted for in this model. In addition to the term $\beta_{12}x_1x_2$, there are three other terms that could be fit to completely account for the four degrees of freedom for interaction, namely $\beta_{112}x_1^2x_2$, $\beta_{122}x_1x_2^2$, and $\beta_{1122}x_1^2x_2^2$.

■ **TABLE 5.18**

JMP Output for the Second-Order Model, Example 5.5

Response Tool Life
Actual by Predicted Plot

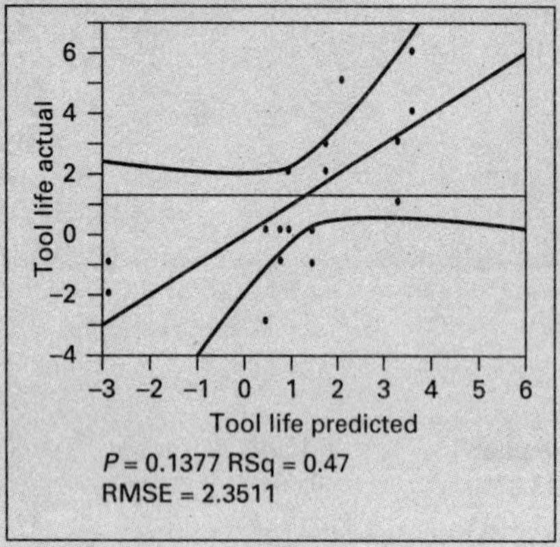

Summary of Fit

RSquare	0.465054
RSquare Adj	0.242159
Root Mean Square Error	2.351123
Mean of Response	1.333333
Observations (or Sum Wgts)	18

Analysis of Variance

Source	DF	Sum of Squares	Mean Square	*F* Ratio
Model	5	57.66667	11.5333	2.0864
Error	12	66.33333	5.5278	Prob > *F*
C. Total	17	124.00000		0.1377

Parameter Estimates

Term	Estimate	Std. Error	*t* Ratio	Prob > |*t*|
Intercept	−8	5.048683	−1.58	0.1390
Angle	0.1666667	0.135742	1.23	0.2431
Speed	0.0533333	0.027148	1.96	0.0731

■ **TABLE 5.18** (*Continued*)

(Angle-20)*(Speed-150)	−0.008	0.00665	−1.20	0.2522
(Angle-20)*(Angle-20)	−0.08	0.047022	−1.70	0.1146
(Speed-150)*(Speed-150)	−0.0016	0.001881	−0.85	0.4116

Prediction Profiler

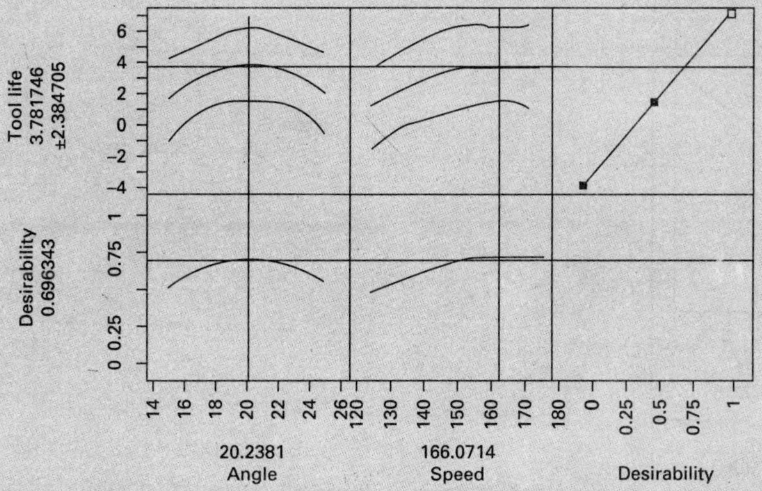

JMP output for the second-order model with the additional higher-order terms is shown in Table 5.19. While these higher-order terms are components of the two-factor interaction, the final model is a reduced quartic. Although there are some large *P*-values, all model terms have been retained to ensure hierarchy. The prediction profiler indicates that maximum tool life is achieved around an angle of 25 degrees and speed of 150 in/min.

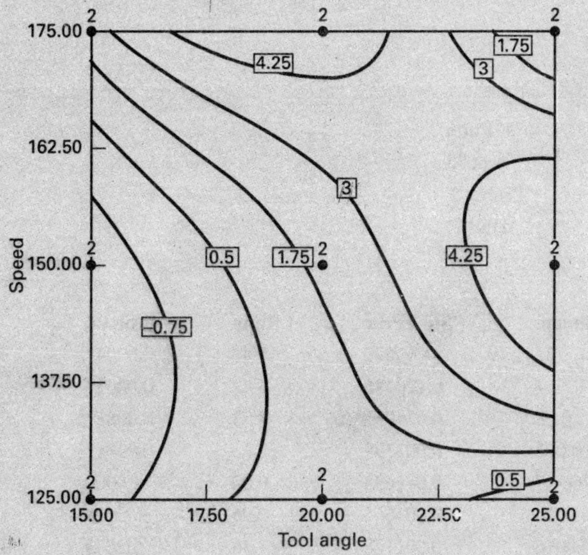

■ **FIGURE 5.19**
Two-dimensional contour plot of the tool life response surface for Example 5.5

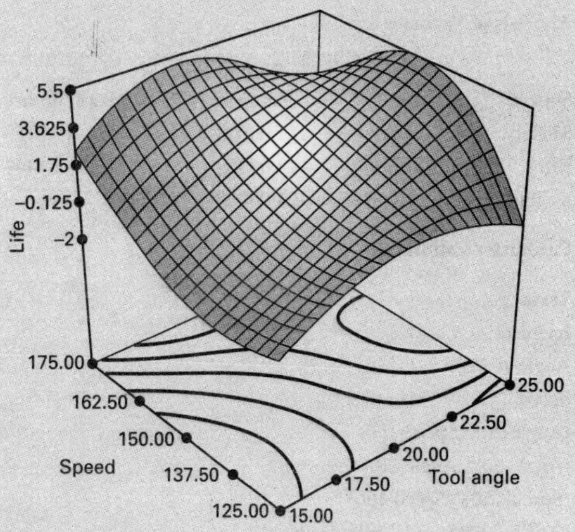

■ **FIGURE 5.20**
Three-dimensional tool life response surface for Example 5.5

Figure 5.19 is the contour plot of tool life for this model and Figure 5.20 is a three-dimensional response surface plot. These plots confirm the estimate of the optimum operating conditions found from the JMP prediction profiler. Exploration of response surfaces is an important use of designed experiments, which we will discuss in more detail in Chapter 11.

■ **TABLE 5.19**
JMP Output for the Expanded Model in Example 5.5

Response Y
Actual by Predicted Plot

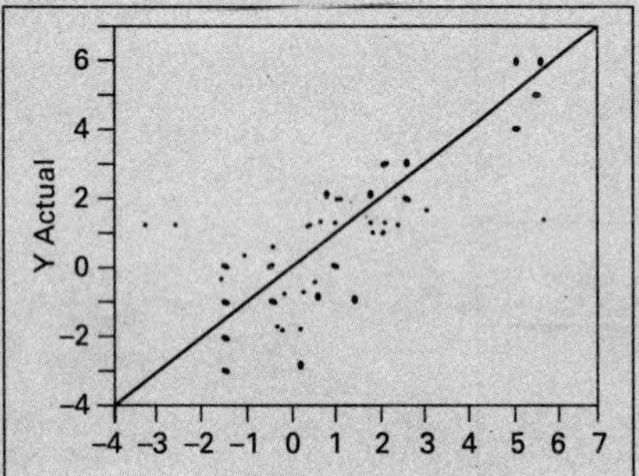

Summary of Fit

RSquare	0.895161
RSquare Adj	0.801971
Root Mean Square Error	1.20185
Mean of Response	1.333333
Observations (or Sum Wgts)	18

Analysis of Variance

Source	DF	Sum of Squares	Mean Square	F Ratio
Model	8	111.00000	13.8750	9.6058
Error	9	13.00000	1.4444	Prob > F
C. Total	17	124.00000		0.0013*

Parameter Estimates

Term	Estimate	Std Error	t Ratio	Prob>|t|
Intercept	−24	4.41588	−5.43	0.0004*
Angle	0.7	0.120185	5.82	0.0003*
Speed	0.08	0.024037	3.33	0.0088*
(Angle-20)*(Speed-150)	−0.008	0.003399	−2.35	0.0431*
(Angle-20)*(Angle-20)	2.776e-17	0.041633	0.00	1.0000
(Speed-150)*(Speed-150)	0.0016	0.001665	0.96	0.3618
(Angle-20)*(Speed-150)*(Angle-20)	−0.0016	0.001178	−1.36	0.2073
(Speed-150)*(Speed-150)*(Angle-20)	−0.00128	0.000236	−5.43	0.0004*
(Angle-20)*(Speed-150)*(Angle-20)*(Speed-150)	−0.000192	8.158a-5	−2.35	0.0431*

■ **TABLE 5.19** *(Continued)*

Effect Tests

Source	Nparm	DF	Sum of Squares	F Ratio	Prob > F
Angle	1	1	49.000000	33.9231	0.0003*
Speed	1	1	16.000000	11.0769	0.0088*
Angle*Speed	1	1	8.000000	5.5385	0.0431*
Angle*Angle	1	1	6.4198e-31	0.0000	1.0000
Speed*Speed	1	1	1.333333	0.9231	0.3618
Angle*Speed*Angle	1	1	2.666667	1.8462	0.2073
Speed*Speed*Angle	1	1	42.666667	29.5385	0.0004*
Angle*Speed*Angle*Speed	1	1	8.000000	5.5385	0.0431*

Sorted Parameter Estimates

Term	Estimate	Std Error	t Ratio	Prob>|t|
Angle	0.7	0.120185	5.82	0.0003*
(Speed-150)*(Speed-150)*(Angle-20)	−0.00128	0.000236	−5.43	0.0004*
Speed	0.08	0.024037	3.33	0.0088*
(Angle-20)*(Speed-150)*(Angle-20)*(Speed-150)	−0.000192	8.158a-5	−2.35	0.0431*
(Angle-20)*(Speed-150)	−0.008	0.003399	−2.35	0.0431*
(Angle-20)*(Speed-150)*(Angle-20)	−0.0016	0.001178	−1.36	0.2073
(Speed-150)*(Speed-150)	0.0016	0.001665	0.96	0.3618
(Angle-20)*(Angle-20)	2.776e-17	0.041633	0.00	1.0000

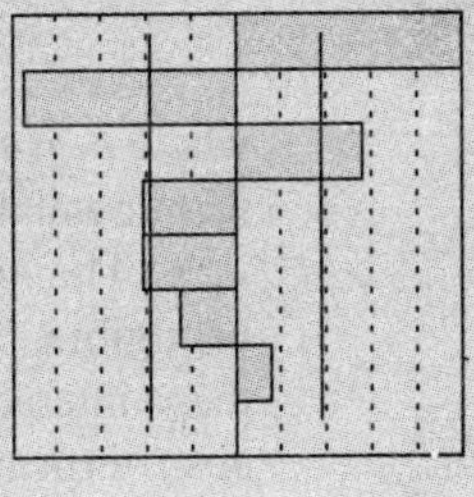

Prediction Profiler

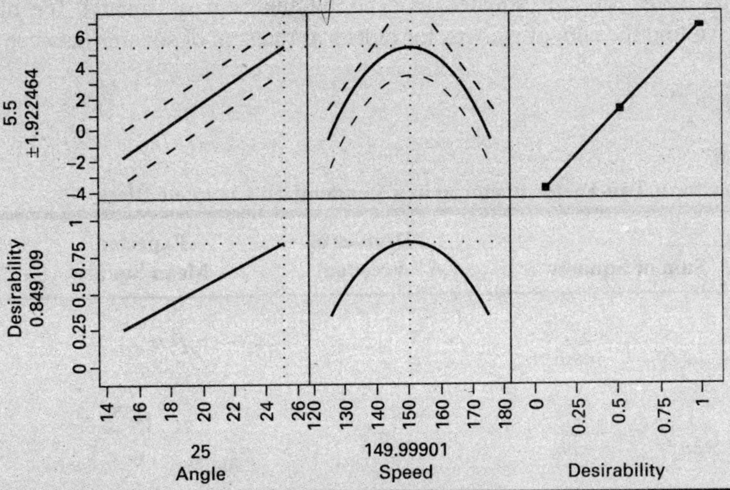

5.6 Blocking in a Factorial Design

We have discussed factorial designs in the context of a completely randomized experiment. Sometimes, it is not feasible or practical to completely randomize all of the runs in a factorial. For example, the presence of a nuisance factor may require that the experiment be run in blocks. We discussed the basic concepts of blocking in the context of a single-factor experiment in Chapter 4. We now show how blocking can be incorporated in a factorial. Some other aspects of blocking in factorial designs are presented in Chapters 7, 8, 9, and 13.

Consider a factorial experiment with two factors (*A* and *B*) and *n* replicates. The linear statistical model for this design is

$$y_{ijk} = \mu + \tau_i + \beta_j + (\tau\beta)_{ij} + \epsilon_{ijk} \qquad \begin{cases} i = 1, 2, \ldots, a \\ j = 1, 2, \ldots, b \\ k = 1, 2, \ldots, n \end{cases} \tag{5.36}$$

where τ_i, β_j, and $(\tau\beta)_{ij}$ represent the effects of factors *A*, *B*, and the *AB* interaction, respectively. Now suppose that to run this experiment a particular raw material is required. This raw material is available in batches that are not large enough to allow *all abn* treatment combinations to be run from the *same* batch. However, if a batch contains enough material for *ab* observations, then an alternative design is to run each of the *n* replicates using a separate batch of raw material. Consequently, the batches of raw material represent a randomization restriction or a **block**, and a single replicate of a complete factorial experiment is run within each block. The effects model for this new design is

$$y_{ijk} = \mu + \tau_i + \beta_j + (\tau\beta)_{ij} + \delta_k + \epsilon_{ijk} \qquad \begin{cases} i = 1, 2, \ldots, a \\ j = 1, 2, \ldots, b \\ k = 1, 2, \ldots, n \end{cases} \tag{5.37}$$

where δ_k is the effect of the *k*th block. Of course, within a block the order in which the treatment combinations are run is completely randomized.

The model (Equation 5.37) assumes that interaction between blocks and treatments is negligible. This was assumed previously in the analysis of randomized block designs. If these interactions do exist, they cannot be separated from the error component. In fact, the error term in this model really consists of the $(\tau\delta)_{ik}$, $(\beta\delta)_{jk}$, and $(\tau\beta\delta)_{ijk}$ interactions. The analysis of variance is outlined in Table 5.20. The layout closely resembles that of a factorial design, with the error sum of squares reduced by the sum of squares for blocks. Computationally, we find the sum of squares for blocks as the sum of squares between the *n*

■ **TABLE 5.20**

Analysis of Variance for a Two-Factor Factorial in a Randomized Complete Block

Source of Variation	Sum of Squares	Degrees of Freedom	Expected Mean Square	F_0
Blocks	$\frac{1}{ab}\sum_k y_{..k}^2 - \frac{y_{...}^2}{abn}$	$n - 1$	$\sigma^2 + ab\sigma_\delta^2$	
A	$\frac{1}{bn}\sum_i y_{i..}^2 - \frac{y_{...}^2}{abn}$	$a - 1$	$\sigma^2 + \frac{bn\sum \tau_i^2}{a-1}$	$\frac{MS_A}{MS_E}$
B	$\frac{1}{an}\sum_j y_{.j.}^2 - \frac{y_{...}^2}{abn}$	$b - 1$	$\sigma^2 + \frac{an\sum \beta_j^2}{b-1}$	$\frac{MS_B}{MS_E}$
AB	$\frac{1}{n}\sum_i\sum_j y_{ij.}^2 - \frac{y_{...}^2}{abn} - SS_A - SS_B$	$(a-1)(b-1)$	$\sigma^2 + \frac{n\sum\sum(\tau\beta)_{ij}^2}{(a-1)(b-1)}$	$\frac{MS_{AB}}{MS_E}$
Error	Subtraction	$(ab-1)(n-1)$	σ^2	
Total	$\sum_i\sum_j\sum_k y_{ijk}^2 - \frac{y_{...}^2}{abn}$	$abn - 1$		

block totals $\{y_{..k}\}$. The ANOVA in Table 5.20 assumes that both factors are fixed and that blocks are random. The ANOVA estimator of the variance component for blocks σ_δ^2, is

$$\sigma_\delta^2 = \frac{MS_{Blocks} - MS_E}{ab}$$

In the previous example, the randomization was restricted to within a batch of raw material. In practice, a variety of phenomena may cause randomization restrictions, such as time and operators. For example, if we could not run the entire factorial experiment on one day, then the experimenter could run a complete replicate on day 1, a second replicate on day 2, and so on. Consequently, each day would be a block.

EXAMPLE 5.6

An engineer is studying methods for improving the ability to detect targets on a radar scope. Two factors she considers to be important are the amount of background noise, or "ground clutter," on the scope and the type of filter placed over the screen. An experiment is designed using three levels of ground clutter and two filter types. We will consider these as fixed-type factors. The experiment is performed by randomly selecting a treatment combination (ground clutter level and filter type) and then introducing a signal representing the target into the scope. The intensity of this target is increased until the operator observes it. The intensity level at detection is then measured as the response variable. Because of operator availability, it is convenient to select an operator and keep him or her at the scope until all the necessary runs have been made. Furthermore, operators differ in their skill and ability to use the scope. Consequently, it seems logical to use the operators as blocks. Four operators are randomly selected. Once an operator is chosen, the order in which the six treatment combinations are run is randomly determined. Thus, we have a 3 × 2 factorial experiment run in a randomized complete block. The data are shown in Table 5.21.

The linear model for this experiment is

$$y_{ijk} = \mu + \tau_i + \beta_j + (\tau\beta)_{ij} + \delta_k + \epsilon_{ijk} \quad \begin{cases} i = 1, 2, 3 \\ j = 1, 2 \\ k = 1, 2, 3, 4 \end{cases}$$

where τ_i represents the ground clutter effect, β_j represents the filter type effect, $(\tau\beta)_{ij}$ is the interaction, δ_k is the block effect, and ϵ_{ijk} is the NID(0, σ^2) error component. The sums of squares for ground clutter, filter type, and their interaction are computed in the usual manner. The sum of squares due to blocks is found from the operator totals $\{y_{..k}\}$ as follows:

$$SS_{\text{Blocks}} = \frac{1}{ab}\sum_{k=1}^{n} y_{..k}^2 - \frac{y_{...}^2}{abn}$$

$$= \frac{1}{(3)(2)}[(572)^2 + (579)^2 + (597)^2 + (530)^2] - \frac{(2278)^2}{(3)(2)(4)}$$

$$= 402.17$$

■ TABLE 5.21
Intensity Level at Target Detection

Operators (blocks)	1		2		3		4	
Filter Type	1	2	1	2	1	2	1	2
Ground clutter								
Low	90	86	96	84	100	92	92	81
Medium	102	87	106	90	105	97	96	80
High	114	93	112	91	108	95	98	83

The complete ANOVA for this experiment is summarized in Table 5.22. The presentation in Table 5.22 indicates that all effects are tested by dividing their mean squares by the mean square error. Both ground clutter level and filter type are significant at the 1 percent level, whereas their interaction is significant only at the 10 percent level. Thus, we conclude that both ground clutter level and the type of scope filter used affect the operator's ability to detect the target, and there is some evidence of mild interaction between these factors. The ANOVA estimate of the variance component for blocks is

$$\hat{\sigma}_\delta^2 = \frac{MS_{Blocks} - MS_E}{ab} = \frac{134.06 - 11.09}{(3)(2)} = 20.50$$

■ **TABLE 5.22**
Analysis of Variance for Example 5.6

Source of Variation	Sum of Squares	Degrees of Freedom	Mean Square	F_0	*P*-Value
Ground clutter (*G*)	335.58	2	167.79	15.13	0.0003
Filter type (*F*)	1066.67	1	1066.67	96.19	<0.0001
GF	77.08	2	38.54	3.48	0.0573
Blocks	402.17	3	134.06		
Error	166.33	15	11.09		
Total	2047.83	23			

The JMP output for this experiment is shown in Table 5.23. The REML estimate of the variance component for blocks is shown in this output, and because this is a balanced design, the REML and ANOVA estimates agree. JMP also provides the confidence intervals on both variance components σ^2 and σ_δ^2.

■ **TABLE 5.23**
JMP Output for Example 5.6

Whole Model
Actual by Predicted Plot

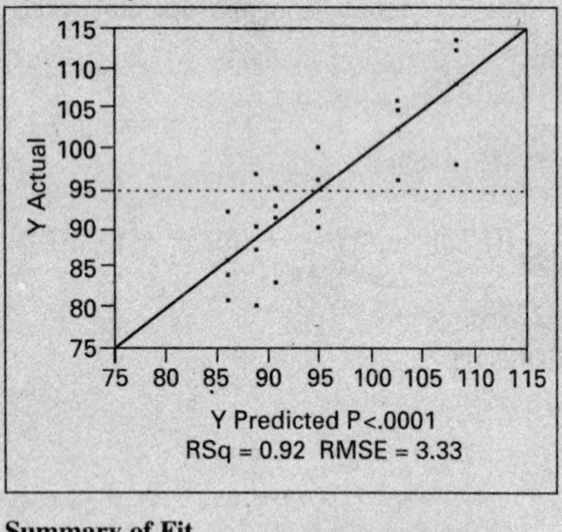

Summary of Fit

RSquare	0.917432
RSquare Adj	0.894497
Root Mean Square Error	3.329998
Mean of Response	94.91667
Observations (or Sum Wgts)	24

REML Variance component Estimates

Random Effect	Var Ratio	Var Component	Std Error	95% Lower	95% Upper	Pct of Total
Operators (Blocks)	1.8481964	20.494444	18.255128	−15.28495	56.273839	64.890
Residual		11.088889	4.0490897	6.0510389	26.561749	35.110
Total		31.583333				100.000

-2 LogLikelihood = 118.73680261

Covariance Matrix of Variance Component Estimates

Random Effect	Operators (Blocks)	Residual
Operators (Blocks)	333.24972	−2.732521
Residual	−2.732521	16.395128

Fixed Effect Tests

Source	Nparm	DF	DFDen	F Ratio	Prob > F
Clutter	2	2	15	15.1315	0.0003*
Filter Type	1	1	15	96.1924	<.0001*
Clutter*Filter Type	2	2	15	3.4757	0.0575

Residual by Predicted Plot

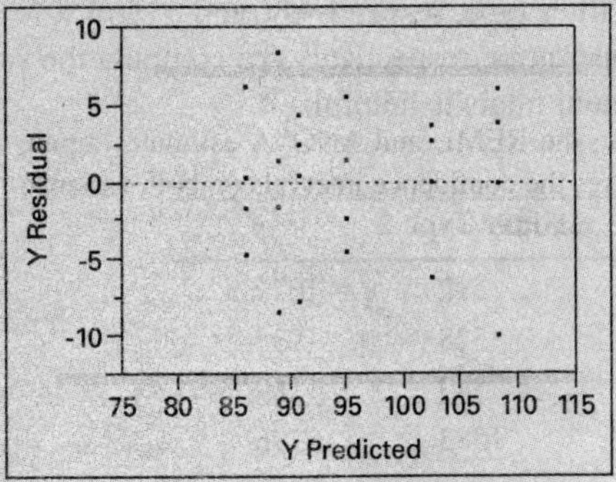

In the case of two randomization restrictions, each with p levels, if the number of treatment combinations in a k-factor factorial design exactly equals the number of restriction levels, that is, if $p = ab \ldots m$, then the factorial design may be run in a $p \times p$ Latin square. For example, consider a modification of the radar target detection experiment of Example 5.6. The factors in this experiment are filter type (two levels) and ground clutter (three levels), and operators are considered as blocks. Suppose now that because of the setup time required, only six runs can be made per day. Thus, days become a second randomization restriction, resulting in the 6×6 Latin square design, as shown in Table 5.24. In this table we have used the lowercase letters f_i and g_j to represent the ith and jth levels of filter type and ground clutter, respectively. That is, f_1g_2 represents filter type 1 and medium ground clutter. Note that now six operators are required, rather than four as in the original experiment, so the number of treatment combinations in the 3×2 factorial design exactly equals the number of restriction levels. Furthermore, in this design, each operator would be used only once on each day. The Latin letters A, B, C, D, E, and F represent the $3 \times 2 = 6$ factorial treatment combinations as follows: $A = f_1g_1$, $B = f_1g_2$, $C = f_1g_3$, $D = f_2g_1$, $E = f_2g_2$, and $F = f_2g_3$.

The five degrees of freedom between the six Latin letters correspond to the main effects of filter type (one degree of freedom), ground clutter (two degrees of freedom), and their interaction (two degrees of freedom). The linear statistical model for this design is

$$y_{ijkl} = \mu + \alpha_i + \tau_j + \beta_k + (\tau\beta)_{jk} + \theta_l + \epsilon_{ijkl} \quad \begin{cases} i = 1, 2, \ldots, 6 \\ j = 1, 2, 3 \\ k = 1, 2 \\ l = 1, 2, \ldots, 6 \end{cases} \tag{5.38}$$

■ **TABLE 5.24**

Radar Detection Experiment Run in a 6 × 6 Latin Square

	Operator					
Day	**1**	**2**	**3**	**4**	**5**	**6**
1	$A(f_1g_1 = 90)$	$B(f_1g_2 = 106)$	$C(f_1g_3 = 108)$	$D(f_2g_1 = 81)$	$F(f_2g_3 = 90)$	$E(f_2g_2 = 88)$
2	$C(f_1g_3 = 114)$	$A(f_1g_1 = 96)$	$B(f_1g_2 = 105)$	$F(f_2g_3 = 83)$	$E(f_2g_2 = 86)$	$D(f_2g_1 = 84)$
3	$B(f_1g_2 = 102)$	$E(f_2g_2 = 90)$	$G(f_2g_3 = 95)$	$A(f_1g_1 = 92)$	$D(f_2g_1 = 85)$	$C(f_1g_3 = 104)$
4	$E(f_2g_2 = 87)$	$D(f_2g_1 = 84)$	$A(f_1g_1 = 100)$	$B(f_1g_2 = 96)$	$C(f_1g_3 = 110)$	$F(f_2g_3 = 91)$
5	$F(f_2g_3 = 93)$	$C(f_1g_3 = 112)$	$D(f_2g_1 = 92)$	$E(f_2g_2 = 80)$	$A(f_1g_1 = 90)$	$B(f_1g_2 = 98)$
6	$D(f_2g_1 = 86)$	$F(f_2g_3 = 91)$	$E(f_2g_2 = 97)$	$C(f_1g_3 = 98)$	$B(f_1g_2 = 100)$	$A(f_1g_1 = 92)$

where τ_j and β_k are effects of ground clutter and filter type, respectively, and α_i and θ_l represent the randomization restrictions of days and operators, respectively. To compute the sums of squares, the following two-way table of treatment totals is helpful:

Ground Clutter	Filter Type 1	Filter Type 2	$y_{.j..}$
Low	560	512	1072
Medium	607	528	1135
High	646	543	1189
$y_{..k.}$	1813	1583	$3396 = y_{....}$

Furthermore, the row and column totals are

Rows ($y_{.jkl}$):	563	568	568	568	565	564
Columns ($y_{ijk.}$):	572	579	597	530	561	557

The ANOVA is summarized in Table 5.25. We have added a column to this table indicating how the number of degrees of freedom for each sum of squares is determined.

■ **TABLE 5.25**

Analysis of Variance for the Radar Detection Experiment Run as a 3 × 2 Factorial in a Latin Square

Source of Variation	Sum of Squares	Degrees of Freedom	General Formula for Degrees of Freedom	Mean Square	F_0	*P*-Value
Ground clutter, *G*	571.50	2	$a - 1$	285.75	28.86	<0.0001
Filter type, *F*	1469.44	1	$b - 1$	1469.44	148.43	<0.0001
GF	126.73	2	$(a - 1)(b - 1)$	63.37	6.40	0.0071
Days (rows)	4.33	5	$ab - 1$	0.87		
Operators (columns)	428.00	5	$ab - 1$	85.60		
Error	198.00	20	$(ab - 1)(ab - 2)$	9.90		
Total	2798.00	35	$(ab)^2 - 1$			

5.7 Problems

5.1. The following output was obtained from a computer program that performed a two-factor ANOVA on a factorial experiment.

Two-way ANOVA: y versus A, B

Source	DF	SS	MS	F	P
A	1	___	0.0002	___	___
B	___	180.378	___	___	___
Interaction	3	8.479	___	___	0.932
Error	8	158.797	___		
Total	15	347.653			

(a) Fill in the blanks in the ANOVA table. You can use bounds on the P-values.

(b) How many levels were used for factor B?

(c) How many replicates of the experiment were performed?

(d) What conclusions would you draw about this experiment?

5.2. The following output was obtained from a computer program that performed a two-factor ANOVA on a factorial experiment.

Two-way ANOVA: y versus, A, B

Source	DF	SS	MS	F	P
A	1	4.250	___	___	___
B	___	80.554	40.2771	4.59	___
Interaction	___	___	___	___	___
Error	12	105.327	8.7773		
Total	17	231.551			

(a) Fill in the blanks in the ANOVA table. You can use bounds on the P-values.

(b) How many levels were used for factor B?

(c) How many replicates of the experiment were performed?

(d) What conclusions would you draw about this experiment?

5.3. An engineer suspects that the surface finish of a metal part is influenced by the feed rate and the depth of cut. He selects three feed rates and four depths of cut. He then conducts a factorial experiment and obtains the following data:

Feed Rate (in/min)	Depth of Cut (in) 0.15	0.18	0.20	0.25
	75	79	82	99
0.20	66	68	88	104
	61	75	92	96
	91	98	99	104
0.25	86	104	108	110
	88	89	95	99
	100	104	108	114
0.30	98	99	110	111
	102	95	98	107

(a) Analyze the data and draw conclusions. Use $\alpha = 0.05$.

(b) Prepare appropriate residual plots and comment on the model's adequacy.

(c) Obtain point estimates of the mean surface finish at each feed rate.

(d) Find the P-values for the tests in part (a).

5.4. The yield of a chemical process is being studied. The two most important variables are thought to be the pressure and the temperature. Three levels of each factor are selected, and a factorial experiment with two replicates is performed. The yield data are as follows:

Temperature (°C)	Pressure (psig) 200	215	230
150	90.4	90.7	90.2
	90.2	90.6	90.4
160	90.1	90.5	89.9
	90.3	90.6	90.1
170	90.5	90.8	90.4
	90.7	90.9	90.1

(a) Analyze the data and draw conclusions. Use $\alpha = 0.05$.

(b) Prepare appropriate residual plots and comment on the model's adequacy.

(c) Under what conditions would you operate this process?

5.5. For the data in Problem 5.3, compute a 95 percent confidence interval estimate of the mean difference in response for feed rates of 0.20 and 0.25 in/min.

5.6. An article in *Industrial Quality Control* (1956, pp. 5–8) describes an experiment to investigate the effect of the type of glass and the type of phosphor on the brightness of a television tube. The response variable is the current necessary (in microamps) to obtain a specified brightness level. The data are as follows:

Glass Type	Phosphor Type 1	2	3
	280	300	290
1	290	310	285
	285	295	290
	230	260	220
2	235	240	225
	240	235	230

(a) Is there any indication that either factor influences brightness? Use $\alpha = 0.05$.

(b) Do the two factors interact? Use $\alpha = 0.05$.

(c) Analyze the residuals from this experiment.

5.7. Johnson and Leone (*Statistics and Experimental Design in Engineering and the Physical Sciences*, Wiley, 1977) describe an experiment to investigate warping of copper plates. The two factors studied were the temperature and the copper content of the plates. The response variable was a measure of the amount of warping. The data were as follows:

	Copper Content (%)			
Temperature (°C)	40	60	80	100
50	17, 20	16, 21	24, 22	28, 27
75	12, 9	18, 13	17, 12	27, 31
100	16, 12	18, 21	25, 23	30, 23
125	21, 17	23, 21	23, 22	29, 31

(a) Is there any indication that either factor affects the amount of warping? Is there any interaction between the factors? Use $\alpha = 0.05$.

(b) Analyze the residuals from this experiment.

(c) Plot the average warping at each level of copper content and compare them to an appropriately scaled t distribution. Describe the differences in the effects of the different levels of copper content on warping. If low warping is desirable, what level of copper content would you specify?

(d) Suppose that temperature cannot be easily controlled in the environment in which the copper plates are to be used. Does this change your answer for part (c)?

5.8. The factors that influence the breaking strength of a synthetic fiber are being studied. Four production machines and three operators are chosen and a factorial experiment is run using fiber from the same production batch. The results are as follows:

	Machine			
Operator	1	2	3	4
1	110	111	108	109
	110	115	109	108
2	111	110	112	114
	112	111	108	112
3	116	112	114	120
	114	115	119	117

(a) Analyze the data and draw conclusions. Use $\alpha = 0.05$.

(b) Prepare appropriate residual plots and comment on the model's adequacy.

5.9. An experiment is conducted to study the influence of operating temperature and three types of faceplate glass in the light output of an oscilloscope tube. The following data are collected:

	Temperature		
Glass Type	100	125	150
	580	1090	1392
1	568	1087	1380
	570	1085	1386
	550	1070	1328
2	530	1035	1312
	579	1000	1299
	546	1045	867
3	575	1053	904
	599	1066	889

(a) Use $\alpha = 0.05$ in the analysis. Is there a significant interaction effect? Does glass type or temperature affect the response? What conclusions can you draw?

(b) Fit an appropriate model relating light output to glass type and temperature.

(c) Analyze the residuals from this experiment. Comment on the adequacy of the models you have considered.

5.10. A mechanical engineer is studying the thrust force developed by a drill press. He suspects that the drilling speed and the feed rate of the material are the most important factors. He selects four feed rates and uses a high and low drill speed chosen to represent the extreme operating conditions. He obtains the following results. Analyze the data and draw conclusions. Use $\alpha = 0.05$.

	Feed Rate			
Drill Speed	0.015	0.030	0.045	0.060
125	2.70	2.45	2.60	2.75
	2.78	2.49	2.72	2.86
200	2.83	2.85	2.86	2.94
	2.86	2.80	2.87	2.88

5.11. Use Tukey's test to determine which levels of the pressure factor are significantly different for the data in Problem 5.4.

5.12. Consider the experiment in Problem 5.4. Fit an appropriate model to the response data. Use this model to provide guidance concerning operating conditions for the process.

5.13. An experiment was conducted to determine whether either firing temperature or furnace position affects the baked density of a carbon anode. The data are shown below:

Position	Temperature (°C) 800	825	850
1	571	1063	565
	564	1080	511
	584	1043	588
2	530	988	525
	548	1026	538
	520	1004	530

Suppose we assume that no interaction exists. Write down the statistical model. Conduct the analysis of variance and test hypotheses on the main effects. What conclusions can be drawn? Comment on the model's adequacy.

5.14. Consider the following data from a two-factor factorial experiment. Analyze the data and draw conclusions. Perform a test for nonadditivity. Use $\alpha = 0.05$.

Row Factor	Column Factor 1	2	3	4
1	36	39	36	32
2	18	20	22	20
3	30	37	33	34

5.15. The shear strength of an adhesive is thought to be affected by the application pressure and temperature. A factorial experiment is performed in which both factors are assumed to be fixed. Analyze the data and draw conclusions. Perform a test for nonadditivity.

Pressure (lb/in^2)	Temperature (°F) 250	260	270
120	9.60	11.28	9.00
130	9.69	10.10	9.57
140	8.43	11.01	9.03
150	9.98	10.44	9.80

5.16. Derive the expected mean squares for a two-factor analysis of variance with one observation per cell, assuming that both factors are fixed.

5.17. Consider the three-factor model

$$y_{ijk} = \mu + \tau_i + \beta_j + \gamma_k + (\tau\beta)_{ij} + (\beta\gamma)_{jk} + \epsilon_{ijk} \quad \begin{cases} i = 1, 2, \ldots, a \\ j = 1, 2, \ldots, b \\ k = 1, 2, \ldots, c \end{cases}$$

Notice that there is only one replicate. Assuming all the factors are fixed, write down the analysis of variance table, including the expected mean squares. What would you use as the "experimental error" to test hypotheses?

5.18. The percentage of hardwood concentration in raw pulp, the vat pressure, and the cooking time of the pulp are being investigated for their effects on the strength of paper. Three levels of hardwood concentration, three levels of pressure, and two cooking times are selected. A factorial experiment with two replicates is conducted, and the following data are obtained:

Percentage of Hardwood Concentration	Cooking Time 3.0 Hours, Pressure 400	500	650
2	196.6	197.7	199.8
	196.0	196.0	199.4
4	198.5	196.0	198.4
	197.2	196.9	197.6
8	197.5	195.6	197.4
	196.6	196.2	198.1

Percentage of Hardwood Concentration	Cooking Time 4.0 Hours, Pressure 400	500	650
2	198.4	199.6	200.6
	198.6	200.4	200.9
4	197.5	198.7	199.6
	198.1	198.0	199.0
8	197.6	197.0	198.5
	198.4	197.8	199.8

(a) Analyze the data and draw conclusions. Use $\alpha = 0.05$.

(b) Prepare appropriate residual plots and comment on the model's adequacy.

(c) Under what set of conditions would you operate this process? Why?

5.19. The quality control department of a fabric finishing plant is studying the effect of several factors on the dyeing of cotton–synthetic cloth used to manufacture men's shirts. Three operators, three cycle times, and two temperatures were selected, and three small specimens of cloth were dyed under each set of conditions. The finished cloth was compared to a standard, and a numerical score was assigned. The results are as follows. Analyze the data and draw conclusions. Comment on the model's adequacy.

	Temperature					
	300°C			350°C		
	Operator			Operator		
Cycle Time	1	2	3	1	2	3
	23	27	31	24	38	34
40	24	28	32	23	36	36
	25	26	29	28	35	39
	36	34	33	37	34	34
50	35	38	34	39	38	36
	36	39	35	35	36	31
	28	35	26	26	36	28
60	24	35	27	29	37	26
	27	34	25	25	34	24

5.20. In Problem 5.4, suppose that we wish to reject the null hypothesis with a high probability if the difference in the true mean yield at any two pressures is as great as 0.5. If a reasonable prior estimate of the standard deviation of yield is 0.1, how many replicates should be run?

5.21. The yield of a chemical process is being studied. The two factors of interest are temperature and pressure. Three levels of each factor are selected; however, only nine runs can be made in one day. The experimenter runs a complete replicate of the design on each day. The data are shown in the following table. Analyze the data, assuming that the days are blocks.

	Day 1			Day 2		
	Pressure			Pressure		
Temperature	250	260	270	250	260	270
Low	86.3	84.0	85.8	86.1	85.2	87.3
Medium	88.5	87.3	89.0	89.4	89.9	90.3
High	89.1	90.2	91.3	91.7	93.2	93.7

5.22. Consider the data in Problem 5.7. Analyze the data, assuming that replicates are blocks.

5.23. Consider the data in Problem 5.8. Analyze the data, assuming that replicates are blocks.

5.24. An article in the *IEEE Transactions on Electron Devices* (Nov. 1986, pp. 1754) describes a study on polysilicon doping. The experiment shown below is a variation of their study. The response variable is base current.

Polysilicon Doping (ions)	Anneal Temperature (°C)		
	900	950	1000
1×10^{20}	4.60	10.15	11.01
	4.40	10.20	10.58
2×10^{20}	3.20	9.38	10.81
	3.50	10.02	10.60

(a) Is there evidence (with $\alpha = 0.05$) indicating that either polysilicon doping level or anneal temperature affects base current?

(b) Prepare graphical displays to assist in interpreting this experiment.

(c) Analyze the residuals and comment on model adequacy.

(d) Is the model

$$y = \beta_0 + \beta_1 x_1 + \beta_2 x_2 + \beta_{22} x_2^2 + \beta_{12} x_1 x_2 + \epsilon$$

supported by this experiment (x_1 = doping level, x_2 = temperature)? Estimate the parameters in this model and plot the response surface.

5.25. An article in the *Journal of Testing and Evaluation* (Vol. 16, No. 2, pp. 508–515) investigated the effects of cyclic loading and environmental conditions on fatigue crack growth at a constant 22 MPa stress for a particular material. The data from this experiment are shown below (the response is crack growth rate):

	Environment		
Frequency	Air	H_2O	Salt H_2O
	2.29	2.06	1.90
10	2.47	2.05	1.93
	2.48	2.23	1.75
	2.12	2.03	2.06
	2.65	3.20	3.10
1	2.68	3.18	3.24
	2.06	3.96	3.98
	2.38	3.64	3.24
	2.24	11.00	9.96
0.1	2.71	11.00	10.01
	2.81	9.06	9.36
	2.08	11.30	10.40

(a) Analyze the data from this experiment (use $\alpha = 0.05$).

(b) Analyze the residuals.

(c) Repeat the analyses from parts (a) and (b) using ln (y) as the response. Comment on the results.

5.26. An experiment was conducted to study the life (in hours) of two different brands of batteries in three different devices (radio, camera, and portable DVD player). A completely randomized two-factor factorial experiment was conducted and the following data resulted.

Brand of Battery	Device: Radio	Camera	DVD Player
A	8.6	7.8	5.5
	8.2	8.4	5.7
B	9.4	8.5	5.8
	8.8	8.9	5.9

(a) Analyze the data and draw conclusions, using $\alpha = 0.05$.

(b) Investigate model adequacy by plotting the residuals.

(c) Which brand of batteries would you recommend?

5.27. I have recently purchased new golf clubs, which I believe will significantly improve my game. Below are the scores of three rounds of golf played at three different golf courses with the old and the new clubs.

Clubs	Course: Ahwatukee	Karsten	Foothills
Old	90	91	88
	87	91	86
	86	90	90
New	88	90	86
	87	91	85
	85	88	88

(a) Conduct an analysis of variance. Using $\alpha = 0.05$, what conclusions can you draw?

(b) Investigate model adequacy by plotting the residuals.

5.28. Bone anchors are used by orthopedic surgeons in repairing torn rotator cuffs (a common shoulder tendon injury among baseball players). The bone anchor is a threaded insert that is screwed into a hole that has been drilled into the shoulder bone near the site of the torn tendon. The torn tendon is then sutured to the anchor. In a successful operation, the tendon is stabilized and reattaches itself to the bone. However, bone anchors can pull out if they are subjected to high loads. An experiment was performed to study the force required to pull out the anchor for three anchor types and two different foam densities (the foam simulates the natural variability found in real bone). Two replicates of the experiment were performed. The experimental design and the pullout force response data are as follows:

Anchor Type	Foam Density: Low	High
A	190, 200	241, 255
B	185, 190	230, 237
C	210, 205	256, 260

(a) Analyze the data from this experiment.

(b) Investigate model adequacy by constructing appropriate residual plots.

(c) What conclusions can you draw?

5.29. An experiment was performed to investigate the keyboard feel on a computer (crisp or mushy) and the size of the keys (small, medium, or large). The response variable is typing speed. Three replicates of the experiment were performed. The experimental design and the data are as follows:

Key Size	Keyboard Feel: Mushy	Crisp
Small	31, 33, 35	36, 40, 41
Medium	36, 35, 33	40, 41, 42
Large	37, 34, 33	38, 36, 39

(a) Analyze the data from this experiment.

(b) Investigate model adequacy by constructing appropriate residual plots.

(c) What conclusions can you draw?

5.30. A manufacturer of laundry products is investigating the performance of a newly formulated stain remover. The new formulation is compared to the original formulation with respect to its ability to remove a standard tomato-like stain in a test article of cotton cloth using a factorial experiment. The other factors in the experiment are the number of times the test article is washed (1 or 2) and whether or not a detergent booster is used. The response variable is the stain shade after washing (12 is the darkest, 0 is the lightest). The data are shown in the following table.

	Number of Washings 1		Number of Washings 2	
Formulation	Booster		Booster	
	Yes	No	Yes	No
New	6, 5	6, 5	3, 2	4, 1
Original	10, 9	11, 11	10, 9	9, 10

(a) Conduct an analysis of variance. Using $\alpha = 0.05$, what conclusions can you draw?

(b) Investigate model adequacy by plotting the residuals.

5.31. An article in *Quality Progress* (May 2011, pp. 42–48) describes the use of factorial experiments to improve a silver powder production process. This product is used in conductive pastes to manufacture a wide variety of products ranging from silicon wafers to elastic membrane switches. Powder density (g/cm^2) and surface area (cm^2/g) are the two critical characteristics of this product. The experiments involved three factors; reaction temperature, ammonium percent, and stirring rate. Each of these factors had two levels and the design was replicated twice. The design is shown below.

Ammonium (%)	Stir Rate (RPM)	Temperature (°C)	Density	Surface Area
2	100	8	14.68	0.40
2	100	8	15.18	0.43
30	100	8	15.12	0.42
30	100	8	17.48	0.41
2	150	8	7.54	0.69
2	150	8	6.66	0.67
30	150	8	12.46	0.52
30	150	8	12.62	0.36
2	100	40	10.95	0.58
2	100	40	17.68	0.43
30	100	40	12.65	0.57
30	100	40	15.96	0.54
2	150	40	8.03	0.68
2	150	40	8.84	0.75
30	150	40	14.96	0.41
30	150	40	14.96	0.41

(a) Analyze the density response. Are any interactions significant? Draw appropriate conclusions about the effects of the significant factors on the response.
(b) Prepare appropriate residual plots and comment on model adequacy.
(c) Construct contour plots to aid in practical interpretation of the density response.
(d) Analyze the surface area response. Are any interactions significant? Draw appropriate conclusions about the effects of the significant factors on the response.
(e) Prepare appropriate residual plots and comment on model adequacy.
(f) Construct contour plots to aid in practical interpretation of the surface area response.

5.32. ***Continuation of Problem 5.31.*** Suppose that the specifications require that surface area must be between 0.3 and 0.6 cm^2/g and that density must be less than 14 g/cm^3. Find a set of operating conditions that will result in a product that meets these requirements.

5.33. An article in *Biotechnology Progress* (2001, Vol. 17, pp. 366–368) described an experiment to investigate nisin extraction in aqueous two-phase solutions. A two-factor factorial experiment was conducted using factors A = concentration of PEG and B = concentration of Na_2SO_4. Data similar to that reported in the paper are shown below.

A	B	Extraction (%)
13	11	62.9
13	11	65.4
15	11	76.1
15	11	72.3
13	13	87.5
13	13	84.2
15	13	102.3
15	13	105.6

(a) Analyze the extraction response. Draw appropriate conclusions about the effects of the significant factors on the response.
(b) Prepare appropriate residual plots and comment on model adequacy.
(c) Construct contour plots to aid in practical interpretation of the density response.

5.34. Reconsider the experiment in Problem 5.3. Suppose that this experiment had been conducted in three blocks, with each replicate a block. Assume that the observations in the data table are given in order, that is, the first observation in each cell comes from the first replicate, and so on. Reanalyze the data as a factorial experiment in blocks and estimate the variance component for blocks. Does it appear that blocking was useful in this experiment?

5.35. Reconsider the experiment in Problem 5.6. Suppose that this experiment had been conducted in three blocks, with each replicate a block. Assume that the observations in the data table are given in order, that is, the first observation in each cell comes from the first replicate, and so on. Reanalyze the data as a factorial experiment in blocks and estimate the variance component for blocks. Does it appear that blocking was useful in this experiment?

5.36. Reconsider the experiment in Problem 5.8. Suppose that this experiment had been conducted in two blocks, with each replicate a block. Assume that the observations in the data table are given in order; that is, the first observation in each cell comes from the first replicate, and so on. Reanalyze the data as a factorial experiment in blocks and estimate the variance component for blocks. Does it appear that blocking was useful in this experiment?

5.37. Reconsider the three-factor factorial experiment in Problem 5.18. Suppose that this experiment had been conducted in two blocks, with each replicate a block. Assume that the observations in the data table are given in order; that is, the first observation in each cell comes from the first replicate, and so on. Reanalyze the data as a factorial experiment in blocks and estimate the variance component for blocks. Does it appear that blocking was useful in this experiment?

5.38. Reconsider the three-factor factorial experiment in Problem 5.19. Suppose that this experiment had been conducted in three blocks, with each replicate a block. Assume that the observations in the data table are given in order; that is, the first observation in each cell comes from the first replicate, and so on. Reanalyze the data as a factorial experiment in blocks and estimate the variance component for blocks. Does it appear that blocking was useful in this experiment?

5.39. Reconsider the bone anchor experiment in Problem 5.28. Suppose that this experiment had been conducted in two blocks, with each replicate a block. Assume that the observations in the data table are given in order; that is, the first observation in each cell comes from the first replicate, and so on. Reanalyze the data as a factorial experiment in blocks and estimate the variance component for blocks. Does it appear that blocking was useful in this experiment?

5.40. Reconsider the keyboard experiment in Problem 5.29. Suppose that this experiment had been conducted in three blocks, with each replicate a block. Assume that the observations in the data table are given in order; that is, the first observation in each cell comes from the first replicate, and so on. Reanalyze the data as a factorial experiment in blocks and estimate the variance component for blocks. Does it appear that blocking was useful in this experiment?

5.41. The C. F. Eye Care company manufactures lenses for transplantation into the eye following cataract surgery. An engineering group has conducted an experiment involving two factors to determine their effect on the lens polishing process. The results of this experiment are summarized in the following ANOVA display:

Source	DF	SS	MS	F	*P*-value
Factor A	—	—	0.0833	0.05	0.952
Factor B	—	96.333	96.3333	57.80	<0.001
Interaction	2	12.167	6.0833	3.65	—
Error	6	10.000	—		
Total	11	118.667			

Answer the following questions about this experiment.

(a) The sum of squares for factor A is______.

(b) The number of degrees of freedom for factor A in the experiment is_____.

(c) The number of degrees of freedom for factor B is_______.

(d) The mean square for error is______.

(e) An upper bound for the *P*-value for the interaction test statistic is______.

(f) The engineers used ______ levels of the factor A in this experiment.

(g) The engineers used ______ levels of the factor B in this experiment.

(h) There are ______ replicates of this experiment.

(i) Would you conclude that the effect of factor B depends on the level of factor A (Yes or No)?

(j) An estimate of the standard deviation of the response variable is ______.

5.42. Reconsider the lens polishing experiment in Problem 5.41. Suppose that this experiment had been conducted as a RCBD. The sum of squares for blocks is 4.00. Reconstruct the ANOVA given this new information. What impact does the blocking have on the conclusions from the original experiment?

5.43. In Problem 4.53 you met physics PhD student Laura Van Ertia who had conducted a single factor experiment in her pursuit of the unified theory. She's at it again, and this time she has moved on to a two-factor factorial conducted as a CRD. From her experiment, Laura has constructed the following incomplete ANOVA display:

Source	SS	DF	MS	F
A	350.00	2		
B	300.00		150	
AB	200.00		50	
Error	150.00	18		
Total	1000.00			

(a) How many levels of factor B did she use in the experiment? ______

(b) How many degrees of freedom are associated with interaction? ______

(c) The error mean square is ______.

(d) The mean square for factor A is ______.

(e) How many replicates of the experiment were conducted? ______

(f) What are your conclusions about interaction and the two main effects?

(g) An estimate of the standard deviation of the response variable is ______.

(h) If this experiment had been run in blocks there would have been ______ degrees of freedom for blocks.

5.44. ***Continuation of Problem 5.43.*** Suppose that Laura did actually conduct the experiment in Problem 5.43 as an RCBD. Assume that the block sum of squares is 60.00. Reconstruct the ANOVA display under this new set of assumptions.

5.45. Consider the following ANOVA for a two-factor factorial experiment:

Source	DF	SS	MS	F	P
A	2	8.0000	4.00000	2.00	0.216
B	1	8.3333	8.33333	4.17	0.087
Interaction	2	10.6667	5.33333	2.67	0.148
Error	6	12.0000	2.00000		
Total	11	39.0000			

In addition to the ANOVA, you are given the following data totals. Row totals (factor A) = 18, 10, 14; column totals (factor B) = 16, 26; cell totals = 10, 8, 2, 8, 4, 10, and replicate totals = 19, 23. The grand total is 42. The original experiment was a completely randomized design. Now suppose that the experiment had been run in two complete blocks. Answer the following questions about the ANOVA for the blocked experiment.

(a) The block sum of squares is ______.

(b) There are______ degrees of freedom for blocks.

(c) The error sum of squares is now ______.

(d) The interaction effect is now significant at 1 percent (Yes or No).

5.46. Consider the following incomplete ANOVA table:

Source	SS	DF	MS	F
A	50.00	1	50.00	
B	80.00	2	40.00	
AB	30.00	2	15.00	
Error		12		
Total	172.00	17		

In addition to the ANOVA table you know that the experiment has been replicated three times and that the totals of the three replicates are 10, 12, and 14 respectively. The original experiment was run as a completely randomized design. Answer the following questions:

(a) The pure error estimate of the standard deviation of the sample observations is 1 (Yes or No).

(b) Suppose that the experiment had been run in blocks, so that it is an RCBD. The number of degrees of freedom for blocks would be ______.

(c) The block sum of squares is ______.

(d) The error sum of squares in the RCBD is now ______.

(e) For the RCBD, what is the estimate of the standard deviation of the sample observations?

CHAPTER 6

Two-Level Factorial Designs

CHAPTER OUTLINE

The supplemental material is on the textbook website www.wiley.com/go/global/montgomery.

6.1 Introduction

Factorial designs are widely used in experiments involving several factors where it is necessary to study the joint effect of the factors on a response. Chapter 5 presented general methods for the analysis of factorial designs. However, several special cases of the general factorial design are important because they are widely used in research work and also because they form the basis of other designs of considerable practical value.

The most important of these special cases is that of k factors, each at only two levels. These levels may be **quantitative**, such as two values of temperature, pressure, or time; or they may be **qualitative**, such as two machines, two operators, the "high" and "low" levels of a factor, or perhaps the presence and absence of a factor. A complete replicate of such a design requires $2 \times 2 \times \ldots \times 2 = 2^k$ observations and is called a **2^k factorial design**.

This chapter focuses on this extremely important class of designs. Throughout this chapter, we assume that (1) the factors are fixed, (2) the designs are completely randomized, and (3) the usual normality assumptions are satisfied.

The 2^k design is particularly useful in the early stages of experimental work when many factors are likely to be investigated. It provides the smallest number of runs with which k factors can be studied in a complete factorial design. Consequently, these designs are widely used in **factor screening experiments**.

Because there are only two levels for each factor, we assume that the response is approximately linear over the range of the factor levels chosen. In many factor screening experiments, when we are just starting to study the process or the system, this is often a reasonable assumption. In Section 6.8, we will present a simple method for checking this assumption and discuss what action to take if it is violated. The book by Mee (2009) is a useful supplement to this chapter and chapters 7 and 8.

6.2 The 2^2 Design

The first design in the 2^k series is one with only two factors, say *A* and *B*, each run at two levels. This design is called a **2^2 factorial design**. The levels of the factors may be arbitrarily called "low" and "high." As an example, consider an investigation into the effect of the concentration of the reactant and the amount of the catalyst on the conversion (yield) in a chemical process. The objective of the experiment was to determine if adjustments to either of these two factors would increase the yield. Let the reactant concentration be factor *A* and let the two levels of interest be 15 and 25 percent. The catalyst is factor *B*, with the high level denoting the use of 2 pounds of the catalyst and the low level denoting the use of only 1 pound. The experiment is replicated three times, so there are 12 runs. The order in which the runs are made is random, so this is a **completely randomized experiment**. The data obtained are as follows:

Factor			Replicate			
A	*B*	Treatment Combination	I	II	III	Total
−	−	*A* low, *B* low	28	25	27	80
+	−	*A* high, *B* low	36	32	32	100
−	+	*A* low, *B* high	18	19	23	60
+	+	*A* high, *B* high	31	30	29	90

The four treatment combinations in this design are shown graphically in Figure 6.1. By convention, we denote the effect of a factor by a capital Latin letter. Thus "*A*" refers to the effect of factor *A*, "*B*" refers to the effect of factor *B*, and "*AB*" refers to the *AB* interaction. In the 2^2 design, the low and high levels of *A* and *B* are denoted by "−" and "+," respectively, on the *A* and *B* axes. Thus, − on the *A* axis represents the low level of concentration (15%), whereas + represents the high level (25%), and − on the *B* axis represents the low level of catalyst, and + denotes the high level.

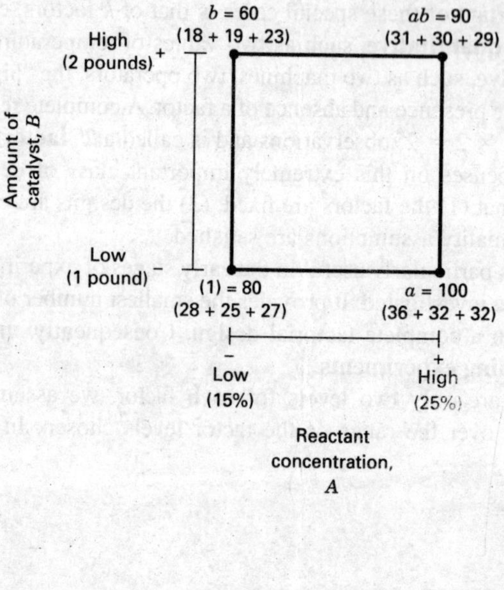

■ **FIGURE 6.1** **Treatment combinations in the 2^2 design**

The four treatment combinations in the design are also represented by lowercase letters, as shown in Figure 6.1. We see from the figure that the high level of any factor in the treatment combination is denoted by the corresponding lowercase letter and that the low level of a factor in the treatment combination is denoted by the absence of the corresponding letter. Thus, a represents the treatment combination of A at the high level and B at the low level, b represents A at the low level and B at the high level, and ab represents both factors at the high level. By convention, (1) is used to denote both factors at the low level. This notation is used throughout the 2^k series.

In a two-level factorial design, we may define the average effect of a factor as the change in response produced by a change in the level of that factor averaged over the levels of the other factor. Also, the symbols (1), a, b, and ab now represent the *total* of the response observation at all n replicates taken at the treatment combination, as illustrated in Figure 6.1. Now the effect of A at the low level of B is $[a - (1)]/n$, and the effect of A at the high level of B is $[ab - b]/n$. Averaging these two quantities yields the **main effect** of A:

$$\begin{aligned} A &= \frac{1}{2n}\{[ab - b] + [a - (1)]\} \\ &= \frac{1}{2n}[ab + a - b - (1)] \end{aligned} \tag{6.1}$$

The average main effect of B is found from the effect of B at the low level of A (i.e., $[b - (1)]/n$) and at the high level of A (i.e., $[ab - a]/n$) as

$$\begin{aligned} B &= \frac{1}{2n}\{[ab - a] + [b - (1)]\} \\ &= \frac{1}{2n}[ab + b - a - (1)] \end{aligned} \tag{6.2}$$

We define the **interaction effect** AB as the average difference between the effect of A at the high level of B and the effect of A at the low level of B. Thus,

$$\begin{aligned} AB &= \frac{1}{2n}\{[ab - b] - [a - (1)]\} \\ &= \frac{1}{2n}[ab + (1) - a - b] \end{aligned} \tag{6.3}$$

Alternatively, we may define AB as the average difference between the effect of B at the high level of A and the effect of B at the low level of A. This will also lead to Equation 6.3.

The formulas for the effects of A, B, and AB may be derived by another method. The effect of A can be found as the difference in the average response of the two treatment combinations on the right-hand side of the square in Figure 6.1 (call this average $\bar{y}_{A^+}$ because it is the average response at the treatment combinations where A is at the high level) and the two treatment combinations on the left-hand side (or $\bar{y}_{A^-}$). That is,

$$\begin{aligned} A &= \bar{y}_{A^+} - \bar{y}_{A^-} \\ &= \frac{ab + a}{2n} - \frac{b + (1)}{2n} \\ &= \frac{1}{2n}[ab + a - b - (1)] \end{aligned}$$

This is exactly the same result as in Equation 6.1. The effect of B, Equation 6.2, is found as the difference between the average of the two treatment combinations on the top of the square ($\bar{y}_{B^+}$) and the average of the two treatment combinations on the bottom ($\bar{y}_{B^-}$), or

$$B = \bar{y}_{B^+} - \bar{y}_{B^-}$$
$$= \frac{ab + b}{2n} - \frac{a + (1)}{2n}$$
$$= \frac{1}{2n}[ab + b - a - (1)]$$

Finally, the interaction effect AB is the average of the right-to-left diagonal treatment combinations in the square [ab and (1)] minus the average of the left-to-right diagonal treatment combinations (a and b), or

$$AB = \frac{ab + (1)}{2n} - \frac{a + b}{2n}$$
$$= \frac{1}{2n}[ab + (1) - a - b]$$

which is identical to Equation 6.3.

Using the experiment in Figure 6.1, we may estimate the average effects as

$$A = \frac{1}{2(3)}(90 + 100 - 60 - 80) = 8.33$$

$$B = \frac{1}{2(3)}(90 + 60 - 100 - 80) = -5.00$$

$$AB = \frac{1}{2(3)}(90 + 80 - 100 - 60) = 1.67$$

The effect of A (reactant concentration) is positive; this suggests that increasing A from the low level (15%) to the high level (25%) will increase the yield. The effect of B (catalyst) is negative; this suggests that increasing the amount of catalyst added to the process will decrease the yield. The interaction effect appears to be small relative to the two main effects.

In experiments involving 2^k designs, it is always important to examine the **magnitude** and **direction** of the factor effects to determine which variables are likely to be important. The **analysis of variance** can generally be used to confirm this interpretation (t-tests could be used too). Effect magnitude and direction should always be considered along with the ANOVA, because the ANOVA alone does not convey this information. There are several excellent statistics software packages that are useful for setting up and analyzing 2^k designs. There are also special time-saving methods for performing the calculations manually.

Consider determining the sums of squares for A, B, and AB. Note from Equation 6.1 that a **contrast** is used in estimating A, namely

$$\text{Contrast}_A = ab + a - b - (1) \tag{6.4}$$

We usually call this contrast the **total effect** of A. From Equations 6.2 and 6.3, we see that contrasts are also used to estimate B and AB. Furthermore, these three contrasts are **orthogonal**. The sum of squares for any contrast can be computed from Equation 3.29, which states that the sum of squares for any contrast is equal to the contrast squared divided by the number of observations in each total in the contrast times the sum of the squares of the contrast coefficients. Consequently, we have

$$SS_A = \frac{[ab + a - b - (1)]^2}{4n} \tag{6.5}$$

$$SS_B = \frac{[ab + b - a - (1)]^2}{4n} \tag{6.6}$$

and

$$SS_{AB} = \frac{[ab + (1) - a - b]^2}{4n} \tag{6.7}$$

as the sums of squares for *A, B*, and *AB*. Notice how simple these equations are. We can compute sums of squares by only squaring one number.

Using the experiment in Figure 6.1, we may find the sums of squares from Equations 6.5, 6.6, and 6.7 as

$$SS_A = \frac{(50)^2}{4(3)} = 208.33$$

$$SS_B = \frac{(-30)^2}{4(3)} = 75.00 \tag{6.8}$$

and

$$SS_{AB} = \frac{(10)^2}{4(3)} = 8.33$$

The total sum of squares is found in the usual way, that is,

$$SS_T = \sum_{i=1}^{2}\sum_{j=1}^{2}\sum_{k=1}^{n} y_{ijk}^2 - \frac{y_{...}^2}{4n} \tag{6.9}$$

In general, SS_T has $4n - 1$ degrees of freedom. The error sum of squares, with $4(n - 1)$ degrees of freedom, is usually computed by subtraction as

$$SS_E = SS_T - SS_A - SS_B - SS_{AB} \tag{6.10}$$

For the experiment in Figure 6.1, we obtain

$$SS_T = \sum_{i=1}^{2}\sum_{j=1}^{2}\sum_{k=1}^{3} y_{ijk}^2 - \frac{y_{...}^2}{4(3)}$$

$$= 9398.00 - 9075.00 = 323.00$$

and

$$SS_E = SS_T - SS_A - SS_B - SS_{AB}$$

$$= 323.00 - 208.33 - 75.00 - 8.33$$

$$= 31.34$$

using SS_A, SS_B, and SS_{AB} from Equations 6.8. The complete ANOVA is summarized in Table 6.1. On the basis of the *P*-values, we conclude that the main effects are statistically significant and that there is no interaction between these factors. This confirms our initial interpretation of the data based on the magnitudes of the factor effects.

It is often convenient to write down the treatment combinations in the order (1), *a*, *b*, *ab*. This is referred to as **standard order** (or Yates's order, for Frank Yates who was one of Fisher coworkers and who made many important contributions to designing and analyzing experiments). Using this standard order, we see that the contrast coefficients used in estimating the effects are

Effects	(1)	*a*	*b*	*ab*
A	−1	+1	−1	+1
B	−1	−1	+1	+1
AB	+1	−1	−1	+1

■ **TABLE 6.1**
Analysis of Variance for the Experiment in Figure 6.1

Source of Variation	Sum of Squares	Degrees of Freedom	Mean Square	F_q	*P*-Value
A	208.33	1	208.33	53.15	0.0001
B	75.00	1	75.00	19.13	0.0024
AB	8.33	1	8.33	2.13	0.1826
Error	31.34	8	3.92		
Total	323.00	11			

Note that the contrast coefficients for estimating the interaction effect are just the product of the corresponding coefficients for the two main effects. The contrast coefficient is always either $+1$ or -1, and a **table of plus and minus signs** such as in Table 6.2 can be used to determine the proper sign for each treatment combination. The column headings in Table 6.2 are the main effects (*A* and *B*), the *AB* interaction, and *I*, which represents the total or average of the entire experiment. Notice that the column corresponding to *I* has only plus signs. The row designators are the treatment combinations. To find the contrast for estimating any effect, simply multiply the signs in the appropriate column of the table by the corresponding treatment combination and add. For example, to estimate *A*, the contrast is $-(1) + a - b + ab$, which agrees with Equation 6.1. Note that the contrasts for the effects *A*, *B*, and *AB* are **orthogonal**. Thus, the 2^2 (and all 2^k designs) is an **orthogonal design**. The ± 1 coding for the low and high levels of the factors is often called the **orthogonal coding** or the **effects coding**.

The Regression Model. In a 2^k factorial design, it is easy to express the results of the experiment in terms of a **regression model**. Because the 2^k is just a factorial design, we could also use either an effects or a means model, but the regression model approach is much more natural and intuitive. For the chemical process experiment in Figure 6.1, the regression model is

$$y = \beta_0 + \beta_1 x_1 + \beta_2 x_2 + \epsilon$$

where x_1 is a **coded variable** that represents the reactant concentration, x_2 is a coded variable that represents the amount of catalyst, and the β's are regression coefficients. The relationship between the **natural variables**, the reactant concentration and the amount of catalyst, and the coded variables is

$$x_1 = \frac{\text{Conc} - (\text{Conc}_{\text{low}} + \text{Conc}_{\text{high}})/2}{(\text{Conc}_{\text{high}} - \text{Conc}_{\text{low}})/2}$$

and

$$x_2 = \frac{\text{Catalyst} - (\text{Catalyst}_{\text{low}} + \text{Catalyst}_{\text{high}})/2}{(\text{Catalyst}_{\text{high}} - \text{Catalyst}_{\text{low}})/2}$$

■ **TABLE 6.2**
Algebraic Signs for Calculating Effects in the 2^2 Design

Treatment Combination	Factorial Effect *I*	*A*	*B*	*AB*
(1)	+	−	−	+
a	+	+	−	−
b	+	−	+	−
ab	+	+	+	+

When the natural variables have only two levels, this coding will produce the familiar ± 1 notation for the levels of the coded variables. To illustrate this for our example, note that

$$x_1 = \frac{\text{Conc} - (15 + 25)/2}{(25 - 15)/2}$$
$$= \frac{\text{Conc} - 20}{5}$$

Thus, if the concentration is at the high level (Conc = 25%), then $x_1 = +1$; if the concentration is at the low level (Conc = 15%), then $x_1 = -1$. Furthermore,

$$x_2 = \frac{\text{Catalyst} - (1 + 2)/2}{(2 - 1)/2}$$
$$= \frac{\text{Catalyst} - 1.5}{0.5}$$

Thus, if the catalyst is at the high level (Catalyst = 2 pounds), then $x_2 = +1$; if the catalyst is at the low level (Catalyst = 1 pound), then $x_2 = -1$.

The fitted regression model is

$$\hat{y} = 27.5 + \left(\frac{8.33}{2}\right) x_1 + \left(\frac{-5.00}{2}\right) x_2$$

where the intercept is the grand average of all 12 observations, and the regression coefficients $\hat{\beta}_1$ and $\hat{\beta}_2$ are one-half the corresponding factor effect estimates. The regression coefficient is one-half the effect estimate because a regression coefficient measures the effect of a one-unit change in x on the mean of y, and the effect estimate is based on a two-unit change (from -1 to $+1$). This simple method of estimating the regression coefficients results in **least squares** parameter estimates. We will return to this topic again in Section 6.7. Also see the **supplemental material** for this chapter.

Residuals and Model Adequacy. The regression model can be used to obtain the predicted or fitted value of y at the four points in the design. The residuals are the differences between the observed and fitted values of y. For example, when the reactant concentration is at the low level ($x_1 = -1$) and the catalyst is at the low level ($x_2 = -1$), the predicted yield is

$$\hat{y} = 27.5 + \left(\frac{8.33}{2}\right)(-1) + \left(\frac{-5.00}{2}\right)(-1) = 25.835$$

There are three observations at this treatment combination, and the residuals are

$$e_1 = 28 - 25.835 = 2.165$$
$$e_2 = 25 - 25.835 = -0.835$$
$$e_3 = 27 - 25.835 = 1.165$$

The remaining predicted values and residuals are calculated similarly. For the high level of the reactant concentration and the low level of the catalyst,

$$\hat{y} = 27.5 + \left(\frac{8.33}{2}\right)(+1) + \left(\frac{-5.00}{2}\right)(-1) = 34.165$$

and

$$e_4 = 36 - 34.165 = 1.835$$
$$e_5 = 32 - 34.165 = -2.165$$
$$e_6 = 32 - 34.165 = -2.165$$

For the low level of the reactant concentration and the high level of the catalyst,

$$\hat{y} = 27.5 + \left(\frac{8.33}{2}\right)(-1) + \left(\frac{-5.00}{2}\right)(+1) = 20.835$$

and

$$e_7 = 18 - 20.835 = -2.835$$
$$e_8 = 19 - 20.835 = -1.835$$
$$e_9 = 23 - 20.835 = 2.165$$

Finally, for the high level of both factors,

$$\hat{y} = 27.5 + \left(\frac{8.33}{2}\right)(+1) + \left(\frac{-5.00}{2}\right)(+1) = 29.165$$

and

$$e_{10} = 31 - 29.165 = 1.835$$
$$e_{11} = 30 - 29.165 = 0.835$$
$$e_{12} = 29 - 29.165 = -0.165$$

Figure 6.2 presents a normal probability plot of these residuals and a plot of the residuals versus the predicted yield. These plots appear satisfactory, so we have no reason to suspect that there are any problems with the validity of our conclusions.

The Response Surface. The regression model

$$\hat{y} = 27.5 + \left(\frac{8.33}{2}\right)x_1 + \left(\frac{-5.00}{2}\right)x_2$$

can be used to generate response surface plots. If it is desirable to construct these plots in terms of the **natural factor levels**, then we simply substitute the relationships between the natural and coded variables that we gave earlier into the regression model, yielding

$$\hat{y} = 27.5 + \left(\frac{8.33}{2}\right)\left(\frac{\text{Conc} - 20}{5}\right) + \left(\frac{-5.00}{2}\right)\left(\frac{\text{Catalyst} - 1.5}{0.5}\right)$$

$$= 18.33 + 0.8333\ \text{Conc} - 5.00\ \text{Catalyst}$$

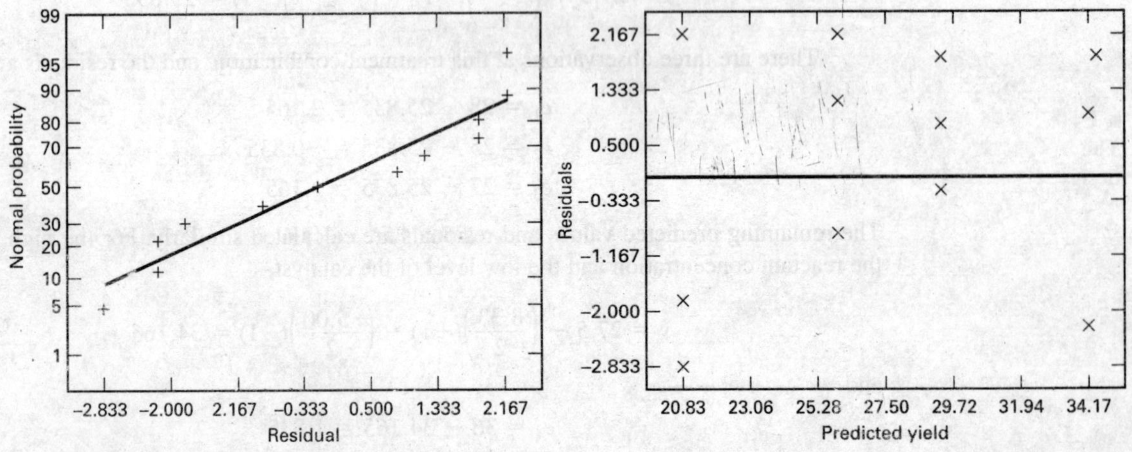

(*a*) Normal probability plot

(*b*) Residuals versus predicted yield

■ **FIGURE 6.2 Residual plots for the chemical process experiment**

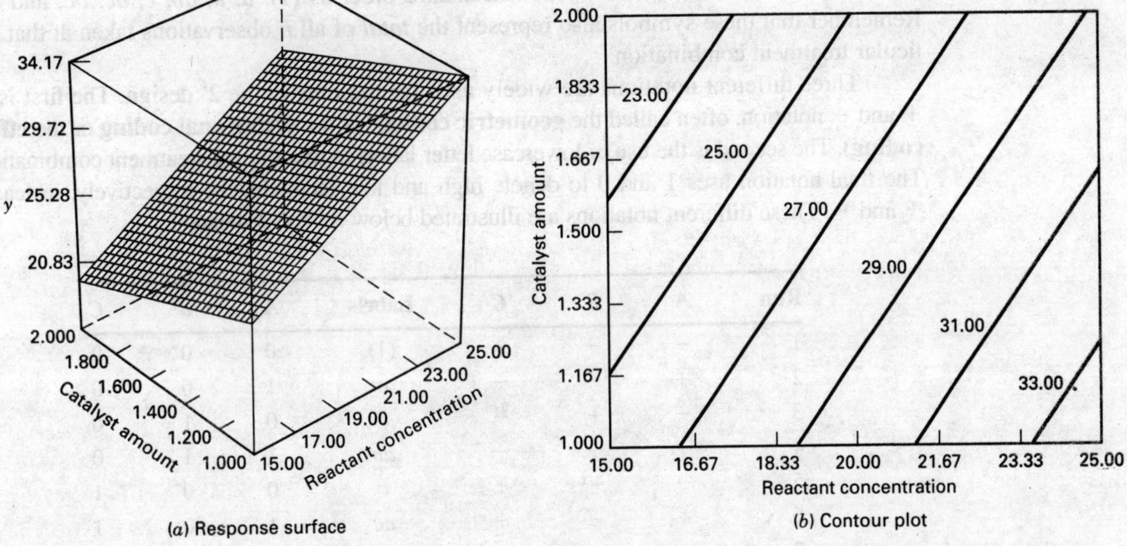

(*a*) Response surface (*b*) Contour plot

■ **FIGURE 6.3** **Response surface plot and contour plot of yield from the chemical process experiment**

Figure 6.3*a* presents the three-dimensional response surface plot of yield from this model, and Figure 6.3*b* is the contour plot. Because the model is **first-order** (that is, it contains only the main effects), the fitted response surface is a plane. From examining the contour plot, we see that yield increases as reactant concentration increases and catalyst amount decreases. Often, we use a fitted surface such as this to find a **direction of potential improvement** for a process. A formal way to do so, called the **method of steepest ascent**, will be presented in Chapter 11 when we discuss methods for systematically exploring response surfaces.

6.3 The 2^3 Design

Suppose that three factors, *A*, *B*, and *C*, each at two levels, are of interest. The design is called a **2^3 factorial design**, and the eight treatment combinations can now be displayed geometrically as a cube, as shown in Figure 6.4*a*. Using the "+ and −" orthogonal coding to represent the low and high levels of the factors, we may list the eight runs in the 2^3 design as in Figure 6.4*b*.

■ **FIGURE 6.4**
The 2^3 factorial design

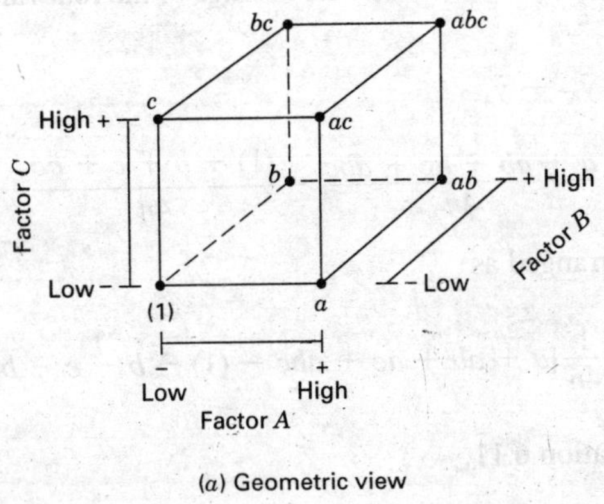

(*a*) Geometric view

Run	Factor A	B	C
1	−	−	−
2	+	−	−
3	−	+	−
4	+	+	−
5	−	−	+
6	+	−	+
7	−	+	+
8	+	+	+

(*b*) Design matrix

This is sometimes called the **design matrix**. Extending the label notation discussed in Section 6.2, we write the treatment combinations in standard order as (1), *a*, *b*, *ab*, *c*, *ac*, *bc*, and *abc*. Remember that these symbols also represent the *total* of all *n* observations taken at that particular treatment combination.

Three different notations are widely used for the runs in the 2^k design. The first is the + and − notation, often called the **geometric coding** (or the **orthogonal coding** or the **effects coding**). The second is the use of lowercase letter labels to identify the treatment combinations. The final notation uses 1 and 0 to denote high and low factor levels, respectively, instead of + and −. These different notations are illustrated below for the 2^3 design:

Run	*A*	*B*	*C*	Labels	*A*	*B*	*C*
1	−	−	−	(1)	0	0	0
2	+	−	−	*a*	1	0	0
3	−	+	−	*b*	0	1	0
4	+	+	−	*ab*	1	1	0
5	−	−	+	*c*	0	0	1
6	+	−	+	*ac*	1	0	1
7	−	+	+	*bc*	0	1	1
8	+	+	+	*abc*	1	1	1

There are seven degrees of freedom between the eight treatment combinations in the 2^3 design. Three degrees of freedom are associated with the main effects of *A*, *B*, and *C*. Four degrees of freedom are associated with interactions; one each with *AB*, *AC*, and *BC* and one with *ABC*.

Consider estimating the main effects. First, consider estimating the main effect *A*. The effect of *A* when *B* and *C* are at the low level is $[a - (1)]/n$. Similarly, the effect of *A* when *B* is at the high level and *C* is at the low level is $[ab - b]/n$. The effect of *A* when *C* is at the high level and *B* is at the low level is $[ac - c]/n$. Finally, the effect of *A* when both *B* and *C* are at the high level is $[abc - bc]/n$. Thus, the average effect of *A* is just the average of these four, or

$$A = \frac{1}{4n}[a - (1) + ab - b + ac - c + abc - bc] \tag{6.11}$$

This equation can also be developed as a contrast between the four treatment combinations in the right face of the cube in Figure 6.5*a* (where *A* is at the high level) and the four in the left face (where *A* is at the low level). That is, the *A* effect is just the average of the four runs where *A* is at the high level ($\bar{y}_{A^+}$) minus the average of the four runs where *A* is at the low level ($\bar{y}_{A^-}$), or

$$A = \bar{y}_{A^+} - \bar{y}_{A^-}$$

$$= \frac{a + ab + ac + abc}{4n} - \frac{(1) + b + c + bc}{4n}$$

This equation can be rearranged as

$$A = \frac{1}{4n}[a + ab + ac + abc - (1) - b - c - bc]$$

which is identical to Equation 6.11.

■ FIGURE 6.5 **Geometric presentation of contrasts corresponding to the main effects and interactions in the 2^3 design**

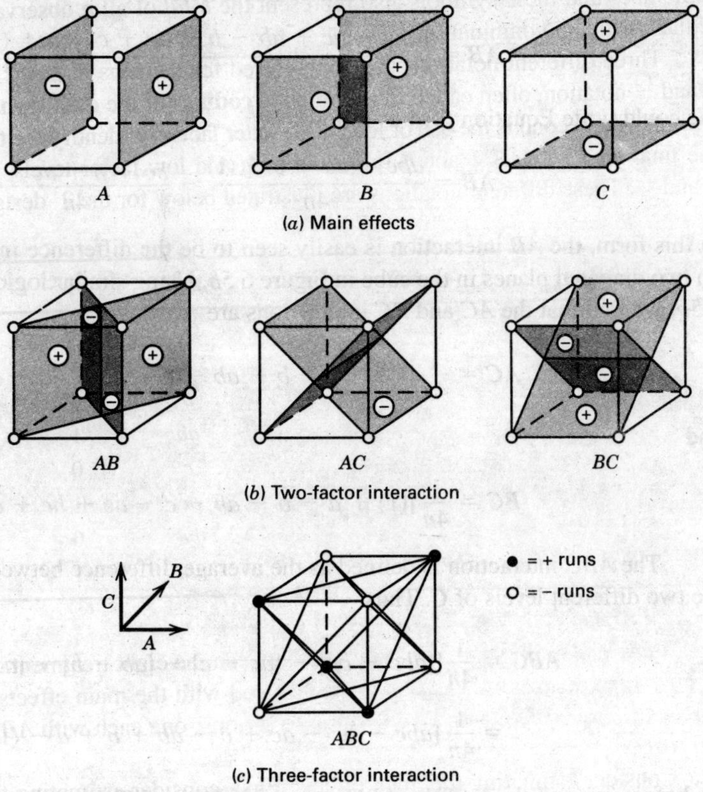

In a similar manner, the effect of B is the difference in averages between the four treatment combinations in the front face of the cube and the four in the back. This yields

$$B = \bar{y}_{B^+} - \bar{y}_{B^-}$$
$$= \frac{1}{4n}[b + ab + bc + abc - (1) - a - c - ac] \tag{6.12}$$

The effect of C is the difference in averages between the four treatment combinations in the top face of the cube and the four in the bottom, that is,

$$C = \bar{y}_{C^+} - \bar{y}_{C^-}$$
$$= \frac{1}{4n}[c + ac + bc + abc - (1) - a - b - ab] \tag{6.13}$$

The two-factor interaction effects may be computed easily. A measure of the AB interaction is the difference between the average A effects at the two levels of B. By convention, one-half of this difference is called the AB interaction. Symbolically,

B	Average A Effect
High (+)	$\frac{[(abc - bc) + (ab - b)]}{2n}$
Low (−)	$\frac{\{(ac - c) + [a - (1)]\}}{2n}$
Difference	$\frac{[abc - bc + ab - b - ac + c - a + (1)]}{2n}$

Because the AB interaction is one-half of this difference,

$$AB = \frac{[abc - bc + ab - b - ac + c - a + (1)]}{4n} \tag{6.14}$$

We could write Equation 6.14 as follows:

$$AB = \frac{abc + ab + c + (1)}{4n} - \frac{bc + b + ac + a}{4n}$$

In this form, the AB interaction is easily seen to be the difference in averages between runs on two diagonal planes in the cube in Figure 6.5*b*. Using similar logic and referring to Figure 6.5*b*, we find that the AC and BC interactions are

$$AC = \frac{1}{4n}[(1) - a + b - ab - c + ac - bc + abc] \tag{6.15}$$

and

$$BC = \frac{1}{4n}[(1) + a - b - ab - c - ac + bc + abc] \tag{6.16}$$

The ABC interaction is defined as the average difference between the AB interaction at the two different levels of C. Thus,

$$\begin{aligned} ABC &= \frac{1}{4n}\{[abc - bc] - [ac - c] - [ab - b] + [a - (1)]\} \\ &= \frac{1}{4n}[abc - bc - ac + c - ab + b + a - (1)] \end{aligned} \tag{6.17}$$

As before, we can think of the ABC interaction as the difference in two averages. If the runs in the two averages are isolated, they define the vertices of the two tetrahedra that comprise the cube in Figure 6.5*c*.

In Equations 6.11 through 6.17, the quantities in brackets are **contrasts** in the treatment combinations. A table of plus and minus signs can be developed from the contrasts, which is shown in Table 6.3. Signs for the main effects are determined by associating a plus with the high level and a minus with the low level. Once the signs for the main effects have been established, the signs for the remaining columns can be obtained by multiplying the appropriate

■ **TABLE 6.3**
Algebraic Signs for Calculating Effects in the 2^3 Design

Treatment Combination	Factorial Effect							
	I	*A*	*B*	*AB*	*C*	*AC*	*BC*	*ABC*
(1)	+	−	−	+	−	+	+	−
a	+	+	−	−	−	−	+	+
b	+	−	+	−	−	+	−	+
ab	+	+	+	+	−	−	−	−
c	+	−	−	+	+	−	−	+
ac	+	+	−	−	+	+	−	−
bc	+	−	+	−	+	−	+	−
abc	+	+	+	+	+	+	+	+

preceding columns row by row. For example, the signs in the *AB* column are the product of the *A* and *B* column signs in each row. The contrast for any effect can be obtained easily from this table.

Table 6.3 has several interesting properties: (1) Except for column *I*, every column has an equal number of plus and minus signs. (2) The sum of the products of the signs in any two columns is zero. (3) Column *I* multiplied times any column leaves that column unchanged. That is, *I* is an **identity element**. (4) The product of any two columns yields a column in the table. For example, $A \times B = AB$, and

$$AB \times B = AB^2 = A$$

We see that the exponents in the products are formed by using **modulus 2** arithmetic. (That is, the exponent can only be 0 or 1; if it is greater than 1, it is reduced by multiples of 2 until it is either 0 or 1.) All of these properties are implied by the **orthogonality** of the 2^3 design and the contrasts used to estimate the effects.

Sums of squares for the effects are easily computed because each effect has a corresponding single-degree-of-freedom contrast. In the 2^3 design with *n* replicates, the sum of squares for any effect is

$$SS = \frac{(\text{Contrast})^2}{8n} \tag{6.18}$$

EXAMPLE 6.1

A 2^3 factorial design was used to develop a nitride etch process on a single-wafer plasma etching tool. The design factors are the gap between the electrodes, the gas flow (C_2F_6 is used as the reactant gas), and the RF power applied to the cathode (see Figure 3.1 for a schematic of the plasma etch tool). Each factor is run at two levels, and the design is replicated twice. The response variable is the etch rate for silicon nitride (Å/m). The etch rate data are shown in Table 6.4, and the design is shown geometrically in Figure 6.6.

Using the totals under the treatment combinations shown in Table 6.4, we may estimate the factor effects as follows:

$$\begin{aligned} A &= \frac{1}{4n}[a - (1) + ab - b + ac - c + abc - bc] \\ &= \frac{1}{8}[1319 - 1154 + 1277 - 1234 \\ &\quad + 1617 - 2089 + 1589 - 2138] \\ &= \frac{1}{8}[-813] = -101.625 \end{aligned}$$

■ **TABLE 6.4**
The Plasma Etch Experiment, Example 6.1

	Coded Factors			Etch Rate			Factor Levels		
Run	*A*	*B*	*C*	Replicate 1	Replicate 2	Total		Low (−1)	High (+1)
1	−1	−1	−1	550	604	(1) = 1154	*A* (Gap, cm)	0.80	1.20
2	1	−1	−1	669	650	*a* = 1319	*B* (C_2F_6 flow, SCCM)	125	200
3	−1	1	−1	633	601	*b* = 1234	*C* (Power, W)	275	325
4	1	1	−1	642	635	*ab* = 1277			
5	−1	−1	1	1037	1052	*c* = 2089			
6	1	−1	1	749	868	*ac* = 1617			
7	−1	1	1	1075	1063	*bc* = 2138			
8	1	1	1	729	860	*abc* = 1589			

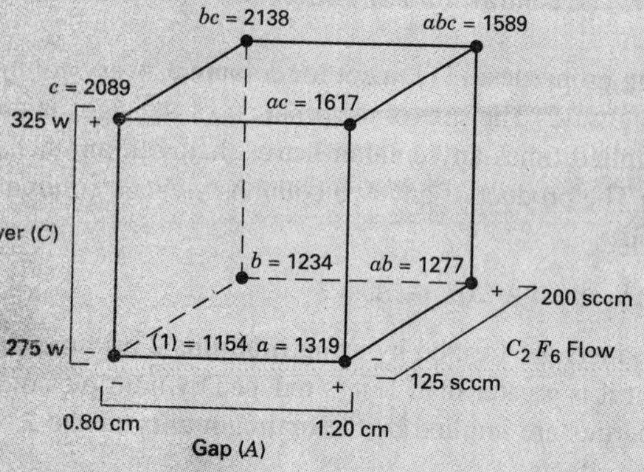

■ **FIGURE 6.6** **The 2^3 design for the plasma etch experiment for Example 6.1**

$$B = \frac{1}{4n}[b + ab + bc + abc - (1) - a - c - ac]$$

$$= \frac{1}{8}[1234 + 1277 + 2138 + 1589 - 1154 - 1319 - 2089 - 1617]$$

$$= \frac{1}{8}[59] = 7.375$$

$$C = \frac{1}{4n}[c + ac + bc + abc - (1) - a - b - ab]$$

$$= \frac{1}{8}[2089 + 1617 + 2138 + 1589 - 1154 - 1319 - 1234 - 1277]$$

$$= \frac{1}{8}[2449] = 306.125$$

$$AB = \frac{1}{4n}[ab - a - b + (1) + abc - bc - ac + c]$$

$$= \frac{1}{8}[1277 - 1319 - 1234 + 1154 + 1589 - 2138 - 1617 + 2089]$$

$$= \frac{1}{8}[-199] = -24.875$$

$$AC = \frac{1}{4n}[(1) - a + b - ab - c + ac - bc + abc]$$

$$= \frac{1}{8}[1154 - 1319 + 1234 - 1277 - 2089 + 1617 - 2138 + 1589]$$

$$= \frac{1}{8}[-1229] = -153.625$$

$$BC = \frac{1}{4n}[(1) + a - b - ab - c - ac + bc + abc]$$

$$= \frac{1}{8}[1154 + 1319 - 1234 - 1277 - 2089 - 1617 + 2138 + 1589]$$

$$= \frac{1}{8}[-17] = -2.125$$

and

$$ABC = \frac{1}{4n}[abc - bc - ac + c - ab + b + a - (1)]$$

$$= \frac{1}{8}[1589 - 2138 - 1617 + 2089 - 1277 + 1234 + 1319 - 1154]$$

$$= \frac{1}{8}[45] = 5.625$$

The largest effects are for power ($C = 306.125$), gap ($A = -101.625$), and the power–gap interaction ($AC = -153.625$).

The sums of squares are calculated from Equation 6.18 as follows:

$$SS_A = \frac{(-813)^2}{16} = 41{,}310.5625$$

$$SS_B = \frac{(59)^2}{16} = 217.5625$$

$$SS_C = \frac{(2449)^2}{16} = 374{,}850.0625$$

$$SS_{AB} = \frac{(-199)^2}{16} = 2475.0625$$

$$SS_{AC} = \frac{(-1229)^2}{16} = 94{,}402.5625$$

$$SS_{BC} = \frac{(-17)^2}{16} = 18.0625$$

and

$$SS_{ABC} = \frac{(45)^2}{16} = 126.5625$$

The total sum of squares is $SS_T = 531{,}420.9375$ and by subtraction $SS_E = 18{,}020.50$. Table 6.5 summarizes the effect estimates and sums of squares. The column labeled "percent contribution" measures the percentage contribution of each model term relative to the total sum of squares. The percentage contribution is often a rough but effective guide to the relative importance of each model term. Note that the main effect of C (Power) really dominates this process, accounting for over 70 percent of the total variability, whereas the main effect of A (Gap) and the AC interaction account for about 8 and 18 percent, respectively.

The ANOVA in Table 6.6 may be used to confirm the magnitude of these effects. We note from Table 6.6 that the main effects of Gap and Power are highly significant (both have very small P-values). The AC interaction is also highly significant; thus, there is a strong interaction between Gap and Power.

■ **TABLE 6.5**
Effect Estimate Summary for Example 6.1

Factor	Effect Estimate	Sum of Squares	Percent Contribution
A	−101.625	41,310.5625	7.7736
B	7.375	217.5625	0.0409
C	306.125	374,850.0625	70.5373
AB	−24.875	2475.0625	0.4657
AC	−153.625	94,402.5625	17.7642
BC	−2.125	18.0625	0.0034
ABC	5.625	126.5625	0.0238

■ **TABLE 6.6**
Analysis of Variance for the Plasma Etching Experiment

Source of Variation	Sum of Squares	Degrees of Freedom	Mean Square	F_0	*P*-Value
Gap (*A*)	41,310.5625	1	41,310.5625	18.34	0.0027
Gas flow (*B*)	217.5625	1	217.5625	0.10	0.7639
Power (*C*)	374,850.0625	1	374,850.0625	166.41	0.0001
AB	2475.0625	1	2475.0625	1.10	0.3252
AC	94,402.5625	1	94,402.5625	41.91	0.0002
BC	18.0625	1	18.0625	0.01	0.9308
ABC	126.5625	1	126.5625	0.06	0.8186
Error	18,020.5000	8	2252.5625		
Total	531,420.9375	15			

The Regression Model and Response Surface. The regression model for predicting etch rate is

$$\hat{y} = \hat{\beta}_0 + \hat{\beta}_1 x_1 + \hat{\beta}_3 x_3 + \hat{\beta}_{13} x_1 x_3$$

$$= 776.0625 + \left(\frac{-101.625}{2}\right) x_1 + \left(\frac{306.125}{2}\right) x_3 + \left(\frac{-153.625}{2}\right) x_1 x_3$$

where the coded variables x_1 and x_3 represent *A* and *C*, respectively. The $x_1 x_3$ term is the *AC* interaction. Residuals can be obtained as the difference between observed and predicted etch rate values. We leave the analysis of these residuals as an exercise for the reader.

Figure 6.7 presents the response surface and contour plot for etch rate obtained from the regression model. Notice that because the model contains interaction, the contour lines of constant etch rate are curved (or the response surface is a "twisted" plane). It is desirable to operate this process so that the etch rate is close to 900 Å/m. The contour plot shows that several combinations of gap and power will satisfy this objective. However, it will be necessary to control both of these variables very precisely.

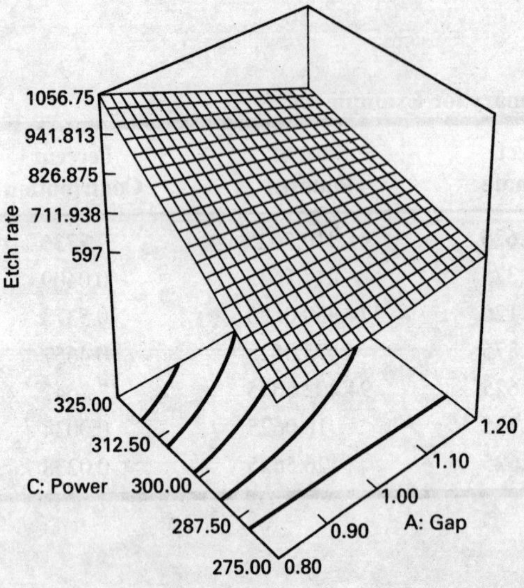

(a) The response surface

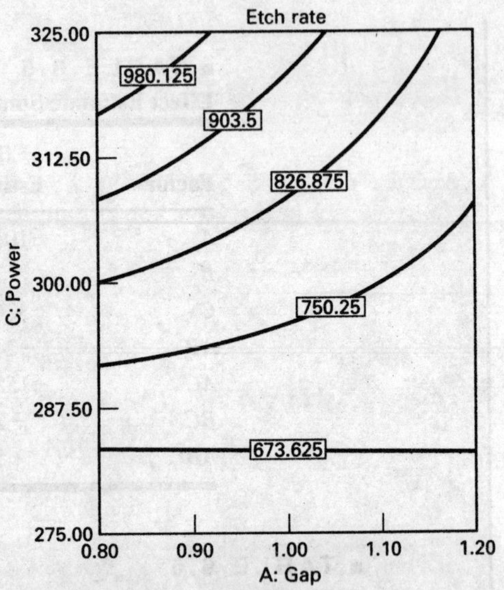

(b) The contour plot

■ **FIGURE 6.7** **Response surface and contour plot of etch rate for Example 6.1**

Computer Solution. Many statistics software packages are available that will set up and analyze two-level factorial designs. The output from one of these computer programs, Design-Expert, is shown in Table 6.7. In the upper part of the table, an ANOVA for the **full model** is presented. The format of this presentation is somewhat different from the ANOVA results given in Table 6.6. Notice that the first line of the ANOVA is an overall summary for the full model (all main effects and interactions), and the model sum of squares is

$$SS_{\text{Model}} = SS_A + SS_B + SS_C + SS_{AB} + SS_{AC} + SS_{BC} + SS_{ABC} = 5.134 \times 10^5$$

Thus the statistic

$$F_0 = \frac{MS_{\text{Model}}}{MS_E} = \frac{73{,}342.92}{2252.56} = 32.56$$

is testing the hypotheses

$$H_0{:}\beta_1 = \beta_2 = \beta_3 = \beta_{12} = \beta_{13} = \beta_{23} = \beta_{123} = 0$$
$$H_1{:}\text{at least one } \beta \neq 0$$

Because F_0 is large, we would conclude that at least one variable has a nonzero effect. Then each individual factorial effect is tested for significance using the F statistic. These results agree with Table 6.6.

Below the full model ANOVA in Table 6.7, several R^2 statistics are presented. The ordinary R^2 is

$$R^2 = \frac{SS_{\text{Model}}}{SS_{\text{Total}}} = \frac{5.134 \times 10^5}{5.314 \times 10^5} = 0.9661$$

and it measures the proportion of total variability explained by the model. A potential problem with this statistic is that it always increases as factors are added to the model, even if these factors are not significant. The **adjusted** R^2 statistic, defined as

$$R^2_{\text{Adj}} = 1 - \frac{SS_E/df_E}{SS_{\text{Total}}/df_{\text{Total}}} = 1 - \frac{18{,}020.50/8}{5.314 \times 10^5/15} = 0.9364$$

■ **TABLE 6.7**

Design-Expert Output for Example 6.1

Response: Etch rate
ANOVA for Selected Factorial Model
Analysis of variance table [Partial sum of squares]

Source	Sum of Squares	DF	Mean Square	F Value	Prob > F
Model	5.134E+005	7	73342.92	32.56	<0.0001
A	*41310.56*	*1*	*41310.56*	*18.34*	*0.0027*
B	*217.56*	*1*	*217.56*	*0.097*	*0.7639*
C	*3.749E+005*	*1*	*3.749E+005*	*166.41*	*<0.0001*
AB	*2475.06*	*1*	*2475.06*	*1.10*	*0.3252*
AC	*94402.56*	*1*	*94402.56*	*41.91*	*0.0002*
BC	*18.06*	*1*	*18.06*	*8.019E-003*	*0.9308*
ABC	*126.56*	*1*	*126.56*	*0.056*	*0.8186*
Pure Error	18020.50	8	2252.56		
Cor Total	5.314E+005	15			

Std. Dev.	47.46	R-Squared	0.9661
Mean	776.06	Adj R-Squared	0.9364
C.V.	6.12	Pred R-Squared	0.8644
PRESS	72082.00	Adeq Precisioin	14.660

Factor	Coefficient Estimated	DF	Standard Error	95% CI Low	95% CI High	VIF
Intercept	776.06	1	11.87	748.70	803.42	
A-Gap	−50.81	1	11.87	−78.17	−23.45	1.00
B-Gas flow	3.69	1	11.87	−23.67	31.05	1.00
C-Power	153.06	1	11.87	125.70	180.42	1.00
AB	−12.44	1	11.87	−39.80	14.92	1.00
AC	−76.81	1	11.87	−104.17	−49.45	1.00
BC	−1.06	1	11.87	−28.42	26.30	1.00
ABC	2.81	1	11.87	−24.55	30.17	1.00

Final Equation in Terms of Coded Factors:

Etch rate	=
+776.06	
−50.81	* *A*
+3.69	* *B*
+153.06	* *C*
−12.44	* *A* * *B*
−76.81	* *A* * *C*
+1.06	* *B* * *C*
+2.81	* *A* * *B* * *C*

Final Equation in Terms of Actual Factors:

Etch rate	=
−6487.33333	
+5355.41667	* Gap
+6.59667	* Gas flow
+24.10667	* Power
−6.15833	* Gap * Gas flow
−17.80000	* Gap * Power
−0.016133	* Gas flow * Power
+0.015000	* Gap * Gas flow * Power

■ **TABLE 6.7** (*Continued*)

Response: Etch rate

ANOVA for Selected Factorial Model

Analysis of variance table [Partial sum of squares]

Source	Sum of Squares	DF	Mean Square	*F* Value	Prob > *F*
Model	5.106E+005	3	1.702E+005	97.91	<0.0001
A	*41310.56*	*1*	*41310.56*	*23.77*	*0.0004*
C	*3.749E+005*	*1*	*3.749E+005*	*215.66*	*<0.0001*
AC	*94402.56*	*1*	*94402.56*	*54.31*	*<0.0001*
Residual	20857.75	12	1738.15		
Lack of Fit	*2837.25*	*4*	*709.31*	*0.31*	*0.8604*
Pure Error	*18020.50*	*8*	*2252.56*		
Cor Total	5.314E+005	15			

Std. Dev.	41.69	R-Squared	0.9608
Mean	776.06	Adj R-Squared	0.9509
C.V.	5.37	Pred R-Squared	0.9302
PRESS	37080.44	Adeq Precision	22.055

Factor	Coefficient Estimate	DF	Standard Error	95% CI Low	95% CI High	VIF
Intercept	776.06	1	10.42	753.35	798.77	
A-Gap	−50.81	1	10.42	−73.52	28.10	1.00
C-Power	153.06	1	10.42	130.35	175.77	1.00
AC	−76.81	1	10.42	−99.52	−54.10	1.00

Final Equation in Terms of Coded Factors:

Etch rate	=	
+776.06		
−50.81	* *A*	
+153.06	* *C*	
−76.81	* *A* * *C*	

Final Equation in Terms of Actual Factors:

Etch rate	=
−5415.37500	
+4354.68750	* Gap
+21.48500	* Power
−15.36250	* Gap * Power

Diagnostics Case Statistics

Standard Order	Actual Value	Predicted Value	Residual	Leverage	Student Residual	Cook's Distance	Outlier *t*	Run Order
1	550.00	597.00	−47.00	0.250	−1.302	0.141	−1.345	9
2	604.00	597.00	7.00	0.250	0.194	0.003	0.186	6
3	669.00	649.00	20.00	0.250	0.554	0.026	0.537	14
4	650.00	649.00	1.00	0.250	0.028	0.000	0.027	1
5	633.00	597.00	36.00	0.250	0.997	0.083	0.997	3
6	601.00	597.00	4.00	0.250	0.111	0.001	0.106	12
7	642.00	649.00	−7.00	0.250	−0.194	0.003	−0.186	13
8	635.00	649.00	−14.00	0.250	−0.388	0.013	−0.374	8
9	1037.00	1056.75	−19.75	0.250	−0.547	0.025	−0.530	5
10	1052.00	1056.75	−4.75	0.250	−0.132	0.001	−0.126	16
11	749.00	801.50	−52.50	0.250	−1.454	0.176	−1.534	2
12	868.00	801.50	66.50	0.250	1.842	0.283	2.082	15
13	1075.00	1056.75	18.25	0.250	0.505	0.021	0.489	4
14	1063.00	1056.75	6.25	0.250	0.173	0.002	0.166	7
15	729.00	801.50	−72.50	0.250	−2.008	0.336	−2.359	10
16	860.00	801.50	58.50	0.250	1.620	0.219	1.755	11

is a statistic that is adjusted for the "size" of the model, that is, the number of factors. The adjusted R^2 can actually decrease if nonsignificant terms are added to a model. The PRESS statistic is a measure of how well the model will predict new data. (PRESS is actually an acronym for prediction error sum of squares, and it is computed as the sum of the squared prediction errors obtained by predicting the *i*th data point with a model that includes all observations *except* the *i*th one.) A model with a small value of PRESS indicates that the model is likely to be a good predictor. The "Prediction R^2" statistic is computed as

$$R^2_{\text{Pred}} = 1 - \frac{\text{PRESS}}{SS_{\text{Total}}} = 1 - \frac{72{,}082.00}{5.314 \times 10^5} = 0.8644$$

This indicates that the full model would be expected to explain about 86 percent of the variability in new data.

The next portion of the output presents the regression coefficient for each model term and the **standard error** of each coefficient, defined as

$$se(\hat{\beta}) = \sqrt{V(\hat{\beta})} = \sqrt{\frac{MS_E}{n2^k}} = \sqrt{\frac{MS_E}{N}} = \sqrt{\frac{2252.56}{2(8)}} = 11.87$$

The standard errors of all model coefficients are equal because the design is **orthogonal**. The 95 percent confidence intervals on each regression coefficient are computed from

$$\hat{\beta} - t_{0.025,N-p}se(\hat{\beta}) \leq \beta \leq \hat{\beta} + t_{0.025,N-p}se(\hat{\beta})$$

where the degrees of freedom on t are the number of degrees of freedom for error; that is, N is the total number of runs in the experiment (16), and p is the number of model parameters (8). The full model in terms of both the coded variables and the natural variables is also presented.

The last part of the display in Table 6.7 illustrates the output following the removal of the nonsignificant interaction terms. This **reduced model** now contains only the main effects *A, C,* and the *AC* interaction. The **error** or **residual** sum of squares is now composed of a **pure error** component arising from the replication of the eight corners of the cube and a **lack-of-fit** component consisting of the sums of squares for the factors that were dropped from the model (*B, AB, BC,* and *ABC*). Once again, the regression model representation of the experimental results is given in terms of both coded and natural variables. The proportion of total variability in etch rate that is explained by this model is

$$R^2 = \frac{SS_{\text{Model}}}{SS_{\text{Total}}} = \frac{5.106 \times 10^5}{5.314 \times 10^5} = 0.9608$$

which is smaller than the R^2 for the full model. Notice, however, that the adjusted R^2 for the reduced model is actually slightly larger than the adjusted R^2 for the full model, and PRESS for the reduced model is considerably smaller, leading to a larger value of R^2_{Pred} for the reduced model. Clearly, removing the nonsignificant terms from the full model has produced a final model that is likely to function more effectively as a predictor of new data. Notice that the confidence intervals on the regression coefficients for the reduced model are shorter than the corresponding confidence intervals for the full model.

The last part of the output presents the residuals from the reduced model. Design-Expert will also construct all of the residual plots that we have previously discussed.

Other Methods for Judging the Significance of Effects. The analysis of variance is a formal way to determine which factor effects are nonzero. Several other methods are useful. Below, we show how to calculate the **standard error of the effects**, and we use these standard errors to construct **confidence intervals** on the effects. Another method, which we will illustrate in Section 6.5, uses **normal probability plots** to assess the importance of the effects.

The standard error of an effect is easy to find. If we assume that there are n replicates at each of the 2^k runs in the design, and if $y_{i1}, y_{i2}, \ldots, y_{in}$ are the observations at the ith run, then

$$S_i^2 = \frac{1}{n-1}\sum_{j=1}^{n}(y_{ij} - \bar{y}_i)^2 \qquad i = 1, 2, \ldots, 2^k$$

is an estimate of the variance at the ith run. The 2^k variance estimates can be combined to give an overall variance estimate:

$$S^2 = \frac{1}{2^k(n-1)}\sum_{i=1}^{2^k}\sum_{j=1}^{n}(y_{ij} - \bar{y}_i)^2 \tag{6.19}$$

This is also the variance estimate given by the error mean square in the analysis of variance. The *variance* of each effect estimate is

$$V(\text{Effect}) = V\left(\frac{\text{Contrast}}{n2^{k-1}}\right) = \frac{1}{(n2^{k-1})^2}V(\text{Contrast})$$

Each contrast is a linear combination of 2^k treatment totals, and each total consists of n observations. Therefore,

$$V(\text{Contrast}) = n2^k\sigma^2$$

and the variance of an effect is

$$V(\text{Effect}) = \frac{1}{(n2^{k-1})^2}n2^k\sigma^2 = \frac{1}{n2^{k-2}}\sigma^2$$

The estimated standard error would be found by replacing σ^2 by its estimate S^2 and taking the square root of this last expression:

$$se(\text{Effect}) = \frac{2S}{\sqrt{n2^k}} \tag{6.20}$$

Notice that the standard error of an effect is twice the standard error of an estimated regression coefficient in the regression model for the 2^k design (see the Design-Expert computer output for Example 6.1). It would be possible to test the significance of any effect by comparing the effect estimates to its standard error:

$$t_0 = \frac{\text{Effect}}{se(\text{Effect})}$$

This is a t statistic with $N - p$ degrees of freedom.

The $100(1 - \alpha)$ percent confidence intervals on the effects are computed from Effect $\pm\, t_{\alpha/2,N-p}se(\text{Effect})$, where the degrees of freedom on t are just the error or residual degrees of freedom ($N - p$ = total number of runs − number of model parameters).

To illustrate this method, consider the plasma etching experiment in Example 6.1. The mean square error for the full model is $MS_E = 2252.56$. Therefore, the standard error of each effect is (using $S^2 = MS_E$)

$$se(\text{Effect}) = \frac{2S}{\sqrt{n2^k}} = \frac{2\sqrt{2252.56}}{\sqrt{2(2^3)}} = 23.73$$

■ **FIGURE 6.8** Ranges of etch rates for Example 6.1

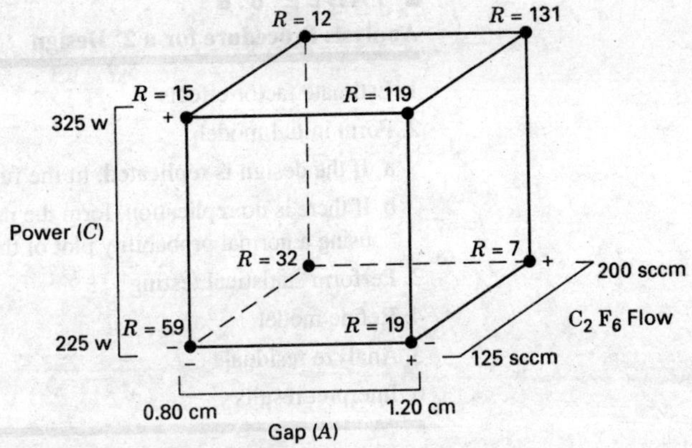

Now $t_{0.025,8} = 2.31$ and $t_{0.025,8}\ se(\text{Effect}) = 2.31(23.73) = 54.82$, so approximate 95 percent confidence intervals on the factor effects are

$$\begin{aligned}
A&: \ -101.625 \pm 54.82\\
B&: \ 7.375 \pm 54.82\\
C&: \ 306.125 \pm 54.82\\
AB&: \ -24.875 \pm 54.82\\
AC&: \ -153.625 \pm 54.82\\
BC&: \ -2.125 \pm 54.82\\
ABC&: \ 5.625 \pm 54.82
\end{aligned}$$

This analysis indicates that A, C, and AC are important factors because they are the only factor effect estimates for which the approximate 95 percent confidence intervals do not include zero.

Dispersion Effects. The process engineer working on the plasma etching tool was also interested in **dispersion effects**; that is, do any of the factors affect variability in etch rate from run to run? One way to answer the question is to look at the **range** of etch rates for each of the eight runs in the 2^3 design. These ranges are plotted on the cube in Figure 6.8. Notice that the ranges in etch rates are much larger when both Gap and Power are at their high levels, indicating that this combination of factor levels may lead to more variability in etch rate than other recipes. Fortunately, etch rates in the desired range of 900 Å/m can be achieved with settings of Gap and Power that avoid this situation.

6.4 The General 2^k Design

The methods of analysis that we have presented thus far may be generalized to the case of a **2^k factorial design**, that is, a design with k factors each at two levels. The statistical model for a 2^k design would include k main effects, $\binom{k}{2}$ two-factor interactions, $\binom{k}{3}$ three-factor interactions, . . . , and one k-factor interaction. That is, the complete model would contain $2^k - 1$ effects for a 2^k design. The notation introduced earlier for treatment combinations is also used here. For example, in a 2^5 design *abd* denotes the treatment combination with factors A, B, and D at the high level and factors C and E at the low level. The treatment combinations may be written in **standard order** by introducing the factors one at a time, with each new factor being successively combined with those that precede it. For example, the standard order for a 2^4 design is (1), *a*, *b*, *ab*, *c*, *ac*, *bc*, *abc*, *d*, *ad*, *bd*, *abd*, *cd*, *acd*, *bcd*, and *abcd*.

The general approach to the statistical analysis of the 2^k design is summarized in Table 6.8. As we have indicated previously, a computer software package is usually employed in this analysis process.

■ **TABLE 6.8**
Analysis Procedure for a 2^k Design

1. Estimate factor effects
2. Form initial model
   a. If the design is replicated, fit the full model
   b. If there is no replication, form the model using a normal probability plot of the effects
3. Perform statistical testing
4. Refine model
5. Analyze residuals
6. Interpret results

The sequence of steps in Table 6.8 should, by now, be familiar. The first step is to estimate factor effects and examine their signs and magnitudes. This gives the experimenter preliminary information regarding which factors and interactions may be important and in which directions these factors should be adjusted to improve the response. In forming the initial model for the experiment, we usually choose the **full model**, that is, all main effects and interactions, provided that at least one of the design points has been replicated (in the next section, we discuss a modification to this step). Then in step 3, we use the analysis of variance to formally test for the significance of main effects and interaction. Table 6.9 shows the general form of an analysis of variance for a 2^k factorial design with n replicates. Step 4, refine the

■ **TABLE 6.9**
Analysis of Variance for a 2^k Design

Source of Variation	Sum of Squares	Degrees of Freedom
k main effects		
A	SS_A	1
B	SS_B	1
$\vdots$	$\vdots$	$\vdots$
K	SS_K	1
$\binom{k}{2}$ two-factor interactions		
AB	SS_{AB}	1
AC	SS_{AC}	1
$\vdots$	$\vdots$	$\vdots$
JK	SS_{JK}	1
$\binom{k}{3}$ three-factor interactions		
ABC	SS_{ABC}	1
ABD	SS_{ABD}	1
$\vdots$	$\vdots$	$\vdots$
IJK	SS_{IJK}	1
$\vdots$	$\vdots$	$\vdots$
$\binom{k}{k}$ k-factor interaction		
$ABC \cdots K$	$SS_{ABC \cdots K}$	1
Error	SS_E	$2^k(n-1)$
Total	SS_T	$n2^k - 1$

model, usually consists of removing any nonsignificant variables from the full model. Step 5 is the usual residual analysis to check for model adequacy and assumptions. Sometimes model refinement will occur after residual analysis if we find that the model is inadequate or assumptions are badly violated. The final step usually consists of graphical analysis—either main effect or interaction plots, or response surface and contour plots.

Although the calculations described above are almost always done with a computer, occasionally it is necessary to manually calculate an effect estimate or sum of squares for an effect. To estimate an effect or to compute the sum of squares for an effect, we must first determine the contrast associated with that effect. This can always be done by using a table of plus and minus signs, such as Table 6.2 or Table 6.3. However, this is awkward for large values of k and we can use an alternate method. In general, we determine the contrast for effect $AB \cdots K$ by expanding the right-hand side of

$$\text{Contrast}_{AB\cdots K} = (a \pm 1)(b \pm 1)\cdots(k \pm 1) \tag{6.21}$$

In expanding Equation 6.21, ordinary algebra is used with "1" being replaced by (1) in the final expression. The sign in each set of parentheses is negative if the factor is included in the effect and positive if the factor is not included.

To illustrate the use of Equation 6.21, consider a 2^3 factorial design. The contrast for AB would be

$$\begin{aligned}\text{Contrast}_{AB} &= (a-1)(b-1)(c+1)\\ &= abc + ab + c + (1) - ac - bc - a - b\end{aligned}$$

As a further example, in a 2^5 design, the contrast for $ABCD$ would be

$$\begin{aligned}\text{Contrast}_{ABCD} &= (a-1)(b-1)(c-1)(d-1)(e+1)\\ &= abcde + cde + bde + ade + bce\\ &\quad + ace + abe + e + abcd + cd + bd\\ &\quad + ad + bc + ac + ab + (1) - a - b - c\\ &\quad - abc - d - abd - acd - bcd - ae\\ &\quad - be - ce - abce - de - abde - acde - bcde\end{aligned}$$

Once the contrasts for the effects have been computed, we may estimate the effects and compute the sums of squares according to

$$AB\cdots K = \frac{2}{n2^k}(\text{Contrast}_{AB\cdots K}) \tag{6.22}$$

and

$$SS_{AB\cdots K} = \frac{1}{n2^k}(\text{Contrast}_{AB\cdots K})^2 \tag{6.23}$$

respectively, where n denotes the number of replicates. There is also a tabular algorithm due to Frank Yates that can occasionally be useful for manual calculation of the effect estimates and the sums of squares. Refer to the **supplemental text material** for this chapter.

6.5 A Single Replicate of the 2^k Design

For even a moderate number of factors, the total number of treatment combinations in a 2^k factorial design is large. For example, a 2^5 design has 32 treatment combinations, a 2^6 design has 64 treatment combinations, and so on. Because resources are usually limited, the number of replicates that the experimenter can employ may be restricted. Frequently, available

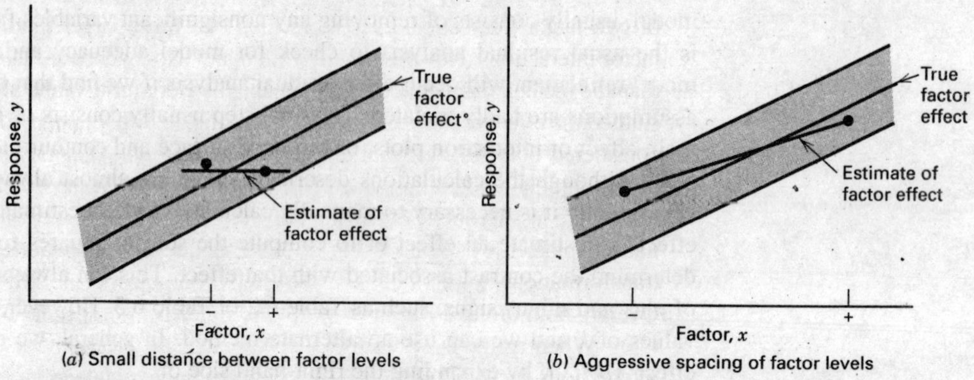

■ **FIGURE 6.9** **The impact of the choice of factor levels in an unreplicated design**

resources only allow a **single replicate** of the design to be run, unless the experimenter is willing to omit some of the original factors.

An obvious risk when conducting an experiment that has only one run at each test combination is that we may be fitting a model to noise. That is, if the response y is highly variable, misleading conclusions may result from the experiment. The situation is illustrated in Figure 6.9*a*. In this figure, the straight line represents the true factor effect. However, because of the random variability present in the response variable (represented by the shaded band), the experimenter actually obtains the two measured responses represented by the dark dots. Consequently, the estimated factor effect is close to zero, and the experimenter has reached an erroneous conclusion concerning this factor. Now if there is less variability in the response, the likelihood of an erroneous conclusion will be smaller. Another way to ensure that reliable effect estimates are obtained is to increase the distance between the low (−) and high (+) levels of the factor, as illustrated in Figure 6.9*b*. Notice that in this figure, the increased distance between the low and high factor levels results in a reasonable estimate of the true factor effect.

The single-replicate strategy is often used in screening experiments when there are relatively many factors under consideration. Because we can never be entirely certain in such cases that the experimental error is small, a good practice in these types of experiments is to spread out the factor levels aggressively. You might find it helpful to reread the guidance on choosing factor levels in Chapter 1.

A single replicate of a 2^k design is sometimes called an **unreplicated factorial**. With only one replicate, there is no internal estimate of error (or "pure error"). One approach to the analysis of an unreplicated factorial is to assume that certain high-order interactions are negligible and combine their mean squares to estimate the error. This is an appeal to the **sparsity of effects principle**; that is, most systems are dominated by some of the main effects and low-order interactions, and most high-order interactions are negligible.

While the effect sparsity principle has been observed by experimenters for many decades, only recently has it been studied more objectively. A paper by Li, Sudarsanam, and Frey (2006) studied 113 response variables obtained from 43 published experiments from a wide range of science and engineering disciplines. All of the experiments were full factorials with between three and seven factors, so no assumptions had to be made about interactions. Most of the experiments had either three or four factors. The authors found that about 40 percent of the main effects in the experiments they studied were significant, while only about 11 percent of the two-factor interactions were significant. Three-factor interactions were very rare, occurring only about 5 percent of the time. The authors also investigated the absolute values of factor effects for main effects, two-factor interactions, and three-factor interactions. The median of main effect strength was about four times larger than the median strength of two-factor interactions. The median strength of two-factor interactions was more than two times larger than the median strength of three-factor interactions. However, there were many two- and three-factor interactions that were larger than the median main effect. Another paper by Bergquist, Vanhatalo, and Nordenvaad (2011) also studied the effect of the sparsity question using 22 different experiments with 35 responses. They considered both full factorial and fractional factorial designs with factors at two levels. Their results largely agree with those of Li et al. (2006), with the exception that three-factor interactions were less frequent, occurring only about 2 percent of the time. This difference may be partially explained by the inclusion of experiments with indications of curvature and the need for transformations in the Li et al. (2006) study. Bergquist et al. (2011) excluded such experiments. Overall, both of these studies confirm the validly of the sparsity of effects principle.

When analyzing data from unreplicated factorial designs, occasionally real high-order interactions occur. The use of an error mean square obtained by pooling high-order interactions is inappropriate in these cases. A method of analysis attributed to Daniel (1959) provides a simple way to overcome this problem. Daniel suggests examining a **normal probability plot** of the estimates of the effects. The effects that are negligible are normally distributed, with mean zero and variance σ^2 and will tend to fall along a straight line on this plot, whereas significant effects will have nonzero means and will not lie along the straight line. Thus, the preliminary model will be specified to contain those effects that are apparently nonzero, based on the normal probability plot. The apparently negligible effects are combined as an estimate of error.

EXAMPLE 6.2 A Single Replicate of the 2^4 Design

A chemical product is produced in a pressure vessel. A factorial experiment is carried out in the pilot plant to study the factors thought to influence the filtration rate of this product. The four factors are temperature (A), pressure (B), concentration of formaldehyde (C), and stirring rate (D). Each factor is present at two levels. The design matrix and the response data obtained from a single replicate of the 2^4 experiment are shown in Table 6.10 and

■ **TABLE 6.10**
Pilot Plant Filtration Rate Experiment

Run Number	Factor A	B	C	D	Run Label	Filtration Rate (gal/h)
1	−	−	−	−	(1)	45
2	+	−	−	−	*a*	71
3	−	+	−	−	*b*	48
4	+	+	−	−	*ab*	65
5	−	−	+	−	*c*	68
6	+	−	+	−	*ac*	60
7	−	+	+	−	*bc*	80
8	+	+	+	−	*abc*	65
9	−	−	−	+	*d*	43
10	+	−	−	+	*ad*	100
11	−	+	−	+	*bd*	45
12	+	+	−	+	*abd*	104
13	−	−	+	+	*cd*	75
14	+	−	+	+	*acd*	86
15	−	+	+	+	*bcd*	70
16	+	+	+	+	*abcd*	96

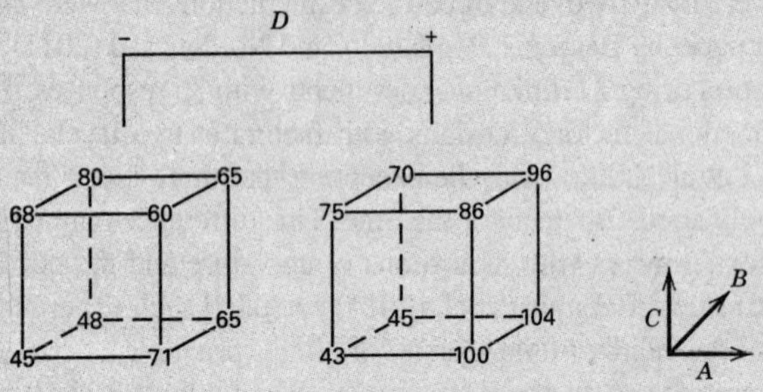

■ **FIGURE 6.10** **Data from the pilot plant filtration rate experiment for Example 6.2**

Figure 6.10. The 16 runs are made in random order. The process engineer is interested in maximizing the filtration rate. Current process conditions give filtration rates of around 75 gal/h. The process also currently uses the concentration of formaldehyde, factor *C*, at the high level. The engineer would like to reduce the formaldehyde concentration as much as possible but has been unable to do so because it always results in lower filtration rates.

We will begin the analysis of these data by constructing a normal probability plot of the effect estimates. The table of plus and minus signs for the contrast constants for the 2^4 design are shown in Table 6.11. From these contrasts, we may estimate the 15 factorial effects and the sums of squares shown in Table 6.12.

■ **TABLE 6.12**
Factor Effect Estimates and Sums of Squares for the 2^4 Factorial in Example 6.2

Model Term	Effect Estimate	Sum of Squares	Percent Contribution
A	21.625	1870.56	32.6397
B	3.125	39.0625	0.681608
C	9.875	390.062	6.80626
D	14.625	855.563	14.9288
AB	0.125	0.0625	0.00109057
AC	−18.125	1314.06	22.9293
AD	16.625	1105.56	19.2911
BC	2.375	22.5625	0.393696
BD	−0.375	0.5625	0.00981515
CD	−1.125	5.0625	0.0883363
ABC	1.875	14.0625	0.245379
ABD	4.125	68.0625	1.18763
ACD	−1.625	10.5625	0.184307
BCD	−2.625	27.5625	0.480942
ABCD	1.375	7.5625	0.131959

■ **TABLE 6.11**
Contrast Constants for the 2^4 Design

	A	*B*	*AB*	*C*	*AC*	*BC*	*ABC*	*D*	*AD*	*BD*	*ABD*	*CD*	*ACD*	*BCD*	*ABCD*
(1)	−	−	+	−	+	+	−	−	+	+	−	+	−	−	+
a	+	−	−	−	−	+	+	−	−	+	+	+	+	−	−
b	−	+	−	−	+	−	+	−	+	−	+	+	−	+	−
ab	+	+	+	−	−	−	−	−	−	−	−	+	+	+	+
c	−	−	+	+	−	−	+	−	+	+	−	−	+	+	−
ac	+	−	−	+	+	−	−	−	−	+	+	−	−	+	+
bc	−	+	−	+	−	+	−	−	+	−	+	−	+	−	+
abc	+	+	+	+	+	+	+	−	−	−	−	−	−	−	−
d	−	−	+	−	+	+	−	+	−	−	+	−	+	+	−
ad	+	−	−	−	−	+	+	+	+	−	−	−	−	+	+
bd	−	+	−	−	+	−	+	+	−	+	−	−	+	−	+
abd	+	+	+	−	−	−	−	+	+	+	+	−	−	−	−
cd	−	−	+	+	−	−	+	+	−	−	+	+	−	−	+
acd	+	−	−	+	+	−	−	+	+	−	−	+	+	−	−
bcd	−	+	−	+	−	+	−	+	−	+	−	+	−	+	−
abcd	+	+	+	+	+	+	+	+	+	+	+	+	+	+	+

The normal probability plot of these effects is shown in Figure 6.11. All of the effects that lie along the line are negligible, whereas the large effects are far from the line. The important effects that emerge from this analysis are the main effects of *A*, *C*, and *D* and the *AC* and *AD* interactions.

The main effects of *A*, *C*, and *D* are plotted in Figure 6.12*a*. All three effects are positive, and if we considered only these main effects, we would run all three factors at the high level to maximize the filtration rate. However, it is always necessary to examine any interactions that are important. Remember that main effects do not have much meaning when they are involved in significant interactions.

The *AC* and *AD* interactions are plotted in Figure 6.12*b*. These interactions are the key to solving the problem. Note from the *AC* interaction that the temperature effect is very small when the concentration is at the high level and very large when the concentration is at the low level, with the best results obtained with low concentration and high temperature. The *AD* interaction indicates that stirring rate *D* has little effect at low temperature but a large positive effect at high temperature. Therefore, the best filtration rates would appear to be obtained when *A* and *D* are at the high level and *C* is at the low level. This would allow the reduction of the formaldehyde concentration to a lower level, another objective of the experimenter.

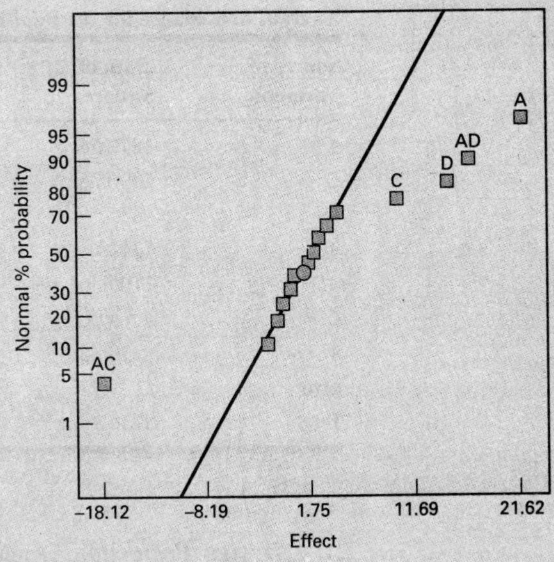

■ **FIGURE 6.11** **Normal probability plot of the effects for the 2^4 factorial in Example 6.2**

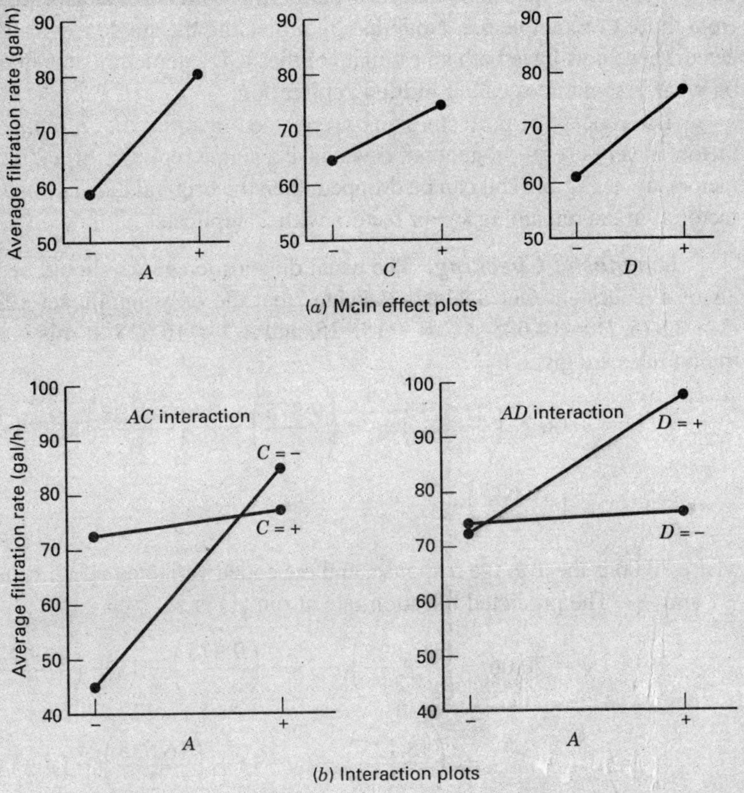

■ **FIGURE 6.12** **Main effect and interaction plots for Example 6.2**

■ **TABLE 6.13**
Analysis of Variance for the Pilot Plant Filtration Rate Experiment in *A*, *C*, and *D*

Source of Variation	Sum of Squares	Degrees of Freedom	Mean Square	F_0	*P*-Value
A	1870.56	1	1870.56	83.36	<0.0001
C	390.06	1	390.06	17.38	<0.0001
D	855.56	1	855.56	38.13	<0.0001
AC	1314.06	1	1314.06	58.56	<0.0001
AD	1105.56	1	1105.56	49.27	<0.0001
CD	5.06	1	5.06	<1	
ACD	10.56	1	10.56	<1	
Error	179.52	8	22.44		
Total	5730.94	15			

Design Projection. Another interpretation of the effects in Figure 6.11 is possible. Because *B* (pressure) is not significant and all interactions involving *B* are negligible, we may discard *B* from the experiment so that the design becomes a 2^3 factorial in *A*, *C*, and *D* with two replicates. This is easily seen from examining only columns *A*, *C*, and *D* in the design matrix shown in Table 6.10 and noting that those columns form two replicates of a 2^3 design. The analysis of variance for the data using this simplifying assumption is summarized in Table 6.13. The conclusions that we would draw from this analysis are essentially unchanged from those of Example 6.2. Note that by projecting the single replicate of the 2^4 into a replicated 2^3, we now have both an estimate of the *ACD* interaction and an estimate of error based on what is sometimes called **hidden replication.**

The concept of projecting an unreplicated factorial into a replicated factorial in fewer factors is very useful. In general, if we have a single replicate of a 2^k design, and if h $(h < k)$ factors are negligible and can be dropped, then the original data correspond to a full two-level factorial in the remaining $k - h$ factors with 2^h replicates.

Diagnostic Checking. The usual diagnostic checks should be applied to the residuals of a 2^k design. Our analysis indicates that the only significant effects are $A = 21.625$, $C = 9.875$, $D = 14.625$, $AC = -18.125$, and $AD = 16.625$. If this is true, the estimated filtration rates are given by

$$\hat{y} = 70.06 + \left(\frac{21.625}{2}\right)x_1 + \left(\frac{9.875}{2}\right)x_3 + \left(\frac{14.625}{2}\right)x_4 - \left(\frac{18.125}{2}\right)x_1x_3 + \left(\frac{16.625}{2}\right)x_1x_4$$

where 70.06 is the average response, and the coded variables x_1, x_3, x_4 take on values between -1 and $+1$. The predicted filtration rate at run (1) is

$$\begin{aligned}\hat{y} &= 70.06 + \left(\frac{21.625}{2}\right)(-1) + \left(\frac{9.875}{2}\right)(-1) + \left(\frac{14.625}{2}\right)(-1)\\ &\quad - \left(\frac{18.125}{2}\right)(-1)(-1) + \left(\frac{16.625}{2}\right)(-1)(-1)\\ &= 46.22\end{aligned}$$

Because the observed value is 45, the residual is $e = y - \hat{y} = 45 - 46.25 = -1.25$. The values of y, $\hat{y}$, and $e = y - \hat{y}$ for all 16 observations are as follows:

	y	$\hat{y}$	$e = y - \hat{y}$
(1)	45	46.25	−1.25
a	71	69.38	1.63
b	48	46.25	1.75
ab	65	69.38	−4.38
c	68	74.25	−6.25
ac	60	61.13	−1.13
bc	80	74.25	5.75
abc	65	61.13	3.88
d	43	44.25	−1.25
ad	100	100.63	−0.63
bd	45	44.25	0.75
abd	104	100.63	3.38
cd	75	72.25	2.75
acd	86	92.38	−6.38
bcd	70	72.25	−2.25
abcd	96	92.38	3.63

A normal probability plot of the residuals is shown in Figure 6.13. The points on this plot lie reasonably close to a straight line, lending support to our conclusion that *A*, *C*, *D*, *AC*, and *AD* are the only significant effects and that the underlying assumptions of the analysis are satisfied.

The Response Surface. We used the interaction plots in Figure 6.12 to provide a practical interpretation of the results of this experiment. Sometimes we find it helpful to use the response surface for this purpose. The response surface is generated by the regression model

$$\hat{y} = 70.06 + \left(\frac{21.625}{2}\right) x_1 + \left(\frac{9.875}{2}\right) x_3 + \left(\frac{14.625}{2}\right) x_4 - \left(\frac{18.125}{2}\right) x_1x_3 + \left(\frac{16.625}{2}\right) x_1x_4$$

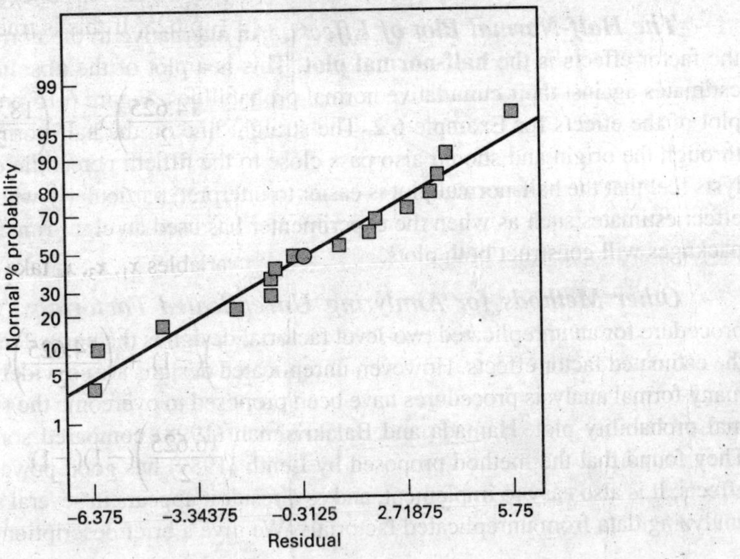

■ **FIGURE 6.13** **Normal probability plot of residuals for Example 6.2**

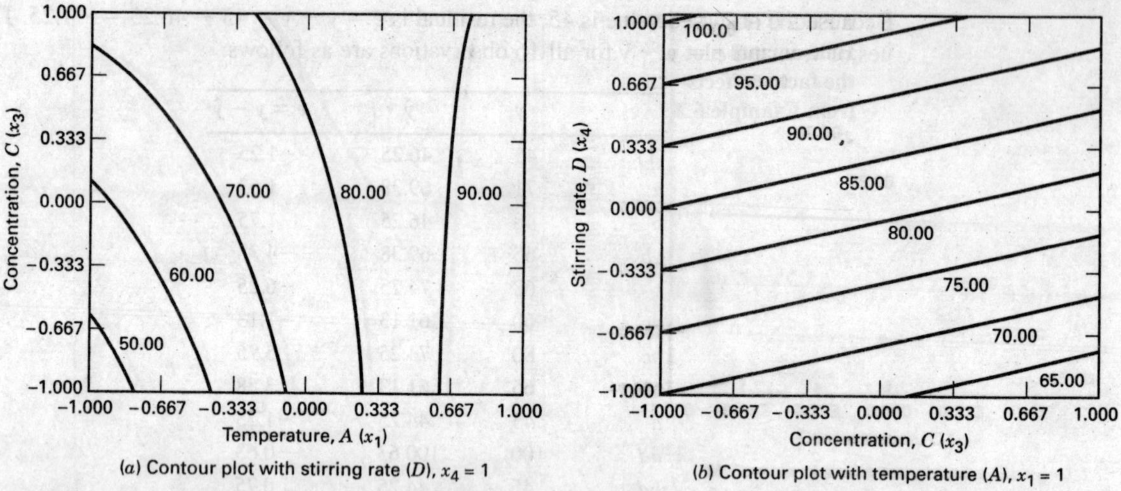

■ **FIGURE 6.14** **Contour plots of filtration rate, Example 6.2**

Figure 6.14*a* shows the response surface contour plot when stirring rate is at the high level (i.e., $x_4 = 1$). The contours are generated from the above model with $x_4 = 1$, or

$$\hat{y} = 77.3725 + \left(\frac{38.25}{2}\right)x_1 + \left(\frac{9.875}{2}\right)x_3 - \left(\frac{18.125}{2}\right)x_1x_3$$

Notice that the contours are curved lines because the model contains an interaction term.

Figure 6.14*b* is the response surface contour plot when temperature is at the high level (i.e., $x_1 = 1$). When we put $x_1 = 1$ in the regression model, we obtain

$$\hat{y} = 80.8725 - \left(\frac{8.25}{2}\right)x_3 + \left(\frac{31.25}{2}\right)x_4$$

These contours are parallel straight lines because the model contains only the main effects of factors C (x_3) and D (x_4).

Both contour plots indicate that if we want to maximize the filtration rate, variables A (x_1) and D (x_4) should be at the high level and that the process is relatively robust to concentration C. We obtained similar conclusions from the interaction graphs.

The Half-Normal Plot of Effects. An alternative to the normal probability plot of the factor effects is the **half-normal plot**. This is a plot of the **absolute value** of the effect estimates against their cumulative normal probabilities. Figure 6.15 presents the half-normal plot of the effects for Example 6.2. The straight line on the half-normal plot always passes through the origin and should also pass close to the fiftieth percentile data value. Many analysts feel that the half-normal plot is easier to interpret, particularly when there are only a few effect estimates such as when the experimenter has used an eight-run design. Some software packages will construct both plots.

Other Methods for Analyzing Unreplicated Factorials. The standard analysis procedure for an unreplicated two-level factorial design is the normal (or half-normal) plot of the estimated factor effects. However, unreplicated designs are so widely used in practice that many formal analysis procedures have been proposed to overcome the subjectivity of the normal probability plot. Hamada and Balakrishnan (1998) compared some of these methods. They found that the method proposed by Lenth (1989) has good power to detect significant effects. It is also easy to implement, and as a result it appears in several software packages for analyzing data from unreplicated factorials. We give a brief description of **Lenth's method**.

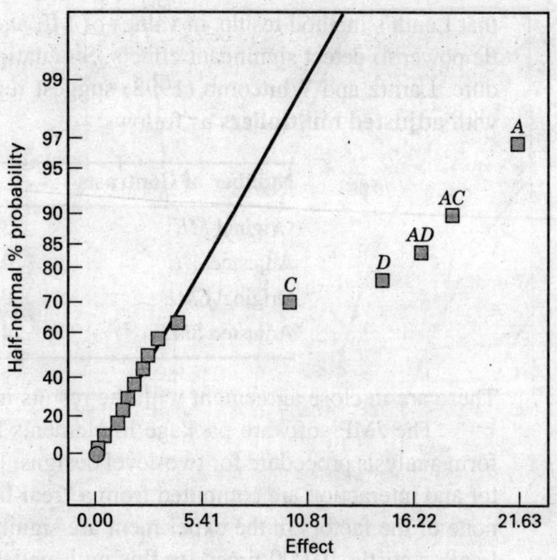

■ **FIGURE 6.15** **Half-normal plot of the factor effects from Example 6.2**

Suppose that we have m contrasts of interest, say $c_1, c_2, \ldots, c_m$. If the design is an unreplicated 2^k factorial design, these contrasts correspond to the $m = 2^k - 1$ factor effect estimates. The basis of Lenth's method is to estimate the variance of a contrast from the smallest (in absolute value) contrast estimates. Let

$$s_0 = 1.5 \times \text{median}(|c_j|)$$

and

$$PSE = 1.5 \times \text{median}(|c_j| : |c_j| < 2.5s_0)$$

PSE is called the "pseudostandard error," and Lenth shows that it is a reasonable estimator of the contrast variance when there are only a few active (significant) effects. The *PSE* is used to judge the significance of contrasts. An individual contrast can be compared to the **margin of error**

$$ME = t_{0.025,d} \times PSE$$

where the degrees of freedom are defined as $d = m/3$. For inference on a group of contrasts, Lenth suggests using the **simultaneous margin of error**

$$SME = t_{\gamma,d} \times PSE$$

where the percentage point of the t distribution used is $\gamma = 1 - (1 + 0.95^{1/m})/2$.

To illustrate Lenth's method, consider the 2^4 experiment in Example 6.2. The calculations result in $s_0 = 1.5 \times |-2.625| = 3.9375$ and $2.5 \times 3.9375 = 9.84375$, so

$$PSE = 1.5 \times |1.75| = 2.625$$

$$ME = 2.571 \times 2.625 = 6.75$$

$$SME = 5.219 \times 2.625 = 13.70$$

Now consider the effect estimates in Table 6.12. The *SME* criterion would indicate that the four largest effects (in magnitude) are significant because their effect estimates exceed *SME*. The main effect of C is significant according to the *ME* criterion, but not with respect to *SME*. However, because the *AC* interaction is clearly important, we would probably include C in the list of significant effects. Notice that in this example, Lenth's method has produced the same answer that we obtained previously from examination of the normal probability plot of effects.

Several authors [see Loughin and Nobel (1997), Hamada and Balakrishnan (1998), Larntz and Whitcomb (1998), Loughin (1998), and Edwards and Mee (2008)] have observed that Lenth's method results in values of ME and SME that are too conservative and have little power to detect significant effects. Simulation methods can be used to calibrate his procedure. Larntz and Whitcomb (1998) suggest replacing the original *ME* and *SME* multipliers with **adjusted multipliers** as follows:

Number of Contrasts	7	15	31
Original *ME*	3.764	2.571	2.218
Adjusted *ME*	2.295	2.140	2.082
Original *SME*	9.008	5.219	4.218
Adjusted *SME*	4.891	4.163	4.030

These are in close agreement with the results in Ye and Hamada (2000).

The JMP software package implements Lenth's method as part of the screening platform analysis procedure for two-level designs. In their implementation, *P*-values for each factor and interaction are computed from a "real-time" simulation. This simulation assumes that none of the factors in the experiment are significant and calculates the observed value of the Lenth statistic 10,000 times for this null model. Then *P*-values are obtained by determining where the observed Lenth statistics fall relative to the tails of these simulation-based reference distributions. These *P*-values can be used as guidance in selecting factors for the model. Table 6.14 shows the JMP output from the screening analysis platform for the resin filtration rate experiment in Example 6.2. Notice that in addition to the Lenth statistics, the JMP output includes a half-normal plot of the effects and a "Pareto" chart of the effect (contrast) magnitudes. When the factors are entered into the model, the Lenth procedure would recommend including the same factors in the model that we identified previously.

The final JMP output for the fitted model is shown in Table 6.15. The **Prediction Profiler** at the bottom of the table has been set to the levels of the factors than maximize filtration rate. These are the same settings that we determined earlier by looking at the contour plots.

In general, the Lenth method is a clever and very useful procedure. However, we recommend using it as a **supplement** to the usual normal probability plot of effects, not as a replacement for it.

Bisgaard (1998–1999) has provided a nice graphical technique, called a **conditional inference chart**, to assist in interpreting the normal probability plot. The purpose of the graph is to help the experimenter in judging significant effects. This would be relatively easy if the standard deviation σ were known, or if it could be estimated from the data. In unreplicated designs, there is no internal estimate of σ, so the conditional inference chart is designed to help the experimenter evaluate effect magnitude for a *range* of standard deviation values. Bisgaard bases the graph on the result that the standard error of an effect in a two-level design with *N* runs (for an unreplicated factorial, $N = 2^k$) is

$$\frac{2\sigma}{\sqrt{N}}$$

where σ is the standard deviation of an individual observation. Then ± 2 times the standard error of an effect is

$$\pm\frac{4\sigma}{\sqrt{N}}$$

Once the effects are estimated, plot a graph as shown in Figure 6.16, with the effect estimates plotted along the vertical or *y*-axis. In this figure, we have used the effect estimates from Example 6.2. The horizontal, or *x*-axis, of Figure 6.16 is a standard deviation (σ) scale. The two lines are at

$$y = +\frac{4\sigma}{\sqrt{N}} \quad \text{and} \quad y = -\frac{4\sigma}{\sqrt{N}}$$

■ **TABLE 6.14**
JMP Screening Platform Output for Example 6.2

Response Y
Summary of Fit

RSquare	1
RSquare Adj	-
Root Mean Square Error	-
Mean of Response	70.0625
Observations (or Sum Wgts)	16

Sorted Parameter Estimates

Term	Estimate	Relative Std Error	Pseudo t-Ratio	Pseudo t-Ratio	Pseudo p-Value
Temp	10.8125	0.25	8.24		0.0004*
Temp*Conc	−9.0625	0.25	−6.90		0.0010*
Temp*StirR	8.3125	0.25	6.33		0.0014*
StirR	7.3125	0.25	5.57		0.0026*
Conc	4.9375	0.25	3.76		0.0131*
Temp*Pressure*StirR	2.0625	0.25	1.57		0.1769
Pressure	1.5625	0.25	1.19		0.2873
Pressure*Conc*StirR	−1.3125	0.25	−1.00		0.3632
Pressure*Conc	1.1875	0.25	0.90		0.4071
Temp*Pressure*Conc	0.9375	0.25	0.71		0.5070
Temp*Conc*StirR	−0.8125	0.25	−0.62		0.5630
Temp*Pressure*Conc*StirR	0.6875	0.25	0.52		0.6228
Conc*StirR	−0.5625	0.25	−0.43		0.6861
Pressure*StirR	−0.1875	0.25	−0.14		0.8920
Temp*Pressure	0.0625	0.25	0.05		0.9639

No error degrees of freedom, so ordinary tests uncomputable. Relative Std Error corresponds to residual standard error of 1.
Pseudo t-Ratio and p-Value calculated using Lenth PSE = 1.3125 and DFE = 5

Effect Screening
The parameter estimates have equal variances.
The parameter estimates are not correlated.

Lenth PSE
1.3125

Orthog t Test used Pseudo Standard Error

Normal Plot

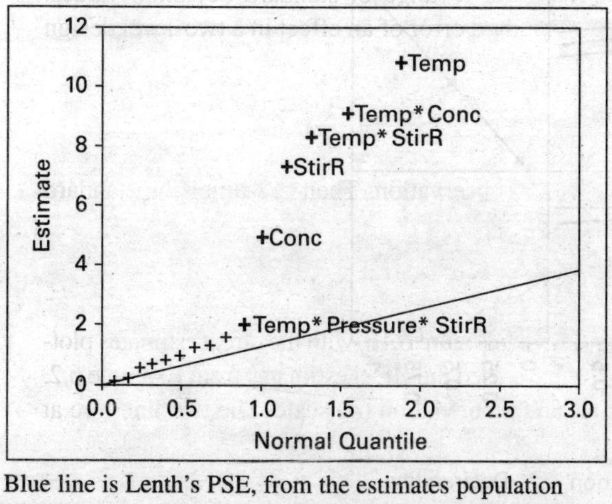

Blue line is Lenth's PSE, from the estimates population

■ **TABLE 6.15**

JMP Output for the Fitted Model Example 6.2

Response Filtration Rate Actual by Predicted Plot

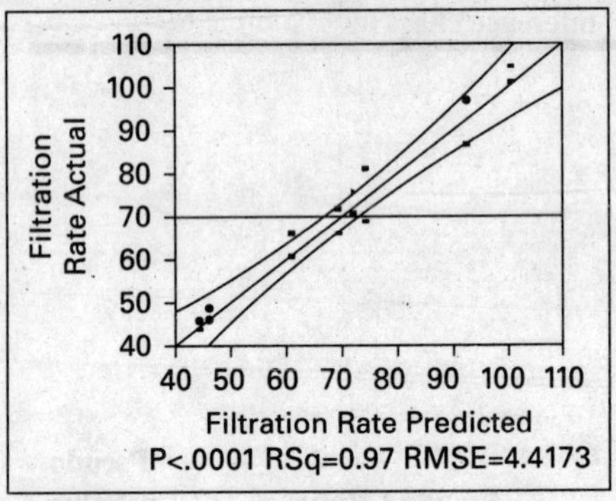

Summary of Fit

RSquare	0.965952
RSquare Adj	0.948929
Root Mean Square Error	4.417296
Mean of Response	70.0625
Observations (or Sum Wgts)	16

Analysis of Variance

Source	DF	Sum of Squares	Mean Square	F Ratio
Model	5	5535.8125	1107.16	56.7412
Error	10	195.1250	19.51	**Prob > F**
C. Total	15	5730.9375		<.0001*

Lack of Fit

Source	DF	Sum of Squares	Mean Square	F Ratio
				0.3482
Lack of Fit	2	15.62500	7.8125	**Prob > F**
Pure Error	8	179.50000	22.4375	0.7162
Total Error	10	195.12500		**Max RSq**
				0.9687

Parameter Estimates

Term	Estimate	Std Error	t Ratio	Prob>|t|
Intercept	70.0625	1.104324	63.44	<.0001*
Temperature	10.8125	1.104324	9.79	<.0001*
Stirring Rate	7.3125	1.104324	6.62	<.0001*
Concentration	4.9375	1.104324	4.47	0.0012*
Temperature *Stirring Rate	8.3125	1.104324	7.53	<.0001*
Temperature *Concentration	−9.0625	1.104324	−8.21	<.0001*

Sorted Parameter Estimates

Term	Estimate	Std Error	t Ratio		Prob < |t|
Temperature	10.8125	1.104324	9.79		<. 0001*
Temperature *Concentration	−9.0625	1.104324	−8.21		<. 0001*
Temperature *Stirring Rate	8.3125	1.104324	7.53		<. 0001*
Stirring Rate	7.3125	1.104324	6.62		<. 0001*
Concentration	4.9375	1.104324	4.47		0.0012*

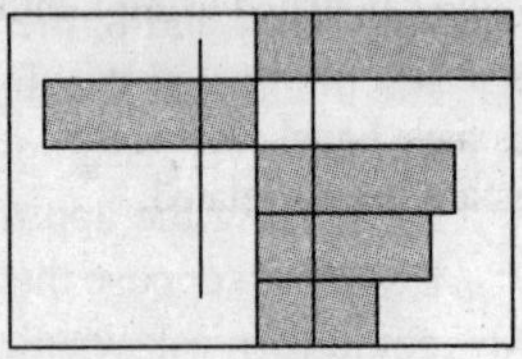

Prediction Profiler

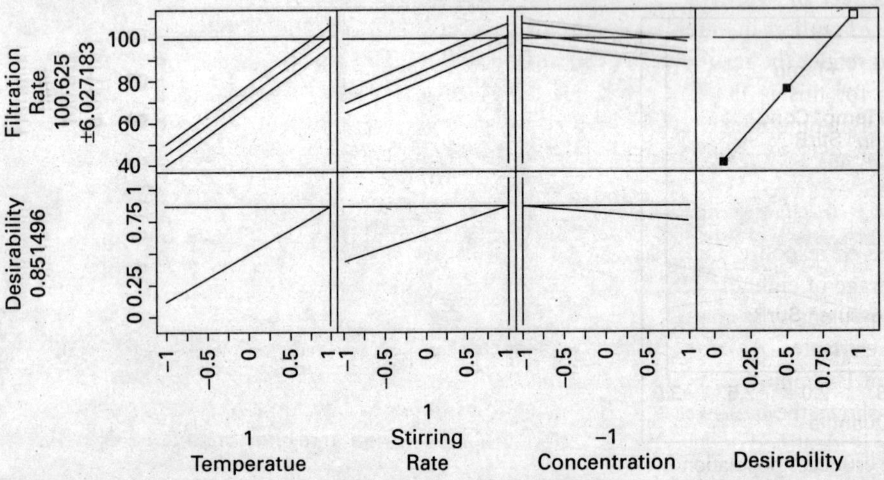

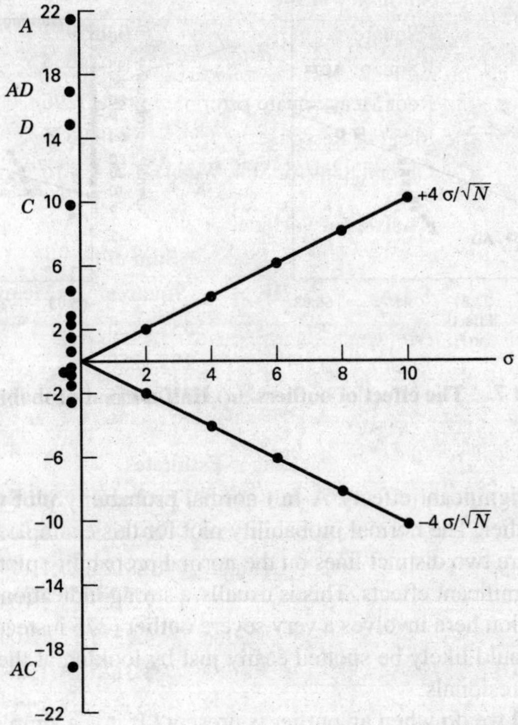

■ **FIGURE 6.16** A conditional inference chart for Example 6.2

In our example, $N = 16$, so the lines are at $y = +\sigma$ and $y = -\sigma$. Thus, for any given value of the standard deviation σ, we can read off the distance between these two lines as an approximate 95 percent confidence interval on the negligible effects.

In Figure 6.16, we observe that if the experimenter thinks that the standard deviation is between 4 and 8, then factors *A, C, D*, and the *AC* and *AD* interactions are significant. If he or she thinks that the standard deviation is as large as 10, factor *C* may not be significant. That is, for any given assumption about the magnitude of σ, the experimenter can construct a "yardstick" for judging the approximate significance of effects. The chart can also be used in reverse. For example, suppose that we were uncertain about whether factor *C* is significant. The experimenter could then ask whether it is reasonable to expect that σ could be as large as 10 or more. If it is unlikely that σ is as large as 10, then we can conclude that *C* is significant.

Effect of Outliers in Unreplicated Designs. Experimenters often worry about the impact of outliers in unreplicated designs, concerned that the outlier will invalidate the analysis and render the results of the experiment useless. This usually isn't a major concern. The reason for this is that the effect estimates are reasonably robust to outliers. To see this, consider an unreplicated 2^4 design with an outlier for (say) the *cd* treatment combination. The effect of any factor, say for example A, is

$$A = \bar{y}_{A^+} - \bar{y}_{A^-}$$

and the *cd* response appears in only one of the averages, in this case $\bar{y}_{A^-}$. The average $\bar{y}_{A^-}$ is an average of eight observations (half of the 16 runs in the 2^4), so the impact of the outlier *cd* is damped out by averaging it with the other seven runs. This will happen with all of the other effect estimates. As an illustration, consider the 2^4 design in the resin filtration rate experiment of Example 6.2. Suppose that the run $cd = 375$ (the correct response was 75). Figure 6.17*a* shows the half-normal plot of the effects. It is obvious that the correct set of important effects is identified on the graph. However, the half-normal plot gives an indication that an outlier may be present. Notice that the straight line identifying the nonsignificant effects does not point toward the origin. In fact, the reference line from the origin is not even close to the

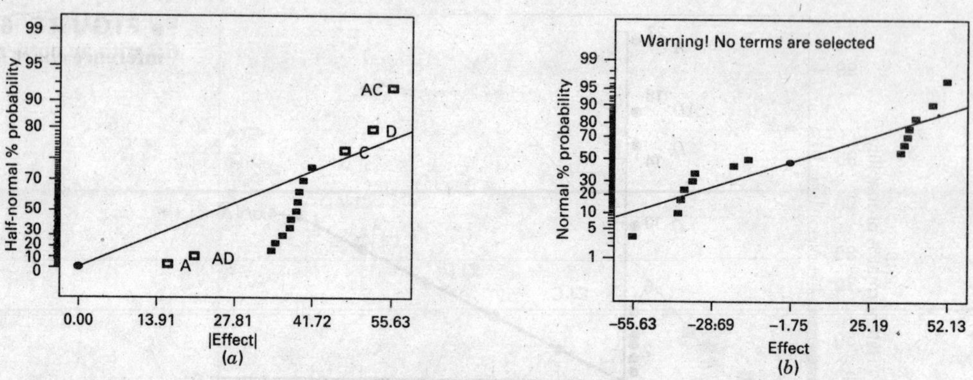

■ **FIGURE 6.17** **The effect of outliers. (*a*) Half-normal probability plot. (*b*) Normal probability plot**

collection of nonsignificant effects. A full normal probability plot would also have provided evidence of an outlier. The normal probability plot for this example is shown in Figure 6.17*b*. Notice that there are two distinct lines on the normal probability plot, not a single line passing through the nonsignificant effects. This is usually a strong indication that on outlier is present.

The illustration here involves a very severe outlier (375 instead of 75). This outlier is so dramatic that it would likely be spotted easily just by looking at the sample data or certainly by examining the residuals.

What should we do when an outlier is present? If it is a simple data recording or transposition error, an experimenter may be able to correct the outlier, replacing it with the right value. One suggestion is to replace it by an estimate (following the tactic introduced in Chapter 4 for blocked designs). This will preserve the orthogonality of the design and make interpretation easy. Replacing the outlier with an estimate that makes the highest order interaction estimate zero (in this case, replacing *cd* with a value that makes ABCD = 0) is one option. Discarding the outlier and analyzing the remaining observations is another option. This same approach would be used if one of the observations from the experiment is missing. Exercise 6.32 asks the reader to follow through with this suggestion for Example 6.2.

Modern computer software can analyze the data from 2^k designs with missing values because they use the method of least squares to estimate the effects, and least squares does not require an orthogonal design. The impact of this is that the effect estimates are no longer uncorrelated as they would be from an orthogonal design. The normal probability plotting technique requires that the effect estimates be uncorrelated with equal variance, but the degree of correlation introduced by a missing observation is relatively small in 2^k designs where the number of factor k is at least four. The correlation between the effect estimates and the model regression coefficients will not usually cause significant problems in interpreting the normal probability plot.

Figure 6.18 presents the half-normal probability plot obtained for the effect estimates if the outlier observation $cd = 375$ in Example 6.2 is omitted. This plot is easy to interpret, and exactly the same significant effects are identified as when the full set of experimental data was used. The correlation between design factors in this situation is ± 0.0714. It can be shown that the correlation between the model regression coefficients is larger, that is ± 0.5, but this still does not lead to any difficulty in interpreting the half-normal probability plot.

6.6 Additional Examples of Unreplicated 2^k Designs

Unreplicated 2^k designs are widely used in practice. They may be the most common variation of the 2^k design. This section presents four interesting applications of these designs, illustrating some additional analysis that can be helpful.

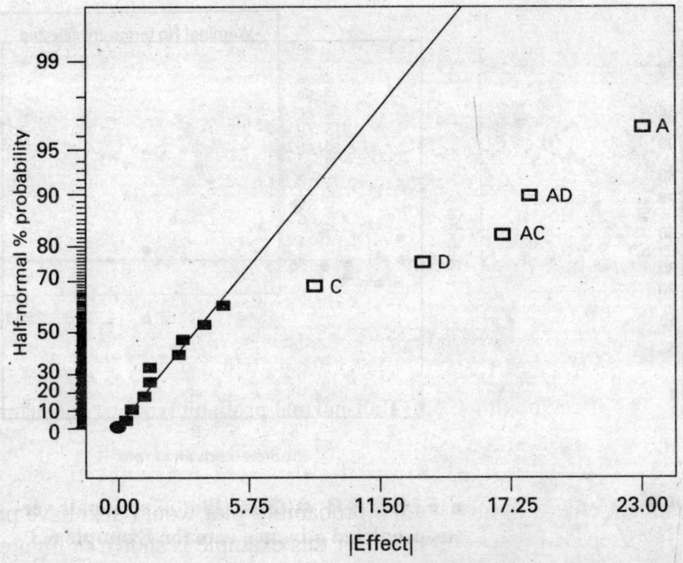

■ **FIGURE 6.18** **Analysis of Example 6.2 with an outlier removed**

EXAMPLE 6.3 Data Transformation in a Factorial Design

Daniel (1976) describes a 2^4 factorial design used to study the advance rate of a drill as a function of four factors: drill load (A), flow rate (B), rotational speed (C), and the type of drilling mud used (D). The data from the experiment are shown in Figure 6.19.

The normal probability plot of the effect estimates from this experiment is shown in Figure 6.20. Based on this plot, factors B, C, and D along with the BC and BD interactions require interpretation. Figure 6.21 is the normal probability plot of the residuals and Figure 6.22 is the plot of the residuals versus the predicted advance rate from the model containing the identified factors. There are clearly problems with normality and equality of variance. A data transformation is often used to deal with such problems. Because the response variable is a rate, the log transformation seems a reasonable candidate.

Figure 6.23 presents a normal probability plot of the effect estimates following the transformation $y^* = \ln y$. Notice that a much simpler interpretation now seems possible because only factors B, C, and D are active. That is, expressing the data in the correct metric has simplified its structure to the point that the two interactions are no longer required in the explanatory model.

Figures 6.24 and 6.25 present, respectively, a normal probability plot of the residuals and a plot of the residuals versus the predicted advance rate for the model in the log

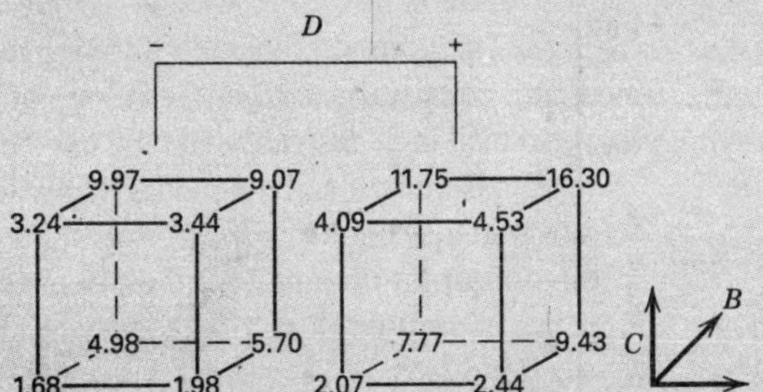

■ **FIGURE 6.19** **Data from the drilling experiment of Example 6.3**

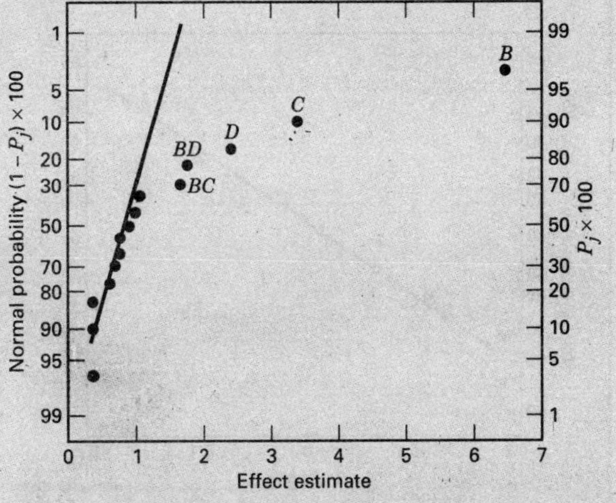

■ **FIGURE 6.20** **Normal probability plot of effects for Example 6.3**

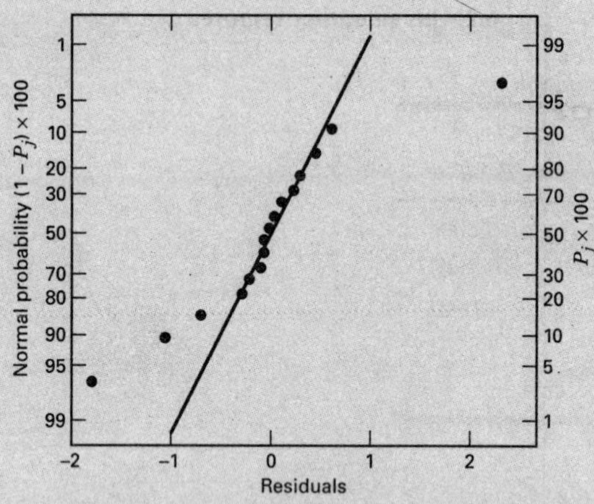

■ **FIGURE 6.21** Normal probability plot of residuals for Example 6.3

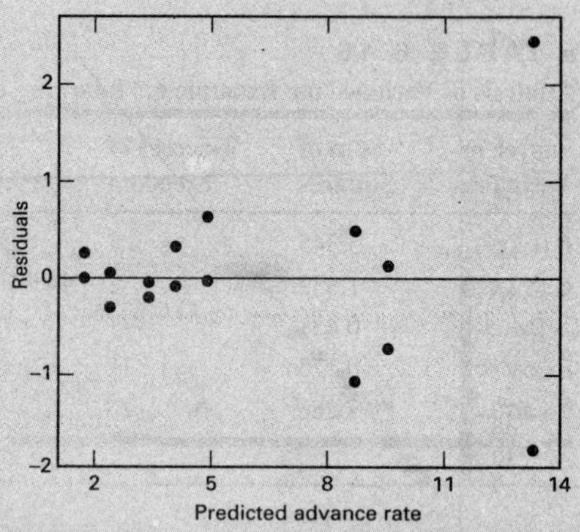

■ **FIGURE 6.22** Plot of residuals versus predicted advance rate for Example 6.3

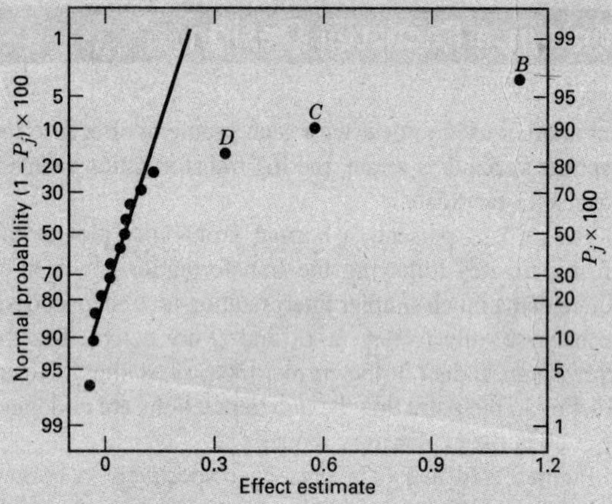

■ **FIGURE 6.23** Normal probability plot of effects for Example 6.3 following log transformation

scale containing *B*, *C*, and *D*. These plots are now satisfactory. We conclude that the model for $y^* = \ln y$ requires only factors *B*, *C*, and *D* for adequate interpretation. The ANOVA for this model is summarized in Table 6.16. The model sum of squares is

$$\begin{aligned} SS_{\text{Model}} &= SS_B + SS_C + SS_D \\ &= 5.345 + 1.339 + 0.431 \\ &= 7.115 \end{aligned}$$

and $R^2 = SS_{\text{Model}}/SS_T = 7.115/7.288 = 0.98$, so the model explains about 98 percent of the variability in the drill advance rate.

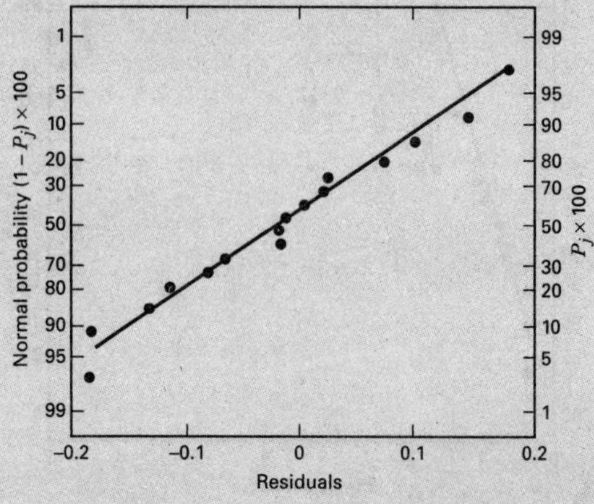

■ **FIGURE 6.24** Normal probability plot of residuals for Example 6.3 following log transformation

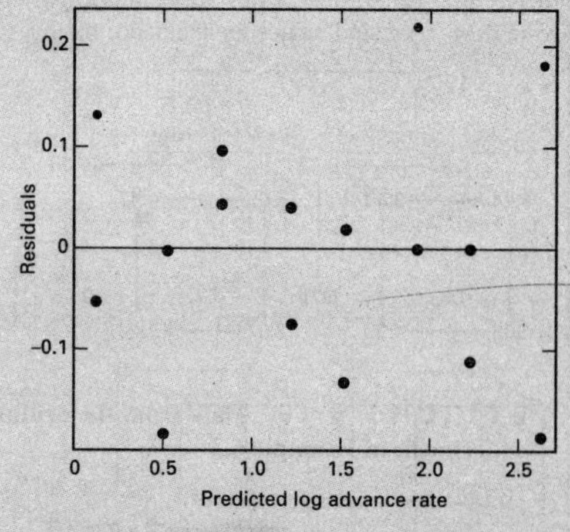

■ **FIGURE 6.25** Plot of residuals versus predicted advance rate for Example 6.3 following log transformation

■ **TABLE 6.16**

Analysis of Variance for Example 6.3 following the Log Transformation

Source of Variation	Sum of Squares	Degrees of Freedom	Mean Square	F_0	*P*-Value
B (Flow)	5.345	1	5.345	381.79	<0.0001
C (Speed)	1.339	1	1.339	95.64	<0.0001
D (Mud)	0.431	1	0.431	30.79	<0.0001
Error	0.173	12	0.014		
Total	7.288	15			

EXAMPLE 6.4 Location and Dispersion Effects in an Unreplicated Factorial

A 2^4 design was run in a manufacturing process producing interior sidewall and window panels for commercial aircraft. The panels are formed in a press, and under present conditions the average number of defects per panel in a press load is much too high. (The current process average is 5.5 defects per panel.) Four factors are investigated using a single replicate of a 2^4 design, with each replicate corresponding to a single press load. The factors are temperature (*A*), clamp time (*B*), resin flow (*C*), and press closing time (*D*). The data for this experiment are shown in Figure 6.26.

A normal probability plot of the factor effects is shown in Figure 6.27. Clearly, the two largest effects are $A = 5.75$ and $C = -4.25$. No other factor effects appear to be large, and *A* and *C* explain about 77 percent of the total variability. We therefore conclude that lower temperature (*A*) and higher resin flow (*C*) would reduce the incidence of panel defects.

Careful residual analysis is an important aspect of any experiment. A normal probability plot of the residuals showed no anomalies, but when the experimenter plotted the residuals versus each of the factors *A* through *D*, the plot of residuals versus *B* (clamp time) presented the pattern shown in Figure 6.28. This factor, which is unimportant insofar as the average number of defects per panel is concerned, is very important in its effect on process variability, with the lower clamp time resulting in less variability in the average number of defects per panel in a press load.

Factors	Low (−)	High (+)
A = Temperature (°F)	295	325
B = Clamp time (min)	7	9
C = Resin flow	10	20
D = Closing time (s)	15	30

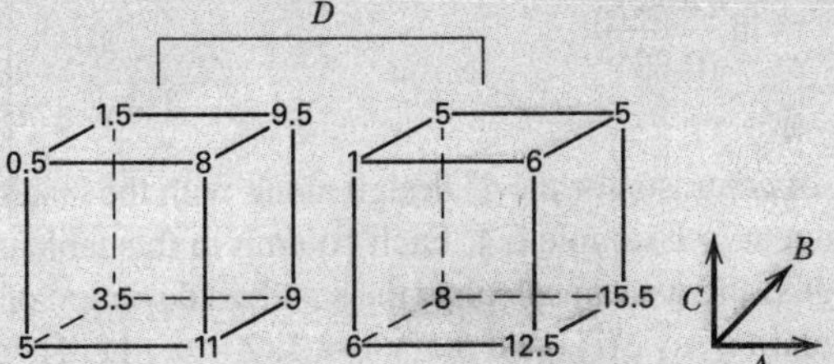

■ **FIGURE 6.26 Data for the panel process experiment of Example 6.4**

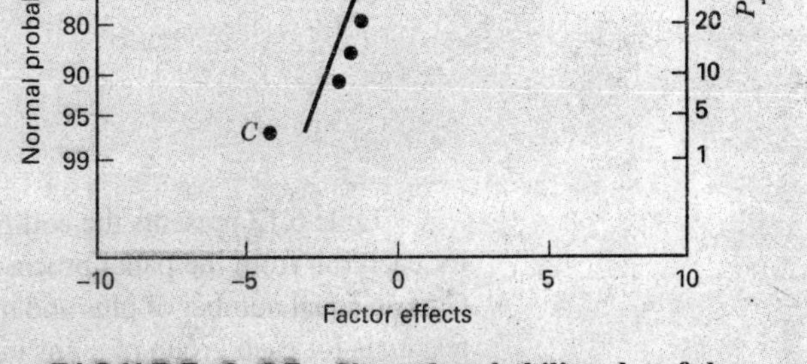

■ **FIGURE 6.27 Normal probability plot of the factor effects for the panel process experiment of Example 6.4**

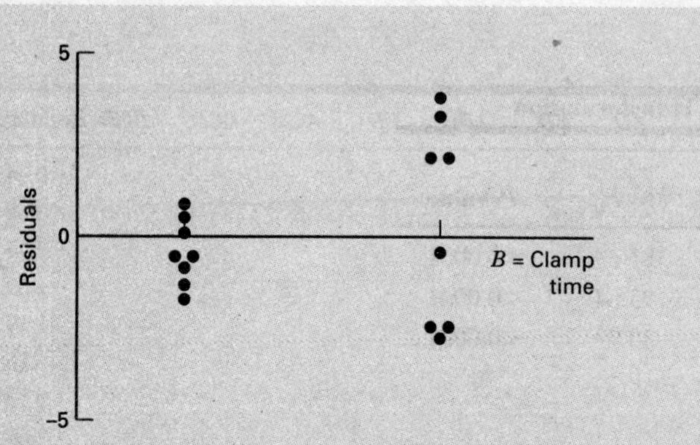

■ **FIGURE 6.28 Plot of residuals versus clamp time for Example 6.4**

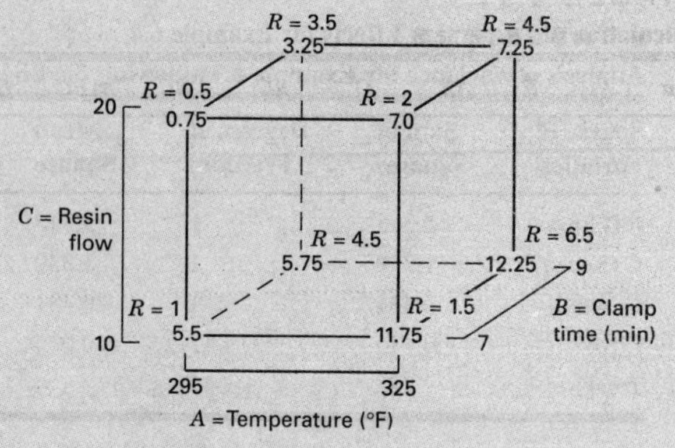

■ **FIGURE 6.29 Cube plot of temperature, clamp time, and resin flow for Example 6.4**

The dispersion effect of clamp time is also very evident from the **cube plot** in Figure 6.29, which plots the average number of defects per panel and the range of the number of defects at each point in the cube defined by factors *A*, *B*, and *C*. The average range when *B* is at the high level (the back face of the cube in Figure 6.29) is $\overline{R}_{B^+} = 4.75$ and when *B* is at the low level, it is $\overline{R}_{B^-} = 1.25$.

As a result of this experiment, the engineer decided to run the process at low temperature and high resin flow to reduce the average number of defects, at low clamp time to reduce the variability in the number of defects per panel, and at low press closing time (which had no effect on either location or dispersion). The new set of operating conditions resulted in a new process average of less than one defect per panel.

The residuals from a 2^k design provide much information about the problem under study. Because residuals can be thought of as observed values of the noise or error, they often give insight into process variability. We can systematically examine the residuals from an unreplicated 2^k design to provide information about process variability.

Consider the residual plot in Figure 6.28. The standard deviation of the eight residuals where *B* is at the low level is $S(B^-) = 0.83$, and the standard deviation of the eight residuals where *B* is at the high level is $S(B^+) = 2.72$. The statistic

$$F_B^* = \ln \frac{S^2(B^+)}{S^2(B^-)} \tag{6.24}$$

has an approximate normal distribution if the two variances $\sigma^2(B^+)$ and $\sigma^2(B^-)$ are equal. To illustrate the calculations, the value of F_B^* is

$$\begin{aligned} F_B^* &= \ln \frac{S^2(B^+)}{S^2(B^-)} \\ &= \ln \frac{(2.72)^2}{(0.83)^2} \\ &= 2.37 \end{aligned}$$

Table 6.17 presents the complete set of contrasts for the 2^4 design along with the residuals for each run from the panel process experiment in Example 6.4. Each column in this table contains an equal number of plus and minus signs, and we can calculate the standard deviation of the residuals for each group of signs in each column, say $S(i^+)$ and $S(i^-)$, $i = 1, 2, \ldots, 15$. Then

$$F_i^* = \ln \frac{S^2(i^+)}{S^2(i^-)} \qquad i = 1, 2, \ldots, 15 \tag{6.25}$$

■ **TABLE 6.17**
Calculation of Dispersion Effects for Example 6.4

Run	*A*	*B*	*AB*	*C*	*AC*	*BC*	*ABC*	*D*	*AD*	*BD*	*ABD*	*CD*	*ACD*	*BCD*	*ABCD*	Residual
1	−	−	+	−	+	+	−	−	+	+	−	+	−	−	+	−0.94
2	+	−	−	−	−	+	+	−	−	+	+	+	+	−	−	−0.69
3	−	+	−	−	+	−	+	−	+	−	+	+	−	+	−	−2.44
4	+	+	+	−	−	−	−	−	−	−	−	+	+	+	+	−2.69
5	−	−	+	+	−	−	+	−	+	+	−	−	+	+	−	−1.19
6	+	−	−	+	+	−	−	−	−	+	+	−	−	+	+	0.56
7	−	+	−	+	−	+	−	−	+	−	+	−	+	−	+	−0.19
8	+	+	+	+	+	+	+	−	−	−	−	−	−	−	−	2.06
9	−	−	+	−	+	+	−	+	−	−	+	−	+	+	−	0.06
10	+	−	−	−	−	+	+	+	+	−	−	−	−	+	+	0.81
11	−	+	−	−	+	−	+	+	−	+	−	−	+	−	+	2.06
12	+	+	+	−	−	−	−	+	+	+	+	−	−	−	−	3.81
13	−	−	+	+	−	−	+	+	−	−	+	+	−	−	+	−0.69
14	+	−	−	+	+	−	−	+	+	−	−	+	+	−	−	−1.44
15	−	+	−	+	−	+	−	+	−	+	−	+	−	+	−	3.31
16	+	+	+	+	+	+	+	+	+	+	+	+	+	+	+	−2.44
$S(i^+)$	2.25	2.72	2.21	1.91	1.81	1.80	1.80	2.24	2.05	2.28	1.97	1.93	1.52	2.09	1.61	
$S(i^-)$	1.85	0.83	1.86	2.20	2.24	2.26	2.24	1.55	1.93	1.61	2.11	1.58	2.16	1.89	2.33	
F_i^*	0.39	2.37	0.34	−0.28	−0.43	−0.46	−0.44	0.74	0.12	0.70	−0.14	0.40	−0.70	0.20	−0.74	

is a statistic that can be used to assess the magnitude of the **dispersion effects** in the experiment. If the variance of the residuals for the runs where factor i is positive equals the variance of the residuals for the runs where factor i is negative, then F_i^* has an approximate normal distribution. The values of F_i^* are shown below each column in Table 6.15.

Figure 6.30 is a normal probability plot of the dispersion effects F_i^*. Clearly, B is an important factor with respect to process dispersion. For more discussion of this procedure, see

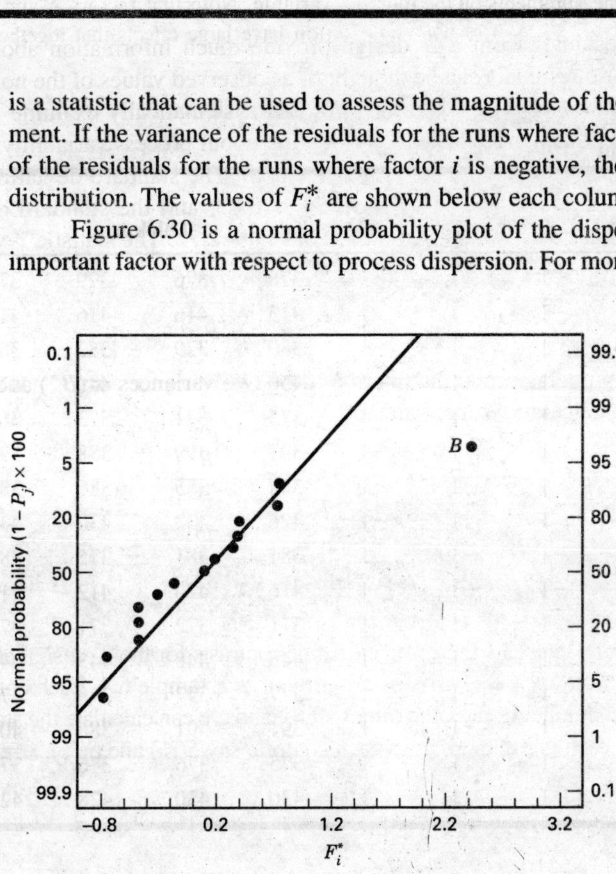

■ **FIGURE 6.30** **Normal probability plot of the dispersion effects F_i^* for Example 6.4**

Box and Meyer (1986) and Myers, Montgomery and Anderson-Cook (2009). Also, in order for the model residuals to properly convey information about dispersion effects, the **location model** must be correctly specified. Refer to the supplemental text material for this chapter for more details and an example.

EXAMPLE 6.5 Duplicate Measurements on the Response

A team of engineers at a semiconductor manufacturer ran a 2^4 factorial design in a vertical oxidation furnace. Four wafers are "stacked" in the furnace, and the response variable of interest is the oxide thickness on the wafers. The four design factors are temperature (A), time (B), pressure (C), and gas flow (D). The experiment is conducted by loading four wafers into the furnace, setting the process variables to the test conditions required by the experimental design, processing the wafers, and then measuring the oxide thickness on all four wafers. Table 6.18 presents the design and the resulting thickness measurements. In this table, the four columns labeled "Thickness" contain the oxide thickness measurements on each individual wafer, and the last two columns contain the sample average and sample variance of the thickness measurements on the four wafers in each run.

The proper analysis of this experiment is to consider the individual wafer thickness measurements as **duplicate measurements** and not as replicates. If they were really replicates, each wafer would have been processed individually on a single run of the furnace. However, because all four wafers were processed together, they received the treatment factors (that is, the levels of the design variables) *simultaneously*, so there is much less variability in the individual wafer thickness measurements than would have been observed if each wafer was a replicate. Therefore, the **average** of the thickness measurements is the correct response variable to initially consider.

Table 6.19 presents the effect estimates for this experiment, using the average oxide thickness $\bar{y}$ as the response variable. Note that factors A and B and the AB interaction have large effects that together account for nearly 90

■ **TABLE 6.18**
The Oxide Thickness Experiment

Standard Order	Run Order	A	B	C	D	Thickness				$\bar{y}$	s^2
1	10	−1	−1	−1	−1	378	376	379	379	378	2
2	7	1	−1	−1	−1	415	416	416	417	416	0.67
3	3	−1	1	−1	−1	380	379	382	383	381	3.33
4	9	1	1	−1	−1	450	446	449	447	448	3.33
5	6	−1	−1	1	−1	375	371	373	369	372	6.67
6	2	1	−1	1	−1	391	390	388	391	390	2
7	5	−1	1	1	−1	384	385	386	385	385	0.67
8	4	1	1	1	−1	426	433	430	431	430	8.67
9	12	−1	−1	−1	1	381	381	375	383	380	12.00
10	16	1	−1	−1	1	416	420	412	412	415	14.67
11	8	−1	1	−1	1	371	372	371	370	371	0.67
12	1	1	1	−1	1	445	448	443	448	446	6
13	14	−1	−1	1	1	377	377	379	379	378	1.33
14	15	1	−1	1	1	391	391	386	400	392	34
15	11	−1	1	1	1	375	376	376	377	376	0.67
16	13	1	1	1	1	430	430	428	428	429	1.33

■ **TABLE 6.19**
Effect Estimates for Example 6.5, Response Variable Is Average Oxide Thickness

Model Term	Effect Estimate	Sum of Squares	Percent Contribution
A	43.125	7439.06	67.9339
B	18.125	1314.06	12.0001
C	−10.375	430.562	3.93192
D	−1.625	10.5625	0.0964573
AB	16.875	1139.06	10.402
AC	−10.625	451.563	4.12369
AD	1.125	5.0625	0.046231
BC	3.875	60.0625	0.548494
BD	−3.875	60.0625	0.548494
CD	1.125	5.0625	0.046231
ABC	−0.375	0.5625	0.00513678
ABD	2.875	33.0625	0.301929
ACD	−0.125	0.0625	0.000570753
BCD	−0.625	1.5625	0.0142688
ABCD	0.125	0.0625	0.000570753

percent of the variability in average oxide thickness. Figure 6.31 is a normal probability plot of the effects. From examination of this display, we would conclude that factors *A*, *B*, and *C* and the *AB* and *AC* interactions are important.

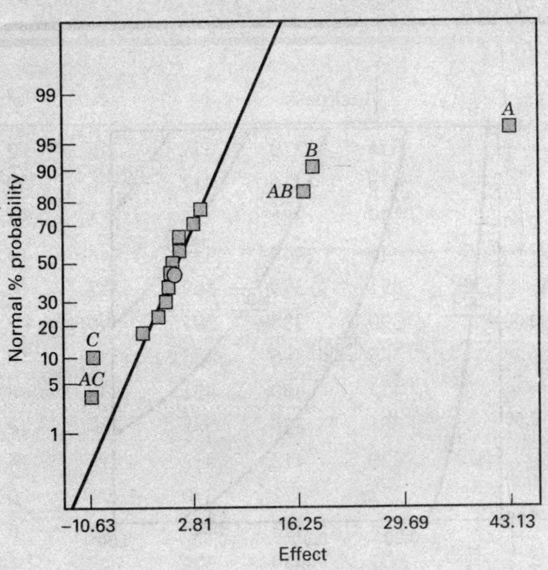

■ **FIGURE 6.31 Normal probability plot of the effects for the average oxide thickness response, Example 6.5**

The analysis of variance display for this model is shown in Table 6.20.

The model for predicting average oxide thickness is

$$\hat{y} = 399.19 + 21.56x_1 + 9.06x_2 - 5.19x_3 + 8.44x_1x_2 - 5.31x_1x_3$$

The residual analysis of this model is satisfactory.

The experimenters are interested in obtaining an average oxide thickness of 400 Å, and product specifications require that the thickness must lie between 390 and 410 Å. Figure 6.32 presents two contour plots of average thickness, one with factor *C* (or x_3), pressure, at the low level (that is, $x_3 = -1$) and the other with *C* (or x_3) at the high level (that is, $x_3 = +1$). From examining these contour plots, it is obvious that there are many combinations of time and temperature (factors *A* and *B*) that will produce acceptable results. However, if pressure is held constant at the low level, the operating "window" is shifted toward the left, or lower, end of the time axis, indicating that lower cycle times will be required to achieve the desired oxide thickness.

It is interesting to observe the results that would be obtained if we **incorrectly** consider the individual wafer oxide thickness measurements as replicates. Table 6.21 presents a full-model ANOVA based on treating the experiment as a replicated 2^4 factorial. Notice that there are many significant factors in this analysis, suggesting a much more complex model than the one that we found when using the average oxide thickness as the response. The reason for this is that the estimate of the error variance in Table 6.21 is too small ($\hat{\sigma}^2 = 6.12$). The residual mean square in Table 6.21 reflects a combination of the variability between wafers *within* a run and variability *between* runs. The estimate of error obtained from Table 6.20 is much larger, $\hat{\sigma}^2 = 17.61$, and it is primarily a measure of the between-run variability. This is the best estimate of error to use in judging the significance of process variables that are changed from run to run.

A logical question to ask is: What harm results from identifying too many factors as important, as the incorrect analysis in Table 6.21 would certainly do. The answer is that trying to manipulate or optimize the unimportant factors would be a waste of resources, and it could result in adding unnecessary variability to *other* responses of interest.

When there are duplicate measurements on the response, these observations almost always contain useful information about some aspect of process variability. For example, if the duplicate measurements are multiple tests by a gauge on the same experimental unit, then the duplicate measurements give some insight about *gauge capability*. If the duplicate measurements are made at different locations on an experimental unit, they may give some information about the *uniformity* of the response variable

■ **TABLE 6.20**
Analysis of Variance (from Design-Expert) for the Average Oxide Thickness Response, Example 6.5

Source	Sum of Squares	DF	Mean Square	F Value	Prob > F
Model	10774.31	5	2154.86	122.35	<0.000
A	*7439.06*	*1*	*7439.06*	*422.37*	*<0.000*
B	*1314.06*	*1*	*1314.06*	*74.61*	*<0.000*
C	*430.56*	*1*	*430.56*	*24.45*	*0.0006*
AB	*1139.06*	*1*	*1139.06*	*64.67*	*<0.000*
AC	*451.46*	*1*	*451.56*	*25.64*	*0.0005*
Residual	176.12	10	17.61		
Cor Total	10950.44	15			
Std. Dev.	4.20	R-Squared	0.9839		
Mean	399.19	Adj R-Squared	0.9759		
C.V.	1.05	Pred R-Squared	0.9588		
PRESS	450.88	Adeq Precision	27.967		

Factor	Coefficient Estimate	DF	Standard Error	95% CI Low	95% CI High
Intercept	399.19	1	1.05	396.85	401.53
A-Time	21.56	1	1.05	19.22	23.90
B-Temp	9.06	1	1.05	6.72	11.40
C-Pressure	−5.19	1	1.05	−7.53	−2.85
AB	8.44	1	1.05	6.10	10.78
AC	−5.31	1	1.05	−7.65	−2.97

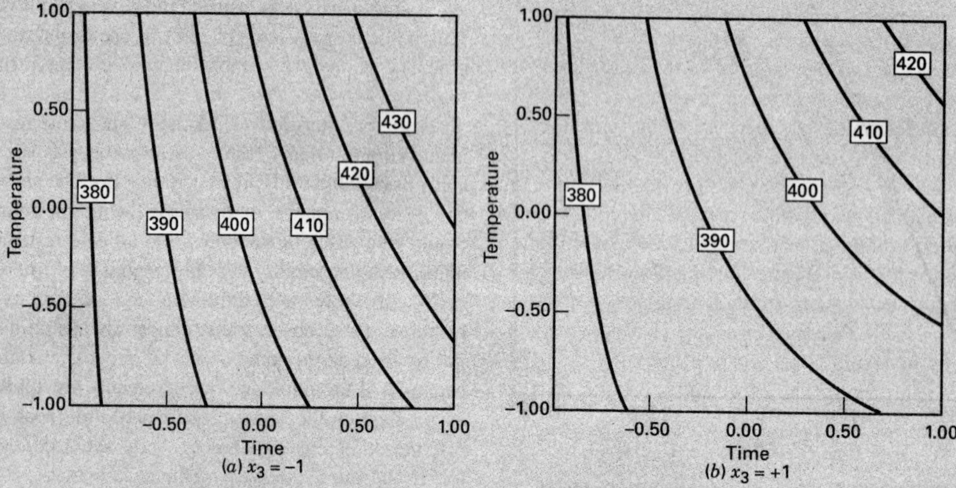

■ **FIGURE 6.32** **Contour plots of average oxide thickness with pressure (x_3) held constant**

■ **TABLE 6.21**
Analysis of Variance (from Design-Expert) of the Individual Wafer Oxide Thickness Response

Source	Sum of Squares	DF	Mean Square	*F* Value	Prob > *F*
Model	43801.75	15	2920.12	476.75	<0.0001
A	*29756.25*	*1*	*29756.25*	*4858.16*	*<0.0001*
B	*5256.25*	*1*	*5256.25*	*858.16*	*<0.0001*
C	*1722.25*	*1*	*1722.25*	*281.18*	*<0.0001*
D	*42.25*	*1*	*42.25*	*6.90*	*0.0115*
AB	*4556.25*	*1*	*4556.25*	*743.88*	*<0.0001*
AC	*1806.25*	*1*	*1806.25*	*294.90*	*<0.0001*
AD	*20.25*	*1*	*20.25*	*3.31*	*0.0753*
BC	*240.25*	*1*	*240.25*	*39.22*	*<0.0001*
BD	*240.25*	*1*	*240.25*	*39.22*	*<0.0001*
CD	*20.25*	*1*	*20.25*	*3.31*	*0.0753*
ABD	*132.25*	*1*	*132.25*	*21.59*	*<0.0001*
ABC	*2.25*	*1*	*2.25*	*0.37*	*0.5473*
ACD	*0.25*	*1*	*0.25*	*0.041*	*0.8407*
BCD	*6.25*	*1*	*6.25*	*1.02*	*0.3175*
ABCD	*0.25*	*1*	*0.25*	*0.041*	*0.8407*
Residual	294.00	48	6.12		
Lack of Fit	*0.000*	*0*			
Pure Error	*294.00*	*48*	*6.13*		
Cor Total	44095.75	63			

across that unit. In our example, because we have one observation on each of the four experimental units that have undergone processing together, we have some information about the *within-run* variability in the process. This information is contained in the variance of the oxide thickness measurements from the four wafers in each run. It would be of interest to determine whether any of the process variables influence the within-run variability.

Figure 6.33 is a normal probability plot of the effect estimates obtained using $\ln(s^2)$ as the response. Recall from Chapter 3 that we indicated that the log transformation is generally appropriate for modeling variability. There are not any strong individual effects, but factor *A* and *BD* interaction are the largest. If we also include the main effects of *B* and *D* to obtain a hierarchical model, then the model for $\ln(s^2)$ is

$$\widehat{\ln(s^2)} = 1.08 + 0.41x_1 - 0.40x_2 + 0.20x_4 - 0.56x_2x_4$$

The model accounts for just slightly less than half of the variability in the $\ln(s^2)$ response, which is certainly not

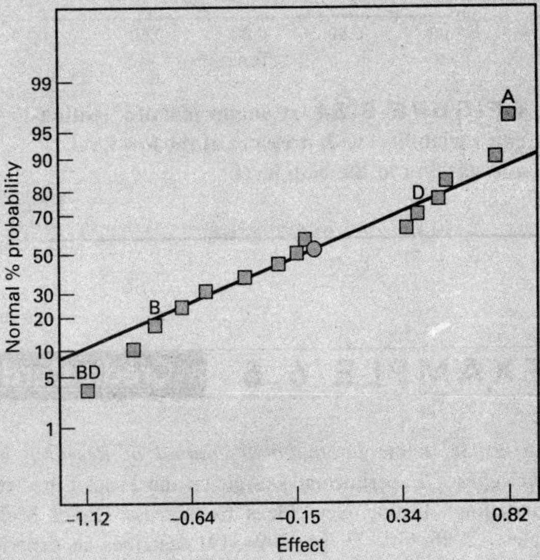

■ **FIGURE 6.33** Normal probability plot of the effects using $\ln(s^2)$ as the response, Example 6.5

spectacular as empirical models go, but it is often difficult to obtain exceptionally good models of variances.

Figure 6.34 is a contour plot of the predicted variance (not the log of the predicted variance) with pressure x_3 at the low level (recall that this minimizes cycle time) and gas flow x_4 at the high level. This choice of gas flow gives the lowest values of predicted variance in the region of the contour plot.

The experimenters here were interested in selecting values of the design variables that gave a mean oxide thickness within the process specifications and as close to 400 Å as possible, while simultaneously making the within-run variability small, say $s^2 \leq 2$. One possible way to find a suitable set of conditions is to overlay the contour plots in Figures 6.32 and 6.34. The overlay plot is shown in Figure 6.35, with the specifications on mean oxide thickness and the constraint $s^2 \leq 2$ shown as contours. In this plot, pressure is held constant at the low level and gas flow is held constant at the high level. The open region near the upper left center of the graph identifies a feasible region for the variables time and temperature.

This is a simple example of using contour plots to study two responses simultaneously. We will discuss this problem in more detail in Chapter 11.

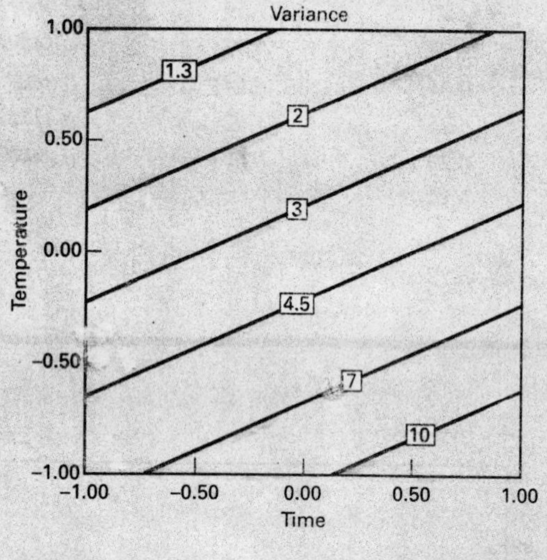

■ **FIGURE 6.34** **Contour plot of s^2 (within-run variability) with pressure at the low level and gas flow at the high level**

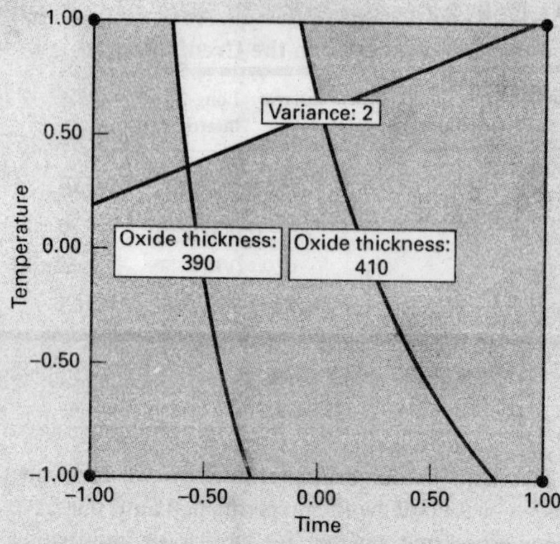

■ **FIGURE 6.35** **Overlay of the average oxide thickness and s^2 responses with pressure at the low level and gas flow at the high level**

EXAMPLE 6.6 Credit Card Marketing

An article in the *International Journal of Research in Marketing* ("Experimental Design on the Front Lines of Marketing: Testing New Ideas to Increase Direct Mail Sales," 2006, Vol. 23, pp. 309–319) describes an experiment to test new ideas to increase direct mail sales by the credit card division of a financial services company. They want to improve the response rate to its credit card offers. They know from experience that the interest rates are an important factor in attracting potential customers, so they have decided to focus on factors involving both interest rates and fees. They want to test changes in both introductory and long-term rates, as well as the effects of adding an account-opening fee and lowering the annual fee. The factors tested in the experiment are as follows:

Factor	(−) Control	(+) New idea
A: Annual fee	Current	Lower
B: Account-opening fee	No	Yes
C: Initial interest rate	Current	Lower
D: Long-term interest rate	Low	High

The marketing team used columns A through D of the 2^4 factorial test matrix shown in Table 6.22 to create 16 mail packages. The +/− sign combinations in the 11 interaction (product) columns are used solely to facilitate the statistical analysis of the results. Each of the 16 test combinations was mailed to 7500 customers, and 2837 customers responded positively to the offers.

■ **TABLE 6.22**

The 2^4 Factorial Design Used in the Credit Card Marketing Experiment, Example 6.6

	Annual-fee	Account-opening fee	Initial interest rate	Long-term interest rate	(interactions)												
Test cell	A	B	C	D	AB	AC	AD	BC	BD	CD	ABC	ABD	ACD	BCD	ABCD	Orders	Response rate
1	−	−	−	−	+	+	+	+	+	+	−	−	−	−	+	184	2.45%
2	+	−	−	−	−	−	−	+	+	+	+	+	+	−	−	252	3.36%
3	−	+	−	−	−	+	+	−	−	+	+	+	−	−	−	162	2.16%
4	+	+	−	−	+	−	−	−	−	+	−	−	+	+	+	172	2.29%
5	−	−	+	−	+	−	+	−	+	−	+	−	+	+	−	187	2.49%
6	+	−	+	−	−	+	−	−	+	−	−	+	−	+	+	254	3.39%
7	−	+	+	−	−	−	+	+	−	−	−	+	+	−	+	174	2.32%
8	+	+	+	−	+	+	−	+	−	−	+	−	−	−	−	183	2.44%
9	−	−	−	+	+	+	−	+	−	−	−	+	+	+	−	138	1.84%
10	+	−	−	+	−	−	+	+	−	−	+	−	−	+	+	168	2.24%
11	−	+	−	+	−	+	−	−	+	−	+	−	+	−	+	127	1.69%
12	+	+	−	+	+	−	+	−	+	−	−	+	−	−	−	140	1.87%
13	−	−	+	+	+	−	−	−	−	+	+	+	−	−	+	172	2.29%
14	+	−	+	+	−	+	+	−	−	+	−	−	+	−	−	219	2.92%
15	−	+	+	+	−	−	−	+	+	+	−	−	−	+	−	153	2.04%
16	+	+	+	+	+	+	+	+	+	+	+	+	+	+	+	152	2.03%

Table 6.23 is the JMP output for the screening analysis. Lenth's method with simulated *P*-values is used to identify significant factors. All four main effects are significant, and one interaction (AB, or Annual Fee × Account Opening Fee). The prediction profiler indicates the settings of the four factors that will result in the maximum response rate. The lower annual fee, no account opening fee, the lower long-term interest rate and either value of the initial interest rate produce the best response, 3.39 percent. The optimum conditions occur at one of the actual test combinations because all four design factors were treated as qualitative. With continuous factors, the optimal conditions are usually not at one of the experimental runs.

■ **TABLE 6.23**
JMP Output for Example 6.6

Response Response Rate

Summary of Fit

RSquare	1
RSquare Adj	.
Root Mean Square Error	.
Mean of Response	2.36375
Observations (or Sum Wgts)	16

Sorted Parameter Estimates

Term	Estimate	Relative Std Error	Pseudo t-Ratio	Pseudo p-Value
Account Opening Fee[No]	0.25875	0.25	3.63	0.0150*
Long-term Interest Rate[Low]	0.24875	0.25	3.49	0.0174*
Annual Fee[Current]	−0.20375	0.25	−2.86	0.0354*
Annual Fee[Current]*Account Opening Fee[No]	−0.15125	0.25	−2.12	0.0872
initial Interest Rate[Current]	−0.12625	0.25	−1.77	0.1366
initial Interest Rate[Current]*Long-term Interest Rate[Low]	0.07875	0.25	1.11	0.3194
Annual Fee[Current]*Long-term Interest Rate[Low]	−0.05375	0.25	−0.75	0.4846
Account Opening Fee[No]*initial Interest Rate[Current]*Long-term Interest Rate[Low]	0.05375	0.25	0.75	0.4846
Account Opening Fee[No]*Long-term Interest Rate[Low]	0.05125	0.25	0.72	0.5042
Annual Fee[Current]*Account Opening Fee[No]*Long-term Interest Rate[Low]	−0.04375	0.25	−0.61	0.5661
Annual Fee[Current]*Account Opening Fee[No]*initial Interest Rate[Current]	0.02625	0.25	0.37	0.7276
Annual Fee[Current]*Account Opening Fee[No]*initial Interest Rate[Current]*Long-term Interest Rate[Low]	−0.02625	0.25	−0.37	0.7276
Account Opening Fee[No]*initial Interest Rate[Current]	−0.02375	0.25	−0.33	0.7524
Annual Fee[Current]*initial Interest Rate[Current]*Long-term Interest Rate[Low]	−0.00375	0.25	−0.05	0.9601
Annual Fee[Current]*initial Interest Rate[Current]	0.00125	0.25	0.02	0.9867

No error degrees of freedom, so ordinary tests uncomputable.
Relative Std Error corresponds to residual standard error of 1.
Pseudo t-Ratio and p-Value calculated using Lenth PSE = 0.07125 and DFE=5

Prediction Profiler

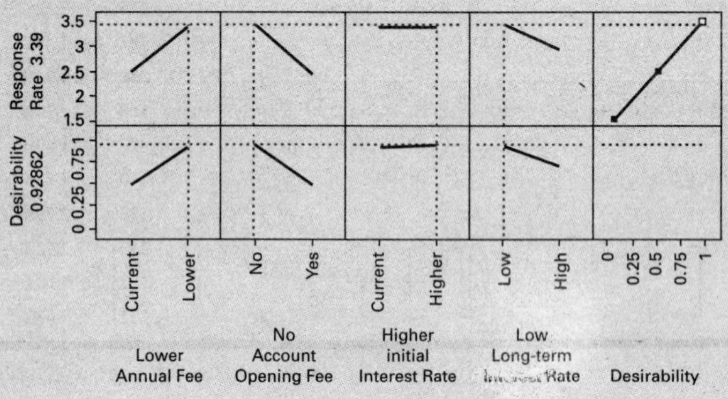

6.7 2^k Designs are Optimal Designs

Two-level factorial designs have many interesting and useful properties. In this section, a brief description of some of these properties is given. We have remarked in previous sections that the model regression coefficients and effect estimates from a 2^k design are least squares estimates. This is discussed in the supplemental text material for this chapter and presented in more detail in Chapter 10, but it is useful to give a proof of this here.

Consider a very simple case the 2^2 design with one replicate. This is a four-run design, with treatment combinations (1), *a*, *b*, and *ab*. The design is shown geometrically in Figure 6.1. The model we fit to the data from this design is

$$y = \beta_0 + \beta_1 x_1 + \beta_2 x_2 + \beta_{12} x_1 x_2 + \varepsilon$$

where x_1 and x_2 are the main effects of the two factors on the ± 1 scale and $x_1 x_2$ is the two-factor interaction. We can write out each one of the four runs in this design in terms of this model as follows:

$$(1) = \beta_0 + \beta_1(-1) + \beta_2(-1) + \beta_{12}(-1)(-1) + \epsilon_1$$
$$a = \beta_0 + \beta_1(1) + \beta_2(-1) + \beta_{12}(1)(-1) + \epsilon_2$$
$$b = \beta_0 + \beta_1(-1) + \beta_2(1) + \beta_2(-1)(1) + \epsilon_3$$
$$ab = \beta_0 + \beta_1(1) + \beta_2(1) + \beta_{12}(1)(1) + \epsilon_4$$

It is much easier if we write these four equations in matrix form:

$$\mathbf{y} = \mathbf{X}\boldsymbol{\beta} + \boldsymbol{\epsilon}, \text{ where } \mathbf{y} = \begin{bmatrix} (1) \\ a \\ b \\ ab \end{bmatrix}, \mathbf{X} = \begin{bmatrix} 1 & -1 & -1 & 1 \\ 1 & 1 & -1 & -1 \\ 1 & -1 & 1 & -1 \\ 1 & 1 & 1 & 1 \end{bmatrix}, \boldsymbol{\beta} = \begin{bmatrix} \beta_0 \\ \beta_1 \\ \beta_2 \\ \beta_{12} \end{bmatrix}, \text{ and } \boldsymbol{\epsilon} = \begin{bmatrix} \varepsilon_1 \\ \varepsilon_2 \\ \varepsilon_3 \\ \varepsilon_4 \end{bmatrix}$$

The least squares estimates of the model parameters are the values of the β's that minimize the sum of the squares of the model errors, ϵ_i, $i = 1, 2, 3, 4$. The least squares estimates are

$$\hat{\boldsymbol{\beta}} = (\mathbf{X}'\mathbf{X})^{-1}\mathbf{X}'\mathbf{y}$$

where the prime (′) denotes a transpose and $(\mathbf{X}'\mathbf{X})^{-1}$ is the inverse of $\mathbf{X}'\mathbf{X}$. We will prove this result later in Chapter 10. For the 2^2 design, the quantities $\mathbf{X}'\mathbf{X}$ and $\mathbf{X}'\mathbf{y}$ are

$$\mathbf{X}'\mathbf{X} = \begin{bmatrix} 1 & 1 & 1 & 1 \\ -1 & 1 & -1 & 1 \\ -1 & -1 & 1 & 1 \\ 1 & -1 & -1 & 1 \end{bmatrix} \begin{bmatrix} 1 & -1 & -1 & 1 \\ 1 & 1 & -1 & -1 \\ 1 & -1 & 1 & -1 \\ 1 & 1 & 1 & 1 \end{bmatrix} = \begin{bmatrix} 4 & 0 & 0 & 0 \\ 0 & 4 & 0 & 0 \\ 0 & 0 & 4 & 0 \\ 0 & 0 & 0 & 4 \end{bmatrix}$$

and

$$\mathbf{X}'\mathbf{y} = \begin{bmatrix} 1 & 1 & 1 & 1 \\ -1 & 1 & -1 & 1 \\ -1 & -1 & 1 & 1 \\ 1 & -1 & -1 & 1 \end{bmatrix} \begin{bmatrix} (1) \\ a \\ b \\ ab \end{bmatrix} = \begin{bmatrix} (1) + a + b + ab \\ -(1) + a - b + ab \\ -(1) - a + b + ab \\ (1) - a - b + ab \end{bmatrix}$$

The $\mathbf{X}'\mathbf{X}$ matrix is diagonal because the 2^2 design is orthogonal. The least squares estimates are as follows:

$$\hat{\boldsymbol{\beta}} = (\mathbf{X}'\mathbf{X})^{-1}\mathbf{X}'\mathbf{y}$$

$$= \begin{bmatrix} 4 & 0 & 0 & 0 \\ 0 & 4 & 0 & 0 \\ 0 & 0 & 4 & 0 \\ 0 & 0 & 0 & 4 \end{bmatrix}^{-1} \begin{bmatrix} (1) + a + b + ab \\ -(1) + a - b + ab \\ -(1) - a + b + ab \\ (1) - a - b + ab \end{bmatrix}$$

$$= \begin{bmatrix} \frac{(1) + a + b + ab}{4} \\ \frac{-(1) + a - b + ab}{4} \\ \frac{-(1) - a + b + ab}{4} \\ \frac{(1) - a - b + ab}{4} \end{bmatrix}$$

The least squares estimates of the model regression coefficients are exactly equal to one-half of the usual effect estimates.

It turns out that the variance of any model regression coefficient is easy to find:

$$V(\hat{\beta}) = \sigma^2(\text{diagonal element of } (\mathbf{X'X})^{-1})$$
$$= \frac{\sigma^2}{4}$$

All model regression coefficients have the same variance. Furthermore, there is no other four-run design on the design space bounded by ± 1 that makes the variance of the model regression coefficients smaller. In general, the variance of any model regression coefficient in a 2^k design where each design point is replicated n times is $V(\hat{\beta}) = \sigma^2/(n2^k) = \sigma^2/N$, where N is the total number of runs in the design. This is the minimum possible variance for the regression coefficient.

For the 2^2 design, the determinant of the $\mathbf{X'X}$ matrix is

$$|(\mathbf{X'X})| = 256$$

This is the maximum possible value of the determinant for a four-run design on the design space bounded by ± 1. It turns out that the volume of the joint confidence region that contains all the model regression coefficients is inversely proportional to the square root of the determinant of $\mathbf{X'X}$. Therefore, to make this joint confidence region as small as possible, we would want to choose a design that makes the determinant of $\mathbf{X'X}$ as large as possible. This is accomplished by choosing the 2^2 design.

In general, a design that minimizes the variance of the model regression coefficients is called a ***D*-optimal design**. The *D* terminology is used because these designs are found by selecting runs in the design to maximize the determinant of $\mathbf{X'X}$. The 2^k design is a *D*-optimal design for fitting the first-order model or the first-order model with interaction. Many computer software packages, such as JMP, Design-Expert, and Minitab, have algorithms for finding *D*-optimal designs. These algorithms can be very useful in constructing experimental designs for many practical situations. We will make use of them in subsequent chapters.

Now consider the variance of the predicted response in the 2^2 design

$$V[\hat{y}(x_1 x_2)] = V(\hat{\beta}_0 + \hat{\beta}_1 x_1 + \hat{\beta}_2 x_2 + \hat{\beta}_{12} x_1 x_2)$$

The variance of the predicted response is a function of the point in the design space where the prediction is made (x_1 and x_2) and the variance of the model regression coefficients. The estimates of the regression coefficients are independent because the 2^2 design is orthogonal and they all have variance $\sigma^2/4$, so

$$V[\hat{y}(x_1,x_2)] = V(\hat{\beta}_0 + \hat{\beta}_1 x_1 + \hat{\beta}_2 x_2 + \hat{\beta}_{12} x_1 x_2)$$
$$= \frac{\sigma^2}{4}(1 + x_1^2 + x_2^2 + x_1^2 x_2^2)$$

The maximum prediction variance occurs when $x_1 = x_2 = \pm 1$ and is equal to σ^2. To determine how good this is, we need to know the best possible value of prediction variance that we can attain. It turns out that the smallest possible value of the maximum prediction variance over the design space is $p\sigma^2/N$, where p is the number of model parameters and N is the number of runs in the design. The 2^2 design has $N = 4$ runs and the model has $p = 4$ parameters, so the model that we fit to the data from this experiment minimizes the maximum prediction variance over the design region. A design that has this property is called a ***G*-optimal design**. In general, 2^k designs are *G*-optimal designs for fitting the first-order model or the first-order model with interaction.

We can evaluate the prediction variance at any point of interest in the design space. For example, when we are at the center of the design where $x_1 = x_2 = 0$, the prediction variance is

$$V[\hat{y}(x_1 = 0, x_2 = 0)] = \frac{\sigma^2}{4}$$

When $x_1 = 1$ and $x_2 = 0$, the prediction variance is

$$V[\hat{y}(x_1 = 1, x_2 = 0)] = \frac{\sigma^2}{2}$$

An alternative to evaluating the prediction variance at a lot of points in the design space is to consider the **average prediction variance** over the design space. One way to calculate this average prediction variance is

$$I = \frac{1}{A}\int_{-1}^{1}\int_{-1}^{1} V[\hat{y}(x_1, x_2)]dx_1dx_2$$

where A is the area (in general the volume) of the design space. To compute the average, we are integrating the variance function over the design space and dividing by the area of the region.

Sometimes I is called the **integrated variance** criterion. Now for a 2^2 design, the area of the design region is $A = 4$, and

$$\begin{aligned} I &= \frac{1}{A}\int_{-1}^{1}\int_{-1}^{1} V[\hat{y}(x_1, x_2)]dx_1dx_2 \\ &= \frac{1}{4}\int_{-1}^{1}\int_{-1}^{1} \sigma^2\frac{1}{4}(1 + x_1^2 + x_2^2 + x_1^2x_2^2)dx_1dx_2 \\ &= \frac{4\sigma^2}{9} \end{aligned}$$

It turns out that this is the smallest possible value of the average prediction variance that can be obtained from a four-run design used to fit a first-order model with interaction on this design space. A design with this property is called an ***I*-optimal design**. In general, 2^k designs are *I*-optimal designs for fitting the first-order model or the first-order model with interaction. The JMP software will construct *I*-optimal designs. This can be very useful in constructing designs when response prediction is the goal of the experiment.

It is also possible to display the prediction variance over the design space graphically. Figure 6.36 is output from JMP illustrating three possible displays of the prediction variance from a 2^2 design. The first graph is the **prediction variance profiler**, which plots the **unscaled prediction variance**

$$UPV = \frac{V[\hat{y}(x_1, x_2)]}{\sigma^2}$$

against the levels of each design factor. The "crosshairs" on the graphs are adjustable, so that the unscaled prediction variance can be displayed at any desired combination of the variables

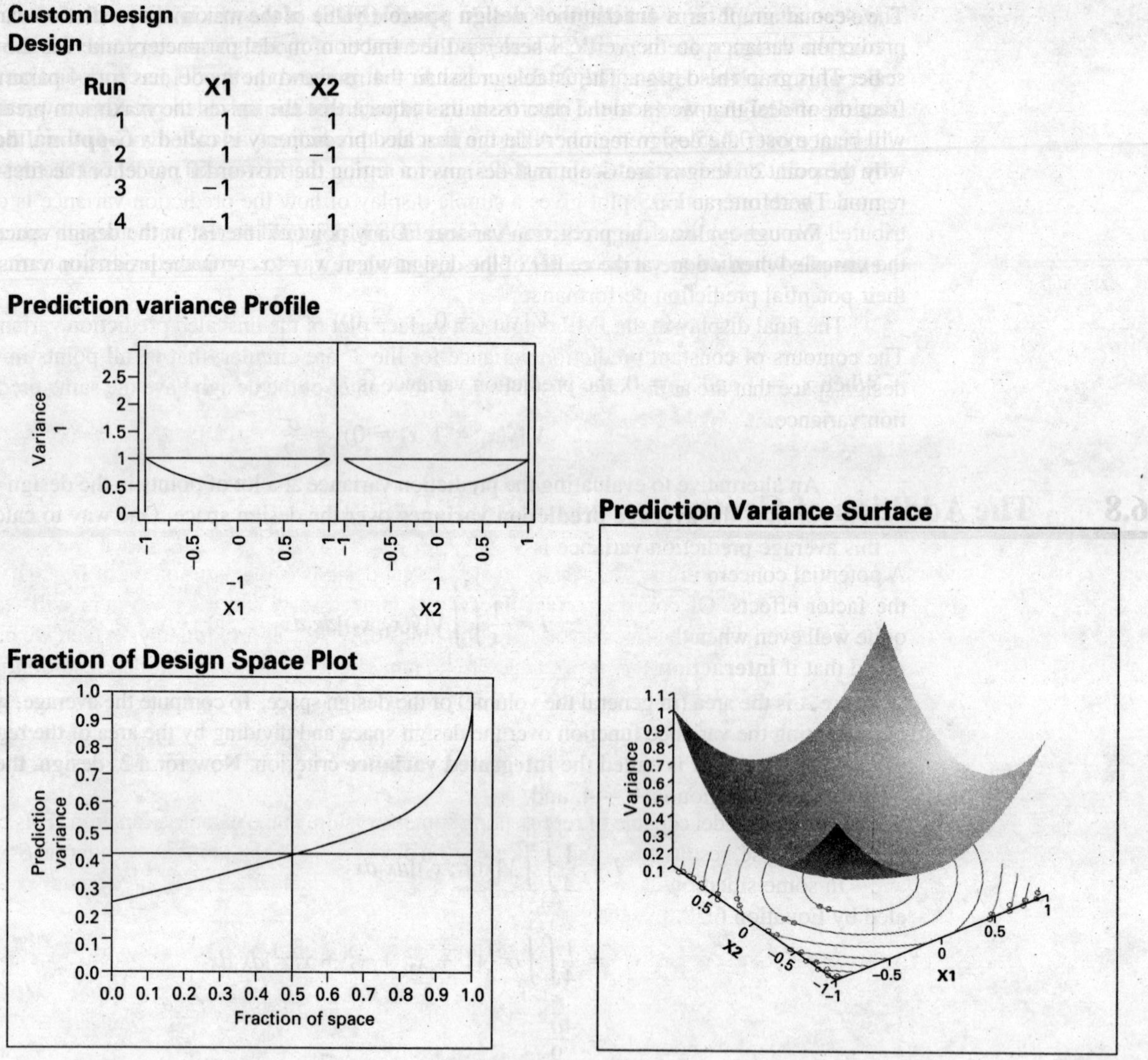

■ **FIGURE 6.36** **JMP prediction variance output for the 2^2 design**

x_1 and x_2. Here, the values chosen are $x_1 = -1$ and $x_2 = +1$, for which the unscaled prediction variance is

$$
\begin{aligned}
UPV &= \frac{V[\hat{y}(x_1, x_2)]}{\sigma^2} \\
&= \frac{\frac{\sigma^2}{4}(1 + x_1^2 + x_2^2 + x_1^2 x_2^2)}{\sigma^2} \\
&= \frac{\frac{\sigma^2}{4}(4)}{\sigma^2} \\
&= 1
\end{aligned}
$$

The second graph is a **fraction of design space (FDS) plot**, which shows the unscaled prediction variance on the vertical scale and the fraction of design space on the horizontal scale. This graph also has an adjustable crosshair that is shown at the 50 percent point on the fraction of design space scale. The crosshairs indicate that the unscaled prediction variance will be at most 0.425 σ^2 (remember that the unscaled prediction variance divides by σ^2, that's why the point on the vertical scale is 0.425) over a region that covers 50 percent of the design region. Therefore, an FDS plot gives a simple display of how the prediction variance is distributed throughout the design region. An ideal FDS plot would be flat with a small value of the unscaled prediction variance. FDS plots are an ideal way to compare designs in terms of their potential prediction performance.

The final display in the JMP output is a surface plot of the unscaled prediction variance. The contours of constant prediction variance for the 2^2 are circular; that is, all points in the design space that are at the same distance from the center of the design have the same prediction variance.

6.8 The Addition of Center Points to the 2^k Design

A potential concern in the use of two-level factorial designs is the assumption of **linearity** in the factor effects. Of course, perfect linearity is unnecessary, and the 2^k system will work quite well even when the linearity assumption holds only very approximately. In fact, we have noted that if **interaction terms** are added to a main effect or first-order model, resulting in

$$y = \beta_0 + \sum_{j=1}^{k} \beta_j x_j + \sum_{i<j}\sum \beta_{ij} x_i x_j + \epsilon \qquad (6.28)$$

then we have a model capable of representing some curvature in the response function. This curvature, of course, results from the twisting of the plane induced by the interaction terms $\beta_{ij} x_i x_j$.

In some situations, the curvature in the response function will not be adequately modeled by Equation 6.28. In such cases, a logical model to consider is

$$y = \beta_0 + \sum_{j=1}^{k} \beta_j x_j + \sum_{i<j}\sum \beta_{ij} x_i x_j + \sum_{j=1}^{k} \beta_{jj} x_j^2 + \epsilon \qquad (6.29)$$

where the β_{jj} represent pure second-order or **quadratic effects**. Equation 6.29 is called a **second-order response surface model.**

In running a two-level factorial experiment, we usually anticipate fitting the first-order model in Equation 6.28, but we should be alert to the possibility that the second-order model in Equation 6.29 is more appropriate. There is a method of replicating certain points in a 2^k factorial that will provide protection against curvature from second-order effects as well as allow an independent estimate of error to be obtained. The method consists of adding **center points** to the 2^k design. These consist of n_C replicates run at the points $x_i = 0$ $(i = 1, 2, \ldots, k)$. One important reason for adding the replicate runs at the design center is that center points do not affect the usual effect estimates in a 2^k design. When we add center points, we assume that the k factors are **quantitative**.

To illustrate the approach, consider a 2^2 design with one observation at each of the factorial points $(-, -)$, $(+, -)$, $(-, +)$, and $(+, +)$ and n_C observations at the center point $(0, 0)$. Figures 6.37 and 6.38 illustrate the situation. Let $\bar{y}_F$ be the average of the four runs at the four factorial points, and $\bar{y}_C$ be the average of the n_C runs at the center point. If the difference $\bar{y}_F - \bar{y}_C$ is small, then the center points lie on or near the plane passing through the factorial points, and there is no quadratic curvature. On the other hand, if $\bar{y}_F - \bar{y}_C$ is large,

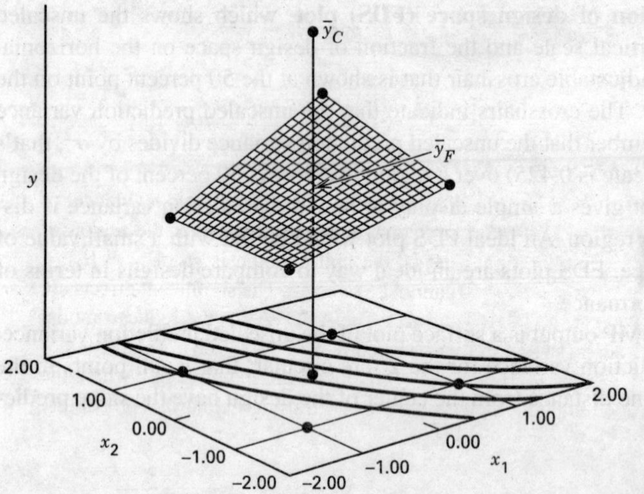

■ **FIGURE 6.37** **A 2^2 design with center points**

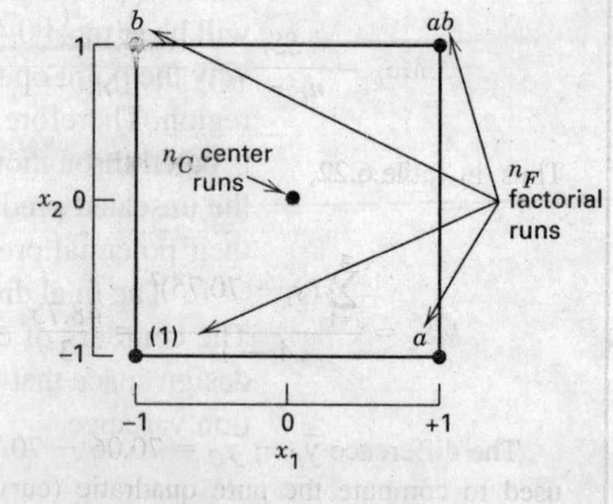

■ **FIGURE 6.38** **A 2^2 design with center points**

then quadratic curvature is present. A single-degree-of-freedom **sum of squares for pure quadratic curvature** is given by

$$SS_{\text{Pure quadratic}} = \frac{n_F n_C(\bar{y}_F - \bar{y}_C)^2}{n_F + n_C} \tag{6.30}$$

where, in general, n_F is the number of factorial design points. This sum of squares may be incorporated into the ANOVA and may be compared to the error mean square to test for pure quadratic curvature. More specifically, when points are added to the center of the 2^k design, the test for curvature (using Equation 6.30) actually tests the hypotheses

$$H_0: \sum_{j=1}^{k} \beta_{jj} = 0$$

$$H_1: \sum_{j=1}^{k} \beta_{jj} \neq 0$$

Furthermore, if the factorial points in the design are unreplicated, one may use the n_C center points to construct an estimate of error with $n_C - 1$ degrees of freedom. A t-test can also be used to test for curvature. Refer to the **supplemental text material** for this chapter.

EXAMPLE 6.7

We will illustrate the addition of center points to a 2^k design by reconsidering the pilot plant experiment in Example 6.2. Recall that this is an unreplicated 2^4 design. Refer to the original experiment shown in Table 6.10. Suppose that four center points are added to this experiment, and at the points $x_1 = x_2 = x_3 = x_4 = 0$ the four observed filtration rates were 73, 75, 66, and 69. The average of these four center points is $\bar{y}_C = 70.75$, and the average of the 16 factorial runs is $\bar{y}_F = 70.06$. Since $\bar{y}_C$ and $\bar{y}_F$ are very similar, we suspect that there is no strong curvature present.

Table 6.24 summarizes the analysis of variance for this experiment. In the upper portion of the table, we have fit the full model. The mean square for pure error is calculated from the center points as follows:

$$MS_E = \frac{SS_E}{n_C - 1} = \frac{\sum_{\text{Center points}} (y_i - \bar{y}_c)^2}{n_C - 1} \qquad (6.29)$$

Thus, in Table 6.22,

$$MS_E = \frac{\sum_{i=1}^{4} (y_i - 70.75)^2}{4 - 1} = \frac{48.75}{3} = 16.25$$

The difference $\bar{y}_F - \bar{y}_C = 70.06 - 70.75 = -0.69$ is used to compute the pure quadratic (curvature) sum of squares in the ANOVA table from Equation 6.30 as follows:

$$SS_{\text{Pure quadratic}} = \frac{n_F n_C(\bar{y}_F - \bar{y}_C)^2}{n_F + n_C} = \frac{(16)(4)(-0.69)^2}{16 + 4} = 1.51$$

The ANOVA indicates that there is no evidence of second-order curvature in the response over the region of exploration. That is, the null hypothesis $H_0: \beta_{11} + \beta_{22} + \beta_{33} + \beta_{44} = 0$ cannot be rejected. The significant effects are *A*, *C*, *D*, *AC*, and *AD*. The ANOVA for the reduced model is shown in the lower portion of Table 6.24. The results of this analysis agree with those from Example 6.2, where the important effects were isolated using the normal probability plotting method.

■ **TABLE 6.24**
Analysis of Variance for Example 6.6

ANOVA for the Full Model

Source of Variation	Sum of Squares	DF	Mean Square	F	Prob > F
Model	5730.94	15	382.06	23.51	0.0121
A	*1870.56*	*1*	*1870.56*	*115.11*	*0.0017*
B	*39.06*	*1*	*39.06*	*2.40*	*0.2188*
C	*390.06*	*1*	*390.06*	*24.00*	*0.0163*
D	*855.56*	*1*	*855.56*	*52.65*	*0.0054*
AB	*0.063*	*1*	*0.063*	*3.846E-003*	*0.9544*
AC	*1314.06*	*1*	*1314.06*	*80.87*	*0.0029*
AD	*1105.56*	*1*	*1105.56*	*68.03*	*0.0037*
BC	*22.56*	*1*	*22.56*	*1.39*	*0.3236*
BD	*0.56*	*1*	*0.56*	*0.035*	*0.8643*
CD	*5.06*	*1*	*5.06*	*0.31*	*0.6157*
ABC	*14.06*	*1*	*14.06*	*0.87*	*0.4209*
ABD	*68.06*	*1*	*68.06*	*4.19*	*0.1332*
ACD	*10.56*	*1*	*10.56*	*0.65*	*0.4791*
BCD	*27.56*	*1*	*27.56*	*1.70*	*0.2838*
ABCD	*7.56*	*1*	*7.56*	*0.47*	*0.5441*
Pure quadratic					
Curvature	1.51	1	1.51	0.093	0.7802
Pure error	48.75	3	16.25		
Cor total	5781.20	19			

■ **TABLE 6.24** *(Continued)*

ANOVA for the Reduced Model

Source of Variation	Sum of Squares	DF	Mean Square	F	Prob > F
Model	5535.81	5	1107.16	59.02	< 0.000
A	*1870.56*	*1*	*1870.56*	*99.71*	*< 0.000*
C	*390.06*	*1*	*390.06*	*20.79*	*0.0005*
D	*855.56*	*1*	*855.56*	*45.61*	*< 0.000*
AC	*1314.06*	*1*	*1314.06*	*70.05*	*< 0.000*
AD	*1105.56*	*1*	*1105.56*	*58.93*	*< 0.000*
Pure quadratic curvature	1.51	1	1.51	0.081	0.7809
Residual	243.87	13	18.76		
Lack of fit	*195.12*	*10*	*19.51*	*1.20*	*0.4942*
Pure error	*48.75*	*3*	*16.25*		
Cor total	5781.20	19			

In Example 6.6, we concluded that there was no indication of quadratic effects; that is, a first-order model in *A*, *C*, *D*, along with the *AC*, and *AD* interaction is appropriate. However, there will be situations where the quadratic terms (x_i^2) will be required. To illustrate for the case of $k = 2$ design factors, suppose that the curvature test is significant so that we will now have to assume a second-order model such as

$$y = \beta_0 + \beta_1 x_1 + \beta_2 x_2 + \beta_{12} x_1 x_2 + \beta_{11} x_1^2 + \beta_{22} x_2^2 + \epsilon$$

Unfortunately, we cannot estimate the unknown parameters (the β's) in this model because there are six parameters to estimate and the 2^2 design and center points in Figure 6.38 have only five independent runs.

A simple and highly effective solution to this problem is to augment the 2^k design with four **axial runs**, as shown in Figure 6.39*a* for the case of $k = 2$. The resulting design, called a **central composite design**, can now be used to fit the second-order model. Figure 6.39*b* shows a central composite design for $k = 3$ factors. This design has $14 + n_C$ runs (usually $3 \leq n_C \leq 5$) and is a very efficient design for fitting the 10-parameter second-order model in $k = 3$ factors.

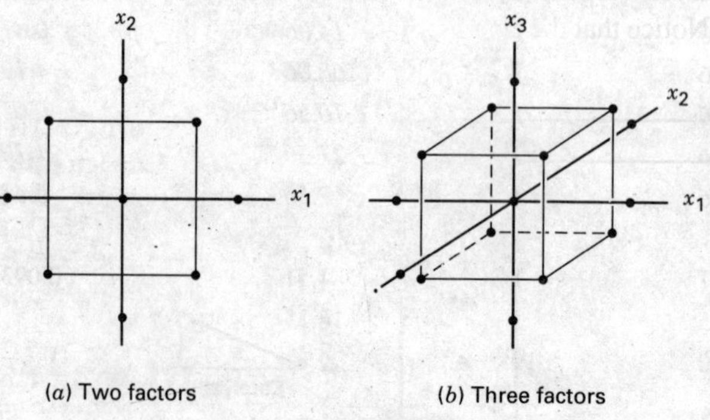

■ **FIGURE 6.39** **Central composite designs**

Central composite designs are used extensively in building second-order response surface models. These designs will be discussed in more detail in Chapter 11.

We conclude this section with a few additional useful suggestions and observations concerning the use of center points.

1. When a factorial experiment is conducted in an ongoing process, consider using the current operating conditions (or recipe) as the center point in the design. This often assures the operating personnel that at least some of the runs in the experiment are going to be performed under familiar conditions, and so the results obtained (at least for these runs) are unlikely to be any worse than are typically obtained.
2. When the center point in a factorial experiment corresponds to the usual operating recipe, the experimenter can use the observed responses at the center point to provide a rough check of whether anything "unusual" occurred during the experiment. That is, the center point responses should be very similar to the responses observed historically in routine process operation. Often operating personnel will maintain a control chart for monitoring process performance. Sometimes the center point responses can be plotted directly on the control chart as a check of the manner in which the process was operating during the experiment.
3. Consider running the replicates at the center point in nonrandom order. Specifically, run one or two center points at or near the beginning of the experiment, one or two near the middle, and one or two near the end. By spreading the center points out in time, the experimenter has a rough check on the stability of the process during the experiment. For example, if a trend has occurred in the response while the experiment was performed, plotting the center point responses versus time order may reveal this.
4. Sometimes experiments must be conducted in situations where there is little or no prior information about process variability. In these cases, running two or three center points as the first few runs in the experiment can be very helpful. These runs can provide a preliminary estimate of variability. If the magnitude of the variability seems reasonable, continue; on the contrary, if larger than anticipated (or reasonable!) variability is observed, stop. Often it will be very profitable to study the question of why the variability is so large before proceeding with the rest of the experiment.
5. Usually, center points are employed when all design factors are **quantitative**. However, sometimes there will be one or more qualitative or categorical variables and several quantitative ones. Center points can still be employed in these cases. To illustrate, consider an experiment with two quantitative factors, time and temperature, each at two levels, and a single qualitative factor, catalyst type, also with two levels (organic and nonorganic). Figure 6.40 shows the 2^3 design for these factors. Notice that the center points are placed in the opposed faces of the cube that involve

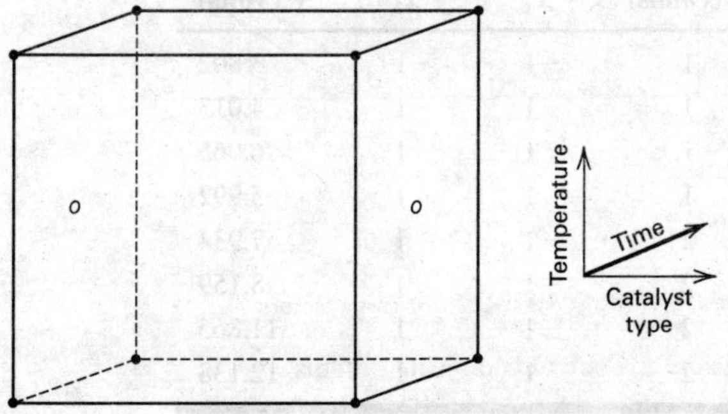

■ FIGURE 6.40 A 2^3 factorial design with one qualitative factor and center points

the quantitative factors. In other words, the center points can be run at the high- and low-level treatment combinations of the qualitative factors as long as those subspaces involve only quantitative factors.

6.9 Why We Work with Coded Design Variables

The reader will have noticed that we have performed all of the analysis and model fitting for a 2^k factorial design in this chapter using **coded** design variables, $-1 \leq x_i \leq +1$, and not the design factors in their **original** units (sometimes called actual, natural, or **engineering** units). When the engineering units are used, we can obtain different numerical results in comparison to the coded unit analysis, and often the results will not be as easy to interpret.

To illustrate some of the differences between the two analyses, consider the following experiment. A simple DC-circuit is constructed in which two different resistors, 1 and 2Ω, can be connected. The circuit also contains an ammeter and a variable-output power supply. With a resistor installed in the circuit, the power supply is adjusted until a current flow of either 4 or 6 amps is obtained. Then the voltage output of the power supply is read from a voltmeter. Two replicates of a 2^2 factorial design are performed, and Table 6.25 presents the results. We know that Ohm's law determines the observed voltage, apart from measurement error. However, the analysis of these data via empirical modeling lends some insight into the value of coded units and the engineering units in designed experiments.

Table 6.26 and 6.27 present the regression models obtained using the design variables in the usual coded variables (x_1 and x_2) and the engineering units, respectively. Minitab was used to perform the calculations. Consider first the coded variable analysis in Table 6.26. The design is orthogonal and the coded variables are also orthogonal. Notice that both main effects (x_1 = current) and (x_2 = resistance) are significant as is the interaction. In the coded variable analysis, the magnitudes of the model coefficients are directly comparable; that is, they all are dimensionless, and they measure the effect of changing each design factor over a one-unit interval. Furthermore, they are all estimated with the same precision (notice that the standard error of all three coefficients is 0.053). The interaction effect is smaller than either main effect, and the effect of current is just slightly more than one-half the resistance effect. This suggests that over the range of the factors studied, resistance is a more important variable. Coded variables are very effective for determining the **relative size** of factor effects.

Now consider the analysis based on the engineering units, as shown in Table 6.27. In this model, only the interaction is significant. The model coefficient for the interaction term

■ **TABLE 6.25**
The Circuit Experiment

I (Amps)	*R* (Ohms)	x_1	x_2	V (Volts)
4	1	−1	−1	3.802
4	1	−1	−1	4.013
6	1	1	−1	6.065
6	1	1	−1	5.992
4	2	−1	1	7.934
4	2	−1	1	8.159
6	2	1	1	11.865
6	2	1	1	12.138

■ **TABLE 6.26**
Regression Analysis for the Circuit Experiment Using Coded Variables

```
The regression equation is
V = 7.50 + 1.52 × 1 + 2.53 × 2 + 0.458 × 1 × 2

Predictor      Coef        StDev        T         P
Constant     7.49600     0.05229     143.35     0.000
x1           1.51900     0.05229      29.05     0.000
x2           2.52800     0.05229      48.34     0.000
x1x2         0.45850     0.05229       8.77     0.001

S = 0.1479     R-Sq = 99.9%     R-Sq(adj) = 99.8%

Analysis of Variance

Source            DF      SS        MS          F        P
Regression         3    71.267    23.756    1085.95    0.000
Residual Error     4     0.088     0.022
Total              7    71.354
```

■ **TABLE 6.27**
Regression Analysis for the Circuit Experiment Using Engineering Units

```
The regression equation is
V = -0.806 + 0.144 I  +  0.471 R  +  0.917  IR

Predictor      Coef        StDev       T         P
Constant     -0.8055      0.8432     -0.96     0.394
I             0.1435      0.1654      0.87     0.434
R             0.4710      0.5333      0.88     0.427
IR            0.9170      0.1046      8.77     0.001

S  =  0.1479    R-Sq  = 99.9%    R-Sq(adj)  =  99.8%

Analysis of Variance

Source            DF      SS        MS          F        P
Regression         3    71.267    23.756    1085.95    0.000
Residual Error     4     0.088     0.022
Total              7    71.354
```

is 0.9170, and the standard error is 0.1046. We can construct a t statistic for testing the hypothesis that the interaction coefficient is unity:

$$t_0 = \frac{\hat{\beta}_{IR} - 1}{se(\hat{\beta}_{IR})} = \frac{0.9170 - 1}{0.1046} = -0.7935$$

The P-value for this test statistic is $P = 0.76$. Therefore, we cannot reject the null hypothesis that the coefficient is unity, which is consistent with Ohm's law. Note that the regression coefficients are not dimensionless and that they are estimated with differing precision. This is because the experimental design, with the factors in the engineering units, is not orthogonal.

Because the intercept and the main effects are not significant, we could consider fitting a model containing only the interaction term IR. The results are shown in Table 6.28. Notice

■ **TABLE 6.28**
Regression Analysis for the Circuit Experiment (Interaction Term Only)

```
The regression equation is
V = 1.00 IR

Predictor         Coef        Std. Dev.          T        P
Noconstant
IR             1.00073          0.00550     181.81    0.000

S = 0.1255
Analysis of Variance

Source            DF        SS          MS          F        P
Regression         3    71.267      23.756    1085.95    0.000
Residual Error     4     0.088       0.022
Total              7    71.354
```

that the estimate of the interaction term regression coefficient is now different from what it was in the previous engineering-units analysis because the design in engineering units is not orthogonal. The coefficient is also virtually unity.

Generally, the engineering units are not directly comparable, but they may have physical meaning as in the present example. This could lead to possible simplification based on the underlying mechanism. In almost all situations, the coded unit analysis is preferable. It is fairly unusual for a simplification based on some underlying mechanism (as in our example) to occur. The fact that coded variables let an experimenter see the relative importance of the design factors is useful in practice.

6.10 Problems

6.1. A router is used to cut locating notches on a printed circuit board. The vibration level at the surface of the board as it is cut is considered to be a major source of dimensional variation in the notches. Two factors are thought to influence vibration: bit size (A) and cutting speed (B). Two bit sizes ($\frac{1}{16}$ and $\frac{1}{8}$ in.) and two speeds (40 and 90 rpm) are selected, and four boards are cut at each set of conditions shown below. The response variable is vibration measured as the resultant vector of three accelerometers (x, y, and z) on each test circuit board.

			Replicate			
A	*B*	Treatment Combination	I	II	III	IV
−	−	(1)	18.2	18.9	12.9	14.4
+	−	*a*	27.2	24.0	22.4	22.5
−	+	*b*	15.9	14.5	15.1	14.2
+	+	*ab*	41.0	43.9	36.3	39.9

(a) Analyze the data from this experiment.

(b) Construct a normal probability plot of the residuals, and plot the residuals versus the predicted vibration level. Interpret these plots.

(c) Draw the *AB* interaction plot. Interpret this plot. What levels of bit size and speed would you recommend for routine operation?

6.2. An engineer is interested in the effects of cutting speed (A), tool geometry (B), and cutting angle (C) on the life (in hours) of a machine tool. Two levels of each factor are chosen, and three replicates of a 2^3 factorial design are run. The results are as follows:

				Replicate		
A	*B*	*C*	Treatment Combination	I	II	III
−	−	−	(1)	21	31	25
+	−	−	*a*	33	43	29
−	+	−	*b*	34	34	50
+	+	−	*ab*	56	47	46
−	−	+	*c*	43	45	38
+	−	+	*ac*	41	37	36
−	+	+	*bc*	61	50	54
+	+	+	*abc*	40	41	47

(a) Estimate the factor effects. Which effects appear to be large?

(b) Use the analysis of variance to confirm your conclusions for part (a).

(c) Write down a regression model for predicting tool life (in hours) based on the results of this experiment.

(d) Analyze the residuals. Are there any obvious problems?

(e) On the basis of an analysis of main effect and interaction plots, what coded factor levels of *A*, *B*, and *C* would you recommend using?

6.3. Reconsider part (c) of Problem 6.2. Use the regression model to generate response surface and contour plots of the tool life response. Interpret these plots. Do they provide insight regarding the desirable operating conditions for this process?

6.4. Find the standard error of the factor effects and approximate 95 percent confidence limits for the factor effects in Problem 6.2. Do the results of this analysis agree with the conclusions from the analysis of variance?

6.5. Plot the factor effects from Problem 6.2 on a graph relative to an appropriately scaled *t* distribution. Does this graphical display adequately identify the important factors? Compare the conclusions from this plot with the results from the analysis of variance.

6.6. Reconsider the experiment described in Problem 6.2. Suppose that the experimenter only performed the eight trials from replicate I. In addition, he ran four center points and obtained the following response values: 36, 40, 43, 45.

(a) Estimate the factor effects. Which effects are large?

(b) Perform an analysis of variance, including a check for pure quadratic curvature. What are your conclusions?

(c) Write down an appropriate model for predicting tool life, based on the results of this experiment. Does this model differ in any substantial way from the model in Problem 6.2, part (c)?

(d) Analyze the residuals.

(e) What conclusions would you draw about the appropriate operating conditions for this process?

6.7. An experiment was performed to improve the yield of a chemical process. Four factors were selected, and two replicates of a completely randomized experiment were run. The results are shown in the following table:

Treatment Combination	Replicate I	Replicate II	Treatment Combination	Replicate I	Replicate II
(1)	91	93	*d*	99	95
a	75	78	*ad*	73	76
b	80	85	*bd*	86	83
ab	84	80	*abd*	84	86
c	76	78	*cd*	98	90
ac	81	80	*acd*	80	75
bc	89	82	*bcd*	86	84
abc	74	70	*abcd*	81	80

(a) Estimate the factor effects.

(b) Prepare an analysis of variance table and determine which factors are important in explaining yield.

(c) Write down a regression model for predicting yield, assuming that all four factors were varied over the range from −1 to +1 (in coded units).

(d) Plot the residuals versus the predicted yield and on a normal probability scale. Does the residual analysis appear satisfactory?

(e) Two three-factor interactions, *ABC* and *ABD*, apparently have large effects. Draw a cube plot in the factors *A*, *B*, and *C* with the average yields shown at each corner. Repeat using the factors *A*, *B*, and *D*. Do these two plots aid in data interpretation? Where would you recommend that the process be run with respect to the four variables?

6.8. An industrial engineer employed by a beverage bottler is interested in the effects of two different types of 32-ounce bottles on the time to deliver 12-bottle cases of the product. The two bottle types are glass and plastic. Two workers are used to perform a task consisting of moving 40 cases of the product 50 feet on a standard type of hand truck and stacking the cases in a display. Four replicates of a 2^2 factorial design are performed, and the times observed are listed in the following table. Analyze the data and draw appropriate conclusions. Analyze the residuals and comment on the model's adequacy.

Bottle Type	Worker 1		Worker 2	
Glass	5.12	4.89	6.65	6.24
	4.98	5.00	5.49	5.55
Plastic	4.95	4.43	5.28	4.91
	4.27	4.25	4.75	4.71

6.9. A bacteriologist is interested in the effects of two different culture media and two different times on the growth of a particular virus. He or she performs six replicates of a 2^2 design, making the runs in random order. Analyze the bacterial growth data that follow and draw appropriate conclusions. Analyze the residuals and comment on the model's adequacy.

Time (h)	Culture Medium 1		Culture Medium 2	
	21	22	25	26
12	23	28	24	25
	20	26	29	27
	37	39	31	34
18	38	38	29	33
	35	36	30	35

6.10. In Problem 6.8, the engineer was also interested in potential fatigue differences resulting from the two types of bottles. As a measure of the amount of effort required, he measured the elevation of the heart rate (pulse) induced by the task. The results follow. Analyze the data and draw conclusions. Analyze the residuals and comment on the model's adequacy.

Bottle Type	Worker 1		Worker 2	
Glass	40	46	20	13
	58	35	16	11
Plastic	44	35	13	10
	42	21	16	15

6.11. Calculate approximate 95 percent confidence limits for the factor effects in Problem 6.10. Do the results of this analysis agree with the analysis of variance performed in Problem 6.10?

6.12. An article in the *AT&T Technical Journal* (March/April 1986, Vol. 65, pp. 39–50) describes the application of two-level factorial designs to integrated circuit manufacturing. A basic processing step is to grow an epitaxial layer on polished silicon wafers. The wafers mounted on a susceptor are positioned inside a bell jar, and chemical vapors are introduced. The susceptor is rotated, and heat is applied until the epitaxial layer is thick enough. An experiment was run using two factors: arsenic flow rate (A) and deposition time (B). Four replicates were run, and the epitaxial layer thickness was measured (μm). The data are shown in Table P6.1.

■ **TABLE P6.1**
The 2^2 Design for Problem 6.12

A	*B*	Replicate I	II	III	IV		Factor Levels Low (−)	High (+)
−	−	14.037	16.165	13.972	13.907	*A*	55%	59%
+	−	13.880	13.860	14.032	13.914			
−	+	14.821	14.757	14.843	14.878	*B*	Short (10 min)	Long (15 min)
+	+	14.888	14.921	14.415	14.932			

(a) Estimate the factor effects.

(b) Conduct an analysis of variance. Which factors are important?

(c) Write down a regression equation that could be used to predict epitaxial layer thickness over the region of arsenic flow rate and deposition time used in this experiment.

(d) Analyze the residuals. Are there any residuals that should cause concern?

(e) Discuss how you might deal with the potential outlier found in part (d).

6.13. ***Continuation of Problem 6.12.*** Use the regression model in part (c) of Problem 6.12 to generate a response surface contour plot for epitaxial layer thickness. Suppose it is critically important to obtain layer thickness of 14.5μm. What settings of arsenic flow rate and decomposition time would you recommend?

6.14. ***Continuation of Problem 6.13.*** How would your answer to Problem 6.13 change if arsenic flow rate was more difficult to control in the process than the deposition time?

6.15. A nickel–titanium alloy is used to make components for jet turbine aircraft engines. Cracking is a potentially serious problem in the final part because it can lead to nonrecoverable failure. A test is run at the parts producer to determine the effect of four factors on cracks. The four factors are pouring temperature (A), titanium content (B), heat treatment method (C), and amount of grain refiner used (D). Two replicates of a 2^4 design are run, and the length of crack (in mm $\times 10^{-2}$) induced in a sample coupon subjected to a standard test is measured. The data are shown in Table P6.2.

■ **TABLE P6.2**
The Experiment for Problem 6.15

A	*B*	*C*	*D*	Treatment Combination	Replicate I	II
−	−	−	−	(1)	7.037	6.376
+	−	−	−	*a*	14.707	15.219
−	+	−	−	*b*	11.635	12.089
+	+	−	−	*ab*	17.273	17.815
−	−	+	−	*c*	10.403	10.151
+	−	+	−	*ac*	4.368	4.098
−	+	+	−	*bc*	9.360	9.253
+	+	+	−	*abc*	13.440	12.923
−	−	−	+	*d*	8.561	8.951
+	−	−	+	*ad*	16.867	17.052
−	+	−	+	*bd*	13.876	13.658
+	+	−	+	*abd*	19.824	19.639
−	−	+	+	*cd*	11.846	12.337
+	−	+	+	*acd*	6.125	5.904
−	+	+	+	*bcd*	11.190	10.935
+	+	+	+	*abcd*	15.653	15.053

(a) Estimate the factor effects. Which factor effects appear to be large?

(b) Conduct an analysis of variance. Do any of the factors affect cracking? Use $\alpha = 0.05$.

(c) Write down a regression model that can be used to predict crack length as a function of the significant main effects and interactions you have identified in part (b).

(d) Analyze the residuals from this experiment.

(e) Is there an indication that any of the factors affect the variability in cracking?

(f) What recommendations would you make regarding process operations? Use interaction and/or main effect plots to assist in drawing conclusions.

6.16. ***Continuation of Problem 6.15.*** One of the variables in the experiment described in Problem 6.15, heat treatment method (C), is a categorical variable. Assume that the remaining factors are continuous.

(a) Write two regression models for predicting crack length, one for each level of the heat treatment method variable. What differences, if any, do you notice in these two equations?

(b) Generate appropriate response surface contour plots for the two regression models in part (a).

(c) What set of conditions would you recommend for the factors A, B, and D if you use heat treatment method $C = +$?

(d) Repeat part (c) assuming that you wish to use heat treatment method $C = -$.

6.17. An experimenter has run a single replicate of a 2^4 design. The following effect estimates have been calculated:

$A = 79.25$	$AB = -51.32$	$ABC = -2.82$
$B = -67.52$	$AC = 11.69$	$ABD = -3.50$
$C = -7.84$	$AD = 9.78$	$ACD = 11.05$
$D = -18.73$	$BC = 20.78$	$BCD = -7.98$
	$BD = 14.74$	$ABCD = -6.25$
	$CD = 1.27$	

(a) Construct a normal probability plot of these effects.

(b) Identify a tentative model, based on the plot of the effects in part (a).

6.18. The effect estimates from a 2^4 factorial design are as follows: ABCD = −1.5000, ABC = −1.2500, ABD = −0.9852, ACD = −0.7500, BCD = −0.4842, CD = −0.0795, BD = −0.0793, AD = 0.5988, BC = 0.9216, AC = 1.1616, AB = 1.3266, D = 4.6744, C = 5.1458, B = 8.2500, and A = 12.7500. Are you comfortable with the conclusions that all main effects are active?

6.19. The effect estimates from a 2^4 factorial experiment are listed here. Are any of the effects significant? ABCD = −2.5251, BCD = 4.4054, ACD = −0.4932, ABD = −5.0842, ABC = −5.7696, CD = 4.6707, BD = −4.6620, BC = −0.7982, AD = −1.6564, AC = 1.1109, AB = −10.5229, D = −6.0275, C = −8.2045, B = −6.5304, and A = −0.7914.

6.20. Consider a variation of the bottle filling experiment from Example 5.3. Suppose that only two levels of carbonation are used so that the experiment is a 2^3 factorial design with two replicates. The data are shown in Table P6.3.

■ TABLE P6.3
Fill Height Experiment from Problem 6.20

	Coded Factors			Fill Height Deviation			Factor Levels	
Run	**A**	**B**	**C**	**Replicate 1**	**Replicate 2**		**Low (−1)**	**High (+1)**
1	−	−	−	−3	−1	A (%)	10	12
2	+	−	−	0	1	B (psi)	25	30
3	−	+	−	−1	0	C (b/m)	200	250
4	+	+	−	2	3			
5	−	−	+	−1	0			
6	+	−	+	2	1			
7	−	+	+	1	1			
8	+	+	+	6	5			

(a) Analyze the data from this experiment. Which factors significantly affect fill height deviation?

(b) Analyze the residuals from this experiment. Are there any indications of model inadequacy?

(c) Obtain a model for predicting fill height deviation in terms of the important process variables. Use this model to construct contour plots to assist in interpreting the results of the experiment.

(d) In part (a), you probably noticed that there was an interaction term that was borderline significant. If you did not include the interaction term in your model, include it now and repeat the analysis. What difference did this make? If you elected to include the interaction term in part (a), remove it and repeat the analysis. What difference does the interaction term make?

6.21. I am always interested in improving my golf scores. Since a typical golfer uses the putter for about 35–45 percent of his or her strokes, it seems reasonable that improving one's putting is a logical and perhaps simple way to improve a golf score ("The man who can putt is a match for any man."—Willie Parks, 1864–1925, two time winner of the British Open). An experiment was conducted to study the effects of four factors on putting accuracy. The design factors are length of putt, type of putter, breaking putt versus straight putt, and level versus downhill putt. The response variable is distance from the ball to the center of the cup after the ball comes to rest. One golfer performs the experiment, a 2^4 factorial design with seven replicates was used, and all putts are made in random order. The results are shown in Table P6.4.

■ **TABLE P6.4**
The Putting Experiment from Problem 6.21

Design Factors				Distance from Cup (replicates)						
Length of putt (ft)	Type of putter	Break of putt	Slope of putt	1	2	3	4	5	6	7
10	Mallet	Straight	Level	10.0	18.0	14.0	12.5	19.0	16.0	18.5
30	Mallet	Straight	Level	0.0	16.5	4.5	17.5	20.5	17.5	33.0
10	Cavity back	Straight	Level	4.0	6.0	1.0	14.5	12.0	14.0	5.0
30	Cavity back	Straight	Level	0.0	10.0	34.0	11.0	25.5	21.5	0.0
10	Mallet	Breaking	Level	0.0	0.0	18.5	19.5	16.0	15.0	11.0
30	Mallet	Breaking	Level	5.0	20.5	18.0	20.0	29.5	19.0	10.0
10	Cavity back	Breaking	Level	6.5	18.5	7.5	6.0	0.0	10.0	0.0
30	Cavity back	Breaking	Level	16.5	4.5	0.0	23.5	8.0	8.0	8.0
10	Mallet	Straight	Downhill	4.5	18.0	14.5	10.0	0.0	17.5	6.0
30	Mallet	Straight	Downhill	19.5	18.0	16.0	5.5	10.0	7.0	36.0
10	Cavity back	Straight	Downhill	15.0	16.0	8.5	0.0	0.5	9.0	3.0
30	Cavity back	Straight	Downhill	41.5	39.0	6.5	3.5	7.0	8.5	36.0
10	Mallet	Breaking	Downhill	8.0	4.5	6.5	10.0	13.0	41.0	14.0
30	Mallet	Breaking	Downhill	21.5	10.5	6.5	0.0	15.5	24.0	16.0
10	Cavity back	Breaking	Downhill	0.0	0.0	0.0	4.5	1.0	4.0	6.5
30	Cavity back	Breaking	Downhill	18.0	5.0	7.0	10.0	32.5	18.5	8.0

(a) Analyze the data from this experiment. Which factors significantly affect putting performance?

(b) Analyze the residuals from this experiment. Are there any indications of model inadequacy?

6.22. Semiconductor manufacturing processes have long and complex assembly flows, so matrix marks and automated 2d-matrix readers are used at several process steps throughout factories. Unreadable matrix marks negatively affect factory run rates because manual entry of part data is required before manufacturing can resume. A 2^4 factorial experiment was conducted to develop a 2d-matrix laser mark on a metal cover that protects a substrate-mounted die. The design factors are A = laser power (9 and 13 W), B = laser pulse frequency (4000 and 12,000 Hz), C = matrix cell size (0.07 and 0.12 in.), and D = writing speed (10 and 20 in./sec), and the response variable is the unused error correction (UEC). This is a measure of the unused portion of the redundant information embedded in the 2d-matrix. A UEC of 0 represents the lowest reading that still results in a decodable matrix, while a value of 1 is the highest reading. A DMX Verifier was used to measure UEC. The data from this experiment are shown in Table P6.5.

(a) Analyze the data from this experiment. Which factors significantly affect UEC?

(b) Analyze the residuals from this experiment. Are there any indications of model inadequacy?

6.23. Reconsider the experiment described in Problem 6.20. Suppose that four center points are available and that the UEC response at these four runs is 0.98, 0.95, 0.93, and 0.96, respectively. Reanalyze the experiment incorporating a test for curvature into the analysis. What conclusions can you draw? What recommendations would you make to the experimenters?

6.24. A company markets its products by direct mail. An experiment was conducted to study the effects of three factors on the customer response rate for a particular product. The three factors are A = type of mail used (3rd class, 1st class), B = type of descriptive brochure (color, black-and-white), and C = offered price ($19.95, $24.95). The mailings are made to two groups of 8000 randomly selected customers, with 1000 customers in each group receiving each treatment combination. Each group of customers is considered as a replicate. The response variable is the number of orders placed. The experimental data are shown in Table P6.6.

■ **TABLE P6.5**
The 2^4 Experiment for Problem 6.22

Standard Order	Run Order	Laser Power	Pulse Frequency	Cell Size	Writing Speed	UEC
8	1	1.00	1.00	1.00	−1.00	0.8
10	2	1.00	−1.00	−1.00	1.00	0.81
12	3	1.00	1.00	−1.00	1.00	0.79
9	4	−1.00	−1.00	−1.00	1.00	0.6
7	5	−1.00	1.00	1.00	−1.00	0.65
15	6	−1.00	1.00	1.00	1.00	0.55
2	7	1.00	−1.00	−1.00	−1.00	0.98
6	8	1.00	−1.00	1.00	−1.00	0.67
16	9	1.00	1.00	1.00	1.00	0.69
13	10	−1.00	−1.00	1.00	1.00	0.56
5	11	−1.00	−1.00	1.00	−1.00	0.63
14	12	1.00	−1.00	1.00	1.00	0.65
1	13	−1.00	−1.00	−1.00	−1.00	0.75
3	14	−1.00	1.00	−1.00	−1.00	0.72
4	15	1.00	1.00	−1.00	−1.00	0.98
11	16	−1.00	1.00	−1.00	1.00	0.63

■ **TABLE P6.6**
The Direct Mail Experiment from Problem 6.24

	Coded Factors			Number of Orders			Factor Levels	
Run	A	B	C	Replicate 1	Replicate 2		Low (−1)	High (+1)
1	−	−	−	52	56	*A* (class)	3rd	1st
2	+	−	−	43	41	*B* (type)	BW	Color
3	−	+	−	47	48	*C* ($)	$19.95	$24.95
4	+	+	−	42	43			
5	−	−	+	49	48			
6	+	−	+	49	45			
7	−	+	+	47	48			
8	+	+	+	58	53			

(a) Analyze the data from this experiment. Which factors significantly affect the customer response rate?

(b) Analyze the residuals from this experiment. Are there any indications of model inadequacy?

(c) What would you recommend to the company?

6.25. An experiment was run in a semiconductor fabrication plant in an effort to increase yield. Five factors, each at two levels, were studied. The factors (and levels) were A = aperture setting (small, large), B = exposure time (20% below nominal, 20% above nominal), C = development time (30 and 45 s), D = mask dimension (small, large), and E = etch time (14.5 and 15.5 min). The unreplicated 2^5 design shown below was run.

(1) = 7	$d = 8$	$e = 8$	$de = 6$
$a = 9$	$ad = 10$	$ae = 12$	$ade = 10$
$b = 34$	$bd = 32$	$be = 35$	$bde = 30$
$ab = 55$	$abd = 50$	$abe = 52$	$abde = 53$
$c = 16$	$cd = 18$	$ce = 15$	$cde = 15$
$ac = 20$	$acd = 21$	$ace = 22$	$acde = 20$
$bc = 40$	$bcd = 44$	$bce = 45$	$bcde = 41$
$abc = 60$	$abcd = 61$	$abce = 65$	$abcde = 63$

(a) Construct a normal probability plot of the effect estimates. Which effects appear to be large?

(b) Conduct an analysis of variance to confirm your findings for part (a).

(c) Write down the regression model relating yield to the significant process variables.

(d) Plot the residuals on normal probability paper. Is the plot satisfactory?

(e) Plot the residuals versus the predicted yields and versus each of the five factors. Comment on the plots.

(f) Interpret any significant interactions.

(g) What are your recommendations regarding process operating conditions?

(h) Project the 2^5 design in this problem into a 2^k design in the important factors. Sketch the design and show the average and range of yields at each run. Does this sketch aid in interpreting the results of this experiment?

6.26. ***Continuation of Problem 6.25.*** Suppose that the experimenter had run four center points in addition to the 32 trials in the original experiment. The yields obtained at the center point runs were 68, 74, 76, and 70.

(a) Reanalyze the experiment, including a test for pure quadratic curvature.

(b) Discuss what your next step would be.

6.27. Consider the single replicate of the 2^4 design in Example 6.2. Suppose that we had arbitrarily decided to analyze the data assuming that all three- and four-factor interactions were negligible. Conduct this analysis and compare your results with those obtained in the example. Do you think that it is a good idea to arbitrarily assume interactions to be negligible even if they are relatively high-order ones?

6.28. In a process development study on yield, four factors were studied, each at two levels: time (*A*), concentration (*B*), pressure (*C*), and temperature (*D*). A single replicate of a 2^4 design was run, and the resulting data are shown in Table P6.7.

■ **TABLE P6.7**
Process Development Experiment from Problem 6.28

Run Number	Actual Run Order	*A*	*B*	*C*	*D*	Yield (lbs)	Factor Levels	Low (−)	High (+)
1	5	−	−	−	−	11	*A* (h)	2.5	3
2	9	+	−	−	−	18	*B* (%)	14	18
3	8	−	+	−	−	12	*C* (psi)	60	80
4	13	+	+	−	−	16	*D* (°C)	225	250
5	3	−	−	+	−	17			
6	7	+	−	+	−	15			
7	14	−	+	+	−	22			
8	1	+	+	+	−	15			
9	6	−	−	−	+	9			
10	11	+	−	−	+	27			
11	2	−	+	−	+	13			
12	15	+	+	−	+	25			
13	4	−	−	+	+	19			
14	16	+	−	+	+	22			
15	10	−	+	+	+	17			
16	12	+	+	+	+	24			

(a) Construct a normal probability plot of the effect estimates. Which factors appear to have large effects?

(b) Conduct an analysis of variance using the normal probability plot in part (a) for guidance in forming an error term. What are your conclusions?

(c) Write down a regression model relating yield to the important process variables.

(d) Analyze the residuals from this experiment. Does your analysis indicate any potential problems?

(e) Can this design be collapsed into a 2^3 design with two replicates? If so, sketch the design with the average and range of yield shown at each point in the cube. Interpret the results.

6.29. ***Continuation of Problem 6.28.*** Use the regression model in part (c) of Problem 6.28 to generate a response surface contour plot of yield. Discuss the practical value of this response surface plot.

6.30. ***The scrumptious brownie experiment.*** The author is an engineer by training and a firm believer in learning by doing. I have taught experimental design for many years to a wide variety of audiences and have always assigned the planning, conduct, and analysis of an actual experiment to the class participants. The participants seem to enjoy this practical experience and always learn a great deal from it. This problem uses the results of an experiment performed by Gretchen Krueger at Arizona State University.

There are many different ways to bake brownies. The purpose of this experiment was to determine how the pan material, the brand of brownie mix, and the stirring method affect the scrumptiousness of brownies. The factor levels were

Factor	Low (−)	High (+)
A = pan material	Glass	Aluminum
B = stirring method	Spoon	Mixer
C = brand of mix	Expensive	Cheap

The response variable was scrumptiousness, a subjective measure derived from a questionnaire given to the subjects who sampled each batch of brownies. (The questionnaire dealt with such issues as taste, appearance, consistency, aroma, and so forth.) An eight-person test panel sampled each batch and filled out the questionnaire. The design matrix and the response data are as follows.

(a) Analyze the data from this experiment as if there were eight replicates of a 2^3 design. Comment on the results.

(b) Is the analysis in part (a) the correct approach? There are only eight batches; do we really have eight replicates of a 2^3 factorial design?

(c) Analyze the average and standard deviation of the scrumptiousness ratings. Comment on the results. Is this analysis more appropriate than the one in part (a)? Why or why not?

Brownie				Test Panel Results							
Batch	A	B	C	1	2	3	4	5	6	7	8
1	−	−	−	11	9	10	10	11	10	8	9
2	+	−	−	15	10	16	14	12	9	6	15
3	−	+	−	9	12	11	11	11	11	11	12
4	+	+	−	16	17	15	12	13	13	11	11
5	−	−	+	10	11	15	8	6	8	9	14
6	+	−	+	12	13	14	13	9	13	14	9
7	−	+	+	10	12	13	10	7	7	17	13
9	+	+	+	15	12	15	13	12	12	9	14

6.31. An experiment was conducted on a chemical process that produces a polymer. The four factors studied were temperature (A), catalyst concentration (B), time (C), and pressure (D). Two responses, molecular weight and viscosity, were observed. The design matrix and response data are shown in Table P6.8.

■ **TABLE P6.8**

The 2^4 Experiment for Problem 6.31

Run Number	Actual Run Order	A	B	C	D	Molecular Weight	Viscosity	Factor Levels	Low (−)	High (+)
1	18	−	−	−	−	2400	1400	A (°C)	100	120
2	9	+	−	−	−	2415	1500	B (%)	4	8
3	13	−	+	−	−	2325	1520	C (min)	20	30
4	8	+	+	−	−	2520	1630	D (psi)	60	75
5	3	−	−	+	−	2625	1380			
6	11	+	−	+	−	2630	1525			
7	14	−	+	+	−	2400	1500			
8	17	+	+	+	−	2775	1620			
9	6	−	−	−	+	2400	1400			
10	7	+	−	−	+	2390	1525			
11	2	−	+	−	+	2300	1500			
12	10	+	+	−	+	2520	1500			
13	4	−	−	+	+	2650	1420			
14	19	+	−	+	+	2635	1490			
15	15	−	+	+	+	2500	1500			
16	20	+	+	+	+	2720	1600			
17	1	0	0	0	0	2515	1500			
18	5	0	0	0	0	2500	1460			
19	16	0	0	0	0	2400	1525			
20	12	0	0	0	0	2475	1500			

(a) Consider only the molecular weight response. Plot the effect estimates on a normal probability scale. What effects appear important?

(b) Use an analysis of variance to confirm the results from part (a). Is there indication of curvature?

(c) Write down a regression model to predict molecular weight as a function of the important variables.

(d) Analyze the residuals and comment on model adequacy.

(e) Repeat parts (a)–(d) using the viscosity response.

6.32. ***Continuation of Problem 6.31.*** Use the regression models for molecular weight and viscosity to answer the following questions.

(a) Construct a response surface contour plot for molecular weight. In what direction would you adjust the process variables to increase molecular weight?

(b) Construct a response surface contour plot for viscosity. In what direction would you adjust the process variables to decrease viscosity?

(c) What operating conditions would you recommend if it was necessary to produce a product with molecular weight between 2400 and 2500 and the lowest possible viscosity?

6.33. Consider the single replicate of the 2^4 design in Example 6.2. Suppose that we ran five points at the center (0, 0, 0, 0) and observed the responses 93, 95, 91, 89, and 96. Test for curvature in this experiment. Interpret the results.

6.34. ***A missing value in a 2^k factorial.*** It is not unusual to find that one of the observations in a 2^k design is missing due to faulty measuring equipment, a spoiled test, or some other reason. If the design is replicated n times ($n > 1$), some of the techniques discussed in Chapter 5 can be employed. However, for an unreplicated factorial ($n = 1$) some other method must be used. One logical approach is to estimate the missing value with a number that makes the highest order interaction contrast zero. Apply this technique to the experiment in Example 6.2 assuming that run ab is missing. Compare the results with the results of Example 6.2.

6.35. An engineer has performed an experiment to study the effect of four factors on the surface roughness of a machined part. The factors (and their levels) are A = tool angle (12, 15°), B = cutting fluid viscosity (300, 400), C = feed rate (10 and 15 in./min), and D = cutting fluid cooler used (no, yes). The data from this experiment (with the factors coded to the usual -1, $+1$ levels) are shown in Table P6.9.

(a) Estimate the factor effects. Plot the effect estimates on a normal probability plot and select a tentative model.

(b) Fit the model identified in part (a) and analyze the residuals. Is there any indication of model inadequacy?

(c) Repeat the analysis from parts (a) and (b) using $1/y$ as the response variable. Is there an indication that the transformation has been useful?

(d) Fit a model in terms of the coded variables that can be used to predict the surface roughness. Convert this prediction equation into a model in the natural variables.

■ **TABLE P6.9**

The Surface Roughness Experiment from Problem 6.35

Run	*A*	*B*	*C*	*D*	Surface Roughness
1	−	−	−	−	0.00340
2	+	−	−	−	0.00362
3	−	+	−	−	0.00301
4	+	+	−	−	0.00182
5	−	−	+	−	0.00280
6	+	−	+	−	0.00290
7	−	+	+	−	0.00252
8	+	+	+	−	0.00160
9	−	−	−	+	0.00336
10	+	−	−	+	0.00344
11	−	+	−	+	0.00308
12	+	+	−	+	0.00184
13	−	−	+	+	0.00269
14	+	−	+	+	0.00284
15	−	+	+	+	0.00253
16	+	+	+	+	0.00163

6.36. Resistivity on a silicon wafer is influenced by several factors. The results of a 2^4 factorial experiment performed during a critical processing step is shown in Table P6.10.

■ **TABLE P6.10**

The Resistivity Experiment from Problem 6.36

Run	*A*	*B*	*C*	*D*	Resistivity
1	−	−	−	−	1.92
2	+	−	−	−	11.28
3	−	+	−	−	1.09
4	+	+	−	−	5.75
5	−	−	+	−	2.13
6	+	−	+	−	9.53
7	−	+	+	−	1.03
8	+	+	+	−	5.35
9	−	−	−	+	1.60
10	+	−	−	+	11.73
11	−	+	−	+	1.16
12	+	+	−	+	4.68
13	−	−	+	+	2.16
14	+	−	+	+	9.11
15	−	+	+	+	1.07
16	+	+	+	+	5.30

(a) Estimate the factor effects. Plot the effect estimates on a normal probability plot and select a tentative model.

(b) Fit the model identified in part (a) and analyze the residuals. Is there any indication of model inadequacy?

(c) Repeat the analysis from parts (a) and (b) using ln (y) as the response variable. Is there an indication that the transformation has been useful?

(d) Fit a model in terms of the coded variables that can be used to predict the resistivity.

6.37. ***Continuation of Problem 6.36.*** Suppose that the experimenter had also run four center points along with the 16 runs in Problem 6.36. The resistivity measurements at the center points are 8.15, 7.63, 8.95, and 6.48. Analyze the experiment again incorporating the center points. What conclusions can you draw now?

6.38. The book by Davies (*Design and Analysis of Industrial Experiments*) describes an experiment to study the yield of isatin. The factors studied and their levels are as follows:

Factor	Low (−)	High (+)
A: Acid strength (%)	87	93
B: Reaction time (min)	15	30
C: Amount of acid (mL)	35	45
D: Reaction temperature (°C)	60	70

The data from the 2^4 factorial is shown in Table P6.11.

(a) Fit a main-effects-only model to the data from this experiment. Are any of the main effects significant?

(b) Analyze the residuals. Are there any indications of model inadequacy or violation of the assumptions?

(c) Find an equation for predicting the yield of isatin over the design space. Express the equation in both coded and engineering units.

(d) Is there any indication that adding interactions to the model would improve the results that you have obtained?

■ TABLE P6.11
The 2^4 Factorial Experiment in Problem 6.38

A	B	C	D	Yield
−1	−1	−1	−1	6.08
1	−1	−1	−1	6.04
−1	1	−1	−1	6.53
1	1	−1	−1	6.43
−1	−1	1	−1	6.31
1	−1	1	−1	6.09
−1	1	1	−1	6.12
1	1	1	−1	6.36
−1	−1	−1	1	6.79
1	−1	−1	1	6.68
−1	1	−1	1	6.73
1	1	−1	1	6.08
−1	−1	1	1	6.77
1	−1	1	1	6.38
−1	1	1	1	6.49
1	1	1	1	6.23

6.39. An article in *Quality and Reliability Engineering International* (2010, Vol. 26, pp. 223–233) presents a 2^5 factorial design. The experiment is shown in Table P6.12.

■ TABLE P6.12
The 2^5 Design in Problem 6.39

A	B	C	D	E	y
−1.00	−1.00	−1.00	−1.00	−1.00	8.11
1.00	−1.00	−1.00	−1.00	−1.00	5.56
−1.00	1.00	−1.00	−1.00	−1.00	5.77
1.00	1.00	−1.00	−1.00	−1.00	5.82
−1.00	−1.00	1.00	−1.00	−1.00	9.17
1.00	−1.00	1.00	−1.00	−1.00	7.8
−1.00	1.00	1.00	−1.00	−1.00	3.23
1.00	1.00	1.00	−1.00	−1.00	5.69
−1.00	−1.00	−1.00	1.00	−1.00	8.82
1.00	−1.00	−1.00	1.00	−1.00	14.23
−1.00	1.00	−1.00	1.00	−1.00	9.2
1.00	1.00	−1.00	1.00	−1.00	8.94
−1.00	−1.00	1.00	1.00	−1.00	8.68
1.00	−1.00	1.00	1.00	−1.00	11.49
−1.00	1.00	1.00	1.00	−1.00	6.25
1.00	1.00	1.00	1.00	−1.00	9.12
−1.00	−1.00	−1.00	−1.00	1.00	7.93
1.00	−1.00	−1.00	−1.00	1.00	5
−1.00	1.00	−1.00	−1.00	1.00	7.47
1.00	1.00	−1.00	−1.00	1.00	12
−1.00	−1.00	1.00	−1.00	1.00	9.86
1.00	−1.00	1.00	−1.00	1.00	3.65
−1.00	1.00	1.00	−1.00	1.00	6.4
1.00	1.00	1.00	−1.00	1.00	11.61
−1.00	−1.00	−1.00	1.00	1.00	12.43
1.00	−1.00	−1.00	1.00	1.00	17.55
−1.00	1.00	−1.00	1.00	1.00	8.87
1.00	1.00	−1.00	1.00	1.00	25.38
−1.00	−1.00	1.00	1.00	1.00	13.06
1.00	−1.00	1.00	1.00	1.00	18.85
−1.00	1.00	1.00	1.00	1.00	11.78
1.00	1.00	1.00	1.00	1.00	26.05

(a) Analyze the data from this experiment. Identify the significant factors and interactions.

(b) Analyze the residuals from this experiment. Are there any indications of model inadequacy or violations of the assumptions?

(c) One of the factors from this experiment does not seem to be important. If you drop this factor, what type of design remains? Analyze the data using the full factorial model for only the four active factors. Compare your results with those obtained in part (a).

(d) Find settings of the active factors that maximize the predicted response.

6.40. An article in the *Journal of Hazardous Materials* ("Feasibility of Using Natural Fishbone Apatite as a Substitute for Hydroxyapatite in Remediating Aqueous Heavy Metals," Vol. 69, Issue 2, 1999, pp. 187–196) describes an experiment to study the suitability of fishbone, a natural, apatite rich substance, as a substitute for hydroxyapatite in the sequestering of aqueous divalent heavy metal ions. Direct comparison of hydroxyapatite and fishbone apatite was performed using a three-factor two-level full factorial design. Apatite (30 or 60 mg) was added to 100 mL deionized water and gently agitated overnight in a shaker. The pH was then adjusted to 5 or 7 using nitric acid. Sufficient concentration of lead nitrate solution was added to each flask to result in a final volume of 200 mL and a lead concentration of 0.483 or 2.41 mM, respectively. The experiment was a 2^3 replicated twice and it was performed for both fishbone and synthetic apatite. Results are shown in Table P6.13.

■ TABLE P6.13

The Experiment for Problem 6.40. For apatite, + is 60 mg and − is 30 mg per 200 mL metal solution. For initial pH, + is 7 and − is 4. For Pb + is 2.41 mM (500 ppm) and − is 0.483 mM (100 ppm)

			Fishbone		Hydroxyapatite	
Apatite	**pH**	**Pb**	**Pb, mM**	**pH**	**Pb, mM**	**pH**
+	+	+	1.82	5.22	0.11	3.49
+	+	+	1.81	5.12	0.12	3.46
+	+	−	0.01	6.84	0.00	5.84
+	+	−	0.00	6.61	0.00	5.90
+	−	+	1.11	3.35	0.80	2.70
+	−	+	1.04	3.34	0.76	2.74
+	−	−	0.00	5.77	0.03	3.36
+	−	−	0.01	6.25	0.05	3.24
−	+	+	2.11	5.29	1.03	3.22
−	+	+	2.18	5.06	1.05	3.22
−	+	−	0.03	5.93	0.00	5.53
−	+	−	0.05	6.02	0.00	5.43
−	−	+	1.70	3.39	1.34	2.82
−	−	+	1.69	3.34	1.26	2.79
−	−	−	0.05	4.50	0.06	3.28
−	−	−	0.05	4.74	0.07	3.28

(a) Analyze the lead response for fishbone apatite. What factors are important?

(b) Analyze the residuals from this response and comment on model adequacy.

(c) Analyze the pH response for fishbone apatite. What factors are important?

(d) Analyze the residuals from this response and comment on model adequacy.

(e) Analyze the lead response for hydroxyapatite apatite. What factors are important?

(f) Analyze the residuals from this response and comment on model adequacy.

(g) Analyze the pH response for hydroxyapatite apatite. What factors are important?

(h) Analyze the residuals from this response and comment on model adequacy.

(i) What differences do you see between fishbone and hydroxyapatite apatite? The authors of this paper concluded that that fishbone apatite was comparable to hydroxyapatite apatite. Because the fishbone apatite is cheaper, it was recommended for adoption. Do you agree with these conclusions?

6.41. A paper in the *Journal of Chemical Technology and Biotechnology* ("Response Surface Optimization of the Critical Media Components for the Production of Surfactin," 1997, Vol. 68, pp. 263–270) describes the use of a designed experiment to maximize surfactin production. A portion of the data from this experiment is shown in Table P6.14. Surfactin was assayed by an indirect method, which involves measurement of surface tensions of the diluted broth samples. Relative surfactin concentrations were determined by serially diluting the broth until the critical micelle concentration (CMC) was reached. The dilution at which the surface tension starts rising abruptly was denoted by CMC^{-1} and was considered proportional to the amount of surfactant present in the original sample.

■ TABLE P6.14

The Factorial Experiment in Problem 6.41

Run	Glucose (g dm^{-3})	NH_4NO_3 (g dm^{-3})	$FeSO_4$ (g dm^{-3} $\times 10^{-4}$)	$MnSO_4$ (g dm^{-3} $\times 10^{-2}$)	y (CMC)$^{-1}$
1	20.00	2.00	6.00	4.00	23
2	60.00	2.00	6.00	4.00	15
3	20.00	6.00	6.00	4.00	16
4	60.00	6.00	6.00	4.00	18
5	20.00	2.00	30.00	4.00	25
6	60.00	2.00	30.00	4.00	16
7	20.00	6.00	30.00	4.00	17
8	60.00	6.00	30.00	4.00	26
9	20.00	2.00	6.00	20.00	28
10	60.00	2.00	6.00	20.00	16
11	20.00	6.00	6.00	20.00	18

12	60.00	6.00	6.00	20.00	21
13	20.00	2.00	30.00	20.00	36
14	60.00	2.00	30.00	20.00	24
15	20.00	6.00	30.00	20.00	33
16	60.00	6.00	30.00	20.00	34

(a) Analyze the data from this experiment. Identify the significant factors and interactions.

(b) Analyze the residuals from this experiment. Are there any indications of model inadequacy or violations of the assumptions?

(c) What conditions would optimize the surfactin production?

6.42. ***Continuation of Problem 6.41.*** The experiment in Problem 6.41 actually included six center points. The responses at these conditions were 35, 35, 35, 36, 36, and 34. Is there any indication of curvature in the response function? Are additional experiments necessary? What would you recommend doing now?

6.43. Often the fitted regression model from a 2^k factorial design is used to make predictions at points of interest in the design space. Assume that the model contains all main effects and two-factor interactions.

(a) Find the variance of the predicted response $\hat{y}$ at a point $x_1, x_2, \ldots, x_k$ in the design space. *Hint:* Remember that the x's are coded variables and assume a 2^k design with an equal number of replicates n at each design point so that the variance of a regression coefficient $\hat{\beta}$ is $\sigma^2/(n2^k)$ and that the covariance between any pair of regression coefficients is zero.

(b) Use the result in part (a) to find an equation for a $100(1-\alpha)$ percent confidence interval on the true mean response at the point $x_1, x_2, \ldots, x_k$ in design space.

6.44. ***Hierarchical models.*** Several times we have used the hierarchy principle in selecting a model; that is, we have included nonsignificant lower order terms in a model because they were factors involved in significant higher order terms. Hierarchy is certainly not an absolute principle that must be followed in all cases. To illustrate, consider the model resulting from Problem 6.2, which required that a nonsignificant main effect be included to achieve hierarchy. Using the data from Problem 6.2.

(a) Fit both the hierarchical and the nonhierarchical models.

(b) Calculate the PRESS statistic, the adjusted R^2, and the mean square error for both models.

(c) Find a 95 percent confidence interval on the estimate of the mean response at a cube corner ($x_1 = x_2 = x_3 = \pm 1$). *Hint:* Use the results of Problem 6.36.

(d) Based on the analyses you have conducted, which model do you prefer?

6.45. Suppose that you want to run a 2^3 factorial design. The variance of an individual observation is expected to be about 4. Suppose that you want the length of a 95 percent confidence interval on any effect to be less than or equal to 1.5. How many replicates of the design do you need to run?

CHAPTER 7

Blocking and Confounding Systems for Two-Level Factorials

CHAPTER OUTLINE

The supplemental material is on the textbook website www.wiley.com/go/global/montgomery.

7.1 Introduction

In many situations it is impossible to perform all of the runs in a 2^k factorial experiment under homogeneous conditions. For example, a single batch of raw material might not be large enough to make all of the required runs. In other cases, it might be desirable to deliberately vary the experimental conditions to ensure that the treatments are equally effective (i.e., robust) across many situations that are likely to be encountered in practice. For example, a chemical engineer may run a pilot plant experiment with several batches of raw material because he knows that different raw material batches of different quality grades are likely to be used in the actual full-scale process.

The design technique used in these situations is **blocking**. Chapter 4 was an introduction to the blocking principle, and you may find it helpful to read the introductory material in that chapter again. We also discussed blocking general factorial experiments in Chapter 5. In this chapter, we will build on the concepts introduced in Chapter 4, focusing on some special techniques for blocking in the 2^k factorial design.

7.2 Blocking a Replicated 2^k Factorial Design

Suppose that the 2^k factorial design has been replicated n times. This is identical to the situation discussed in Chapter 5, where we showed how to run a general factorial design in blocks. If there are n replicates, then each set of nonhomogeneous conditions defines a block, and each replicate is run in one of the blocks. The runs in each block (or replicate) would be made in random order. The analysis of the design is similar to that of any blocked factorial experiment; for example, see the discussion in Section 5.6. If blocks are random, the REML method can be used to find both a point estimate and an approximate confidance interval for the variance component for blocks.

EXAMPLE 7.1

Consider the chemical process experiment first described in Section 6.2. Suppose that only four experimental trials can be made from a single batch of raw material. Therefore, three batches of raw material will be required to run all three replicates of this design. Table 7.1 shows the design, where each batch of raw material corresponds to a block.

The ANOVA for this blocked design is shown in Table 7.2. All of the sums of squares are calculated exactly as in a standard, unblocked 2^k design. The sum of squares for blocks is calculated from the block totals. Let B_1, B_2, and B_3 represent the block totals (see Table 7.1). Then

$$SS_{\text{Blocks}} = \sum_{i=1}^{3} \frac{B_i^2}{4} - \frac{y_{...}^2}{12}$$

$$= \frac{(113)^2 + (106)^2 + (111)^2}{4} - \frac{(330)^2}{12}$$

$$= 6.50$$

There are two degrees of freedom among the three blocks. Table 7.2 indicates that the conclusions from this analysis, had the design been run in blocks, are identical to those in Section 6.2 and that the block effect is relatively small.

■ **TABLE 7.1**
Chemical Process Experiment in Three Blocks

	Block 1	Block 2	Block 3
	(1) = 28	(1) = 25	(1) = 27
	a = 36	a = 32	a = 32
	b = 18	b = 19	b = 23
	ab = 31	ab = 30	ab = 29
Block totals:	B_1 = 113	B_2 = 106	B_3 = 111

■ **TABLE 7.2**
Analysis of Variance for the Chemical Process Experiment in Three Blocks

Source of Variation	Sum of Squares	Degrees of Freedom	Mean Square	F_0	P-Value
Blocks	6.50	2	3.25		
A (concentration)	208.33	1	208.33	50.32	0.0004
B (catalyst)	75.00	1	75.00	18.12	0.0053
AB	8.33	1	8.33	2.01	0.2060
Error	24.84	6	4.14		
Total	323.00	11			

7.3 Confounding in the 2^k Factorial Design

In many problems it is impossible to perform a complete replicate of a factorial design in one block. **Confounding** is a design technique for arranging a complete factorial experiment in blocks, where the block size is smaller than the number of treatment combinations in one replicate. The technique causes information about certain treatment effects (usually high-order interactions) to be **indistinguishable from**, or **confounded with, blocks**. In this chapter we concentrate on confounding systems for the 2^k factorial design. Note that even though the designs presented are **incomplete block designs** because each block does not contain all the treatments or treatment combinations, the special structure of the 2^k factorial system allows a simplified method of analysis.

We consider the construction and analysis of the 2^k factorial design in 2^p incomplete blocks, where $p < k$. Consequently, these designs can be run in two blocks ($p = 1$), four blocks ($p = 2$), eight blocks ($p = 3$), and so on.

7.4 Confounding the 2^k Factorial Design in Two Blocks

Suppose that we wish to run a single replicate of the 2^2 design. Each of the $2^2 = 4$ treatment combinations requires a quantity of raw material, for example, and each batch of raw material is only large enough for two treatment combinations to be tested. Thus, two batches of raw material are required. If batches of raw material are considered as blocks, then we must assign two of the four treatment combinations to each block.

Figure 7.1 shows one possible design for this problem. The geometric view, Figure 7.1*a*, indicates that treatment combinations on opposing diagonals are assigned to different blocks. Notice from Figure 7.1*b* that block 1 contains the treatment combinations (1) and *ab* and that block 2 contains *a* and *b*. Of course, the *order* in which the treatment combinations are run within a block is randomly determined. We would also randomly decide which block to run first. Suppose we estimate the main effects of *A* and *B* just as if no blocking had occurred. From Equations 6.1 and 6.2, we obtain

$$A = \tfrac{1}{2}[ab + a - b - (1)]$$
$$B = \tfrac{1}{2}[ab + b - a - (1)]$$

Note that both *A* and *B* are unaffected by blocking because in each estimate there is one plus and one minus treatment combination from each block. That is, any difference between block 1 and block 2 will cancel out.

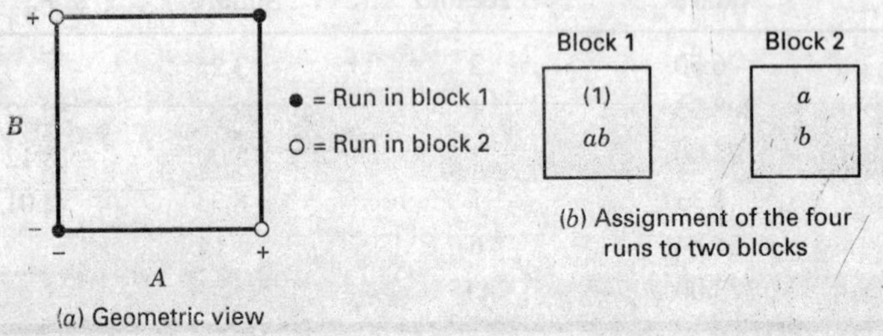

■ FIGURE 7.1 A 2^2 design in two blocks

■ **TABLE 7.3**
Table of Plus and Minus Signs for the 2^2 Design

Treatment Combination	Factorial Effect				
	I	A	B	AB	Block
(1)	+	−	−	+	1
a	+	+	−	−	2
b	+	−	+	−	2
ab	+	+	+	+	1

Now consider the AB interaction

$$AB = \tfrac{1}{2}[ab + (1) - a - b]$$

Because the two treatment combinations with the plus sign [ab and (1)] are in block 1 and the two with the minus sign (a and b) are in block 2, the block effect and the AB interaction are identical. That is, AB is **confounded** with blocks.

The reason for this is apparent from the table of plus and minus signs for the 2^2 design. This was originally given as Table 6.2, but for convenience it is reproduced as Table 7.3 here. From this table, we see that all treatment combinations that have a plus sign on AB are assigned to block 1, whereas all treatment combinations that have a minus sign on AB are assigned to block 2. This approach can be used to confound any effect (A, B, or AB) with blocks. For example, if (1) and b had been assigned to block 1 and a and ab to block 2, the main effect A would have been confounded with blocks. The usual practice is to confound the highest order interaction with blocks.

This scheme can be used to confound any 2^k design in two blocks. As a second example, consider a 2^3 design run in two blocks. Suppose we wish to confound the three-factor interaction ABC with blocks. From the table of plus and minus signs shown in Table 7.4, we assign the treatment combinations that are minus on ABC to block 1 and those that are plus on ABC to block 2. The resulting design is shown in Figure 7.2. Once again, we emphasize that the treatment combinations *within* a block are run in random order.

Other Methods for Constructing the Blocks. There is another method for constructing these designs. The method uses the linear combination

$$L = \alpha_1 x_1 + \alpha_2 x_2 + \cdots + \alpha_k x_k \tag{7.1}$$

■ **TABLE 7.4**
Table of Plus and Minus Signs for the 2^3 Design

Treatment Combination	Factorial Effect								
	I	A	B	AB	C	AC	BC	ABC	Block
(1)	+	−	−	+	−	+	+	−	1
a	+	+	−	−	−	−	+	+	2
b	+	−	+	−	−	+	−	+	2
ab	+	+	+	+	−	−	−	−	1
c	+	−	−	+	+	−	−	+	2
ac	+	+	−	−	+	+	−	−	1
bc	+	−	+	−	+	−	+	−	1
abc	+	+	+	+	+	+	+	+	2

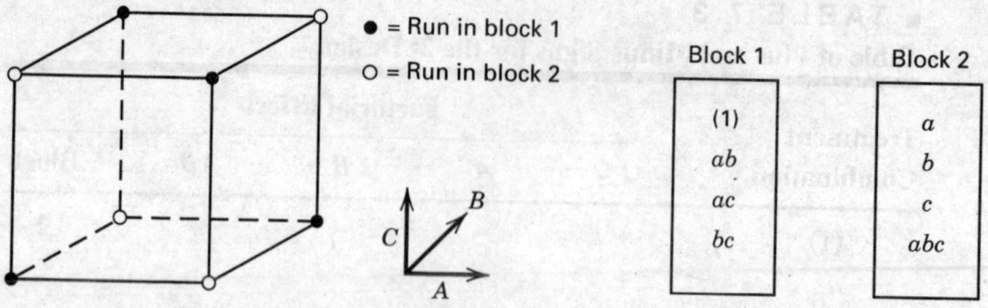

(*a*) Geometric view

(*b*) Assignment of the eight runs to two blocks

■ **FIGURE 7.2** **The 2^3 design in two blocks with *ABC* confounded**

where x_i is the level of the ith factor appearing in a particular treatment combination and α_i is the exponent appearing on the ith factor in the effect to be confounded. For the 2^k system, we have $\alpha_i = 0$ or 1 and $x_i = 0$ (low level) or $x_i = 1$ (high level). Equation 7.1 is called a **defining contrast**. Treatment combinations that produce the same value of L (mod 2) will be placed in the same block. Because the only possible values of L (mod 2) are 0 and 1, this will assign the 2^k treatment combinations to exactly two blocks.

To illustrate the approach, consider a 2^3 design with *ABC* confounded with blocks. Here x_1 corresponds to A, x_2 to B, x_3 to C, and $\alpha_1 = \alpha_2 = \alpha_3 = 1$. Thus, the defining contrast corresponding to *ABC* is

$$L = x_1 + x_2 + x_3$$

The treatment combination (1) is written 000 in the (0, 1) notation; therefore,

$$L = 1(0) + 1(0) + 1(0) = 0 = 0 \text{ (mod 2)}$$

Similarly, the treatment combination a is 100, yielding

$$L = 1(1) + 1(0) + 1(0) = 1 = 1 \text{ (mod 2)}$$

Thus, (1) and a would be run in different blocks. For the remaining treatment combinations, we have

$$b\colon L = 1(0) + 1(1) + 1(0) = 1 = 1 \text{ (mod 2)}$$
$$ab\colon L = 1(1) + 1(1) + 1(0) = 2 = 0 \text{ (mod 2)}$$
$$c\colon L = 1(0) + 1(0) + 1(1) = 1 = 1 \text{ (mod 2)}$$
$$ac\colon L = 1(1) + 1(0) + 1(1) = 2 = 0 \text{ (mod 2)}$$
$$bc\colon L = 1(0) + 1(1) + 1(1) = 2 = 0 \text{ (mod 2)}$$
$$abc\colon L = 1(1) + 1(1) + 1(1) = 3 = 1 \text{ (mod 2)}$$

Thus (1), *ab*, *ac*, and *bc* are run in block 1 and *a*, *b*, *c*, and *abc* are run in block 2. This is the same design shown in Figure 7.2, which was generated from the table of plus and minus signs.

Another method may be used to construct these designs. The block containing the treatment combination (1) is called the **principal block**. The treatment combinations in this block have a useful group-theoretic property; namely, they form a group with respect to multiplication modulus 2. This implies that any element [except (1)] in the principal block may be generated by multiplying two other elements in the principal block modulus 2. For example, consider the principal block of the 2^3 design with *ABC* confounded, as shown in Figure 7.2.

Note that

$$ab \cdot ac = a^2bc = bc$$

$$ab \cdot bc = ab^2c = ac$$

$$ac \cdot bc = abc^2 = ab$$

Treatment combinations in the other block (or blocks) may be generated by multiplying one element in the new block by each element in the principal block modulus 2. For the 2^3 with *ABC* confounded, because the principal block is (1), *ab*, *ac*, and *bc*, we know that *b* is in the other block. Thus, the elements of this second block are

$$b \cdot (1) \qquad = b$$

$$b \cdot ab = ab^2 \quad = a$$

$$b \cdot ac \qquad = abc$$

$$b \cdot bc = b^2c \quad = c$$

This agrees with the results obtained previously.

Estimation of Error. When the number of variables is small, say $k = 2$ or 3, it is usually necessary to replicate the experiment to obtain an estimate of error. For example, suppose that a 2^3 factorial must be run in two blocks with *ABC* confounded, and the experimenter decides to replicate the design four times. The resulting design might appear as in Figure 7.3. Note that *ABC* is confounded in each replicate.

The analysis of variance for this design is shown in Table 7.5. There are 32 observations and 31 total degrees of freedom. Furthermore, because there are eight blocks, seven degrees of freedom must be associated with these blocks. One breakdown of those seven degrees of freedom is shown in Table 7.5. The error sum of squares actually consists of the interactions between replicates and each of the effects (*A*, *B*, *C*, *AB*, *AC*, *BC*). It is usually safe to consider these interactions to be zero and to treat the resulting mean square as an estimate of error. Main effects and two-factor interactions are tested against the mean square error. Cochran and Cox (1957) observe that the block or *ABC* mean square could be compared to the error for the *ABC* mean square, which is really replicates × blocks. This test is usually very insensitive.

If resources are sufficient to allow the replication of confounded designs, it is generally better to use a slightly different method of designing the blocks in each replicate. This approach consists of confounding a different effect in each replicate so that some information on all effects is obtained. Such a procedure is called **partial confounding** and is discussed in Section 7.8.

If *k* is moderately large, say $k \geq 4$, we can frequently afford only a single replicate. The experimenter usually assumes higher order interactions to be negligible and combines their sums of squares as error. The normal probability plot of factor effects can be very helpful in this regard.

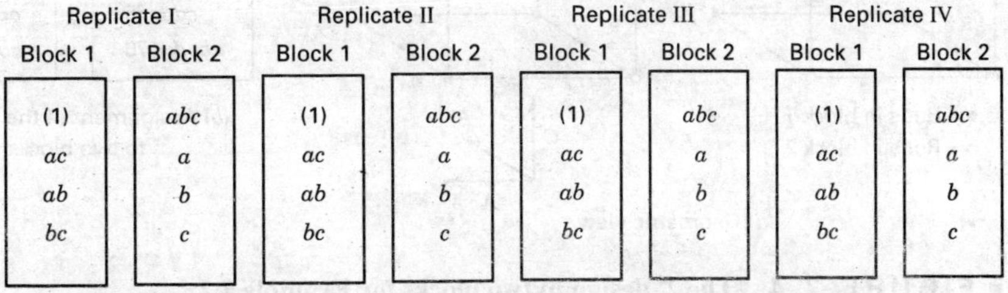

■ **FIGURE 7.3** **Four replicates of the 2^3 design with *ABC* confounded**

■ **TABLE 7.5**
Analysis of Variance for Four Replicates of a 2^3 Design with ABC Confounded

Source of Variation	Degrees of Freedom
Replicates	3
Blocks (ABC)	1
Error for ABC (replicates × blocks)	3
A	1
B	1
C	1
AB	1
AC	1
BC	1
Error (or replicates × effects)	18
Total	31

EXAMPLE 7.2

Consider the situation described in Example 6.2. Recall that four factors—temperature (A), pressure (B), concentration of formaldehyde (C), and stirring rate (D)—are studied in a pilot plant to determine their effect on product filtration rate. We will use this experiment to illustrate the ideas of blocking and confounding in an unreplicated design. We will make two modifications to the original experiment. First, suppose that the $2^4 = 16$ treatment combinations cannot all be run using one batch of raw material. The experimenter can run eight treatment combinations from a single batch of material, so a 2^4 design confounded in two blocks seems appropriate. It is logical to confound the highest order interaction $ABCD$ with blocks. The defining contrast is

$$L = x_1 + x_2 + x_3 + x_4$$

and it is easy to verify that the design is as shown in Figure 7.4. Alternatively, one may examine Table 6.11 and observe that the treatment combinations that are + in the $ABCD$ column are assigned to block 1 and those that are − in $ABCD$ column are in block 2.

The second modification that we will make is to introduce a **block effect** so that the utility of blocking can be demonstrated. Suppose that when we select the two batches of raw material required to run the experiment, one of them is of much poorer quality and, as a result, all responses will be 20 units lower in this material batch than in the other. The

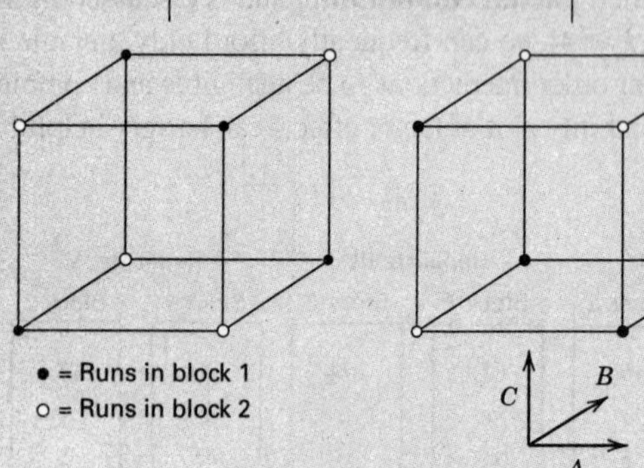

(a) Geometric view

Block 1	Block 2
(1) = 25	a = 71
ab = 45	b = 48
ac = 40	c = 68
bc = 60	d = 43
ad = 80	abc = 65
bd = 25	bcd = 70
cd = 55	acd = 86
$abcd$ = 76	abd = 104

(b) Assignment of the 16 runs to two blocks

■ **FIGURE 7.4** The 2^4 design in two blocks for Example 7.2

poor quality batch becomes block 1 and the good quality batch becomes block 2 (it doesn't matter which batch is called block 1 or which batch is called block 2). Now all the tests in block 1 are performed first (the eight runs in the block are, of course, performed in random order), but the responses are 20 units lower than they would have been if good quality material had been used. Figure 7.4*b* shows the resulting responses—note that these have been found by subtracting the block effect from the original observations given in Example 6.2. That is, the original response for treatment combination (1) was 45, and in Figure 7.4*b* it is reported as (1) = 25 (= 45 − 20). The other responses in this block are obtained similarly. After the tests in block 1 are performed, the eight tests in block 2 follow. There is no problem with the raw material in this batch, so the responses are exactly as they were originally in Example 6.2.

The effect estimates for this "modified" version of Example 6.2 are shown in Table 7.6. Note that the estimates of the four main effects, the six two-factor interactions, and the four three-factor interactions are identical to the effect estimates obtained in Example 6.2 where there was *no block effect*. When a normal probability of these effect estimates is constructed, factors *A*, *C*, *D*, and the *AC* and *AD* interactions emerge as the important effects, just as in the original experiment. (The reader should verify this.)

What about the *ABCD* interaction effect? The estimate of this effect in the original experiment (Example 6.2) was *ABCD* = 1.375. In the present example, the estimate of the *ABCD* interaction effect is *ABCD* = −18.625. Because *ABCD* is confounded with blocks, the *ABCD* interaction estimates the *original interaction effect* (1.375) plus the *block effect* (−20), so *ABCD* = 1.375 + (−20) = −18.625. (Do you see why the block effect is −20?) The block effect may also be calculated directly as the difference in average response between the two blocks, or

$$\begin{aligned}\text{Block effect} &= \bar{y}_{\text{Block 1}} - \bar{y}_{\text{Block 2}} \\ &= \frac{406}{8} - \frac{555}{8} \\ &= \frac{-149}{8} \\ &= -18.625\end{aligned}$$

Of course, this effect really estimates Blocks + *ABCD*.

Table 7.7 summarizes the ANOVA for this experiment. The effects with large estimates are included in the model, and the block sum of squares is

$$SS_{\text{Blocks}} = \frac{(406)^2 + (555)^2}{8} - \frac{(961)^2}{16} = 1387.5625$$

The conclusions from this experiment exactly match those from Example 6.2, where no block effect was present. Notice that if the experiment had not been run in blocks, and if an effect of magnitude −20 had affected the first 8 trials (which would have been selected in a random fashion, because the 16 trials would be run in random order in an unblocked design), the results could have been very different.

■ **TABLE 7.6**
Effect Estimates for the Blocked 2^4 Design in Example 7.2

Model Term	Regression Coefficient	Effect Estimate	Sum of Squares	Percent Contribution
A	10.81	21.625	1870.5625	26.30
B	1.56	3.125	39.0625	0.55
C	4.94	9.875	390.0625	5.49
D	7.31	14.625	855.5625	12.03
AB	0.062	0.125	0.0625	<0.01
AC	−9.06	−18.125	1314.0625	18.48
AD	8.31	16.625	1105.5625	15.55
BC	1.19	2.375	22.5625	0.32
BD	−0.19	−0.375	0.5625	<0.01
CD	−0.56	−1.125	5.0625	0.07
ABC	0.94	1.875	14.0625	0.20
ABD	2.06	4.125	68.0625	0.96
ACD	−0.81	−1.625	10.5625	0.15
BCD	−1.31	−2.625	27.5625	0.39
Block (*ABCD*)		−18.625	1387.5625	19.51

■ **TABLE 7.7**
Analysis of Variance for Example 7.2

Source of Variation	Sum of Squares	Degrees of Freedom	Mean Square	F_0	*P*-Value
Blocks (*ABCD*)	1387.5625	1			
A	1870.5625	1	1870.5625	89.76	<0.0001
C	390.0625	1	390.0625	18.72	0.0019
D	855.5625	1	855.5625	41.05	0.0001
AC	1314.0625	1	1314.0625	63.05	<0.0001
AD	1105.5625	1	1105.5625	53.05	<0.0001
Error	187.5625	9	20.8403		
Total	7110.9375	15			

7.5 Another Illustration of Why Blocking Is Important

Blocking is a very useful and important design technique. In Chapter 4 we pointed out that blocking has such dramatic potential to reduce the noise in an experiment that an experimenter should always consider the potential impact of nuisance factors, and when in doubt, block.

To illustrate what can happen if an experimenter doesn't block when he or she should have, consider a variation of Example 7.2 from the previous section. In this example we utilized a 2^4 unreplicated factorial experiment originally presented as Example 6.2. We constructed the design in two blocks of eight runs each, and we inserted a "block effect" or nuisance factor effect of magnitude -20 that affects all of the observations in block 1 (refer to Figure 7.4). Now suppose that we had not run this design in blocks and that the -20 nuisance factor effect impacted the first eight observations that were taken (in random or run order). The modified data are shown in Table 7.8.

Figure 7.5 is a normal probability plot of the factor effects from this modified version of the experiment. Notice that although the appearance of this plot is not too dissimilar from

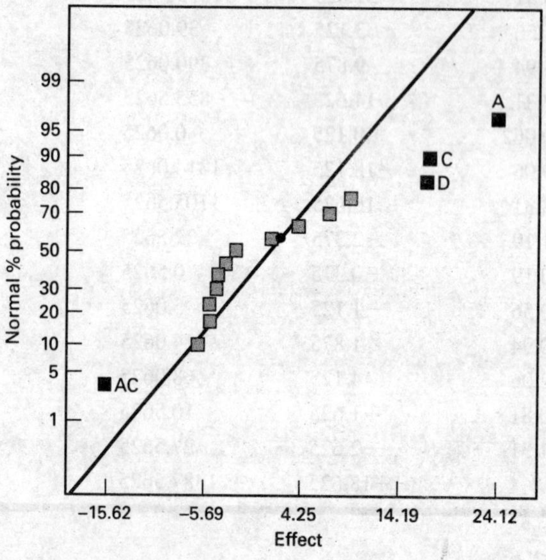

■ **FIGURE 7.5** **Normal probability plot for the data in Table 7.8**

■ **TABLE 7.8**
The Modified Data from Example 7.2

Run Order	Std. Order	Factor *A*: Temperature	Factor *B*: Pressure	Factor *C*: Concentration	Factor *D*: Stirring Rate	Response Filtration Rate
8	1	−1	−1	−1	−1	25
11	2	1	−1	−1	−1	71
1	3	−1	1	−1	−1	28
3	4	1	1	−1	−1	45
9	5	−1	−1	1	−1	68
12	6	1	−1	1	−1	60
2	7	−1	1	1	−1	60
13	8	1	1	1	−1	65
7	9	−1	−1	−1	1	23
6	10	1	−1	−1	1	80
16	11	−1	1	−1	1	45
5	12	1	1	−1	1	84
14	13	−1	−1	1	1	75
15	14	1	−1	1	1	86
10	15	−1	1	1	1	70
4	16	1	1	1	1	76

the one given with the original analysis of the experiment in Chapter 6 (refer to Figure 6.11), one of the important interactions, *AD*, is not identified. Consequently, we will not discover this important effect that turns out to be one of the keys to solving the original problem. We remarked in Chapter 4 that blocking is a **noise reduction technique**. If we don't block, then the added variability from the nuisance variable effect ends up getting distributed across the other design factors.

Some of the nuisance variability also ends up in the error estimate as well. The residual mean square for the model based on the data in Table 7.8 is about 109, which is several times larger than the residual mean square based on the original data (see Table 6.13).

7.6 Confounding the 2^k Factorial Design in Four Blocks

It is possible to construct 2^k factorial designs confounded in four blocks of 2^{k-2} observations each. These designs are particularly useful in situations where the number of factors is moderately large, say $k \geq 4$, and block sizes are relatively small.

As an example, consider the 2^5 design. If each block will hold only eight runs, then four blocks must be used. The construction of this design is relatively straightforward. Select *two* effects to be confounded with blocks, say *ADE* and *BCE*. These effects have the two defining contrasts

$$L_1 = x_1 + x_4 + x_5$$

$$L_2 = x_2 + x_3 + x_5$$

Block 1 $L_1 = 0$ $L_2 = 0$	Block 2 $L_1 = 1$ $L_2 = 0$	Block 3 $L_1 = 0$ $L_2 = 1$	Block 4 $L_1 = 1$ $L_2 = 1$
(1) *abe*	*a* *be*	*b* *abce*	*e* *abcde*
ad *ace*	*d* *abde*	*abd* *ae*	*ade* *bd*
bc *cde*	*abc* *ce*	*c* *bcde*	*bce* *ac*
abcd *bde*	*bcd* *acde*	*acd* *de*	*ab* *cd*

■ **FIGURE 7.6** **The 2^5 design in four blocks with *ADE, BCE*, and *ABCD* confounded**

associated with them. Now every treatment combination will yield a particular pair of values of L_1 (mod 2) and L_2 (mod 2), that is, either $(L_1, L_2) = (0, 0), (0, 1), (1, 0)$, or $(1, 1)$. Treatment combinations yielding the same values of (L_1, L_2) are assigned to the same block. In our example we find

$L_1 = 0, L_2 = 0$ for (1), *ad, bc, abcd, abe, ace, cde, bde*

$L_1 = 1, L_2 = 0$ for *a, d, abc, bcd, be, abde, ce, acde*

$L_1 = 0, L_2 = 1$ for *b, abd, c, acd, ae, de abce, bcde*

$L_1 = 1, L_2 = 1$ for *e, ade, bce, abcde, ab, bd, ac, cd*

These treatment combinations would be assigned to different blocks. The complete design is as shown in Figure 7.6.

With a little reflection we realize that another effect in addition to *ADE* and *BCE* must be confounded with blocks. Because there are four blocks with three degrees of freedom between them, and because *ADE* and *BCE* have only one degree of freedom each, clearly an additional effect with one degree of freedom must be confounded. This effect is the **generalized interaction** of *ADE* and *BCE*, which is defined as the product of *ADE* and *BCE* modulus 2. Thus, in our example the generalized interaction $(ADE)(BCE) = ABCDE^2 = ABCD$ is also confounded with blocks. It is easy to verify this by referring to a table of plus and minus signs for the 2^5 design, such as in Davies (1956). Inspection of such a table reveals that the treatment combinations are assigned to the blocks as follows:

Treatment Combinations in	Sign on *ADE*	Sign on *BCE*	Sign on *ABCD*
Block 1	−	−	+
Block 2	+	−	−
Block 3	−	+	−
Block 4	+	+	+

Notice that the product of signs of any two effects for a particular block (e.g., *ADE* and *BCE*) yields the sign of the other effect for that block (in this case, *ABCD*). Thus, *ADE, BCE*, and *ABCD* are all confounded with blocks.

The group-theoretic properties of the principal block mentioned in Section 7.4 still hold. For example, we see that the product of two treatment combinations in the principal block yields another element of the principal block. That is,

$$ad \cdot bc = abcd \quad \text{and} \quad abe \cdot bde = ab^2de^2 = ad$$

and so forth. To construct another block, select a treatment combination that is not in the principal block (e.g., b) and multiply b by all the treatment combinations in the principal block. This yields

$$b \cdot (1) = b \quad b \cdot ad = abd \quad b \cdot bc = b^2c = c \quad b \cdot abcd = ab^2cd = acd$$

and so forth, which will produce the eight treatment combinations in block 3. In practice, the principal block can be obtained from the defining contrasts and the group-theoretic property, and the remaining blocks can be determined from these treatment combinations by the method shown above.

The general procedure for constructing a 2^k design confounded in four blocks is to choose two effects to generate the blocks, automatically confounding a third effect that is the generalized interaction of the first two. Then, the design is constructed by using the two defining contrasts (L_1, L_2) and the group-theoretic properties of the principal block. In selecting effects to be confounded with blocks, care must be exercised to obtain a design that does not confound effects that may be of interest. For example, in a 2^5 design we might choose to confound *ABCDE* and *ABD*, which automatically confounds *CE*, an effect that is probably of interest. A better choice is to confound *ADE* and *BCE*, which automatically confounds *ABCD*. It is preferable to sacrifice information on the three-factor interactions *ADE* and *BCE* instead of the two-factor interaction *CE*.

7.7 Confounding the 2^k Factorial Design in 2^p Blocks

The methods described above may be extended to the construction of a 2^k factorial design confounded in 2^p blocks ($p < k$), where each block contains exactly 2^{k-p} runs. We select p independent effects to be confounded, where by "independent" we mean that no effect chosen is the generalized interaction of the others. The blocks may be generated by use of the p defining contrasts $L_1, L_2, \ldots, L_p$ associated with these effects. In addition, exactly $2^p - p - 1$ other effects will be confounded with blocks, these being the generalized interactions of those p independent effects initially chosen. Care should be exercised in selecting effects to be confounded so that information on effects that may be of potential interest is not sacrificed.

The statistical analysis of these designs is straightforward. Sums of squares for all the effects are computed as if no blocking had occurred. Then, the block sum of squares is found by adding the sums of squares for all the effects confounded with blocks.

Obviously, the choice of the p effects used to generate the block is critical because the confounding structure of the design directly depends on them. Table 7.9 presents a list of useful designs. To illustrate the use of this table, suppose we wish to construct a 2^6 design confounded in $2^3 = 8$ blocks of $2^3 = 8$ runs each. Table 7.9 indicates that we would choose *ABEF*, *ABCD*, and *ACE* as the $p = 3$ independent effects to generate the blocks. The remaining $2^p - p - 1 = 2^3 - 3 - 1 = 4$ effects that are confounded are the generalized interactions of these three; that is,

$$(ABEF)(ABCD) = A^2B^2CDEF = CDEF$$

$$(ABEF)(ACE) = A^2BCE^2F = BCF$$

$$(ABCD)(ACE) = A^2BC^2ED = BDE$$

$$(ABEF)(ABCD)(ACE) = A^3B^2C^2DE^2F = ADF$$

The reader is asked to generate the eight blocks for this design in Problem 7.11.

■ TABLE 7.9
Suggested Blocking Arrangements for the 2^k Factorial Design

Number of Factors, k	Number of Blocks, 2^p	Block Size, 2^{k-p}	Effects Chosen to Generate the Blocks	Interactions Confounded with Blocks
3	2	4	*ABC*	*ABC*
	4	2	*AB, AC*	*AB, AC, BC*
4	2	8	*ABCD*	*ABCD*
	4	4	*ABC, ACD*	*ABC, ACD, BD*
	8	2	*AB, BC, CD*	*AB, BC, CD, AC, BD, AD, ABCD*
5	2	16	*ABCDE*	*ABCDE*
	4	8	*ABC, CDE*	*ABC, CDE, ABDE*
	8	4	*ABE, BCE, CDE*	*ABE, BCE, CDE, AC, ABCD, BD, ADE*
	16	2	*AB, AC, CD, DE*	All two- and four-factor interactions (15 effects)
6	2	32	*ABCDEF*	*ABCDEF*
	4	16	*ABCF, CDEF*	*ABCF, CDEF, ABDE*
	8	8	*ABEF, ABCD, ACE*	*ABEF, ABCD, ACE, BCF, BDE, CDEF, ADF*
	16	4	*ABF, ACF, BDF, DEF*	*ABF, ACF, BDF, DEF, BC, ABCD, ABDE, AD, ACDE, CE, CDF, BCDEF, ABCEF, AEF, BE*
	32	2	*AB, BC, CD, DE, EF*	All two-, four-, and six-factor interactions (31 effects)
7	2	64	*ABCDEFG*	*ABCDEFG*
	4	32	*ABCFG, CDEFG*	*ABCFG, CDEFG, ABDE*
	8	16	*ABCD, CDEF, ADFG*	*ABC, DEF, AFG, ABCDEF, BCFG, ADEG, BCDEG*
	16	8	*ABCD, EFG, CDE, ADG*	*ABCD, EFG, CDE, ADG, ABCDEFG, ABE, BCG, CDFG, ADEF, ACEG, ABFG, BCEF, BDEG, ACF, BDF*
	32	4	*ABG, BCG, CDG, DEG, EFG*	*ABG, BCG, CDG, DEG, EFG, AC, BD, CE, DF, AE, BF, ABCD, ABDE, ABEF, BCDE, BCEF, CDEF, ABCDEFG, ADG, ACDEG, ACEFG, ABDFG, ABCEG, BEG, BDEFG, CFG, ADEF, ACDF, ABCF, AFG, BCDFG*
	64	2	*AB, BC, CD, DE, EF, FG*	All two-, four-, and six-factor interactions (63 effects)

7.8 Partial Confounding

We remarked in Section 7.4 that, unless experimenters have a prior estimate of error or are willing to assume certain interactions to be negligible, they must replicate the design to obtain an estimate of error. Figure 7.3 shows a 2^3 factorial in two blocks with *ABC* confounded, replicated four times. From the analysis of variance for this design, shown in Table 7.5, we note that information on the *ABC* interaction cannot be retrieved because *ABC* is confounded with blocks *in each replicate*. This design is said to be **completely confounded**.

Consider the alternative shown in Figure 7.7. Once again, there are four replicates of the 2^3 design, but a *different* interaction has been confounded in each replicate. That is, *ABC* is confounded in replicate I, *AB* is confounded in replicate II, *BC* is confounded in replicate III,

Replicate I, *ABC* Confounded		Replicate II, *AB* Confounded		Replicate III, *BC* Confounded		Replicate IV, *AC* Confounded	
(1)	*a*	(1)	*a*	(1)	*b*	(1)	*a*
ab	*b*	*c*	*b*	*a*	*c*	*b*	*c*
ac	*c*	*ab*	*ac*	*bc*	*ab*	*ac*	*ab*
bc	*abc*	*abc*	*bc*	*abc*	*ac*	*abc*	*bc*

■ **FIGURE 7.7 Partial confounding in the 2^3 design**

and *AC* is confounded in replicate IV. As a result, information on *ABC* can be obtained from the data in replicates II, III, and IV; information on *AB* can be obtained from replicates I, III, and IV; information on *AC* can be obtained from replicates I, II, and III; and information on *BC* can be obtained from replicates I, II, and IV. We say that three-quarters information can be obtained on the interactions because they are unconfounded in only three replicates. Yates (1937) calls the ratio 3/4 the **relative information for the confounded effects**. This design is said to be **partially confounded**.

The analysis of variance for this design is shown in Table 7.10. In calculating the interaction sums of squares, only data from the replicates in which an interaction is unconfounded are used. The error sum of squares consists of replicates × main effect sums of squares plus replicates × interaction sums of squares for each replicate in which that interaction is unconfounded (e.g., replicates × *ABC* for replicates II, III, and IV). Furthermore, there are seven degrees of freedom among the eight blocks. This is usually partitioned into three degrees of freedom for replicates and four degrees of freedom for blocks within replicates. The composition of the sum of squares for blocks is shown in Table 7.10 and follows directly from the choice of the effect confounded in each replicate.

■ **TABLE 7.10**
Analysis of Variance for a Partially Confounded 2^3 Design

Source of Variation	Degrees of Freedom
Replicates	3
Blocks within replicates [or *ABC* (rep. I) + *AB* (rep. II) + *BC* (rep. III) + *AC* (rep. IV)]	4
A	1
B	1
C	1
AB (from replicates I, III, and IV)	1
AC (from replicates I, II, and III)	1
BC (from replicates I, II, and IV)	1
ABC (from replicates II, III, and IV)	1
Error	17
Total	31

EXAMPLE 7.3 A 2^3 Design with Partial Confounding

Consider Example 6.1, in which an experiment was conducted to develop a plasma etching process. There were three factors, A = gap, B = gas flow, and C = RF power, and the response variable is the etch rate. Suppose that only four treatment combinations can be tested during a shift, and because there could be shift-to-shift differences in etching tool performance, the experimenters decide to use shifts as a blocking factor. Thus, each replicate of the 2^3 design must be run in two blocks. Two replicates are run, with ABC confounded in replicate I and AB confounded in replicate II. The data are as follows:

Replicate I, *ABC* Confounded: Block 1	Replicate I, *ABC* Confounded: Block 2	Replicate II, *AB* Confounded: Block 1	Replicate II, *AB* Confounded: Block 2
(1) = 550	a = 669	(1) = 604	a = 650
ab = 642	b = 633	c = 1052	b = 601
ac = 749	c = 1037	ab = 635	ac = 868
bc = 1075	abc = 729	abc = 860	bc = 1063

The sums of squares for A, B, C, AC, and BC may be calculated in the usual manner, using all 16 observations. However, we must find SS_{ABC} using only the data in replicate II and SS_{AB} using only the data in replicate I as follows:

$$SS_{ABC} = \frac{[a + b + c + abc - ab - ac - bc - (1)]^2}{n2^k}$$

$$= \frac{[650 + 601 + 1052 + 860 - 635 - 868 - 1063 - 604]^2}{(1)(8)} = 6.1250$$

$$SS_{AB} = \frac{[(1) + abc - ac + c - a - b + ab - bc]^2}{n2^k}$$

$$= \frac{[550 + 729 - 749 + 1037 - 669 - 633 + 642 - 1075]^2}{(1)(8)} = 3528.0$$

The sum of squares for the replicates is, in general,

$$SS_{\text{Rep}} = \sum_{h=1}^{n} \frac{R_h^2}{2^k} - \frac{y_{...}^2}{N}$$

$$= \frac{(6084)^2 + (6333)^2}{8} - \frac{(12{,}417)^2}{16} = 3875.0625$$

where R_h is the total of the observations in the hth replicate. The block sum of squares is the sum of SS_{ABC} from replicate I and SS_{AB} from replicate II, or $SS_{\text{Blocks}} = 458.1250$.

The analysis of variance is summarized in Table 7.11. The main effects of A and C and the AC interaction are important.

■ **TABLE 7.11**
Analysis of Variance for Example 7.3

Source of Variation	Sum of Squares	Degrees of Freedom	Mean Square	F_0	*P*-Value
Replicates	3875.0625	1	3875.0625	—	
Blocks within replicates	458.1250	2	229.0625	—	
A	41,310.5625	1	41,310.5625	16.20	0.01
B	217.5625	1	217.5625	0.08	0.78
C	374,850.5625	1	374,850.5625	146.97	<0.001
AB (rep. I only)	3528.0000	1	3528.0000	1.38	0.29
AC	94,404.5625	1	94,404.5625	37.01	<0.001
BC	18.0625	1	18.0625	0.007	0.94
ABC (rep. II only)	6.1250	1	6.1250	0.002	0.96
Error	12,752.3125	5	2550.4625		
Total	531,420.9375	15			

7.9 Problems

7.1. Consider the experiment described in Problem 6.2. Analyze this experiment assuming that each replicate represents a block of a single production shift.

7.2. Consider the experiment described in Problem 6.1. Analyze this experiment assuming that each one of the four replicates represents a block.

7.3. Consider the alloy cracking experiment described in Problem 6.15. Suppose that only 16 runs could be made on a single day, so each replicate was treated as a block. Analyze the experiment and draw conclusions.

7.4. Consider the data from the first replicate of Problem 6.2. Suppose that these observations could not all be run using the same bar stock. Set up a design to run these observations in two blocks of four observations each with *ABC* confounded. Analyze the data.

7.5. Consider the data from the first replicate of Problem 6.7. Construct a design with two blocks of eight observations each with *ABCD* confounded. Analyze the data.

7.6. Repeat Problem 7.5 assuming that four blocks are required. Confound *ABD* and *ABC* (and consequently *CD*) with blocks.

7.7. Using the data from the 2^5 design in Problem 6.25, construct and analyze a design in two blocks with *ABCDE* confounded with blocks.

7.8. Repeat Problem 7.7 assuming that four blocks are necessary. Suggest a reasonable confounding scheme.

7.9. Consider the data from the 2^5 design in Problem 6.25. Suppose that it was necessary to run this design in four blocks with *ACDE* and *BCD* (and consequently *ABE*) confounded. Analyze the data from this design.

7.10. Consider the fill height deviation experiment in Problem 6.20. Suppose that each replicate was run on a separate day. Analyze the data assuming that days are blocks.

7.11. Consider the fill height deviation experiment in Problem 6.20. Suppose that only four runs could be made on each shift. Set up a design with *ABC* confounded in replicate 1 and *AC* confounded in replicate 2. Analyze the data and comment on your findings.

7.12. Consider the putting experiment in Problem 6.21. Analyze the data considering each replicate as a block.

7.13. Using the data from the 2^4 design in Problem 6.22, construct and analyze a design in two blocks with *ABCD* confounded with blocks.

7.14. Consider the direct mail experiment in Problem 6.24. Suppose that each group of customers is in a different part of the country. Suggest an appropriate analysis for the experiment.

7.15. Consider the isatin yield experiment in Problem 6.38. Set up the 2^4 experiment in this problem in two blocks with ABCD confounded. Analyze the data from this design. Is the block effect large?

7.16. The experiment in Problem 6.39 is a 2^5 factorial. Suppose that this design had been run in four blocks of eight runs each.

(a) Recommend a blocking scheme and set up the design.

(b) Analyze the data from this blocked design. Is blocking important?

7.17. Repeat Problem 6.16 using a design in two blocks.

7.18. The design in Problem 6.41 is a 2^4 factorial. Set up this experiment in two blocks with ABCD confounded. Analyze the data from this design. Is the block effect large?

7.19. The design in Problem 6.40 is a 2^3 factorial replicated twice. Suppose that each replicate was a block. Analyze all of the responses from this blocked design. Are the results comparable to those from Problem 6.40? Is the block effect large?

7.20. Design an experiment for confounding a 2^6 factorial in four blocks. Suggest an appropriate confounding scheme, different from the one shown in Table 7.8.

7.21. Consider the 2^6 design in eight blocks of eight runs each with *ABCD*, *ACE*, and *ABEF* as the independent effects chosen to be confounded with blocks. Generate the design. Find the other effects confounded with blocks.

7.22. Consider the 2^2 design in two blocks with *AB* confounded. Prove algebraically that $SS_{AB} = SS_{\text{Blocks}}$.

7.23. Consider the data in Example 7.2. Suppose that all the observations in block 2 are increased by 20. Analyze the data that would result. Estimate the block effect. Can you explain its magnitude? Do blocks now appear to be an important factor? Are any other effect estimates impacted by the change you made to the data?

7.24. Suppose that in Problem 6.2 we had confounded *ABC* in replicate I, *AB* in replicate II, and *BC* in replicate III. Calculate the factor effect estimates. Construct the analysis of variance table.

7.25. Repeat the analysis of Problem 6.2 assuming that *ABC* was confounded with blocks in each replicate.

7.26. Suppose that in Problem 6.7 *ABCD* was confounded in replicate I and *ABC* was confounded in replicate II. Perform the statistical analysis of this design.

7.27. Construct a 2^3 design with *ABC* confounded in the first two replicates and *BC* confounded in the third. Outline the analysis of variance and comment on the information obtained.

CHAPTER 8

Two-Level Fractional Factorial Designs

CHAPTER OUTLINE

The supplemental material is on the textbook website www.wiley.com/go/global/montgomery.

8.1 Introduction

As the number of factors in a 2^k factorial design increases, the number of runs required for a complete replicate of the design rapidly outgrows the resources of most experimenters. For example, a complete replicate of the 2^6 design requires 64 runs. In this design only 6 of the 63 degrees of freedom correspond to main effects, and only 15 degrees of freedom correspond to two-factor interactions. There are only 21 degrees of freedom associated with effects that are likely to be of major interest. The remaining 42 degrees of freedom are associated with three-factor and higher interactions.

If the experimenter can reasonably assume that certain high-order interactions are negligible, information on the main effects and low-order interactions may be obtained by running only a fraction of the complete factorial experiment. These **fractional factorial designs** are among the most widely used types of designs for product and process design, process improvement, and industrial/business experimentation.

A major use of fractional factorials is in **screening experiments**—experiments in which many factors are considered and the objective is to identify those factors (if any) that have large effects. Screening experiments are usually performed in the early stages of a project when many of the factors initially considered likely have little or no effect on the response. The factors identified as important are then investigated more thoroughly in subsequent experiments.

The successful use of fractional factorial designs is based on three key ideas:

1. ***The sparsity of effects principle.*** When there are several variables, the system or process is likely to be driven primarily by some of the main effects and low-order interactions.
2. ***The projection property.*** Fractional factorial designs can be projected into stronger (larger) designs in the subset of significant factors.
3. ***Sequential experimentation.*** It is possible to combine the runs of two (or more) fractional factorials to construct sequentially a larger design to estimate the factor effects and interactions of interest.

We will focus on these principles in this chapter and illustrate them with several examples.

8.2 The One-Half Fraction of the 2^k Design

8.2.1 Definitions and Basic Principles

Consider a situation in which three factors, each at two levels, are of interest, but the experimenters cannot afford to run all $2^3 = 8$ treatment combinations. They can, however, afford four runs. This suggests a **one-half fraction** of a 2^3 design. Because the design contains $2^{3-1} = 4$ treatment combinations, a one-half fraction of the 2^3 design is often called a $\mathbf{2^{3-1}}$ **design.**

The table of plus and minus signs for the 2^3 design is shown in Table 8.1. Suppose we select the four treatment combinations *a, b, c,* and *abc* as our one-half fraction. These runs are shown in the top half of Table 8.1 and in Figure 8.1*a*.

Notice that the 2^{3-1} design is formed by selecting only those treatment combinations that have a plus in the *ABC* column. Thus, *ABC* is called the **generator** of this particular

■ **TABLE 8.1**
Plus and Minus Signs for the 2^3 Factorial Design

Treatment Combination	Factorial Effect							
	I	*A*	*B*	*C*	*AB*	*AC*	*BC*	*ABC*
a	+	+	−	−	−	−	+	+
b	+	−	+	−	−	+	−	+
c	+	−	−	+	+	−	−	+
abc	+	+	+	+	+	+	+	+
ab	+	+	+	−	+	−	−	−
ac	+	+	−	+	−	+	−	−
bc	+	−	+	+	−	−	+	−
(1)	+	−	−	−	+	+	+	−

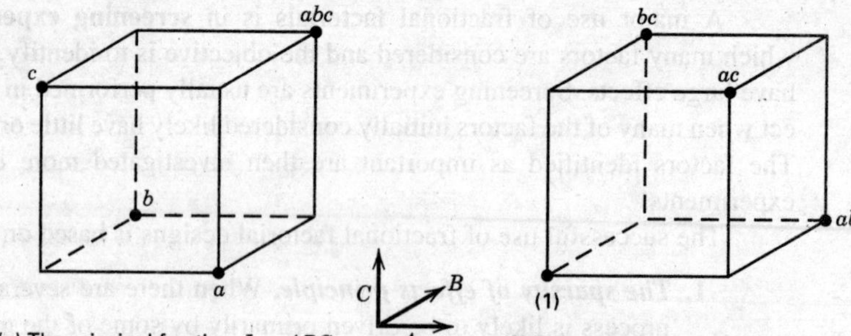

(a) The principal fraction, $I = +ABC$ (b) The alternate fraction, $I = -ABC$

■ **FIGURE 8.1 The two one-half fractions of the 2^3 design**

fraction. Usually we will refer to a generator such as ABC as a **word**. Furthermore, the identity column I is also always plus, so we call

$$I = ABC$$

the **defining relation** for our design. In general, the defining relation for a fractional factorial will always be the set of all columns that are equal to the identity column I.

The treatment combinations in the 2^{3-1} design yield three degrees of freedom that we may use to estimate the main effects. Referring to Table 8.1, we note that the linear combinations of the observations used to estimate the main effects of A, B, and C are

$$[A] = \tfrac{1}{2}(a - b - c + abc)$$
$$[B] = \tfrac{1}{2}(-a + b - c + abc)$$
$$[C] = \tfrac{1}{2}(-a - b + c + abc)$$

Where the notation $[A]$, $[B]$, and $[C]$ is used to indicate the linear combinations associated with the main effects. It is also easy to verify that the linear combinations of the observations used to estimate the two-factor interactions are

$$[BC] = \tfrac{1}{2}(a - b - c + abc)$$
$$[AC] = \tfrac{1}{2}(-a + b - c + abc)$$
$$[AB] = \tfrac{1}{2}(-a - b + c + abc)$$

Thus, $[A] = [BC]$, $[B] = [AC]$, and $[C] = [AB]$; consequently, it is impossible to differentiate between A and BC, B and AC, and C and AB. In fact, when we estimate A, B, and C we are *really* estimating $A + BC$, $B + AC$, and $C + AB$. Two or more effects that have this property are called **aliases**. In our example, A and BC are aliases, B and AC are aliases, and C and AB are aliases. We indicate this by the notation $[A] \rightarrow A + BC$, $[B] \rightarrow B + AC$, and $[C] \rightarrow C + AB$.

The alias structure for this design may be easily determined by using the defining relation $I = ABC$. Multiplying any column (or effect) by the defining relation yields the aliases for that column (or effect). In our example, this yields as the alias of A

$$A \cdot I = A \cdot ABC = A^2BC$$

or, because the square of any column is just the identity I,

$$A = BC$$

Similarly, we find the aliases of B and C as

$$B \cdot I = B \cdot ABC$$
$$B = AB^2C = AC$$

and

$$C \cdot I = C \cdot ABC$$
$$C = ABC^2 = AB$$

This one-half fraction, with $I = +ABC$, is usually called the **principal fraction**.

Now suppose that we had chosen the *other* one-half fraction, that is, the treatment combinations in Table 8.1 associated with minus in the ABC column. This **alternate**, or **complementary**, one-half fraction (consisting of the runs (1), *ab*, *ac*, and *bc*) is shown in Figure 8.1*b*. The defining relation for this design is

$$I = -ABC$$

The linear combination of the observations, say $[A]'$, $[B]'$, and $[C]'$, from the alternate fraction gives us

$$[A]' \rightarrow A - BC$$
$$[B]' \rightarrow B - AC$$
$$[C]' \rightarrow C - AB$$

Thus, when we estimate A, B, and C with this particular fraction, we are really estimating $A - BC$, $B - AC$, and $C - AB$.

In practice, it does not matter which fraction is actually used. Both fractions belong to the same **family**; that is, the two one-half fractions form a complete 2^3 design. This is easily seen by reference to parts *a* and *b* of Figure 8.1.

Suppose that after running one of the one-half fractions of the 2^3 design, the other fraction was also run. Thus, all eight runs associated with the full 2^3 are now available. We may now obtain de-aliased estimates of all the effects by analyzing the eight runs as a full 2^3 design in two blocks of four runs each. This could also be done by adding and subtracting the linear combination of effects from the two individual fractions. For example, consider $[A] \rightarrow A + BC$ and $[A]' \rightarrow A - BC$. This implies that

$$\tfrac{1}{2}([A] + [A]') = \tfrac{1}{2}(A + BC + A - BC) \rightarrow A$$

and that

$$\tfrac{1}{2}([A] - [A]') = \tfrac{1}{2}(A + BC - A + BC) \rightarrow BC$$

Thus, for all three pairs of linear combinations, we would obtain the following:

i	From $\frac{1}{2}([i] + [i]')$	From $\frac{1}{2}([i] - [i]')$
A	A	BC
B	B	AC
C	C	AB

Furthermore, by assembling the full 2^3 in this fashion with $I = +ABC$ in the first group of runs and $I = -ABC$ in the second, the 2^3 confounds ABC with blocks.

8.2.2 Design Resolution

The preceding 2^{3-1} design is called a **resolution III design**. In such a design, main effects are aliased with two-factor interactions. A design is of resolution R if no p-factor effect is aliased with another effect containing less than $R - p$ factors. We usually employ a Roman numeral subscript to denote design resolution; thus, the one-half fraction of the 2^3 design with the defining relation $I = ABC$ (or $I = -ABC$) is a 2^{3-1}_{III} design.

Designs of resolution III, IV, and V are particularly important. The definitions of these designs and an example of each follow:

1. ***Resolution III designs.*** These are designs in which no main effects are aliased with any other main effect, but main effects are aliased with two-factor interactions and some two-factor interactions may be aliased with each other. The 2^{3-1} design in Table 8.1 is of resolution III (2^{3-1}_{III}).
2. ***Resolution IV designs.*** These are designs in which no main effect is aliased with any other main effect *or* with any two-factor interaction, but two-factor interactions are aliased with each other. A 2^{4-1} design with $I = ABCD$ is a resolution IV design (2^{4-1}_{IV}).
3. ***Resolution V designs.*** These are designs in which no main effect or two-factor interaction is aliased with any other main effect or two-factor interaction, but two-factor interactions are aliased with three-factor interactions. A 2^{5-1} design with $I = ABCDE$ is a resolution V design (2^{5-1}_{V}).

In general, the resolution of a two-level fractional factorial design is equal to the number of letters in the shortest word in the defining relation. Consequently, we could call the preceding design types three-, four-, and five-letter designs, respectively. We usually like to employ fractional designs that have the highest possible resolution consistent with the degree of fractionation required. The higher the resolution, the less restrictive the assumptions that are required regarding which interactions are negligible to obtain a unique interpretation of the results.

8.2.3 Construction and Analysis of the One-Half Fraction

A one-half fraction of the 2^k design of the highest resolution may be constructed by writing down a **basic design** consisting of the runs for a *full* 2^{k-1} factorial and then adding the kth factor by identifying its plus and minus levels with the plus and minus signs of the highest order interaction $ABC \cdots (K - 1)$. Therefore, the 2^{3-1}_{III} fractional factorial is obtained by writing down the full 2^2 factorial as the basic design and then equating factor C to the AB interaction. The alternate fraction would be obtained by equating factor C to the $-AB$ interaction. This approach is illustrated in Table 8.2. Notice that the basic design always has the right number

■ **TABLE 8.2**
The Two One-Half Fractions of the 2^3 Design

	Full 2^2 Factorial (Basic Design)		$2^{3-1}_{\text{III}}, I = ABC$			$2^{3-1}_{\text{III}}, I = -ABC$		
Run	A	B	A	B	$C = AB$	A	B	$C = -AB$
1	−	−	−	−	+	−	−	−
2	+	−	+	−	−	+	−	+
3	−	+	−	+	−	−	+	+
4	+	+	+	+	+	+	+	−

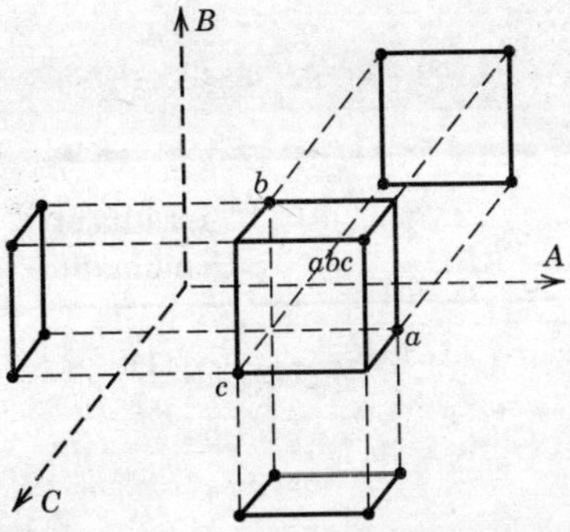

■ **FIGURE 8.2** Projection of a 2^{3-1}_{III} design into three 2^2 designs

of runs (rows), but it is missing one column. The generator $I = ABC \cdots K$ is then solved for the missing column (K) so that $K = ABC \cdots (K - 1)$ defines the product of plus and minus signs to use in each row to produce the levels for the kth factor.

Note that *any* interaction effect could be used to generate the column for the kth factor. However, using any effect other than $ABC \cdots (K - 1)$ will not produce a design of the highest possible resolution.

Another way to view the construction of a one-half fraction is to partition the runs into two blocks with the highest order interaction $ABC \cdots K$ confounded. Each block is a 2^{k-1} fractional factorial design of the highest resolution.

Projection of Fractions into Factorials. Any fractional factorial design of resolution R contains complete factorial designs (possibly replicated factorials) in any subset of $R - 1$ factors. This is an important and useful concept. For example, if an experimenter has several factors of potential interest but believes that only $R - 1$ of them have important effects, then a fractional factorial design of resolution R is the appropriate choice of design. If the experimenter is correct, the fractional factorial design of resolution R will project into a full factorial in the $R - 1$ significant factors. This property is illustrated in Figure 8.2 for the 2^{3-1}_{III} design, which projects into a 2^2 design in every subset of two factors.

Because the maximum possible resolution of a one-half fraction of the 2^k design is $R = k$, every 2^{k-1} design will project into a full factorial in any $(k - 1)$ of the original k factors. Furthermore, a 2^{k-1} design may be projected into two replicates of a full factorial in any subset of $k - 2$ factors, four replicates of a full factorial in any subset of $k - 3$ factors, and so on.

EXAMPLE 8.1

Consider the filtration rate experiment in Example 6.2. The original design, shown in Table 6.10, is a single replicate of the 2^4 design. In that example, we found that the main effects A, C, and D and the interactions AC and AD were different from zero. We will now return to this experiment and simulate what would have happened if a half-fraction of the 2^4 design had been run instead of the full factorial.

We will use the 2^{4-1} design with $I = ABCD$, because this choice of generator will result in a design of the highest possible resolution (IV). To construct the design, we first write down the basic design, which is a 2^3 design, as shown in the first three columns of Table 8.3. This basic design has the necessary number of runs (eight) but only three columns (factors). To find the fourth factor levels, solve $I = ABCD$ for D, or $D = ABC$. Thus, the level of D in each run is the product of the plus and minus signs in columns A, B, and C. The process is illustrated in Table 8.3. Because the generator $ABCD$ is positive, this 2^{4-1}_{IV} design is the principal fraction. The design is shown graphically in Figure 8.3.

Using the defining relation, we note that each main effect is aliased with a three-factor interaction; that is, $A = A^2BCD = BCD$, $B = AB^2CD = ACD$, $C = ABC^2D = ABD$, and $D = ABCD^2 = ABC$. Furthermore, every two-factor

■ **TABLE 8.3**
The 2_{IV}^{4-1} Design with the Defining Relation $I = ABCD$

Run	Basic Design A	B	C	$D = ABC$	Treatment Combination	Filtration Rate
1	−	−	−	−	(1)	45
2	+	−	−	+	*ad*	100
3	−	+	−	+	*bd*	45
4	+	+	−	−	*ab*	65
5	−	−	+	+	*cd*	75
6	+	−	+	−	*ac*	60
7	−	+	+	−	*bc*	80
8	+	+	+	+	*abcd*	96

interaction is aliased with another two-factor interaction. These alias relationships are $AB = CD$, $AC = BD$, and $BC = AD$. The four main effects plus the three two-factor interaction alias pairs account for the seven degrees of freedom for the design.

At this point, we would normally randomize the eight runs and perform the experiment. Because we have already run the full 2^4 design, we will simply select the eight observed filtration rates from Example 6.2 that correspond to the runs in the 2_{IV}^{4-1} design. These observations are shown in the last column of Table 8.3 as well as in Figure 8.3.

The estimates of the effects obtained from this 2_{IV}^{4-1} design are shown in Table 8.4. To illustrate the calculations, the linear combination of observations associated with the *A* effect is

$$[A] = \tfrac{1}{4}(-45 + 100 - 45 + 65 - 75 + 60 - 80 + 96) = 19.00 \rightarrow A + BCD$$

whereas for the *AB* effect, we would obtain

$$[AB] = \tfrac{1}{4}(45 - 100 - 45 + 65 + 75 - 60 - 80 + 96) = -1.00 \rightarrow AB + CD$$

From inspection of the information in Table 8.4, it is not unreasonable to conclude that the main effects *A*, *C*, and *D* are large. The $AB + CD$ alias chain has a small estimate, so the simplest interpretation is that both the *AB* and *CD* interactions are negligible (otherwise, both *AB* and *CD* are large, but they have nearly identical magnitudes and opposite signs—this is fairly unlikely). Furthermore, if *A*, *C*, and *D* are the important main effects, then it is logical to conclude that the two interaction alias chains $AC + BD$ and $AD + BC$ have large effects because the *AC* and *AD* interactions are also significant. In other words, if *A*, *C*, and *D* are significant then the significant interactions are most likely *AC* and *AD*. This is an application of **Ockham's razor** (after William of Ockham), a scientific principle that when one is confronted with several different possible interpretations of a phenomena, the simplest interpretation is usually the correct one. Note that this interpretation agrees with the conclusions from the analysis of the complete 2^4 design in Example 6.2.

Another way to view this interpretation is from the standpoint of **effect heredity**. Suppose that *AB* is significant and that both main effects *A* and *B* are significant. This is called **strong heredity**, and it is the usual situation (if an intraction is significant and only one of the main

■ **FIGURE 8.3** **The 2_{IV}^{4-1} design for the filtration rate experiment of Example 8.1**

■ **TABLE 8.4**
Estimates of Effects and Aliases from Example 8.1[a]

Estimate	Alias Structure
[A] = 19.00	$[A] \rightarrow A + BCD$
[B] = 1.50	$[B] \rightarrow B + ACD$
[C] = 14.00	$[C] \rightarrow C + ABD$
[D] = 16.50	$[D] \rightarrow D + ABC$
[AB] = −1.00	$[AB] \rightarrow AB + CD$
[AC] = −18.50	$[AC] \rightarrow AC + BD$
[AD] = 19.00	$[AD] \rightarrow AD + BC$

[a] Significant effects are shown in boldface type.

effects is significant this is called **weak heredity**; and this is relatively less common). So in this example, with A significant and B not significant this support the assumption that AB is not significant.

Because factor B is not significant, we may drop it from consideration. Consequently, we may project this 2_{IV}^{4-1} design into a single replicate of the 2^3 design in factors A, C, and D, as shown in Figure 8.4. Visual examination of this cube plot makes us more comfortable with the conclusions reached above. Notice that if the temperature (A) is at the low level, the concentration (C) has a large positive effect, whereas if the temperature is at the high level, the concentration has a very small effect. This is probably due to an AC interaction. Furthermore, if the temperature is at the low level, the effect of the stirring rate (D) is negligible, whereas if the temperature is at the high level, the stirring rate has a large positive effect. This is probably due to the AD interaction tentatively identified previously.

Based on the above analysis, we can now obtain a model to predict filtration rate over the experimental region. This model is

$$\hat{y} = \hat{\beta}_0 + \hat{\beta}_1 x_1 + \hat{\beta}_3 x_3 + \hat{\beta}_4 x_4 + \hat{\beta}_{13} x_1 x_3 + \hat{\beta}_{14} x_1 x_4$$

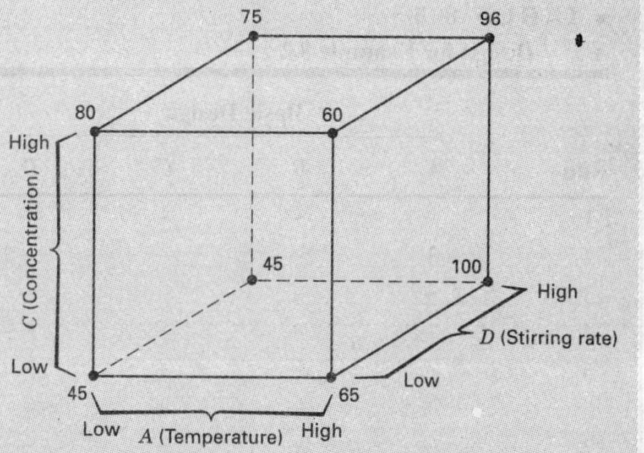

■ **FIGURE 8.4** **Projection of the 2_{IV}^{4-1} design into a 2^3 design in A, C, and D for Example 8.1**

where x_1, x_3, and x_4 are coded variables ($-1 \leq x_i \leq +1$) that represent A, C, and D, and the $\hat{\beta}$'s are regression coefficients that can be obtained from the effect estimates as we did previously. Therefore, the prediction equation is

$$\hat{y} = 70.75 + \left(\frac{19.00}{2}\right) x_1 + \left(\frac{14.00}{2}\right) x_3 + \left(\frac{16.50}{2}\right) x_4 + \left(\frac{-18.50}{2}\right) x_1 x_3 + \left(\frac{19.00}{2}\right) x_1 x_4$$

Remember that the intercept $\hat{\beta}_0$ is the average of all responses at the eight runs in the design. This model is very similar to the one that resulted from the full 2^k factorial design in Example 6.2.

EXAMPLE 8.2 A 2^{5-1} Design Used for Process Improvement

Five factors in a manufacturing process for an integrated circuit were investigated in a 2^{5-1} design with the objective of improving the process yield. The five factors were A = aperture setting (small, large), B = exposure time (20 percent below nominal, 20 percent above nominal), C = develop time (30 and 45 sec), D = mask dimension (small, large), and E = etch time (14.5 and 15.5 min). The construction of the 2^{5-1} design is shown in Table 8.5. Notice that the design was constructed by writing down the basic design having 16 runs (a 2^4 design in A, B, C, and D), selecting $ABCDE$ as the generator, and then setting the levels of the fifth factor $E = ABCD$. Figure 8.5 gives a pictorial representation of the design.

The defining relation for the design is $I = ABCDE$. Consequently, every main effect is aliased with a four-factor interaction (for example, $[A] \rightarrow A + BCDE$), and every two-factor interaction is aliased with a three-factor interaction (e.g., $[AB] \rightarrow AB + CDE$). Thus, the design is of resolution V. We would expect this 2^{5-1} design to provide excellent information concerning the main effects and two-factor interactions.

■ **TABLE 8.5**
A 2^{5-1} Design for Example 8.2

Run	Basic Design A	B	C	D	E = ABCD	Treatment Combination	Yield
1	−	−	−	−	+	*e*	8
2	+	−	−	−	−	*a*	9
3	−	+	−	−	−	*b*	34
4	+	+	−	−	+	*abe*	52
5	−	−	+	−	−	*c*	16
6	+	−	+	−	+	*ace*	22
7	−	+	+	−	+	*bce*	45
8	+	+	+	−	−	*abc*	60
9	−	−	−	+	−	*d*	6
10	+	−	−	+	+	*ade*	10
11	−	+	−	+	+	*bde*	30
12	+	+	−	+	−	*abd*	50
13	−	−	+	+	+	*cde*	15
14	+	−	+	+	−	*acd*	21
15	−	+	+	+	−	*bcd*	44
16	+	+	+	+	+	*abcde*	63

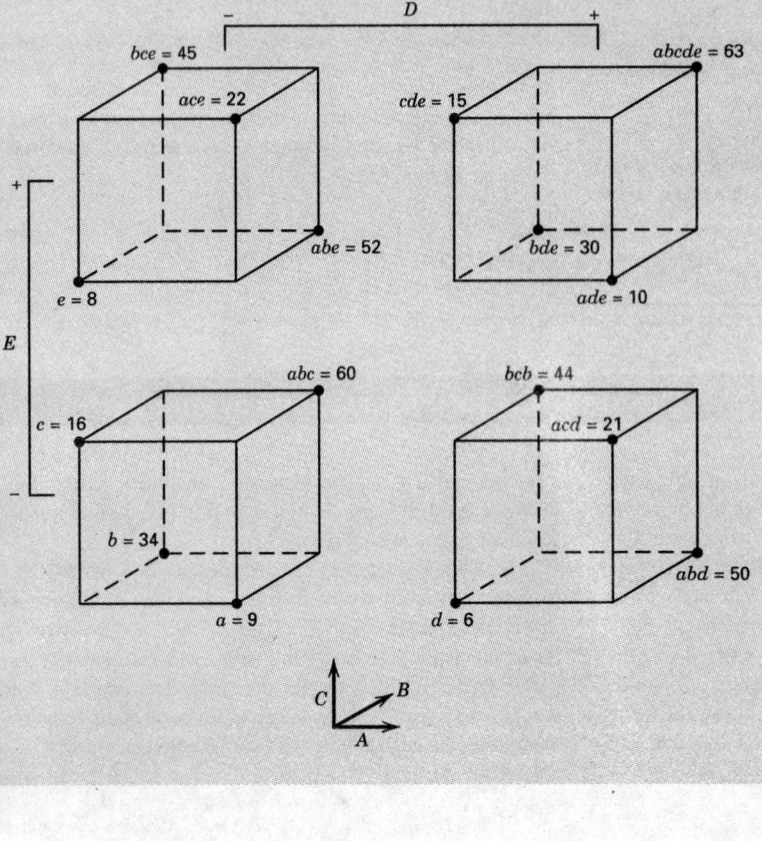

■ **FIGURE 8.5** The 2_V^{5-1} design for Example 8.2

■ **TABLE 8.6**
Effects, Regression Coefficients, and Sums of Squares for Example 8.2

Variable	Name	− 1 Level	+ 1 Level
A	Aperture	Small	Large
B	Exposure time	−20%	+20%
C	Develop time	30 sec	40 sec
D	Mask dimension	Small	Large
E	Etch time	14.5 min	15.5 min

Variable	Regression Coefficient	Estimated Effect	Sum of Squares
Overall Average	30.3125		
A	5.5625	11.1250	495.062
B	16.9375	33.8750	4590.062
C	5.4375	10.8750	473.062
D	−0.4375	−0.8750	3.063
E	0.3125	0.6250	1.563
AB	3.4375	6.8750	189.063
AC	0.1875	0.3750	0.563
AD	0.5625	1.1250	5.063
AE	0.5625	1.1250	5.063
BC	0.3125	0.6250	1.563
BD	−0.0625	−0.1250	0.063
BE	−0.0625	−0.1250	0.063
CD	0.4375	0.8750	3.063
CE	0.1875	0.3750	0.563
DE	−0.6875	−1.3750	7.563

Table 8.6 contains the effect estimates, sums of squares, and model regression coefficients for the 15 effects from this experiment. Figure 8.6 presents a normal probability plot of the effect estimates from this experiment. The main effects of *A*, *B*, and *C* and the *AB* interaction are large. Remember that, because of aliasing, these effects are really $A + BCDE$, $B + ACDE$, $C + ABDE$, and $AB + CDE$. However, because it seems plausible that three-factor and higher interactions are negligible, we feel safe in concluding that only *A*, *B*, *C*, and *AB* are important effects.

Table 8.7 summarizes the analysis of variance for this experiment. The model sum of squares is $SS_{\text{Model}} = SS_A + SS_B + SS_C + SS_{AB} = 5747.25$, and this accounts for over 99 percent of the total variability in yield. Figure 8.7 presents a normal probability plot of the residuals, and Figure 8.8 is a plot of the residuals versus the predicted values. Both plots are satisfactory.

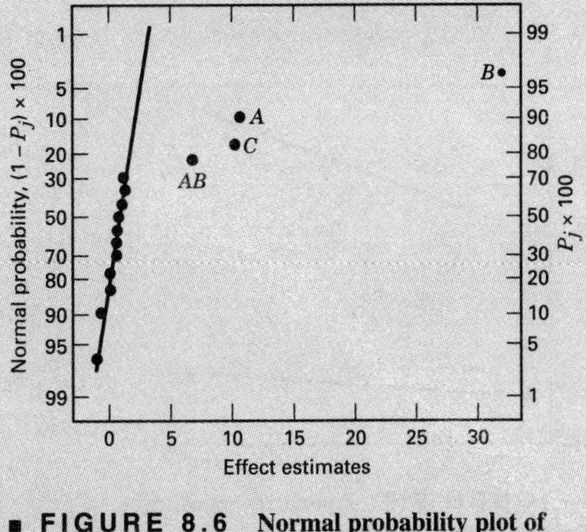

■ **FIGURE 8.6** Normal probability plot of effects for Example 8.2

■ **TABLE 8.7**
Analysis of Variance for Example 8.2

Source of Variation	Sum of Squares	Degrees of Freedom	Mean Square	F_0	*P*-Value
A (Aperture)	495.0625	1	495.0625	193.20	<0.0001
B (Exposure time)	4590.0625	1	4590.0625	1791.24	<0.0001
C (Develop time)	473.0625	1	473.0625	184.61	<0.0001
AB	189.0625	1	189.0625	73.78	<0.0001
Error	28.1875	11	2.5625		
Total	5775.4375	15			

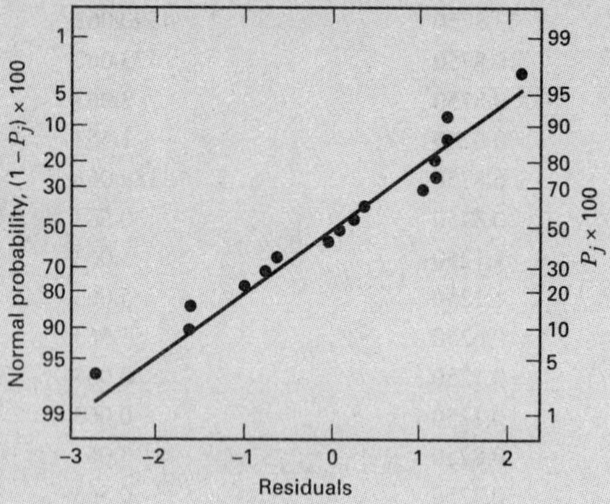

■ **FIGURE 8.7 Normal probability plot of the residuals for Example 8.2**

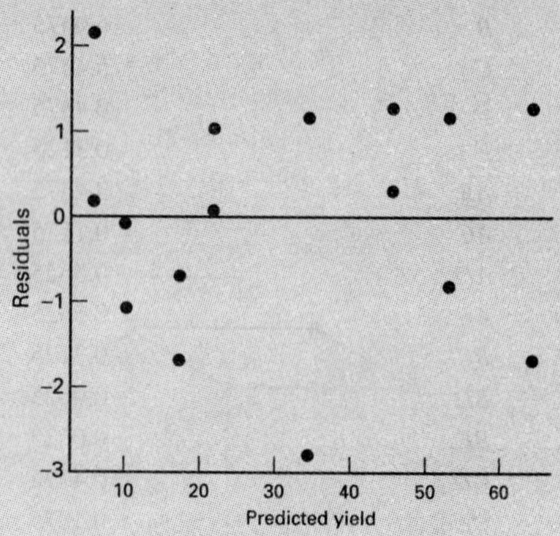

■ **FIGURE 8.8 Plot of residuals versus predicted yield for Example 8.2**

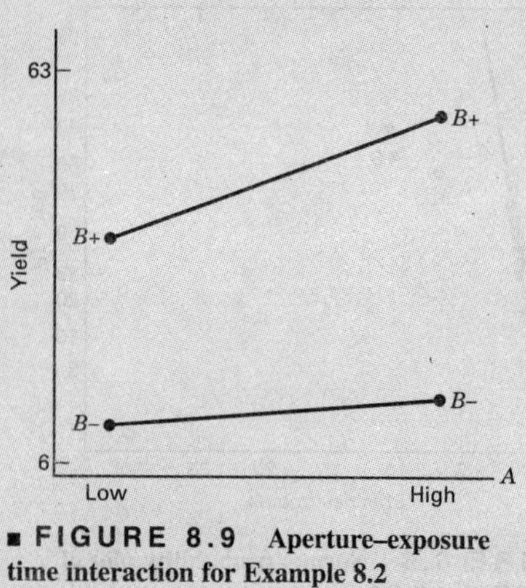

■ **FIGURE 8.9 Aperture–exposure time interaction for Example 8.2**

The three factors *A*, *B*, and *C* have large positive effects. The *AB* or aperture–exposure time interaction is plotted in Figure 8.9. This plot confirms that the yields are higher when both *A* and *B* are at the high level.

The 2^{5-1} design will collapse into two replicates of a 2^3 design in any three of the original five factors. (Looking at Figure 8.5 will help you visualize this.) Figure 8.10 is a cubę plot in the factors *A*, *B*, and *C* with the average yields superimposed on the eight corners. It is clear from inspection of the cube plot that highest yields are achieved with *A*, *B*, and *C* all at the high level. Factors *D* and *E* have little effect on average process yield and may be set to values that optimize other objectives (such as cost).

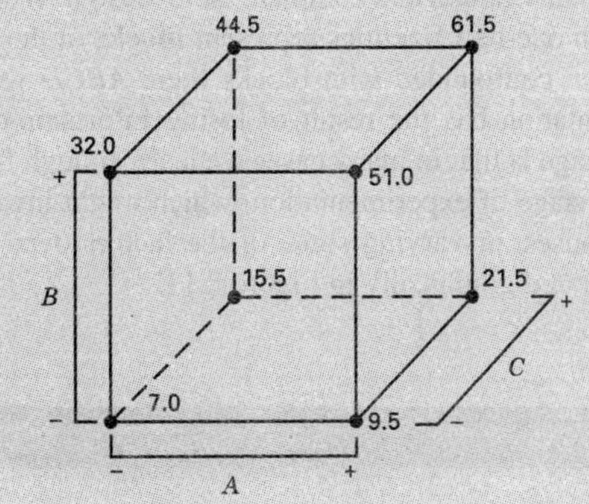

■ **FIGURE 8.10** **Projection of the 2_V^{5-1} design in Example 8.2 into two replicates of a 2^3 design in the factors *A*, *B*, and *C***

Sequences of Fractional Factorials. Using fractional factorial designs often leads to great economy and efficiency in experimentation, particularly if the runs can be made **sequentially**. For example, suppose that we are investigating $k = 4$ factors ($2^4 = 16$ runs). It is almost always preferable to run a 2_{IV}^{4-1} fractional design (eight runs), analyze the results, and then decide on the best set of runs to perform next. If it is necessary to resolve ambiguities,

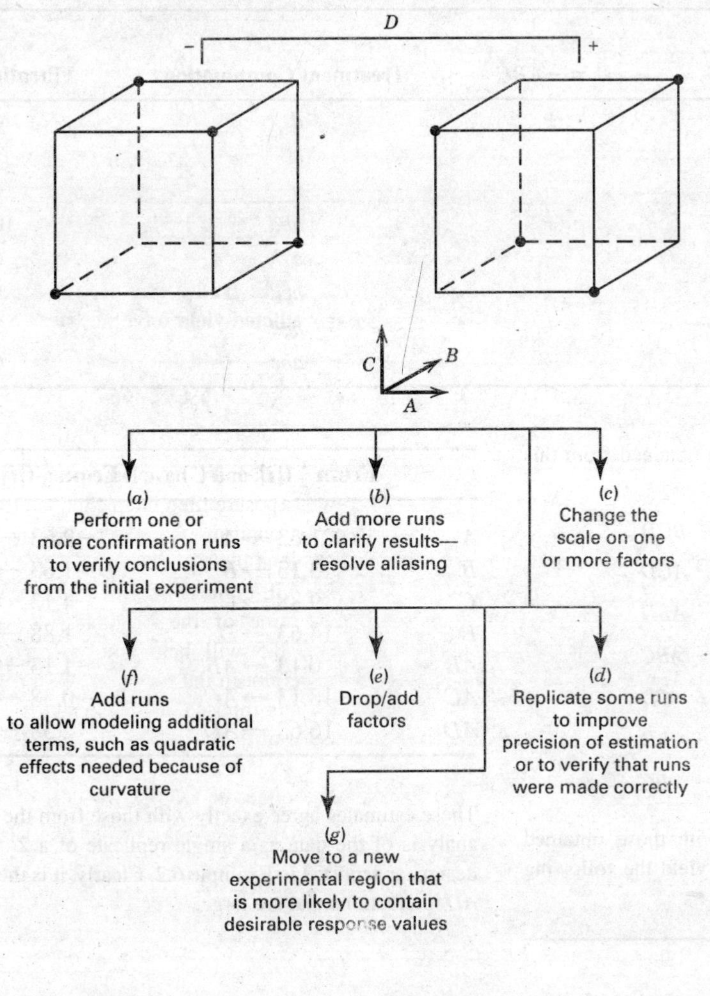

■ **FIGURE 8.11** **Possibilities for follow-up experimentation after an initial fractional factorial experiment**

we can always run the alternate fraction and complete the 2^4 design. When this method is used to complete the design, both one-half fractions represent **blocks** of the complete design with the highest order interaction confounded with blocks (here *ABCD* would be confounded). Thus, sequential experimentation has the result of losing information only on the highest order interaction. Its advantage is that in many cases we learn enough from the one-half fraction to proceed to the next stage of experimentation, which might involve adding or removing factors, changing responses, or varying some of the factors over new ranges. Some of these possibilities are illustrated graphically in Figure 8.11.

EXAMPLE 8.3

Reconsider the experiment in Example 8.1. We have used a 2_{IV}^{4-1} design and tentatively identified three large main effects—*A*, *C*, and *D*. There are two large effects associated with two-factor interactions, $AC + BD$ and $AD + BC$. In Example 8.2, we used the fact that the main effect of *B* was negligible to tentatively conclude that the important interactions were *AC* and *AD*. Sometimes the experimenter will have process knowledge that can assist in discriminating between interactions likely to be important. However, we can always isolate the significant interaction by running the alternate fraction, given by $I = -ABCD$. It is straightforward to show that the design and the responses are as follows:

	Basic Design					
Run	**A**	**B**	**C**	$D = -ABC$	**Treatment Combination**	**Filtration Rate**
1	−	−	−	+	*d*	43
2	+	−	−	−	*a*	71
3	−	+	−	−	*b*	48
4	+	+	−	+	*abd*	104
5	−	−	+	−	*c*	68
6	+	−	+	+	*acd*	86
7	−	+	+	+	*bcd*	70
8	+	+	+	−	*abc*	65

The effect estimates (and their aliases) obtained from this alternate fraction are

$$[A]' = 24.25 \rightarrow A - BCD$$
$$[B]' = 4.75 \rightarrow B - ACD$$
$$[C]' = 5.75 \rightarrow C - ABD$$
$$[D]' = 12.75 \rightarrow D - ABC$$
$$[AB]' = 1.25 \rightarrow AB - CD$$
$$[AC]' = -17.75 \rightarrow AC - BD$$
$$[AD]' = 14.25 \rightarrow AD - BC$$

These estimates may be combined with those obtained from the original one-half fraction to yield the following estimates of the effects:

i	From $\frac{1}{2}([i] + [i]')$	From $\frac{1}{2}([i] - [i]')$
A	21.63 → *A*	−2.63 → *BCD*
B	3.13 → *B*	−1.63 → *ACD*
C	9.88 → *C*	4.13 → *ABD*
D	14.63 → *D*	1.88 → *ABC*
AB	0.13 → *AB*	−1.13 → *CD*
AC	−18.13 → *AC*	−0.38 → *BD*
AD	16.63 → *AD*	2.38 → *BC*

These estimates agree exactly with those from the original analysis of the data as a single replicate of a 2^4 factorial design, as reported in Example 6.2. Clearly, it is the *AC* and *AD* interactions that are large.

Confirmation Experiments. Adding the alternate fraction to the principal fraction may be thought of as a type of **confirmation experiment** in that it provides information that will allow us to strengthen our initial conclusions about the two-factor interaction effects. We will investigate some other aspects of combining fractional factorials to isolate interactions in Sections 8.5 and 8.6.

A confirmation experiment need not be this elaborate. A very simple confirmation experiment is to use the model equation to predict the response at a point of interest in the design space (this should not be one of the runs in the current design) and then actually run that treatment combination (perhaps several times), comparing the predicted and observed responses. Reasonably close agreement indicates that the interpretation of the fractional factorial was correct, whereas serious discrepancies mean that the interpretation was problematic. This would be an indication that additional experimentation is required to resolve ambiguities.

To illustrate, consider the 2^{4-1} fractional factorial in Example 8.1. The experimenters are interested in finding a set of conditions where the response variable filtration rate is high, but low concentrations of formaldehyde (factor C) are desirable. This would suggest that factors A and D should be at the high level and factor C should be at the low level. Examining Figure 8.3, we note that when B is at the low level, this treatment combination was run in the fractional factorial, producing an observed response of 100. The treatment combination with B at the high level was not in the original fraction, so this would be a reasonable confirmation run. With A, B, and D at the high level and C at the low level, we use the model equation from Example 8.1 to calculate the predicted response as follows:

$$\begin{aligned}\hat{y} &= 70.75 + \left(\frac{19.00}{2}\right)x_1 + \left(\frac{14.00}{2}\right)x_3 + \left(\frac{16.50}{2}\right)x_4 + \left(\frac{-18.50}{2}\right)x_1x_3 + \left(\frac{19.00}{2}\right)x_1x_4 \\ &= 70.75 + \left(\frac{19.00}{2}\right)(1) + \left(\frac{14.00}{2}\right)(-1) + \left(\frac{16.50}{2}\right)(1) + \left(\frac{-18.50}{2}\right)(1)(-1) \\ &\quad + \left(\frac{19.00}{2}\right)(1)(1) \\ &= 100.25\end{aligned}$$

The observed response at this treatment combination is 104 (refer to Figure 6.10 where the response data for the complete 2^4 factorial design are presented). Since the observed and predicted values of filtration rate are very similar, we have a successful confirmation run. This is additional evidence that our interpretation of the fractional factorial was correct.

There will be situations where the predicted and observed values in a confirmation experiment will not be this close together, and it will be necessary to answer the question of whether the two values are sufficiently close to reasonably conclude that the interpretation of the fractional design was correct. One way to answer this question is to construct a **prediction interval** on the future observation for the confirmation run and then see if the actual observation falls inside the prediction interval. We show how to do this using this example in Section 10.6, where prediction intervals for a regression model are introduced.

8.3 The One-Quarter Fraction of the 2^k Design

For a moderately large number of factors, smaller fractions of the 2^k design are frequently useful. Consider a one-quarter fraction of the 2^k design. This design contains 2^{k-2} runs and is usually called a $\mathbf{2^{k-2}}$ **fractional factorial**.

The 2^{k-2} design may be constructed by first writing down a **basic design** consisting of the runs associated with a full factorial in $k - 2$ factors and then associating the two additional columns with appropriately chosen interactions involving the first $k - 2$ factors. Thus, a one-quarter fraction of the 2^k design has two generators. If P and Q represent the generators chosen, then $I = P$ and $I = Q$ are called the **generating relations** for the design. The signs of P and Q (either + or −) determine which one of the one-quarter fractions is produced. All four fractions associated with the choice of **generators** $\pm P$ and $\pm Q$ are members of the same **family**. The fraction for which both P and Q are positive is the principal fraction.

The **complete defining relation** for the design consists of all the columns that are equal to the identity column I. These will consist of P, Q, and their **generalized interaction** PQ; that is, the defining relation is $I = P = Q = PQ$. We call the elements P, Q, and PQ in the defining relation **words**. The aliases of any effect are produced by the multiplication of the column for that effect by each word in the defining relation. Clearly, each effect has *three* aliases. The experimenter should be careful in choosing the generators so that potentially important effects are not aliased with each other.

As an example, consider the 2^{6-2} design. Suppose we choose $I = ABCE$ and $I = BCDF$ as the design generators. Now the generalized interaction of the generators $ABCE$ and $BCDF$ is $ADEF$; therefore, the complete defining relation for this design is

$$I = ABCE = BCDF = ADEF$$

Consequently, this is a resolution IV design. To find the aliases of any effect (e.g., A), multiply that effect by each word in the defining relation. For A, this produces

$$A = BCE = ABCDF = DEF$$

It is easy to verify that every main effect is aliased by three- and five-factor interactions, whereas two-factor interactions are aliased with each other and with higher order interactions. Thus, when we estimate A, for example, we are really estimating $A + BCE + DEF + ABCDF$. The complete alias structure of this design is shown in Table 8.8. If three-factor and higher interactions are negligible, this design gives clear estimates of the main effects.

To construct the design, first write down the **basic design**, which consists of the 16 runs for a full $2^{6-2} = 2^4$ design in A, B, C, and D. Then the two factors E and F are added by associating their plus and minus levels with the plus and minus signs of the interactions ABC and BCD, respectively. This procedure is shown in Table 8.9.

Another way to construct this design is to derive the four blocks of the 2^6 design with $ABCE$ and $BCDF$ confounded and then choose the block with treatment combinations that are positive on $ABCE$ and $BCDF$. This would be a 2^{6-2} fractional factorial with generating relations $I = ABCE$ and $I = BCDF$, and because both generators $ABCE$ and $BCDF$ are positive, this is the principal fraction.

■ **TABLE 8.8**
Alias Structure for the 2^{6-2}_{IV} Design with $I = ABCE = BCDF = ADEF$

$A = BCE = DEF = ABCDF$	$AB = CE = ACDF = BDEF$
$B = ACE = CDF = ABDEF$	$AC = BE = ABDF = CDEF$
$C = ABE = BDF = ACDEF$	$AD = EF = BCDE = ABCF$
$D = BCF = AEF = ABCDE$	$AE = BC = DF = ABCDEF$
$E = ABC = ADF = BCDEF$	$AF = DE = BCEF = ABCD$
$F = BCD = ADE = ABCEF$	$BD = CF = ACDE = ABEF$
	$BF = CD = ACEF = ABDE$
$ABD = CDE = ACF = BEF$	
$ACD = BDE = ABF = CEF$	

■ **TABLE 8.9**
Construction of the 2^{6-2}_{IV} Design with the Generators $I = ABCE$ and $I = BCDF$

	Basic Design					
Run	A	B	C	D	$E = ABC$	$F = BCD$
1	−	−	−	−	−	−
2	+	−	−	−	+	−
3	−	+	−	−	+	+
4	+	+	−	−	−	+
5	−	−	+	−	+	+
6	+	−	+	−	−	+
7	−	+	+	−	−	−
8	+	+	+	−	+	−
9	−	−	−	+	−	+
10	+	−	−	+	+	+
11	−	+	−	+	+	−
12	+	+	−	+	−	−
13	−	−	+	+	+	−
14	+	−	+	+	−	−
15	−	+	+	+	−	+
16	+	+	+	+	+	+

There are, of course, three **alternate fractions** of this particular 2^{6-2}_{IV} design. They are the fractions with generating relationships $I = ABCE$ and $I = -BCDF$; $I = -ABCE$ and $I = BCDF$; and $I = -ABCE$ and $I = -BCDF$. These fractions may be easily constructed by the method shown in Table 8.9. For example, if we wish to find the fraction for which $I = ABCE$ and $I = -BCDF$, then in the last column of Table 8.9 we set $F = -BCD$, and the column of levels for factor F becomes

$$+ + - - - - + + - - + + + + - -$$

The complete defining relation for this alternate fraction is $I = ABCE = -BCDF = -ADEF$. Certain signs in the alias structure in Table 8.9 are now changed; for instance, the aliases of A are $A = BCE = -DEF = -ABCDF$. Thus, the linear combination of the observations $[A]$ actually estimates $A + BCE - DEF - ABCDF$.

Finally, note that the 2^{6-2}_{IV} fractional factorial will **project** into a single replicate of a 2^4 design in any subset of four factors that is not a word in the defining relation. It also collapses to a replicated one-half fraction of a 2^4 in any subset of four factors that is a word in the defining relation. Thus, the design in Table 8.9 becomes two replicates of a 2^{4-1} in the factors $ABCE$, $BCDF$, and $ADEF$, because these are the words in the defining relation. There are 12 other combinations of the six factors, such as $ABCD$, $ABCF$, for which the design projects to a single replicate of the 2^4. This design also collapses to two replicates of a 2^3 in *any* subset of three of the six factors or four replicates of a 2^2 in any subset of two factors.

In general, any 2^{k-2} fractional factorial design can be collapsed into either a full factorial or a fractional factorial in some subset of $r \leq k - 2$ of the original factors. Those subsets of variables that form full factorials are not words in the complete defining relation.

EXAMPLE 8.4

Parts manufactured in an injection molding process are showing excessive shrinkage. This is causing problems in assembly operations downstream from the injection molding area. A quality improvement team has decided to use a designed experiment to study the injection molding process so that shrinkage can be reduced. The team decides to investigate six factors—mold temperature (A), screw speed (B), holding time (C), cycle time (D), gate size (E), and holding pressure (F)—each at two levels, with the objective of learning how each factor affects shrinkage and also, something about how the factors interact.

The team decides to use the 16-run two-level fractional factorial design in Table 8.9. The design is shown again in Table 8.10, along with the observed shrinkage ($\times 10$) for the test part produced at each of the 16 runs in the design. Table 8.11 shows the effect estimates, sums of squares, and the regression coefficients for this experiment.

A normal probability plot of the effect estimates from this experiment is shown in Figure 8.12. The only large effects are A (mold temperature), B (screw speed), and the AB interaction. In light of the alias relationships in Table 8.8, it seems reasonable to adopt these conclusions tentatively. The plot of the AB interaction in Figure 8.13 shows that the process is very insensitive to temperature if the screw speed is at the low level but very sensitive to temperature if the screw speed is at the high level. With the screw speed at the low level, the process should produce an average shrinkage of around 10 percent regardless of the temperature level chosen.

Based on this initial analysis, the team decides to set both the mold temperature and the screw speed at the low level. This set of conditions will reduce the *mean* shrinkage of parts to around 10 percent. However, the variability in shrinkage from part to part is still a potential problem. In effect, the mean shrinkage can be adequately reduced by the above modifications; however, the part-to-part variability in shrinkage over a production run could still cause problems in assembly. One way to address this issue is to see if any of the process factors affect the *variability* in parts shrinkage.

■ **TABLE 8.10**

A 2^{6-2}_{IV} Design for the Injection Molding Experiment in Example 8.4

	Basic Design						Observed Shrinkage	
Run	A	B	C	D	$E = ABC$	$F = BCD$	($\times$ 10)	
1	−	−	−	−	−	−	6	(1)
2	+	−	−	−	+	−	10	*ae*
3	−	+	−	−	+	+	32	*bef*
4	+	+	−	−	−	+	60	*abf*
5	−	−	+	−	+	+	4	*cef*
6	+	−	+	−	−	+	15	*acf*
7	−	+	+	−	−	−	26	*bc*
8	+	+	+	−	+	−	60	*abce*
9	−	−	−	+	−	+	8	*df*
10	+	−	−	+	+	+	12	*adef*
11	−	+	−	+	+	−	34	*bde*
12	+	+	−	+	−	−	60	*abd*
13	−	−	+	+	+	−	16	*cde*
14	+	−	+	+	−	−	5	*acd*
15	−	+	+	+	−	+	37	*bcdf*
16	+	+	+	+	+	+	52	*abcdef*

■ **TABLE 8.11**
Effects, Sums of Squares, and Regression Coefficients for Example 8.4

Variable	Name	−1 Level	+1 Level
A	Mold temperature	−1.000	1.000
B	Screw speed	−1.000	1.000
C	Holding time	−1.000	1.000
D	Cycle time	−1.000	1.000
E	Gate size	−1.000	1.000
F	Hold pressure	−1.000	1.000

Variable[a]	Regression Coefficient	Estimated Effect	Sum of Squares
Overall Average	27.3125		
A	6.9375	13.8750	770.062
B	17.8125	35.6250	5076.562
C	−0.4375	−0.8750	3.063
D	0.6875	1.3750	7.563
E	0.1875	0.3750	0.563
F	0.1875	0.3750	0.563
AB + *CE*	5.9375	11.8750	564.063
AC + *BE*	−0.8125	−1.6250	10.562
AD + *EF*	−2.6875	−5.3750	115.562
AE + *BC* + *DF*	−0.9375	−1.8750	14.063
AF + *DE*	0.3125	0.6250	1.563
BD + *CF*	−0.0625	−0.1250	0.063
BF + *CD*	−0.0625	−0.1250	0.063
ABD	0.0625	0.1250	0.063
ABF	−2.4375	−4.8750	95.063

[a]Only main effects and two-factor interactions.

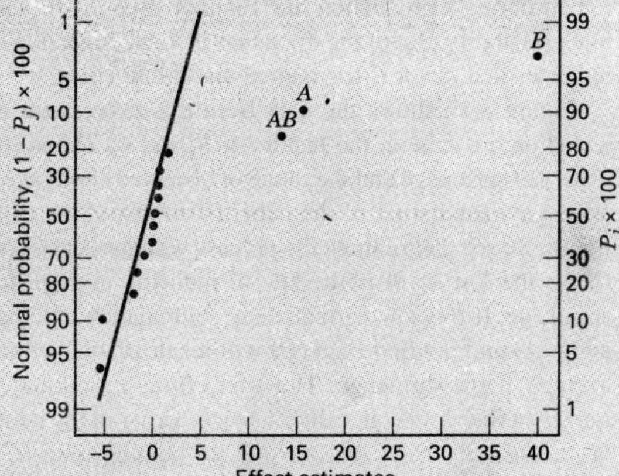

■ **FIGURE 8.12** **Normal probability plot of effects for Example 8.4**

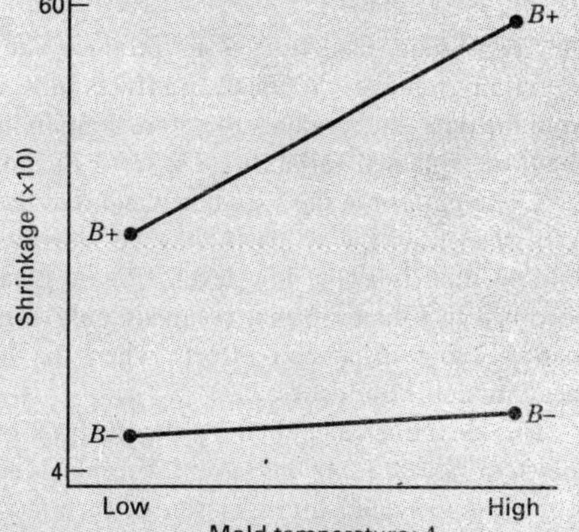

■ **FIGURE 8.13** **Plot of *AB* (mold temperature–screw speed) interaction for Example 8.4**

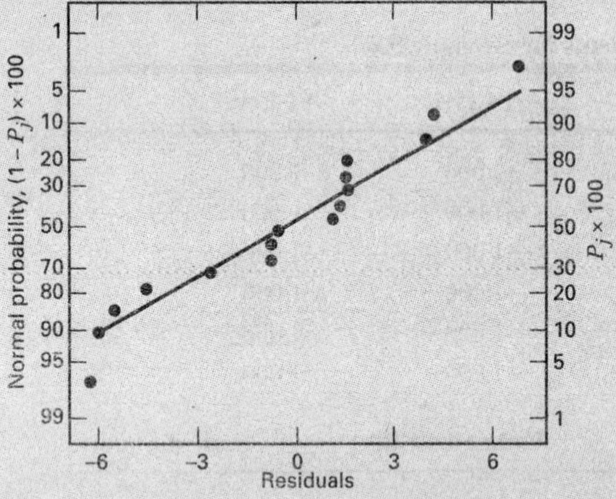

■ **FIGURE 8.14 Normal probability plot of residuals for Example 8.4**

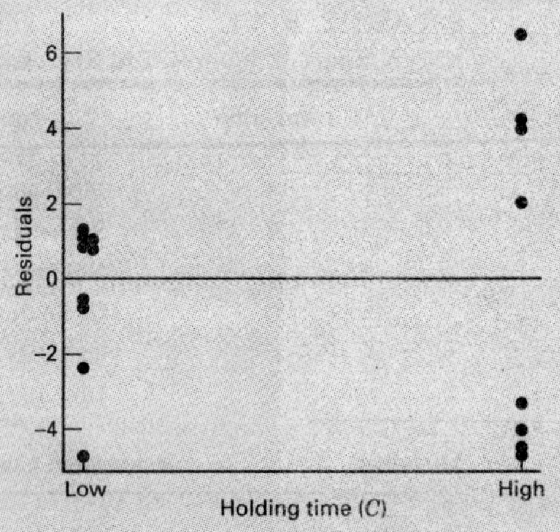

■ **FIGURE 8.15 Residuals versus holding time (*C*) for Example 8.4**

Figure 8.14 presents the normal probability plot of the residuals. This plot appears satisfactory. The plots of residuals versus each factor were then constructed. One of these plots, that for residuals versus factor *C* (holding time), is shown in Figure 8.15. The plot reveals that there is much less scatter in the residuals at the low holding time than at the high holding time. These residuals were obtained in the usual way from a model for predicted shrinkage:

$$\begin{aligned}\hat{y} &= \hat{\beta}_0 + \hat{\beta}_1 x_1 + \hat{\beta}_2 x_2 + \hat{\beta}_{12} x_1 x_2 \\ &= 27.3125 + 6.9375x_1 + 17.8125x_2 + 5.9375x_1x_2\end{aligned}$$

where x_1, x_2, and x_1x_2 are coded variables that correspond to the factors *A* and *B* and the *AB* interaction. The residuals are then

$$e = y - \hat{y}$$

The regression model used to produce the residuals essentially removes the **location effects** of *A*, *B*, and *AB* from the data; the residuals therefore contain information about unexplained variability. Figure 8.15 indicates that there is a *pattern* in the variability and that the variability in the shrinkage of parts may be smaller when the holding time is at the low level. (Please recall that we observed in Chapter 6 that residuals only convey information about dispersion effects when the location or mean model is correct.)

This is further amplified by the analysis of residuals shown in Table 8.12. In this table, the residuals are arranged at the low (−) and high (+) levels of each factor, and the standard deviations of the residuals at the low and high levels of each factor have been calculated. Note that the standard deviation of the residuals with *C* at the low level $[S(C^-) = 1.63]$ is considerably smaller than the standard deviation of the residuals with *C* at the high level $[S(C^+) = 5.70]$.

The bottom line of Table 8.12 presents the statistic

$$F_i^* = \ln \frac{S^2(i^+)}{S^2(i^-)}$$

Recall that if the variances of the residuals at the high (+) and low (−) levels of factor *i* are equal, then this ratio is approximately normally distributed with mean zero, and it can be used to judge the difference in the response variability at the two levels of factor *i*. Because the ratio F_C^* is relatively large, we would conclude that the apparent **dispersion** or **variability effect** observed in Figure 8.15 is real. Thus, setting the holding time at its low level would contribute to reducing the variability in shrinkage from part to part during a production run. Figure 8.16 presents a normal probability plot of the F_i^* values in Table 8.12; this also indicates that factor *C* has a large dispersion effect.

Figure 8.17 shows the data from this experiment projected onto a cube in the factors *A*, *B*, and *C*. The average observed shrinkage and the range of observed shrinkage are shown at each corner of the cube. From inspection of this figure, we see that running the process with the screw speed (*B*) at the low level is the key to reducing average parts shrinkage. If *B* is low, virtually any combination of temperature (*A*) and holding time (*C*) will result in low values of average parts shrinkage. However, from examining the ranges of the shrinkage values at each corner of the cube, it is immediately clear that setting the holding time (*C*) at the low level is the only reasonable choice if we wish to keep the part-to-part variability in shrinkage low during a production run.

■ TABLE 8.12

Calculation of Dispersion Effects for Example 8.4

Run	A	B	$AB = CE$	C	$AC = BE$	$AE = BC = DF$	E	D	$AD = EF$	$BD = CE$	ABD	$BF = CD$	ACD	F	$AF = DE$	Residual
1	−	−	+	−	+	+	−	−	+	+	−	+	−	−	+	−2.50
2	+	−	−	−	−	+	+	−	−	+	+	+	+	−	−	−0.50
3	−	+	−	−	+	−	+	−	+	−	+	+	−	+	−	−0.25
4	+	+	+	−	−	−	−	−	−	−	−	+	+	+	+	2.00
5	−	−	+	+	−	−	+	−	+	+	−	−	+	+	−	−4.50
6	+	−	−	+	+	−	−	−	−	+	+	−	−	+	+	4.50
7	−	+	−	+	−	+	−	−	+	−	+	−	+	−	+	−6.25
8	+	+	+	+	+	+	+	−	−	−	−	−	−	−	−	2.00
9	−	−	+	−	+	+	−	+	−	−	+	−	+	+	−	−0.50
10	+	−	−	−	−	+	+	+	+	−	−	−	−	+	+	1.50
11	−	+	−	−	+	−	+	+	−	+	−	−	+	−	+	1.75
12	+	+	+	−	−	−	−	+	+	+	+	−	−	−	−	2.00
13	−	−	+	+	−	−	+	+	−	−	+	+	−	−	+	7.50
14	+	−	−	+	+	−	−	+	+	−	−	+	+	−	−	−5.50
15	−	+	−	+	−	+	−	+	−	+	−	+	−	+	−	4.75
16	+	+	+	+	+	+	+	+	+	+	+	+	+	+	+	−6.00
$S(i^+)$	3.80	4.01	4.33	5.70	3.68	3.85	4.17	4.64	3.39	4.01	4.72	4.71	3.50	3.88	4.87	
$S(i^-)$	4.60	4.41	4.10	1.63	4.53	4.33	4.25	3.59	2.75	4.41	3.64	3.65	3.12	4.52	3.40	
F_i^*	−0.38	−0.19	0.11	2.50	−0.42	−0.23	−0.04	0.51	0.42	−0.19	0.52	0.51	0.23	−0.31	0.72	

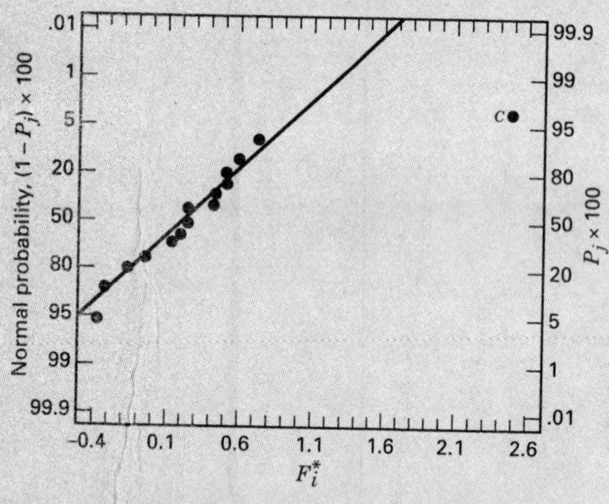

■ **FIGURE 8.16** **Normal probability plot of the dispersion effects F_i^* for Example 8.4**

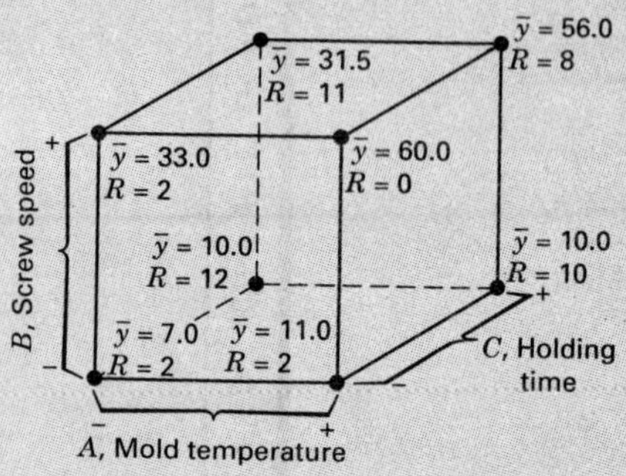

■ **FIGURE 8.17** **Average shrinkage and range of shrinkage in factors A, B, and C for Example 8.4**

8.4 The General 2^{k-p} Fractional Factorial Design

8.4.1 Choosing a Design

A 2^k fractional factorial design containing 2^{k-p} runs is called a $1/2^p$ fraction of the 2^k design or, more simply, a **2^{k-p} fractional factorial design**. These designs require the selection of p independent generators. The defining relation for the design consists of the p generators initially chosen and their $2^p - p - 1$ generalized interactions. In this section, we discuss the construction and analysis of these designs.

The alias structure may be found by multiplying each effect column by the defining relation. Care should be exercised in choosing the generators so that effects of potential interest are not aliased with each other. Each effect has $2^p - 1$ aliases. For moderately large values of k, we usually assume higher order interactions (say, third- or fourth-order and higher) to be negligible, and this greatly simplifies the alias structure.

It is important to select the p generators for a 2^{k-p} fractional factorial design in such a way that we obtain the best possible alias relationships. A reasonable criterion is to select the generators such that the resulting 2^{k-p} design has the **highest possible resolution.** To illustrate, consider the 2_{IV}^{6-2} design in Table 8.9, where we used the generators $E = ABC$ and $F = BCD$, thereby producing a design of resolution IV. This is the maximum resolution design. If we had selected $E = ABC$ and $F = ABCD$, the complete defining relation would have been $I = ABCE = ABCDF = DEF$, and the design would be of resolution III. Clearly, this is an inferior choice because it needlessly sacrifices information about interactions.

Sometimes resolution alone is insufficient to distinguish between designs. For example, consider the three 2_{IV}^{7-2} designs in Table 8.13. All of these designs are of resolution IV, but they have rather different alias structures (we have assumed that three-factor and higher interactions are negligible) with respect to the two-factor interactions. Clearly, design A has more extensive aliasing and design C the least, so design C would be the best choice for a 2_{IV}^{7-2}.

The three word lengths in design A are all 4; that is, the word length **pattern** is {4, 4, 4}. For design B it is {4, 4, 6}, and for design C it is {4, 5, 5}. Notice that the defining relation for design C has only one four-letter word, whereas the other designs have two or three.

■ **TABLE 8.13**
Three Choices of Generators for the 2_{IV}^{7-2} Design

Design *A* Generators: $F = ABC$, $G = BCD$ $I = ABCF = BCDG = ADFG$	Design *B* Generators: $F = ABC$, $G = ADE$ $I = ABCF = ADEG = BCDEFG$	Design *C* Generators: $F = ABCD$, $G = ABDE$ $I = ABCDF = ABDEG = CEFG$
Aliases (two-factor interactions)	Aliases (two-factor interactions)	Aliases (two-factor interactions)
$AB = CF$	$AB = CF$	$CE = FG$
$AC = BF$	$AC = BF$	$CF = EG$
$AD = FG$	$AD = EG$	$CG = EF$
$AG = DF$	$AE = DG$	
$BD = CG$	$AF = BC$	
$BG = CD$	$AG = DE$	
$AF = BC = DG$		

Thus, design *C* minimizes the number of words in the defining relation that are of minimum length. We call such a design a **minimum aberration design**. Minimizing aberration in a design of resolution *R* ensures that the design has the minimum number of main effects aliased with interactions of order $R - 1$, the minimum number of two-factor interactions aliased with interactions of order $R - 2$, and so forth. Refer to Fries and Hunter (1980) for more details.

Table 8.14 presents a selection of 2^{k-p} fractional factorial designs for $k \leq 15$ factors and up to $n \leq 128$ runs. The suggested generators in this table will result in a design of the highest possible resolution. These are also the minimum aberration designs.

The alias relationships for all of the designs in Table 8.14 for which $n \leq 64$ are given in Appendix Table V(a–w). The alias relationships presented in this table focus on main effects and two- and three-factor interactions. The complete defining relation is given for each design. This appendix table makes it very easy to select a design of sufficient resolution to ensure that any interactions of potential interest can be estimated.

EXAMPLE 8.5

To illustrate the use of Table 8.14, suppose that we have seven factors and that we are interested in estimating the seven main effects and getting some insight regarding the two-factor interactions. We are willing to assume that three-factor and higher interactions are negligible. This information suggests that a resolution IV design would be appropriate.

Table 8.14 shows that there are two resolution IV fractions available: the 2_{IV}^{7-2} with 32 runs and the 2_{IV}^{7-3} with 16 runs. Appendix Table V contains the complete alias relationships for these two designs. The aliases for the 2_{IV}^{7-3} 16-run design are in Appendix Table V(i). Notice that all seven main effects are aliased with three-factor interactions. The two-factor interactions are all aliased in groups of three. Therefore, this design will satisfy our objectives; that is, it will allow the estimation of the main effects, and it will give some insight regarding two-factor interactions. It is not necessary to run the 2_{IV}^{7-2} design, which would require 32 runs. Appendix Table V(j) shows that this design would allow the estimation of all seven main effects and that 15 of the 21 two-factor interactions could also be uniquely estimated. (Recall that three-factor and higher interactions are negligible.) This is probably more information about interactions than is necessary. The complete 2_{IV}^{7-3} design is shown in Table 8.15. Notice that it was constructed by starting with the 16-run 2^4 design in *A*, *B*, *C*, and *D* as the basic design and then adding the three columns $E = ABC$, $F = BCD$, and $G = ACD$. The generators are $I = ABCE$, $I = BCDF$, and $I = ACDG$ (Table 8.14). The complete defining relation is $I = ABCE = BCDF = ADEF = ACDG = BDEG = CEFG = ABFG$.

(*Continued on* p. 343)

■ **TABLE 8.14**

Selected 2^{k-p} Fractional Factorial Designs

Number of Factors, k	Fraction	Number of Runs	Design Generators
3	2^{3-1}_{III}	4	$C = \pm AB$
4	2^{4-1}_{IV}	8	$D = \pm ABC$
5	2^{5-1}_{V}	16	$E = \pm ABCD$
	2^{5-2}_{III}	8	$D = \pm AB$
			$E = \pm AC$
6	2^{6-1}_{VI}	32	$F = \pm ABCDE$
	2^{6-2}_{IV}	16	$E = \pm ABC$
			$F = \pm BCD$
	2^{6-3}_{III}	8	$D = \pm AB$
			$E = \pm AC$
			$F = \pm BC$
7	2^{7-1}_{VII}	64	$G = \pm ABCDEF$
	2^{7-2}_{IV}	32	$F = \pm ABCD$
			$G = \pm ABDE$
	2^{7-3}_{IV}	16	$E = \pm ABC$
			$F = \pm BCD$
			$G = \pm ACD$
	2^{7-4}_{III}	8	$D = \pm AB$
			$E = \pm AC$
			$F = \pm BC$
			$G = \pm ABC$
8	2^{8-2}_{V}	64	$G = \pm ABCD$
			$H = \pm ABEF$
	2^{8-3}_{IV}	32	$F = \pm ABC$
			$G = \pm ABD$
			$H = \pm BCDE$
	2^{8-4}_{IV}	16	$E = \pm BCD$
			$F = \pm ACD$
			$G = \pm ABC$
			$H = \pm ABD$
9	2^{9-2}_{VI}	128	$H = \pm ACDFG$
			$J = \pm BCEFG$
	2^{9-3}_{IV}	64	$G = \pm ABCD$
			$H = \pm ACEF$
			$J = \pm CDEF$
	2^{9-4}_{IV}	32	$F = \pm BCDE$
			$G = \pm ACDE$
			$H = \pm ABDE$
			$J = \pm ABCE$

Number of Factors, k	Fraction	Number of Runs	Design Generators
	2^{9-5}_{III}	16	$E = \pm ABC$
			$F = \pm BCD$
			$G = \pm ACD$
			$H = \pm ABD$
			$J = \pm ABCD$
10	2^{10-3}_{V}	128	$H = \pm ABCG$
			$J = \pm BCDE$
			$K = \pm ACDF$
	2^{10-4}_{IV}	64	$G = \pm BCDF$
			$H = \pm ACDF$
			$J = \pm ABDE$
			$K = \pm ABCE$
	2^{10-5}_{IV}	32	$F = \pm ABCD$
			$G = \pm ABCE$
			$H = \pm ABDE$
			$J = \pm ACDE$
			$K = \pm BCDE$
	2^{10-6}_{III}	16	$E = \pm ABC$
			$F = \pm BCD$
			$G = \pm ACD$
			$H = \pm ABD$
			$J = \pm ABCD$
			$K = \pm AB$
11	2^{11-5}_{IV}	64	$G = \pm CDE$
			$H = \pm ABCD$
			$J = \pm ABF$
			$K = \pm BDEF$
			$L = \pm ADEF$
	2^{11-6}_{IV}	32	$F = \pm ABC$
			$G = \pm BCD$
			$H = \pm CDE$
			$J = \pm ACD$
			$K = \pm ADE$
			$L = \pm BDE$
	2^{11-7}_{III}	16	$E = \pm ABC$
			$F = \pm BCD$
			$G = \pm ACD$
			$H = \pm ABD$
			$J = \pm ABCD$
			$K = \pm AB$

Number of Factors, k	Fraction	Number of Runs	Design Generators
			$L = \pm AC$
12	2^{12-8}_{III}	16	$E = \pm ABC$
			$F = \pm ABD$
			$G = \pm ACD$
			$H = \pm BCD$
			$J = \pm ABCD$
			$K = \pm AB$
			$L = \pm AC$
			$M = \pm AD$
13	2^{13-9}_{III}	16	$E = \pm ABC$
			$F = \pm ABD$
			$G = \pm ACD$
			$H = \pm BCD$
			$J = \pm ABCD$
			$K = \pm AB$
			$L = \pm AC$
			$M = \pm AD$
			$N = \pm BC$
14	2^{14-10}_{III}	16	$E = \pm ABC$
			$F = \pm ABD$
			$G = \pm ACD$
			$H = \pm BCD$
			$J = \pm ABCD$
			$K = \pm AB$
			$L = \pm AC$
			$M = \pm AD$
			$N = \pm BC$
			$O = \pm BD$
15	2^{15-11}_{III}	16	$E = \pm ABC$
			$F = \pm ABD$
			$G = \pm ACD$
			$H = \pm BCD$
			$J = \pm ABCD$
			$K = \pm AB$
			$L = \pm AC$
			$M = \pm AD$
			$N = \pm BC$
			$O = \pm BD$
			$P = \pm CD$

■ **TABLE 8.15**

A 2_{IV}^{7-3} Fractional Factorial Design

	Basic Design						
Run	A	B	C	D	$E = ABC$	$F = BCD$	$G = ACD$
1	−	−	−	−	−	−	−
2	+	−	−	−	+	−	+
3	−	+	−	−	+	+	−
4	+	+	−	−	−	+	+
5	−	−	+	−	+	+	+
6	+	−	+	−	−	+	−
7	−	+	+	−	−	−	+
8	+	+	+	−	+	−	−
9	−	−	−	+	−	+	+
10	+	−	−	+	+	+	−
11	−	+	−	+	+	−	+
12	+	+	−	+	−	−	−
13	−	−	+	+	+	−	−
14	+	−	+	+	−	−	+
15	−	+	+	+	−	+	−
16	+	+	+	+	+	+	+

8.4.2 Analysis of 2^{k-p} Fractional Factorials

There are many computer programs that can be used to analyze the 2^{k-p} fractional factorial design. For example, Design-Expert JMP, and Minitab all have this capability.

The design may also be analyzed by resorting to first principles; the ith effect is estimated by

$$\text{Effect}_i = \frac{2(\text{Contrast}_i)}{N} = \frac{\text{Contrast}_i}{(N/2)}$$

where the Contrast_i is found using the plus and minus signs in column i and $N = 2^{k-p}$ is the total number of observations. The 2^{k-p} design allows only $2^{k-p} - 1$ effects (and their aliases) to be estimated. Normal probability plots of the effect estimates and Lenth's method are very useful analysis tools.

Projection of the 2^{k-p} Fractional Factorial. The 2^{k-p} design collapses into either a full factorial or a fractional factorial in any subset of $r \leq k - p$ of the original factors. Those subsets of factors providing fractional factorials are subsets appearing as words in the complete defining relation. This is particularly useful in screening experiments when we suspect at the outset of the experiment that most of the original factors will have small effects. The original 2^{k-p} fractional factorial can then be projected into a full factorial, say, in the most interesting factors. Conclusions drawn from designs of this type should be considered tentative and subject to further analysis. It is usually possible to find alternative explanations of the data involving higher order interactions.

As an example, consider the 2_{IV}^{7-3} design from Example 8.5. This is a 16-run design involving seven factors. It will project into a full factorial in any four of the original seven factors that is not a word in the defining relation. There are 35 subsets of four factors, seven of which appear in the complete defining relation (see Table 8.15). Thus, there are 28 subsets of four factors that would form 2^4 designs. One combination that is obvious upon inspecting Table 8.15 is *A, B, C,* and *D*.

To illustrate the usefulness of this projection properly, suppose that we are conducting an experiment to improve the efficiency of a ball mill and the seven factors are as follows:

1. Motor speed
2. Gain
3. Feed mode
4. Feed sizing
5. Material type
6. Screen angle
7. Screen vibration level

We are fairly certain that motor speed, feed mode, feed sizing, and material type will affect efficiency and that these factors may interact. The role of the other three factors is less well known, but it is likely that they are negligible. A reasonable strategy would be to assign motor speed, feed mode, feed sizing, and material type to columns *A, B, C,* and *D*, respectively, in Table 8.15. Gain, screen angle, and screen vibration level would be assigned to columns *E, F,* and *G*, respectively. If we are correct and the "minor variables" *E, F,* and *G* are negligible, we will be left with a full 2^4 design in the key process variables.

8.4.3 Blocking Fractional Factorials

Occasionally, a fractional factorial design requires so many runs that all of them cannot be made under homogeneous conditions. In these situations, fractional factorials may be confounded in blocks. Appendix Table V contains recommended blocking arrangements for many of the fractional factorial designs in Table 8.14. The minimum block size for these designs is eight runs.

To illustrate the general procedure, consider the 2_{IV}^{6-2} fractional factorial design with the defining relation $I = ABCE = BCDF = ADEF$ shown in Table 8.10. This fractional design contains 16 treatment combinations. Suppose we wish to run the design in two blocks of eight treatment combinations each. In selecting an interaction to confound with blocks, we note from examining the alias structure in Appendix Table V(f) that there are two alias sets involving only three-factor interactions. The table suggests selecting *ABD* (and its aliases) to be confounded with blocks. This would give the two blocks shown in Figure 8.18. Notice that the

Block 1	Block 2
(1)	ae
abf	acf
cef	bef
abce	bc
adef	df
bde	abd
acd	cde
bcdf	abcdef

■ FIGURE 8.18 The 2_{IV}^{6-2} design in two blocks with *ABD* confounded principal block contains those treatment combinations that have an even number of letters in common with *ABD*. These are also the treatment combinations for which $L = x_1 + x_2 + x_4 = 0 \pmod 2$.

EXAMPLE 8.6

A five-axis CNC machine is used to produce an impeller for a jet turbine engine. The blade profiles are an important quality characteristic. Specifically, the deviation of the blade profile from the profile specified on the engineering drawing is of interest. An experiment is run to determine which machine parameters affect profile deviation. The eight factors selected for the design are as follows:

Factor	Low Level (−)	High Level (+)
A = x-Axis shift (0.001 in.)	0	15
B = y-Axis shift (0.001 in.)	0	15
C = z-Axis shift (0.001 in.)	0	15
D = Tool supplier	1	2
E = a-Axis shift (0.001 deg)	0	30
F = Spindle speed (%)	90	110
G = Fixture height (0.001 in.)	0	15
H = Feed rate (%)	90	110

One test blade on each part is selected for inspection. The profile deviation is measured using a coordinate measuring machine, and the standard deviation of the difference between the actual profile and the specified profile is used as the response variable.

The machine has four spindles. Because there may be differences in the spindles, the process engineers feel that the spindles should be treated as blocks.

The engineers feel confident that three-factor and higher interactions are not too important, but they are reluctant to ignore the two-factor interactions. From Table 8.14, two designs initially appear appropriate: the 2_{IV}^{8-4} design with 16 runs and the 2_{IV}^{8-3} design with 32 runs. Appendix Table V(1) indicates that if the 16-run design is used, there will be fairly extensive aliasing of two-factor interactions. Furthermore, this design cannot be run in four blocks without confounding four two-factor interactions with blocks. Therefore, the experimenters decide to use the 2_{IV}^{8-3} design in four blocks. This confounds one three-factor interaction alias chain and one two-factor interaction (*EH*) and its three-factor interaction aliases with blocks. The *EH* interaction is the interaction between the a-axis shift and the feed rate, and the engineers consider an interaction between these two variables to be fairly unlikely.

Table 8.16 contains the design and the resulting responses as standard deviation $\times$ 10^3 in.. Because the response variable is a standard deviation, it is often best to perform the analysis following a log transformation. The effect estimates are shown in Table 8.17. Figure 8.19 is a normal probability plot of the effect estimates, using ln (standard deviation $\times$ 10^3) as the response variable. The only large effects are A = x-axis shift, B = y-axis shift, and the alias chain involving $AD + BG$. Now AD is the x-axis shift-tool supplier interaction, and BG is the y-axis shift-fixture height interaction, and since these two interactions are aliased it is impossible to separate them based on the data from the current experiment. Since both interactions involve one large main effect it is also difficult to apply any "obvious" simplifying logic such as effect heredity to the situation either. If there is some engineering knowledge or process knowledge available that sheds light on the situation, then perhaps a choice could be made between the two interactions; otherwise, more data will be required to separate these two effects. (The problem of adding runs to a fractional factorial to de-alias interactions is discussed in Sections 8.6 and 8.7.)

Suppose that process knowledge suggests that the appropriate interaction is likely to be *AD*. Table 8.18 is the resulting analysis of variance for the model with factors *A*, *B*, *D*, and *AD* (factor *D* was included to preserve the hierarchy principle). Notice that the block effect is small, suggesting that the machine spindles are not very different.

Figure 8.20 is a normal probability plot of the residuals from this experiment. This plot is suggestive of slightly heavier than normal tails, so possibly other transformations should be considered. The *AD* interaction plot is in Figure 8.21. Notice that tool supplier (*D*) and the magnitude of the x-axis shift (*A*) have a profound impact on the variability of the blade profile from design specifications. Running *A* at the low level (0 offset) and buying tools from supplier 1 gives the best results. Figure 8.22 shows the projection of this 2_{IV}^{8-3} design into four replicates of a 2^3 design in factors *A*, *B*, and *D*. The best combination of operating conditions is *A* at the low level (0 offset), *B* at the high level (0.015 in offset), and *D* at the low level (tool supplier 1).

■ **TABLE 8.16**
The 2^{8-3} Design in Four Blocks for Example 8.6

	Basic Design									Actual Run	Standard Deviation
Run	A	B	C	D	E	$F = ABC$	$G = ABD$	$H = BCDE$	Block	Order	($\times 10^3$ in)
1	−	−	−	−	−	−	−	+	3	18	2.76
2	+	−	−	−	−	+	+	+	2	16	6.18
3	−	+	−	−	−	+	+	−	4	29	2.43
4	+	+	−	−	−	−	−	−	1	4	4.01
5	−	−	+	−	−	+	−	−	1	6	2.48
6	+	−	+	−	−	−	+	−	4	26	5.91
7	−	+	+	−	−	−	+	+	2	14	2.39
8	+	+	+	−	−	+	−	+	3	22	3.35
9	−	−	−	+	−	−	+	−	1	8	4.40
10	+	−	−	+	−	+	−	−	4	32	4.10
11	−	+	−	+	−	+	−	+	2	15	3.22
12	+	+	−	+	−	−	+	+	3	19	3.78
13	−	−	+	+	−	+	+	+	3	24	5.32
14	+	−	+	+	−	−	−	+	2	11	3.87
15	−	+	+	+	−	−	−	−	4	27	3.03
16	+	+	+	+	−	+	+	−	1	3	2.95
17	−	−	−	−	+	−	−	−	2	10	2.64
18	+	−	−	−	+	+	+	−	3	21	5.50
19	−	+	−	−	+	+	+	+	1	7	2.24
20	+	+	−	−	+	−	−	+	4	28	4.28
21	−	−	+	−	+	+	−	+	4	30	2.57
22	+	−	+	−	+	−	+	+	1	2	5.37
23	−	+	+	−	+	−	+	−	3	17	2.11
24	+	+	+	−	+	+	−	−	2	13	4.18
25	−	−	−	+	+	−	+	+	4	25	3.96
26	+	−	−	+	+	+	−	+	1	1	3.27
27	−	+	−	+	+	+	−	−	3	23	3.41
28	+	+	−	+	+	−	+	−	2	12	4.30
29	−	−	+	+	+	+	+	−	2	9	4.44
30	+	−	+	+	+	−	−	−	3	20	3.65
31	−	+	+	+	+	−	−	+	1	5	4.41
32	+	+	+	+	+	+	+	+	4	31	3.40

■ **TABLE 8.17**

Effect Estimates, Regression Coefficients, and Sums of Squares for Example 8.6

Variable	Name	−1 Level	+1 Level
A	*x*-Axis shift	0	15
B	*y*-Axis shift	0	15
C	*z*-Axis shift	0	15
D	Tool supplier	1	2
E	*a*-Axis shift	0	30
F	Spindle speed	90	110
G	Fixture height	0	15
H	Feed rate	90	110

Variable	Regression Coefficient	Estimated Effect	Sum of Squares
Overall average	1.28007		
A	0.14513	0.29026	0.674020
B	−0.10027	−0.20054	0.321729
C	−0.01288	−0.02576	0.005310
D	0.05407	0.10813	0.093540
E	−2.531E-04	−5.063E-04	2.050E-06
F	−0.01936	−0.03871	0.011988
G	0.05804	0.11608	0.107799
H	0.00708	0.01417	0.001606
AB + *CF* + *DG*	−0.00294	−0.00588	2.767E-04
AC + *BF*	−0.03103	−0.06206	0.030815
AD + *BG*	−0.18706	−0.37412	1.119705
AE	0.00402	0.00804	5.170E-04
AF + *BC*	−0.02251	−0.04502	0.016214
AG + *BD*	0.02644	0.05288	0.022370
AH	−0.02521	−0.05042	0.020339
BE	0.04925	0.09851	0.077627
BH	0.00654	0.01309	0.001371
CD + *FG*	0.01726	0.03452	0.009535
CE	0.01991	0.03982	0.012685
CG + *DF*	−0.00733	−0.01467	0.001721
CH	0.03040	0.06080	0.029568
DE	0.00854	0.01708	0.002334
DH	0.00784	0.01569	0.001969
EF	−0.00904	−0.01808	0.002616
EG	−0.02685	−0.05371	0.023078
EH	−0.01767	−0.03534	0.009993
FH	−0.01404	−0.02808	0.006308
GH	0.00245	0.00489	1.914E-04
ABE	0.01665	0.03331	0.008874
ABH	−0.00631	−0.01261	0.001273
ACD	−0.02717	−0.05433	0.023617

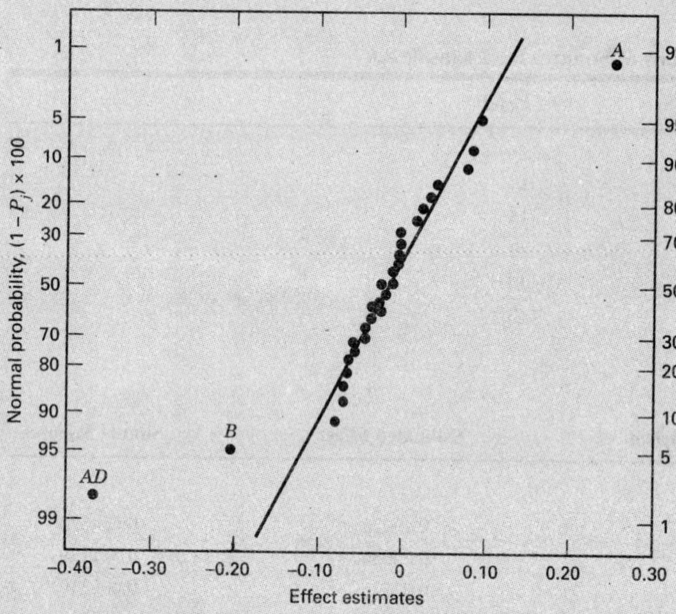

■ **FIGURE 8.19** **Normal probability plot of the effect estimates for Example 8.6**

■ **TABLE 8.18**
Analysis of Variance for Example 8.6

Source of Variation	Sum of Squares	Degrees of Freedom	Mean Square	F_0	*P*-Value
A	0.6740	1	0.6740	39.42	<0.0001
B	0.3217	1	0.3217	18.81	0.0002
D	0.0935	1	0.0935	5.47	0.0280
AD	1.1197	1	1.1197	65.48	<0.0001
Blocks	0.0201	3	0.0067		
Error	0.4099	24	0.0171		
Total	2.6389	31			

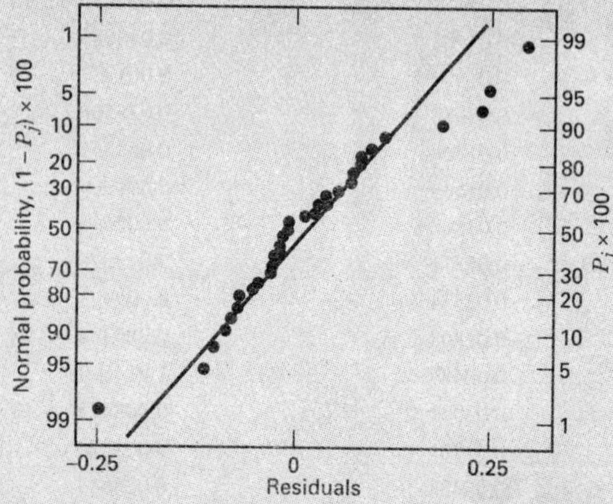

■ **FIGURE 8.20** **Normal probability plot of the residuals for Example 8.6**

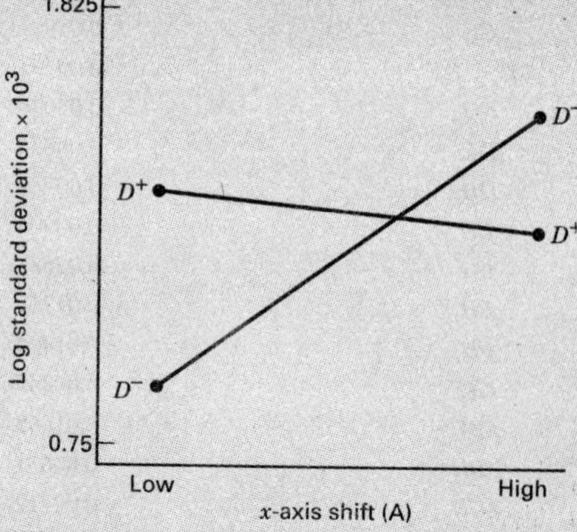

■ **FIGURE 8.21** **Plot of *AD* interaction for Example 8.6**

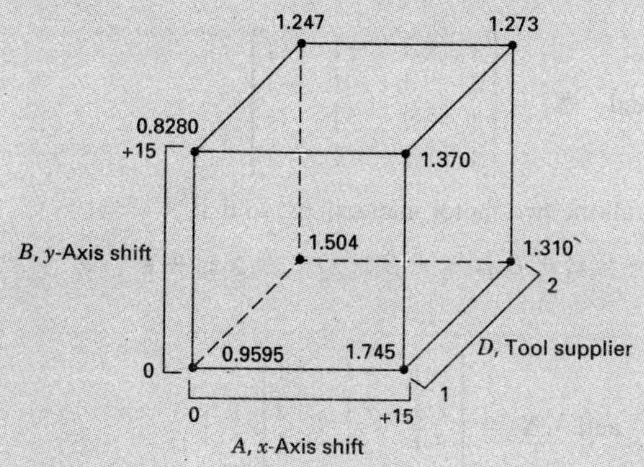

■ **FIGURE 8.22** **The 2_{IV}^{8-3} design in Example 8.6 projected into four replicates of a 2^3 design in factors A, B, and D**

8.5 Alias Structures in Fractional Factorials and Other Designs

In this chapter, we show how to find the alias relationships in a 2^{k-p} fractional factorial design by use of the complete defining relation. This method works well in simple designs, such as the regular fractions we use most frequently, but it does not work as well in more complex settings, such as some of the nonregular fractions and partial fold-over designs that we will discuss subsequently. Furthermore, there are some fractional factorials that do not have defining relations, such as the Plackett–Burman designs in Section 8.6.3, so the defining relation method will not work for these types of designs at all.

Fortunately, there is a general method available that works satisfactorily in many situations. The method uses the polynomial or regression model representation of the model, say

$$\mathbf{y} = \mathbf{X}_1\boldsymbol{\beta}_1 + \boldsymbol{\epsilon}$$

where $\mathbf{y}$ is an $n \times 1$ vector of the responses, $\mathbf{X}_1$ is an $n \times p_1$ matrix containing the design matrix expanded to the form of the model that the experimenter is fitting, $\boldsymbol{\beta}_1$ is a $p_1 \times 1$ vector of the model parameters, and $\boldsymbol{\epsilon}$ is an $n \times 1$ vector of errors. The least squares estimate of $\boldsymbol{\beta}_1$ is

$$\hat{\boldsymbol{\beta}}_1 = (\mathbf{X}_1'\mathbf{X}_1)^{-1}\mathbf{X}_1'\mathbf{y}$$

Suppose that the **true** model is

$$\mathbf{y} = \mathbf{X}_1\boldsymbol{\beta}_1 + \mathbf{X}_2\boldsymbol{\beta}_2 + \boldsymbol{\epsilon}$$

where $\mathbf{X}_2$ is an $n \times p_2$ matrix containing additional variables that are not in the fitted model and $\boldsymbol{\beta}_2$ is a $p_2 \times 1$ vector of the parameters associated with these variables. It can be shown that

$$\begin{aligned} E(\hat{\boldsymbol{\beta}}_1) &= \boldsymbol{\beta}_1 + (\mathbf{X}_1'\mathbf{X}_1)^{-1}\mathbf{X}_1'\mathbf{X}_2\boldsymbol{\beta}_2 \\ &= \boldsymbol{\beta}_1 + \mathbf{A}\boldsymbol{\beta}_2 \end{aligned} \tag{8.1}$$

The matrix $\mathbf{A} = (\mathbf{X}_1'\mathbf{X}_1)^{-1}\mathbf{X}_1'\mathbf{X}_2$ is called the **alias matrix**. The elements of this matrix operating on $\boldsymbol{\beta}_2$ identify the alias relationships for the parameters in the vector $\boldsymbol{\beta}_1$.

We illustrate the application of this procedure with a familiar example. Suppose that we have conducted a 2^{3-1} design with defining relation $I = ABC$ or $I = x_1x_2x_3$. The model that the experimenter plans to fit is the main-effects-only model

$$y = \beta_0 + \beta_1x_1 + \beta_2x_2 + \beta_3x_3 + \epsilon$$

In the notation defined above

$$\boldsymbol{\beta}_1 = \begin{bmatrix} \beta_0 \\ \beta_1 \\ \beta_2 \\ \beta_3 \end{bmatrix} \quad \text{and} \quad \mathbf{X}_1 = \begin{bmatrix} 1 & -1 & -1 & 1 \\ 1 & 1 & -1 & -1 \\ 1 & -1 & 1 & -1 \\ 1 & 1 & 1 & 1 \end{bmatrix}$$

Suppose that the true model contains all the two-factor interactions, so that

$$y = \beta_0 + \beta_1 x_1 + \beta_2 x_2 + \beta_3 x_3 + \beta_{12} x_1 x_2 + \beta_{13} x_1 x_3 + \beta_{23} x_2 x_3 + \epsilon$$

and

$$\boldsymbol{\beta}_2 = \begin{bmatrix} \beta_{12} \\ \beta_{13} \\ \beta_{23} \end{bmatrix}, \quad \text{and} \quad \mathbf{X}_2 = \begin{bmatrix} 1 & -1 & -1 \\ -1 & -1 & 1 \\ -1 & 1 & -1 \\ 1 & 1 & 1 \end{bmatrix}$$

Now

$$\mathbf{X}_1'\mathbf{X}_1 = 4\,\mathbf{I}_4 \quad \text{and} \quad \mathbf{X}_1'\mathbf{X}_2 = \begin{bmatrix} 0 & 0 & 0 \\ 0 & 0 & 4 \\ 0 & 4 & 0 \\ 4 & 0 & 0 \end{bmatrix}$$

Therefore,

$$(\mathbf{X}_1'\mathbf{X}_1)^{-1} = \frac{1}{4}\mathbf{I}_4$$

and

$$E(\hat{\boldsymbol{\beta}}_1) = \boldsymbol{\beta}_1 + \mathbf{A}\,\boldsymbol{\beta}_2$$

$$E\begin{bmatrix} \hat{\beta}_0 \\ \hat{\beta}_1 \\ \hat{\beta}_2 \\ \hat{\beta}_3 \end{bmatrix} = \begin{bmatrix} \beta_0 \\ \beta_1 \\ \beta_2 \\ \beta_3 \end{bmatrix} + \frac{1}{4}\mathbf{I}_4 \begin{bmatrix} 0 & 0 & 0 \\ 0 & 0 & 4 \\ 0 & 4 & 0 \\ 4 & 0 & 0 \end{bmatrix} \begin{bmatrix} \beta_{12} \\ \beta_{13} \\ \beta_{23} \end{bmatrix}$$

$$= \begin{bmatrix} \beta_0 \\ \beta_1 \\ \beta_2 \\ \beta_3 \end{bmatrix} + \begin{bmatrix} 0 & 0 & 0 \\ 0 & 0 & 1 \\ 0 & 1 & 0 \\ 1 & 0 & 0 \end{bmatrix} \begin{bmatrix} \beta_{12} \\ \beta_{13} \\ \beta_{23} \end{bmatrix}$$

$$= \begin{bmatrix} \beta_0 \\ \beta_1 \\ \beta_2 \\ \beta_3 \end{bmatrix} + \begin{bmatrix} 0 \\ \beta_{23} \\ \beta_{13} \\ \beta_{12} \end{bmatrix}$$

$$= \begin{bmatrix} \beta_0 \\ \beta_1 + \beta_{23} \\ \beta_2 + \beta_{13} \\ \beta_3 + \beta_{12} \end{bmatrix}$$

The interpretation of this, of course, is that each of the main effects is aliased with one of the two-factor interactions, which we know to be the case for this design. Notice that every row of the alias matrix represents one of the factors in $\boldsymbol{\beta}_1$ and every column represents one of the factors in $\boldsymbol{\beta}_2$. While this is a very simple example, the method is very general and can be applied to much more complex designs.

8.6 Resolution III Designs

8.6.1 Constructing Resolution III Designs

As indicated earlier, the sequential use of fractional factorial designs is very useful, often leading to great economy and efficiency in experimentation. This application of fractional factorials occurs frequently in situations of pure factor screening; that is, there are relatively many factors but only a few of them are expected to be important. Resolution III designs can be very useful in these situations.

It is possible to construct resolution III designs for investigating up to $k = N - 1$ factors in only N runs, where N is a multiple of 4. These designs are frequently useful in industrial experimentation. Designs in which N is a power of 2 can be constructed by the methods presented earlier in this chapter, and these are presented first. Of particular importance are designs requiring 4 runs for up to 3 factors, 8 runs for up to 7 factors, and 16 runs for up to 15 factors. If $k = N - 1$, the fractional factorial design is said to be **saturated**.

A design for analyzing up to three factors in four runs is the 2_{III}^{3-1} design, presented in Section 8.2. Another very useful saturated fractional factorial is a design for studying seven factors in eight runs, that is, the 2_{III}^{7-4} design. This design is a one-sixteenth fraction of the 2^7. It may be constructed by first writing down as the basic design the plus and minus levels for a full 2^3 design in A, B, and C and then associating the levels of four additional factors with the interactions of the original three as follows: $D = AB$, $E = AC$, $F = BC$, and $G = ABC$. Thus, the generators for this design are $I = ABD$, $I = ACE$, $I = BCF$, and $I = ABCG$. The design is shown in Table 8.19.

The complete defining relation for this design is obtained by multiplying the four generators ABD, ACE, BCF, and $ABCG$ together two at a time, three at a time, and four at a time, yielding

$$I = ABD = ACE = BCF = ABCG = BCDE = ACDF = CDG$$

$$= ABEF = BEG = AFG = DEF = ADEG = CEFG = BDFG = ABCDEFG$$

■ **TABLE 8.19**

The 2_{III}^{7-4} Design with the Generators $I = ABD$, $I = ACE$, $I = BCF$, and $I = ABCG$

	Basic Design							
Run	A	B	C	$D = AB$	$E = AC$	$F = BC$	$G = ABC$	
1	−	−	−	+	+	+	−	*def*
2	+	−	−	−	−	+	+	*afg*
3	−	+	−	−	+	−	+	*beg*
4	+	+	−	+	−	−	−	*abd*
5	−	−	+	+	−	−	+	*cdg*
6	+	−	+	−	+	−	−	*ace*
7	−	+	+	−	−	+	−	*bcf*
8	+	+	+	+	+	+	+	*abcdefg*

To find the aliases of any effect, simply multiply the effect by each word in the defining relation. For example, the aliases of B are

$$B = AD = ABCE = CF = ACG = CDE = ABCDF = BCDG = AEF = EG$$
$$= ABFG = BDEF = ABDEG = BCEFG = DFG = ACDEFG$$

This design is a one-sixteenth fraction, and because the signs chosen for the generators are positive, this is the principal fraction. It is also a resolution III design because the smallest number of letters in any word of the defining contrast is three. Any one of the 16 different 2_{III}^{7-4} designs in this family could be constructed by using the generators with one of the 16 possible arrangements of signs in $I = \pm ABD, I = \pm ACE, I = \pm BCF, I = \pm ABCG$.

The seven degrees of freedom in this design may be used to estimate the seven main effects. Each of these effects has 15 aliases; however, if we assume that three-factor and higher interactions are negligible, then considerable simplification in the alias structure results. Making this assumption, each of the linear combinations associated with the seven main effects in this design actually estimates the main effect and three two-factor interactions:

$$\begin{aligned}
[A] &\rightarrow A + BD + CE + FG \\
[B] &\rightarrow B + AD + CF + EG \\
[C] &\rightarrow C + AE + BF + DG \\
[D] &\rightarrow D + AB + CG + EF \\
[E] &\rightarrow E + AC + BG + DF \\
[F] &\rightarrow F + BC + AG + DE \\
[G] &\rightarrow G + CD + BE + AF
\end{aligned} \tag{8.2}$$

These aliases are found in Appendix Table V(h), ignoring three-factor and higher interactions.

The saturated 2_{III}^{7-4} design in Table 8.19 can be used to obtain resolution III designs for studying fewer than seven factors in eight runs. For example, to generate a design for six factors in eight runs, simply drop any one column in Table 8.19, for example, column G. This produces the design shown in Table 8.20.

■ **TABLE 8.20**
A 2_{III}^{6-3} Design with the Generators $I = ABD$, $I = ACE$, and $I = BCF$

Run	Basic Design A	Basic Design B	Basic Design C	$D = AB$	$E = AC$	$F = BC$	
1	−	−	−	+	+	+	*def*
2	+	−	−	−	−	+	*af*
3	−	+	−	−	+	−	*be*
4	+	+	−	+	−	−	*abd*
5	−	−	+	+	−	−	*cd*
6	+	−	+	−	+	−	*ace*
7	−	+	+	−	−	+	*bcf*
8	+	+	+	+	+	+	*abcdef*

It is easy to verify that this design is also of resolution III; in fact, it is a 2_{III}^{6-3}, or a one-eighth fraction, of the 2^6 design. The defining relation for the 2_{III}^{6-3} design is equal to the defining relation for the original 2_{III}^{7-4} design with any words containing the letter G deleted. Thus, the defining relation for our new design is

$$I = ABD = ACE = BCF = BCDE = ACDF = ABEF = DEF$$

In general, when d factors are dropped to produce a new design, the new defining relation is obtained as those words in the original defining relation that do not contain any dropped letters. When constructing designs by this method, care should be exercised to obtain the best arrangement possible. If we drop columns B, D, F, and G from Table 8.19, we obtain a design for three factors in eight runs, yet the treatment combinations correspond to two replicates of a 2^{3-1} design. The experimenter would probably prefer to run a full 2^3 design in A, C, and E.

It is also possible to obtain a resolution III design for studying up to 15 factors in 16 runs. This saturated 2_{III}^{15-11} design can be generated by first writing down the 16 treatment combinations associated with a 2^4 design in A, B, C, and D and then equating 11 new factors with the two-, three-, and four-factor interactions of the original four. In this design, each of the 15 main effects is aliased with seven two-factor interactions. A similar procedure can be used for the 2_{III}^{31-26} design, which allows up to 31 factors to be studied in 32 runs.

8.6.2 Fold Over of Resolution III Fractions to Separate Aliased Effects

By combining fractional factorial designs in which certain signs are switched, we can systematically isolate effects of potential interest. This type of sequential experiment is called a **fold over** of the original design. The alias structure for any fraction with the signs for one or more factors reversed is obtained by making changes of sign on the appropriate factors in the alias structure of the original fraction.

Consider the 2_{III}^{7-4} design in Table 8.19. Suppose that along with this principal fraction a second fractional design with the signs reversed in the column for factor D is also run. That is, the column for D in the second fraction is

$$- + + - - + + -$$

The effects that may be estimated from the first fraction are shown in Equation 8.2, and from the second fraction we obtain

$$
\begin{aligned}
[A]' &\rightarrow A - BD + CE + FG \\
[B]' &\rightarrow B - AD + CF + EG \\
[C]' &\rightarrow C + AE + BF - DG \\
[D]' &\rightarrow D - AB - CG - EF \\
[-D]' &\rightarrow -D + AB + CG + EF \\
[E]' &\rightarrow E + AC + BG - DF \\
[F]' &\rightarrow F + BC + AG - DE \\
[G]' &\rightarrow G - CD + BE + AF
\end{aligned}
\tag{8.3}
$$

assuming that three-factor and higher interactions are insignificant. Now from the two linear combinations of effects $\frac{1}{2}([i] + [i]')$ and $\frac{1}{2}([i] - [i]')$ we obtain

i	From $\frac{1}{2}([i] + [i]')$	From $\frac{1}{2}([i] - [i]')$
A	$A + CE + FG$	BD
B	$B + CF + EG$	AD
C	$C + AE + BF$	DG
D	D	$AB + CG + EF$
E	$E + AC + BG$	DF
F	$F + BC + AG$	DE
G	$G + BE + AF$	CD

Thus, we have isolated the main effect of D and all of its two-factor interactions. In general, if we add to a fractional design of resolution III or higher a further fraction with the signs of a *single factor* reversed, then the combined design will provide estimates of the main effect of that factor and its two-factor interactions. This is sometimes called a **single-factor fold over**.

Now suppose we add to a resolution III fractional a second fraction in which the *signs for all the factors are reversed*. This type of fold over (sometimes called a **full fold over** or a reflection) breaks the alias links between all main effects and their two-factor interactions. That is, we may use the **combined design** to estimate all of the main effects clear of any two-factor interactions. The following example illustrates the full fold-over technique.

EXAMPLE 8.7

A human performance analyst is conducting an experiment to study eye focus time and has built an apparatus in which several factors can be controlled during the test. The factors he initially regards as important are acuity or sharpness of vision (A), distance from target to eye (B), target shape (C), illumination level (D), target size (E), target density (F), and subject (G). Two levels of each factor are considered. He suspects that only a few of these seven factors are of major importance and that high-order interactions between the factors can be neglected. On the basis of this assumption, the analyst decides to run a screening experiment to identify the most important factors and then to concentrate further study on those. To screen these seven factors, he runs the treatment combinations from the 2_{III}^{7-4} design in Table 8.19 in random order, obtaining the focus times in milliseconds, as shown in Table 8.21.

■ **TABLE 8.21**
A 2_{III}^{7-4} Design for the Eye Focus Time Experiment

	Basic Design								
Run	A	B	C	$D = AB$	$E = AC$	$F = BC$	$G = ABC$		Time
1	−	−	−	+	+	+	−	*def*	85.5
2	+	−	−	−	−	+	+	*afg*	75.1
3	−	+	−	−	+	−	+	*beg*	93.2
4	+	+	−	+	−	−	−	*abd*	145.4
5	−	−	+	+	−	−	+	*cdg*	83.7
6	+	−	+	−	+	−	−	*ace*	77.6
7	−	+	+	−	−	+	−	*bcf*	95.0
8	+	+	+	+	+	+	+	*abcdefg*	141.8

Seven main effects and their aliases may be estimated from these data. From Equation 8.2, we see that the effects and their aliases are

$$[A] = 20.63 \rightarrow A + BD + CE + FG$$

$$[B] = 38.38 \rightarrow B + AD + CF + EG$$

$$[C] = -0.28 \rightarrow C + AE + BF + DG$$

$$[D] = 28.88 \rightarrow D + AB + CG + EF$$

$$[E] = -0.28 \rightarrow E + AC + BG + DF$$

$$[F] = -0.63 \rightarrow F + BC + AG + DE$$

$$[G] = -2.43 \rightarrow G + CD + BE + AF$$

For example, the estimate of the main effect of *A* and its aliases is

$$[A] = \tfrac{1}{4}(-85.5 + 75.1 - 93.2 + 145.4 - 83.7 + 77.6 - 95.0 + 141.8) = 20.63$$

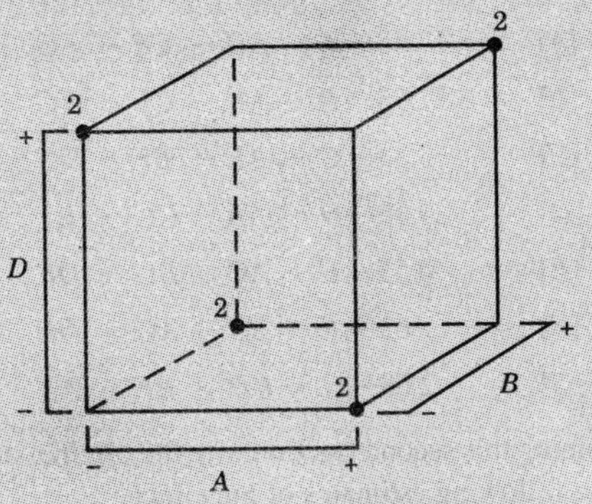

■ **FIGURE 8.23 The 2_{III}^{7-4} design projected into two replicates of a 2_{III}^{3-1} design in *A*, *B*, and *D***

The three largest effects are [*A*], [*B*], and [*D*]. The simplest interpretation of the results of this experiment is that the main effects of *A*, *B*, and *D* are all significant. However, this interpretation is not unique, because one could also logically conclude that *A*, *B*, and the *AB* interaction, or perhaps *B*, *D*, and the *BD* interaction, or perhaps *A*, *D*, and the *AD* interaction are the true effects.

Notice that *ABD* is a word in the defining relation for this design. Therefore, this 2_{III}^{7-4} design does not project into a full 2^3 factorial in *ABD*; instead, it projects into two replicates of a 2^{3-1} design, as shown in Figure 8.23. Because the 2^{3-1} design is a resolution III design, *A* will be aliased with *BD*, *B* will be aliased with *AD*, and *D* will be aliased with *AB*, so the interactions cannot be separated from the main effects. The experimenter here may have been unlucky. If he had assigned the factor illumination level to *C* instead of *D*, the design would have projected into a full 2^3 design, and the interpretation could have been simpler.

To separate the main effects and the two-factor interactions, the full fold-over technique is used, and a second fraction is run with all the signs reversed. This fold-over design is shown in Table 8.22 along with the observed responses. Notice that when we construct a full fold over of a resolution III design, we (in effect) change the signs on the generators that have an odd number of letters. The effects estimated by this fraction are

■ **TABLE 8.22**
A Fold-Over 2_{III}^{7-4} Design for the Eye Focus Experiment

	Basic Design								
Run	*A*	*B*	*C*	*D* = −*AB*	*E* = −*AC*	*F* = −*BC*	*G* = *ABC*		Time
1	+	+	+	−	−	−	+	*abcg*	91.3
2	−	+	+	+	+	−	−	*bcde*	136.7
3	+	−	+	+	−	+	−	*acdf*	82.4
4	−	−	+	−	+	+	+	*cefg*	73.4
5	+	+	−	−	+	+	−	*abef*	94.1
6	−	+	−	+	−	+	+	*bdfg*	143.8
7	+	−	−	+	+	−	+	*adeg*	87.3
8	−	−	−	−	−	−	−	(1)	71.9

$[A]' = -17.68 \rightarrow A - BD - CE - FG$

$[B]' = 37.73 \rightarrow B - AD - CF - EG$

$[C]' = -3.33 \rightarrow C - AE - BF - DG$

$[D]' = 29.88 \rightarrow D - AB - CG - EF$

$[E]' = 0.53 \rightarrow E - AC - BG - DF$

$[F]' = 1.63 \rightarrow F - BC - AG - DE$

$[G]' = 2.68 \rightarrow G - CD - BE - AF$

By combining this second fraction with the original one, we obtain the following estimates of the effects:

i	From $\frac{1}{2}([i] + [i]')$	From $\frac{1}{2}([i] - [i]')$
A	$A = 1.48$	$BD + CE + FG = 19.15$
B	$B = 38.05$	$AD + CF + EG = 0.33$
C	$C = -1.80$	$AE + BF + DG = 1.53$
D	$D = 29.38$	$AB + CG + EF = -0.50$
E	$E = 0.13$	$AC + BG + DF = -0.40$
F	$F = 0.50$	$BC + AG + DE = -1.13$
G	$G = 0.13$	$CD + BE + AF = -2.55$

The two largest effects are B and D. Furthermore, the third largest effect is $BD + CE + FG$, so it seems reasonable to attribute this to the BD interaction. The experimenter used the two factors distance (B) and illumination level (D) in subsequent experiments with the other factors A, C, E, and F at standard settings and verified the results obtained here. He decided to use subjects as blocks in these new experiments rather than ignore a potential subject effect because several different subjects had to be used to complete the experiment.

The Defining Relation for a Fold-Over Design. Combining fractional factorial designs via fold over as demonstrated in Example 8.7 is a very useful technique. It is often of interest to know the defining relation for the combined design. It can be easily determined. Each separate fraction will have $L + U$ words used as generators: L words of like sign and U words of unlike sign. The combined design will have $L + U - 1$ words used as generators. These will be the L words of like sign and the $U - 1$ words consisting of independent even products of the words of unlike sign. (**Even products** are words taken two at a time, four at a time, and so forth.)

To illustrate this procedure, consider the design in Example 8.7. For the first fraction, the generators are

$$I = ABD, \quad I = ACE, \quad I = BCF, \quad \text{and} \quad I = ABCG$$

and for the second fraction, they are

$$I = -ABD, \quad I = -ACE, \quad I = -BCF, \quad \text{and} \quad I = ABCG$$

Notice that in the second fraction we have switched the signs on the generators with an odd number of letters. Also, notice that $L + U = 1 + 3 = 4$. The combined design will have $I = ABCG$ (the like sign word) as a generator and two words that are independent even products of the words of unlike sign. For example, take $I = ABD$ and $I = ACE$; then $I = (ABD)(ACE) = BCDE$ is a generator of the combined design. Also, take $I = ABD$ and $I = BCF$; then $I = (ABD)(BCF) = ACDF$ is a generator of the combined design. The complete defining relation for the combined design is

$$I = ABCG = BCDE = ACDF = ADEG = BDFG = ABEF = CEFG$$

Blocking in a Fold-Over Design. Usually a fold-over design is conducted in two distinct time periods. Following the initial fraction, some time usually elapses while the data are analyzed and the fold-over runs are planned. Then the second set of runs is made, often on a different day, or different shift, or using different operating personnel, or perhaps material from a different source. This leads to a situation where **blocking** to eliminate potential nuisance effects between the two time periods is of interest. Fortunately, blocking in the combined experiment is easily accomplished.

To illustrate, consider the fold-over experiment in Example 8.7. In the initial group of eight runs shown in Table 8.21, the generators are $D = AB$, $E = AC$, $F = BC$, and $G = ABC$. In the fold-over set of runs, Table 8.22, the signs are changed on three of the generators so that $D = -AB$, $E = -AC$, and $F = -BC$. Thus, in the first group of eight runs the signs on the effects ABD, ACE, and BCF are positive, and in the second group of eight runs the signs on ABD, ACE, and BCF are negative; therefore, these effects are confounded with blocks. Actually, there is a single-degree-of-freedom alias chain confounded with blocks (remember that there are two blocks, so there must be one degree of freedom for blocks), and the effects in this alias chain may be found by multiplying any one of the effects ABD, ACE, and BCF through the defining relation for the design. This yields

$$ABD = CDG = ACE = BCF = BEG = AFG = DEF = ABCDEFG$$

as the complete set of effects that are confounded with blocks. In general, a completed fold-over experiment will always form two blocks with the effects whose signs are positive in one block and negative in the other (and their aliases) confounded with blocks. These effects can always be determined from the generators whose signs have been switched to form the fold over.

8.6.3 Plackett–Burman Designs

These are two-level fractional factorial designs developed by Plackett and Burman (1946) for studying $k = N - 1$ variables in N runs, where N is a multiple of 4. If N is a power of 2, these designs are identical to those presented earlier in this section. However, for $N = 12, 20, 24, 28$, and 36, the Plackett–Burman designs are sometimes of interest. Because these designs cannot be represented as cubes, they are sometimes called **nongeometric designs.**

The upper half of Table 8.23 presents rows of plus and minus signs that are used to construct the Plackett–Burman designs for $N = 12, 20, 24$, and 36, whereas the lower half of the

■ **TABLE 8.23**
Plus and Minus Signs for the Plackett–Burman Designs

$k = 11, N = 12$	+ + − + + + − − − + −
$k = 19, N = 20$	+ + − − + + + + − + − + − − − − + + −
$k = 23, N = 24$	+ + + + + − + − + + − − + + − − + − + − − − −
$k = 35, N = 36$	− + − + + + − − − + + + + + − + + + − − + − − − − + − + − + + − − + −

$k = 27, N = 28$		
+ − + + + + − − −	− + − − − + − − +	+ + − + − + + − +
+ + − + + + − − −	− − + + − − + − −	− + + + + − + + −
− + + + + + − − −	+ − − − + − − + −	+ − + − + + − + +
− − − + − + + + +	− − + − + − − − +	+ − + + + − + − +
− − − + + − + + +	+ − − − − + + − −	+ + − − + + + + −
− − − − + + + + +	− + − + − − − + −	− + + + − + − + +
+ + + − − − + − +	− − + − − + − + −	+ − + + − + + + −
+ + + − − − + + −	+ − − + − − − − +	+ + − + + − − + +
+ + + − − − − + +	− + − − + − + − −	− + + − + + + − +

table presents blocks of plus and minus signs for constructing the design for $N = 28$. The designs for $N = 12$, 20, 24, and 36 are obtained by writing the appropriate row in Table 8.23 as a column (or row). A second column (or row) is then generated from this first one by moving the elements of the column (or row) down (or to the right) one position and placing the last element in the first position. A third column (or row) is produced from the second similarly, and the process is continued until column (or row) k is generated. A row of minus signs is then added, completing the design. For $N = 28$, the three blocks X, Y, and Z are written down in the order

$$\begin{matrix} X & Y & Z \\ Z & X & Y \\ Y & Z & X \end{matrix}$$

and a row of minus signs is added to these 27 rows. The design for $N = 12$ runs and $k = 11$ factors is shown in Table 8.24.

The nongeometric Plackett–Burman designs for $N = 12$, 20, 24, 28, and 36 have complex alias structures. For example, in the 12-run design every main effect is **partially aliased** with every two-factor interaction not involving itself. For example, the AB interaction is aliased with the nine main effects $C, D, \ldots, K$ and the AC interaction is aliased with the nine main effects $B, D, \ldots, K$. Furthermore, each main effect is **partially aliased** with 45 two-factor interactions. As an example, consider the aliases of the main effect of factor A:

$$[A] = A - \frac{1}{3}BC - \frac{1}{3}BD - \frac{1}{3}BE + \frac{1}{3}BF + \cdots - \frac{1}{3}KL$$

Each one of the 45 two-factor interactions in the alias chain in weighed by the constant $\pm\frac{1}{3}$. This weighting of the two-factor interactions occurs throughout the Plackett–Burman series of nongeometric designs. In other Plackett–Burman designs, the constant will be different than $\pm\frac{1}{3}$.

■ **TABLE 8.24**

Plackett–Burman Design for $N = 12$, $k = 11$

Run	A	B	C	D	E	F	G	H	I	J	K
1	+	−	+	−	−	−	+	+	+	−	+
2	+	+	−	+	−	−	−	+	+	+	−
3	−	+	+	−	+	−	−	−	+	+	+
4	+	−	+	+	−	+	−	−	−	+	+
5	+	+	−	+	+	−	+	−	−	−	+
6	+	+	+	−	+	+	−	+	−	−	−
7	−	+	+	+	−	+	+	−	+	−	−
8	−	−	+	+	+	−	+	+	−	+	−
9	−	−	−	+	+	+	−	+	+	−	+
10	+	−	−	−	+	+	+	−	+	+	−
11	−	+	−	−	−	+	+	+	−	+	+
12	−	−	−	−	−	−	−	−	−	−	−

Plackett–Burman designs are examples of **nonregular designs**. This term appears frequently in the experimental design literature. Basically, a **regular** design is one in which all effects can be estimated independently of the other effects and in the case of a fractional factorial, the effects that cannot be estimated are completely aliased with the other effects. Obviously, a full factorial such as the 2^k is a regular design, and so are the 2^{k-p} fractional factorials because while all of the effects cannot be estimated the "constants" in the alias chains for these designs are always either zero or plus or minus unity. That is, the effects that are not estimable because of the fractionation are completely aliased (some say completely confounded) with the effects that can be estimated. In nonregular designs, because some of the nonzero constants in the alias chains are not equal to ± 1, there is always at least a chance that some information on the aliased effects may be available.

The projection properties of the nongeometric Plackett–Burman designs are interesting, and in many cases, useful. For example, consider the 12-run design in Table 8.24. This design will project into three replicates of a full 2^2 design in any two of the original 11 factors. In three factors, the projected design is a full 2^3 factorial plus a 2^{3-1}_{III} fractional factorial (see Figure 8.24*a*). All Plackett–Burman designs will project into a full factorial plus some additional runs in any three factors. Thus, the resolution III Plackett–Burman design has **projectivity** 3, meaning it will collapse into a full factorial in any subset of three factors (actually, some of the larger Plackett–Burman designs, such as those with 68, 72, 80, and 84 runs, have projectivity 4). In contrast, the 2^{k-p}_{III} design only has projectivity 2. The four-dimensional projections of the 12-run design are shown in Figure 8.24*b*. Notice that there are 11 distinct runs. This design can fit all four of the main effects and all 6 two-factor interactions, assuming that all other main effects and interactions are negligible. The design in Fig. 8.24b needs 5 additional runs to form a complete 2^4 (with one additional run) and only a single run to form a 2^{4-1} (with 5 additional runs). Regression methods can be used to fit models involving main effects and interactions using those projected designs.

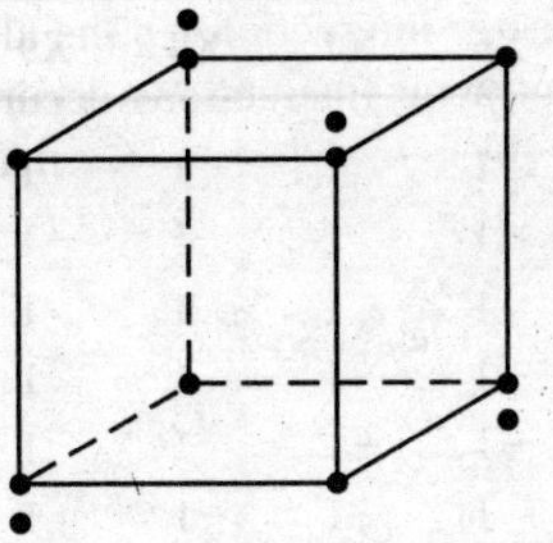

(*a*) Projection into three factors

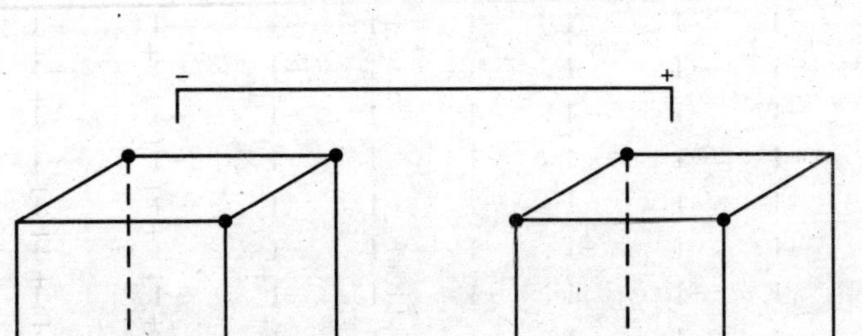

(*b*) Projection into four factors

■ **FIGURE 8.24 Projection of the 12-run Plackett–Burman design into three- and four-factor designs**

EXAMPLE 8.8

We will illustrate the analysis of a Plackett–Burman design with an example involving 12 factors. The smallest regular fractional factorial for 12 factors is a 16-run 2^{12-8} fractional factorial design. In this design, all 12 main effects are aliased with four two-factor interactions and three chains of two-factor interactions each containing six two-factor interactions (refer to Appendix V, design w). If there are significant two-factor interactions along with the main effects it is very possible that additional runs will be required to dealias some of these effects.

Suppose that we decide to use a 20-run Plackett–Burman design for this problem. Now this has more runs that the smallest regular fraction, but it contains fewer runs than would be required by either a full fold-over or a partial fold-over of the 16 run regular fraction. This design was created in JMP and is shown in Table 8.25, along with the observed response data obtained when the experiment was conducted. The alias matrix for this design, also produced from JMP is in Table 8.26. Note that the coefficients of the aliased two-factor interactions are not either 0, -1 or $+1$ because this is a nonregular design). Hopefully this will provide some flexibility with which to estimate interactions if necessary.

Table 8.27 shows the JMP analysis of this design, using a forward-stepwise regression procedure to fit the model. In forward-stepwise regression, variables are entered into the model one at a time, beginning with those that appear most important, until no variables remain that are reasonable candidates for entry. In this analysis, we consider all main effects and two-factor interactions as possible variables of interest for the model.

Considering the P-values for the variables in Table 8.27, the most important factor is x_2, so this factor is entered into the model first. JMP then recalculates the P-values and the next variable entered would be x_4. Then the x_1x_4 interaction is entered along with the main effect of x_1 to preserve the hierarchy of the model. This is followed by the x_1x_4 interactions. The JMP output for these steps is not shown but is summarized at the bottom of Table 8.28. Finally, the last variable entered is x_5. Table 8.28 summarizes the final model.

■ **TABLE 8.25**
Plackett–Burman Design for Example 8.8

Run	X1	X2	X3	X4	X5	X6	X7	X8	X9	X10	X11	X12	y
1	1	1	1	1	1	1	1	1	1	1	1	1	221.5032
2	−1	1	−1	−1	1	1	1	1	−1	1	−1	1	213.8037
3	−1	−1	1	−1	−1	1	1	1	1	−1	1	−1	167.5424
4	1	−1	−1	1	−1	−1	1	1	1	1	−1	1	232.2071
5	1	1	−1	−1	1	−1	−1	1	1	1	1	−1	186.3883
6	−1	1	1	−1	−1	1	−1	−1	1	1	1	1	210.6819
7	−1	−1	1	1	−1	−1	1	−1	−1	1	1	1	168.4163
8	−1	−1	−1	1	1	−1	−1	1	−1	−1	1	1	180.9365
9	−1	−1	−1	−1	1	1	−1	−1	1	−1	−1	1	172.5698
10	1	−1	−1	−1	−1	1	1	−1	−1	1	−1	−1	181.8605
11	−1	1	−1	−1	−1	−1	1	1	−1	−1	1	−1	202.4022
12	1	−1	1	−1	−1	−1	−1	1	1	−1	−1	1	186.0079
13	−1	1	−1	1	−1	−1	−1	−1	1	1	−1	−1	216.4375
14	1	−1	1	−1	1	−1	−1	−1	−1	1	1	−1	192.4121
15	1	1	−1	1	−1	1	−1	−1	−1	−1	1	1	224.4362
16	1	1	1	−1	1	−1	1	−1	−1	−1	−1	1	190.3312
17	1	1	1	1	−1	1	−1	1	−1	−1	−1	−1	228.3411
18	−1	1	1	1	1	−1	1	−1	1	−1	−1	−1	223.6747
19	−1	−1	1	1	1	1	−1	1	−1	1	−1	−1	163.5351
20	1	−1	−1	1	1	1	1	−1	1	−1	1	−1	236.5124

■ TABLE 8.26
The Alias Matrix

Effect	1 2	1 3	1 4	1 5	1 6	1 7	1 8	1 9	1 10	1 11	1 12	2 3	2 4	2 5	2 6	2 7	2 8
Intercept	0	0	0	0	0	0	0	0	0	0	0	0	0	0	0	0	0
X1	0	0	0	0	0	0	0	0	0	0	0	0.2	0.2	0.2	0.2	−0.2	0.2
X2	0	0.2	0.2	0.2	0.2	−0.2	0.2	−0.2	−0.2	0.2	0.2	0	0	0	0	0	0
X3	0.2	0	−0.2	0.2	−0.2	−0.2	0.2	−0.2	−0.2	−0.2	0.2	0	0.2	0.2	0.2	0.2	−0.2
X4	0.2	−0.2	0	−0.2	0.6	0.2	0.2	0.2	−0.2	0.2	0.2	0.2	0	−0.2	0.2	−0.2	−0.2
X5	0.2	0.2	−0.2	0	−0.2	0.2	−0.2	0.2	0.2	0.6	−0.2	0.2	−0.2	0	−0.2	0.6	0.2
X6	0.2	−0.2	0.6	−0.2	0	0.2	−0.2	−0.2	−0.2	0.2	−0.2	0.2	0.2	−0.2	0	−0.2	0.2
X7	−0.2	−0.2	0.2	0.2	0.2	0	−0.2	0.2	0.2	−0.2	0.2	0.2	−0.2	0.6	−0.2	0	0.2
X8	0.2	0.2	0.2	−0.2	−0.2	−0.2	0	0.6	0.2	−0.2	0.2	−0.2	−0.2	0.2	0.2	0.2	0
X9	−0.2	−0.2	0.2	0.2	−0.2	0.2	0.6	0	0.2	0.2	0.2	0.2	0.2	0.2	−0.2	−0.2	−0.2
X10	−0.2	−0.2	−0.2	0.2	−0.2	0.2	0.2	0.2	0	0.2	−0.2	−0.2	−0.2	0.2	0.2	−0.2	0.2
X11	0.2	−0.2	0.2	0.6	0.2	−0.2	−0.2	0.2	0.2	0	−0.2	−0.2	−0.2	−0.2	0.2	−0.2	0.2
X12	0.2	0.2	0.2	−0.2	−0.2	0.2	0.2	0.2	−0.2	−0.2	0	0.2	−0.2	0.2	0.6	0.2	−0.2

Effect	2 9	2 10	2 11	2 12	3 4	3 5	3 6	3 7	3 8	3 9	3 10	3 11	3 12	4 5	4 6	4 7	4 8
Intercept	0	0	0	0	0	0	0	0	0	0	0	0	0	0	0	0	0
X1	−0.2	−0.2	0.2	0.2	−0.2	0.2	−0.2	−0.2	0.2	−0.2	−0.2	−0.2	0.2	−0.2	0.6	0.2	0.2
X2	0	0	0	0	0.2	0.2	0.2	0.2	−0.2	0.2	−0.2	−0.2	0.2	−0.2	0.2	−0.2	−0.2
X3	0.2	−0.2	−0.2	0.2	0	0	0	0	0	0	0	0	0	0.2	0.2	0.2	0.2
X4	0.2	−0.2	−0.2	−0.2	0	0.2	0.2	0.2	0.2	−0.2	0.2	−0.2	−0.2	0	0	0	0
X5	0.2	0.2	−0.2	0.2	0	0	−0.2	0.2	−0.2	−0.2	0.2	−0.2	−0.2	0	0.2	0.2	0.2
X6	−0.2	0.2	0.2	0.6	0.2	−0.2	0	−0.2	0.6	0.2	0.2	0.2	−0.2	0.2	0	−0.2	0.2
X7	−0.2	−0.2	−0.2	0.2	0.2	0.2	−0.2	0	−0.2	0.2	−0.2	0.2	0.2	0.2	−0.2	0	−0.2
X8	−0.2	0.2	0.2	−0.2	0.2	−0.2	0.6	−0.2	0	0.2	−0.2	−0.2	−0.2	0.2	0.2	−0.2	0
X9	0	0.6	0.2	−0.2	−0.2	−0.2	0.2	0.2	0.2	0	−0.2	0.2	0.2	0.2	−0.2	0.6	−0.2
X10	0.6	0	0.2	0.2	0.2	0.2	0.2	−0.2	−0.2	−0.2	0	0.6	0.2	−0.2	−0.2	0.2	0.2
X11	0.2	0.2	0	0.2	−0.2	−0.2	0.2	0.2	−0.2	0.2	0.6	0	0.2	0.2	0.2	0.2	−0.2
X12	−0.2	0.2	0.2	0	−0.2	−0.2	−0.2	0.2	−0.2	0.2	0.2	0.2	0	−0.2	−0.2	0.2	0.2

Effect	4 9	4 10	4 11	4 12	5 6	5 7	5 8	5 9	5 10	5 11	5 12	6 7	6 8	6 9	6 10	6 11
Intercept	0	0	0	0	0	0	0	0	0	0	0	0	0	0	0	0
X1	0.2	−0.2	0.2	0.2	−0.2	0.2	−0.2	0.2	0.2	0.6	−0.2	0.2	−0.2	−0.2	−0.2	0.2
X2	0.2	−0.2	−0.2	−0.2	−0.2	0.6	0.2	0.2	0.2	−0.2	0.2	−0.2	0.2	−0.2	0.2	0.2
X3	−0.2	0.2	−0.2	−0.2	−0.2	0.2	−0.2	−0.2	0.2	−0.2	−0.2	−0.2	0.6	0.2	0.2	0.2
X4	0	0	0	0	0.2	0.2	0.2	0.2	−0.2	0.2	−0.2	−0.2	0.2	−0.2	−0.2	0.2
X5	0.2	−0.2	0.2	−0.2	0	0	0	0	0	0	0	0.2	0.2	0.2	0.2	−0.2
X6	−0.2	−0.2	0.2	−0.2	0	0.2	0.2	0.2	0.2	−0.2	0.2	0	0	0	0	0
X7	0.6	0.2	0.2	0.2	0.2	0	−0.2	0.2	−0.2	−0.2	0.2	0	0.2	0.2	0.2	0.2
X8	−0.2	0.2	−0.2	0.2	0.2	−0.2	0	−0.2	0.6	0.2	0.2	0.2	0	−0.2	0.2	−0.2
X9	0	0.2	−0.2	−0.2	0.2	0.2	−0.2	0	−0.2	0.2	−0.2	0.2	−0.2	0	−0.2	0.6
X10	0.2	0	−0.2	0.2	0.2	−0.2	0.6	−0.2	0	0.2	−0.2	0.2	0.2	−0.2	0	−0.2
X11	−0.2	−0.2	0	0.6	−0.2	−0.2	0.2	0.2	0.2	0	−0.2	0.2	−0.2	0.6	−0.2	0
X12	−0.2	0.2	0.6	0	0.2	0.2	0.2	−0.2	−0.2	−0.2	0	−0.2	−0.2	0.2	0.2	0.2

Effect	6 12	7 8	7 9	7 10	7 11	7 12	8 9	8 10	8 11	8 12	9 10	9 11	9 12	10 11	10 12	11 12
Intercept	0	0	0	0	0	0	0	0	0	0	0	0	0	0	0	0
X1	−0.2	−0.2	0.2	0.2	−0.2	0.2	0.6	0.2	−0.2	0.2	0.2	0.2	0.2	0.2	−0.2	−0.2
X2	0.6	0.2	−0.2	−0.2	−0.2	0.2	−0.2	0.2	0.2	−0.2	0.6	0.2	−0.2	0.2	0.2	0.2
X3	−0.2	−0.2	0.2	−0.2	0.2	0.2	0.2	−0.2	−0.2	−0.2	−0.2	0.2	0.2	0.6	0.2	0.2
X4	−0.2	−0.2	0.6	0.2	0.2	0.2	−0.2	0.2	−0.2	0.2	0.2	−0.2	−0.2	−0.2	0.2	0.6
X5	0.2	−0.2	0.2	−0.2	−0.2	0.2	−0.2	0.6	0.2	0.2	−0.2	0.2	−0.2	0.2	−0.2	−0.2
X6	0	0.2	0.2	0.2	0.2	−0.2	−0.2	0.2	−0.2	−0.2	−0.2	0.6	0.2	−0.2	0.2	0.2
X7	−0.2	0	0	0	0	0	0.2	0.2	0.2	0.2	−0.2	0.2	−0.2	−0.2	0.6	−0.2
X8	−0.2	0	0.2	0.2	0.2	0.2	0	0	0	0	0.2	0.2	0.2	−0.2	0.2	−0.2
X9	0.2	0.2	0	−0.2	0.2	−0.2	0	0.2	0.2	0.2	0	0	0	0.2	0.2	−0.2
X10	0.2	0.2	−0.2	0	−0.2	0.6	0.2	0	−0.2	0.2	0	0.2	0.2	0	0	0.2
X11	0.2	0.2	0.2	−0.2	0	−0.2	0.2	−0.2	0	−0.2	0.2	0	−0.2	0	0.2	0
X12	0	0.2	−0.2	0.6	−0.2	0	0.2	0.2	−0.2	0	0.2	−0.2	0	0.2	0	0

■ **TABLE 8.27**
JMP Stepwise Regression Analysis of Example 8.8, Initial Solution

Stepwise Fit

Response:

Y

Stepwise Regression Control

Prob to Enter 0.250

Prob to Leave 0.100

Current Estimates

SSE	DFE	MSE	RSquare	RSquare Adj	Cp
10732	19	564.84211	0.0000	0.0000	12

Parameter	Estimate	nDF	SS	"F Ratio"	"Prob>F"
Intercept	200	1	0	0.000	1.0000
X1	0	1	1280	2.438	0.1359
X2	0	1	2784.8	6.307	0.0218
X3	0	1	452.279	0.792	0.3853
X4	0	1	1843.2	3.733	0.0693
X5	0	1	67.21943	0.113	0.7401
X6	0	1	86.41367	0.146	0.7068
X7	0	1	292.6697	0.505	0.4866
X8	0	1	60.08353	0.101	0.7539
X9	0	1	572.9881	1.015	0.3270
X10	0	1	32.53443	0.055	0.8177
X11	0	1	15.37763	0.026	0.8741
X12	0	1	0.159759	0.000	0.9871
X1*X2	0	3	5908	6.532	0.0043
X1*X3	0	3	1736.782	1.030	0.4058
X1*X4	0	3	5543.2	5.698	0.0075
X1*X5	0	3	1358.09	0.773	0.5261
X1*X6	0	3	2795.154	1.878	0.1740
X1*X7	0	3	1581.316	0.922	0.4528
X1*X8	0	3	1767.483	1.052	0.3970
X1*X9	0	3	1866.724	1.123	0.3692
X1*X10	0	3	1609.033	0.941	0.4441
X1*X11	0	3	1821.162	1.090	0.3818
X1*X12	0	3	1437.829	0.825	0.4991
X2*X3	0	3	4473.249	3.812	0.0309
X2*X4	0	3	4671.721	4.111	0.0243
X2*X5	0	3	3011.798	2.081	0.1431
X2*X6	0	3	3561.431	2.649	0.0842
X2*X7	0	3	3635.536	2.732	0.0781
X2*X8	0	3	2848.428	1.927	0.1659
X2*X9	0	3	3944.319	3.099	0.0564
X2*X10	0	3	2828.937	1.909	0.1688
X2*X11	0	3	2867.948	1.945	0.1631
X2*X12	0	3	2786.331	1.870	0.1753
X3*X4	0	3	2576.807	1.685	0.2102

■ **TABLE 8.27** *(Continued)*

Parameter	Estimate	nDF	SS	"F Ratio"	"Prob>F"
X3*X5	0	3	995.7837	0.545	0.6582
X3*X6	0	3	558.5936	0.293	0.8300
X3*X7	0	3	1201.228	0.672	0.5815
X3*X8	0	3	512.677	0.268	0.8478
X3*X9	0	3	1058.287	0.583	0.6344
X3*X10	0	3	626.2659	0.331	0.8034
X3*X11	0	3	569.497	0.299	0.8257
X3*X12	0	3	452.4973	0.235	0.8708
X4*X5	0	3	2038.876	1.251	0.3244
X4*X6	0	3	2132.749	1.323	0.3017
X4*X7	0	3	2320.382	1.471	0.2599
X4*X8	0	3	2034.576	1.248	0.3255
X4*X9	0	3	4886.816	4.459	0.0185
X4*X10	0	3	3125.433	2.191	0.1288
X4*X11	0	3	1970.181	1.199	0.3418
X4*X12	0	3	2194.402	1.371	0.2875
X5*X6	0	3	189.5188	0.096	0.9612
X5*X7	0	3	4964.273	4.590	0.0168
X5*X8	0	3	332.1148	0.170	0.9149
X5*X9	0	3	1065.334	0.588	0.6318
X5*X10	0	3	136.8974	0.069	0.9757
X5*X11	0	3	866.5116	0.468	0.7084
X5*X12	0	3	185.205	0.094	0.9625
X6*X7	0	3	434.1661	0.225	0.8777
X6*X8	0	3	185.7122	0.094	0.9623
X6*X9	0	3	1302.2	0.737	0.5455
X6*X10	0	3	246.5934	0.125	0.9437
X6*X11	0	3	2492.598	1.613	0.2256
X6*X12	0	3	913.7187	0.496	0.6900
X7*X8	0	3	935.8699	0.510	0.6813
X7*X9	0	3	1876.723	1.130	0.3665
X7*X10	0	3	345.5343	0.177	0.9101
X7*X11	0	3	577.8999	0.304	0.8224
X7*X12	0	3	328.611	0.168	0.9161
X8*X9	0	3	1111.212	0.616	0.6146
X8*X10	0	3	936.6248	0.510	0.6811
X8*X11	0	3	710.6107	0.378	0.7700
X8*X12	0	3	1517.358	0.878	0.4731
X9*X10	0	3	2360.154	1.504	0.2517
X9*X11	0	3	588.4157	0.309	0.8183
X9*X12	0	3	587.527	0.309	0.8186
X10*X11	0	3	125.3218	0.063	0.9786
X10*X12	0	3	2241.266	1.408	0.2770
X11*X12	0	3	94.12651	0.047	0.9859

■ **TABLE 8.28**
JMP Final Stepwise Regression Solution, Example 8.8

Stepwise Fit

Response:

Y

Stepwise Regression Control

Prob to Enter	0.250
Prob to Leave	0.100
Direction:	
Rules:	

Current Estimates

SSE	DFE	MSE	RSquare	RSquare Adj	Cp
381.79001	13	29.368462	0.9644	0.9480	72

Parameter	Estimate	nDF	SS	"F Ratio"	"Prob>F"
Intercept	200	1	0	0.000	1.0000
X1	8	3	5654.991	64.184	0.0000
X2	9.89242251	2	4804.208	81.792	0.0000
X3	0	1	2.547056	0.081	0.7813
X4	12.1075775	2	4442.053	75.626	0.0000
X5	2.581897	1	122.21	4.161	0.0622
X6	0	1	44.86956	1.598	0.2302
X7	0	1	7.652516	0.245	0.6292
X8	0	1	28.02042	0.950	0.3488
X9	0	1	19.33012	0.640	0.4393
X10	0	1	76.73973	3.019	0.1079
X11	0	1	1.672382	0.053	0.8221
X12	0	1	10.36884	0.335	0.5734
X1*X2	−12.537887	1	2886.987	98.302	0.0000
X1*X3	0	2	6.20474	0.091	0.9138
X1*X4	9.53788744	1	1670.708	56.888	0.0000
X1*X5	0	1	1.889388	0.060	0.8111
X1*X6	0	2	45.6286	0.747	0.4966
X1*X7	0	2	10.10477	0.150	0.8628
X1*X8	0	2	41.24821	0.666	0.5332
X1*X9	0	2	90.27392	1.703	0.2268
X1*X10	0	2	76.84386	1.386	0.2905
X1*X11	0	2	27.15307	0.421	0.6665
X1*X12	0	2	37.51692	0.599	0.5662
X2*X3	0	2	54.47309	0.915	0.4288
X2*X4	0	1	3.403658	0.108	0.7482
X2*X5	0	1	0.216992	0.007	0.9355
X2*X6	0	2	46.47256	0.762	0.4897
X2*X7	0	2	37.44377	0.598	0.5668
X2*X8	0	2	65.97489	1.149	0.3522
X2*X9	0	2	69.32501	1.220	0.3322
X2*X10	0	2	98.35266	1.908	0.1943

■ **TABLE 8.28** (*Continued*)

Parameter	Estimate	nDF	SS	"F Ratio"	"Prob>F"
X2*X11	0	2	141.1503	3.226	0.0790
X2*X12	0	2	52.05325	0.868	0.4466
X3*X4	0	2	111.3687	2.265	0.1500
X3*X5	0	2	80.40096	1.467	0.2724
X3*X6	0	3	67.40344	0.715	0.5653
X3*X7	0	3	99.64513	1.177	0.3667
X3*X8	0	3	66.19013	0.699	0.5737
X3*X9	0	3	29.41242	0.278	0.8399
X3*X10	0	3	120.8801	1.544	0.2632
X3*X11	0	3	4.678496	0.041	0.9881
X3*X12	0	3	56.41798	0.578	0.6426
X4*X5	0	1	49.01055	1.767	0.2084
X4*X6	0	2	148.7678	3.511	0.0662
X4*X7	0	2	10.61344	0.157	0.8564
X4*X8	0	2	29.55318	0.461	0.6420
X4*X9	0	2	25.40367	0.392	0.6847
X4*X10	0	2	112.0974	2.286	0.1478
X4*X11	0	2	1.673771	0.024	0.9761
X4*X12	0	2	24.16136	0.372	0.6980
X5*X6	0	2	169.9083	4.410	0.0392
X5*X7	0	2	31.18914	0.489	0.6258
X5*X8	0	2	90.33176	1.705	0.2265
X5*X9	0	2	34.4118	0.545	0.5948
X5*X10	0	2	154.654	3.745	0.0575
X5*X11	0	2	10.09686	0.149	0.8629
X5*X12	0	2	12.34385	0.184	0.8346
X6*X7	0	3	59.7591	0.619	0.6187
X6*X8	0	3	94.11651	1.091	0.3974
X6*X9	0	3	57.73503	0.594	0.6331
X6*X10	0	3	165.7402	2.557	0.1139
X6*X11	0	3	77.11154	0.844	0.5007
X6*X12	0	3	58.58914	0.604	0.6270
X7*X8	0	3	44.58254	0.441	0.7290
X7*X9	0	3	29.92824	0.284	0.8362
X7*X10	0	3	86.08846	0.970	0.4445
X7*X11	0	3	63.54514	0.666	0.5920
X7*X12	0	3	31.78299	0.303	0.8229
X8*X9	0	3	60.30138	0.625	0.6148
X8*X10	0	3	104.4506	1.255	0.3414
X8*X11	0	3	33.70238	0.323	0.8089
X8*X12	0	3	51.03759	0.514	0.6816
X9*X10	0	3	110.8786	1.364	0.3092
X9*X11	0	3	50.35583	0.506	0.6865
X9*X12	0	3	119.2043	1.513	0.2706
X10*X11	0	3	93.00237	1.073	0.4037
X10*X12	0	3	94.6634	1.099	0.3943

■ **TABLE 8.28** (*Continued*)

Parameter	Estimate	nDF	SS	"F Ratio"	"Prob>F"	
X11*X12	0	3	38.30184	0.372	0.7753	
Step History						
Step	**Parameter**	**Action**	**"Sig Prob"**	**Seq SS**	**RSquare**	**Cp**
1	X2	Entered	0.0218	2784.8	0.2595	.
2	X4	Entered	0.0368	1843.2	0.4312	.
3	X1*X2	Entered	0.0003	4044.8	0.8081	.
4	X1*X4	Entered	0.0000	1555.2	0.9530	.
5	X5	Entered	0.0622	122.21	0.9644	.

The final model for this experiment contains the main effects of factors x_1, x_2, x_4, and x_5, plus the two-factor interactions x_1x_2 and x_1x_4. Now, it turns out that the data for this experiment were simulated from a model. The model used was

$$y = 200 + 8x_1 + 10x_2 + 12x_4 - 12x_1x_2 + 9x_1x_4 + \epsilon$$

where the random error term was normal with mean zero and standard deviation 5. The Plackett–Burman design was able to correctly identify all of the significant main effects and the two significant two-factor interactions. From Table 8.28 we observe that the model parameter estimates are actually very close to the values chosen for the model.

The partial aliasing structure of the Plackett–Burman design has been very helpful in identifying the significant interactions. Another approach to the analysis would be to realize that this design could be used to fit the main effects in any four factors and all of their two factor interactions, than use a normal probability plot to identify the four largest main effects, and finally fit the four factorial model in those four factors.

Notice that there is the main effect x_5 is identified as significant that was not in the simulation model used to generate the data. A type I error has been committed with respect to this factor. In screening experiments type I errors are not as serious as type II errors. A type I error results in a nonsignificant factor being identified as important and retained for subsequent experimentation and analysis. Eventually, we will likely discover that this factor really isn't important. However, a type II error means that an important factor has not been discovered. This variable will be dropped from subsequent studies and if it really turns out to be a critical factor, product or process performance can be negatively impacted. It is highly likely that the effect of this factor will never be discovered because it was discarded early in the research. In our example, all important factors were discovered, including the interactions, and that is the key point.

8.7 Resolution IV and V Designs

8.7.1 Resolution IV Designs

A 2^{k-p} fractional factorial design is of resolution IV if the main effects are clear of two-factor interactions and some two-factor interactions are aliased with each other. Thus, if three-factor and higher interactions are suppressed, the main effects may be estimated directly in a 2_{IV}^{k-p} design. An example is the 2_{IV}^{6-2} design in Table 8.10. Furthermore, the two combined fractions of the 2_{III}^{7-4} design in Example 8.7 yields a 2_{IV}^{7-3} design. Resolution IV designs are used extensively as screening experiments. The 2^{4-1} with eight runs and the 16 run fractions with 6, 7, and 8 factors are very popular.

Any 2_{IV}^{k-p} design must contain at least $2k$ runs. Resolution IV designs that contain exactly $2k$ runs are called **minimal designs**. Resolution IV designs may be obtained from resolution III designs by the process of **fold over**. Recall that to fold over a 2_{III}^{k-p} design, simply add to the original fraction a second fraction with all the signs reversed. Then the plus signs in the identity column I in the first fraction could be switched in the second fraction, and a $(k + 1)$st

■ **TABLE 8.29**

A 2_{IV}^{4-1} Design Obtained by Fold Over

D			
I	*A*	*B*	*C*
Original 2_{III}^{3-1} $I = ABC$			
+	−	−	+
+	+	−	−
+	−	+	−
+	+	+	+
Second 2_{III}^{3-1} with Signs Switched			
−	+	+	−
−	−	+	+
−	+	−	+
−	−	−	−

factor could be associated with this column. The result is a 2_{IV}^{k+1-p} fractional factorial design. The process is demonstrated in Table 8.29 for the 2_{III}^{3-1} design. It is easy to verify that the resulting design is a 2_{IV}^{4-1} design with defining relation $I = ABCD$.

Table 8.30 provides a convenient summary of 2^{k-p} fractional factorial designs with $N =$ 4, 8, 16, and 32 runs. Notice that although 16-run resolution IV designs are available for $6 \leq k \leq 8$ factors, if there are nine or more factors the smallest resolution IV design in the 2^{9-p} family is the 2^{9-4}, which requires 32 runs. Since this is a rather large number of runs, many experimenters are interested in smaller designs. Recall that a resolution IV design must contain at least $2k$ runs, so for example, a nine-factor resolution IV design must have at least 18 runs. A design with exactly $N = 18$ runs can be created by using an algorithm for constructing "optimal" designs. This design is a nonregular design, and it will be illustrated in chapter 9 as part of a broader discussion of nonregular designs.

8.7.2 Sequential Experimentation with Resolution IV Designs

Because resolution IV designs are used as screening experiments, it is not unusual to find that upon conducting and analyzing the original experiment, additional experimentation is necessary

■ **TABLE 8.30**

Useful Factorial and Fractional Factorial Designs from the 2^{k-p} System. The Numbers in the Cells Are the Numbers of Factors in the Experiment

	Number of Runs			
Design Type	**4**	**8**	**16**	**32**
Full factorial	2	3	4	5
Half-fraction	3	4	5	6
Resolution IV fraction	—	4	6–8	7–16
Resolution III fraction	3	5–7	9–15	17–31

to completely resolve all of the effects. We discussed this in Section 8.6.2 for the case of resolution III designs and introduced **fold over** as a sequential experimentation strategy. In the resolution III situation, main effects are aliased with two-factor interaction, so the purpose of the fold over is to separate the main effects from the two-factor interactions. It is also possible to fold over resolution IV designs to separate two-factor interactions that are aliased with each other.

Montgomery and Runger (1996) observe that an experimenter may have several objectives in folding over a resolution IV design, such as

1. breaking as many two-factor interaction alias chains as possible,
2. breaking the two-factor interactions on a specific alias chain, or
3. breaking the two-factor interaction aliases involving a specific factor.

However, one has to be careful in folding over a resolution IV design. The full fold-over rule that we used for resolution III designs, simply run another fraction with all of the signs reversed, will not work for the resolution IV case. If this rule is applied to a resolution IV design, the result will be to produce exactly the same design with the runs in a different order. Try it! Use the 2_{IV}^{6-2} in Table 8.9 and see what happens when you reverse all of the signs in the test matrix.

The simplest way to fold over a resolution IV design is to switch the signs on a single variable of the original design matrix. This single-factor fold over allows all the two-factor interactions involving the factor whose signs are switched to be separated and accomplishes the third objective listed above.

To illustrate how a single-factor fold over is accomplished for a resolution IV design, consider the 2_{IV}^{6-2} design in Table 8.31 (the runs are in standard order, not run order). This experiment was conducted to study the effects of six factors on the thickness of photoresist coating applied to a silicon wafer. The design factors are A = spin speed, B = acceleration,

■ **TABLE 8.31**
The Initial 2_{IV}^{6-2} Design for the Spin Coater Experiment

A Speed RPM	*B* Acceleration	*C* Vol (cc)	*D* Time (sec)	*E* Resist Viscosity	*F* Exhaust Rate	Thickness (mil)
−	−	−	−	−	−	4524
+	−	−	−	+	−	4657
−	+	−	−	+	+	4293
+	+	−	−	−	+	4516
−	−	+	−	+	+	4508
+	−	+	−	−	+	4432
−	+	+	−	−	−	4197
+	+	+	−	+	−	4515
−	−	−	+	−	+	4521
+	−	−	+	+	+	4610
−	+	−	+	+	−	4295
+	+	−	+	−	−	4560
−	−	+	+	+	−	4487
+	−	+	+	−	−	4485
−	+	+	+	−	+	4195
+	+	+	+	+	+	4510

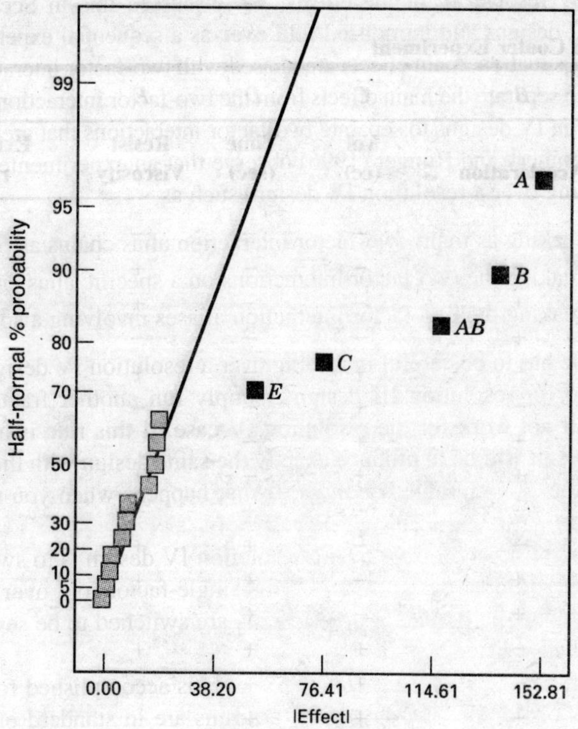

■ **FIGURE 8.25** **Half-normal plot of effects for the initial spin coater experiment in Table 8.31**

C = volume of resist applied, D = spin time, E = resist viscosity, and F = exhaust rate. The alias relationships for this design are given in Table 8.8. The half-normal probability plot of the effects is shown in Figure 8.25. Notice that the largest main effects are A, B, C, and E, and since these effects are aliased with three-factor or higher interactions, it is logical to assume that these are real effects. However, the effect estimate for the $AB + CE$ alias chain is also large. Unless other process knowledge or engineering information is available, we do not know whether this is AB, CE, or both of the interaction effects.

The fold-over design is constructed by setting up a new 2_{IV}^{6-2} fractional factorial design and changing the signs on factor A. The complete design following the addition of the fold-over runs is shown (in standard order) in Table 8.32. Notice that the runs have been assigned to two blocks; the runs from the initial 2_{IV}^{6-2} design in Table 8.32 are in block 1, and the fold-over runs are in block 2. The effects that are estimated from the combined set of runs are (ignoring interactions involving three or more factors)

$$[A] = A \qquad [AE] = AE$$

$$[B] = B \qquad [AF] = AF$$

$$[C] = C \qquad [BC] = BC + DF$$

$$[D] = D \qquad [BD] = BD + CF$$

$$[E] = E \qquad [BE] = BE$$

$$[F] = F \qquad [BF] = BF + CD$$

$$[AB] = AB \qquad [CE] = CE$$

$$[AC] = AC \qquad [DE] = DE$$

$$[AD] = AD \qquad [EF] = EF$$

■ **TABLE 8.32**
The Completed Fold Over for the Spin Coater Experiment

		A	*B*	*C*	*D*	*E*	*F*	
Std. Order	**Block**	**Speed (RPM)**	**Acceleration**	**Vol (cc)**	**Time (sec)**	**Resist Viscosity**	**Exhaust rate**	**Thickness (mil)**
1	1	−	−	−	−	−	−	4524
2	1	+	−	−	−	+	−	4657
3	1	−	+	−	−	+	+	4293
4	1	+	+	−	−	−	+	4516
5	1	−	−	+	−	+	+	4508
6	1	+	−	+	−	−	+	4432
7	1	−	+	+	−	−	−	4197
8	1	+	+	+	−	+	−	4515
9	1	−	−	−	+	−	+	4521
10	1	+	−	−	+	+	+	4610
11	1	−	+	−	+	+	−	4295
12	1	+	+	−	+	−	−	4560
13	1	−	−	+	+	+	−	4487
14	1	+	−	+	+	−	−	4485
15	1	−	+	+	+	−	+	4195
16	1	+	+	+	+	+	+	4510
17	2	+	−	−	−	−	−	4615
18	2	−	−	−	−	+	−	4445
19	2	+	+	−	−	+	+	4475
20	2	−	+	−	−	−	+	4285
21	2	+	−	+	−	+	+	4610
22	2	−	−	+	−	−	+	4325
23	2	+	+	+	−	−	−	4330
24	2	−	+	+	−	+	−	4425
25	2	+	−	−	+	−	+	4655
26	2	−	−	−	+	+	+	4525
27	2	+	+	−	+	+	−	4485
28	2	−	+	−	+	−	−	4310
29	2	+	−	+	+	+	−	4620
30	2	−	−	+	+	−	−	4335
31	2	+	+	+	+	−	+	4345
32	2	−	+	+	+	+	+	4305

Notice that all of the two-factor interactions involving factor *A* are now clear of other two-factor interactions. Also, *AB* is no longer aliased with *CE*. The half-normal probability plot of the effects from the combined design is shown in Figure 8.26. Clearly it is the CE interaction that is significant.

It is easy to show that the completed fold-over design in Table 8.32 allows estimation of the 6 main effects and 12 two-factor interaction alias chains shown previously, along with estimation of 12 other alias chains involving higher order interactions and the block effect.

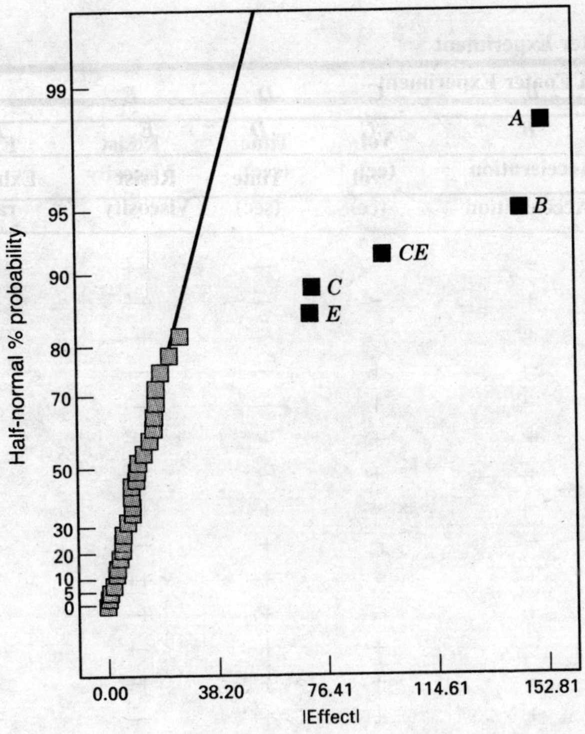

■ **FIGURE 8.26 Half-normal plot of effects for the spin coater experiment in Table 8.32**

The generators for the original fractions are $E = ABC$ and $F = BCD$, and because we changed the signs in column A to create the fold over, the generators for the second group of 16 runs are $E = -ABC$ and $F = BCD$. Since there is only one word of like sign ($L = 1, U = 1$) and the combined design has only one generator (it is a one-half fraction), the generator for the combined design is $F = BCD$. Furthermore, since $ABCE$ is positive in block 1 and $ABCE$ is negative in block 2, $ABCE$ plus its alias $ADEF$ are confounded with blocks.

Examination of the alias chains involving the two-factor interactions for the original 16-run design and the completed fold over reveals some troubling information. In the original resolution IV fraction, every two-factor interaction was aliased with another two-factor interaction in six alias chains, and in one alias chain there were three two-factor interactions (refer to Table 8.8). Thus, seven degrees of freedom were available to estimate two-factor interactions. In the completed fold over, there are nine two-factor interactions that are estimated free of other two-factor interactions and three alias chains involving two two-factor interactions, resulting in 12 degrees of freedom for estimating two-factor interactions. Put another way, we used 16 additional runs but only gained five additional degrees of freedom for estimating two-factor interactions. This is not a terribly efficient use of experimental resources.

Fortunately, there is another alternative to using a complete fold over. In a **partial fold over** (or **semifold**) we make only half of the runs required for a complete fold over, which for the spin coater experiment would be eight runs. The following steps will produce a partial fold-over design:

1. Construct a single-factor fold over from the original design in the usual way by changing the signs on a factor that is involved in a two-factor interaction of interest.
2. Select only half of the fold-over runs by choosing those runs where the chosen factor is either at its high or low level. Selecting the level that you believe will generate the most desirable response is usually a good idea.

Table 8.33 is the partial fold-over design for the spin coater experiment. Notice that we selected the runs where A is at its low level because in the original set of 16 runs (Table 8.31),

■ **TABLE 8.33**
The Partial Fold Over for the Spin Coater Experiment

		A	*B*	*C*	*D*	*E*	*F*	
Std. Order	**Block**	**Speed (RPM)**	**Acceleration**	**Vol (cc)**	**Time (sec)**	**Resist Viscosity**	**Exhaust rate**	**Thickness (mil)**
1	1	−	−	−	−	−	−	4524
2	1	+	−	−	−	+	−	4657
3	1	−	+	−	−	+	+	4293
4	1	+	+	−	−	−	+	4516
5	1	−	−	+	−	+	+	4508
6	1	+	−	+	−	−	+	4432
7	1	−	+	+	−	−	−	4197
8	1	+	+	+	−	+	−	4515
9	1	−	−	−	+	−	+	4521
10	1	+	−	−	+	+	+	4610
11	1	−	+	−	+	+	−	4295
12	1	+	+	−	+	−	−	4560
13	1	−	−	+	+	+	−	4487
14	1	+	−	+	+	−	−	4485
15	1	−	+	+	+	−	+	4195
16	1	+	+	+	+	+	+	4510
17	2	−	−	−	−	+	−	4445
18	2	−	+	−	−	−	+	4285
19	2	−	−	+	−	−	+	4325
20	2	−	+	+	−	+	−	4425
21	2	−	−	−	+	+	+	4525
22	2	−	+	−	+	−	−	4310
23	2	−	−	+	+	−	−	4335
24	2	−	+	+	+	+	+	4305

thinner coatings of photoresist (which are desirable in this case) were obtained with *A* at the low level. (The estimate of the *A* effect is positive in the analysis of the original 16 runs, also suggesting that *A* at the low level produces the desired results.)

The alias relations from the partial fold over (ignoring interactions involving three or more factors) are

$$[A] = A \qquad [AE] = AE$$
$$[B] = B \qquad [AF] = AF$$
$$[C] = C \qquad [BC] = BC + DF$$
$$[D] = D \qquad [BD] = BD + CF$$
$$[E] = E \qquad [BE] = BE$$
$$[F] = F \qquad [BF] = BF + CD$$
$$[AB] = AB \qquad [CE] = CE$$
$$[AC] = AC \qquad [DE] = DE$$
$$[AD] = AD \qquad [EF] = EF$$

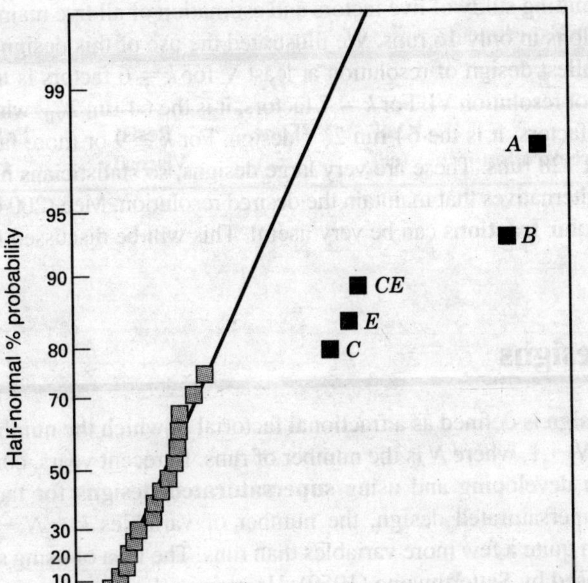

■ **FIGURE 8.27 Half-normal plot of effects from the partial fold over of the spin coater experiment in Table 8.33**

Notice that there are 12 degrees of freedom available to estimate two-factor interactions, exactly as in the complete fold over. Furthermore, AB is no longer aliased with CE. The half-normal plot of the effects from the partial fold over is shown in Figure 8.27. As in the complete fold over, CE is identified as the significant two-factor interaction.

The partial fold-over technique is very useful with resolution IV designs and usually leads to an efficient use of experimental resources. Resolution IV designs always provide good estimates of main effects (assuming that three-factor interactions are negligible), and usually the number of possible two-factor interaction that need to be de-aliased is not large. A partial fold over of a resolution IV design will usually support estimation of as many two-factor interactions as a full fold over. One disadvantage of the partial fold over is that it is not orthogonal. This causes parameter estimates to be correlated and leads to inflation in the standard errors of the effects or regression model coefficients. For example, in the partial fold over of the spin coater experiment, the standard errors of the regression model coefficients range from 0.20σ to 0.25σ, while in the complete fold over, which is orthogonal, the standard errors of the model coefficients are 0.18σ. For more information on partial fold overs, see Mee and Peralta (2000) and the supplemental material for this chapter.

8.7.3 Resolution V Designs

Resolution V designs are fractional factorials in which the main effects and the two-factor interactions do not have other main effects and two-factor interactions as their aliases. Consequently, these are very powerful designs, allowing unique estimation of all main effects and two-factor interactions, provided of course that all interactions involving three or more factors are negligible. The shortest word in the defining relation of a resolution V design must have five letters. The 2^{5-1} design with $I = ABCDE$ is perhaps the most widely used resolution

V design, permitting study of five factors and estimation of all five main effects and all 10 two-factor interactions in only 16 runs. We illustrated the use of this design in Example 8.2.

The smallest design of resolution at least V for $k = 6$ factors is the 2_{VI}^{6-1} design with 32 runs, which is of resolution VI. For $k = 7$ factors, it is the 64-run 2_{VII}^{7-1} which is of resolution VII, and for $k = 8$ factors, it is the 64 run 2_{V}^{8-2} design. For $k \geq 9$ or more factors, all these designs require at least 128 runs. These are very large designs, so statisticians have long been interested in smaller alternatives that maintain the desired resolution. Mee (2004) gives a survey of this topic. **Nonregular fractions** can be very useful. This will be discussed further in chapter 9.

8.8 Supersaturated Designs

A saturated design is defined as a fractional factorial in which the number of factors or design variables $k = N - 1$, where N is the number of runs. In recent years, considerable interest has been shown in developing and using **supersaturated designs** for factor screening experiments. In a supersaturated design, the number of variables $k > N - 1$, and usually these designs contain quite a few more variables than runs. The idea of using supersaturated designs was first proposed by Satterthwaite (1959). He proposed generating these designs at random. In an extensive discussion of this paper, some of the leading authorities in experimental design of the day, including Jack Youden, George Box, J. Stuart Hunter, William Cochran, John Tukey, Oscar Kempthorne, and Frank Anscombe, criticized random balanced designs. As a result, supersaturated designs received little attention for the next 30 years. A notable exception is the systematic supersaturated design developed by Booth and Cox (1962). Their designs were not randomly generated, which was a significant departure from Satterthwaite's proposal. They generated their designs with elementary computer search methods. They also developed the basic criteria by which supersaturated designs are judged.

Lin (1993) revisited the supersaturated design concept and stimulated much additional research on the topic. Many authors have proposed methods to construct supersaturated designs. A good survey is in Lin (2000). Most design construction techniques are limited computer search procedures based on simple heuristics [see Lin (1995), Li and Wu (1997), and Holcomb and Carlyle (2002), for example]. Others have proposed methods based on optimal design construction techniques (we will discuss optimal designs in Chapter 11).

Another construction method for supersaturated designs is based on the structure of existing orthogonal designs. These include using the half-fraction of Hadamard matrices [Lin (1993)] and enumerating the two-factor interactions of certain Hadamard matrices [Wu (1997)]. A Hadamard matrix is a square orthogonal matrix whose elements are either -1 or $+1$. When the number of factors in the experiment exceeds the number of runs, the design matrix cannot be orthogonal. Consequently, the factor effect estimates are not independent. An experiment with one dominant factor may contaminate and obscure the contribution of another factor. Supersaturated designs are created to minimize this amount of nonorthogonality between factors. Supersaturated designs can also be constructed using the optimal design approach. The custom designer in JMP uses this approach to constructing supersaturated designs.

The supersaturated designs that are based on the half fraction of a Hadamard matrix are very easy to construct. Table 8.34 is the Plackett–Burman design for $N = 12$ runs and $k = 11$ factors. It is also a Hadamard matrix design. In the table, the design has been sorted by the signs in the last column (Factor 11 or L). This is sometimes called the **branching column**. Now retain only the runs that are positive (say) in column L from the design and delete column L from this group of runs. The resulting design is a supersaturated design for $k = 10$ factors in $N = 6$ runs. We could have used the runs that are negative in column L equally well.

■ **TABLE 8.34**

A Supersaturated Design Derived from a 12-Run Hadamard Matrix (Plackett–Burman) Design

Run	I	Factor 1 (A)	Factor 2 (B)	Factor 3 (C)	Factor 4 (D)	Factor 5 (E)	Factor 6 (F)	Factor 7 (G)	Factor 8 (H)	Factor 9 (J)	Factor 10 (K)	Factor 11 (L)
1	+	−	+	+	+	−	+	+	−	+	−	−
2	+	−	−	−	−	−	−	−	−	−	−	−
3	+	+	−	−	−	+	+	+	−	+	+	−
4	+	−	−	+	+	+	−	+	+	−	+	−
5	+	+	+	+	−	+	+	−	+	−	−	−
6	+	+	+	−	+	−	−	−	+	+	+	−
7	+	−	+	+	−	+	−	−	−	+	+	+
8	+	+	+	−	+	+	−	+	−	−	−	+
9	+	+	−	+	−	−	−	+	+	+	−	+
10	+	+	−	+	+	−	+	−	−	−	+	+
11	+	−	−	−	+	+	+	−	+	+	−	+
12	+	−	+	−	−	−	+	+	+	−	+	+

This procedure will always produce a supersaturated design for $k = N - 2$ factors in $N/2$ runs. If there are fewer than $N - 2$ factors of interest, additional columns can be removed from the complete design.

Supersaturated designs are typically analyzed by regression model-fitting methods, such as the forward selection method we have illustrated previously. In this procedure, variables are selected one at a time for inclusion in the model until no other variables appear useful in explaining the response. Abraham, Chipman, and Vijayan (1999) and Holcomb, Montgomery, and Carlyle (2003) have studied analysis methods for supersaturated designs. Generally, these designs can experience large type I and type II errors, but some analysis methods can be tuned to emphasize type I errors so that the type II error rate will be moderate. In a factor screening situation, it is usually more important not to exclude an active factor than it is to conclude that inactive factors are important, so type I errors are less critical than type II errors. However, because both error rates can be large, the philosophy in using a supersaturated design should be to eliminate a large portion of the inactive factors, and not to clearly identify the few important or active factors. Holcomb, Montgomery, and Carlyle (2003) found that some types of supersaturated designs perform better than others with respect to type I and type II errors. Generally, the designs produced by search algorithms were outperformed by designs constructed from standard orthogonal designs. Supersaturated designs created using the D-optimality criterion also usually work well.

Supersaturated designs have not had widespread use. However, they are an interesting and potentially useful method for experimentation with systems where there are many variables and only a very few of these are expected to produce large effects.

8.9 Summary

This chapter has introduced the 2^{k-p} fractional factorial design. We have emphasized the use of these designs in screening experiments to quickly and efficiently identify the subset of factors that are active and to provide some information on interaction. The projective property of these designs makes it possible in many cases to examine the active factors in more detail. Sequential assembly of these designs via fold over is a very effective way to gain additional information about interactions that an initial experiment may identify as possibly important.

In practice, 2^{k-p} fractional factorial designs with $N = 4$, 8, 16, and 32 runs are highly useful. Table 8.28 summarizes these designs, identifying how many factors can be used with each design to obtain various types of screening experiments. For example, the 16-run design is a full factorial for 4 factors, a one-half fraction for 5 factors, a resolution IV fraction for 6 to 8 factors, and a resolution III fraction for 9 to 15 factors. All of these designs may be constructed using the methods discussed in this chapter, and many of their alias structures are shown in Appendix Table V.

8.10 Problems

8.1. Suppose that in the chemical process development experiment described in Problem 6.7, it was only possible to run a one-half fraction of the 2^4 design. Construct the design and perform the statistical analysis, using the data from replicate I.

8.2. Suppose that in Problem 6.15, only a one-half fraction of the 2^4 design could be run. Construct the design and perform the analysis, using the data from replicate I.

8.3. Consider the plasma etch experiment described in Example 6.1. Suppose that only a one-half fraction of the design could be run. Set up the design and analyze the data.

8.4. Problem 6.25 describes a process improvement study in the manufacturing process of an integrated circuit. Suppose that only eight runs could be made in this process. Set up an appropriate 2^{5-2} design and find the alias structure. Use the appropriate observations from Problem 6.24 as the observations in this design and estimate the factor effects. What conclusions can you draw?

8.5. ***Continuation of Problem 8.4.*** Suppose you have made the eight runs in the 2^{5-2} design in Problem 8.4. What additional runs would be required to identify the factor effects that are of interest? What are the alias relationships in the combined design?

8.6. In Example 6.6, a 2^4 factorial design was used to improve the response rate to a credit card mail marketing offer. Suppose that the researchers had used the 2^{4-1} fractional factorial design with $I = ABCD$ instead. Set up the design and select the responses for the runs from the full factorial data in Example 6.6. Analyze the data and draw conclusions. Compare your findings with those from the full factorial in Example 6.6.

8.7. ***Continuation of Problem 8.6.*** In Example 6.6 we found that all four main effects and the two-factor AB interaction were significant. Show that if the alternate fraction ($I = -ABCD$) is added to the 2^{4-1} design in Problem 8.6 that the analysis of the results from the combined design produce results identical to those found in Exercise 6.6.

8.8. ***Continuation of Problem 8.6.*** Reconsider the 2^{4-1} fractional factorial design with $I = ABCD$ from Problem 8.6. Set a partial fold-over of this fraction to isolate the AB interaction. Select the appropriate set of responses from the full factorial data in Example 6.6 and analyze the resulting data.

8.9. R. D. Snee ("Experimenting with a Large Number of Variables," in *Experiments in Industry: Design, Analysis and Interpretation of Results*, by R. D. Snee, L. B. Hare, and J. B. Trout, Editors, ASQC, 1985) describes an experiment in which a 2^{5-1} design with $I = ABCDE$ was used to investigate the effects of five factors on the color of a chemical product. The factors are A = solvent/reactant, B = catalyst/reactant, C = temperature, D = reactant purity, and E = reactant pH. The responses obtained were as follows:

$e = -0.60$	$d = 6.85$
$a = 2.55$	$ade = 5.50$
$b = -2.68$	$bde = 3.48$
$abe = 1.60$	$abd = 5.75$
$c = 2.06$	$cde = 5.25$
$ace = 1.22$	$acd = 4.40$
$bce = -2.09$	$bcd = 4.32$
$abc = 1.93$	$abcde = 4.00$

(a) Prepare a normal probability plot of the effects. Which effects seem active?

(b) Calculate the residuals. Construct a normal probability plot of the residuals and plot the residuals versus the fitted values. Comment on the plots.

(c) If any factors are negligible, collapse the 2^{5-1} design into a full factorial in the active factors. Comment on the resulting design, and interpret the results.

8.10. An article by J. J. Pignatiello Jr. and J. S. Ramberg in the *Journal of Quality Technology* (Vol. 17, 1985, pp. 198–206) describes the use of a replicated fractional factorial to investigate the effect of five factors on the free height of leaf springs used in an automotive application. The factors are A = furnace temperature, B = heating time, C = transfer time, D = hold down time, and E = quench oil temperature. The data are shown in Table P8.1.

■ **TABLE P8.1**
Leaf Spring Experiment

A	*B*	*C*	*D*	*E*	Free Height		
−	−	−	−	−	7.78	7.78	7.81
+	−	−	+	−	8.15	8.18	7.88
−	+	−	+	−	7.50	7.56	7.50
+	+	−	−	−	7.59	7.56	7.75
−	−	+	+	−	7.54	8.00	7.88
+	−	+	−	−	7.69	8.09	8.06
−	+	+	−	−	7.56	7.52	7.44
+	+	+	+	−	7.56	7.81	7.69
−	−	−	−	+	7.50	7.25	7.12
+	−	−	+	+	7.88	7.88	7.44
−	+	−	+	+	7.50	7.56	7.50
+	+	−	−	+	7.63	7.75	7.56
−	−	+	+	+	7.32	7.44	7.44
+	−	+	−	+	7.56	7.69	7.62
−	+	+	−	+	7.18	7.18	7.25
+	+	+	+	+	7.81	7.50	7.59

(a) Write out the alias structure for this design. What is the resolution of this design?

(b) Analyze the data. What factors influence the mean free height?

(c) Calculate the range and standard deviation of the free height for each run. Is there any indication that any of these factors affects variability in the free height?

(d) Analyze the residuals from this experiment, and comment on your findings.

(e) Is this the best possible design for five factors in 16 runs? Specifically, can you find a fractional design for five factors in 16 runs with a higher resolution than this one?

8.11. Consider the leaf spring experiment in Problem 8.7. Suppose that factor *E* (quench oil temperature) is very difficult to control during manufacturing. Where would you set factors *A*, *B*, *C*, and *D* to reduce variability in the free height as much as possible regardless of the quench oil temperature used?

8.12. An article in *Industrial and Engineering Chemistry* ("More on Planning Experiments to Increase Research Efficiency," 1970, pp. 60–65) uses a 2^{5-2} design to investigate the effect of A = condensation temperature, B = amount of material 1, C = solvent volume, D = condensation time, and E = amount of material 2 on yield. The results obtained are as follows:

$$e = 23.2 \quad ad = 16.9 \quad cd = 23.8 \quad bde = 16.8$$
$$ab = 15.5 \quad bc = 16.2 \quad ace = 23.4 \quad abcde = 18.1$$

(a) Verify that the design generators used were $I = ACE$ and $I = BDE$.

(b) Write down the complete defining relation and the aliases for this design.

(c) Estimate the main effects.

(d) Prepare an analysis of variance table. Verify that the *AB* and *AD* interactions are available to use as error.

(e) Plot the residuals versus the fitted values. Also construct a normal probability plot of the residuals. Comment on the results.

8.13. Construct a 2^{7-2} design by choosing two four-factor interactions as the independent generators. Write down the complete alias structure for this design. Outline the analysis of variance table. What is the resolution of this design?

8.14. Consider the 2^5 design in Problem 6.25. Suppose that only a one-half fraction could be run. Furthermore, two days were required to take the 16 observations, and it was necessary to confound the 2^{5-1} design in two blocks. Construct the design and analyze the data.

8.15. Analyze the data in Problem 6.28 as if it came from a 2^{4-1}_{IV} design with $I = ABCD$. Project the design into a full factorial in the subset of the original four factors that appear to be significant.

8.16. Repeat Problem 8.15 using $I = -ABCD$. Does the use of the alternate fraction change your interpretation of the data?

8.17. Project the 2^{4-1}_{IV} design in Example 8.1 into two replicates of a 2^2 design in the factors *A* and *B*. Analyze the data and draw conclusions.

8.18. Construct a 2^{5-2}_{III} design. Determine the effects that may be estimated if a full fold over of this design is performed.

8.19. Construct a 2^{6-3}_{III} design. Determine the effects that may be estimated if a full fold over of this design is performed.

8.20. Consider the 2^{6-3}_{III} design in Problem 8.18. Determine the effects that may be estimated if a single factor fold over of this design is run with the signs for factor *A* reversed.

8.21. Fold over the 2^{7-4}_{III} design in Table 8.19 to produce an eight-factor design. Verify that the resulting design is a 2^{8-4}_{IV} design. Is this a minimal design?

8.22. Fold over a 2^{5-2}_{III} design to produce a six-factor design. Verify that the resulting design is a 2^{6-2}_{IV} design. Compare this design to the 2^{6-2}_{IV} design in Table 8.10.

8.23. Construct a 2^{5-1} design. Show how the design may be run in two blocks of eight observations each. Are any main effects or two-factor interactions confounded with blocks?

8.24. Construct a 2^{7-2} design. Show how the design may be run in four blocks of eight observations each. Are any main effects or two-factor interactions confounded with blocks?

8.25. An industrial engineer is conducting an experiment using a Monte Carlo simulation model of an inventory system. The independent variables in her model are the order quantity (*A*), the reorder point (*B*), the setup cost (*C*), the backorder cost (*D*), and the carrying cost rate (*E*). The response variable is average annual cost. To conserve computer time, she decides to investigate these factors using a 2^{5-2}_{III} design with $I = ABD$ and $I = BCE$. The results she obtains are $de = 95$,

$ae = 134$, $b = 158$, $abd = 190$, $cd = 92$, $ac = 187$, $bce = 155$, and $abcde = 185$.

(a) Verify that the treatment combinations given are correct. Estimate the effects, assuming three-factor and higher interactions are negligible.

(b) Suppose that a second fraction is added to the first, for example, $ade = 136$, $e = 93$, $ab = 187$, $bd = 153$, $acd = 139$, $c = 99$, $abce = 191$, and $bcde = 150$. How was this second fraction obtained? Add this data to the original fraction, and estimate the effects.

(c) Suppose that the fraction $abc = 189$, $ce = 96$, $bcd = 154$, $acde = 135$, $abe = 193$, $bde = 152$, $ad = 137$, and $(1) = 98$ was run. How was this fraction obtained? Add this data to the original fraction and estimate the effects.

8.26. ***Nonregular fractions of the 2^k [John (1971)].*** Consider a 2^4 design. We must estimate the four main effects and the six two-factor interactions, but the full 2^4 factorial cannot be run. The largest possible block size contains 12 runs. These 12 runs can be obtained from the four one-quarter replicates defined by $I = \pm AB = \pm ACD = \pm BCD$ by omitting the principal fraction. Show how the remaining three 2^{4-2} fractions can be combined to estimate the required effects, assuming three-factor and higher interactions are negligible. This design could be thought of as a three-quarter fraction.

8.27. Carbon anodes used in a smelting process are baked in a ring furnace. An experiment is run in the furnace to determine which factors influence the weight of packing material that is stuck to the anodes after baking. Six variables are of interest, each at two levels: A = pitch/fines ratio (0.45, 0.55), B = packing material type (1, 2), C = packing material temperature (ambient, 325°C), D = flue location (inside, outside), E = pit temperature (ambient, 195°C), and F = delay time before packing (zero, 24 hours). A 2^{6-3} design is run, and three replicates are obtained at each of the design points. The weight of packing material stuck to the anodes is measured in grams. The data in run order are as follows: abd = (984, 826, 936); $abcdef$ = (1275, 976, 1457); be = (1217, 1201, 890); af = (1474, 1164, 1541); def = (1320, 1156, 913); cd = (765, 705, 821); ace = (1338, 1254, 1294); and bcf = (1325, 1299, 1253). We wish to minimize the amount of stuck packing material.

(a) Verify that the eight runs correspond to a 2_{III}^{6-3} design. What is the alias structure?

(b) Use the average weight as a response. What factors appear to be influential?

(c) Use the range of the weights as a response. What factors appear to be influential?

(d) What recommendations would you make to the process engineers?

8.28. A 16-run experiment was performed in a semiconductor manufacturing plant to study the effects of six factors on the curvature or camber of the substrate devices produced. The six variables and their levels are shown in Table P8.2.

TABLE P8.2
Factor Levels for the Experiment in Problem 8.28

Run	Lamination Temperature (°C)	Lamination Time (sec)	Lamination Pressure (tn)	Firing Temperature (°C)	Firing Cycle Time (h)	Firing Dew Point (°C)
1	55	10	5	1580	17.5	20
2	75	10	5	1580	29	26
3	55	25	5	1580	29	20
4	75	25	5	1580	17.5	26
5	55	10	10	1580	29	26
6	75	10	10	1580	17.5	20
7	55	25	10	1580	17.5	26
8	75	25	10	1580	29	20
9	55	10	5	1620	17.5	26
10	75	10	5	1620	29	20
11	55	25	5	1620	29	26
12	75	25	5	1620	17.5	20
13	55	10	10	1620	29	20
14	75	10	10	1620	17.5	26
15	55	25	10	1620	17.5	20
16	75	25	10	1620	29	26

■ **TABLE P8.3**
Data from the Experiment in Problem 8.28

Run	Camber for Replicate (in./in.) 1	2	3	4	Total (10^{-4} in./in.)	Mean (10^{-4} in./in.)	Standard Deviation
1	0.0167	0.0128	0.0149	0.0185	629	157.25	24.418
2	0.0062	0.0066	0.0044	0.0020	192	48.00	20.976
3	0.0041	0.0043	0.0042	0.0050	176	44.00	4.083
4	0.0073	0.0081	0.0039	0.0030	223	55.75	25.025
5	0.0047	0.0047	0.0040	0.0089	223	55.75	22.410
6	0.0219	0.0258	0.0147	0.0296	920	230.00	63.639
7	0.0121	0.0090	0.0092	0.0086	389	97.25	16.029
8	0.0255	0.0250	0.0226	0.0169	900	225.00	39.42
9	0.0032	0.0023	0.0077	0.0069	201	50.25	26.725
10	0.0078	0.0158	0.0060	0.0045	341	85.25	50.341
11	0.0043	0.0027	0.0028	0.0028	126	31.50	7.681
12	0.0186	0.0137	0.0158	0.0159	640	160.00	20.083
13	0.0110	0.0086	0.0101	0.0158	455	113.75	31.12
14	0.0065	0.0109	0.0126	0.0071	371	92.75	29.51
15	0.0155	0.0158	0.0145	0.0145	603	150.75	6.75
16	0.0093	0.0124	0.0110	0.0133	460	115.00	17.45

Each run was replicated four times, and a camber measurement was taken on the substrate. The data are shown in Table P8.3.

(a) What type of design did the experimenters use?

(b) What are the alias relationships in this design?

(c) Do any of the process variables affect average camber?

(d) Do any of the process variables affect the variability in camber measurements?

(e) If it is important to reduce camber as much as possible, what recommendations would you make?

8.29. A spin coater is used to apply photoresist to a bare silicon wafer. This operation usually occurs early in the semiconductor manufacturing process, and the average coating thickness and the variability in the coating thickness have an important impact on downstream manufacturing steps. Six variables are used in the experiment. The variables and their high and low levels are as follows:

Factor	Low Level	High Level
Final spin speed	7350 rpm	6650 rpm
Acceleration rate	5	20
Volume of resist applied	3 cc	5 cc
Time of spin	14 sec	6 sec
Resist batch variation	Batch 1	Batch 2
Exhaust pressure	Cover off	Cover on

The experimenter decides to use a 2^{6-1} design and to make three readings on resist thickness on each test wafer. The data are shown in Table P8.4.

(a) Verify that this is a 2^{6-1} design. Discuss the alias relationships in this design.

(b) What factors appear to affect average resist thickness?

(c) Because the volume of resist applied has little effect on average thickness, does this have any important practical implications for the process engineers?

(d) Project this design into a smaller design involving only the significant factors. Graphically display the results. Does this aid in interpretation?

(e) Use the range of resist thickness as a response variable. Is there any indication that any of these factors affect the variability in resist thickness?

(f) Where would you recommend that the engineers run the process?

8.30. Harry Peterson-Nedry (a friend of the author) owns a vineyard and winery in Newberg, Oregon. He grows several varieties of grapes and produces wine. Harry has used factorial designs for process and product development in the winemaking segment of the business. This problem describes the experiment conducted for the 1985 Pinot Noir. Eight variables, shown in Table P8.5, were originally studied in this experiment:

■ **TABLE P8.4**
Data for Problem 8.29

	A	*B*	*C*	*D*	*E*	*F*	Resist Thickness				
Run	**Volume**	**Batch**	**Time (sec)**	**Speed**	**Acc.**	**Cover**	**Left**	**Center**	**Right**	**Avg.**	**Range**
1	5	Batch 2	14	7350	5	Off	4531	4531	4515	4525.7	16
2	5	Batch 1	6	7350	5	Off	4446	4464	4428	4446	36
3	3	Batch 1	6	6650	5	Off	4452	4490	4452	4464.7	38
4	3	Batch 2	14	7350	20	Off	4316	4328	4308	4317.3	20
5	3	Batch 1	14	7350	5	Off	4307	4295	4289	4297	18
6	5	Batch 1	6	6650	20	Off	4470	4492	4495	4485.7	25
7	3	Batch 1	6	7350	5	On	4496	4502	4482	4493.3	20
8	5	Batch 2	14	6650	20	Off	4542	4547	4538	4542.3	9
9	5	Batch 1	14	6650	5	Off	4621	4643	4613	4625.7	30
10	3	Batch 1	14	6650	5	On	4653	4670	4645	4656	25
11	3	Batch 2	14	6650	20	On	4480	4486	4470	4478.7	16
12	3	Batch 1	6	7350	20	Off	4221	4233	4217	4223.7	16
13	5	Batch 1	6	6650	5	On	4620	4641	4619	4626.7	22
14	3	Batch 1	6	6650	20	On	4455	4480	4466	4467	25
15	5	Batch 2	14	7350	20	On	4255	4288	4243	4262	45
16	5	Batch 2	6	7350	5	On	4490	4534	4523	4515.7	44
17	3	Batch 2	14	7350	5	On	4514	4551	4540	4535	37
18	3	Batch 1	14	6650	20	Off	4494	4503	4496	4497.7	9
19	5	Batch 2	6	7350	20	Off	4293	4306	4302	4300.3	13
20	3	Batch 2	6	7350	5	Off	4534	4545	4512	4530.3	33
21	5	Batch 1	14	6650	20	On	4460	4457	4436	4451	24
22	3	Batch 2	6	6650	5	On	4650	4688	4656	4664.7	38
23	5	Batch 1	14	7350	20	Off	4231	4244	4230	4235	14
24	3	Batch 2	6	7350	20	On	4225	4228	4208	4220.3	20
25	5	Batch 1	14	7350	5	On	4381	4391	4376	4382.7	15
26	3	Batch 2	6	6650	20	Off	4533	4521	4511	4521.7	22
27	3	Batch 1	14	7350	20	On	4194	4230	4172	4198.7	58
28	5	Batch 2	6	6650	5	Off	4666	4695	4672	4677.7	29
29	5	Batch 1	6	7350	20	On	4180	4213	4197	4196.7	33
30	5	Batch 2	6	6650	20	On	4465	4496	4463	4474.7	33
31	5	Batch 2	14	6650	5	On	4653	4685	4665	4667.7	32
32	3	Batch 2	14	6650	5	Off	4683	4712	4677	4690.7	35

■ **TABLE P8.5**
Factors and Levels for the Winemaking Experiment

Variable	**Low Level (−)**	**High Level (+)**
A = Pinot Noir clone	Pommard	Wadenswil
B = Oak type	Allier	Troncais
C = Age of barrel	Old	New
D = Yeast/skin contact	Champagne	Montrachet
E = Stems	None	All
F = Barrel toast	Light	Medium
G = Whole cluster	None	10%
H = Fermentation temperature	Low (75°F max)	High (92°F max)

Harry decided to use a 2_{IV}^{8-4} design with 16 runs. The wine was tastetested by a panel of experts on March 8, 1986. Each expert ranked the 16 samples of wine tasted, with rank 1 being the best. The design and the taste-test panel results are shown in Table P8.6.

(a) What are the alias relationships in the design selected by Harry?

(b) Use the average ranks ($\overline{y}$) as a response variable. Analyze the data and draw conclusions. You will find it helpful to examine a normal probability plot of the effect estimates.

(c) Use the standard deviation of the ranks (or some appropriate transformation such as log s) as a response variable. What conclusions can you draw about the effects of the eight variables on variability in wine quality?

(d) After looking at the results, Harry decides that one of the panel members (DCM) knows more about beer than he does about wine, so they decide to delete his ranking. What effect would this have on the results and conclusions from parts (b) and (c)?

(e) Suppose that just before the start of the experiment, Harry and Judy discovered that the eight new barrels they ordered from France for use in the experiment would not arrive in time, and all 16 runs would have to be made with old barrels. If Harry just drops column C from their design, what does this do to the alias relationships? Does he need to start over and construct a new design?

(f) Harry knows from experience that some treatment combinations are unlikely to produce good results. For example, the run with all eight variables at the high level generally results in a poorly rated wine. This was confirmed in the March 8, 1986 taste test. He wants to set up a new design for their 1986 Pinot Noir using these same eight variables, but he does not want to make the run with all eight factors at the high level. What design would you suggest?

■ **TABLE P8.6**
Design and Results for Wine Tasting Experiment

	Variable								Panel Rankings					Summary	
Run	A	B	C	D	E	F	G	H	HPN	JPN	CAL	DCM	RGB	$\overline{y}$	s
1	−	−	−	−	−	−	−	−	12	6	13	10	7	9.6	3.05
2	+	−	−	−	−	+	+	+	10	7	14	14	9	10.8	3.11
3	−	+	−	−	+	−	+	+	14	13	10	11	15	12.6	2.07
4	+	+	−	−	+	+	−	−	9	9	7	9	12	9.2	1.79
5	−	−	+	−	+	+	+	−	8	8	11	8	10	9.0	1.41
6	+	−	+	−	+	−	−	+	16	12	15	16	16	15.0	1.73
7	−	+	+	−	−	+	−	+	6	5	6	5	3	5.0	1.22
8	+	+	+	−	−	−	+	−	15	16	16	15	14	15.2	0.84
9	−	−	−	+	+	+	−	+	1	2	3	3	2	2.2	0.84
10	+	−	−	+	+	−	+	−	7	11	4	7	6	7.0	2.55
11	−	+	−	+	−	+	+	−	13	3	8	12	8	8.8	3.96
12	+	+	−	+	−	−	−	+	3	1	5	1	4	2.8	1.79
13	−	−	+	+	−	−	+	+	2	10	2	4	5	4.6	3.29
14	+	−	+	+	−	+	−	−	4	4	1	2	1	2.4	1.52
15	−	+	+	+	+	−	−	−	5	15	9	6	11	9.2	4.02
16	+	+	+	+	+	+	+	+	11	14	12	13	13	12.6	1.14

8.31. Consider the isatin yield data from the experiment described in Problem 6.38. The original experiment was a 2^4 full factorial. Suppose that the original experimenters could only afford eight runs. Set up the 2^{4-1} fractional factorial design with $I = ABCD$ and select the responses for the runs from the full factorial data in Example 6.38. Analyze the data and draw conclusions. Compare your findings with those from the full factorial in Problem 6.38.

8.32. Consider the 2^5 factorial in Problem 6.39. Suppose that the experimenters could only afford 16 runs. Set up the 2^{5-1} fractional factorial design with $I = ABCDE$ and select the responses for the runs from the full factorial data in Problem 6.39.

(a) Analyze the data and draw conclusions.

(b) Compare your findings with those from the full factorial in Problem 6.39.

(c) Are there any potential interactions that need further study? What additional runs do you recommend? Select these runs from the full factorial design in Problem 6.39 and analyze the new design. Discuss your conclusions.

8.33. Consider the 2^4 factorial experiment for surfactin production in Problem 6.41. Suppose that the experimenters could only afford eight runs. Set up the 2^{4-1} fractional factorial design with $I = ABCD$ and select the responses for the runs from the full factorial data in Problem 6.40.

(a) Analyze the data and draw conclusions.

(b) Compare your findings with those from the full factorial in Problem 6.40.

8.34. Consider the replicated 2^3 factorial experiment in Problem 6.42. Suppose that the experimenters could only afford eight runs. Set up the 2^{3-1} fractional factorial design with $I = ABC$ and select the responses for the runs from the full factorial data in Problem 6.42.

(a) Analyze the data for all of the responses and draw conclusions.

(b) Compare your findings with those from the full factorial in Problem 6.42.

8.35. An article in the *Journal of Chromatography A* ("Simultaneous supercritical fluid derivatization and extraction of formaldehyde by the Hantzsch reaction," 2000, Vol. 896, pp. 51–59) describes an experiment where the Hantzsch reaction is used to produce the chemical derivatization of formaldehyde in a supercritical medium. Pressure, temperature, and other parameters such as static and dynamic extraction time must be optimized to increase the yield of this kinetically controlled reaction. A 2^{5-1} fractional factorial design with one center run was used to study the significant parameters affecting the supercritical process in terms of resolution and sensitivity. Ultraviolet–visible spectrophotometry was used as the detection technique. The experimental design and the responses are shown in Table P8.7.

■ **TABLE P8.7**
The 2^{5-1} Fractional Factorial Design for Problem 8.35

Experiment	*P* (MPa)	*T* (°C)	*s* (min)	*d* (min)	*c* (μl)	Resolution	Sensitivity
1	13.8	50	2	2	100	0.00025	0.057
2	55.1	50	2	2	10	0.33333	0.094
3	13.8	120	2	2	10	0.02857	0.017
4	55.1	120	2	2	100	0.20362	1.561
5	13.8	50	15	2	10	0.00027	0.010
6	55.1	50	15	2	100	0 52632	0.673
7	13.8	120	15	2	100	0.00026	0.028
8	55.1	120	15	2	10	0.52632	1.144
9	13.8	50	2	15	10	0 42568	0.142
10	55.1	50	2	15	100	0.60150	0.399
11	13.8	120	2	15	100	0.06098	0.767
12	55.1	120	2	15	10	0.74165	1.086
13	13.8	50	15	15	100	0.08780	0.252
14	55.1	50	15	15	10	0.40000	0.379
15	13.8	120	15	15	10	0.00026	0.028
16	55.1	120	15	15	100	0.28091	3.105
Central	34.5	85	8.5	8.5	55	0.75000	1.836

(a) Analyze the data from this experiment and draw conclusions.

(b) Analyze the residuals. Are there any concerns about model adequacy or violations of assumptions?

(c) Does the single center point cause any concerns about curvature or the possible need for second-order terms?

(d) Do you think that running one center point was a good choice in this design?

8.36. An article in *Thin Solid Films* (504, "A Study of Si/SiGe selective epitaxial growth by experimental design approach," 2006, Vol. 504, pp. 95–100) describes the use of a fractional factorial design to investigate the sensitivity of low-temperature (740–760 °C) Si/SiGe selective epitaxial growth to changes in five factors and their two-factor interactions. The five factors are SiH_2Cl_2, GeH_4, HCl, B_2H_6 and temperature. The factor levels studied are:

Factors	Levels (−)	Levels (+)
SiH_2Cl_2 (sccm)	8	12
GeH_4 (sccm)	7.2	10.8
HCl (sccm)	3.2	4.8
B_2H_6 (sccm)	4.4	6.6
Temperature (°C)	740	760

Table P8.8 contains the design matrix and the three measured responses. Bede RADS Mercury software based on the Takagi–Taupin dynamical scattering theory was used to extract the Si cap thickness, SiGe thickness, and Ge concentration of each sample.

(a) What design did the experimenters use? What is the defining relation?

(b) Will the experimenters be able to estimate all main effects and two-factor interactions with this experimental design?

(c) Analyze all three responses and draw conclusions.

(d) Is there any indication of curvature in the responses?

(e) Analyze the residuals and comment on model adequacy.

8.37. An article in *Soldering & Surface Mount Technology* ("Characterization of a solder paste printing process and its optimization," 1999, Vol. 11, No. 3, pp. 23–26) describes the use of a 2^{8-3} fractional factorial experiment to study the effect of eight factors on two responses; percentage volume matching (PVM)—the ratio of the actual printed solder paste volume to the designed volume; and non-conformities per unit (NPU)—the number of solder paste printing defects determined by visual inspection (20′ magnification) after printing according to an industry workmanship standard. The factor levels are shown below and the test matrix and response data are shown in Table P8.9.

■ TABLE P8.8

The Epitaxial Growth Experiment in Problem 8.36

Run order	Factors A	B	C	D	E	Si cap thickness (Å)	SiGe thickness (Å)	Ge concentration (at.%)
7	−	−	−	−	+	371.18	475.05	8.53
17	−	−	−	+	−	152.36	325.21	9.74
6	−	−	+	−	−	91.69	258.60	9.78
10	−	−	+	+	+	234.48	392.27	9.14
16	−	+	−	−	−	151.36	440.37	12.13
2	−	+	−	+	+	324.49	623.60	10.68
15	−	+	+	−	+	215.91	518.50	11.42
4	−	+	+	+	−	97.91	356.67	12.96
9	+	−	−	−	−	186.07	320.95	7.87
13	+	−	−	+	+	388.69	487.16	7.14
18	+	−	+	−	+	277.39	422.35	6.40
5	+	−	+	+	−	131.25	241.51	8.54
14	+	+	−	−	+	378.41	630.90	9.17
3	+	+	−	+	−	192.65	437.53	10.35
1	+	+	+	−	−	128.99	346.22	10.95
12	+	+	+	+	+	298.40	526.69	9.73
8	0	0	0	0	0	215.70	416.44	9.78
11	0	0	0	0	0	212.21	419.24	9.80

Parameters	Levels	
	Low (−)	High (+)
A. Squeegee pressure, *MPa*	0.1	0.3
B. Printing speed, *mm/s*	24	32
C. Squeegee angle, *deg*	45	65
D. Temperature, °*C*	20	28
E. Viscosity, *kCps*	1100–1150	1250–1300
F. Cleaning interval, *stroke*	8	15
G. Separation speed, *mm/s*	0.4	0.8
H. Relative humidity, *%*	30	70

(a) Verify that the generators are I = ABCF, I = ABDG, and I = BCDEH for this design.

(b) What are the aliases for the main effects and two-factor interactions? You can ignore all interactions of order three and higher.

(c) Analyze both the PVM and NPU responses.

(d) Analyze the residual for both responses. Are there any problems with model adequacy?

(e) The ideal value of PVM is unity and the NPU response should be as small as possible. Recommend suitable operating conditions for the process based on the experimental results.

■ **TABLE P8.9**
The Solder Paste Experiment

Run order	Parameters A	B	C	D	E	F	G	H	PVM	NPU (%)
4	−	−	−	−	−	−	−	+	1.00	5
13	+	−	−	−	−	+	+	+	1.04	13
6	−	+	−	−	−	+	+	−	1.02	16
3	+	+	−	−	−	−	−	−	0.99	12
19	−	−	+	−	−	+	−	−	1.02	15
25	+	−	+	−	−	−	+	−	1.01	9
21	−	+	+	−	−	−	+	+	1.01	12
14	+	+	+	−	−	+	−	+	1.03	17
10	−	−	−	+	−	−	+	−	1.04	21
22	+	−	−	+	−	+	−	−	1.14	20
1	−	+	−	+	−	+	−	+	1.20	25
2	+	+	−	+	−	−	+	+	1.13	21
30	−	−	+	+	−	+	+	+	1.14	25
8	+	−	+	+	−	−	−	+	1.07	13
9	−	+	+	+	−	−	−	−	1.06	20
20	+	+	+	+	−	+	+	−	1.13	26
17	−	−	−	−	+	−	−	−	1.02	10
18	+	−	−	−	+	+	+	−	1.10	13
5	−	+	−	−	+	+	+	+	1.09	17
26	+	+	−	−	+	−	−	+	0.96	13
31	−	−	+	−	+	+	−	+	1.02	14
11	+	−	+	−	+	−	+	+	1.07	11
29	−	+	+	−	+	−	+	−	0.98	10
23	+	+	+	−	+	+	−	−	0.95	14
32	−	−	−	+	+	−	+	+	1.10	28
7	+	−	−	+	+	+	−	+	1.12	24
15	−	+	−	+	+	+	−	−	1.19	22
27	+	+	−	+	+	−	+	−	1.13	15
12	−	−	+	+	+	+	+	−	1.20	21
28	+	−	+	+	+	−	−	−	1.07	19
24	−	+	+	+	+	−	−	+	1.12	21
16	+	+	+	+	+	+	+	+	1.21	27

8.38. An article in the *International Journal of Research in Marketing* ("Experimental design on the front lines of marketing: Testing new ideas to increase direct mail sales," 2006, Vol. 23, pp. 309–319) describes the use of a 20-run Plackett–Burman design to investigate the effects of 19 factors to improve the response rate to a direct mail sales campaign to attract new customers to a credit card. The 19 factors are as follows:

Factor	(−) Control	(+) New idea
A: Envelope teaser	General offer	Product-specific offer
B: Return address	Blind	Add company name
C: "Official" ink-stamp on envelope	Yes	No
D: Postage	Pre-printed	Stamp
E: Additional graphic on envelope	Yes	No
F: Price graphic on letter	Small	Large
G: Sticker	Yes	No
H: Personalize letter copy	No	Yes
I: Copy message	Targeted	Generic
J: Letter headline	Headline 1	Headline 2
K: List of benefits	Standard layout	Creative layout
L: Postscript on letter	Control version	New P.S.
M: Signature	Manager	Senior executive
N: Product selection	Many	Few
O: Value of free gift	High	Low
P: Reply envelope	Control	New style
Q: Information on buckslip	Product info	Free gift info
R: 2nd buckslip	No	Yes
S: Interest rate	Low	High

The 20-run Plackett–Burman design is shown in Table P8.10. Each test combination in Table P8.10 was mailed to 5000 potential customers, and the response rate is the percentage of customers who responded positively to the offer.

(a) Verify that in this design each main effect is aliased with all two-factor interactions except those that involve that main effect. That is, in the 19 factor design, the main effect for each factor is aliased with all two-factor interactions involving the other 18 factors, or 153 two-factor interactions (18!/2!16!).

(b) Show that for the 20-run Plackett–Burman design in Table P8.10, the weights (or correlations) that multiple the two-factor interactions in each alias chain are either −0.2, +0.2, or −0.6. Of the 153 interactions that are aliased with each main effect, 144 have weights of −0.2 or +0.2, while nine interactions have weights of −0.6.

(c) Verify that the five largest main effects are S, G, R, I, and J.

(d) Factors S (interest rate) and G (presence of a sticker) are by far the largest main effects. The correlation between the main effect of R (2nd buckslip) and the SG interaction is −0.6. This means that a significant SG interaction would bias the estimate of the main effect of R by −0.6 times the value of the interaction. This suggests that it may not be the main effect of factor R that is important, but the two-factor interaction between S and G.

(e) Since this design projects into a full factorial in any three factors, obtain the projection in factors S, G, and R and verify that it is a full factorial with some runs replicated. Fit a full factorial model involving all three of these factors and the interactions (you will need to use a regression program to do this). Show that S, G, and the SG interaction are significant.

■ **TABLE P8.10**

The Plackett–Burman Design for the Direct Mail Experiment in Problem 8.38

	Envelope teaser	Return address	"Official" ink-stamp on envelope	Postage	Additional graphic on envelope	Price graphic on letter	Sticker	Personalize letter copy	Copy message	Letter headline
Test cell	**A**	**B**	**C**	**D**	**E**	**F**	**G**	**H**	**I**	**J**
1	+	+	−	−	+	+	+	+	−	+
2	−	+	+	−	−	+	+	+	+	−
3	+	−	+	+	−	−	+	+	+	+
4	+	+	−	+	+	−	−	+	+	+
5	−	+	+	−	+	+	−	−	+	+
6	−	−	+	+	−	+	+	−	−	+
7	−	−	−	+	+	−	+	+	−	−
8	−	−	−	−	+	+	−	+	+	−
9	+	−	−	−	−	+	+	−	+	+
10	−	+	−	−	−	−	+	+	−	+
11	+	−	+	−	−	−	−	+	+	−
12	−	+	−	+	−	−	−	−	+	+
13	+	−	+	−	+	−	−	−	−	+
14	+	+	−	+	−	+	−	−	−	−
15	+	+	+	−	+	−	+	−	−	−
16	+	+	+	+	−	+	−	+	−	−
17	−	+	+	+	+	−	+	−	+	−
18	−	−	+	+	+	+	−	+	−	+
19	+	−	−	+	+	+	+	−	+	−
20	−	−	−	−	−	−	−	−	−	−

List of benefits	Postscript on letter	Signature	Product selection	Value of free gift	Reply envelope	Information on buckslip	2nd buckslip	Interest rate	Orders	Response rate
K	**L**	**M**	**N**	**O**	**P**	**Q**	**R**	**S**		
−	+	−	−	−	−	+	+	−	52	1.04%
+	−	+	−	−	−	−	+	+	38	0.76%
−	+	−	+	−	−	−	−	+	42	0.84%
+	−	+	−	+	−	−	−	−	134	2.68%
+	+	−	+	−	+	−	−	−	104	2.08%
+	+	+	−	+	−	+	−	−	60	1.20%
+	+	+	+	−	+	−	+	−	61	1.22%
−	+	+	+	+	−	+	−	+	68	1.36%
−	−	+	+	+	+	−	+	−	57	1.14%
+	−	−	+	+	+	+	−	+	30	0.60%
+	+	−	−	+	+	+	+	−	108	2.16%
−	+	+	−	−	+	+	+	+	39	0.78%
+	−	+	+	−	−	+	+	+	40	0.80%
+	+	−	+	+	−	−	+	+	49	0.98%
−	+	+	−	+	+	−	−	+	37	0.74%
−	−	+	+	−	+	+	−	−	99	1.98%
−	−	−	+	+	−	+	+	−	86	1.72%
−	−	−	−	+	+	−	+	+	43	0.86%
+	−	−	−	−	+	+	−	+	47	0.94%
−	−	−	−	−	−	−	−	−	104	2.08%

8.39. Consider the following experiment:

Run	Treatment combination
1	d
2	ae
3	b
4	abde
5	cde
6	ac
7	bce
8	abcd

Answer the following questions about this experiment:

(a) How many factors did this experiment investigate?

(b) How many factors are in the basic design?

(c) Assume that the factors in the experiment are represented by the initial letters of the alphabet (i.e., A, B, etc.), what are the design generators for the factors beyond the basic design?

(d) Is this design a principal fraction?

(e) What is the complete defining relation?

(f) What is the resolution of this design?

8.40. Consider the following experiment:

Run	Treatment combination	y
1	(1)	8
2	ad	10
3	bd	12
4	ab	7
5	cd	13
6	ac	6
7	bc	5
8	abcd	11

Answer the following questions about this experiment:

(a) How many factors did this experiment investigate?

(b) What is the resolution of this design?

(c) Calculate the estimates of the main effects.

(d) What is the complete defining relation for this design?

8.41. An unreplicated 2^{5-1} fractional factorial experiment with four center points has been run in a chemical process. The response variable is molecular weight. The experimenter has used the following factors:

Factor	Natural levels	Coded levels (x's)
A - time	20, 40 (minutes)	−1, 1
B - temperature	160, 180 (deg C)	−1, 1
C - concentration	30, 60 (percent)	−1, 1
D - stirring rate	100, 150 (RPM)	−1, 1
E - catalyst type	1, 2 (Type)	−1. 1

Suppose that the prediction equation that results from this experiment is $\hat{y} = 10 + 3x_1 + 2x_2 - 1x_1x_2$. What is the predicted response at A = 30, B = 165, C = 50, D = 135, and E = 1?

8.42. An unreplicated 2^{4-1} fractional factorial experiment with four center points has been run. The experimenter has used the following factors:

Factor	Natural levels	Coded levels (x's)
A - time	10, 50 (minutes)	−1, 1
B - temperature	200, 300 (deg C)	−1, 1
C - concentration	70, 90 (percent)	−1, 1
D - pressure	260, 300 (psi)	−1, 1

(a) Suppose that the average of the 16 factorial design points is 100 and the average of the center points is 120, what is the sum of squares for pure quadratic curvature?

(b) Suppose that the prediction equation that results from this experiment is $\hat{y} = 50 + 5x_1 + 2x_2 - 2x_1x_2$. Find the predicted response at A = 20, B = 250, C = 80, and D = 275.

8.43. An unreplicated 2^{4-1} fractional factorial experiment has been run. The experimenter has used the following factors:

Factor	Natural levels	Coded levels (x's)
A	20, 50	−1, 1
B	200, 280	−1, 1
C	50, 100	−1, 1
D	150, 200	−1, 1

(a) Suppose that this design has four center runs that average 100. The average of the 16 factorial design points is 95. What is the sum of squares for pure quadratic curvature?

(b) Suppose that the prediction equation that results from this experiment is $\hat{y} = 100 + -2x_1 + 10x_2 - 4x_1x_2$. What is the predicted response at A = 41, B = 280, C = 60, and D = 185?

8.44. A 2^{6-2} factorial experiment with three replicates has been run in a pharmaceutical drug manufacturing process. The experimenter has used the following factors:

Factor	Natural levels	Coded levels (x's)
A	50, 100	−1, 1
B	20, 60	−1, 1
C	10, 30	−1, 1
D	12, 18	−1, 1
E	15, 30	−1, 1
F	60, 100	−1, 1

(a) If two main effects and one two-factor interaction are included in the final model, how many degrees of freedom for error will be available?

(b) Suppose that the significant factors are A, C, AB, and AC. What other effects need to be included to obtain a hierarchical model?

8.45. Consider the following design:

Run	A	B	C	D	E	y
1	−1	−1	−1	−1	−1	63
2	1	−1	−1	−1	1	21
3	−1	1	−1	−1	1	36
4	1	1	−1	−1	−1	99
5	−1	−1	1	−1	1	24
6	1	−1	1	−1	−1	66
7	−1	1	1	−1	−1	71
8	1	1	1	−1	1	54
9	−1	−1	−1	1	−1	23
10	1	−1	−1	1	1	74
11	−1	1	−1	1	1	80
12	1	1	−1	1	−1	33
13	−1	−1	1	1	1	63
14	1	−1	1	1	−1	21
15	−1	1	1	1	−1	44
16	1	1	1	1	1	96

(a) What is the generator for column E?

(b) If ABC is confounded with blocks, run 1 above goes in the ______ block. Answer either + or −.

(c) What is the resolution of this design?

(d) Find the estimates of the main effects and their aliases.

8.46. Consider the following design:

Run	A	B	C	D	E	y
1	−1	−1	−1	−1	−1	65
2	1	−1	−1	−1	1	25
3	−1	1	−1	−1	1	30
4	1	1	−1	−1	−1	89
5	−1	−1	1	−1	1	25
6	1	−1	1	−1	−1	60
7	−1	1	1	−1	−1	70
8	1	1	1	−1	1	50
9	−1	−1	−1	1	1	20
10	1	−1	−1	1	−1	70
11	−1	1	−1	1	−1	80
12	1	1	−1	1	1	30
13	−1	−1	1	1	−1	60
14	1	−1	1	1	1	20
15	−1	1	1	1	1	40
16	1	1	1	1	−1	90

(a) What is the generator for column E?

(b) If ABE is confounded with blocks, run 16 goes in the ______ block. Answer either − or +.

(c) The resolution of this design is ______.

(d) Find the estimates of the main effects and their aliases.

8.47. Consider the following design:

Run	A	B	C	D	E	y
1	−1	−1	−1	1	−1	50
2	1	−1	−1	−1	−1	20
3	−1	1	−1	−1	1	40
4	1	1	−1	1	1	25
5	−1	−1	1	−1	1	45
6	1	−1	1	1	1	30
7	−1	1	1	1	−1	40
8	1	1	1	−1	−1	30

(a) What is the generator for column D?

(b) What is the generator for column E?

(c) If this design were run in two blocks with the AB interaction confounded with blocks, the run *d* would be in the block where the sign on AB is ______. Answer either − or +.

8.48. Consider the following design:

Std	A	B	C	D	E	y
1	−1	−1	−1	1	1	40
2	1	−1	−1	−1	1	10
3	−1	1	−1	−1	−1	30
4	1	1	−1	1	−1	20
5	−1	−1	1	−1	−1	40
6	1	−1	1	1	−1	30
7	−1	1	1	1	1	20
8	1	1	1	−1	1	30

(a) What is the generator for column D?

(b) What is the generator for column E?

(c) If this design were folded over, what is the resolution of the combined design?

8.49. In an article in *Quality Engineering* ("An Application of Fractional Factorial Experimental Designs," 1988, Vol. 1, pp. 19–23), M. B. Kilgo describes an experiment to determine the effect of CO_2 pressure (A), CO_2 temperature (B), peanut moisture (C), CO_2 flow rate (D), and peanut particle size (E) on the total yield of oil per batch of peanuts (y). The levels that she used for these factors are shown in Table P8.11. She conducted the 16-run fractional factorial experiment shown in Table P8.12.

■ **TABLE P8.11**
Factor Levels for the Experiment in Problem 8.49

Coded Level	*A*, Pressure (bar)	*B*, Temp, (°C)	*C*, Moisture (% by weight)	*D*, Flow (liters/min)	*E*, Part. Size (mm)
−1	415	25	5	40	1.28
1	550	95	15	60	4.05

■ **TABLE P8.12**
The Peanut Oil Experiment

A	*B*	*C*	*D*	*E*	*y*
415	25	5	40	1.28	63
550	25	5	40	4.05	21
415	95	5	40	4.05	36
550	95	5	40	1.28	99
415	25	15	40	4.05	24
550	25	15	40	1.28	66
415	95	15	40	1.28	71
550	95	15	40	4.05	54
415	25	5	60	4.05	23
550	25	5	60	1.28	74
415	95	5	60	1.28	80
550	95	5	60	4.05	33
415	25	15	60	1.28	63
550	25	15	60	4.05	21
415	95	15	60	4.05	44
550	95	15	60	1.28	96

(a) What type of design has been used? Identify the defining relation and the alias relationships.

(b) Estimate the factor effects and use a normal probability plot to tentatively identify the important factors.

(c) Perform an appropriate statistical analysis to test the hypotheses that the factors identified in part (b) above have a significant effect on the yield of peanut oil.

(d) Fit a model that could be used to predict peanut oil yield in terms of the factors that you have identified as important.

(e) Analyze the residuals from this experiment and comment on model adequacy.

8.50. A 16-run fractional factorial experiment in 10 factors on sand-casting of engine manifolds was conducted by engineers at the Essex Aluminum Plant of the Ford Motor Company and described in the article "Evaporative Cast Process 3.0 Liter Intake Manifold Poor Sandfill Study," by D. Becknell (*Fourth Symposium on Taguchi Methods*, American Supplier Institute, Dearborn, MI, 1986, pp. 120–130). The purpose was to determine which of 10 factors has an effect on the proportion of defective castings. The design and the resulting proportion of nondefective castings $\hat{p}$ observed on each run are shown in Table P8.13. This is a resolution III fraction with generators $E = CD$, $F = BD$, $G = BC$, $H = AC$, $J = AB$, and $K = ABC$. Assume that the number of castings made at each run in the design is 1000.

(a) Find the defining relation and the alias relationships in this design.

(b) Estimate the factor effects and use a normal probability plot to tentatively identify the important factors.

(c) Fit an appropriate model using the factors identified in part (b) above.

(d) Plot the residuals from this model versus the predicted proportion of nondefective castings. Also prepare a normal probability plot of the residuals. Comment on the adequacy of these plots.

(e) In part (d) you should have noticed an indication that the variance of the response is not constant. (Considering that the response is a proportion, you should have expected this.) The previous table also shows a transformation on $\hat{p}$, the arcsin square root, that is a widely used *variance stabilizing transformation* for proportion data (refer to the discussion of variance stabilizing transformations in Chapter 3). Repeat parts (a) through (d) above using the transformed response and comment on your results. Specifically, are the residual plots improved?

(f) There is a modification to the arcsin square root transformation, proposed by Freeman and Tukey ("Transformations Related to the Angular and the Square Root," *Annals of Mathematical Statistics*, Vol. 21, 1950, pp. 607–611), that improves its performance in the tails. F&T's modification is

$$[\arcsin\sqrt{n\hat{p}/(n+1)} + \arcsin\sqrt{(n\hat{p}+1)/(n+1)}]/2$$

■ **TABLE P8.13**
The Sand-Casting Experiment

Run	*A*	*B*	*C*	*D*	*E*	*F*	*G*	*H*	*J*	*K*	$\hat{p}$	Arcsin $\sqrt{\hat{p}}$	F&T's Modification
1	−	−	−	−	+	+	+	+	+	−	0.958	1.364	1.363
2	+	−	−	−	+	+	+	−	−	+	1.000	1.571	1.555
3	−	+	−	−	+	−	−	+	−	+	0.977	1.419	1.417
4	+	+	−	−	+	−	−	−	+	−	0.775	1.077	1.076
5	−	−	+	−	−	+	−	−	+	+	0.958	1.364	1.363
6	+	−	+	−	−	+	−	+	−	−	0.958	1.364	1.363
7	−	+	+	−	−	−	+	−	−	−	0.813	1.124	1.123
8	+	+	+	−	−	−	+	+	+	+	0.906	1.259	1.259
9	−	−	−	+	−	−	+	+	+	−	0.679	0.969	0.968
10	+	−	−	+	−	−	+	−	−	+	0.781	1.081	1.083
11	−	+	−	+	−	+	−	+	−	+	1.000	1.571	1.556
12	+	+	−	+	−	+	−	−	+	−	0.896	1.241	1.242
13	−	−	+	+	+	−	−	−	+	+	0.958	1.364	1.363
14	+	−	+	+	+	−	−	+	−	−	0.818	1.130	1.130
15	−	+	+	+	+	+	+	−	−	−	0.841	1.161	1.160
16	+	+	+	+	+	+	+	+	+	+	0.955	1.357	1.356

Rework parts (a) through (d) using this transformation and comment on the results. (For an interesting discussion and analysis of this experiment, refer to "Analysis of Factorial Experiments with Defects or Defectives as the Response," by S. Bisgaard and H. T. Fuller, *Quality Engineering*, Vol. 7, 1994–95, pp. 429–443.)

8.51. A 16-run fractional factorial experiment in nine factors was conducted by Chrysler Motors Engineering and described in the article "Sheet Molded Compound Process Improvement," by P. I. Hsieh and D. E. Goodwin (*Fourth Symposium on Taguchi Methods*, American Supplier Institute, Dearborn, MI, 1986, pp. 13–21). The purpose was to reduce the number of defects in the finish of sheet-molded grill opening panels. The design, and the resulting number of defects, *c*, observed on each run, is shown in Table P8.14. This is a resolution III fraction with generators $E = BD$, $F = BCD$, $G = AC$, $H = ACD$, and $J = AB$.

(a) Find the defining relation and the alias relationships in this design.

(b) Estimate the factor effects and use a normal probability plot to tentatively identify the important factors.

(c) Fit an appropriate model using the factors identified in part (b) above.

(d) Plot the residuals from this model versus the predicted number of defects. Also, prepare a normal probability plot of the residuals. Comment on the adequacy of these plots.

(e) In part (d) you should have noticed an indication that the variance of the response is not constant. (Considering that the response is a count, you should have expected this.) The previous table also shows a transformation on *c*, the square root, that is a widely used *variance stabilizing transformation* for count data. (Refer to the discussion of variance stabilizing transformations in Chapter 3.) Repeat parts (a) through (d) using the transformed response and comment on your results. Specifically, are the residual plots improved?

(f) There is a modification to the square root transformation, proposed by Freeman and Tukey ("Transformations Related to the Angular and the Square Root," *Annals of Mathematical Statistics*, Vol. 21, 1950, pp. 607–611) that improves its performance. F&T's modification to the square root transformation is

$$[\sqrt{c} + \sqrt{(c+1)}]/2$$

Rework parts (a) through (d) using this transformation and comment on the results. (For an interesting discussion and analysis of this experiment, refer to "Analysis of Factorial Experiments with Defects or

■ **TABLE P8.14**
The Grill Defects Experiment

Run	A	B	C	D	E	F	G	H	J	c	$\sqrt{c}$	F&T's Modification
1	−	−	−	−	+	−	+	−	+	56	7.48	7.52
2	+	−	−	−	+	−	−	+	−	17	4.12	4.18
3	−	+	−	−	−	+	+	−	−	2	1.41	1.57
4	+	+	−	−	−	+	−	+	+	4	2.00	2.12
5	−	−	+	−	+	+	−	+	+	3	1.73	1.87
6	+	−	+	−	+	+	+	−	−	4	2.00	2.12
7	−	+	+	−	−	−	−	+	−	50	7.07	7.12
8	+	+	+	−	−	−	+	−	+	2	1.41	1.57
9	−	−	−	+	−	+	+	+	+	1	1.00	1.21
10	+	−	−	+	−	+	−	−	−	0	0.00	0.50
11	−	+	−	+	+	−	+	+	−	3	1.73	1.87
12	+	+	−	+	+	−	−	−	+	12	3.46	3.54
13	−	−	+	+	−	−	−	−	+	3	1.73	1.87
14	+	−	+	+	−	−	+	+	−	4	2.00	2.12
15	−	+	+	+	+	+	−	−	−	0	0.00	0.50
16	+	+	+	+	+	+	+	+	+	0	0.00	0.50

Defectives as the Response," by S. Bisgaard and H. T. Fuller, *Quality Engineering*, Vol. 7, 1994–95, pp. 429–443.)

8.52. An experiment is run in a semiconductor factory to investigate the effect of six factors on transistor gain. The design selected is the 2_{IV}^{6-2} shown in Table P8.15.

■ **TABLE P8.15**
The Transistor Gain Experiment

Standard Order	Run Order	A	B	C	D	E	F	Gain
1	2	−	−	−	−	−	−	1455
2	8	+	−	−	−	+	−	1511
3	5	−	+	−	−	+	+	1485
4	9	+	+	−	−	−	+	1600
5	3	−	−	+	−	+	+	1430
6	14	+	−	+	−	−	+	1481
7	11	−	+	+	−	−	−	1450
8	10	+	+	+	−	+	−	1549
9	15	−	−	−	+	−	+	1454
10	13	+	−	−	+	+	+	1512
11	1	−	+	−	+	+	−	1487
12	6	+	+	−	+	−	−	1596
13	12	−	−	+	+	+	−	1450
14	4	+	−	+	+	−	−	1473
15	7	−	+	+	+	−	+	1461
16	16	+	+	+	+	+	+	1565

(a) Use a normal plot of the effects to identify the significant factors.

(b) Conduct appropriate statistical tests for the model identified in part (a).

(c) Analyze the residuals and comment on your findings.

(d) Can you find a set of operating conditions that produce gain of 1500 ± 25?

8.53. An article by L. B. Hare ("In the Soup: A Case Study to Identify Contributors to Filling Variability," *Journal of Quality Technology*, Vol. 20, pp. 36–43) describes a factorial experiment used to study the filling variability of dry soup mix packages. The factors are A = number of mixing ports through which the vegetable oil was added (1, 2), B = temperature surrounding the mixer (cooled, ambient), C = mixing time (60, 80 sec), D = batch weight (1500, 2000 lb), and

E = number of days of delay between mixing and packaging (1, 7). Between 125 and 150 packages of soup were sampled over an 8-hour period for each run in the design, and the standard deviation of package weight was used as the response variable. The design and resulting data are shown in Table P8.16.

■ **TABLE P8.16**
The Soup Experiment

Std. Order	*A* Mixer Ports	*B* Temp.	*C* Time	*D* Batch Weight	*E* Delay	*y* Std. Dev
1	−	−	−	−	−	1.13
2	+	−	−	−	+	1.25
3	−	+	−	−	+	0.97
4	+	+	−	−	−	1.7
5	−	−	+	−	+	1.47
6	+	−	+	−	−	1.28
7	−	+	+	−	−	1.18
8	+	+	+	−	+	0.98
9	−	−	−	+	+	0.78
10	+	−	−	+	−	1.36
11	−	+	−	+	−	1.85
12	+	+	−	+	+	0.62
13	−	−	+	+	−	1.09
14	+	−	+	+	+	1.1
15	−	+	+	+	+	0.76
16	+	+	+	+	−	2.1

(a) What is the generator for this design?
(b) What is the resolution of this design?
(c) Estimate the factor effects. Which effects are large?
(d) Does a residual analysis indicate any problems with the underlying assumptions?
(e) Draw conclusions about this filling process.

8.54. Heat treating is often used to carbonize metal parts, such as gears. The thickness of the carbonized layer is a critical output variable from this process, and it is usually measured by performing a carbon analysis on the gear pitch (the top of the gear tooth). Six factors were studied in a 2_{IV}^{6-2} design: A = furnace temperature, B = cycle time, C = carbon concentration, D = duration of the carbonizing cycle, E = carbon concentration of the diffuse cycle, and F = duration of the diffuse cycle. The experiment is shown in Table P8.17.

■ **TABLE P8.17**
The Heat Treating Experiment

Standard Order	Run Order	*A*	*B*	*C*	*D*	*E*	*F*	Pitch
1	5	−	−	−	−	−	−	74
2	7	+	−	−	−	+	−	190
3	8	−	+	−	−	+	+	133
4	2	+	+	−	−	−	+	127
5	10	−	−	+	−	+	+	115
6	12	+	−	+	−	−	+	101
7	16	−	+	+	−	−	−	54
8	1	+	+	+	−	+	−	144
9	6	−	−	−	+	−	+	121
10	9	+	−	−	+	+	+	188
11	14	−	+	−	+	+	−	135
12	13	+	+	−	+	−	−	170
13	11	−	−	+	+	+	−	126
14	3	+	−	+	+	−	−	175
15	15	−	+	+	+	−	+	126
16	4	+	+	+	+	+	+	193

(a) Estimate the factor effects and plot them on a normal probability plot. Select a tentative model.
(b) Perform appropriate statistical tests on the model.
(c) Analyze the residuals and comment on model adequacy.
(d) Interpret the results of this experiment. Assume that a layer thickness of between 140 and 160 is desirable.

8.55. Consider the 2_{IV}^{6-2} design.
(a) Suppose that the design had been folded over by changing the signs in column B instead of column A. What changes would have resulted in the effects that can be estimated from the combined design?
(b) Suppose that the design had been folded over by changing the signs in column E instead of column A. What changes would have resulted in the effects that can be estimated from the combined design?

8.56. Consider the 2_{IV}^{7-3} design. Suppose that a fold over of this design is run by changing the signs in column A. Determine the alias relationships in the combined design.

8.57. Reconsider the 2_{IV}^{7-3} design in Problem 8.56.
(a) Suppose that a fold over of this design is run by changing the signs in column B. Determine the alias relationships in the combined design.

(b) Compare the aliases from this combined design to those from the combined design from Problem 8.35. What differences resulted by changing the signs in a different column?

8.58. Consider the 2_{IV}^{7-3} design.

(a) Suppose that a partial fold over of this design is run using column *A* (+ signs only). Determine the alias relationships in the combined design.

(b) Rework part (a) using the negative signs to define the partial fold over. Does it make any difference which signs are used to define the partial fold over?

8.59. Consider a partial fold over for the 2_{IV}^{6-2} design. Suppose that the signs are reversed in column *A*, but the eight runs that are retained are the runs that have positive signs in column *C*. Determine the alias relationships in the combined design.

8.60. Consider a partial fold over for the 2_{III}^{7-4} design. Suppose that the partial fold over of this design is constructed using column *A* (+ signs only). Determine the alias relationships in the combined design.

8.61. Consider a partial fold over for the 2_{III}^{5-2} design. Suppose that the partial fold over of this design is constructed using column *A* (+ signs only). Determine the alias relationships in the combined design.

8.62. Reconsider the 2^{4-1} design in Example 8.1. The significant factors are *A, C, D, AC + BD*, and *AD + BC*. Find a partial fold-over design that will allow the *AC, BD, AD*, and *BC* interactions to be estimated.

8.63. Construct a supersaturated design for $k = 8$ factors in $P = 6$ runs.

8.64. Consider the 2^{8-3} design in Problem 8.37. Suppose that the alias chain involving the AB interaction was large. Recommend a partial fold-one design to resolve the ambiguity about this interaction.

8.65. Construct a supersaturated design for $h = 12$ factors in $N = 10$ runs.

8.66. How could an "optimal design" approach be used to augment a fractional factorial design to de-alias effects of potential interest?

CHAPTER 9

Other Topics on Factorial and Fractional Factorial Designs

CHAPTER OUTLINE

The supplemental material is on the textbook website www.wiley.com/go/global/montgomery.

The two-level series of factorial and fractional factorial designs discussed in Chapters 6, 7, and 8 are widely used in industrial research and development. This chapter discusses some extensions and variations of these designs one important case is the situation where all the factors are present at three levels. These 3^k designs and their fractions are discussed in this chapter. We will also consider cases where some factors have two levels and other factors have either three or four levels. In chapter 8 use introduced Plackett–Burman designs and observed that they are nonregular fractions. The general case of nonregular fractions with all factors at two levels is discussed in more detail here. We also illustrate how optimal design tools can be useful for constructing designs in many important situations.

9.1 The 3^k Factorial Design

9.1.1 Notation and Motivation for the 3^k Design

We now discuss the 3^k factorial design—that is, a factorial arrangement with k factors, each at three levels. Factors and interactions will be denoted by capital letters. We will refer to the three levels of the factors as low, intermediate, and high. Several different notations may be used to represent these factor levels; one possibility is to represent the factor levels by the digits 0 (low), 1 (intermediate), and 2 (high). Each treatment combination in the 3^k design will be denoted by k digits, where the first digit indicates the level of factor A, the second digit indicates the level of factor B, . . . , and the kth digit indicates the level of factor K. For example, in a 3^2 design, 00 denotes the treatment combination corresponding to A and B both at the low level, and 01 denotes the treatment combination corresponding to A at the low level and B at the intermediate level. Figures 9.1 and 9.2 show the geometry of the 3^2 and the 3^3 design, respectively, using this notation.

This system of notation could have been used for the 2^k designs presented previously, with 0 and 1 used in place of the ± 1s, respectively. In the 2^k design, we prefer the ± 1 notation because it facilitates the geometric view of the design and because it is directly applicable to regression modeling, blocking, and the construction of fractional factorials.

In the 3^k system of designs, when the factors are **quantitative**, we often denote the low, intermediate, and high levels by -1, 0, and $+1$, respectively. This facilitates fitting a **regression model** relating the response to the factor levels. For example, consider the 3^2 design in Figure 9.1, and let x_1 represent factor A and x_2 represent factor B. A regression model relating the response y to x_1 and x_2 that is supported by this design is

$$y = \beta_0 + \beta_1 x_1 + \beta_2 x_2 + \beta_{12} x_1 x_2 + \beta_{11} x_1^2 + \beta_{22} x_2^2 + \epsilon \tag{9.1}$$

Notice that the addition of a third factor level allows the relationship between the response and design factors to be modeled as a quadratic.

The 3^k design is certainly a possible choice by an experimenter who is concerned about curvature in the response function. However, two points need to be considered:

1. The 3^k design is not the most efficient way to model a quadratic relationship; the **response surface designs** discussed in Chapter 11 are superior alternatives.
2. The 2^k design augmented with center points, as discussed in Chapter 6, is an excellent way to obtain an indication of curvature. It allows one to keep the size and

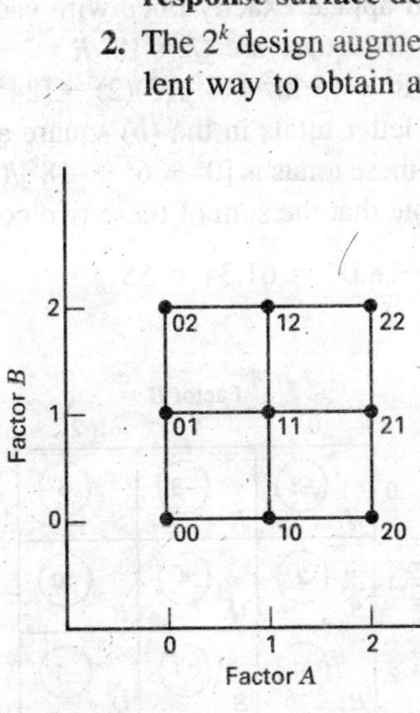

■ **FIGURE 9.1** Treatment combinations in a 3^2 design

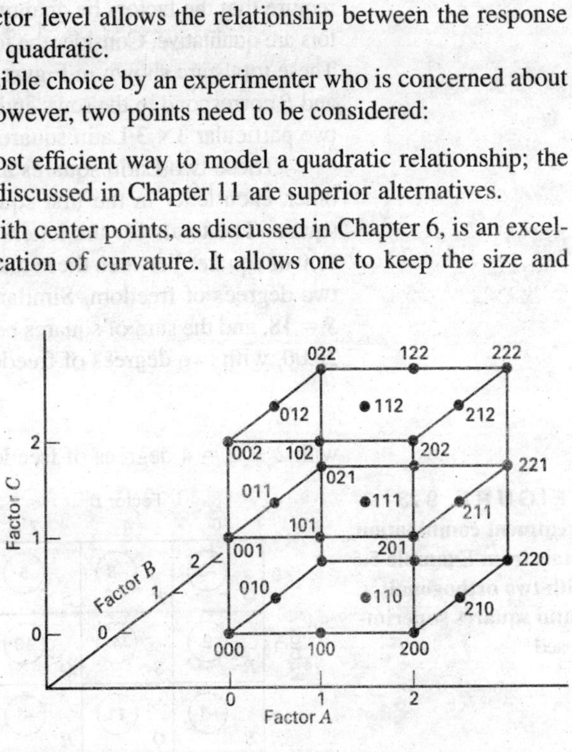

■ **FIGURE 9.2** Treatment combinations in a 3^3 design

complexity of the design low and simultaneously obtain some protection against curvature. Then, if curvature is important, the two-level design can be augmented with **axial runs** to obtain a **central composite design**, as shown in Figure 6.37. This **sequential strategy** of experimentation is far more efficient than running a 3^k factorial design with quantitative factors.

9.1.2 The 3^2 Design

The simplest design in the 3^k system is the 3^2 design, which has two factors, each at three levels. The treatment combinations for this design are shown in Figure 9.1. Because there are $3^2 = 9$ treatment combinations, there are eight degrees of freedom between these treatment combinations. The main effects of A and B each have two degrees of freedom, and the AB interaction has four degrees of freedom. If there are n replicates, there will be $n3^2 - 1$ total degrees of freedom and $3^2(n - 1)$ degrees of freedom for error.

The sums of squares for A, B, and AB may be computed by the usual methods for factorial designs discussed in Chapter 5. Each main effect can be represented by a linear and a quadratic component, each with a single degree of freedom, as demonstrated in Equation 9.1. Of course, this is meaningful only if the factor is quantitative.

The two-factor interaction AB may be partitioned in two ways. Suppose that both factors A and B are quantitative. The first method consists of subdividing AB into the four single-degree-of-freedom components corresponding to $AB_{L\times L}$, $AB_{L\times Q}$, $AB_{Q\times L}$, and $AB_{Q\times Q}$. This can be done by fitting the terms $\beta_{12}x_1x_2$, $\beta_{122}x_1x_2^2$, $\beta_{112}x_1^2x_2$, and $\beta_{1122}x_1^2x_2^2$, respectively, as demonstrated in Example 5.5. For the tool life data, this yields $SS_{AB_{L\times L}} = 8.00$, $SS_{AB_{L\times Q}} = 42.67$, $SS_{AB_{Q\times L}} = 2.67$, and $SS_{AB_{Q\times Q}} = 8.00$. Because this is an orthogonal partitioning of AB, note that $SS_{AB} = SS_{AB_{L\times L}} + SS_{AB_{L\times Q}} + SS_{AB_{Q\times L}} + SS_{AB_{Q\times Q}} = 61.34$.

The second method is based on **orthogonal Latin squares**. This method does not require that the factors be quantitative, and it is usually associated with the case where all factors are qualitative. Consider the totals of the treatment combinations for the data in Example 5.5. These totals are shown in Figure 9.3 as the circled numbers in the squares. The two factors A and B correspond to the rows and columns, respectively, of a 3×3 Latin square. In Figure 9.3, two particular 3×3 Latin squares are shown superimposed on the cell totals.

These two Latin squares are **orthogonal**; that is, if one square is superimposed on the other, each letter in the first square will appear exactly once with each letter in the second square. The totals for the letters in the (a) square are $Q = 18$, $R = -2$, and $S = 8$, and the sum of squares between these totals is $[18^2 + (-2)^2 + 8^2]/(3)(2) - [24^2/(9)(2)] = 33.34$, with two degrees of freedom. Similarly, the letter totals in the (b) square are $Q = 0$, $R = 6$, and $S = 18$, and the sum of squares between these totals is $[0^2 + 6^2 + 18^2]/(3)(2) - [24^2/(9)(2)] = 28.00$, with two degrees of freedom. Note that the sum of these two components is

$$33.34 + 28.00 = 61.34 = SS_{AB}$$

with $2 + 2 = 4$ degrees of freedom.

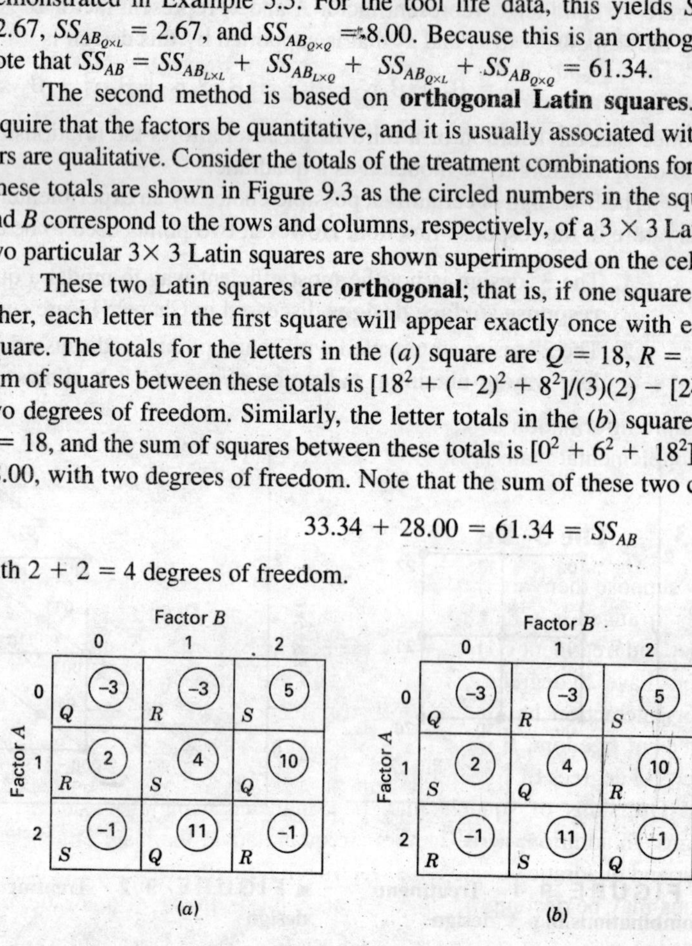

■ **FIGURE 9.3**
Treatment combination totals from Example 5.5 with two orthogonal Latin squares superimposed

In general, the sum of squares computed from square (*a*) is called the ***AB* component of interaction**, and the sum of squares computed from square (*b*) is called the ***AB*² component of interaction**. The components AB and AB^2 each have two degrees of freedom. This terminology is used because if we denote the levels (0, 1, 2) for A and B by x_1 and x_2, respectively, then we find that the letters occupy cells according to the following pattern:

Square (*a*)	Square (*b*)
Q: $x_1 + x_2 = 0 \pmod 3$	Q: $x_1 + 2x_2 = 0 \pmod 3$
R: $x_1 + x_2 = 1 \pmod 3$	S: $x_1 + 2x_2 = 1 \pmod 3$
S: $x_1 + x_2 = 2 \pmod 3$	R: $x_1 + 2x_2 = 2 \pmod 3$

For example, in square (*b*), note that the middle cell corresponds to $x_1 = 1$ and $x_2 = 1$; thus, $x_1 + 2x_2 = 1 + (2)(1) = 3 = 0 \pmod 3$, and Q would occupy the middle cell. When considering expressions of the form A^pB^q, we establish the convention that the only exponent allowed on the first letter is 1. If the first letter exponent is not 1, the entire expression is squared and the exponents are reduced modulus 3. For example, A^2B is the same as AB^2 because

$$A^2B = (A^2B)^2 = A^4B^2 = AB^2$$

The AB and AB^2 components of the AB interaction have no actual meaning and are usually not displayed in the analysis of variance table. However, this rather arbitrary partitioning of the AB interaction into two orthogonal two-degree-of-freedom components is very useful in constructing more complex designs. Also, there is no connection between the AB and AB^2 components of interaction and the sums of squares for $AB_{L\times L}$, $AB_{L\times Q}$, $AB_{Q\times L}$, and $AB_{Q\times Q}$.

The AB and AB^2 components of interaction may be computed another way. Consider the treatment combination totals in either square in Figure 9.3. If we add the data by diagonals downward from left to right, we obtain the totals $-3 + 4 - 1 = 0$, $-3 + 10 - 1 = 6$, and $5 + 11 + 2 = 18$. The sum of squares between these totals is 28.00 (AB^2). Similarly, the diagonal totals downward from right to left are $5 + 4 - 1 = 8$, $-3 + 2 - 1 = -2$, and $-3 + 11 + 10 = 18$. The sum of squares between these totals is 33.34 (AB). Yates called these components of interaction as the ***I* and *J* components of interaction**, respectively. We use both notations interchangeably; that is,

$$I(AB) = AB^2$$

$$J(AB) = AB$$

For more information about decomposing the sums of squares in three-level designs, refer to the supplemental material for this chapter.

9.1.3 The 3^3 Design

Now suppose there are three factors (A, B, and C) under study and that each factor is at three levels arranged in a factorial experiment. This is a 3^3 factorial design, and the experimental layout and treatment combination notation are shown in Figure 9.2. The 27 treatment combinations have 26 degrees of freedom. Each main effect has two degrees of freedom, each two-factor interaction has four degrees of freedom, and the three-factor interaction has eight degrees of freedom. If there are n replicates, there are $n3^3 - 1$ total degrees of freedom and $3^3(n - 1)$ degrees of freedom for error.

The sums of squares may be calculated using the standard methods for factorial designs. In addition, if the factors are quantitative, the main effects may be partitioned into linear and quadratic components, each with a single degree of freedom. The two-factor interactions may be decomposed into linear × linear, linear × quadratic, quadratic × linear, and quadratic × quadratic effects. Finally, the three-factor interaction ABC can be partitioned into eight single-degree-of-freedom components corresponding to linear × linear × linear, linear × linear × quadratic, and so on. Such a breakdown for the three-factor interaction is generally not very useful.

It is also possible to partition the two-factor interactions into their I and J components. These would be designated AB, AB^2, AC, AC^2, BC, and BC^2, and each component would have two degrees of freedom. As in the 3^2 design, these components have no physical significance.

The three-factor interaction ABC may be partitioned into four orthogonal two-degrees-of-freedom components, which are usually called the W, X, Y, and Z components of the interaction. They are also referred to as the AB^2C^2, AB^2C, ABC^2, and ABC components of the ABC interaction, respectively. The two notations are used interchangeably; that is,

$$W(ABC) = AB^2C^2$$

$$X(ABC) = AB^2C$$

$$Y(ABC) = ABC^2$$

$$Z(ABC) = ABC$$

Note that no first letter can have an exponent other than 1. Like the I and J components, the W, X, Y, and Z components have no practical interpretation. They are, however, useful in constructing more complex designs.

EXAMPLE 9.1

A machine is used to fill 5-gallon metal containers with soft drink syrup. The variable of interest is the amount of syrup loss due to frothing. Three factors are thought to influence frothing: the nozzle design (A), the filling speed (B), and the operating pressure (C). Three nozzles, three filling speeds, and three pressures are chosen, and two replicates of a 3^3 factorial experiment are run. The coded data are shown in Table 9.1.

The analysis of variance for the syrup loss data is shown in Table 9.2. The sums of squares have been computed by the usual methods. We see that the filling speed and operating pressure are statistically significant. All three two-factor interactions are also significant. The two-factor interactions are analyzed graphically in Figure 9.4. The middle level of speed gives the best performance, nozzle types 2 and 3, and either the low (10 psi) or high (20 psi) pressure seems most effective in reducing syrup loss.

■ **TABLE 9.1**
Syrup Loss Data for Example 9.1 (units are cubic centimeters −70)

	Nozzle Type (A)								
	1			2			3		
Pressure (in psi)	Speed (in RPM) (B)								
(C)	**100**	**120**	**140**	**100**	**120**	**140**	**100**	**120**	**140**
10	−35	−45	−40	17	−65	20	−39	−55	15
	−25	−60	15	24	−58	4	−35	−67	−30
15	110	−10	80	55	−55	110	90	−28	110
	75	30	54	120	−44	44	113	−26	135
20	4	−40	31	−23	−64	−20	−30	−61	54
	5	−30	36	−5	−62	−31	−55	−52	4

■ **TABLE 9.2**

Analysis of Variance for Syrup Loss Data

Source of Variation	Sum of Squares	Degrees of Freedom	Mean Square	F_0	*P*-Value
A, nozzle	993.77	2	496.89	1.17	0.3256
B, speed	61,190.33	2	30,595.17	71.74	<0.0001
C, pressure	69,105.33	2	34,552.67	81.01	<0.0001
AB	6,300.90	4	1,575.22	3.69	0.0160
AC	7,513.90	4	1,878.47	4.40	0.0072
BC	12,854.34	4	3,213.58	7.53	0.0003
ABC	4,628.76	8	578.60	1.36	0.2580
Error	11,515.50	27	426.50		
Total	174,102.83	53			

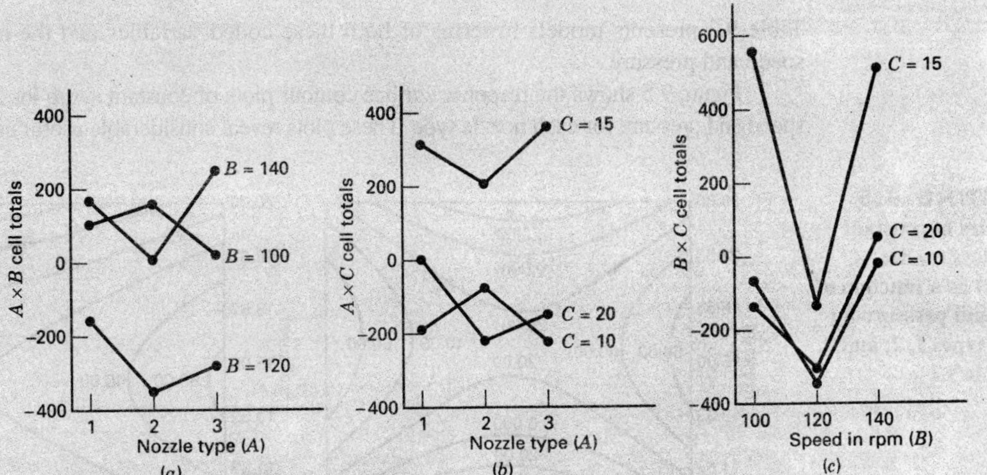

■ **FIGURE 9.4** **Two-factor interactions for Example 9.1**

Example 9.1 illustrates a situation where the three-level design often finds some application; one or more of the factors are **qualitative**, naturally taking on three levels, and the remaining factors are **quantitative**. In this example, suppose only three nozzle designs are of interest. This is clearly, then, a qualitative factor that requires three levels. The filling speed and the operating pressure are quantitative factors. Therefore, we could fit a quadratic model such as Equation 9.1 in the two factors speed and pressure at each level of the nozzle factor.

Table 9.3 shows these quadratic regression models. The β's in these models were estimated using a standard linear regression computer program. (We will discuss least squares regression in more detail in Chapter 10.) In these models, the variables x_1 and x_2 are coded to the levels $-1, 0, +1$ as discussed previously, and we assumed the following natural levels for pressure and speed:

■ **TABLE 9.3**
Regression Models for Example 9.1

Nozzle Type	x_1 = Speed (S), x_2 = Pressure (P) in Coded Units
1	$\hat{y} = 22.1 + 3.5x_1 + 16.3x_2 + 51.7x_1^2 - 71.8x_2^2 + 2.9x_1x_2$
	$\hat{y} = 1217.3 - 31.256S + 86.017P + 0.12917S^2 - 2.8733P^2 + 0.02875SP$
2	$\hat{y} = 25.6 - 22.8x_1 - 12.3x_2 + 14.1x_1^2 - 56.9x_2^2 - 0.7x_1x_2$
	$\hat{y} = 180.1 - 9.475S + 66.75P + 0.035S^2 - 2.2767P^2 - 0.0075SP$
3	$\hat{y} = 15.1 + 20.3x_1 + 5.9x_2 + 75.8x_1^2 - 94.9x_2^2 + 10.5x_1x_2$
	$\hat{y} = 1940.1 - 46.058S + 102.48P + 0.18958S^2 - 3.7967P^2 + 0.105SP$

Coded Level	Speed (psi)	Pressure (rpm)
−1	100	10
0	120	15
+1	140	20

Table 9.3 presents models in terms of both these coded variables and the natural levels of speed and pressure.

Figure 9.5 shows the response surface contour plots of constant syrup loss as a function of speed and pressure for each nozzle type. These plots reveal considerable useful information about

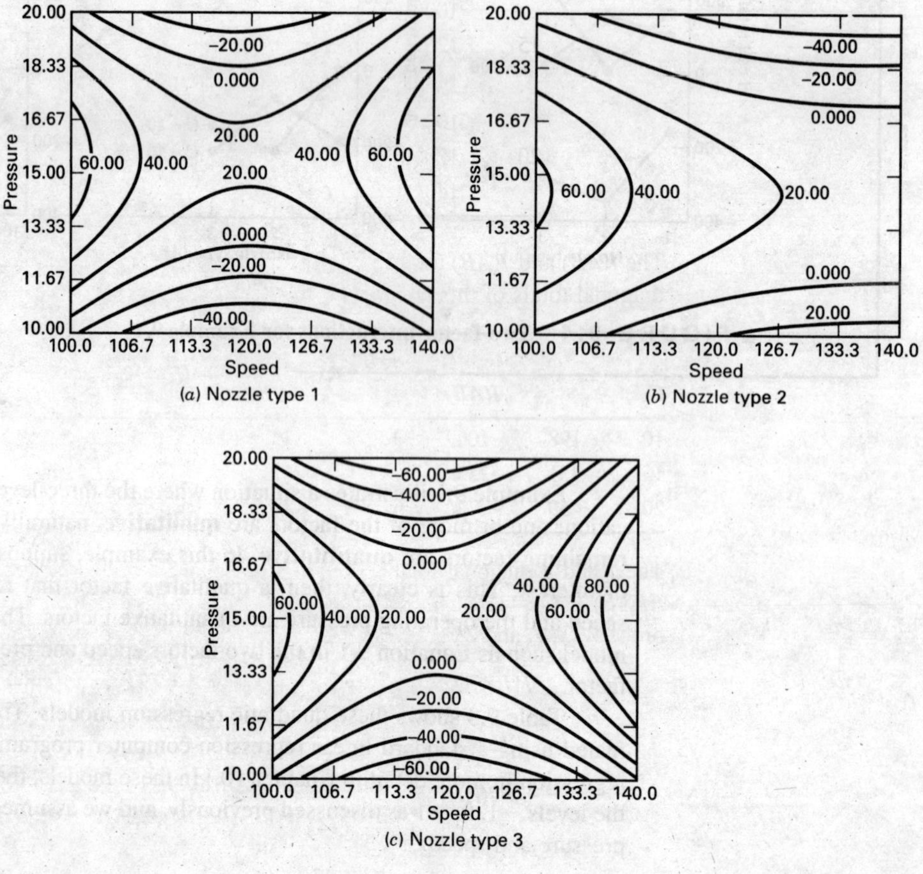

■ **FIGURE 9.5** **Contours of constant syrup loss (units: cc −70) as a function of speed and pressure for nozzle types 1, 2, and 3, Example 9.1**

the performance of this filling system. Because the objective is to minimize syrup loss, nozzle type 3 would be preferred, as the smallest observed contours (−60) appear only on this plot. Filling speed near the middle level of 120 rpm and the either low or high pressure levels should be used.

When constructing contour plots for an experiment that has a mixture of quantitative and qualitative factors, it is not unusual to find that the shapes of the surfaces in the quantitative factors are very different at each level of the qualitative factors. This is noticeable to some degree in Figure 9.5, where the shape of the surface for nozzle type 2 is considerably elongated in comparison to the surfaces for nozzle types 1 and 3. When this occurs, it implies that there are interactions present between the quantitative and qualitative factors, and as a result, the optimum operating conditions (and other important conclusions) in terms of the quantitative factors are very different at each level of the qualitative factors.

We can easily show the numerical partitioning of the ABC interaction into its four orthogonal two-degrees-of-freedom components using the data in Example 9.1. The general procedure has been described by Davies (1956) and Cochran and Cox (1957). First, select any two of the three factors, say AB, and compute the I and J totals of the AB interaction at each level of the third factor C. These calculations are as follows:

		A			Totals	
C	*B*	1	2	3	*I*	*J*
	100	−60	41	−74	−198	−222
10	120	−105	−123	−122	−106	−79
	140	−25	24	−15	−155	−158
	100	185	175	203	331	238
15	120	20	−99	−54	255	440
	140	134	154	245	377	285
	100	9	−28	−85	−59	−144
20	120	−70	−126	−113	−74	−40
	140	67	−51	58	−206	−155

The $I(AB)$ and $J(AB)$ totals are now arranged in a two-way table with factor C, and the I and J diagonal totals of this new display are computed as follows:

				Totals						Totals	
C		*I(AB)*		*I*	*J*	*C*		*J(AB)*		*I*	*J*
10	−198	−106	−155	−149	41	10	−222	−79	−158	63	138
15	331	255	377	212	19	15	238	440	285	62	4
20	−59	−74	−206	102	105	20	−144	−40	−155	40	23

The I and J diagonal totals computed above are actually the totals representing the quantities $I[I(AB) \times C] = AB^2C^2$, $J[I(AB) \times C] = AB^2C$, $I[J(AB) \times C] = ABC^2$, and $J[J(AB) \times C] = ABC$ or the W, X, Y, and Z components of ABC. The sums of squares are found in the usual way; that is,

$$I[I(AB) \times C)] = AB^2C^2 = W(ABC)$$
$$= \frac{(-149)^2 + (212)^2 + (102)^2}{18} - \frac{(165)^2}{54} = 3804.11$$
$$J[I(AB) \times C] = AB^2C = X(ABC)$$
$$= \frac{(41)^2 + (19)^2 + (105)^2}{18} - \frac{(165)^2}{54} = 221.77$$

$$J[J(AB) \times C] = ABC^2 = Y(ABC)$$
$$= \frac{(63)^2 + (62)^2 + (40)^2}{18} - \frac{(165)^2}{54} = 18.77$$
$$J[J(AB) \times C] = ABC = Z(ABC)$$
$$= \frac{(138)^2 + (4)^2 + (23)^2}{18} - \frac{(165)^2}{54} = 584.11$$

Although this is an orthogonal partitioning of SS_{ABC}, we point out again that it is not customarily displayed in the analysis of variance table. In subsequent sections, we discuss the occasional need for the computation of one or more of these components.

9.1.4 The General 3^k Design

The concepts utilized in the 3^2 and 3^3 designs can be readily extended to the case of k factors, each at three levels, that is, to a 3^k factorial design. The usual digital notation is employed for the treatment combinations, so 0120 represents a treatment combination in a 3^4 design with A and D at the low levels, B at the intermediate level, and C at the high level. There are 3^k treatment combinations, with $3^k - 1$ degrees of freedom between them. These treatment combinations allow sums of squares to be determined for k main effects, each with two degrees of freedom; $\binom{k}{2}$ two-factor interactions, each with four degrees of freedom; . . . ; and one k-factor interaction with 2^k degrees of freedom. In general, an h-factor interaction has 2^h degrees of freedom. If there are n replicates, there are $n3^k - 1$ total degrees of freedom and $3^k(n - 1)$ degrees of freedom for error.

Sums of squares for effects and interactions are computed by the usual methods for factorial designs. Typically, three-factor and higher interactions are not broken down any further. However, any h-factor interaction has 2^{h-1} orthogonal two-degrees-of-freedom components. For example, the four-factor interaction $ABCD$ has $2^{4-1} = 8$ orthogonal two-degrees-of-freedom components, denoted by $ABCD^2$, ABC^2D, AB^2CD, $ABCD$, ABC^2D^2, AB^2C^2D, AB^2CD^2, and $AB^2C^2D^2$. In writing these components, note that the only exponent allowed on the first letter is 1. If the exponent on the first letter is not 1, then the entire expression must be squared and the exponents reduced modulus 3. To demonstrate this, consider

$$A^2BCD = (A^2BCD)^2 = A^4B^2C^2D^2 = AB^2C^2D^2$$

These interaction components have no physical interpretation, but they are useful in constructing more complex designs.

The size of the design increases rapidly with k. For example, a 3^3 design has 27 treatment combinations per replication, a 3^4 design has 81, a 3^5 design has 243, and so on. Therefore, only a single replicate of the 3^k design is frequently considered, and higher order interactions are combined to provide an estimate of error. As an illustration, if three-factor and higher interactions are negligible, then a single replicate of the 3^3 design provides 8 degrees of freedom for error, and a single replicate of the 3^4 design provides 48 degrees of freedom for error. These are still large designs for $k \geq 3$ factors and, consequently, not too useful.

9.2 Confounding in the 3^k Factorial Design

Even when a single replicate of the 3^k design is considered, the design requires so many runs that it is unlikely that all 3^k runs can be made under uniform conditions. Thus, confounding in blocks is often necessary. The 3^k design may be confounded in 3^p incomplete blocks, where $p < k$. Thus, these designs may be confounded in three blocks, nine blocks, and so on.

9.2.1 The 3^k Factorial Design in Three Blocks

Suppose that we wish to confound the 3^k design in three incomplete blocks. These three blocks have two degrees of freedom among them; thus, there must be two degrees of freedom confounded with blocks. Recall that in the 3^k factorial series each main effect has two degrees of freedom. Furthermore, every two-factor interaction has four degrees of freedom and can be decomposed into two components of interaction (e.g., AB and AB^2), each with two degrees of freedom; every three-factor interaction has eight degrees of freedom and can be decomposed into four components of interaction (e.g., ABC, ABC^2, AB^2C, and AB^2C^2), each with two degrees of freedom; and so on. Therefore, it is convenient to confound a component of interaction with blocks.

The general procedure is to construct a **defining contrast**

$$L = \alpha_1 x + \alpha_2 x_2 + \cdots + \alpha_k x_k \tag{9.2}$$

where α_i represents the exponent on the ith factor in the effect to be confounded and x_i is the level of the ith factor in a particular treatment combination. For the 3^k series, we have $\alpha_i = 0$, 1, or 2 with the first nonzero α_i being unity, and $x_i = 0$ (low level), 1 (intermediate level), or 2 (high level). The treatment combinations in the 3^k design are assigned to blocks based on the value of L (mod 3). Because L (mod 3) can take on only the values 0, 1, or 2, three blocks are uniquely defined. The treatment combinations satisfying $L = 0$ (mod 3) constitute the **principal block**. This block will always contain the treatment combination 00 . . . 0.

For example, suppose we wish to construct a 3^2 factorial design in three blocks. Either component of the AB interaction, AB or AB^2, may be confounded with blocks. Arbitrarily choosing AB^2, we obtain the defining contrast

$$L = x_1 + 2x_2$$

The value of L (mod 3) of each treatment combination may be found as follows:

$$00: \quad L = 1(0) + 2(0) = 0 = 0 \text{ (mod 3)} \qquad 11: \quad L = 1(1) + 2(1) = 3 = 0 \text{ (mod 3)}$$
$$01: \quad L = 1(0) + 2(1) = 2 = 2 \text{ (mod 3)} \qquad 21: \quad L = 1(2) + 2(1) = 4 = 1 \text{ (mod 3)}$$
$$02: \quad L = 1(0) + 2(2) = 4 = 1 \text{ (mod 3)} \qquad 12: \quad L = 1(1) + 2(2) = 5 = 2 \text{ (mod 3)}$$
$$10: \quad L = 1(1) + 2(0) = 1 = 1 \text{ (mod 3)} \qquad 22: \quad L = 1(2) + 2(2) = 6 = 0 \text{ (mod 3)}$$
$$20: \quad L = 1(2) + 2(0) = 2 = 2 \text{ (mod 3)}$$

The blocks are shown in Figure 9.6.

The elements in the principal block form a group with respect to addition modulus 3. Referring to Figure 9.6, we see that 11 + 11 = 22 and 11 + 22 = 00. Treatment combinations

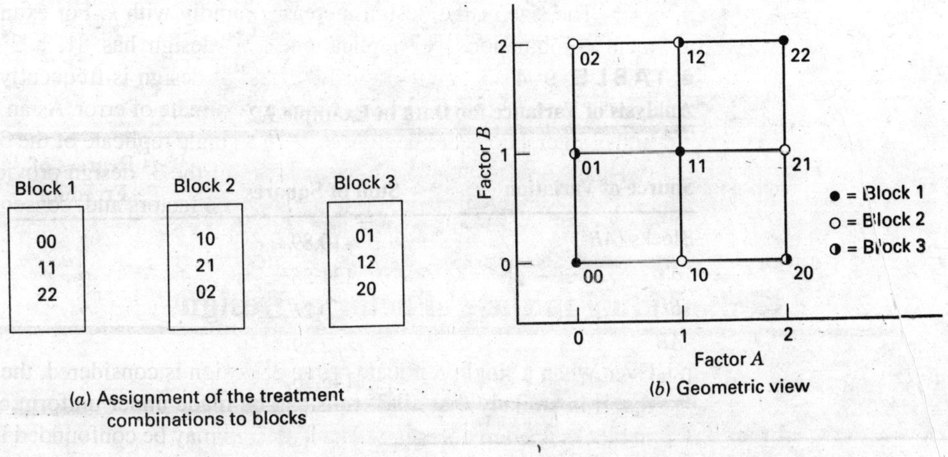

■ FIGURE 9.6 The 3^2 design in three blocks with AB^2 confounded

(a) Assignment of the treatment combinations to blocks

(b) Geometric view

in the other two blocks may be generated by adding, modulus 3, any element in the new block to the elements of the principal block. Thus, we use 10 for block 2 and obtain

$$10 + 00 = 10 \quad 10 + 11 = 21 \quad \text{and} \quad 10 + 22 = 02$$

To generate block 3, we find using 01

$$01 + 00 = 01 \quad 01 + 11 = 12 \quad \text{and} \quad 01 + 22 = 20$$

EXAMPLE 9.2

We illustrate the statistical analysis of the 3^2 design confounded in three blocks by using the following data, which come from the single replicate of the 3^2 design shown in Figure 9.6.

	Block 1	Block 2	Block 3
	00 = 4	10 = −2	01 = 5
	11 = −4	21 = 1	12 = −5
	22 = 0	02 = 8	20 = 0
Block totals =	0	7	0

Using conventional methods for the analysis of factorials, we find that $SS_A = 131.56$ and $SS_B = 0.22$.

We also find that

$$SS_{\text{Blocks}} = \frac{(0)^2 + (7)^2 + (0)^2}{3} - \frac{(7)^2}{9} = 10.89$$

However, SS_{Blocks} is exactly equal to the AB^2 component of interaction. To see this, write the observations as follows:

		Factor *B*		
		0	1	2
	0	4	5	8
Factor *A*	1	−2	−4	−5
	2	0	1	0

Recall from Section 9.1.2 that the *I* or AB^2 component of the *AB* interaction may be found by computing the sum of squares between the left-to-right diagonal totals in the above layout. This yields

$$SS_{AB^2} = \frac{(0)^2 + (0)^2 + (7)^2}{3} - \frac{(7)^2}{9} = 10.89$$

which is identical to SS_{Blocks}.

The analysis of variance is shown in Table 9.4. Because there is only one replicate, no formal tests can be performed. It is not a good idea to use the *AB* component of interaction as an estimate of error.

■ **TABLE 9.4**
Analysis of Variance for Data in Example 9.2

Source of Variation	Sum of Squares	Degrees of Freedom
Blocks (AB^2)	10.89	2
A	131.56	2
B	0.22	2
AB	2.89	2
Total	145.56	8

We now look at a slightly more complicated design—a 3^3 factorial confounded in three blocks of nine runs each. The AB^2C^2 component of the three-factor interaction will be confounded with blocks. The defining contrast is

$$L = x_1 + 2x_2 + 2x_3$$

It is easy to verify that the treatment combinations 000, 012, and 101 belong in the principal block. The remaining runs in the principal block are generated as follows:

(1) 000	(4) 101 + 101 = 202	(7) 101 + 021 = 122
(2) 012	(5) 012 + 012 = 021	(8) 012 + 202 = 211
(3) 101	(6) 101 + 012 = 110	(9) 021 + 202 = 220

To find the runs in another block, note that the treatment combination 200 is not in the principal block. Thus, the elements of block 2 are

(1) 200 + 000 = 200	(4) 200 + 202 = 102	(7) 200 + 122 = 022
(2) 200 + 012 = 212	(5) 200 + 021 = 221	(8) 200 + 211 = 111
(3) 200 + 101 = 001	(6) 200 + 110 = 010	(9) 200 + 220 = 120

Notice that all these runs satisfy $L = 2 \pmod 3$. The final block is found by observing that 100 does not belong in block 1 or 2. Using 100 as above yields

(1) 100 + 000 = 100	(4) 100 + 202 = 002	(7) 100 + 122 = 222
(2) 100 + 012 = 112	(5) 100 + 021 = 121	(8) 100 + 211 = 011
(3) 100 + 101 = 201	(6) 100 + 110 = 210	(9) 100 + 220 = 020

The blocks are shown in Figure 9.7.

The analysis of variance for this design is shown in Table 9.5. Through the use of this confounding scheme, information on all the main effects and two-factor interactions is available. The remaining components of the three-factor interaction (ABC, AB^2C, and ABC^2) are combined as an estimate of error. The sum of squares for those three components could be obtained by subtraction. In general, for the 3^k design in three blocks, we would always select a component of the highest order interaction to confound with blocks. The remaining unconfounded components of this interaction could be obtained by computing the k-factor interaction in the usual way and subtracting from this quantity the sum of squares for blocks.

■ FIGURE 9.7 **The 3^3 design in three blocks with AB^2C^2 confounded**

Block 1	Block 2	Block 3
000	200	100
012	212	112
101	001	201
202	102	002
021	221	121
110	010	210
122	022	222
211	111	011
220	120	020

(a) Assignment of the treatment combinations to blocks

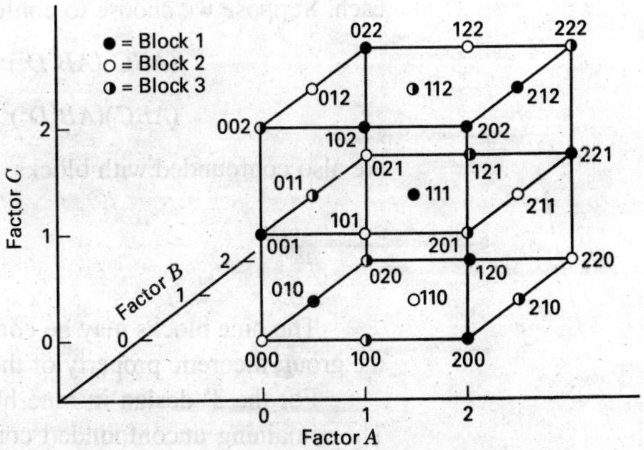

(b) Geometric view

■ **TABLE 9.5**
Analysis of Variance for a 3^3 Design with AB^2C^2 Confounded

Source of Variation	Degrees of Freedom
Blocks (AB^2C^2)	2
A	2
B	2
C	2
AB	4
AC	4
BC	4
Error ($ABC + AB^2C + ABC^2$)	6
Total	26

9.2.2 The 3^k Factorial Design in Nine Blocks

In some experimental situations, it may be necessary to confound the 3^k design in nine blocks. Thus, eight degrees of freedom will be confounded with blocks. To construct these designs, we choose *two* components of interaction, and, as a result, two more will be confounded automatically, yielding the required eight degrees of freedom. These two are the generalized interactions of the two effects originally chosen. In the 3^k system, the **generalized interactions** of two effects (e.g., P and Q) are defined as PQ and PQ^2 (or P^2Q).

The two components of interaction initially chosen yield *two* defining contrasts

$$L_1 = \alpha_1 x_1 + \alpha_2 x_2 + \cdots + \alpha_k x_k = u \pmod 3 \qquad u = 0, 1, 2$$

$$L_2 = \beta_1 x_1 + \beta_2 x_2 + \cdots + \beta_k x_k = h \pmod 3 \qquad h = 0, 1, 2 \qquad (9.3)$$

where $\{\alpha_i\}$ and $\{\beta_j\}$ are the exponents in the first and second generalized interactions, respectively, with the convention that the first nonzero α_i and β_j are unity. The defining contrasts in Equation 9.3 imply nine simultaneous equations specified by the pair of values for L_1 and L_2. Treatment combinations having the same pair of values for (L_1, L_2) are assigned to the same block.

The principal block consists of treatment combinations satisfying $L_1 = L_2 = 0 \pmod 3$. The elements of this block form a group with respect to addition modulus 3; thus, the scheme given in Section 9.2.1 can be used to generate the blocks.

As an example, consider the 3^4 factorial design confounded in nine blocks of nine runs each. Suppose we choose to confound ABC and AB^2D^2. Their generalized interactions

$$(ABC)(AB^2D^2) = A^2B^3CD^2 = (A^2B^3CD^2)^2 = AC^2D$$

$$(ABC)(AB^2D^2)^2 = A^3B^5CD^4 = B^2CD = (B^2CD)^2 = BC^2D^2$$

are also confounded with blocks. The defining contrasts for ABC and AB^2D^2 are

$$L_1 = x_1 + x_2 + x_3$$

$$L_2 = x_1 + 2x_2 + 2x_4 \qquad (9.4)$$

The nine blocks may be constructed by using the defining contrasts (Equation 9.4) and the group-theoretic property of the principal block. The design is shown in Figure 9.8.

For the 3^k design in nine blocks, four components of interaction will be confounded. The remaining unconfounded components of these interactions can be determined by subtracting the sum of squares for the confounded component from the sum of squares for the

Block 1	Block 2	Block 3	Block 4	Block 5	Block 6	Block 7	Block 8	Block 9
0000	0001	2000	0200	0020	0010	1000	0100	0002
0122	0120	2122	0022	0112	0102	1122	0222	0121
0211	0212	2211	0111	0201	0221	1211	0011	0210
1021	1022	0021	1221	1011	1001	2021	1121	1020
1110	1111	0110	1010	1100	1120	2110	1210	1112
1202	1200	0202	1102	1222	1212	2202	1002	1201
2012	2010	1012	2212	2002	2022	0012	2112	2011
2101	2102	1101	2001	2121	2111	0101	2201	2100
2220	2221	1220	2120	2210	2200	0220	2020	2222
$(L_1, L_2) = (0,0)$	(0,2)	(2,2)	(2,1)	(2,0)	(1,6)	(1,1)	(1,2)	(0,1)

■ **FIGURE 9.8 The 3^4 design in nine blocks with ABC, AB^2D^2, AC^2D, and BC^2D^2 confounded**

entire interaction. The method described in Section 9.1.3 may be useful in computing the components of interaction.

9.2.3 The 3^k Factorial Design in 3^p Blocks

The 3^k factorial design may be confounded in 3^p blocks of 3^{k-p} observations each, where $p < k$. The procedure is to select p independent effects to be confounded with blocks. As a result, exactly $(3^p - 2p - 1)/2$ other effects are automatically confounded. These effects are the generalized interactions of those effects originally chosen.

As an illustration, consider a 3^7 design to be confounded in 27 blocks. Because $p = 3$, we would select three independent components of interaction and automatically confound $[3^3 - 2(3) - 1]/2 = 10$ others. Suppose we choose ABC^2DG, BCE^2F^2G, and $BDEFG$. Three defining contrasts can be constructed from these effects, and the 27 blocks can be generated by the methods previously described. The other 10 effects confounded with blocks are

$$(ABC^2DG)(BCE^2F^2G) = AB^2DE^2F^2G^2$$

$$(ABC^2DG)(BCE^2F^2G)^2 = AB^3C^4DE^4F^4G^3 = ACDEF$$

$$(ABC^2DG)(BDEFG) = AB^2C^2D^2EFG^2$$

$$(ABC^2DG)(BDEFG)^2 = AB^3C^2D^3E^2F^2G^3 = AC^2E^2F^2$$

$$(BCE^2F^2G)(BDEFG) = B^2CDE^3F^3G^2 = BC^2D^2G$$

$$(BCE^2F^2G)(BDEFG)^2 = B^3CD^2E^4F^4G^3 = CD^2EF$$

$$(ABC^2DG)(BCE^2F^2G)(BDEFG) = AB^3C^3D^2E^3F^3G^3 = AD^2$$

$$(ABC^2DG)^2(BCE^2F^3G)(BDEFG) = A^2B^4C^5D^3G^4 = AB^2CG^2$$

$$(ABC^2DG)(BCE^2F^2G)^2(BDEFG) = ABCD^2E^2F^2G$$

$$(ABC^2DG)(BCE^2F^2G)(BDEFG)^2 = ABC^3D^3E^4F^4G^4 = ABEFG$$

This is a huge design requiring $3^7 = 2187$ observations arranged in 27 blocks of 81 observations each.

9.3 Fractional Replication of the 3^k Factorial Design

The concept of fractional replication can be extended to the 3^k factorial designs. Because a complete replicate of the 3^k design can require a rather large number of runs even for moderate values of k, fractional replication of these designs is of interest. As we shall see, however, some of these designs have complex alias structures.

9.3.1 The One-Third Fraction of the 3^k Factorial Design

The largest fraction of the 3^k design is a one-third fraction containing 3^{k-1} runs. Consequently, we refer to this as a 3^{k-1} fractional factorial design. To construct a 3^{k-1} fractional factorial design, select a two-degrees-of-freedom component of interaction (generally, the highest order interaction) and partition the *full* 3^k design into three blocks. Each of the three resulting blocks is a 3^{k-1} fractional design, and any one of the blocks may be selected for use. If $AB^{\alpha_2}C^{\alpha_3}\cdots K^{\alpha_k}$ is the component of interaction used to define the blocks, then $I = AB^{\alpha_2}C^{\alpha_3}\cdots K^{\alpha_k}$ is called the **defining relation** of the fractional factorial design. Each main effect or component of interaction estimated from the 3^{k-1} design has two aliases, which may be found by multiplying the effect by *both* I and I^2 modulus 3.

As an example, consider a one-third fraction of the 3^3 design. We may select any component of the ABC interaction to construct the design, that is, ABC, AB^2C, ABC^2, or AB^2C^2. Thus, there are actually 12 *different* one-third fractions of the 3^3 design defined by

$$x_1 + \alpha_2 x_2 + \alpha_3 x_3 = u \ (\text{mod } 3)$$

where $\alpha = 1$ or 2 and $u = 0$, 1, or 2. Suppose we select the component of AB^2C^2. Each fraction of the resulting 3^{3-1} design will contain exactly $3^2 = 9$ treatment combinations that must satisfy

$$x_1 + 2x_2 + 2x_3 = u \ (\text{mod } 3)$$

where $u = 0$, 1, or 2. It is easy to verify that the three one-third fractions are as shown in Figure 9.9.

If any one of the 3^{3-1} designs in Figure 9.9 is run, the resulting alias structure is

$$A = A(AB^2C^2) = A^2B^2C^2 = ABC$$
$$A = A(AB^2C^2)^2 = A^3B^4C^4 = BC$$
$$B = B(AB^2C^2) = AB^3C^2 = AC^2$$
$$B = B(AB^2C^2)^2 = A^2B^5C^4 = ABC^2$$
$$C = C(AB^2C^2) = AB^2C^3 = AB^2$$
$$C = C(AB^2C^2)^2 = A^2B^4C^5 = AB^2C$$
$$AB = AB(AB^2C^2) = A^2B^3C^2 = AC$$
$$AB = AB(AB^2C^2)^2 = A^3B^5C^4 = BC^2$$

Consequently, the four effects that are actually estimated from the eight degrees of freedom in the design are $A + BC + ABC$, $B + AC^2 + ABC^2$, $C + AB^2 + AB^2C$, and $AB + AC + BC^2$. This design would be of practical value only if all the interactions were small relative to the main effects. Because the main effects are aliased with two-factor interactions, this is a resolution III design. Notice how complex the alias relationships are in this design. Each main effect is aliased with a *component* of interaction. If, for example, the two-factor interaction BC is large, this will potentially distort the estimate of the main effect of A and make the

Design 1 $u = 0$	Design 1 $u = 1$	Design 1 $u = 2$
000	100	200
012	112	212
101	201	001
202	002	102
021	121	221
110	210	010
122	222	022
211	011	111
220	020	120

(*a*) Treatment combinations

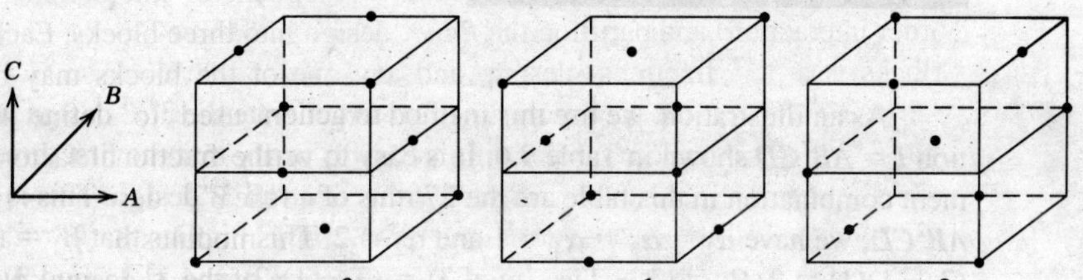

(*b*) Geometric view

■ **FIGURE 9.9 The three one-third fractions of the 3^3 design with defining relation $I = AB^2C^2$**

$AB + AC + BC^2$ effect very difficult to interpret. It is very difficult to see how this design could be useful unless we assume that all interactions are negligible.

Before leaving the 3_{III}^{3-1} design, note that for the design with $u = 0$ (see Figure 9.9) if we let A denote the row and B denote the column, then the design can be written as

000	012	021
101	110	122
202	211	220

which is a 3×3 **Latin square**. The assumption of negligible interactions required for unique interpretations of the 3_{III}^{3-1} design is paralleled in the Latin square design. However, the two designs arise from different motives, one as a consequence of fractional replication and the other from randomization restrictions. From Table 4.13, we observe that there are only twelve 3×3 Latin squares and that each one corresponds to one of the twelve different 3^{3-1} fractional factorial designs.

The treatment combinations in a 3^{k-1} design with the defining relation $I = AB^{\alpha_2}C^{\alpha_3}\cdots K^{\alpha_k}$ can be constructed using a method similar to that employed in the 2^{k-p} series. First, write down the 3^{k-1} runs for a **full** three-level factorial design in $k-1$ factors, with the usual 0, 1, 2 notation. This is the **basic design** in the terminology of Chapter 8. Then introduce the kth factor by equating its levels x_k to the appropriate component of the highest order interaction, say $AB^{\alpha_2}C^{\alpha_3}\cdots(K-1)^{\alpha_{k-1}}$, through the relationship

$$x_k = \beta_1 x_1 + \beta_2 x_2 + \cdots + \beta_{k-1}x_{k-1} \tag{9.5}$$

where $\beta_i = (3 - \alpha_k)\alpha_i \pmod 3$ for $1 \le i \le k-1$. This yields a design of the highest possible resolution.

■ **TABLE 9.6**
A 3_{IV}^{4-1} Design with $I = AB^2CD$

0000	0012	2221
0101	0110	0021
1100	0211	0122
1002	1011	0220
0202	1112	1020
1201	1210	1121
2001	2010	1222
2102	2111	2022
2200	2212	2120

As an illustration, we use this method to generate the 3_{IV}^{4-1} design with the defining relation $I = AB^2CD$ shown in Table 9.6. It is easy to verify that the first three digits of each treatment combination in this table are the 27 runs of a full 3^3 design. This is the basic design. For AB^2CD, we have $\alpha_1 = \alpha_3 = \alpha_4 = 1$ and $\alpha_2 = 2$. This implies that $\beta_1 = (3 - 1)\alpha_1 \pmod 3 = (3 - 1)(1) = 2$, $\beta_2 = (3 - 1)\alpha_2 \pmod 3 = (3 - 1)(2) = 4 = 1 \pmod 3$, and $\beta_3 = (3 - 1)\alpha_3 \pmod 3 = (3 - 1)(1) = 2$. Thus, Equation 9.5 becomes

$$x_4 = 2x_1 + x_2 + 2x_3 \tag{9.6}$$

The levels of the fourth factor satisfy Equation 9.6. For example, we have $2(0) + 1(0) + 2(0) = 0$, $2(0) + 1(1) + 2(0) = 1$, $2(1) + 1(1) + 2(0) = 3 = 0$, and so on.

The resulting 3_{IV}^{4-1} design has 26 degrees of freedom that may be used to compute the sums of squares for the 13 main effects and components of interactions (and their aliases). The aliases of any effect are found in the usual manner; for example, the aliases of A are $A(AB^2CD) = ABC^2D^2$ and $A(AB^2CD)^2 = BC^2D^2$. One may verify that the four main effects are clear of any two-factor interaction components, but that some two-factor interaction components are aliased with each other. Once again, we notice the complexity of the alias structure. If any two-factor interactions are large, it will likely be very difficult to isolate them with this design.

The statistical analysis of a 3^{k-1} design is accomplished by the usual analysis of variance procedures for factorial experiments. The sums of squares for the components of interaction may be computed as in Section 9.1. Remember when interpreting results that the components of interactions have no practical interpretation.

9.3.2 Other 3^{k-p} Fractional Factorial Designs

For moderate to large values of k, even further fractionation of the 3^k design is potentially desirable. In general, we may construct a $(\frac{1}{3})^p$ fraction of the 3^k design for $p < k$, where the fraction contains 3^{k-p} runs. Such a design is called a 3^{k-p} fractional factorial design. Thus, a 3^{k-2} design is a one-ninth fraction, a 3^{k-3} design is a one-twenty-seventh fraction, and so on.

The procedure for constructing a 3^{k-p} fractional factorial design is to select p components of interaction and use these effects to partition the 3^k treatment combinations into 3^p blocks. Each block is then a 3^{k-p} fractional factorial design. The defining relation I of any fraction consists of the p effects initially chosen and their $(3^p - 2p - 1)/2$ generalized interactions. The alias of any main effect or component of interaction is produced by multiplication modulus 3 of the effect by I and I^2.

■ **TABLE 9.7**

A 3_{III}^{4-2} Design with $I = AB^2C$ and $I = BCD$

0000	0111	0222
1021	1102	1210
2012	2120	2201

We may also generate the runs defining a 3^{k-p} fractional factorial design by first writing down the treatment combinations of a full 3^{k-p} factorial design and then introducing the additional p factors by equating them to components of interaction, as we did in Section 9.3.1.

We illustrate the procedure by constructing a 3^{4-2} design, that is, a one-ninth fraction of the 3^4 design. Let AB^2C and BCD be the two components of interaction chosen to construct the design. Their generalized interactions are $(AB^2C)(BCD) = AC^2D$ and $(AB^2C)(BCD)^2 = ABD^2$. Thus, the defining relation for this design is $I = AB^2C = BCD = AC^2D = ABD^2$, and the design is of resolution III. The nine treatment combinations in the design are found by writing down a 3^2 design in the factors A and B, and then adding two new factors by setting

$$x_3 = 2x_1 + x_2$$
$$x_4 = 2x_2 + 2x_3$$

This is equivalent to using AB^2C and BCD to partition the full 3^4 design into nine blocks and then selecting one of these blocks as the desired fraction. The complete design is shown in Table 9.7.

This design has eight degrees of freedom that may be used to estimate four main effects and their aliases. The aliases of any effect may be found by multiplying the effect modulus 3 by AB^2C, BCD, AC^2D, ABD^2, and their squares. The complete alias structure for the design is given in Table 9.8.

From the alias structure, we see that this design is useful only in the absence of interaction. Furthermore, if A denotes the rows and B denotes the columns, then from examining Table 9.7 we see that the 3_{III}^{4-2} design is also a Graeco–Latin square.

The publication by Connor and Zelen (1959) contains an extensive selection of designs for $4 \leq k \leq 10$. This pamphlet was prepared for the National Bureau of Standards and is the most complete table of fractional 3^{k-p} plans available.

In this section, we have noted several times the complexity of the alias relationships in 3^{k-p} fractional factorial designs. In general, if k is moderately large, say $k \geq 4$ or 5, the size of the 3^k design will drive most experimenters to consider fairly small fractions. These designs have alias relationships that involve the **partial aliasing** of two-degrees-of-freedom components of interaction. This, in turn, results in a design that can be difficult and in many cases impossible to interpret if interactions are not negligible. Furthermore,

■ **TABLE 9.8**

Alias Structure for the 3_{III}^{4-2} Design in Table 9.7

Effect	Aliases							
	I				I^2			
A	ABC^2	$ABCD$	ACD^2	AB^2D	BC^2	$AB^2C^2D^2$	CD^2	BD^2
B	AC	BC^2D^2	ABC^2D	AB^2D^2	ABC	CD	AB^2C^2D	AD^2
C	AB^2C^2	BC^2D	AD	$ABCD^2$	AB^2	BD	ACD	ABC^2D^2
D	AB^2CD	BCD^2	AC^2D^2	AB	AB^2CD^2	BC	AC^2	ABD

there are no simple augmentation schemes (such as fold over) that can be used to combine two or more fractions to isolate significant interactions. The 3^k design is often suggested as appropriate when curvature is present. However, more efficient alternatives (see Chapter 11) are possible.

9.4 Factorials with Mixed Levels

We have emphasized factorial and fractional factorial designs in which all the factors have the same number of levels. The two-level system discussed in Chapters 6, 7, and 8 is particularly useful. The three-level system presented earlier in this chapter is much less useful because the designs are relatively large even for a modest number of factors, and most of the small fractions have complex alias relationships that would require very restrictive assumptions regarding interactions to be useful.

It is our belief that the two-level factorial and fractional factorial designs should be the cornerstone of industrial experimentation for product and process development, troubleshooting, and improvement. In some situations, however, it is necessary to include a factor (or a few factors) that has more than two levels. This usually occurs when there are both quantitative and qualitative factors in the experiment, and the qualitative factor has (say) three levels. If all factors are quantitative, then two-level designs with center points should be employed. In this section, we show how some three- and four-level factors can be accommodated in a 2^k design.

9.4.1 Factors at Two and Three Levels

Occasionally, there is interest in a design that has some factors at two levels and some factors at three levels. If these are full factorials, then construction and analysis of these designs presents no new challenges. However, interest in these designs can occur when a fractional factorial design is being contemplated. If all of the factors are quantitative, mixed-level fractions are usually poor alternatives to a 2^{k-p} fractional factorial with center points. Usually when these designs are considered, the experimenter has a mix of qualitative and quantitative factors, with the qualitative factors taking on three levels. The complex aliasing we observed in the 3^{k-p} design with qualitative factors carries over to a great extent in the mixed-level fractional system. Thus, mixed-level fractional designs with all qualitative factors or a mix of qualitative and quantitative factors should be used very carefully. This section gives a brief discussion of some of these designs.

Designs in which some factors have two levels and other factors have three levels can be derived from the table of plus and minus signs for the usual 2^k design. The general procedure is best illustrated with an example. Suppose we have two variables, with A at two levels and X at three levels. Consider a table of plus and minus signs for the usual eight-run 2^3 design. The signs in columns B and C have the pattern shown on the left side of Table 9.9. Let

■ **TABLE 9.9**
Use of Two-Level Factors to Form a Three-Level Factor

Two-Level Factors		Three-Level Factor
B	C	X
−	−	x_1
+	−	x_2
−	+	x_2
+	+	x_3

■ **TABLE 9.10**
One 2-Level and One 3-Level Factor in a 2^3 Design

	A	X_L	X_L	$A \times X_L$	$A \times X_L$	X_Q	$A \times X_Q$	Actual Treatment Combinations	
Run	A	B	C	AB	AC	BC	ABC	A	X
1	−	−	−	+	+	+	−	Low	Low
2	+	−	−	−	−	+	+	High	Low
3	−	+	−	−	+	−	+	Low	Med
4	+	+	−	+	−	−	−	High	Med
5	−	−	+	+	−	−	+	Low	Med
6	+	−	+	−	+	−	−	High	Med
7	−	+	+	−	−	+	−	Low	High
8	+	+	+	+	+	+	+	High	High

the levels of X be represented by x_1, x_2, and x_3. The right side of Table 9.9 shows how the sign patterns for B and C are combined to form the levels of the three-level factor.

Now factor X has two degrees of freedom and if the factor is quantitative, it can be partitioned into a linear and a quadratic component, each component having one degree of freedom. Table 9.10 shows a 2^3 design with the columns labeled to show the actual effects that they estimate, with X_L and X_Q denoting the linear and quadratic effects of X, respectively. Note that the linear effect of X is the sum of the two effect estimates computed from the columns usually associated with B and C and that the effect of A can only be computed from the runs where X is at either the low or high levels, namely, runs 1, 2, 7, and 8. Similarly, the $A \times X_L$ effect is the sum of the two effects that would be computed from the columns usually labeled AB and AC. Furthermore, note that runs 3 and 5 are replicates. Therefore, a one-degree-of-freedom estimate of error can be made using these two runs. Similarly, runs 4 and 6 are replicates, and this would lead to a second one-degree-of-freedom estimate of error. The average variance at these two pairs of runs could be used as a mean square for error with two degrees of freedom. The complete analysis of variance is summarized in Table 9.11.

If we are willing to assume that the two-factor and higher interactions are negligible, we can convert the design in Table 9.10 into a resolution III fraction with up to four two-level factors and a single three-level factor. This would be accomplished by associating the two-level factors with columns A, AB, AC, and ABC. Column BC cannot be used for a two-level factor because it contains the quadratic effect of the three-level factor X.

■ **TABLE 9.11**
Analysis of Variance for the Design in Table 9.10

Source of Variation	Sum of Squares	Degrees of Freedom	Mean Square
A	SS_A	1	MS_A
X ($X_L + X_Q$)	SS_X	2	MS_{X_1}
AX ($A \times X_L + A \times X_Q$)	SS_{AX}	2	MS_{AX}
Error (from runs 3 and 5 and runs 4 and 6)	SS_E	2	MS_E
Total	SS_T	7	

■ **TABLE 9.12**
Four-Level Factor A Expressed as Two Two-Level Factors

	Two-Level Factors		Four-Level Factor
Run	***P***	***Q***	***A***
1	−	−	a_1
2	+	−	a_2
3	−	+	a_3
4	+	+	a_4

This same procedure can be applied to the 16-, 32-, and 64-run 2^k designs. For 16 runs, it is possible to construct resolution V fractional factorials with two two-level factors and either two or three factors at three levels. A 16-run resolution V fraction can also be obtained with three two-level factors and one three-level factor. If we include 4 two-level factors and a single three-level factor in 16 runs, the design will be of resolution III. The 32- and 64-run designs allow similar arrangements. For additional discussion of some of these designs, see Addelman (1962).

9.4.2 Factors at Two and Four Levels

It is very easy to accommodate a four-level factor in a 2^k design. The procedure for doing this involves using two two-level factors to represent the four-level factor. For example, suppose that A is a four-level factor with levels a_1, a_2, a_3, and a_4. Consider two columns of the usual table of plus and minus signs, say columns P and Q. The pattern of signs in these two columns is as shown on the left side of Table 9.12. The right side of this table shows how these four sign patterns would correspond to the four levels of factor A. The effects represented by columns P and Q and the PQ interaction are mutually orthogonal and correspond to the three-degrees-of-freedom A effect. This method of constructing a four-level factor from two two-level factors is called the method of **replacement**.

To illustrate this idea more completely, suppose that we have one four-level factor and two two-level factors and that we need to estimate all the main effects and interactions involving these factors. This can be done with a 16-run design. Table 9.13 shows the usual table of plus and minus signs for the 16-run 2^4 design, with columns A and B used to form the four-level factor, say X, with levels x_1, x_2, x_3, and x_4. Sums of squares would be calculated for each column $A, B, \ldots, ABCD$ just as in the usual 2^k system. Then the sums of squares for all factors X, C, D, and their interactions are formed as follows:

$$SS_X = SS_A + SS_B + SS_{AB} \quad \text{(3 degrees of freedom)}$$

$$SS_C = SS_C \quad \text{(1 degree of freedom)}$$

$$SS_D = SS_D \quad \text{(1 degree of freedom)}$$

$$SS_{CD} = SS_{CD} \quad \text{(1 degree of freedom)}$$

$$SS_{XC} = SS_{AC} + SS_{BC} + SS_{ABC} \quad \text{(3 degrees of freedom)}$$

$$SS_{XD} = SS_{AD} + SS_{BD} + SS_{ABD} \quad \text{(3 degrees of freedom)}$$

$$SS_{XCD} = SS_{ACD} + SS_{BCD} + SS_{ABCD} \quad \text{(3 degrees of freedom)}$$

■ **TABLE 9.13**
A Single Four-Level Factor and Two Two-Level Factors in 16 Runs

Run	(*A*	*B*)	= *X*	*C*	*D*	*AB*	*AC*	*BC*	*ABC*	*AD*	*BD*	*ABD*	*CD*	*ACD*	*BCD*	*ABCD*
1	−	−	x_1	−	−	+	+	+	−	+	+	−	+	−	−	+
2	+	−	x_2	−	−	−	−	+	+	−	+	+	+	+	−	−
3	−	+	x_3	−	−	−	+	−	+	+	−	+	+	−	+	−
4	+	+	x_4	−	−	+	−	−	−	−	−	−	+	+	+	+
5	−	−	x_1	+	−	+	−	−	+	+	+	−	−	+	+	−
6	+	−	x_2	+	−	−	+	−	−	−	+	+	−	−	+	+
7	−	+	x_3	+	−	−	−	+	−	+	−	+	−	+	−	+
8	+	+	x_4	+	−	+	+	+	+	−	−	−	−	−	−	−
9	−	−	x_1	−	+	+	+	+	−	−	−	+	−	+	+	−
10	+	−	x_2	−	+	−	−	+	+	+	−	−	−	−	+	+
11	−	+	x_3	−	+	−	+	−	+	−	+	−	−	+	−	+
12	+	+	x_4	−	+	+	−	−	−	+	+	+	−	−	−	−
13	−	−	x_1	+	+	+	−	−	+	−	−	+	+	−	−	+
14	+	−	x_2	+	+	−	+	−	−	+	−	−	+	+	−	−
15	−	+	x_3	+	+	−	−	+	−	−	+	−	+	−	+	−
16	+	+	x_4	+	+	+	+	+	+	+	+	+	+	+	+	+

This could be called a 4×2^2 design. If we are willing to ignore two-factor interactions, up to nine additional two-level factors can be associated with the two-factor interaction (except *AB*), three-factor interaction, and four-factor interaction columns.

There are a wide range of fractional factorial designs with a mix of two- and four-level factors available. However, we recommend using these designs cautiously. If all factors are quantitative, the 2^{k-p} system with center points will usually be a superior alternative. Designs with factors at two and four levels that are of resolution IV or higher, which would usually be necessary if there are both quantitative and qualitative factors present, and typically rather large, requiring $n \geq 32$ runs in many cases.

9.5 Nonregular Fractional Factorial Designs

The regular two-level fractional factorial designs in Chapter 8 are a staple for factor screening in modern industrial applications. Resolution IV designs are particularly popular because they avoid the confounding of main effects and two-factor interactions found in resolution III designs while avoiding the larger sample size requirements of resolution V designs. However, when the number of factors is relatively large, say $k = 9$ or more, resolution III designs are widely used. In Chapter 8 we discussed the regular minimum aberration versions of the 2^{k-p} fractional factorials of resolutions III, IV, and V.

The two-factor interaction aliasing in the resolution III and IV designs can result in experiments whose outcomes have ambiguous conclusions. For example, in Chapter 8 we illustrated a 2^{6-2} design used in a spin coating process applying photoresist where four main effects A, B, C, and E were found to be important along with one two-factor interaction alias chain AB + CE. Without external process knowledge, the experimenter could not decide whether the AB interaction, the CE interaction, or some linear combination of them represents the true state of nature.

To resolve this ambiguity requires additional runs. In Chapter 8 we illustrated the use of both a fold-over and a partial fold-over strategy to resolve the aliasing. We also saw an example of a resolution III 2^{7-4} fractional factorial in an eye focus time experiment where fold-over was required to identify a large two-factor interaction effect.

While strong two-factor interactions may be less likely than strong main effects, there are likely to be many more interactions than main effects in screening situations. As a result, the likelihood of at least one significant interaction effect is quite high. There is often substantial reluctance to commit additional time and material to a study with unclear results. Consequently, experimenters often want to avoid the need for a follow-up study. In this section we show how specific choices of nonregular two-level fractional factorial designs can be used in experiments with between 6 and 14 factors and potentially avoid subsequent experimentation when two-factor interactions are active. Section 9.5.1 presents designs for 6, 7, and 8 factors in 16 runs. These designs have no complete confounding of pairs of two-factor interactions. These designs are excellent alternatives for the regular minimum aberration resolution IV fractional factorials. In Section 9.5.2 we present nonregular designs for between 9 and 14 factors in 16 runs that have no complete aliasing of main effects and two-factor interactions. These designs are alternative to the regular minimum aberration resolution III fractional factorials. We also present metrics to evaluate these fractional factorial designs, show how the recommended nonregular 16-run designs were obtained, and discuss analysis methods.

Screening designs are primarily concerned with the discovery of active factors. This factor activity generally expresses itself through a main effect or a factor's involvement in a two-factor interaction. Consider the model

$$\mathbf{y} = \mathbf{X}\boldsymbol{\beta} + \boldsymbol{\varepsilon} \tag{9.7}$$

where $\mathbf{X}$ contains columns for the intercept, main effects and all two-factor interactions, $\boldsymbol{\beta}$ is the vector of model parameters, and $\boldsymbol{\varepsilon}$ is the usual vector of NID(0, σ^2) random errors. Consider the case of six factors in 16 runs and a model with all main effects and two-factor interactions. For this situation the $\mathbf{X}$ matrix has more columns than rows. Thus, it is not of full rank and the usual least squares estimate for $\boldsymbol{\beta}$ does not exist because the matrix $\mathbf{X'X}$ is singular. With respect to this model, every 16 run design is supersaturated. Booth and Cox (1962) introduced the $E(s^2)$ criterion as a diagnostic measure for comparing supersaturated designs, where

$$E(s^2) = \Sigma_{i<j}(\mathbf{X}_i'\mathbf{X}_j)^2/(k(k-1))$$

and k is the number of columns in $\mathbf{X}$.

Minimizing the $E(s^2)$ criterion is equivalent to minimizing the sum of squared off-diagonal elements of the correlation matrix of $\mathbf{X}$. Removing the constant column from $\mathbf{X}$, the correlation matrix of the regular resolution IV 16-run six-factor design is 21×21 with one row and column for each of the six main effects and 15 two-factor interactions. Figure 9.10 shows the **cell plot** of the **correlation matrix** for the principal fraction of this design. In Figure 9.10 we note that the correlation is zero between all main effects and two-factor interactions (because the design is resolution IV) and that the correlation is $+1$ between every two-factor interaction and at least one other two-factor interaction. These two-factor interactions are completely confounded. If another member of the same design family had been used, at least one of the generators would gave been used with a negative sign in design construction and some of the entries of the correlation matrix would have been -1. There still would be complete confounding of two-factor interactions in the design.

Jones and Montgomery (2010) introduced the cell plot of the correlation matrix as a useful graphical way to show the aliasing relationships in fractional factorials and to compare nonregular designs to their regular fractional factorial counterparts. In Figure 9.10 it is a display

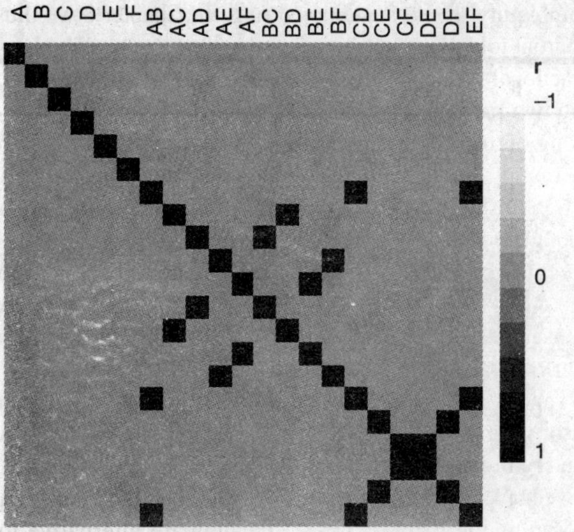

■ **FIGURE 9.10 The correlation matrix for the regular 2^{6-2} resolution IV fractional factorial design**

of the confounding pattern, much like what can be seen in the alias matrix. We introduced the alias matrix in Chapter 8. Recall that we plan to fit the model

$$\mathbf{y} = \mathbf{X}_1\boldsymbol{\beta}_1 + \boldsymbol{\varepsilon}$$

where $\mathbf{X}_1$ is the design matrix for the experiment that has been conducted expanded to model form, $\boldsymbol{\beta}_1$ is the vector of model parameters, and $\boldsymbol{\varepsilon}$ is the usual vector of NID(0, σ^2) errors but that the true model is

$$\mathbf{y} = \mathbf{X}_1\boldsymbol{\beta}_1 + \mathbf{X}_2\boldsymbol{\beta}_2 + \boldsymbol{\varepsilon}$$

where the columns of $\mathbf{X}_2$ contain additional factors not included in the original model (such as interactions) and $\boldsymbol{\beta}_2$ is the corresponding vector of model parameters. In Chapter 8 we observed that the expected value of $\hat{\boldsymbol{\beta}}_1$, the least squares estimate of $\boldsymbol{\beta}_1$, is

$$E(\hat{\boldsymbol{\beta}}_1) = \boldsymbol{\beta}_1 + (\mathbf{X}_1'\mathbf{X}_1)^{-1}\mathbf{X}'_1\mathbf{X}_2\boldsymbol{\beta}_2 = \boldsymbol{\beta}_1 + \mathbf{A}\boldsymbol{\beta}_2$$

The **alias matrix** $\mathbf{A} = (\mathbf{X}'_1\mathbf{X}_1)^{-1}\,\mathbf{X}'_1\,\mathbf{X}_2$ shows how estimates of terms in the fitted model are biased by active terms that are not in the fitted model. Each row of **A** is associated with a parameter in the fitted model. Nonzero elements in a row of **A** show the degree of biasing of the fitted model parameter due to terms associated with the columns of $\mathbf{X}_2$.

In a regular design, an arbitrary entry in the alias matrix, say A_{ij}, is either 0 or $\pm$ 1. If A_{ij} is 0 then the ith column of $\mathbf{X}_1$ is orthogonal to the jth column of $\mathbf{X}_2$. Otherwise if A_{ij} is $\pm$ 1, then the ith column of $\mathbf{X}_1$ and the jth column of $\mathbf{X}_2$ are perfectly correlated.

For nonregular designs, the aliasing is more complex. If $\mathbf{X}_1$ is the design matrix for the main effects model and $\mathbf{X}_2$ is the design matrix for the two-factor interactions, then the entries of the alias matrix for orthogonal nonregular designs for 16 runs take the values 0, $\pm$1, or $\pm$ 0.5. A small subset of these designs have no entries of $\pm$1.

Bursztyn and Steinberg (2006) propose using the trace of **AA′** (or equivalently the trace of **A′A**) as a scalar measure of the total bias in a design. They use this as a means for comparing designs for computer simulations but this measure works equally well for ranking competitive screening designs.

■ **TABLE 9.14**
The Hall I Design

Run	A	B	C	D	E	F	G	H	J	K	L	M	N	P	Q
1	−1	−1	1	−1	1	1	−1	−1	1	1	−1	1	−1	−1	1
2	1	−1	−1	−1	−1	1	1	−1	−1	1	1	1	1	−1	−1
3	−1	1	−1	−1	1	−1	1	−1	1	−1	1	1	−1	1	−1
4	1	1	1	−1	−1	−1	−1	−1	−1	−1	−1	1	1	1	1
5	−1	−1	1	1	−1	−1	1	−1	1	1	−1	−1	1	1	−1
6	1	−1	−1	1	1	−1	−1	−1	−1	1	1	−1	−1	1	1
7	−1	1	−1	1	−1	1	−1	−1	1	−1	1	−1	1	−1	1
8	1	1	1	1	1	1	1	−1	−1	−1	−1	−1	−1	−1	−1
9	−1	−1	1	−1	1	1	−1	1	−1	−1	1	−1	1	1	−1
10	1	−1	−1	−1	−1	1	1	1	1	−1	−1	−1	−1	1	1
11	−1	1	−1	−1	1	−1	1	1	−1	1	−1	−1	1	−1	1
12	1	1	1	−1	−1	−1	−1	1	1	1	1	−1	−1	−1	−1
13	−1	−1	1	1	−1	−1	1	1	−1	−1	1	1	−1	−1	1
14	1	−1	−1	1	1	−1	−1	1	1	−1	−1	1	1	−1	−1
15	−1	1	−1	1	−1	1	−1	1	−1	1	−1	1	−1	1	−1
16	1	1	1	1	1	1	1	1	1	1	1	1	1	1	1

9.5.1 Nonregular Fractional Factorial Designs for 6, 7, and 8 Factors in 16 Runs

These designs were introduced by Jones and Montgomery (2010) as alternatives to the usual regular minimum aberration fraction. Hall (1961) identified five **nonisomorphic** orthogonal designs for 15 factors in 16 runs. By nonisomorphic, we mean that one cannot obtain one of these designs from another one by permuting the rows or columns or by changing the labels of the factor. The Jones and Montgomery designs are projections of the Hall designs created by selecting the specific sets of columns that minimize the $E(s^2)$ and trace $\mathbf{AA'}$ criteria. They searched all of the non-isomorphic orthogonal projections of the Hall designs. Tables 9.14 through 9.18 show the Hall designs. Table 9.19 shows the number of nonisomorphic orthogonal 16 run designs.

■ **TABLE 9.15**
The Hall II Design

Run	A	B	C	D	E	F	G	H	J	K	L	M	N	P	Q
1	1	1	1	1	1	1	1	1	1	1	1	1	1	1	1
2	1	1	1	1	1	1	1	−1	−1	−1	−1	−1	−1	−1	−1
3	1	1	1	−1	−1	−1	−1	1	1	1	1	−1	−1	−1	−1
4	1	1	1	−1	−1	−1	−1	−1	−1	−1	−1	1	1	1	1
5	1	−1	−1	1	1	−1	−1	1	1	−1	−1	1	1	−1	−1
6	1	−1	−1	1	1	−1	1	−1	−1	1	1	−1	−1	1	1
7	1	−1	−1	−1	−1	1	1	1	1	−1	−1	−1	−1	1	1
8	1	−1	−1	−1	−1	1	1	−1	−1	1	1	1	1	−1	−1
9	−1	1	−1	1	−1	1	−1	1	−1	1	−1	1	−1	1	−1
10	−1	1	−1	1	−1	1	−1	−1	1	−1	1	−1	1	−1	1
11	−1	1	−1	−1	1	−1	1	1	−1	1	−1	−1	1	−1	1
12	−1	1	−1	−1	1	−1	1	−1	1	−1	1	1	−1	1	−1

13	−1	−1	1	1	−1	−1	1	1	−1	−1	1	−1	1	1	−1
14	−1	−1	1	1	−1	−1	1	−1	1	1	−1	1	−1	−1	1
15	−1	−1	1	−1	1	1	−1	1	−1	−1	1	1	−1	−1	1
16	−1	−1	1	−1	1	1	−1	−1	1	1	−1	−1	1	1	−1

■ **TABLE 9.16**
The Hall III Design

Run	A	B	C	D	E	F	G	H	J	K	L	M	N	P	Q
1	1	1	1	1	1	1	1	1	1	1	1	1	1	1	1
2	1	1	1	1	1	1	1	−1	−1	−1	−1	−1	−1	−1	−1
3	1	1	1	−1	−1	−1	−1	1	1	1	1	−1	−1	−1	−1
4	1	1	1	−1	−1	−1	−1	−1	−1	−1	−1	1	1	1	1
5	1	−1	−1	1	1	−1	−1	1	1	−1	−1	1	1	−1	−1
6	1	−1	−1	1	1	−1	−1	−1	−1	1	1	−1	−1	1	1
7	1	−1	−1	−1	−1	1	1	1	1	−1	−1	−1	−1	1	1
8	1	−1	−1	−1	−1	1	1	−1	−1	1	1	1	1	−1	−1
9	−1	1	−1	1	−1	1	−1	1	−1	1	−1	1	−1	1	−1
10	−1	1	−1	1	−1	1	−1	−1	1	−1	1	−1	1	−1	1
11	−1	1	−1	−1	1	−1	1	1	−1	−1	1	1	−1	−1	1
12	−1	1	−1	−1	1	−1	1	−1	1	1	−1	−1	1	1	−1
13	−1	−1	1	1	−1	−1	1	1	−1	−1	1	−1	1	1	−1
14	−1	−1	1	1	−1	−1	1	−1	1	1	−1	1	−1	−1	1
15	−1	−1	1	−1	1	1	−1	1	−1	1	−1	−1	1	−1	1
16	−1	−1	1	−1	1	1	−1	−1	1	−1	1	1	1	1	−1

■ **TABLE 9.17**
The Hall IV Design

Run	A	B	C	D	E	F	G	H	J	K	L	M	N	P	Q
1	1	1	1	1	1	1	1	1	1	1	1	1	1	1	1
2	1	1	1	1	1	1	1	−1	−1	−1	−1	−1	−1	−1	−1
3	1	1	1	−1	−1	−1	−1	1	1	1	1	−1	−1	−1	−1
4	1	1	1	−1	−1	−1	−1	−1	−1	−1	−1	1	1	1	1
5	1	−1	−1	1	1	−1	−1	1	1	−1	−1	1	1	−1	−1
6	1	−1	−1	1	1	−1	−1	−1	−1	1	1	−1	−1	1	1
7	1	−1	−1	−1	−1	1	1	1	1	−1	−1	−1	−1	1	1
8	1	−1	−1	−1	−1	1	1	−1	−1	1	1	1	1	−1	−1
9	−1	1	−1	1	−1	1	−1	1	−1	1	−1	1	−1	1	−1
10	−1	1	−1	1	−1	−1	1	1	−1	−1	1	−1	1	−1	1
11	−1	1	−1	−1	1	1	−1	1	1	−1	1	1	−1	−1	1
12	−1	1	−1	−1	1	−1	1	−1	1	1	−1	−1	1	1	−1
13	−1	−1	1	1	−1	1	−1	−1	1	−1	1	−1	1	1	−1
14	−1	−1	1	1	−1	−1	1	−1	1	1	−1	1	−1	−1	1
15	−1	−1	1	−1	1	1	−1	1	−1	1	−1	−1	1	−1	1
16	−1	−1	1	−1	1	−1	1	1	−1	−1	1	1	−1	1	−1

■ **TABLE 9.18**
The Hall V Design

Run	A	B	C	D	E	F	G	H	J	K	L	M	N	P	Q
1	1	1	1	1	1	1	1	1	1	1	1	1	1	1	1
2	1	1	1	1	1	1	1	−1	−1	−1	−1	1	−1	−1	−1
3	1	1	1	−1	−1	−1	−1	1	1	1	1	−1	−1	−1	−1
4	1	1	1	−1	−1	−1	−1	−1	−1	−1	−1	1	1	1	1
5	1	−1	−1	1	1	−1	−1	1	1	−1	−1	1	1	−1	−1
6	1	−1	−1	1	1	−1	−1	−1	−1	1	1	−1	−1	1	1
7	1	−1	−1	−1	−1	1	1	1	−1	1	−1	1	−1	1	−1
8	1	−1	−1	−1	−1	1	1	−1	1	−1	1	−1	1	−1	1
9	−1	1	−1	1	−1	1	−1	1	1	−1	−1	−1	−1	1	1
10	−1	1	−1	1	−1	1	−1	−1	−1	1	1	1	1	−1	−1
11	−1	1	−1	−1	1	−1	1	1	−1	−1	1	−1	1	1	−1
12	−1	1	−1	−1	1	−1	1	−1	1	1	−1	1	−1	−1	1
13	−1	−1	1	1	−1	−1	1	1	−1	1	−1	−1	1	−1	1
14	−1	−1	1	1	−1	−1	1	−1	1	−1	1	1	−1	1	−1
15	−1	−1	1	−1	1	1	−1	1	−1	−1	1	1	−1	−1	1
16	−1	−1	1	−1	1	1	−1	−1	1	1	−1	−1	1	1	−1

■ **TABLE 9.19**
Number of 16-Run Orthogonal Non-Isomorphic Designs

Number of Factors	Number of Designs
6	27
7	55
8	80

The nonregular designs that Jones and Montgomery recommended are shown in Tables 9.20, 9.21, and 9.22. The six-factor design in Table 9.20 is found from columns D, E, H, K, M, and Q of Hall II. The correlation matrix for this design along with the correlation matrix for the corresponding regular fraction is in Figure 9.11. Notice that like the regular 2^{6-2} design the design in Table 9.20 is first-order orthogonal but unlike the regular design, there are no two-factor interactions that are aliased with each other. All of the off-diagonal entries in the correlation matrix are either zero, -0.5, or $+0.5$. Because there is no complete confounding of two-factor interactions, Jones and Montgomery called this nonregular fraction a **no-confounding design**.

Table 9.21 presents the recommended seven-factor 16-run design. This design was constructed by selecting columns A, B, D, H, J, M, and Q from Hall III. The correlation matrix for this design and the regular 2^{7-3} fraction is shown in Figure 9.12. The no-confounding design is first-order orthogonal and there is no complete confounding of two-factor interactions. All off-diagonal elements of the correlation matrix are either zero, -0.5, or $+0.5$.

Table 9.22 presents the recommended eight-factor 16-run design. This design was constructed by choosing columns A, B, D, F, H, J, M, and O from Hall IV. The correlation matrix for this design and the regular 2^{8-4} fraction is shown in Figure 9.13. The no-confounding design is orthogonal for the first-order model and there is no complete confounding of two-factor interactions. All off-diagonal elements of the correlation matrix are either zero, -0.5, or $+0.5$.

■ **TABLE 9.20**

A Nonregular Orthogonal Design for $k = 6$ Factors in 16 Runs

Run	A	B	C	D	E	F
1	1	1	1	1	1	1
2	1	1	−1	−1	−1	−1
3	−1	−1	1	1	−1	−1
4	−1	−1	−1	−1	1	1
5	1	1	1	−1	1	−1
6	1	1	−1	1	−1	1
7	−1	−1	1	−1	−1	1
8	−1	−1	−1	1	1	−1
9	1	−1	1	1	1	−1
10	1	−1	−1	−1	−1	1
11	−1	1	1	1	−1	1
12	−1	1	−1	−1	1	−1
13	1	−1	1	−1	−1	−1
14	1	−1	−1	1	1	1
15	−1	1	1	−1	1	1
16	−1	1	−1	1	−1	−1

■ **TABLE 9.21**

A Nonregular Orthogonal Design for $k = 7$ Factors in 16 Runs

Run	A	B	C	D	E	F	G
1	1	1	1	1	1	1	1
2	1	1	1	−1	−1	−1	−1
3	1	1	−1	1	1	−1	−1
4	1	1	−1	−1	−1	1	1
5	1	−1	1	1	−1	1	−1
6	1	−1	1	−1	1	−1	1
7	1	−1	−1	1	−1	−1	1
8	1	−1	−1	−1	1	1	−1
9	−1	1	1	1	1	1	−1
10	−1	1	1	−1	−1	−1	1
11	−1	1	−1	1	−1	1	1
12	−1	1	−1	−1	1	−1	−1
13	−1	−1	1	1	−1	−1	−1
14	−1	−1	1	−1	1	1	1
15	−1	−1	−1	1	1	−1	1
16	−1	−1	−1	−1	−1	1	−1

Table 9.23 compares the popular minimum aberration resolution IV designs to the nonregular alternatives designs on the metrics described previously. As shown in the cell plots of the correlation matrices, the recommended designs outperform the minimum aberration designs for the number of confounded pairs of effects. They also are substantially better with

■ **TABLE 9.22**
A Nonregular Orthogonal Design for $k = 7$ Factors in 16 Runs

Run	A	B	C	D	E	F	G	H
1	1	1	1	1	1	1	1	1
2	1	1	1	1	−1	−1	−1	−1
3	1	1	−1	−1	1	1	−1	−1
4	1	1	−1	−1	−1	−1	1	1
5	1	−1	1	−1	1	−1	1	−1
6	1	−1	1	−1	−1	1	−1	1
7	1	−1	−1	1	1	−1	−1	1
8	1	−1	−1	1	−1	1	1	−1
9	−1	1	1	1	1	1	1	1
10	−1	1	1	−1	1	−1	−1	−1
11	−1	1	−1	1	−1	−1	1	−1
12	−1	1	−1	−1	−1	1	−1	1
13	−1	−1	1	1	−1	−1	−1	1
14	−1	−1	1	−1	−1	1	1	−1
15	−1	−1	−1	1	1	1	−1	−1
16	−1	−1	−1	−1	1	−1	1	1

■ **TABLE 9.23**
Design Comparison on Metrics

N Factors	Design	Confounded Effect Pairs	$E(s^2)$	Trace(AA')
6	Recommended	0	7.31	6
	Resolution IV	9	10.97	0
7	Recommended	0	10.16	6
	Resolution IV	21	14.20	0
8	Recommended	0	12.80	10.5
	Resolution IV	42	17.07	0

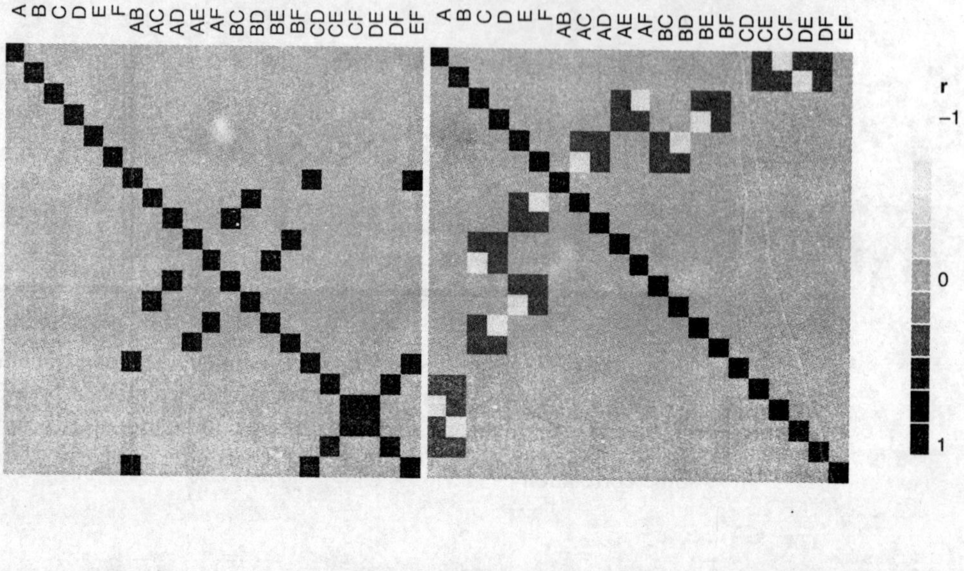

■ **FIGURE 9.11** Correlation matrix (a) regular 2^{6-2} fractional factorial (b) the non-regular no-confounding design

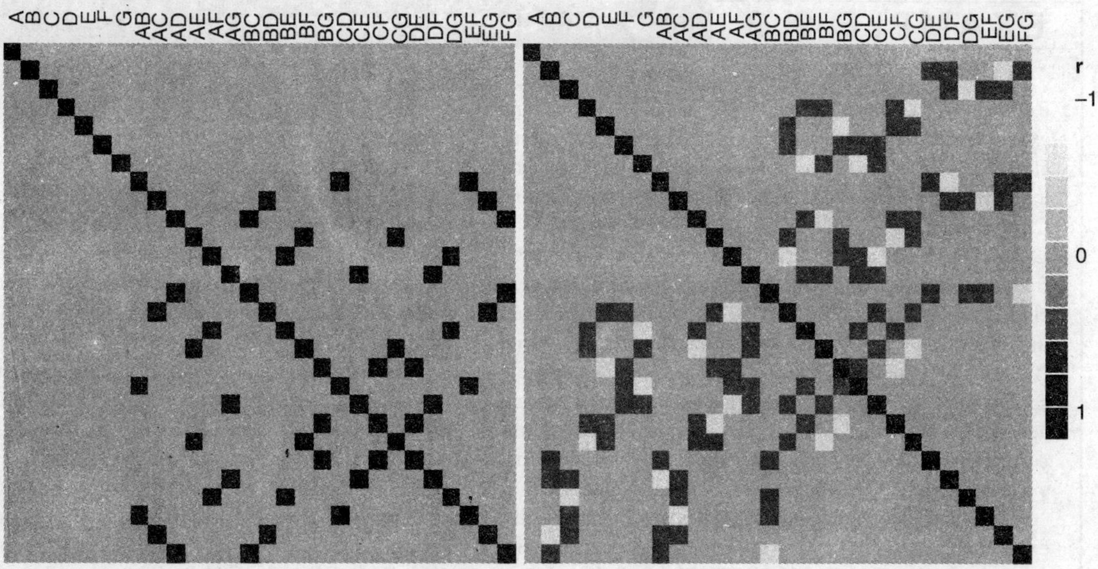

■ **FIGURE 9.12 Correlation matrix (a) Regular 2^{7-3} fractional factorial (b) the nonregular no-confounding design**

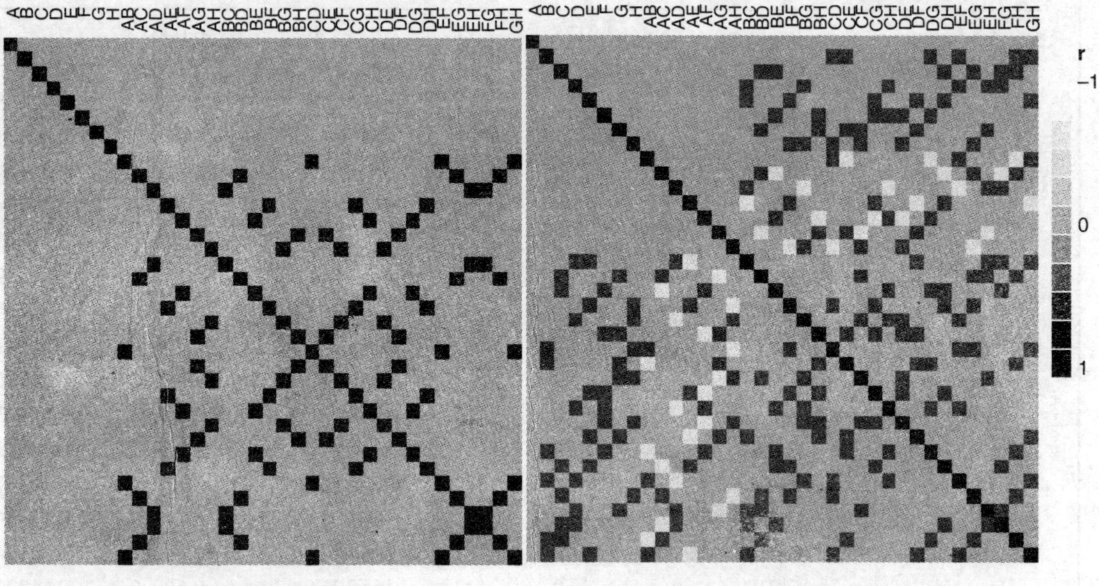

■ **FIGURE 9.13 Correlation matrix (a) regular 2^{8-4} fractional factorial (b) the nonregular no-confounding design**

respect to the $E(s^2)$ criterion. The recommended designs all achieve the minimum value of the trace criterion for all of the possible nonregular designs. The price that the Jones and Montgomery recommended designs pay for avoiding any pure confounding is that there is some correlation between main effects and two-factor interactions.

EXAMPLE 9.3 The Spin Coating Experiment

Recall from Chapter 8 (Section 8.7.2) the 2^{6-2} spin coating experiment that involved application of a photoresist material to silicon wafers. The response variable is thickness and the design factors are A = Speed RPM, B = acceleration, C = Volume, D = Time, E = Resist Viscosity, and F = Exhaust Rate. The design is the regular minimum aberration fraction. From the original analysis in Chapter 8, we concluded that the main effects of factors A, B, C, and E are important and that the two-factor interaction alias chain AB + CE is important. Because AB and CE are completely confounded, either additional information or assumptions are necessary to analytical ambiguity. A complete fold-over was performed to resolve this ambiguity and this additional experimentation indicated that the CE interaction was active.

Jones and Montgomery (2010) considered an alternative experimental design for this problem, the no-confounding six-variable design from Table 9.20. Table 9.24 presents this design with a set of simulated response data. In constructing the simulation, they assumed that the main effects that were important were A, B, C, and E, and that the CE interaction was the true source of the AB + CE effect observed in the actual study. They added normal random noise in the simulated data to match the RMSE of the fitted model in the original data. They also matched the model parameter estimates to those from the original experiment. The intent is to create a fair realization of the data that might have been observed if the no-confounding design had been used.

Jones and Montgomery analyzed this experiment using forward stepwise regression with all main effect and two-factor interactions as candidate effects. The reason that all two-factor interactions can be considered as candidate effects is that none of these interactions are completely confounded. The JMP stepwise regression output is in Figure 9.14. Stepwise regression selects the main effects of A, B, C, E, along with the CE interaction.

The no-confounding design correctly identifies the model unambiguously and without requiring additional runs.

Lock	Entered	Parameter	Estimate	nDF	SS	"F Ratio"	"Prob>F"
×	×	Intercept	4462.8125	1	0	0.000	1
	×	A	85.3125	1	77634.37	53.976	2.46e-5
	×	B	−77.6825	1	64368.76	44.753	5.43e-5
	×	C	−34.1875	2	42735.84	14.856	0.00101
		D	0	1	31.19857	0.020	0.89184
	×	E	21.5625	2	31474.34	10.941	0.00304
		F	0	1	2024.045	1.474	0.25562
		A*B	0	1	395.8518	0.255	0.6259
		A*C	0	1	476.1781	0.308	0.59234
		A*D	0	2	3601.749	1.336	0.31571
		A*E	0	1	119.4661	0.075	0.78986
		A*F	0	2	4961.283	2.106	0.18413
		B*C	0	1	60.91511	0.038	0.84923
		B*D	0	2	938.8809	0.279	0.76337
		B*E	0	1	3677.931	3.092	0.11254
		B*F	0	2	2044.119	0.663	0.54164
		C*D	0	2	1655.264	0.520	0.61321
	×	C*E	54.8125	1	24035.28	16.711	0.00219
		C*F	0	2	2072.497	0.673	0.53667
		D*E	0	2	79.65054	0.022	0.97803
		D*F	0	0	0	.	.
		E*F	0	2	5511.275	2.485	0.14476

■ **FIGURE 9.14** JMP stepwise regression output for the no-confounding design in Table 9.24

■ **TABLE 9.24**
The No-Confounding Design for the Photoresist Application Experiment

Run	A	B	C	D	E	F	Thickness
1	1	1	1	1	1	1	4494
2	1	1	−1	−1	−1	−1	4592
3	−1	−1	1	1	−1	−1	4357
4	−1	−1	−1	−1	1	1	4489
5	1	1	1	−1	1	−1	4513
6	1	1	−1	1	−1	1	4483
7	−1	−1	1	−1	−1	1	4288
8	−1	−1	−1	1	1	−1	4448
9	1	−1	1	1	1	−1	4691
10	1	−1	−1	−1	−1	1	4671
11	−1	1	1	1	−1	1	4219
12	−1	1	−1	−1	1	−1	4271
13	1	−1	1	−1	−1	−1	4530
14	1	−1	−1	1	1	1	4632
15	−1	1	1	−1	1	1	4337
16	−1	1	−1	1	−1	−1	4391

9.5.2 Nonregular Fractional Factorial Designs for 9 Through 14 Factors in 16 Runs

Resolution III fractional factorial designs are popular for factor screening problems in situations where there are a moderate-to-large number of factors because these designs contain a relatively small number of runs and they are effective in identifying the unimportant factors and elevating potentially important factors for further experimentation. Designs in 16 runs are extremely popular because the number of runs is usually within the resources available to most experimenters.

Because the regular resolution III designs alias main effects and two-factor interactions, and the aliased effects are completely confounded, experimenters often end up with ambiguous conclusions about which main effects and two-factor interactions are important. Resolving these ambiguities requires either additional experimentation (such as use of a fold-over design to augment the original fraction) or assumptions about which effects are important or external process knowledge. This is very similar to the situation encountered in the previous section, except now main effects are completely confounded with two-factor interactions. Just as in that section, it is possible to develop no-confounding designs for 9–14 factors in 16 runs that are good alternatives to the usual minimum aberration resolution III designs when there are only a few main effects and two-factor interactions that are important. Table 9.25 is an extension of Table 9.19, showing all possible nonisomorphic nonregular 16 run designs with from 6 to 15 factors. The recommended designs in Tables 9.26 through 9.31 are chosen from the designs in this table. They are projections of the Hall designs. The correlation matrices of the designs are shown in Figures 9.15 through 9.20. All recommended designs are first-order orthogonal (100% D-efficient) and the correlations between main effects and two-factor interactions are $\pm .5$.

■ **TABLE 9.25**
Number of Non-Isomorphic Nonregular 16-Run Designs

Number of Factors	Number of Non-Isomorphic Designs
6	27
7	55
8	80
9	87
10	78
11	58
12	36
13	18
14	10
15	5

■ **TABLE 9.26**
Recommended 16-Run 9-Factor No-Confounding Design

Run	A	B	C	D	E	F	G	H	J
1	−1	−1	−1	−1	−1	−1	1	−1	1
2	−1	−1	−1	1	−1	1	−1	1	−1
3	−1	−1	1	−1	1	1	1	1	−1
4	−1	−1	1	1	1	−1	−1	−1	1
5	−1	1	−1	−1	1	1	−1	1	1
6	−1	1	−1	1	1	−1	1	−1	−1
7	−1	1	1	−1	−1	−1	−1	1	−1
8	−1	1	1	1	−1	1	1	−1	1
9	1	−1	−1	−1	1	−1	−1	−1	−1
10	1	−1	−1	1	1	1	1	1	1
11	1	−1	1	−1	−1	1	−1	−1	1
12	1	−1	1	1	−1	−1	1	1	−1
13	1	1	−1	−1	−1	1	1	−1	−1
14	1	1	−1	1	−1	−1	−1	1	1
15	1	1	1	−1	1	−1	1	1	1
16	1	1	1	1	1	1	−1	−1	−1

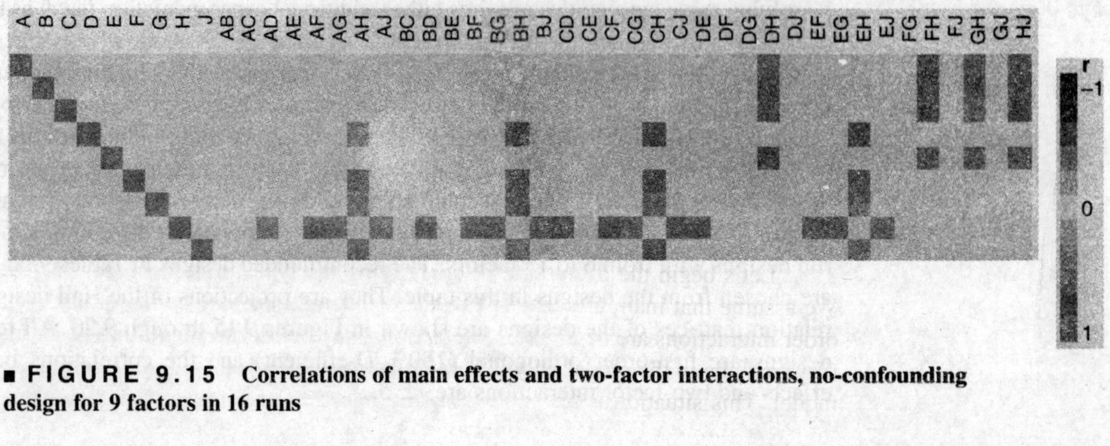

■ **FIGURE 9.15** **Correlations of main effects and two-factor interactions, no-confounding design for 9 factors in 16 runs**

■ **TABLE 9.27**

Recommended 16-Run 10-Factor No-Confounding Design

Run	A	B	C	D	E	F	G	H	J	K
1	−1	−1	−1	−1	1	−1	−1	1	−1	1
2	−1	−1	−1	1	1	1	−1	−1	1	1
3	−1	−1	1	−1	−1	1	1	1	1	1
4	−1	−1	1	−1	1	−1	1	−1	1	−1
5	−1	1	−1	1	−1	1	1	1	−1	1
6	−1	1	−1	1	1	−1	1	−1	−1	−1
7	−1	1	1	−1	−1	−1	−1	1	−1	−1
8	−1	1	1	1	−1	1	−1	−1	1	−1
9	1	−1	−1	−1	−1	1	1	−1	−1	−1
10	1	−1	−1	1	−1	−1	−1	1	1	−1
11	1	−1	1	1	−1	−1	1	−1	−1	1
12	1	−1	1	1	1	1	−1	1	−1	−1
13	1	1	−1	−1	−1	−1	−1	−1	1	1
14	1	1	−1	−1	1	1	1	1	1	−1
15	1	1	1	−1	1	1	−1	−1	−1	1
16	1	1	1	1	1	−1	1	1	1	1

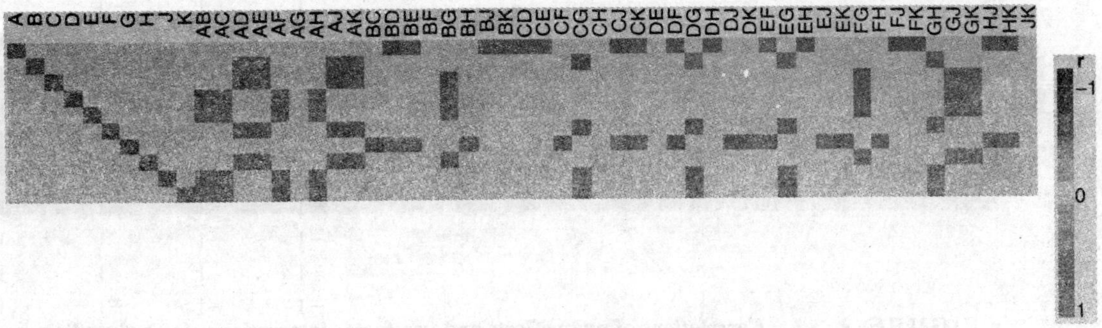

■ **FIGURE 9.16** **Correlations of main effects and two-factor interactions, no-confounding design for 10 factors in 16 runs**

9.5.3 Analysis of Nonregular Fractional Factorial Designs

In Section 9.5.1 we illustrated the use of forward selection regression to analyze a nonregular design, the 16-run no-confounding design with $k = 6$ factors. This approach was very successful as the correct model was identified. Generally, forward selection regression is a very useful approach for analyzing nonregular designs. There are variations of the procedure that are useful in some situations.

Let's begin the discussion by identifying the types of models that may be of interest. We assume that main effects and two-factor interactions may be important, and that higher-order interactions are negligible. Interactions may be **hierarchical**; that is, an interaction AB (say) may be in the model only if both of the main effects (here both A and B) are also in the model. This situation is also called **strong heredity** and it occurs frequently in practice so

■ **TABLE 9.28**
Recommended 16-Run 11-Factor No-Confounding Design

Run	A	B	C	D	E	F	G	H	J	K	L
1	−1	−1	−1	1	1	−1	−1	−1	1	−1	1
2	−1	−1	1	−1	−1	−1	1	−1	1	−1	−1
3	−1	−1	1	−1	1	1	−1	1	−1	−1	−1
4	−1	−1	1	1	−1	1	1	1	−1	1	1
5	−1	1	−1	−1	−1	−1	−1	−1	−1	1	1
6	−1	1	−1	1	−1	1	1	−1	−1	−1	−1
7	−1	1	−1	1	1	1	−1	1	1	1	−1
8	−1	1	1	−1	1	−1	1	1	1	1	1
9	1	−1	−1	−1	−1	1	−1	1	1	−1	1
10	1	−1	−1	−1	1	1	1	−1	−1	1	1
11	1	−1	−1	1	−1	−1	1	1	1	1	−1
12	1	−1	1	1	1	−1	−1	−1	−1	1	−1
13	1	1	−1	−1	1	−1	1	1	−1	−1	−1
14	1	1	1	−1	−1	1	−1	−1	1	1	−1
15	1	1	1	1	−1	−1	−1	1	−1	−1	1
16	1	1	1	1	1	1	1	−1	1	−1	1

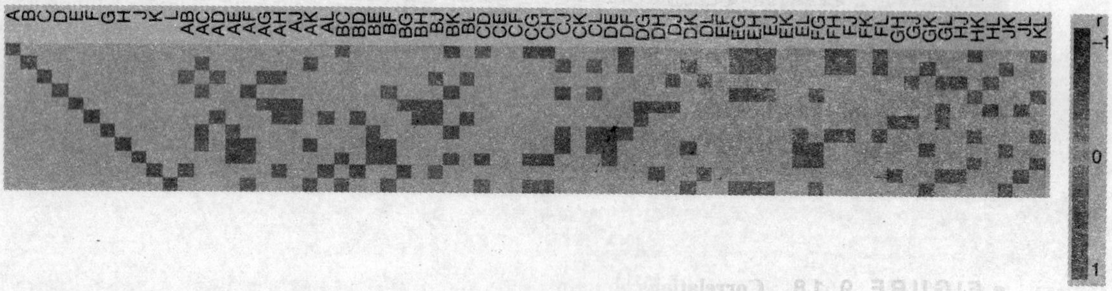

■ **FIGURE 9.17** **Correlations of main effects and two-factor interactions, no-confounding design for 11 factors in 16 runs**

assuming hierarchy (strong heredity) in analyzing data from a nonregular design is usually not a bad assumption. Interactions may also obey only the **weak heredity** principle; this means that AB can be in the model if either A or B is in the model. This situation also occurs fairly often, although not as often as hierarchy, but ignoring this possibility could result in the experimenter failing to identify all of the large interactions effects. Finally, there can be situations where an interaction such as AB is active but neither main effect A or B is active. This case is relatively uncommon.

There are several variations of forward selection regression that are useful. They are briefly described as follows:

1. Use forward selection without concern about model hierarchy. This can result in including too many interaction terms.

■ **TABLE 9.29**
Recommended 16-Run 12-Factor No-Confounding Design

Run	A	B	C	D	E	F	G	H	J	K	L	M
1	−1	−1	−1	−1	1	−1	−1	1	1	−1	1	1
2	−1	−1	−1	1	−1	1	1	1	−1	−1	1	−1
3	−1	−1	1	−1	−1	−1	1	−1	1	1	−1	1
4	−1	−1	1	1	1	1	−1	−1	−1	1	−1	−1
5	−1	1	−1	1	−1	−1	−1	−1	−1	−1	−1	1
6	−1	1	−1	1	1	1	1	−1	1	1	1	1
7	−1	1	1	−1	−1	1	−1	1	1	−1	−1	−1
8	−1	1	1	−1	1	−1	1	1	−1	1	1	−1
9	1	−1	−1	−1	−1	1	−1	−1	1	1	1	−1
10	1	−1	−1	−1	1	−1	1	−1	−1	−1	−1	−1
11	1	−1	1	1	−1	−1	−1	1	−1	1	1	1
12	1	−1	1	1	1	1	1	1	1	−1	−1	1
13	1	1	−1	−1	−1	1	1	1	−1	1	−1	1
14	1	1	−1	1	1	−1	−1	1	1	1	−1	−1
15	1	1	1	−1	1	1	−1	−1	−1	−1	1	1
16	1	1	1	1	−1	−1	1	−1	1	−1	1	−1

■ **FIGURE 9.18** **Correlations of main effects and two-factor interactions, no-confounding design for 12 factors in 16 runs**

2. Use forward selection restricted to hierarchy. This means that if AB is selected for entry, then the entire group of terms A, B, and AB are entered in the model if A and B are not already included.
3. Consider using larger than "usual" P-values for entering factors. Many stepwise regression computer programs have "default" values for entering factors, such as $P = 0.05$. This may be too restrictive. Values of 0.10 or even 0.15 may work better. The big danger in screening designs is not identifying important effects (type II errors), so type I errors are usually not of too much concern.
4. Use forward selection in two steps. First, select terms from all the main effects. Then run forward selection a second time using all two-factor interactions that satisfy the weak heredity assumption based on the main effects identified in step 1.
5. You could also include any two-factor interactions that experience or process knowledge suggests should be considered.

■ **TABLE 9.30**
Recommended 16-Run 13-Factor No-Confounding Design

Run	A	B	C	D	E	F	G	H	J	K	L	M	N
1	−1	−1	−1	1	1	−1	−1	1	−1	1	1	−1	1
2	−1	−1	1	−1	−1	−1	−1	−1	1	1	−1	−1	1
3	−1	−1	1	−1	1	1	1	1	1	−1	1	−1	−1
4	−1	−1	1	1	−1	1	1	1	−1	1	−1	1	−1
5	−1	1	−1	−1	−1	1	−1	−1	−1	−1	1	−1	−1
6	−1	1	−1	−1	1	1	1	−1	−1	1	−1	1	1
7	−1	1	−1	1	−1	−1	1	1	1	−1	1	1	1
8	−1	1	1	1	1	−1	−1	−1	1	−1	−1	1	−1
9	1	−1	−1	−1	−1	−1	1	−1	1	1	1	1	−1
10	1	−1	−1	−1	1	−1	−1	1	−1	−1	−1	1	−1
11	1	−1	−1	1	1	1	1	−1	1	−1	−1	−1	1
12	1	−1	1	1	−1	1	−1	−1	−1	−1	1	1	1
13	1	1	−1	1	−1	1	−1	1	1	1	−1	−1	−1
14	1	1	1	−1	−1	−1	1	1	−1	−1	−1	−1	1
15	1	1	1	−1	1	1	−1	1	1	1	1	1	1
16	1	1	1	1	1	−1	1	−1	−1	1	1	−1	−1

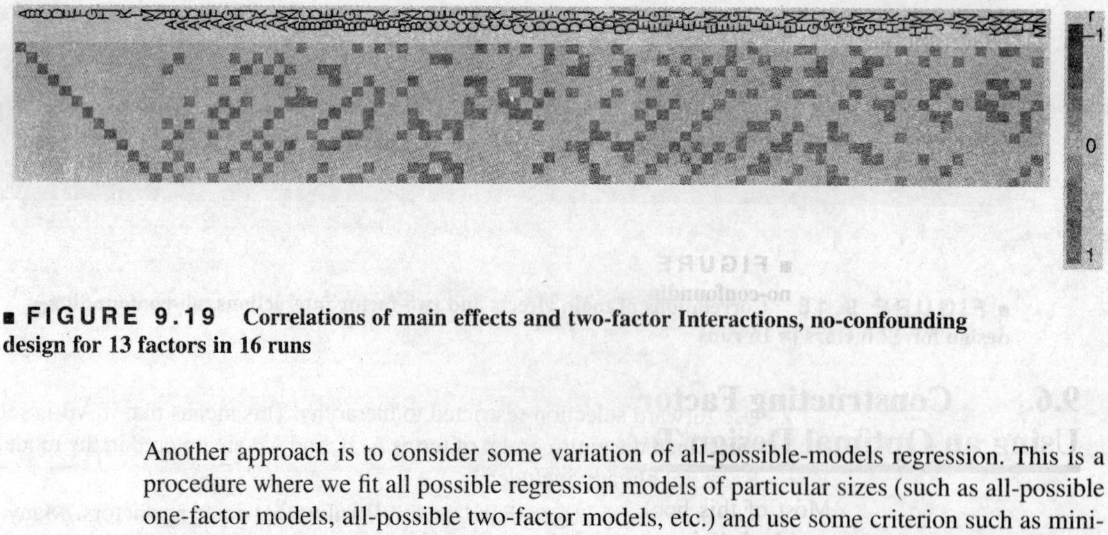

■ **FIGURE 9.19 Correlations of main effects and two-factor Interactions, no-confounding design for 13 factors in 16 runs**

Another approach is to consider some variation of all-possible-models regression. This is a procedure where we fit all possible regression models of particular sizes (such as all-possible one-factor models, all-possible two-factor models, etc.) and use some criterion such as minimum mean square error, or restrict attention to models with either strong or weak heredity, or models associated with a large increase in adjusted R^2, to narrow down the set of possible models for further consideration.

In some cases, nonregular designs have useful projection properties, and this could suggest an appropriate analysis. For example, the 12-run Plackett–Burman design will support a model will all main effects and all two-factor interactions in any $k = 4$ factors. So if up to four main effects appear large, we could analyze this design simply by fitting the main effects plus two-factor interaction model to the four apparently active effects. In such situations, it still may be useful to consider other possible interaction terms for inclusion in the final model.

■ **TABLE 9.31**
Recommended 16-Run 14-Factor No-Confounding Design

Run	A	B	C	D	E	F	G	H	J	K	L	M	N	P
1	−1	−1	−1	−1	1	−1	1	−1	1	1	−1	−1	1	
2	−1	−1	−1	1	−1	−1	1	−1	1	1	−1	1	1	−1
3	−1	−1	1	−1	−1	1	−1	1	1	1	−1	−1	1	1
4	−1	−1	1	1	1	1	1	−1	−1	−1	1	−1	1	−1
5	−1	1	−1	−1	−1	1	1	−1	1	−1	1	1	−1	1
6	−1	1	−1	1	1	1	−1	1	−1	1	−1	1	−1	−1
7	−1	1	1	−1	−1	−1	−1	1	−1	−1	1	1	1	−1
8	−1	1	1	1	1	−1	−1	−1	1	−1	−1	−1	−1	1
9	1	−1	−1	−1	1	1	−1	−1	−1	−1	−1	1	1	1
10	1	−1	−1	1	−1	1	−1	1	1	−1	1	−1	−1	−1
11	1	−1	1	−1	1	−1	−1	−1	1	1	1	1	−1	−1
12	1	−1	1	1	−1	−1	1	1	−1	−1	−1	1	−1	1
13	1	1	−1	−1	1	−1	1	1	1	−1	−1	−1	1	−1
14	1	1	−1	1	−1	−1	−1	−1	−1	1	1	−1	1	1
15	1	1	1	−1	−1	1	1	−1	−1	1	−1	−1	−1	−1
16	1	1	1	1	1	1	1	1	1	1	1	1	1	1

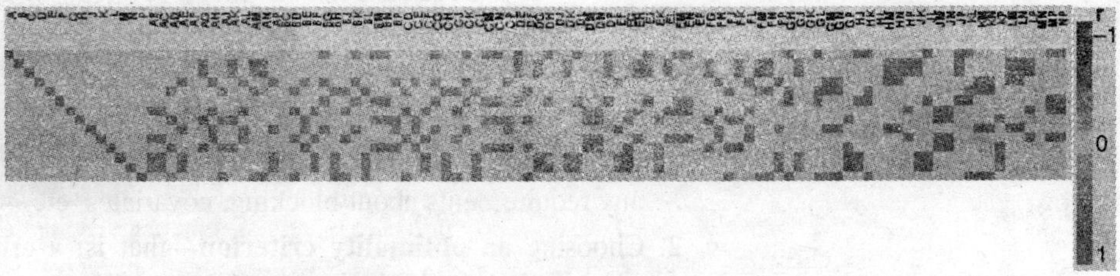

■ **FIGURE 9.20** **Correlations of main effects and two-factor interactions, no-confounding design for 14 factors in 16 runs**

9.6. Constructing Factorial and Fractional Factorial Designs Using an Optimal Design Tool

Most of this book has focused on standard factorial and fractional factorial design. These standard designs work well when the experimental research problem and the design are a good match. But there are many situations where the requirements of a standard design and the research problem are not a good fit. Some of these include:

1. The experimenter has unusual resource restrictions, so either the number of runs that can be made in the experiment or the size of the blocks required are different from the sample size and/or block sizes required by a standard design. We will see an example of this situation in Chapter 11.
2. There are restrictions or constraints on the design region. That is, the standard cuboidal regions for factorial and fractional factorial designs and spherical or cuboidal regions for response surface designs are not appropriate either because it

is impossible to experiment in some portions of the factor space (such as temperatures and pressures that are simultaneously beyond certain boundaries leading to unsafe operating conditions) or there are infeasible combinations of some factors. An example of an optimal design for a problem with a constrained region will be given in Chapter 11.

3. The experimenter needs to fit a nonstandard model. Models containing a mix of factors of different types. For example, suppose that the experimenter is interested in fitting a full quadratic model in two variables x_1 and x_2, but there is a third two-level categorical factor z that is also of interest. The model that the experimenter wants to entertain is

$$y = \beta_0 + \beta_1 x_1 + \beta_2 x_2 + \beta_{12} x_1 x_2 + \beta_{11} x_1^2 + \beta_{22} x_2^2 + yz + \delta_1 z x_1 + \delta_2 z x_2$$
$$+\delta_{12} z x_1 x_2 + \delta_{11} z x_1^2 + \delta_{22} z x_2^2 + \varepsilon$$

This is a full quadratic model in the two continuous factors and it also contains the main effect of the categorical factor plus all interactions between the categorical factor and the linear, interaction, and pure quadratic effects of the continuous factors. If this full 12-parameter model is the final model for the experiment, then the model describes two completely different response functions at the two different levels of the categorical factor. Assuming that the experimenter can only conduct 15 runs, there is not a standard response surface design for this problem. The closest standard design that would work for this problem would be the 3 × 3 × 2 factorial, which requires 18 runs.

Designing experiments for these types of problems requires a different approach. We can't look in the textbook or course notes and try to match the designs we find there to the problem. Instead we need to create a **custom design** that fits our specific problem. Creating this custom design requires:

1. Information about the problem—specifically the model that the experimenter wants to entertain, the region of interest, the number of runs that can be performed, and any requirements about blocking, covariates, etc.
2. Choosing an **optimality criterion**—that is, a criterion for selecting the design points to be run. In the next section we will give a brief review of optimality criterion for design experiments.
3. A software package to construct the design. Sometimes optimal designs are called **computer-generated** designs. Several standard software packages do a good job of finding optimal designs.

It is always better to create a custom design for the actual problem that you want to solve than to force your problem to fit a standard design. Fortunately, it has been relatively easy to construct optimal design for about the last 15 years. The early research work on the theory of design optimality began with Kiefer (1959, 1961) and Kiefer and Wolfowitz (1959). The first practical algorithm for construction of optimal designs was developed by Mitchell (1974). This was a point exchange method, in which runs from a candidate set of all possible runs that the experimenter would consider running were systematically exchanged with the runs in a current design until no further improvement in the optimality criterion could be achieved. Several variations of the point exchange approach were developed and implemented over the next 20 years. Meyer and Nachtsheim (1995) developed a coordinate exchange algorithm in which individual design coordinates were systematically searched to find the optimal settings. No candidate set of runs was required. This approach quickly became the standard one and today almost all efficient optimal design software makes use of the coordinate exchange approach.

9.6.1 Design Optimality Criterion

In Chapter 6 (Section 6.7) we introduced three design optimality criteria—***D*-optimality**, ***G*-optimality**, and ***I*-optimality**. The ***D*-optimal design** employs a criterion on the selection of design points that results in the minimization of the volume of the joint confidence region of the regression coefficients. This is achieved by maximizing the determinant (hence, the "*D*" in *D*-optimal) of the $\mathbf{X'X}$ matrix. That is, the quantity $|\mathbf{X'X}|$ is maximized over all possible designs with *N* runs. The covariance or dispersion matrix, $(\mathbf{X'X})^{-1}$, contains the variances and covariance of the regression coefficients and the square of the volume of the confidence region is inversely proportional to $|\mathbf{X'X}|$. Controlling the volume of the confidence region is related to the precision of the regression coefficients; a smaller confidence region, for the same level of confidence, means more precise estimates. The **G-optimal design** minimizes the maximum value of prediction variance in the design region *R*. The *I*-optimal design minimizes the integrated or average prediction variance of the regression model over the design region *R*.

The *D*-optimal and *I*-optimal are the two most widely used. Experimental designs that are created with respect to both *D*-optimal and *I*-optimal criteria are available in many commercially available software packages. Creating these designs requires an optimization algorithm. Techniques such as the coordinate exchange method of Meyer and Nachtsheim (1995) have been developed that minimize the computational burden and reduce the time required to find the optimal design. These techniques do not always guarantee a global optimal, but the efficiency, a metric that quantifies the quality of an optimal design, in terms of the best possible design is reported by the software programs.

9.6.2 Examples of Optimal Designs

Many of the standard designs this book are optimal designs. To illustrate, recall the 2^k factorial design. The 2^k factorial and its many variants are probably the most widely used family of designs in industrial research and development. In Section 6.7 of Chapter 6 we showed that these designs are optimal designs with respect to the *D, G,* and *I* criteria.

The fact that many widely used standard designs are optimal designs suggests that the optimal design approach is applicable in any design situation. If the problem turns out to be a standard one, an optimal design algorithm will generate the required standard design. But if not, then the optimal design approach will be necessary to construct the appropriate design for this specific research problem.

As an example, consider a situation in which there are two categorical factors with three levels each. A scenario where this situation might occur is in missile testing. For example, assume that the White Sands missile range testing center wants to compare distance from target for three different types of missiles (x_1=L1, L2, L3), each containing a slightly different metal alloy encasing, and three different launching mechanisms (x_2 = L1, L2, L3). Using this example, both *D*-optimal and *I*-optimal designs can be created for the main effect only model and main effects plus two-factor interaction model. Let us assume that the experimenter is interested in a 9-run design. The *D*-optimal and *I*–optimal designs for the main effects only model were found using JMP and are presented in Tables 9.32 and 9.33, respectively. Notice that both designs are identical; in fact, they are both 3^2 factorial designs. If we augment the model to include the two-factor interaction term and construct 9-run *D*-optimal and *I*-optimal designs we get the same results.

Both of the designs in Tables 9.32 and 9.33 are unreplicated factorials. Usually the experimenter would like to replicate the design in order to obtain an estimate of experimental error that would support statistical testing. If the design is replicated twice, this would require a total of 18 runs. We would still have a standard design. However, since each run requires firing a missile, and these runs are likely very expensive, the experimenter would probably be interested in a design with fewer runs. Suppose that the experimenter wants to fit

■ **TABLE 9.32**
The 9-Run *D*-Optimal Design for Two 3-Level Categorical Factors

Run	X1	X2
1	L1	L1
2	L2	L1
3	L3	L1
4	L1	L2
5	L2	L2
6	L3	L2
7	L1	L3
8	L2	L3
9	L3	L3

■ **TABLE 9.33**
The 9-Run *I*-Optimal Design for Two 3-Level Categorical Factors

Run	X1	X2
1	L1	L1
2	L2	L1
3	L3	L1
4	L1	L2
5	L2	L2
6	L3	L2
7	L1	L3
8	L2	L3
9	L3	L3

the main effects plus interaction model and can afford a total of 12 runs. Where should these replicate runs made? An optimal design approach can be used to determine the best place to allocate replicate runs to an existing design.

Tables 9.34 and 9.35 present the *D*-optimal and *I*-optimal designs obtained from JMP, respectively. From inspection of the tables we see that both designs are full 3^2 full factorials with three replicated runs.

The *D*-optimal design replicates the treatment combinations (L1, L2), (L2, L2), and (L3, L2). Thus, L2 appears six times in column X2 while L1 and L3 appear only three times. By contrast, the *I*-optimal design replicates combinations (L1, L1), (L2, L2), and (L3, L2). In column X2 of this design, L1 appears three times, L2 appears five times, and L3 appears four times. The only difference between the two designs is that one of the replicated runs is different in each design. The average scaled prediction variance is 0.833 for both designs.

■ **TABLE 9.34**
The 12-Run *D*-Optimal Design for Two 3-Level Categorical Factors

Run	X1	X2
1	L1	L1
2	L2	L1
3	L3	L1
4	L1	L2
5	L1	L2
6	L2	L2
7	L2	L2
8	L3	L2
9	L3	L2
10	L1	L3
11	L2	L3
12	L3	L3

■ **TABLE 9.35**
The 12-Run *I*-Optimal Design for Two 3-Level Categorical Factors

Run	X1	X2
1	L3	L3
2	L2	L2
3	L2	L1
4	L1	L1
5	L3	L1
6	L3	L2
7	L2	L3
8	L3	L2
9	L1	L3
10	L2	L2
11	L1	L1
12	L1	L2

EXAMPLE 9.4 An Experiment with Unusual Blocking Requirements

Suppose an investigator wishes to run a screening experiment with six continuous factors and can perform three runs in one day. The budget for the experiment allows for 12 runs. So, the experiment will involve four days of experimentation with three runs per day. There could be significant day-to-day variation, so days should be treated as block.

The 12-run Plackett–Burman design is a natural choice for a main effects model but there is no blocking scheme for these designs that accommodates blocks of three runs. So, no textbook design quite matches this problem description.

Table 9.36 shows the factor settings and block assignments for the 12-run *I*-optimal design for a main effects model tailored to fit the problem. This design has a *D*-efficiency of 95.4 percent. The most notable feature of this design is that each factor has one setting at the middle of its range. This means that if any factor has a strong quadratic effect, there is a good chance both of detecting the curvature and identifying the active factor. By contrast, a two-level design has no way to detect strong curvature if it exists without adding center points.

■ **TABLE 9.36**
An *I*-Optimal Design for 6 Factors in 4 Blocks of Size 3

A	B	C	D	E	F	Day
−1	−1	−1	−1	1	1	1
1	1	1	−1	1	−1	1
1	−1	1	1	−1	1	1
0	−1	1	−1	1	1	2
−1	1	−1	1	−1	−1	2
1	−1	1	0	1	0	2
1	−1	−1	1	1	−1	3
−1	−1	1	−1	−1	−1	3
−1	1	1	1	1	1	3
1	1	0	−1	−1	1	4
−1	−1	1	1	1	−1	4
1	0	−1	−1	0	1	4

Optimal design construction methodology can also be used to create specialized fractional factorial designs. We saw examples of this in the presentation of no-confounding design earlier in this chapter. To illustrate the usefulness of optimal design techniques, consider the case of resolution IV designs. We know that there are regular resolution IV designs available for 6, 7, and 8 factors in 16 runs, and no-confounding designs that are good competitors for these regular designs (the no-confounding designs can be thought of as resolution III.5 designs). However, for 9 factors the smallest regular resolution IV fraction is the 2^{9-4}, which has 32 runs. This is a large number of runs. Since a resolution IV design must have at least $2k$ runs, an 18-run optimal design could be a good competitor to the regular design. Table 9.37 presents the 18-run design constructed using the *D*-optimality criterion.

The alias relationships (for only the main effects and two-factor interactions) for the design in Table 9.37 are

$$[A] = A,\ [B] = B,\ [C] = C,\ [D] = D,\ [E] = E,\ [F] = F,\ [G] = G,\ [H] = H,\ [J] = J$$

$$[AB] = AB - 0.429\ BC - 0.429\ BD - 0.429\ BE + 0.429\ BF - 0.143\ BG + 0.429\ BH - 0.429\ BJ + 0.571\ CG - 0.571\ CH + 0.571\ DG + 0.571\ DJ - 0.571\ EF + 0.571\ EG - 0.571\ FG - 0.571\ GH + 0.571\ GJ$$

■ **TABLE 9.37**
An 18-run Minimum Run Resolution IV Design in $k = 9$ Factors

A	B	C	D	E	F	G	H	J
−	−	−	+	−	+	−	+	+
+	+	+	+	−	+	+	−	+
−	+	+	+	+	−	−	−	+
+	−	+	−	−	−	−	+	+
+	−	+	+	+	+	−	+	−
−	−	+	−	−	+	−	−	−
+	−	−	−	+	+	−	−	+
+	+	+	−	+	−	+	−	−
+	+	−	−	−	+	−	+	−
−	+	+	−	+	+	+	+	+
−	−	−	−	+	−	−	+	−
−	+	−	+	+	+	+	−	−
+	−	−	−	−	+	+	+	−
−	+	−	−	−	−	+	−	+
+	+	−	+	+	−	+	+	+
−	+	+	+	−	−	+	+	−
−	−	+	+	+	−	+	−	+
+	−	−	+	−	−	−	−	−

$$[AC] = AC - 0.143\,BC - 0.143\,BD - 0.143\,BE + 0.143\,BF + 0.286\,BG - 0.857\,BH$$
$$- 0.143\,BJ - 0.143\,CG + 0.143\,CH + DF - 0.143\,DG - 0.143\,DJ + 0.143\,EF$$
$$- 0.143\,EG - EJ + 0.143\,FG - 0.857\,GH - 0.143\,GJ$$

$$[AD] = AD - 0.143\,BC - 0.143\,BD - 0.143\,BE + 0.143\,BF + 0.286\,BG + 0.143\,BH$$
$$+ 0.857\,BJ + CF - 0.143\,CG + 0.143\,CH - 0.143\,DG - 0.143\,DJ + 0.143\,EF$$
$$- 0.143\,EG + EH + 0.143\,FG + 0.143\,GH + 0.857\,GJ$$

$$[AE] = AE - 0.143\,BC - 0.143\,BD - 0.143\,BE - 0.857\,BF + 0.286\,BG + 0.143\,BH$$
$$- 0.143\,BJ - 0.143\,CG + 0.143\,CH - CJ - 0.143\,DG + DH - 0.143\,DJ$$
$$+ 0.143\,EF - 0.143\,EG - 0.857\,FG + 0.143\,GH - 0.143\,GJ$$

$$[AF] = AF + 0.143\,BC + 0.143\,BD - 0.857\,BE - 0.143\,BF - 0.286\,BG - 0.143\,BH$$
$$+ 0.143\,BJ + CD + 0.143\,CG - 0.143\,CH + 0.143\,DG + 0.143\,DJ - 0.143\,EF$$
$$- 0.857\,EG - 0.143\,FG - 0.143\,GH + 0.143\,GJ - HJ$$

$$[AG] = AG + 0.571\,BC + 0.571\,BD + 0.571\,BE - 0.571\,BF - 0.143\,BG - 0.571\,BH$$
$$+ 0.571\,BJ - 0.429\,CG - 0.571\,CH - 0.429\,DG + 0.571\,DJ - 0.571\,EF$$
$$- 0.429\,EG + 0.429\,FG + 0.429\,GH - 0.429\,GJ$$

$$[AH] = AH - 0.857\,BC + 0.143\,BD + 0.143\,BE - 0.143\,BF - 0.286\,BG - 0.143\,BH$$
$$+ 0.143\,BJ - 0.857\,CG - 0.143\,CH + DE + 0.143\,DG + 0.143\,DJ - 0.143\,EF$$
$$+ 0.143\,EG - 0.143\,FG - FJ - 0.143\,GH + 0.143\,GJ$$

$$[AJ] = AJ - 0.143\,BC + 0.857\,BD - 0.143\,BE + 0.143\,BF + 0.286\,BG + 0.143\,BH$$
$$- 0.143\,BJ - CE - 0.143\,CG + 0.143\,CH + 0.857\,DG - 0.143\,DJ + 0.143\,EF$$
$$- 0.143\,EG + 0.143\,FG - FH + 0.143\,GH - 0.143\,GJ$$

We see that, as in any resolution IV, design the main effects are estimated free of any two-factor interactions, and the two-factor interactions are aliased with each other. However, note that there is partial aliasing of the two-factor interaction effects (for example, BC appears in more than one alias chain and the constants in the alias chains are not all either zero or ± 1). Therefore, this is a **nonregular** design. The two-factor interaction alias relationships in the 18-run design are much more complicated than they are in the standard 32-run 2^{9-4} design. Because of partial aliasing of the two-factor interactions it may be possible to estimate some of these effects. Furthermore, the standard errors of the main effects and interaction regression model coefficients are 0.24σ while in the standard 32-run 2^{9-4} design they are 0.18σ, so the 18-run design does not provide as much precision in parameter estimation as the standard 32-run design. Finally, the standard 2^{9-4} design is an orthogonal design whereas the 18-run design is not. This results in correlation between the model coefficients and contributes to the inflation of the standard errors of the model coefficients for the 18-run design.

It is also of interest to construct minimum-run resolution IV designs as alternatives to the standard resolution IV designs for $k = 6$ or 7 factors. The 12-run resolution IV design for six factors is shown in Table 9.38. The alias relationships for this design (ignoring three-factor and higher order interactions) are

$$[A] = A, [B] = B, [C] = C, [D] = D, [E] = E, [F] = F$$

$$[AB] = AB - 0.2\,BC + 0.6\,BD - 0.2\,BE - 0.6\,BF + 0.4\,CD - 0.8\,CE - 0.4\,CF + 0.4\,DE - 0.4\,DF - 0.4\,EF$$

$$[AC] = AC + 0.2\,BC + 0.4\,BD - 0.8\,BE - 0.4\,BF + 0.6\,CD - 0.2\,CE - 0.6\,CF - 0.4\,DE + 0.4\,DF + 0.4\,EF$$

$$[AD] = AD + 0.4\,BC - 0.2\,BD + 0.4\,BE - 0.8\,BF + 0.2\,CD - 0.4\,CE + 0.8\,CF + 0.2\,DE - 0.2\,DF + 0.8\,EF$$

$$[AE] = AE - 0.8\,BC + 0.4\,BD + 0.2\,BE - 0.4\,BF - 0.4\,CD - 0.2\,CE + 0.4\,CF + 0.6\,DE + 0.4\,DF - 0.6\,EF$$

$$[AF] = AF - 0.4\,BC - 0.8\,BD - 0.4\,BE - 0.2\,BF + 0.8\,CD + 0.4\,CE + 0.2\,CF + 0.8\,DE + 0.2\,DF + 0.2\,EF$$

Once again, notice that the price an experimenter is paying to reduce the number of runs from 16 to 12 is to introduce more complication into the alias relationships for the two-factor interactions. There is also a loss in the precision of estimation for model coefficients in comparison to the standard design. However, because we do not have complete confounding between two-factor interactions it may be possible to estimate some of these effects.

TABLE 9.38

A 12-run Resolution IV Design in $k = 6$ Factors

A	*B*	*C*	*D*	*E*	*F*
−	+	−	−	−	−
−	−	+	−	−	+
+	+	−	+	+	−
+	+	−	−	−	+
−	−	−	−	+	+
+	−	−	+	−	−
−	+	+	−	+	+
−	−	+	+	+	−
+	−	+	−	+	−
+	−	+	+	+	+
+	+	+	+	−	−
−	+	−	+	−	+

These minimum-run resolution IV designs are additional examples of nonregular fractional factorial designs. Design-Expert contains a selection of these designs for $5 \leq k \leq 50$ factors. Similar design can be created using the "custom designer" feature in JMP. Generally, these will be nonregular designs and there is no guarantee that they will be orthogonal. Like the no-confounding designs, these designs can be very useful alternatives to the standard 2_{IV}^{k-p} fractional factorial designs in screening problems where main effects are of primary interest but two-factor interactions cannot be completely ignored. If two-factor interactions prove to be important, in many cases these interactions can be estimated by using stepwise regression methods. In other cases follow-on experimentation will be necessary to determine which interaction effects are important.

Small resolution V designs can be constructed similarly. Table 9.39 contains a nonregular two-level fraction for $k = 6$ factors and $N = 22$ runs. Since the two-factor interaction model has 22 parameters, this is a minimum-run design. This design supports estimation of all main effect and two-factor interactions, just as the 2_{VI}^{6-1} will, but with 10 fewer runs. However, the design in Table 9.39 is not orthogonal, and this impacts the precision of estimation for effects and regression coefficients. The standard error of the regression model coefficients ranges from 0.26σ to 0.29σ, while in the 2_{VI}^{6-1}, the corresponding standard errors are 0.18σ.

As a final example, Table 9.40 presents a nonregular two-level fraction for $k = 8$ factors in $N = 38$ runs. This design supports estimation of all main effect and two-factor interactions, just as the 2_{V}^{8-2} will, but with 26 fewer runs. However, the nonorthogonality of the

■ **TABLE 9.39**
A Resolution V Two-Level Fraction in $k = 6$ Factors

Run	*A*	*B*	*C*	*D*	*E*	*F*
1	+	−	−	−	+	−
2	+	−	+	−	+	+
3	−	+	+	−	−	−
4	−	−	−	−	+	+
5	+	+	−	+	+	+
6	+	+	−	+	−	+
7	+	−	−	+	−	+
8	+	+	−	−	−	+
9	−	−	−	+	+	−
10	−	−	−	−	−	−
11	+	−	+	+	+	−
12	−	+	−	−	+	−
13	+	+	+	+	−	+
14	+	−	+	−	−	−
15	−	−	+	−	−	+
16	+	+	−	+	−	−
17	−	−	+	+	+	+
18	−	−	+	+	−	−
19	−	+	−	+	−	+
20	+	+	+	−	+	−
21	−	+	+	+	+	−
22	−	+	+	−	+	+

■ **TABLE 9.40**
A Resolution V Two-Level Fraction in $k = 8$ Factors

Run	A	B	C	D	E	F	G	H
1	−	−	+	−	+	−	+	−
2	−	+	+	−	+	+	−	+
3	−	+	+	+	−	−	−	−
4	−	−	−	−	−	−	+	−
5	+	+	−	−	+	−	+	−
6	+	−	+	+	+	+	+	−
7	−	−	−	+	+	+	−	−
8	+	−	−	+	+	−	−	−
9	−	−	+	−	−	+	+	+
10	+	+	+	−	+	−	−	+
11	+	+	+	−	−	+	−	+
12	+	+	−	+	−	−	+	+
13	−	+	−	+	+	−	+	−
14	+	−	−	−	−	−	+	+
15	+	+	−	+	+	+	+	−
16	−	+	−	−	−	+	+	+
17	+	−	+	+	−	+	−	−
18	+	−	+	−	+	+	−	−
19	−	−	−	−	−	+	−	−
20	−	+	+	+	−	+	+	+
21	−	+	+	−	−	−	+	+
22	−	+	−	+	+	−	−	+
23	+	+	+	+	−	−	+	−
24	−	−	−	+	+	−	+	+
25	+	−	+	+	−	−	+	+
26	+	−	+	−	−	−	−	−
27	+	−	−	+	−	+	+	+
28	+	+	−	−	+	+	−	+
29	+	+	+	−	+	+	+	+
30	+	+	−	−	−	−	−	−
31	−	+	+	−	−	+	+	−
32	−	−	−	−	+	+	+	−
33	−	−	−	+	−	−	−	+
34	+	−	+	+	+	+	−	+
35	+	+	+	+	+	+	−	−
36	−	−	+	+	+	−	−	+
37	−	−	−	−	+	−	−	+
38	−	+	−	+	−	+	−	−

design has some modest impact on the precision of estimation for effects and regression coefficients. For the design in Table 9.40, the standard error of the regression model coefficients ranges from 0.18σ to 0.26σ, while in the 2_{V}^{8-2}, the corresponding standard error is 0.13σ.

Despite the loss in precision of estimation, these nonregular fractions can be of value when experimental resources are scarce. Design-Expert contains a selection of these designs for $6 \leq k \leq 50$ factors. These designs were constructed using a D-optimal design construction tool. The custom designer capability in JMP can also be very useful in constructing small resolution V fractions.

■ **TABLE 9.41**
A 24-Run *D*-optimal Design for Two Three-level Factors and One Two-level Factor

Run	Factor *A*	Factor *B*	Factor *C*
1	L1	L1	−1
2	L1	L1	−1
3	L1	L1	1
4	L1	L2	−1
5	L1	L2	1
6	L1	L3	−1
7	L1	L3	1
8	L1	L3	1
9	L2	L1	−1
10	L2	L1	−1
11	L2	L1	1
12	L2	L2	−1
13	L2	L2	−1
14	L2	L2	1
15	L2	L3	−1
16	L2	L3	1
17	L3	L1	−1
18	L3	L1	1
19	L3	L2	−1
20	L3	L2	1
21	L3	L2	1
22	L3	L3	−1
23	L3	L3	1
24	L3	L3	1

The optimal design approach is also an excellent way to create mixed-level designs. The *D*-optimality criterion discussed earlier usually produces good designs. The custom design tool in JMP is an excellent way to construct *D*-optimal mixed-level designs. For example, suppose that we have two three-level factors (categorical) and a single quantitative two-level factor. We want to estimate all main effects and all two-factor interactions. The JMP custom designer recommends a 24-run design. The design is shown in Table 9.41. This design is nearly orthogonal; notice that each level of the three-level factors *A* and *B* appears eight times, but the design is not balanced with respect to these two factors. Also, while there are exactly 12 runs with factor *C* at the low and high levels, the levels of *C* are not exactly balanced against the levels of factors *A* and *B*. Table 9.42 shows the relative variances (that is, variances divided by σ^2) of the individual single-degree-of-freedom model components from this design. Notice that all of the relative variances are almost identical, illustrating the near-orthogonality of the design. In an orthogonal design, all of the relative variances would be equal.

It is possible to construct a smaller design for this problem. The minimum number of runs for this situation is $N = 14$, and the *D*-optimal design, constructed using the JMP custom design tool, is shown in Table 9.43. This design is not orthogonal, but it does permit unique estimates of all main effects and two-factor interactions. The relative variances of the model parameters, shown in Table 9.44, are both larger than they were in the 24-run design

■ TABLE 9.42

Relative Variances for the Individual Model Effects for the 24-Run *D*-optimal Design in Table 9.41

Effect	Relative Variance
Intercept	0.046
A1	0.045
A2	0.045
B1	0.046
B2	0.044
C	0.045
A*B1	0.044
A*B2	0.047
A*B3	0.046
A*B4	0.044
A*C1	0.046
A*C2	0.046
B*C1	0.046
B*C2	0.046

(this should not be a surprise—a larger sample size gives smaller variances of the estimates) and more uneven, indicating that this design is much further from orthogonal than was the 24-run design.

As a final illustration of the power and flexibility of optimal designs for mixed-level fractional factorials, suppose that an experimenter has five factors: *A* is categorical with five levels, *B* is categorical with four levels, *C* is categorical with three levels, and *B* and *C* are continuous with two levels. The experimenter is interested in estimating all of the main effects

■ TABLE 9.43

A 14-Run *D*-optimal Design for Two Three-level Factors and One Two-level Factor

Run	Factor *A*	Factor *B*	Factor *C*
1	L1	L3	−1
2	L3	L2	1
3	L1	L1	1
4	L1	L2	1
5	L3	L3	1
6	L2	L3	1
7	L1	L1	−1
8	L3	L1	1
9	L2	L3	−1
10	L3	L3	−1
11	L2	L2	1
12	L2	L1	1
13	L1	L2	−1
14	L3	L1	−1

■ **TABLE 9.44**

Relative Variances for the Individual Model Effects for the 14-Run *D*-optimal Design in Table 9.43

Effect	Relative Variance
Intercept	0.340
A1	0.093
A2	0.179
B1	0.093
B2	0.179
C	0.167
A*B1	0.111
A*B2	0.185
A*B3	0.185
A*B4	0.136
A*C1	0.167
A*C2	0.278
B*C1	0.167
B*C2	0.278

of these factors. The full factorial has $N = 5 \times 4 \times 3 \times 2 \times 2 = 240$ runs and is an orthogonal design. However, it is not necessary to use 240 runs to estimate the main effects, as only 11 degrees of freedom are required. A design with 120 runs would be a one-half fraction. This design is almost orthogonal, but probably too large for practical use. Both the one-quarter and one-eighth fractions with 60 and 30 runs, respectively, are nearly orthogonal but still too large. Let's see what can be done with a 15-run design.

Table 9.45 shows the 15-run *D*-optimal design constructed using the optimal design tool in JMP. This design is not perfectly balanced; this isn't possible with 15 runs and a four-level

■ **TABLE 9.45**

A 15-Run *D*-optimal Mixed-level Design for Five Factors

Run	Factor *A*	Factor *B*	Factor *C*	Factor *D*	Factor *E*
1	L4	L2	L1	1	1
2	L1	L1	L3	1	1
3	L5	L4	L2	1	1
4	L3	L3	L2	1	−1
5	L4	L1	L2	−1	−1
6	L2	L4	L3	1	−1
7	L1	L4	L1	−1	−1
8	L5	L2	L3	−1	−1
9	L3	L2	L3	1	−1
10	L3	L1	L1	−1	1
11	L2	L2	L2	−1	1
12	L4	L3	L3	−1	1
13	L5	L3	L1	1	−1
14	L1	L2	L2	1	−1
15	L2	L1	L1	1	−1

■ **TABLE 9.46**
Relative Variances for the Individual Model Effects for the 15-Run *D*-optimal Design in Table 9.16

Effect	Relative Variance
Intercept	0.077
A1	0.075
A2	0.069
A3	0.078
A4	0.084
B1	0.087
B2	0.063
B3	0.100
C1	0.070
C2	0.068
D	0.077
E	0.077

factor. However, it is nearly orthogonal. To see this, consider the relative variances of the model coefficients shown in Table 9.46, and notice that all of the relative variances are very similar.

9.6.3 Extensions of the Optimal Design Approach

We have discussed the use of design optimality for relatively simple situations. There are several extensions of these ideas that practitioners should be aware of.

A criticism often leveled at the optimal design approach is that the final design depends on the model chosen by the experimenter. DuMouchel and Jones (1994) introduce a Bayesian modification of the *D*-optimal design that affords protection to experimenters against terms that are not in the assumed model. They assume that the model contains p primary terms but they want to obtain protection against q potential model terms. These potential terms are typically of higher order than those in the primary model. Their Bayesian *D*-optimal designs have N runs, where $p < N < p + q$. These designs allow some of the potential model terms to be fit if necessary. Jones, Lin, and Nachtsheim (2008) use the Bayesian *D*-optimality approach to construct supersaturated fractional factorial designs. Andere-Rendon, Montgomery, and Rollier (1997) use this approach to design mixture experiments in the presence of model uncertainty. Mixture problems are introduced in Chapter 11.

We noted earlier that the *G*-optimal design criterion makes use of an objective function that involves minimizes the maximum value of the prediction variance in the design region. Constructing *G*-optimal designs has historically proved difficult because two optimization problems must be solved—finding the best coordinate value to change in the current design and determining the maximum value of the scaled prediction variance for each new design that is evaluated. Rodriguez et al. (2010) describe a commercially viable algorithm for constructing *G*-optimal designs and compare the performance of several *G*-optimal designs to their *I*-optimal and *D*-optimal counterparts.

It is also possible to construct optimal designs for nonlinear models. In linear models the optimal design problem is relatively simple because the model covariance matrix $\mathbf{X}'\mathbf{X}$ does not contain any of the unknown parameters $\boldsymbol{\beta}$. However, if the model is nonlinear, this is not the case. To find a *D*-optimal design for a nonlinear model, we must find design points that maximize the determinant of $\mathbf{D}'\mathbf{D}$, where $\mathbf{D}$ is a matrix of partial derivatives of the nonlinear model expectation function with respect to each model parameter evaluated at each design point. This matrix is a function of the unknown parameters, so finding a *D*-optimal design would require knowledge of the model parameters. One possible approach to this problem is to assume values for the unknown β's. This would produce a **conditional** *D*-optimal design.

An alternative is to use a Bayesian approach employing a prior distribution $f(\boldsymbol{\beta})$ to specify the uncertainty in the *p* parameter values. This leads to a design criterion

$$\phi(\mathbf{D}) = \int \log |\mathbf{D}'\mathbf{D}| f(\boldsymbol{\beta})d\boldsymbol{\beta} \tag{9.9}$$

This is the expectation of the logarithm of the information matrix. This criterion was proposed by Chaloner and Larntz (1989) for the single-factor logistic regression model. The difficulty in using Equation 9.9 as a design criterion is that the *p*-dimensional integral must be evaluated a very large number of times. Gotwalt, Jones, and Steinberg (2009) have recently developed a clever quadrature scheme that greatly reduces the computing time to evaluate the integral in Equation 9.9 with excellent accuracy. This procedure is implemented in the nonlinear design routine of JMP, and allows computationally efficient construction of *D*-optimal designs for nonlinear models. Gotwalt et al. (2009) present examples of the use of this technique. Also see Johnson and Montgomery (2010).

A very important type of nonlinear model that occurs frequently in industrial experimentation is the generalized linear model. This is a family of models that unify linear and nonlinear regression models with response distributions that are a member of the exponential family (which includes the binomial, Poisson, normal, exponential, and gamma distributions). Important special cases include logistic regression, Poisson regression, and regression with exponential responses. Often an experimenter will know in advance that the response distribution is binomial (for example). Then a design for a logistic regression model would be appropriate. The method described in Gotwalt et al. (2009) can be used to construct *D*-optimal designs for this experiment. For examples of designed experiments for generalized linear models, also see Johnson and Montgomery (2009) and Myers et al. (2010).

9.7 Problems

9.1. The effects of developer strength (*A*) and development time (*B*) on the density of photographic plate film are being studied. Three strengths and three times are used, and four replicates of a 3^2 factorial experiment are run. The data from this experiment follow. Analyze the data using the standard methods for factorial experiments.

Developer Strength	Development Time (minutes) 10		14		18	
1	1	2	0	3	2	5
	6	5	4	2	4	6
2	5	7	6	8	9	11
	8	5	7	7	8	5
3	7	11	10	10	13	10
	7	8	8	7	9	8

9.2. Compute the *I* and *J* components of the two-factor interaction in Problem 9.1.

9.3. An experiment was performed to study the effect of three different types of 32-ounce bottles (*A*) and three different shelf types (*B*)—smooth permanent shelves, end-aisle displays with grilled shelves, and beverage coolers—on the time it takes to stock ten 12-bottle cases on the shelves. Three workers (factor *C*) were employed in the experiment, and two replicates of a 3^3 factorial design were run. The observed time data are shown in the following table. Analyze the data and draw conclusions.

		Replicate I		
Worker	Bottle Type	Permanent	End Aisle	Cooler
1	Plastic	3.55	4.14	5.80
	28-mm glass	4.01	4.38	5.48
	38-mm glass	4.20	4.25	5.75
2	Plastic	4.80	5.22	6.21
	28-mm glass	4.52	5.15	6.28
	38-mm glass	4.96	5.17	6.05
3	Plastic	4.10	3.94	5.14
	28-mm glass	4.30	4.53	4.99
	38-mm glass	4.17	4.86	4.85

		Replicate II		
		Per-manent	**End Aisle**	**Cooler**
	Plastic	3.36	4.19	5.25
1	28-mm glass	3.52	4.26	4.90
	38-mm glass	3.68	4.37	5.58
	Plastic	4.40	4.75	5.88
2	28-mm glass	4.44	4.65	6.20
	38-mm glass	4.39	4.75	6.38
	Plastic	3.65	4.08	4.49
3	28-mm glass	4.04	4.08	4.59
	38-mm glass	3.88	4.48	4.93

9.4. A medical researcher is studying the effect of lidocaine on the enzyme level in the heart muscle of beagle dogs. Three different commercial brands of lidocaine (A), three dosage levels (B), and three dogs (C) are used in the experiment, and two replicates of a 3^3 factorial design are run. The observed enzyme levels follow. Analyze the data from this experiment.

		Replicate I		
		Dog		
Lidocaine Brand	**Dosage Strength**	**1**	**2**	**3**
	1	96	84	85
1	2	94	99	98
	3	101	106	98
	1	85	84	86
2	2	95	98	97
	3	108	114	109
	1	84	83	81
3	2	95	97	93
	3	105	100	106

		Replicate II		
		Dog		
		1	**2**	**3**
	1	84	85	86
1	2	95	97	90
	3	105	104	103
	1	80	82	84
2	2	93	99	95
	3	110	102	100
	1	83	80	79
3	2	92	96	93
	3	102	111	108

9.5. Compute the I and J components of the two-factor interactions for Problem 9.4.

9.6. An experiment is run in a chemical process using a 3^2 factorial design. The design factors are temperature and pressure, and the response variable is yield. The data that result from this experiment are as follows.

	Pressure, psig		
Temperature, °C	**100**	**120**	**140**
80	47.58, 48.77	64.97, 69.22	80.92, 72.60
90	51.86, 82.43	88.47, 84.23	93.95, 88.54
100	71.18, 92.77	96.57, 88.72	76.58, 83.04

(a) Analyze the data from this experiment by conducting an analysis of variance. What conclusions can you draw?

(b) Graphically analyze the residuals. Are there any concerns about underlying assumptions or model adequacy?

(c) Verify that if we let the low, medium, and high levels of both factors in this design take on the levels −1, 0, and +1, then a least squares fit to a second-order model for yield is

$$\hat{y} = 86.81 + 10.4x_1 + 8.42x_2 - 7.17x_1^2 - 7.84x_2^2 - 7.69x_1x_2$$

(d) Confirm that the model in part (c) can be written in terms of the natural variables temperature (T) and pressure (P) as

$$\hat{y} = -1335.63 + 18.56T + 8.59P - 0.072T^2 - 0.0196P^2 - 0.0384TP$$

(e) Construct a contour plot for yield as a function of pressure and temperature. Based on examination of this plot, where would you recommend running this process?

9.7. (a) Confound a 3^3 design in three blocks using the ABC^2 component of the three-factor interaction. Compare your results with the design in Figure 9.7.

(b) Confound a 3^3 design in three blocks using the AB^2C component of the three-factor interaction. Compare your results with the design in Figure 9.7.

(c) Confound a 3^3 design in three blocks using the ABC component of the three-factor interaction. Compare your results with the design in Figure 9.7.

(d) After looking at the designs in parts (a), (b), and (c) and Figure 9.7, what conclusions can you draw?

9.8. Confound a 3^4 design in three blocks using the AB^2CD component of the four-factor interaction.

9.9. Consider the data from the first replicate of Problem 9.3. Assuming that not all 27 observations could be run on the same day, set up a design for conducting the experiment over three days with AB^2C confounded with blocks. Analyze the data.

9.10. Outline the analysis of variance table for the 3^4 design in nine blocks. Is this a practical design?

9.11. Consider the data in Problem 9.3. If ABC is confounded in replicate I and ABC^2 is confounded in replicate II, perform the analysis of variance.

9.12. Consider the data from replicate I of Problem 9.3. Suppose that only a one-third fraction of this design with $I = ABC$ is run. Construct the design, determine the alias structure, and analyze the data.

9.13. From examining Figure 9.9, what type of design would remain if after completing the first nine runs, one of the three factors could be dropped?

9.14. Construct a 3_{IV}^{4-1} design with $I = ABCD$. Write out the alias structure for this design.

9.15. Verify that the design in Problem 9.14 is a resolution IV design.

9.16. Construct a 3^{5-2} design with $I = ABC$ and $I = CDE$. Write out the alias structure for this design. What is the resolution of this design?

9.17. Construct a 3^{9-6} design and verify that it is a resolution III design.

9.18. Construct a 4×2^3 design confounded in two blocks of 16 observations each. Outline the analysis of variance for this design.

9.19. Outline the analysis of variance table for a $2^2 3^2$ factorial design. Discuss how this design may be confounded in blocks.

9.20. Starting with a 16-run 2^4 design, show how two three-level factors can be incorporated in this experiment. How many two-level factors can be included if we want some information on two-factor interactions?

9.21. Starting with a 16-run 2^4 design, show how one three-level factor and three two-level factors can be accommodated and still allow the estimation of two-factor interactions.

9.22. In Problem 8.30, you met Harry Peterson-Nedry, a friend of the author who has a winery and vineyard in Newberg, Oregon. That problem described the application of two-level fractional factorial designs to their 1985 Pinot Noir product. In 1987, he wanted to conduct another Pinot Noir experiment. The variables for this experiment were

Variable	Levels
Clone of Pinot Noir	Wadenswil, Pommard
Berry size	Small, large
Fermentation temperature	80°, 85°, 90/80°, and 90°F
Whole berry	None, 10%
Maceration time	10 and 21 days
Yeast type	Assmanhau, Champagne
Oak type	Tronçais, Allier

Harry decided to use a 16-run two-level fractional factorial design, treating the four levels of fermentation temperature as two two-level variables. As he has done previously, Harry used the rankings from a taste-test panel as the response variable. The design and the resulting average ranks are as follows.

Run	Clone	Berry Size	Ferm. Temp.	Whole Berry	Macer. Time	Yeast Type	Oak Type	Average Rank
1	−	−	− −	−	−	−	−	4
2	+	−	− −	−	+	+	+	10
3	−	+	− −	+	−	+	+	6
4	+	+	− −	+	+	−	−	9
5	−	−	+ −	+	+	+	−	11
6	+	−	+ −	+	−	−	+	1
7	−	+	+ −	−	+	−	+	15
8	+	+	+ −	−	−	+	−	5
9	−	−	− +	+	+	−	+	12
10	+	−	− +	+	−	+	−	2
11	−	+	− +	−	+	+	−	16
12	+	+	− +	−	−	−	+	3
13	−	−	+ +	−	−	+	+	8
14	+	−	+ +	−	+	−	−	14
15	−	+	+ +	+	−	−	−	7
16	+	+	+ +	+	+	+	+	13

(a) Describe the aliasing in this design.

(b) Analyze the data and draw conclusions.

9.23. An article by W. D. Baten in the 1956 volume of *Industrial Quality Control* described an experiment to study the effect of three factors on the lengths of steel bars. Each bar was subjected to one of two heat treatment processes and was cut on one of four machines at one of three times during the day (8 A.M., 11 A.M., or 3 P.M.). The coded length data are as follows:

Time of Day	Heat Treatment Process	Machine 1		Machine 2		Machine 3		Machine 4	
8 A.M.	1	6	9	7	9	1	2	6	6
		1	3	5	5	0	4	7	3
	2	4	6	6	5	−1	0	4	5
		0	1	3	4	0	1	5	4
11 A.M.	1	6	3	8	7	3	2	7	9
		1	−1	4	8	1	0	11	6
	2	3	1	6	4	2	0	9	4
		1	−2	1	3	−1	1	6	3
3 P.M.	1	5	4	10	11	−1	2	10	5
		9	6	6	4	6	1	4	8
	2	6	0	8	7	0	−2	4	3
		3	7	10	0	4	−4	7	0

(a) Analyze the data from this experiment, assuming that the four observations in each cell are replicates.

(b) Analyze the residuals from this experiment. Is there any indication that there is an outlier in one cell? If you find an outlier, remove it and repeat the analysis from part (a). What are your conclusions?

(c) Suppose that the observations in the cells are the lengths (coded) of bars processed together in heat treatment and then cut sequentially (that is, in order) on the four machines. Analyze the data to determine the effects of the three factors on mean length.

(d) Calculate the log variance of the observations in each cell. Analyze this response. What conclusions can you draw?

(e) Suppose the time at which a bar is cut really cannot be controlled during routine production. Analyze the average length and the log variance of the length for each of the 12 bars cut at each machine/heat treatment process combination. What conclusions can you draw?

9.24. Reconsider the experiment in Problem 9.23. Suppose that it was necessary to estimate all main effects and two-factor interactions, but the full factorial with 24 runs (not counting replication) was too expensive. Recommend an alternative design.

9.25. Suppose there are four three-level categorical factor and a single two-level continuous factor. What is the minimum number of runs required to estimate all main effects and two-factor interactions? Construct this design.

9.26. Reconsider the experiment in Problem 9.25. Construct a design with $N = 48$ runs and compare it to the design you constructed in Problem 9.25.

9.27. Reconsider the experiment in Problem 9.25. Suppose that you are only interested in main effects. Construct a design with $N = 12$ runs for this experiment.

9.28. An article in the *Journal of Chemical Technology and Biotechnology* ("A study of antifungal antibiotic production by *Thermomonospora* sp MTCC 3340 using full factorial design," 2003, Vol. 78, pp. 605–610) investigated three independent variables, concentration of carbon source (glucose),

Run no	Percent carbon (glucose)	Percent nitrogen (soybean meal)	Temperature (°C)	Activity against *C albicans*[a]
1	2	0.5		25.84
2	2	1.0		51.86
3	2	3.0		32.59
4	5	0.5		20.48
5	5	1.0	25	25.84
6	5	3.0		12.87
7	7.5	0.5		20.48
8	7.5	1.0		25.84
9	7.5	3.0		10.20
10	2	0.5		51.86
11	2	1.0		131.33
12	2	3.0		41.11
13	5	0.5		41.11
14	5	1.0	30	104.11
15	5	3.0		32.59
16	7.5	0.5		65.42
17	7.5	1.0		82.53
18	7.5	3.0		51.86
19	2	0.5		41.11
20	2	1.0		104.11
21	2	3.0		32.59
22	5	0.5		32.59
23	5	1.0	35	82.53
24	5	3.0		25.84
25	7.5	0.5		51.86
26	7.5	1.0		65.42
27	7.5	3.0		41.11

concentration of nitrogen source (soybean meal), and temperature of incubation for their effects on the production of antifungal antibiotic by the isolate *Thermomonospora* sp MTCC 3340. A 3^3 factorial design was conducted and the results are shown in the table on the previous page.

(a) Analyze the data from this experiment.

(b) Fit a second-order model to the activity response. Construct contour plots and response surface plots to assist in interpreting the results of this experiment.

(c) What operating conditions would you recommend to optimize this process?

9.29. Construct a minimum-run D-optimal resolution IV design for 10 factors. Find the alias relationships. What approach would you recommend for analyzing the data from this experiment?

9.30. Construct a minimum-run D-optimal resolution IV design for 12 factors. Find the alias relationships. What approach would you recommend for analyzing the data from this experiment?

9.31. Suppose that you must design an experiment to investigate 9 continuous factors. It is thought that running all factors at two levels is adequate but that all two-factor interactions are of interest.

(a) How many runs are required to estimate all main effects and two-factor interactions?

(b) Find a minimum-run *D*-optimal design that is suitable for this problem.

9.32. Suppose that you must design an experiment to investigate 7 continuous factors. Running all factors at two levels is thought to be appropriate but that only the two-factor interactions involving factor A are of interest.

(a) How many runs are required to estimate all of the relevant effects?

(b) Find a minimum-run *D*-optimal design that is suitable for this problem.

9.33. Suppose that you must design an experiment to investigate 6 continuous factors. It is thought that running all factors at two levels is adequate but that only the AB, AC, and AD two-factor interactions are of interest.

(a) How many runs are required to estimate all of the effects that are of interest?

(b) Find a minimum-run *D*-optimal design that is suitable for this problem.

9.34. Suppose that you must design an experiment with six categorical factors. Factor A has 6 levels, factor B has 5 levels, factor C has 5 levels, factor D has 3 levels, and factors E and F have 2 levels. You are interested in main effects and two-factor interactions.

(a) How many runs are required to estimate all of the effects that are of interest?

(b) Find a *D*-optimal design that is suitable for this problem.

(c) Suppose that the experimenter decides that this design requires too many runs. What strategy would you recommend?

CHAPTER 10

Regression Modeling

CHAPTER OUTLINE

The supplemental material is on the textbook website www.wiley.com/go/global/montgomery.

10.1 Introduction

In many problems two or more variables are related, and it is of interest to model and explore this relationship. For example, in a chemical process the yield of product is related to the operating temperature. The chemical engineer may want to build a model relating yield to temperature and then use the model for prediction, process optimization, or process control.

In general, suppose that there is a single **dependent variable** or **response** y that depends on k **independent** or **regressor variables**, for example, $x_1, x_2, \ldots, x_k$. The relationship between these variables is characterized by a mathematical model called a **regression model**. The regression model is fit to a set of sample data. In some instances, the experimenter knows the exact form of the true functional relationship between y and $x_1, x_2, \ldots, x_k$, say $y = \phi(x_1, x_2, \ldots, x_k)$. However, in most cases, the true functional relationship is unknown, and the experimenter chooses an appropriate function to approximate ϕ. Low-order polynomial models are widely used as approximating functions.

There is a strong interplay between design of experiments and regression analysis. Throughout this book we have emphasized the importance of expressing the results of an experiment quantitatively, in terms of an **empirical model**, to facilitate understanding, interpretation, and implementation. Regression models are the basis for this. On numerous occasions we have shown the regression model that represented the results of an experiment. In this chapter, we present some aspects of fitting these models. More complete presentations of regression are available in Montgomery, Peck, and Vining (2006) and Myers (1990).

Regression methods are frequently used to analyze data from **unplanned experiments**, such as might arise from observation of uncontrolled phenomena or historical records. Regression methods are also very useful in designed experiments where something has "gone wrong." We will illustrate some of these situations in this chapter.

10.2 Linear Regression Models

We will focus on fitting linear regression models. To illustrate, suppose that we wish to develop an empirical model relating the viscosity of a polymer to the temperature and the catalyst feed rate. A model that might describe this relationship is

$$y = \beta_0 + \beta_1 x_1 + \beta_2 x_2 + \epsilon \tag{10.1}$$

where y represents the viscosity, x_1 represents the temperature, and x_2 represents the catalyst feed rate. This is a **multiple linear regression model** with two independent variables. We often call the independent variables **predictor variables** or **regressors**. The term **linear** is used because Equation 10.1 is a linear function of the unknown parameters β_0, β_1, and β_2. The model describes a plane in the two-dimensional x_1, x_2 space. The parameter β_0 defines the intercept of the plane. We sometimes call β_1 and β_2 *partial regression coefficients* because β_1 measures the expected change in y per unit change in x_1 when x_2 is held constant and β_2 measures the expected change in y per unit change in x_2 when x_1 is held constant.

In general, the response variable y may be related to k regressor variables. The model

$$y = \beta_0 + \beta_1 x_1 + \beta_2 x_2 + \cdots + \beta_k x_k + \epsilon \tag{10.2}$$

is called a *multiple linear regression model* with k regressor variables. The parameters β_j, $j = 0, 1, \ldots, k$, are called the **regression coefficients**. This model describes a hyperplane in the k-dimensional space of the regressor variables $\{x_j\}$. The parameter β_j represents the expected change in response y per unit change in x_j when all the remaining independent variables x_i $(i \neq j)$ are held constant.

Models that are more complex in appearance than Equation 10.2 may often still be analyzed by multiple linear regression techniques. For example, consider adding an interaction term to the first-order model in two variables, say

$$y = \beta_0 + \beta_1 x_1 + \beta_2 x_2 + \beta_{12} x_1 x_2 + \epsilon \tag{10.3}$$

If we let $x_3 = x_1 x_2$ and $\beta_3 = \beta_{12}$, then Equation 10.3 can be written as

$$y = \beta_0 + \beta_1 x_1 + \beta_2 x_2 + \beta_3 x_3 + \epsilon \tag{10.4}$$

which is a standard multiple linear regression model with three regressors. Recall that we presented empirical models like Equations 10.2 and 10.4 in several examples in Chapters 6, 7, and 8 to quantitatively express the results of a two-level factorial design. As another example, consider the second-order **response surface model** in two variables:

$$y = \beta_0 + \beta_1 x_1 + \beta_2 x_x + \beta_{11} x_1^2 + \beta_{22} x_2^2 + \beta_{12} x_1 x_2 + \epsilon \tag{10.5}$$

If we let $x_3 = x_1^2$, $x_4 = x_2^2$, $x_5 = x_1x_2$, $\beta_3 = \beta_{11}$, $\beta_4 = \beta_{22}$, and $\beta_5 = \beta_{12}$, then this becomes

$$y = \beta_0 + \beta_1 x_1 + \beta_2 x_2 + \beta_3 x_3 + \beta_4 x_4 + \beta_5 x_5 + \epsilon \tag{10.6}$$

which is a linear regression model. We have also seen this model in examples earlier in the text. In general, any regression model that is linear in the parameters (the β's) is a linear regression model, regardless of the shape of the response surface that it generates.

In this chapter we will summarize methods for estimating the parameters in multiple linear regression models. This is often called **model fitting**. We have used some of these results in previous chapters, but here we give the developments. We will also discuss methods for testing hypotheses and constructing confidence intervals for these models as well as for checking the adequacy of the model fit. Our focus is primarily on those aspects of regression analysis useful in designed experiments. For more complete presentations of regression, refer to Montgomery, Peck, and Vining (2006) and Myers (1990).

10.3 Estimation of the Parameters in Linear Regression Models

The method of least squares is typically used to estimate the regression coefficients in a multiple linear regression model. Suppose that $n > k$ observations on the response variable are available, say $y_1, y_2, \ldots, y_n$. Along with each observed response y_i, we will have an observation on each regressor variable and let x_{ij} denote the ith observation or level of variable x_j. The data will appear as in Table 10.1. We assume that the error term ϵ in the model has $E(\epsilon) = 0$ and $V(\epsilon) = \sigma^2$ and that the $\{\epsilon_i\}$ are uncorrelated random variables.

We may write the model equation (Equation 10.2) in terms of the observations in Table 10.1 as

$$\begin{aligned} y_i &= \beta_0 + \beta_1 x_{i1} + \beta_2 x_{i2} + \cdots + \beta_k x_{ik} + \epsilon_i \\ &= \beta_0 + \sum_{j=1}^{k} \beta_j x_{ij} + \epsilon_i \quad i = 1, 2, \ldots, n \end{aligned} \tag{10.7}$$

The method of least squares chooses the β's in Equation 10.7 so that the sum of the squares of the errors, ϵ_i, is minimized. The least squares function is

$$L = \sum_{i=1}^{n} \epsilon_i^2 = \sum_{i=1}^{n} \left(y_i - \beta_0 - \sum_{j=1}^{k} \beta_j x_{ij} \right)^2 \tag{10.8}$$

The function L is to be minimized with respect to $\beta_0, \beta_1, \ldots, \beta_k$. The least squares estimators, say $\hat{\beta}_0, \hat{\beta}_1, \ldots, \hat{\beta}_k$, must satisfy

$$\left.\frac{\partial L}{\partial \beta_0}\right|_{\hat{\beta}_0, \hat{\beta}_1, \ldots, \hat{\beta}_k} = -2 \sum_{i=1}^{n} \left(y_i - \hat{\beta}_0 - \sum_{j=1}^{k} \hat{\beta}_j x_{ij} \right) = 0 \tag{10.9a}$$

■ **TABLE 10.1**
Data for Multiple Linear Regression

y	x_1	x_2	...	x_k
y_1	x_{11}	x_{12}	...	x_{1k}
y_2	x_{21}	x_{22}	...	x_{2k}
$\vdots$	$\vdots$	$\vdots$		$\vdots$
y_n	x_{n1}	x_{n2}	...	x_{nk}

and

$$\left.\frac{\partial L}{\partial \beta_j}\right|_{\hat{\beta}_0, \hat{\beta}_1, \ldots, \hat{\beta}_k} = -2 \sum_{i=1}^{n} \left(y_i - \hat{\beta}_0 - \sum_{j=1}^{k} \hat{\beta}_j x_{ij} \right) x_{ij} = 0 \quad j = 1, 2, \ldots, k \qquad \textbf{(10.9b)}$$

Simplifying Equation 10.9, we obtain

$$\begin{aligned}
n\hat{\beta}_0 + \hat{\beta}_1 \sum_{i=1}^{n} x_{i1} + \hat{\beta}_2 \sum_{i=1}^{n} x_{i2} + \cdots + \hat{\beta}_k \sum_{i=1}^{n} x_{ik} &= \sum_{i=1}^{n} y_i \\
\hat{\beta}_0 \sum_{i=1}^{n} x_{i1} + \hat{\beta}_1 \sum_{i=1}^{n} x_{i1}^2 + \hat{\beta}_2 \sum_{i=1}^{n} x_{i1}x_{i2} + \cdots + \hat{\beta}_k \sum_{i=1}^{n} x_{i1}x_{ik} &= \sum_{i=1}^{n} x_{i1}y_i \\
&\vdots \\
\hat{\beta}_0 \sum_{i=1}^{n} x_{ik} + \hat{\beta}_1 \sum_{i=1}^{n} x_{ik}x_{i1} + \hat{\beta}_2 \sum_{i=1}^{n} x_{ik}x_{i2} + \cdots + \hat{\beta}_k \sum_{i=1}^{n} x_{ik}^2 &= \sum_{i=1}^{n} x_{ik}y_i
\end{aligned} \qquad \textbf{(10.10)}$$

These equations are called the **least squares normal equations.** Note that there are $p = k + 1$ normal equations, one for each of the unknown regression coefficients. The solution to the normal equations will be the least squares estimators of the regression coefficients $\hat{\beta}_0, \hat{\beta}_1, \ldots, \hat{\beta}_k$.

It is simpler to solve the normal equations if they are expressed in matrix notation. We now give a matrix development of the normal equations that parallels the development of Equation 10.10. The model in terms of the observations, Equation 10.7, may be written in matrix notation as

$$\mathbf{y} = \mathbf{X}\boldsymbol{\beta} + \boldsymbol{\epsilon}$$

where

$$\mathbf{y} = \begin{bmatrix} y_1 \\ y_2 \\ \vdots \\ y_n \end{bmatrix}, \quad \mathbf{X} = \begin{bmatrix} 1 & x_{11} & x_{12} & \cdots & x_{1k} \\ 1 & x_{21} & x_{22} & \cdots & x_{2k} \\ \vdots & \vdots & \vdots & & \vdots \\ 1 & x_{n1} & x_{n2} & \cdots & x_{nk} \end{bmatrix}, \quad \boldsymbol{\beta} = \begin{bmatrix} \beta_0 \\ \beta_1 \\ \vdots \\ \beta_k \end{bmatrix}, \text{ and } \quad \boldsymbol{\epsilon} = \begin{bmatrix} \epsilon_1 \\ \epsilon_2 \\ \vdots \\ \epsilon_n \end{bmatrix}$$

In general, $\mathbf{y}$ is an $(n \times 1)$ vector of the observations, $\mathbf{X}$ is an $(n \times p)$ matrix of the levels of the independent variables, $\boldsymbol{\beta}$ is a $(p \times 1)$ vector of the regression coefficients, and $\boldsymbol{\epsilon}$ is an $(n \times 1)$ vector of random errors.

We wish to find the vector of least squares estimators, $\hat{\boldsymbol{\beta}}$, that minimizes

$$L = \sum_{i=1}^{n} \epsilon_i^2 = \boldsymbol{\epsilon}'\boldsymbol{\epsilon} = (\mathbf{y} - \mathbf{X}\boldsymbol{\beta})'(\mathbf{y} - \mathbf{X}\boldsymbol{\beta})$$

Note that L may be expressed as

$$\begin{aligned}
L &= \mathbf{y}'\mathbf{y} - \boldsymbol{\beta}'\mathbf{X}'\mathbf{y} - \mathbf{y}'\mathbf{X}\boldsymbol{\beta} + \boldsymbol{\beta}'\mathbf{X}'\mathbf{X}\boldsymbol{\beta} \\
&= \mathbf{y}'\mathbf{y} - 2\boldsymbol{\beta}'\mathbf{X}'\mathbf{y} + \boldsymbol{\beta}'\mathbf{X}'\mathbf{X}\boldsymbol{\beta}
\end{aligned} \qquad \textbf{(10.11)}$$

because $\boldsymbol{\beta}'\mathbf{X}'\mathbf{y}$ is a (1×1) matrix, or a scalar, and its transpose $(\boldsymbol{\beta}'\mathbf{X}'\mathbf{y})' = \mathbf{y}'\mathbf{X}\boldsymbol{\beta}$ is the same scalar. The least squares estimators must satisfy

$$\left.\frac{\partial L}{\partial \boldsymbol{\beta}}\right|_{\hat{\boldsymbol{\beta}}} = -2\mathbf{X}'\mathbf{y} + 2\mathbf{X}'\mathbf{X}\hat{\boldsymbol{\beta}} = \mathbf{0}$$

which simplifies to

$$\mathbf{X}'\mathbf{X}\hat{\boldsymbol{\beta}} = \mathbf{X}'\mathbf{y} \qquad \textbf{(10.12)}$$

Equation 10.12 is the matrix form of the least squares normal equations. It is identical to Equation 10.10. To solve the normal equations, multiply both sides of Equation 10.12 by the inverse of $\mathbf{X}'\mathbf{X}$. Thus, the least squares estimator of $\boldsymbol{\beta}$ is

$$\hat{\boldsymbol{\beta}} = (\mathbf{X}'\mathbf{X})^{-1}\mathbf{X}'\mathbf{y} \qquad \textbf{(10.13)}$$

It is easy to see that the matrix form of the normal equations is identical to the scalar form. Writing out Equation 10.12 in detail, we obtain

$$\begin{bmatrix} n & \sum_{i=1}^{n} x_{i1} & \sum_{i=1}^{n} x_{i2} & \cdots & \sum_{i=1}^{n} x_{ik} \\ \sum_{i=1}^{n} x_{i1} & \sum_{i=1}^{n} x_{i1}^2 & \sum_{i=1}^{n} x_{i1}x_{i2} & \cdots & \sum_{i=1}^{n} x_{i1}x_{ik} \\ \vdots & \vdots & \vdots & & \vdots \\ \sum_{i=1}^{n} x_{ik} & \sum_{i=1}^{n} x_{ik}x_{i1} & \sum_{i=1}^{n} x_{ik}x_{i2} & \cdots & \sum_{i=1}^{n} x_{ik}^2 \end{bmatrix} \begin{bmatrix} \hat{\beta}_0 \\ \hat{\beta}_1 \\ \vdots \\ \hat{\beta}_k \end{bmatrix} = \begin{bmatrix} \sum_{i=1}^{n} y_i \\ \sum_{i=1}^{n} x_{i1}y_i \\ \vdots \\ \sum_{i=1}^{n} x_{ik}y_i \end{bmatrix}$$

If the indicated matrix multiplication is performed, the scalar form of the normal equations (i.e., Equation 10.10) will result. In this form it is easy to see that $\mathbf{X}'\mathbf{X}$ is a $(p \times p)$ symmetric matrix and $\mathbf{X}'\mathbf{y}$ is a $(p \times 1)$ column vector. Note the special structure of the $\mathbf{X}'\mathbf{X}$ matrix. The diagonal elements of $\mathbf{X}'\mathbf{X}$ are the sums of squares of the elements in the columns of $\mathbf{X}$, and the off-diagonal elements are the sums of cross products of the elements in the columns of $\mathbf{X}$. Furthermore, note that the elements of $\mathbf{X}'\mathbf{y}$ are the sums of cross products of the columns of $\mathbf{X}$ and the observations $\{y_i\}$.

The fitted regression model is

$$\hat{\mathbf{y}} = \mathbf{X}\hat{\boldsymbol{\beta}} \tag{10.14}$$

In scalar notation, the fitted model is

$$\hat{y}_i = \hat{\beta}_0 + \sum_{j=1}^{k} \hat{\beta}_j x_{ij} \qquad i = 1, 2, \ldots, n$$

The difference between the actual observation y_i and the corresponding fitted value $\hat{y}_i$ is the **residual**, say $e_i = y_i - \hat{y}_i$. The $(n \times 1)$ vector of residuals is denoted by

$$\mathbf{e} = \mathbf{y} - \hat{\mathbf{y}} \tag{10.15}$$

Estimating σ^2. It is also usually necessary to estimate σ^2. To develop an estimator of this parameter, consider the sum of squares of the residuals, say

$$SS_E = \sum_{i=1}^{n} (y_i - \hat{y}_i)^2 = \sum_{i=1}^{n} e_i^2 = \mathbf{e}'\mathbf{e}$$

Substituting $\mathbf{e} = \mathbf{y} - \hat{\mathbf{y}} = \mathbf{y} - \mathbf{X}\hat{\boldsymbol{\beta}}$, we have

$$\begin{aligned} SS_E &= (\mathbf{y} - \mathbf{X}\hat{\boldsymbol{\beta}})'(\mathbf{y} - \mathbf{X}\hat{\boldsymbol{\beta}}) \\ &= \mathbf{y}'\mathbf{y} - \hat{\boldsymbol{\beta}}'\mathbf{X}'\mathbf{y} - \mathbf{y}'\mathbf{X}\hat{\boldsymbol{\beta}} + \hat{\boldsymbol{\beta}}'\mathbf{X}'\mathbf{X}\hat{\boldsymbol{\beta}} \\ &= \mathbf{y}'\mathbf{y} - 2\hat{\boldsymbol{\beta}}'\mathbf{X}'\mathbf{y} + \hat{\boldsymbol{\beta}}'\mathbf{X}'\mathbf{X}\hat{\boldsymbol{\beta}} \end{aligned}$$

Because $\mathbf{X}'\mathbf{X}\hat{\boldsymbol{\beta}} = \mathbf{X}'\mathbf{y}$, this last equation becomes

$$SS_E = \mathbf{y}'\mathbf{y} - \hat{\boldsymbol{\beta}}'\mathbf{X}'\mathbf{y} \tag{10.16}$$

Equation 10.16 is called the **error** or **residual sum of squares**, and it has $n - p$ degrees of freedom associated with it. It can be shown that

$$E(SS_E) = \sigma^2(n - p)$$

so an unbiased estimator of σ^2 is given by

$$\hat{\sigma}^2 = \frac{SS_E}{n - p} \tag{10.17}$$

Properties of the Estimators. The method of least squares produces an unbiased estimator of the parameter $\boldsymbol{\beta}$ in the linear regression model. This may be easily demonstrated by taking the expected value of $\hat{\boldsymbol{\beta}}$ as follows:

$$E(\hat{\boldsymbol{\beta}}) = E[(\mathbf{X}'\mathbf{X})^{-1}\mathbf{X}'\mathbf{y}] = E[(\mathbf{X}'\mathbf{X})^{-1}\mathbf{X}'(\mathbf{X}\boldsymbol{\beta} + \boldsymbol{\epsilon})]$$
$$= E[(\mathbf{X}'\mathbf{X})^{-1}\mathbf{X}'\mathbf{X}\boldsymbol{\beta} + (\mathbf{X}'\mathbf{X})^{-1}\mathbf{X}'\boldsymbol{\epsilon}] = \boldsymbol{\beta}$$

because $E(\boldsymbol{\epsilon}) = \mathbf{0}$ and $(\mathbf{X}'\mathbf{X})^{-1}\mathbf{X}'\mathbf{X} = \mathbf{I}$. Thus, $\hat{\boldsymbol{\beta}}$ is an unbiased estimator of $\boldsymbol{\beta}$.

The variance property of $\hat{\boldsymbol{\beta}}$ is expressed in the **covariance matrix**:

$$\text{Cov}(\hat{\boldsymbol{\beta}}) \equiv E\{[\hat{\boldsymbol{\beta}} - E(\hat{\boldsymbol{\beta}})][\hat{\boldsymbol{\beta}} - E(\hat{\boldsymbol{\beta}})]'\} \tag{10.18}$$

which is just a symmetric matrix whose ith main diagonal element is the variance of the individual regression coefficient $\hat{\beta}_i$ and whose (ij)th element is the covariance between $\hat{\beta}_i$ and $\hat{\beta}_j$. The covariance matrix of $\hat{\boldsymbol{\beta}}$ is

$$\text{Cov}(\hat{\boldsymbol{\beta}}) = \sigma^2(\mathbf{X}'\mathbf{X})^{-1} \tag{10.19}$$

If σ^2 in Equation 10.19 is replaced with the estimate $\hat{\sigma}^2$ from Equation 10.12, we obtain an estimate of the covariance matrix of $\hat{\boldsymbol{\beta}}$. The square roots of the main diagonal elements of this matrix are the **standard errors** of the model parameters.

EXAMPLE 10.1

Sixteen observations on the viscosity of a polymer (y) and two process variables—reaction temperature (x_1) and catalyst feed rate (x_2)—are shown in Table 10.2. We will fit a multiple linear regression model

$$y = \beta_0 + \beta_1 x_1 + \beta_2 x_2 + \epsilon$$

to these data. The $\mathbf{X}$ matrix and $\mathbf{y}$ vector are

$$\mathbf{X} = \begin{bmatrix} 1 & 80 & 8 \\ 1 & 93 & 9 \\ 1 & 100 & 10 \\ 1 & 82 & 12 \\ 1 & 90 & 11 \\ 1 & 99 & 8 \\ 1 & 81 & 8 \\ 1 & 96 & 10 \\ 1 & 94 & 12 \\ 1 & 93 & 11 \\ 1 & 97 & 13 \\ 1 & 95 & 11 \\ 1 & 100 & 8 \\ 1 & 85 & 12 \\ 1 & 86 & 9 \\ 1 & 87 & 12 \end{bmatrix} \quad \mathbf{y} = \begin{bmatrix} 2256 \\ 2340 \\ 2426 \\ 2293 \\ 2330 \\ 2368 \\ 2250 \\ 2409 \\ 2364 \\ 2379 \\ 2440 \\ 2364 \\ 2404 \\ 2317 \\ 2309 \\ 2328 \end{bmatrix}$$

The $\mathbf{X}'\mathbf{X}$ matrix is

$$\mathbf{X}'\mathbf{X} = \begin{bmatrix} 1 & 1 & \cdots & 1 \\ 80 & 93 & \cdots & 87 \\ 8 & 9 & \cdots & 12 \end{bmatrix} \begin{bmatrix} 1 & 80 & 8 \\ 1 & 93 & 9 \\ \vdots & \vdots & \vdots \\ 1 & 87 & 12 \end{bmatrix}$$
$$= \begin{bmatrix} 16 & 1458 & 164 \\ 1458 & 133{,}560 & 14{,}946 \\ 164 & 14{,}946 & 1{,}726 \end{bmatrix}$$

and the $\mathbf{X}'\mathbf{y}$ vector is

$$\mathbf{X}'\mathbf{y} = \begin{bmatrix} 1 & 1 & \cdots & 1 \\ 80 & 93 & \cdots & 87 \\ 8 & 9 & \cdots & 12 \end{bmatrix} \begin{bmatrix} 2256 \\ 2340 \\ \vdots \\ 2328 \end{bmatrix} = \begin{bmatrix} 37{,}577 \\ 3{,}429{,}550 \\ 385{,}562 \end{bmatrix}$$

The least squares estimate of $\boldsymbol{\beta}$ is

$$\hat{\boldsymbol{\beta}} = (\mathbf{X}'\mathbf{X})^{-1}\mathbf{X}'\mathbf{y}$$

or

$$\hat{\boldsymbol{\beta}} = \begin{bmatrix} 14.176004 & -0.129746 & -0.223453 \\ -0.129746 & 1.429184 \times 10^{-3} & -4.763947 \times 10^{-5} \\ -0.223453 & -4{,}763947 \times 10^{-5} & 2.222381 \times 10^{-2} \end{bmatrix} \begin{bmatrix} 37{,}577 \\ 3{,}429{,}550 \\ 385{,}562 \end{bmatrix} = \begin{bmatrix} 1566.07777 \\ 7.62129 \\ 8.58485 \end{bmatrix}$$

■ **TABLE 10.2**
Viscosity Data for Example 10.1 (viscosity in centistokes @ 100°c)

Observation	Temperature (x_1, °C)	Catalyst Feed Rate (x_2, lb/h)	Viscosity
1	80	8	2256
2	93	9	2340
3	100	10	2426
4	82	12	2293
5	90	11	2330
6	99	8	2368
7	81	8	2250
8	96	10	2409
9	94	12	2364
10	93	11	2379
11	97	13	2440
12	95	11	2364
13	100	8	2404
14	85	12	2317
15	86	9	2309
16	87	12	2328

The least squares fit, with the regression coefficients reported to two decimal places, is

$$\hat{y} = 1566.08 + 7.62x_1 + 8.58x_2$$

The first three columns of Table 10.3 present the actual observations y_i, the predicted or fitted values $\hat{y}_i$, and the residuals. Figure 10.1 is a normal probability plot of the residuals. Plots of the residuals versus the predicted values $\hat{y}_i$ and versus the two variables x_1 and x_2 are shown in Figures 10.2, 10.3, and 10.4, respectively. Just as in designed experiments, residual plotting is an integral part of regression model building. These plots indicate that the variance of the observed viscosity tends to increase with the magnitude of viscosity. Figure 10.3 suggests that the variability in viscosity is increasing as temperature increases.

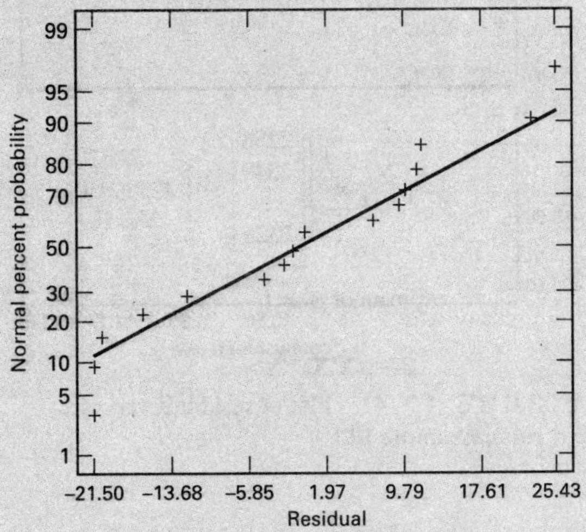

■ **FIGURE 10.1** **Normal probability plot of residuals, Example 10.1**

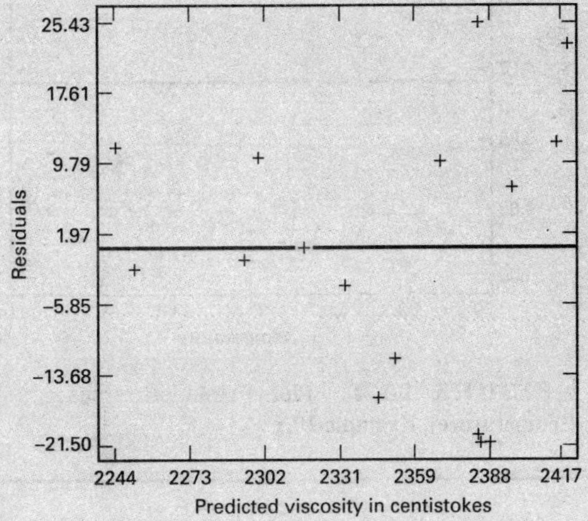

■ **FIGURE 10.2** **Plot of residuals versus predicted viscosity, Example 10.1**

■ **TABLE 10.3**
Predicted Values, Residuals, and Other Diagnostics from Example 10.1

Observation i	y_i	Predicted Value $\hat{y}_i$	Residual e_i	h_{ii}	Studentized Residual	D_i	R-Student
1	2256	2244.5	11.5	0.350	0.87	0.137	0.87
2	2340	2352.1	−12.1	0.102	−0.78	0.023	−0.77
3	2426	2414.1	11.9	0.177	0.80	0.046	0.79
4	2293	2294.0	−1.0	0.251	−0.07	0.001	−0.07
5	2330	2346.4	−16.4	0.077	−1.05	0.030	−1.05
6	2368	2389.3	−21.3	0.265	−1.52	0.277	−1.61
7	2250	2252.1	−2.1	0.319	−0.15	0.004	−0.15
8	2409	2383.6	25.4	0.098	1.64	0.097	1.76
9	2364	2385.5	−21.5	0.142	−1.42	0.111	−1.48
10	2379	2369.3	9.7	0.080	0.62	0.011	0.60
11	2440	2416.9	23.1	0.278	1.66	0.354	1.80
12	2364	2384.5	−20.5	0.096	−1.32	0.062	−1.36
13	2404	2396.9	7.1	0.289	0.52	0.036	0.50
14	2317	2316.9	0.1	0.185	0.01	0.000	<0.01
15	2309	2298.8	10.2	0.134	0.67	0.023	0.66
16	2328	2332.1	−4.1	0.156	−0.28	0.005	−0.27

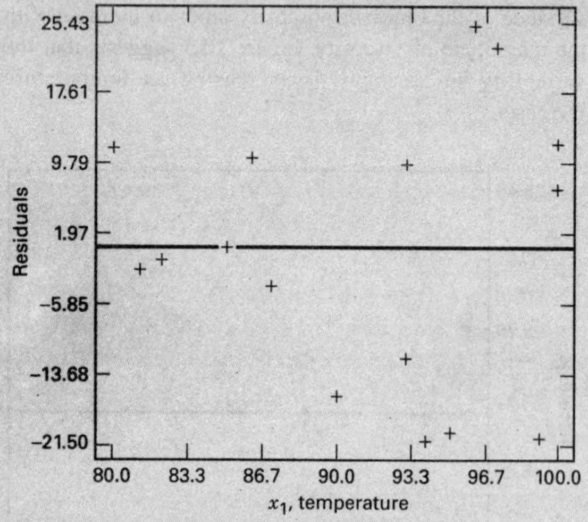

■ **FIGURE 10.3** **Plot of residuals versus x_1 (temperature), Example 10.1**

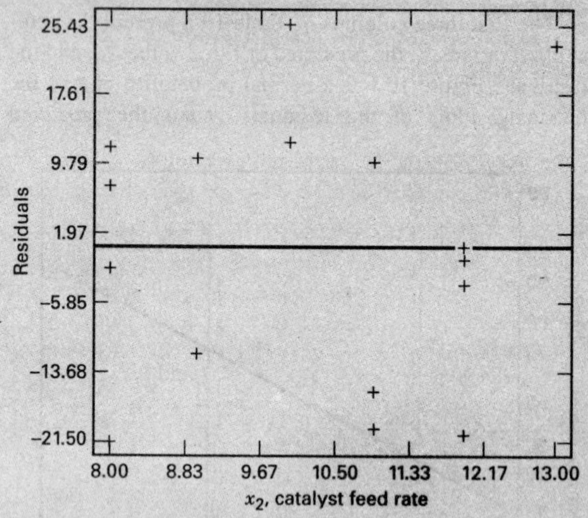

■ **FIGURE 10.4** **Plot of residuals versus x_2 (feed rate), Example 10.1**

Using the Computer. Regression model fitting is almost always done using a statistical software package, such as Minitab or JMP. Table 10.4 shows some of the output obtained when Minitab is used to fit the viscosity regression model in Example 10.1. Many of the quantities in this output should be familiar because they have similar meanings to the quantities in the output displays for computer analysis of data from designed experiments. We have seen many such computer outputs previously in the book. In subsequent sections, we will discuss the analysis of variance and t-test information in Table 10.4 in detail and will show exactly how these quantities were computed.

Fitting Regression Models in Designed Experiments. We have often used a regression model to present the results of a designed experiment in a quantitative form. We now give a complete illustrative example. This is followed by three other brief examples that illustrative other useful applications of regression analysis in designed experiments.

EXAMPLE 10.2 Regression Analysis of a 2^3 Factorial Design

A chemical engineer is investigating the yield of a process. Three process variables are of interest: temperature, pressure, and catalyst concentration. Each variable can be run at a low and a high level, and the engineer decides to run a 2^3 design with four center points. The design and the resulting yields are shown in Figure 10.5, where we have shown both the natural levels of the design factor and the $+1, -1$ coded variable notation normally employed in 2^k factorial designs to represent the factor levels.

Suppose that the engineer decides to fit a main effects only model, say

$$y = \beta_0 + \beta_1 x_1 + \beta_2 x_2 + \beta_3 x_3 + \epsilon$$

For this model the $\mathbf{X}$ matrix and $\mathbf{y}$ vector are

$$\mathbf{X} = \begin{bmatrix} 1 & -1 & -1 & -1 \\ 1 & 1 & -1 & -1 \\ 1 & -1 & 1 & -1 \\ 1 & 1 & 1 & -1 \\ 1 & -1 & -1 & 1 \\ 1 & 1 & -1 & 1 \\ 1 & -1 & 1 & 1 \\ 1 & 1 & 1 & 1 \\ 1 & 0 & 0 & 0 \\ 1 & 0 & 0 & 0 \\ 1 & 0 & 0 & 0 \\ 1 & 0 & 0 & 0 \end{bmatrix} \quad \text{and } \mathbf{y} = \begin{bmatrix} 32 \\ 46 \\ 57 \\ 65 \\ 36 \\ 48 \\ 57 \\ 68 \\ 50 \\ 44 \\ 53 \\ 56 \end{bmatrix}$$

The 2^3 is an orthogonal design, and even with the added center runs it is still orthogonal. Therefore

$$\mathbf{X}'\mathbf{X} = \begin{bmatrix} 12 & 0 & 0 & 0 \\ 0 & 8 & 0 & 0 \\ 0 & 0 & 8 & 0 \\ 0 & 0 & 0 & 8 \end{bmatrix} \quad \text{and} \quad \mathbf{X}'\mathbf{y} = \begin{bmatrix} 612 \\ 45 \\ 85 \\ 9 \end{bmatrix}$$

Because the design is orthogonal, the $\mathbf{X}'\mathbf{X}$ matrix is *diagonal*, the required inverse is also diagonal, and the vector of least squares estimates of the regression coefficients is

$$\hat{\boldsymbol{\beta}} = (\mathbf{X}'\mathbf{X})^{-1}\mathbf{X}'\mathbf{y} = \begin{bmatrix} 1/12 & 0 & 0 & 0 \\ 0 & 1/8 & 0 & 0 \\ 0 & 0 & 1/8 & 0 \\ 0 & 0 & 0 & 1/8 \end{bmatrix} \begin{bmatrix} 612 \\ 45 \\ 85 \\ 9 \end{bmatrix}$$

$$= \begin{bmatrix} 51.000 \\ 5.625 \\ 10.625 \\ 1.125 \end{bmatrix}$$

The fitted regression model is

$$\hat{y} = 51.000 + 5.625x_1 + 10.625x_2 + 1.125x_3$$

As we have made use of on many occasions, the regression coefficients are closely connected to the effect estimates that would be obtained from the usual analysis of a 2^3 design. For example, the effect of temperature is (refer to Figure 10.5)

$$T = \bar{y}_{T^+} - \bar{y}_{T^-}$$

$$= 56.75 - 45.50 = 11.25$$

Notice that the regression coefficient for x_1 is

$$(11.25)/2 = 5.625$$

That is, the regression coefficient is exactly one-half the usual effect estimate. This will always be true for a 2^k design. As noted above, we used this result in Chapters 6 through 8 to produce regression models, fitted values, and residuals for several two-level experiments. This example demonstrates that the effect estimates from a 2^k design are least squares estimates.

The variance of the regression model parameter are found from the diagonal elements of $(\mathbf{X}'\mathbf{X})^{-1}$. That is,

$$V(\hat{\beta}_0) = \frac{\sigma^2}{12}, \quad \text{and} \quad V(\hat{\beta}_i) = \frac{\sigma^2}{8}, i = 1,2,3.$$

The **relative variance** are

$$\frac{V(\hat{\beta}_0)}{\sigma^2} = \frac{1}{12} \quad \text{and} \quad \frac{V(\hat{\beta}_i)}{\sigma^2} = \frac{1}{8}, i = 1,2,3.$$

	Process Variables			Coded Variables			
Run	Temp (°C)	Pressure (psig)	Conc (g/l)	x_1	x_2	x_3	Yield, y
1	120	40	15	−1	−1	−1	32
2	160	80	15	1	−1	−1	46
3	120	40	15	−1	1	−1	57
4	160	80	15	1	1	−1	65
5	120	40	30	−1	−1	1	36
6	160	80	30	1	−1	1	48
7	120	40	30	−1	1	1	57
8	160	80	30	1	1	1	68
9	140	60	22.5	0	0	0	50
10	140	60	22.5	0	0	0	44
11	140	60	22.5	0	0	0	53
12	140	60	22.5	0	0	0	56

$$x_1 = \frac{\text{Temp} - 140}{20}, \quad x_2 = \frac{\text{Pressure} - 60}{20}, \quad x_3 = \frac{\text{Conc} - 22.5}{7.5}$$

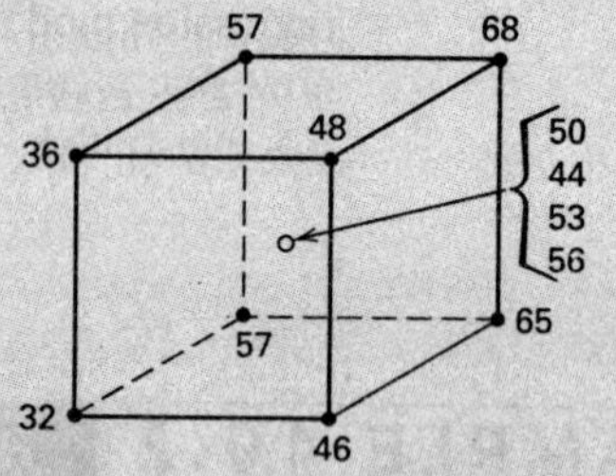

■ **FIGURE 10.5** **Experimental design for Example 10.2**

■ **TABLE 10.4**
Minitab Output for the Viscosity Regression Model, Example 10.1

Regression Analysis

```
The regression equation is
Viscosity = 1566 + 7.62 Temp + 8.58 Feed Rate

Predictor        Coef      Std. Dev.        T        P
Constant      1566.08        61.59      25.43    0.000
Temp           7.6213       0.6184      12.32    0.000
Feed Rat        8.585        2.439       3.52    0.004

S = 16.36     R-Sq = 92.7%     R-Sq (adj) = 91.6%

Analysis of Variance

Source            DF        SS        MS        F        P
Regression         2     44157     22079    82.50    0.000
Residual Error    13      3479       268
Total             15     47636

Source       DF      Seq SS
Temp          1       40841
Feed Rat      1        3316
```

In Example 10.2, the inverse matrix is easy to obtain because $\mathbf{X}'\mathbf{X}$ is diagonal. Intuitively, this seems to be advantageous, not only because of the computational simplicity but also because the estimators of all the regression coefficients are uncorrelated; that is, $\text{Cov}(\hat{\beta}_i, \hat{\beta}_j) = 0$. If we can choose the levels of the x variables before the data are collected, we might wish to design the experiment so that a diagonal $\mathbf{X}'\mathbf{X}$ will result.

In practice, it can be relatively easy to do this. We know that the off-diagonal elements in $\mathbf{X}'\mathbf{X}$ are the sums of cross products of the columns in $\mathbf{X}$. Therefore, we must make the inner product of the columns of $\mathbf{X}$ equal to zero; that is, these columns must be **orthogonal**. As we have noted before, experimental designs that have this property for fitting a regression model are called **orthogonal designs**. In general, the 2^k factorial design is an orthogonal design for fitting the multiple linear regression model.

Regression methods are extremely useful when something "goes wrong" in a designed experiment. This is illustrated in the next two examples.

EXAMPLE 10.3 A 2^3 Factorial Design with a Missing Observation

Consider the 2^3 factorial design with four center points from Example 10.2. Suppose that when this experiment was performed, the run with all variables at the high level (run 8 in Figure 10.5) was missing. This can happen for a variety of reasons; the measurement system can produce a faulty reading, the combination of factor levels may prove infeasible, the experimental unit may be damaged, and so forth.

We will fit the main effects model

$$y = \beta_0 + \beta_1 x_1 + \beta_2 x_2 + \beta_3 x_3 + \epsilon$$

using the 11 remaining observations. The $\mathbf{X}$ matrix and $\mathbf{y}$ vector are

$$\mathbf{X} = \begin{bmatrix} 1 & -1 & -1 & -1 \\ 1 & 1 & -1 & -1 \\ 1 & -1 & 1 & -1 \\ 1 & 1 & 1 & -1 \\ 1 & -1 & -1 & 1 \\ 1 & 1 & -1 & 1 \\ 1 & -1 & 1 & 1 \\ 1 & 0 & 0 & 0 \\ 1 & 0 & 0 & 0 \\ 1 & 0 & 0 & 0 \\ 1 & 0 & 0 & 0 \end{bmatrix} \quad \text{and} \quad \mathbf{y} = \begin{bmatrix} 32 \\ 46 \\ 57 \\ 65 \\ 36 \\ 48 \\ 57 \\ 50 \\ 44 \\ 53 \\ 56 \end{bmatrix}$$

To estimate the model parameters, we form

$$\mathbf{X}'\mathbf{X} = \begin{bmatrix} 11 & -1 & -1 & -1 \\ -1 & 7 & -1 & -1 \\ -1 & -1 & 7 & -1 \\ -1 & -1 & -1 & 7 \end{bmatrix} \quad \text{and } \mathbf{X}'\mathbf{y} = \begin{bmatrix} 544 \\ -23 \\ 17 \\ -59 \end{bmatrix}$$

Because there is a missing observation, the design is no longer orthogonal. Now

$$\hat{\boldsymbol{\beta}} = (\mathbf{X}'\mathbf{X})^{-1}\mathbf{X}'\mathbf{y}$$

$$= \begin{bmatrix} 9.61538 \times 10^{-2} & 1.92307 \times 10^{-2} & 1.92307 \times 10^{-2} & 1.92307 \times 10^{-2} \\ 1.92307 \times 10^{-2} & 0.15385 & 2.88462 \times 10^{-2} & 2.88462 \times 10^{-2} \\ 1.92307 \times 10^{-2} & 2.88462 \times 10^{-2} & 0.15385 & 2.88462 \times 10^{-2} \\ 1.92307 \times 10^{-2} & 2.88462 \times 10^{-2} & 2.88462 \times 10^{-2} & 0.15385 \end{bmatrix} \begin{bmatrix} 544 \\ -23 \\ 17 \\ -59 \end{bmatrix}$$

$$= \begin{bmatrix} 51.25 \\ 5.75 \\ 10.75 \\ 1.25 \end{bmatrix}$$

Therefore, the fitted model is

$$\hat{y} = 51.25 + 5.75x_1 + 10.75x_2 + 1.25x_3$$

Compare this model to the one obtained in Example 10.2, where all 12 observations were used. The regression coefficients are very similar. Because the regression coefficients are closely related to the factor effects, our conclusions would not be seriously affected by the missing observation. However, notice that the effect estimates are no longer orthogonal because $\mathbf{X}'\mathbf{X}$ and its inverse are no longer diagonal. Further more the variances of the regression coefficients are larger than they were in the original orthogonal design with no missing data.

EXAMPLE 10.4 Inaccurate Levels in Design Factors

When running a designed experiment, it is sometimes difficult to reach and hold the precise factor levels required by the design. Small discrepancies are not important, but large ones are potentially of more concern. Regression methods are useful in the analysis of a designed experiment where the experimenter has been unable to obtain the required factor levels.

To illustrate, the experiment presented in Table 10.5 shows a variation of the 2^3 design from Example 10.2, where many of the test combinations are not exactly the ones specified in the design. Most of the difficulty seems to have occurred with the temperature variable.

We will fit the main effects model

$$y = \beta_0 + \beta_1 x_1 + \beta_2 x_2 + \beta_3 x_3 + \epsilon$$

The **X** matrix and **y** vector are

$$\mathbf{X} = \begin{bmatrix} 1 & -0.75 & -0.95 & -1.133 \\ 1 & 0.90 & -1 & -1 \\ 1 & -0.95 & 1.1 & -1 \\ 1 & 1 & 0 & -1 \\ 1 & -1.10 & -1.05 & 1.4 \\ 1 & 1.15 & -1 & 1 \\ 1 & -0.90 & 1 & 1 \\ 1 & 1.25 & 1.15 & 1 \\ 1 & 0 & 0 & 0 \\ 1 & 0 & 0 & 0 \\ 1 & 0 & 0 & 0 \\ 1 & 0 & 0 & 0 \end{bmatrix} \qquad \mathbf{y} = \begin{bmatrix} 32 \\ 46 \\ 57 \\ 65 \\ 36 \\ 48 \\ 57 \\ 68 \\ 50 \\ 44 \\ 53 \\ 56 \end{bmatrix}$$

To estimate the model parameters, we need

$$\mathbf{X}'\mathbf{X} = \begin{bmatrix} 12 & 0.60 & 0.25 & 0.2670 \\ 0.60 & 8.18 & 0.31 & -0.1403 \\ 0.25 & 0.31 & 8.5375 & -0.3437 \\ 0.2670 & -0.1403 & -0.3437 & 9.2437 \end{bmatrix}$$

$$\mathbf{X}'\mathbf{y} = \begin{bmatrix} 612 \\ 77.55 \\ 100.7 \\ 19.144 \end{bmatrix}$$

■ **TABLE 10.5**
Experimental Design for Example 10.4

	Process Variables			Coded Variables			Yield
Run	Temp (°C)	Pressure (psig)	Conc (g/l)	x_1	x_2	x_3	y
1	125	41	14	−0.75	−0.95	−1.133	32
2	158	40	15	0.90	−1	−1	46
3	121	82	15	−0.95	1.1	−1	57
4	160	80	15	1	1	−1	65
5	118	39	33	−1.10	−1.05	1.14	36
6	163	40	30	1.15	−1	1	48
7	122	80	30	−0.90	1	1	57
8	165	83	30	1.25	1.15	1	68
9	140	60	22.5	0	0	0	50
10	140	60	22.5	0	0	0	44
11	140	60	22.5	0	0	0	53
12	140	60	22.5	0	0	0	56

Then

$\hat{\boldsymbol{\beta}} = (\mathbf{X}'\mathbf{X})^{-1}\mathbf{X}'\mathbf{y}$

$$= \begin{bmatrix} 8.37447 \times 10^{-2} & -6.09871 \times 10^{-3} & -2.33542 \times 10^{-3} & -2.59833 \times 10^{-3} \\ -6.09871 \times 10^{-3} & 0.12289 & -4.20766 \times 10^{-3} & 1.88490 \times 10^{-3} \\ -2.33542 \times 10^{-3} & -4.20766 \times 10^{-3} & 0.11753 & 4.37851 \times 10^{-3} \\ -2.59833 \times 10^{-3} & 1.88490 \times 10^{-3} & 4.37851 \times 10^{-3} & 0.10845 \end{bmatrix} \begin{bmatrix} 612 \\ 77.55 \\ 100.7 \\ 19.144 \end{bmatrix} = \begin{bmatrix} 50.49391 \\ 5.40996 \\ 10.16316 \\ 1.07245 \end{bmatrix}$$

The fitted regression model, with the coefficients reported to two decimal places, is

$$\hat{y} = 50.49 + 5.41x_1 + 10.16x_2 + 1.07x_3$$

Comparing this to the original model in Example 10.2, where the factor levels were exactly those specified by the design, we note very little difference. The practical interpretation of the results of this experiment would not be seriously affected by the inability of the experimenter to achieve the desired factor levels exactly.

EXAMPLE 10.5 De-aliasing Interactions in a Fractional Factorial

We observed in Chapter 8 that it is possible to de-alias interactions in a fractional factorial design by a process called fold over. For a resolution III design, a full fold over is constructed by running a second fraction in which the signs are reversed from those in the original fraction. Then the combined design can be used to de-alias all main effects from the two-factor interactions.

A difficulty with a full fold over is that it requires a second group of runs of identical size as the original design. It is usually possible to de-alias certain interactions of interest by augmenting the original design with fewer runs than required in a full fold over. The partial fold-over technique was used to solve this problem. Regression methods are an easy way to see how the partial fold-over technique works and, in some cases, find even more efficient fold-over designs.

To illustrate, suppose that we have run a 2_{IV}^{4-1} design. Table 8.3 shows the principal fraction of this design, in which $I = ABCD$. Suppose that after the data from the first eight trials were observed, the largest effects were A, B, C, D (we ignore the three-factor interactions that are aliased with these main effects) and the $AB + CD$ alias chain. The other two alias chains can be ignored, but clearly either AB, CD, or both two-factor interactions are large. To find out which interactions are important, we could, of course, run the alternate fraction, which would require another eight trials. Then all 16 runs could be used to estimate the main effects and the two-factor interactions. An alternative would be to use a partial fold over involving four additional runs.

It is possible to de-alias AB and CD in fewer than four additional trials. Suppose that we wish to fit the model

$$y = \beta_0 + \beta_1 x_1 + \beta_2 x_2 + \beta_3 x_3 + \beta_4 x_4 + \beta_{12} x_1 x_2 + \beta_{34} x_3 x_4 + \epsilon$$

where x_1, x_2, x_3, and x_4 are the coded variables representing A, B, C, and D. Using the design in Table 8.3, the $\mathbf{X}$ matrix for this model is

$$\mathbf{X} = \begin{array}{c} \\ \end{array}\overset{\begin{array}{ccccccc} & x_1 & x_2 & x_3 & x_4 & x_1x_2 & x_3x_4 \end{array}}{\begin{bmatrix} 1 & -1 & -1 & -1 & -1 & 1 & 1 \\ 1 & 1 & -1 & -1 & 1 & -1 & -1 \\ 1 & -1 & 1 & -1 & 1 & -1 & -1 \\ 1 & 1 & 1 & -1 & -1 & 1 & 1 \\ 1 & -1 & -1 & 1 & 1 & 1 & 1 \\ 1 & 1 & -1 & 1 & -1 & -1 & -1 \\ 1 & -1 & 1 & 1 & -1 & -1 & -1 \\ 1 & 1 & 1 & 1 & 1 & 1 & 1 \end{bmatrix}}$$

where we have written the variables above the columns to facilitate understanding. Notice that the x_1x_2 column is identical to the x_3x_4 column (as anticipated, because AB or x_1x_2 is aliased with CD or x_3x_4), implying a linear dependency in the columns of $\mathbf{X}$. Therefore, we cannot estimate both β_{12} and β_{34} in the model. However, suppose that we add a single run $x_1 = -1$, $x_2 = -1$, $x_3 = -1$, and $x_4 = 1$

from the alternate fraction to the original eight runs. The **X** matrix for the model now becomes

$$\mathbf{X} = \begin{bmatrix} 1 & -1 & -1 & -1 & -1 & 1 & 1 \\ 1 & 1 & -1 & -1 & 1 & -1 & -1 \\ 1 & -1 & 1 & -1 & 1 & -1 & -1 \\ 1 & 1 & 1 & -1 & -1 & 1 & 1 \\ 1 & -1 & -1 & 1 & 1 & 1 & 1 \\ 1 & 1 & -1 & 1 & -1 & -1 & -1 \\ 1 & -1 & 1 & 1 & -1 & -1 & -1 \\ 1 & 1 & 1 & 1 & 1 & 1 & 1 \\ 1 & -1 & -1 & -1 & 1 & 1 & -1 \end{bmatrix} \quad (\text{columns: } x_1, x_2, x_3, x_4, x_1x_2, x_3x_4)$$

Notice that the columns x_1x_2 and x_3x_4 are now no longer identical, and we can fit the model including both the x_1x_2 (AB) and x_3x_4 (CD) interactions. The magnitudes of the regression coefficients will give insight regarding which interactions are important.

Although adding a single run will de-alias the AB and CD interactions, this approach does have a disadvantage. Suppose that there is a time effect (or a block effect) between the first eight runs and the last run added above. Add a column to the **X** matrix for blocks, and you obtain the following:

$$\mathbf{X} = \begin{bmatrix} 1 & -1 & -1 & -1 & -1 & 1 & 1 & -1 \\ 1 & 1 & -1 & -1 & 1 & -1 & -1 & -1 \\ 1 & -1 & 1 & -1 & 1 & -1 & -1 & -1 \\ 1 & 1 & 1 & -1 & -1 & 1 & 1 & -1 \\ 1 & -1 & -1 & 1 & 1 & 1 & 1 & -1 \\ 1 & 1 & -1 & 1 & -1 & -1 & -1 & -1 \\ 1 & -1 & 1 & 1 & -1 & -1 & -1 & -1 \\ 1 & 1 & 1 & 1 & 1 & 1 & 1 & -1 \\ 1 & -1 & -1 & -1 & 1 & 1 & -1 & 1 \end{bmatrix} \quad (\text{columns: } x_1, x_2, x_3, x_4, x_1x_2, x_3x_4, \text{block})$$

We have assumed the block factor was at the low or " − " level during the first eight runs, and at the high or " + " level during the ninth run. It is easy to see that the sum of the cross products of every column with the block column does not sum to zero, meaning that blocks are no longer orthogonal to treatments, or that the block effect now affects the estimates of the model regression coefficients. To block orthogonally, you must add an even number of runs. For example, the four runs

x_1	x_2	x_3	x_4
−1	−1	−1	1
1	−1	−1	−1
−1	1	1	1
1	1	1	−1

will de-alias AB from CD and allow orthogonal blocking (you can see this by writing out the **X** matrix as we did previously). This is equivalent to a partial fold over, in terms of the number of runs that are required.

In general, it is usually straightforward to examine the **X** matrix for the reduced model obtained from a fractional factorial and determine which runs to augment the original design with to de-alias interactions of potential interest. Furthermore, the impact of specific augmentation strategies can be evaluated using the general results for regression models given later in this chapter. There are also computer-based optimal design methods for constructing designs that can be useful for **design augmentation** to de-alias effects (refer to the supplemental material for Chapter 8).

10.4 Hypothesis Testing in Multiple Regression

In multiple linear regression problems, certain tests of hypotheses about the model parameters are helpful in measuring the usefulness of the model. In this section, we describe several important hypothesis-testing procedures. These procedures require that the errors ϵ_i in the model be normally and independently distributed with mean zero and variance σ^2, abbreviated $\epsilon \sim$ NID$(0, \sigma^2)$. As a result of this assumption, the observations y_i are normally and independently distributed with mean $\beta_0 + \sum_{j=1}^{k}\beta_j x_{ij}$ and variance σ^2

10.4.1 Test for Significance of Regression

The test for significance of regression is a test to determine whether a linear relationship exists between the response variable y and a subset of the regressor variables $x_1, x_2, \ldots, x_k$. The appropriate hypotheses are

$$H_0\colon \beta_1 = \beta_2 = \cdots = \beta_k = 0 \qquad \textbf{(10.20)}$$
$$H_1\colon \beta_j \neq 0 \quad \text{for at least one } j$$

Rejection of H_0 in Equation 10.20 implies that at least one of the regressor variables x_1, $x_2, \ldots, x_k$ contributes significantly to the model. The test procedure involves an analysis of variance partitioning of the total sum of squares SS_T into a sum of squares due to the model (or to regression) and a sum of squares due to residual (or error), say

$$SS_T = SS_R + SS_E \tag{10.21}$$

Now if the null hypothesis $H_0: \beta_1 = \beta_2 = \cdots = \beta_k = 0$ is true, then SS_R/σ^2 is distributed as χ^2_k, where the number of degrees of freedom for χ^2 is equal to the number of regressor variables in the model. Also, we can show that SS_E/σ^2 is distributed as χ^2_{n-k-1} and that SS_E and SS_R are independent. The test procedure for $H_0: \beta_1 = \beta_2 = \cdots = \beta_k = 0$ is to compute

$$F_0 = \frac{SS_R/k}{SS_E/(n-k-1)} = \frac{MS_R}{MS_E} \tag{10.22}$$

and to reject H_0 if F_0 exceeds $F_{\alpha,k,n-k-1}$. Alternatively, we could use the *P*-value approach to hypothesis testing and, reject H_0 if the *P*-value for the statistic F_0 is less than α. The test is usually summarized in an analysis of variance table such as Table 10.6.

A computational formula for SS_R may be found easily. We have derived a computational formula for SS_E in Equation 10.16—that is,

$$SS_E = \mathbf{y}'\mathbf{y} - \hat{\boldsymbol{\beta}}'\mathbf{X}'\mathbf{y}$$

Now, because $SS_T = \Sigma_{i=1}^n y_i^2 - (\Sigma_{i=1}^n y_i)^2/n = \mathbf{y}'\mathbf{y} - (\Sigma_{i=1}^n y_i)^2/n$, we may rewrite the foregoing equation as

$$SS_E = \mathbf{y}'\mathbf{y} - \frac{\left(\sum_{i=1}^n y_i\right)^2}{n} - \left[\hat{\boldsymbol{\beta}}'\mathbf{X}'\mathbf{y} - \frac{\left(\sum_{i=1}^n y_i\right)^2}{n}\right]$$

or

$$SS_E = SS_T - SS_R$$

Therefore, the regression sum of squares is

$$SS_R = \hat{\boldsymbol{\beta}}'\mathbf{X}'\mathbf{y} - \frac{\left(\sum_{i=1}^n y_i\right)^2}{n} \tag{10.23}$$

and the error sum of squares is

$$SS_E = \mathbf{y}'\mathbf{y} - \hat{\boldsymbol{\beta}}'\mathbf{X}'\mathbf{y} \tag{10.24}$$

and the total sum of squares is

$$SS_T = \mathbf{y}'\mathbf{y} - \frac{\left(\sum_{i=1}^n y_i\right)^2}{n} \tag{10.25}$$

■ **TABLE 10.6**

Analysis of Variance for Significance of Regression in Multiple Regression

Source of Variation	Sum of Squares	Degrees of Freedom	Mean Square	F_0
Regression	SS_R	k	MS_R	MS_R/MS_E
Error or residual	SS_E	$n-k-1$	MS_E	
Total	SS_T	$n-1$		

These computations are almost always performed with regression software. For instance, Table 10.4 shows some of the output from Minitab for the viscosity regression model in Example 10.1. The upper portion in this display is the analysis of variance for the model. The test of significance of regression in this example involves the hypotheses

$$H_0: \beta_1 = \beta_2 = 0$$
$$H_1: \beta_j \neq 0 \quad \text{for at least one } j$$

The P-value in Table 10.4 for the F statistic (Equation 10.22) is very small, so we would conclude that at least one of the two variables—temperature (x_1) and feed rate (x_2)—has a nonzero regression coefficient.

Table 10.4 also reports the coefficient of multiple determination R^2, where

$$R^2 = \frac{SS_R}{SS_T} = 1 - \frac{SS_E}{SS_T} \tag{10.26}$$

Just as in designed experiments, R^2 is a measure of the amount of reduction in the variability of y obtained by using the regressor variables $x_1, x_2, \ldots, x_k$ in the model. However, as we have noted previously, a large value of R^2 does not necessarily imply that the regression model is a good one. Adding a variable to the model will always increase R^2, regardless of whether the additional variable is statistically significant or not. Thus, it is possible for models that have large values of R^2 to yield poor predictions of new observations or estimates of the mean response.

Because R^2 always increases as we add terms to the model, some regression model builders prefer to use an **adjusted R^2 statistic** defined as

$$R^2_{\text{adj}} = 1 - \frac{SS_E/(n-p)}{SS_T/(n-1)} = 1 - \left(\frac{n-1}{n-p}\right)(1 - R^2) \tag{10.27}$$

In general, the adjusted R^2 statistic will not always increase as variables are added to the model. In fact, if unnecessary terms are added, the value of R^2_{adj} will often decrease.

For example, consider the viscosity regression model. The adjusted R^2 for the model is shown in Table 10.4. It is computed as

$$\begin{aligned} R^2_{\text{adj}} &= 1 - \left(\frac{n-1}{n-p}\right)(1 - R^2) \\ &= 1 - \left(\frac{15}{13}\right)(1 - 0.92697) = 0.915735 \end{aligned}$$

which is very close to the ordinary R^2. When R^2 and R^2_{adj} differ dramatically, there is a good chance that nonsignificant terms have been included in the model.

10.4.2 Tests on Individual Regression Coefficients and Groups of Coefficients

We are frequently interested in testing hypotheses on the individual regression coefficients. Such tests would be useful in determining the value of each regressor variable in the regression model. For example, the model might be more effective with the inclusion of additional variables or perhaps with the deletion of one or more of the variables already in the model.

Adding a variable to the regression model always causes the sum of squares for regression to increase and the error sum of squares to decrease. We must decide whether the increase in the regression sum of squares is sufficient to warrant using the additional variable in the model. Furthermore, adding an unimportant variable to the model can actually increase the mean square error, thereby decreasing the usefulness of the model.

The hypotheses for testing the significance of any individual regression coefficient, say β_j, are

$$H_0: \beta_j = 0$$
$$H_1: \beta_j \neq 0$$

If $H_0: \beta_j = 0$ is not rejected, then this indicates that x_j can be deleted from the model. The test statistic for this hypothesis is

$$t_0 = \frac{\hat{\beta}_j}{\sqrt{\hat{\sigma}^2 C_{jj}}} \tag{10.28}$$

where C_{jj} is the diagonal element of $(\mathbf{X}'\mathbf{X})^{-1}$ corresponding to $\hat{\beta}_j$. The null hypothesis H_0: $\beta_j = 0$ is rejected if $|t_0| > t_{\alpha/2, n-k-1}$. Note that this is really a partial or marginal test because the regression coefficient $\hat{\beta}_j$ depends on all the other regressor variables x_i $(i \neq j)$ that are in the model.

The denominator of Equation 10.28, $\sqrt{\hat{\sigma}^2 C_{jj}}$, is often called the **standard error** of the regression coefficient $\hat{\beta}_j$. That is,

$$se(\hat{\beta}_j) = \sqrt{\hat{\sigma}^2 C_{jj}} \tag{10.29}$$

Therefore, an equivalent way to write the test statistic in Equation (10.28) is

$$t_0 = \frac{\hat{\beta}_j}{se(\hat{\beta}_j)} \tag{10.30}$$

Most regression computer programs provide the *t*-test for each model parameter. For example, consider Table 10.4, which contains the Minitab output for Example 10.1. The upper portion of this table gives the least squares estimate of each parameter, the standard error, the *t* statistic, and the corresponding *P*-value. We would conclude that both variables, temperature and feed rate, contribute significantly to the model.

We may also directly examine the contribution to the regression sum of squares for a particular variable, say x_j, given that other variables x_i $(i \neq j)$ are included in the model. The procedure for doing this is the general regression significance test or, as it is often called, the **extra sum of squares method**. This procedure can also be used to investigate the contribution of a *subset* of the regressor variables to the model. Consider the regression model with k regressor variables:

$$\mathbf{y} = \mathbf{X}\boldsymbol{\beta} + \boldsymbol{\epsilon}$$

where $\mathbf{y}$ is $(n \times 1)$, $\mathbf{X}$ is $(n \times p)$, $\boldsymbol{\beta}$ is $(p \times 1)$, $\boldsymbol{\epsilon}$ is $(n \times 1)$, and $p = k + 1$. We would like to determine if the subset of regressor variables $x_1, x_2, \ldots, x_r$ $(r < k)$ contribute significantly to the regression model. Let the vector of regression coefficients be partitioned as follows:

$$\boldsymbol{\beta} = \begin{bmatrix} \boldsymbol{\beta}_1 \\ \boldsymbol{\beta}_2 \end{bmatrix}$$

where $\boldsymbol{\beta}_1$ is $(r \times 1)$ and $\boldsymbol{\beta}_2$ is $[(p - r) \times 1]$. We wish to test the hypotheses

$$H_0: \boldsymbol{\beta}_1 = \mathbf{0}$$
$$H_1: \boldsymbol{\beta}_1 \neq \mathbf{0} \tag{10.31}$$

The model may be written as

$$\mathbf{y} = \mathbf{X}\boldsymbol{\beta} + \boldsymbol{\epsilon} = \mathbf{X}_1\boldsymbol{\beta}_1 + \mathbf{X}_2\boldsymbol{\beta}_2 + \boldsymbol{\epsilon} \tag{10.32}$$

where $\mathbf{X}_1$ represents the columns of $\mathbf{X}$ associated with $\boldsymbol{\beta}_1$ and $\mathbf{X}_2$ represents the columns of $\mathbf{X}$ associated with $\boldsymbol{\beta}_2$.

For the **full model** (including both $\boldsymbol{\beta}_1$ and $\boldsymbol{\beta}_2$), we know that $\hat{\boldsymbol{\beta}} = (\mathbf{X}'\mathbf{X})^{-1}\mathbf{X}'\mathbf{y}$. Also, the regression sum of squares for all variables including the intercept is

$$SS_R(\boldsymbol{\beta}) = \hat{\boldsymbol{\beta}}'\mathbf{X}'\mathbf{y} \qquad (p \text{ degrees of freedom})$$

and

$$MS_E = \frac{\mathbf{y}'\mathbf{y} - \hat{\boldsymbol{\beta}}\mathbf{X}'\mathbf{y}}{n - p}$$

$SS_R(\boldsymbol{\beta})$ is called the regression sum of squares due to $\boldsymbol{\beta}$. To find the contribution of the terms in $\boldsymbol{\beta}_1$ to the regression, we fit the model assuming the null hypothesis $H_0: \boldsymbol{\beta}_1 = \mathbf{0}$ to be true. The **reduced model** is found from Equation 10.32 with $\boldsymbol{\beta}_1 = \mathbf{0}$:

$$\mathbf{y} = \mathbf{X}_2\boldsymbol{\beta}_2 + \boldsymbol{\epsilon} \tag{10.33}$$

The least squares estimator of $\boldsymbol{\beta}_2$ is $\hat{\boldsymbol{\beta}}_2 = (\mathbf{X}_2'\mathbf{X}_2)^{-1}\mathbf{X}_2'\mathbf{y}$, and

$$SS_R(\boldsymbol{\beta}_2) = \hat{\boldsymbol{\beta}}_2'\mathbf{X}_2'\mathbf{y} \qquad (p - r \text{ degrees of freedom}) \tag{10.34}$$

The regression sum of squares due to $\boldsymbol{\beta}_1$ given that $\boldsymbol{\beta}_2$ is already in the model is

$$SS_R(\boldsymbol{\beta}_1|\boldsymbol{\beta}_2) = SS_R(\boldsymbol{\beta}) - SS_R(\boldsymbol{\beta}_2) \tag{10.35}$$

This sum of squares has r degrees of freedom. It is the "extra sum of squares" due to $\boldsymbol{\beta}_1$. Note that $SS_R(\boldsymbol{\beta}_1|\boldsymbol{\beta}_2)$ is the increase in the regression sum of squares due to inclusion of variables $x_1, x_2, \ldots, x_r$ in the model.

Now, $SS_R(\boldsymbol{\beta}_1|\boldsymbol{\beta}_2)$ is independent of MS_E, and the null hypothesis $\boldsymbol{\beta}_1 = \mathbf{0}$ may be tested by the statistic

$$F_0 = \frac{Ss_R(\boldsymbol{\beta}_1|\boldsymbol{\beta}_2)/r}{MS_E} \tag{10.36}$$

If $F_0 > F_{\alpha,r,n-p}$, we reject H_0, concluding that at least one of the parameters in $\boldsymbol{\beta}_1$ is not zero, and, consequently, at least one of the variables $x_1, x_2, \ldots, x_r$ in $\mathbf{X}_1$ contributes significantly to the regression model. Some authors call the test in Equation 10.36 a **partial *F* test**.

The partial F test is very useful. We can use it to measure the contribution of x_j as if it were the last variable added to the model by computing

$$SS_R(\beta_j|\beta_0, \beta_1, \ldots, \beta_{j-1}, \beta_{j+1}, \ldots, \beta_k)$$

This is the increase in the regression sum of squares due to adding x_j to a model that already includes $x_1, \ldots, x_{j-1}, x_{j+1}, \ldots, x_k$. Note that the partial F test on a single variable x_j is equivalent to the t test in Equation 10.28. However, the partial F test is a more general procedure in that we can measure the effect of sets of variables.

EXAMPLE 10.6

Consider the viscosity data in Example 10.1. Suppose that we wish to investigate the contribution of the variable x_2 (feed rate) to the model. That is, the hypotheses we wish to test are

$$H_0: \beta_2 = 0$$

$$H_1: \beta_2 \neq 0$$

This will require the extra sum of squares due to β_2, or

$$\begin{aligned} SS_R(\beta_2|\beta_1, \beta_0) &= SS_R(\beta_0, \beta_1, \beta_2) - SS_R(\beta_0, \beta_1) \\ &= Ss_R(\beta_1, \beta_2|\beta_0) - SS_R(\beta_2|\beta_0) \end{aligned}$$

Now from Table 10.4, where we tested for significance of regression, we have

$$SS_R(\beta_1, \beta_2|\beta_0) = 44{,}157.1$$

which was called the model sum of squares in the table. This sum of squares has two degrees of freedom.

The reduced model is

$$y = \beta_0 + \beta_1 x_1 + \epsilon$$

The least squares fit for this model is

$$\hat{y} = 1652.3955 + 7.6397 x_1$$

and the regression sum of squares for this model (with one degree of freedom) is

$$SS_R(\beta_1|\beta_0) = 40{,}840.8$$

Note that $SS_R(\beta_1|\beta_0)$ is shown at the bottom of the Minitab output in Table 10.4 under the heading "Seq SS." Therefore,

$$SS_R(\beta_2|\beta_0, \beta_1) = 44{,}157.1 - 40{,}840.8$$
$$= 3316.3$$

with $2 - 1 = 1$ degree of freedom. This is the increase in the regression sum of squares that results from adding x_2 to a model already containing x_1, and it is shown at the bottom of the Minitab output on Table 10.4. To test $H_0: \beta_2 = 0$, from the test statistic we obtain

$$F_0 = \frac{Ss_R(\beta_2|\beta_0, \beta_1)/1}{MS_E} = \frac{3316.3/1}{267.604} = 12.3926$$

Note that MS_E from the full model (Table 10.4) is used in the denominator of F_0. Now, because $F_{0.05,1,13} = 4.67$, we would reject $H_0: \beta_2 = 0$ and conclude that x_2 (feed rate) contributes significantly to the model.

Because this partial F test involves only a single regressor, it is equivalent to the t-test because the square of a t random variable with v degrees of freedom is an F random variable with 1 and v degrees of freedom. To see this, note from Table 10.4 that the t-statistic for $H_0: \beta_2 = 0$ resulted in $t_0 = 3.5203$ and that $t_0^2 = (3.5203)^2 = 12.3925 \simeq F_0$.

10.5 Confidence Intervals in Multiple Regression

It is often necessary to construct confidence interval estimates for the regression coefficients $\{\beta_j\}$ and for other quantities of interest from the regression model. The development of a procedure for obtaining these confidence intervals requires that we assume the errors $\{\epsilon_i\}$ to be normally and independently distributed with mean zero and variance σ^2, the same assumption made in the section on hypothesis testing in Section 10.4.

10.5.1 Confidence Intervals on the Individual Regression Coefficients

Because the least squares estimator $\hat{\boldsymbol{\beta}}$ is a linear combination of the observations, it follows that $\hat{\boldsymbol{\beta}}$ is normally distributed with mean vector $\boldsymbol{\beta}$ and covariance matrix $\sigma^2(\mathbf{X}'\mathbf{X})^{-1}$. Then each of the statistics

$$\frac{\hat{\beta}_j - \beta_j}{\sqrt{\hat{\sigma}^2 C_{jj}}} \qquad j = 0, 1, \ldots, k \tag{10.37}$$

is distributed as t with $n - p$ degrees of freedom, where C_{jj} is the (jj)th element of the $(\mathbf{X}'\mathbf{X})^{-1}$ matrix, and $\hat{\sigma}^2$ is the estimate of the error variance, obtained from Equation 10.17. Therefore, a $100(1 - \alpha)$ percent confidence interval for the regression coefficient β_j, $j = 0, 1, \ldots, k$, is

$$\hat{\beta}_j - t_{\alpha/2,n-p}\sqrt{\hat{\sigma}^2 C_{jj}} \le \beta_j \le \hat{\beta}_j + t_{\alpha/2,n-p}\sqrt{\hat{\sigma}^2 C_{jj}} \tag{10.38}$$

Note that this confidence interval could also be written as

$$\hat{\beta}_j - t_{\alpha/2,n-p} se(\hat{\beta}_j) \le \beta_j \le \hat{\beta}_j + t_{\alpha/2,n-p} se(\hat{\beta}_j)$$

because $se(\hat{\beta}_j) = \sqrt{\hat{\sigma}^2 C_{jj}}$.

EXAMPLE 10.7

We will construct a 95 percent confidence interval for the parameter β_1 in Example 10.1. Now $\hat{\beta}_1 = 7.62129$, and because $\hat{\sigma}^2 = 267.604$ and $C_{11} = 1.429184 \times 10^{-3}$, we find that

$$\hat{\beta}_1 - t_{0.025,13}\sqrt{\hat{\sigma}^2 C_{11}} \leq \beta_1 \leq \hat{\beta}_1 + t_{0.025,13}\sqrt{\hat{\sigma}^2 C_{11}}$$

$$7.62129 - 2.16\sqrt{(267.604)(1.429184 \times 10^{-3})} \leq \beta_1$$

$$\leq 7.62129 + 2.16\sqrt{(267.604)(1.429184 \times 10^{-3})}$$

$$7.62129 - 2.16(0.6184) \leq \beta_1 \leq 7.62129 + 2.16(0.6184)$$

and the 95 percent confidence interval on β_1 is

$$6.2855 \leq \beta_1 \leq 8.9570$$

10.5.2 Confidence Interval on the Mean Response

We may also obtain a confidence interval on the mean response at a particular point, say, x_{01}, $x_{02}, \ldots, x_{0k}$. We first define the vector

$$\mathbf{x}_0 = \begin{bmatrix} 1 \\ x_{01} \\ x_{02} \\ \vdots \\ x_{0k} \end{bmatrix}$$

The mean response at this point is

$$\mu_{y|\mathbf{x}_0} = \beta_0 + \beta_1 x_{01} + \beta_2 x_{02} + \cdots + \beta_k x_{0k} = \mathbf{x}_0'\boldsymbol{\beta}$$

The estimated mean response at this point is

$$\hat{y}(\mathbf{x}_0) = \mathbf{x}_0'\hat{\boldsymbol{\beta}} \qquad \textbf{(10.39)}$$

This estimator is unbiased because $E[\hat{y}(\mathbf{x}_0)] = E(\mathbf{x}_0'\hat{\boldsymbol{\beta}}) = \mathbf{x}_0'\boldsymbol{\beta} = \mu_{y|\mathbf{x}_0}$, and the variance of $\hat{y}(\mathbf{x}_0)$ is

$$V[\hat{y}(\mathbf{x}_0)] = \sigma^2 \mathbf{x}_0'(\mathbf{X}'\mathbf{X})^{-1}\mathbf{x}_0 \qquad \textbf{(10.40)}$$

Therefore, a $100(1-\alpha)$ percent confidence interval on the mean response at the point x_{01}, $x_{02}, \ldots, x_{0k}$ is

$$\hat{y}(\mathbf{x}_0) - t_{\alpha/2,n-p}\sqrt{\hat{\sigma}^2 \mathbf{x}_0'(\mathbf{X}'\mathbf{X})^{-1}\mathbf{x}_0} \leq \mu_{y|\mathbf{x}_0} \leq \hat{y}(\mathbf{x}_0) + t_{\alpha/2,n-p}\sqrt{\hat{\sigma}^2 \mathbf{x}_0'(\mathbf{X}'\mathbf{X})^{-1}\mathbf{x}_0} \qquad \textbf{(10.41)}$$

10.6 Prediction of New Response Observations

A regression model can be used to predict future observations on the response y corresponding to particular values of the regressor variables, say $x_{01}, x_{02}, \ldots, x_{0k}$. If $\mathbf{x}_0' = [1, x_{01}, x_{02}, \ldots, x_{0k}]$, then a point estimate for the future observation y_0 at the point $x_{01}, x_{02}, \ldots, x_{0k}$ is computed from Equation 10.39:

$$\hat{y}(\mathbf{x}_0) = \mathbf{x}_0'\hat{\boldsymbol{\beta}}$$

A 100(1 − α) **percent prediction interval** for this future observation is

$$\hat{y}(\mathbf{x}_0) - t_{\alpha/2,n-p}\sqrt{\hat{\sigma}^2(1 + \mathbf{x}_0'(\mathbf{X}'\mathbf{X})^{-1}\mathbf{x}_0)} \leq y_0 \leq \hat{y}(\mathbf{x}_0) + t_{\alpha/2,n-p}\sqrt{\hat{\sigma}^2(1 + \mathbf{x}_0'(\mathbf{X}'\mathbf{X})^{-1}\mathbf{x}_0)} \quad \textbf{(10.42)}$$

In predicting new observations and in estimating the mean response at a given point x_{01}, $x_{02}, \ldots, x_{0k}$, we must be careful about extrapolating beyond the region containing the original observations. It is very possible that a model that fits well in the region of the original data will no longer fit well outside of that region.

The prediction interval in Equation 10.42 has many useful applications. One of these is in confirmation experiments following a factorial or fractional factorial experiment. In a confirmation experiment, we are usually testing the model developed from the original experiment to determine if our interpretation was correct. Often we will do this by using the model to predict the response at some point of interest in the design space and then comparing the predicted response with an actual observation obtained by conducting another trial at that point. We illustrated this in Chapter 8, using the 2^{4-1} fractional factorial design in Example 8.1. A useful measure of confirmation is to see if the new observation falls inside the prediction interval on the response at that point.

To illustrate, reconsider the situation in Example 8.1. The interpretation of this experiment indicated that three of the four main effects (A, C, and D) and two of the two-factor interactions (AC and AD) were important. The point with A, B, and D at the high level and C at the low level was considered to be a reasonable confirmation run, and the predicted value of the response at that point was 100.25. If the fractional factorial has been interpreted correctly and the model for the response is valid, we would expect the observed value at this point to fall inside the prediction interval computed from Equation 10.42. This interval is easy to calculate. Since the 2^{4-1} is an orthogonal design, and the model contains six terms (the intercept, the three main effects, and the two two-factor interactions), the $(\mathbf{X}'\mathbf{X})^{-1}$ matrix has a particularly simple form, namely $(\mathbf{X}'\mathbf{X})^{-1} = \frac{1}{8}\mathbf{I}_6$. Furthermore, the coordinates of the point of interest are $x_1 = 1$, $x_2 = 1$, $x_3 = -1$, and $x_4 = 1$, but since B (or x_2) isn't in the model and the two interactions AC and AD (or x_1x_3 and $x_1x_4 = 1$) are in the model, the coordinates of the point of interest $\mathbf{x}_0$ are given by $\mathbf{x}_0' = [1, x_1, x_3, x_4, x_1x_3, x_1x_4] = [1, 1, -1, 1, -1, 1]$. It is also easy to show that the estimate of σ^2 (with two degrees of freedom) for this model is $\hat{\sigma}^2 = 3.25$. Therefore, using Equation 10.42, a 95 percent prediction interval on the observation at this point is

$$\hat{y}(\mathbf{x}_0) - t_{0.025,2}\sqrt{\hat{\sigma}^2(1 + \mathbf{x}_0'(\mathbf{X}'\mathbf{X})^{-1}\mathbf{x}_0)} \leq y_0 \leq \hat{y}(\mathbf{x}_0) + t_{0.025,2}\sqrt{\hat{\sigma}^2(1 + \mathbf{x}_0'(\mathbf{X}'\mathbf{X})^{-1}\mathbf{x}_0)}$$

$$100.25 - 4.30\sqrt{3.25\left(1 + \mathbf{x}_0'\frac{1}{8}\mathbf{I}_6\mathbf{x}_0\right)} \leq y_0 \leq 100.25 + 4.30\sqrt{3.25\left(1 + \mathbf{x}_0'\frac{1}{8}\mathbf{I}_6\mathbf{x}_0\right)}$$

$$100.25 - 4.30\sqrt{3.25(1 + 0.75)} \leq y_0 \leq 100.25 + 4.30\sqrt{3.25(1 + 0.75)}$$

$$100.25 - 10.25 \leq y_0 \leq 100.25 + 10.25$$

$$90 \leq y_0 \leq 110.50$$

Therefore, we would expect the confirmation run with A, B, and D at the high level and C at the low level to result in an observation on the filtration rate response that falls between 90 and 110.50. The actual observation was 104. The successful confirmation run provides some assurance that the fractional factorial was interpreted correctly.

10.7 Regression Model Diagnostics

As we emphasized in designed experiments, **model adequacy checking** is an important part of the data analysis procedure. This is equally important in building regression models, and as we illustrated in Example 10.1, the **residual plots** that we used with designed experiments should always be examined for a regression model. In general, it is always necessary to (1) examine the fitted model to ensure that it provides an adequate approximation to the true system and (2) verify that none of the least squares regression assumptions are violated. The regression model will probably give poor or misleading results unless it is an adequate fit.

In addition to residual plots, other model diagnostics are frequently useful in regression. This section briefly summarizes some of these procedures. For more complete presentations, see Montgomery, Peck, and Vining (2006) and Myers (1990).

10.7.1 Scaled Residuals and PRESS

Standardized and Studentized Residuals. Many model builders prefer to work with **scaled residuals** in contrast to the ordinary least squares residuals. These scaled residuals often convey more information than do the ordinary residuals.

One type of scaled residual is the **standardized residual**:

$$d_i = \frac{e_i}{\hat{\sigma}} \quad i = 1, 2, \ldots, n \tag{10.43}$$

where we generally use $\hat{\sigma} = \sqrt{MS_E}$ in the computation. These standardized residuals have mean zero and approximately unit variance; consequently, they are useful in looking for **outliers**. Most of the standardized residuals should lie in the interval $-3 \le d_i \le 3$, and any observation with a standardized residual outside of this interval is potentially unusual with respect to its observed response. These outliers should be carefully examined because they may represent something as simple as a data-recording error or something of more serious concern, such as a region of the regressor variable space where the fitted model is a poor approximation to the true response surface.

The standardizing process in Equation 10.43 scales the residuals by dividing them by their approximate average standard deviation. In some data sets, residuals may have standard deviations that differ greatly. We now present a scaling that takes this into account.

The vector of fitted values $\hat{y}_i$ corresponding to the observed values y_i is

$$\begin{aligned} \hat{\mathbf{y}} &= \mathbf{X}\hat{\boldsymbol{\beta}} \\ &= \mathbf{X}(\mathbf{X}'\mathbf{X})^{-1}\mathbf{X}'\mathbf{y} \\ &= \mathbf{H}\mathbf{y} \end{aligned} \tag{10.44}$$

The $n \times n$ matrix $\mathbf{H} = \mathbf{X}(\mathbf{X}'\mathbf{X})^{-1}\mathbf{X}'$ is usually called the "hat" matrix because it maps the vector of observed values into a vector of fitted values. The hat matrix and its properties play a central role in regression analysis.

The residuals from the fitted model may be conveniently written in matrix notation as

$$\mathbf{e} = \mathbf{y} - \hat{\mathbf{y}}$$

and it turns out that the covariance matrix of the residuals is

$$\text{Cov}(\mathbf{e}) = \sigma^2(\mathbf{I} - \mathbf{H}) \tag{10.45}$$

The matrix $\mathbf{I} - \mathbf{H}$ is generally not diagonal, so the residuals have different variances and they are correlated.

Thus, the variance of the ith residual is

$$V(e_i) = \sigma^2(1 - h_{ii}) \tag{10.46}$$

where h_{ii} is the ith diagonal element of **H**. Because $0 \le h_{ii} \le 1$, using the residual mean square MS_E to estimate the variance of the residuals actually overestimates $V(e_i)$. Furthermore, because h_{ii} is a measure of the location of the ith point in x-space, the variance of e_i depends on where the point x_i lies. Generally, residuals near the center of the x space have larger variance than do residuals at more remote locations. Violations of model assumptions are more likely at remote points, and these violations may be hard to detect from inspection of e_i (or d_i) because their residuals will usually be smaller.

We recommend taking this inequality of variance into account when scaling the residuals. We suggest plotting the **studentized residuals**:

$$r_i = \frac{e_i}{\sqrt{\hat{\sigma}^2(1 - h_{ii})}} \qquad i = 1, 2, \ldots, n \tag{10.47}$$

with $\hat{\sigma}^2 = MS_E$ instead of e_i (or d_i). The studentized residuals have constant variance $V(r_i) = 1$ regardless of the location of $\mathbf{x}_i$ when the form of the model is correct. In many situations the variance of the residuals stabilizes, particularly for large data sets. In these cases, there may be little difference between the standardized and studentized residuals. Thus standardized and studentized residuals often convey equivalent information. However, because any point with a large residual and a large h_{ii} is potentially highly influential on the least squares fit, examination of the studentized residuals is generally recommended. Table 10.3 displays the hat diagonals h_{ii} and the studentized residuals for the viscosity regression model in Example 10.1.

PRESS Residuals. The prediction error sum of squares (PRESS) provides a useful residual scaling. To calculate PRESS, we select an observation—for example, i. We fit the regression model to the remaining $n - 1$ observations and use this equation to predict the withheld observation y_i. Denoting this predicted value $\hat{y}_{(i)}$, we may find the prediction error for point i as $e_{(i)} = y_i - \hat{y}_{(i)}$. The prediction error is often called the ith PRESS residual. This procedure is repeated for each observation $i = 1, 2, \ldots, n$, producing a set of n PRESS residuals $e_{(1)}, e_{(2)}, \ldots, e_{(n)}$. Then the PRESS statistic is defined as the sum of squares of the n PRESS residuals as in

$$\text{PRESS} = \sum_{i=1}^{n} e_{(i)}^2 = \sum_{i=1}^{n} [y_i - \hat{y}_{(i)}]^2 \tag{10.48}$$

Thus PRESS uses each possible subset of $n - 1$ observations as an estimation data set, and every observation in turn is used to form a prediction data set.

It would initially seem that calculating PRESS requires fitting n different regressions. However, it is possible to calculate PRESS from the results of a single least squares fit to all n observations. It turns out that the ith PRESS residual is

$$e_{(i)} = \frac{e_i}{1 - h_{ii}} \tag{10.49}$$

Thus because PRESS is just the sum of the squares of the PRESS residuals, a simple computing formula is

$$\text{PRESS} = \sum_{i=1}^{n} \left(\frac{e_i}{1 - h_{ii}} \right)^2 \tag{10.50}$$

From Equation 10.49, it is easy to see that the PRESS residual is just the ordinary residual weighted according to the diagonal elements of the hat matrix h_{ii}. Data points for which h_{ii} are large will have large PRESS residuals. These observations will generally be **high influence** points. Generally, a large difference between the ordinary residual and the PRESS residuals will indicate a point where the model fits the data well, but a model built without that point predicts poorly. In the next section we will discuss some other measures of influence.

Finally, we note that PRESS can be used to compute an approximate R^2 for prediction, say

$$R^2_{\text{Prediction}} = 1 - \frac{\text{PRESS}}{SS_T} \tag{10.51}$$

This statistic gives some indication of the predictive capability of the regression model. For the viscosity regression model from Example 10.1, we can compute the PRESS residuals using the ordinary residuals and the values of h_{ii} found in Table 10.3. The corresponding value of the PRESS statistic is PRESS = 5207.7. Then

$$\begin{aligned} R^2_{\text{Prediction}} &= 1 - \frac{\text{PRESS}}{SS_T} \\ &= 1 - \frac{5207.7}{47{,}635.9} = 0.8907 \end{aligned}$$

Therefore, we could expect this model to "explain" about 89 percent of the variability in predicting new observations, as compared to the approximately 93 percent of the variability in the original data explained by the least squares fit. The overall predictive capability of the model based on this criterion seems very satisfactory.

***R*-Student.** The studentized residual r_i discussed above is often considered an outlier diagnostic. It is customary to use MS_E as an estimate of σ^2 in computing r_i. This is referred to as internal scaling of the residual because MS_E is an internally generated estimate of σ^2 obtained from fitting the model to all n observations. Another approach would be to use an estimate of σ^2 based on a data set with the ith observation removed. We denote the estimate of σ^2 so obtained by $S^2_{(i)}$. We can show that

$$S^2_{(i)} = \frac{(n-p)MS_E - e_i^2/(1-h_{ii})}{n-p-1} \tag{10.52}$$

The estimate of σ^2 in Equation 10.52 is used instead of MS_E to produce an externally studentized residual, usually called R-student, given by

$$t_i = \frac{e_i}{\sqrt{S^2_{(i)}(1-h_{ii})}} \qquad i = 1, 2, \ldots, n \tag{10.53}$$

In many situations, t_i will differ little from the studentized residual r_i. However, if the ith observation is influential, then $S^2_{(i)}$ can differ significantly from MS_E, and thus the R-student will be more sensitive to this point. Furthermore, under the standard assumptions, t_i has a t_{n-p-1} distribution. Thus R-student offers a more formal procedure for outlier detection via hypothesis testing. Table 10.3 displays the values of R-student for the viscosity regression model in Example 10.1. None of those values are unusually large.

10.7.2 Influence Diagnostics

We occasionally find that a small subset of the data exerts a disproportionate influence on the fitted regression model. That is, parameter estimates or predictions may depend more on the influential subset than on the majority of the data. We would like to locate these influential points and assess their impact on the model. If these influential points are "bad" values, they should be eliminated. On the contrary, there may be nothing wrong with these points. But if they control key model properties, we would like to know it because it could affect the use of the model. In this section we describe and illustrate some useful measures of influence.

Leverage Points. The disposition of points in x space is important in determining model properties. In particular, remote observations potentially have disproportionate leverage on the parameter estimates, predicted values, and the usual summary statistics.

The hat matrix $\mathbf{H} = \mathbf{X}(\mathbf{X}'\mathbf{X})^{-1}\mathbf{X}'$ is very useful in identifying influential observations. As noted earlier, $\mathbf{H}$ determines the variances and covariances of $\hat{\mathbf{y}}$ and $\mathbf{e}$ because $V(\hat{\mathbf{y}}) = \sigma^2\mathbf{H}$ and $V(\mathbf{e}) = \sigma^2(\mathbf{I} - \mathbf{H})$. The elements h_{ij} of $\mathbf{H}$ may be interpreted as the amount of leverage exerted by y_j on $\hat{y}_i$. Thus, inspection of the elements of $\mathbf{H}$ can reveal points that are potentially influential by virtue of their location in x space. Attention is usually focused on the diagonal elements h_{ii}. Because $\Sigma_{i=1}^{n} h_{ii} = \text{rank}(\mathbf{H}) = \text{rank}(\mathbf{X}) = p$, the average size of the diagonal element of the $\mathbf{H}$ matrix is p/n. As a rough guideline, then, if a diagonal element h_{ii} is greater than $2p/n$, observation i is a high-leverage point. To apply this to the viscosity model in Example 10.1, note that $2p/n = 2(3)/16 = 0.375$. Table 10.3 gives the hat diagonals h_{ii} for the first-order model; because none of the h_{ii} exceeds 0.375, we would conclude that there are no leverage points in these data.

Influence on Regression Coefficients. The hat diagonals will identify points that are potentially influential due to their location in x space. It is desirable to consider both the location of the point and the response variable in measuring influence. Cook (1977, 1979) has suggested using a measure of the squared distance between the least squares estimate based on all n points $\hat{\boldsymbol{\beta}}$ and the estimate obtained by deleting the i point, say $\hat{\boldsymbol{\beta}}_{(i)}$. This distance measure can be expressed as

$$D_1 = \frac{(\hat{\boldsymbol{\beta}}_{(i)} - \hat{\boldsymbol{\beta}})'\mathbf{X}'\mathbf{X}(\hat{\boldsymbol{\beta}}_{(i)} - \hat{\boldsymbol{\beta}})}{pMS_E} \qquad i = 1, 2, \ldots, n \tag{10.54}$$

A reasonable cutoff for D_i is unity. That is, we usually consider observations for which $D_i > 1$ to be influential.

The D_i statistic is actually calculated from

$$D_i = \frac{r_i^2}{p}\frac{V[\hat{y}(\mathbf{x}_i)]}{V(e_i)} = \frac{r_i^2}{p}\frac{h_{ii}}{(1 - h_{ii})} \qquad i = 1, 2, \ldots, n \tag{10.55}$$

Note that, apart from the constant p, D_i is the product of the square of the ith studentized residual and $h_{ii}/(1 - h_{ii})$. This ratio can be shown to be the distance from the vector $\mathbf{x}_i$ to the centroid of the remaining data. Thus, D_i is made up of a component that reflects how well the model fits the ith observation y_i and a component that measures how far that point is from the rest of the data. Either component (or both) may contribute to a large value of D_i.

Table 10.3 presents the values of D_i for the regression model fit to the viscosity data in Example 10.1. None of these values of D_i exceeds 1, so there is no strong evidence of influential observations in these data.

10.8 Testing for Lack of Fit

In Section 6.8 we showed how adding center points to a 2^k factorial design allows the experimenter to obtain an estimate of pure experimental error. This allows the partitioning of the residual sum of squares SS_E into two components; that is

$$SS_E = SS_{PE} + SS_{LOF}$$

where SS_{PE} is the sum of squares due to pure error and SS_{LOF} is the sum of squares due to lack of fit.

We may give a general development of this partitioning in the context of a regression model. Suppose that we have n_i observations on the response at the ith level of the regressors $\mathbf{x}_i$, $i = 1, 2, \ldots, m$. Let y_{ij} denote the jth observation on the response at $\mathbf{x}_i$, $i = 1, 2, \ldots, m$ and $j = 1, 2, \ldots, n_i$. There are $n = \Sigma_{i=1}^{m} n_i$ total observations. We may write the (ij)th residual as

$$y_{ij} - \hat{y}_i = (y_{ij} - \bar{y}_i) + (\bar{y}_i - \hat{y}_i) \tag{10.56}$$

where $\bar{y}_i$ is the average of the n_i observations at $\mathbf{x}_i$. Squaring both sides of Equation 10.56 and summing over i and j yields

$$\sum_{i=1}^{m}\sum_{j=1}^{n_i}(y_{ij} - \hat{y}_i)^2 = \sum_{i=1}^{m}\sum_{j=1}^{n_i}(y_{ij} - \bar{y}_i)^2 + \sum_{i=1}^{m} n_i(\bar{y}_i - \hat{y}_i)^2 \tag{10.57}$$

The left-hand side of Equation 10.57 is the usual residual sum of squares. The two components on the right-hand side measure pure error and lack of fit. We see that the pure error sum of squares

$$SS_{PE} = \sum_{i=1}^{m}\sum_{j=1}^{n_i}(y_{ij} - \bar{y}_i)^2 \tag{10.58}$$

is obtained by computing the corrected sum of squares of the repeat observations at each level of $\mathbf{x}$ and then pooling over the m levels of $\mathbf{x}$. If the assumption of constant variance is satisfied, this is a **model-independent** measure of pure error because only the variability of the y's at each $\mathbf{x}_i$ level is used to compute SS_{PE}. Because there are $n_i - 1$ degrees of freedom for pure error at each level $\mathbf{x}_i$, the total number of degrees of freedom associated with the pure error sum of squares is

$$\sum_{i=1}^{m}(n_i - 1) = n - m \tag{10.59}$$

The sum of squares for lack of fit

$$SS_{LOF} = \sum_{i=1}^{m} n_i(\bar{y}_i - \hat{y}_i)^2 \tag{10.60}$$

is a weighted sum of squared deviations between the mean response $\bar{y}_i$ at each $\mathbf{x}_i$ level and the corresponding fitted value. If the fitted values $\hat{y}_i$ are close to the corresponding average responses $\bar{y}_i$, then there is a strong indication that the regression function is linear. If the $\hat{y}_i$ deviate greatly from the $\bar{y}_i$, then it is likely that the regression function is not linear. There are $m - p$ degrees of freedom associated with SS_{LOF} because there are m levels of $\mathbf{x}$, and p degrees of freedom are lost because p parameters must be estimated for the model. Computationally we usually obtain SS_{LOF} by subtracting SS_{PE} from SS_E.

The test statistic for lack of fit is

$$F_0 = \frac{SS_{LOF}/(m - p)}{SS_{PE}/(n - m)} = \frac{MS_{LOF}}{MS_{PE}} \tag{10.61}$$

The expected value of MS_{PE} is σ^2, and the expected value of MS_{LOF} is

$$E(MS_{LOF}) = \sigma^2 + \frac{\sum_{i=1}^{m} n_i\left[E(y_i) - \beta_0 - \sum_{j=1}^{k}\beta_j x_{ij}\right]^2}{m - 2} \tag{10.62}$$

If the true regression function is linear, then $E(y_i) = \beta_0 + \Sigma_{j=1}^{k} \beta_j x_{ij}$, and the second term of Equation 10.62 is zero, resulting in $E(MS_{LOF}) = \sigma^2$. However, if the true regression function is not linear, then $E(y_i) \neq \beta_0 + \Sigma_{j=1}^{k} \beta_j x_{ij}$, and $E(MS_{LOF}) > \sigma^2$. Furthermore, if the true regression function is linear, then the statistic F_0 follows the $F_{m-p,n-m}$ distribution. Therefore, to test for lack of fit, we would compute the test statistic F_0 and conclude that the regression function is not linear if $F_0 > F_{\alpha,m-p,n-m}$.

This test procedure may be easily incorporated into the analysis of variance. If we conclude that the regression function is not linear, then the tentative model must be abandoned and attempts made to find a more appropriate equation. Alternatively, if F_0 does not exceed $F_{\alpha,m-p,n-m}$, there is no strong evidence of lack of fit and MS_{PE} and MS_{LOF} are often combined to estimate σ^2. Example 6.7 is a very complete illustration of this procedure, where the replicate runs are center points in a 2^4 factorial design.

10.9 Problems

10.1. The tensile strength of a paper product is related to the amount of hardwood in the pulp. Ten samples are produced in the pilot plant, and the data obtained are shown in the following table.

Strength	Percent Hardwood	Strength	Percent Hardwood
160	10	181	20
171	15	188	25
175	15	193	25
182	20	195	28
184	20	200	30

(a) Fit a linear regression model relating strength to percent hardwood.

(b) Test the model in part (a) for significance of regression.

(c) Find a 95 percent confidence interval on the parameter β_1.

10.2. A plant distills liquid air to produce oxygen, nitrogen, and argon. The percentage of impurity in the oxygen is thought to be linearly related to the amount of impurities in the air as measured by the "pollution count" in parts per million (ppm). A sample of plant operating data is shown below:

Purity (%)	93.3	92.0	92.4	91.7	94.0	94.6	93.6	
Pollution count (ppm)	1.10	1.45	1.36	1.59	1.08	0.75	1.20	
Purity (%)	93.1	93.2	92.9	92.2	91.3	90.1	91.6	91.9
Pollution count (ppm)	0.99	0.83	1.22	1.47	1.81	2.03	1.75	1.68

(a) Fit a linear regression model to the data.

(b) Test for significance of regression.

(c) Find a 95 percent confidence interval on β_1.

10.3. Plot the residuals from Problem 10.1 and comment on model adequacy.

10.4. Plot the residuals from Problem 10.2 and comment on model adequacy.

10.5. Using the results of Problem 10.1, test the regression model for lack of fit.

10.6. A study was performed on wear of a bearing y and its relationship to x_1 = oil viscosity and x_2 = load. The following data were obtained:

y	x_1	x_2
195	1.6	851
230	15.5	816
172	22.0	1058
93	43.0	1201
115	33.0	1357
125	40.0	1115

(a) Fit a multiple linear regression model to the data.

(b) Test for significance of regression.

(c) Compute t statistics for each model parameter. What conclusions can you draw?

10.7. The brake horsepower developed by an automobile engine on a dynamometer is thought to be a function of the engine speed in revolutions per minute (rpm), the road octane number of the fuel, and the engine compression. An experiment is run in the laboratory and the data that follow are collected:

Brake Horsepower	rpm	Road Octane Number	Compression
225	2000	90	100
212	1800	94	95
229	2400	88	110
222	1900	91	96
219	1600	86	100
278	2500	96	110

246	3000	94	98
237	3200	90	100
233	2800	88	105
224	3400	86	97
223	1800	90	100
230	2500	89	104

(a) Fit a multiple regression model to these data.

(b) Test for significance of regression. What conclusions can you draw?

(c) Based on t-tests, do you need all three regressor variables in the model?

10.8. Analyze the residuals from the regression model in Problem 10.7. Comment on model adequacy.

10.9. The yield of a chemical process is related to the concentration of the reactant and the operating temperature. An experiment has been conducted with the following results.

Yield	Concentration	Temperature
83	1.00	150
90	1.00	180
85	2.00	150
92	2.00	180
80	1.00	150
88	1.00	180
84	2.00	150
91	2.00	180

(a) Suppose we wish to fit a main effects model to this data. Set up the $\mathbf{X'X}$ matrix using the data exactly as it appears in the table.

(b) Is the matrix you obtained in part (a) diagonal? Discuss your response.

(c) Suppose we write our model in terms of the "usual" coded variables

$$x_1 = \frac{\text{Conc} - 1.5}{0.5} \quad x_2 = \frac{\text{Temp} - 165}{15}$$

Set up the $\mathbf{X'X}$ matrix for the model in terms of these coded variables. Is this matrix diagonal? Discuss your response.

(d) Define a new set of coded variables

$$x_1 = \frac{\text{Conc} - 1.0}{1.0} \quad x_2 = \frac{\text{Temp} - 150}{30}$$

Set up the $\mathbf{X'X}$ matrix for the model in terms of this set of coded variables. Is this matrix diagonal? Discuss your response.

(e) Summarize what you have learned from this problem about coding the variables.

10.10. Consider the 2^4 factorial experiment in Example 6.2. Suppose that the last observation is missing. Reanalyze the data and draw conclusions. How do these conclusions compare with those from the original example?

10.11. Consider the 2^4 factorial experiment in Example 6.2. Suppose that the last two observations are missing. Reanalyze the data and draw conclusions. How do these conclusions compare with those from the original example?

10.12. Given the following data, fit the second-order polynomial regression model

$$y = \beta_0 + \beta_1 x_1 + \beta_2 x_2 + \beta_{11} x_1^2 + \beta_{22} x_2^2 + \beta_{12} x_1 x_2 + \epsilon$$

y	x_1	x_2
26	1.0	1.0
24	1.0	1.0
175	1.5	4.0
160	1.5	4.0
163	1.5	4.0
55	0.5	2.0
62	1.5	2.0
100	0.5	3.0
26	1.0	1.5
30	0.5	1.5
70	1.0	2.5
71	0.5	2.5

After you have fit the model, test for significance of regression.

10.13.

(a) Consider the quadratic regression model from Problem 10.12. Compute t statistics for each model parameter and comment on the conclusions that follow from these quantities.

(b) Use the extra sum of squares method to evaluate the value of the quadratic terms x_1^2, x_2^2, and x_1x_2 to the model.

10.14. ***Relationship between analysis of variance and regression.*** Any analysis of variance model can be expressed in terms of the general linear model $\mathbf{y} = \mathbf{x}\boldsymbol{\beta} + \boldsymbol{\epsilon}$, where the $\mathbf{X}$ matrix consists of 0s and 1s. Show that the single-factor model $y_{ij} = \mu + \tau_i + \epsilon_{ij}$, $i = 1, 2, 3$, $j = 1, 2, 3, 4$ can be written in general linear model form. Then,

(a) Write the normal equations $(\mathbf{X'X})\hat{\boldsymbol{\beta}} = \mathbf{X'y}$ and compare them with the normal equations found for this model in Chapter 3.

(b) Find the rank of $\mathbf{X'X}$. Can $(\mathbf{X'X})^{-1}$ be obtained?

(c) Suppose the first normal equation is deleted and the restriction $\Sigma_{i=1}^{3} n\hat{\tau}_i = 0$ is added. Can the resulting system of equations be solved? If so, find the solution. Find the regression sum of squares $\hat{\boldsymbol{\beta}}'\mathbf{X'y}$, and compare it to the treatment sum of squares in the single-factor model.

10.15. Suppose that we are fitting a straight line and we desire to make the variance of $\hat{\beta}_1$ as small as possible. Restricting ourselves to an even number of experimental points, where should we place these points so as to minimize $V(\hat{\beta}_1)$? [*Note:* Use the design called for in this exercise with *great* caution because, even though it minimizes $V(\hat{\beta}_1)$, it has some undesirable properties; for example, see Myers, Montgomery and Anderson-Cook (2009). Only if you are *very sure* the true functional relationship is linear should you consider using this design.]

10.16. *Weighted least squares.* Suppose that we are fitting the straight line $y = \beta_0 + \beta_1 x + \epsilon$, but the variance of the y's now depends on the level of x; that is,

$$V(y|x_i) = \sigma_i^2 = \frac{\sigma^2}{w_i} \qquad i = 1, 2, \ldots, n$$

where the w_i are known constants, often called weights. Show that if we choose estimates of the regression coefficients to minimize the weighted sum of squared errors given by $\sum_{i=1}^{n} w_i(y_i - \beta_0 - \beta_1 x_i)^2$, the resulting least squares normal equations are

$$\hat{\beta}_0 \sum_{i=1}^{n} w_i + \hat{\beta}_1 \sum_{i=1}^{n} w_i x_i = \sum_{i=1}^{n} w_i y_i$$

$$\hat{\beta}_0 \sum_{i=1}^{n} w_i x_i + \hat{\beta}_1 \sum_{i=1}^{n} w_i x_i^2 = \sum_{i=1}^{n} w_i x_i y_i$$

10.17. Consider the 2_{IV}^{4-1} design discussed in Example 10.5.

(a) Suppose you elect to augment the design with the single run selected in that example. Find the variances and covariances of the regression coefficients in the model (ignoring blocks):

$$y = \beta_0 + \beta_1 x_1 + \beta_2 x_2 + \beta_3 x_3 + \beta_4 x_4 + \beta_{12} x_1 x_2 + \beta_{34} x_3 x_4 + \epsilon$$

(b) Are there any other runs in the alternate fraction that would de-alias *AB* from *CD*?

(c) Suppose you augment the design with the four runs suggested in Example 10.5. Find the variances and covariances of the regression coefficients (ignoring blocks) for the model in part (a).

(d) Considering parts (a) and (c), which augmentation strategy would you prefer, and why?

10.18. Consider a 2_{III}^{7-4} design. Suppose after running the experiment, the largest observed effects are $A + BD$, $B + AD$, and $D + AB$. You wish to augment the original design with a group of four runs to de-alias these effects.

(a) Which four runs would you make?

(b) Find the variances and covariances of the regression coefficients in the model

$$y = \beta_0 + \beta_1 x_1 + \beta_2 x_2 + \beta_4 x_4 + \beta_{12} x_1 x_2 + \beta_{14} x_1 x_4 + \beta_{24} x_2 x_4 + \epsilon.$$

(c) Is it possible to de-alias these effects with fewer than four additional runs?

CHAPTER 11

Response Surface Methodology

CHAPTER OUTLINE

The supplemental material is on the textbook website www.wiley.com/go/global/montgomery.

11.1 Introduction to Response Surface Methodology

Response surface methodology, or **RSM**, is a collection of mathematical and statistical techniques useful for the modeling and analysis of problems in which a response of interest is influenced by several variables and the objective is to optimize this response. For example, suppose that a chemical engineer wishes to find the levels of temperature (x_1) and pressure (x_2) that maximize the yield (y) of a process. The process yield is a function of the levels of temperature and pressure, say

$$y = f(x_1, x_2) + \epsilon$$

where ϵ represents the noise or error observed in the response y. If we denote the expected response by $E(y) = f(x_1, x_2) = \eta$, then the surface represented by

$$\eta = f(x_1, x_2)$$

is called a **response surface**.

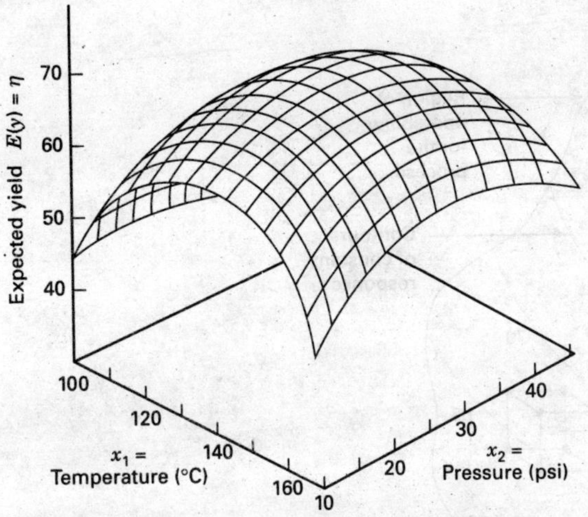

■ **FIGURE 11.1 A three-dimensional response surface showing the expected yield (η) as a function of temperature (x_1) and pressure (x_2)**

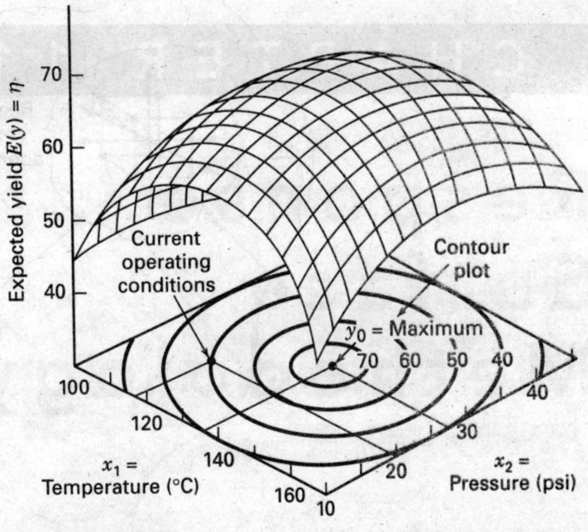

■ **FIGURE 11.2 A contour plot of a response surface**

We usually represent the response surface graphically, such as in Figure 11.1, where η is plotted versus the levels of x_1 and x_2. We have seen such response surface plots before, particularly in the chapters on factorial designs. To help visualize the shape of a response surface, we often plot the contours of the response surface as shown in Figure 11.2. In the contour plot, lines of constant response are drawn in the x_1, x_2 plane. Each contour corresponds to a particular height of the response surface. We have also previously seen the utility of contour plots.

In most RSM problems, the form of the relationship between the response and the independent variables is unknown. Thus, the first step in RSM is to find a suitable approximation for the true functional relationship between y and the set of independent variables. Usually, a low-order polynomial in some region of the independent variables is employed. If the response is well modeled by a linear function of the independent variables, then the approximating function is the **first-order model**

$$y = \beta_0 + \beta_1 x_1 + \beta_2 x_2 + \cdots + \beta_k x_k + \epsilon \tag{11.1}$$

If there is curvature in the system, then a polynomial of higher degree must be used, such as the **second-order model**

$$y = \beta_0 + \sum_{i=1}^{k} \beta_i x_i + \sum_{i=1}^{k} \beta_{ii} x_i^2 + \sum_{i<j}\sum \beta_{ij} x_i x_j + \epsilon \tag{11.2}$$

Almost all RSM problems use one or both of these models. Of course, it is unlikely that a polynomial model will be a reasonable approximation of the true functional relationship over the entire space of the independent variables, but for a relatively small region they usually work quite well.

The method of least squares, discussed in Chapter 10, is used to estimate the parameters in the approximating polynomials. The response surface analysis is then performed using the fitted surface. If the fitted surface is an adequate approximation of the true response function, then analysis of the fitted surface will be approximately equivalent to analysis of the actual system. The model parameters can be estimated most effectively if proper experimental designs are used to collect the data. Designs for fitting response surfaces are called **response surface designs**. These designs are discussed in Section 11.4.

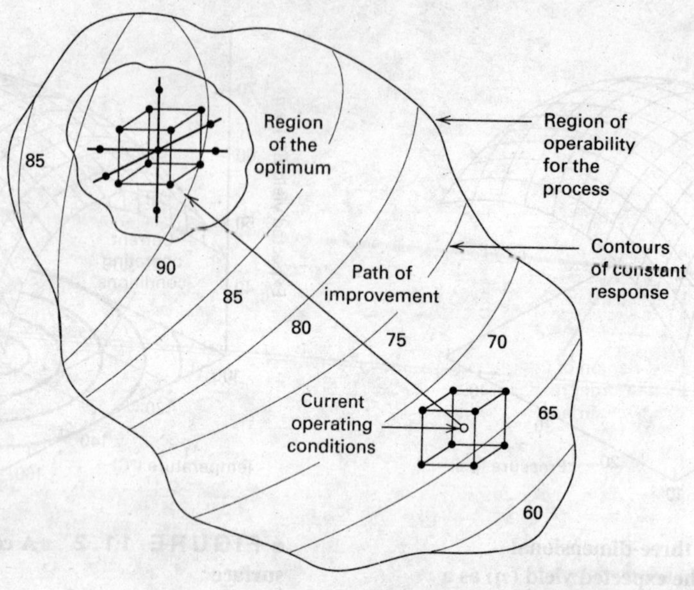

■ **FIGURE 11.3** The sequential nature of RSM

RSM is a **sequential procedure**. Often, when we are at a point on the response surface that is remote from the optimum, such as the current operating conditions in Figure 11.3, there is little curvature in the system and the first-order model will be appropriate. Our objective here is to lead the experimenter rapidly and efficiently along a path of improvement toward the general vicinity of the **optimum**. Once the region of the optimum has been found, a more elaborate model, such as the second-order model, may be employed, and an analysis may be performed to locate the optimum. From Figure 11.3, we see that the analysis of a response surface can be thought of as "climbing a hill," where the top of the hill represents the point of maximum response. If the true optimum is a point of minimum response, then we may think of "descending into a valley."

The eventual objective of RSM is to determine the optimum operating conditions for the system or to determine a region of the factor space in which operating requirements are satisfied. More extensive presentations of RSM are in Khuri and Cornell (1996), Myers, Montgomery and Anderson-Cook (2009), and Box and Draper (2007). The review paper by Myers et al. (2004) is also a useful reference.

11.2 The Method of Steepest Ascent

Frequently, the initial estimate of the optimum operating conditions for the system will be far from the actual optimum. In such circumstances, the objective of the experimenter is to move rapidly to the general vicinity of the optimum. We wish to use a simple and economically efficient experimental procedure. When we are remote from the optimum, we usually assume that a first-order model is an adequate approximation to the true surface in a small region of the x's.

The **method of steepest ascent** is a procedure for moving sequentially in the direction of the maximum increase in the response. Of course, if minimization is desired, then we call this technique the **method of steepest descent**. The fitted first-order model is

$$\hat{y} = \hat{\beta}_0 + \sum_{i=1}^{k} \hat{\beta}_i x_i \tag{11.3}$$

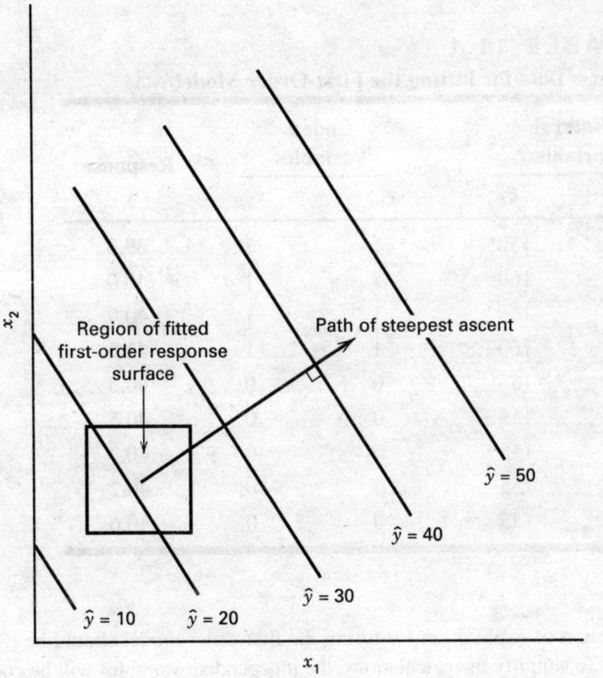

■ **FIGURE 11.4 First-order response surface and path of steepest ascent**

and the first-order response surface, that is, the contours of $\hat{y}$, is a series of parallel lines such as shown in Figure 11.4. The direction of steepest ascent is the direction in which $\hat{y}$ increases most rapidly. This direction is normal to the fitted response surface. We usually take as the **path of steepest ascent** the line through the center of the region of interest and normal to the fitted surface. Thus, the steps along the path are proportional to the regression coefficients $\{\hat{\beta}_i\}$. The actual step size is determined by the experimenter based on process knowledge or other practical considerations.

Experiments are conducted along the path of steepest ascent until no further increase in response is observed. Then a new first-order model may be fit, a new path of steepest ascent determined, and the procedure continued. Eventually, the experimenter will arrive in the vicinity of the optimum. This is usually indicated by lack of fit of a first-order model. At that time, additional experiments are conducted to obtain a more precise estimate of the optimum.

EXAMPLE 11.1

A chemical engineer is interested in determining the operating conditions that maximize the yield of a process. Two controllable variables influence process yield: reaction time and reaction temperature. The engineer is currently operating the process with a reaction time of 35 minutes and a temperature of 155°F, which result in yields of around 40 percent. Because it is unlikely that this region contains the optimum, she fits a first-order model and applies the method of steepest ascent.

■ **TABLE 11.1**
Process Data for Fitting the First-Order Model

Natural Variables		Coded Variables		Response
ξ_1	ξ_2	x_1	x_2	y
30	150	−1	−1	39.3
30	160	−1	1	40.0
40	150	1	−1	40.9
40	160	1	1	41.5
35	155	0	0	40.3
35	155	0	0	40.5
35	155	0	0	40.7
35	155	0	0	40.2
35	155	0	0	40.6

The engineer decides that the region of exploration for fitting the first-order model should be (30, 40) minutes of reaction time and (150, 160) Fahrenheit. To simplify the calculations, the independent variables will be coded to the usual (−1, 1) interval. Thus, if ξ_1 denotes the **natural variable** time and ξ_2 denotes the **natural variable** temperature, then the **coded variables** are

$$x_1 = \frac{\xi_1 - 35}{5} \quad \text{and} \quad x_2 = \frac{\xi_2 - 155}{5}$$

The experimental design is shown in Table 11.1. Note that the design used to collect these data is a 2^2 factorial augmented by five center points. Replicates at the center are used to estimate the experimental error and to allow for checking the adequacy of the first-order model. Also, the design is centered about the current operating conditions for the process.

A first-order model may be fit to these data by least squares. Employing the methods for two-level designs, we obtain the following model in the coded variables:

$$\hat{y} = 40.44 + 0.775x_1 + 0.325x_2$$

Before exploring along the path of steepest ascent, the adequacy of the first-order model should be investigated. The 2^2 design with center points allows the experimenter to

1. Obtain an estimate of error.
2. Check for interactions (cross-product terms) in the model.
3. Check for quadratic effects (curvature).

The replicates at the center can be used to calculate an estimate of error as follows:

$$\hat{\sigma}^2 = \frac{(40.3)^2 + (40.5)^2 + (40.7)^2 + (40.2)^2 + (40.6)^2 - (202.3)^2/5}{4} = 0.0430$$

The first-order model assumes that the variables x_1 and x_2 have an **additive effect** on the response. Interaction between the variables would be represented by the coefficient β_{12} of a cross-product term x_1x_2 added to the model. The least squares estimate of this coefficient is just one-half the interaction effect calculated as in an ordinary 2^2 factorial design, or

$$\hat{\beta}_{12} = \tfrac{1}{4}[(1 \times 39.3) + (1 \times 41.5) + (-1 \times 40.0) + (-1 \times 40.9)] = \tfrac{1}{4}(-0.1) = -0.025$$

The single-degree-of-freedom sum of squares for interaction is

$$SS_{\text{Interaction}} = \frac{(-0.1)^2}{4} = 0.0025$$

Comparing $SS_{\text{Interaction}}$ to $\hat{\sigma}^2$ gives a lack-of-fit statistic

$$F = \frac{SS_{\text{Interaction}}}{\hat{\sigma}^2} = \frac{0.0025}{0.0430} = 0.058$$

which is small, indicating that interaction is negligible.

Another check of the adequacy of the straight-line model is obtained by applying the check for a pure quadratic curvature effect described in Section 6.8. Recall that this consists of comparing the average response at the four points in the factorial portion of the design, say $\bar{y}_F = 40.425$, with the average response at the design center, say $\bar{y}_C = 40.46$. If there is quadratic curvature in the true response function, then $\bar{y}_F - \bar{y}_C$ is a measure of this curvature. If β_{11} and β_{22} are the coefficients of the "pure quadratic" terms x_1^2 and x_2^2, then $\bar{y}_F - \bar{y}_C$ is an estimate of $\beta_{11} + \beta_{22}$. In our example, an estimate of the pure quadratic term is

$$\hat{\beta}_{11} + \hat{\beta}_{22} = \bar{y}_F - \bar{y}_C = 40.425 - 40.46 = -0.035$$

The single-degree-of-freedom sum of squares associated with the null hypothesis, H_0: $\beta_{11} + \beta_{22} = 0$, is

$$SS_{\text{Pure Quadratic}} = \frac{n_F n_C(\bar{y}_F - \bar{y}_C)^2}{n_F + n_C} = \frac{(4)(5)(-0.035)^2}{4+5} = 0.0027$$

where n_F and n_C are the number of points in the factorial portion and the number of center points, respectively. Because

$$F = \frac{SS_{\text{Pure Quadratic}}}{\hat{\sigma}^2} = \frac{0.0027}{0.0430} = 0.063$$

is small, there is no indication of a pure quadratic effect.

The analysis of variance for this model is summarized in Table 11.2. Both the interaction and curvature checks are not significant, whereas the F test for the overall regression is significant. Furthermore, the standard error of $\hat{\beta}_1$ and $\hat{\beta}_2$ is

$$se(\hat{\beta}_i) = \sqrt{\frac{MS_E}{4}} = \sqrt{\frac{\hat{\sigma}^2}{4}} = \sqrt{\frac{0.0430}{4}} = 0.10 \qquad i = 1, 2$$

Both regression coefficients $\hat{\beta}_1$ and $\hat{\beta}_2$ are large relative to their standard errors. At this point, we have no reason to question the adequacy of the first-order model.

To move away from the design center—the point ($x_1 = 0$, $x_2 = 0$)—along the path of steepest ascent, we would move 0.775 units in the x_1 direction for every 0.325 units in the x_2 direction. Thus, the path of steepest ascent passes through the point ($x_1 = 0$, $x_2 = 0$) and has a slope 0.325/0.775. The engineer decides to use 5 minutes of reaction time as the basic step size. Using the relationship between ξ_1 and x_1, we see that 5 minutes of reaction time is equivalent to a step in the *coded* variable x_1 of $\Delta x_1 = 1$. Therefore, the steps along the path of steepest ascent are $\Delta x_1 = 1.0000$ and $\Delta x_2 = (0.325/0.775) = 0.42$.

■ TABLE 11.2
Analysis of Variance for the First-Order Model

Source of Variation	Sum of Squares	Degrees of Freedom	Mean Square	F_0	*P*-Value
Model (β_1, β_2)	2.8250	2	1.4125	47.83	0.0002
Residual	0.1772	6			
(Interaction)	(0.0025)	1	0.0025	0.058	0.8215
(Pure quadratic)	(0.0027)	1	0.0027	0.063	0.8142
(Pure error)	(0.1720)	4	0.0430		
Total	3.0022	8			

■ **TABLE 11.3**
Steepest Ascent Experiment for Example 11.1

Steps	Coded Variables x_1	Coded Variables x_2	Natural Variables ξ_1	Natural Variables ξ_2	Response y
Origin	0	0	35	155	
Δ	1.00	0.42	5	2	
Origin + Δ	1.00	0.42	40	157	41.0
Origin + 2Δ	2.00	0.84	45	159	42.9
Origin + 3Δ	3.00	1.26	50	161	47.1
Origin + 4Δ	4.00	1.68	55	163	49.7
Origin + 5Δ	5.00	2.10	60	165	53.8
Origin + 6Δ	6.00	2.52	65	167	59.9
Origin + 7Δ	7.00	2.94	70	169	65.0
Origin + 8Δ	8.00	3.36	75	171	70.4
Origin + 9Δ	9.00	3.78	80	173	77.6
Origin + 10Δ	10.00	4.20	85	175	80.3
Origin + 11Δ	11.00	4.62	90	179	76.2
Origin + 12Δ	12.00	5.04	95	181	75.1

The engineer computes points along this path and observes the yields at these points until a decrease in response is noted. The results are shown in Table 11.3 in both coded and natural variables. Although the coded variables are easier to manipulate mathematically, the natural variables must be used in running the process. Figure 11.5 plots the yield at each step along the path of steepest ascent. Increases in response are observed through the tenth step; however, all steps beyond this point result in a decrease in yield. Therefore, another first-order model should be fit in the general vicinity of the point ($\xi_1 = 85$, $\xi_2 = 175$).

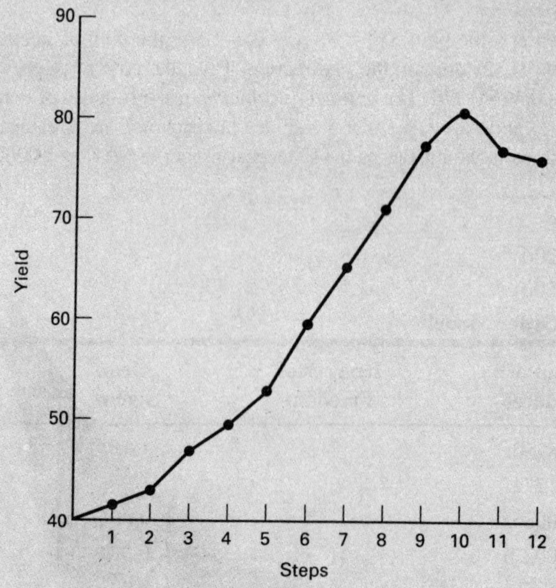

■ **FIGURE 11.5 Yield versus steps along the path of steepest ascent for Example 11.1**

■ **TABLE 11.4**
Data for Second First-Order Model

Natural Variables		Coded Variables		Response
ξ_1	ξ_2	x_1	x_2	y
80	170	−1	−1	76.5
80	180	−1	1	77.0
90	170	1	−1	78.0
90	180	1	1	79.5
85	175	0	0	79.9
85	175	0	0	80.3
85	175	0	0	80.0
85	175	0	0	79.7
85	175	0	0	79.8

A new first-order model is fit around the point ($\xi_1 = 85$, $\xi_2 = 175$). The region of exploration for ξ_1 is [80, 90], and it is [170, 180] for ξ_2. Thus, the coded variables are

$$x_1 = \frac{\xi_1 - 85}{5} \quad \text{and} \quad x_2 = \frac{\xi_2 - 175}{5}$$

Once again, a 2^2 design with five center points is used. The experimental design is shown in Table 11.4.

The first-order model fit to the coded variables in Table 11.4 is

$$\hat{y} = 78.97 + 1.00x_1 + 0.50x_2$$

The analysis of variance for this model, including the interaction and pure quadratic term checks, is shown in Table 11.5. The interaction and pure quadratic checks imply that the first-order model is not an adequate approximation. This curvature in the true surface may indicate that we are near the optimum. At this point, additional analysis must be done to locate the optimum more precisely.

■ **TABLE 11.5**
Analysis of Variance for the Second First-Order Model

Source of Variation	Sum of Squares	Degrees of Freedom	Mean Square	F_0	*P*-Value
Regression	5.00	2			
Residual	11.1200	6			
(Interaction)	(0.2500)	1	0.2500	4.72	0.0955
(Pure quadratic)	(10.6580)	1	10.6580	201.09	0.0001
(Pure error)	(0.2120)	4	0.0530		
Total	16.1200	8			

We notice from Example 11.1 that the *path of steepest ascent is proportional to the signs and magnitudes of the regression coefficients* in the fitted first-order model

$$\hat{y} = \hat{\beta}_0 + \sum_{i=1}^{k} \hat{\beta}_i x_i$$

It is easy to give a general algorithm for determining the coordinates of a point on the path of steepest ascent. Assume that the point $x_1 = x_2 = \cdots = x_k = 0$ is the base or origin point. Then

1. Choose a step size in one of the process variables, say Δx_j. Usually, we would select the variable we know the most about, or we would select the variable that has the largest absolute regression coefficient $|\hat{\beta}_j|$.
2. The step size in the other variables is

$$\Delta x_i = \frac{\hat{\beta}_i}{\hat{\beta}_j/\Delta x_j} \qquad i = 1, 2, \ldots, k \qquad i \neq j$$

3. Convert the Δx_i from coded variables to the natural variables.

To illustrate, consider the path of steepest ascent computed in Example 11.1. Because x_1 has the largest regression coefficient, we select reaction time as the variable in step 1 of the above procedure. Five minutes of reaction time is the step size (based on process knowledge). In terms of the coded variables, this is $\Delta x_1 = 1.0$. Therefore, from guideline 2, the step size in temperature is

$$\Delta x_2 = \frac{\hat{\beta}_2}{\hat{\beta}_1/\Delta x_1} = \frac{0.325}{(0.775/1.0)} = 0.42$$

To convert the coded step sizes ($\Delta x_1 = 1.0$ and $\Delta x_2 = 0.42$) to the natural units of time and temperature, we use the relationships

$$\Delta x_1 = \frac{\Delta \xi_1}{5} \qquad \text{and} \qquad \Delta x_2 = \frac{\Delta \xi_2}{5}$$

which results in

$$\Delta \xi_1 = \Delta x_1(5) = 1.0(5) = 5 \text{ min}$$

and

$$\Delta \xi_2 = \Delta x_2(5) = 0.42(5) = 2°\text{F}$$

11.3 Analysis of a Second-Order Response Surface

When the experimenter is relatively close to the optimum, a model that incorporates curvature is usually required to approximate the response. In most cases, the second-order model

$$y = \beta_0 + \sum_{i=1}^{k} \beta_i x_i + \sum_{i=1}^{k} \beta_{ii} x_i^2 + \sum_{i<j}\sum \beta_{ij} x_i x_j + \epsilon \tag{11.4}$$

is adequate. In this section, we will show how to use this fitted model to find the optimum set of operating conditions for the x's and to characterize the nature of the response surface.

11.3.1 Location of the Stationary Point

Suppose we wish to find the levels of $x_1, x_2, \ldots, x_k$ that optimize the predicted response. This point, if it exists, will be the set of $x_1, x_2, \ldots, x_k$ for which the partial derivatives $\partial\hat{y}/\partial x_1 = \partial\hat{y}/\partial x_2 = \cdots = \partial\hat{y}/\partial x_k = 0$. This point, say $x_{1,s}, x_{2,s}, \ldots, x_{k,s}$, is called the **stationary point**. The stationary point could represent a point of **maximum response**, a point of

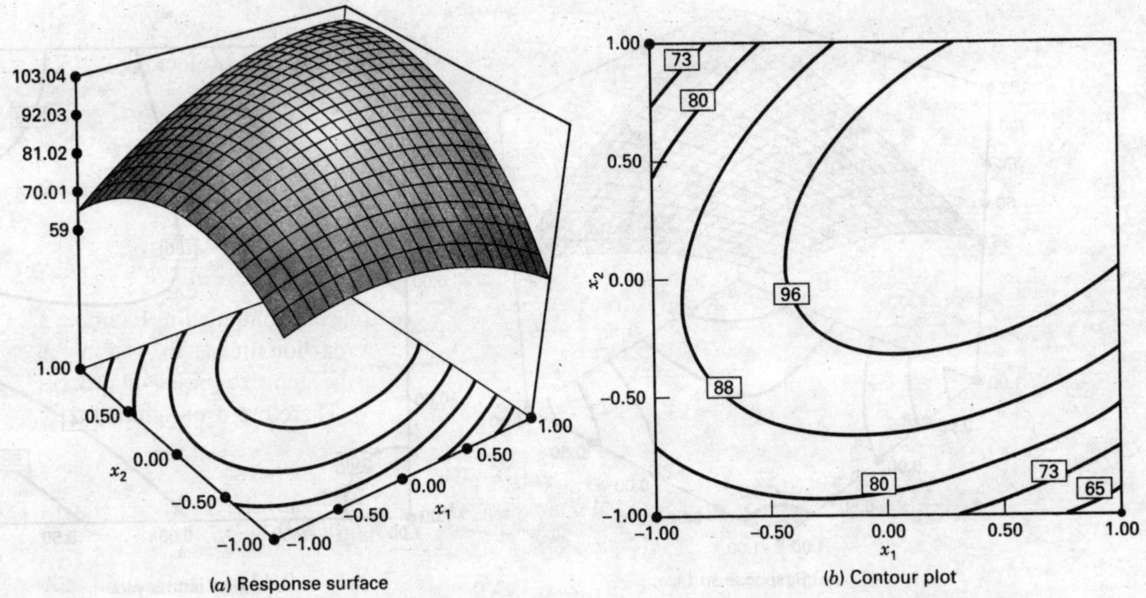

■ **FIGURE 11.6** **Response surface and contour plot illustrating a surface with a maximum**

minimum response, or a **saddle point**. These three possibilities are shown in Figures 11.6, 11.7, and 11.8.

Contour plots play a very important role in the study of the response surface. By generating contour plots using computer software for response surface analysis, the experimenter can usually characterize the shape of the surface and locate the optimum with reasonable precision.

We may obtain a general mathematical solution for the location of the stationary point. Writing the fitted second-order model in matrix notation, we have

$$\hat{y} = \hat{\beta}_0 + \mathbf{x}'\mathbf{b} + \mathbf{x}'\mathbf{B}\mathbf{x} \tag{11.5}$$

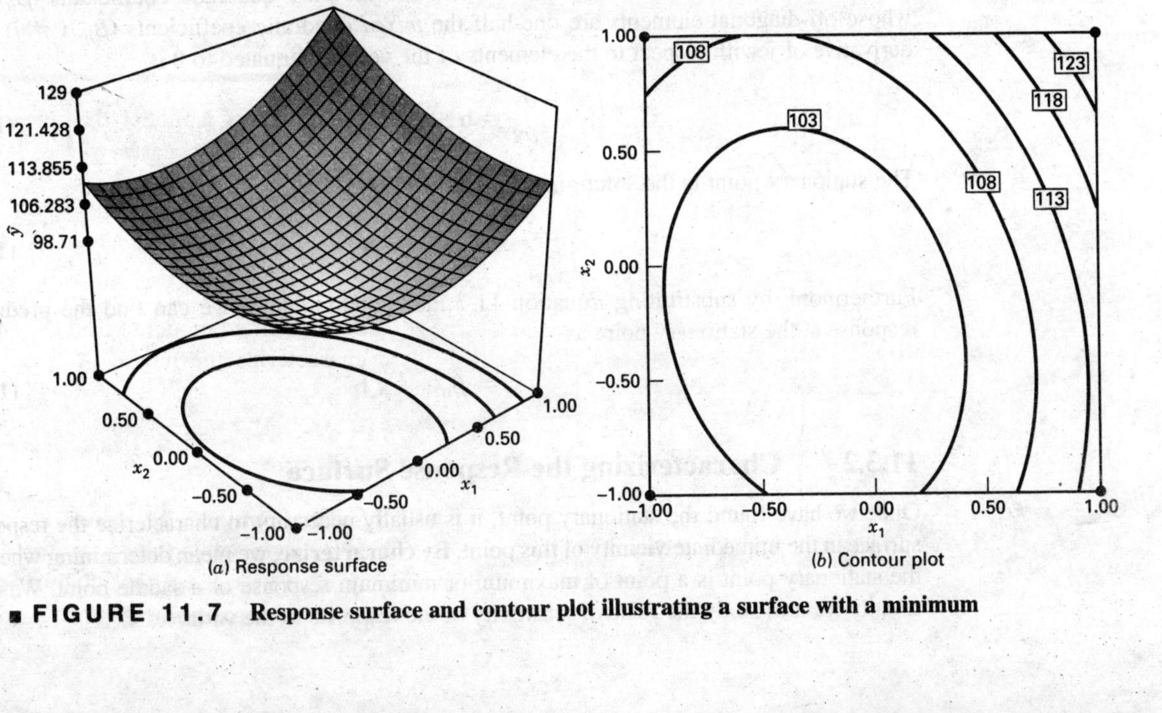

■ **FIGURE 11.7** **Response surface and contour plot illustrating a surface with a minimum**

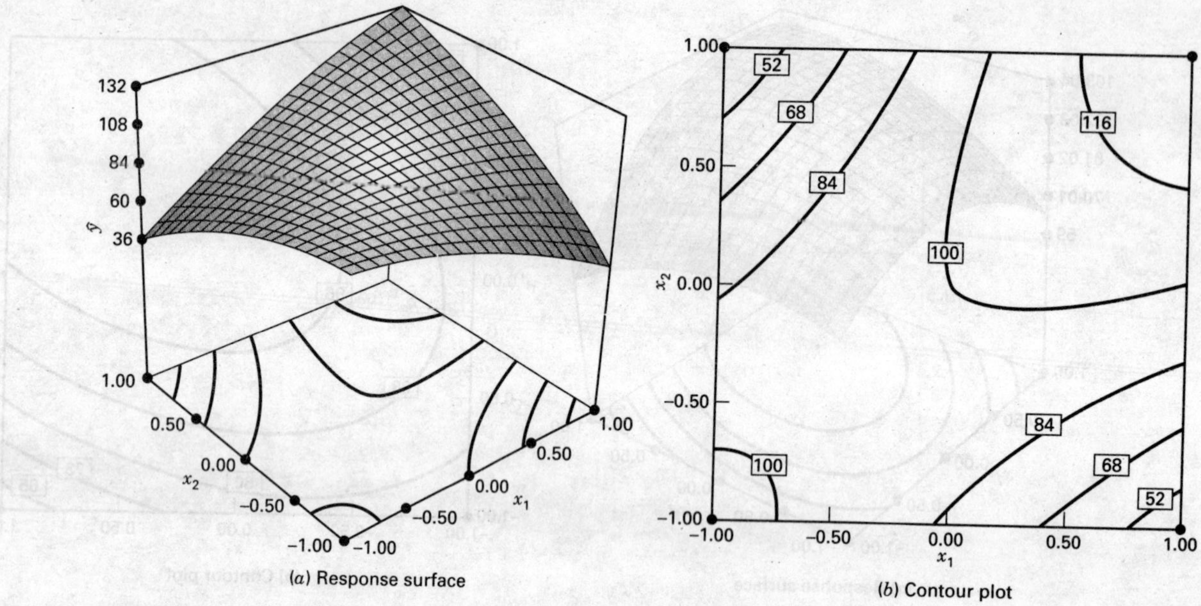

■ **FIGURE 11.8** **Response surface and contour plot illustrating a saddle point (or minimax)**

where

$$\mathbf{x} = \begin{bmatrix} x_1 \\ x_2 \\ \vdots \\ x_k \end{bmatrix} \quad \mathbf{b} = \begin{bmatrix} \hat{\beta}_1 \\ \hat{\beta}_2 \\ \vdots \\ \hat{\beta}_k \end{bmatrix} \quad \text{and} \quad \mathbf{B} = \begin{bmatrix} \hat{\beta}_{11}, \hat{\beta}_{12}/2, \ldots, \hat{\beta}_{1k}/2 \\ \hat{\beta}_{22}, \ldots, \hat{\beta}_{2k}/2 \\ \ddots \\ \text{sym.} \qquad \hat{\beta}_{kk} \end{bmatrix}$$

That is, $\mathbf{b}$ is a $(k \times 1)$ vector of the first-order regression coefficients and $\mathbf{B}$ is a $(k \times k)$ symmetric matrix whose main diagonal elements are the *pure* quadratic coefficients ($\hat{\beta}_{ii}$) and whose off-diagonal elements are one-half the *mixed* quadratic coefficients ($\hat{\beta}_{ij}$, $i \neq j$). The derivative of $\hat{y}$ with respect to the elements of the vector $\mathbf{x}$ equated to $\mathbf{0}$ is

$$\frac{\partial \hat{y}}{\partial \mathbf{x}} = \mathbf{b} + 2\mathbf{B}\mathbf{x} = \mathbf{0} \tag{11.6}$$

The stationary point is the solution to Equation 11.6, or

$$\mathbf{x}_s = -\tfrac{1}{2}\mathbf{B}^{-1}\mathbf{b} \tag{11.7}$$

Furthermore, by substituting Equation 11.7 into Equation 11.5, we can find the predicted response at the stationary point as

$$\hat{y}_s = \hat{\beta}_0 + \tfrac{1}{2}\mathbf{x}_s'\mathbf{b} \tag{11.8}$$

11.3.2 Characterizing the Response Surface

Once we have found the stationary point, it is usually necessary to characterize the response surface in the immediate vicinity of this point. By **characterize**, we mean determining whether the stationary point is a point of maximum or minimum response or a saddle point. We also usually want to study the relative sensitivity of the response to the variables $x_1, x_2, \ldots, x_k$.

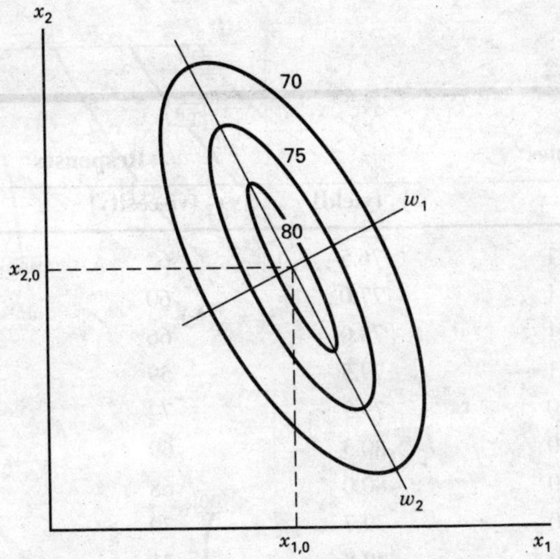

■ **FIGURE 11.9** **Canonical form of the second-order model**

As we mentioned previously, the most straightforward way to do this is to examine a **contour plot** of the fitted model. If there are only two or three process variables (the x's), the construction and interpretation of this contour plot is relatively easy. However, even when there are relatively few variables, a more formal analysis, called the **canonical analysis**, can be useful.

It is helpful first to transform the model into a new coordinate system with the origin at the stationary point $\mathbf{x}_s$ and then to rotate the axes of this system until they are parallel to the principal axes of the fitted response surface. This transformation is illustrated in Figure 11.9. We can show that this results in the fitted model

$$\hat{y} = \hat{y}_s + \lambda_1 w_1^2 + \lambda_2 w_2^2 + \cdots + \lambda_k w_k^2 \tag{11.9}$$

where the $\{w_i\}$ are the transformed independent variables and the $\{\lambda_i\}$ are constants. Equation 11.9 is called the **canonical form** of the model. Furthermore, the $\{\lambda_i\}$ are just the **eigenvalues** or **characteristic roots** of the matrix $\mathbf{B}$.

The nature of the response surface can be determined from the stationary point and the *signs* and *magnitudes* of the $\{\lambda_i\}$. First, suppose that the stationary point is within the region of exploration for fitting the second-order model. If the $\{\lambda_i\}$ are all positive, $\mathbf{x}_s$ is a point of minimum response; if the $\{\lambda_i\}$ are all negative, $\mathbf{x}_s$ is a point of maximum response; and if the $\{\lambda_i\}$ have different signs, $\mathbf{x}_s$ is a saddle point. Furthermore, the surface is steepest in the w_i direction for which $|\lambda_i|$ is the greatest. For example, Figure 11.9 depicts a system for which $\mathbf{x}_s$ is a maximum (λ_1 and λ_2 are negative) with $|\lambda_1| > |\lambda_2|$.

EXAMPLE 11.2

We will continue the analysis of the chemical process in Example 11.1. A second-order model in the variables x_1 and x_2 cannot be fit using the design in Table 11.4. The experimenter decides to augment this design with enough points to fit a second-order model.[1] She obtains four observations at ($x_1 = 0, x_2 = \pm 1.414$) and ($x_1 = \pm 1.414, x_2 = 0$). The complete experiment is shown in Table 11.6, and the design is displayed in Figure 11.10. This design is called a **central composite design** (or a CCD) and will be discussed in more detail in Section 11.4.2. In this second phase of the study, two

[1] The engineer ran the additional four observations at about the same time he or she ran the original nine observations. If substantial time had elapsed between the two sets of runs, blocking would have been necessary. Blocking in response surface designs is discussed in Section 11.4.3.

■ **TABLE 11.6**
Central Composite Design for Example 11.2

Natural Variables		Coded Variables		Responses		
ξ_1	ξ_2	x_1	x_2	y_1 (yield)	y_2 (viscosity)	y_3 (molecular weight)
80	170	−1	−1	76.5	62	2940
80	180	−1	1	77.0	60	3470
90	170	1	−1	78.0	66	3680
90	180	1	1	79.5	59	3890
85	175	0	0	79.9	72	3480
85	175	0	0	80.3	69	3200
85	175	0	0	80.0	68	3410
85	175	0	0	79.7	70	3290
85	175	0	0	79.8	71	3500
92.07	175	1.414	0	78.4	68	3360
77.93	175	−1.414	0	75.6	71	3020
85	182.07	0	1.414	78.5	58	3630
85	167.93	0	−1.414	77.0	57	3150

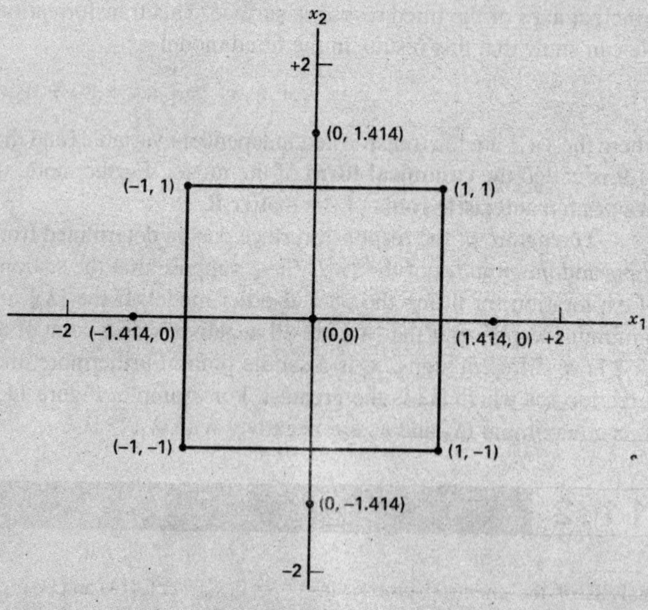

■ **FIGURE 11.10** **Central composite design for Example 11.2**

additional responses were of interest: the viscosity and the molecular weight of the product. The responses are also shown in Table 11.6.

We will focus on fitting a quadratic model to the yield response y_1 (the other responses will be discussed in Section 11.3.4). We generally use computer software to fit a response surface and to construct the contour plots. Table 11.7 contains the output from Design-Expert. From examining this table, we notice that this software package first computes the "sequential or extra sums of squares" for the

■ **TABLE 11.7**

Computer Output from Design-Expert for Fitting a Model to the Yield Response in Example 11.2

Response: yield

*****WARNING: The Cubic Model is Aliased!*****

Sequential Model Sum of Squares

Source	Sum of Squares	DF	Mean Square	F Value	Prob > F	
Mean	80062.16	1	80062.16			
Linear	10.04	2	5.02	2.69	0.1166	
2FI	0.25	1	0.25	0.12	0.7350	
Quadratic	*17.95*	*2*	*8.98*	*126.88*	*<0.001*	*Suggested*
Cubic	2.042E-003	2	1.021E-003	0.010	0.9897	Aliased
Residual	0.49	5	0.099			
Total	80090.90	13	6160.84			

"Sequential Model Sum of Squares": Select the highest order polynomial where the additional terms are significant.

Lack-of-Fit Tests

Source	Sum of Squares	DF	Mean Square	F Value	Prob > F	
Linear	18.49	6	3.08	58.14	0.0008	
2FI	18.24	5	3.65	68.82	0.0006	
Quadratic	*0.28*	*3*	*0.094*	*1.78*	*0.2897*	*Suggested*
Cubic	0.28	1	0.28	5.31	0.0826	Aliased
Pure Error	0.21	4	0.053			

"Lack-of-Fit Tests": Want the selected model to have insignificant lack-of-fit.

Model summary Statistics

Source	Std. Dev.	R-Squared	Adjusted R-Squared	Predicted R-Squared	PRESS	
Linear	1.37	0.3494	0.2193	−0.0435	29.99	
2FI	1.43	0.3581	0.1441	−0.2730	36.59	
Quadratic	*0.27*	*0.9828*	*0.9705*	*0.9184*	*2.35*	*Suggested*
Cubic	0.31	0.9828	0.9588	0.3622	18.33	Aliased

"Model Summary Statistics": Focus on the model minimizing the "PRESS," or equivalently maximizing the "PRED R-SQR."

Response: yield
ANOVA for Response Surface Quadratic Model
Analysis of variance table [Partial sum of squares]

Source	Sum of Squares	DF	Mean Square	F Value	Prob > F
Model	28.25	5	5.65	79.85	<0.0001
A	*7.92*	*1*	*7.92*	*111.93*	*<0.0001*
B	*2.12*	*1*	*2.12*	*30.01*	*0.0009*
A^2	*13.18*	*1*	*13.18*	*186.22*	*<0.0001*
B^2	*6.97*	*1*	*6.97*	*98.56*	*<0.0001*
AB	*0.25*	*1*	*0.25*	*3.53*	*0.1022*
Residual	0.50	7	0.071		
Lack of Fit	*0.28*	*3*	*0.094*	*1.78*	*0.2897*
Pure Error	*0.21*	*4*	*0.053*		
Cor Total	28.74	12			

■ **TABLE 11.7** *(Continued)*

Std. Dev.	0.27	R-Squared	0.9828
Mean	78.48	Adj R-Squared	0.9705
C.V.	0.34	Pred R-Squared	0.9184
PRESS	2.35	Adeq Precision	23.018

Factor	Coefficient Estimate	DF	Standard Error	95% CI Low	95% CI High	VIF
Intercept	79.94	1	0.12	79.66	80.22	
A-time	0.99	1	0.094	0.77	1.22	1.00
B-temp	0.52	1	0.094	0.29	0.74	1.00
A^2	−1.38	1	0.10	−1.61	−1.14	1.02
B^2	−1.00	1	0.10	−1.24	−0.76	1.02
AB	0.25	1	0.13	−0.064	0.56	1.00

Final Equation in Terms of Coded Factors:

yield =
+79.94
+0.99 * A
+0.52 * B
−1.38 * A^2
−1.00 * B^2
+0.25 * A * B

Final Equation in Terms of Actual Factors:

yield =
−1430.52285
+7.80749 * time
+13.27053 * temp
−0.055050 * time^2
−0.040050 * temp^2
+0.010000 * time * temp

Diagnostics Case Statistics

Run Order	Standard Order	Actual Value	Predicted Value	Residual	Leverage	Student Residual	Cook's Distance	Outlier t
8	1	76.50	76.30	0.20	0.625	1.213	0.409	1.264
6	2	78.00	77.79	0.21	0.625	1.275	0.452	1.347
9	3	77.00	76.83	0.17	0.625	1.027	0.293	1.032
11	4	79.50	79.32	0.18	0.625	1.089	0.329	1.106
12	5	75.60	75.78	−0.18	0.625	−1.107	0.341	−1.129
10	6	78.40	78.59	−0.19	0.625	−1.195	0.396	−1.240
7	7	77.00	77.21	−0.21	0.625	−1.283	0.457	−1.358
1	8	78.50	78.67	−0.17	0.625	−1.019	0.289	−1.023
5	9	79.90	79.94	−0.040	0.200	−0.168	0.001	−0.156
3	10	80.30	79.94	0.36	0.200	1.513	0.095	1.708
13	11	80.00	79.94	0.060	0.200	0.252	0.003	0.235
2	12	79.70	79.94	−0.24	0.200	−1.009	0.042	−1.010
4	13	79.80	79.94	−0.14	0.200	−0.588	0.014	−0.559

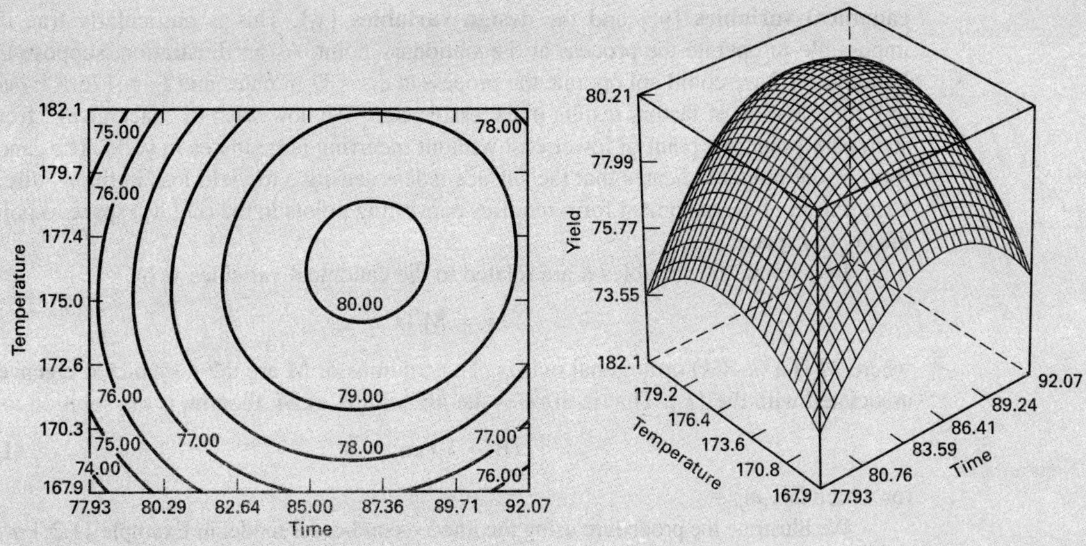

(*a*) The contour plot (*b*) The response surface plot

■ **FIGURE 11.11** **Contour and response surface plots of the yield response, Example 11.2**

linear, quadratic, and cubic terms in the model (there is a warning message concerning aliasing in the cubic model because the CCD does not contain enough runs to support a full cubic model). On the basis of the small *P*-value for the quadratic terms, we decided to fit the second-order model to the yield response. The computer output shows the final model in terms of both the coded variables and the natural or actual factor levels.

Figure 11.11 shows the three-dimensional response surface plot and the contour plot for the yield response in terms of the process variables time and temperature. It is relatively easy to see from examining these figures that the optimum is very near 175°F and 85 minutes of reaction time and that the response is at a maximum at this point. From examination of the contour plot, we note that the process may be slightly more sensitive to changes in reaction time than to changes in temperature.

We could also find the location of the stationary point using the general solution in Equation 11.7. Note that

$$\mathbf{b} = \begin{bmatrix} 0.995 \\ 0.515 \end{bmatrix} \quad \mathbf{B} = \begin{bmatrix} -1.376 & 0.1250 \\ 0.1250 & -1.001 \end{bmatrix}$$

and from Equation 11.7, the stationary point is

$$\mathbf{x}_s = -\tfrac{1}{2}\mathbf{B}^{-1}\mathbf{b}$$

$$= -\frac{1}{2}\begin{bmatrix} -0.7345 & -0.0917 \\ -0.0917 & -1.0096 \end{bmatrix}\begin{bmatrix} 0.995 \\ 0.515 \end{bmatrix} = \begin{bmatrix} 0.389 \\ 0.306 \end{bmatrix}$$

That is, $x_{1,s} = 0.389$ and $x_{2,s} = 0.306$. In terms of the natural variables, the stationary point is

$$0.389 = \frac{\xi_1 - 85}{5} \qquad 0.306 = \frac{\xi_2 - 175}{5}$$

which yields $\xi_1 = 86.95 \simeq 87$ minutes of reaction time and $\xi_2 = 176.53 \simeq 176.5$°F. This is very close to the stationary point found by visual examination of the contour plot in Figure 11.11. Using Equation 11.8, we may find the predicted response at the stationary point as $\hat{y}_s = 80.21$.

We may also use the canonical analysis described in this section to characterize the response surface. First, it is necessary to express the fitted model in canonical form (Equation 11.9). The eigenvalues λ_1 and λ_2 are the roots of the determinantal equation

$$|\mathbf{B} - \lambda\mathbf{I}| = 0$$

$$\begin{vmatrix} -1.376 - \lambda & 0.1250 \\ 0.1250 & -1.001 - \lambda \end{vmatrix} = 0$$

which reduces to

$$\lambda^2 + 2.3788\lambda + 1.3639 = 0$$

The roots of this quadratic equation are $\lambda_1 = -0.9634$ and $\lambda_2 = -1.4141$. Thus, the canonical form of the fitted model is

$$\hat{y} = 80.21 - 0.9634w_1^2 - 1.4141w_2^2$$

Because both λ_1 and λ_2 are negative and the stationary point is within the region of exploration, we conclude that the stationary point is a maximum.

In some RSM problems, it may be necessary to find the relationship between the **canonical variables** $\{w_i\}$ and the **design variables** $\{x_i\}$. This is particularly true if it is impossible to operate the process at the stationary point. As an illustration, suppose that in Example 11.2 we could not operate the process at $\xi_1 = 87$ minutes and $\xi_2 = 176.5°F$ because this combination of factors results in excessive cost. We now wish to "back away" from the stationary point to a point of lower cost without incurring large losses in yield. The canonical form of the model indicates that the surface is less sensitive to yield loss in the w_1 direction. Exploration of the canonical form requires converting points in the (w_1, w_2) space to points in the (x_1, x_2) space.

In general, the variables **x** are related to the canonical variables **w** by

$$\mathbf{w} = \mathbf{M}'(\mathbf{x} - \mathbf{x}_s)$$

where **M** is a $(k \times k)$ orthogonal matrix. The columns of **M** are the normalized eigenvectors associated with the $\{\lambda_i\}$. That is, if $\mathbf{m}_i$ is the ith column of **M**, then $\mathbf{m}_i$ is the solution to

$$(\mathbf{B} - \lambda_i \mathbf{I})\mathbf{m}_i = \mathbf{0} \qquad \textbf{(11.10)}$$

for which $\Sigma_{j=1}^{k} m_{ji}^2 = 1$.

We illustrate the procedure using the fitted second-order model in Example 11.2. For $\lambda_1 = -0.9634$, Equation 11.10 becomes

$$\begin{bmatrix} (-1.376 + 0.9634) & 0.1250 \\ 0.1250 & (-1.001 + 0.9634) \end{bmatrix} \begin{bmatrix} m_{11} \\ m_{21} \end{bmatrix} = \begin{bmatrix} 0 \\ 0 \end{bmatrix}$$

or

$$-0.4129m_{11} + 0.1250m_{21} = 0$$
$$0.1250m_{11} + 0.0377m_{21} = 0$$

We wish to obtain the normalized solution to these equations, that is, the one for which $m_{11}^2 + m_{21}^2 = 1$. There is no unique solution to these equations, so it is most convenient to assign an arbitrary value to one unknown, solve the system, and normalize the solution. Letting $m_{21}^* = 1$, we find $m_{11}^* = 0.3027$. To normalize this solution, we divide m_{11}^* and m_{21}^* by

$$\sqrt{(m_{11}^*)^2 + (m_{21}^*)^2} = \sqrt{(0.3027)^2 + (1)^2} = 1.0448$$

This yields the normalized solution

$$m_{11} = \frac{m_{11}^*}{1.0448} = \frac{0.3027}{1.0448} = 0.2898$$

and

$$m_{21} = \frac{m_{21}^*}{1.0448} = \frac{1}{1.0448} = 0.9571$$

which is the first column of the **M** matrix.

Using $\lambda_2 = -1.4141$, we can repeat the above procedure, obtaining $m_{12} = -0.9571$ and $m_{22} = 0.2898$ as the second column of **M**. Thus, we have

$$\mathbf{M} = \begin{bmatrix} 0.2898 & -0.9571 \\ 0.9571 & 0.2898 \end{bmatrix}$$

The relationship between the **w** and **x** variables is

$$\begin{bmatrix} w_1 \\ w_2 \end{bmatrix} = \begin{bmatrix} 0.2898 & 0.9571 \\ -0.9574 & 0.2898 \end{bmatrix} \begin{bmatrix} x_1 - 0.389 \\ x_2 - 0.306 \end{bmatrix}$$

or

$$w_1 = 0.2897(x_1 - 0.389) + 0.9571(x_2 - 0.306)$$
$$w_2 = -0.9574(x_1 - 0.389) + 0.2888(x_2 - 0.306)$$

If we wished to explore the response surface in the vicinity of the stationary point, we could determine appropriate points at which to take observations in the (w_1, w_2) space and then use the above relationship to convert these points into the (x_1, x_2) space so that the runs may be made.

11.3.3 Ridge Systems

It is not unusual to encounter variations of the pure maximum, minimum, or saddle point response surfaces discussed in the previous section. Ridge systems, in particular, are fairly common. Consider the canonical form of the second-order model given previously in Equation 11.9:

$$\hat{y} = \hat{y}_s + \lambda_1 w_1^2 + \lambda_2 w_2^2 + \cdots + \lambda_k w_k^2$$

Now suppose that the stationary point $\mathbf{x}_s$ is within the region of experimentation; furthermore, let one or more of the λ_i be very small (e.g., $\lambda_i = 0$). The response variable is then very insensitive to the variables w_i multiplied by the small λ_i.

A contour plot illustrating this situation is shown in Figure 11.12 for $k = 2$ variables with $\lambda_1 = 0$. (In practice, λ_1 would be close to but not exactly equal to zero.) The canonical model for this response surface is theoretically

$$\hat{y} = \hat{y}_s + \lambda_2 w_2^2$$

with λ_2 negative. Notice that the severe elongation in the w_1 direction has resulted in a line of centers at $\hat{y} = 70$ and the optimum may be taken anywhere along that line. This type of response surface is called a **stationary ridge system**.

If the stationary point is far outside the region of exploration for fitting the second-order model and one (or more) λ_i is near zero, then the surface may be a **rising ridge**. Figure 11.13 illustrates a rising ridge for $k = 2$ variables with λ_1 near zero and λ_2 negative. In this type of ridge system, we cannot draw inferences about the true surface or the stationary point because $\mathbf{x}_s$ is outside the region where we have fit the model. However, further exploration is warranted in the w_1 direction. If λ_2 had been positive, we would call this system a falling ridge.

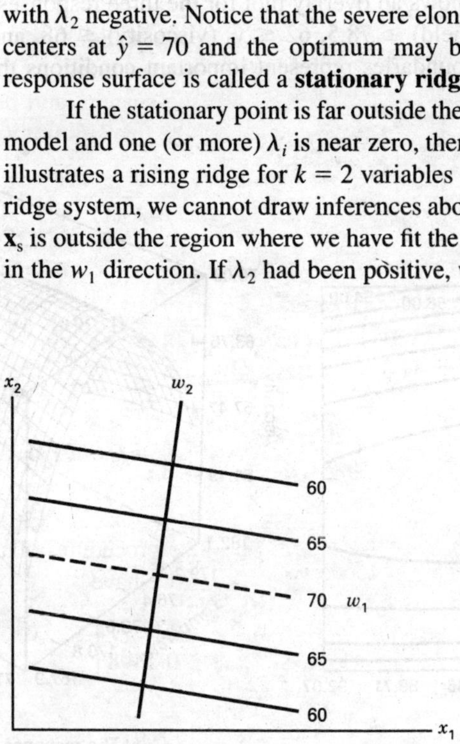

■ **FIGURE 11.12** **A contour plot of a stationary ridge system**

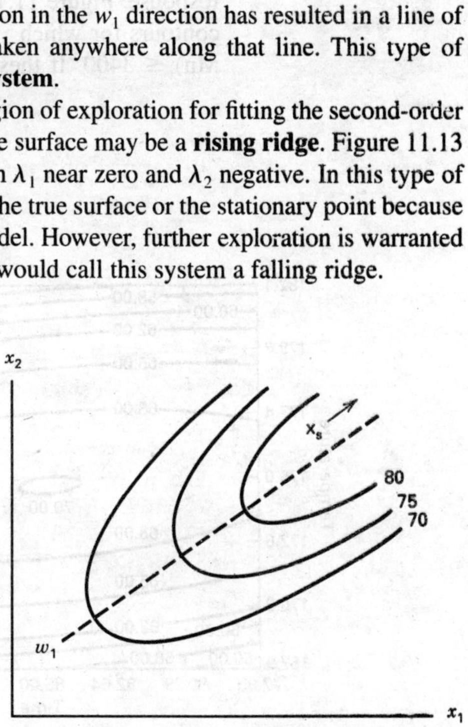

■ **FIGURE 11.13** **A contour plot of a rising ridge system**

11.3.4 Multiple Responses

Many response surface problems involve the analysis of several responses. For instance, in Example 11.2, the experimenter measured three responses. In this example, we optimized the process with respect to only the yield response y_1.

Simultaneous consideration of multiple responses involves first building an appropriate response surface model for each response and then trying to find a set of operating conditions that in some sense optimizes all responses or at least keeps them in desired ranges. An extensive treatment of the multiple response problem is given in Myers, Montgomery and Anderson-Cook (2009).

We may obtain models for the viscosity and molecular weight responses (y_2 and y_3, respectively) in Example 11.2 as follows:

$$\hat{y}_2 = 70.00 - 0.16x_2 - 0.95x_2 - 0.69x_1^2 - 6.69x_2^2 - 1.25x_1x_2$$

$$\hat{y}_3 = 3386.2 + 205.1x_1 + 177.4x_2$$

In terms of the natural levels of time (ξ_1) and temperature (ξ_2), these models are

$$\hat{y}_2 = -9030.74 + 13.393\xi_1 + 97.708\xi_2$$
$$-2.75 \times 10^{-2}\xi_1^2 - 0.26757\xi_2^2 - 5 \times 10^{-2}\xi_1\xi_2$$

and

$$\hat{y}_3 = -6308.8 + 41.025\xi_1 + 35.473\xi_2$$

Figures 11.14 and 11.15 present the contour and response surface plots for these models.

A relatively straightforward approach to optimizing several responses that works well when there are only a few process variables is to **overlay the contour plots** for each response. Figure 11.16 shows an overlay plot for the three responses in Example 11.2, with contours for which y_1 (yield) ≥ 78.5, $62 \leq y_2$ (viscosity) ≤ 68, and y_3 (molecular weight Mn) ≤ 3400. If these boundaries represent important conditions that must be met by the

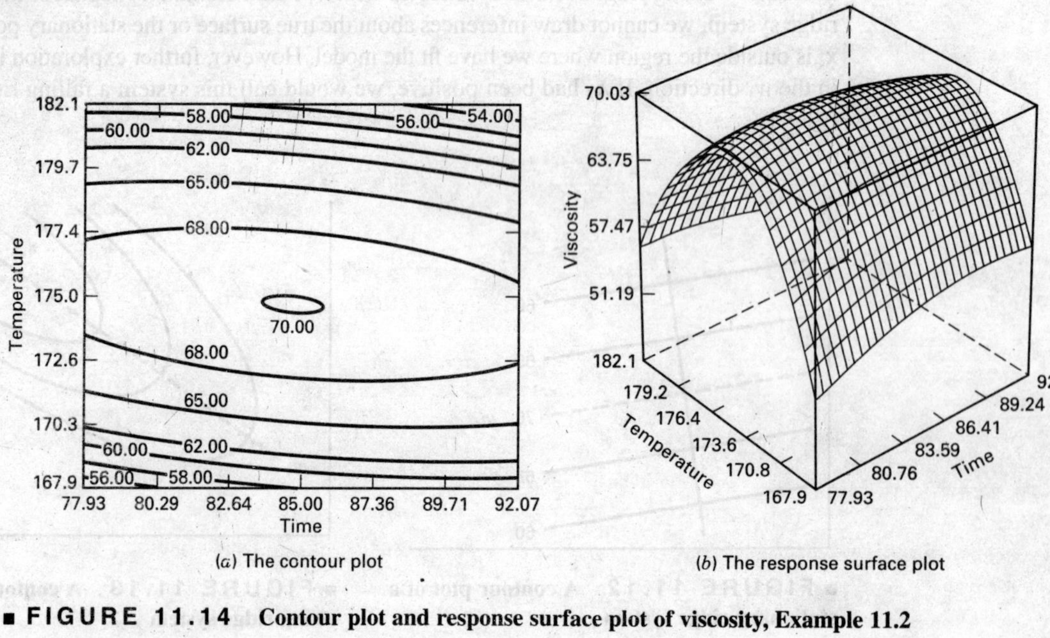

(a) The contour plot (b) The response surface plot

■ **FIGURE 11.14 Contour plot and response surface plot of viscosity, Example 11.2**

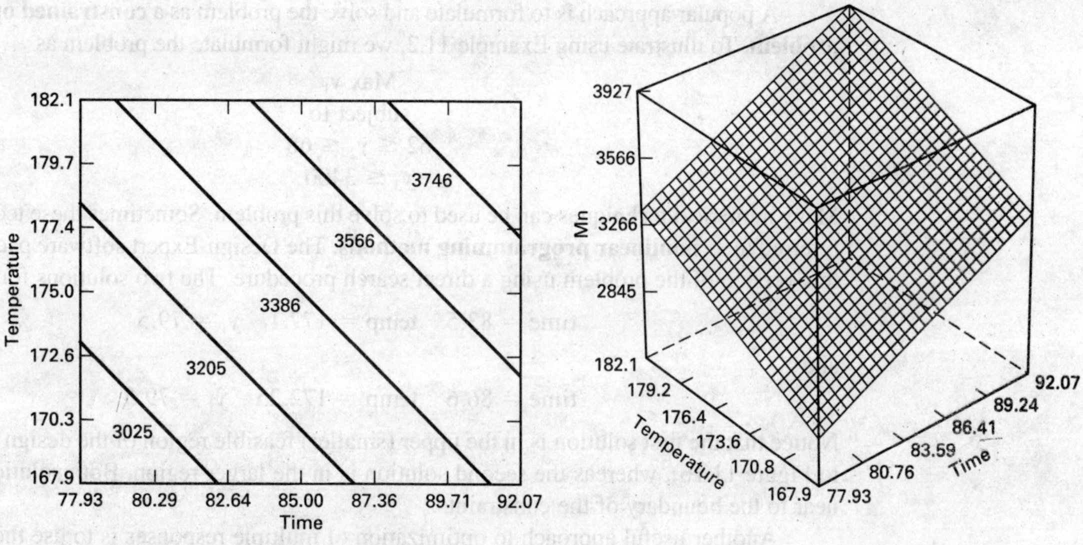

(a) The contour plot (b) The response surface plot

■ **FIGURE 11.15** **Contour plot and response surface plot of molecular weight, Example 11.2**

process, then as the unshaded portion of Figure 11.16 shows, a number of combinations of time and temperature will result in a satisfactory process. The experimenter can visually examine the contour plot to determine appropriate operating conditions. For example, it is likely that the experimenter would be most interested in the larger of the two feasible operating regions shown in Figure 11.16.

When there are more than three design variables, overlaying contour plots becomes awkward because the contour plot is two dimensional, and $k - 2$ of the design variables must be held constant to construct the graph. Often a lot of trial and error is required to determine which factors to hold constant and what levels to select to obtain the best view of the surface. Therefore, there is practical interest in more formal optimization methods for multiple responses.

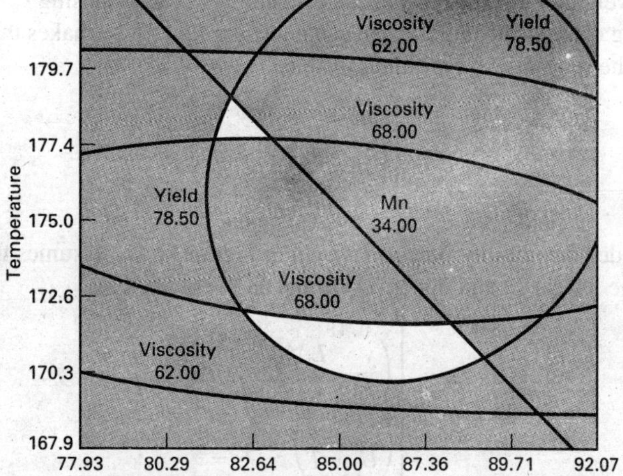

■ **FIGURE 11.16**
Region of the optimum found by overlaying yield, viscosity, and molecular weight response surfaces, Example 11.2

A popular approach is to formulate and solve the problem as a **constrained optimization problem**. To illustrate using Example 11.2, we might formulate the problem as

$$\begin{gathered} \text{Max } y_1 \\ \text{subject to} \\ 62 \le y_2 \le 68 \\ y_3 \le 3400 \end{gathered}$$

Many numerical techniques can be used to solve this problem. Sometimes these techniques are referred to as **nonlinear programming methods**. The Design-Expert software package solves this version of the problem using a direct search procedure. The two solutions found are

$$\text{time} = 83.5 \quad \text{temp} = 177.1 \quad \hat{y}_1 = 79.5$$

and

$$\text{time} = 86.6 \quad \text{temp} = 172.25 \quad \hat{y}_1 = 79.5$$

Notice that the first solution is in the upper (smaller) feasible region of the design space (refer to Figure 11.16), whereas the second solution is in the larger region. Both solutions are very near to the boundary of the constraints.

Another useful approach to optimization of multiple responses is to use the simultaneous optimization technique popularized by Derringer and Suich (1980). Their procedure makes use of **desirability functions**. The general approach is to first convert each response y_i into an individual desirability function d_i that varies over the range

$$0 \le d_i \le 1$$

where if the response y_i is at its goal or target, then $d_i = 1$ and if the response is outside an acceptable region, $d_i = 0$. Then the design variables are chosen to maximize the overall desirability

$$D = (d_1 \cdot d_2 \cdot \cdots \cdot d_m)^{1/m}$$

where there are m responses. The overall desirability will be zero if any of the individual responses is undesirable.

The individual desirability functions are structured as shown in Figure 11.17. If the objective or target T for the response y is a maximum value,

$$d = \begin{cases} 0 & y < L \\ \left(\dfrac{y - L}{T - L}\right)^r & L \le y \le T \\ 1 & y > T \end{cases} \tag{11.11}$$

when the weight $r = 1$, the desirability function is linear. Choosing $r > 1$ places more emphasis on being close to the target value and choosing $0 < r < 1$ makes this less important. If the target for the response is a minimum value,

$$d = \begin{cases} 1 & y < T \\ \left(\dfrac{U - y}{U - T}\right)^r & T \le y \le U \\ 0 & y > U \end{cases} \tag{11.12}$$

The two-sided desirability function shown in Figure 11.17c assumes that the target is located between the lower (L) and upper (U) limits and is defined as

$$d = \begin{cases} 0 & y < L \\ \left(\dfrac{y - L}{T - L}\right)^{r_1} & L \le y \le T \\ \left(\dfrac{U - y}{U - T}\right)^{r_2} & T \le y \le U \\ 0 & y > U \end{cases} \tag{11.13}$$

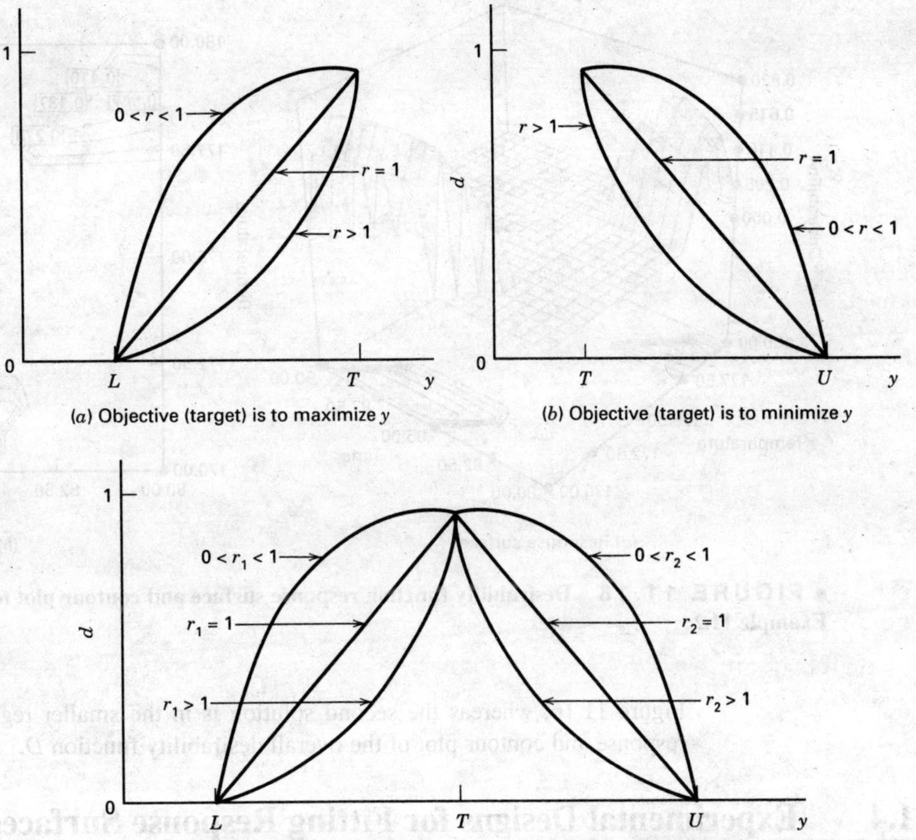

(a) Objective (target) is to maximize y

(b) Objective (target) is to minimize y

(c) Objective is for y to be as close as posible to the target

■ **FIGURE 11.17** **Individual desirability functions for simultaneous optimization**

The Design-Expert software package was used to solve Example 11.2 using the desirability function approach. We chose $T = 80$ as the target for the yield response with $U = 70$ and set the weight for this individual desirability equal to unity. We set $T = 65$ for the viscosity response with $L = 62$ and $U = 68$ (to be consistent with specifications), with both weights $r_1 = r_2 = 1$. Finally, we indicated that any molecular weight between 3200 and 3400 was acceptable. Two solutions were found.

Solution 1

Time = 86.5	Temp = 170.5	$D = 0.822$
$\hat{y}_1 = 78.8$	$\hat{y}_2 = 65$	$\hat{y}_3 = 3287$

Solution 2

Time = 82	Temp = 178.8	$D = 0.792$
$\hat{y}_1 = 78.5$	$\hat{y}_2 = 65$	$\hat{y}_3 = 3400$

Solution 1 has the highest overall desirability. Notice that it results in on-target viscosity and acceptable molecular weight. This solution is in the larger of the two operating regions in

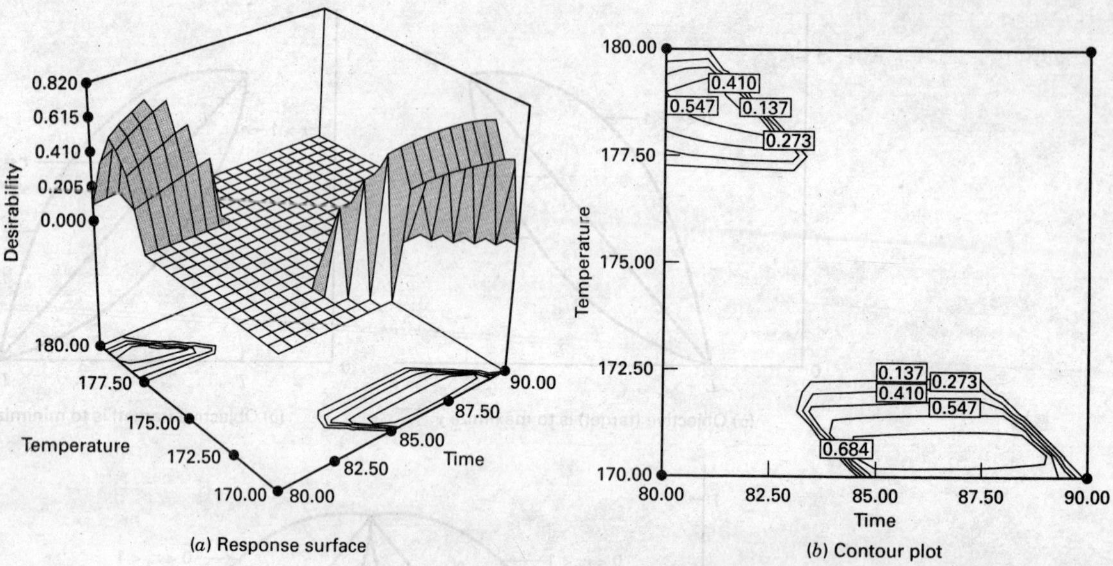

■ **FIGURE 11.18 Desirability function response surface and contour plot for the problem in Example 11.2**

Figure 11.16, whereas the second solution is in the smaller region. Figure 11.18 shows a response and contour plot of the overall desirability function D.

11.4 Experimental Designs for Fitting Response Surfaces

Fitting and analyzing response surfaces is greatly facilitated by the proper choice of an experimental design. In this section, we discuss some aspects of selecting appropriate designs for fitting response surfaces.

When selecting a response surface design, some of the features of a desirable design are as follows:

1. Provides a reasonable distribution of data points (and hence information) throughout the region of interest
2. Allows model adequacy, including lack of fit, to be investigated
3. Allows experiments to be performed in blocks
4. Allows designs of higher order to be built up sequentially
5. Provides an internal estimate of error
6. Provides precise estimates of the model coefficients
7. Provides a good profile of the prediction variance throughout the experimental region
8. Provides reasonable robustness against outliers or missing values
9. Does not require a large number of runs
10. Does not require too many levels of the independent variables
11. Ensures simplicity of calculation of the model parameters

These features are sometimes conflicting, so judgment must often be applied in design selection. For more information on the choice of a response surface design, refer to Khuri and Cornell (1996), Myers, Montgomery and Anderson-Cook (2009), and Box and Draper (2007).

11.4.1 Designs for Fitting the First-Order Model

Suppose we wish to fit the first-order model in k variables

$$y = \beta_0 + \sum_{i=1}^{k} \beta_i x_i + \epsilon \tag{11.14}$$

There is a unique class of designs that minimizes the variance of the regression coefficients $\{\hat{\beta}_i\}$. These are the **orthogonal first-order designs**. A first-order design is orthogonal if the off-diagonal elements of the $(\mathbf{X}'\mathbf{X})$ matrix are all zero. This implies that the cross products of the columns of the $\mathbf{X}$ matrix sum to zero.

The class of orthogonal first-order designs includes the 2^k factorial and fractions of the 2^k series in which main effects are not aliased with each other. In using these designs, we assume that the low and high levels of the k factors are coded to the usual ± 1 levels.

The 2^k design does not afford an estimate of the experimental error unless some runs are replicated. A common method of including replication in the 2^k design is to augment the design with several observations at the center (the point $x_i = 0$, $i = 1, 2, \ldots, k$). The addition of center points to the 2^k design does not influence the $\{\hat{\beta}_i\}$ for $i \geq 1$, but the estimate of β_0 becomes the grand average of all observations. Furthermore, the addition of center points does not alter the orthogonality property of the design. Example 11.1 illustrates the use of a 2^2 design augmented with five center points to fit a first-order model.

Another orthogonal first-order design is the **simplex**. The simplex is a regularly sided figure with $k + 1$ vertices in k dimensions. Thus, the simplex design for $k = 2$ is an equilateral triangle, and it is a regular tetrahedron for $k = 3$. Simplex designs in two and three dimensions are shown in Figure 11.19.

11.4.2 Designs for Fitting the Second-Order Model

We have informally introduced in Example 11.2 (and even earlier, in Example 6.6) the **central composite design** or **CCD** for fitting a second-order model. This is the most popular class of designs used for fitting these models. Generally, the CCD consists of a 2^k factorial (or fractional factorial of resolution V) with n_F factorial runs, $2k$ axial or star runs, and n_C center runs. Figure 11.20 shows the CCD for $k = 2$ and $k = 3$ factors.

The practical deployment of a CCD often arises through **sequential experimentation**, as in Examples 11.1 and 11.2. That is, a 2^k has been used to fit a first-order model, this model has exhibited lack of fit, and the axial runs are then added to allow the quadratic terms to be incorporated into the model. The CCD is a very efficient design for fitting the second-order

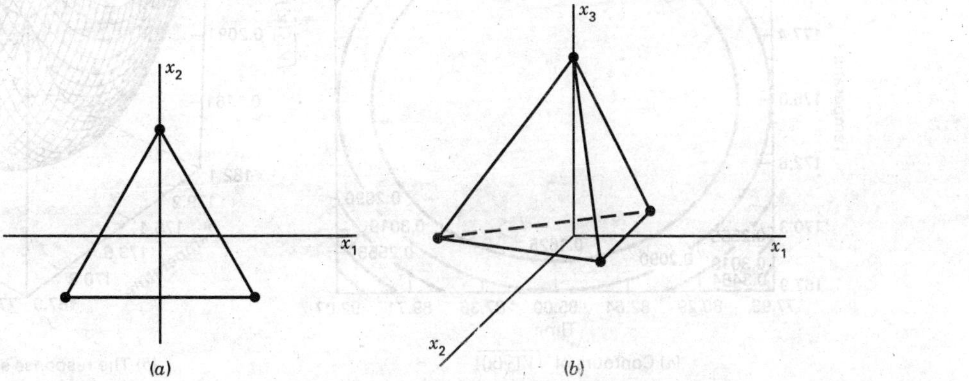

■ **FIGURE 11.19** **The simplex design for (a) $k = 2$ variables and (b) $k = 3$ variables**

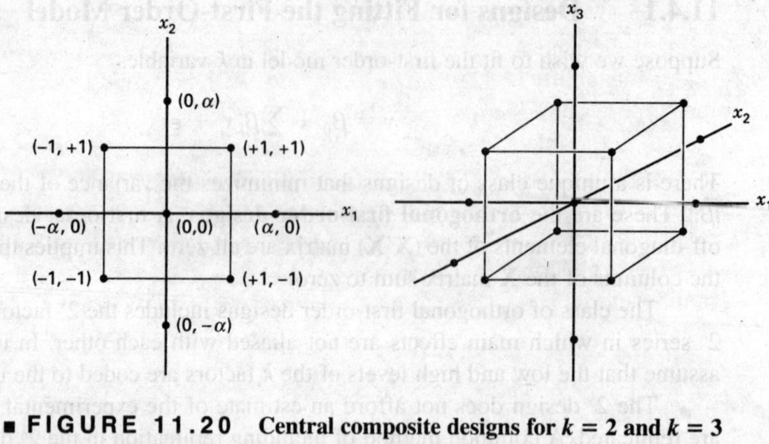

■ **FIGURE 11.20** **Central composite designs for $k = 2$ and $k = 3$**

model. There are two parameters in the design that must be specified: the distance α of the axial runs from the design center and the number of center points n_C. We now discuss the choice of these two parameters.

Rotatability. It is important for the second-order model to provide good predictions throughout the region of interest. One way to define "good" is to require that the model should have a reasonably consistent and stable variance of the predicted response at points of interest **x**. Recall from Equation 10.40 that the variance of the predicted response at some point **x** is

$$V[\hat{y}(\mathbf{x})] = \sigma^2 \mathbf{x}'(\mathbf{X}'\mathbf{X})^{-1}\mathbf{x} \tag{11.15}$$

Box and Hunter (1957) suggested that a second-order response surface design should be **rotatable**. This means that the $V[\hat{y}(\mathbf{x})]$ is the same at all points **x** that are at the same distance from the design center. That is, the variance of predicted response is constant on spheres.

Figure 11.21 shows contours of constant $\sqrt{V[\hat{y}(\mathbf{x})]}$ for the second-order model fit using the CCD in Example 11.2. Notice that the contours of constant standard deviation of predicted

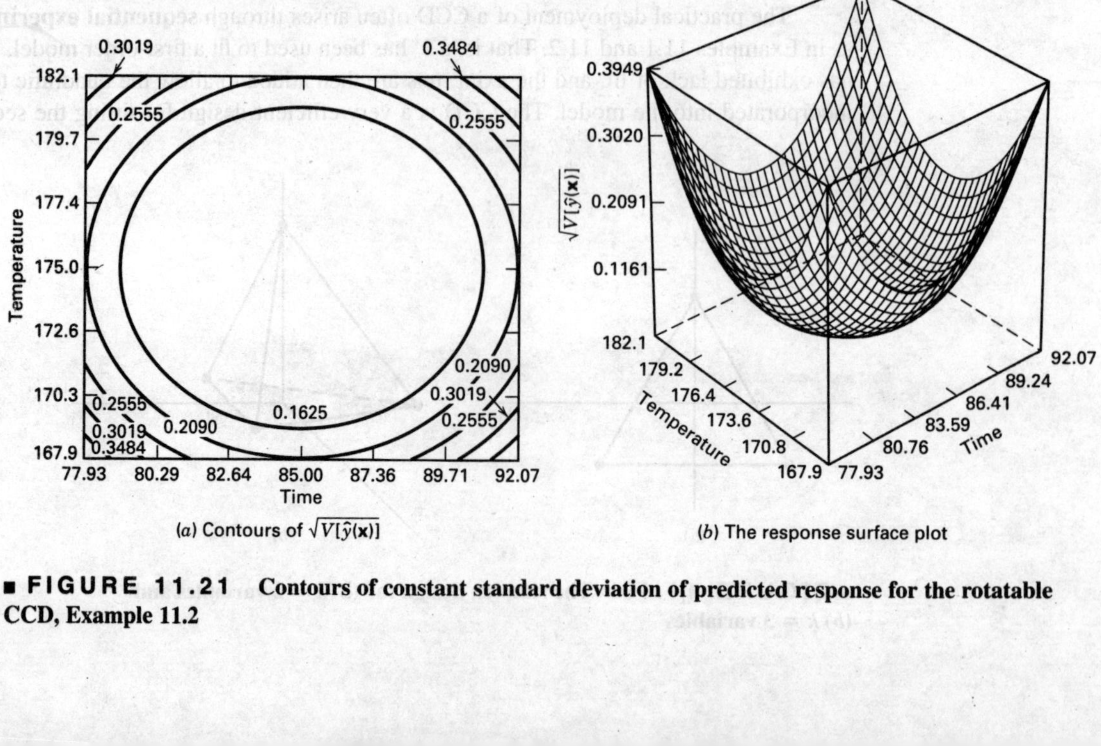

(*a*) Contours of $\sqrt{V[\hat{y}(\mathbf{x})]}$

(*b*) The response surface plot

■ **FIGURE 11.21** **Contours of constant standard deviation of predicted response for the rotatable CCD, Example 11.2** response are concentric circles. A design with this property will leave the variance of $\hat{y}$ unchanged when the design is rotated about the center (0, 0, . . . , 0), hence the name *rotatable* design.

Rotatability is a reasonable basis for the selection of a response surface design. Because the purpose of RSM is optimization and the location of the optimum is unknown prior to running the experiment, it makes sense to use a design that provides equal precision of estimation in all directions. (It can be shown that any first-order orthogonal design is rotatable.)

A central composite design is made rotatable by the choice of α. The value of α for rotatability depends on the number of points in the factorial portion of the design; in fact, $\alpha = (n_F)^{1/4}$ yields a rotatable central composite design where n_F is the number of points used in the factorial portion of the design.

The Spherical CCD. Rotatability is a **spherical property**; that is, it makes the most sense as a design criterion when the region of interest is a sphere. However, it is not important to have exact rotatability to have a good design. For a spherical region of interest, the best choice of α from a prediction variance viewpoint for the CCD is to set $\alpha = \sqrt{k}$. This design, called a **spherical CCD**, puts all the factorial and axial design points on the surface of a sphere of radius $\sqrt{k}$. For more discussion of this, see Myers, Montgomery and Anderson-Cook (2009).

Center Runs in the CCD. The choice of α in the CCD is dictated primarily by the region of interest. When this region is a sphere, the design must include center runs to provide reasonably stable variance of the predicted response. Generally, three to five center runs are recommended.

The Box–Behnken Design. Box and Behnken (1960) have proposed some three-level designs for fitting response surfaces. These designs are formed by combining 2^k factorials with incomplete block designs. The resulting designs are usually very efficient in terms of the number of required runs, and they are either rotatable or nearly rotatable.

Table 11.8 shows a three-variable Box–Behnken design. The design is also shown geometrically in Figure 11.22. Notice that the Box–Behnken design is a spherical design, with all points

■ **TABLE 11.8**
A Three-Variable Box–Behnken Design

Run	x_1	x_2	x_3
1	−1	−1	0
2	−1	1	0
3	1	−1	0
4	1	1	0
5	−1	0	−1
6	−1	0	1
7	1	0	−1
8	1	0	1
9	0	−1	−1
10	0	−1	1
11	0	1	−1
12	0	1	1
13	0	0	0
14	0	0	0
15	0	0	0

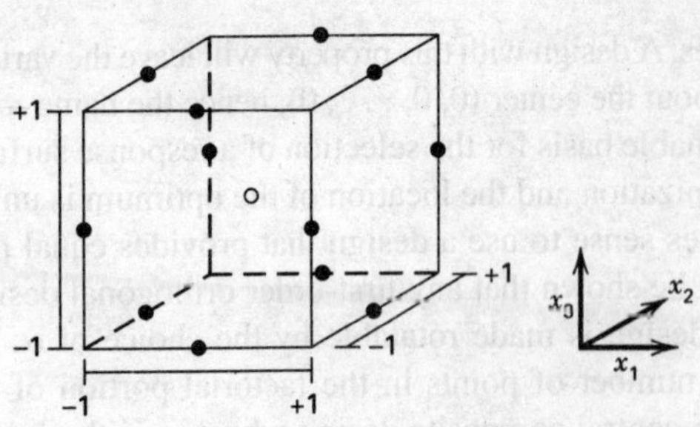

■ **FIGURE 11.22 A Box–Behnken design for three factors**

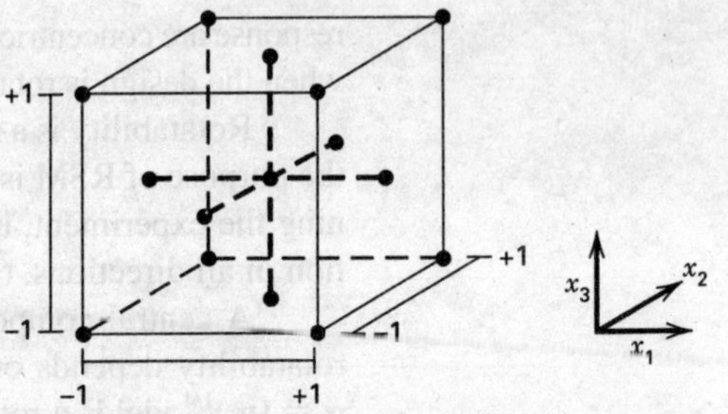

■ **FIGURE 11.23 A face-centered central composite design for $k = 3$**

lying on a sphere of radius $\sqrt{2}$. Also, the Box–Behnken design does not contain any points at the vertices of the cubic region created by the upper and lower limits for each variable. This could be advantageous when the points on the corners of the cube represent factor-level combinations that are prohibitively expensive or impossible to test because of physical process constraints.

Cuboidal Region of Interest. In many situations, the region of interest is cuboidal rather than spherical. In these cases, a useful variation of the central composite design is the **face-centered central composite design** or the **face-centered cube**, in which $\alpha = 1$. This design locates the star or axial points on the centers of the faces of the cube, as shown in Figure 11.23 for $k = 3$. This variation of the central composite design is also sometimes used because it requires only three levels of each factor, and in practice it is frequently difficult to change factor levels. However, note that face-centered central composite designs are not rotatable.

The face-centered cube does not require as many center points as the spherical CCD. In practice, $n_C = 2$ or 3 is sufficient to provide good variance of prediction throughout the experimental region. It should be noted that sometimes more center runs will be employed to give a reasonable estimate of experimental error. Figure 11.24 shows the square root of prediction variance $\sqrt{V[\hat{y}(\mathbf{x})]}$ for the face-centered cube for $k = 3$ with $n_C = 3$ center points. Notice

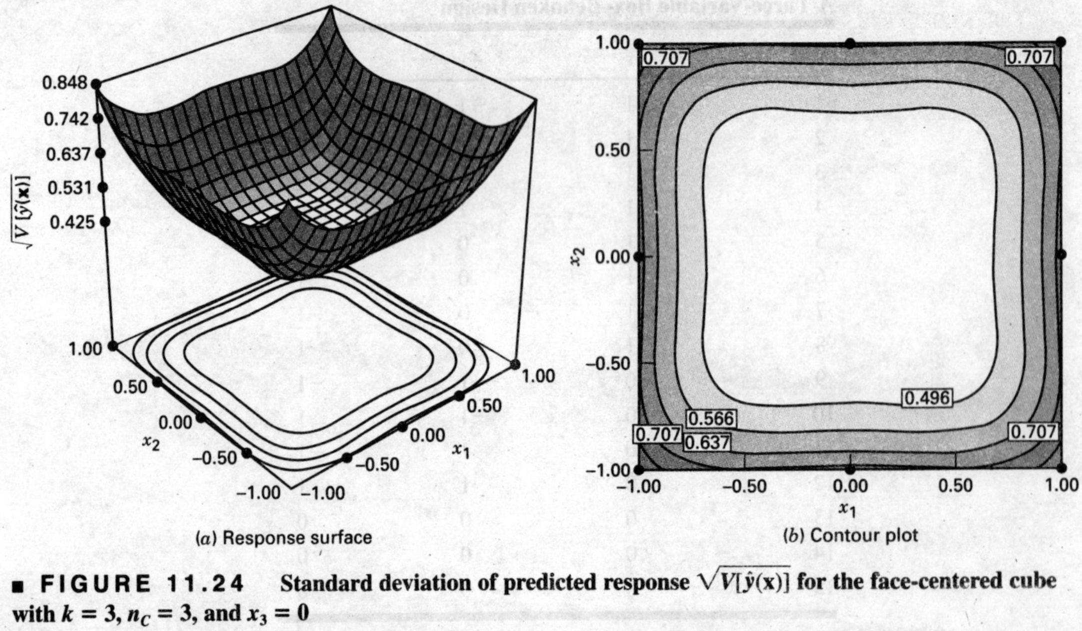

(*a*) Response surface (*b*) Contour plot

■ **FIGURE 11.24 Standard deviation of predicted response $\sqrt{V[\hat{y}(\mathbf{x})]}$ for the face-centered cube with $k = 3$, $n_C = 3$, and $x_3 = 0$**

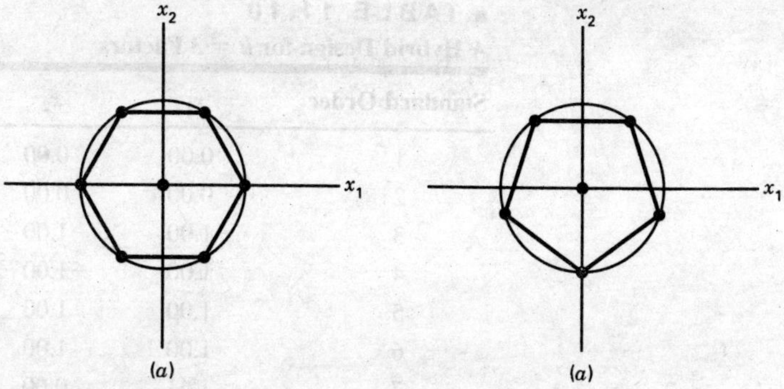

■ **FIGURE 11.25** **Equiradial designs for two variables. (*a*) Hexagon, (*b*) Pentagon**

that the standard deviation of predicted response is reasonably uniform over a relatively large portion of the design space.

Other Designs. Many other response surface designs are occasionally useful in practice. For two variables, we could use designs consisting of points that are equally spaced on a circle and that form regular polygons. Because the design points are equidistant from the origin, these arrangements are often called **equiradial designs.**

For $k = 2$, a rotatable equiradial design is obtained by combining $n_2 \geq 5$ points equally spaced on a circle with $n_1 \geq 1$ points at the center of the circle. Particularly useful designs for $k = 2$ are the pentagon and the hexagon. These designs are shown in Figure 11.25. **The small composite design** is another alternative. The small composite design consists of a fractional factorial in the cube of resolution III* (main effects aliased with two-factor interactions and no two-factor interactions aliased with each other) and the usual axial and center runs. While the small composite design may be of interest when it is important to reduce the number of runs these design do not enjoy good prediction variance properties relative to those of the CCD.

A small composite design for $k = 3$ factors is shown in Table 11.9. This design uses the standard one-half fraction of the 2^3 in the cube because it meets the resolution III* criteria.

■ **TABLE 11.9**
A Small Composite Design for $k = 3$ Factors

Standard Order	x_1	x_2	x_3
1	1.00	1.00	−1.00
2	1.00	−1.00	1.00
3	−1.00	1.00	1.00
4	−1.00	−1.00	−1.00
5	−1.73	0.00	0.00
6	1.73	0.00	0.00
7	0.00	−1.73	0.00
8	0.00	1.73	0.00
9	0.00	0.00	−1.73
10	0.00	0.00	1.73
11	0.00	0.00	0.00
12	0.00	0.00	0.00
13	0.00	0.00	0.00
14	0.00	0.00	0.00

■ **TABLE 11.10**
A Hybrid Design for $k = 3$ Factors

Standard Order	x_1	x_2	x_3
1	0.00	0.00	1.41
2	0.00	0.00	−1.41
3	−1.00	−1.00	0.71
4	1.00	−1.00	0.71
5	−1.00	1.00	0.71
6	1.00	1.00	0.71
7	1.41	0.00	−0.71
8	−1.41	0.00	−0.71
9	0.00	1.41	−0.71
10	0.00	−1.41	−0.71
11	0.00	0.00	0.00

The design has four runs in the cube and six axial runs, and it must have at least one center point. Thus the design has a minimum of $N = 11$ trials, and the second-order model in $k = 3$ variables has $p = 10$ parameters to estimate, so this is a very efficient design with respect to the number of runs. The design in Table 11.9 has $n_C = 4$ center runs. We selected $\alpha = 1.73$ to give a spherical design because the small composite design cannot be made rotatable.

The **hybrid design** is another alternative when it is important to reduce the number of runs. A hybrid design for $k = 3$ is shown in Table 11.10. Some of these designs have irregular levels, and this can be a limiting factor in their application. However, they are very small designs, and they have excellent prediction variance properties. For more details about small composite and hybrid designs, refer to Myers, Montgomery and Anderson-Cook (2009).

Graphical Evaluation of Response Surface Designs. Response surface designs are most often used to build models for making predictions. Therefore, the prediction variance (defined in Equation 11.15) is of considerable importance in evaluating or comparing designs. Two-dimensional contour plots or three-dimensional response surface plots of prediction variance (or its square root, prediction standard deviation) such as Figures 11.21 and 11.24 can be of value in this. However, for a design in k factors, these plots allow only two design factors to be displayed on the plot. Because all remaining $k - 2$ factors are held constant, these plots give an incomplete picture of how the prediction variance is distributed over the design space. Both the **fraction of design space (FDS)** plot introduced in Section 6.7 and the **variance dispersion graph (VDG)** developed by Giovannitti-Jensen and Myers (1989) can be used to solve this problem.

A VDG is a graph displaying the minimum, maximum, and average prediction variance for a specific design and response model versus the distance of the design point from the center of the region. The distance or radius usually varies from zero (the design center) to $\sqrt{k}$, which for a spherical design is the distance of the most remote point in the design from the center. It is customary to plot the **scaled prediction variance (SPV)**

$$\frac{NV[\hat{y}(\mathbf{x})]}{\sigma^2} = N\mathbf{x}'(\mathbf{X}'\mathbf{X})^{-1}\mathbf{x} \tag{11.16}$$

on a VDG. Notice that the SPV is the prediction variance in Equation 11.15 multiplied by the number of runs in the design (N) and divided by the error variance σ^2. Dividing by σ^2 eliminates an unknown parameter and multiplying by N often serves to facilitate comparing designs of different sizes.

Figure 11.26*a* is a VDG for the rotatable CCD with $k = 3$ variables and four center runs. Because the design is rotatable, the minimum, maximum, and average SPV are identical for all points that are at the same distance from the center of the design, so there is only one line on the VDG. Notice how the graph displays the behavior of the SPV over the design space, with nearly constant variance out to a radius of approximately 1.2, and then increasing steadily from there out to the boundary of the design. Figure 11.26*b* is the VDG for a spherical CCD with $k = 3$ variables and four center runs. Notice that there is very little difference between the three lines for minimum, maximum, and average SPV, leading us to conclude that any practical difference between the rotatable and spherical versions of this design is very minimal.

Figure 11.27 is the VDG for the rotatable CCD with $k = 4$ factors. In this VDG, the number of center points in the design varies from $n_C = 1$ to $n_C = 5$. The VDG shows clearly that a design with too few center points will have a very unstable distribution of prediction variance but that prediction variance quickly stabilizes with increasing values of n_C. Using either four or five center runs will give reasonably stable prediction variance over the design region. VDGs have been used to study the effect of changing the number of center runs in response surface design, and the recommendations given earlier in the chapter are based on some of these studies.

11.4.3 Blocking in Response Surface Designs

When using response surface designs, it is often necessary to consider blocking to eliminate nuisance variables. For example, this problem may occur when a second-order design is assembled sequentially from a first-order design, as was illustrated in Examples 11.1 and 11.2. Considerable time may elapse between the running of the first-order design and the running of the supplemental experiments required to build up a second-order design, and test conditions may change during this time, thus necessitating blocking.

A response surface design is said to **block orthogonally** if it is divided into blocks such that block effects do not affect the parameter estimates of the response surface model. If a 2^k or 2^{k-p} design is used as a first-order response surface design, the methods of Chapter 7 may be used to arrange the runs in 2^r blocks. The center points in these designs should be allocated equally among the blocks.

For a second-order design to block orthogonally, two conditions must be satisfied. If there are n_b observations in the bth block, then these conditions are

1. Each block must be a first-order orthogonal design; that is,

$$\sum_{u=1}^{n_b} x_{iu}x_{ju} = 0 \qquad i \neq j = 0, 1, \ldots, k \qquad \text{for all } b$$

where x_{iu} and x_{ju} are the levels of ith and jth variables in the uth run of the experiment with $x_{0u} = 1$ for all u.

2. The fraction of the total sum of squares for each variable contributed by every block must be equal to the fraction of the total observations that occur in the block; that is,

$$\frac{\sum_{u=1}^{n_b} x_{iu}^2}{\sum_{u=1}^{N} x_{iu}^2} = \frac{n_b}{N} \qquad i = 1, 2, \ldots, k \qquad \text{for all } b$$

where N is the number of runs in the design.

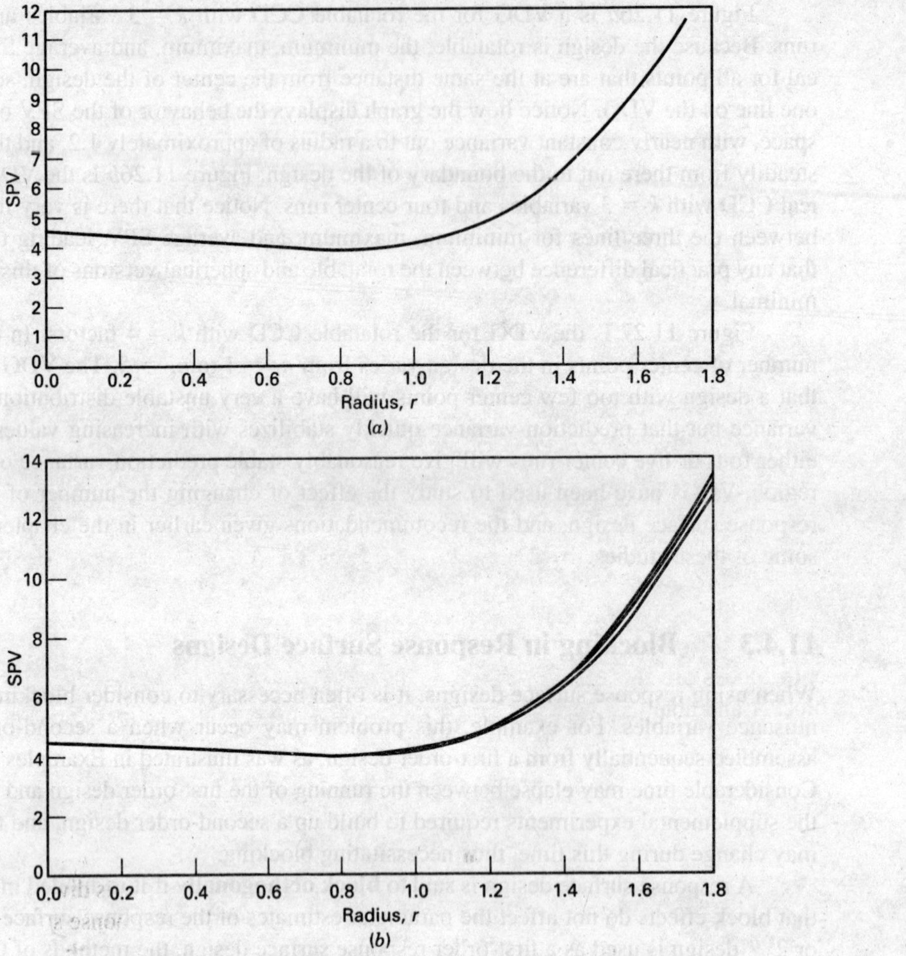

■ **FIGURE 11.26** **Variance dispersion graphs. (*a*) the CCD with $k = 3$ and $\alpha = 1.68$ (four center runs). (*b*) The CCD with $k = 3$ and $\alpha = 1.732$ (four center runs)**

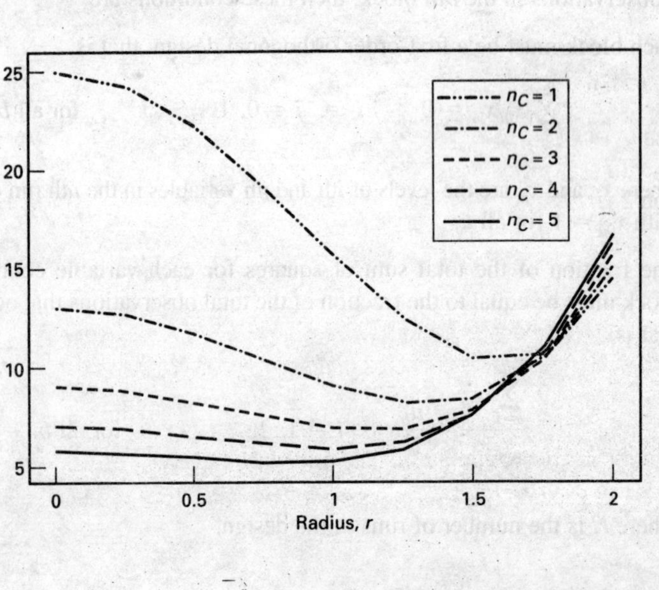

■ **FIGURE 11.27** **Variance dispersion graph for CCD with $k = 4$ and $\alpha = 2$**

As an example of applying these conditions, consider a rotatable central composite design in $k = 2$ variables with $N = 12$ runs. We may write the levels of x_1 and x_2 for this design in the design matrix

$$\mathbf{D} = \begin{bmatrix} x_1 & x_2 \\ -1 & -1 \\ 1 & -1 \\ -1 & 1 \\ 1 & 1 \\ 0 & 0 \\ 0 & 0 \\ 1.414 & 0 \\ -1.414 & 0 \\ 0 & 1.414 \\ 0 & -1.414 \\ 0 & 0 \\ 0 & 0 \end{bmatrix} \begin{matrix} \\ \left.\begin{matrix} \\ \\ \\ \\ \\ \\ \end{matrix}\right\} \text{Block 1} \\ \left.\begin{matrix} \\ \\ \\ \\ \\ \\ \end{matrix}\right\} \text{Block 2} \end{matrix}$$

Notice that the design has been arranged in two blocks, with the first block consisting of the factorial portion of the design plus two center points and the second block consisting of the axial points plus two additional center points. It is clear that condition 1 is met; that is, both blocks are first-order orthogonal designs. To investigate condition two, consider first block 1 and note that

$$\sum_{u=1}^{n_1} x_{1u}^2 = \sum_{u=1}^{n_1} x_{2u}^2 = 4$$

$$\sum_{u=1}^{N} x_{1u}^2 = \sum_{u=1}^{N} x_{2u}^2 = 8 \qquad \text{and} \qquad n_1 = 6$$

Therefore,

$$\frac{\sum_{u=1}^{n_1} x_{iu}^2}{\sum_{u=1}^{n} x_{iu}^2} = \frac{n_1}{N}$$

or

$$\frac{4}{8} = \frac{6}{12}$$

Thus, condition 2 is satisfied in block 1. For block 2, we have

$$\sum_{u=1}^{n_2} x_{1u}^2 = \sum_{u=1}^{n_2} x_{2u}^2 = 4 \qquad \text{and} \qquad n_2 = 6$$

Therefore,

$$\frac{\sum_{u=1}^{n_2} x_{iu}^2}{\sum_{u=1}^{N} x_{iu}^2} = \frac{n_2}{N}$$

or

$$\frac{4}{8} = \frac{6}{12}$$

Because condition 2 is also satisfied in block 2, this design blocks orthogonally.

In general, the central composite design can always be constructed to block orthogonally in two blocks, with the first block consisting of n_F factorial points plus n_{CF} center points and the second block consisting of $n_A = 2k$ axial points plus n_{CA} center points. The first condition for orthogonal blocking will always hold regardless of the value used for α in the design. For the second condition to hold,

$$\frac{\sum_{u}^{n_2} x_{iu}^2}{\sum_{u}^{n_1} x_{iu}^2} = \frac{n_A + n_{CA}}{n_F + n_{CF}} \tag{11.17}$$

The left-hand side of Equation 11.17 is $2\alpha^2/n_F$, and after substituting in this quantity, we may solve the equation for the value of α that will result in orthogonal blocking as

$$\alpha = \left[\frac{n_F(n_A + n_{CA})}{2(n_F + n_{CF})}\right]^{1/2} \tag{11.18}$$

This value of α does not, in general, result in a rotatable or spherical design. If the design is also required to be rotatable, then $\alpha = (n_F)^{1/4}$ and

$$(n_F)^{1/2} = \frac{n_F(n_A + n_{CA})}{2(n_F + n_{CF})} \tag{11.19}$$

It is not always possible to find a design that exactly satisfies Equation 11.19. For example if $k = 3$, $n_F = 8$, and $n_A = 6$, Equation 11.19 reduces to

$$(8)^{1/2} = \frac{8(6 + n_{CA})}{2(8 + n_{CF})}$$

$$2.83 = \frac{48 + 8n_{CA}}{16 + 2n_{CF}}$$

It is impossible to find values of n_{CA} and n_{CF} that exactly satisfy this last equation. However, note that if $n_{CF} = 3$ and $n_{CA} = 2$, then the right-hand side is

$$\frac{48 + 8(2)}{16 + 2(3)} = 2.91$$

so the design nearly blocks orthogonally. In practice, one could relax somewhat the requirement of either rotatability or orthogonal blocking without any major loss of information.

The central composite design is very versatile in its ability to accommodate blocking. If k is large enough, the factorial portion of the design can be divided into two or more blocks. (The number of factorial blocks must be a power of 2, with the axial portion forming a single block.) Table 11.11 presents several useful blocking arrangements for the central composite design.

There are two important points about the analysis of variance when the response surface design has been run in blocks. The first concerns the use of center points to calculate an estimate of pure error. Only center points that are run in the same block can be considered to be replicates, so the pure error term can only be calculated within each block. If the variability is consistent across blocks, then these pure error estimates could be pooled. The second point concerns the block effect. If the design blocks orthogonally in m blocks, the sum of squares for blocks is

$$SS_{\text{Blocks}} = \sum_{b=1}^{m} \frac{B_b^2}{n_b} - \frac{G^2}{N} \tag{11.20}$$

■ **TABLE 11.11**
Some Rotatable and Near-Rotatable Central Composite Designs That Block Orthogonally

k	2	3	4	5	5 $\frac{1}{2}$ Rep.	6	6 $\frac{1}{2}$ Rep.	7	7 $\frac{1}{2}$ Rep.
Factorial Block(s)									
n_F	4	8	16	32	16	64	32	128	64
Number of blocks	1	2	2	4	1	8	2	16	8
Number of points in each block	4	4	8	8	16	8	16	8	8
Number of center points in each block	3	2	2	2	6	1	4	1	1
Total number of points in each block	7	6	10	10	22	9	20	9	9
Axial Block									
n_A	4	6	8	10	10	12	12	14	14
n_{CA}	3	2	2	4	1	6	2	11	4
Total number of points in the axial block	7	8	10	14	11	18	14	25	18
Total number of points N in the design	14	20	30	54	33	90	54	169	80
Values of α									
Orthogonal blocking	1.4142	1.6330	2.0000	2.3664	2.0000	2.8284	2.3664	3.3333	2.8284
Rotatability	1.4142	1.6818	2.0000	2.3784	2.0000	2.8284	2.3784	3.3636	2.8284

where B_b is the total of the n_b observations in the bth block and G is the grand total of all N observations in all m blocks. When blocks are not exactly orthogonal, the general regression significance test (the "extra sum of squares" method) described in Chapter 10 can be used.

11.4.4 Optimal Designs for Response Surfaces

The standard response surface designs discussed in the previous sections, such as the central composite design, the Box–Behnken design, and their variations (such as the face-centered cube), are widely used because they are quite general and flexible designs. If the experimental region is either a cube or a sphere, typically a standard response surface design will be applicable to the problem. However, occasionally an experimenter encounters a situation where a standard response surface design may not be the obvious choice. **Optimal designs** are an alternative to consider in these cases.

As we have noted before, there are several situations where some type of computer-generated design may be appropriate.

1. ***An irregular experimental region.*** If the region of interest for the experiment is not a cube or a sphere, standard designs may not be the best choice. Irregular regions of interest occur fairly often. For example, an experimenter is investigating the properties of a particular adhesive. The adhesive is applied to two parts and then cured at an elevated temperature. The two factors of interest are the amount of adhesive applied and the cure temperature. Over the ranges of these two factors, taken as -1 to $+1$ on the usual coded variable scale, the experimenter knows that if too little adhesive is applied and the cure temperature is too low, the parts will not bond

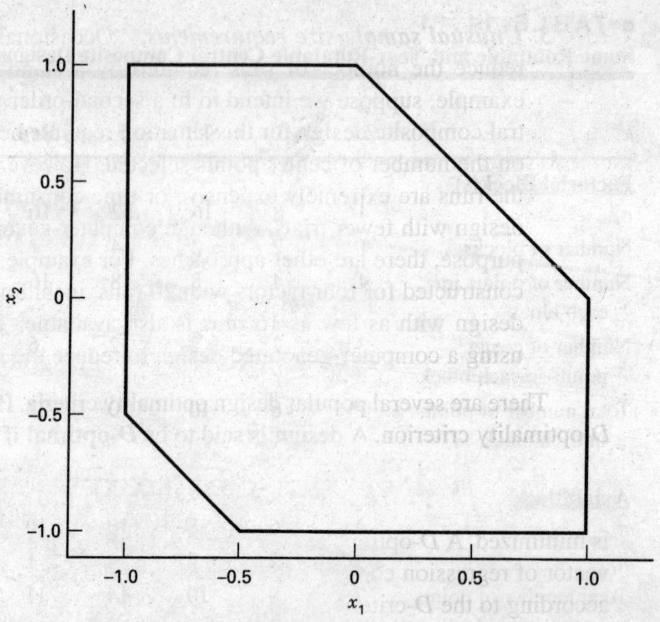

■ **FIGURE 11.28** **A constrained design region in two variables**

satisfactorily. In terms of the coded variables, this leads to a **constraint** on the design variables, say

$$-1.5 \leq x_1 + x_2$$

where x_1 represents the application amount of adhesive and x_2 represents the temperature. Furthermore, if the temperature is too high and too much adhesive is applied, the parts will be either damaged by heat stress or an inadequate bond will result. Thus, there is another constraint on the factor levels

$$x_1 + x_2 \leq 1$$

Figure 11.28 shows the experimental region that results from applying these constraints. Notice that the constraints effectively remove two corners of the square, producing an irregular experimental region (sometimes these irregular regions are called "dented cans"). There is no standard response surface design that will exactly fit into this region.

2. ***A nonstandard model.*** Usually an experimenter elects a first- or second-order response surface model, realizing that this **empirical model** is an approximation to the true underlying mechanism. However, sometimes the experimenter may have some special knowledge or insight about the process being studied that may suggest a nonstandard model. For example, the model

$$y = \beta_0 + \beta_1 x_1 + \beta_2 x_2 + \beta_{12} x_1 x_2 + \beta_{11} x_1^2 + \beta_{22} x_2^2 + \beta_{112} x_1^2 x_2 + \beta_{1112} x_1^3 x_2 + \epsilon$$

may be of interest. The experimenter would be interested in obtaining an efficient design for fitting this reduced quartic model. As another illustration, sometimes we encounter response surface problems where some of the design factors are categorical variables. There are no standard response surface designs for this situation [refer to Myers, Montgomery and Anderson-Cook (2009) for a discussion of categorical variables in response surface problems].

3. ***Unusual sample size requirements.*** Occasionally, an experimenter may need to reduce the number of runs required by a standard response surface design. For example, suppose we intend to fit a second-order model in four variables. The central composite design for this situation requires between 28 and 30 runs, depending on the number of center points selected. However, the model has only 15 terms. If the runs are extremely expensive or time-consuming, the experimenter will want a design with fewer trials. Although computer-generated designs can be used for this purpose, there are other approaches. For example, a small composite design can be constructed for four factors with 20 runs, including four center points, and a hybrid design with as few as 16 runs is also available. These may be superior choices to using a computer-generated design to reduce the number of trials.

There are several popular design optimality criteria. Perhaps the most widely used is the *D*-optimality criterion. A design is said to be ***D*-optimal** if

$$|(\mathbf{X}'\mathbf{X})^{-1}|$$

is minimized. A *D*-optimal design minimizes the volume of the joint confidence region on the vector of regression coefficients. A measure of the relative efficiency of design 1 to design 2 according to the *D*-criterion is given by

$$D_e = \left(\frac{|(\mathbf{X}_2'\mathbf{X}_2)^{-1}|}{|(\mathbf{X}_1'\mathbf{X}_1)^{-1}|} \right)^{1/p} \tag{11.21}$$

where $\mathbf{X}_1$ and $\mathbf{X}_2$ are the $\mathbf{X}$ matrices for the two designs and p is the number of model parameters. Many popular software packages including JMP, Design-Expert, and Minitab will construct D-optimal designs.

The ***A*-optimality** criterion deals with only the variances of the regression coefficients. A design is *A*-optimal if it minimizes the sum of the main diagonal elements of $(\mathbf{X}'\mathbf{X})^{-1}$. (This is called the **trace** of $(\mathbf{X}'\mathbf{X})^{-1}$, usually denoted $\text{tr}(\mathbf{X}'\mathbf{X})^{-1}$.) Thus, an *A*-optimal design minimizes the sum of the variances of the regression coefficients.

Because many response surface experiments are concerned with the prediction of the response, **prediction variance criteria** are of considerable practical interest. Perhaps the most popular of these is the ***G*-optimality criterion**. A design is said to be *G*-optimal if it minimizes the maximum scaled prediction variance over the design region; that is, if the maximum value of

$$\frac{NV[\hat{y}(\mathbf{x})]}{\sigma^2}$$

over the design region is a minimum, where N is the number of points in the design. If the model has p parameters, the *G*-efficiency of a design is just

$$G_e = \frac{p}{\max \dfrac{NV[\hat{y}(\mathbf{x})]}{\sigma^2}} \tag{11.22}$$

The ***V*-criterion** considers the prediction variance at a *set* of points of interest in the design region, say $\mathbf{x}_1, \mathbf{x}_2, \ldots, \mathbf{x}_m$. The set of points could be the candidate set from which the design was selected, or it could be some other collection of points that have specific meaning to the experimenter. A design that minimizes the *average* prediction variance over this set of m points is a ***V*-optimal design**.

As we observed in Chapter 6 (Section 6.7), an alterative to calculating the prediction variance at a finite set of points in the design space is to compute an **average or integrated**

variance over the design space, say

$$I = \frac{1}{A}\int_R V[\hat{y}(\mathbf{x})]d\mathbf{x}$$

where R is the design region and A is the volume of the region. Note that this is a more general form of the I-criterion discussed in Chapter 6. The I criterion is also sometimes called the IV or Q criterion. JMP can construct I-optimal designs.

Generally, we think of the D criteria as the most appropriate for first-oder designs, as they are associated with parameter estimation, which is very important in screening situations where first-order model are most often used. The G and I criteria are prediction-oriented criteria, so they would be most likely used for second-order models, as second-order models are often used for optimization, and good prediction properties are essential for optimization. The I criteria is much easier to implement than G, and is available in several software packages.

One of the design construction methods is based on a **point exchange algorithm**. In the simplest form of this algorithm, a grid of candidate points is selected by the experimenter, and an initial design is selected (perhaps by random) from this set of points. Then the algorithm exchanges points that are in the grid but not in the design with points currently in the design in an effort to improve the selected optimality criterion. Because not every possible design is explicitly evaluated, there is no guarantee that an optimal design has been found, but the exchange procedure usually ensures that a design that is "close" to optimal results. The procedure is also sensitive to the grid of candidate points that have been specified. Some implementations repeat the design construction process several times, starting from different initial designs, to increase the likelihood that a final design that is very near the optimal will result.

Another way to construct optimal design is with a **coordinate exchange** algorithm. This method searches over each coordinate of every point in the initial design recursively until no improvement in the optimality criterion is found. The procedure is usually repeated several times with each cycle starting with a randomly generated initial design. Coordinate exchange is usually much more efficient than point exchange and is the standard method in many software packages.

To illustrate some of these ideas, consider the adhesive experiment discussed previously that led to the irregular experimental region in Figure 11.28. Suppose that the response of interest is pull-off force and that we wish to fit a second-order model to this response. In Figure 11.29*a*, we show a central composite design with four center points (12 runs total) inscribed inside this region. This is not a rotatable design, but it is the largest CCD that we can fit inside the design space. For this design, $|(\mathbf{X}'\mathbf{X})^{-1}| = 1.852$ E-2 and the trace of $(\mathbf{X}'\mathbf{X})^{-1}$ is 6.375. Also shown in Figure 11.29*a* are the contours of constant standard deviation of the predicted response, calculated assuming that $\sigma = 1$. Figure 11.29*b* shows the corresponding response surface plot.

Figure 11.30*a* and Table 11.12 show a 12-run D-optimal design for this problem, generated with the Design-Expert software package. For this design, $|(\mathbf{X}'\mathbf{X})^{-1}| = 2.153$ E-4. Notice that the D-criterion is considerably better for this design than for the inscribed CCD. The relative efficiency of the inscribed CCD to the D-optimal design is

$$D_e = \left(\frac{|(\mathbf{X}_2'\mathbf{X}_2)^{-1}|}{|(\mathbf{X}_1'\mathbf{X}_1)^{-1}|}\right)^{1/p} = \left(\frac{0.0002153}{0.01852}\right)^{1/6} = 0.476$$

That is, the inscribed CCD is only 47.6 percent as efficient as the D-optimal design. This implies that the CCD would have to be replicated 1/0.476 = 2.1 times (or approximately twice) to have the same precision of estimation for the regression coefficients as achieved with the D-optimal design. The trace of $(\mathbf{X}'\mathbf{X})^{-1}$ is 2.516 for the D-optimal design, indicating that the sum of the variances of the regression coefficients is considerably smaller for this

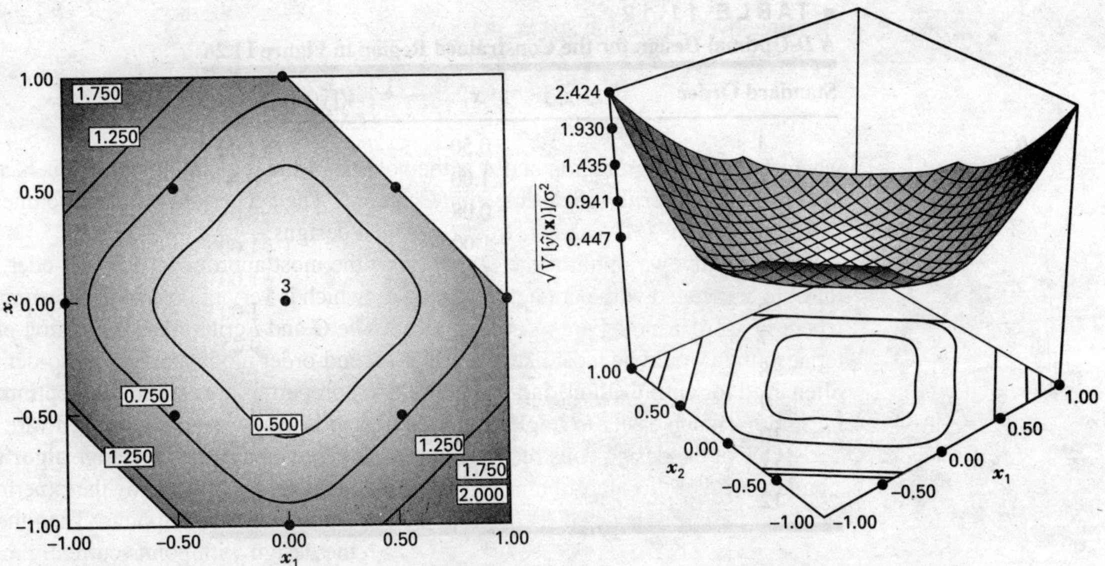

(a) The design and contours of constant $\sqrt{V[\hat{y}(\mathbf{x})]/\sigma^2}$

(b) The response surface plot

■ **FIGURE 11.29** **An inscribed central composite design for the constrained design region in Figure 11.28**

design than for the CCD. Figure 11.30*a* and *b* also shows the contours of constant standard deviation of predicted response and the associated response surface plot (assuming that $\sigma = 1$). Generally, the prediction standard deviation contours are lower for the *D*-optimal design than for the inscribed CCD, particularly near the boundaries of the region of interest where the inscribed CCD does not have any design points.

Figure 11.31*a* shows a third design, created by taking the two replicates at the corners of the region in the *D*-optimal design and moving them to the design center. This could be a useful idea because Figure 11.30*b* shows that the standard deviation of predicted response increases

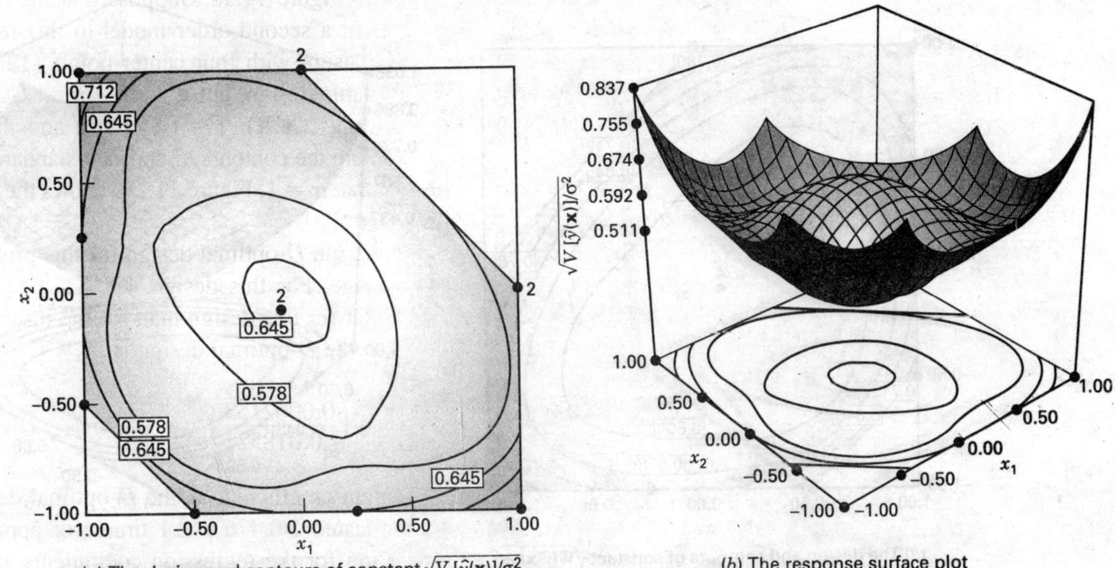

(a) The design and contours of constant $\sqrt{V[\hat{y}(\mathbf{x})]/\sigma^2}$

(b) The response surface plot

■ **FIGURE 11.30** **A *D*-optimal design for the constrained design region in Figure 11.28**

■ **TABLE 11.12**
A *D*-Optimal Design for the Constrained Region in Figure 11.26

Standard Order	x_1	x_2
1	−0.50	−1.00
2	1.00	0.00
3	−0.08	−0.08
4	−1.00	1.00
5	1.00	−1.00
6	0.00	1.00
7	−1.00	0.25
8	0.25	−1.00
9	−1.00	−0.50
10	1.00	0.00
11	0.00	1.00
12	−0.08	−0.08

slightly near the center of the design region for the *D*-optimal design. Figure 11.31*a* also shows the contours of constant standard deviation of prediction for this *modified D*-optimal design, and Figure 11.31*b* shows the response surface plot. The *D*-criterion for this design is $|(\mathbf{X'X})^{-1}| = 3.71$ E-4, and the relative efficiency is

$$D_e = \left(\frac{|(\mathbf{X}_2'\mathbf{X}_2)^{-1}|}{|(\mathbf{X}_1'\mathbf{X}_1)^{-1}|}\right)^{1/p} = \left(\frac{0.0002153}{0.000371}\right)^{1/6} = 0.91$$

That is, this design is almost as efficient as the *D*-optimal design. The trace of $(\mathbf{X'X})^{-1}$ is 2.448 for this design, a slightly smaller value than was obtained for the *D*-optimal design. The contours

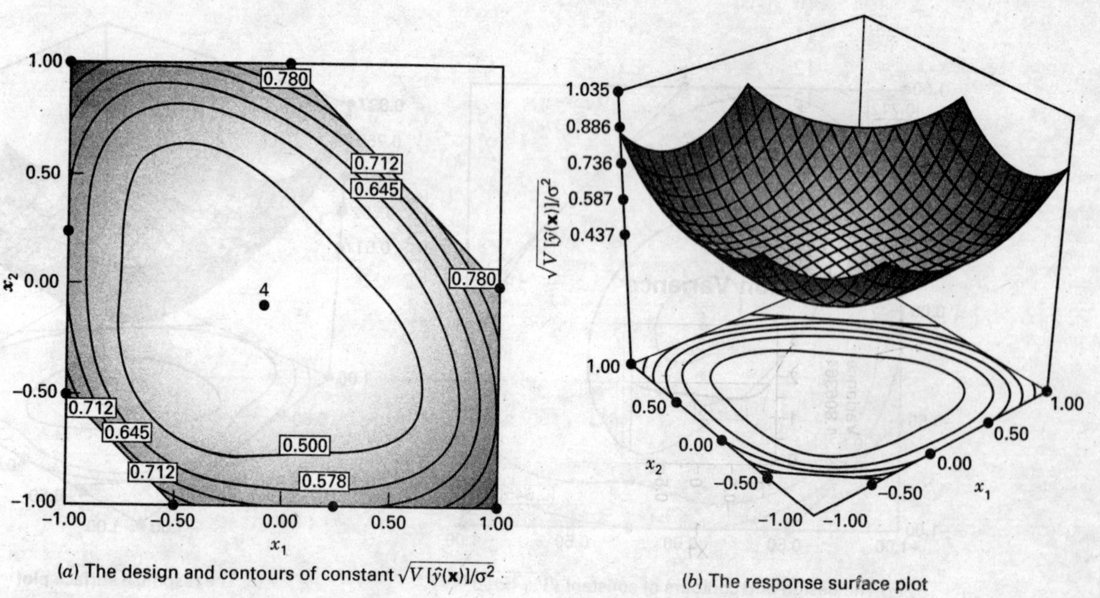

(*a*) The design and contours of constant $\sqrt{V[\hat{y}(\mathbf{x})]/\sigma^2}$ (*b*) The response surface plot

■ **FIGURE 11.31** **A modified *D*-optimal design for the constrained design region in Figure 11.28** of constant prediction standard deviation for this design visually look at least as good as those for the D-optimal design, particularly at the center of the region. This points out the necessity of **design evaluation**; that is, carefully examine the properties of a computer-generated design before you decide to use it.

EXAMPLE 11.3

As an illustration of the different designs that can be constructed using both the D and I optimality criteria, suppose that we want to fit a second-order model in four factors on a cubic region. The standard design for this problem would be a face-centered cube, a design with 24 factorial and axial runs plus 2 or 3 center points, or a total of 26 or 27 runs. The second-order model in $k = 4$ factors has 15 parameters, so a **minimal** design must have 15 runs. Suppose that we want to employ a design with 16 runs. Since there is not a standard design available with 16 runs, we will consider using an optimal design.

Table 11.13 is the output from the JMP custom design tool for this problem, where a D-optimal design has been

■ **TABLE 11.13**
The D-Optimal Design

Design Matrix

Run	X1	X2	X3	X4
1	1	1	1	1
2	1	−1	−1	1
3	1	1	−1	−1
4	−1	−1	1	−1
5	1	−1	1	−1
6	0	0	0	−1
7	0	0	1	0
8	0	1	−1	1
9	−1	1	−1	−1
10	−1	−1	−1	1
11	0	1	1	−1
12	0	−1	1	1
13	0	−1	−1	−1
14	1	1	0	0
15	1	0	−1	1
16	−1	1	1	1

Prediction Variance Profile

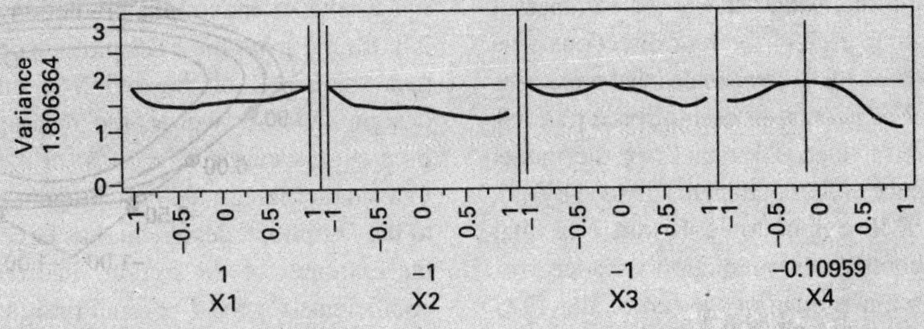

■ **TABLE 11.13** *(Continued)*

Fraction of Design Space Plot

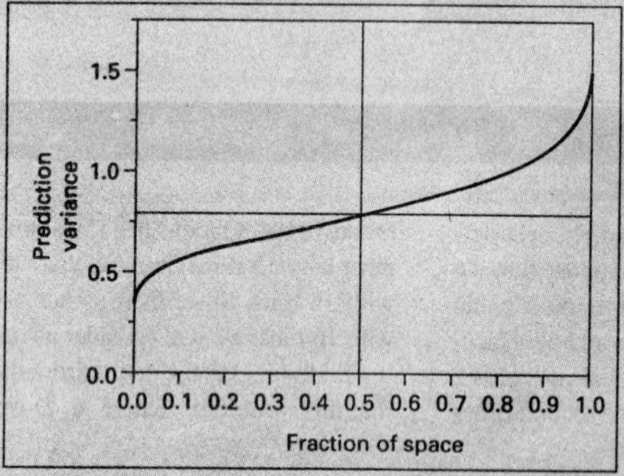

Relative Variance of Coefficients

Effect	Variance
Intercept	0.909
X1	0.115
X2	0.092
X3	0.088
X4	0.088
X1*X1	0.319
X1*X2	0.124
X2*X2	0.591
X1*X3	0.120
X2*X3	0.093
X3*X3	0.839
X1*X4	0.120
X2*X4	0.093
X3*X4	0.090
X4*X4	0.839

requested. A coordinate exchange algorithm was used to construct the design. Immediately below the design matrix is the prediction variance profile, which shows the variance of the predicted response along each of the four directions. The crosshair on the plot has been set to coordinates that maximize the prediction variance. The fraction of design space plot follows, along with a table of relative variances of the model coefficients (that is, variance of the coefficients divided by σ^2).

Table 11.14 is the JMP output for a 16-run *I*-optimal design. This table also contains the prediction variance profile showing the maximum prediction variance, the FDS plot, and the table of relative variance of the model coefficients. Several important differences between the *D* and *I* optimal designs can be observed. First, the *D*-optimal design has a smaller maximum prediction variance (1.806 versus 2.818), but from the FDS plot we observe that the variance near the center of the region is smaller for the *I*-optimal design. In other words, the *I*-optimal design has smaller prediction variance over most of the design space (leading to a smaller integrated or average variance) when compared to the *D*-optimal design but has larger prediction variance at the extremes of the region. The relative variances of the coefficients for the *I*-optimal design are in almost all cases smaller for the *D*-optimal design. This is not unexpected as the D criterion focuses on minimizing the variances of the model coefficients while the I criterion focuses on minimizing a measure of average prediction variance. This comparison also reveals why the I-criterion is generally preferable for second-order models or situations where prediction and/or optimization is required because it results in a design having small prediction variances over most of the design space and performs only poorly at the extremes.

■ **TABLE 11.14**
The *I*-Optimal Design

Design Matrix

Run	X1	X2	X3	X4
1	0	1	1	1
2	0	1	−1	−1
3	−1	−1	0	−1
4	1	−1	−1	0
5	−1	−1	−1	1
6	−1	1	0	1
7	−1	1	1	−1
8	−1	0	1	1
9	1	1	−1	1
10	1	−1	1	1
11	1	1	0	0
12	−1	0	−1	0
13	0	0	0	0
14	0	−1	1	0
15	0	0	0	1
16	1	0	1	−1

Prediction Variance Profile

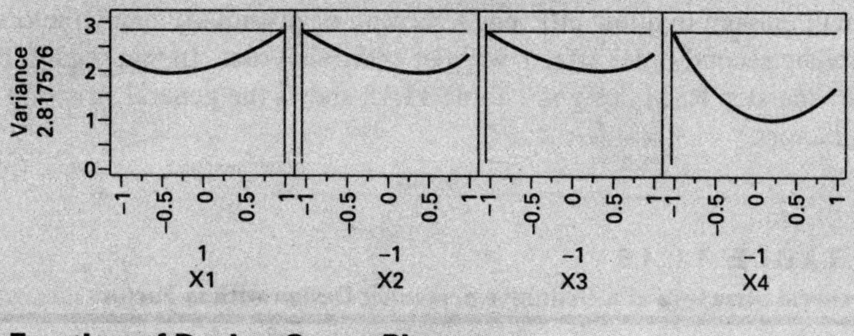

Fraction of Design Space Plot

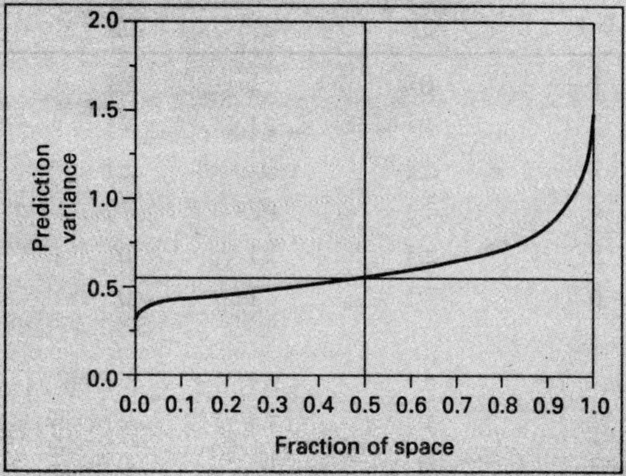

■ **TABLE 11.14** *(Continued)*

Relative Variance of Coefficients

Effect	Variance
Intercept	0.508
X1	0.118
X2	0.118
X3	0.118
X4	0.121
X1*X1	0.379
X1*X2	0.174
X2*X2	0.379
X1*X3	0.174
X2*X3	0.174
X3*X3	0.379
X1*X4	0.186
X2*X4	0.186
X3*X4	0.186
X4*X4	0.399

Jones and Nachtsheim (2011b) have introduced a potentially useful class of response surface designs for quantitative factors whose construction can be facilitated by an optimal design algorithm. They refer to these as "definitive screening designs" because they are small enough to allow efficient screening of potentially many factors yet they can accommodate second-order effects without additional runs. In that regard, they can be thought of as "one-step RSM designs". Table 11.15 shows the general structure of these designs with m factors.

■ **TABLE 11.15**
General Structure of a Definitive Screening Design with m Factors

Foldover Pair	Run (i)	Factor Levels $x_{i,1}$	$x_{i,2}$	$x_{i,3}$	...	$x_{i,m}$
1	1	0	±1	±1	...	±1
	2	0	∓1	∓1	...	∓1
2	3	±1	0	±1	...	±1
	4	∓1	0	∓1	...	∓1
3	5	±1	±1	0	...	±1
	6	∓1	∓1	0	...	∓1
$\vdots$	$\vdots$	$\vdots$	$\vdots$	$\vdots$	$\ddots$	$\vdots$
m	$2m-1$	±1	±1	±1	...	0
	$2m$	∓1	∓1	∓1	...	0
Centerpoint	$m+1$	0	0	0	...	0

Notice that for m factors, there are only $2m+1$ runs based on m fold-over pairs and an overall center point. Each run (excluding the center run) has exactly one factor level at its center point and all others levels at the extremes. These designs have the following desirable properties:

1. The number of required runs is only one more than twice the number of factors. Consequently, these are very small designs.
2. Unlike resolution III designs, main effects are completely independent of two-factor interactions. As a result, estimates of main effects are not biased by the presence of active two-factor interactions, regardless of whether the interactions are included in the model.
3. Unlike resolution IV designs, two-factor interactions are not completely confounded with other two-factor interactions, although they may be correlated.
4. Unlike resolution III, IV and V designs with added center points, all quadratic effects can be estimated in models comprised of any number of linear and quadratic main effect terms.
5. Quadratic effects are orthogonal to main effects and not completely confounded (though correlated) with interaction effects.
6. With six through or more factors, the designs are capable of estimating all possible full quadratic models involving three or fewer factors with very high levels of statistical efficiency.

These designs are an excellent compromise between Resolution III fractions for screening and small RSM designs. They also admit the possibility of moving directly from screening to optimization using the results of a single experiment. Jones and Nachtsheim found these designs using an optimization technique they had previously developed for finding minimum aliasing designs [see Jones and Nachtsheim (2011a)] Their algorithm minimizes the sum of the squares of the elements of the alias matrix subject to a constraint on the D-efficiency of the resulting design. Figure 11.32 shows these designs for the cases of 4 through 12 factors.

These designs can also be constructed from **conference matrices** [see Xiao, Lin and Bai (2012)]. A conference matrix $\mathbf{C}$ is an $n \times n$ matrix that has diagonal elements equal to zero and all off-diagonal elements equal to ± 1. They have the property that $\mathbf{C}'\mathbf{C}$ is a multiple of the identity matrix. For the $n \times n$ conference matrix $\mathbf{C}$, $\mathbf{C}'\mathbf{C} = (n - 1)\mathbf{I}$. Conference matrices first arose in connection with a problem in telephony. They were used in constructing ideal telephone conference networks from ideal transformers. These networks were represented by conference matrices. There are other applications.

The conference matrix of order 6 is given by:

$$\begin{pmatrix} 0 & +1 & +1 & +1 & +1 & +1 \\ +1 & 0 & +1 & -1 & -1 & +1 \\ +1 & +1 & 0 & +1 & -1 & -1 \\ +1 & -1 & +1 & 0 & +1 & -1 \\ +1 & -1 & -1 & +1 & 0 & +1 \\ +1 & +1 & -1 & -1 & +1 & 0 \end{pmatrix}$$

The 13-run 6-factor definitive screening design can be found by folding over each row of this conference matrix and adding a row of zeros at the bottom. In general, if $\mathbf{C}$ is the conference matrix of order n the m-factor definitive screening design matrix can be found as follows:

$$\mathbf{D} = \begin{bmatrix} \mathbf{C} \\ -\mathbf{C} \\ \mathbf{0} \end{bmatrix}$$

where $\mathbf{0}$ denotes the $1 \times n$ row vector of zeros and $m = 2n+1$.

```
m = 4

1  0+--
2  0-++
3  -0-+
4  +0+-
5  --0-
6  ++0+
7  -++0
8  +--0
9  0000
```

```
m = 5

1  0++--
2  0--++
3  +0--+
4  -0++-
5  +-0+-
6  -+0-+
7  +-+0+
8  -+-0-
9  ++++0
10 ----0
11 00000
```

```
m = 6

1  0+----
2  0-++++
3  +0-++-
4  -0+--+
5  --0+--
6  ++0-++
7  -++0+-
8  +--0-+
9  +-+-0-
10 -+-+0+
11 ++++-0
12 ----+0
13 000000
```

```
m = 7

1  0+-+-+-
2  0-+-+-+
3  -0+-++-
4  +0-+--+
5  +-0++++
6  -+0----
7  +--0+--
8  -++0-++
9  --++0--
10 ++--0++
11 -+-++0+
12 +-+--0-
13 +++++-0
14 -----+0
15 0000000
```

```
m = 8

1  0-++-+++
2  0+--+---
3  -0-++++-
4  +0+----+
5  --0++--+
6  ++0--++-
7  +-+0++--
8  -+-0--++
9  --+-0-+-
10 ++-+0+-+
11 +---+0++
12 -+++-0--
13 -++-++0+
14 +--+--0-
15 +++++-+0
16 -----+-0
17 00000000
```

```
m = 9

1  0++++++++
2  0--------
3  +0+-+--+-
4  -0-+-++-+
5  -+0-+-+--
6  +-0+-+-++
7  --+0+---+
8  ++-0-+++-
9  +-+-0++--
10 -+-+0--++
11 ----+0+++
12 ++++-0---
13 ++--++0-+
14 --++--0+-
15 ---+++-0-
16 +++---+0+
17 -++--+-+0
18 +--++-+-0
19 000000000
```

```
m = 10

1  0++-++++-+
2  0--+----+-
3  +0-++-++--
4  -0+--+--++
5  -+0---+---
6  +-0+++-+++
7  -++0+--++-
8  +--0-++--+
9  ----0++++-
10 ++++0----+
11 -+-++0+-++
12 +-+--0-+--
13 ++----0+++
14 --++++0---
15 ++++-++0+-
16 ----+--0-+
17 ++--++--0-
18 --++--++0+
19 +-+-+-+-+0
20 -+-+-+-+-0
21 0000000000
```

```
m = 11

1  0-+-----+-+
2  0+-+++++-+-
3  -0--+----++
4  +0++-++++--
5  --0++++---+
6  ++0----+++-
7  ---0-++-++-
8  +++0+--+--+
9  +--+0+-++++
10 -++-0-+----
11 --++-0-+-+-
12 ++--+0+-+-+
13 ---++-0++--
14 +++--+0--++
15 -+++--+0+++
16 +---++-0---
17 -+---+-+0-+
18 +-+++-+-0+-
19 +-----++-0+
20 -+++++--+0-
21 ++-+-------0
22 --+-++++++0
23 00000000000
```

```
m = 12

1  0--+-+-+++-+
2  0++-+-+---+-
3  -0+++++++---
4  +0--------+++
5  ++0-++-++-++
6  --0+--+--+--
7  +--0+-+-+--+
8  -++0-+-+-++-
9  ++++0-++++++
10 ----0+------
11 +-+-+0++-+-+
12 -+-+-0--+-+-
13 ++++-+0----+
14 ----+-0++++-
15 --+++--0--++
16 ++---++0++--
17 +-++++--0++-
18 -+----++0--+
19 ++-++--+-0--
20 --+--++-+0++
21 -+-++++--+0+
22 +-+----++-0-
23 +--+-+++--+0
24 -++-+---++-0
25 000000000000
```

■ **FIGURE 11.32** **Definitive Screening Designs for 4 Through 12 Factors**

11.5 Experiments with Computer Models

We customarily think of applying designed experiments to a *physical* process, such as chemical vapor deposition in semiconductor manufacturing, wave soldering or machining. However, designed experiments can also be successfully applied to *computer simulation models* of physical systems. In such applications, the data from the experimental design is used to build a model of the system being modeled by the computer simulation—a **metamodel**—and optimization is carried out on the metamodel. The assumption is that if the computer simulation model is a faithful representation of the real system, then optimization of the model will result in adequate determination of the optimum conditions for the real system.

Generally, there are two types of simulation models, **stochastic** and **deterministic**. In a stochastic simulation model, the output responses are random variables. Examples include systems simulations such as the factory planning and scheduling models used in the semiconductor industry and traffic flow simulators employed by civil engineers, and Monte Carlo simulations that sample from probability distributions to study complex mathematical phenomena that do not have direct analytical solutions. Sometimes the output from a stochastic simulation model will be in the form of a time series. Often standard experimental design techniques can be applied to the output from a stochastic simulation model, although a number of specialized techniques have been developed. Sometimes polynomials of higher-order than the usual quadratic response surface model are used.

In a **deterministic simulation** model the output responses are not random variables; they are entirely deterministic quantities whose values are determined by the (often highly complex) mathematical models upon which the computer model is based. Deterministic simulation models are often used by engineers and scientists as computer-based design tools. Typical examples are circuit simulators used for designing electronic circuits and semiconductor devices, finite element analysis models for mechanical and structural design and computational models for physical phenomena such as fluid dynamics. These are often very complex models, requiring considerable computer resources and time to run.

As an example of a situation where a finite element analysis model may be employed, consider the problem of designing a turbine engine to contain a failed compressor rotor. Many factors may influence the design, such as engine operating conditions as well as the location, size, and material properties of surrounding parts. Figure 11.33 shows a cutaway view of a typical compressor containment model. Many parameters for each component are potentially important. The thickness, material type, and geometric feature (bend radius, bolt hole size and location, stiffening ribs or gussets, etc.) are engineering design parameters and, potentially, experimental factors that could be included in a response surface model. One can see that large numbers of factors are potentially important in the design of such a product. Furthermore, the sign or direction of the effect of many of these factors is unknown. For instance, setting factors that increase the axial stiffness of a backface (such as increasing the thickness of the transition duct) may help align a rotor fragment, centering the impact on the containment structure. On the other hand, the increased stiffness may nudge the fragment too much, causing it to miss the primary containment structure. From experience the design engineers may confidently assume that only a small number of these potentially important factors have a significant effect on the performance of the design in containing a failed part. Detailed analysis or testing of the turbine engine is needed to understand which factors are important and to quantify their effect on the design. The cost of building a prototype turbine engine frequently exceeds one million dollars, so studying the effects of these factors using a computer model is very attractive. The type of model used is called a **finite element analysis** model. Simulating a containment event with a finite element analysis model is very computationally intensive. The model shown in Figure 11.32 has over 100,000 elements and takes about 90 hr of computer time to model 2 ms

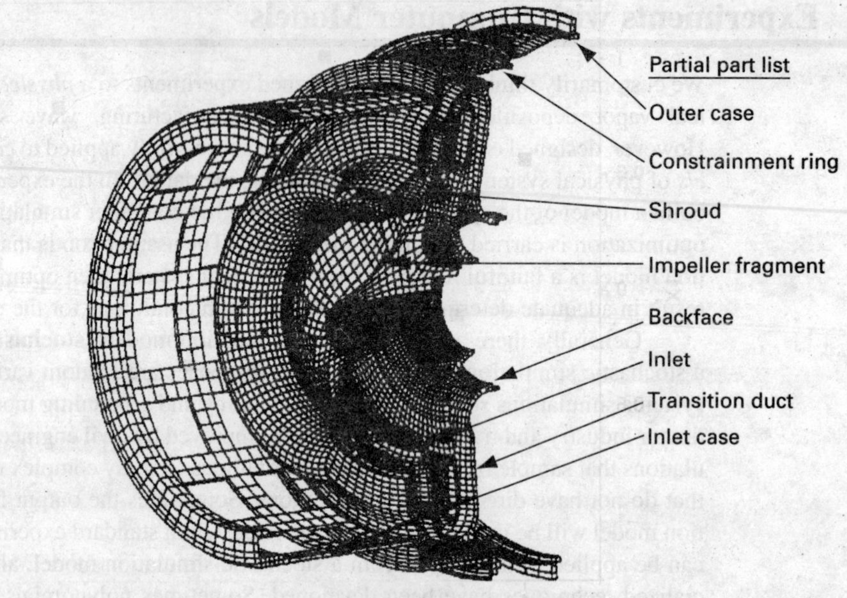

■ **FIGURE 11.33 Finite element model for compressor containment analysis of a turbine engine and partial parts list**

of event time. Frequently as much as 10 ms of event time must be modeled. Clearly the need to limit experimentation or simulation is great. Therefore the typical approach of factor screening followed by optimization might well be applied to this scenario.

Remember that the response surface approach is based on a philosophy of **sequential experimentation**, with the objective of approximating the response with a low-order polynomial in a relatively small region of interest that contains the optimum solution. Some computer experimenters advocate a somewhat different philosophy. They seek to find a model that approximates the true response surface over a much wider range of the design variables, sometimes extending over the entire region of operability. As mentioned earlier in this section, this can lead to situations where the model that is considered is much more complex than the first- and second-order polynomials typically employed in response surface methodology [see, for example, Barton (1992, 1994), Mitchell and Morris (1992), and Simpson and Peplinski (1997)].

The choice of a design for a computer simulation experiment presents some interesting alternatives. If the experimenter is considering a polynomial model, then an optimal design such as a *D*-optimal or *I*-optimal design is a possible choice. In recent years, various types of **space-filling designs** have been suggested for computer experiments. Space-filling designs are often thought to be particularly appropriate for deterministic computer models because in general they spread the design points out nearly evenly or uniformly (in some sense) throughout the region of experimentation. This is a desirable feature if the experimenter doesn't know the form of the model that is required, and believes that interesting phenomena are likely to be found in different regions of the experimental space. Furthermore, most space-filling designs do not contain any replicate runs. For a **deterministic** computer model this is desirable, because a single run of the computer model at a design point provides all of the information about the response at that point. Many space-filling designs do not contain replicates even if some factors are dropped and they are projected into lower dimensions.

The first space-filling design proposed was the Latin hypercube design [McKay, Conover and Beckman (1979)]. A Latin hypercube in n runs for k factors in an $n \times k$ matrix

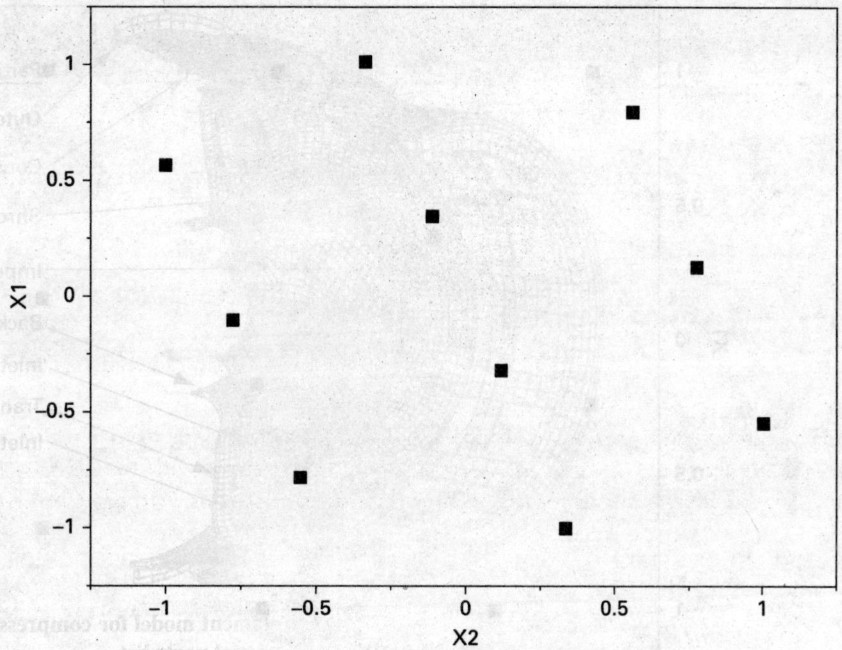

■ **FIGURE 11.34** A 10-run Latin hypercube design

where each column is a random permutation of the levels 1, 2, . . . , *n*. JMP can create Latin hypercube designs. An example of a 10-run Latin hypercube design in two factors from JMP on the interval −1 to +1 is shown in Figure 11.34.

The sphere-packing design in chosen so that the minimum distance between pairs of points is maximized. These designs were proposed by Johnson, Moore and Ylvisaker (1990) and are also called maximin designs. An example of a 10-run sphere-packing design in two factors constructed using JMP is shown in Figure 11.35.

Uniform designs were proposed by Fang (1980). These designs attempt to place the design points so that they are uniformly scattered through the regions as would a sample from a uniform distribution. There are a number of algorithms for creating these designs and several measures of uniformity. See the book by Fang, Li and Sudjianto (2006). An example of a 10-run uniform design in two factors constructed using JMP is in Figure 11.36.

Maximum entropy designs were proposed by Shewry and Wynn (1987). Entropy can be thought of as a measure of the amount of information contained in the distribution of a data set. Suppose that the data comes from a normal distribution with mean vector $\boldsymbol{\mu}$ and covariance matrix $\sigma^2\mathbf{R}(\boldsymbol{\theta})$, where $\mathbf{R}(\boldsymbol{\theta})$ is a correlation matrix having elements

$$r_{ij} = e^{-\sum_{s=1}^{k}\theta_s(x_{is}-x_{js})^2} \tag{11.23}$$

The quantities r_{ij} are the correlations between the responses at two design points. The maximum entropy design maximizes the determinant of $\mathbf{R}(\boldsymbol{\theta})$. Figure 11.37 shows a 10-run maximum entropy design in two factors created using JMP.

The **Gaussian process model** is often used to fit the data from a deterministic computer experiment. These models were introduced as models for computer experiments by Sacks, Welch, Mitchell and Wynn (1989). They are desirable because they provide an exact fit to the observations from the experiment. Now this is no assurance that they will interpolate well at locations in the region of interest where there is no data, and no one seriously believes that

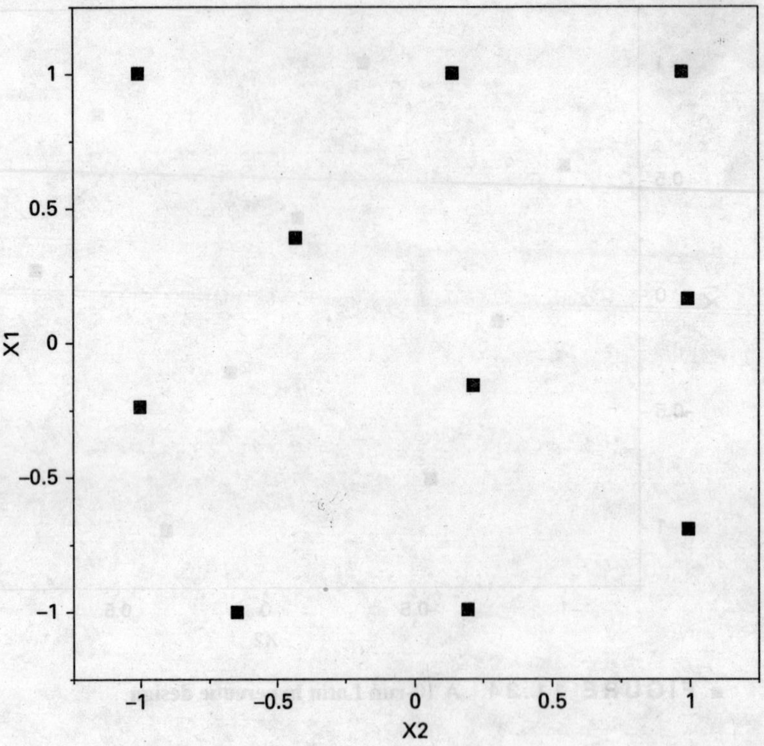

■ **FIGURE 11.35** A 10-run sphere-packing design

■ **FIGURE 11.36** A 10-run uniform design

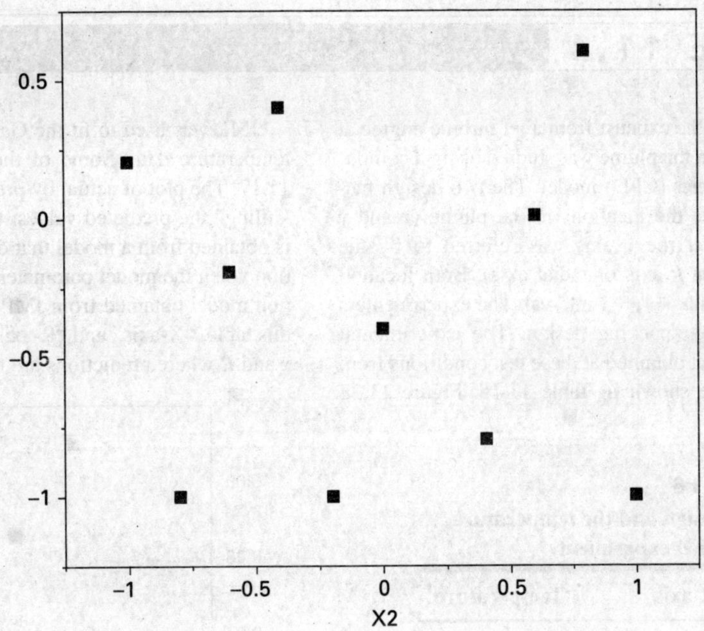

■ **FIGURE 11.37** **A 10-run maximum entropy design**

the Gaussian process model is the correct model for the relationship between the response and the design variables. However, the "exact fit" nature of the model and the fact that it only requires one parameter for each factor considered in the experiment has made it quite popular. The Gaussian process model is

$$y = \mu + z(\mathbf{x})$$

where $z(\mathbf{x})$ is a Gaussian stochastic process with covariance matrix $\sigma^2\mathbf{R}(\boldsymbol{\theta})$, and the elements of $\mathbf{R}(\boldsymbol{\theta})$ are defined in Equation (11.23). The Gaussian process model is essentially a spatial correlation model, where the correlation of the response between two observations decreases as the values of the design factors become further apart. When design points are close together, this causes ill-conditioning in the data for the Gaussian process model, much like multicollinearity resulting from predictors that are nearly linearly dependent in linear regression models. The parameters μ and θ_s, $s = 1, 2, \ldots, k$ are estimated using the method of maximum likelihood. Predicted values of the response at the point $\mathbf{x}$ are computed from

$$\hat{y}(\mathbf{x}) = \hat{\mu} + \mathbf{r}'(\mathbf{x})\mathbf{R}(\hat{\boldsymbol{\theta}})^{-1}(\mathbf{y} - \mathbf{j}\hat{\mu})$$

where $\hat{\mu}$ and $\hat{\boldsymbol{\theta}}$ are the maximum likelihood estimates of the model parameters μ and $\boldsymbol{\theta}$, and $\mathbf{r}'(\mathbf{x}) = [\mathrm{r}(\mathbf{x}_1,\mathbf{x}), \mathrm{r}(\mathbf{x}_2, \mathbf{x}) \ldots, \mathrm{r}(\mathbf{x}_n, \mathbf{x})]$. The prediction equation contains one model term for each design point in the original experiment. JMP will fit and provide predictions from the Gaussian process model. More details about the Gaussian process model are in Santner, Williams, and Notz (2003). A good review of designs for computer experiments and the Gaussian process model is Jones and Johnson (2009).

EXAMPLE 11.4

The temperature in the exhaust from a jet turbine engine at different locations in the plume was studied using a computational fluid dynamics (CFD) model. The two design factors of interest were the locations in the plume (x and y coordinates, however the y axis was referred to by the experimenters as the R-axis or radial axis). Both location axes were coded to the -1, $+1$ interval. The experimenters used a 10-run sphere-packing design. The experimental design and the output obtained at these test conditions from the CFD model are shown in Table 11.16. Figure 11.38 shows the design.

JMP was used to fit the Gaussian process model to the temperature data. Some of the output is shown in Table 11.17. The plot of actual by predicted is obtained by "jackknifing" the predicted values; that is, each predicted value is obtained from a model that doesn't contain that observation when the model parameters are estimated. The prediction model obtained from JMP is shown in Table 11.18. In this table, "X-axis" and "R-axis" refer to the coordinates in x and R where predictions are to be made.

■ **TABLE 11.16**
Sphere-packing design and the temperature responses in the CFD experiment

x-axis	*R*-axis	Temperature
0.056	0.062	338.07
0.095	0.013	1613.04
0.077	0.062	335.91
0.095	0.061	327.82
0.090	0.037	449.23
0.072	0.038	440.58
0.064	0.015	1173.82
0.050	0.000	1140.36
0.050	0.035	453.83
0.079	0.000	1261.39

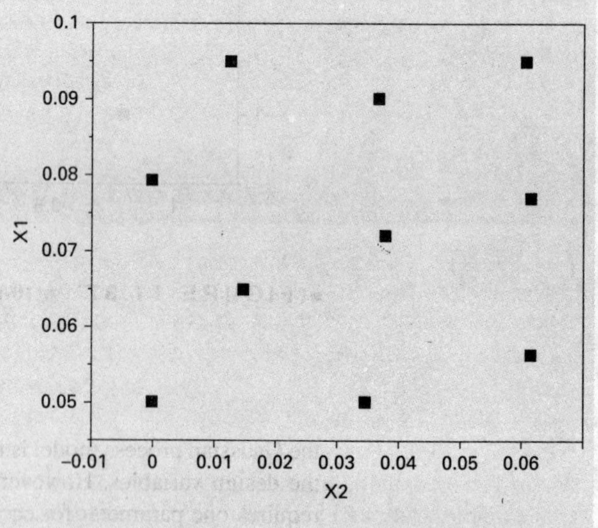

■ **FIGURE 11.38** **The sphere-packing design for the CFD experiment**

■ **TABLE 11.17**
JMP output for the Gaussian process model for the CFD experiment in Table 11.16

Gaussian Process
Actual by Predicted Plot

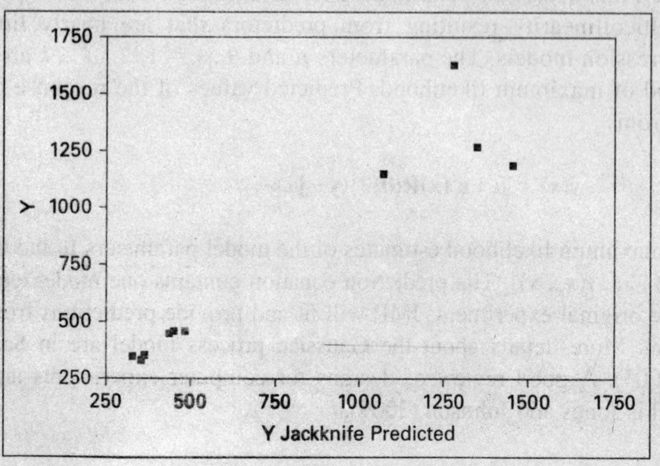

■ **TABLE 11.17** *(Continued)*

Model Report

Column	Theta	Total Sensitivity	Main Effect	X-Axis Interaction	R-axis Interaction
X-Axis	65.40254	0.0349982	0.0141129		0.0208852
R-axis	3603.2483	0.9858871	0.9650018	0.0208852	

Mu	Sigma
734.54584	212205.18

−2*LogLikelihood
132.98004

Contour Profiler

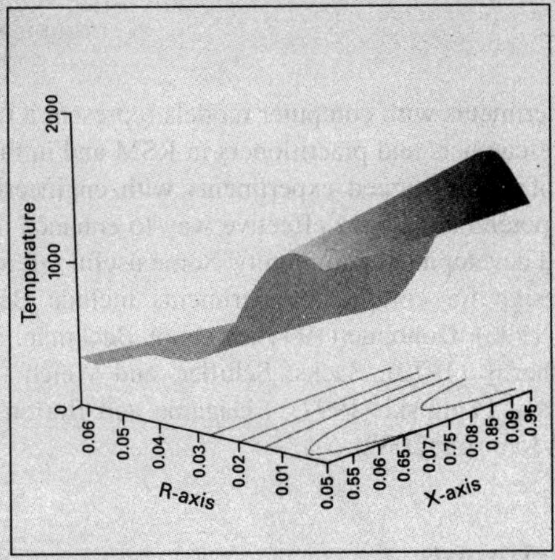

■ **TABLE 11.18**
The JMP Gaussian process prediction model for the CFD experiment

$\hat{y}$ = 734.545842514493 +
(−1943.3447961328 * Exp(−(65.4025404276544 * (("X-Axis") − 0.0560573769818389)^2 +
3603.24827558717 * ("R-axis") − 0.0618)^2)) +
3941.78888206788 * Exp(−(65.4025404276544 * (("X-Axis") − 0.0947)^2 + 3603.24827558717
* (("R-axis") − 0.0126487944665913)^2)) +
3488.57543918861 * Exp(−(65.4025404276544 * (("X-Axis") − 0.0765974898313444)^2 +
3603.24827558717 * (("R-axis") − 0.0618)^2)) + −
2040.39522592773 * Exp(−(65.4025404276544 * (("X-Axis") − 0.0947)^2 + 3603.24827558717
* (("R-axis") − 0.0608005210868486)^2)) +
−742.642897583584 * Exp(−(65.4025404276544 * (("X-Axis") − 0.898402482375096)^2 +
3603.24827558717 * ("R-axis") − 0.0367246615426894)^2)) +

■ **TABLE 11.18** *(Continued)*

519.91871208163 * Exp(−(65.4025404276544 * ((“X-Axis”) − 0.0717377150616494)^2 +
3603.24827558717 * ((“R-axis”) − 0.377241897055609)^2)) +
−3082.85411601115 * Exp(−(65.4025404276544 * ((“X-Axis”) − 0.0644873310121405)^2 +
3603.24827558717 * (“R-axis”) −0.0148210408248663)^2)) +
958.926988711818 * Exp(−(65.4025404276544 * ((“X-Axis”) − 0.0499)^2 + 3603.24827558717
* (“R-axis”) ^2)) +
80.468182554262 * Exp(−(65.4025404276544 * ((“X-Axis”) − 0.0499)^2 + 3603.24827558717
* (“R-axis”) − 0.0347687447931648)^2)) +
−1180.44117607546 * Exp(−(65.4025404276544 * ((“X-Axis”) − 0.0790747191607881)^2 +
3603.24827558717 * (“R-axis”) ^2)))

Experiments with computer models represent a relatively new and challenging area for both researchers and practitioners in RSM and in the broader engineering community. The use of well-designed experiments with engineering computer models for product design is potentially a very effective way to enhance the productivity of the engineering design and development community. Some useful references on the general subject of statistical design for computer experiments include Barton (1992, 1994), Bettonvil and Kleijnen (1996), Donohue (1994), McKay, Beckman, and Conover (1979), Montgomery and Weatherby (1979), Sacks, Schiller, and Welch (1989), Sacks and Welch (1989), Simpson and Peplinski (1997), Slaagame and Barton (1997), Welch et al. (1992), and Jones and Johnson (2009).

11.6 Mixture Experiments

In previous sections, we have presented response surface designs for those situations in which the levels of each factor are independent of the levels of other factors. In **mixture experiments**, the factors are the components or ingredients of a mixture, and consequently their levels are not independent. For example, if $x_1, x_2, \ldots, x_p$ denote the proportions of p components of a mixture, then

$$0 \leq x_i \leq 1 \qquad i = 1, 2, \ldots, p$$

and

$$x_1 + x_2 + \cdots + x_p = 1 \qquad \text{(i.e., 100 percent)}$$

These restrictions are illustrated graphically in Figure 11.39 for $p = 2$ and $p = 3$ components. For two components, the factor space for the design includes all values of the two components that lie on the line segment $x_1 + x_2 = 1$, with each component being bounded by 0 and 1. With three components, the mixture space is a triangle with vertices corresponding to formulations that are **pure blends** (mixtures that are 100 percent of a single component).

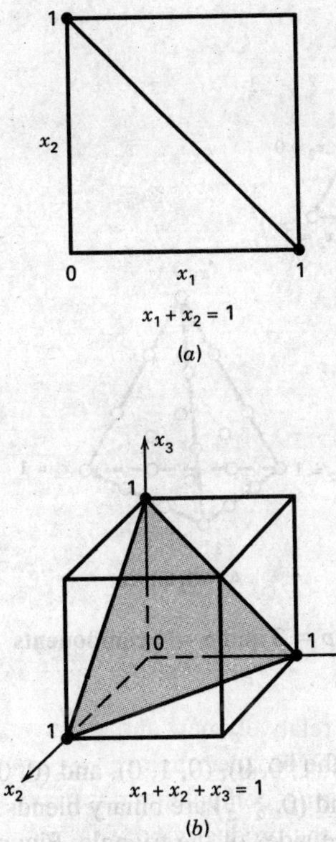

■ **FIGURE 11.39** **Constrained factor space for mixtures with (*a*) $p = 2$ components and (*b*) $p = 3$ components**

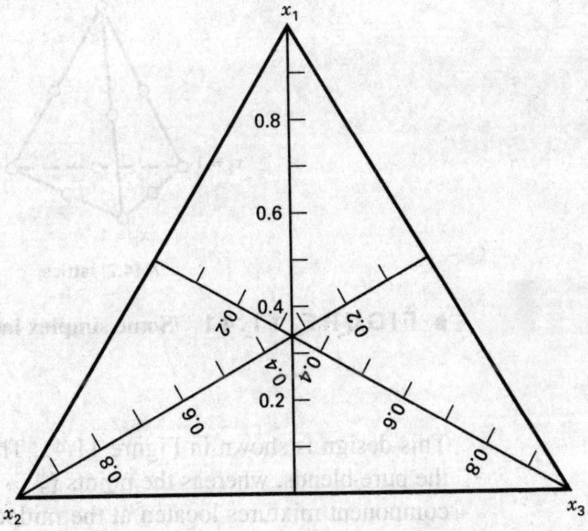

■ **FIGURE 11.40** **Trilinear coordinate system**

When there are three components of the mixture, the constrained experimental region can be conveniently represented on **trilinear coordinate paper** as shown in Figure 11.40. Each of the three sides of the graph in Figure 11.40 represents a mixture that has none of the three components (the component labeled on the opposite vertex). The nine grid lines in each direction mark off 10 percent increments in the respective components.

Simplex designs are used to study the effects of mixture components on the response variable. A $\{p, m\}$ **simplex lattice design** for p components consists of points defined by the following coordinate settings: the proportions assumed by each component take the $m + 1$ equally spaced values from 0 to 1,

$$x_i = 0, \frac{1}{m}, \frac{2}{m}, \ldots, 1 \qquad i = 1, 2, \ldots, p \tag{11.24}$$

and all possible combinations (mixtures) of the proportions from Equation 11.24 are used. As an example, let $p = 3$ and $m = 2$. Then

$$x_i = 0, \tfrac{1}{2}, 1 \qquad i = 1, 2, 3$$

and the simplex lattice consists of the following six runs:

$$(x_1, x_2, x_3) = (1, 0, 0), (0, 1, 0), (0, 0, 1), (\tfrac{1}{2}, \tfrac{1}{2}, 0), (\tfrac{1}{2}, 0, \tfrac{1}{2}), (0, \tfrac{1}{2}, \tfrac{1}{2})$$

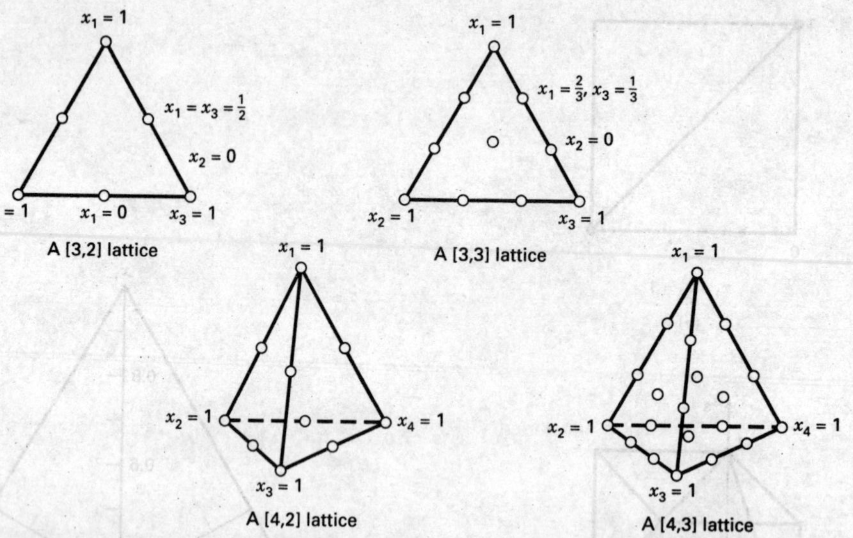

■ **FIGURE 11.41** Some simplex lattice designs for $p = 3$ and $p = 4$ components

This design is shown in Figure 11.41. The three vertices (1, 0, 0), (0, 1, 0), and (0, 0, 1) are the pure blends, whereas the points $(\frac{1}{2}, \frac{1}{2}, 0)$, $(\frac{1}{2}, 0, \frac{1}{2})$, and $(0, \frac{1}{2}, \frac{1}{2})$ are binary blends or two-component mixtures located at the midpoints of the three sides of the triangle. Figure 11.41 also shows the {3, 3}, {4, 2}, and {4, 3} simplex lattice designs. In general, the number of points in a $\{p, m\}$ simplex lattice design is

$$N = \frac{(p + m - 1)!}{m!(p - 1)!}$$

An alternative to the simplex lattice design is the **simplex centroid design**. In a p-component simplex centroid design, there are $2^p - 1$ points, corresponding to the p permutations of $(1, 0, 0, \ldots, 0)$, the $\binom{p}{2}$ permutations of $(\frac{1}{2}, \frac{1}{2}, 0, \ldots, 0)$, the $\binom{p}{3}$ permutations of $(\frac{1}{3}, \frac{1}{3}, \frac{1}{3}, 0, \ldots, 0)$, . . . , and the overall centroid $(\frac{1}{p}, \frac{1}{p}, \ldots, \frac{1}{p})$. Figure 11.42 shows some simplex centroid designs.

A criticism of the simplex designs described above is that most of the experimental runs occur on the boundary of the region and, consequently, include only $p - 1$ of the p components. It is usually desirable to augment the simplex lattice or simplex centroid with additional points in the interior of the region where the blends will consist of all p mixture components. For more discussion, see Cornell (2002) and Myers, Montgomery and Anderson-Cook (2009).

Mixture models differ from the usual polynomials employed in response surface work because of the constraint $\Sigma x_i = 1$. The standard forms of the mixture models that are in widespread use are

Linear

$$E(y) = \sum_{i=1}^{p} \beta_i x_i \tag{11.25}$$

Quadratic

$$E(y) = \sum_{i=1}^{p} \beta_i x_i + \sum_{i<j}^{p} \sum \beta_{ij} x_i x_j \tag{11.26}$$

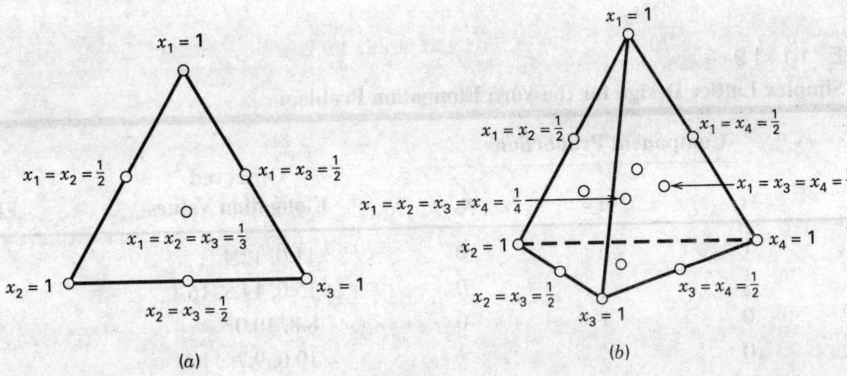

■ **FIGURE 11.42** **Simplex centroid designs with (*a*) $p = 3$ components and (*b*) $p = 4$ components**

Full cubic

$$E(y) = \sum_{i=1}^{p} \beta_i x_i + \sum\sum_{i<j}^{p} \beta_{ij} x_i x_j + \sum\sum_{i<j}^{p} \delta_{ij} x_i x_j (x_i - x_j) + \sum\sum\sum_{i<j<k} \beta_{ijk} x_i x_j x_k \tag{11.27}$$

Special cubic

$$E(y) = \sum_{i=1}^{p} \beta_i x_i + \sum\sum_{i<j}^{p} \beta_{ij} x_i x_j + \sum\sum\sum_{i<j<k} \beta_{ijk} x_i x_j x_k \tag{11.28}$$

The terms in these models have relatively simple interpretations. In Equations 11.25 through 11.28, the parameter β_i represents the expected response to the pure blend $x_i = 1$ and $x_j = 0$ when $j \neq i$. The portion $\Sigma_{i=1}^{p} \beta_i x_i$ is called the **linear blending portion**. When curvature arises from nonlinear blending between component pairs, the parameters β_{ij} represent either **synergistic** or **antagonistic blending**. Higher order terms are frequently necessary in mixture models because (1) the phenomena studied may be complex and (2) the experimental region is frequently the entire operability region and therefore large, requiring an elaborate model.

EXAMPLE 11.5 A Three-Component Mixture

Cornell (2002) describes a mixture experiment in which three components—polyethylene (x_1), polystyrene (x_2), and polypropylene (x_3)—were blended to form fiber that will be spun into yarn for draperies. The response variable of interest is yarn elongation in kilograms of force applied. A {3, 2} simplex lattice design is used to study the product. The design and the observed responses are shown in Table 11.19. Notice that all of the design points involve either **pure** or **binary**

■ TABLE 11.19
The {3, 2} Simplex Lattice Design for the Yarn Elongation Problem

Design Point	Component Proportions x_1	x_2	x_3	Observed Elongation Values	Average Elongation Value ($\bar{y}$)
1	1	0	0	11.0, 12.4	11.7
2	$\frac{1}{2}$	$\frac{1}{2}$	0	15.0, 14.8, 16.1	15.3
3	0	1	0	8.8, 10.0	9.4
4	0	$\frac{1}{2}$	$\frac{1}{2}$	10.0, 9.7, 11.8	10.5
5	0	0	1	16.8, 16.0	16.4
6	$\frac{1}{2}$	0	$\frac{1}{2}$	17.7, 16.4, 16.6	16.9

blends; that is, at most only two of the three components are used in any formulation of the product. Replicate observations are also run, with two replicates at each of the pure blends and three replicates at each of the binary blends. The error standard deviation can be estimated from these replicate observations as = 0.85. Cornell fits the second-order mixture polynomial to the data, resulting in

$$\hat{y} = 11.7x_1 + 9.4x_2 + 16.4x_3 + 19.0x_1x_2 + 11.4x_1x_3 - 9.6x_2x_3$$

This model can be shown to be an adequate representation of the response. Note that because $\hat{\beta}_3 > \hat{\beta}_1 > \hat{\beta}_2$ we would conclude that component 3 (polypropylene) produces yarn with the highest elongation. Furthermore, because $\hat{\beta}_{12}$ and $\hat{\beta}_{13}$ are positive, blending components 1 and 2 or components 1 and 3 produces higher elongation values than would be expected just by averaging the elongations of the pure blends. This is an example of "synergistic" blending effects. Components 2 and 3 have antagonistic blending effects because $\hat{\beta}_{23}$ is negative.

Figure 11.43 plots the contours of elongation, and this may be helpful in interpreting the results. From examining the figure, we note that if maximum elongation is desired, a blend of components 1 and 3 should be chosen consisting of about 80 percent component 3 and 20 percent component 1.

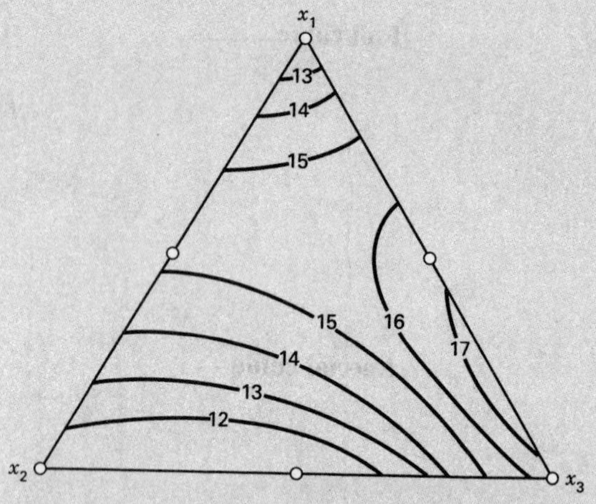

■ FIGURE 11.43 Contours of constant estimated yarn elongation from the second-order mixture model for Example 11.5

We noted previously that the simplex lattice and simplex centroid designs are **boundary point designs**. If the experimenter wants to make predictions about the properties of complete mixtures, it would be highly desirable to have more runs in the interior of the simplex. We recommend augmenting the usual simplex designs with **axial runs** and the overall centroid (if the centroid is not already a design point).

The **axis of component *i*** is the line or ray extending from the base point $x_i = 0, x_j = 1/(p - 1)$ for all $j \neq i$ to the opposite vertex where $x_i = 1, x_j = 0$ for all $j \neq i$. The base point will always lie at the centroid of the $(p - 2)$-dimensional boundary of the simplex that is opposite the vertex $x_i = 1, x_j = 0$ for all $j \neq i$. [the boundary is sometimes called a $(p - 2)$-flat.] The length of the component axis is one unit. **Axial points** are positioned along the component axes at a distance Δ from the centroid. The maximum value for Δ is $(p - 1)/p$. We recommend that

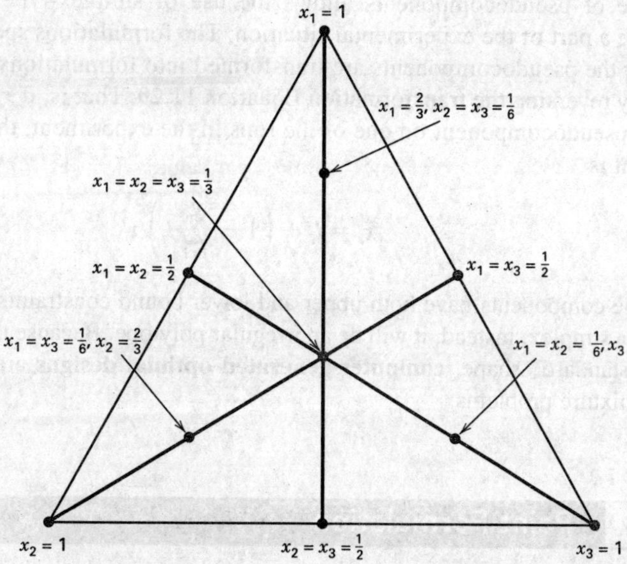

■ **FIGURE 11.44**
An augmented simplex-lattice design

axial runs be placed midway between the centroid of the simplex and each vertex so that $\Delta = (p - 1)/2p$. Sometimes these points are called **axial check blends** because a fairly common practice is to exclude them when fitting the preliminary mixture model and then use the responses at these axial points to check the adequacy of the fit of the preliminary model.

Figure 11.44 shows the {3, 2} simplex lattice design augmented with the axial points. This design has 10 points, with four of these points in the interior of the simplex. The {3, 3} simplex lattice will support fitting the full cubic model, whereas the augmented simplex lattice will not; however, the augmented simplex lattice will allow the experimenter to fit the special cubic model or to add special quartic terms such as $\beta_{1233}x_1x_2x_3^2$ to the quadratic model. The augmented simplex lattice is superior for studying the response of complete mixtures in the sense that it can detect and model curvature in the interior of the triangle that cannot be accounted for by the terms in the full cubic model. The augmented simplex lattice has more power for detecting lack of fit than does the {3, 3} lattice. This is particularly useful when the experimenter is unsure about the proper model to use and also plans to sequentially build a model by starting with a simple polynomial (perhaps first order), test the model for lack of fit, and then augment the model with higher order terms, test the new model for lack of fit, and so forth.

In some mixture problems, **constraints** on the individual components arise. Lower bound constraints of the form

$$l_i \leq x_i \leq 1 \qquad i = 1, 2, \ldots, p$$

are fairly common. When only lower bound constraints are present, the feasible design region is still a simplex, but it is inscribed inside the original simplex region. This situation may be simplified by the introduction of **pseudocomponents**, defined as

$$x_i' = \frac{x_i - l_i}{\left(1 - \sum_{j=1}^{p} l_j\right)} \tag{11.29}$$

with $\Sigma_{j=1}^{p} l_j < 1$. Now

$$x_1' + x_2' + \cdots + x_p' = 1$$

so the use of pseudocomponents allows the use of simplex-type designs when lower bounds are a part of the experimental situation. The formulations specified by the simplex design for the pseudocomponents are transformed into formulations for the original components by reversing the transformation Equation 11.29. That is, if x_i' is the value assigned to the ith pseudocomponent on one of the runs in the experiment, the ith original mixture component is

$$x_i = l_i + \left(1 - \sum_{j=1}^{p} l_j\right) x_i' \qquad (11.30)$$

If the components have both upper and lower bound constraints, the feasible region is no longer a simplex; instead, it will be an irregular polytope. Because the experimental region is not a "standard" shape, **computer-generated optimal designs** are very useful for these types of mixture problems.

EXAMPLE 11.6 Paint Formulation

An experimenter is trying to optimize the formulation of automotive clear coat paint. These are complex products that have very specific performance requirements. Specifically, the customer wants the Knoop hardness to exceed 25 and the percentage of solids to be below 30. The clear coat is a three-component mixture, consisting of a monomer (x_1), a crosslinker (x_2), and a resin (x_3). There are constraints on the component proportions:

$$x_1 + x_2 + x_3 = 100$$
$$5 \le x_1 \le 25$$
$$25 \le x_2 \le 40$$
$$50 \le x_3 \le 70$$

The result is the constrained region of experimentation shown in Figure 11.44. Because the region of interest is not a simplex, we will use a D-optimal design for this problem. Assuming that both responses are likely to be modeled with a quadratic mixture model, we can generate the D-optimal design shown in Figure 11.39 using Design-Expert. We assumed that in addition to the six runs required to fit the quadratic mixture model, four additional distinct runs would be made to check for lack of fit and that four of these runs would be replicated to provide an estimate of pure error. Design-Expert used the vertices, the edge centers, the overall centroid, and the check runs (points located halfway between the centroid and the vertices) as the candidate points.

The 14-run design is shown in Table 11.20, along with the hardness and solids responses. The results of fitting quadratic models to both responses are summarized in Tables 11.21 and 11.22. Notice that quadratic models fit nicely to both the hardness and the solids responses. The fitted equations for both responses (in terms of the pseudocomponents) are shown in these tables. Contour plots of the responses are shown in Figures 11.46 and 11.47.

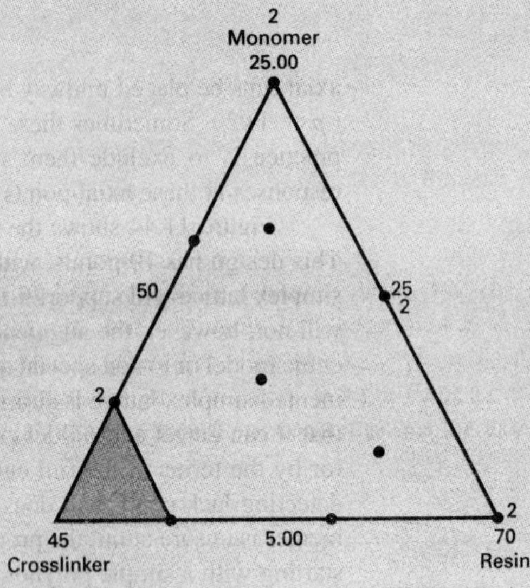

■ **FIGURE 11.45 The constrained experimental region for the paint formulation problem in Example 11.6 (shown in the actual component scale)**

Figure 11.48 is an overlay plot of the two response surfaces, showing the Knoop hardness contour of 25 and the 30 percent contour for solids. The feasible region for this product is the unshaded region near the center of the plot. Obviously, there are a number of choices for the proportions of monomer, crosslinker, and resin for the clear coat that will give a product satisfying the performance requirements.

■ **TABLE 11.20**
A *D*-Optimal Design for the Paint Formulation Problem in Example 11.5

Standard Order	Run	Monomer x_1	Crosslinker x_2	Resin x_3	Hardness y_1	Solids y_2
1	2	17.50	32.50	50.00	29	9.539
2	1	10.00	40.00	50.00	26	27.33
3	4	15.00	25.00	60.00	17	29.21
4	13	25.00	25.00	50.00	28	30.46
5	7	5.00	25.00	70.00	35	74.98
6	3	5.00	32.50	62.50	31	31.5
7	6	11.25	32.50	56.25	21	15.59
8	11	5.00	40.00	55.00	20	19.2
9	10	18.13	28.75	53.13	29	23.44
10	14	8.13	28.75	63.13	25	32.49
11	12	25.00	25.00	50.00	19	23.01
12	9	15.00	25.00	60.00	14	41.46
13	5	10.00	40.00	50.00	30	32.98
14	8	5.00	25.00	70.00	23	70.95

■ **TABLE 11.21**
Model Fitting for the Hardness Response

Response: hardness
ANOVA for Mixture Quadratic Model
Analysis of variance table [Partial sum of squares]

Source	Sum of Squares	DF	Mean Square	F Value	Prob > F
Model	279.73	5	55.95	2.37	0.1329
Linear Mixture	*29.13*	*2*	*14.56*	*0.62*	*0.5630*
AB	*72.61*	*1*	*72.61*	*3.08*	*0.1174*
AC	*179.67*	*1*	*179.67*	*7.62*	*0.0247*
BC	*8.26*	*1*	*8.26*	*0.35*	*0.5703*
Residual	188.63	8	23.58		
Lack of Fit	*63.63*	*4*	*15.91*	*0.51*	*0.7354*
Pure Error	*125.00*	*4*	*31.25*		
Cor Total	468.36	13			

Std. Dev.	4.86	R-Squared	0.5973
Mean	24.79	Adj R-Squared	0.3455
C.V.	19.59	Pred R-Squared	−0.3635
PRESS	638.60	Adeq Precision	4.975

■ **TABLE 11.21** *(Continued)*

Component	Coefficient Estimate	DF	Standard Error	95% CI Low	95% CI High
A-Monomer	23.81	1	3.36	16.07	31.55
B-Crosslinker	16.40	1	7.68	−1.32	34.12
C-Reşin	29.45	1	3.36	21.71	37.19
AB	44.42	1	25.31	−13.95	102.80
AC	−44.01	1	15.94	−80.78	−7.25
BC	13.80	1	23.32	−39.97	67.57

Final Equation in Terms of Pseudocomponents:

hardness =
+23.81 * A
+16.40 * B
+29.45 * C
+44.42 * A * B
−44.01 * A * C
+13.80 * B * C

■ **TABLE 11.22**
Model Fitting for the Solids Response

Response: solids
ANOVA for Mixture Quadratic Model
Analysis of variance table [Partial sum of squares]

Source	Sum of Squares	DF	Mean Square	F Value	Prob > F
Model	4297.94	5	859.59	25.78	<0.0001
Linear Mixture	*2931.09*	*2*	*1465.66*	*43.95*	*<0.0001*
AB	*211.20*	*1*	*211.20*	*6.33*	*0.0360*
AC	*285.67*	*1*	*285.67*	*8.57*	*0.0191*
BC	*1036.72*	*1*	*1036.72*	*31.09*	*0.0005*
Residual	266.79	8	33.35		
Lack of Fit	*139.92*	*4*	*34.98*	*1.10*	*0.4633*
Pure Error	*126.86*	*4*	*31.72*		
Cor Total	4564.73	13			

Std. Dev.	5.77	R-Squared	0.9416
Mean	33.01	Adj R-Squared	0.9050
C.V.	17.49	Pred R-Squared	0.7827
PRESS	991.86	Adeq Precision	15.075

■ **TABLE 11.22** ***(Continued)***

Component	Coefficient Estimate	DF	Standard Error	95% CI Low	95% CI High
A-Monomer	26.53	1	3.99	17.32	35.74
B-Crosslinker	46.60	1	9.14	25.53	67.68
C-Resin	73.23	1	3.99	64.02	82.43
AB	−75.76	1	30.11	−145.19	−6.34
AC	−55.50	1	18.96	−99.22	−11.77
BC	−154.61	1	27.73	−218.56	−90.67

Final Equation in Terms of Pseudocomponents:

$$\begin{aligned} \text{solids} = &\ +26.53 * A \\ &+46.60 * B \\ &+73.23 * C \\ &-75.76 * A * B \\ &-55.50 * A * C \\ &-154.61 * B * C \end{aligned}$$

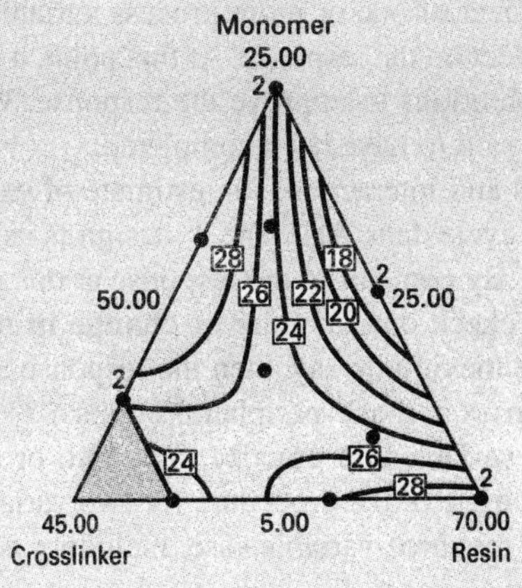

■ **FIGURE 11.46** **Contour plot of the Knoop hardness response, Example 11.6**

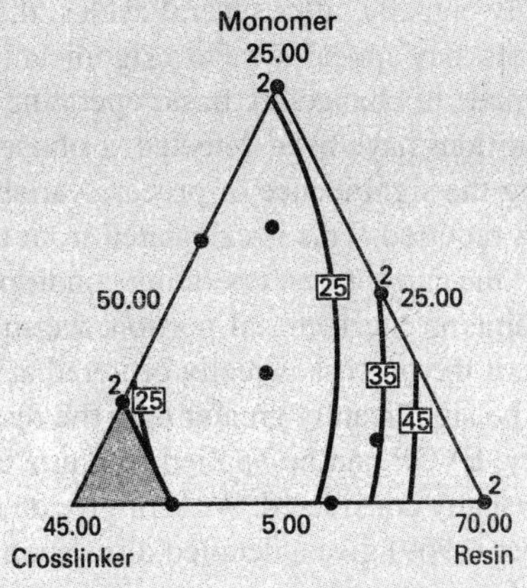

■ **FIGURE 11.47** **Contour plot of the percentage of solids response, Example 11.6**

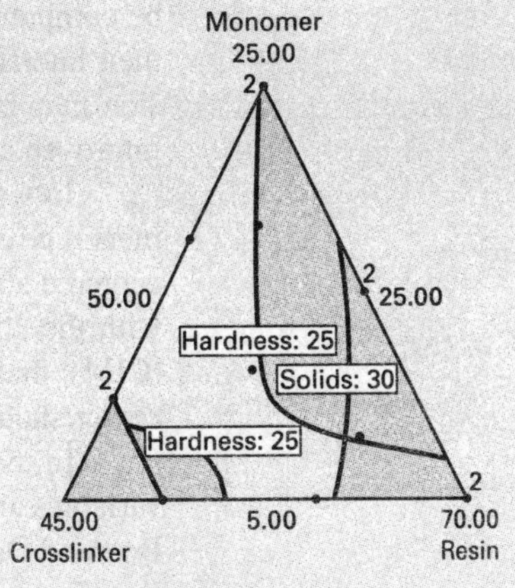

■ **FIGURE 11.48** **Overlay plot of the Knoop hardness and percentage of solids response, showing the feasible region for the paint formulation**

11.7 Evolutionary Operation

Response surface methodology is often applied to pilot plant operations by research and development personnel. When it is applied to a full-scale production process, it is usually done only once (or very infrequently) because the experimental procedure is relatively elaborate. However, conditions that were optimum for the pilot plant may not be optimum for the full-scale process. The pilot plant may produce 2 pounds of product per day, whereas the full-scale process may produce 2000 pounds per day. This "scale-up" of the pilot plant to the full-scale production process usually results in distortion of the optimum conditions. Even if the full-scale plant begins operation at the optimum, it will eventually "drift" away from that point because of variations in raw materials, environmental changes, and operating personnel.

A method is needed for the continuous monitoring and improvement of a full-scale process with the goal of moving the operating conditions toward the optimum or following a "drift." The method should not require large or sudden changes in operating conditions that might disrupt production. **Evolutionary operation (EVOP)** was proposed by Box (1957) as such an operating procedure. It is designed as a method of routine plant operation that is carried out by manufacturing personnel with minimum assistance from the research and development staff.

EVOP consists of systematically introducing small changes in the levels of the operating variables under consideration. Usually, a 2^k design is employed to do this. The changes in the variables are assumed to be so small enough that serious disturbances in yield, quality, or quantity will not occur, yet large enough that potential improvements in process performance will eventually be discovered. Data are collected on the response variables of interest at each point of the 2^k design. When one observation has been taken at each design point, a cycle is said to have been completed. The effects and interactions of the process variables may then be computed. Eventually, after several cycles, the effect of one or more process variables or their interactions may appear to have a significant effect on the response. At this point, a decision may be made to change the basic operating conditions to improve the response. When improved conditions have been detected, a **phase** is said to have been completed.

In testing the significance of process variables and interactions, an estimate of experimental error is required. This is calculated from the cycle data. Also, the 2^k design is usually centered about the current best operating conditions. By comparing the response at this point with the 2^k points in the factorial portion, we may check on curvature or **change in mean** (CIM); that is, if the process is really centered at the maximum, say, then the response at the center should be significantly greater than the responses at the 2^k-peripheral points.

In theory, EVOP can be applied to k process variables. In practice, only two or three variables are usually considered. We will give an example of the procedure for two variables. Box and Draper (1969) give a detailed discussion of the three-variable case, including necessary forms and worksheets.

EXAMPLE 11.7

Consider a chemical process whose yield is a function of temperature (x_1) and pressure (x_2). The current operating conditions are $x_1 = 250°F$ and $x_2 = 145$ psi. The EVOP procedure uses the 2^2 design plus the center point shown in Figure 11.49. The cycle is completed by running each design point in numerical order (1, 2, 3, 4, 5). The yields in the first cycle are also shown in Figure 11.49.

The yields from the first cycle are entered in the EVOP calculation sheet, as shown in Table 11.23. At the end of the first cycle, no estimate of the standard deviation can be

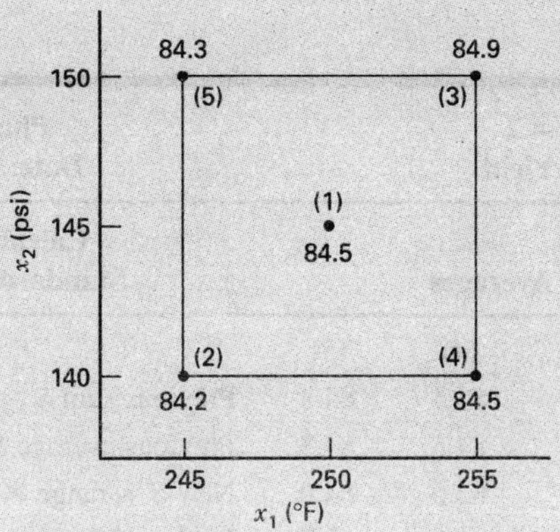

■ **FIGURE 11.49** A 2^2 design for EVOP

■ **TABLE 11.23**
EVOP Calculation Sheet for Example 11.7, $n = 1$

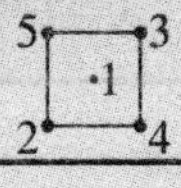

Cycle: $n = 1$ **Response: Yield** **Phase: 1** **Date: 1/11/04**

	Calculation of Averages					Calculation of Standard Deviation
Operating Conditions	**(1)**	**(2)**	**(3)**	**(4)**	**(5)**	
(i) Previous cycle sum						Previous sum S =
(ii) Previous cycle average						Previous average S =
(iii) New observations	84.5	84.2	84.9	84.5	84.3	New S = range $\times f_{5,n}$ =
(iv) Differences [(ii) − (iii)]						Range of (iv) =
(v) New sums [(i) + (iii)]	84.5	84.2	84.9	84.5	84.3	New sum S =
(vi) New averages [$\bar{y}_i$ = (v)/n]	84.5	84.2	84.9	84.5	84.3	New average $S = \dfrac{\text{New sum } S}{n-1}$

Calculation of Effects	Calculation of Error Limits
Temperature effect $= \frac{1}{2}(\bar{y}_3 + \bar{y}_4 - \bar{y}_2 - \bar{y}_5) = 0.45$	For new average $= \dfrac{2}{\sqrt{n}} S =$
Pressure effect $= \frac{1}{2}(\bar{y}_3 + \bar{y}_5 - \bar{y}_2 - \bar{y}_4) = 0.25$	For new effects $\dfrac{2}{\sqrt{n}} S =$
$T \times P$ interaction effect $= \frac{1}{2}(\bar{y}_2 + \bar{y}_3 - \bar{y}_4 - \bar{y}_5) = 0.15$	
Change-in-mean effect $= \frac{1}{5}(\bar{y}_2 + \bar{y}_3 + \bar{y}_4 + \bar{y}_5 - 4\bar{y}_1) = 0.02$	For change in mean $\dfrac{1.78}{\sqrt{n}} S =$

made. The effects and interactions for temperature and pressure are calculated in the usual manner for a 2^2 design.

A second cycle is then run and the yield data entered in another EVOP calculation sheet, shown in Table 11.24. At the end of the second cycle, the experimental error can be estimated and the estimates of the effects compared to approximate 95 percent (two standard deviation) limits. Note that the range refers to the range of the differences in row (iv); thus, the range is $+1.0 - (-1.0) = 2.0$. Because none of the effects in Table 11.24 exceed their error limits, the true effect is probably zero, and no changes in operating conditions are contemplated.

The results of a third cycle are shown in Table 11.25. The effect of pressure now exceeds its error limit and the temperature effect is equal to the error limit. A change in operating conditions is now probably justified.

■ **TABLE 11.24**
EVOP Calculation Sheet for Example 11.7, $n = 2$

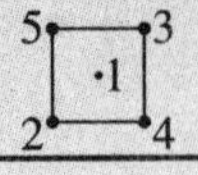

Cycle: $n = 2$
Response: Yield

Phase: 1
Date: 1/11/04

Operating Conditions	**(1)**	**(2)**	**(3)**	**(4)**	**(5)**	**Calculation of Standard Deviation**
			Calculation of Averages			
(i) Previous cycle sum	84.5	84.2	84.9	84.5	84.3	Previous sum S =
(ii) Previous cycle average	84.5	84.2	84.9	84.5	84.3	Previous average S =
(iii) New observations	84.9	84.6	85.9	83.5	84.0	New S = range $\times f_{5,n}$ = 0.60
(iv) Differences [(ii) − (iii)]	−0.4	−0.4	−1.0	+1.0	0.3	Range of (iv) = 2.0
(v) New sums [(i) + (iii)]	169.4	168.8	170.8	168.0	168.3	New sum S = 0.60
(vi) New average [$\bar{y}_i = (\text{v})/n$]	84.70	84.40	85.40	84.00	84.15	New average $S = \dfrac{\text{New sum } S}{n-1} = 0.60$

Calculation of Effects

Temperature effect $= \frac{1}{2}(\bar{y}_3 + \bar{y}_4 - \bar{y}_2 - \bar{y}_5) = 0.43$

Pressure effect $= \frac{1}{2}(\bar{y}_3 + \bar{y}_5 - \bar{y}_2 - \bar{y}_4) = 0.58$

$T \times P$ interaction effect $= \frac{1}{2}(\bar{y}_2 + \bar{y}_3 - \bar{y}_4 - \bar{y}_5) = 0.83$

Change-in-mean effect $= \frac{1}{5}(\bar{y}_2 + \bar{y}_3 - \bar{y}_4 - \bar{y}_5 - 4\bar{y}_1) = -0.17$

Calculation of Error Limits

For new average $= \dfrac{2}{\sqrt{n}} S = 0.85$

For new effects $\dfrac{2}{\sqrt{n}} S = 0.85$

For change in mean $\dfrac{1.78}{\sqrt{n}} S = 0.76$

■ **TABLE 11.25**
EVOP Calculation Sheet for Example 11.7, $n = 3$

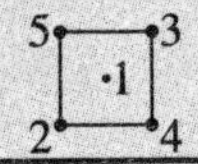

Cycle: $n = 1$
Response: Yield

Phase: 1
Date: 1/11/04

Operating Conditions	**(1)**	**(2)**	**(3)**	**(4)**	**(5)**	**Calculation of Standard Deviation**
			Calculation of Averages			
(i) Previous cycle sum	169.4	168.8	170.8	168.0	168.3	Previous sum S = 0.60
(ii) Previous cycle average	84.70	84.40	85.40	84.00	84.15	Previous average S = 0.60
(iii) New observations	85.0	84.0	86.6	84.9	85.2	New S = range $\times f_{5,n}$ = 0.56
(iv) Differences [(ii) − (iii)]	−0.30	+0.40	−1.20	−0.90	−1.05	Range of (iv) = 1.60
(v) New sums [(i) + (iii)]	254.4	252.8	257.4	252.9	253.5	New sum S = 1.16
(vi) New average [$\bar{y}_i = (\text{v})/n$]	84.80	84.27	85.80	84.30	84.50	New average $S = \dfrac{\text{New sum } S}{n-1} = 0.58$

■ **TABLE 11.25** *(Continued)*

Calculation of Effects	Calculation of Error Limits
Temperature effect $= \frac{1}{2}(\bar{y}_3 + \bar{y}_4 - \bar{y}_2 - \bar{y}_5) = 0.67$	For new average $= \frac{2}{\sqrt{n}}S = 0.67$
Pressure effect $= \frac{1}{2}(\bar{y}_3 + \bar{y}_5 - \bar{y}_2 - \bar{y}_4) = 0.87$	For new effects $\frac{2}{\sqrt{n}}S = 0.67$
$T \times P$ interaction effect $= \frac{1}{2}(\bar{y}_2 + \bar{y}_3 - \bar{y}_4 - \bar{y}_5) = 0.64$	
Change-in-mean effect $= \frac{1}{5}(\bar{y}_2 + \bar{y}_3 + \bar{y}_4 + \bar{y}_5 - 4\bar{y}_1) = -0.07$	For change in mean $\frac{1.78}{\sqrt{n}}S = 0.60$

■ **TABLE 11.26**
EVOP Information Board, Cycle 3

Response: Percent Yield

Requirement: Maximize

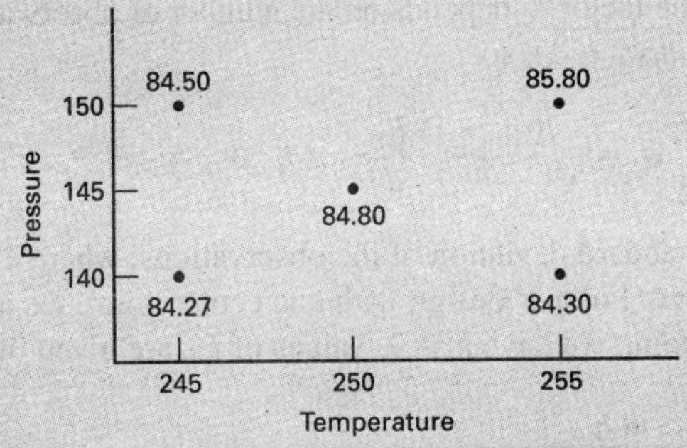

Error Limits for Averages: ±0.67		
Effects with	Temperature	0.67 ± 0.67
95 percent error	Pressure	0.87 ± 0.67
Limits	$T \times P$	0.64 ± 0.67
	Change in mean	0.07 ± 0.60
Standard deviation	0.58	

In light of the results, it seems reasonable to begin a new EVOP phase about point (3). Thus, $x_1 = 225°F$ and $x_2 = 150$ psi would become the center of the 2^2 design in the second phase.

An important aspect of EVOP is feeding the information generated back to the process operators and supervisors. This is accomplished by a prominently displayed EVOP information board. The information board for this example at the end of cycle 3 is shown in Table 11.26.

Most of the quantities on the EVOP calculation sheet follow directly from the analysis of the 2^k factorial design. For example, the variance of any effect, such as $\frac{1}{2}(\bar{y}_3 + \bar{y}_5 - \bar{y}_2 - \bar{y}_4)$, is simply σ^2/n where σ^2 is the variance of the observations (y). Thus, two standard deviation (corresponding to 95 percent) error limits on any effect would be $\pm 2\sigma/\sqrt{n}$. The variance of the change in mean is

$$V(\text{CIM}) = V\left[\frac{1}{5}(\bar{y}_2 + \bar{y}_3 + \bar{y}_4 + \bar{y}_5 - 4\bar{y}_1)\right]$$

$$= \frac{1}{25}(4\sigma^2_{\bar{y}} + 16\sigma^2_{\bar{y}}) = \left(\frac{20}{25}\right)\frac{\sigma^2}{n}$$

Thus, two standard deviation error limits on the CIM are $\pm(2\sqrt{20/25})\sigma/\sqrt{n} = \pm 1.78\sigma/\sqrt{n}$.

The standard deviation σ is estimated by the range method. Let $y_i(n)$ denote the observation at the ith design point in cycle n and $\bar{y}_i(n)$ the corresponding average of $y_i(n)$ after n cycles. The quantities in row (iv) of the EVOP calculation sheet are the differences $y_i(n) - \bar{y}_i(n-1)$. The variance of these differences is

$$V[y_i(n) - \bar{y}_i(n-1)] \equiv \sigma^2_D = \sigma^2\left[1 + \frac{1}{(n-1)}\right] = \sigma^2\frac{n}{(n-1)}$$

The range of the differences, say R_D, is related to the estimate of the standard deviation of the differences by $\hat{\sigma}_D = R_D/d_2$. The factor d_2 depends on the number of observations used in computing R_D. Now $R_D/d_2 = \hat{\sigma}\sqrt{n/(n-1)}$, so

$$\hat{\sigma} = \sqrt{\frac{(n-1)}{n}}\frac{R_D}{d_2} = (f_{k,n})R_D \equiv s$$

can be used to estimate the standard deviation of the observations, where k denotes the number of points used in the design. For a 2^2 design with one center point, we have $k = 5$ and for a 2^3 design with one center point, we have $k = 9$. Values of $f_{k,n}$ are given in Table 11.27.

■ **TABLE 11.27** Values of $f_{k,n}$

$n =$	2	3	4	5	6	7	8	9	10
$k = 5$	0.30	0.35	0.37	0.38	0.39	0.40	0.40	0.40	0.41
9	0.24	0.27	0.29	0.30	0.31	0.31	0.31	0.32	0.32
10	0.23	0.26	0.28	0.29	0.30	0.30	0.30	0.31	0.31

11.8 Problems

11.1. An industrial engineer has developed a computer simulation model of a two-item inventory system. The decision variables are the order quantity and the reorder point for each item. The response to be minimized is total inventory cost. The simulation model is used to produce the data shown in Table P11.1. Identify the experimental design. Find the path of steepest descent.

■ **TABLE P11.1**
The Inventory Experiment, Problem 11.1

Item 1	
Order Quantity (ξ_1)	**Reorder Point (ξ_2)**
100	25
140	45
140	25
140	25
100	45

100	45
100	25
140	45
120	35
120	35
120	35

Item 2

Order Quantity (ξ_3)	Reorder Point (ξ_4)	Total Cost
250	40	625
250	40	670
300	40	663
250	80	654
300	40	648
250	80	634
300	80	692
300	80	686
275	60	680
275	60	674
275	60	681

11.2. A chemical plant produces oxygen by liquifying air and separating it into its component gases by fractional distillation. The purity of the oxygen is a function of the main condenser temperature and the pressure ratio between the upper and lower columns. Current operating conditions are temperature (ξ_1) = −220°C and pressure ratio (ξ_2) = 1.2. Using the following data, find the path of steepest ascent:

Temperature (ξ_1)	Pressure Ratio (ξ_2)	Purity
−225	1.1	82.8
−225	1.3	83.5
−215	1.1	84.7
−215	1.3	85.0
−220	1.2	84.1
−220	1.2	84.5
−220	1.2	83.9
−220	1.2	84.3

11.3. Verify that the following design is a simplex. Fit the first-order model and find the path of steepest ascent.

x_1	x_2	x_3	y
0	$\sqrt{2}$	−1	18.5
$-\sqrt{2}$	0	1	19.8
0	$-\sqrt{2}$	−1	17.4
$\sqrt{2}$	0	1	22.5

11.4. The region of experimentation for three factors are time ($40 \le T_1 \le 80$ min), temperature ($200 \le T_2 \le 300$°C), and pressure ($20 \le P \le 50$ psig). A first-order model in coded variables has been fit to yield data from a 2^3 design. The model is

$$\hat{y} = 30 + 5x_1 + 2.5x_2 + 3.5x_3$$

Is the point $T_1 = 85$, $T_2 = 325$, $P = 60$ on the path of steepest ascent?

11.5. For the first-order model

$$\hat{y} = 60 + 1.5x_1 - 0.8x_2 + 2.0x_3$$

find the path of steepest ascent. The variables are coded as $-1 \le x_i \le 1$.

11.6. The region of experimentation for two factors are temperature ($100 \le T \le 300$°F) and catalyst feed rate ($10 \le C \le 30$ lb/in). A first-order model in the usual ±1 coded variables has been fit to a molecular weight response, yielding the following model:

$$\hat{y} = 2000 + 125x_1 + 40x_2$$

(a) Find the path of steepest ascent.

(b) It is desired to move to a region where molecular weights are above 2500. On the basis of the information you have from experimentation in this region, about how many steps along the path of steepest ascent might be required to move to the region of interest?

11.7. The path of steepest ascent is usually computed assuming that the model is truly first order; that is, there is no interaction. However, even if there is interaction, steepest ascent ignoring the interaction still usually produces good results. To illustrate, suppose that we have fit the model

$$\hat{y} = 20 + 5x_1 - 8x_2 + 3x_1x_2$$

using coded variables ($-1 \le x_i \le +1$).

(a) Draw the path of steepest ascent that you would obtain if the interaction were ignored.

(b) Draw the path of steepest ascent that you would obtain with the interaction included in the model. Compare this with the path found in part (a).

11.8. The data shown in the Table P11.2 were collected in an experiment to optimize crystal growth as a function of three variables x_1, x_2, and x_3. Large values of y (yield in grams) are desirable. Fit a second-order model and analyze the fitted surface. Under what set of conditions is maximum growth achieved?

■ **TABLE P11.2**
The Crystal Growth Experiment, Problem 11.8

x_1	x_2	x_3	y
−1	−1	−1	65
−1	−1	1	71
−1	1	−1	79
−1	1	1	61
1	−1	−1	82
1	−1	1	71
1	1	−1	100
1	1	1	75
−1.682	0	0	103
1.682	0	0	80
0	−1.682	0	68
0	1.682	0	63
0	0	−1.682	65
0	0	1.682	82
0	0	0	115
0	0	0	101
0	0	0	118
0	0	0	88
0	0	0	100
0	0	0	85

11.9. The data in Table P11.3 were collected by a chemical engineer. The response y is filtration time, x_1 is temperature, and x_2 is pressure. Fit a second-order model.

(a) What operating conditions would you recommend if the objective is to minimize the filtration time?

(b) What operating conditions would you recommend if the objective is to operate the process at a mean filtration rate very close to 46?

■ **TABLE P11.3**
The Experiment for Problem 11.9

x_1	x_2	y
−1	−1	54
−1	1	45
1	−1	32
1	1	47
−1.414	0	50
1.414	0	53
0	−1.414	47
0	1.414	51
0	0	41
0	0	39
0	0	44
0	0	42
0	0	40

11.10. The hexagon design in Table P11.4 is used in an experiment that has the objective of fitting a second-order model:

■ **TABLE P11.4**
A Hexagon Design

x_1	x_2	y
1	0	68
0.5	$\sqrt{0.75}$	74
−0.5	$\sqrt{0.75}$	65
−1	0	60
−0.5	$-\sqrt{0.75}$	63
0.5	$-\sqrt{0.75}$	70
0	0	58
0	0	60
0	0	57
0	0	55
0	0	69

(a) Fit the second-order model.

(b) Perform the canonical analysis. What type of surface has been found?

(c) What operating conditions on x_1 and x_2 lead to the stationary point?

(d) Where would you run this process if the objective is to obtain a response that is as close to 65 as possible?

11.11. Consider the three-variable central composite design shown in Table P11.5. Analyze the data and draw conclusions, assuming that we wish to maximize conversion (y_1) with activity (y_2) between 55 and 60.

■ **TABLE P11.5**
A Three Variable CCD

Run	Time (min)	Temperature (°C)	Catalyst (%)	Conversion (%) y_1	Activity y_2
1	−1.000	−1.000	−1.000	74.00	53.20
2	1.000	−1.000	−1.000	51.00	62.90
3	−1.000	1.000	−1.000	88.00	53.40
4	1.000	1.000	−1.000	70.00	62.60
5	−1.000	−1.000	1.000	71.00	57.30
6	1.000	−1.000	1.000	90.00	67.90
7	−1.000	1.000	1.000	66.00	59.80
8	1.000	1.000	1.000	97.00	67.80
9	0.000	0.000	0.000	81.00	59.20
10	0.000	0.000	0.000	75.00	60.40
11	0.000	0.000	0.000	76.00	59.10
12	0.000	0.000	0.000	83.00	60.60
13	−1.682	0.000	0.000	76.00	59.10
14	1.682	0.000	0.000	79.00	65.90
15	0.000	−1.682	0.000	85.00	60.00
16	0.000	1.682	0.000	97.00	60.70
17	0.000	0.000	−1.682	55.00	57.40
18	0.000	0.000	1.682	81.00	63.20
19	0.000	0.000	0.000	80.00	60.80
20	0.000	0.000	0.000	91.00	58.90

11.12. An experimenter has run a Box–Behnken design and obtained the results as shown in Table P11.6, where the response variable is the viscosity of a polymer:

■ **TABLE P11.6**
The Box–Behnken Design for Problem 11.12

Level	Temp.	Agitation Rate	Pressure	x_1	x_2	x_3
High	200	10.0	25	+1	+1	+1
Middle	175	7.5	20	0	0	0
Low	150	5.0	15	−1	−1	−1

Run	x_1	x_2	x_3	y_1
1	−1	−1	0	540
2	+1	−1	0	581
3	−1	+1	0	596
4	+1	+1	0	563
5	−1	0	−1	645
6	+1	0	−1	458
7	−1	0	+1	350
8	+1	0	+1	600
9	0	−1	−1	595
10	0	+1	−1	648
11	0	−1	+1	532
12	0	+1	+1	656
13	0	0	0	653
14	0	0	0	599
15	0	0	0	620

(a) Fit the second-order model.

(b) Perform the canonical analysis. What type of surface has been found?

(c) What operating conditions on x_1, x_2, and x_3 lead to the stationary point?

(d) What operating conditions would you recommend if it is important to obtain a viscosity that is as close to 600 as possible?

11.13. A manufacturer of cutting tools has developed two empirical equations for tool life in hours (y_1) and for tool cost in dollars (y_2). Both models are linear functions of steel hardness (x_1) and manufacturing time (x_2). The two equations are

$$\hat{y} = 10 + 5x_1 + 2x_2$$
$$\hat{y}_2 = 23 + 3x_1 + 4x_2$$

and both equations are valid over the range $-1.5 \le x_i \le 1.5$. Unit tool cost must be below \$27.50 and life must exceed 12 hours for the product to be competitive. Is there a feasible set of operating conditions for this process? Where would you recommend that the process be run?

11.14. A central composite design is run in a chemical vapor deposition process, resulting in the experimental data shown in Table P11.7. Four experimental units were processed simultaneously on each run of the design, and the responses are the mean and the variance of thickness, computed across the four units.

(a) Fit a model to the mean response. Analyze the residuals.

(b) Fit a model to the variance response. Analyze the residuals.

(c) Fit a model to $\ln(s^2)$. Is this model superior to the one you found in part (b)?

■ **TABLE P11.7**
The CCD for Problem 11.14

x_1	x_2	$\bar{y}$	s^2
−1	−1	360.6	6.689
1	−1	445.2	14.230
−1	1	412.1	7.088
1	1	601.7	8.586
1.414	0	518.0	13.130
−1.414	0	411.4	6.644
0	1.414	497.6	7.649
0	−1.414	397.6	11.740
0	0	530.6	7.836
0	0	495.4	9.306
0	0	510.2	7.956
0	0	487.3	9.127

(d) Suppose you want the mean thickness to be in the interval 450 ± 25. Find a set of operating conditions that achieves this objective and simultaneously minimizes the variance.

(e) Discuss the variance minimization aspects of part (d). Have you minimized the *total* process variance?

11.15. Verify that an orthogonal first-order design is also first-order rotatable.

11.16. Show that augmenting a 2^k design with n_C center points does not affect the estimates of the β_i ($i = 1, 2, \ldots, k$) but that the estimate of the intercept β_0 is the average of all $2^k + n_c$ observations.

11.17. ***The rotatable central composite design.*** It can be shown that a second-order design is rotatable if $\Sigma_{u=1}^{n} x_{iu}^{a} x_{ju}^{b} = 0$, if a or b (or both) are odd, and if $\Sigma_{u=1}^{n} x_{iu}^{4} = 3\Sigma_{u=1}^{n} x_{iu}^{2} x_{ju}^{2}$. Show that for the central composite design these conditions lead to $\alpha = (n_F)^{1/4}$ for rotatability, where n_F is the number of points in the factorial portion.

11.18. Verify that the central composite design shown in Table P11.8 blocks orthogonally:

■ **TABLE P11.8**
A CCD in Three Blocks

Block 1			Block 2		
x_1	x_2	x_3	x_1	x_2	x_3
0	0	0	0	0	0
0	0	0	0	0	0
1	1	1	1	1	−1
1	−1	−1	1	−1	1
−1	−1	1	−1	1	1
−1	1	−1	−1	−1	−1

Block 3		
x_1	x_2	x_3
−1.633	0	0
1.633	0	0
0	−1.633	0
0	1.633	0
0	0	−1.633
0	0	1.633
0	0	0
0	0	0

11.19. ***Blocking in the central composite design.*** Consider a central composite design for $k = 4$ variables in two blocks. Can a rotatable design always be found that blocks orthogonally?

11.20. How could a hexagon design be run in two orthogonal blocks?

11.21. Yield during the first four cycles of a chemical process is shown in the following table. The variables are percentage of concentration (x_1) at levels 30, 31, and 32 and temperature (x_2) at 140, 142, and 144°F. Analyze by EVOP methods.

	Conditions				
Cycle	(1)	(2)	(3)	(4)	(5)
1	60.7	59.8	60.2	64.2	57.5
2	59.1	62.8	62.5	64.6	58.3
3	56.6	59.1	59.0	62.3	61.1
4	60.5	59.8	64.5	61.0	60.1

11.22. Suppose that we approximate a response surface with a model of order d_1, such as $\mathbf{y} = \mathbf{X}_1\boldsymbol{\beta}_1 + \boldsymbol{\epsilon}$, when the true surface is described by a model of order $d_2 > d_1$; that is, $E(y) = \mathbf{X}_1\boldsymbol{\beta}_1 + \mathbf{X}_1\boldsymbol{\beta}_2$.

(a) Show that the regression coefficients are biased, that is, $E(\hat{\boldsymbol{\beta}}_1) = \boldsymbol{\beta}_1 + \mathbf{A}\boldsymbol{\beta}_2$, where $\mathbf{A} = (\mathbf{X}_1'\mathbf{X}_1)^{-1}\mathbf{X}_1'\mathbf{X}_2$. **A** is usually called the alias matrix.

(b) If $d_1 = 1$ and $d_2 = 2$, and a full 2^k is used to fit the model, use the result in part (a) to determine the alias structure.

(c) If $d_1 = 1$, $d_2 = 2$, and $k = 3$, find the alias structure assuming that a 2^{3-1} design is used to fit the model.

(d) If $d_1 = 1$, $d_2 = 2$, and $k = 3$, and the simplex design in Problem 11.3 is used to fit the model, determine the alias structure and compare the results with part (c).

11.23. Suppose that you need to design an experiment to fit a quadratic model over the region $-1 \leq x_i \leq +1$, $i = 1, 2$ subject to the constraint $x_1 + x_2 \leq 1$. If the constraint is violated, the process will not work properly. You can afford to make no more than $n = 12$ runs. Set up the following designs:

(a) An "inscribed" CCD with center point at $x_1 = x_2 = 0$.

(b) An "inscribed" 3^2 factorial with center point at $x_1 = x_2 = -0.25$.

(c) A *D*-optimal design.

(d) A modified *D*-optimal design that is identical to the one in part (c), but with all replicate runs at the design center.

(e) Evaluate the $|(\mathbf{X}'\mathbf{X})^{-1}|$ criterion for each design.

(f) Evaluate the *D*-efficiency for each design relative to the *D*-optimal design in part (c).

(g) Which design would you prefer? Why?

11.24. Consider a 2^3 design for fitting a first-order model.

(a) Evaluate the *D*-criterion $|(\mathbf{X}'\mathbf{X})^{-1}|$ for this design.

(b) Evaluate the *A*-criterion $\text{tr}(\mathbf{X}'\mathbf{X})^{-1}$ for this design.

(c) Find the maximum scaled prediction variance for this design. Is this design G-optimal?

11.25. Repeat Problem 11.24 using a first-order model with the two-factor interactions.

11.26. A chemical engineer wishes to fit a calibration curve for a new procedure used to measure the concentration of a particular ingredient in a product manufactured in his facility. Twelve samples can be prepared, having known concentration. The engineer wants to build a model for the measured concentrations. He or she suspects that a linear calibration curve will be adequate to model the measured concentration as a function of the known concentrations; that is, $y = \beta_0 + \beta_1 x + \epsilon$, where x is the actual concentration. Four experimental designs are under consideration. Design 1 consists of six runs at known concentration 1 and six runs at known concentration 10. Design 2 consists of four runs at concentrations 1, 5.5, and 10. Design 3 consists of three runs at concentration 1, 4, 7, and 10. Finally, design 4 consists of three runs at concentrations 1 and 10 and 6 runs at concentration 5.5.

(a) Plot the scaled variance of prediction for all four designs on the same graph over the concentration range $1 \leq x \leq 10$. Which design would be preferable?

(b) For each design, calculate the determinant of $(\mathbf{X}'\mathbf{X})^{-1}$. Which design would be preferred according to the *D*-criterion?

(c) Calculate the *D*-efficiency of each design relative to the "best" design that you found in part (b).

(d) For each design, calculate the average variance of prediction over the set of points given by $x = 1, 1.5, 2, 2.5, \ldots, 10$. Which design would you prefer according to the *V*-criterion?

(e) Calculate the *V*-efficiency of each design relative to the best design that you found in part (d).

(f) What is the *G*-efficiency of each design?

11.27. Rework Problem 11.26 assuming that the model the engineer wishes to fit is a quadratic. Obviously, only designs 2, 3, and 4 can now be considered.

11.28. Suppose that you want to fit a second-order model in $k = 5$ factors. You cannot afford more than 25 runs. Construct both a *D*-optimal and on I-optimal design for this situation. Compare the prediction variance properties of the designs. Which design would you prefer?

11.29. Suppose that you want to fit a *second*-order response surface model in a situation where there are $k = 4$ factors; however, one of the factors is categorical with two levels. What model should you consider for this experiment? Suggest an appropriate design for this situation.

11.30. An experimenter wishes to run a three-component mixture experiment. The constraints in the component proportions are as follows:

$$0.2 \leq x_1 \leq 0.4$$
$$0.1 \leq x_2 \leq 0.3$$
$$0.4 \leq x_2 \leq 0.7$$

(a) Set up an experiment to fit a quadratic mixture model. Use $n = 14$ runs, with four replicates. Use the *D*-criterion.

(b) Draw the experimental region.

(c) Set up an experiment to fit a quadratic mixture model with $n = 12$ runs, assuming that three of these runs are replicates. Use the *D*-criterion.

(d) Comment on the two designs you have found.

11.31. Myers, Montgomery and Anderson-Cook (2009) describe a gasoline blending experiment involving three mixture components. There are no constraints on the mixture proportions, and the following 10-run design is used:

Design Point	x_1	x_2	x_3	y (mi/gal)
1	1	0	0	24.5, 25.1
2	0	1	0	24.8, 23.9
3	0	0	1	22.7, 23.6
4	$\frac{1}{2}$	$\frac{1}{2}$	0	25.1
5	$\frac{1}{2}$	0	$\frac{1}{2}$	24.3
6	0	$\frac{1}{2}$	$\frac{1}{2}$	23.5
7	$\frac{1}{3}$	$\frac{1}{3}$	$\frac{1}{3}$	24.8, 24.1
8	$\frac{2}{3}$	$\frac{1}{6}$	$\frac{1}{6}$	24.2
9	$\frac{1}{6}$	$\frac{2}{3}$	$\frac{1}{6}$	23.9
10	$\frac{1}{6}$	$\frac{1}{6}$	$\frac{2}{3}$	23.7

(a) What type of design did the experimenters use?

(b) Fit a quadratic mixture model to the data. Is this model adequate?

(c) Plot the response surface contours. What blend would you recommend to maximize the miles per gallon?

11.32. An article in *Quality Progress* ("For Starbucks, It's in the Bag," March 2011, pp. 18–23) describes using a central composite design to improve the packaging of one-pound coffee. The objective is to produce an airtight seal that is easy to open without damaging the top of the coffee bag. The experimenters studied three factors; x_1 = plastic viscosity (300–400 centipoise), x_2 = clamp pressure (170–190 psi), and x_3 = plate gap (−3, +3 mm) and two responses, y_1 = tear and y_2 = leakage. The design is shown in Table P11.9. The tear response was measure on a scale from 0–9 (good to bad) and leakage was proportion failing. Each run used a sample of 20 bags for response measurement.

(a) Build a second-order model for the tear response.

(b) Build a second-order model for the leakage response.

(c) Analyze the residuals for both models. Do transformations seem necessary for either response? If so, refit the models in the transformed metric.

(d) Construct response surface plots and contour plots for both responses. Provide interpretations for the fitted surfaces.

(e) What conditions would you recommend for process operation to minimize leakage and keep tear below 0.75?

■ **TABLE P11.9**
The Coffee Bag Experiment in Problem 11.32

Run	Viscosity	Pressure	Plate gap	Tear	Leakage
Center	350	180	0	0	0.15
Axial	350	170	0	0	0.5
Factorial	319	186	1.8	0.45	0.15
Factorial	380	174	1.8	0.85	0.05
Center	350	180	0	0.35	0.15
Axial	300	180	0	0.3	0.45
Axial	400	180	0	0.7	0.25
Axial	350	190	0	1.9	0
Center	350	180	0	0.25	0.05
Factorial	319	186	−1.8	0.1	0.35
Factorial	380	186	−1.8	0.15	0.4
Axial	350	180	3	3.9	0
Factorial	380	174	−1.8	0	0.45
Center	350	180	0	0.55	0.2
Axial	350	180	−3	0	1
Factorial	319	174	−1.8	0.05	0.2
Factorial	319	174	1.8	0.4	0.25
Factorial	380	186	1.8	4.3	0.05
Center	350	180	0	0	0

11.33. Table P11.10 shows a six-variable RSM design from Jones and Nachtsheim (2011). Analyze the response data from this experiment.

■ **TABLE P11.10**
The Design for Problem 11.31

Run (i)	$x_{i.1}$	$x_{i.2}$	$x_{i.3}$	$x_{i.4}$	$x_{i.5}$	$x_{i.6}$	y_i
1	0	1	−1	−1	−1	−1	21.04
2	0	−1	1	1	1	1	10.48
3	1	0	−1	1	1	−1	17.89
4	−1	0	1	−1	−1	1	10.07
5	−1	−1	0	1	−1	−1	7.74
6	1	1	0	−1	1	1	21.01
7	−1	1	1	0	1	−1	16.53
8	1	−1	−1	0	−1	1	20.38
9	1	−1	1	−1	0	−1	8.62
10	−1	1	−1	1	0	1	7.80
11	1	1	1	1	−1	0	23.56
12	−1	−1	−1	−1	1	0	15.24
13	0	0	0	0	0	0	19.91

11.34. Box and Liu (1999) describe an experiment flying paper helicopters where the objective is to maximize flight time. They used the central composite design shown in Table P11.11. Each run involved a single helicopter made to the following specifications: x_1 = wing area (in^2), −1 = 11.80 and +1 = 13.00; x_2 = wing-length to width ratio, −1 = 2.25 and +1 = 2.78; x_3 = base width (in), −1 = 1.00 and +1 = 1.50; and x_4 = base length (in), −1 = 1.50 and +1 = 2.50. Each helicopter was flown four times and the average flight time and the standard deviation of flight time was recorded.

(a) Fit a second-order model to the average flight time response.

(b) Fit a second-order model to the standard deviation of flight time response.

(c) Analyze the residuals for both models from parts (a) and (b). Are transformations on the response(s) necessary? If so, fit the appropriate models.

(d) What design would you recommend to maximize the flight time?

(e) What design would you recommend to maximize the flight time while simultaneously minimizing the standard deviation of flight time?

■ **TABLE P11.11**
The Paper Helicopter Experiment

Std. order	Run order	Wing area	Wing ratio	Base width	Base length	Avg. flight time	Std. Dev flight time
1	9	−1	−1	−1	−1	3.67	0.052
2	21	1	−1	−1	−1	3.69	0.052
3	14	−1	1	−1	−1	3.74	0.055
4	4	1	1	−1	−1	3.7	0.062
5	2	−1	−1	1	−1	3.72	0.052
6	19	1	−1	1	−1	3.55	0.065
7	22	−1	1	1	−1	3.97	0.052
8	25	1	1	1	−1	3.77	0.098
9	27	−1	−1	−1	1	3.5	0.079
10	13	1	−1	−1	1	3.73	0.072
11	20	−1	1	−1	1	3.58	0.083
12	6	1	1	−1	1	3.63	0.132
13	12	−1	−1	1	1	3.44	0.058
14	17	1	−1	1	1	3.55	0.049
15	26	−1	1	1	1	3.7	0.081
16	1	1	1	1	1	3.62	0.051
17	8	−2	0	0	0	3.61	0.129
18	15	2	0	0	0	3.64	0.085
19	7	0	−2	0	0	3.55	0.1
20	5	0	2	0	0	3.73	0.063
21	29	0	0	−2	0	3.61	0.051
22	28	0	0	2	0	3.6	0.095
23	16	0	0	0	−2	3.8	0.049
24	18	0	0	0	2	3.6	0.055
25	24	0	0	0	0	3.77	0.032
26	10	0	0	0	0	3.75	0.055
27	23	0	0	0	0	3.7	0.072
28	11	0	0	0	0	3.68	0.055
29	3	0	0	0	0	3.69	0.078
30	30	0	0	0	0	3.66	0.058

11.35. The Paper Helicopter Experiment Revisited. Reconsider the paper helicopter experiment in Problem 11.34. This experiment was actually run in two blocks. Block 1 consisted of the first 16 runs in Table P11.11 (standard order runs 1–16) and two center points (standard order runs 25 and 26).

(a) Fit main-effects plus two-factor interaction models to the block 1 data, using both responses.

(b) For the models in part (a) use the two center points to test for lack of fit. Is there an indication that second-order terms are needed?

(c) Now use the data from block 2 (standard order runs 17–24 and the remaining center points, standard order runs 27–30) to augment block 1 and fit second-order models to both responses. Check the adequacy of the fit for both models. Does blocking seem to have been important in this experiment?

(d) What design would you recommend to maximize the flight time while simultaneously minimizing the standard deviation of flight time?

11.36. An article in the *Electronic Journal of Biotechnology* ("Optimization of Medium Composition for Transglutaminase Production by a Brazilian Soil *Streptomyces* sp," available at http://www.ejbiotechnology.info/content/vol10/issue4/full/10.index.html) describes the use of designed experiments to improve the medium for cells used in a new microbial source of transglutaminase (MTGase), an enzyme that catalyzes an acyl transfer reaction using peptide-bond glutamine residues as acyl donors and some primary amines as acceptors. Reactions catalyzed by MTGase can be used in food processing. The article describes two phases of experimentation; screening with a fractional factorial and optimization. We will use only the optimization experiment. The design was a central composite design in four factors; x_1 = KH_2PO_4, x_2 = MgSO4.7H₂O, x_3 = soybean flower, and x_4 = peptone. MTGase activity is the response, which should be maximized. Table P11.12 contains the design and the response data.

■ TABLE P11.12
The MTGase Optimization Experiment

Standard Order	X1	X2	X3	X4	MTGase activity
1	−1	−1	−1	−1	0.87
2	1	−1	−1	−1	0.74
3	−1	1	−1	−1	0.51
4	1	1	−1	−1	0.99
5	−1	−1	1	−1	0.67
6	1	−1	1	−1	0.72
7	−1	1	1	−1	0.81
8	1	1	1	−1	1.01
9	−1	−1	−1	1	1.33
10	1	−1	−1	1	0.7
11	−1	1	−1	1	0.82
12	1	1	−1	1	0.78
13	−1	−1	1	1	0.36
14	1	−1	1	1	0.23
15	−1	1	1	1	0.21
16	1	1	1	1	0.44
17	−2	0	0	0	0.56
18	2	0	0	0	0.49
19	0	−2	0	0	0.57
20	0	2	0	0	0.81
21	0	0	−2	0	0.9
22	0	0	2	0	0.65
23	0	0	0	−2	0.91
24	0	0	0	2	0.49
25	0	0	0	0	1.43
26	0	0	0	0	1.17
27	0	0	0	0	1.5

(a) Fit a second-order model to the MTGase activity response.

(b) Analyze the residuals from this model.

(c) Recommend operating conditions that maximize MTGase activity.

11.37. An article in the *Journal of Chromatography A* ("Optimization of the Capillary Electrophoresis Separation of Ranitidine and Related Compounds," Vol. 766, pp. 245–254) describes an experiment to optimize the production of ranitidine, a compound that is the primary active ingredient of Zantac, a pharmaceutical product used to treat ulcers, gastroesophageal reflux disease (a condition in which backward flow of acid from the stomach causes heartburn and injury of the esophagus), and other conditions where the stomach produces too much acid, such as Zollinger–Ellison syndrome. The authors used three factors (x_1 = pH of the buffer solution, x_2 = the electrophoresis voltage, and the concentration of one component of the buffer solution) in a central composite design. The response is chromatographic exponential function (CEF), which should be minimized. Table P11.13 shows the design.

(a) Fit a second-order model to the CEF response. Analyze the residuals from this model. Does it seem that all model terms are necessary?

(b) Reduce the model from part (a) as necessary. Did model reduction improve the fit?

(c) Does transformation of the CEF response seem like a useful idea? What aspect of either the data or the residual analysis suggests that transformation would be helpful?

(d) Fit a second-order model to the transformed CEF response. Analyze the residuals from this model. Does it seem that all model terms are necessary? What would you choose as the final model?

(e) Suppose that you had some information that suggests that the separation process malfunctioned during run 7. Delete this run and analyze the data from this experiment again.

(f) What conditions would you recommend to minimize CEF?

■ **TABLE P11.13**
The Ranitidine Separation Experiment

Standard Order	X1	X2	X3	CEF
1	−1	−1	−1	17.3
2	1	−1	−1	45.5
3	−1	1	−1	10.3
4	1	1	−1	11757.1
5	−1	−1	1	16.942
6	1	−1	1	25.4
7	−1	1	1	31697.2
8	1	1	1	12039.2
9	−1.68	0	0	7.5
10	1.68	0	0	6.3
11	0	−1.68	0	11.1
12	0	1.68	0	6.664
13	0	0	−1.68	16548.7
14	0	0	1.68	26351.8
15	0	0	0	9.9
16	0	0	0	9.6
17	0	0	0	8.9
18	0	0	0	8.8
19	0	0	0	8.013
20	0	0	0	8.059

CHAPTER 12

Robust Design

CHAPTER OUTLINE

The supplemental material is on the textbook website www.wiley.com/go/global/montgomery.

12.1 Introduction

Robust parameter design (RPD) is an approach to product realization activities that focuses on choosing the levels of controllable factors (or parameters) in a process or a product to achieve two objectives: (1) to ensure that the mean of the output response is at a desired level or target and (2) to ensure that the variability around this target value is as small as possible. When an RPD study is conducted on a process, it is usually called a **process robustness study**. The general RPD problem was developed by a Japanese engineer, Genichi Taguchi, and introduced in the United States in the 1980s (see Taguchi and Wu, 1980; Taguchi, 1987). Taguchi proposed an approach to solving the RPD problem based on designed experiments and some novel methods for analysis of the resulting data. His philosophy and technical methods generated widespread interest among engineers and statisticians, and during the 1980s his methodology was used at many large corporations, including AT&T Bell Laboratories, Ford Motor Company, and Xerox. These techniques generated controversy and debate in the statistical and engineering communities. The controversy was not about the basic RPD problem, which is an extremely important one, but rather about the experimental procedures and the data analysis methods that Taguchi advocated. Extensive analysis revealed that Taguchi's technical methods were usually inefficient and, in many cases, ineffective. Consequently, a period of extensive research and development on new approaches to the RPD problem followed. From these efforts, response surface methodology (RSM) emerged as an approach to the RPD problem that not only allows us to employ Taguchi's robust design concept but also provides a sounder and more efficient approach to design and analysis.

This chapter is about the RSM approach to the RPD problem. More information about the original Taguchi approach, including discussion that identifies the pitfalls and inefficiencies of his methods, is presented in the supplemental text material for this chapter. Other useful references include Hunter (1985, 1989), Box (1988), Box, Bisgaard, and Fung (1988), Pignatiello and Ramberg (1992), Montgomery (1999), Myers, Montgomery and Anderson-Cook (2009), and the panel discussion edited by Nair (1992).

In a robust design problem, the focus is usually on one or more of the following:

1. Designing systems that are insensitive to environmental factors that can affect performance once the system is deployed in the field. An example is the development of an exterior paint that should exhibit long life when exposed to a variety of weather conditions. Because the weather conditions are not entirely predictable, and certainly not constant, the product formulator wants the paint to be robust against or withstand a wide range of temperature, humidity, and precipitation factors that affect the wear and finish of the paint.
2. Designing products so that they are insensitive to variability transmitted by the components of the system. An example is designing an electronic amplifier so that the output voltage is as close as possible to the desired target regardless of the variability in the electrical parameters of the transistors, resistors, and power supplies that are the components of the system.
3. Designing processes so that the manufactured product will be as close as possible to the desired target specifications, even though some process variables (such as temperature) or raw material properties are impossible to control precisely.
4. Determining the operating conditions for a process so that the critical process characteristics are as close as possible to the desired target values and the variability around this target is minimized. Examples of this type of problem occur frequently. For example, in semiconductor manufacturing we want the oxide thickness on a wafer to be as close as possible to the target mean thickness, and we want the variability in thickness across the wafer (a measure of uniformity) to be as small as possible.

RPD problems are not new. Product and process designers/developers have been concerned about robustness issues for decades, and efforts to solve the problem long predate Taguchi's contributions. One of the classical approaches used to achieve robustness is to redesign the product using stronger components, or components with tighter tolerances, or to use different materials. However, this may lead to problems with **overdesign**, resulting in a product that is more expensive, more difficult to manufacture, or suffers a weight and subsequent performance penalty. Sometimes different design methods or incorporation of new technology into the design can be exploited. For example, for many years automobile speedometers were driven by a metal cable, and over time the lubricant in the cable deteriorated, which could lead to operating noise in cold weather or erratic measurement of vehicle speed. Sometimes the cable would break, resulting in an expensive repair. This is an example of robustness problems caused by product aging. Modern automobiles use electronic speedometers that are not subject to these problems. In a process environment, older equipment may be replaced with new tools, which may improve process robustness but usually at a significant cost. Another possibility may be to exercise tighter control over the variables that impact robustness. For example, if variations in environmental conditions cause problems with robustness, then those conditions may have to be controlled more tightly. The use of clean rooms in semiconductor manufacturing is a result of efforts to control environmental conditions. In some cases, effort will be directed to controlling raw material properties or process variables more tightly if these factors impact robustness. These classical approaches are still useful, but Taguchi's principal contribution was the recognition that experimental design and other statistical tools could be applied to the problem in many cases.

An important aspect of Taguchi's approach was his notion that certain types of variables cause variability in the important system response variables. We refer to these types of variables as **noise variables** or **uncontrollable variables**. We have discussed this concept before—for example, see Figure 1.1. These noise factors are often functions of environmental conditions such as temperature or relative humidity. They may be properties of raw materials that vary from batch to batch or over time in the process. They may be process variables that are difficult to control or to keep at specified targets. In some cases, they may involve the way the consumer handles or uses the product. Noise variables may often be controlled at the research or development level, but they cannot be controlled at the production or product use level. An integral part of the RPD problem is identifying the controllable variables and the noise variables that affect process or product performance and then finding the settings for the controllable variables that minimize the variability transmitted from the noise variables.

As an illustration of controllable and noise variables, consider a product developer who is formulating a cake mix. The developer must specify the ingredients and composition of the cake mix, including the amounts of flour, sugar, dry milk, hydrogenated oils, corn starch, and flavorings. These variables can be controlled reasonably easily when the cake mix is manufactured. When the consumer bakes the cake, water and eggs are added, the mixture of wet and dry ingredients is blended into cake batter, and the cake is baked in an oven at a specified temperature for a specified time. The product formulator cannot control exactly how much water is added to the dry cake mix, how well the wet and dry ingredients are blended, or the exact baking time or oven temperature. Target values for these variables can be and usually are specified, but they are really noise variables, as there will be variation (perhaps considerable variation) in the levels of these factors that are used by different customers. Therefore, the product formulator has a robust design problem. The objective is to formulate a cake mix that will perform well and meet or exceed customer expectations regardless of the variability transmitted into the final cake by the noise variables.

12.2 Crossed Array Designs

The original Taguchi methodology for the RPD problem revolved around the use of a statistical design for the controllable variables and another statistical design for the noise variables. Then these two designs were "crossed"; that is, every treatment combination in the design for the controllable variables was run in combination with every treatment combination in the noise variable design. This type of experimental design was called a **crossed array design**.

We will illustrate the crossed array design approach using the leaf spring experiment originally introduced as Problem 8.10. In this experiment, five factors were studied to determine their effect on the free height of a leaf spring used in an automotive application. There were five factors in the experiment; A = furnace temperature, B = heating time, C = transfer time, D = hold down time, and E = quench oil temperature. This was originally an RPD problem, and quench oil temperature was the noise variable. The data from this experiment are shown in Table 12.1. The design for the controllable factors is a 2^{4-1} fractional factorial design with generator $D = ABC$. This is called the **inner array** design. The design for the single noise factor is a 2^1 design, and it is called the **outer array** design. Notice how each run in the outer array is performed for all eight treatment combinations in the inner array, producing the crossed array structure. In the leaf spring experiment, each of the 16 distinct design points was replicated three times, resulting in 48 observations on free height.

An important point about the crossed array design is that it provides information about interactions between controllable factors and noise factors. These interactions are crucial to the

■ **TABLE 12.1**
The Leaf Spring Experiment

A	*B*	*C*	*D*	*E* = −	*E* = +	$\bar{y}$	s^2
−	−	−	−	7.78, 7.78, 7.81	7.50, 7.25, 7.12	7.54	0.090
+	−	−	+	8.15, 8.18, 7.88	7.88, 7.88, 7.44	7.90	0.071
−	+	−	+	7.50, 7.56, 7.50	7.50, 7.56, 7.50	7.52	0.001
+	+	−	−	7.59, 7.56, 7.75	7.63, 7.75, 7.56	7.64	0.008
−	−	+	+	7.54, 8.00, 7.88	7.32, 7.44, 7.44	7.60	0.074
+	−	+	−	7.69, 8.09, 8.06	7.56, 7.69, 7.62	7.79	0.053
−	+	+	−	7.56, 7.52, 7.44	7.18, 7.18, 7.25	7.36	0.030
+	+	+	+	7.56, 7.81, 7.69	7.81, 7.50, 7.59	7.66	0.017

solution of an RPD problem. For example, consider the two-factor interaction graphs in Figure 12.1, where x is the controllable factor and z is the noise factor. In Figure 12.1*a*, there is no interaction between x and z; therefore, there is no setting for the controllable variable x that will affect the variability transmitted to the response by the variability in the noise factor z. However, in Figure 12.1*b*, there is a strong interaction between x and z. Note that when x is set to its low level, there is much less variability in the response variable than when x is at the high level. Thus, unless there is at least one controllable factor—noise factor interaction—there is no robust design problem. As we will subsequently see, focusing on identifying and modeling these interactions is one of the keys to an efficient and effective approach to solving the RPD problem.

Table 12.2 presents another example of an RPD problem, taken from Byrne and Taguchi (1987). This problem involved the development of an elastometric connector that would deliver the required pull-off force when assembled with a nylon tube. There are four controllable factors, each at three levels (A = interference, B = connector wall thickness, C = insertion depth, and D = percent adhesive), and three noise or uncontrollable factors, each at two levels (E = conditioning time, F = conditioning temperature, and G = conditioning relative humidity). Panel (*a*) of Table 12.2 contains the inner array design for the controllable factors. Notice that the design is a three-level fractional factorial, and specifically, it is a 3^{4-2} design. Panel (*b*) of Table 12.2 contains a 2^3 outer array design for the noise factors. Now as before, each run in the inner

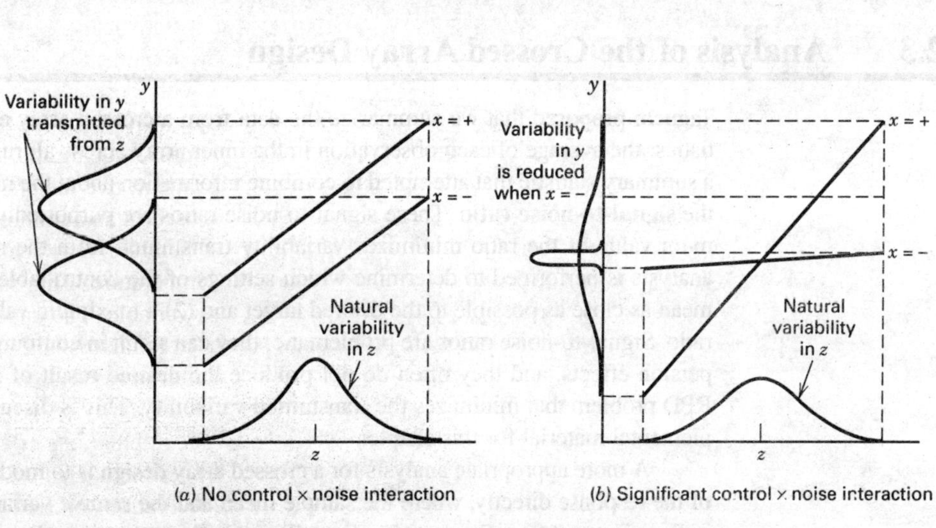

■ **FIGURE 12.1** **The role of the control × noise interaction in robust design**

■ **TABLE 12.2**
The Design for the Connector Pull-Off Force Experiment

(a) Inner Array					(b) Outer Array							
				E	−	−	−	−	+	+	+	+
				F	−	−	+	+	−	−	+	+
				G	−	+	−	+	−	−	−	+
Run	*A*	*B*	*C*	*D*								
1	−1	−1	−1	−1	15.6	9.5	16.9	19.9	19.6	19.6	20.0	19.1
2	−1	0	0	0	15.0	16.2	19.4	19.2	19.7	19.8	24.2	21.9
3	−1	+1	+1	+1	16.3	16.7	19.1	15.6	22.6	18.2	23.3	20.4
4	0	−1	0	+1	18.3	17.4	18.9	18.6	21.0	18.9	23.2	24.7
5	0	0	+1	−1	19.7	18.6	19.4	25.1	25.6	21.4	27.5	25.3
6	0	+1	−1	0	16.2	16.3	20.0	19.8	14.7	19.6	22.5	24.7
7	+1	−1	+1	0	16.4	19.1	18.4	23.6	16.8	18.6	24.3	21.6
8	+1	0	−1	+1	14.2	15.6	15.1	16.8	17.8	19.6	23.2	24.2
9	+1	+1	0	−1	16.1	19.9	19.3	17.3	23.1	22.7	22.6	28.6

array is performed for all treatment combinations in the outer array, producing the crossed array design with 72 observations on pull-off force shown in the table.

Examination of the crossed array design in Table 12.2 reveals a major problem with the Taguchi design strategy; namely, the crossed array approach can lead to a very large experiment. In our example, there are only seven factors, yet the design has 72 runs. Furthermore, the inner array design is a 3^{4-2} resolution III design (see Chapter 9 for discussion of this design), so in spite of the large number of runs, we cannot obtain any information about interactions among the controllable variables. Indeed, even information about the main effects is potentially tainted because the main effects are heavily aliased with the two-factor interactions. In Section 12.4, we will introduce the **combined array** design, which is generally much more efficient than the crossed array.

12.3 Analysis of the Crossed Array Design

Taguchi proposed that we summarize the data from a crossed array experiment with two statistics: the average of each observation in the inner array across all runs in the outer array and a summary statistic that attempted to combine information about the mean and variance, called the **signal-to-noise ratio**. These signal-to-noise ratios are purportedly defined so that a maximum value of the ratio minimizes variability transmitted from the noise variables. Then an analysis is performed to determine which settings of the controllable factors result in (1) the mean as close as possible to the desired target and (2) a maximum value of the signal-to-noise ratio. Signal-to-noise ratios are problematic; they can result in confounding of location and dispersion effects, and they often do not produce the desired result of finding a solution to the RPD problem that minimizes the transmitted variability. This is discussed in detail in the supplemental material for this chapter.

A more appropriate analysis for a crossed array design is to model the mean and variance of the response directly, where the sample mean and the sample variance for each observation in the inner array are computed across all runs in the outer array. Because of the crossed array structure, the sample means $\bar{y}_i$ and variances s_i^2 are computed over the same levels of the noise variables, so any differences between these quantities are due to differences in the levels of the controllable variables. Consequently, choosing the levels of the controllable variables to optimize the mean and simultaneously minimize the variability is a valid approach.

To illustrate this approach, consider the leaf spring experiment in Table 12.1. The last two columns of this table show the sample means $\bar{y}_i$ and variances s_i^2 for each run in the inner array. Figure 12.2 is the half-normal probability plot of the effects for the mean free height response. Clearly, factors *A*, *B*, and *D* have important effects. Since these factors are aliased with three-factor interactions, it seems reasonable to conclude that these effects are real. The model for the mean free height response is

$$\hat{\bar{y}}_i = 7.63 + 0.12x_1 - 0.081x_2 + 0.044x_4$$

where the *x*'s represent the original design factors *A*, *B*, and *D*. Because the sample variance does not have a normal distribution (it is scaled chi-square), it is usually best to analyze the natural log of the variance. Figure 12.3 is the half-normal probability plot of the effects of the $\ln(s_i^2)$ response. The only significant effect is factor *B*. The model for the $\ln(s_i^2)$ response is

$$\widehat{\ln(s_i^2)} = -3.74 - 1.09x_2$$

Figure 12.4 is a contour plot of the mean free height in terms of factors *A* and *B* with factor $D = 0$, and Figure 12.5 is a plot of the variance response in the original scale. Clearly, the variance of the free height decreases as the heating time (factor *B*) increases.

Suppose that the objective of the experimenter is to find a set of conditions that results in a mean free height between 7.74 and 7.76 inches, with minimum variability. This is a standard multiple response optimization problem and can be solved by any of the methods for solving these problems described in Chapter 11. Figure 12.6 is an overlay plot of the two responses, with factor D = hold down time held constant at the high level. By also selecting

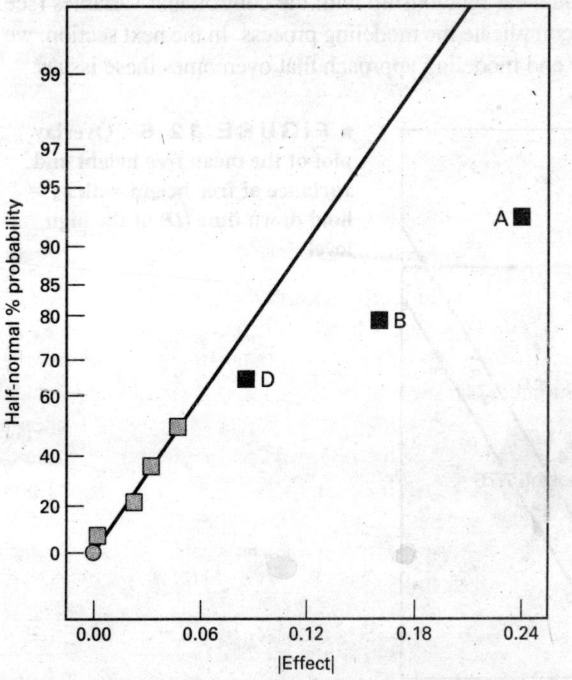

■ **FIGURE 12.2 Half-normal plot of effect, mean free height response**

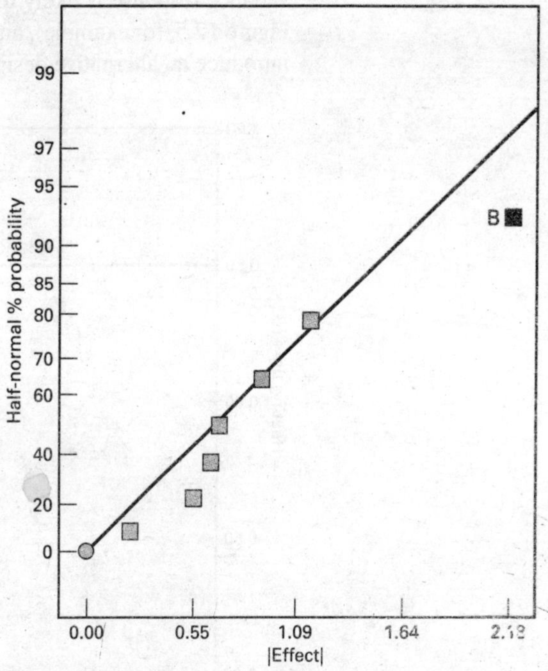

■ **FIGURE 12.3 Half-normal plot of effects, $\ln(s_i^2)$ response**

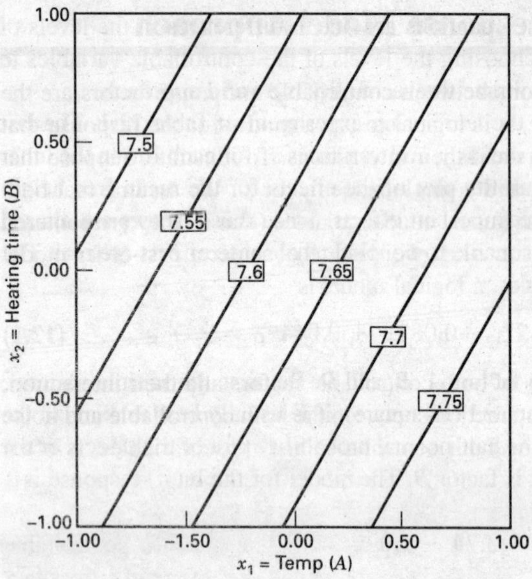

■ **FIGURE 12.4 Contour plot of the mean free height response with D = hold down time = 0**

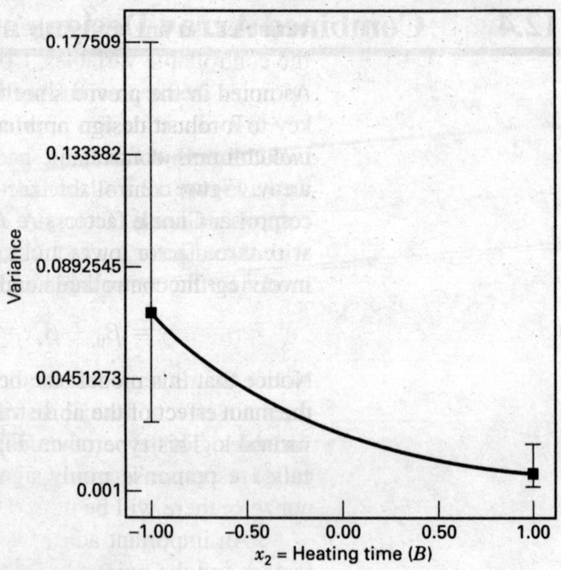

■ **FIGURE 12.5 Plot of the variance of free height versus x = heating time (B)**

A = temperature at the high level and B = heating time at 0.50 (in coded units), we can achieve a mean free height between the desired limits with variance of approximately 0.0138.

A disadvantage of the mean and variance modeling approach using the crossed array design is that it does not take direct advantage of the interactions between controllable variables and noise variables. In some instances, it can even mask these relationships. Furthermore, the variance response is likely to have a nonlinear relationship with the controllable variables (see Figure 12.5, for example), and this can complicate the modeling process. In the next section, we introduce an alternative design strategy and modeling approach that overcomes these issues.

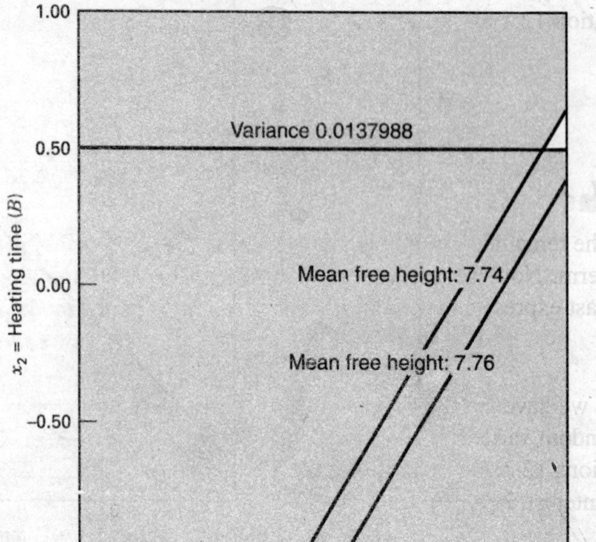

■ **FIGURE 12.6 Overlay plot of the mean free height and variance of free height with x_4 = hold down time (D) at the high level**

12.4 Combined Array Designs and the Response Model Approach

As noted in the previous section, interactions between controllable and noise factors are the key to a robust design problem. Therefore, it is logical to use a **model** for the response that includes both controllable and noise factors and their interactions. To illustrate, suppose that we have two controllable factors x_1 and x_2 and a single noise factor z_1. We assume that both control and noise factors are expressed as the usual coded variables (that is, they are centered at zero and have lower and upper limits at $\pm a$). If we wish to consider a first-order model involving the controllable and noise variables, a logical model is

$$y = \beta_0 + \beta_1 x_1 + \beta_2 x_2 + \beta_{12} x_1 x_2 + \gamma_1 z_1 + \delta_{11} x_1 z_1 + \delta_{21} x_2 z_1 + \epsilon \quad (12.1)$$

Notice that this model has the main effects of both controllable factors and their interaction, the main effect of the noise variable, and interactions between the both controllable and noise variables. This type of model, incorporating both controllable and noise variables, is often called a **response model**. Unless at least one of the regression coefficients δ_{11} and δ_{21} is nonzero, there will be no robust design problem.

An important advantage of the response model approach is that both the controllable factors and the noise factors can be placed in a single experimental design; that is, the inner and outer array structure of the Taguchi approach can be avoided. We usually call the design containing both controllable and noise factors a **combined array design**.

As mentioned previously, we assume that noise variables are random variables, although they are controllable for purposes of an experiment. Specifically, we assume that the noise variables are expressed in coded units, they have expected value zero, and variance σ_z^2, and if there are several noise variables, they have zero covariances. Under these assumptions, it is easy to find a model for the mean response just by taking the expected value of y in Equation 12.1. This yields

$$E_z(y) = \beta_0 + \beta_1 x_1 + \beta_2 x_2 + \beta_{12} x_1 x_2 \quad (12.2)$$

where the z subscript on the expectation operator is a reminder to take expected value with respect to *both* random variables in Equation 12.1, z_1 *and* ϵ. To find a model for the variance of the response y, we **use the transmission of error approach**. First, expand the response model Equation 12.1 in a first-order Taylor series around $z_1 = 0$. This gives

$$\begin{aligned} y &\cong y_{z=0} + \frac{dy}{dz_1}(z_1 - 0) + R + \epsilon \\ &\cong \beta_0 + \beta_1 x_1 + \beta_2 x_2 + \beta_{12} x_1 x_2 \\ &\quad + (\gamma_1 + \delta_{11} x_1 + \delta_{21} x_2) z_1 + R + \epsilon \end{aligned}$$

where R is the remainder term in the Taylor series. As is the usual practice, we will ignore the remainder term. Now the variance of y can be obtained by applying the variance operator across this last expression (without R). The resulting variance model is

$$V_z(y) = \sigma_z^2(\gamma_1 + \delta_{11} x_1 + \delta_{21} x_2)^2 + \sigma^2 \quad (12.3)$$

Once again, we have used the z subscript on the variance operator as a reminder that *both* z_1 *and* ϵ are random variables.

Equations 12.2 and 12.3 are simple models for the mean and variance of the response variable of interest. Note the following:

1. The mean and variance models involve **only the controllable variables**. This means that we can potentially set the controllable variables to achieve a target value of the mean and minimize the variability transmitted by the noise variable.

2. Although the variance model involves only the controllable variables, it also involves the *interaction regression coefficients* between the controllable and noise variables. This is how the noise variable influences the response.
3. The variance model is a **quadratic function** of the controllable variables.
4. The variance model (apart from σ^2) is just the square of the **slope** of the fitted response model in the direction of the noise variable.

To use these models operationally, we would

1. Perform an experiment and fit an appropriate response model, such as Equation 12.1.
2. Replace the unknown regression coefficients in the mean and variance models with their least squares estimates from the response model and replace σ^2 in the variance model by the residual mean square found when fitting the response model.
3. Optimize the mean and variance model using the standard multiple response optimization methods discussed in Section 11.3.4.

It is very easy to generalize these results. Suppose that there are k controllable variables and r noise variables. We will write the general response model involving these variables as

$$y(\mathbf{x}, \mathbf{z}) = f(\mathbf{x}) + h(\mathbf{x}, \mathbf{z}) + \epsilon \tag{12.4}$$

where $f(\mathbf{x})$ is the portion of the model that involves only the controllable variables and $h(\mathbf{x}, \mathbf{z})$ are the terms that involve the main effects of the noise factors and the interactions between the controllable and noise factors. Typically, the structure for $h(\mathbf{x}, \mathbf{z})$ is

$$h(\mathbf{x}, \mathbf{z}) = \sum_{i=1}^{r} \gamma_i z_i + \sum_{i=1}^{k} \sum_{j=1}^{r} \delta_{ij} x_i z_j$$

The structure for $f(\mathbf{x})$ will depend on what type of model for the controllable variables the experimenter thinks is appropriate. The logical choices are the first-order model with interaction and the second-order model. If we assume that the noise variables have mean zero, variances $\sigma^2_{z_i}$, and zero covariances and that the noise variables and the random errors ϵ have zero covariances, then the mean model for the response is just

$$E_z[y(\mathbf{x}, \mathbf{z})] = f(\mathbf{x}) \tag{12.5}$$

and the variance model for the response is

$$V_z[y(\mathbf{x}, \mathbf{z})] = \sum_{i=1}^{r} \left[\frac{\partial y(\mathbf{x}, \mathbf{z})}{\partial z_i}\right]^2 \sigma^2_{z_i} + \sigma^2 \tag{12.6}$$

Myers, Montgomery and Anderson-Cook (2009) give a slightly more general form for Equation (12.6) based on applying a conditional variance operator directly to the response model.

EXAMPLE 12.1

To illustrate the foregoing procedure, reconsider Example 6.2 in which four factors were studied in a 2^4 factorial design to investigate their effect on the filtration rate of a chemical product. We will assume that factor A, temperature, is potentially difficult to control in the full-scale process, but it can be controlled during the experiment (which was performed in a pilot plant). The other three factors, pressure (B), concentration (C), and stirring rate (D), are easy to control. Thus, the noise factor z_1 is temperature, and the controllable variables x_1, x_2, and x_3 are pressure, concentration, and stirring rate, respectively. Because both the controllable factors and the noise factor are in the same design, the 2^4 factorial design used in this experiment is an example of a **combined array design**.

Using the results from Example 6.2, the response model is

$$\begin{aligned}\hat{y}(\mathbf{x}, z_1) &= 70.06 + \left(\frac{21.625}{2}\right) z_1 \\ &+ \left(\frac{9.875}{2}\right) x_2 + \left(\frac{14.625}{2}\right) x_3 \\ &- \left(\frac{18.125}{2}\right) x_2 z_1 + \left(\frac{16.625}{2}\right) x_3 z_1 \\ &= 70.06 + 10.81 z_1 + 4.94 x_2 + 7.31 x_3 \\ &\quad - 9.06 x_2 z_1 + 8.31 x_3 z_1\end{aligned}$$

Using Equations (12.5) and (12.6), we can find the mean and variance models as

$$E_z[y(\mathbf{x}, z_1)] = 70.06 + 4.94 x_2 + 7.31 x_3$$

and

$$\begin{aligned}V_z[y(\mathbf{x}, z_1)] &= \sigma_z^2 (10.81 - 9.06 x_2 + 8.31 x_3)^2 + \sigma^2 \\ &= \sigma_z^2 (116.91 + 82.08 x_2^2 + 69.06 x_3^2 \\ &\quad - 195.88 x_2 + 179.66 x_3 - 150.58 x_2 x_3) + \sigma^2\end{aligned}$$

respectively. Now assume that the low and high levels of the noise variable temperature have been run at one standard deviation on either side of its typical or average value, so that $\sigma_z^2 = 1$ and use $\hat{\sigma}^2 = 19.51$ (this is the residual mean square obtained by fitting the response model). Therefore, the variance model becomes

$$\begin{aligned}V_z[y(\mathbf{x}, z_1)] &= 136.42 - 195.88 x_2 + 179.66 x_3 \\ &\quad - 150.58 x_2 x_3 + 82.08 x_2^2 + 69.06 x_3^2\end{aligned}$$

Figure 12.7 presents a contour plot from the Design-Expert software package of the response contours from the mean model. To construct this plot, we held the noise factor (temperature) at zero and the nonsignificant controllable factor (pressure) at zero. Notice that mean filtration rate increases as both concentration and stirring rate increase. Design-Expert will also automatically construct plots of the **square root** of the variance contours, which it labels **propagation of error**, or **POE**. Obviously, the POE is just the standard deviation of the transmitted variability in the response as a function of the controllable variables. Figure 12.8 shows a contour plot and a three-dimensional response surface plot of the POE, obtained from Design-Expert. (In this plot, the noise variable is held constant at zero, as explained previously.)

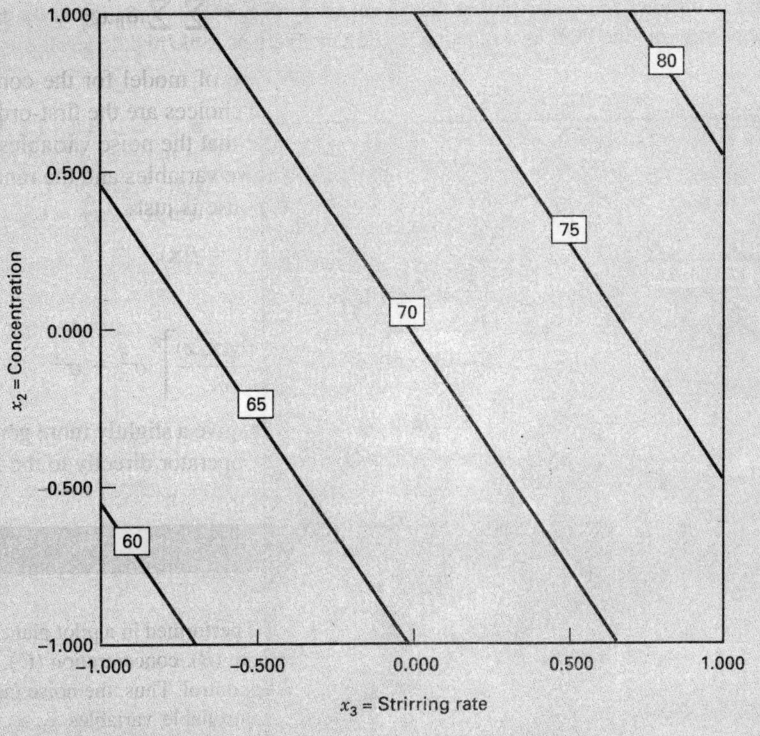

■ **FIGURE 12.7** **Contours of constant mean filtration rate, Example 12.1, with x_1 = temperature = 0**

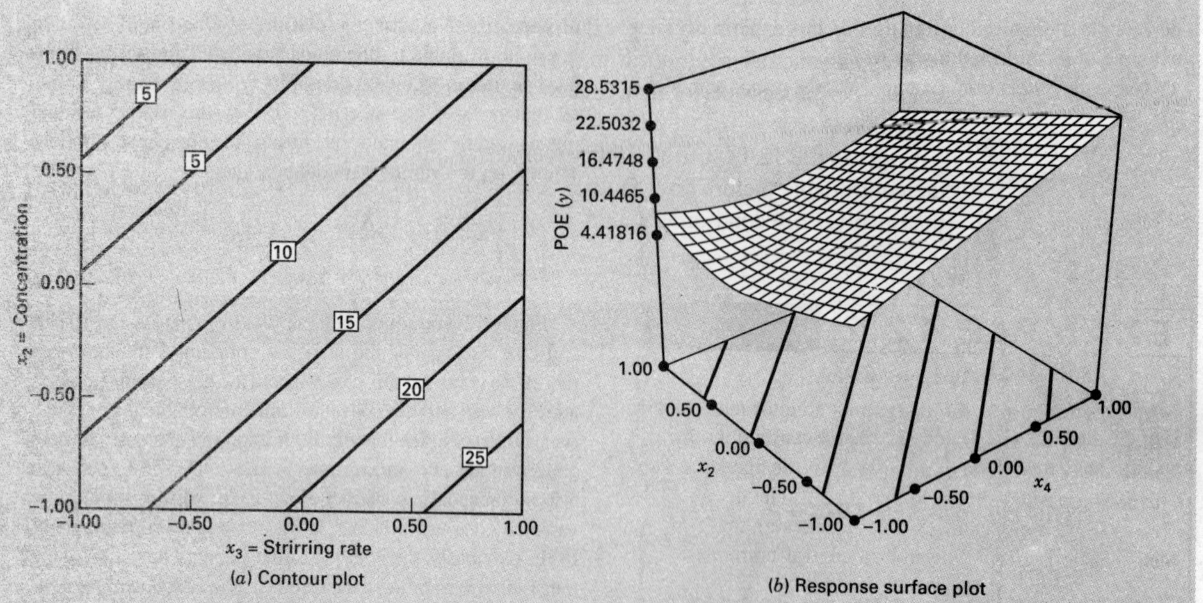

■ **FIGURE 12.8 Contour plot and response surface of propagation of error for Example 12.1, with x_1 = temperature = 0**

Suppose that the experimenter wants to maintain a mean filtration rate of about 75 and minimize the variability around this value. Figure 12.9 shows an overlay plot of the contours of mean filtration rate and the POE as a function of concentration and stirring rate, the significant controllable variables. To achieve the desired objectives, it will be necessary to hold concentration at the high level and stirring rate very near the middle level.

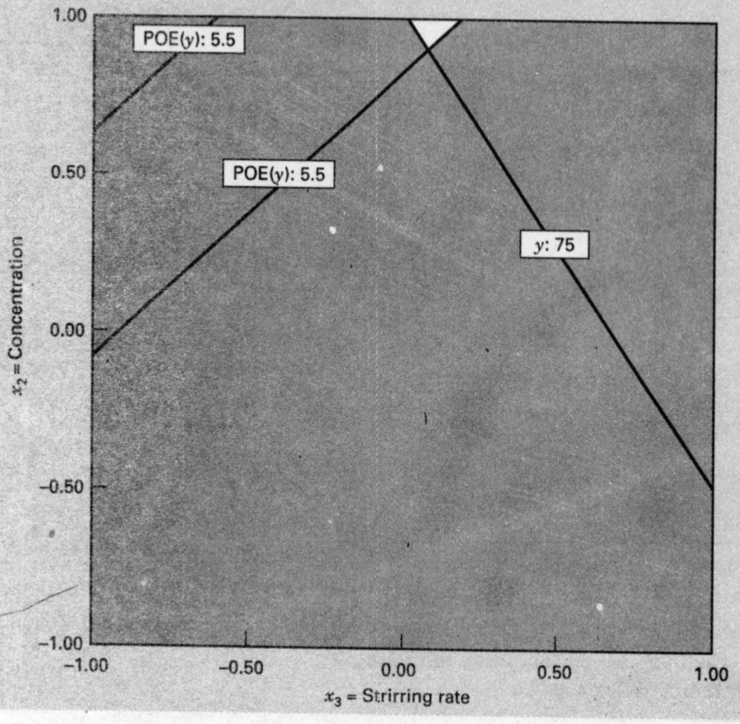

■ **FIGURE 12.9 Overlay plot of mean and POE contours for filtration rate, Example 12.1, with x_1 = temperature = 0**

We observe that the standard deviation of the filtration rate response in Example 12.1 is still very large. This illustrates that sometimes a process robustness study may not yield an entirely satisfactory solution. It may still be necessary to employ other measures to achieve satisfactory process performance, such as controlling temperature more precisely in the full-scale process.

Example 12.1 illustrates the use of a first-order model with interaction as the model for the controllable factors $f(\mathbf{x})$. We now present an example adapted from Montgomery (1999) that involves a second-order model.

EXAMPLE 12.2

An experiment was run in a semiconductor manufacturing facility involving two controllable variables and three noise variables. The combined array design used by the experimenters is shown in Table 12.3. The design is a 23-run variation of a central composite design that was created by starting with a standard central composite design (CCD) for five factors (the cube portion is a 2^{5-1}) and deleting the axial runs associated with the three noise variables. This design will support a response model that has a second-order model in the controllable variables, the main effects of the three noise variables, and the interactions between the control and noise factors. The fitted

TABLE 12.3
Combined Array Experiment with Two Controllable Variables and Three Noise Variables, Example 12.2

Run Number	x_1	x_2	z_1	z_2	z_3	y
1	−1.00	−1.00	−1.00	−1.00	1.00	44.2
2	1.00	−1.00	−1.00	−1.00	−1.00	30.0
3	−1.00	1.00	−1.00	−1.00	−1.00	30.0
4	1.00	1.00	−1.00	−1.00	1.00	35.4
5	−1.00	−1.00	1.00	−1.00	−1.00	49.8
6	1.00	−1.00	1.00	−1.00	1.00	36.3
7	−1.00	1.00	1.00	−1.00	1.00	41.3
8	1.00	1.00	1.00	−1.00	−1.00	31.4
9	−1.00	−1.00	−1.00	1.00	−1.00	43.5
10	1.00	−1.00	−1.00	1.00	1.00	36.1
11	−1.00	1.00	−1.00	1.00	1.00	22.7
12	1.00	1.00	−1.00	1.00	−1.00	16.0
13	−1.00	−1.00	1.00	1.00	1.00	43.2
14	1.00	−1.00	1.00	1.00	−1.00	30.3
15	−1.00	1.00	1.00	1.00	−1.00	30.1
16	1.00	1.00	1.00	1.00	1.00	39.2
17	−2.00	0.00	0.00	0.00	0.00	46.1
18	2.00	0.00	0.00	0.00	0.00	36.1
19	0.00	−2.00	0.00	0.00	0.00	47.4
20	0.00	2.00	0.00	0.00	0.00	31.5
21	0.00	0.00	0.00	0.00	0.00	30.8
22	0.00	0.00	0.00	0.00	0.00	30.7
23	0.00	0.00	0.00	0.00	0.00	31.0

response model is

$$\begin{aligned}\hat{y}(\mathbf{x}, \mathbf{z}) = {} & 30.37 - 2.92x_1 - 4.13x_2 \\ & + 2.60x_1^2 + 2.18x_2^2 + 2.87x_1x_2 \\ & + 2.73z_1 - 2.33z_2 + 2.33z_3 - 0.27x_1z_1 \\ & + 0.89x_1z_2 + 2.58x_1z_3 \\ & + 2.01x_2z_1 - 1.43x_2z_2 + 1.56x_2z_3\end{aligned}$$

The mean and variance models are

$$\begin{aligned}E_z[y(\mathbf{x}, \mathbf{z})] = {} & 30.37 - 2.92x_1 - 4.13x_2 \\ & + 2.60x_1^2 + 2.18x_2^2 + 2.87x_1x_2\end{aligned}$$

and

$$\begin{aligned}V_z[y(\mathbf{x}, \mathbf{z})] = {} & 19.26 + 6.40x_1 + 24.91x_2 + 7.52x_1^2 \\ & + 8.52x_2^2 + 4.42x_1x_2\end{aligned}$$

where we have substituted parameter estimates from the fitted response model into the equations for the mean and variance models and, as in the previous example, assumed that $\sigma_z^2 = 1$. Figures 12.10 and 12.11 (from Design-Expert) present contour plots of the process mean and POE (remember POE is the square root of the variance response surface) generated from these models.

In this problem, it is desirable to keep the process mean below 30. From inspection of Figures 12.10 and 12.11, it is clear that some trade-off will be necessary if we wish to make the process variance small. Because there are only two controllable variables, a logical way to accomplish this trade-off is to overlay the contours of constant mean response and constant variance, as shown in Figure 12.12. This plot shows the contours for which the process mean is less than or equal to 30 and the process standard deviation is less than or equal to 5. The region bounded by these contours would represent a typical operating region of low mean response and low process variance.

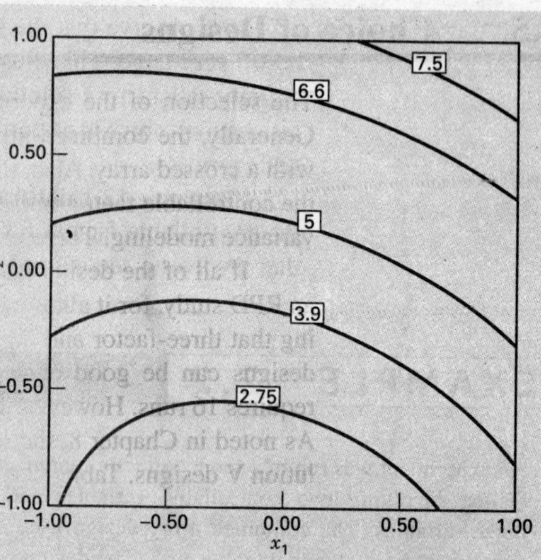

■ **FIGURE 12.11** **Contour plot of the POE, Example 12.2**

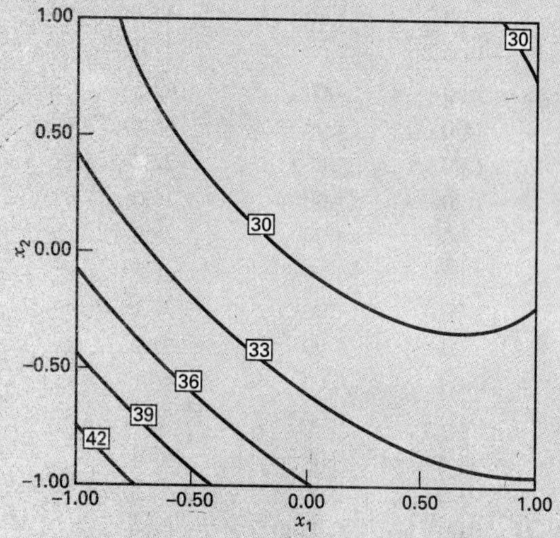

■ **FIGURE 12.10** **Contour plot of the mean model, Example 12.2**

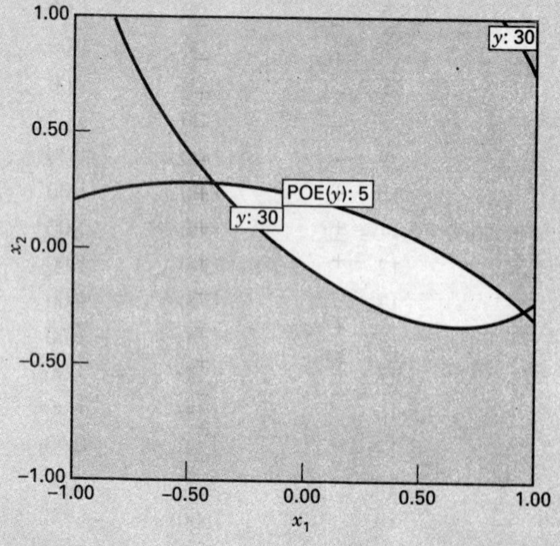

■ **FIGURE 12.12** **Overlay of the mean and POE contours for Example 12.2, with the open region indicating satisfactory operating conditions for process mean and variance**

12.5 Choice of Designs

The selection of the experimental design is a very important aspect of an RPD problem. Generally, the combined array approach will result in smaller designs that will be obtained with a crossed array. Also, the response modeling approach allows the direct incorporation of the controllable factor–noise factor interactions, which is usually superior to direct mean and variance modeling. Therefore, our comments in this section are confined to combined arrays.

If all of the design factors are at two levels, a resolution V design is a good choice for an RPD study, for it allows all main effect and two-factor interactions to be estimated, assuming that three-factor and higher interactions are negligible. Standard 2_V^{k-p} fractional factorial designs can be good choices in some cases. For example, with five factors, this design requires 16 runs. However, with six or more factors, the standard 2_V^{k-p} designs are rather large. As noted in Chapter 8, the software package Design-Expert contains smaller two-level resolution V designs. Table 12.4 is the design from this package for seven factors, which requires

■ TABLE 12.4
A Resolution V Design in Seven Factors and 30 Runs

A	*B*	*C*	*D*	*E*	*F*	*G*
−	+	−	−	−	+	−
+	−	−	−	−	−	−
−	+	+	+	−	+	−
+	−	+	−	+	−	−
−	+	−	+	−	−	+
+	−	+	+	−	−	−
+	+	+	+	−	+	+
−	+	−	+	+	−	−
+	−	+	−	−	−	+
−	−	−	+	+	−	+
+	−	+	−	−	+	−
−	+	+	+	+	−	+
−	−	+	−	−	−	−
−	−	+	+	−	+	+
+	+	−	+	−	−	−
+	+	−	+	+	−	+
+	+	−	−	+	+	−
−	−	+	+	+	−	−
+	+	+	−	−	−	−
−	+	+	−	−	+	+
−	−	−	−	−	−	+
−	+	+	−	+	+	−
−	+	−	−	+	−	+
−	+	−	+	+	+	+
−	−	−	−	+	+	−
+	−	−	+	−	+	+
+	−	−	−	+	+	+
+	+	+	−	+	+	+
+	−	−	+	+	+	−
+	−	+	+	+	+	+

30 runs. This design will accommodate any combination of controllable and noise variables totaling seven and allow all seven main effects and all two-factor interactions between these factors to be estimated.

Sometimes a design with fewer runs can be employed. For example, suppose that there are three controllable variables (*A*, *B*, and *C*) and four noise variables (*D*, *E*, *F*, and *G*). It is only necessary to estimate the main effects and two-factor interactions of the controllable variables (six parameters), the main effects of the noise variables (four parameters), and the interactions between the controllable and noise variables (12 parameters). Including the intercept, only 23 parameters must be estimated. Often very nice designs for these problems can be constructed using either the *D* or *I*-optimality criterion.

Table 12.5 is a *D*-optimal design with 23 runs for this situation. In this design, there are no two-factor interactions involving the controllable factors aliased with each other or with two-factor interactions involving control and noise variables. However, these main effects and two-factor interactions are aliased with the two-factor interactions involving the noise factors, so the usefulness of this design depends on the assumption that two-factor interactions involving only the noise factors are negligible.

When it is of interest to fit a complete second-order model in the controllable variables, the CCD is a logical basis for selecting the experimental design. The CCD can be modified

■ **TABLE 12.5**
A *D*-Optimal Design with 23 Runs for Three Controllable and Four Noise Variables

A	*B*	*C*	*D*	*E*	*F*	*G*
−	+	−	+	+	+	+
+	−	−	+	+	+	+
−	−	+	−	−	+	+
+	+	−	−	−	−	+
−	+	+	−	−	−	+
+	+	−	+	−	+	−
+	+	+	−	−	−	−
+	−	−	−	−	+	−
+	−	+	−	−	−	+
−	−	+	−	−	−	−
−	+	+	−	+	+	−
−	−	+	+	+	−	+
−	−	−	−	+	−	−
−	+	−	+	+	−	−
−	−	−	+	−	+	−
+	−	+	+	−	+	−
−	+	+	+	−	+	−
−	−	−	−	−	−	+
+	−	+	−	+	+	−
+	−	−	+	+	−	−
+	+	−	−	+	+	+
−	+	−	−	−	+	−
+	+	+	+	+	−	+

■ **TABLE 12.6**
A *D*-Optimal Design for Fitting a Second-Order Response Model with Three Control and Two Noise Variables

A	*B*	*C*	*D*	*E*
+	+	+	+	−
+	+	−	+	+
+	−	−	+	−
0	−	+	−	−
+	+	+	−	+
−	−	−	+	+
−	+	−	+	−
−	−	+	+	−
+	+	−	−	−
+	−	−	−	+
+	−	0	−	−
−	−	−	−	−
−	+	−	−	+
−	−	+	−	+
+	0	+	−	−
0	0	0	0	+
−	+	+	+	+
−	+	+	−	−

as in Example 12.2 by using only the axial runs in the directions of the controllable variables. For example, if there were three controllable variables and four noise variables, adding six axial runs for factors *A*, *B*, and *C* along with four center runs to the 30-run design in Table 12.4 would produce a very nice design for fitting the response model. The resulting design would have 40 runs, and the response model would have 26 parameters.

Other methods can be used to construct designs for the second-order case. For example, suppose that there are three controllable factors and two noise factors. A modified CCD would have 16 runs (a 2^{5-1}) in the cube, six axial runs in the directions of the controllable variables, and (say) four center runs. This yields a design with 26 runs to estimate a model with 18 parameters. Another alternative would be to use a small composite design in the cube (11 runs), along with the six axial runs in the directions of the controllable variables and the four center runs. This results in a design with only 21 runs. A *D*-optimal or *I*-optimal approach could also be used. The 18-run design in Table 12.6 was constructed using Design-Expert. Note that this is a saturated design. Remember that as the design gets smaller, in general the parameters in the response model may not be estimated as well as they would have been with a larger design, and the variance of the predicted response may also be larger. For more information on designs for RPD and process robustness studies, see Myers, Montgomery and Anderson-Cook (2009) and the references therein.

12.6 Problems

12.1. Reconsider the leaf spring experiment in Table 12.1. Suppose that the objective is to find a set of conditions where the mean free height is as close as possible to 7.6 inches, with the variance of free height as small as possible. What conditions would you recommend to achieve these objectives?

12.2. Consider the bottle-filling experiment in Problem 6.20. Suppose that the percentage of carbonation (A) is a noise variable ($\sigma_z^2 = 1$ in coded units).

(a) Fit the response model to these data. Is there a robust design problem?

(b) Find the mean model and either the variance model or the POE.

(c) Find a set of conditions that result in mean fill deviation as close to zero as possible with minimum transmitted variance.

12.3. Consider the experiment in Problem 11.12. Suppose that temperature is a noise variable ($\sigma_z^2 = 1$ in coded units). Fit response models for both responses. Is there a robust design problem with respect to both responses? Find a set of conditions that maximize conversion with activity between 55 and 60 and that minimize the variability transmitted from temperature.

12.4. Reconsider the leaf spring experiment from Table 12.1. Suppose that factors A, B, and C are controllable variables and that factors D and E are noise factors. Set up a crossed array design to investigate this problem, assuming that all of the two-factor interactions involving the controllable variables are thought to be important. What type of design have you obtained?

12.5. ***Continuation of Problem 12.5.*** Reconsider the leaf spring experiment from Table 12.1. Suppose that factors A, B and C are controllable variables and that factors D and E are noise factors. Show how a combined array design can be employed to investigate this problem that allows all two-factor interaction to be estimated and only requires 16 runs. Compare this with the crossed array design from Problem 12.5. Can you see how in general combined array designs have fewer runs than crossed array designs?

12.6. Consider the connector pull-off force experiment shown in Table 12.2. What main effects and interaction involving the controllable variables can be estimated with this design? Remember that all of the controllable variables are quantitative factors.

12.7. Consider the connector pull-off force experiment shown in Table 12.2. Show how an experiment can be designed for this problem that will allow a full quadratic model to be fit in the controllable variables along all main effects of the noise variables and their interactions with the controllable variables. How many runs will be required in this design? How does this compare with the design in Table 12.2?

12.8. Consider the experiment in Problem 11.11. Suppose that pressure is a noise variable ($\sigma_z^2 = 1$ in coded units). Fit the response model for the viscosity response. Find a set of conditions that result in viscosity as close as possible to 600 and that minimize the variability transmitted from the noise variable pressure.

12.9. ***A variation of Example 12.1.*** In Example 12.1 (which utilized data from Example 6.2), we found that one of the process variables (B = pressure) was not important. Dropping this variable produces two replicates of a 2^3 design. The data are as follows:

C	D	$A(+)$	$A(-)$	$\bar{y}$	s^2
−	−	45, 48	71, 65	57.75	121.19
+	−	68, 80	60, 65	68.25	72.25
−	+	43, 45	100, 104	73.00	1124.67
+	+	75, 70	86, 96	81.75	134.92

Assume that C and D are controllable factors and that A is a noise variable.

(a) Fit a model to the mean response.

(b) Fit a model to the $\ln(s^2)$ response.

(c) Find operating conditions that result in the mean filtration rate response exceeding 75 with minimum variance.

(d) Compare your results with those from Example 12.1, which used the transmission of error approach. How similar are the two answers?

12.10. In an article ("Let's All Beware the Latin Square," *Quality Engineering*, Vol. 1, 1989, pp. 453–465), J. S. Hunter illustrates some of the problems associated with 3^{k-p} fractional factorial designs. Factor A is the amount of ethanol added to a standard fuel, and factor B represents the air/fuel ratio. The response variable is carbon monoxide (CO) emission in g/m^3. The design is as follows,

Design				Observations	
A	B	x_1	x_2	y	
0	0	−1	−1	66	62
1	0	0	−1	78	81
2	0	+1	−1	90	94
0	1	−1	0	72	67
1	1	0	0	80	81
2	1	+1	0	75	78
0	2	−1	+1	68	66
1	2	0	+1	66	69
2	2	+1	+1	60	58

Notice that we have used the notation system of 0, 1, and 2 to represent the low, medium, and high levels for the factors. We have also used a "geometric notation" of −1, 0, and +1. Each run in the design is replicated twice.

(a) Verify that the second-order model

$$\hat{y} = 78.5 + 4.5x_1 - 7.0x_2 - 4.5x_1^2 - 4.0x_2^2 - 9.0x_1x_2$$

is a reasonable model for this experiment. Sketch the CO concentration contours in the x_1, x_2 space.

(b) Now suppose that instead of only two factors, we had used *four* factors in a 3^{4-2} fractional factorial design and obtained *exactly* the same data as in part (a). The design would be as follows:

Design								Observations y	
A	**B**	**C**	**D**	x_1	x_2	x_3	x_4		
0	0	0	0	−1	−1	−1	−1	66	62
1	0	1	1	0	−1	0	0	78	81
2	0	2	2	+1	−1	+1	+1	90	94
0	1	2	1	−1	0	+1	0	72	67
1	1	0	2	0	0	−1	+1	80	81
2	1	1	0	+1	0	0	−1	75	78
0	2	1	2	−1	+1	0	+1	68	66
1	2	2	0	0	+1	+1	−1	66	69
2	2	0	1	+1	+1	−1	0	60	58

Calculate the marginal averages of the CO response at each level of the four factors *A*, *B*, *C*, and *D*. Construct plots of these marginal averages and interpret the results. Do factors *C* and *D* appear to have strong effects? Do these factors *really* have any effect on CO emission? Why is their apparent effect strong?

(c) The design in part (b) allows the model

$$y = \beta_0 + \sum_{i=1}^{4} \beta_i x_i + \sum_{i=1}^{4} \beta_{ii} x_i^2 + \epsilon$$

to be fitted. Suppose that the *true* model is

$$y = \beta_0 + \sum_{i=1}^{4} \beta_i x_i + \sum_{i=1}^{4} \beta_{ii} x_i^2 + \sum\sum_{i<j} \beta_{ij} x_i x_j + \epsilon$$

Show that if $\hat{\beta}_i$ represents the least squares estimate of the coefficients in the fitted model, then

$$E(\hat{\beta}_0) = \beta_0 - \beta_{13} - \beta_{14} - \beta_{34}$$
$$E(\hat{\beta}_1) = \beta_1 - (\beta_{23} + \beta_{24})/2$$
$$E(\hat{\beta}_2) = \beta_2 - (\beta_{13} + \beta_{14} + \beta_{34})/2$$
$$E(\hat{\beta}_3) = \beta_3 - (\beta_{12} + \beta_{24})/2$$
$$E(\hat{\beta}_4) = \beta_4 - (\beta_{12} + \beta_{23})/2$$
$$E(\hat{\beta}_{11}) = \beta_{11} - (\beta_{23} - \beta_{24})/2$$
$$E(\hat{\beta}_{22}) = \beta_{22} + (\beta_{13} + \beta_{14} + \beta_{34})/2$$
$$E(\hat{\beta}_{33}) = \beta_{33} - (\beta_{24} - \beta_{12})/2 + \beta_{14}$$
$$E(\hat{\beta}_{44}) = \beta_{44} - (\beta_{12} - \beta_{23})/2 + \beta_{13}$$

Does this help explain the strong effects for factors *C* and *D* observed graphically in part (*b*)?

12.11. An experiment has been run in a process that applies a coating material to a wafer. Each run in the experiment produced a wafer, and the coating thickness was measured several times at different locations on the wafer. Then the mean y_1 and the standard deviation y_2 of the thickness measurement were obtained. The data [adapted from Box and Draper (2007)] are shown in the Table P12.1:

■ TABLE P12.1
The Coating Experiment in Problem 12.11

Run	Speed	Pressure	Distance	Mean y_1	Std. Dev. y_2
1	−1	−1	−1	24.0	12.5
2	0	−1	−1	120.3	8.4
3	+1	−1	−1	213.7	42.8
4	−1	0	−1	86.0	3.5
5	0	0	−1	136.6	80.4
6	+1	0	−1	340.7	16.2
7	−1	+1	−1	112.3	27.6
8	0	+1	−1	256.3	4.6
9	+1	+1	−1	271.7	23.6
10	−1	−1	0	81.0	0.0
11	0	−1	0	101.7	17.7
12	+1	−1	0	357.0	32.9
13	−1	0	0	171.3	15.0
14	0	0	0	372.0	0.0
15	+1	0	0	501.7	92.5
16	−1	+1	0	264.0	63.5
17	0	+1	0	427.0	88.6
18	+1	+1	0	730.7	21.1
19	−1	−1	+1	220.7	133.8
20	0	−1	+1	239.7	23.5
21	+1	−1	+1	422.0	18.5
22	−1	0	+1	199.0	29.4
23	0	0	+1	485.3	44.7
24	+1	0	+1	673.7	158.2
25	−1	+1	+1	176.7	55.5
26	0	+1	+1	501.0	138.9
27	+1	+1	+1	1010.0	142.4

(a) What type of design did the experimenters use? Is this a good choice of design for fitting a quadratic model?
(b) Build models of both responses.

(c) Find a set of optimum conditions that result in the mean as large as possible with the standard deviation less than 60.

12.12. Suppose that there are four controllable variables and two noise variables. It is necessary to estimate the main effects and two-factor interactions of all of the controllable variables, the main effects of the noise variables, and the two-factor interactions between all controllable and noise factors. If all factors are at two levels, what is the minimum number of runs that can be used to estimate all of the model parameters using a combined array design? Use a *D*-optimal algorithm to find a design.

12.13. Suppose that there are four controllable variables and two noise variables. It is necessary to fit a complete quadratic model in the controllable variables, the main effects of the noise variables, and the two-factor interactions between all controllable and noise factors. Set up a combined array design for this by modifying a central composite design.

12.14. Reconsider the situation in Problem 12.13. Could a modified small composite design be used for this problem? Are any disadvantages associated with the use of the small composite design?

12.15. Reconsider the situation in Problem 12.13. What is the minimum number of runs that can be used to estimate all of the model parameters using a combined array design? Use a *D*-optimal algorithm to find a reasonable design for this problem.

12.16. Rework Problem 12.15 using the I-criterion to construct the design. Compare this design to the *D*-optimal design in Problem 12.15. Which design would you prefer?

12.17. Rework Problem 12.12 using the I-criterion. Compare this design to the *D*-optimal design in Problem 12.12. Which design would you prefer?

12.18. An experiment was run in a wave soldering process. There are five controllable variables and three noise variables. The response variable is the number of solder defects per million opportunities. The experimental design employed was the following crossed array.

						Outer Array			
					F	−1	1	1	−1
Inner Array					*G*	−1	1	−1	1
A	*B*	*C*	*D*	*E*	*H*	−1	−1	1	1
1	1	1	−1	−1		194	197	193	275
1	1	−1	1	1		136	136	132	136
1	−1	1	−1	1		185	261	264	264
1	−1	−1	1	−1		47	125	127	42
−1	1	1	1	−1		295	216	204	293
−1	1	−1	−1	1		234	159	231	157
−1	−1	1	1	1		328	326	247	322
−1	−1	−1	−1	−1		186	187	105	104

(a) What types of designs were used for the inner and outer arrays? What are the alias relationships in these designs?

(b) Develop models for the mean and variance of solder defects. What set of operating conditions would you recommend?

12.19. Reconsider the wave soldering experiment in Problem 12.16. Find a combined array design for this experiment that requires fewer runs.

12.20. Reconsider the wave soldering experiment in Problem 12.16. Suppose that it was necessary to fit a complete quadratic model in the controllable variables, all main effects of the noise variables, and all controllable variable–noise variable interactions. What design would you recommend?

12.21. Consider the alloy cracking experiment in Problem 6.15. Suppose that temperature (A) is a noise variable. Find the response model, and the model for the mean response, and the model for the transmitted variability. Can you find settings for the controllable factors that minimize crack length and make the transmitted variability small?

CHAPTER 13

Random Effects Models

CHAPTER OUTLINE

The supplemental material is on the textbook website www.wiley.com/go/global/montgomery.

13.1 Random Effects Models

Throughout most of this book we have assumed that the factors in an experiment were **fixed factors**, that is, the levels of the factors used by the experimenter were the specific levels of interest. The implication of this, of course, is that the statistical inferences made about these factors are confined to the specific levels studied. That is, if three material types are investigated as in the battery life experiment of Example 5.1, our conclusions are valid only about those specific material types. A variation of this occurs when the factor or factors are **quantitative**. In these situations, we often use a regression model relating the response to the factors to predict the response over the region spanned by the factor levels used in the experimental design. Several examples of this were presented in Chapters 5 through 9. In general, with a fixed effect, we say that the **inference space** of the experiment is the specific set of factor levels investigated.

In some experimental situations, the factor levels are chosen at random from a larger population of possible levels, and the experimenter wishes to draw conclusions about the entire population of levels, not just those that were used in the experimental design. In this situation, the factor is said to be a **random factor**. We introduced a simple situation in Chapter 3, a single-factor experiment where the factor is random, and we used this to introduce the **random effects model** for the analysis of variance and **components of variance**. We have also discussed experiments where **blocks** are random. However, random factors also occur regularly in factorial experiments as well as in other types of experiments. In this chapter, we focus on methods for the design and analysis of factorial experiments with random factors. In Chapter 14, we will present **nested** and **split-plot designs**, two situations where random factors are frequently encountered in practice.

13.2 The Two-Factor Factorial with Random Factors

Suppose that we have two factors, A and B, and that both factors have a large number of levels that are of interest (as in Chapter 3, we will assume that the number of levels is infinite). We will choose at random a levels of factor A and b levels of factor B and arrange these factor levels in a factorial experimental design. If the experiment is replicated n times, we may represent the observations by the linear model

$$y_{ijk} = \mu + \tau_i + \beta_j + (\tau\beta)_{ij} + \epsilon_{ijk} \quad \begin{cases} i = 1, 2, \ldots, a \\ j = 1, 2, \ldots, b \\ k = 1, 2, \ldots, n \end{cases} \tag{13.1}$$

where the model parameters τ_i, β_j, $(\tau\beta)_{ij}$, and ϵ_{ijk} are all independent random variables. We are also going to assume that the random variables τ_i, β_j, $(\tau\beta)_{ij}$, and ϵ_{ijk} are normally distributed with mean zero and variances given by $V(\tau_i) = \sigma_\tau^2$, $V(\beta_j) = \sigma_\beta^2$, $V[(\tau\beta)_{ij}] = \sigma_{\tau\beta}^2$, and $V(\epsilon_{ijk}) = \sigma^2$. Therefore the variance of any observation is

$$V(y_{ijk}) = \sigma_\tau^2 + \sigma_\beta^2 + \sigma_{\tau\beta}^2 + \sigma^2 \tag{13.2}$$

and σ_τ^2, σ_β^2, $\sigma_{\tau\beta}^2$, and σ^2 are the **variance components**. The hypotheses that we are interested in testing are $H_0\colon \sigma_\tau^2 = 0$, $H_0\colon \sigma_\beta^2 = 0$, and $H_0\colon \sigma_{\tau\beta}^2 = 0$. Notice the similarity to the single-factor random effects model.

The numerical calculations in the analysis of variance remain unchanged; that is, SS_A, SS_B, SS_{AB}, SS_T, and SS_E are all calculated as in the fixed effects case. However, to form the test statistics, we must examine the **expected mean squares**. It may be shown that

$$\begin{aligned} E(MS_A) &= \sigma^2 + n\sigma_{\tau\beta}^2 + bn\sigma_\tau^2 \\ E(MS_B) &= \sigma^2 + n\sigma_{\tau\beta}^2 + an\sigma_\beta^2 \\ E(MS_{AB}) &= \sigma^2 + n\sigma_{\tau\beta}^2 \end{aligned} \tag{13.3}$$

and

$$E(MS_E) = \sigma^2$$

From the expected mean squares, we see that the appropriate statistic for testing the no-interaction hypothesis $H_0\colon \sigma_{\tau\beta}^2 = 0$ is

$$F_0 = \frac{MS_{AB}}{MS_E} \tag{13.4}$$

because under H_0 both numerator and denominator of F_0 have expectation σ^2, and only if H_0 is false is $E(MS_{AB})$ greater than $E(MS_E)$. The ratio F_0 is distributed as $F_{(a-1)(b-1),ab(n-1)}$. Similarly, for testing H_0: $\sigma_\tau^2 = 0$ we would use

$$F_0 = \frac{MS_A}{MS_{AB}} \tag{13.5}$$

which is distributed as $F_{a-1,(a-1)(b-1)}$, and for testing H_0: $\sigma_\beta^2 = 0$ the statistic is

$$F_0 = \frac{MS_B}{MS_{AB}} \tag{13.6}$$

which is distributed as $F_{b-1,(a-1)(b-1)}$. These are all upper-tail, one-tail tests. Notice that these test statistics are not the same as those used if both factors A and B are fixed. The expected mean squares are always used as a guide to test statistic construction.

In many experiments involving random factors, interest centers at least as much on estimating the variance components as on hypothesis testing. Recall from Chapter 3 that there are two approaches to variance component estimation. The variance components may be estimated by the **analysis of variance method**, that is, by equating the observed mean squares in the lines of the analysis of variance table to their expected values and solving for the variance components. This yields

$$\begin{aligned}
\hat{\sigma}^2 &= MS_E \\
\hat{\sigma}_{\tau\beta}^2 &= \frac{MS_{AB} - MS_E}{n} \\
\hat{\sigma}_\beta^2 &= \frac{MS_B - MS_{AB}}{an} \\
\hat{\sigma}_\tau^2 &= \frac{MS_A - MS_{AB}}{bn}
\end{aligned} \tag{13.7}$$

as the point estimates of the variance components in the two-factor random effects model. These are **moment estimators**. Some computer programs use this method. This will be illustrated in the following example.

EXAMPLE 13.1 A Measurement Systems Capability Study

Statistically designed experiments are frequently used to investigate the sources of variability that affect a system. A common industrial application is to use a designed experiment to study the components of variability in a measurement system. These studies are often called **gauge capability studies** or **gauge repeatability and reproducibility (R&R) studies** because these are the components of variability that are of interest (for more discussion of gauge R&R studies, see the supplemental material for this chapter).

A typical gauge R&R experiment from Montgomery (2009) is shown in Table 13.1. An instrument or gauge is used to measure a critical dimension on a part. Twenty parts have been selected from the production process, and three randomly selected operators measure each part twice with this gauge. The order in which the measurements are made is completely randomized, so this is a two-factor factorial experiment with design factors parts and operators, with two replications. Both parts and operators are random factors. The variance component identity in Equation 13.1 applies; namely,

$$\sigma_y^2 = \sigma_\tau^2 + \sigma_\beta^2 + \sigma_{\tau\beta}^2 + \sigma^2$$

■ **TABLE 13.1**
The Measurement Systems Capability Experiment in Example 13.2

Part Number	Operator 1		Operator 2		Operator 3	
1	21	20	20	20	19	21
2	24	23	24	24	23	24
3	20	21	19	21	20	22
4	27	27	28	26	27	28
5	19	18	19	18	18	21
6	23	21	24	21	23	22
7	22	21	22	24	22	20
8	19	17	18	20	19	18
9	24	23	25	23	24	24
10	25	23	26	25	24	25
11	21	20	20	20	21	20
12	18	19	17	19	18	19
13	23	25	25	25	25	25
14	24	24	23	25	24	25
15	29	30	30	28	31	30
16	26	26	25	26	25	27
17	20	20	19	20	20	20
18	19	21	19	19	21	23
19	25	26	25	24	25	25
20	19	19	18	17	19	17

where σ_y^2 is the total variability (including variability due to the different parts, variability due to the different operators, and variability due to the gauge), σ_τ^2 is the variance component for parts, σ_β^2 is the variance component for operators, $\sigma_{\tau\beta}^2$ is the variance component that represents interaction between parts and operators, and σ^2 is the random experimental error. Typically, the variance component σ^2 is called the gauge repeatability because σ^2 can be thought of as reflecting the variation observed when the same part is measured by the same operator, and

$$\sigma_\beta^2 + \sigma_{\tau\beta}^2$$

is usually called the reproducibility of the gauge because it reflects the additional variability in the measurement system resulting from use of the instrument by the operator. These experiments are usually performed with the objective of estimating the variance components.

Table 13.2 shows the ANOVA for this experiment. The computations were performed using the Balanced ANOVA routine in Minitab. Based on the *P*-values, we conclude that the effect of parts is large, operators may have a small effect, and no significant part–operator interaction takes place. We may use Equation 13.7 to estimate the variance components as follows:

$$\hat{\sigma}_\tau^2 = \frac{62.39 - 0.71}{(3)(2)} = 10.28$$

$$\hat{\sigma}_\beta^2 = \frac{1.31 - 0.71}{(20)(2)} = 0.015$$

$$\hat{\sigma}_{\tau\beta}^2 = \frac{0.71 - 0.99}{2} = -0.14$$

and

$$\hat{\sigma}^2 = 0.99$$

The bottom portion of the Minitab output in Table 13.2 contains the expected mean squares for the random model, with numbers in parentheses representing the variance components [(4) represents σ^2, (3) represents $\sigma_{\tau\beta}^2$, etc.]. The estimates of the variance components are also given, along with the error term that was used in testing that variance

■ **TABLE 13.2**
Analysis of Variance (Minitab Balanced ANOVA) for Example 13.1

```
Analysis of Variance (Balanced Designs)

Factor    Type Levels Values
part     random     20        1        2        3        4        5        6        7
                              8        9       10       11       12       13       14
                             15       16       17       18       19       20
operator random      3        1        2        3

Analysis of Variance for y

Source              DF          SS          MS          F          P
part                19    1185.425      62.391      87.65      0.000
operator             2       2.617       1.308       1.84      0.173
part*operator       38      27.050       0.712       0.72      0.861
Error               60      59.500       0.992
Total              119    1274.592

Source              Variance      Error    Expected Mean Square for Each Term
                    component     term     (using unrestricted model)
1 part               10.2798        3      (4) + 2(3) + 6(1)
2 operator            0.0149        3      (4) + 2(3) + 40(2)
3 part*operator      -0.1399        4      (4) + 2(3)
4 Error               0.9917               (4)
```

component in the analysis of variance. We will discuss the terminology **unrestricted model** later; it has no relevance in random models.

Notice that the estimate of one of the variance components, $\sigma^2_{\tau\beta}$, is negative. This is certainly not reasonable because by definition variances are nonnegative. Unfortunately, negative estimates of variance components can result when we use the analysis of variance method of estimation (this is considered one of its drawbacks). We can deal with this negative result in a variety of ways. One possibility is to assume that the negative estimate means that the variance component is really zero and just set it to zero, leaving the other nonnegative estimates unchanged. Another approach is to estimate the variance components with a method that assures nonnegative estimates (this can be done with the maximum likelihood approach). Finally, we could note that the P-value for the interaction term in Table 13.2 is very large, take this as evidence that $\sigma^2_{\tau\beta}$ really is zero and that there is no interaction effect, and then fit a **reduced model** of the form

$$y_{ijk} = \mu + \tau_i + \beta_j + \epsilon_{ijk}$$

that does not include the interaction term. This is a relatively easy approach and one that often works nearly as well as more sophisticated methods.

Table 13.3 shows the analysis of variance for the reduced model. Because there is no interaction term in the model, both main effects are tested against the error term, and the estimates of the variance components are

$$\hat{\sigma}^2_\tau = \frac{62.39 - 0.88}{(3)(2)} = 10.25$$

$$\hat{\sigma}^2_\beta = \frac{1.31 - 0.88}{(20)(2)} = 0.0108$$

$$\hat{\sigma}^2 = 0.88$$

Finally, we could estimate the variance of the gauge as the sum of the variance component estimates $\hat{\sigma}^2$ and $\hat{\sigma}^2_\beta$ as

$$\begin{aligned}\hat{\sigma}^2_{\text{gauge}} &= \hat{\sigma}^2 + \hat{\sigma}^2_\beta \\ &= 0.88 + 0.0108 \\ &= 0.8908\end{aligned}$$

The variability in the gauge appears small relative to the variability in the product. This is generally a desirable situation, implying that the gauge is capable of distinguishing among different grades of product.

■ **TABLE 13.3**
Analysis of Variance for the Reduced Model, Example 13.1

```
Analysis of Variance (Balanced Designs)

Factor       Type     Levels    Values
part         random       20         1      2      3      4      5      6      7
                                     8      9     10     11     12     13     14
                                    15     16     17     18     19     20
operator random            3         1      2      3

Analysis of Variance for y

Source          DF           SS           MS          F          P
part            19     1185.425       62.391      70.64      0.000
operator         2        2.617        1.308       1.48      0.232
Error           98       86.550        0.883
Total          119     1274.592

Source          Variance      Error    Expected Mean Square for Each Term
                component     term     (using unrestricted model)
1 part            10.2513        3     (3) + 6(1)
2 operator         0.0106        3     (3) + 40(2)
3 error            0.8832              (3)
```

Measurement system capability studies are a very common application of designed experiments. These experiments almost always involve random effects. For more information about measurement systems experiments and a bibliography, see Burdick, Borror, and Montgomery (2003).

The other method for variance component estimation is the method of maximum likelihood, which was introduced in Chapter 3. This method is superior to the method of moments approach, because it produces estimators that are approximately normally distributed and it is easy to obtain their standard errors. Therefore, finding confidence intervals on the variance components is straightforward.

To illustrate how this method applies to an experimental design model with random effects, consider a two-factor model where both factors are random and $a = b = n = 2$. The model is

$$y_{ijk} = \mu + \tau_i + \beta_j + (\tau\beta)_{ij} + \epsilon_{ijk}$$

with $i = 1, 2, j = 1, 2$, and $k = 1, 2$. The variance of any observation is

$$V(y_{ijk}) = \sigma_y^2 = \sigma_\tau^2 + \sigma_\beta^2 + \sigma_{\tau\beta}^2 + \sigma^2$$

and the covariances are

$$\begin{aligned} \text{Cov}(y_{ijk}, y_{i'j'k'}) &= \sigma_\tau^2 + \sigma_\beta^2 + \sigma_{\tau\beta}^2 & i = i', j = j', k \neq k' \\ &= \sigma_\tau^2 & i = i', j \neq j' \\ &= \sigma_\beta^2 & i \neq i', j = j' \\ &= 0 & i \neq i', j \neq j' \end{aligned}$$

It is convenient to think of the observations as an 8 × 1 vector, say

$$\mathbf{y} = \begin{bmatrix} y_{111} \\ y_{112} \\ y_{211} \\ y_{212} \\ y_{121} \\ y_{122} \\ y_{221} \\ y_{222} \end{bmatrix}$$

and the variances and covariances can be expressed as an 8 × 8 **covariance matrix**

$$\boldsymbol{\Sigma} = \begin{bmatrix} \boldsymbol{\Sigma}_{11} & \boldsymbol{\Sigma}_{12} \\ \boldsymbol{\Sigma}_{21} & \boldsymbol{\Sigma}_{22} \end{bmatrix}$$

where $\boldsymbol{\Sigma}_{11}$, $\boldsymbol{\Sigma}_{22}$, $\boldsymbol{\Sigma}_{12}$, and $\boldsymbol{\Sigma}_{21} = \boldsymbol{\Sigma}'_{12}$ are 4 × 4 matrices defined as follows:

$$\boldsymbol{\Sigma}_{11} = \boldsymbol{\Sigma}_{22} = \begin{bmatrix} \sigma^2 & \sigma_\tau^2 + \sigma_\beta^2 + \sigma_{\tau\beta}^2 & \sigma_\tau^2 & \sigma_\tau^2 \\ \sigma_\tau^2 + \sigma_\beta^2 + \sigma_{\tau\beta}^2 & \sigma_\tau^2 & \sigma_\tau^2 & \sigma_\tau^2 \\ \sigma_\tau^2 & \sigma_\tau^2 & \sigma_\tau^2 & \sigma_\tau^2 + \sigma_\beta^2 + \sigma_{\tau\beta}^2 \\ \sigma_\tau^2 & \sigma_\tau^2 & \sigma_\tau^2 + \sigma_\beta^2 + \sigma_{\tau\beta}^2 & \sigma_y^2 \end{bmatrix}$$

$$\boldsymbol{\Sigma}_{12} = \begin{bmatrix} \sigma_\beta^2 & \sigma_\beta^2 & 0 & 0 \\ \sigma_\beta^2 & \sigma_\beta^2 & 0 & 0 \\ 0 & 0 & \sigma_\beta^2 & \sigma_\beta^2 \\ 0 & 0 & \sigma_\beta^2 & \sigma_\beta^2 \end{bmatrix}$$

and $\boldsymbol{\Sigma}_{21}$ is just the transpose of $\boldsymbol{\Sigma}_{12}$. Now each observation is normally distributed with variance σ_y^2, and if we assume that all $N = abn$ observations have a joint normal distribution, then the likelihood function for the random model becomes

$$L(\mu, \sigma_\tau^2, \sigma_\beta^2, \sigma_{\tau\beta}^2, \sigma^2) = \frac{1}{(2\pi)^{n/2}|\boldsymbol{\Sigma}|^{1/2}} \exp\left[-\frac{1}{2}(\mathbf{y} - \mathbf{j}_N\mu)'\boldsymbol{\Sigma}^{-1}(\mathbf{y} - \mathbf{j}_N\mu)\right]$$

where $\mathbf{j}_N$ is an $N \times 1$ vector of 1s. The maximum likelihood estimates of μ, σ_τ^2, σ_β^2, $\sigma_{\tau\beta}^2$, and σ^2 are those values of these parameters that maximize the likelihood function. In some situations, it would also be desirable to restrict the variance component estimates to nonnegative values.

Estimating variance components by maximum likelihood requires specialized computer software. JMP computes maximum likelihood estimates of the variance components in random or mixed models using the **residual maximum likelihood (REML) method**.

Table 13.3 is the output from JMP for the two-factor random effects experiment in Example 13.1. The output contains some model summary statistics, and the estimates of the individual variance components, which agree with those obtained via the ANOVA method in Example 13.1 (REML and the ANOVA method will agree for point estimation in balanced designs). Other information includes the ratio of each variance component to the estimated residual error variance, the standard error of each variance component, upper and lower bounds of a large-sample 95 percent confidence interval on each variance component, the percent of total variability accounted for by each variance component and the covariance

■ **TABLE 13.3**
JMP REML Analysis for the Two-Factor Random Model in Example 13.1

Response Y

Whole Model

Summary of Fit

RSquare	0.910717
RSquare Adj	0.910717
Root Mean Square Error	0.995825
Mean of Response	22.39167
Observations (or Sum Wgts)	120

Parameter Estimates

Term	Estimate	Std Error	DFDen	t Ratio	Prob>|t|
Intercept	22.391667	0.724496	19.28	30.91	<.0001*

REML Variance Component Estimates

Random Effect	Var Ratio	Var Component	Std Error	95% Lower	95% Upper	Pct of Total
Parts	10.36621	10.279825	3.3738173	3.6672642	16.892385	92.225
Operators	0.0150376	0.0149123	0.0329622	−0.049692	0.0795169	0.134
Parts*Operators	−0.141088	−0.139912	0.1219114	−0.378854	0.0990296	−1.255
Residual		0.9916667	0.1810527	0.7143057	1.4697982	8.897
Total		11.146491				100.000

−2 LogLikelihood = 408.14904346

Covariance Matrix of Variance Component Estimates

Random Effect	Parts	Operators	Parts*Operators	Residual
Parts	11.382643	0.0001111	−0.002222	3.125e-14
Operators	0.0001111	0.0010865	−0.000333	6.126e-17
Parts*Operators	−0.002222	−0.000333	0.0148624	−0.01639
Residual	3.125e-14	6.126e-17	−0.01639	0.0327801

matrix of the variance component estimates. The square roots of the diagonal elements of the matrix are the standard errors. The lower and upper bounds on the large-sample CI are found from

$$L = \hat{\sigma}_i^2 - Z_{\alpha/2}se(\hat{\sigma}_i^2) \quad \text{and} \quad U = \hat{\sigma}_i^2 + Z_{\alpha/2}se(\hat{\sigma}_i^2)$$

The 95% CI on the interaction variance component includes zero, evidence that this variance component is likely zero. Furthermore, the CI on the operator variance component also includes zero, and although its point estimate is positive, it would not be unreasonable to assume that this variance component is also zero.

13.3 The Two-Factor Mixed Model

We now consider the situation where one of the factors A is fixed and the other factor B is random. This is called the **mixed model** analysis of variance. The linear statistical model is

$$y_{ijk} = \mu + \tau_i + \beta_j + (\tau\beta)_{ij} + \epsilon_{ijk} \begin{cases} i = 1, 2, \ldots, a \\ j = 1, 2, \ldots, b \\ k = 1, 2, \ldots, n \end{cases} \quad \textbf{(13.8)}$$

Here τ_i is a fixed effect, β_j is a random effect, the interaction $(\tau\beta)_{ij}$ is assumed to be a random effect, and ϵ_{ijk} is a random error. We also assume that the $\{\tau_i\}$ are fixed effects such that $\sum_{i=1}^{a}\tau_i = 0$ and β_j is a NID$(0, \sigma_\beta^2)$ random variable. The interaction effect, $(\tau\beta)_{ij}$, is a normal random variable with mean 0 and variance $[(a-1)/a]\sigma_{\tau\beta}^2$; however, summing the interaction component over the fixed factor equals zero. That is,

$$\sum_{i=1}^{a}(\tau\beta)_{ij} = (\tau\beta)_{.j} = 0 \quad j = 1, 2, \ldots, b$$

This restriction implies that certain interaction elements at different levels of the fixed factor are not independent. In fact, we may show that

$$\text{Cov}[(\tau\beta)_{ij}, (\tau\beta)_{i'j}] = -\frac{1}{a}\sigma_{\tau\beta}^2 \qquad i \neq i'$$

The covariance between $(\tau\beta)_{ij'}$ and $(\tau\beta)_{ij'}$ for $j \neq j'$ is zero, and the random error ϵ_{ijk} is NID $(0, \sigma^2)$. Because the sum of the interaction effects over the levels of the fixed factor equals zero, this version of the mixed model is often called the **restricted model**.

In this model, the variance of $(\tau\beta)_{ij}$ is defined as $[(a-1)/a]\sigma_{\tau\beta}^2$ rather than $\sigma_{\tau\beta}^2$ to simplify the expected mean squares. The assumption $(\tau\beta)_{.j} = 0$ also has an effect on the expected mean squares, which we may show are

$$\begin{aligned} E(MS_A) &= \sigma^2 + n\sigma_{\tau\beta}^2 + \frac{bn\sum_{i=1}^{a}\tau_i^2}{a-1} \\ E(MS_B) &= \sigma^2 + an\sigma_\beta^2 \\ E(MS_{AB}) &= \sigma^2 + n\sigma_{\tau\beta}^2 \end{aligned} \quad \textbf{(13.9)}$$

and

$$E(MS_E) = \sigma^2$$

Therefore, the appropriate test statistic for testing that the means of the fixed factor effects are equal, or $H_0\colon \tau_i = 0$, is

$$F_0 = \frac{MS_A}{MS_{AB}}$$

for which the reference distribution is $F_{a-1,(a-1)(b-1)}$. For testing $H_0\colon \sigma_\beta^2 = 0$, the test statistic is

$$F_0 = \frac{MS_B}{MS_E}$$

with reference distribution $F_{b-1,ab(n-1)}$. Finally, for testing the interaction hypothesis $H_0: \sigma^2_{\tau\beta} = 0$, we would use

$$F_0 = \frac{MS_{AB}}{MS_E}$$

which has reference distribution $F_{(a-1)(b-1),ab(n-1)}$.

In the mixed model, it is possible to estimate the fixed factor effects as

$$\hat{\mu} = \bar{y}_{...}$$
$$\hat{\tau}_i = \bar{y}_{i..} - \bar{y}_{...} \qquad i = 1, 2, \ldots, a \tag{13.10}$$

The variance components σ^2_β, $\sigma^2_{\tau\beta}$, and σ^2 may be estimated using the analysis of variance method. Eliminating the first Equation from Equations 13.9 leaves three equations in three unknowns, whose solutions are

$$\hat{\sigma}^2_\beta = \frac{MS_B - MS_E}{an}$$
$$\hat{\sigma}^2_{\tau\beta} = \frac{MS_{AB} - MS_E}{n} \tag{13.11}$$

and

$$\hat{\sigma}^2 = MS_E$$

This general approach can be used to estimate the variance components in *any* mixed model. After eliminating the mean squares containing fixed factors, there will always be a set of equations remaining that can be solved for the variance components.

In mixed models the experimenter may be interested in testing hypotheses or constructing confidence intervals about individual treatment means for the fixed factor. In using such procedures, care must be exercised to use the proper standard error of the treatment mean. The standard error of the fixed effect treatment mean is

$$\left[\frac{\text{Mean square for testing the fixed effect}}{\text{Number of observations in each treatment mean}}\right]^{1/2} = \sqrt{\frac{MS_{AB}}{bn}}$$

Notice that this is just the standard error that we would use if this was a fixed effects model, except that MS_E has been replaced by the mean square used for hypothesis testing.

EXAMPLE 13.2 The Measurement Systems Capability Experiment Revisited

Reconsider the gauge R&R experiment described in Example 13.1. Suppose now that only three operators use this gauge, so the operators are a fixed factor. However, because the parts are chosen at random, the experiment now involves a mixed model.

The ANOVA for the mixed model is shown in Table 13.4. The computations were performed using the Balanced ANOVA routine in Minitab. We specified that the restricted model be used in the Minitab analysis. Minitab also generated the expected mean squares for this model. In the Minitab output, the quantity $Q[2]$ indicates a quadratic expression involving the fixed factor effect operator.

That is, $Q[2] = \Sigma_{j=1}^{b}\beta_j^2/(b-1)$. The conclusions are similar to Example 13.1. The variance components may be estimated from Equation (13.11) as

$$\hat{\sigma}^2_{\text{Parts}} = \frac{MS_{\text{Parts}} - MS_E}{an} = \frac{62.39 - 0.99}{(3)(2)} = 10.23$$

$$\hat{\sigma}^2_{\text{Parts}\times\text{operators}} = \frac{MS_{\text{Parts}\times\text{operators}} - MS_E}{n} = \frac{0.71 - 0.99}{2} = -0.14$$

$$\hat{\sigma}^2 = MS_E = 0.99$$

These results are also given in the Minitab output. Once again, a negative estimate of the interaction variance component results. An appropriate course of action would be to fit a reduced model, as we did in Example 13.1. In the case of a mixed model with two factors, this leads to the same results as in Example 13.1.

■ **TABLE 13.4**
Analysis of Variance (Minitab) for the Mixed Model in Example 13.2. The Restricted Model is Assumed

```
Analysis of Variance (Balanced Designs)

Factor      Type Levels Values
part      random     20      1      2      3      4      5      6      7
                             8      9     10     11     12     13     14
                            15     16     17     18     19     20
operator fixed        3      1      2      3

Analysis of Variance for y

Source              DF          SS          MS          F          P
part                19    1185.425      62.391      62.92      0.000
operator             2       2.617       1.308       1.84      0.173
part*operator       38      27.050       0.712       0.72      0.861
Error               60      59.500       0.992
Total              119    1274.592

Source            Variance    Error   Expected Mean Square for Each Term
                  component   term    (using restricted model)
1 part             10.2332      4     (4) + 6(1)
2 operator                      3     (4) + 2(3) + 40Q[2]
3 part*operator    -0.1399      4     (4) + 2(3)
4 Error             0.9917            (4)
```

Alternate Mixed Models. Several different versions of the mixed model have been proposed. These models differ from the restricted version of the mixed model discussed previously in the assumptions made about the random components. One of these alternate models is now briefly discussed.

Consider the model

$$y_{ijk} = \mu + \alpha_i + \gamma_j + (\alpha\gamma)_{ij} + \epsilon_{ijk}$$

where the α_i ($i = 1, 2, \ldots, a$) are fixed effects such that $\sum_{i=1}^{a}\alpha_i = 0$ and γ_j, $(\alpha\gamma)_{ij}$, and ϵ_{ijk} are uncorrelated random variables having zero means and variances $V(\gamma_j) = \sigma_\gamma^2$, $V[(\alpha\gamma)_{ij}] = \sigma_{\alpha\gamma}^2$, and $V(\epsilon_{ijk}) = \sigma^2$. Note that the restriction imposed previously on the interaction effect is not used here; consequently, this version of the mixed model is often called the **unrestricted mixed model**.

We can show that expected mean squares for this model are (refer to the supplemental text material for this chapter)

$$\begin{aligned} E(MS_A) &= \sigma^2 + n\sigma_{\alpha\gamma}^2 + \frac{bn\sum_{i=1}^{a}\alpha_i^2}{a-1} \\ E(MS_B) &= \sigma^2 + n\sigma_{\alpha\gamma}^2 + an\sigma_\gamma^2 \\ E(MS_{AB}) &= \sigma^2 + n\sigma_{\alpha\gamma}^2 \end{aligned} \tag{13.12}$$

and

$$E(MS_E) = \sigma^2$$

Comparing these expected mean squares with those in Equation 13.9, we note that the only obvious difference is the presence of the variance component $\sigma^2_{\alpha\gamma}$ in the expected mean square for the random effect. (Actually, there are other differences because of the different definitions of the variance of the interaction effect in the two models.) Consequently, we would test the hypothesis that the variance component for the random effect equals zero (H_0: $\sigma^2_\gamma = 0$) using the statistic

$$F_0 = \frac{MS_B}{MS_{AB}}$$

as contrasted with testing H_0: $\sigma^2_\beta = 0$ with $F_0 = MS_B/MS_E$ in the restricted model.

The parameters in the two models are closely related. In fact, we may show that

$$\begin{aligned} \tau_i &= \alpha_i \\ \beta_j &= \gamma_j + \overline{(\alpha\gamma)}_{.j} \\ (\tau\beta)_{ij} &= (\alpha\gamma)_{ij} - \overline{(\alpha\gamma)}_{.j} \\ \sigma^2_\gamma &= \sigma^2_\beta + \frac{1}{a}\sigma^2_{\alpha\gamma} \end{aligned}$$

and

$$\sigma^2_{\tau\beta} = \sigma^2_{\alpha\gamma}$$

The analysis of variance method may be used to estimate the variance components. Referring to the expected mean squares, we find that the only change from Equations 13.11 is that

$$\hat{\sigma}^2_\gamma = \frac{MS_B - MS_{AB}}{an} \quad \textbf{(13.13)}$$

Both of these models are special cases of the mixed model proposed by Scheffé (1956a, 1959). This model assumes that the observations may be represented by

$$y_{ijk} = m_{ij} + \epsilon_{ijk} \begin{cases} i = 1, 2, \ldots, a \\ j = 1, 2, \ldots, b \\ k = 1, 2, \ldots, n \end{cases}$$

where m_{ij} and ϵ_{ijk} are independent random variables. The structure of m_{ij} is

$$\begin{aligned} m_{ij} &= \mu + \tau_i + b_j + c_{ij} \\ E(m_{ij}) &= \mu + \tau_i \\ \sum_{i=1}^{a} \tau_i &= 0 \end{aligned}$$

and

$$c_{.j} = 0 \qquad j = 1, 2, \ldots, b$$

The variances and covariances of b_j and c_{ij} are expressed through the covariances of the m_{ij}. Furthermore, the random effect parameters in other formulations of the mixed model can be related to b_j and c_{ij}. The statistical analysis of Scheffé's model is identical to that for our restricted model, except that in general the statistic MS_A/MS_{AB} is not always distributed as F when H_0: $\tau_i = 0$ is true.

In light of this multiplicity of mixed models, a logical question is: Which model should one use? This author prefers the restricted model, although both restricted and unrestricted models are widely encountered in the literature. The restricted model is actually slightly more general than the unrestricted model, because in the restricted model the covariance between two observations from the same level of the random factor can be either positive or negative, whereas this covariance can only be positive in the unrestricted model. If the correlative structure of the random components is not large, then either mixed model is appropriate, and there are only minor differences between these models. On the contrary, the unrestricted form of the mixed model is preferred when the design is unbalanced, because it is easier to work with, and some computer packages always assume the unrestricted model when displaying expected mean squares. (SAS is an example, JMP uses the unrestricted model, and the default in Minitab is the unrestricted model, although that can be easily changed.) When we subsequently refer to mixed models, we assume the restricted model structure. However, if there are large correlations in the data, then Scheffé's model may have to be employed. The choice of model should always be dictated by the data. The article by Hocking (1973) is a clear summary of various mixed models.

EXAMPLE 13.3 The Unrestricted Model

Some computer software packages support only one mixed model. Minitab supports both the restricted and unrestricted model, although as noted above the default is to the unrestricted model. Table 13.5 shows the Minitab output for the experiment in Example 13.2 using the unrestricted model. Note that the expected mean squares are in agreement with those in Equation 13.12. The conclusions are identical to those from the restricted model analysis, and the variance component estimates are very similar.

TABLE 13.5

Analysis of the Experiment in Example 13.2 Using the Unrestricted Model

```
Analysis of Variance (Balanced Designs)

Factor    Type   Levels  Values
Part    random     20      1      2      3      4      5      6      7
                           8      9     10     11     12     13     14
                          15     16     17     18     19     20
operator fixed      3      1      2      3

Analysis of Variance for y

Source            DF         SS         MS        F        P
part              19   1185.425     62.391    87.65    0.000
operator           2      2.617      1.308     1.84    0.173
part*operator     38     27.050      0.712     0.72    0.861
Error             60     59.500      0.992
Total            119   1274.592

Source           Variance   Error   Expected Mean Square for Each Term
                 component  term    (using unrestricted model)
1 part            10.2798     3     (4) + 2(3) + 6(1)
2 operator                    3     (4) + 2(3) + Q[2]
3 part*operator   -0.1399     4     (4) + 2(3)
4 Error            0.9917           (4)
```

JMP can also analyze the mixed model. JMP uses the REML method for variance component estimation. Table 13.6 is the JMP output for the two-factor mixed model in Example 13.2. Recall that this is a measurement systems capability study, where now the parts are random but the operators are fixed. The JMP output includes both variance components estimates and tests for the fixed effects. JMP assumes the unrestricted form of the mixed model, so the results differ slightly from the previous analysis of this experiment given in Table 13.4 where the restricted form of the mixed model was employed.

■ **TABLE 13.6**
JMP Output for the Two-Factor Mixed Model in Example 13.2

Response Y

Summary of Fit

RSquare	0.911896
RSquare Adj	0.91039
Root Mean Square Error	0.995825
Mean of Response	22.39167
Observations (or Sum Wgts)	120

REML Variance Component Estimates

Random Effect	Var Ratio	Var Component	Std Error	95% Lower	9% Upper	Pct of Total
Parts	10.36621	10.279825	3.3738173	3.6672642	16.892385	92.348
Parts*Operators	−0.141088	−0.139912	0.1219114	−0.378854	0.0990296	−1.257
Residual		0.9916667	0.1810527	0.7143057	1.4697982	8.909
Total	11.131579					100.000

−2 LogLikelihood = 410.4121524

Covariance Matrix of Variance Component Estimates

Random Effect	Parts	Parts*Operators	Residual
Parts	11.382643	−0.002222	2.659e-14
Parts*Operators	−0.002222	0.0148624	−0.01639
Residual	2.659e-14	−0.01639	0.0327801

Fixed Effects Tests

Source	Nparm	DF	DFDen	F Ratio	Prob > F
Operators	2	2	38	1.8380	0.1730

13.4 Sample Size Determination with Random Effects

When designing experiments with one or more random factors it is useful to have a quantitative approach to determining sample sizes. We begin with the single-factor random effects model of Chapter 3. The probability of type II error for this random model is

$$\beta = 1 - P\{\text{Reject } H_0 | H_0 \text{ is false}\}$$

$$= 1 - P\{F_0 > F_{\alpha,a-1,N-a} | \sigma_\tau^2 > 0\} \tag{13.14}$$

$$\beta = 0.20$$

The distribution of the test statistic

$$F_0 = \frac{MS_{\text{Treatments}}}{MS_E} \tag{13.15}$$

under the alternative hypothesis (that is, when the alternative hypothesis is true) is needed. It can be shown that if H_1 is true ($\sigma_\tau^2 > 0$) the distribution of F_0 is the central F distribution with $a - 1$ numerator and $N - a$ denominator degrees of freedom. While it is possible to use tables of the F distribution to evaluate Equation 13.14, it is awkward to do so because only a fixed number of tail areas or probabilities are usually available. For example, the Appendix tables in this book only provide probabilities of 0.25, 0.10, 0.05, 0.025, and 0.01. It is simpler to evaluate the power of the test using a computer software program. Programs to calculate probabilities from common distributions are widely available. This feature is provided in JMP.

Recall that the expected values of the mean squares in the test statistic Equation 13.15 are

$$E(MS_{\text{Treatments}}) = \sigma^2 + n\sigma_\tau^2$$

and

$$E(MS_E) = \sigma^2.$$

So when the null hypothesis is true and $\sigma_\tau^2 = 0$ the value of the test statistic F_0 is expected to be unity and when the null hypothesis is false the magnitude is expected to be $n\sigma_\tau^2/\sigma^2$. Therefore, for a given ample size n two unknown parameters are required to calculate the power of the test. We may be able to estimate σ_τ^2 if we have an idea about how much variability in the population of treatments it is important to detect. An estimate of σ^2 may sometimes be chosen based on prior experience or judgment of the experimenter. Often we define the ratio σ_τ^2/σ^2 that is of interest. Small values of this ratio (such as $\sigma_\tau^2/\sigma^2 = 1$) would be appropriate when we are interested in detecting small differences between treatments, and larger values (such as $\sigma_\tau^2/\sigma^2 = 2$) would be selected when we want to detect larger differences between treatments in the population.

EXAMPLE 13.4

Suppose that we have $a = 5$ treatments selected at random from a large population of treatments and the experiment has been replicated $n = 6$ times. We want to find the power of the test if $\alpha = 0.05$ and σ_τ^2 is equal to the experimental error variance σ^2. Therefore, the ratio $\sigma_\tau^2/\sigma^2 = 1$ and $F_{\alpha,a-1,N-a} = F_{0.05,4,25} = 2.76$. Computer software was used to evaluate Equation 13.14 resulting in $\beta = 0.20$, so the power of the test is 0.80. If we had been interested in larger differences between treatments, say $\sigma_\tau^2/\sigma^2 = 2$, then $\beta = 0.038$ and the power would have been much higher, $1 - 0.038$ or 0.962.

We can also investigate the choice of sample size to give a specified power of the test. Consider the situation in Example 13.4, where there are $a = 5$ treatments and we want to determine the sample size if the difference in treatments is expected to be large, say $\sigma_\tau^2/\sigma^2 = 2$, for the case where $\alpha = 0.05$ and the desired power is at least 0.85. We can evaluate Equation 13.14 for different sample sizes to obtain the following results:

$$n = 2: \beta = 0.55, \text{Power} = 1 - 0.55 = 0.45$$

$$n = 3: \beta = 0.20, \text{Power} = 1 - 0.20 = 0.80$$

$$n = 4: \beta = 0.08, \text{Power} = 1 - 0.08 = 0.92$$

Therefore, $n = 4$ replicates would be adequate to give a power that is at least equal to the desired value of 0.85.

It is also possible to perform similar sample size and power calculations for multifactor experiments with random effects. For example, consider the two-factor random model. Recall that the expected mean squares for this model are:

$$E(MS_A) = \sigma^2 + n\sigma_{\tau\beta}^2 + bn\sigma_\tau^2$$

$$E(MS_B) = \sigma^2 + n\sigma_{\tau\beta}^2 + an\sigma_\beta^2$$

$$E(MS_{AB}) = \sigma^2 + n\sigma_{\tau\beta}^2$$

$$E(MS_E) = \sigma^2$$

So for testing hypotheses about the variance component σ_τ^2 associated with the random effect factor A, it is the magnitude of the ratio $bn\sigma_\tau^2/(\sigma^2 + n\sigma_{\tau\beta}^2)$ that must be taken into account in using Equation 13.14 in computing the type II error probability. This could be done by assuming specific values for the individual variance components or by assuming a value for the ratio. For example suppose that we have $b = 3$ levels of factor B and that the variance component for the interaction is twice the size of the error variance ($\sigma^2 = 1$, $\sigma_{\tau\beta}^2 = 2$). There are 3 levels of factor A and we think that the variance component for factor A is as large as $\sigma_\tau^2 = 2$. If we use $n = 3$ replicates of this design then the type II error probability is approximately $\beta = 0.55$. If a smaller type II error rate or a larger power is required, more replication will be necessary.

13.5 Rules for Expected Mean Squares

An important part of any experimental design problem is conducting the analysis of variance. This involves determining the sum of squares for each component in the model and the number of degrees of freedom associated with each sum of squares. Then, to construct appropriate test statistics, the expected mean squares must be determined. In complex design situations, particularly those involving random or mixed models, it is frequently helpful to have a formal procedure for this process.

We will present a set of rules for writing down the number of degrees of freedom for each model term and the expected mean squares for any balanced factorial, nested[1], or nested factorial experiment. (Note that partially balanced arrangements, such as Latin squares and incomplete block designs, are specifically excluded.) Other rules are available; for example, see Scheffé (1959), Bennett and Franklin (1954), Cornfield and Tukey (1956), and Searle (1971a, 1971b). By examining the expected mean squares, one may develop the appropriate statistic for testing hypotheses about any model parameter. The test statistic is a ratio of mean squares that is chosen such that the expected value of the **numerator** mean square differs from the expected value of the **denominator** mean square only by the variance component or the fixed factor in which we are interested.

It is always possible to determine the expected mean squares in any model as we did in Chapter 3—that is, by the direct application of the expectation operator. This **brute force** method, as it is often called, can be very tedious. The rules that follow always produce the expected mean squares without resorting to the brute force approach, and they are relatively simple to use. We illustrate the rules using the two-factor fixed effects factorial model assuming that there are n replicates.

Rule 1. The error term in the model is $\epsilon_{ij\ldots m}$, where the subscript m denotes the replication subscript. For the two-factor model, this rule implies that the error term is ϵ_{ijk}. The variance component associated with $\epsilon_{ij\ldots m}$ is σ^2.

Rule 2. In addition to an overall mean (μ) and an error term $\epsilon_{ij\ldots m}$, the model contains all the main effects and any interactions that the experimenter assumes exist. If all possible interactions between k factors exist, then there are $\binom{k}{2}$ two-factor interactions, $\binom{k}{3}$ three-factor interactions, . . . , 1 k-factor interaction. If one of the factors in a term appears in parentheses, then there is no interaction between that factor and the other factors in that term.

Rule 3. For each term in the model other than μ and the error term, divide the subscripts into three classes: (a) live—those subscripts that are present in the term and are not in parentheses; (b) dead—those subscripts that are present in the term and are in parentheses; and (c) absent—those subscripts that are present in the model but not in that particular term. Note that the two-factor fixed effects model has no dead subscripts, but we will encounter such models later. Thus, in the two-factor model, for the term $(\tau\beta)_{ij}$, i and j are live and k is absent.

Rule 4. Degrees of freedom. The number of degrees of freedom for any effect in the model is the product of the number of levels associated with each dead subscript and the number of levels minus 1 associated with each live subscript. For example, the number of degrees of freedom associated with $(\tau\beta)_{ij}$ is $(a-1)(b-1)$. The number of degrees of freedom for error is obtained by subtracting the sum of all other degrees of freedom from $N-1$, where N is the total number of observations.

Rule 5. Each term in the model has either a variance component (random effect) or a fixed factor (fixed effect) associated with it. If an interaction contains at least one random effect, the entire interaction is considered as random. A variance component has Greek letters as subscripts to identify the particular random effect. Thus, in a two-factor mixed model with factor A fixed and factor B random, the variance component for B is σ^2_β, and the variance component for AB is $\sigma^2_{\tau\beta}$. A fixed effect is always

[1]Nested designs are discussed in Chapter 14.

represented by the sum of squares of the model components associated with that factor divided by its degrees of freedom. In our example, the fixed effect for A is

$$\frac{\sum_{i=1}^{a} \tau_i^2}{a-1}$$

Rule 6. Expected Mean Squares. There is an expected mean square for each model component. The expected mean square for error is $E(MS_E) = \sigma^2$. In the case of the **restricted model,** for every other model term, the expected mean square contains σ^2 plus either the variance component or the fixed effect component for that term, plus those components for all other model terms that contain the effect in question and that involve no interactions with other fixed effects. The coefficient of each variance component or fixed effect is the number of observations at each distinct value of that component.

To illustrate for the case of the two-factor fixed effects model, consider finding the interaction expected mean square, $E(MS_{AB})$. The expected mean square will contain only the fixed effect for the AB interaction (because no other model terms contain AB) plus σ^2, and the fixed effect for AB will be multiplied by n because there are n observations at each distinct value of the interaction component (the n observations in each cell). Thus, the expected mean square for AB is

$$E(MS_{AB}) = \sigma^2 + \frac{n \sum_{i=1}^{a} \sum_{j=1}^{b} (\tau\beta)_{ij}^2}{(a-1)(b-1)}$$

As another illustration of the two-factor fixed effects model, the expected mean square for the main effect of A would be

$$E(MS_A) = \sigma^2 + \frac{bn \sum_{i=1}^{a} \tau_i^2}{(a-1)}$$

The multiplier in the numerator is bn because there are bn observations at each level of A. The AB interaction term is not included in the expected mean square because while it does include the effect in question (A), factor B is a fixed effect.

To illustrate how Rule 6 applies to a model with random effects, consider the two-factor **random** model. The expected mean square for the AB interaction would be

$$E(MS_{AB}) = \sigma^2 + n\sigma_{\tau\beta}^2$$

and the expected mean square for the main effect of A would be

$$E(MS_A) = \sigma^2 + n\sigma_{\tau\beta}^2 + bn\sigma_\tau^2$$

Note that the variance component for the AB interaction term is included because A is included in AB and B is a random effect.

Now consider the restricted form of the two-factor mixed model. Once again, the expected mean square for the AB interaction term is

$$E(MS_{AB}) = \sigma^2 + n\sigma_{\tau\beta}^2$$

For the main effect of A, the fixed factor, the expected mean square is

$$E(MS_A) = \sigma^2 + n\sigma_{\tau\beta}^2 + \frac{bn \sum_{i=1}^{a} \tau_i^2}{a-1}$$

The interaction variance component is included because A is included in AB and B is a random effect. For the main effect of B, the expected mean square is

$$E(MS_B) = \sigma^2 + an\sigma_\beta^2$$

Here the interaction variance component is not included, because while B is included in AB, A is a fixed effect. Please note that these expected mean squares agree with those given previously for the two-factor mixed model in Equation 13.9.

Rule 6 can be easily modified to give expected mean squares for the **unrestricted** form of the mixed model. Simply include the term for the effect in question, plus all the terms that contain this effect as long as there is at least one random factor. To illustrate, consider the unrestricted form of the two-factor mixed model. The expected mean square for the two-factor interaction term is

$$E(MS_{AB}) = \sigma^2 + n\sigma^2_{\tau\beta}$$

(Please recall the difference in notation for model components between the restricted and unrestricted models.) For the main effect of A, the fixed factor, the expected mean square is

$$E(MS_A) = \sigma^2 + n\sigma^2_{\tau\beta} + \frac{bn\sum_{i=1}^{a}\tau_i^2}{a-1}$$

and for the main effect of the random factor B, the expected mean square would be

$$E(MS_B) = \sigma^2 + n\sigma^2_{\tau\beta} + an\sigma^2_{\beta}$$

Note that these are the expected mean squares given previously in Equation 13.12 for the unrestricted mixed model.

EXAMPLE 13.5

Consider a three-factor factorial experiment with a levels of factor A, b levels of factor B, c levels of factor C, and n replicates. The analysis of this design, assuming that all the factors are fixed effects, is given in Section 5.4. We now determine the expected mean squares assuming that all the factors are *random*. The appropriate statistical model is

$$y_{ijkl} = \mu + \tau_i + \beta_j + \gamma_k + (\tau\beta)_{ij} + (\tau\gamma)_{ik} + (\beta\gamma)_{jk} + (\tau\beta\gamma)_{ijk} + \epsilon_{ijkl}$$

Using the rules previously described, the expected mean squares are shown in Table 13.7.

We notice, by examining the expected mean squares in Table 13.8, that if A, B, and C are all random factors, then no exact test exists for the main effects. That is, if we wish to test the hypothesis $\sigma^2_\tau = 0$, we cannot form a ratio of two expected mean squares such that the only term in the numerator that is not in the denominator is $bcn\sigma^2_\tau$. The same phenomenon occurs for the main effects of B and C. Notice that proper tests do exist for the two- and three-factor interactions. However, it is likely that tests on the main effects are of central importance to the experimenter. Therefore, how should the main effects be tested? This problem is considered in the next section.

■ **TABLE 13.7**
Expected Mean Squares for the Three-Factor Random Effects Model

Model Term	Factor	Expected Mean Squares
τ_i	A, main effect	$\sigma^2 + cn\sigma^2_{\tau\beta} + bn\sigma^2_{\tau\gamma} + n\sigma^2_{\tau\beta\gamma} + bcn\sigma^2_\tau$
β_j	B, main effect	$\sigma^2 + cn\sigma^2_{\tau\beta} + an\sigma^2_{\beta\gamma} + n\sigma^2_{\tau\beta\gamma} + acn\sigma^2_\beta$
γ_k	C, main effect	$\sigma^2 + bn\sigma^2_{\tau\gamma} + an\sigma^2_{\beta\gamma} + n\sigma^2_{\tau\beta\gamma} + abn\sigma^2_\gamma$
$(\tau\beta)_{ij}$	AB, two-factor interaction	$\sigma^2 + n\sigma^2_{\tau\beta\gamma} + cn\sigma^2_{\tau\beta}$
$(\tau\gamma)_{ik}$	AC, two-factor interaction	$\sigma^2 + n\sigma^2_{\tau\beta\gamma} + bn\sigma^2_{\tau\gamma}$
$(\beta\gamma)_{jk}$	BC, two-factor interaction	$\sigma^2 + n\sigma^2_{\tau\beta\gamma} + an\sigma^2_{\beta\gamma}$
$(\tau\beta\gamma)_{ijk}$	ABC, three-factor interaction	$\sigma^2 + n\sigma^2_{\tau\beta\gamma}$
ϵ_{ijkl}	Error	σ^2

13.6 Approximate *F* Tests

In factorial experiments with three or more factors involving a random or mixed model and certain other, more complex designs, there are frequently no exact test statistics for certain effects in the models. One possible solution to this dilemma is to assume that certain interactions are negligible. To illustrate, if we could reasonably assume that all the two-factor interactions in Example 13.5 are negligible, then we could set $\sigma^2_{\tau\beta} = \sigma^2_{\tau\gamma} = \sigma^2_{\beta\gamma} = 0$, and tests for main effects could be conducted.

Although this seems to be an attractive possibility, we must point out that there must be something in the nature of the process—or some strong prior knowledge—for us to assume that one or more of the interactions are negligible. In general, this assumption is not easily made, nor should it be taken lightly. We should not eliminate certain interactions from the model without conclusive evidence that it is appropriate to do so. A procedure advocated by some experimenters is to test the interactions first, then set at zero those interactions found to be insignificant, and then assume that these interactions are zero when testing other effects in the same experiment. Although sometimes done in practice, this procedure can be dangerous because any decision regarding an interaction is subject to both type I and type II errors.

A variation of this idea is to **pool** certain mean squares in the analysis of variance to obtain an estimate of error with more degrees of freedom. For instance, suppose that in Example 13.6 the test statistic $F_0 = MS_{ABC}/MS_E$ was not significant. Thus, $H_0\colon \sigma^2_{\tau\beta\gamma} = 0$ is not rejected, and *both* MS_{ABC} and MS_E estimate the error variance σ^2. The experimenter might consider pooling or combining MS_{ABC} and MS_E according to

$$MS_{E'} = \frac{abc(n-1)MS_E + (a-1)(b-1)(c-1)MS_{ABC}}{abc(n-1) + (a-1)(b-1)(c-1)}$$

so that $E(MS_{E'}) = \sigma^2$. Note that $MS_{E'}$ has $abc(n-1) + (a-1)(b-1)(c-1)$ degrees of freedom, compared to $abc(n-1)$ degrees of freedom for the original MS_E.

The danger of pooling is that one may make a type II error and combine the mean square for a factor that really *is* significant with error, thus obtaining a new residual mean square ($MS_{E'}$) that is too large. This will make other significant effects more difficult to detect. On the contrary, if the original error mean square has a very small number of degrees of freedom (e.g., less than six), the experimenter may have much to gain by pooling because it could potentially increase the precision of further tests considerably. A reasonably practical procedure is as follows. If the original error mean square has six or more degrees of freedom, do not pool. If the original error mean square has fewer than six degrees of freedom, pool only if the *F* statistic for the mean square to be pooled is not significant at a large value of α, such as $\alpha = 0.25$.

If we cannot assume that certain interactions are negligible and we still need to make inferences about those effects for which exact tests do not exist, a procedure attributed to Satterthwaite (1946) can be employed. Satterthwaite's method uses **linear combinations of mean squares**, for example,

$$MS' = MS_r + \cdots + MS_s \tag{13.17}$$

and

$$MS'' = MS_u + \cdots + MS_v \tag{13.18}$$

where the mean squares in Equations 13.31 and 13.32 are chosen so that $E(MS') - E(MS'')$ is equal to a multiple of the effect (the model parameter or variance component) considered in the null hypothesis. Then the test statistic would be

$$F = \frac{MS'}{MS''} \tag{13.19}$$

which is distributed approximately as $F_{p,q}$, where

$$p = \frac{(MS_r + \cdots + MS_s)^2}{MS_r^2/f_r + \cdots + MS_s^2/f_s} \quad (13.20)$$

and

$$q = \frac{(MS_u + \cdots + MS_v)^2}{MS_u^2/f_u + \cdots + MS_v^2/f_v} \quad (13.21)$$

In p and q, f_i is the number of degrees of freedom associated with the mean square MS_i. There is no assurance that p and q will be integers, so it may be necessary to interpolate in the tables of the F distribution. For example, in the three-factor random effects model (Table 13.9), it is relatively easy to see that an appropriate test statistic for $H_0: \sigma_\tau^2 = 0$ would be $F = MS'/MS''$, with

$$MS' = MS_A + MS_{ABC}$$

and

$$MS'' = MS_{AB} + MS_{AC}$$

The degrees of freedom for F would be computed from Equations 13.20 and 13.21.

The theory underlying this test is that both the numerator and the denominator of the test statistic (Equation 13.19) are distributed approximately as multiples of chi-square random variables, and because no mean square appears in both the numerator or denominator of Equation 13.19, the numerator and denominator are independent. Thus F in Equation 13.19 is distributed approximately as $F_{p,q}$. Satterthwaite remarks that caution should be used in applying the procedure when some of the mean squares in MS' and MS'' are involved negatively. Gaylor and Hopper (1969) report that if $MS' = MS_1 - MS_2$, then Satterthwaite's approximation holds reasonably well if

$$\frac{MS_1}{MS_2} > F_{0.025,f_2,f_1} \times F_{0.50,f_2,f_2}$$

and if $f_1 \leq 100$ and $f_2 \geq f_1/2$.

EXAMPLE 13.6

The pressure drop measured across an expansion valve in a turbine is being studied. The design engineer considers the important variables that influence pressure drop reading to be gas temperature on the inlet side (A), operator (B), and the specific pressure gauge used by the operator (C). These three factors are arranged in a factorial design, with gas temperature fixed, and operator and pressure gauge random. The coded data for two replicates are shown in Table 13.8. The linear model for this design is

$$y_{ijkl} = \mu + \tau_i + \beta_j + \gamma_k + (\tau\beta)_{ij} + (\tau\gamma)_{ik} + (\beta\gamma)_{jk} + (\tau\beta\gamma)_{ijk} + \epsilon_{ijkl}$$

where τ_i is the effect of the gas temperature (A), β_j is the operator effect (B), and γ_k is the effect of the pressure gauge (C).

The analysis of variance is shown in Table 13.9. A column entitled Expected Mean Squares has been added to this table, and the entries in this column are derived using the rules discussed in Section 13.5. From the Expected Mean Squares column, we observe that exact tests exist for all effects except the main effect A. Results for these tests are shown in Table 13.10. To test the gas temperature effect, or $H_0: \tau_i = 0$, we could use the statistic

$$F = \frac{MS'}{MS''}$$

where

$$MS' = MS_A + MS_{ABC}$$

and

$$MS'' = MS_{AB} + MS_{AC}$$

because

$$E(MS') - E(MS'') = \frac{bcn\Sigma\tau_i^2}{a-1}$$

■ **TABLE 13.8**

Coded Pressure Drop Data for the Turbine Experiment

	Gas Temperature (A)											
Pressure Gauge	60°F				75°F				90°F			
	Operator (B)				Operator (B)				Operator (B)			
(C)	1	2	3	4	1	2	3	4	1	2	3	4
1	−2	0	−1	4	14	6	1	−7	−8	−2	−1	−2
	−3	−9	−8	4	14	0	2	6	−8	20	−2	1
2	−6	−5	−8	−3	22	8	6	−5	−8	1	−9	−8
	4	−1	−2	−7	24	6	2	2	3	−7	−8	3
3	−1	−4	0	−2	20	2	3	−5	−2	−1	−4	1
	−2	−8	−7	4	16	0	0	−1	−1	−2	−7	3

■ **TABLE 13.9**

Analysis of Variance for the Pressure Drop Data

Source of Variation	Sum of Squares	Degrees of Freedom	Expected Mean Squares	Mean Square	F_0	P-value
Temperature, A	1023.36	2	$\sigma^2 + bn\sigma^2_{\tau\gamma} + cn\sigma^2_{\tau\beta} + n\sigma^2_{\tau\beta\gamma} + \frac{bcn\sum\tau_i^2}{a-1}$	511.68	2.22	0.17
Operator, B	423.82	3	$\sigma^2 + an\sigma^2_{\beta\gamma} + acn\sigma^2_{\beta}$	141.27	4.05	0.07
Pressure gauge, C	7.19	2	$\sigma^2 + an\sigma^2_{\beta\gamma} + abn\sigma^2_{\gamma}$	3.60	0.10	0.90
AB	1211.97	6	$\sigma^2 + n\sigma^2_{\tau\beta\gamma} + cn\sigma^2_{\tau\beta}$	202.00	14.59	<0.01
AC	137.89	4	$\sigma^2 + n\sigma^2_{\tau\beta\gamma} + bn\sigma^2_{\tau\gamma}$	34.47	2.49	0.10
BC	209.47	6	$\sigma^2 + an\sigma^2_{\beta\gamma}$	34.91	1.63	0.17
ABC	166.11	12	$\sigma^2 + n\sigma^2_{\tau\beta\gamma}$	13.84	0.65	0.79
Error	770.50	36	σ^2	21.40		
Total	3950.32	71				

To determine the test statistic for $H_0: \tau_i = 0$, we compute

$$MS' = MS_A + MS_{ABC} = 511.68 + 13.84 = 525.52$$
$$MS'' = MS_{AB} + MS_{AC} = 202.00 + 34.47 = 236.47$$

and

$$F = \frac{MS'}{MS''} = \frac{525.52}{236.47} = 2.22$$

The degrees of freedom for this statistic are found from Equations 13.20 and 13.21 as follows:

$$p = \frac{(MS_A + MS_{ABC})^2}{(MS_A^2/2) + (MS_{ABC}^2/12)} = \frac{(525.52)^2}{[(511.68)^2/2] + [(13.84)^2/12]} = 2.11 \simeq 2$$

and

$$q = \frac{(MS_{AB} + MS_{AC})^2}{(MS_{AB}^2/6) + (MS_{AC}^2/4)} = \frac{(236.47)^2}{[(202.00)^2/6] + [(34.47)^2/4]} = 7.88 \simeq 8$$

Comparing $F = 2.22$ to $F_{0.05,2,8} = 4.46$, we cannot reject H_0. The P-value is approximately 0.17.

The AB, or temperature–operator, interaction is large, and there is some indication of an AC, or temperature–gauge, interaction. The graphical analysis of the AB and AC interactions, shown in Figure 13.2, indicates that the effect of temperature may be large when operator 1 and gauge 3 are used. Thus, it seems possible that the main effects of temperature and operator are masked by the large AB interaction.

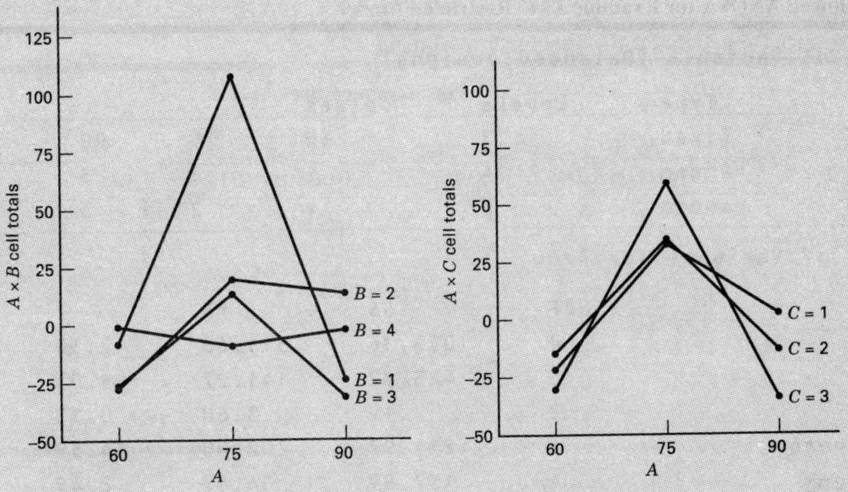

■ **FIGURE 13.2** **Interactions in pressure drop experiment**

Table 13.10 presents the Minitab Balanced ANOVA output for the experiment in Example 13.6. We have specified the **restricted model**. *Q*[1] represents the fixed effect of gas pressure. Notice that the entries in the analysis of variance table are in general agreement with those in Table 13.9, except for the *F* test on gas temperature (factor *A*). Minitab notes that the test is not an exact test (which we see from the expected mean squares). The Synthesized Test constructed by Minitab is actually Satterthwaite's procedure, but it uses a different test statistic than we did. Note that, from the Minitab output, the error mean square for testing factor *A* is

$$(4) + (5) - (7) = MS_{AB} + MS_{AC} - MS_{ABC}$$

for which the expected value is

$$\begin{aligned} E[(4) + (5) - (7)] &= \sigma^2 + n\sigma^2_{\tau\beta\gamma} + cn\sigma^2_{\tau\beta} + \sigma^2 + n\sigma^2_{\tau\beta\gamma} \\ &\quad + bn\sigma^2_{\tau\gamma} - (\sigma^2 + n\sigma^2_{\tau\beta\gamma}) \\ &= \sigma^2 + n\sigma^2_{\tau\beta\gamma} + cn\sigma^2_{\tau\beta} + bn\sigma^2_{\tau\gamma} \end{aligned}$$

which is an appropriate error mean square for testing the mean effect of *A*. This nicely illustrates that there can be more than one way to construct the synthetic mean squares used in Satterthwaite's procedure. However, we would generally prefer the linear combination of mean squares we selected instead of the one chosen by Minitab because it does not have any mean squares involved negatively in the linear combinations.

The analysis of Example 13.6, assuming the unrestricted model, is presented in Table 13.11. The principal difference from the restricted model is that now the expected values of the mean squares for all three mean effects are such that no exact test exists. In the restricted model, the two random mean effects could be tested against their interaction, but now the expected mean square for *B* involves $\sigma^2_{\tau\beta\gamma}$ and $\sigma^2_{\tau\beta}$, and the expected mean square for *C* involves $\sigma^2_{\tau\beta\gamma}$ and $\sigma^2_{\tau\gamma}$. Once again, Minitab constructs synthetic mean squares and tests these effects with Satterthwaite's procedure. The overall conclusions are not radically different from the restricted model analysis, other than the large change in the estimate of the operator variance component. The unrestricted model produces a negative estimate of σ^2_{β}. Because the gauge factor is not significant in either analysis, it is possible that some model reduction is in order.

■ TABLE 13.10
Minitab Balanced ANOVA for Example 13.6, Restricted Model

```
Analysis of Variance (Balanced Designs)

Factor        Type      Levels    Values
GasT          fixed          3       60      75      90
Operator     random          4        1       2       3       4
Gauge        random          3        1       2       3

Analysis of Variance for Drop

Source                  DF        SS          MS        F         P
GasT                     2     1023.36      511.68     2.30     0.171 x
Operator                 3      423.82      141.27     4.05     0.069
Gauge                    2        7.19        3.60     0.10     0.904
GasT*Operator            6     1211.97      202.00    14.59     0.000
GasT*Gauge               4      137.89       34.47     2.49     0.099
Operator*Gauge           6      209.47       34.91     1.63     0.167
GasT*Operator*Gauge     12      166.11       13.84     0.65     0.788
Error                   36      770.50       21.40
Total                   71     3950.32

x Not an exact F test

Source                  Variance   Error  Expected Mean Square for Each Term
                        component  term   (using restricted model)
1 GasT                                *   (8) + 2(7) + 8(5) + 6(4) + 24Q[1]
2 Operator                5.909       6   (8) + 6(6) + 18(2)
3 Gauge                  -1.305       6   (8) + 6(6) + 24(3)
4 GasT*Operator          31.359       7   (8) + 2(7) + 6(4)
5 GasT*Gauge              2.579       7   (8) + 2(7) + 8(5)
6 Operator*Gauge          2.252       8   (8) + 6(6)
7 GasT*Operator*Gauge    -3.780       8   (8) + 2(7)
8 Error                  21.403           (8)

* Synthesized Test

Error Terms for Synthesized Tests

Source          Error DF     Error MS      Synthesis of Error MS
1 GasT            6.97        222.63       (4) + (5) - (7)
```

13.7 Some Additional Topics on Estimation of Variance Components

As we have previously observed, estimating the variance components in a random or mixed model is frequently a subject of considerable importance to the experimenter. In this section, we present some further results and techniques useful in estimating variance components. We concentrate on procedures for finding confidence intervals on variance components.

■ **TABLE 13.11**
Minitab Balanced ANOVA for Example 13.6, Unrestricted Model

Analysis of Variance (Balanced Designs)

Factor	Type	Levels	Values			
GasT	fixed	3	60	75	90	
Operator	random	4	1	2	3	4
Gauge	random	3	1	2	3	

Analysis of Variance for Drop

Source	DF	SS	MS	F	P
GasT	2	1023.36	511.68	2.30	0.171 ×
Operator	3	423.82	141.27	0.63	0.616 ×
Gauge	2	7.19	3.60	0.06	0.938 ×
GasT*Operator	6	1211.97	202.00	14.59	0.000
GasT*Gauge	4	137.89	34.47	2.49	0.099
Operator*Gauge	6	209.47	34.91	2.52	0.081
GasT*Operator*Gauge	12	166.11	13.84	0.65	0.788
Error	36	770.50	21.40		
Total	71	3950.32			

× Not an exact F test

Source	Variance component	Error term	Expected Mean Square for Each Term (using unrestricted model)
1 GasT		*	(8) + 2(7) + 8(5) + 6(4) + Q[1]
2 Operator	−4.544	*	(8) + 2(7) + 6(6) + 6(4) + 18(2)
3 Gauge	−2.164	*	(8) + 2(7) + 6(6) + 8(5) + 24(3)
4 GasT*Operator	31.359	7	(8) + 2(7) + 6(4)
5 GasT*Gauge	2.579	7	(8) + 2(7) + 8(5)
6 Operator*Gauge	3.512	7	(8) + 2(7) + 6(6)
7 GasT*Operator*Gauge	−3.780	8	(8) + 2(7)
8 Error	21.403		(8)

* Synthesized Test

Error Terms for Synthesized Tests

Source	Error DF	Error MS	Synthesis of Error MS
1 GasT	6.97	222.63	(4) + (5) − (7)
2 Operator	7.09	223.06	(4) + (6) − (7)
3 Gauge	5.98	55.54	(5) + (6) − (7)

13.7.1 Approximate Confidence Intervals on Variance Components

When the single-factor random effects model was introduced in Chapter 1, we presented exact $100(1-\alpha)$ percent confidence intervals for σ^2 and for other functions of the variance components in that simple experimental design. It is always possible to find an exact confidence interval on any function of the variance components that is the expected value of one of the mean squares in the analysis of variance. For example, consider the error mean square. Because $E(MS_E) = \sigma^2$, we can always find an exact confidence interval on σ^2 because the quantity

$$f_E MS_E/\sigma^2 = f_E\hat{\sigma}^2/\sigma^2$$

has a chi-square distribution with f_E degrees of freedom. The exact $100(1 - \alpha)$ percent confidence interval is

$$\frac{f_E MS_E}{\chi^2_{\alpha/2, f_E}} \leq \sigma^2 \leq \frac{f_e MS_E}{\chi^2_{1-\alpha/2, f_E}} \tag{13.22}$$

Unfortunately, in more complex experiments involving several design factors, it is generally not possible to find exact confidence intervals on the variance components of interest because these variances are not the expected value of a single mean square from the analysis of variance. The advantage of the REML method is that it produces standard errors of the variance component estimates, so finding approximate confidence intervals is easy. The concepts underlying Satterthwaite's approximate "pseudo" F tests, introduced in Section 13.6, can also be employed to construct approximate confidence intervals on variance components for which no exact CI is available and as an alternative to REML if it is not available.

Recall that Satterthwaite's method uses two linear combinations of mean squares

$$MS' = MS_r + \cdots + MS_s$$

and

$$MS'' = MS_u + \cdots + MS_v$$

with the test statistic

$$F = \frac{MS'}{MS''}$$

having an approximate F distribution. Using appropriate degrees of freedom for MS' and MS'', defined in Equations 13.20 and 13.21, we can use this F statistic in an approximate test of the significance of the parameter or variance component of interest.

For testing the significance of a variance component, say σ_0^2, the two linear combinations, MS' and MS'', are chosen such that the difference in their expected values is equal to a multiple of the component, say

$$E(MS') - E(MS'') = k\sigma_0^2$$

or

$$\sigma_0^2 = \frac{E(MS') - E(MS'')}{k}. \tag{13.23}$$

Equation 13.23 provides a basis for a point estimate of σ_0^2:

$$\hat{\sigma}_0^2 = \frac{MS' - MS''}{k}$$

$$= \frac{1}{k}MS_r + \cdots + \frac{1}{k}MS_s - \frac{1}{k}MS_u - \cdots - \frac{1}{k}MS_v \tag{13.24}$$

The mean squares (MS_i) in Equation 13.24 are independent with $f_i MS_i/\sigma_i^2 = SS_i/\sigma_i^2$ having chi-square distributions with f_i degrees of freedom. The estimate of the variance component, $\hat{\sigma}_0^2$, is a linear combination of multiples of the mean squares, and $r\hat{\sigma}_0^2/\sigma_0^2$ has an approximate chi-square distribution with r degrees of freedom, where

$$r = \frac{(\hat{\sigma}_0^2)^2}{\sum_{i=1}^{m} \frac{1}{k^2}\frac{MS_i^2}{f_i}}$$

$$= \frac{(MS_r + \cdots + MS_s - MS_u - \cdots - MS_v)^2}{\frac{MS_r^2}{f_r} + \cdots + \frac{MS_s^2}{f_s} + \frac{MS_u^2}{f_u} + \cdots + \frac{MS_v^2}{f_v}} \tag{13.25}$$

This result can only be used if $\hat{\sigma}_0^2 > 0$. As r will usually not be an integer, interpolation from the chi-square tables will generally be required. Graybill (1961) derives a general result for r.

Now because $r\hat{\sigma}_0^2/\sigma^2$ has an approximate chi-square distribution with r degrees of freedom,

$$P\left\{\chi^2_{1-\alpha/2,r} \leq \frac{r\hat{\sigma}_0^2}{\sigma_0^2} \leq \chi^2_{\alpha/2,r}\right\} = 1 - \alpha$$

and

$$P\left\{\frac{r\hat{\sigma}_0^2}{\chi^2_{\alpha/2,r}} \leq \sigma_0^2 \leq \frac{r\hat{\sigma}_0^2}{\chi^2_{1-\alpha/2,r}}\right\} = 1 - \alpha$$

Therefore, an approximate $100(1 - \alpha)$ percent confidence interval on σ_0^2 is

$$\frac{r\hat{\sigma}_0^2}{\chi^2_{\alpha/2,r}} \leq \sigma_0^2 \leq \frac{r\hat{\sigma}_0^2}{\chi^2_{1-\alpha/2,r}} \tag{13.26}$$

EXAMPLE 13.8

To illustrate this procedure, reconsider the experiment in Example 13.6, where a three-factor mixed model is used on a study of the pressure drop across an expansion valve of a turbine. The model is

$$y_{ijkl} = \mu + \tau_i + \beta_j + \gamma_k + (\tau\beta)_{ij} + (\tau\gamma)_{ik} + (\beta\gamma)_{jk} + (\tau\beta\gamma)_{ijk} + \epsilon_{ijkl}$$

where τ_i is a fixed effect and all other effects are random. We will find an approximate confidence interval on $\sigma^2_{\tau\beta}$. Using the expected mean squares in Table 13.10, we note that the difference in the expected values of the mean squares for the two-way interaction effect AB and the three-way interaction effect ABC is a multiple of the variance component of interest, $\sigma^2_{\tau\beta}$:

$$E(MS_{AB}) - E(MS_{ABC}) = \sigma^2 + n\sigma^2_{\tau\beta\gamma} + cn\sigma^2_{\tau\beta} - (\sigma^2 + n\sigma^2_{\tau\beta\gamma}) = cn\sigma^2_{\tau\beta}$$

Therefore, the point estimate of $\sigma^2_{\tau\beta}$ is

$$\hat{\sigma}^2_{\tau\beta} = \frac{MS_{AB} - MS_{ABC}}{cn} = \frac{202.00 - 13.84}{(3)(2)} = 31.36$$

and

$$r = \frac{(MS_{AB} - MS_{ABC})^2}{\dfrac{MS^2_{AB}}{(a-1)(b-1)} + \dfrac{MS^2_{ABC}}{(a-1)(b-1)(c-1)}} = \frac{(202.00 - 13.84)^2}{\dfrac{(202.00)^2}{(2)(3)} + \dfrac{(13.84)^2}{(2)(3)(2)}} = 5.19$$

The approximate 95 percent confidence interval on $\sigma^2_{\tau\beta}$ is then found from Equation 13.40 as follows:

$$\frac{r\hat{\sigma}^2_{\tau\beta}}{\chi^2_{0.025,r}} \leq \sigma^2_{\tau\beta} \leq \frac{r\hat{\sigma}^2_{\tau\beta}}{\chi^2_{0.975,r}}$$

$$\frac{(5.19)(31.36)}{13.14} \leq \sigma^2_{\tau\beta} \leq \frac{(5.19)(31.36)}{0.90}$$

$$12.39 \leq \sigma^2_{\tau\beta} \leq 180.84$$

This result is consistent with the results of the exact F test on $\sigma^2_{\tau\beta}$, in that there is strong evidence that this variance component is not zero.

13.7.2 The Modified Large-Sample Method

The Satterthwaite method in the previous section is a relatively simple way to find an approximate confidence interval on a variance component that can be expressed as a linear combination of mean squares, say

$$\hat{\sigma}_0^2 = \sum_{i=1}^{Q} c_i MS_i \tag{13.27}$$

The Satterthwaite method works well when the degrees of freedom on each mean square MS_i are all relatively large and when the constants c_i in Equation 13.27 are all positive. However, often some of the c_i are negative. Graybill and Wang (1980) proposed a procedure called the **modified large-sample method**, which can be a very useful alternative to Satterthwaite's method. If all of the constants c_i in Equation 13.27 are positive, then the approximate 100 $(1 - \alpha)$ percent modified large-sample confidence interval on σ_0^2 is

$$\hat{\sigma}_0^2 - \sqrt{\sum_{i=1}^{Q} G_i^2 c_i^2 MS_i^2} \le \sigma_0^2 \le \hat{\sigma}_0^2 + \sqrt{\sum_{i=1}^{Q} H_i^2 c_i^2 MS_i^2} \tag{13.28}$$

where

$$G_i = 1 - \frac{1}{f_{\alpha, f_i, \infty}} \quad \text{and} \quad H_i = \frac{1}{F_{1-\alpha, f_i, \infty}} - 1$$

Note that an F random variable with an infinite number of denominator degrees of freedom is equivalent to a chi-square random variable divided by its degrees of freedom.

Now consider the more general case of Equation 13.27, where the constants c_i are unrestricted in sign. This may be written as

$$\hat{\sigma}_0^2 = \sum_{i=1}^{P} c_i MS_i - \sum_{j=P+1}^{Q} c_j MS_j \quad c_i, c_j \ge 0 \tag{13.29}$$

Ting et al. (1990) give an approximate $100(1 - \alpha)$ percent lower confidence limit on σ_0^2 as

$$L = \hat{\sigma}_0^2 - \sqrt{V_L} \tag{13.30}$$

where

$$V_L = \sum_{i=1}^{P} G_i^2 c_i^2 MS_i^2 + \sum_{j=P+1}^{Q} H_j^2 c_j^2 MS_j^2 + \sum_{i=1}^{P} \sum_{j=P+1}^{Q} G_{ij}^2 c_i c_j MS_i MS_j$$

$$+ \sum_{i=1}^{P-1} \sum_{t>1}^{P} G_{it}^* c_i c_t MS_i MS_t$$

$$G_i = 1 - \frac{1}{F_{\alpha, f_i, \infty}}$$

$$H_j = \frac{1}{F_{1-\alpha, f_j, \infty}} - 1$$

$$G_{ij} = \frac{(F_{\alpha, f_i, f_j} - 1)^2 - G_i^2 F_{\alpha, f_i, f_j}^2 - H_j^2}{F_{\alpha, f_i, f_j}}$$

$$G_{it}^* = \left[\left(\frac{1}{F_{\alpha, f_i + f_t, \infty}}\right)^2 \frac{(f_i + f_t)^2}{f_i f_t} - \frac{G_i^2 f_i}{f_t} - \frac{G_t^2 f_t}{f_i}\right] (P - 1),$$

$$\text{if } P > 1 \text{ and } G_{it}^* = 0 \text{ if } P = 1$$

These results can also be extended to include approximate confidence intervals on ratios of variance components. For a complete account of these methods, refer to the excellent book by Burdick and Graybill (1992). Also see the supplemental material for this chapter.

EXAMPLE 13.8

To illustrate the modified large-sample method, reconsider the three-factor mixed model in Example 13.6. We will find an approximate 95 percent lower confidence interval on $\sigma^2_{\tau\beta}$. Recall that the point estimate of $\sigma^2_{\tau\beta}$ is

$$\hat{\sigma}^2_{\tau\beta} = \frac{MS_{AB} - MS_{ABC}}{cn} = \frac{202.00 - 13.84}{(3)(2)} = 31.359$$

Therefore, in the notation of Equation 13.29, $c_1 = c_2 = \frac{1}{6}$, and

$$G_1 = 1 - \frac{1}{F_{0.05,6,\infty}} = 1 - \frac{1}{2.1} = 0.524$$

$$H_2 = \frac{1}{F_{0.95,12,\infty}} - 1 = \frac{1}{0.435} - 1 = 1.30$$

$$G_{12} = \frac{(F_{0.05,6,12} - 1)^2 - (G_1)^2 F^2_{0.05,6,12} - (H_2)^2}{F_{0.05,6,12}}$$

$$= \frac{(3.00 - 1)^2 - (0.524)^2(3.00)^2 - (1.3)^2}{3.00} = -0.054$$

$$G^*_{1l} = 0$$

From Equation 13.30

$$V_L = G_1^2 c_1^2 MS^2_{AB} + H_2^2 c_2^2 MS^2_{ABC} + G_{12} c_1 c_2 MS_{AB} MS_{ABC}$$

$$= (0.524)^2(1/6)^2(202.00)^2 + (1.3)^2(1/6)^2(13.84)^2 + (-0.054)(1/6)(1/6)(202.00)(13.84)$$

$$= 316.02$$

So an approximate 95 percent lower confidence limit on $\sigma^2_{\tau\beta}$ is

$$L = \hat{\sigma}^2_{\tau\beta} - \sqrt{V_L} = 31.36 - \sqrt{316.02} = 13.58$$

This result is consistent with the results of the exact F test for this effect.

13.8 Problems

13.1. An experiment was performed to investigate the capability of a measurement system. Ten parts were randomly selected, and two randomly selected operators measured each part three times. The tests were made in random order, and the data are shown in Table P13.1.

■ TABLE P13.1

Measurement Systems Data for Problem 13.1

	Operator 1 Measurements			Operator 2 Measurements		
Part No.	**1**	**2**	**3**	**1**	**2**	**3**
1	49	49	50	50	48	51
2	51	52	51	51	51	51
3	53	51	50	54	52	51
4	50	51	50	47	50	51
5	50	49	48	49	49	48
6	52	50	50	52	50	50
7	50	51	51	51	50	50
8	52	50	51	53	48	50
9	50	51	50	51	48	49
10	48	46	49	46	47	48

(a) Analyze the data from this experiment.

(b) Estimate the variance components using the ANOVA method.

13.2. An article by Hoof and Berman ("Statistical Analysis of Power Module Thermal Test Equipment Performance," *IEEE Transactions on Components, Hybrids, and Manufacturing Technology* Vol. 11, pp. 516–520, 1988) describes an experiment conducted to investigate the capability of measurements in thermal impedance (C°/w × 100) on a power module for an induction motor starter. There are

10 parts, three operators, and three replicates. The data are shown in Table P13.2.

■ **TABLE P13.2**
Power Module Thermal Test Equipment Data for Problem 13.2

	Inspector 1			Inspector 2			Inspector 3		
	Test			Test			Test		
Part No.	1	2	3	1	2	3	1	2	3
1	37	38	37	41	41	40	41	42	41
2	42	41	43	42	42	42	43	42	43
3	30	31	31	31	31	31	29	30	28
4	42	43	42	43	43	43	42	42	42
5	28	30	29	29	30	29	31	29	29
6	42	42	43	45	45	45	44	46	45
7	25	26	27	28	28	30	29	27	27
8	40	40	40	43	42	42	43	43	41
9	25	25	25	27	29	28	26	26	26
10	35	34	34	35	35	34	35	34	35

(a) Analyze the data from this experiment, assuming that both parts and operators are random effects.

(b) Estimate the variance components using the analysis of variance method.

(c) Estimate the variance components using the REML method. Use the confidence intervals on the variance components to assist in drawing conclusions.

13.3. Reconsider the data in Problem 5.8. Suppose that both factors, machines and operators, are chosen at random.

(a) Analyze the data from this experiment.

(b) Find point estimates of the variance components using the analysis of variance method.

13.4. Reconsider the data in Problem 5.15. Suppose that both factors are random.

(a) Analyze the data from this experiment.

(b) Estimate the variance components using the ANOVA method.

13.5. Suppose that in Problem 5.13 the furnace positions were randomly selected, resulting in a mixed model experiment. Reanalyze the data from this experiment under this new assumption. Estimate the appropriate model components using the ANOVA method.

13.6. Reanalyze the measurement systems experiment in Problem 13.1, assuming that operators are a fixed factor. Estimate the appropriate model components using the ANOVA method.

13.7. Reanalyze the measurement system experiment in Problem 13.2, assuming that operators are a fixed factor. Estimate the appropriate model components using the ANOVA method.

13.8. In Problem 5.8, suppose that there are only four machines of interest, but the operators were selected at random.

(a) What type of model is appropriate?

(b) Perform the analysis and estimate the model components using the ANOVA method.

13.9. Rework Problem 13.5 using the REML method.

13.10. Rework Problem 13.6 using the REML method.

13.11. Rework Problem 13.7 using the REML method.

13.12. Rework Problem 13.8 using the REML method.

13.13. By application of the expectation operator, develop the expected mean squares for the two-factor factorial, mixed model. Use the restricted model assumptions. Check your results with the expected mean squares given in Equation 13.9 to see that they agree.

13.14. Consider the three-factor factorial design in Example 13.5. Propose appropriate test statistics for all main effects and interactions. Repeat for the case where A and B are fixed and C is random.

13.15. Consider the experiment in Example 13.6. Analyze the data for the case where A, B, and C are random.

13.16. Derive the expected mean squares shown in Table 13.11.

13.17. Consider a four-factor factorial experiment where factor A is at a levels, factor B is at b levels, factor C is at c levels, factor D is at d levels, and there are n replicates. Write down the sums of squares, the degrees of freedom, and the expected mean squares for the following cases. Assume the restricted model for all mixed models. You may use a computer package such as Minitab.

(a) A, B, C, and D are fixed factors.

(b) A, B, C, and D are random factors.

(c) A is fixed and B, C, and D are random.

(d) A and B are fixed and C and D are random.

(e) A, B, and C are fixed and D is random.

Do exact tests exist for all effects? If not, propose test statistics for those effects that cannot be directly tested.

13.18. Reconsider cases (c), (d), and (e) of Problem 13.17. Obtain the expected mean squares assuming the unrestricted model. You may use a computer package such as Minitab. Compare your results with those for the restricted model.

13.19. In Problem 5.19, assume that the three operators were selected at random. Analyze the data under these conditions and draw conclusions. Estimate the variance components.

13.20. Consider the three-factor factorial model

$$y_{ijk} = \mu + \tau_i + \beta_j + \gamma_k + (\tau\beta)_{ij} + (\beta\gamma)_{jk} + \epsilon_{ijk} \begin{cases} i = 1, 2, \ldots, a \\ j = 1, 2, \ldots, b \\ k = 1, 2, \ldots, c \end{cases}$$

Assuming that all the factors are random, develop the analysis of variance table, including the expected mean squares. Propose appropriate test statistics for all effects.

13.21. The three-factor factorial model for a single replicate is

$$y_{ijk} = \mu + \tau_i + \beta_j + \gamma_k + (\tau\beta)_{ij} + (\beta\gamma)_{jk} + (\tau\gamma)_{ik} + (\tau\beta\gamma)_{ijk} + \epsilon_{ijk}$$

If all the factors are random, can any effects be tested? If the three-factor and $(\tau\beta)_{ij}$ interactions do not exist, can all the remaining effects be tested?

13.22. In Problem 5.8, assume that both machines and operators were chosen randomly. Determine the power of the test for detecting a machine effect such that $\sigma_\beta^2 = \sigma^2$, where σ_β^2 is the variance component for the machine factor. Are two replicates sufficient?

13.23. In the two-factor mixed model analysis of variance, show that $\text{Cov}[(\tau\beta)_{ij}, (\tau\beta)_{i'j}] = -(1/a)\sigma_{\tau\beta}^2$ for $i \neq i'$.

13.24. Show that the method of analysis of variance always produces unbiased point estimates of the variance components in any random or mixed model.

13.25. Invoking the usual normality assumptions, find an expression for the probability that a negative estimate of a variance component will be obtained by the analysis of variance method. Using this result, write a statement giving the probability that $\hat{\sigma}_\tau^2 > 0$ in a one-factor analysis of variance. Comment on the usefulness of this probability statement.

13.26. Analyze the data in Problem 13.1, assuming that operators are fixed, using both the unrestricted and the restricted forms of the mixed models. Compare the results obtained from the two models.

13.27. Consider the two-factor mixed model. Show that the standard error of the fixed factor mean (e.g., A) is $[MS_{AB}/bn]^{1/2}$.

13.28. Consider the variance components in the random model from Problem 13.1.

(a) Find an exact 95 percent confidence interval on σ^2.

(b) Find approximate 95 percent confidence intervals on the other variance components using the Satterthwaite method.

13.29. Use the experiment described in Problem 5.8 and assume that both factors are random. Find an exact 95 percent confidence interval on σ^2. Construct approximate 95 percent confidence intervals on the other variance components using the Satterthwaite method.

13.30. Consider the three-factor experiment in Problem 5.19 and assume that operators were selected at random. Find an approximate 95 percent confidence interval on the operator variance component.

13.31. Rework Problem 13.26 using the modified large-sample approach described in Section 13.7.2. Compare the two sets of confidence intervals obtained and discuss.

13.32. Rework Problem 13.28 using the modified large-sample method described in Section 13.7.2. Compare this confidence interval with the one obtained previously and discuss.

13.33. Consider the experiment described in Problem 5.8. Estimate the variance components using the REML method. Compare the confidence intervals to the approximate CIs found in Problem 13.29.

13.34. Consider the experiment in Problem 13.1. Analyze the data using REML. Compare the CIs to those obtained in Problem 13.28.

13.35. Rework Problem 13.31 using REML. Compare all sets of CIs for the variance components.

CHAPTER 14

Experiments with Nested Factors and Hard-to-Change Factors

CHAPTER OUTLINE

The supplemental material is on the textbook website www.wiley.com/go/global/montgomery.

This chapter introduces two important types of experimental designs, the **nested design** and the **split-plot design**. Both of these designs find reasonably widespread application in the industrial use of designed experiments. They also frequently involve one or more **random factors**, and so some of the concepts introduced in Chapter 13 will find application here.

14.1 The Two-Stage Nested Design

In certain multifactor experiments, the levels of one factor (e.g., factor *B*) are similar but not identical for different levels of another factor (e.g., *A*). Such an arrangement is called a **nested**, or **hierarchical, design**, with the levels of factor *B* nested under the levels of factor *A*. For example, consider a company that purchases its raw material from three different suppliers. The company wishes to determine whether the purity of the raw material is the same from each supplier. There are four batches of raw material available from each supplier, and three determinations of purity are to be taken from each batch. The situation is depicted in Figure 14.1.

This is a **two-stage nested design**, with batches nested under suppliers. At first glance, you may ask why this is not a factorial experiment. If this were a factorial, then batch 1 would always refer to the same batch, batch 2 would always refer to the same batch, and so on. This is clearly not the case because the batches from each supplier are **unique** for that particular supplier. That is, batch 1 from supplier 1 has no connection with batch 1 from any other

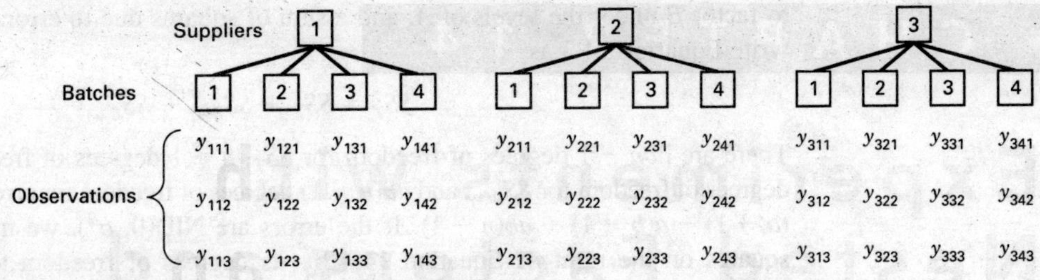

■ **FIGURE 14.1** A two-stage nested design

supplier, batch 2 from supplier 1 has no connection with batch 2 from any other supplier, and so forth. To emphasize the fact that the batches from each supplier are different batches, we may renumber the batches as 1, 2, 3, and 4 from supplier 1; 5, 6, 7, and 8 from supplier 2; and 9, 10, 11, and 12 from supplier 3, as shown in Figure 14.2.

Sometimes we may not know whether a factor is crossed in a factorial arrangement or nested. If the levels of the factor can be renumbered arbitrarily as in Figure 14.2, then the factor is nested.

14.1.1 Statistical Analysis

The linear statistical model for the two-stage nested design is

$$y_{ijk} = \mu + \tau_i + \beta_{j(i)} + \epsilon_{(ij)k} \quad \begin{cases} i = 1, 2, \ldots, a \\ j = 1, 2, \ldots, b \\ k = 1, 2, \ldots, n \end{cases} \tag{14.1}$$

That is, there are a levels of factor A, b levels of factor B nested under each level of A, and n replicates. The subscript $j(i)$ indicates that the jth level of factor B is nested under the ith level of factor A. It is convenient to think of the replicates as being nested within the combination of levels of A and B; thus, the subscript $(ij)k$ is used for the error term. This is a **balanced nested design** because there are an equal number of levels of B within each level of A and an equal number of replicates. Because not every level of factor B appears with every level of factor A, there can be no interaction between A and B.

We may write the total corrected sum of squares as

$$\sum_{i=1}^{a}\sum_{j=1}^{b}\sum_{k=1}^{n}(y_{ijk} - \bar{y}_{...})^2 = \sum_{i=1}^{a}\sum_{j=1}^{b}\sum_{k=1}^{n}[(\bar{y}_{i..} - \bar{y}_{...}) + (\bar{y}_{ij.} - \bar{y}_{i..}) + (y_{ijk} - \bar{y}_{ij.})]^2 \tag{14.2}$$

Expanding the right-hand side of Equation 14.2 yields

$$\sum_{i=1}^{a}\sum_{j=1}^{b}\sum_{k=1}^{n}(y_{ijk} - \bar{y}_{...})^2 = bn\sum_{i=1}^{a}(\bar{y}_{i..} - \bar{y}_{...})^2 + n\sum_{i=1}^{a}\sum_{j=1}^{b}(\bar{y}_{ij.} - \bar{y}_{i..})^2 + \sum_{i=1}^{a}\sum_{j=1}^{b}\sum_{k=1}^{n}(y_{ijk} - \bar{y}_{ij.})^2 \tag{14.3}$$

because the three cross-product terms are zero. Equation 14.3 indicates that the total sum of squares can be partitioned into a sum of squares due to factor A, a sum of squares due

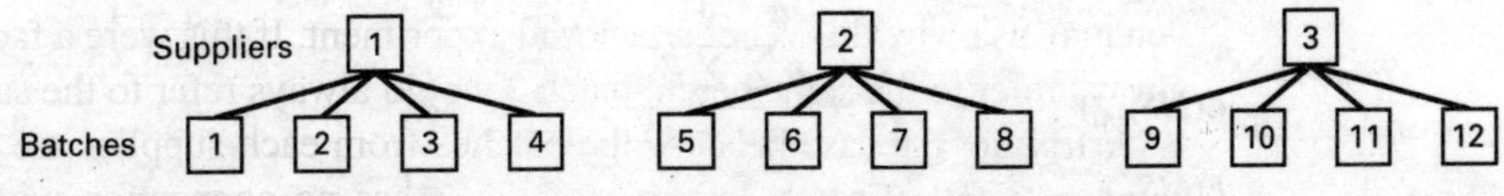

■ **FIGURE 14.2** Alternate layout for the two-stage nested design

to factor B under the levels of A, and a sum of squares due to error. Symbolically, we may write Equation 14.3 as

$$SS_T = SS_A + SS_{B(A)} + SS_E \tag{14.4}$$

There are $abn - 1$ degrees of freedom for SS_T, $a - 1$ degrees of freedom for SS_A, $a(b - 1)$ degrees of freedom for $SS_{B(A)}$, and $ab(n - 1)$ degrees of freedom for error. Note that $abn - 1 = (a - 1) + a(b - 1) + ab(n - 1)$. If the errors are NID$(0, \sigma^2)$, we may divide each sum of squares on the right of Equation 14.4 by its degrees of freedom to obtain independently distributed mean squares such that the ratio of any two mean squares is distributed as F.

The appropriate statistics for testing the effects of factors A and B depend on whether A and B are **fixed or random**. If factors A and B are fixed, we assume that $\sum_{i=1}^{a} \tau_i = 0$ and $\sum_{j=1}^{b} \beta_{j(i)} = 0$ $(i = 1, 2, \ldots, a)$. That is, the A treatment effects sum to zero, and the B treatment effects sum to zero within each level of A. Alternatively, if A and B are random, we assume that τ_i is NID$(0, \sigma_\tau^2)$ and $\beta_{j(i)}$ is NID$(0, \sigma_\beta^2)$. Mixed models with A fixed and B random are also widely encountered. The expected mean squares can be determined by a straightforward application of the rules in Chapter 13. Table 14.1 gives the expected mean squares for these situations.

Table 14.1 indicates that if the levels of A and B are fixed, H_0: $\tau_i = 0$ is tested by MS_A/MS_E and H_0: $\beta_{j(i)} = 0$ is tested by $MS_{B(A)}/MS_E$. If A is a fixed factor and B is random, then H_0: $\tau_i = 0$ is tested by $MS_A/MS_{B(A)}$ and H_0: $\sigma_\beta^2 = 0$ is tested by $MS_{B(A)}/MS_E$. Finally, if both A and B are random factors, we test H_0: $\sigma_\tau^2 = 0$ by $MS_A/MS_{B(A)}$ and H_0: $\sigma_\beta^2 = 0$ by $MS_{B(A)}/MS_E$. The test procedure is summarized in an analysis of variance table as shown in Table 14.2. Computing formulas for the sums of squares may be obtained by expanding the quantities in Equation 14.3 and simplifying. They are

$$SS_A = \frac{1}{bn}\sum_{i=1}^{a} y_{i..}^2 - \frac{y_{...}^2}{abn} \tag{14.5}$$

$$SS_{B(A)} = \frac{1}{n}\sum_{i=1}^{a}\sum_{j=1}^{b} y_{ij.}^2 - \frac{1}{bn}\sum_{i=1}^{a} y_{i..}^2 \tag{14.6}$$

$$SS_E = \sum_{i=1}^{a}\sum_{j=1}^{b}\sum_{k=1}^{n} y_{ijk}^2 - \frac{1}{n}\sum_{i=1}^{a}\sum_{j=1}^{b} y_{ij.}^2 \tag{14.7}$$

$$SS_T = \sum_{i=1}^{a}\sum_{j=1}^{b}\sum_{k=1}^{n} y_{ijk}^2 - \frac{y_{...}^2}{abn} \tag{14.8}$$

We see that Equation 14.6 for $SS_{B(A)}$ can be written as

$$SS_{B(A)} = \sum_{i=1}^{a}\left[\frac{1}{n}\sum_{j=1}^{b} y_{ij.}^2 - \frac{y_{i..}^2}{bn}\right]$$

■ **TABLE 14.1**
Expected Mean Squares in the Two-Stage Nested Design

$E(MS)$	A Fixed B Fixed	A Fixed B Random	A Random B Random
$E(MS_A)$	$\sigma^2 + \frac{bn\sum \tau_i^2}{a-1}$	$\sigma^2 + n\sigma_\beta^2 + \frac{bn\sum \tau_i^2}{a-1}$	$\sigma^2 + n\sigma_\beta^2 + bn\sigma_\tau^2$
$E(MS_{B(A)})$	$\sigma^2 + \frac{n\sum\sum \beta_{j(i)}^2}{a(b-1)}$	$\sigma^2 + n\sigma_\beta^2$	$\sigma^2 + n\sigma_\beta^2$
$E(MS_E)$	σ^2	σ^2	σ^2

■ TABLE 14.2
Analysis of Variance Table for the Two-Stage Nested Design

Source of Variation	Sum of Squares	Degrees of Freedom	Mean Square
A	$bn\sum(\bar{y}_{i..}-\bar{y}_{...})^2$	$a-1$	MS_A
B within A	$n\sum\sum(\bar{y}_{ij.}-\bar{y}_{i..})^2$	$a(b-1)$	$MS_{B(A)}$
Error	$\sum\sum\sum(y_{ijk}-\bar{y}_{ij.})^2$	$ab(n-1)$	MS_E
Total	$\sum\sum\sum(y_{ijk}-\bar{y}_{...})^2$	$abn-1$	

This expresses the idea that $SS_{B(A)}$ is the sum of squares between levels of B for each level of A, summed over all the levels of A.

EXAMPLE 14.1

Consider a company that buys raw material in batches from three different suppliers. The purity of this raw material varies considerably, which causes problems in manufacturing the finished product. We wish to determine whether the variability in purity is attributable to differences between the suppliers. Four batches of raw material are selected at random from each supplier, and three determinations of purity are made on each batch. This is, of course, a two-stage nested design. The data, after coding by subtracting 93, are shown in Table 14.3. The sums of squares are computed as follows:

$$SS_T = \sum_{i=1}^{a}\sum_{j=1}^{b}\sum_{k=1}^{n} y_{ijk}^2 - \frac{y_{...}^2}{abn}$$

$$= 153.00 - \frac{(13)^2}{36} = 148.31$$

$$SS_A = \frac{1}{bn}\sum_{i=1}^{a} y_{i..}^2 - \frac{y_{...}^2}{abn}$$

$$= \frac{1}{(4)(3)}[(-5)^2 + (4)^2 + (14)^2] - \frac{(13)^2}{36} = 15.06$$

$$SS_{B(A)} = \frac{1}{n}\sum_{i=1}^{a}\sum_{j=1}^{b} y_{ij.}^2 - \frac{1}{bn}\sum_{i=1}^{a} y_{i..}^2$$

$$= \frac{1}{3}[(0)^2 + (-9)^2 + (-1)^2 + \cdots + (2)^2 + (6)^2]$$

$$- 19.75 = 69.92$$

and

$$SS_E = \sum_{i=1}^{a}\sum_{j=1}^{b}\sum_{k=1}^{n} y_{ijk}^2 - \frac{1}{n}\sum_{i=1}^{a}\sum_{j=1}^{b} y_{ij.}^2$$

$$= 153.00 - 89.67 = 63.33$$

The analysis of variance is summarized in Table 14.4. Suppliers are fixed and batches are random, so the expected mean squares are obtained from the middle column of Table 14.1. They are repeated for convenience in Table 14.4. From examining the *P*-values, we would conclude that there is no significant effect on purity due to suppliers, but the purity of batches of raw material from the same supplier does differ significantly.

■ TABLE 14.3
Coded Purity Data for Example 14.1 (Code: y_{ijk} = Purity − 93)

		Supplier 1				Supplier 2				Supplier 3			
Batches		1	2	3	4	1	2	3	4	1	2	3	4
		1	−2	−2	1	1	0	−1	0	2	−2	1	3
		−1	−3	0	4	−2	4	0	3	4	0	−1	2
		0	−4	1	0	−3	2	−2	2	0	2	2	1
Batch totals	$y_{ij.}$	0	−9	−1	5	−4	6	−3	5	6	0	2	6
Supplier totals	$y_{i..}$	−5				4				14			

■ **TABLE 14.4**

Analysis of Variance for the Data in Example 14.1

Source of Variation	Sum of Squares	Degrees of Freedom	Mean Square	Expected Mean Square	F_0	*P*-Value
Suppliers	15.06	2	7.53	$\sigma^2 + 3\sigma_\beta^2 + 6\sum\tau_i^2$	0.97	0.42
Batches (within suppliers)	69.92	9	7.77	$\sigma^2 + 3\sigma_\beta^2$	2.94	0.02
Error	63.33	24	2.64	σ^2		
Total	148.31	35				

The practical implications of this experiment and the analysis are very important. The objective of the experimenter is to find the source of the variability in raw material purity. If it results from differences among suppliers, we may be able to solve the problem by selecting the "best" supplier. However, that solution is not applicable here because the major source of variability is the batch-to-batch purity variation *within* suppliers. Therefore, we must attack the problem by working with the suppliers to reduce their batch-to-batch variability. This may involve modifications to the suppliers' production processes or their internal quality assurance system.

Notice what would have happened if we had incorrectly analyzed this design as a two-factor factorial experiment. If batches are considered to be crossed with suppliers, we obtain batch totals of 2, −3, −2, and 16, with each batch × suppliers cell containing three replicates. Thus, a sum of squares due to batches and an interaction sum of squares can be computed. The complete factorial analysis of variance is shown in Table 14.5, assuming the mixed model.

This analysis indicates that batches differ significantly and that there is a significant interaction between batches and suppliers. However, it is difficult to give a practical interpretation of the batches × suppliers interaction. For example, does this significant interaction mean that the supplier effect is not constant from batch to batch? Furthermore, the significant interaction coupled with the nonsignificant supplier effect could lead the analyst to conclude that suppliers really differ but their effect is masked by the significant interaction.

Computing. Some statistical software packages will perform the analysis for a nested design. Table 14.6 presents the output from the Balanced ANOVA procedure in Minitab (using the restricted model). The numerical results are in agreement with the manual calculations reported in Table 14.4. Minitab also reports the expected mean squares in the lower

■ **TABLE 14.5**

Incorrect Analysis of the Two-Stage Nested Design in Example 14.1 as a Factorial (Suppliers Fixed, Batches Random)

Source of Variation	Sum of Squares	Degrees of Freedom	Mean Square	F_0	*P*-Value
Suppliers (*S*)	15.06	2	7.53	1.02	0.42
Batches (*B*)	25 64	3	8.55	3.24	0.04
S × *B* interaction	44.28	6	7.38	2.80	0.03
Error	63.33	24	2.64		
Total	148.31	35			

■ **TABLE 14.6**
Minitab Output (Balanced ANOVA) for Example 14.1

Analysis of Variance (Balanced Designs)

Factor	Type	Levels	Values			
Supplier	fixed	3	1	2	3	
Batch(Supplier)	random	4	1	2	3	4

Analysis of Variance for Purity

Source	DF	SS	MS	F	P
Supplier	2	15.056	7.528	0.97	0.416
Batch(Supplier)	9	69.917	7.769	2.94	0.017
Error	24	63.333	2.639		
Total	35	148.306			

Source	Variance component	Error term	Expected Mean Square for Each Term (using restricted model)
1 Supplier		2	(3) + 3(2) + 12Q[1]
2 Batch(Supplier)	1.710	3	(3) + 3(2)
3 Error	2.639		(3)

portion of Table 14.6. Remember that the symbol $Q[1]$ is a quadratic term that represents the fixed effect of suppliers, so in our notation

$$Q[1] = \frac{\sum_{i=1}^{a} \tau_i^2}{a-1}$$

Therefore, the fixed effect term in the Minitab expected mean square for suppliers $12Q[1] = 12\sum_{i=1}^{3} \tau_i^2/(3-1) = 6\sum_{i=1}^{3} \tau_i^2$, which matches the result given by the algorithm in Table 14.4.

Sometimes a specialized computer program for analyzing nested designs is not available. However, notice from comparing Tables 14.4 and 14.5 that

$$SS_B + SS_{S\times B} = 25.64 + 44.28 = 69.92 \equiv SS_{B(S)}$$

That is, the sum of squares for batches within suppliers consists of the sum of squares of the batches plus the sum of squares for the batches $\times$ suppliers interaction. The degrees of freedom have a similar property; that is,

$$\underset{3}{\text{Batches}} + \underset{6}{\text{Batches} \times \text{Suppliers}} = \underset{9}{\text{Batches within Suppliers}}$$

Therefore, a computer program for analyzing factorial designs could also be used for the analysis of nested designs by pooling the "main effect" of the nested factor and interactions of that factor with the factor under which it is nested.

14.1.2 Diagnostic Checking

The major tool used in diagnostic checking is **residual analysis**. For the two-stage nested design, the residuals are

$$e_{ijk} = y_{ijk} - \hat{y}_{ijk}$$

The fitted value is

$$\hat{y}_{ijk} = \hat{\mu} + \hat{\tau}_i + \hat{\beta}_{j(i)}$$

and if we make the usual restrictions on the model parameters ($\Sigma_i\ \hat{\tau}_i = 0$ and $\Sigma_j\ \hat{\beta}_{j(i)} = 0$, $i = 1, 2, \ldots, a$), then $\hat{\mu} = \bar{y}_{...}$, $\hat{\tau}_i = \bar{y}_{i..} - \bar{y}_{...}$, and $\hat{\beta}_{j(i)} = \bar{y}_{ij.} - \bar{y}_{i..}$. Consequently, the fitted value is

$$\hat{y}_{ijk} = \bar{y}_{...} + (\bar{y}_{i..} - \bar{y}_{...}) + (\bar{y}_{ij.} - \bar{y}_{i..}) = \bar{y}_{ij.}$$

Thus, the residuals from the two-stage nested design are

$$e_{ijk} = y_{ijk} - \bar{y}_{ij.} \qquad \textbf{(14.9)}$$

where $\bar{y}_{ij.}$ are the individual batch averages.

The observations, fitted values, and residuals for the purity data in Example 14.1 follow:

Observed Value y_{ijk}	Fitted Value $\hat{y}_{ijk} = \bar{y}_{ij.}$	$e_{ijk} = y_{ijk} - \bar{y}_{ij.}$
1	0.00	1.00
−1	0.00	−1.00
0	0.00	0.00
−2	−3.00	1.00
−3	−3.00	0.00
−4	−3.00	−1.00
−2	−0.33	−1.67
0	−0.33	0.33
1	−0.33	1.33
1	1.67	−0.67
4	1.67	2.33
0	1.67	−1.67
1	−1.33	2.33
−2	−1.33	−0.67
−3	−1.33	−1.67
0	2.00	−2.00
4	2.00	2.00
2	2.00	0.00
−1	−1.00	0.00
0	−1.00	1.00
−2	−1.00	−1.00
0	1.67	−1.67
3	1.67	1.33
2	1.67	0.33
2	2.00	0.00
4	2.00	2.00
0	2.00	−2.00
−2	0.00	−2.00
0	0.00	0.00
2	0.00	2.00
1	0.67	0.33

Observed Value y_{ijk}	Fitted Value $\hat{y}_{ijk} = \bar{y}_{ij.}$	$e_{ijk} = y_{ijk} - \bar{y}_{ij.}$
−1	0.67	−1.67
2	0.67	1.33
3	2.00	1.00
2	2.00	0.00
1	2.00	−1.00

The usual diagnostic checks—including normal probability plots, checking for outliers, and plotting the residuals versus fitted values—may now be performed. As an illustration, the residuals are plotted versus the fitted values and against the levels of the supplier factor in Figure 14.3.

In a problem situation such as that described in Example 14.1, the residual plots are particularly useful because of the additional diagnostic information they contain. For instance, the analysis of variance has indicated that the mean purity of all three suppliers does not differ but that there is statistically significant batch-to-batch variability (that is, $\sigma_\beta^2 > 0$). But is the variability within batches the same for all suppliers? In effect, we have assumed this to be the case and if it's not true, we would certainly like to know it because it has considerable practical impact on our interpretation of the results of the experiment. The plot of residuals versus suppliers in Figure 14.3*b* is a simple but effective way to check this assumption. Because the spread of the residuals is about the same for all three suppliers, we would conclude that the batch-to-batch variability in purity is about the same for all three suppliers.

14.1.3 Variance Components

For the random effects case, the analysis of variance method can be used to estimate the variance components σ^2, σ_β^2, and σ_τ^2. The maximum likelihood (REML) procedure could also be used. Applying the ANOVA method and using the expected mean squares in the last column of Table 14.1, we obtain

$$\hat{\sigma}^2 = MS_E \tag{14.10}$$

$$\hat{\sigma}_\beta^2 = \frac{MS_{B(A)} - MS_E}{n} \tag{14.11}$$

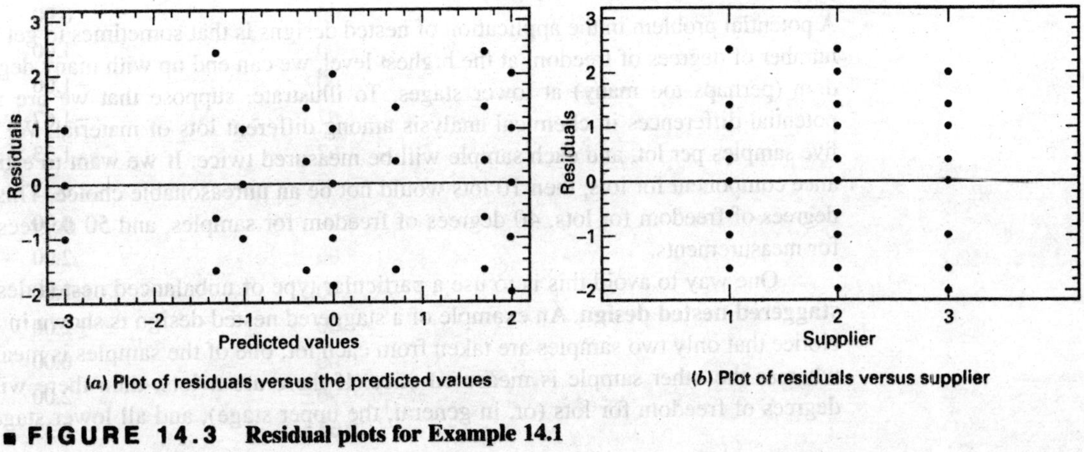

(*a*) Plot of residuals versus the predicted values (*b*) Plot of residuals versus supplier

■ **FIGURE 14.3 Residual plots for Example 14.1**

and

$$\hat{\sigma}_\tau^2 = \frac{MS_A - MS_{B(A)}}{bn} \tag{14.12}$$

Many applications of nested designs involve a **mixed model**, with the main factor (A) fixed and the nested factor (B) random. This is the case for the problem described in Example 14.1, where suppliers (factor A) are fixed, and batches of raw material (factor B) are random. The effects of the suppliers may be estimated by

$$\hat{\tau}_1 = \bar{y}_{1..} - \bar{y}_{...} = \frac{-5}{12} - \frac{13}{36} = \frac{-28}{36}$$

$$\hat{\tau}_2 = \bar{y}_{2..} - \bar{y}_{...} = \frac{4}{12} - \frac{13}{36} = \frac{-1}{36}$$

$$\hat{\tau}_3 = \bar{y}_{3..} - \bar{y}_{...} = \frac{14}{12} - \frac{13}{36} = \frac{29}{36}$$

To estimate the variance components σ^2 and σ_β^2, we eliminate the line in the analysis of variance table pertaining to suppliers and apply the analysis of variance estimation method to the next two lines. This yields

$$\hat{\sigma}^2 = MS_E = 2.64$$

and

$$\hat{\sigma}_\beta^2 = \frac{MS_{B(A)} - MS_E}{n} = \frac{7.77 - 2.64}{3} = 1.71$$

These results are also shown in the lower portion of the Minitab output in Table 14.6. From the analysis in Example 14.1, we know that the τ_i does not differ significantly from zero, whereas the variance component σ_β^2 is greater than zero.

To illustrate the REML method for a nested design, reconsider the experiment in Example 14.1 with suppliers fixed and batches random. The REML output from JMP is shown in Table 14.7. The REML estimates of the variance components agree with the ANOVA estimates, but the REML procedure provides confidence intervals. The fixed effects test on suppliers indicates that there is no significant difference in mean purity among the three suppliers. The 95 percent confidence interval on batches within suppliers has a lower bound that is just less than zero, but the batches within suppliers variance component accounts for about 40 percent of the total variability so there is some evidence that batches within suppliers exhibit some meaningful variability.

14.1.4 Staggered Nested Designs

A potential problem in the application of nested designs is that sometimes to get a reasonable number of degrees of freedom at the highest level, we can end up with many degrees of freedom (perhaps too many) at lower stages. To illustrate, suppose that we are investigating potential differences in chemical analysis among different lots of material. We plan to take five samples per lot, and each sample will be measured twice. If we want to estimate a variance component for lots, then 10 lots would not be an unreasonable choice. This results in 9 degrees of freedom for lots, 40 degrees of freedom for samples, and 50 degrees of freedom for measurements.

One way to avoid this is to use a particular type of unbalanced nested design called a **staggered nested design**. An example of a staggered nested design is shown in Figure 14.4. Notice that only two samples are taken from each lot; one of the samples is measured twice, whereas the other sample is measured once. If there are a lots, then there will be $a - 1$ degrees of freedom for lots (or, in general, the upper stage), and all lower stages will have

■ **TABLE 14.7**

JMP Output for the Nested Design in Example 14.1, Suppliers Fixed and Batches Random

Response Y

Summary of Fit

RSquare	0.518555
RSquare Adj	0.489376
Root Mean Square Error	1.624466
Mean of Response	0.361111
Observations (or Sum Wgts)	36

REML Variance Component Estimates

Random Effect	Var Ratio	Var Component	Std Error	95% Lower	95% Upper	Pct of Total
Supplier [Batches]	0.6479532	1.7098765	1.2468358	−0.733922	4.1536747	39.319
Residual		2.6388889	0.7617816	1.6089119	5.1070532	60.681
Total		4.3487654				100.000

−2 LogLikelihood = 145.04119391

Covariance Matrix of Variance Component Estimates

Random Effect	Supplier [Batches]	Residual
Supplier [Batches]	1.5545995	−0.193437
Residual	−0.193437	0.5803112

Fixed Effect Tests

Source	Nparm	DF	DFDen	F Ratio	Prob > F
Supplier	2	2	9	0.9690	0.4158

exactly a degrees of freedom. For more information on the use and analysis of these designs, see Bainbridge (1965), Smith and Beverly (1981), and Nelson (1983, 1995a, 1995b). The **supplemental text material** for this chapter contains a complete example of a staggered nested design.

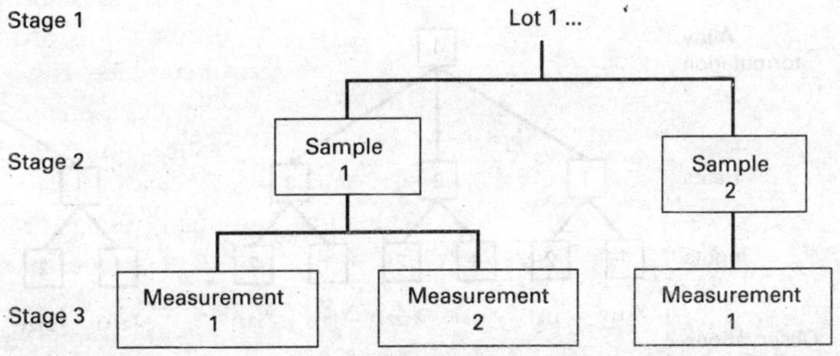

■ **FIGURE 14.4** A three-stage staggered nested design

14.2 The General *m*-Stage Nested Design

The results of Section 14.1 can be easily extended to the case of m completely nested factors. Such a design would be called an ***m*-stage nested design**. As an example, suppose a foundry wishes to investigate the hardness of two different formulations of a metal alloy. Three heats of each alloy formulation are prepared, two ingots are selected at random from each heat for testing, and two hardness measurements are made on each ingot. The situation is illustrated in Figure 14.5.

In this experiment, heats are nested under the levels of the factor alloy formulation, and ingots are nested under the levels of the factor heats. Thus, this is a three-stage nested design with two replicates.

The **model** for the general three stage nested design is

$$y_{ijkl} = \mu + \tau_i + \beta_{j(i)} + \gamma_{k(ij)} + \epsilon_{(ijk)l} \quad \begin{cases} i = 1, 2, \ldots, a \\ j = 1, 2, \ldots, b \\ k = 1, 2, \ldots, c \\ l = 1, 2, \ldots, n \end{cases} \tag{14.13}$$

For our example, τ_i is the effect of the ith alloy formulation, $\beta_{j(i)}$ is the effect of the jth heat within the ith alloy, $\gamma_{k(ij)}$ is the effect of the kth ingot within the jth heat and ith alloy, and $\epsilon_{(ijk)l}$ is the usual NID(0, σ^2) error term. Extension of this model to m factors is straightforward.

Notice that in the above example the overall variability in hardness consists of three components: one that results from alloy formulations, one that results from heats, and one that results from analytical test error. These components of the variability in overall hardness are illustrated in Figure 14.6.

This example demonstrates how the nested design is often used in analyzing processes to identify the major sources of variability in the output. For instance, if the alloy formulation variance component is large, then this implies that overall hardness variability could be reduced by using only one alloy formulation.

The calculation of the sums of squares and the analysis of variance for the m-stage nested design are similar to the analysis presented in Section 14.1. For example, the analysis of variance for the three-stage nested design is summarized in Table 14.8. Definitions of the sums of squares are also shown in this table. Notice that they are a simple extension of the formulas for the two-stage nested design. Many statistics software packages will perform the calculations.

To determine the proper test statistics, we must find the expected mean squares using the methods of Chapter 13. For example, if factors A and B are fixed and factor C is random, then the expected mean squares are as shown in Table 14.9. This table indicates the proper test statistics for this situation.

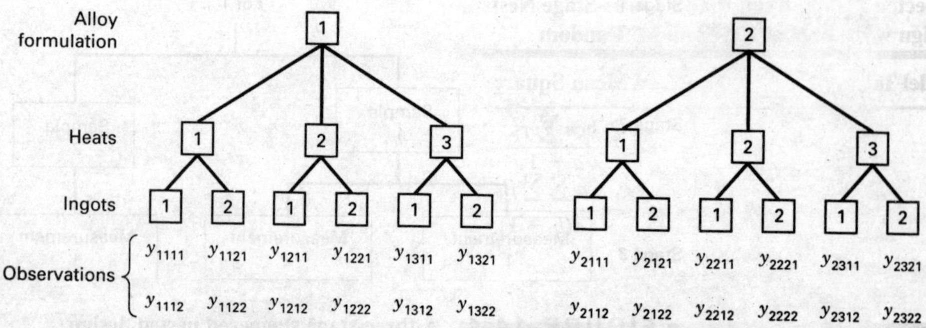

■ **FIGURE 14.5** A three-stage nested design

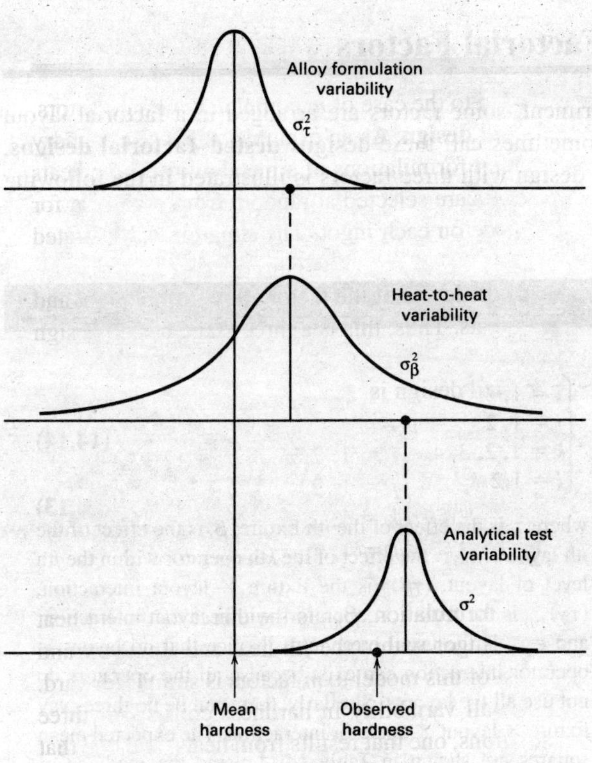

■ **FIGURE 14.6** **Sources of variation in the three-stage nested design example**

■ **TABLE 14.8**
Analysis of Variance for the Three-Stage Nested Design

Source of Variation	Sum of Squares	Degrees of Freedom	Mean Square
A	$bcn \sum_i (\bar{y}_{i...} - \bar{y}_{....})^2$	$a - 1$	MS_A
B (within A)	$cn \sum_i \sum_j (\bar{y}_{ij..} - \bar{y}_{i...})^2$	$a(b - 1)$	$MS_{B(A)}$
C (within B)	$n \sum_i \sum_j \sum_k (\bar{y}_{ijk.} - \bar{y}_{ij..})^2$	$ab(c - 1)$	$MS_{C(B)}$
Error	$\sum_i \sum_j \sum_k \sum_l (y_{ijkl} - \bar{y}_{ijk.})^2$	$abc(n - 1)$	MS_E
Total	$\sum_i \sum_j \sum_k \sum_l (y_{ijkl} - \bar{y}_{....})^2$	$abcn - 1$	

■ **TABLE 14.9**
Expected Mean Squares for a Three-Stage Nested Design with *A* and *B* Fixed and *C* Random

Model Term	Expected Mean Square
τ_i	$\sigma^2 + n\sigma_\gamma^2 + \dfrac{bcn \sum \tau_i^2}{a - 1}$
$\beta_{j(i)}$	$\sigma^2 + n\sigma_\gamma^2 + \dfrac{cn \sum \sum \beta_{j(i)}^2}{a(b - 1)}$
$\gamma_{k(ij)}$	$\sigma^2 + n\sigma_\gamma^2$
$\epsilon_{l(ijk)}$	σ^2

14.3 Designs with Both Nested and Factorial Factors

Occasionally in a multifactor experiment, some factors are arranged in a factorial layout and other factors are nested. We sometimes call these designs **nested–factorial designs**. The statistical analysis of one such design with three factors is illustrated in the following example.

EXAMPLE 14.2

An industrial engineer is studying the hand insertion of electronic components on printed circuit boards to improve the speed of the assembly operation. He has designed three assembly fixtures and two workplace layouts that seem promising. Operators are required to perform the assembly, and it is decided to randomly select four operators for each fixture–layout combination. However, because the workplaces are in different locations within the plant, it is difficult to use the *same* four operators for each layout. Therefore, the four operators chosen for layout 1 are different individuals from the four operators chosen for layout 2. Because there are only three fixtures and two layouts, but the operators are chosen at random, this is a **mixed model**. The treatment combinations in this design are run in random order, and two replicates are obtained. The assembly times are measured in seconds and are shown in Table 14.10.

In this experiment, operators are nested within the levels of layouts, whereas fixtures and layouts are arranged in a factorial. Thus, this design has both nested and factorial factors. The **linear model** for this design is

$$y_{ijkl} = \mu + \tau_i + \beta_j + \gamma_{k(j)} + (\tau\beta)_{ij} + (\tau\gamma)_{ik(j)} + \epsilon_{(ijk)l} \qquad \begin{cases} i = 1, 2, 3 \\ j = 1, 2 \\ k = 1, 2, 3, 4 \\ l = 1, 2 \end{cases} \tag{14.14}$$

where τ_i is the effect of the ith fixture, β_j is the effect of the jth layout, $\gamma_{k(j)}$ is the effect of the kth operator within the jth level of layout, $(\tau\beta)_{ij}$ is the fixture × layout interaction, $(\tau\gamma)_{ik(j)}$ is the fixture × operators within layout interaction, and $\epsilon_{(ijk)l}$ is the usual error term. Notice that no layout × operator interaction can exist because all the operators do not use all the layouts. Similarly, there can be no three-way fixture × layout × operator interaction. The expected mean squares are shown in Table 14.11 using the methods of Chapter 13 and assuming a **restricted** mixed model. The proper test statistic for any effect or interaction can be found from the inspection of this table.

The complete analysis of variance is shown in Table 14.12. We see that assembly fixtures are significant and that operators within layouts also differ significantly. There is also a significant interaction between fixtures and operators within layouts, indicating that the effects of the different fixtures are not the same for all operators. The workplace layouts

■ **TABLE 14.10**
Assembly Time Data for Example 14.2

	Layout 1				Layout 2				
Operator	1	2	3	4	1	2	3	4	$y_{i...}$
Fixture 1	22	23	28	25	26	27	28	24	404
	24	24	29	23	28	25	25	23	
Fixture 2	30	29	30	27	29	30	24	28	447
	27	28	32	25	28	27	23	30	
Fixture 3	25	24	27	26	27	26	24	28	401
	21	22	25	23	25	24	27	27	
Operator totals, $y_{.jk.}$	149	150	171	149	163	159	151	160	
Layout totals, $y_{.j..}$		619				633			$1252 = y_{....}$

■ **TABLE 14.11**
Expected Mean Squares for Example 14.2

Model Term	Expected Mean Square
τ_i	$\sigma^2 + 2\sigma^2_{\tau\gamma} + 8\sum \tau_i^2$
β_j	$\sigma^2 + 6\sigma^2_{\gamma} + 24\sum \beta_j^2$
$\gamma_{k(j)}$	$\sigma^2 + 6\sigma^2_{\gamma}$
$(\tau\beta)_{ij}$	$\sigma^2 + 2\sigma^2_{\tau\gamma} + 4\sum\sum (\tau\beta)^2_{ij}$
$(\tau\gamma)_{ik(j)}$	$\sigma^2 + 2\sigma^2_{\tau\gamma}$
$\epsilon_{(ijk)l}$	σ^2

seem to have little effect on the assembly time. Therefore, to minimize assembly time, we should concentrate on fixture types 1 and 3. (Note that the fixture totals in Table 14.9 are smaller for fixture types 1 and 3 than for type 2. This difference in fixture type means could be formally tested using multiple comparisons.) Furthermore, the interaction between operators and fixtures implies that some operators are more effective than others using the same fixtures. Perhaps these operator–fixture effects could be isolated and the less effective operators' performance improved by retraining them.

■ **TABLE 14.12**
Analysis of Variance for Example 14.2

Source of Variation	Sum of Squares	Degrees of Freedom	Mean Square	F_0	*P*-Value
Fixtures (*F*)	82.80	2	41.40	7.54	0.01
Layouts (*L*)	4.08	1	4.09	0.34	0.58
Operators (within layouts), *O*(*L*)	71.91	6	11.99	5.15	<0.01
FL	19.04	2	9.52	1.73	0.22
FO(*L*)	65.84	12	5.49	2.36	0.04
Error	56.00	24	2.33		
Total	299.67	47			

Computing. A number of statistical software packages can easily analyze nested–factorial designs, including both Minitab and JMP. Table 14.13 presents the output from Minitab (Balanced ANOVA), assuming the restricted form of the mixed model, for Example 14.2. The expected mean squares in the bottom portion of Table 14.13 agree with those shown in Table 14.10. $Q[1]$, $Q[3]$, and $Q[4]$ are the fixed-factor effects for layouts, fixtures, and layouts × fixtures, respectively. The estimates of the variance components are

Operator (layout):	$\sigma^2_{\gamma} = 1.609$
Fixture × Operator (layout):	$\sigma^2_{\tau\gamma} = 1.576$
Error:	$\sigma^2 = 2.333$

■ **TABLE 14.13**
Minitab Balanced ANOVA Analysis of Example 14.2 Using the Restricted Model

Analysis of Variance (Balanced Designs)

Factor	Type	Levels	Values			
Layout	fixed	2	1	2		
Operator(Layout)	random	4	1	2	3	4
Fixture	fixed	3	1	2	3	

Analysis of Variance for Time

Source	DF	SS	MS	F	P
Layout	1	4.083	4.083	0.34	0.581
Operator(Layout)	6	71.917	11.986	5.14	0.002
Fixture	2	82.792	41.396	7.55	0.008
Layout*Fixture	2	19.042	9.521	1.74	0.218
Fixture*Operator(Layout)	12	65.833	5.486	2.35	0.036
Error	24	56.000	2.333		
Total	47	299.667			

Source	Variance component	Error term	Expected Mean Square for Each Term (using restricted model)
1 Layout		2	(6) + 6(2) + 24Q[1]
2 Operator(Layout)	1.609	6	(6) + 6(2)
3 Fixture		5	(6) + 2(5) + 16Q[3]
4 Layout*Fixture		5	(6) + 2(5) + 8Q[4]
5 Fixture*Operator(Layout)	1.576	6	(6) + 2(5)
6 Error	2.333		(6)

Table 14.14 presents the Minitab analysis of Example 14.2 using the **unrestricted** form of the mixed model. The expected mean squares in the lower portion of this table are slightly different from those reported for the restricted model, and so the construction of the test statistic will be slightly different for the operators (layout) factor. Specifically, the *F* ratio denominator for operators (layout) is the fixtures × operators (layout) interaction in the restricted model (12 degrees of freedom for error), and it is the layout × fixtures interaction in the unrestricted model (2 degrees of freedom for error). Because $MS_{\text{layout} \times \text{fixture}} > MS_{\text{fixture} \times \text{operator(layout)}}$ and it has fewer degrees of freedom, we now find that the operator within layout effect is only significant at about the 12 percent level (the *P*-value was 0.002 in the restricted model analysis). Furthermore, the variance component estimate $\hat{\sigma}_\gamma^2 = 1.083$ is smaller. However, because there is a large fixture effect and a significant fixture × operator (layout) interaction, we would still suspect an operator effect, and so the practical conclusions from this experiment are not greatly affected by choosing either the restricted or the unrestricted form of the mixed model. The quantities $Q[1, 4]$ and $Q[3, 4]$ are fixed-type quadratic terms containing the interaction effect of layouts × fixtures.

■ **TABLE 14.14**
Minitab Balanced ANOVA Analysis of Example 14.2 Using the Unrestricted Model

Analysis of Variance (Balanced Designs)

Factor	Type	Levels	Values			
Layout	fixed	2	1	2		
Operator(Layout)	random	4	1	2	3	4
Fixture	fixed	3	1	2	3	

Analysis of Variance for Time

Source	DF	SS	MS	F	P
Layout	1	4.083	4.083	0.34	0.581
Operator(Layout)	6	71.917	11.986	2.18	0.117
Fixture	2	82.792	41.396	7.55	0.008
Layout*Fixture	2	19.042	9.521	1.74	0.218
Fixture*Operator(Layout)	12	65.833	5.486	2.35	0.036
Error	24	56.000	2.333		
Total	47	299.667			

Source	Variance component	Error term	Expected Mean Square for Each Term (using unrestricted model)
1 Layout		2	(6) + 2(5) + 6(2) + Q[1,4]
2 Operator(Layout)	1.083	5	(6) + 2(5) + 6(2)
3 Fixture		5	(6) + 2(5) + Q[3,4]
4 Layout*Fixture		5	(6) + 2(5) + Q[4]
5 Fixture*Operator (Layout)	1.576	6	(6) + 2(5)
6 Error	2.333		(6)

Table 14.15 presents the JMP output for Example 14.3. Because JMP uses the unrestricted form of the mixed model, estimates of the variance components agree with the Minitab ANOVA estimates in Table 14.14, but the REML procedure is a preferred analysis because it provides confidence intervals. The fixed effects tests indicate that there is a strong fixture effect, and even though the confidence intervals on the operators (layout) and the fixture × operators (layout) interaction variance components includes zero, we would be reluctant to discount an operator effect and an interaction because these two variance components account for over 50 percent of the total variability.

If no specialized software such as JMP or Minitab is available, then a program for analyzing factorial experiments can be used to analyze experiments with nested and factorial factors. For instance, the experiment in Example 14.2 could be considered as a three-factor factorial, with fixtures (F), operators (O), and layouts (L) as the factors. Then certain sums of squares and degrees of freedom from the factorial analysis would be

■ **TABLE 14.15**
JMP Output for Example 14.2

Response Time

Summary of Fit

RSquare	0.764291
RSquare Adj	0.73623
Root Mean Square Error	1.527525
Mean of Response	26.08333
Observations for (or Sum Wgts)	48

REML Variance Component Estimates

Random Effect	Var Ratio	Var Component	Std Error	95% Lower	95% Upper	Pct of Total
Operator[Layout]	0.4642857	1.0833333	1.2122659	−1.292708	3.4593745	21.697
Operator*Fixture[Layout]	0.6755952	1.5763889	1.1693951	−0.715625	3.8684033	31.572
Residual		2.3333333	0.6735753	1.4226169	4.5157102	46.732
Total		4.9930556				100.000

−2 LogLikelihood = 195.88509411

Covariance Matrix of Variance Component Estimates

Random Effect	Operator[Layout]	Operator*Fixture[Layout]	Residual
Operator[Layout]	1.4695886	−0.41802	3.608e-15
Operator*Fixture[Layout]	−0.41802	1.3674849	−0.226852
Residual	3.608e-15	−0.226852	0.4537037

Fixed Effect Tests

Source	Nparm	DF	DFDen	F Ratio	Prob > F
Layout	1	1	6	0.3407	0.5807
Fixture	2	2	12	7.5456	0.0076*
Layout*Fixture	2	2	12	1.7354	0.2178

pooled to form the appropriate quantities required for the design with nested and factorial factors as follows:

Factorial Analysis		Nested–Factorial Analysis	
Sum of Squares	Degrees of Freedom	Sum of Squares	Degrees of Freedom
SS_F	2	SS_F	2
SS_L	1	SS_L	1
SS_{FL}	2	SS_{FL}	2
SS_O	3		
SS_{LO}	3	$SS_{O(L)} = SS_O + SS_{LO}$	6
SS_{FO}	6		
SS_{FOL}	6	$SS_{FO(L)} = SS_{FO} + SS_{FOL}$	12
SS_E	24	SS_E	24
SS_T	47	SS_T	47

14.4 The Split-Plot Design

In some multifactor factorial experiments, we may be unable to completely randomize the order of the runs. This often results in a generalization of the factorial design called a **split-plot design.**

As an example, consider a paper manufacturer who is interested in three different pulp preparation methods (the methods differ in the amount of hardwood in the pulp mixture) and four different cooking temperatures for the pulp and who wishes to study the effect of these two factors on the tensile strength of the paper. Each replicate of a factorial experiment requires 12 observations, and the experimenter has decided to run three replicates. This will require a total of 36 runs. The experimenter decides to conduct the experiment as follows. A batch of pulp is produced by one of the three methods under study. Then this batch is divided into four samples, and each sample is cooked at one of the four temperatures. Then a second batch of pulp is made up using another of the three methods. This second batch is also divided into four samples that are tested at the four temperatures. The process is then repeated, until all three replicates (36 runs) of the experiment are obtained. The data are shown in Table 14.16.

Initially, we might consider this to be a factorial experiment with three levels of preparation method (factor *A*) and four levels of temperature (factor *B*). If this is the case, then the order of experimentation within each replicate should be completely randomized. That is, we should randomly select a treatment combination (a preparation method and a temperature) and obtain an observation, then we should randomly select another treatment combination and obtain a second observation, and so on, until all 36 observations have been taken. However, the experimenter did not collect the data this way. He made up a batch of pulp and obtained observations for all four temperatures from that batch. Because of the economics of preparing the batches and the size of the batches, this is the only feasible way to run this experiment. A completely randomized factorial experiment would require 36 batches of pulp, which is completely unrealistic. The split-plot design requires only 9 batches total. Obviously, the split-plot design has resulted in considerable experimental efficiency.

The design used in our example is a split-plot design. In this split-plot design we have 9 **whole plots**, and the preparation methods are called the **whole plot** or **main treatments**. Each whole plot is divided into four parts called **subplots** (or **split-plots**), and one temperature is assigned to each. Temperature is called the **subplot treatment**. Note that if other uncontrolled or undesigned factors are present and if these uncontrolled factors vary as the pulp preparation methods are changed, then any effect of the undesigned factors on the

■ TABLE 14.16
The Experiment on the Tensile Strength of Paper

	Replicate 1			Replicate 2			Replicate 3		
Pulp Preparation Method	**1**	**2**	**3**	**1**	**2**	**3**	**1**	**2**	**3**
Temperature (°F)									
200	30	34	29	28	31	31	31	35	32
225	35	41	26	32	36	30	37	40	34
250	37	38	33	40	42	32	41	39	39
275	36	42	36	41	40	40	40	44	45

response will be completely confounded with the effect of the pulp preparation methods. Because the whole-plot treatments in a split-plot design are confounded with the whole-plots and the subplot treatments are not confounded, it is best to assign the factor we are most interested in to the subplots, if possible.

This example is fairly typical of how the split-plot design is used in an industrial setting. Notice that the two factors were essentially "applied" at different times. Consequently, a split-plot design can be viewed as two experiments "combined" or superimposed on each other. One "experiment" has the whole-plot factor applied to the large experimental units (or it is a factor whose levels are hard to change) and the other "experiment" has the subplot factor applied to the smaller experimental units (or it is a factor whose levels are easy to change).

The **linear model** for the split-plot design is

$$y_{ijk} = \mu + \tau_i + \beta_j + (\tau\beta)_{ij} + \gamma_k + (\tau\gamma)_{ik} + (\beta\gamma)_{jk} + (\tau\beta\gamma)_{ijk} + \epsilon_{ijk} \quad \begin{cases} i = 1, 2, \ldots, r \\ j = 1, 2, \ldots, a \\ k = 1, 2, \ldots, b \end{cases} \tag{14.15}$$

where τ_i, β_j, and $(\tau\beta)_{ij}$ represent the whole plot and correspond, respectively, to replicates, main treatments (factor A), and **whole-plot error** (replicates $\times$ A), and γ_k, $(\tau\gamma)_{ik}$, $(\beta\gamma)_{jk}$, and $(\tau\beta\gamma)_{ijk}$ represent the subplot and correspond, respectively, to the subplot treatment (factor B), the replicates $\times$ B and AB interactions, and the **subplot error** (replicates $\times$ AB). Note that the whole-plot error is the replicates $\times$ A interaction and the subplot error is the three-factor interaction replicates $\times$ AB. The sums of squares for these factors are computed as in the three-way analysis of variance without replication.

The expected mean squares for the split-plot design, with replicates random and main treatments and subplot treatments fixed, are shown in Table 14.17. Note that the main factor (A) in the whole plot is tested against the whole-plot error, whereas the subtreatment (B) is tested against the replicates $\times$ subtreatment interaction. The AB interaction is tested against the subplot error. Notice that there are no tests for the replicate effect (A) or the replicate $\times$ subtreatment (AC) interaction.

■ **TABLE 14.17**
Expected Mean Squares for Split-Plot Design

	Model Term	Expected Mean Square
	τ_i	$\sigma^2 + ab\sigma_\tau^2$
Whole plot	β_j	$\sigma^2 + b\sigma_{\tau\beta}^2 + \dfrac{rb\sum\beta_j^2}{a-1}$
	$(\tau\beta)_{ij}$	$\sigma^2 + b\sigma_{\tau\beta}^2$
	γ_k	$\sigma^2 + a\sigma_{\tau\gamma}^2 + \dfrac{ra\sum\gamma_k^2}{(b-1)}$
	$(\tau\gamma)_{ik}$	$\sigma^2 + a\sigma_{\tau\gamma}^2$
Subplot	$(\beta\gamma)_{jk}$	$\sigma^2 + \sigma_{\tau\beta\gamma}^2 + \dfrac{r\sum\sum(\beta\gamma)_{jk}^2}{(a-1)(b-1)}$
	$(\tau\beta\gamma)_{ijk}$	$\sigma^2 + \sigma_{\tau\beta\gamma}^2$
	$\epsilon_{(ijk)h}$	σ^2 (not estimable)

■ **TABLE 14.18**
Analysis of Variance for the Split-Plot Design Using the Tensile Strength Data from Table 14.14

Source of Variation	Sum of Squares	Degrees of Freedom	Mean Square	F_0	*P*-Value
Replicates	77.55	2	38.78		
Preparation method (*A*)	128.39	2	64.20	7.08	0.05
Whole plot error (replicates × *A*)	36.28	4	9.07		
Temperature (*B*)	434.08	3	144.69	41.94	<0.01
Replicates × *B*	20.67	6	3.45		
AB	75.17	6	12.53	2.96	0.05
Subplot error (replicates × *AB*)	50.83	12	4.24		
Total	822.97	35			

The analysis of variance for the tensile strength data in Table 14.16 is summarized in Table 14.18. Because both preparation methods and temperatures are fixed and replicates are random, the expected mean squares in Table 14.17 apply. The mean square for preparation methods is compared to the whole-plot error mean square, and the mean square for temperatures is compared to the replicate × temperature (*AC*) mean square. Finally, the preparation method × temperature mean square is tested against the subplot error. Both preparation methods and temperature have a significant effect on strength, and their interaction is significant.

Note from Table 14.18 that the subplot error (4.24) is less than the whole-plot error (9.07). This is the usual case in split-plot designs because the subplots are generally more homogeneous than the whole plots. This results in **two different error structures** for the experiment. Because the subplot treatments are compared with greater precision, it is preferable to assign the treatment we are most interested in to the subplots, if possible.

Some authors propose a slightly different statistical model for the split-plot design, say

$$y_{ijk} = \mu + \tau_i + \beta_j + (\tau\beta)_{ij} + \gamma_k + (\beta\gamma)_{jk} + \epsilon_{ijk} \quad \begin{cases} i = 1, 2, \ldots, r \\ j = 1, 2, \ldots, a \\ k = 1, 2, \ldots, b \end{cases} \tag{14.16}$$

In this model, $(\tau\beta)_{ij}$ is still the whole-plot error, but the replicates × *B* and replicates × *AB* interactions have essentially been pooled with ϵ_{ijk} to form the subplot error. If we denote the variance of the subplot error term ϵ_{ijk} by σ_ϵ^2 and make the same assumptions as for model (Equation 14.15), the expected mean squares become

Factor	*E*(*MS*)
τ_i (Replicates)	$\sigma_\epsilon^2 + ab\sigma_\tau^2$
β_j (*A*)	$\sigma_\epsilon^2 + b\sigma_{\tau\beta}^2 + \dfrac{rb\sum \beta_j^2}{a-1}$
$(\tau\beta)_{ij}$	$\sigma_\epsilon^2 + b\sigma_{\tau\beta}^2$ (whole-plot error)
γ_k (*B*)	$\sigma_\epsilon^2 + \dfrac{ra\sum \gamma_k^2}{ab-1}$
$(\beta\gamma)_{jk}$ (*AB*)	$\sigma_\epsilon^2 + \dfrac{r\sum\sum (\beta\gamma)_{jk}^2}{(a-1)(b-1)}$
ϵ_{ijk}	σ_ϵ^2 (subplot error)

Notice that now both the subplot treatment (B) and the AB interaction are tested against the subplot error mean square. If one is reasonably comfortable with the assumption that the interactions of replicates × B and replicates × AB interactions are negligible, then this alternative model is entirely satisfactory.

Because there are two variance components in the split-plot design, REML can be used to estimate them. JMP implements the REML method for the split-plot design using the model in Equation 14.16. Table 14.19 is the JMP output for the split-plot design in Table 14.16. The advantage of the REML method is that explicit estimates of the whole-plot and subplot variance components are provided along with standard errors and approximate confidence intervals.

Sometimes experimenters do not recognize the very specific structure of split-plot designs. They know that there is one (or perhaps more) hard-to-change factor involved in the experiment but they do not design the experiment as a split-plot. They set up a standard

■ **TABLE 14.19**
JMP Output for the Split-Plot Design in Table 14.16

Response Strength

Summary of Fit

RSquare	0.903675
RSquare Adj	0.859526
Root Mean Square Error	1.993043
Mean of Response	36.02778
Observations (or Sum Wgts)	36

REML Variance Component Estimates

Random Effect	Var Ratio	Var Component	Std Error	95% Lower	95% Upper	Pct of Total
Whole Plots	0.6232517	2.4756944	3.2753747	−3.94404	8.8954289	32.059
Subplots	0.3208042	1.2743056	1.6370817	−1.934375	4.4829857	16.502
Residual		3.9722222	1.3240741	2.2679421	8.6869402	51.439
Total		7.7222222				100.000

−2 LogLikelihood = 139.36226272

Covariance Matrix of Variance Component Estimates

Random Effect	Whole Plots	Subplots	Residual
Whole Plots	10.72808	−0.856821	−9.84e-14
Subplots	−0.856821	2.6800365	−0.438293
Residual	−9.84e-14	-0.438293	1.7531722

Fixed Effect Tests

Source	Nparm	DF	DFDen	F Ratio	Prob > F
Method	2	2	4	7.0781	0.0485*
Temp	3	3	18	36.4266	<.0001*
Temp*Method	6	6	18	3.1538	0.0271*

factorial and then reorder the runs to minimize the number of times that the hard-to-change factor must be changed. Then they run the experiment creating an "inadvertent" split-plot and analyze the data as if it were a standard factorial (that is, a completely randomized design or a CRD).

Suppose that this happened with the experiment in Table 14.16. The experimenters made up three batches of pulp in each replicate and then randomized the levels of temperature within each batch, resulting in the split-plot design in Table 14.16. The standard factorial analysis of this experiment as a two-factor completely randomized factorial design is shown in Table 14.20. Recall that in the correct split-plot analysis both the whole-plot and subplot factors and their interaction were significant. However, in the incorrect analysis the interaction is not significant.

Generally, in "inadvertent" split-plots we will tend to make too many type I errors for the whole-plot factors (concluding that unimportant factors are important) and too many type II errors for the subplot factors (failing to identify significant effects). In our example, we did properly identify the pulp preparation methods (the whole-plot factor) as important, but we did not identify the significant interaction, which is a subplot factor. This emphasizes the importance of correctly designing and analyzing experiments involving split-plots.

The split-plot design has an **agricultural heritage**, with the whole plots usually being large areas of land and the subplots being smaller areas of land within the large areas. For example, several varieties of a crop could be planted in different fields (whole plots), one variety to a field. Then each field could be divided into, say, four subplots, and each subplot could be treated with a different type of fertilizer. Here the crop varieties are the main treatments and the different fertilizers are the subtreatments.

Despite its agricultural basis, the split-plot design is useful in many scientific and industrial experiments. In these experimental settings, it is not unusual to find that some factors require large experimental units whereas other factors require small experimental units, such

■ TABLE 14.20
Analysis of the Split-Plot Design in Table 14.16 as a CRD

Summary of Fit

RSquare	0.7748
RSquare Adj	0.671583
Root Mean Square Error	2.778889
Mean of Response	36.02778
Observations (or Sum Wgts)	36

Analysis of Variance

Source	DF	Sum of Squares	Mean Square	F Ratio
Model	11	637.63889	57.9672	7.5065
Error	24	185.33333	7.7222	Prob > F
C. Total	35	822.97222		<.0001*

Effect Tests

Source	Nparm	DF	Sum of Squares	F Ratio	Prob > F
Method	2	2	128.38889	8.3129	0.0018*
Temp	3	3	434.08333	18.7374	<.0001
Temp*Method	6	6	75.16667	1.6223	0.1843

as in the tensile strength problem described above. Alternatively, we sometimes find that complete randomization is not feasible because it is more difficult to change the levels of some factors than others. The hard-to-vary factors form the whole plots whereas the easy-to-vary factors are run in the subplots. The review paper by Jones and Nachtsheim (2009) is an excellent source of information and key references on split-plot designs.

In principle, we must carefully consider how the experiment must be conducted and incorporate all restrictions on randomization into the analysis. We illustrate this point using a modification of the eye focus time experiment in Chapter 6. Suppose there are only two factors: visual acuity (A) and illumination level (B). A factorial experiment with a levels of acuity, b levels of illumination, and n replicates would require that all abn observations be taken in random order. However, in the test apparatus, it is fairly difficult to adjust these two factors to different levels, so the experimenter decides to obtain the n replicates by adjusting the device to one of the a acuities and one of the b illumination levels and running all n observations at once. In the factorial design, the error actually represents the scatter or noise in the system plus the ability of the subject to reproduce the same focus time. The model for the factorial design could be written as

$$y_{ijk} = \mu + \tau_i + \beta_j + (\tau\beta)_{ij} + \phi_{ijk} + \theta_{ijk} \begin{cases} i = 1, 2, \ldots, a \\ j = 1, 2, \ldots, b \\ k = 1, 2, \ldots, n \end{cases} \tag{14.17}$$

where ϕ_{ijk} represents the scatter or noise in the system that results from "experimental error" (that is, our failure to duplicate exactly the same levels of acuity and illumination on different runs, variability in environmental conditions, and the like), and θ_{ijk} represents the "reproducibility error" of the subject. Usually, we combine these components into one overall error term, say $\epsilon_{ijk} = \phi_{ijk} + \theta_{ijk}$. Assume that $V(\epsilon_{ijk}) = \sigma^2 = \sigma^2_\phi + \sigma^2_\theta$. Now, in the factorial design, the error mean square has expectation $\sigma^2 = \sigma^2_\phi + \sigma^2_\theta$, with $ab(n-1)$ degrees of freedom.

If we restrict the randomization as in the second design above, then the "error" mean square in the analysis of variance provides an estimate of the "reproducibility error" σ^2_θ with $ab(n - 1)$ degrees of freedom, but it yields no information on the "experimental error" σ^2_ϕ. Thus, the mean square for error in this second design is too small; consequently, we will wrongly reject the null hypothesis very frequently. As pointed out by John (1971), this design is similar to a split-plot design with ab whole plots, each divided into n subplots, and no subtreatment. The situation is also similar to **subsampling**, as described by Ostle (1963). Assuming that A and B are fixed, we find that the expected mean squares in this case are

$$\begin{aligned} E(MS_A) &= \sigma^2_\theta + n\sigma^2_\phi + \frac{bn \sum \tau_i^2}{a - 1} \\ E(MS_B) &= \sigma^2_\theta + n\sigma^2_\phi + \frac{an \sum \beta_j^2}{b - 1} \\ E(MS_{AB}) &= \sigma^2_\theta + n\sigma^2_\phi + \frac{n \sum \sum (\tau\beta)^2_{ij}}{(a-1)(b-1)} \\ E(MS_E) &= \sigma^2_\theta \end{aligned} \tag{14.18}$$

Thus, there are no tests on the main effects unless interaction is negligible. The situation is exactly that of a two-way analysis of variance with one observation per cell. If both factors are random, then the main effects may be tested against the AB interaction. If only one factor is random, then the fixed factor can be tested against the AB interaction.

In general, if one analyzes a factorial design and all the main effects and interactions are significant, then one should examine carefully **how the experiment was actually conducted**. There may be randomization restrictions in the model not accounted for in the analysis, and consequently, the data should not be analyzed as a factorial.

14.5 Other Variations of the Split-Plot Design

14.5.1 Split-Plot Designs with More Than Two Factors

Sometimes we find that either the whole plot or the subplot will contain two or more factors, arranged in a factorial structure. As an example, consider an experiment conducted on a furnace to grow an oxide on a silicon wafer. The response variables of interest are oxide layer thickness and layer uniformity. There are four design factors: temperature (*A*), gas flow (*B*), time (*C*), and wafer position in the furnace (*D*). The experimenter plans to run a 2^4 factorial design with two replicates (32 trials). Now factors *A* and *B* (temperature and gas flow) are difficult to change, whereas *C* and *D* (time and wafer position) are easy to change. This leads to the split-plot design shown in Figure 14.7. Notice that both replicates of the experiment are split into four whole plots, each containing one combination of the settings of temperature and gas flow. Once these levels are chosen, each whole plot is split into four subplots and a 2^2 factorial in the factors time and wafer position is conducted, where the treatment combinations in the subplot are tested in random order. Only four changes in temperature and gas flow are made in each replicate, whereas the levels of time and wafer position are completely randomized.

A model for this experiment, consistent with Equation 14.16, is

$$\begin{aligned} y_{ijklm} = \mu &+ \tau_i + \beta_j + \gamma_k + (\beta\gamma)_{jk} + \theta_{ijk} + \delta_l + \lambda_m + (\delta\lambda)_{lm} \\ &+ (\beta\delta)_{jl} + (\beta\lambda)_{jm} + (\gamma\delta)_{kl} + (\delta\lambda)_{lm} + (\beta\gamma\delta)_{jkl} + (\beta\gamma\lambda)_{jkm} \\ &+ (\beta\delta\lambda)_{jlm} + (\gamma\delta\lambda)_{klm} + (\beta\gamma\delta\lambda)_{jklm} + \epsilon_{ijklm} \quad \begin{cases} i = 1, 2 \\ j = 1, 2 \\ k = 1, 2 \\ l = 1, 2 \\ m = 1, 2 \end{cases} \end{aligned} \tag{14.19}$$

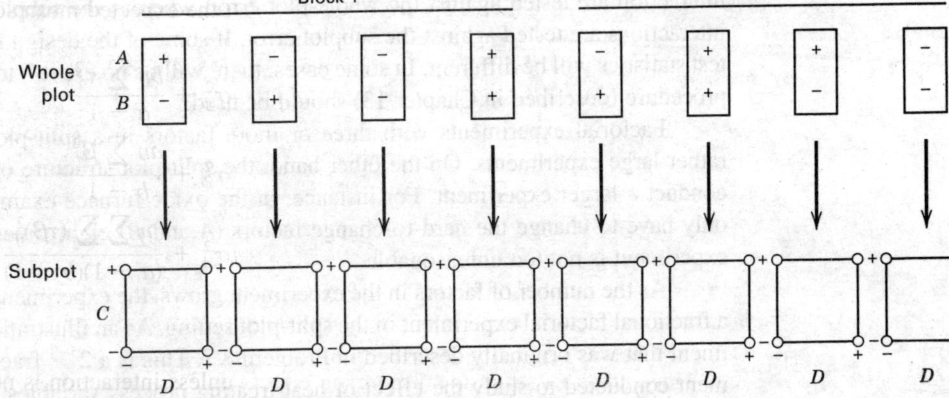

■ **FIGURE 14.7 A split-plot design with four design factors, two in the whole plot and two in the subplot**

■ **TABLE 14.21**

An Abbreviated Analysis for a Split-Plot Design with Factors *A* and *B* in the Whole Plots and Factors *C* and *D* in the Subplots (Refer to Figure 14.7)

Source of Variation	Sum of Squares	Degrees of Freedom	Expected Mean Square
Replicates (τ_i)	$SS_{\text{Replicates}}$	1	$\sigma_\epsilon^2 + 16\sigma_\tau^2$
A (β_j)	SS_A	1	$\sigma_\epsilon^2 + 8\sigma_\theta^2 + A$
B (γ_k)	SS_B	1	$\sigma_\epsilon^2 + 8\sigma_\theta^2 + B$
AB	SS_{AB}	1	$\sigma_\epsilon^2 + 8\sigma_\theta^2 + AB$
Whole-Plot Error (θ_{ijk})	SS_{WP}	3	$\sigma_\epsilon^2 + 8\sigma_\theta^2$
C (δ_l)	SS_C	1	$\sigma_\epsilon^2 + C$
D (λ_m)	SS_D	1	$\sigma_\epsilon^2 + D$
CD	SS_{CD}	1	$\sigma_\epsilon^2 + CD$
AC	SS_{AC}	1	$\sigma_\epsilon^2 + AC$
BC	SS_{BC}	1	$\sigma_\epsilon^2 + BC$
AD	SS_{AD}	1	$\sigma_\epsilon^2 + AD$
BD	SS_{BD}	1	$\sigma_\epsilon^2 + BD$
ABC	SS_{ABC}	1	$\sigma_\epsilon^2 + ABC$
ABD	SS_{ABD}	1	$\sigma_\epsilon^2 + ABD$
ACD	SS_{ACD}	1	$\sigma_\epsilon^2 + ACD$
BCD	SS_{BCD}	1	$\sigma_\epsilon^2 + BCD$
$ABCD$	SS_{ABCD}	1	$\sigma_\epsilon^2 + ABCD$
Subplot Error (ϵ_{ijklm})	SS_{SP}	12	σ_ϵ^2
Total	SS_T	31	

where τ_i represents the replicate effect, β_j and γ_k represent the whole-plot main effects, θ_{ijk} is the whole-plot error, δ_l and λ_m represent the subplot main effects, and ϵ_{ijklm} is the subplot error. We have included all interactions between the four design factors. Table 14.21 presents the analysis of variance for this design, assuming that replicates are random and all design factors are fixed effects. In this table, σ_θ^2 and σ_ϵ^2 represent the variances of the whole-plot and subplot errors, respectively, σ_τ^2 is the variance of the block effects, and (for simplicity) we have used capital Latin letters to denote fixed-type effects. The whole-plot main effects and interaction are tested against the whole-plot error, whereas the subplot factors and all other interactions are tested against the subplot error. If some of the design factors are random, the test statistics will be different. In some cases, there will be no exact F tests and Satterthwaite's procedure (described in Chapter 13) should be used.

Factorial experiments with three or more factors in a split-plot structure tend to be rather large experiments. On the other hand, the split-plot structure often makes it easier to conduct a larger experiment. For instance, in the oxide furnace example, the experimenters only have to change the hard-to-change factors (A and B) eight times, so perhaps a 32-run experiment is not too unreasonable.

As the number of factors in the experiment grows, the experimenter may consider using a fractional factorial experiment in the split-plot setting. As an illustration, consider the experiment that was originally described in Problem 8.7. This is a 2^{5-1} fractional factorial experiment conducted to study the effect of heat-treating process variables on the height of truck springs. The factors are A = transfer time, B = heating time, C = oil quench temperature, D = temperature, and E = hold down time. Suppose that factors A, B, and C are very hard to change and that the other two factors D and E are easy to change. For example, it might be necessary to first manufacture the springs by varying factors A, B, and C, and then hold those factors fixed while varying factors D and E in a subsequent experiment.

We consider a modification of that experiment because the original experimenters may not have run it as a split plot and because they did not use the design generator that we are going to use. Let the whole-plot factors be denoted by A, B, and C and the split-plot factors be denoted by $\boldsymbol{D}$ and $\boldsymbol{E}$ (the bold symbol is used to help us identify the easy-to-change factors). We will select the design generator $\boldsymbol{E} = ABC\boldsymbol{D}$. The layout of this design has eight whole plots (the eight combinations of factors A, B, and C each at two levels). Each whole plot is divided into two subplots, and a combination of the factors $\boldsymbol{D}$ and $\boldsymbol{E}$ are tested in each split plot. (The exact treatment combinations depend on the signs on the treatment combinations for the whole-plot factors through the generator.)

Assume that all three-, four-, and five-factor interactions are negligible. If this assumption is reasonable, then all whole-plot factors A, B, and C and their two-factor interactions can be estimated in the whole plot . If the design is replicated, these effects would be tested against the whole-plot error. Alternatively, if the design is unreplicated, their effects could be assessed via a normal probability plot (or possibly by Lenth's method). The subplot factors $\boldsymbol{D}$ and $\boldsymbol{E}$ and their two-factor interaction $\boldsymbol{DE}$ can also be estimated. However, since $\boldsymbol{DE} = ABC$, the $\boldsymbol{DE}$ interaction needs to be treated as a whole-plot term. There are six two-factor interactions of whole-plot and split-plot factors that can also be estimated: $A\boldsymbol{D}$, $A\boldsymbol{E}$, $B\boldsymbol{D}$, $B\boldsymbol{E}$, $C\boldsymbol{D}$, and $C\boldsymbol{E}$. In general, it turns out that any split-plot main effect or interaction that is aliased with whole-plot main effects or interactions involving only whole-plot factors would be compared to the whole-plot error. Furthermore, split-plot main effects or interactions involving at least one split-plot factor that are not aliased with whole-plot main effects or interactions involving only whole-plot factors are compared to the split-plot error. See Bisgaard (2000) for a thorough discussion. Therefore, in our problem, all of the effects $\boldsymbol{D}$, $\boldsymbol{E}$, $A\boldsymbol{D}$, $A\boldsymbol{E}$, $B\boldsymbol{D}$, $B\boldsymbol{E}$, $C\boldsymbol{D}$, and $C\boldsymbol{E}$ are compared to the split-plot error. Alternatively, they could be assessed via a normal probability plot.

Recently, several papers have appeared on the subject of fractional factorials and response surface experiments in split plots. The papers by Vining, Kowalski and Montgomery (2005), Vining and Kowalski (2008), Macharia and Goos (2010), Bisgaard (2000), Bingham and Sitter (1999), Huang, Chen, and Voelkel (1999), and Almimi, Kulahci and Montgomery (2008a)(2008b)(2009) are recommended.

EXAMPLE 14.3

The factors affecting uniformity in a single-wafer plasma etching process are being investigated. Three factors on the etching tool are relatively difficult to change from run to run: A = electrode gap, B = gas flow, and C = pressure. Two other factors are easy to change from run to run: $\boldsymbol{D}$ = time and $\boldsymbol{E}$ = RF (radio frequency) power. The experimenters want to use a fractional factorial experiment to investigate these five factors because the number of test wafers available is limited. The hard-to-change factors also indicate that a split-plot design should be considered. The experimenters decide to use the strategy discussed above: a 2^{5-1} design with factors A, B, and C in the whole plots and factors $\boldsymbol{D}$ and $\boldsymbol{E}$ in the subplots. The design generator is $\boldsymbol{E} = ABC\boldsymbol{D}$. This produces a 16-run fractional factorial with eight whole plots. Every whole plot contains one of the eight treatment combinations from a complete 2^3 factorial design in factors A, B, and C. Each whole plot is divided into two subplots, with one of the treatment combinations for factors $\boldsymbol{D}$ and $\boldsymbol{E}$ in each subplot. The design and the resulting uniformity data are shown in Table 14.22. The eight whole plots were run in random order, but once a whole plot configuration for factors A, B, and C was set up on the etching tool, both subplot runs were made (also in random order).

The statistical analysis of this experiment involves keeping the whole-plot and subplot factors separate. We assume that all interactions beyond order two are negligible. Table 14.23 lists the effect estimates separated into whole-plot and subplot terms. Furthermore, available degrees of freedom are used to estimate effects, so we cannot estimate either the whole-plot or the subplot error. Therefore, we must use

■ **TABLE 14.22**
The 2^{5-1} Split-Plot Experiment for the Plasma Etching Tool

Whole Plots	Whole-Plot Factors *A*	*B*	*C*	Subplot Factors *D*	*E*	Uniformity
	−	−	−	−	+	40.85
1	−	−	−	+	−	41.07
	+	−	−	−	−	35.67
2	+	−	−	+	+	51.15
	−	+	−	−	−	41.80
3	−	+	−	+	+	37.01
	+	+	−	−	+	91.09
4	+	+	−	+	−	48.67
	−	−	+	−	−	40.32
5	−	−	+	+	+	43.34
	+	−	+	−	+	62.46
6	+	−	+	+	−	38.08
	−	+	+	−	+	31.99
7	−	+	+	+	−	41.03
	+	+	+	−	−	70.31
8	+	+	+	+	+	81.03

■ **TABLE 14.23**
Effects for Plasma Etching Experiment Separated into Whole-Plot and Subplot Effects

Term	Parameter Estimates	Type of Term
Intercept	49.73875	
Gap (*A*)	10.0625	Whole
Gas flow (*B*)	5.6275	Whole
Pressure (*C*)	1.325	Whole
Time (*D*)	−2.0725	Subplot
RF power (*E*)	5.12625	Subplot
AB	7.34625	Whole
AC	1.83125	Whole
AD	−3.00875	Subplot
AE	6.505	Subplot
BC	−0.60125	Whole
BD	−1.35875	Subplot
BE	−0.2125	Subplot
CD	1.86625	Subplot
CE	−1.485	Subplot
DE	0.34	Whole

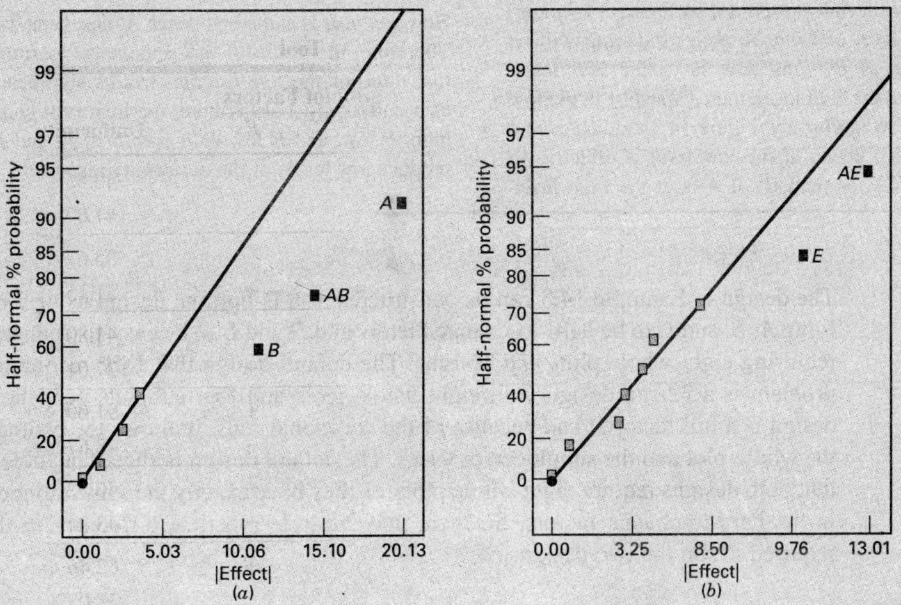

■ **FIGURE 14.8 Half-normal plots of the effects from the 2^{5-1} split-plot experiment in Example 14.3. (*a*) Whole-plot effects. (*b*) Subplot effects**

normal probability plots to evaluate the effects. Figure 14.8*a* is a half-normal probability plot of the effect estimates for only the whole-plot factors, ignoring the factors in the subplots. Notice that factors *A*, *B*, and the *AB* interaction have large effects. Figure 14.8*b* is a half-normal probability plot of the subplot effects ***D*** and ***E***, and the interactions involving those factors, ***DE***, ***AD***, ***AE***, ***BD***, ***BE***, ***CD***, and ***CE***. Only the main effect of ***E*** and the ***AE*** interaction are large.

Figure 14.9*a* and *b* are the two-factor interaction plots of the intersections *AB* and *AE*. The experimenter's objective is

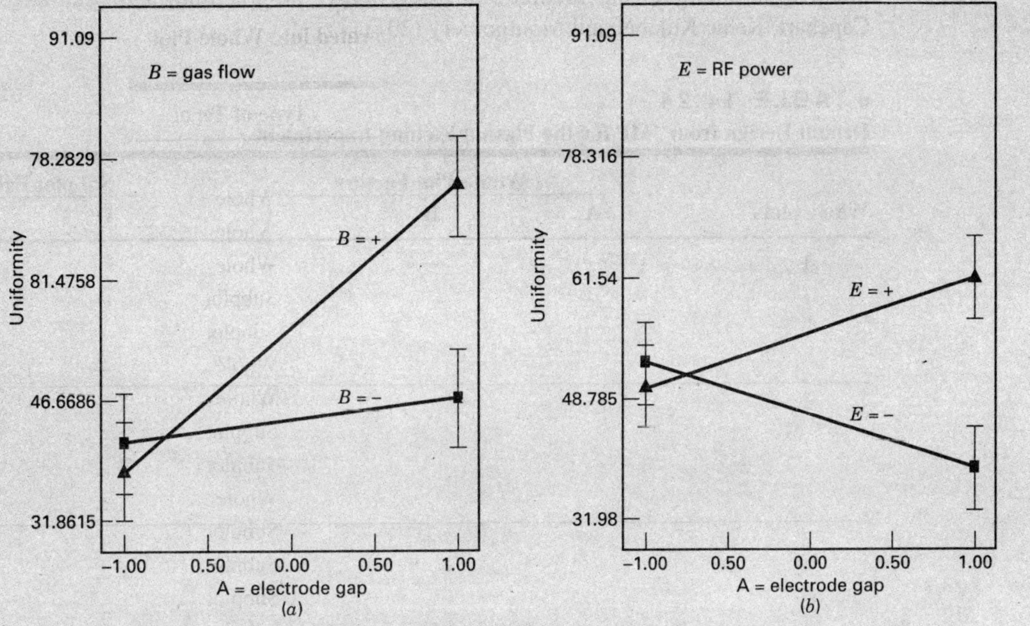

■ **FIGURE 14.9 Two-factor interaction graphs for the 2^{5-1} split-plot experiment in Example 14.3. (*a*) *AB* interaction. (*b*) *AE* interaction**

to minimize the uniformity response, so Figure 14.9*a* suggests that either level of factor A = electrode gap will be effective as long as B = gas flow is at the low level. However, if B is at the high level, then A must be at the low level to achieve low uniformity. Figure 14.9*b* indicates that controlling E = RF power at the low level is effective in reducing uniformity, particularly if A is at the high level. However, if E is at the high level, A must be at the low level. Therefore, the results of this screening experiment indicate that three of the five original factors significantly impact etch uniformity. Furthermore, the treatment combination A high, B low, and E low or A low, B high, and E high will produce low levels of the uniformity response.

The design in Example 14.3 can be constructed as a D-optimal design using JMP, by specifying A, B, and C to be hard-to-change factors and D and E to be easy-to-change factors, and requiring eight whole plots and 16 runs. The default design that JMP recommends for this problem is a 32-run design with eight whole plots and four subplots per whole plot. This design is a full factorial and because of the additional runs, it allows the estimation of both the whole-plot and the subplot error terms. The default design is shown in Table 14.24. Note that both designs require eight whole plots, as they have exactly the same number of changes in the hard-to-change factors. So there may be little practical difference in the resources required to run the two designs.

14.5.2 The Split-Split-Plot Design

The concept of split-plot designs can be extended to situations in which randomization restrictions may occur at any number of levels within the experiment. If there are two levels of randomization restrictions, the layout is called a **split-split-plot design**. The following example illustrates such a design.

Optimal design tools are an important and useful way to construct split-plot designs. Optimal design tools using the coordinate exchange algorithm are very efficient and available in JMP. An alternate construction tool using integer programming has been developed by Capehart, Keha, Kulahci and Montgomery (2011).

■ **TABLE 14.24**
Default Design from JMP for the Plasma Etching Experiment

	Whole-Plot Factors			Subplot Factors	
Whole plots	**A**	**B**	**C**	**D**	**E**
1	−	−	−	−	−
				−	+
				+	−
				+	+
2	−	−	+	−	−
				−	+
				+	−
				+	+
3	−	+	−	−	−
				−	+
				+	−
				+	+

■ **TABLE 14.24** (*Continued*)

4	−	+	+	−	−
				−	+
				+	−
				+	+
5	+	−	−	−	−
				−	+
				+	−
				+	+
6	+	−	+	−	−
				−	+
				+	−
				+	+
7	+	+	−	−	−
				−	+
				+	−
				+	+
8	+	+	+	−	−
				−	+
				+	−
				+	+

EXAMPLE 14.4

A researcher is studying the absorption times of a particular type of antibiotic capsule. There are three technicians, three dosage strengths, and four capsule wall thicknesses of interest to the researcher. Each replicate of a factorial experiment would require 36 observations. The experimenter has decided on four replicates, and it is necessary to run each replicate on a different day. Note that the days can be considered as blocks. Within a replicate (or a block) (day), the experiment is performed by assigning a unit of antibiotic to a technician who conducts the experiment on the three dosage strengths and the four wall thicknesses. Once a particular dosage strength is formulated, all four wall thicknesses are tested at that strength. Then another dosage strength is selected and all four wall thicknesses are tested. Finally, the third dosage strength and the four wall thicknesses are tested. Meanwhile, two other laboratory technicians also follow this plan, each starting with a unit of antibiotic.

Note that there are two randomization restrictions within each replicate (or block): technician and dosage strength. The whole plots correspond to the technician. The order in which the technicians are assigned the units of antibiotic is randomly determined. The dosage strengths form three subplots. Dosage strength may be randomly assigned to a subplot. Finally, within a particular dosage strength, the four capsule wall thicknesses are tested in random order, forming four sub-subplots. The wall thicknesses are usually called sub-subtreatments. Because there are two randomization restrictions in the experiment (some authors say two "splits" in the design), the design is called a split-split-plot design. Figure 14.10 illustrates the randomization restrictions and experimental layout in this design.

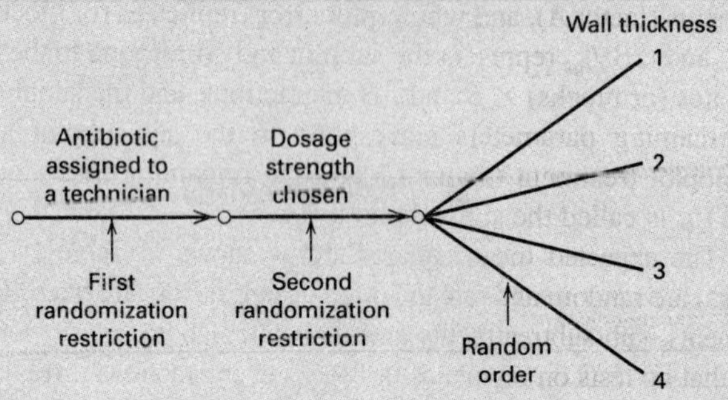

		Technician								
		1			2			3		
Blocks	Dosage strength	1	2	3	1	2	3	1	2	3
1	Wall thicknesses	1 2 3 4	1 2 3 4	1 2 3 4	1 2 3 4	1 2 3 4	1 2 3 4	1 2 3 4	1 2 3 4	1 2 3 4
2	Wall thicknesses	1 2 3 4	1 2 3 4	1 2 3 4	1 2 3 4	1 2 3 4	1 2 3 4	1 2 3 4	1 2 3 4	1 2 3 4
3	Wall thicknesses	1 2 3 4	1 2 3 4	1 2 3 4	1 2 3 4	1 2 3 4	1 2 3 4	1 2 3 4	1 2 3 4	1 2 3 4
4	Wall thicknesses	1 2 3 4	1 2 3 4	1 2 3 4	1 2 3 4	1 2 3 4	1 2 3 4	1 2 3 4	1 2 3 4	1 2 3 4

■ **FIGURE 14.10 A split-split-plot design**

A linear statistical model for the split-split-plot design is

$$
\begin{aligned}
y_{ijkh} = \mu &+ \tau_i + \beta_j + (\tau\beta)_{ij} + \gamma_k + (\tau\gamma)_{ik} + (\beta\gamma)_{jk} + (\tau\beta\gamma)_{ijk} \\
&+ \delta_h + (\tau\delta)_{ih} + (\beta\delta)_{jh} \\
&+ (\tau\beta\delta)_{ijh} + (\gamma\delta)_{kh} + (\tau\gamma\delta)_{ikh} + (\beta\gamma\delta)_{jkh} \\
&+ (\tau\beta\gamma\delta)_{ijkh} + \epsilon_{ijkh} \quad
\begin{cases} i = 1, 2, \ldots, r \\ j = 1, 2, \ldots, a \\ k = 1, 2, \ldots, b \\ h = 1, 2, \ldots, c \end{cases}
\end{aligned}
\qquad (14.20)
$$

where τ_i, β_j, and $(\tau\beta)_{ij}$ represent the whole plot and correspond to replicates or blocks, main treatments (factor A), and whole-plot error (replicates (or blocks) $\times$ A), respectively; γ_k, $(\tau\gamma)_{ik}$, $(\beta\gamma)_{jk}$, and $(\tau\beta\gamma)_{ijk}$ represent the subplot and correspond to the subplot treatment (factor B), the replicates (or blocks) $\times$ B and AB interactions, and the subplot error, respectively; and δ_h and the remaining parameters correspond to the sub-subplot and represent, respectively, the sub-subplot treatment (factor C) and the remaining interactions. The four-factor interaction $(\tau\beta\gamma\delta)_{ijkh}$ is called the sub-subplot error.

The expected mean squares are as shown in Table 14.25, assuming that the replicates (blocks) are random and that the other design factors are fixed. Tests on the main treatments, subtreatments, sub-subtreatments and their interactions are obvious from inspection of this table. Note that no tests on replicates or blocks or interactions involving replicates or blocks exist.

The statistical analysis of a split-split-plot design is like that of a single replicate of a four-factor factorial. The number of degrees of freedom for each test is determined in the usual manner. To illustrate, in Example 14.4, where we had four replicates, three technicians, three dosage strengths, and four wall thicknesses, we would have only $(r - 1)(a - 1) = (4 - 1)(3 - 1) = 6$ whole-plot error degrees of freedom for testing technicians. This is a

■ **TABLE 14.25**

Expected Mean Squares for the Split-Split-Plot Design

	Model Term	Expected Mean Square
	τ_i	$\sigma^2 + abc\sigma_\tau^2$
Whole plot	β_j	$\sigma^2 + bc\sigma_{\tau\beta}^2 + \dfrac{rbc\sum\beta_j^2}{(a-1)}$
	$(\tau\beta)_{ij}$	$\sigma^2 + bc\sigma_{\tau\beta}^2$
	γ_k	$\sigma^2 + ac\sigma_{\tau\gamma}^2 + \dfrac{rac\sum\gamma_k^2}{(b-1)}$
Subplot	$(\tau\gamma)_{ik}$	$\sigma^2 + ac\sigma_{\tau\gamma}^2$
	$(\beta\gamma)_{jk}$	$\sigma^2 + c\sigma_{\tau\beta\gamma}^2 + \dfrac{rc\sum\sum(\beta\gamma)_{jh}^2}{(a-1)(b-1)}$
	$(\tau\beta\gamma)_{ijk}$	$\sigma^2 + c\sigma_{\tau\beta\gamma}^2$
	δ_h	$\sigma^2 + ab\sigma_{\tau\delta}^2 + \dfrac{rab\sum\delta_k^2}{(c-1)}$
	$(\tau\delta)_{ih}$	$\sigma^2 + ab\sigma_{\tau\delta}^2$
	$(\beta\delta)_{jh}$	$\sigma^2 + b\sigma_{\tau\beta\delta}^2 + \dfrac{rb\sum\sum(\beta\delta)_{jh}^2}{(a-1)(c-1)}$
	$(\tau\beta\delta)_{ijh}$	$\sigma^2 + b\sigma_{\tau\beta\delta}^2$
Sub-subplot	$(\gamma\delta)_{kh}$	$\sigma^2 + a\sigma_{\tau\gamma\delta}^2 + \dfrac{ra\sum\sum(\gamma\delta)_{kh}^2}{(b-1)(c-1)}$
	$(\tau\gamma\delta)_{ikh}$	$\sigma^2 + a\sigma_{\tau\gamma\delta}^2$
	$(\beta\gamma\delta)_{jkh}$	$\sigma^2 + \sigma_{\tau\beta\gamma\delta}^2 + \dfrac{r\sum\sum\sum(\beta\gamma\delta)_{ijk}^2}{(a-1)(b-1)(c-1)}$
	$(\tau\beta\gamma\delta)_{ijkh}$	$\sigma^2 + \sigma_{\tau\beta\gamma\delta}^2$
	$\epsilon_{l(ijkh)}$	σ^2 (not estimable)

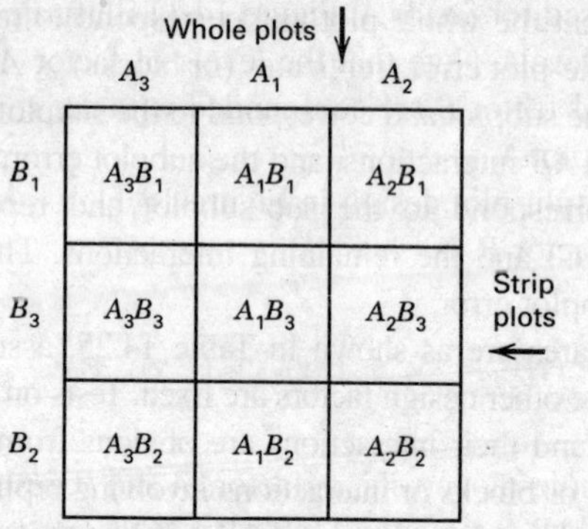

■ **FIGURE 14.11** **One replicate (block) of a strip-split-plot design**

relatively small number of degrees of freedom, and the experimenter might consider using additional replicates to increase the precision of the test. If there are a replicates, we will have $2(r - 1)$ degrees of freedom for whole-plot error. Thus, five replicates yield $2(5 - 1) = 8$ error degrees of freedom, six replicates yield $2(6 - 1) = 10$ error degrees of freedom, seven replicates yield $2(7 - 1) = 12$ error degrees of freedom, and so on. Consequently, we would probably not want to run fewer than four replicates because this would yield only four error degrees of freedom. Each additional replicate allows us to gain two degrees of freedom for error. If we could afford to run five replicates, we could increase the precision of the test by one-third (from six to eight degrees of freedom). Also, in going from five to six replicates, there is an additional 25 percent gain in precision. If resources permit, the experimenter should run five or six replicates.

14.5.3 The Strip-Split-Plot Design

The strip-split-plot design has had an extensive application in the agricultural sciences, but it finds occasional use in industrial experimentation. In the simplest case, we have two factors A and B. Factor A is applied to whole plots just as in the standard split-plot design. Then factor B is applied to **strips** (which are really just another set of whole plots) that are orthogonal

■ **TABLE 14.26**
An Abbreviated Analysis of Variance for a Strip-Split-Plot Design

Source of Variation	Sum of Squares	Degrees of Freedom	Expected Mean Square
Replicates (or blocks)	$SS_{\text{Replicates}}$	$r - 1$	$\sigma_\epsilon^2 + ab\sigma_\tau^2$
A	SS_A	$a - 1$	$\sigma_\epsilon^2 + b\sigma_{\tau\beta}^2 + \frac{rb\sum \beta_j^2}{a-1}$
Whole-plot error$_A$	SS_{WP_A}	$(r-1)(a-1)$	$\sigma_\epsilon^2 + b\sigma_{\tau\beta}^2$
B	SS_B	$b - 1$	$\sigma_\epsilon^2 + a\sigma_{\tau\gamma}^2 + \frac{ra\sum \gamma_k^2}{b-1}$
Whole-plot error$_B$	SS_{WP_B}	$(r-1)(b-1)$	$\sigma_\epsilon^2 + a\sigma_{\tau\gamma}^2$
AB	SS_{AB}	$(a-1)(b-1)$	$\sigma_\epsilon^2 + \frac{r\sum\sum(\tau\beta)_{jk}^2}{(a-1)(b-1)}$
Subplot error	SS_{SP}	$(r-1)(a-1)(b-1)$	σ_ϵ^2
Total	SS_T	$rab - 1$	

to the original whole plots used for factor A. Figure 14.11 illustrates a situation in which both factors A and B have three levels. Note that the levels of factor A are confounded with the whole plots, and the levels of factor B are confounded with the strips (which can be thought of as a **second** set of whole plots).

A model for the strip-split plot design in Figure 14.11, assuming r replicates, a levels of factor A, and b levels of factor B, is

$$y_{ijk} = \mu + \tau_i + \beta_j + (\tau\beta)_{ij} + \gamma_k + (\tau\gamma)_{ik} + (\beta\gamma)_{jk} + \epsilon_{ijk} \quad \begin{cases} i = 1, 2, \ldots, r \\ j = 1, 2, \ldots, a \\ k = 1, 2, \ldots, b \end{cases}$$

where $(\tau\beta)_{ij}$ and $(\tau\gamma)_{ik}$ are whole-plot errors for factors A and B, respectively, and ϵ_{ijk} is a "subplot" error used to test the AB interaction. Table 14.26 shows an abbreviated analysis of variance assuming A and B are fixed factors and replicates are random. The replicates are sometimes considered as blocks.

14.6 Problems

14.1. The surface finish of metal parts made on four machines is being studied. An experiment is conducted in which each machine is run by three different operators and two specimens from each operator are collected and tested. Because of the location of the machines, different operators are used on each machine, and the operators are chosen at random. The data are shown in the following table. Analyze the data and draw conclusions.

	Machine 1			**Machine 2**			**Machine 3**			**Machine 4**		
Operator	**1**	**2**	**3**	**1**	**2**	**3**	**1**	**2**	**3**	**1**	**2**	**3**
	79	94	46	92	85	76	88	53	46	36	40	62
	62	74	57	99	79	68	75	56	57	53	56	47

14.2. A rocket propellant manufacturer is studying the burning rate of propellant from three production processes. Four batches of propellant are randomly selected from the output of each process, and three determinations of burning rate are made on each batch. The results follow. Analyze the data and draw conclusions.

	Process 1				**Process 2**				**Process 3**			
Batch	**1**	**2**	**3**	**4**	**1**	**2**	**3**	**4**	**1**	**2**	**3**	**4**
	26	19	15	15	19	23	18	35	15	35	38	26
	31	28	17	16	17	24	21	27	14	21	54	29
	26	20	14	13	14	21	17	25	20	24	50	34

14.3. A manufacturing engineer is studying the dimensional variability of a particular component that is produced on three machines. Each machine has two spindles, and four components are randomly selected from each spindle. The results follow. Analyze the data, assuming that machines and spindles are fixed factors.

	Machine 1		**Machine 2**		**Machine 3**	
Spindle	**1**	**2**	**1**	**2**	**1**	**2**
	12	8	14	12	14	16
	9	9	15	10	10	15
	11	10	13	11	12	15
	12	8	14	13	11	14

14.4. To simplify production scheduling, an industrial engineer is studying the possibility of assigning one time standard to a particular class of jobs, believing that differences between jobs are negligible. To see if this simplification is possible, six jobs are randomly selected. Each job is given to a different group of three operators. Each operator completes the job twice at different times during the week, and the following results are obtained. What are your conclusions about the use of a common time standard for all jobs in this class? What value would you use for the standard?

Job	**Operator 1**		**Operator 2**		**Operator 3**	
1	158.3	159.4	159.2	159.6	158.9	157.8
2	154.6	154.9	157.7	156.8	154.8	156.3
3	162.5	162.6	161.0	158.9	160.5	159.5
4	160.0	158.7	157.5	158.9	161.1	158.5
5	156.3	158.1	158.3	156.9	157.7	156.9
6	163.7	161.0	162.3	160.3	162.6	161.8

14.5. Consider the three-stage nested design shown in Figure 14.5 to investigate alloy hardness. Using the data that follow, analyze the design, assuming that alloy chemistry and heats are fixed factors and ingots are random. Use the restricted form of the mixed model.

Alloy Chemistry 1						
Heats	1		2		3	
Ingots	1 .	2	1	2	1	2
	40	27	95	69	65	78
	63	30	67	47	54	45

Alloy Chemistry 2						
Heats	1		2		3	
Ingots	1	2	1	2	1	2
	22	23	83	75	61	35
	10	39	62	64	77	42

14.6. Reanalyze the experiment in Problem 14.5 using the unrestricted form of the mixed model. Comment on any differences you observe between the restricted and the unrestricted model results. You may use a computer software package.

14.7. Derive the expected mean squares for a balanced three-stage nested design, assuming that A is fixed and that B and C are random. Obtain formulas for estimating the variance components. Assume the restricted form of the mixed model.

14.8. Repeat Problem 14.7 assuming the unrestricted form of the mixed model. You may use a computer software package to do this. Comment on any differences between the restricted and unrestricted model analysis and conclusions.

14.9. Derive the expected mean squares for a balanced three-stage nested design if all three factors are random. Obtain formulas for estimating the variance components.

14.10. Verify the expected mean squares given in Table 14.1.

14.11. ***Unbalanced nested designs.*** Consider an unbalanced two-stage nested design with b_j levels of B under the ith level of A and n_{ij} replicates in the ijth cell.

(a) Write down the least squares normal equations for this situation. Solve the normal equations.

(b) Construct the analysis of variance table for the unbalanced two-stage nested design.

(c) Analyze the following data, using the results in part (b).

Factor A	1		2		
Factor B	1	2	1	2	3
	6	−3	5	2	1
	4	1	7	4	0
	8		9	3	−3
			6		

14.12. ***Variance components in the unbalanced two-stage nested design.*** Consider the model

$$y_{ijk} = \mu + \tau_i + \beta_{j(i)} + \epsilon_{k(ij)} \quad \begin{cases} i = 1, 2, \ldots, a \\ j = 1, 2, \ldots, b_i \\ k = 1, 2, \ldots, n_{ij} \end{cases}$$

where A and B are random factors. Show that

$$E(MS_A) = \sigma^2 + c_1\sigma_\beta^2 + c_2\sigma_\tau^2$$
$$E(MS_{B(A)}) = \sigma^2 + c_0\sigma_\beta^2$$
$$E(MS_E) = \sigma^2$$

where

$$c_0 = \frac{N - \sum_{i=1}^{a}\left(\sum_{j=1}^{b_i} n_{ij}^2/n_{i.}\right)}{b - a}$$

$$c_1 = \frac{\sum_{i=1}^{a}\left(\sum_{j=1}^{b_i} n_{ij}^2/n_{i.}\right) - \sum_{i=1}^{a}\sum_{j=1}^{b_i} n_{ij}^2/N}{a - 1}$$

$$c_2 = \frac{N - \dfrac{\sum_{i=1}^{a} n_{i.}^2}{N}}{a - 1}$$

14.13. A process engineer is testing the yield of a product manufactured on three machines. Each machine can be operated at two power settings. Furthermore, a machine has three stations on which the product is formed. An experiment is conducted in which each machine is tested at both power settings, and three observations on yield are taken from each station. The runs are made in random order, and the results are shown in Table P14.1. Analyze this experiment, assuming that all three factors are fixed.

■ **TABLE P14.1**
Yield Experiment in Problem 14.13

	Machine 1			Machine 2		
Station	**1**	**2**	**3**	**1**	**2**	**3**
Power setting 1	34.1	33.7	36.2	31.1	33.1	32.8
	30.3	34.9	36.8	33.5	34.7	35.1
	31.6	35.0	37.1	34.0	33.9	34.3
Power setting 2	24.3	28.1	25.7	24.1	24.1	26.0
	26.3	29.3	26.1	25.0	25.1	27.1
	27.1	28.6	24.9	26.3	27.9	23.9

	Machine 3		
Station	**1**	**2**	**3**
Power setting 1	32.9	33.8	33.6
	33.0	33.4	32.8
	33.1	32.8	31.7
Power setting 2	24.2	23.2	24.7
	26.1	27.4	22.0
	25.3	28.0	24.8

14.14. Suppose that in Problem 14.13 a large number of power settings could have been used and that the two selected for the experiment were chosen randomly. Obtain the expected mean squares for this situation assuming the restricted form of the mixed model and modify the previous analysis appropriately.

14.15. Reanalyze the experiment in Problem 14.14 assuming the unrestricted form of the mixed model. You may use a computer software package to do this. Comment on any differences between the restricted and unrestricted model analysis and conclusions.

14.16. A structural engineer is studying the strength of aluminum alloy purchased from three vendors. Each vendor submits the alloy in standard-sized bars of 1.0, 1.5, or 2.0 inches. The processing of different sizes of bar stock from a common ingot involves different forging techniques, and so this factor may be important. Furthermore, the bar stock is forged from ingots made in different heats. Each vendor submits two test specimens of each size bar stock from three heats. The resulting strength data is shown in Table P14.2. Analyze the data, assuming that vendors and bar size are fixed and heats are random. Use the restricted form of the mixed model.

■ **TABLE P14.2**
Strength Data in Problem P14.16

	Vendor 1			Vendor 2		
Heat	**1**	**2**	**3**	**1**	**2**	**3**
Bar size:						
1 in.	1.250	1.350	1.235	1.301	1.350	1.301
	1.259	1.400	1.206	1.263	1.392	1.320
$1\frac{1}{2}$ in.	1.316	1.329	1.250	1.274	1.384	1.356
	1.300	1.362	1.239	1.268	1.375	1.357
2 in.	1.287	1.346	1.273	1.247	1.362	1.336
	1.292	1.382	1.215	1.215	1.328	1.342

	Vendor 3		
Heat	**1**	**2**	**3**
Bar size:			
1 in.	1.247	1.275	1.325
	1.296	1.268	1.310
$1\frac{1}{2}$ in.	1.273	1.260	1.395
	1.264	1.265	1.364
2 in.	1.301	1.280	1.319
	1.262	1.271	1.323

14.17. Rework Problem 14.16 using the unrestricted form of the mixed model. You may use a computer software package to do this. Comment on any differences between the restricted and unrestricted model analysis and conclusions.

14.18. Suppose that in Problem 14.16 the bar stock may be purchased in many sizes and that the three sizes actually used in the experiment were selected randomly. Obtain the expected mean squares for this situation and modify the previous analysis appropriately. Use the restricted form of the mixed model.

14.19. Steel is normalized by heating above the critical temperature, soaking, and then air cooling. This process increases the strength of the steel, refines the grain, and homogenizes the structure. An experiment is performed to determine the effect of temperature and heat treatment time on the strength of normalized steel. Two temperatures and three times are selected. The experiment is performed by heating the oven to a randomly selected temperature and inserting three specimens. After 10 minutes one specimen is removed, after 20 minutes the second is removed, and after 30 minutes the final specimen is removed. Then the temperature is changed to the other level and the process is repeated. Four shifts are required to collect the data, which are shown below. Analyze the data and draw conclusions, assuming both factors are fixed.

		Temperature (°F)	
Shift	**Time (min)**	**1500**	**1600**
	10	63	89
1	20	54	91
	30	61	62
	10	50	80
2	20	52	72
	30	59	69
	10	48	73
3	20	74	81
	30	71	69
	10	54	88
4	20	48	92
	30	59	64

14.20. An experiment is designed to study pigment dispersion in paint. Four different mixes of a particular pigment are studied. The procedure consists of preparing a particular mix and then applying that mix to a panel by three application methods (brushing, spraying, and rolling). The response measured is the percentage reflectance of pigment. Three days are required to run the experiment, and the data obtained follow. Analyze the data and draw conclusions, assuming that mixes and application methods are fixed.

Day	Application Method	Mix 1	2	3	4
1	1	64.5	66.3	74.1	66.5
	2	68.3	69.5	73.8	70.0
	3	70.3	73.1	78.0	72.3
2	1	65.2	65.0	73.8	64.8
	2	69.2	70.3	74.5	68.3
	3	71.2	72.8	79.1	71.5
3	1	66.2	66.5	72.3	67.7
	2	69.0	69.0	75.4	68.6
	3	70.8	74.2	80.1	72.4

14.21. Repeat Problem 14.20, assuming that the mixes are random and the application methods are fixed.

14.22. Consider the split-split-plot design described in Example 14.4. Suppose that this experiment is conducted as described and that the data shown in Table P14.3 are obtained. Analyze the data and draw conclusions.

14.23. Rework Problem 14.22, assuming that the technicians are chosen at random. Use the restricted form of the mixed model.

14.24. Suppose that in Problem 14.22 four technicians had been used. Assuming that all the factors are fixed, how many blocks should be run to obtain an adequate number of degrees of freedom on the test for differences among technicians?

■ TABLE P14.3
The Absorption Time Experiment

Replicates (or Blocks)	Dosage Strengths	Technician 1: 1	1: 2	1: 3	2: 1	2: 2	2: 3	3: 1	3: 2	3: 3
	Wall Thickness									
1	1	95	71	108	96	70	108	95	70	100
	2	104	82	115	99	84	100	102	81	106
	3	101	85	117	95	83	105	105	84	113
	4	108	85	116	97	85	109	107	87	115
2	1	95	78	110	100	72	104	92	69	101
	2	106	84	109	101	79	102	100	76	104
	3	103	86	116	99	80	108	101	80	109
	4	109	84	110	112	86	109	108	86	113
3	1	96	70	107	94	66	100	90	73	98
	2	105	81	106	100	84	101	97	75	100
	3	106	88	112	104	87	109	100	82	104
	4	113	90	117	121	90	117	110	91	112
4	1	90	68	109	98	68	106	98	72	101
	2	100	84	112	102	81	103	102	78	105
	3	102	85	115	100	85	110	105	80	110
	4	114	88	118	118	85	116	110	95	120

14.25. Consider the experiment described in Example 14.4. Demonstrate how the order in which the treatment combinations are run would be determined if this experiment were run as (a) a split-split-plot, (b) a split-plot, (c) a factorial design in a randomized block, and (d) a completely randomized factorial design.

14.26. An article in *Quality Engineering* ("Quality Quandries: Two-Level Factorials Run as Split-Plot Experiments," Bisgaard et al., Vol. 8, No. 4, pp. 705–708, 1996) describes a 2^5 factorial experiment in a plasma process focused on making paper more susceptible to ink. Four of the factors (A–D) are difficult to change from run to run, so the experimenters set up the reactor at the eight sets of conditions specified by the low and high levels of those factors, and then processed the two paper types (factor E) together. The placement of the paper specimens in the reactor (right versus left) was randomized. This produces a split-plot design with *A–D* as the whole-plot factors and factor *E* as the subplot factor. The data from this experiment are shown in Table P14.4. Analyze the data from this experiment and draw conclusions.

14.27. Reconsider the experiment in Problem 14.26. This is a rather large experiment, so suppose that the experimenter had used a 2^{5-1} design instead. Set up the 2^{5-1} design in a split-plot, using the principal fraction. Then select the response data using the information from the full factorial. Analyze the data and draw conclusions. Do they agree with the results of Problem 14.26?

■ **TABLE P14.4**
The Paper-Making Experiment in Problem 14.26

Standard Order	Run Number	A = Pressure	B = Power	C = Gas Flow	D = Gas Type	E = Paper Type	*y* Contact Angle
5	1	−1	−1	+1	Oxygen	E1	37.6
21	2	−1	−1	+1	Oxygen	E2	43.5
2	3	+1	−1	−1	Oxygen	E1	41.2
18	4	+1	−1	−1	Oxygen	E2	38.2
10	5	+1	−1	−1	SiCl4	E1	56.8
26	6	+1	−1	−1	SiCl4	E2	56.2
14	7	+1	−1	+1	SiCl4	E1	47.5
30	8	+1	−1	+1	SiCl4	E2	43.2
11	9	−1	+1	−1	SiCl4	E1	25.6
27	10	−1	+1	−1	SiCl4	E2	33.0
3	11	−1	+1	−1	Oxygen	E1	55.8
19	12	−1	+1	−1	Oxygen	E2	62.9
13	13	−1	−1	+1	SiCl4	E1	13.3
29	14	−1	−1	+1	SiCl4	E2	23.7
6	15	+1	−1	+1	Oxygen	E1	47.2
22	16	+1	−1	+1	Oxygen	E2	44.8
16	17	+1	+1	+1	SiCl4	E1	49.5
32	18	+1	+1	+1	SiCl4	E2	48.2
9	19	−1	−1	−1	SiCl4	E1	5.0
25	20	−1	−1	−1	SiCl4	E2	18.1
15	21	−1	+1	+1	SiCl4	E1	11.3
31	22	−1	+1	+1	SiCl4	E2	23.9
1	23	−1	−1	−1	Oxygen	E1	48.6
17	24	−1	−1	−1	Oxygen	E2	57.0
8	25	+1	+1	+1	Oxygen	E1	48.7
24	26	+1	+1	+1	Oxygen	E2	44.4
7	27	−1	+1	+1	Oxygen	E1	47.2
23	28	−1	+1	+1	Oxygen	E2	54.6
4	29	+1	+1	−1	Oxygen	E1	53.5
20	30	+1	+1	−1	Oxygen	E2	51.3
12	31	+1	+1	−1	SiCl4	E1	41.8
28	32	+1	+1	−1	SiCl4	E2	37.8

CHAPTER 15

Other Topics

CHAPTER OUTLINE

The supplemental material is on the textbook website www.wiley.com/go/global/montgomery.

The subject of statistically designed experiments is an extensive one. The previous chapters have provided an introductory presentation of many of the basic concepts and methods, yet in some cases we have only been able to provide an overview. For example, there are book-length presentations of topics such as response surface methodology, mixture experiments, variance component estimation, and optimal design. In this chapter, we provide an overview of several other topics that the experimenter may potentially find useful.

15.1 Nonnormal Responses and Transformations

15.1.1 Selecting a Transformation: The Box–Cox Method

In Section 3.4.3, we discussed the problem of nonconstant variance in the response variable y from a designed experiment and noted that this is a departure from the standard analysis of variance assumptions. This inequality of variance problem occurs relatively often in practice, often in conjunction with a nonnormal response variable. Examples would include a count of defects or particles, proportion data such as yield or fraction defective, or a response variable that follows some skewed distribution (one "tail" of the response distribution is longer than the other). We introduced **transformation of the response variable** as an appropriate method for stabilizing the variance of the response. Two methods for selecting the **form** of the transformation were discussed—an empirical graphical technique and essentially trial and error in which the experimenter simply tries one or more transformations, selecting the one that produces the most satisfactory or pleasing plot of residuals versus the fitted response.

Generally, transformations are used for three purposes: stabilizing response variance, making the distribution of the response variable closer to the normal distribution, and improving the fit of the model to the data. This last objective could include model simplification, say by eliminating interaction terms. Sometimes a transformation will be reasonably effective in simultaneously accomplishing more than one of these objectives.

We have noted that the **power family** of transformations $y^* = y^\lambda$ is very useful, where λ is the parameter of the transformation to be determined (e.g., $\lambda = \frac{1}{2}$ means use the square root of the original response). Box and Cox (1964) have shown how the transformation parameter λ may be estimated simultaneously with the other model parameters (overall mean and treatment effects). The theory underlying their method uses the method of maximum likelihood. The actual computational procedure consists of performing, for various values of λ, a standard analysis of variance on

$$y^{(\lambda)} = \begin{cases} \dfrac{y^\lambda - 1}{\lambda \dot{y}^{\lambda-1}} & \lambda \neq 0 \\ \dot{y} \ln y & \lambda = 0 \end{cases} \qquad \textbf{(15.1)}$$

where $\dot{y} = \ln^{-1}[(1/n) \Sigma \ln y]$ is the geometric mean of the observations. The maximum likelihood estimate of λ is the value for which the error sum of squares, say $SS_E(\lambda)$, is a minimum. This value of λ is usually found by plotting a graph of $SS_E(\lambda)$ versus λ and then reading the value of λ that minimizes $SS_E(\lambda)$ from the graph. Usually between 10 and 20 values of λ are sufficient for estimating the optimum value. A second iteration using a finer mesh of values can be performed if a more accurate estimate of λ is necessary.

Notice that we *cannot* select the value of λ by *directly* comparing the error sums of squares from analysis of variance on y^λ because for each value of λ, the error sum of squares is measured on a different scale. Furthermore, a problem arises in y when $\lambda = 0$, namely, as λ approaches zero, y^λ approaches unity. That is, when $\lambda = 0$, all the response values are a constant. The component $(y^\lambda - 1)/\lambda$ of Equation 15.1 alleviates this problem because as λ tends to zero, $(y^\lambda - 1)/\lambda$ goes to a limit of $\ln y$. The divisor component $\dot{y}^{\lambda-1}$ in Equation 15.1 rescales the responses so that the error sums of squares are directly comparable.

In using the Box–Cox method, we recommend that the experimenter use simple choices for λ because the practical difference between $\lambda = 0.5$ and $\lambda = 0.58$ is likely to be small, but the square root transformation ($\lambda = 0.5$) is much easier to interpret. Obviously, values of λ close to unity would suggest that no transformation is necessary.

Once a value of λ is selected by the Box–Cox method, the experimenter can analyze the data using y^λ as the response unless of course $\lambda = 0$, in which case he can use $\ln y$. It is perfectly acceptable to use $y^{(\lambda)}$ as the actual response, although the model parameter estimates will have a scale difference and origin shift in comparison to the results obtained using y^λ (or ln y).

An approximate $100(1 - \alpha)$ percent confidence interval for λ can be found by computing

$$SS^* = SS_E(\lambda)\left(1 + \frac{t^2_{\alpha/2,\nu}}{\nu}\right) \tag{15.2}$$

where ν is the number of degrees of freedom, and plotting a line parallel to the λ axis at height SS^* on the graph of $SS_E(\lambda)$ versus λ. Then, by locating the points on the λ axis where SS^* cuts the curve $SS_E(\lambda)$, we can read confidence limits on λ directly from the graph. If this confidence interval includes the value $\lambda = 1$, this implies (as noted above) that the data do not support the need for transformation.

EXAMPLE 15.1

We will illustrate the Box–Cox procedure using the peak discharge data originally presented in Example 3.5. Recall that this is a single-factor experiment (see Table 3.7 for the original data). Using Equation 15.1, we computed values of $SS_E(\lambda)$ for various values of λ:

λ	$SS_E(\lambda)$
−1.00	7922.11
−0.50	687.10
−0.25	232.52
0.00	91.96
0.25	46.99
0.50	35.42
0.75	40.61
1.00	62.08
1.25	109.82
1.50	208.12

A graph of values close to the minimum is shown in Figure 15.1, from which it is seen that $\lambda = 0.52$ gives a minimum value of approximately $SS_E(\lambda) = 35.00$. An approximate 95 percent confidence interval on λ is found by calculating the quantity SS^* from Equation 15.2 as follows:

$$\begin{aligned} SS^* &= SS_E(\lambda)\left(1 + \frac{t^2_{0.025,20}}{20}\right) \\ &= 35.00\left[1 + \frac{(2.086)^2}{20}\right] \\ &= 42.61 \end{aligned}$$

By plotting SS^* on the graph in Figure 15.1 and reading the points on the λ scale where this line intersects the curve, we obtain lower and upper confidence limits on λ of $\lambda^- = 0.27$ and $\lambda^+ = 0.77$. Because these confidence limits do not include the value 1, use of a transformation is indicated, and the square root transformation ($\lambda = 0.50$) actually used is easily justified.

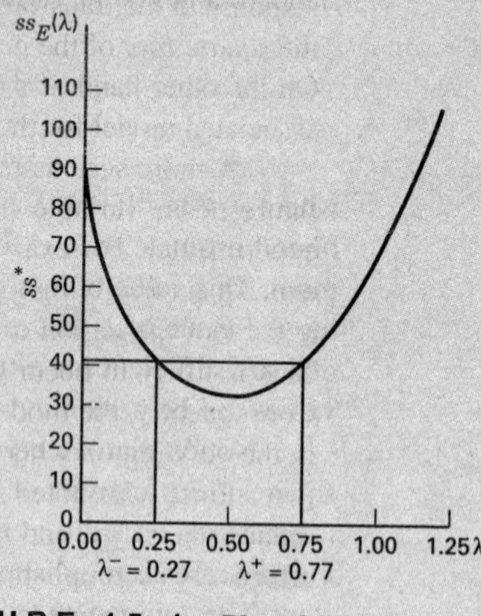

■ **FIGURE 15.1** **Plot of $SS_E(\lambda)$ versus λ for Example 15.1**

Some computer programs include the Box–Cox procedure for selecting a power family transformation. Figure 15.2 presents the output from this procedure as implemented in Design-Expert for the peak discharge data. The results agree closely with the manual calculations summarized in Example 15.1. Notice that the vertical scale of the graph in Figure 15.2 is $\ln[SS_E(\lambda)]$.

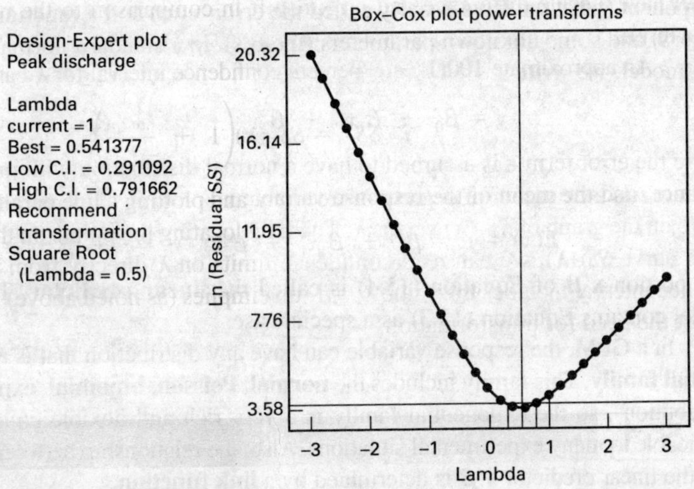

■ **FIGURE 15.2 Output from Design-Expert for the Box–Cox procedure**

15.1.2 The Generalized Linear Model

Data transformations are often a very effective way to deal with the problem of nonnormal responses and the associated inequality of variance. As we have seen in the previous section, the Box–Cox method is an easy and effective way to select the form of the transformation. However, the use of a data transformation poses some problems.

One problem is that the experimenter may be uncomfortable in working with the response in the transformed scale. That is, he or she is interested in the number of defects, not the *square root* of the number of defects, or in resistivity instead of the *logarithm* of resistivity. On the other hand, if a transformation is really successful and improves the analysis and the associated model for the response, experimenters will usually quickly adopt the new metric.

A more serious problem is that a transformation can result in a **nonsensical value** for the response variable over some portion of the design factor space that is of interest to the experimenter. For example, suppose that we have used the square root transformation in an experiment involving the number of defects observed on semiconductor wafers, and for some portion of the region of interest the predicted square root of the count of defects is negative. This is likely to occur for situations where the actual number of defects observed is small. Consequently, the model for the experiment has produced an obviously unreliable prediction in the very region where we would like this model to have good predictive performance.

Finally, as noted in Section 15.1.1, we often use transformations to stabilize variance, induce normality, and simplify the model. There is no assurance that a transformation will effectively accomplish all of these objectives simultaneously.

An alternative to the typical approach of data transformation followed by standard least squares analysis of the transformed response is to use the **generalized linear model (GLM)**. This is an approach developed by Nelder and Wedderburn (1972) that essentially unifies linear and nonlinear models with both normal and nonnormal responses. McCullagh and Nelder (1989) and Myers, Montgomery, Vining and Robinson (2010) give comprehensive presentations of generalized linear models, and Myers and Montgomery (1997) provide a tutorial. More details are also given in the supplemental text material for this chapter. We will provide an overview of the concepts and illustrate them with three short examples.

A generalized linear model is basically a **regression model** (an experimental design model is also a regression model). Like all regression models, it is made up of a random component (what we have usually called the error term) and a function of the design factors (the x's) and some unknown parameters (the β's). In a standard normal-theory linear regression model, we write

$$y = \beta_0 + \beta_1 x_1 + \beta_2 x_2 + \cdots + \beta_k x_k + \epsilon \tag{15.3}$$

where the error term ϵ is assumed to have a normal distribution with mean zero and constant variance, and the mean of the response variable y is

$$E(y) = \mu = \beta_0 + \beta_1 x_1 + \beta_2 x_2 + \cdots + \beta_k x_k = \mathbf{x}'\boldsymbol{\beta} \tag{15.4}$$

The portion $\mathbf{x}'\boldsymbol{\beta}$ of Equation (15.4) is called the **linear predictor**. The generalized linear model contains Equation (15.3) as a special case.

In a GLM, the response variable can have any distribution that is a member of the **exponential family**. This family includes the **normal**, **Poisson**, **binomial**, **exponential**, and **gamma** distributions, so the exponential family is a very rich and flexible collection of distributions applicable to many experimental situations. Also, the relationship between the response mean μ and the linear predictor $\mathbf{x}'\boldsymbol{\beta}$ is determined by a **link function**.

$$g(\mu) = \mathbf{x}'\boldsymbol{\beta} \tag{15.5}$$

The regression model that represents the mean response is then given by

$$E(y) = \mu = g^{-1}(\mathbf{x}'\boldsymbol{\beta}) \tag{15.6}$$

For example, the link function leading to the ordinary linear regression model in Equation 15.3 is called the **identity link** because $\mu = g^{-1}(\mathbf{x}'\boldsymbol{\beta}) = \mathbf{x}'\boldsymbol{\beta}$. As another example, the **log link**

$$\ln(\mu) = \mathbf{x}'\boldsymbol{\beta} \tag{15.7}$$

produces the model

$$\mu = e^{\mathbf{x}'\boldsymbol{\beta}} \tag{15.8}$$

The log link is often used with count data (Poisson response) and with continuous responses that have a distribution that has a long tail to the right (the exponential or gamma distribution). Another important link function used with binomial data is the **logit link**

$$\ln\left(\frac{\mu}{1-\mu}\right) = \mathbf{x}'\boldsymbol{\beta} \tag{15.9}$$

This choice of link function leads to the model

$$\mu = \frac{1}{1 + e^{-\mathbf{x}'\boldsymbol{\beta}}} \tag{15.10}$$

Many choices of link function are possible, but it must always be monotonic and differentiable. Also, note that in a generalized linear model, the variance of the response variable does not have to be a constant; it can be a function of the mean (and the predictor variables through the link function). For example, if the response is Poisson, the variance of the response is exactly equal to the mean.

To use a generalized linear model in practice, the experimenter must specify a response distribution and a link function. Then the model fitting or parameter estimation is done by the method of maximum likelihood, which for the exponential family turns out to be an iterative version of weighted least squares. For ordinary linear regression or experimental design models with a normal response variable, this reduces to standard least squares. Using an approach that is analogous to the analysis of variance for normal-theory data, we can perform inference and diagnostic checking for a GLM. Refer to Myers and Montgomery (1997) and Myers, Montgomery, Vining and Robinson (2010) for the details and examples. Two software packages that support the generalized linear model nicely are SAS (PROC GENMOD) and JMP.

EXAMPLE 15.2

A consumer products company is studying the factors that impact the chances that a customer will redeem a coupon for one of its personal care products. A 2^3 factorial experiment was conducted to investigate the following variables: A = coupon value (low, high), B = length of time for which the coupon is valid, and C = ease of use (easy, hard). A total of 1000 customers were randomly selected for each of the eight cells of the 2^3 design, and the response is the number of coupons redeemed. The experimental results are shown in Table 15.1.

We can think of the response as the number of successes out of 1000 Bernoulli trials in each cell of the design, so a reasonable model for the response is a generalized linear model with a binomial response distribution and a logit link. This particular form of the GLM is usually called **logistic regression**.

Minitab and JMP will fit logistic regression models. The experimenters decided to fit a model involving only the main effects and two-factor interactions. Therefore, the model for the expected response is

$$E(y) = \frac{1}{1 + \exp\left[-\left(\beta_0 + \sum_{i=1}^{3} \beta_i x_i + \sum_{i<}^{2} \sum_{j=2}^{3} \beta_{ij} x_i x_j\right)\right]}$$

Table 15.2 presents a portion of the Minitab output for the data in Table 15.1. The upper portion of the table fits the full model involving all three main effects and the three two-factor interactions. Notice that the output contains a display of the estimated model coefficients and their standard errors. It turns out that the ratio of the estimated regression coefficient to its standard error (a t-like ratio) has an asymptotically normal distribution under the null hypothesis that the regression coefficient is equal to zero. Thus, the ratios $Z = \hat{\beta}/se(\hat{\beta})$ can be used to test the contribution that each main effect and interaction term is significant. Here, the word "asymptotic" means when the sample size is large. Now the sample size here is certainly not large, and we should be careful about interpreting the P-values associated with these t-like ratios, but these statistics can be used as a guide to the analysis of the data. The P-values in the table indicate that the intercept, the main effects of A and B, and the BC interaction are significant.

■ **TABLE 15.1**
Design and Data for the Coupon Redemption Experiment

A	B	C	Number of Coupons Redeemed
−	−	−	200
+	−	−	250
−	+	−	265
+	+	−	347
−	−	+	210
+	−	+	286
−	+	+	271
+	+	+	326

The goodness-of-fit section of the table presents three different test statistics (Pearson, Deviance, and Hosmer–Lemeshow) that measure the overall adequacy of the model. All of the P-values for these goodness-of-fit statistics are large implying that the model is satisfactory. The bottom portion of the table presents the analysis for a reduced model containing the three main effects and the BC interaction (factor C was included to maintain model hierarchy). The fitted model is

$$\hat{y} = \frac{1}{1 + \exp[-(-1.01 + 0.169x_1 + 0.169x_2 + 0.023x_3 - 0.041x_2x_3)]}$$
$$= \frac{1}{1 + \exp(+1.01 - 0.169x_1 - 0.169x_2 - 0.023x_3 + 0.041x_2x_3)}$$

Since the effects of C and the BC interaction are very small, these terms could likely be dropped from the model with no major consequences.

Minitab reports an **odds ratio** for each regression model coefficient. The odds ratio follows directly from the logit link in Equation 15.9 and is interpreted much like factor effect estimates in standard two-level designs. For factor A, it can be interpreted as the ratio of the odds of redeeming a coupon of high value ($x_1 = +1$) to the odds of redeeming a coupon of value $x_1 = 0$. The computed value of the odds ratio is $e^{\hat{\beta}_1} = e^{0.168765} = 1.18$. The quantity $e^{2\hat{\beta}_1} = e^{2(0.168765)}$ $= 1.40$ is the ratio of the odds of redeeming a coupon of high value ($x_1 = +1$) to the odds of redeeming a coupon of low value ($x_1 = -1$). Thus, a high-value coupon increases the odds of redemption by about 40 percent.

■ **TABLE 15.2**

Minitab Output for the Coupon Redemption Experiment

Binary Logistic Regression: Full Model

Link Function: Logit

Response Information

Variable	Value	Count
C5	Success	2155
	Failure	5845
C6	Total	8000

Logistic Regression Table

					Odds	95%	CI
Predictor	Coef	SE Coef	Z	P	Ratio	Lower	Upper
Constant	-1.01154	0.0255150	-39.65	0.000			
A	0.169208	0.0255092	6.63	0.000	1.18	1.13	1.25
B	0.169622	0.0255150	6.65	0.000	1.18	1.13	1.25
C	0.0233173	0.0255099	0.91	0.361	1.02	0.97	1.08
A*B	-0.0062854	0.0255122	-0.25	0.805	0.99	0.95	1.04
A*C	-0.0027726	0.0254324	-0.11	0.913	1.00	0.95	1.05
B*C	-0.0410198	0.0254339	-1.61	0.107	0.96	0.91	1.01

Log-Likelihood = -4615.310

Goodness-of-Fit Tests

Method	Chi-Square	DF	P
Pearson	1.46458	1	0.226
Deviance	1.46451	1	0.226
Hosmer-Lemeshow	1.46458	6	0.962

Binary Logistic Regression: Reduced Model Involving *A, B, C*, and *BC*

Link Function: Logit

Response Information

Variable	Value	Count
C5	Success	2155
	Failure	5845
C6	Total	8000

Logistic Regression Table

					Odds	95%	CI
Predictor	Coef	SE Coef	Z	P	Ratio	Lower	Upper
Constant	-1.01142	0.0255076	-39.65	0.000			
A	0.168675	0.0254235	6.63	0.000	1.18	1.13	1.24
B	0.169116	0.0254321	6.65	0.000	1.18	1.13	1.24
C	0.0230825	0.0254306	0.91	0.364	1.02	0.97	1.08
B*C	-0.0409711	0.0254307	-1.61	0.107	0.96	0.91	1.01

Log-Likelihood = -4615.346

Goodness-of-Fit Tests

Method	Chi-Square	DF	P
Pearson	1.53593	3	0.674
Deviance	1.53602	3	0.674
Hosmer-Lemeshow	1.53593	6	0.957

Logistic regression is widely used and may be the most common application of the GLM. It finds wide application in the biomedical field with dose-response studies (where the design factor is the dose of a particular therapeutic treatment and the response is whether or not the patient responds successfully to the treatment). Many reliability engineering experiments involve binary (success–failure) data, such as when units of products or components are subjected to a stress or load and the response is whether or not the unit fails.

EXAMPLE 15.3 The Grill Defects Experiment

Problem 8.29 introduced an experiment to study the effects of nine factors on defects in sheet-molded grill opening panels. Bisgaard and Fuller (1994–95) performed an interesting and useful analysis of these data to illustrate the value of data transformation in a designed experiment. As we observed in Problem 8.29 part (f), they used a modification of the square root transformation that led to the model

$$\widehat{(\sqrt{y}+\sqrt{y+1})/2} = 2.513 - 0.996x_4 - 1.21x_6 - 0.772x_2x_7$$

where, as usual, the x's represent the coded design factors. This transformation does an excellent job of stabilizing the variance of the number of defectives. The first two panels of Table 15.3 present some information about this model. Under the "transformed" heading, the first column contains the predicted response. Notice that there are two negative predicted values. The "untransformed" heading presents the untransformed predicted values, along with 95 percent confidence intervals on the mean response at each of the 16 design points. Because there were some negative predicted

■ **TABLE 15.3**
Least Squares and Generalized Linear Model Analysis for the Grill Opening Panels Experiment

	Using Least Squares Methods with Freeman and Tukey Modified Square Root Data Transformation				Generalized Linear Model (Poisson Response, Log Link)		Length of the 95% Confidence Interval	
	Transformed		Untransformed					
Observation	Predicted Value	95% Confidence Interval	Predicted Value	95% Confidence Interval	Predicted Value	95% Confidence Interval	Least Squares	GLM
1	5.50	(4.14, 6.85)	29.70	(16.65, 46.41)	51.26	(42.45, 61.90)	29.76	19.45
2	3.95	(2.60, 5.31)	15.12	(6.25, 27.65)	11.74	(8.14, 16.94)	21.39	8.80
3	1.52	(0.17, 2.88)	1.84	(1.69, 7.78)	1.12	(0.60, 2.08)	6.09	1.47
4	3.07	(1.71, 4.42)	8.91	(2.45, 19.04)	4.88	(2.87, 8.32)	16.59	5.45
5	1.52	(0.17, 2.88)	1.84	(1.69, 7.78)	1.12	(0.60, 2.08)	6.09	1.47
6	3.07	(1.71, 4.42)	8.91	(2.45, 19.04)	4.88	(2.87, 8.32)	16.59	5.45
7	5.50	(4.14, 6.85)	29.70	(16.65, 46.41)	51.26	(42.45, 61.90)	29.76	19.45
8	3.95	(2.60, 5.31)	15.12	(6.25, 27.65)	11.74	(8.14, 16.94)	21.39	8.80
9	1.08	(−0.28, 2.43)	0.71	(*, 5.41)	0.81	(0.42, 1.56)	*	1.13
10	−0.47	(−1.82, 0.89)	*	(*, 0.36)	0.19	(0.09, 0.38)	*	0.29
11	1.96	(0.61, 3.31)	3.36	(0.04, 10.49)	1.96	(1.16, 3.30)	10.45	2.14
12	3.50	(2.15, 4.86)	11.78	(4.13, 23.10)	8.54	(5.62, 12.98)	18.96	7.35
13	1.96	(0.61, 3.31)	3.36	(0.04, 10.49)	1.96	(1.16, 3.30)	10.45	2.14
14	3.50	(2.15, 4.86)	11.78	(4.13, 23.10)	8.54	(5.62, 12.98)	18.97	7.35
15	1.08	(−0.28, 2.43)	0.71	(*, 5.41)	0.81	(0.42, 1.56)	*	1.13
16	−0.47	(−1.82, 0.89)	*	(*, 0.36)	0.19	(0.09, 0.38)	*	0.29

values and negative lower confidence limits, we were unable to compute values for all of the entries in this panel of the table.

The response is essentially a square root of the count of defects. A negative predicted value is clearly illogical. Notice that this is occurring where the observed counts were small. If it is important to use the model to predict performance in this region, the model may be unreliable. This should *not* be taken as a criticism of either the original experimenters or Bisgaard and Fuller's analysis. This was an extremely successful screening experiment that clearly defined the important processing variables. Prediction was not the original goal, nor was it the goal of the analysis done by Bisgaard and Fuller.

If obtaining a prediction model *had* been important, however, a generalized linear model would probably have been a good alternative to the transformation approach. Myers and Montgomery use a log link (Equation 15.7) and Poisson response to fit exactly the same linear predictor as given by Bisgaard and Fuller. This produces the model

$$\hat{y} = e^{(1.128 - 0.896x_4 - 1.176x_6 - 0.737x_2x_7)}$$

The third panel of Table 15.3 contains the predicted values from this model and the 95 percent confidence intervals on the mean response at each point in the design. These results were obtained from SAS PROC GENMOD. JMP will also fit the Poisson regression model. There are no negative predicted values (assured by the choice of link function) and no negative lower confidence limits. The last panel of the table compares the lengths of the 95 percent confidence intervals for the untransformed response and the GLM. Notice that the confidence intervals for the generalized linear model are uniformly *shorter* than their least squares counterparts. This is a strong indication that the generalized linear model approach has explained more variability and produced a superior model compared to the transformation approach.

EXAMPLE 15.4 The Worsted Yarn Experiment

Table 15.4 presents a 3^3 factorial design conducted to investigate the performance of worsted yarn under cycles of repeated loading. The experiment is described thoroughly by Box and Draper (2007). The response is the number of cycles to failure. Reliability data such as this is typically nonnegative and continuous and often follows a distribution with a long right tail.

■ **TABLE 15.4**
The Worsted Yarn Experiment

Run	x_1	x_2	x_3	Cycles to Failure	Natural Log of Cycles to Failure
1	−1	−1	−1	674	6.51
2	−1	−1	0	370	5.91
3	−1	−1	1	292	5.68
4	−1	0	−1	338	5.82
5	−1	0	0	266	5.58
6	−1	0	1	210	5.35
7	−1	1	−1	170	5.14
8	−1	1	0	118	4.77
9	−1	1	1	90	4.50
10	0	−1	−1	1414	7.25
11	0	−1	0	1198	7.09
12	0	−1	1	634	6.45
13	0	0	−1	1022	6.93
14	0	0	0	620	6.43
15	0	0	1	438	6.08
16	0	1	−1	442	6.09
17	0	1	0	332	5.81
18	0	1	1	220	5.39
19	1	−1	−1	3636	8.20
20	1	−1	0	3184	8.07
21	1	−1	1	2000	7.60
22	1	0	−1	1568	7.36
23	1	0	0	1070	6.98
24	1	0	1	566	6.34
25	1	1	−1	1140	7.04
26	1	1	0	884	6.78
27	1	1	1	360	5.89

The data were initially analyzed using the standard (least squares) approach, and data transformation was necessary to stabilize the variance. The natural log of the cycles to failure data is found to yield an adequate model in terms of overall model fit and satisfactory residual plots. The model is

$$\ln \hat{y} = 6.33 + 0.82x_1 - 0.63x_2 - 0.38x_3$$

or in terms of the original response, cycles to failure,

$$\hat{y} = e^{6.33 + 0.82x_1 - 0.63x_2 - 0.38x_3}$$

This experiment was also analyzed using the generalized linear model and selecting the gamma response distribution and the log link. We used exactly the same model form found by least squares analysis of the log-transformed response. The model that resulted is

$$\hat{y} = e^{6.35+0.84x_1-0.63x_2-0.39x_3}$$

Table 15.5 presents the predicted values from the least squares model and the generalized linear model, along with 95 percent confidence intervals on the mean response at each of the 27 points in the design. A comparison of the lengths of the confidence intervals reveals that the generalized linear model is likely to be a better predictor than the least squares model.

■ **TABLE 15.5**

Least Squares and Generalized Linear Model Analysis for the Worsted Yarn Experiment

	Least Squares Methods with Log Data Transformation				Generalized Linear Model		Length of the 95% Confidence Interval	
	Transformed		Untransformed					
Obs.	Predicted Value	95% Confidence Interval	Predicted Value	95% Confidence Interval	Predicted Value	95% Confidence Interval	Least Squares	GLM
1	2.83	(2.76, 2.91)	682.50	(573.85, 811.52)	680.52	(583.83, 793.22)	237.67	209.39
2	2.66	(2.60, 2.73)	460.26	(397.01, 533.46)	463.00	(407.05, 526.64)	136.45	119.59
3	2.49	(2.42, 2.57)	310.38	(260.98, 369.06)	315.01	(271.49, 365.49)	108.09	94.00
4	2.56	(2.50, 2.62)	363.25	(313.33, 421.11)	361.96	(317.75, 412.33)	107.79	94.58
5	2.39	(2.34, 2.44)	244.96	(217.92, 275.30)	246.26	(222.55, 272.51)	57.37	49.96
6	2.22	(2.15, 2.28)	165.20	(142.50, 191.47)	167.55	(147.67, 190.10)	48.97	42.42
7	2.29	(2.21, 2.36)	193.33	(162.55, 229.93)	192.52	(165.69, 223.70)	67.38	58.01
8	2.12	(2.05, 2.18)	130.38	(112.46, 151.15)	130.98	(115.43, 148.64)	38.69	33.22
9	1.94	(1.87, 2.02)	87.92	(73.93, 104.54)	89.12	(76.87, 103.32)	30.62	26.45
10	3.20	(3.13, 3.26)	1569.28	(1353.94, 1819.28)	1580.00	(1390.00, 1797.00)	465.34	407.00
11	3.02	(2.97, 3.08)	1058.28	(941.67, 1189.60)	1075.00	(972.52, 1189.00)	247.92	216.48
12	2.85	(2.79, 2.92)	713.67	(615.60, 827.37)	731.50	(644.35, 830.44)	211.77	186.09
13	2.92	(2.87, 2.97)	835.41	(743.19, 938.86)	840.54	(759.65, 930.04)	195.67	170.39
14	2.75	(2.72, 2.78)	563.25	(523.24, 606.46)	571.87	(536.67, 609.38)	83.22	72.70
15	2.58	(2.53, 2.63)	379.84	(337.99, 426.97)	389.08	(351.64, 430.51)	88.99	78.87
16	2.65	(2.58, 2.71)	444.63	(383.53, 515.35)	447.07	(393.81, 507.54)	131.82	113.74
17	2.48	(2.43, 2.53)	299.85	(266.75, 336.98)	304.17	(275.13, 336.28)	70.23	61.15
18	2.31	(2.24, 2.37)	202.16	(174.42, 234.37)	206.95	(182.03, 235.27)	59.95	53.23
19	3.56	(3.48, 3.63)	3609.11	(3034.59, 4292.40)	3670.00	(3165.00, 4254.00)	1257.81	1089.00
20	3.39	(3.32, 3.45)	2433.88	(2099.42, 2821.63)	2497.00	(2200.00, 2833.00)	722.21	633.00
21	3.22	(3.14, 3.29)	1641.35	(1380.07, 1951.64)	1699.00	(1462.00, 1974.00)	571.57	512.00
22	3.28	(3.22, 3.35)	1920.88	(1656.91, 2226.90)	1952.00	(1720.00, 2215.00)	569.98	495.00
23	3.11	(3.06, 3.16)	1295.39	(1152.66, 1455.79)	1328.00	(1200.00, 1470.00)	303.14	270.00
24	2.94	(2.88, 3.01)	873.57	(753.53, 1012.74)	903.51	(793.15, 1029.00)	259.22	235.85
25	3.01	(2.93, 3.08)	1022.35	(859.81, 1215.91)	1038.00	(894.79, 1205.00)	356.10	310.21
26	2.84	(2.77, 2.90)	689.45	(594.70, 799.28)	706.34	(620.99, 803.43)	204.58	182.44
27	2.67	(2.59, 2.74)	464.94	(390.93, 552.97)	480.57	(412.29, 560.15)	162.04	147.86

Generalized linear models have found extensive application in biomedical and pharmaceutical research and development. As more software packages incorporate this capability, it will find widespread application in the general industrial research and development environment. The examples in this section used standard experimental designs in conjuction with a response distribution from the exponential family. When the experimenter knows in advance that a GLM analysis will be required it is possible to design the experiment with this in mind. Optimal designs for GLMs are a special type of optimal design for a nonlinear model. For more discussion and examples, see Johnson and Montgomery (2009) (2010).

15.2 Unbalanced Data in a Factorial Design

The primary focus of this book has been the analysis of **balanced factorial designs**—that is, cases where there are an equal number of observations n in each cell. However, it is not unusual to encounter situations where the number of observations in the cells is unequal. These **unbalanced factorial designs** occur for various reasons. For example, the experimenter may have designed a balanced experiment initially, but because of unforeseen problems in running the experiment, resulting in the loss of some observations, he or she ends up with unbalanced data. On the other hand, some unbalanced experiments are deliberately designed that way. For instance, certain treatment combinations may be more expensive or more difficult to run than others, so fewer observations may be taken in those cells. Alternatively, some treatment combinations may be of greater interest to the experimenter because they represent new or unexplored conditions, and so the experimenter may elect to obtain additional replication in those cells.

The orthogonality property of main effects and interactions present in balanced data does not carry over to the unbalanced case. This means that the usual analysis of variance techniques do not apply. Consequently, the analysis of unbalanced factorials is much more difficult than that for balanced designs.

In this section, we give a brief overview of methods for dealing with unbalanced factorials, concentrating on the case of the two-factor fixed effects model. Suppose that the number of observations in the ijth cell is n_{ij}. Furthermore, let $n_{i.} = \sum_{j=1}^{b} n_{ij}$ be the number of observations in the ith row (the ith level of factor A), $n_{.j} = \sum_{i=1}^{a} n_{ij}$ be the number of observations in the jth column (the jth level of factor B), and $n_{..} = \sum_{i=1}^{a} \sum_{j=1}^{b} n_{ij}$ be the total number of observations.

15.2.1 Proportional Data: An Easy Case

One situation involving unbalanced data presents little difficulty in analysis; this is the case of **proportional data**. That is, the number of observations in the ijth cell is

$$n_{ij} = \frac{n_{i.}n_{.j}}{n_{..}} \tag{15.11}$$

This condition implies that the number of observations in any two rows or columns is proportional. When proportional data occur, the standard analysis of variance can be employed. Only minor modifications are required in the manual computing formulas for the sums of squares, which become

$$SS_T = \sum_{i=1}^{a} \sum_{j=1}^{b} \sum_{k=1}^{n_{ij}} y_{ijk}^2 - \frac{y_{...}^2}{n_{..}}$$

$$SS_A = \sum_{i=1}^{a} \frac{y_{i..}^2}{n_{i.}} - \frac{y_{...}^2}{n_{..}}$$

■ TABLE 15.6
Battery Design Experiment with Proportional Data

Material Type	Temperature (°F) 15	70	125	
1	$n_{11} = 4$: 130, 155, 74, 180	$n_{12} = 4$: 34, 40, 80, 75	$n_{13} = 2$: 70, 58	$n_{1.} = 10$, $y_{1..} = 896$
2	$n_{21} = 2$: 159, 126	$n_{22} = 2$: 136, 115	$n_{23} = 1$: 45	$n_{2.} = 5$, $y_{2..} = 581$
3	$n_{31} = 2$: 138, 160	$n_{32} = 2$: 150, 139	$n_{33} = 1$: 96	$n_{3.} = 5$, $y_{3..} = 683$
	$n_{.1} = 8$, $y_{.1.} = 1122$	$n_{.2} = 8$, $y_{.2.} = 769$	$n_{.3} = 4$, $y_{.3.} = 269$	$n_{..} = 20$, $y_{...} = 2160$

$$SS_B = \sum_{j=1}^{b} \frac{y_{.j.}^2}{n_{.j}} - \frac{y_{...}^2}{n_{..}}$$

$$SS_{AB} = \sum_{i=1}^{a} \sum_{j=1}^{b} \frac{y_{ij.}^2}{n_{ij}} - \frac{y_{...}^2}{n_{..}} - SS_A - SS_B$$

$$SS_E = SS_T - SS_A - SS_B - SS_{AB} = \sum_{i=1}^{a} \sum_{j=1}^{b} \sum_{k=1}^{n_{ij}} y_{ijk}^2 - \sum_{i=1}^{a} \sum_{j=1}^{b} \frac{y_{ij.}^2}{n_{ij}}$$

This produces an ANOVA based on a sequential model fitting analysis, with factor A fit before factor B (an alternative would be to use an "adjusted" model fitting strategy similar to the one used with incomplete block designs in Chapter 4—both procedures can be implemented using the Minitab Balanced ANOVA routine).

As an example of proportional data, consider the battery design experiment in Example 5.1. A modified version of the original data is shown in Table 15.6. Clearly, the data are proportional; for example, in cell 1,1 we have

$$n_{11} = \frac{n_{1.}n_{.1}}{n_{..}} = \frac{10(8)}{20} = 4$$

observations. The results of applying the usual analysis of variance to these data are shown in Table 15.7. Both material type and temperature are significant, and the interaction is only significant at about $\alpha = 0.17$. Therefore, the conclusions mostly agree with the analysis of the full data set in Example 5.1, except that the interaction effect is not significant.

■ TABLE 15.7
Analysis of Variance for Battery Design Data in Table 15.6

Source of Variance	Sum of Squares	Degrees of Freedom	Mean Square	F_0
Material types	7,811.6	2	3,905.8	4.78
Temperature	16,090.9	2	8,045.5	9.85
Interaction	6,266.5	4	1,566.6	1.92
Error	8,981.0	11	816.5	
Total	39,150.0	19		

15.2.2 Approximate Methods

When unbalanced data are not too far from the balanced case, it is sometimes possible to use **approximate procedures** that convert the unbalanced problem into a balanced one. This, of course, makes the analysis only approximate, but the analysis of balanced data is so easy that we are frequently tempted to use this approach. In practice we must decide when the data are not sufficiently different from the balanced case to make the degree of approximation introduced relatively unimportant. We now briefly describe some of these approximate methods. We assume that every cell has at least one observation (i.e., $n_{ij} \geq 1$).

Estimating Missing Observations. If only a few n_{ij} are different, a reasonable procedure is to estimate the missing values. For example, consider the unbalanced design in Table 15.8. Clearly, estimating the single missing value in cell 2,2 is a reasonable approach. For a model with interaction, the estimate of the missing value in the ijth cell that minimizes the error sum of squares is $\bar{y}_{ij.}$. That is, we estimate the missing value by taking the average of the observations that are available in that cell.

The estimated value is treated just like actual data. The only modification to the analysis of variance is to reduce the error degrees of freedom by the number of missing observations that have been estimated. For example, if we estimate the missing value in cell 2,2 in Table 15.8, we would use 26 error degrees of freedom instead of 27.

Setting Data Aside. Consider the data in Table 15.9. Note that cell 2,2 has only one more observation than the others. Estimating missing values for the remaining eight cells is probably not a good idea here because this would result in estimates constituting about 18 percent of the final data. An alternative is to set aside one of the observations in cell 2,2, giving a balanced design with $n = 4$ replicates.

The observation that is set aside should be chosen randomly. Furthermore, rather than completely discarding the observation, we could return it to the design, and then randomly choose another observation to set aside and repeat the analysis. And, we hope, these two analyses will not lead to conflicting interpretations of the data. If they do, we suspect that the observation that was set aside is an outlier or a wild value and should be handled accordingly. In practice, this confusion is unlikely to occur when only small numbers of observations are set aside and the variability within the cells is small.

Method of Unweighted Means. In this approach, introduced by Yates (1934), the cell averages are treated as data and are subjected to a standard balanced data analysis to obtain sums of squares for rows, columns, and interaction. The error mean square is found as

$$MS_E = \frac{\sum_{i=1}^{a}\sum_{j=1}^{b}\sum_{k=1}^{n_{ij}} (y_{ijk} - \bar{y}_{ij.})^2}{n_{..} - ab} \tag{15.12}$$

■ **TABLE 15.8**
The n_{ij} Values for an Unbalanced Design

Rows	Columns 1	2	3
1	4	4	4
2	4	3	4
3	4	4	4

■ **TABLE 15.9**
The n_{ij} Values for an Unbalanced Design

Rows	Columns 1	2	3
1	4	4	4
2	4	5	4
3	4	4	4

Now MS_E estimates σ^2, the variance of y_{ijk}, an individual observation. However, we have done an analysis of variance on the cell *averages*, and because the variance of the average in the ijth cell is σ^2/n_{ij}, the error mean square actually used in the analysis of variance should be an estimate of the average variance of the $\bar{y}_{ij.}$, say

$$\bar{V}(\bar{y}_{ij.}) = \frac{\sum_{i=1}^{a}\sum_{j=1}^{b} \sigma^2/n_{ij}}{ab} = \frac{\sigma^2}{ab}\sum_{i=1}^{a}\sum_{j=1}^{b}\frac{1}{n_{ij}} \quad \textbf{(15.13)}$$

Using MS_E from Equation 15.12 to estimate σ^2 in Equation 15.13, we obtain

$$MS'_E = \frac{MS_E}{ab}\sum_{i=1}^{a}\sum_{j=1}^{b}\frac{1}{n_{ij}} \quad \textbf{(15.14)}$$

as the error mean square (with $n_{..} - ab$ degrees of freedom) to use in the analysis of variance.

The method of unweighted means is an approximate procedure because the sums of squares for rows, columns, and intefaction are not distributed as chi-square random variables. The primary advantage of the method seems to be its computational simplicity. When the n_{ij} are not dramatically different, the method of unweighted means often works reasonably well.

A related technique is the **weighted squares of means method**, also proposed by Yates (1934). This technique is also based on the sums of squares of the cell means, but the terms in the sums of squares are weighted in inverse proportion to their variances. For further details of the procedure, see Searle (1971a) and Speed, Hocking, and Hackney (1978).

15.2.3 The Exact Method

In situations where approximate methods are inappropriate, such as when empty cells occur (some $n_{ij} = 0$) or when the n_{ij} are dramatically different, the experimenter must use an exact analysis. The approach used to develop sums of squares for testing main effects and interactions is to represent the analysis of variance model as a **regression model**, fit that model to the data, and use the general regression significance test approach. However, this may be done in several ways, and these methods may result in different values for the sums of squares. Furthermore, the hypotheses that are being tested are not always direct analogs of those for the balanced case, nor are they always easily interpretable. For further reading on the subject, see the supplemental text material for this chapter. Other good references are Searle (1971a); Hocking and Speed (1975); Hocking, Hackney, and Speed (1978); Speed, Hocking, and Hackney (1978); Searle, Speed, and Henderson (1981); Milliken and Johnson (1984); and Searle (1987). The SAS system of statistical software provides an excellent approach to the analysis of unbalanced data through PROC GLM.

15.3 The Analysis of Covariance

In Chapters 2 and 4, we introduced the use of the blocking principle to improve the precision with which comparisons between treatments are made. The paired t-test was the procedure illustrated in Chapter 2, and the randomized block design was presented in Chapter 4. In general, the **blocking principle** can be used to eliminate the effect of controllable nuisance factors. The **analysis of covariance (ANCOVA)** is another technique that is occasionally useful for improving the precision of an experiment. Suppose that in an experiment with a response variable y there is another variable, say x, and that y is linearly related to x. Furthermore, suppose that x cannot be controlled by the experimenter but can be observed along with y. The variable x is called a **covariate** or **concomitant variable**. The analysis of covariance involves adjusting the observed response variable for the effect of the concomitant

■ **TABLE 15.10**

Breaking Strength Data (y = strength in pounds and x = diameter in 10^{-3} in.)

Machine 1		Machine 2		Machine 3	
y	x	y	x	y	x
36	20	40	22	35	21
41	25	48	28	37	23
39	24	39	22	42	26
42	25	45	30	34	21
49	32	44	28	32	15
207	126	216	130	180	106

variable. If such an adjustment is not performed, the concomitant variable could inflate the error mean square and make true differences in the response due to treatments harder to detect. Thus, the analysis of covariance is a method of adjusting for the effects of an uncontrollable nuisance variable. As we will see, the procedure is a combination of analysis of variance and regression analysis.

As an example of an experiment in which the analysis of covariance may be employed, consider a study performed to determine if there is a difference in the strength of a monofilament fiber produced by three different machines. The data from this experiment are shown in Table 15.10. Figure 15.3 presents a scatter diagram of strength (y) versus the diameter (or thickness) of the sample. Clearly, the strength of the fiber is also affected by its thickness; consequently, a thicker fiber will generally be stronger than a thinner one. The analysis of covariance could be used to remove the effect of thickness (x) on strength (y) when testing for differences in strength between machines.

15.3.1 Description of the Procedure

The basic procedure for the analysis of covariance is now described and illustrated for a single-factor experiment with one covariate. Assuming that there is a linear relationship between the

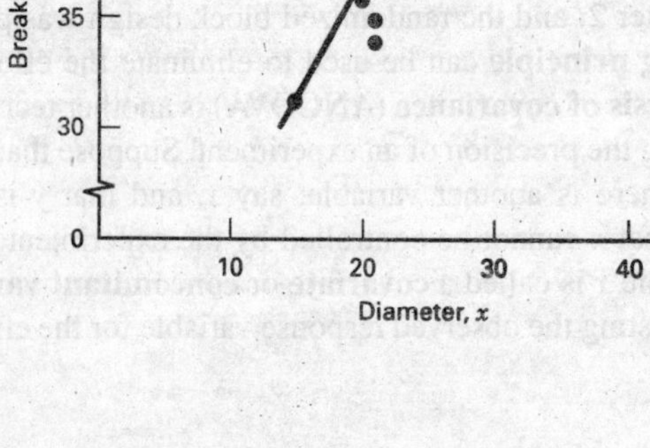

■ **FIGURE 15.3**
Breaking strength (y) versus fiber diameter (x)

response and the covariate, we find that an appropriate statistical model is

$$y_{ij} = \mu + \tau_i + \beta(x_{ij} - \bar{x}_{..}) + \epsilon_{ij} \begin{cases} i = 1, 2, \ldots, a \\ j = 1, 2, \ldots, n \end{cases} \tag{15.15}$$

where y_{ij} is the jth observation on the response variable taken under the ith treatment or level of the single factor, x_{ij} is the measurement made on the covariate or concomitant variable corresponding to y_{ij} (i.e., the ijth run), $\bar{x}_{..}$ is the mean of the x_{ij} values, μ is an overall mean, τ_i is the effect of the ith treatment, β is a linear regression coefficient indicating the dependency of y_{ij} on x_{ij}, and ϵ_{ij} is a random error component. We assume that the errors ϵ_{ij} are NID(0, σ^2), that the slope $\beta \neq 0$ and the true relationship between y_{ij} and x_{ij} is linear, that the regression coefficients for each treatment are identical, that the treatment effects sum to zero ($\Sigma_{i=1}^{a} \tau_i = 0$), and that the concomitant variable x_{ij} is not affected by the treatments.

This model assumes that all treatment regression lines have identical slopes. If the treatments interact with the covariates this can result in non-identical slopes. Covariance analysis is not appropriate in these cases. Estimating and comparing different regression models is the correct approach.

Equation 15.15 assumes a linear relationship between y and x. However, any other relationship, such as a quadratic (for example) could be used.

Notice from Equation 15.15 that the analysis of covariance model is a combination of the linear models employed in analysis of variance and regression. That is, we have treatment effects $\{\tau_i\}$ as in a single-factor analysis of variance and a regression coefficient β as in a regression equation. The concomitant variable in Equation 15.15 is expressed as $(x_{ij} - \bar{x}_{..})$ instead of x_{ij} so that the parameter μ is preserved as the overall mean. The model could have been written as

$$y_{ij} = \mu' + \tau_i + \beta x_{ij} + \epsilon_{ij} \begin{cases} i = 1, 2, \ldots, a \\ j = 1, 2, \ldots, n \end{cases} \tag{15.16}$$

where μ' is a constant not equal to the overall mean, which for this model is $\mu' + \beta\bar{x}_{..}$. Equation 15.15 is more widely found in the literature.

To describe the analysis, we introduce the following notation:

$$S_{yy} = \sum_{i=1}^{a}\sum_{j=1}^{n}(y_{ij} - \bar{y}_{..})^2 = \sum_{i=1}^{a}\sum_{j=1}^{n} y_{ij}^2 - \frac{y_{..}^2}{an} \tag{15.17}$$

$$S_{xx} = \sum_{i=1}^{a}\sum_{j=1}^{n}(x_{ij} - \bar{x}_{..})^2 = \sum_{i=1}^{a}\sum_{j=1}^{n} x_{ij}^2 - \frac{x_{..}^2}{an} \tag{15.18}$$

$$S_{xy} = \sum_{i=1}^{a}\sum_{j=1}^{n}(x_{ij} - \bar{x}_{..})(y_{ij} - \bar{y}_{..}) = \sum_{i=1}^{a}\sum_{j=1}^{n} x_{ij}y_{ij} - \frac{(x_{..})(y_{..})}{an} \tag{15.19}$$

$$T_{yy} = n\sum_{i=1}^{a}(\bar{y}_{i.} - \bar{y}_{..})^2 = \frac{1}{n}\sum_{i=1}^{a} y_{i.}^2 - \frac{y_{..}^2}{an} \tag{15.20}$$

$$T_{xx} = n\sum_{i=1}^{a}(\bar{x}_{i.} - \bar{x}_{..})^2 = \frac{1}{n}\sum_{i=1}^{a} x_{i.}^2 - \frac{x_{..}^2}{an} \tag{15.21}$$

$$T_{xy} = n\sum_{i=1}^{a}(\bar{x}_{i.} - \bar{x}_{..})(\bar{y}_{i.} - \bar{y}_{..}) = \frac{1}{n}\sum_{i=1}^{a}(x_{i.})(y_{i.}) - \frac{(x_{..})(y_{..})}{an} \tag{15.22}$$

$$E_{yy} = \sum_{i=1}^{a}\sum_{j=1}^{n}(y_{ij} - \bar{y}_{i.})^2 = S_{yy} - T_{yy} \tag{15.23}$$

$$E_{xx} = \sum_{i=1}^{a}\sum_{j=1}^{n}(x_{ij} - \bar{x}_{i.})^2 = S_{xx} - T_{xx} \tag{15.24}$$

$$E_{xy} = \sum_{i=1}^{a}\sum_{j=1}^{n}(x_{ij} - \bar{x}_{i.})(y_{ij} - \bar{y}_{i.}) = S_{xy} - T_{xy} \tag{15.25}$$

Note that, in general, $S = T + E$, where the symbols S, T, and E are used to denote sums of squares and cross products for total, treatments, and error, respectively. The sums of squares for x and y must be nonnegative; however, the sums of cross products (xy) may be negative.

We now show how the analysis of covariance adjusts the response variable for the effect of the covariate. Consider the full model (Equation 15.15). The least squares estimators of μ, τ_i, and β are $\hat{\mu} = \bar{y}_{..}$, $\hat{\tau}_i = \bar{y}_{i.} - \bar{y}_{..} - \hat{\beta}(\bar{x}_{i.} - \bar{x}_{..})$, and

$$\hat{\beta} = \frac{E_{xy}}{E_{xx}} \tag{15.26}$$

The error sum of squares in this model is

$$SS_E = E_{yy} - (E_{xy})^2/E_{xx} \tag{15.27}$$

with $a(n - 1) - 1$ degrees of freedom. The experimental error variance is estimated by

$$MS_E = \frac{SS_E}{a(n-1) - 1}$$

Now suppose that there is no treatment effect. The model (Equation 15.15) would then be

$$y_{ij} = \mu + \beta(x_{ij} - \bar{x}_{..}) + \epsilon_{ij} \tag{15.28}$$

and it can be shown that the least squares estimators of μ and β are $\hat{\mu} = \bar{y}_{..}$ and $\hat{\beta} = S_{xy}/S_{xx}$. The sum of squares for error in this reduced model is

$$SS'_E = S_{yy} - (S_{xy})^2/S_{xx} \tag{15.29}$$

with $an - 2$ degrees of freedom. In Equation 15.29, the quantity $(S_{xy})^2/S_{xx}$ is the reduction in the sum of squares of y obtained through the linear regression of y on x. Furthermore, note that SS_E is smaller than SS'_E [because the model (Equation 15.15) contains additional parameters $\{\tau_i\}$] and that the quantity $SS'_E - SS_E$ is a reduction in sum of squares due to the $\{\tau_i\}$. Therefore, the difference between SS'_E and SS_E, that is, $SS'_E - SS_E$, provides a sum of squares with $a - 1$ degrees of freedom for testing the hypothesis of no treatment effects. Consequently, to test H_0: $\tau_i = 0$, compute

$$F_0 = \frac{(SS'_E - SS_E)/(a-1)}{SS_E/[a(n-1) - 1]} \tag{15.30}$$

which, if the null hypothesis is true, is distributed as $F_{a-1,a(n-1)-1}$. Thus, we reject H_0: $\tau_i = 0$ if $F_0 > F_{\alpha,a-1,a(n-1)-1}$. The P-value approach could also be used.

It is instructive to examine the display in Table 15.11. In this table we have presented the analysis of covariance as an "adjusted" analysis of variance. In the source of variation column, the total variability is measured by S_{yy} with $an - 1$ degrees of freedom. The source of variation "regression" has the sum of squares $(S_{xy})^2/S_{xx}$ with one degree of freedom. If there were no concomitant variable, we would have $S_{xy} = S_{xx} = E_{xy} = E_{xx} = 0$. Then the sum of squares for error would be simply E_{yy} and the sum of squares for treatments would be $S_{yy} - E_{yy} = T_{yy}$. However, because of the presence of the concomitant variable, we must "adjust" S_{yy} and E_{yy} for the regression of y on x as shown in Table 15.11. The adjusted error sum of squares

■ **TABLE 15.11**
Analysis of Covariance as an "Adjusted" Analysis of Variance

Source of Variation	Sum of Squares	Degrees of Freedom	Mean Square	F_0
Regression	$(S_{xy})^2/S_{xx}$	1		
Treatments	$SS'_E - SS_E = S_{yy} - (S_{xy})^2/S_{xx} - [E_{yy} - (E_{xy})^2/E_{xx}]$	$a - 1$	$\frac{SS'_E - SS_E}{a-1}$	$\frac{(SS'_E - SS_E)/(a-1)}{MS_E}$
Error	$SS_E = E_{yy} - (E_{xy})^2/E_{xx}$	$a(n-1) - 1$	$MS_E = \frac{SS_E}{a(n-1)-1}$	
Total	S_{yy}	$an - 1$		

has $a(n-1) - 1$ degrees of freedom instead of $a(n-1)$ degrees of freedom because an additional parameter (the slope β) is fitted to the data.

Manual computations are usually displayed in an analysis of covariance table such as Table 15.12. This layout is employed because it conveniently summarizes all the required sums of squares and cross products as well as the sums of squares for testing hypotheses about treatment effects. In addition to testing the hypothesis that there are no differences in the treatment effects, we frequently find it useful in interpreting the data to present the adjusted treatment means. These adjusted means are computed according to

$$\text{Adjusted } \bar{y}_{i.} = \bar{y}_{i.} - \hat{\beta}(\bar{x}_{i.} - \bar{x}_{..}) \qquad i = 1, 2, \ldots, a \tag{15.31}$$

where $\hat{\beta} = E_{xy}/E_{xx}$. This adjusted treatment mean is the least squares estimator of $\mu + \tau_i$, $i = 1, 2, \ldots, a$, in the model (Equation 15.15). The standard error of any adjusted treatment mean is

$$S_{\text{adj}\bar{y}_{i.}} = \left[MS_E\left(\frac{1}{n} + \frac{(\bar{x}_{i.} - \bar{x}_{..})^2}{E_{xx}}\right)\right]^{1/2} \tag{15.32}$$

Finally, we recall that the regression coefficient β in the model (Equation 15.15) has been assumed to be nonzero. We may test the hypothesis $H_0: \beta = 0$ by using the test statistic

$$F_0 = \frac{(E_{xy})^2/E_{xx}}{MS_E} \tag{15.33}$$

which under the null hypothesis is distributed as $F_{1,a(n-1)-1}$. Thus, we reject $H_0: \beta = 0$ if $F_0 > F_{\alpha,1,a(n-1)-1}$.

■ **TABLE 15.12**
Analysis of Covariance for a Single-Factor Experiment with One Covariate

Source of Variation	Degrees of Freedom	Sums of Squares and Products: x	xy	y	Adjusted for Regression: y	Degrees of Freedom	Mean Square
Treatments	$a - 1$	T_{xx}	T_{xy}	T_{yy}			
Error	$a(n-1)$	E_{xx}	E_{xy}	E_{yy}	$SS_E = E_{yy} - (E_{xy})^2/E_{xx}$	$a(n-1) - 1$	$MS_E = \frac{SS_E}{a(n-1)-1}$
Total	$an - 1$	S_{xx}	S_{xy}	S_{yy}	$SS'_E = S_{yy} - (S_{xy})^2/S_{xx}$	$an - 2$	
Adjusted Treatments					$SS'_E - SS_E$	$a - 1$	$\frac{SS'_E - SS_E}{a-1}$

EXAMPLE 15.5

Consider the experiment described at the beginning of Section 15.3. Three different machines produce a monofilament fiber for a textile company. The process engineer is interested in determining if there is a difference in the breaking strength of the fiber produced by the three machines. However, the strength of a fiber is related to its diameter, with thicker fibers being generally stronger than thinner ones. A random sample of five fiber specimens is selected from each machine. The fiber strength (y) and the corresponding diameter (x) for each specimen are shown in Table 15.10.

The scatter diagram of breaking strength versus the fiber diameter (Figure 15.3) shows a strong suggestion of a linear relationship between breaking strength and diameter, and it seems appropriate to remove the effect of diameter on strength by an analysis of covariance. Assuming that a linear relationship between breaking strength and diameter is appropriate, we see that the model is

$$y_{ij} = \mu + \tau_i + \beta(x_{ij} - \bar{x}_{..}) + \epsilon_{ij} \quad \begin{cases} i = 1, 2, 3 \\ j = 1, 2, \ldots, 5 \end{cases}$$

Using Equations 15.17 through 15.25, we may compute

$$S_{yy} = \sum_{i=1}^{3}\sum_{j=1}^{5} y_{ij}^2 - \frac{y_{..}^2}{an} = (36)^2 + (41)^2 + \cdots + (32)^2 - \frac{(603)^2}{(3)(5)} = 346.40$$

$$S_{xx} = \sum_{i=1}^{3}\sum_{j=1}^{5} x_{ij}^2 - \frac{x_{..}^2}{an} = (20)^2 + (25)^2 + \cdots + (15)^2 - \frac{(362)^2}{(3)(5)} = 261.73$$

$$S_{xy} = \sum_{i=1}^{3}\sum_{j=1}^{5} x_{ij}y_{ij} - \frac{(x_{..})(y_{..})}{an} = (20)(36) + (25)(41) + \cdots + (15)(32) - \frac{(362)(603)}{(3)(5)} = 282.60$$

$$T_{yy} = \frac{1}{n}\sum_{i=1}^{3} y_{i.}^2 - \frac{y_{..}^2}{an} = \frac{1}{5}[(207)^2 + (216)^2 + (180)^2] - \frac{(603)^2}{(3)(5)} = 140.40$$

$$T_{xx} = \frac{1}{n}\sum_{i=1}^{3} x_{i.}^2 - \frac{x_{..}^2}{an} = \frac{1}{5}[(126)^2 + (130)^2 + (106)^2] - \frac{(362)^2}{(3)(5)} = 66.13$$

$$T_{xy} = \frac{1}{n}\sum_{i=1}^{3} x_{i.}y_{i.} - \frac{(x_{..})(y_{..})}{an} = \frac{1}{5}[(126)(207) + (130)(216) + (106)(184)] - \frac{(362)(603)}{(3)(5)} = 96.00$$

$$E_{yy} = S_{yy} - T_{yy} = 346.40 - 140.40 = 206.00$$

$$E_{xx} = S_{xx} - T_{xx} = 261.73 - 66.13 = 195.60$$

$$E_{xy} = S_{xy} - T_{xy} = 282.60 - 96.00 = 186.60$$

From Equation 15.29, we find

$$\begin{aligned} SS_E' &= S_{yy} - (S_{xy})^2/S_{xx} \\ &= 346.40 - (282.60)^2/261.73 \\ &= 41.27 \end{aligned}$$

with $an - 2 = (3)(5) - 2 = 13$ degrees of freedom; and from Equation 15.27, we find

$$\begin{aligned} SS_E &= E_{yy} - (E_{xy})^2/E_{xx} \\ &= 206.00 - (186.60)^2/195.60 \\ &= 27.99 \end{aligned}$$

with $a(n - 1) - 1 = 3(5 - 1) - 1 = 11$ degrees of freedom.

■ **TABLE 15.13**
Analysis of Covariance for the Breaking Strength Data

Source of Variation	Degrees of Freedom	Sums of Squares and Products			Adjusted for Regression				
		x	xy	y	y	Degrees of Freedom	Mean Square	F_0	P-Value
Machines	2	66.13	96.00	140.40					
Error	12	195.60	186.60	206.00	27.99	11	2.54		
Total	14	261.73	282.60	346.40	41.27	13			
Adjusted Machines					13.28	2	6.64	2.61	0.1181

The sum of squares for testing $H_0: \tau_1 = \tau_2 = \tau_3 = 0$ is

$$SS'_E - SS_E = 41.27 - 27.99 = 13.28$$

with $a - 1 = 3 - 1 = 2$ degrees of freedom. These calculations are summarized in Table 15.13.

To test the hypothesis that machines differ in the breaking strength of fiber produced, that is, $H_0: \tau_i = 0$, we compute the test statistic from Equation 15.30 as

$$F_0 = \frac{(SS'_E - SS_E)/(a-1)}{SS_E/[a(n-1)-1]} = \frac{13.28/2}{27.99/11} = \frac{6.64}{2.54} = 2.61$$

Comparing this to $F_{0.10,2,11} = 2.86$, we find that the null hypothesis cannot be rejected. The P-value of this test statistic is 0.1181. Thus, there is no strong evidence that the fibers produced by the three machines differ in breaking strength.

The estimate of the regression coefficient is computed from Equation 15.26 as

$$\hat{\beta} = \frac{E_{xy}}{E_{xx}} = \frac{186.60}{195.60} = 0.9540$$

We may test the hypothesis $H_0: \beta = 0$ by using Equation 15.33. The test statistic is

$$F_0 = \frac{(E_{xy})^2/E_{xx}}{MS_E} = \frac{(186.60)^2/195.60}{2.54} = 70.08$$

and because $F_{0.01,1,11} = 9.65$, we reject the hypothesis that $\beta = 0$. Therefore, there is a linear relationship between breaking strength and diameter, and the adjustment provided by the analysis of covariance was necessary.

The adjusted treatment means may be computed from Equation 15.31. These adjusted means are

$$\text{Adjusted } \bar{y}_{1.} = \bar{y}_{1.} - \hat{\beta}(\bar{x}_{1.} - \bar{x}_{..}) = 41.40 - (0.9540)(25.20 - 24.13) = 40.38$$

$$\text{Adjusted } \bar{y}_{2.} = \bar{y}_{2.} - \hat{\beta}(\bar{x}_{2.} - \bar{x}_{..}) = 43.20 - (0.9540)(26.00 - 24.13) = 41.42$$

and

$$\text{Adjusted } \bar{y}_{3.} = \bar{y}_{3.} - \hat{\beta}(\bar{x}_{3.} - \bar{x}_{..}) = 36.00 - (0.9540)(21.20 - 24.13) = 38.80$$

Comparing the adjusted treatment means with the unadjusted treatment means (the $\bar{y}_{i.}$), we note that the adjusted means are much closer together, another indication that the covariance analysis was necessary.

A basic assumption in the analysis of covariance is that the treatments do not influence the covariate x because the technique removes the effect of variations in the $\bar{x}_{i.}$. However, if the variability in the $\bar{x}_{i.}$ is due in part to the treatments, then analysis of covariance removes part of the treatment effect. Thus, we must be reasonably sure that the treatments do not affect the values x_{ij}. In some experiments this may be obvious from the nature of the covariate, whereas in others it may be more doubtful. In our example, there may be a difference in fiber diameter (x_{ij}) between the three machines. In such cases, Cochran and Cox (1957) suggest that an analysis of variance on the x_{ij} values may be helpful in determining the validity of this assumption. For our problem, this procedure yields

$$F_0 = \frac{66.13/2}{195.60/12} = \frac{33.07}{16.30} = 2.03$$

which is less than $F_{0.10,2,12} = 2.81$, so there is no reason to believe that machines produce fibers of different diameters.

Diagnostic checking of the covariance model is based on residual analysis. For the covariance model, the residuals are

$$e_{ij} = y_{ij} - \hat{y}_{ij}$$

where the fitted values are

$$\begin{aligned}\hat{y}_{ij} &= \hat{\mu} + \hat{\tau}_i + \hat{\beta}(x_{ij} - \bar{x}_{..}) = \bar{y}_{..} + [\bar{y}_{i.} - \bar{y}_{..} - \hat{\beta}(\bar{x}_{i.} - \bar{x}_{..})] \\ &\quad + \hat{\beta}(x_{ij} - \bar{x}_{..}) = \bar{y}_{i.} + \hat{\beta}(x_{ij} - \bar{x}_{i.})\end{aligned}$$

Thus,

$$e_{ij} = y_{ij} - \bar{y}_{i.} - \hat{\beta}(x_{ij} - \bar{x}_{i.}) \tag{15.34}$$

To illustrate the use of Equation 15.34, the residual for the first observation from the first machine in Example 15.5 is

$$\begin{aligned}e_{11} &= y_{11} - \bar{y}_{1.} - \hat{\beta}(x_{11} - \bar{x}_{1.}) = 36 - 41.4 - (0.9540)(20 - 25.2) \\ &= 36 - 36.4392 = -0.4392\end{aligned}$$

A complete listing of observations, fitted values, and residuals is given in the following table:

Observed Value, y_{ij}	Fitted Value, $\hat{y}_{ij}$	Residual, $e_{ij} = y_{ij} - \hat{y}_{ij}$
36	36.4392	−0.4392
41	41.2092	−0.2092
39	40.2552	−1.2552
42	41.2092	0.7908
49	47.8871	1.1129
40	39.3840	0.6160
48	45.1079	2.8921
39	39.3840	−0.3840
45	47.0159	−2.0159
44	45.1079	−1.1079
35	35.8092	−0.8092
37	37.7171	−0.7171
42	40.5791	1.4209
34	35.8092	−1.8092
32	30.0852	1.9148

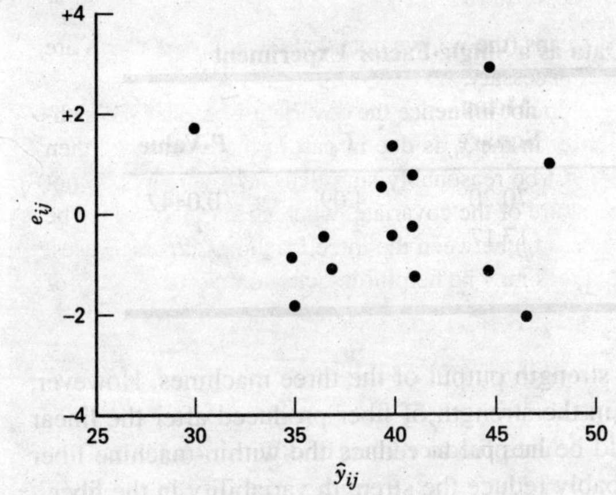

■ **FIGURE 15.4 Plot of residuals versus fitted values for Example 15.5**

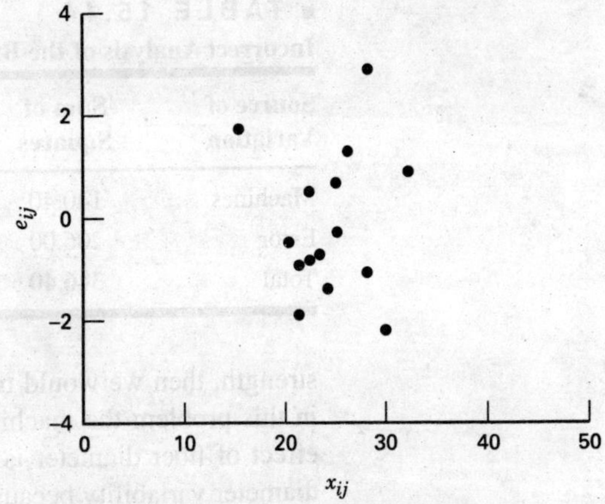

■ **FIGURE 15.5 Plot of residuals versus fiber diameter *x* for Example 15.5**

The residuals are plotted versus the fitted values $\hat{y}_{ij}$ in Figure 15.4, versus the covariate x_{ij} in Figure 15.5 and versus the machines in Figure 15.6. A normal probability plot of the residuals is shown in Figure 15.7. These plots do not reveal any major departures from the assumptions, so we conclude that the covariance model (Equation 15.15) is appropriate for the breaking strength data.

It is interesting to note what would have happened in this experiment if an analysis of covariance had not been performed, that is, if the breaking strength data (y) had been analyzed as a completely randomized single-factor experiment in which the covariate x was ignored. The analysis of variance of the breaking strength data is shown in Table 15.14. We immediately notice that the error estimate is much longer in the CRD analysis (17.17 versus 2.54). This is a reflection of the effectiveness of analysis of covariance in reducing error variability. We would also conclude, based on the CRD analysis, that machines differ significantly in the strength of fiber produced. This is exactly *opposite the conclusion* reached by the covariance analysis. If we suspected that the machines differed significantly in their effect on fiber

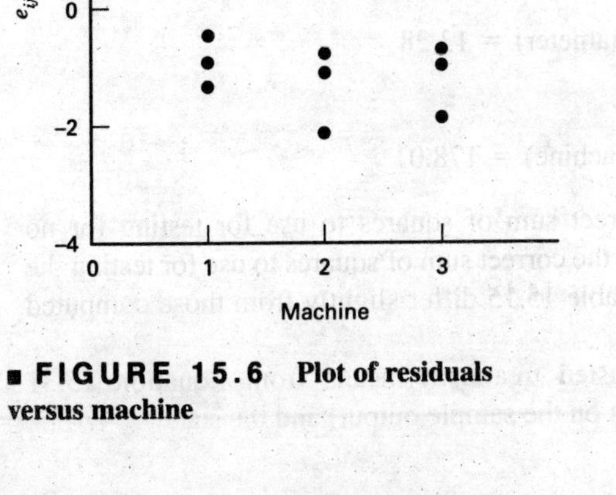

■ **FIGURE 15.6 Plot of residuals versus machine**

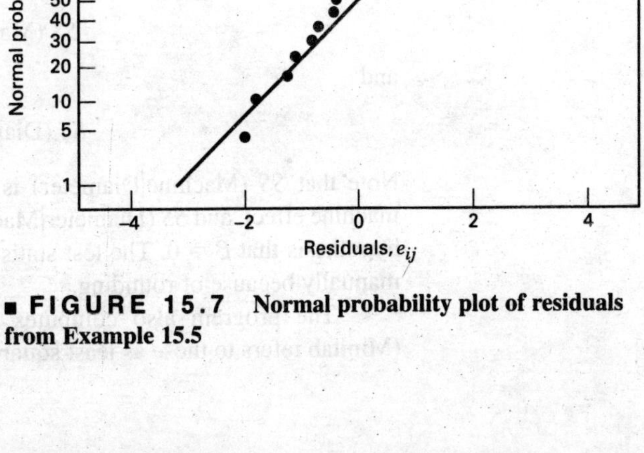

■ **FIGURE 15.7 Normal probability plot of residuals from Example 15.5**

■ **TABLE 15.14**
Incorrect Analysis of the Breaking Strength Data as a Single-Factor Experiment

Source of Variation	Sum of Squares	Degrees of Freedom	Mean Square	F_0	*P*-Value
Machines	140.40	2	70.20	4.09	0.0442
Error	206.00	12	17.17		
Total	346.40	14			

strength, then we would try to equalize the strength output of the three machines. However, in this problem the machines do not differ in the strength of fiber produced after the linear effect of fiber diameter is removed. It would be helpful to reduce the within-machine fiber diameter variability because this would probably reduce the strength variability in the fiber.

Analysis of Covariance as an Alternative to Blocking In some situations, the experimenter may have a choice between either running a completely randomized design with a covariate or running a randomized block design with the covariate used in some fashion to form the blocks. If the relationship between the covariate and the response is really well-approximated by a straight line and that is the form of the covariance model that the experimenter chooses, then either of these approaches is about equally effective. However, if the relationship isn't linear and a linear model is assumed, the analysis of covariance will be outperformed in error reduction by the randomized block design. Randomized block designs do not make any explicit assumptions about relationships between the nuisance variables (covariates) and the response. Generally, a randomized block design will have fewer degrees if freedom for error than a covariance model but the resulting loss of statistical power is usually quite small.

15.3.2 Computer Solution

Several computer software packages now available can perform the analysis of covariance. The output from the Minitab General Linear Models procedure for the data in Example 15.4 is shown in Table 15.15. This output is very similar to those presented previously. In the section of the output entitled "Analysis of Variance," the "Seq SS" correspond to a "sequential" partitioning of the overall model sum of squares, say

$$\begin{aligned} SS\,(\text{Model}) &= SS\,(\text{Diameter}) + SS\,(\text{Machine}|\text{Diameter}) \\ &= 305.13 + 13.28 \\ &= 318.41 \end{aligned}$$

whereas the "Adj SS" corresponds to the "extra" sum of squares for each factor, that is,

$$SS\,(\text{Machine}|\text{Diameter}) = 13.28$$

and

$$SS\,(\text{Diameter}|\text{Machine}) = 178.01$$

Note that SS (Machine|Diameter) is the correct sum of squares to use for testing for no machine effect, and SS (Diameter|Machine) is the correct sum of squares to use for testing the hypothesis that $\beta = 0$. The test statistics in Table 15.15 differ slightly from those computed manually because of rounding.

The program also computes the adjusted treatment means from Equation 15.31 (Minitab refers to those as least squares means on the sample output) and the standard errors.

■ **TABLE 15.15**

Minitab Output (Analysis of Covariance) for Example 15.5

```
General Linear Model

Factor      Type      Levels      Values
Machine     fixed          3       1 2 3

Analysis of Variance for Strength, using Adjusted SS for Tests

Source      DF     Seq SS     Adj SS     Adj MS       F       P
Diameter     1     305.13     178.01     178.01   69.97   0.000
Machine      2      13.28      13.28       6.64    2.61   0.118
Error       11      27.99      27.99       2.54
Total       14     346.40

Term         Coef       Std. Dev.      T       P
Constant   17.177        2.783       6.17    0.000
Diameter   0.9540        0.1140      8.36    0.000
Machine
1          0.1824        0.5950      0.31    0.765
2          1.2192        0.6201      1.97    0.075

Means for Covariates

Covariate     Mean   Std. Dev.
Diameter     24.13     4.324

Least Squares Means for Strength

Machine      Mean      Std. Dev.
1           40.38       0.7236
2           41.42       0.7444
3           38.80       0.7879
```

The program will also compare all pairs of treatment means using the pairwise multiple comparison procedures discussed in Chapter 3.

15.3.3 Development by the General Regression Significance Test

It is possible to develop formally the ANCOVA procedure for testing $H_0: \tau_i = 0$ in the covariance model

$$y_{ij} = \mu + \tau_i + \beta(x_{ij} - \bar{x}_{..}) + \epsilon_{ij} \quad \begin{cases} i = 1, 2, \ldots, a \\ j = 1, 2, \ldots, n \end{cases} \tag{15.35}$$

using the general regression significance test. Consider estimating the parameters in the model (Equation 15.15) by least squares. The least squares function is

$$L = \sum_{i=1}^{a} \sum_{j=1}^{n} [y_{ij} - \mu - \tau_i - \beta(x_{ij} - \bar{x}_{..})]^2 \tag{15.36}$$

and from $\partial L/\partial\mu = \partial L/\partial\tau_i = \partial L/\partial\beta = 0$, we obtain the normal equations

$$\mu\text{: } an\hat{\mu} + n\sum_{i=1}^{a}\hat{\tau}_i = y_{..} \tag{15.37a}$$

$$\tau_i\text{: } n\hat{\mu} + n\hat{\tau}_i + \hat{\beta}\sum_{j=1}^{n}(x_{ij} - \bar{x}_{..}) = y_{i.} \qquad i = 1, 2, \ldots, a \tag{15.37b}$$

$$\beta\text{: } \sum_{i=1}^{a}\hat{\tau}_i\sum_{j=1}^{n}(x_{ij} - \bar{x}_{..}) + \hat{\beta}S_{xx} = S_{xy} \tag{15.37c}$$

Adding the a equations in Equation 15.37b, we obtain Equation 15.37a because $\Sigma_{i=1}^{a}\Sigma_{j=1}^{n}(x_{ij} - \bar{x}_{..}) = 0$, so there is one linear dependency in the normal equations. Therefore, it is necessary to augment Equations 15.37 with a linearly independent equation to obtain a solution. A logical side condition is $\Sigma_{i=1}^{a}\hat{\tau}_i = 0$.

Using this condition, we obtain from Equation 15.37a

$$\hat{\mu} = \bar{y}_{..} \tag{15.38a}$$

and from Equation 15.37b

$$\hat{\tau}_i = \bar{y}_{i.} - \bar{y}_{..} - \hat{\beta}(\bar{x}_{i.} - \bar{x}_{..}) \tag{15.38b}$$

Equation 15.37c may be rewritten as

$$\sum_{i=1}^{a}(\bar{y}_{i.} - \bar{y}_{..})\sum_{j=1}^{n}(x_{ij} - \bar{x}_{..}) - \hat{\beta}\sum_{i=1}^{a}(\bar{x}_{i.} - \bar{x}_{..})\sum_{j=1}^{n}(x_{ij} - \bar{x}_{..}) + \hat{\beta}S_{xx} = S_{xy}$$

after substituting for $\hat{\tau}_i$. But we see that

$$\sum_{i=1}^{a}(\bar{y}_{i.} - \bar{y}_{..})\sum_{j=1}^{n}(x_{ij} - \bar{x}_{..}) = T_{xy}$$

and

$$\sum_{i=1}^{a}(\bar{x}_{i.} - \bar{x}_{..})\sum_{j=1}^{n}(x_{ij} - \bar{x}_{..}) = T_{xx}$$

Therefore, the solution to Equation 15.37c is

$$\hat{\beta} = \frac{S_{xy} - T_{xy}}{S_{xx} - T_{xx}} = \frac{E_{xy}}{E_{xx}}$$

which was the result given previously in Section 15.3.1, Equation 15.26.

We may express the reduction in the total sum of squares due to fitting the full model (Equation 15.15) as

$$\begin{aligned} R(\mu, \tau, \beta) &= \hat{\mu}y_{..} + \sum_{i=1}^{a}\hat{\tau}_i y_{i.} + \hat{\beta}S_{xy} \\ &= (\bar{y}_{..})y_{..} + \sum_{i=1}^{a}[\bar{y}_{i.} - \bar{y}_{..} - (E_{xy}/E_{xx})(\bar{x}_{i.} - \bar{x}_{..})]y_{i.} + (E_{xy}/E_{xx})S_{xy} \\ &= y_{..}^2/an + \sum_{i=1}^{a}(\bar{y}_{i.} - \bar{y}_{..})y_{i.} - (E_{xy}/E_{xx})\sum_{i=1}^{a}(\bar{x}_{i.} - \bar{x}_{..})y_{i.} + (E_{xy}/E_{xx})S_{xy} \\ &= y_{..}^2/an + T_{yy} - (E_{xy}/E_{xx})(T_{xy} - S_{xy}) \\ &= y_{..}^2/an + T_{yy} + (E_{xy})^2/E_{xx} \end{aligned}$$

This sum of squares has $a + 1$ degrees of freedom because the rank of the normal equations is $a + 1$. The error sum of squares for this model is

$$\begin{aligned} SS_E &= \sum_{i=1}^{a}\sum_{j=1}^{n} y_{ij}^2 - R(\mu, \tau, \beta) \\ &= \sum_{i=1}^{a}\sum_{j=1}^{n} y_{ij}^2 - y_{..}^2/an - T_{yy} - (E_{xy})^2/E_{xx} \\ &= S_{yy} - T_{yy} - (E_{xy})^2/E_{xx} \\ &= E_{yy} - (E_{xy})^2/E_{xx} \end{aligned} \tag{15.39}$$

with $an - (a + 1) = a(n - 1) - 1$ degrees of freedom. This quantity was obtained previously as Equation 15.27.

Now consider the model restricted to the null hypothesis, that is, to H_0: $\tau_1 = \tau_2 = \cdots = \tau_a = 0$. This reduced model is

$$y_{ij} = \mu + \beta(x_{ij} - \bar{x}_{..}) + \epsilon_{ij} \quad \begin{cases} i = 1, 2, \ldots, a \\ j = 1, 2, \ldots, n \end{cases} \tag{15.40}$$

This is a simple linear regression model, and the least squares normal equations for this model are

$$an\hat{\mu} = y_{..} \tag{15.41a}$$

$$\hat{\beta}S_{xx} = S_{xy} \tag{15.41b}$$

The solutions to these equations are $\hat{\mu} = \bar{y}_{..}$ and $\hat{\beta} = S_{xy}/S_{xx}$, and the reduction in the total sum of squares due to fitting the reduced model is

$$\begin{aligned} R(\mu, \beta) &= \hat{\mu}y_{..} + \hat{\beta}S_{xy} \\ &= (\bar{y}_{..})y_{..} + (S_{xy}/S_{xx})S_{xy} \\ &= y_{..}^2/an + (S_{xy})^2/S_{xx} \end{aligned} \tag{15.42}$$

This sum of squares has two degrees of freedom.

We may find the appropriate sum of squares for testing H_0: $\tau_1 = \tau_2 = \cdots = \tau_a = 0$ as

$$\begin{aligned} R(\tau|\mu, \beta) &= R(\mu, \tau, \beta) - R(\mu, \beta) \\ &= y_{..}^2/an + T_{yy} + (E_{xy})^2/E_{xx} - y_{..}^2/an - (S_{xy})^2/S_{xx} \\ &= S_{yy} - (S_{xy})^2/S_{xx} - [E_{yy} - (E_{xy})^2/E_{xx}] \end{aligned} \tag{15.43}$$

using $T_{yy} = S_{yy} - E_{yy}$. Note that $R(\tau|\mu, \beta)$ has $a + 1 - 2 = a - 1$ degrees of freedom and is identical to the sum of squares given by $SS_E' - SS_E$ in Section 15.3.1. Thus, the test statistic for H_0: $\tau_i = 0$ is

$$F_0 = \frac{R(\tau|\mu, \beta)/(a - 1)}{SS_E/[a(n - 1) - 1]} = \frac{(SS_E' - SS_E)/(a - 1)}{SS_E/[a(n - 1) - 1]} \tag{15.44}$$

which we gave previously as Equation 15.30. Therefore, by using the general regression significance test, we have justified the heuristic development of the analysis of covariance in Section 15.3.1.

15.3.4 Factorial Experiments with Covariates

Analysis of covariance can be applied to more complex treatment structures, such as factorial designs. Provided enough data exists for every treatment combination, nearly any complex treatment structure can be analyzed through the analysis of covariance approach. We now show how the analysis of covariance could be used in the most common family of factorial designs used in industrial experimentation, the 2^k factorials.

Imposing the assumption that the covariate affects the response variable identically across all treatment combinations, an analysis of covariance table similar to the procedure given in Section 15.3.1 could be performed. The only difference would be the treatment sum of squares. For a 2^2 factorial with n replicates, the treatment sum of squares (T_{yy}) would be $(1/n) \sum_{i=1}^{2} \sum_{j=1}^{2} y_{ij.}^2 - y_{...}^2/(2)(2)n$. This quantity is the sum of the sums of squares for factors A, B, and the AB interaction. The adjusted treatment sum of squares could then be partitioned into individual effect components, that is, adjusted main effects sum of squares SS_A and SS_B, and an adjusted interaction sum of squares, SS_{AB}.

The amount of replication is a key issue when broadening the design structure of the treatments. Consider a 2^3 factorial arrangement. A minimum of two replicates is needed to evaluate all treatment combinations with a separate covariate for each treatment combination (covariate by treatment interaction). This is equivalent to fitting a simple regression model to each treatment combination or design cell. With two observations per cell, one degree of freedom is used to estimate the intercept (the treatment effect), and the other is used to estimate the slope (the covariate effect). With this saturated model, no degrees of freedom are available to estimate the error. Thus, at least three replicates are needed for a complete analysis of covariance, assuming the most general case. This problem becomes more pronounced as the number of distinct design cells (treatment combinations) and covariates increases.

If the amount of replication is limited, various assumptions can be made to allow some useful analysis. The simplest assumption (and typically the worst) that can be made is that the covariate has no effect. If the covariate is erroneously not considered, the entire analysis and subsequent conclusions could be dramatically in error. Another choice is to assume that there is no treatment by covariate interaction. Even if this assumption is incorrect, the average affect of the covariate across all treatments will still increase the precision of the estimation and testing of the treatment effects. One disadvantage of this assumption is that if several treatment levels interact with the covariate, the various terms may cancel one another out and the covariate term, if estimated alone with no interaction, may be insignificant. A third choice would be to assume some of the factors (such as some two-factor and higher interactions) are insignificant. This allows some degrees of freedom to be used to estimate error. This course of action, however, should be undertaken carefully and the subsequent models evaluated thoroughly because the estimation of error will be relatively imprecise unless enough degrees of freedom are allocated for it. With two replicates, each of these assumptions will free some degrees of freedom to estimate error and allow useful hypothesis tests to be performed. Which assumption to enforce should be dictated by the experimental situation and how much risk the experimenter is willing to assume. We caution that in the effects model-building strategy if a treatment factor is eliminated, then the resulting two "replicates" of each original 2^3 are not truly replicates. These "hidden replicates" do free $\nu 1$ degrees of freedom for parameter estimation but should not be used as replicates to estimate pure error because the execution of the original design may not have been randomized that way.

To illustrate some of these ideas, consider the 2^3 factorial design with two replicates and a covariate shown in Table 15.16. If the response variable y is analyzed without accounting for the covariate, the following model results:

$$\hat{y} = 25.03 + 11.20A + 18.05B + 7.24C - 18.91AB + 14.80AC$$

The overall model is significant at the $\alpha = 0.01$ level with $R^2 = 0.786$ and $MS_E = 470.82$. The residual analysis indicates no problem with this model except the observation with $y = 103.01$ is unusual.

If the second assumption, common slopes with no treatment by covariate interaction, is chosen, the full effects model and the covariate effect can be estimated. The JMP output is shown in Table 15.17. Notice that the MS_E has been reduced considerably by considering the covariate. The final resulting analysis after sequentially removing each nonsignificant interaction and the

■ **TABLE 15.16**

Response and Covariate Data for a 2^3 with 2 Replicates

A	*B*	*C*	*x*	*y*
−1	−1	−1	4.05	−30.73
1	−1	−1	0.36	9.07
−1	1	−1	5.03	39.72
1	1	−1	1.96	16.30
−1	−1	1	5.38	−26.39
1	−1	1	8.63	54.58
−1	1	1	4.10	44.54
1	1	1	11.44	66.20
−1	−1	−1	3.58	−26.46
1	−1	−1	1.06	10.94
−1	1	−1	15.53	103.01
1	1	−1	2.92	20.44
−1	−1	1	2.48	−8.94
1	−1	1	13.64	73.72
−1	1	1	−0.67	15.89
1	1	1	5.13	38.57

■ **TABLE 15.17**

JMP Analysis of Covariance for the Experiment in Table 15.16, Assuming a Common Slope

Response Y

Summary of Fit

RSquare	0.971437
RSquare Adj	0.938795
Root Mean Square Error	9.47287
Mean of Response	25.02875
Observations for (or Sum Wgts)	16

Analysis of Variance

Source	**DF**	**Sum of Squares**	**Mean Square**	**F Ratio**
Model	8	21363.834	2670.48	29.7595
Error	7	628.147	89.74	**Prob > F**
C. Total	15	21991.981		<.0001*

Parameter Estimates

Term	**Estimate**	**Std Error**	**t Ratio**	**Prob>|t|**
Intercept	−1.015872	5.454106	−0.19	0.8575
X	4.9245327	0.928977	5.30	0.0011*
X1	9.4566966	2.39091	3.96	0.0055*

■ **TABLE 15.17** *(Continued)*

X2	16.128277	2.395946	6.73	0.0003*
X3	2.4287693	2.536347	0.96	0.3702
X1*X2	−15.59941	2.448939	−6.37	0.0004*
X1*X3	−0.419306	3.72135	−0.11	0.9135
X2*X3	−0.863837	2.824779	−0.31	0.7686
X1*X2*X3	1.469927	2.4156	0.61	0.5621

Sorted Parameter Estimates

Term	**Estimate**	**Std Error**	**t Ratio**	**Prob>|t|**
X2	16.128277	2.395946	6.73	0.0003*
X1*X2	−15.59941	2.448939	−6.37	0.0004*
x	4.9245327	0.928977	5.30	0.0011*
X1	9.4566966	2.39091	3.96	0.0055*
X3	2.4287693	2.536347	0.96	0.3702
X1*X2*X3	1.469927	2.4156	0.61	0.5621
X2*X3	−0.863837	2.824779	−0.31	0.7686
X1*X3	−0.419306	3.72135	−0.11	0.9135

main effect C is shown in Table 15.18. This reduced model provides an even smaller MS_E than does the full model with the covariate in Table 15.17.

Finally, we could consider a third course of action, assuming certain interaction terms are negligible. We consider the full model that allows for different slopes between treatments and treatment by covariate interaction. We assume that the three-factor interactions (both ABC and $ABCx$) are not significant and use their associated degrees of freedom to estimate error in the most general effects model that can be fit. This is often a practical assumption. Three-factor and higher interactions are usually negligible in most experimental settings. We used JMP for the analysis, and the results are shown in Table 15.19. The type III sums of squares are the adjusted sums of squares that we require.

With a near-saturated model, the estimate of error will be fairly imprecise. Even with only a few terms being individually significant at the $\alpha = 0.05$ level, the overall sense is that this model is better than the two previous scenarios (based on R^2 and the mean square for error). Because the treatment effects aspect of the model is of more interest, we sequentially remove terms from the covariate portion of the model to add degrees of freedom to the estimate of error. If we sequentially remove the ACx term followed by BCx, the MS_E decreases and several terms are insignificant. The final model is shown in Table 15.20 after sequentially removing Cx, AC, and BC.

This example emphasizes the need to have degrees of freedom available to estimate experimental error in order to increase the precision of the hypothesis tests associated with the individual terms in the model. This process should be done sequentially to avoid eliminating significant terms masked by a poor estimate of error.

Reviewing the results obtained from the three approaches, we note that each method successively improves the model fit in this example. If there is a strong reason to believe that the covariate does not interact with the factors, it may be best to make that assumption at the outset of the analysis. This choice may also be dictated by software. Although experimental design software packages may only be able to model covariates that do not interact with treatments, the analyst may have a reasonable chance of identifying the major factors influencing the process, even if there is some covariate by treatment interaction. We also note that

■ **TABLE 15.18**

JMP Analysis of Covariance, Reduced Model for the Experiment in Table 15.16

Response Y

Whole Model

Summary of Fit

RSquare	0.96529
RSquare Adj	0.952668
Root Mean Square Error	8.330324
Mean of Response	25.02875
Observations (or Sum Wgts)	16

Analysis of Variance

Source	DF	Sum of Squares	Mean Square	F Ratio
Model	4	21228.644	5307.16	76.4783
Error	11	763.337	69.39	Prob > F
C. Total	15	21991.981		<.0001*

Parameter Estimates

Term	Estimate	Std Error	t Ratio	Prob>|t|
Intercept	−1.878361	3.224761	−0.58	0.5720
X	5.0876125	0.465535	10.93	<.0001*
X1	9.3990071	2.089082	4.50	0.0009*
X2	16.064472	2.090531	7.68	<.0001*
X1*X2	−15.48994	2.105895	−7.36	<.0001*

■ **TABLE 15.19**

JMP Output for the Experiment in Table 15.16

Response Y

Summary of Fit

RSquare	0.999872
RSquare Adj	0.999044
Root Mean Square Error	1.18415
Mean of Response	25.02875
Observations (or Sum Wgts)	16

Analysis of Variance

Source	DF	Sum of Squares	Mean Square	F Ratio
Model	13	21989.177	1691.48	1206.292
Error	2	2.804	1.40	Prob > F
C. Total	15	21991.981		0.0008*

■ **TABLE 15.19** *(Continued)*

Parameter Estimates

Term	Estimate	Std Error	t Ratio	Prob>|t|
Intercept	10.063551	1.018886	9.88	0.0101*
x	2.1420242	0.361866	5.92	0.0274*
X1	13.887593	1.29103	10.76	0.0085*
X2	19.4443	1.12642	17.26	0.0033*
X3	5.0908008	1.114861	4.57	0.0448*
X1*X2	−19.59556	0.701412	−27.94	0.0013*
X1*X3	−0.23434	1.259161	−0.19	0.8695
X2*X3	−0.368071	0.887678	−0.41	0.7186
X1*(x-5.28875)	2.1171372	0.429771	4.93	0.0388*
X2*(x-5.28875)	3.0175357	0.364713	8.27	0.0143*
X3*(x-5.28875)	−0.096716	0.356887	−0.27	0.8118
X1*X2*(x-5.28875)	−2.949107	0.190284	−15.50	0.0041*
X1*X3*(x-5.28875)	−0.012116	0.415868	−0.03	0.9794
X2*X3*(x-5.28875)	0.0848196	0.401999	0.21	0.8524

Sorted Parameter Estimates

Term	Estimate	Std Error	t Ratio		Prob>|t|
X1*X2	−19.59556	0.701412	−27.94		0.0013*
X2	19.4443	1.12642	17.26		0.0033*
X1*X2*(x-5.28875)	−2.949107	0.190284	−15.50		0.0041*
X1	13.887593	1.29103	10.76		0.0085*
X2*(x-5.28875)	3.0175357	0.364713	8.27		0.0143*
x	2.1420242	0.361866	5.92		0.0274*
X1*(x-5.28875)	2.1171372	0.429771	4.93		0.0388*
X3	5.0908008	1.114861	4.57		0.0448*
X2*X3	−0.368071	0.887678	−0.41		0.7186
X3*(x-5.28875)	−0.096716	0.356887	−0.27		0.8118
X2*X3*(x-5.28875)	0.0848196	0.401999	0.21		0.8524
X1*X3*	−0.23434	1.259161	−0.19		0.8695
X1*X3*(x-5.28875)	−0.012116	0.415868	−0.03		0.9794

all the usual tests of model adequacy are still appropriate and are strongly recommended as part of the ANCOVA model building process.

Another situation involving covariates arises often in practice. The experimenter has available a collection of experimental units, and these units can be described or characterized by some measured quantities than have the potential to affect the outcome of the experiment in which they are used. This happens frequently in clinical studies where the experimental units are patients and they are characterized in terms of factors such as gender, age, blood pressure, weight, or other parameters relevant to the specific study. The design factors for the experiment can include type of pharmacological agent, dosage, and how frequently the dose is administered. The experimenter wants to select a subset of the experimental units that is optimal with respect to the design factors that are to be studied. In this type of problem the

■ **TABLE 15.20**
JMP Output for the Experiment in Table 15.16, Reduced Model

Response Y

Summary of Fit

RSquare	0.999743
RSquare Adj	0.99945
Root Mean Square Error	0.898184
Mean of Response	25.02875
Observations (or Sum Wgts)	16

Analysis of Variance

Source	DF	Sum of Squares	Mean Square	F Ratio
Model	8	21986.334	2748.29	3406.688
Error	7	5.647	0.81	Prob > F
C. Total	15	21991.981		<.0001*

Parameter Estimates

Term	Estimate	Std Error	t Ratio	Prob>|t|
Intercept	10.234044	0.544546	18.79	<.0001*
x	2.05036	0.118305	17.33	<.0001*
X1	13.694768	0.273254	50.12	<.0001*
X2	19.709839	0.272411	72.35	<.0001*
X3	5.4433644	0.321496	16.93	<.0001*
X1*X2	−19.40839	0.27236	−71.26	<.0001*
X1*(x-5.28875)	2.0628522	0.121193	17.02	<.0001*
X2*(x-5.28875)	3.0321079	0.116027	26.13	<.0001*
X1*X2*(x-5.28875)	−3.031387	0.11686	−25.94	<.0001*

Sorted Parameter Estimates

Term	Estimate	Std Error	t Ratio		Prob>|t|
X2	19.709839	0.272411	72.35		<.0001*
X1*X2	−19.40839	0.27236	−71.26		<.0001*
X1	13.694768	0.273254	50.12		<.0001*
X2(x-5.28875)	3.0321079	0.116027	26.13		<.0001*
X1*X2*(x-5.28875)	−3.031387	0.11686	−25.94		<.0001*
x	2.05036	0.118305	17.33		<.0001*
X1*(x-5.28875)	2.0628522	0.121193	17.02		<.0001*
X3	5.4433644	0.321496	16.93		<.0001*

experimenter knows the values of the covariates in advance of running the experiment and wants to obtain the best possible design taking the values of the covariates into account. This design problem can be solved using the *D*-optimality criterion 9 the mathematical details are beyond the scope of this book, but see the supplemental material for this chapter).

To illustrate, suppose that a chemical manufacturer is producing an additive for motor oil. He knows that final properties of his additive depend on two design factors that he can control and also to some extent on the viscosity and molecular weight of one of the basic raw materials. There are 40 samples of this raw material available for use in his experiment. Table 15.21 contains the viscosity and molecular weight measurements for the 40 raw material samples.

Figure 15.8 is a scatter plot of the viscosity and molecular weight data in Table 15.21. As the manufacturer suspects, there is a relationship between the two quantities.

■ **TABLE 15.21**
Viscosity and Molecular Weight Data

Sample	Viscosity	Molecular Weight
1	32	1264
2*	33.5	1250
3	37	1290
4*	30.9	1250
5*	50	1325
6	37	1296
7	55	1340
8	58	1359
9*	46	1278
10	44	1260
11*	58.3	1329
12	31.9	1250
13*	32.5	1246
14*	59	1304
15*	57.8	1303
16*	60	1336
17	33	1275
18*	36	1290
19	57.1	1326
20	58.3	1330
21*	32	1264
22*	33.5	1250
23*	37	1290
24*	30.9	1250
25*	50	1325
26*	37	1296
27*	55	1340
28	58	1359
29*	46	1278
30	44	1260
31	58.3	1329
32	31.9	1250
33	32.5	1246
34	59	1304
35	57.8	1303
36	60	1336
37	33	1275
38*	36	1290
39*	57.1	1326
40	58.3	1330

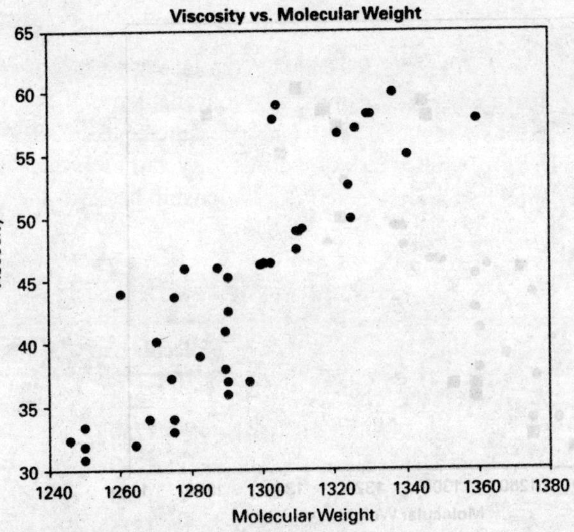

■ **FIGURE 15.8 Scatter Plot of the Viscosity versus Molecular Weight Data from Table 15.21.**

There are two quantitative factors x_3 and x_4 that the manufacturer suspects affect the final properties of his product. He wants to conduct a factorial experiment to study these two factors. Each run of the experiment will require one sample of the available 40 samples of raw material. He feels that a 20-run experiment will be adequate to investigate the effects of the two factors. Table 15.22 is the *D*-optimal design from JMP assuming a main effects plus

■ **TABLE 15.22**
The 20-Run *D*-optimal Design from JMP

Run	Viscosity	Molecular Weight	X_3	X_4
1	44	1260	1	1
2	32.5	1246	1	−1
3	36	1290	−1	−1
4	60	1336	−1	1
5	30.9	1250	1	1
6	33.5	1250	−1	1
7	59	1304	−1	−1
8	58	1359	1	1
9	37	1290	−1	1
10	33	1275	1	−1
11	58.3	1330	1	−1
12	32	1264	−1	−1
13	57.1	1326	1	−1
14	58.3	1329	−1	1
15	46	1278	−1	−1
16	57.8	1303	1	1
17	31.9	1250	−1	1
18	37	1296	1	1
19	55	1340	−1	−1
20	50	1325	1	−1

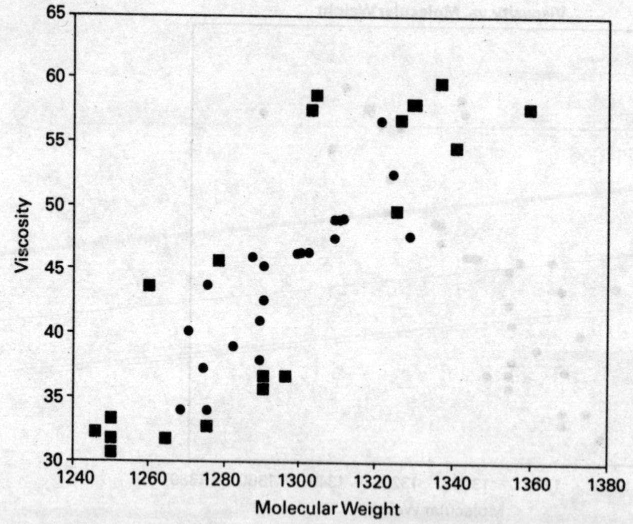

■ **FIGURE 15.9 Scatter Plot of the Viscosity versus Molecular Weight Data from Table 15.21, with the Selected Design Points Shown as Larger Dots**

two-factor interaction in the design factors and a sample size of 20. The 20 raw material samples that are selected by the *D*-optimal algorithm are indicated in Table 15.21 with asterisks in the sample column. Figure 15.9 is the scatter plot of viscosity versus molecular weight with the selected samples shown in larger dots. Notice that the selected samples "spread out" over the boundary of the convex hull of points. This is consistent with the tendency of *D*-optimal designs to spread observations to the boundaries of the experimental region. The design points selected in the two factors x_3 and x_4 correspond to a 2^2 factorial experiment with $n = 5$ replicates.

Table 15.23 presents the relative variance of the coefficients for the model for the *D*-optimal design. Notice that the design factors are all estimated with the best possible precision (relative variance $= 1/N = 1/20 = 0.05$) while the covariates are estimated with different precisions that depends on their spread in the original set of 40 samples. The alias matrix is shown in Table 15.24. The numbers in the table are the correlations between model terms. All correlations are small, a consequence of the *D*-optimality criterion spreading the design points out as much as possible.

■ **TABLE 15.23**
Relative Variances of the Coefficients for the *D*-Optimal Design in Table 15.22

Effect	Relative Variance
Intercept	0.058
Viscosity	0.314
Molecular Weight	0.516
X_3	0.050
X_4	0.050
X_3X_4	0.050

■ **TABLE 15.24**
The Alias Matrix

Effect	12	13	14	23	24	34
Intercept	0.417	0.066	0.007	0.084	0.005	0
Viscosity	−0.19	−0.2	−0.15	−0.24	−0.18	
Molecular weight	0.094	0.252	0.33	0.387	0.415	0
X_3	0.011	−0.01	0.008	−0.14	−0.02	0
X_4	0.063	0.019	0.005	0	−0.12	0
X_3X_4	−0.09	−0.03	−0.04	−0.04	−0.042	1

15.4 Repeated Measures

In experimental work in the social and behavioral sciences and some aspects of engineering the physical sciences, and business, the experimental units are frequently people. Because of differences in experience, training, or background, the differences in the responses of different people to the same treatment may be very large in some experimental situations. Unless it is controlled, this variability between people would become part of the experimental error, and in some cases, it would significantly inflate the error mean square, making it more difficult to detect real differences between treatments. In many repeated measures the experimental units are not necessarily people; they could be different stores in a marketing study, or plants, or experimental animals, and so on. We typically think of these experimental units as **subjects**.

It is possible to control this variability between people or "projects" by using a design in which each of the a treatments is used on each person. Such a design is called a **repeated measures design**. In this section, we give a brief introduction to repeated measures experiments with a single factor.

Suppose that an experiment involves a treatments and every treatment is to be used exactly once on each of n subjects. The data would appear as in Table 15.25. Note that the observation y_{ij} represents the response of subject j to treatment i and that only n subjects are used. The model that we use for this design is

$$y_{ij} = \mu + \tau_i + \beta_j + \epsilon_{ij} \tag{15.45}$$

■ **TABLE 15.25**
Data for a Single-Factor Repeated Measures Design

	Subject				Treatment
Treatment	1	2	...	n	Totals
1	y_{11}	y_{12}	...	y_{1n}	$y_{1.}$
2	y_{21}	y_{22}	...	y_{2n}	$y_{2.}$
⋮	⋮	⋮		⋮	⋮
a	y_{a1}	y_{a2}		y_{an}	$y_{a.}$
Subject Totals	$y_{.1}$	$y_{.2}$	...	$y_{.n}$	$y_{..}$

where τ_i is the effect of the ith treatment and β_j is a parameter associated with the jth subject. We assume that treatments are fixed (so $\sum_{i=1}^{a}\tau_i = 0$) and that the subjects employed are a random sample of subjects from some larger population of potential subjects. Thus, the subjects collectively represent a random effect, so we assume that the mean of β_j is zero and that the variance of β_j is σ_β^2. Because the term β_j is common to all a measurements on the same subject, the covariance between y_{ij} and $y_{i'j}$ is not, in general, zero. It is customary to assume that the covariance between y_{ij} and $y_{i'j}$ is constant across all treatments and subjects.

Consider an analysis of variance partitioning of the total sum of squares, say

$$\sum_{i=1}^{a}\sum_{j=1}^{n}(y_{ij}-\bar{y}_{..})^2 = a\sum_{j=1}^{n}(\bar{y}_{.j}-\bar{y}_{..})^2 + \sum_{i=1}^{a}\sum_{j=1}^{n}(y_{ij}-\bar{y}_{.j})^2 \tag{15.46}$$

We may view the first term on the right-hand side of Equation 15.46 as a sum of squares that results from differences *between subjects* and the second term as a sum of squares of differences *within subjects*. That is,

$$SS_T = SS_{\text{Between Subjects}} + SS_{\text{Within Subjects}}$$

The sums of squares $SS_{\text{Between Subjects}}$ and $SS_{\text{Within Subjects}}$ are statistically independent, with degrees of freedom

$$an - 1 = (n - 1) + n(a - 1)$$

The differences within subjects depend on both differences in treatment effects and uncontrolled variability (noise or error). Therefore, we may decompose the sum of squares resulting from differences within subjects as follows:

$$\sum_{i=1}^{a}\sum_{j=1}^{n}(y_{ij}-\bar{y}_{.j})^2 = n\sum_{i=1}^{a}(\bar{y}_{i.}-\bar{y}_{..})^2 + \sum_{i=1}^{a}\sum_{j=1}^{n}(y_{ij}-\bar{y}_{i.}-\bar{y}_{.j}+\bar{y}_{..})^2 \tag{15.47}$$

The first term on the right-hand side of Equation 15.47 measures the contribution of the difference between treatment means to $SS_{\text{Within Subjects}}$, and the second term is the residual variation due to error. Both components of $SS_{\text{Within Subjects}}$ are independent. Thus,

$$SS_{\text{Within Subjects}} = SS_{\text{Treatments}} + SS_E$$

with the degrees of freedom given by

$$n(a - 1) = (a - 1) + (a - 1)(n - 1)$$

respectively.

To test the hypothesis of no treatment effect, that is,

$$H_0: \tau_1 = \tau_2 = \cdots = \tau_a = 0$$
$$H_1: \text{At least one } \tau_i \neq 0$$

we would use the ratio

$$F_0 = \frac{SS_{\text{Treatment}}/(a-1)}{SS_E/(a-1)(n-1)} = \frac{MS_{\text{Treatments}}}{MS_E} \tag{15.48}$$

If the model errors are normally distributed, then under the null hypothesis, H_0: $\tau_i = 0$, the statistic F_0 follows an $F_{a-1,(a-1)(n-1)}$ distribution. The null hypothesis would be rejected if $F_0 > F_{\alpha,a-1,(a-1)(n-1)}$.

■ **TABLE 15.26**
Analysis of Variance for a Single-Factor Repeated Measures Design

Source of Variation	Sums of Squares	Degrees of Freedom	Mean Square	F_0
1. Between subjects	$\sum_{j=1}^{n} \frac{y_{.j}^2}{a} - \frac{y_{..}^2}{an}$	$n - 1$		
2. Within subjects	$\sum_{i=1}^{a}\sum_{j=1}^{n} y_{ij}^2 - \sum_{j=1}^{n} \frac{y_{.j}^2}{a}$	$n(a - 1)$		
3. Treatments	$\sum_{i=1}^{a} \frac{y_{i.}^2}{n} - \frac{y_{..}^2}{an}$	$a - 1$	$MS_{\text{Treatment}} = \frac{SS_{\text{Treatment}}}{a-1}$	$\frac{MS_{\text{Treatment}}}{MS_E}$
4. Error	Subtraction: line (2) − line (3)	$(a - 1)(n - 1)$	$MS_E = \frac{SS_E}{(a-1)(n-1)}$	
5. Total	$\sum_{i=1}^{a}\sum_{j=1}^{n} y_{ij}^2 - \frac{y_{..}^2}{an}$	$an - 1$		

The analysis of variance procedure is summarized in Table 15.26, which also gives convenient computing formulas for the sums of squares. Readers should recognize the analysis of variance for a single-factor design with repeated measures as equivalent to the analysis for a randomized complete block design, with subjects considered to be the blocks. Residual analysis and model adequacy checking is done exactly as in the RCBO.

15.5 Problems

15.1. Reconsider the experiment in Problem 5.25. Use the Box–Cox procedure to determine whether a transformation on the response is appropriate (or useful) in the analysis of the data from this experiment.

15.2. In Example 6.3 we selected a log transformation for the drill advance rate response. Use the Box–Cox procedure to demonstrate that this is an appropriate data transformation.

15.3. Reconsider the smelting process experiment in Problem 8.27, where a 2^{6-3} fractional factorial design was used to study the weight of packing material that is stuck to carbon anodes after baking. Each of the eight runs in the design was replicated three times, and both the average weight and the range of the weights at each test combination were treated as response variables. Is there any indication that a transformation is required for either response?

15.4. In Problem 8.28 a replicated fractional factorial design was used to study substrate camber in semiconductor manufacturing. Both the mean and standard deviation of the camber measurements were used as response variables. Is there any indication that a transformation is required for either response?

15.5. Reconsider the photoresist experiment in Problem 8.29. Use the variance of the resist thickness at each test combination as the response variable. Is there any indication that a transformation is required?

15.6. In the grill defects experiment described in Problem 8.51, a variation of the square root transformation was employed in the analysis of the data. Use the Box–Cox method to determine whether this is the appropriate transformation.

15.7. In the experimental design of Problem 12.11, two responses were obtained, the mean and variance of an oxide thickness. Use the Box–Cox method to investigate the potential usefulness of transformation for both of these responses. Is the log transformation suggested in part (c) of that problem appropriate?

15.8. In the 3^3 factorial design of Problem 12.11 one of the responses is a standard deviation. Use the Box–Cox method to investigate the usefulness of transformations for this response. Would your answer change if we used the variance as the response?

15.9. Problem 12.9 suggests using ln (s^2) as the response [refer to part (b)]. Does the Box–Cox method indicate that a transformation is appropriate?

15.10. Myers, Montgomery, Vining and Robinson (2010) describe an experiment to study spermatozoa survival. The design factors are the amount of sodium citrate, the amount of glycerol, and equilibrium time, each at two levels. The response variable is the number of spermatozoa that survive out of 50 that were tested at each set of conditions. The data are shown in the following table:

Sodium Citrate	Glycerol	Equilibrium Time	Number Survived
−	−	−	34
+	−	−	20
−	+	−	8
+	+	−	21
−	−	+	30
+	−	+	20
−	+	+	10
+	+	+	25

Analyze the data from this experiment with logistic regression.

15.11. A soft drink distributor is studying the effectiveness of delivery methods. Three different types of hand trucks have been developed, and an experiment is performed in the company's methods engineering laboratory. The variable of interest is the delivery time in minutes (y); however, delivery time is also strongly related to the case volume delivered (x). Each hand truck is used four times and the data that follow are obtained. Analyze these data and draw appropriate conclusions. Use $\alpha = 0.05$.

Hand Truck Type					
1		2		3	
y	x	y	x	y	x
28	23	24	26	40	37
45	40	35	31	22	25
33	36	46	42	52	50
41	42	26	25	17	20

15.12. Compute the adjusted treatment means and the standard errors of the adjusted treatment means for the data in Problem 15.11.

15.13. The sums of squares and products for a single-factor analysis of covariance follow. Complete the analysis and draw appropriate conclusions. Use $\alpha = 0.05$.

Source of Variation	Degrees of Freedom	Sums of Squares and Products x	xy	y
Treatment	3	1500	1000	650
Error	12	6000	1200	550
Total	15	7500	2200	1200

15.14. Find the standard errors of the adjusted treatment means in Example 15.5.

15.15. Four different formulations of an industrial glue are being tested. The tensile strength of the glue when it is applied to join parts is also related to the application thickness. Five observations on strength (y) in pounds and thickness (x) in 0.01 inches are obtained for each formulation. The data are shown in the following table. Analyze these data and draw appropriate conclusions.

Glue Formulation							
1		2		3		4	
y	x	y	x	y	x	y	x
46.5	13	48.7	12	46.3	15	44.7	16
45.9	14	49.0	10	47.1	14	43.0	15
49.8	12	50.1	11	48.9	11	51.0	10
46.1	12	48.5	12	48.2	11	48.1	12
44.3	14	45.2	14	50.3	10	48.6	11

15.16. Compute the adjusted treatment means and their standard errors using the data in Problem 15.15.

15.17. An engineer is studying the effect of cutting speed on the rate of metal removal in a machining operation. However, the rate of metal removal is also related to the hardness of the test specimen. Five observations are taken at each cutting speed. The amount of metal removed (y) and the hardness of the specimen (x) are shown in the following table. Analyze the data using an analysis of covariance. Use $\alpha = 0.05$.

Cutting Speed (rpm)					
1000		1200		1400	
y	x	y	x	y	x
68	120	112	165	118	175
90	140	94	140	82	132
98	150	65	120	73	124
77	125	74	125	92	141
88	136	85	133	80	130

15.18. Show that in a single-factor analysis of covariance with a single covariate a $100(1-\alpha)$ percent confidence interval on the ith adjusted treatment mean is

$$\bar{y}_{i.} - \hat{\beta}(\bar{x}_{i.} - \bar{x}_{..}) \pm t_{\alpha/2, a(n-1)-1}$$

$$\left[MS_E\left(\frac{1}{n} + \frac{(\bar{x}_{i.} - \bar{x}_{..})^2}{E_{xx}}\right)\right]^{1/2}$$

Using this formula, calculate a 95 percent confidence interval on the adjusted mean of machine 1 in Example 15.5.

15.19. Show that in a single-factor analysis of covariance with a single covariate, the standard error of the difference between any two adjusted treatment means is

$$S_{\text{Adj}\bar{y}_{i.} - \text{Adj}\bar{y}_{j.}} = \left[MS_E\left(\frac{2}{n} + \frac{(\bar{x}_{i.} - \bar{x}_{j.})^2}{E_{xx}}\right)\right]^{1/2}$$

15.20. Discuss how the operating characteristic curves for the analysis of variance can be used in the analysis of covariance.

15.21. Three different Pinot Noir wines were evaluated by a panel of eight judges. The judges are considered a random panel of all possible judges. The wines are evaluated on a 100-point scale. The wines were presented in random order to each judge, and the following results obtained.

	Wine		
Judge	**1**	**2**	**3**
1	85	88	93
2	90	89	94
3	88	90	98
4	91	93	96
5	92	92	95
6	89	90	95
7	90	91	97
8	91	89	98

Analyze the data from this experiment. Is there a difference in wine quality? Analyze the residuals and comment on model adequacy.

Appendix

I Cumulative Standard Normal Distribution[a]

$$\Phi(z) = \int_{-\infty}^{z} \frac{1}{\sqrt{2\pi}} e^{-u^2/2}\, du$$

z	0.00	0.01	0.02	0.03	0.04	z
0.0	0.50000	0.50399	0.50798	0.51197	0.51595	0.0
0.1	0.53983	0.54379	0.54776	0.55172	0.55567	0.1
0.2	0.57926	0.58317	0.58706	0.59095	0.59483	0.2
0.3	0.61791	0.62172	0.62551	0.62930	0.63307	0.3
0.4	0.65542	0.65910	0.66276	0.66640	0.67003	0.4
0.5	0.69146	0.69497	0.69847	0.70194	0.70540	0.5
0.6	0.72575	0.72907	0.73237	0.73565	0.73891	0.6
0.7	0.75803	0.76115	0.76424	0.76730	0.77035	0.7
0.8	0.78814	0.79103	0.79389	0.79673	0.79954	0.8
0.9	0.81594	0.81859	0.82121	0.82381	0.82639	0.9
1.0	0.84134	0.84375	0.84613	0.84849	0.85083	1.0
1.1	0.86433	0.86650	0.86864	0.87076	0.87285	1.1
1.2	0.88493	0.88686	0.88877	0.89065	0.89251	1.2
1.3	0.90320	0.90490	0.90658	0.90824	0.90988	1.3
1.4	0.91924	0.92073	0.92219	0.92364	0.92506	1.4
1.5	0.93319	0.93448	0.93574	0.93699	0.93822	1.5
1.6	0.94520	0.94630	0.94738	0.94845	0.94950	1.6
1.7	0.95543	0.95637	0.95728	0.95818	0.95907	1.7
1.8	0.96407	0.96485	0.96562	0.96637	0.96711	1.8
1.9	0.97128	0.97193	0.97257	0.97320	0.97381	1.9
2.0	0.97725	0.97778	0.97831	0.97882	0.97932	2.0
2.1	0.98214	0.98257	0.98300	0.98341	0.93882	2.1
2.2	0.98610	0.98645	0.98679	0.98713	0.98745	2.2
2.3	0.98928	0.98956	0.98983	0.99010	0.99036	2.3
2.4	0.99180	0.99202	0.99224	0.99245	0.99266	2.4
2.5	0.99379	0.99396	0.99413	0.99430	0.99446	2.5
2.6	0.99534	0.99547	0.99560	0.99573	0.99585	2.6
2.7	0.99653	0.99664	0.99674	0.99683	0.99693	2.7
2.8	0.99744	0.99752	0.99760	0.99767	0.99774	2.8
2.9	0.99813	0.99819	0.99825	0.99831	0.99836	2.9
3.0	0.99865	0.99869	0.99874	0.99878	0.99882	3.0
3.1	0.99903	0.99906	0.99910	0.99913	0.99916	3.1
3.2	0.99931	0.99934	0.99936	0.99938	0.99940	3.2
3.3	0.99952	0.99953	0.99955	0.99957	0.99958	3.3
3.4	0.99966	0.99968	0.99969	0.99970	0.99971	3.4
3.5	0.99977	0.99978	0.99978	0.99979	0.99980	3.5
3.6	0.99984	0.99985	0.99985	0.99986	0.99986	3.6
3.7	0.99989	0.99990	0.99990	0.99990	0.99991	3.7
3.8	0.99993	0.99993	0.99993	0.99994	0.99994	3.8
3.9	0.99995	0.99995	0.99996	0.99996	0.99996	3.9

[a]Reproduced with permission from *Probability and Statistics in Engineering and Management Science*, 3rd edition, by W. W. Hines and D. C. Montgomery, Wiley, New York, 1990.

I Cumulative Standard Normal Distribution (*Continued*)

$$\Phi(z) = \int_{-\infty}^{z} \frac{1}{\sqrt{2\pi}} e^{-u^2/2}\, du$$

z	**0.05**	**0.06**	**0.07**	**0.08**	**0.09**	*z*
0.0	0.51994	0.52392	0.52790	0.53188	0.53586	0.0
0.1	0.55962	0.56356	0.56749	0.57142	0.57534	0.1
0.2	0.59871	0.60257	0.60642	0.61026	0.61409	0.2
0.3	0.63683	0.64058	0.64431	0.64803	0.65173	0.3
0.4	0.67364	0.67724	0.68082	0.68438	0.68793	0.4
0.5	0.70884	0.71226	0.71566	0.71904	0.72240	0.5
0.6	0.74215	0.74537	0.74857	0.75175	0.75490	0.6
0.7	0.77337	0.77637	0.77935	0.78230	0.78523	0.7
0.8	0.80234	0.80510	0.80785	0.81057	0.81327	0.8
0.9	0.82894	0.83147	0.83397	0.83646	0.83891	0.9
1.0	0.85314	0.85543	0.85769	0.85993	0.86214	1.0
1.1	0.87493	0.87697	0.87900	0.88100	0.88297	1.1
1.2	0.89435	0.89616	0.89796	0.89973	0.90147	1.2
1.3	0.91149	0.91308	0.91465	0.91621	0.91773	1.3
1.4	0.92647	0.92785	0.92922	0.93056	0.93189	1.4
1.5	0.93943	0.90462	0.94179	0.94295	0.94408	1.5
1.6	0.95053	0.95154	0.95254	0.95352	0.95448	1.6
1.7	0.95994	0.96080	0.96164	0.96246	0.96327	1.7
1.8	0.96784	0.96856	0.96926	0.96995	0.97062	1.8
1.9	0.97441	0.97500	0.97558	0.97615	0.97670	1.9
2.0	0.97982	0.98030	0.98077	0.98124	0.98169	2.0
2.1	0.98422	0.98461	0.98500	0.98537	0.98574	2.1
2.2	0.98778	0.98809	0.98840	0.98870	0.98899	2.2
2.3	0.99061	0.99086	0.99111	0.99134	0.99158	2.3
2.4	0.99286	0.99305	0.99324	0.99343	0.99361	2.4
2.5	0.99461	0.99477	0.99492	0.99506	0.99520	2.5
2.6	0.99598	0.99609	0.99621	0.99632	0.99643	2.6
2.7	0.99702	0.99711	0.99720	0.99728	0.99736	2.7
2.8	0.99781	0.99788	0.99795	0.99801	0.99807	2.8
2.9	0.99841	0.99846	0.99851	0.99856	0.99861	2.9
3.0	0.99886	0.99889	0.99893	0.99897	0.99900	3.0
3.1	0.99918	0.99921	0.99924	0.99926	0.99929	3.1
3.2	0.99942	0.99944	0.99946	0.99948	0.99950	3.2
3.3	0.99960	0.99961	0.99962	0.99964	0.99965	3.3
3.4	0.99972	0.99973	0.99974	0.99975	0.99976	3.4
3.5	0.99981	0.99981	0.99982	0.99983	0.99983	3.5
3.6	0.99987	0.99987	0.99988	0.99988	0.99989	3.6
3.7	0.99991	0.99992	0.99992	0.99992	0.99992	3.7
3.8	0.99994	0.99994	0.99995	0.99995	0.99995	3.8
3.9	0.99996	0.99996	0.99996	0.99997	0.99997	3.9

II Percentage Points of the t Distribution[a]

ν \ α	0.40	0.25	0.10	0.05	0.025	0.01	0.005	0.0025	0.001	0.0005
1	0.325	1.000	3.078	6.314	12.706	31.821	63.657	127.32	318.31	636.62
2	0.289	0.816	1.886	2.920	4.303	6.965	9.925	14.089	23.326	31.598
3	0.277	0.765	1.638	2.353	3.182	4.541	5.841	7.453	10.213	12.924
4	0.271	0.741	1.533	2.132	2.776	3.747	4.604	5.598	7.173	8.610
5	0.267	0.727	1.476	2.015	2.571	3.365	4.032	4.773	5.893	6.869
6	0.265	0.727	1.440	1.943	2.447	3.143	3.707	4.317	5.208	5.959
7	0.263	0.711	1.415	1.895	2.365	2.998	3.499	4.019	4.785	5.408
8	0.262	0.706	1.397	1.860	2.306	2.896	3.355	3.833	4.501	5.041
9	0.261	0.703	1.383	1.833	2.262	2.821	3.250	3.690	4.297	4.781
10	0.260	0.700	1.372	1.812	2.228	2.764	3.169	3.581	4.144	4.587
11	0.260	0.697	1.363	1.796	2.201	2.718	3.106	3.497	4.025	4.437
12	0.259	0.695	1.356	1.782	2.179	2.681	3.055	3.428	3.930	4.318
13	0.259	0.694	1.350	1.771	2.160	2.650	3.012	3.372	3.852	4.221
14	0.258	0.692	1.345	1.761	2.145	2.624	2.977	3.326	3.787	4.140
15	0.258	0.691	1.341	1.753	2.131	2.602	2.947	3.286	3.733	4.073
16	0.258	0.690	1.337	1.746	2.120	2.583	2.921	3.252	3.686	4.015
17	0.257	0.689	1.333	1.740	2.110	2.567	2.898	3.222	3.646	3.965
18	0.257	0.688	1.330	1.734	2.101	2.552	2.878	3.197	3.610	3.922
19	0.257	0.688	1.328	1.729	2.093	2.539	2.861	3.174	3.579	3.883
20	0.257	0.687	1.325	1.725	2.086	2.528	2.845	3.153	3.552	3.850
21	0.257	0.686	1.323	1.721	2.080	2.518	2.831	3.135	3.527	3.819
22	0.256	0.686	1.321	1.717	2.074	2.508	2.819	3.119	3.505	3.792
23	0.256	0.685	1.319	1.714	2.069	2.500	2.807	3.104	3.485	3.767
24	0.256	0.685	1.318	1.711	2.064	2.492	2.797	3.091	3.467	3.745
25	0.256	0.684	1.316	1.708	2.060	2.485	2.787	3.078	3.450	3.725
26	0.256	0.684	1.315	1.706	2.056	2.479	2.779	3.067	3.435	3.707
27	0.256	0.684	1.314	1.703	2.052	2.473	2.771	3.057	3.421	3.690
28	0.256	0.683	1.313	1.701	2.048	2.467	2.763	3.047	3.408	3.674
29	0.256	0.683	1.311	1.699	2.045	2.462	2.756	3.038	3.396	3.659
30	0.256	0.683	1.310	1.697	2.042	2.457	2.750	3.030	3.385	3.646
40	0.255	0.681	1.303	1.684	2.021	2.423	2.704	2.971	3.307	3.551
60	0.254	0.679	1.296	1.671	2.000	2.390	2.660	2.915	3.232	3.460
120	0.254	0.677	1.289	1.658	1.980	2.358	2.617	2.860	3.160	3.373
∞	0.253	0.674	1.282	1.645	1.960	2.326	2.576	2.807	3.090	3.291

ν = Degrees of freedom.

[a]Adapted with permission from *Biometrika Tables for Statisticians*, Vol. 1, 3rd edition, by E. S. Pearson and H. O. Hartley, Cambridge University Press, Cambridge, 1966.

III Percentage Points of the χ^2 Distribution[a]

ν \ α	0.995	0.990	0.975	0.950	0.500	0.050	0.025	0.010	0.005
1	0.00 +	0.00 +	0.00 +	0.00 +	0.45	3.84	5.02	6.63	7.88
2	0.01	0.02	0.05	0.10	1.39	5.99	7.38	9.21	10.60
3	0.07	0.11	0.22	0.35	2.37	7.81	9.35	11.34	12.84
4	0.21	0.30	0.48	0.71	3.36	9.49	11.14	13.28	14.86
5	0.41	0.55	0.83	1.15	4.35	11.07	12.38	15.09	16.75
6	0.68	0.87	1.24	1.64	5.35	12.59	14.45	16.81	18.55
7	0.99	1.24	1.69	2.17	6.35	14.07	16.01	18.48	20.28
8	1.34	1.65	2.18	2.73	7.34	15.51	17.53	20.09	21.96
9	1.73	2.09	2.70	3.33	8.34	16.92	19.02	21.67	23.59
10	2.16	2.56	3.25	3.94	9.34	18.31	20.48	23.21	25.19
11	2.60	3.05	3.82	4.57	10.34	19.68	21.92	24.72	26.76
12	3.07	3.57	4.40	5.23	11.34	21.03	23.34	26.22	28.30
13	3.57	4.11	5.01	5.89	12.34	22.36	24.74	27.69	29.82
14	4.07	4.66	5.63	6.57	13.34	23.68	26.12	29.14	31.32
15	4.60	5.23	6.27	7.26	14.34	25.00	27.49	30.58	32.80
16	5.14	5.81	6.91	7.96	15.34	26.30	28.85	32.00	34.27
17	5.70	6.41	7.56	8.67	16.34	27.59	30.19	33.41	35.72
18	6.26	7.01	8.23	9.39	17.34	28.87	31.53	34.81	37.16
19	6.84	7.63	8.91	10.12	18.34	30.14	32.85	36.19	38.58
20	7.43	8.26	9.59	10.85	19.34	31.41	34.17	37.57	40.00
25	10.52	11.52	13.12	14.61	24.34	37.65	40.65	44.31	46.93
30	13.79	14.95	16.79	18.49	29.34	43.77	46.98	50.89	53.67
40	20.71	22.16	24.43	26.51	39.34	55.76	59.34	63.69	66.77
50	27.99	29.71	32.36	34.76	49.33	67.50	71.42	76.15	79.49
60	35.53	37.48	40.48	43.19	59.33	79.08	83.30	88.38	91.95
70	43.28	45.44	48.76	51.74	69.33	90.53	95.02	100.42	104.22
80	51.17	53.54	57.15	60.39	79.33	101.88	106.63	112.33	116.32
90	59.20	61.75	65.65	69.13	89.33	113.14	118.14	124.12	128.30
100	67.33	70.06	74.22	77.93	99.33	124.34	129.56	135.81	140.17

ν = Degrees of freedom.

[a]Adapted with permission from *Biometrika Tables for Statisticians*, Vol. 1, 3rd edition by E. S. Pearson and H. O. Hartley, Cambridge University Press, Cambridge, 1966.

IV Percentage Points of the F Distribution[a]

$F_{0.25,\nu_1,\nu_2}$

Degrees of Freedom for the Numerator (ν_1); rows: Degrees of Freedom for the Denominator (ν_2)

ν_2 \ ν_1	1	2	3	4	5	6	7	8	9	10	12	15	20	24	30	40	60	120	∞
1	5.83	7.50	8.20	8.58	8.82	8.98	9.10	9.19	9.26	9.32	9.41	9.49	9.58	9.63	9.67	9.71	9.76	9.80	9.85
2	2.57	3.00	3.15	3.23	3.28	3.31	3.34	3.35	3.37	3.38	3.39	3.41	3.43	3.43	3.44	3.45	3.46	3.47	3.48
3	2.02	2.28	2.36	2.39	2.41	2.42	2.43	2.44	2.44	2.44	2.45	2.46	2.46	2.46	2.47	2.47	2.47	2.47	2.47
4	1.81	2.00	2.05	2.06	2.07	2.08	2.08	2.08	2.08	2.08	2.08	2.08	2.08	2.08	2.08	2.08	2.08	2.08	2.08
5	1.69	1.85	1.88	1.89	1.89	1.89	1.89	1.89	1.89	1.89	1.89	1.89	1.88	1.88	1.88	1.88	1.87	1.87	1.87
6	1.62	1.76	1.78	1.79	1.79	1.78	1.78	1.78	1.77	1.77	1.77	1.76	1.76	1.75	1.75	1.75	1.74	1.74	1.74
7	1.57	1.70	1.72	1.72	1.71	1.71	1.70	1.70	1.70	1.69	1.68	1.68	1.67	1.67	1.66	1.66	1.65	1.65	1.65
8	1.54	1.66	1.67	1.66	1.66	1.65	1.64	1.64	1.63	1.63	1.62	1.62	1.61	1.60	1.60	1.59	1.59	1.58	1.58
9	1.51	1.62	1.63	1.63	1.62	1.61	1.60	1.60	1.59	1.59	1.58	1.57	1.56	1.56	1.55	1.54	1.54	1.53	1.53
10	1.49	1.60	1.60	1.59	1.59	1.58	1.57	1.56	1.56	1.55	1.54	1.53	1.52	1.52	1.51	1.51	1.50	1.49	1.48
11	1.47	1.58	1.58	1.57	1.56	1.55	1.54	1.53	1.53	1.52	1.51	1.50	1.49	1.49	1.48	1.47	1.47	1.46	1.45
12	1.46	1.56	1.56	1.55	1.54	1.53	1.52	1.51	1.51	1.50	1.49	1.48	1.47	1.46	1.45	1.45	1.44	1.43	1.42
13	1.45	1.55	1.55	1.53	1.52	1.51	1.50	1.49	1.49	1.48	1.47	1.46	1.45	1.44	1.43	1.42	1.42	1.41	1.40
14	1.44	1.53	1.53	1.52	1.51	1.50	1.49	1.48	1.47	1.46	1.45	1.44	1.43	1.42	1.41	1.41	1.40	1.39	1.38
15	1.43	1.52	1.52	1.51	1.49	1.48	1.47	1.46	1.46	1.45	1.44	1.43	1.41	1.41	1.40	1.39	1.38	1.37	1.36
16	1.42	1.51	1.51	1.50	1.48	1.47	1.46	1.45	1.44	1.44	1.43	1.41	1.40	1.39	1.38	1.37	1.36	1.35	1.34
17	1.42	1.51	1.50	1.49	1.47	1.46	1.45	1.44	1.43	1.43	1.41	1.40	1.39	1.38	1.37	1.36	1.35	1.34	1.33
18	1.41	1.50	1.49	1.48	1.46	1.45	1.44	1.43	1.42	1.42	1.40	1.39	1.38	1.37	1.36	1.35	1.34	1.33	1.32
19	1.41	1.49	1.49	1.47	1.46	1.44	1.43	1.42	1.41	1.41	1.40	1.38	1.37	1.36	1.35	1.34	1.33	1.32	1.30
20	1.40	1.49	1.48	1.47	1.45	1.44	1.43	1.42	1.41	1.40	1.39	1.37	1.36	1.35	1.34	1.33	1.32	1.31	1.29
21	1.40	1.48	1.48	1.46	1.44	1.43	1.42	1.41	1.40	1.39	1.38	1.37	1.35	1.34	1.33	1.32	1.31	1.30	1.28
22	1.40	1.48	1.47	1.45	1.44	1.42	1.41	1.40	1.39	1.39	1.37	1.36	1.34	1.33	1.32	1.31	1.30	1.29	1.28
23	1.39	1.47	1.47	1.45	1.43	1.42	1.41	1.40	1.39	1.38	1.37	1.35	1.34	1.33	1.32	1.31	1.30	1.28	1.27
24	1.39	1.47	1.46	1.44	1.43	1.41	1.40	1.39	1.38	1.38	1.36	1.35	1.33	1.32	1.31	1.30	1.29	1.28	1.26
25	1.39	1.47	1.46	1.44	1.42	1.41	1.40	1.39	1.38	1.37	1.36	1.34	1.33	1.32	1.31	1.29	1.28	1.27	1.25
26	1.38	1.46	1.45	1.44	1.42	1.41	1.39	1.38	1.37	1.37	1.35	1.34	1.32	1.31	1.30	1.29	1.28	1.26	1.25
27	1.38	1.46	1.45	1.43	1.42	1.40	1.39	1.38	1.37	1.36	1.35	1.33	1.32	1.31	1.30	1.28	1.27	1.26	1.24
28	1.38	1.46	1.45	1.43	1.41	1.40	1.39	1.38	1.37	1.36	1.34	1.33	1.31	1.30	1.29	1.28	1.27	1.25	1.24
29	1.38	1.45	1.45	1.43	1.41	1.40	1.38	1.37	1.36	1.35	1.34	1.32	1.31	1.30	1.29	1.27	1.26	1.25	1.23
30	1.38	1.45	1.44	1.42	1.41	1.39	1.38	1.37	1.36	1.35	1.34	1.32	1.30	1.29	1.28	1.27	1.26	1.24	1.23
40	1.36	1.44	1.42	1.40	1.39	1.37	1.36	1.35	1.34	1.33	1.31	1.30	1.28	1.26	1.25	1.24	1.22	1.21	1.19
60	1.35	1.42	1.41	1.38	1.37	1.35	1.33	1.32	1.31	1.30	1.29	1.27	1.25	1.24	1.22	1.21	1.19	1.17	1.15
120	1.34	1.40	1.39	1.37	1.35	1.33	1.31	1.30	1.29	1.28	1.26	1.24	1.22	1.21	1.19	1.18	1.16	1.13	1.10
∞	1.32	1.39	1.37	1.35	1.33	1.31	1.29	1.28	1.27	1.25	1.24	1.22	1.19	1.18	1.16	1.14	1.12	1.08	1.00

ν = Degrees of freedom.

[a] Adapted from Biometrika. E.S. Pearson and H.O. Hartley. *Biometrika Tables for Statisticians*, Vol. 1, 3rd edition (1966) by permission of Oxford University Press.

IV Percentage Points of the F Distribution (*Continued*)

$F_{0.10,\nu_1,\nu_2}$

Degrees of Freedom for the Numerator (ν_1); Degrees of Freedom for the Denominator (ν_2)

ν_2 \ ν_1	1	2	3	4	5	6	7	8	9	10	12	15	20	24	30	40	60	120	∞
1	39.86	49.50	53.59	55.83	57.24	58.20	58.91	59.44	59.86	60.19	60.71	61.22	61.74	62.00	62.26	62.53	62.79	63.06	63.33
2	8.53	9.00	9.16	9.24	9.29	9.33	9.35	9.37	9.38	9.39	9.41	9.42	9.44	9.45	9.46	9.47	9.47	9.48	9.49
3	5.54	5.46	5.39	5.34	5.31	5.28	5.27	5.25	5.24	5.23	5.22	5.20	5.18	5.18	5.17	5.16	5.15	5.14	5.13
4	4.54	4.32	4.19	4.11	4.05	4.01	3.98	3.95	3.94	3.92	3.90	3.87	3.84	3.83	3.82	3.80	3.79	3.78	3.76
5	4.06	3.78	3.62	3.52	3.45	3.40	3.37	3.34	3.32	3.30	3.27	3.24	3.21	3.19	3.17	3.16	3.14	3.12	3.10
6	3.78	3.46	3.29	3.18	3.11	3.05	3.01	2.98	2.96	2.94	2.90	2.87	2.84	2.82	2.80	2.78	2.76	2.74	2.72
7	3.59	3.26	3.07	2.96	2.88	2.83	2.78	2.75	2.72	2.70	2.67	2.63	2.59	2.58	2.56	2.54	2.51	2.49	2.47
8	3.46	3.11	2.92	2.81	2.73	2.67	2.62	2.59	2.56	2.54	2.50	2.46	2.42	2.40	2.38	2.36	2.34	2.32	2.29
9	3.36	3.01	2.81	2.69	2.61	2.55	2.51	2.47	2.44	2.42	2.38	2.34	2.30	2.28	2.25	2.23	2.21	2.18	2.16
10	3.29	2.92	2.73	2.61	2.52	2.46	2.41	2.38	2.35	2.32	2.28	2.24	2.20	2.18	2.16	2.13	2.11	2.08	2.06
11	3.23	2.86	2.66	2.54	2.45	2.39	2.34	2.30	2.27	2.25	2.21	2.17	2.12	2.10	2.08	2.05	2.03	2.00	1.97
12	3.18	2.81	2.61	2.48	2.39	2.33	2.28	2.24	2.21	2.19	2.15	2.10	2.06	2.04	2.01	1.99	1.96	1.93	1.90
13	3.14	2.76	2.56	2.43	2.35	2.28	2.23	2.20	2.16	2.14	2.10	2.05	2.01	1.98	1.96	1.93	1.90	1.88	1.85
14	3.10	2.73	2.52	2.39	2.31	2.24	2.19	2.15	2.12	2.10	2.05	2.01	1.96	1.94	1.91	1.89	1.86	1.83	1.80
15	3.07	2.70	2.49	2.36	2.27	2.21	2.16	2.12	2.09	2.06	2.02	1.97	1.92	1.90	1.87	1.85	1.82	1.79	1.76
16	3.05	2.67	2.46	2.33	2.24	2.18	2.13	2.09	2.06	2.03	1.99	1.94	1.89	1.87	1.84	1.81	1.78	1.75	1.72
17	3.03	2.64	2.44	2.31	2.22	2.15	2.10	2.06	2.03	2.00	1.96	1.91	1.86	1.84	1.81	1.78	1.75	1.72	1.69
18	3.01	2.62	2.42	2.29	2.20	2.13	2.08	2.04	2.00	1.98	1.93	1.89	1.84	1.81	1.78	1.75	1.72	1.69	1.66
19	2.99	2.61	2.40	2.27	2.18	2.11	2.06	2.02	1.98	1.96	1.91	1.86	1.81	1.79	1.76	1.73	1.70	1.67	1.63
20	2.97	2.59	2.38	2.25	2.16	2.09	2.04	2.00	1.96	1.94	1.89	1.84	1.79	1.77	1.74	1.71	1.68	1.64	1.61
21	2.96	2.57	2.36	2.23	2.14	2.08	2.02	1.98	1.95	1.92	1.87	1.83	1.78	1.75	1.72	1.69	1.66	1.62	1.59
22	2.95	2.56	2.35	2.22	2.13	2.06	2.01	1.97	1.93	1.90	1.86	1.81	1.76	1.73	1.70	1.67	1.64	1.60	1.57
23	2.94	2.55	2.34	2.21	2.11	2.05	1.99	1.96	1.92	1.89	1.84	1.80	1.74	1.72	1.69	1.66	1.62	1.59	1.55
24	2.93	2.54	2.33	2.19	2.10	2.04	1.98	1.94	1.91	1.88	1.83	1.78	1.73	1.70	1.67	1.64	1.61	1.57	1.53
25	2.92	2.53	2.32	2.18	2.09	2.02	1.97	1.93	1.89	1.87	1.82	1.77	1.72	1.69	1.66	1.63	1.59	1.56	1.52
26	2.91	2.52	2.31	2.17	2.08	2.01	1.96	1.92	1.88	1.86	1.81	1.76	1.71	1.68	1.65	1.61	1.58	1.54	1.50
27	2.90	2.51	2.30	2.17	2.07	2.00	1.95	1.91	1.87	1.85	1.80	1.75	1.70	1.67	1.64	1.60	1.57	1.53	1.49
28	2.89	2.50	2.29	2.16	2.06	2.00	1.94	1.90	1.87	1.84	1.79	1.74	1.69	1.66	1.63	1.59	1.56	1.52	1.48
29	2.89	2.50	2.28	2.15	2.06	1.99	1.93	1.89	1.86	1.83	1.78	1.73	1.68	1.65	1.62	1.58	1.55	1.51	1.47
30	2.88	2.49	2.28	2.14	2.03	1.98	1.93	1.88	1.85	1.82	1.77	1.72	1.67	1.64	1.61	1.57	1.54	1.50	1.46
40	2.84	2.44	2.23	2.09	2.00	1.93	1.87	1.83	1.79	1.76	1.71	1.66	1.61	1.57	1.54	1.51	1.47	1.42	1.38
60	2.79	2.39	2.18	2.04	1.95	1.87	1.82	1.77	1.74	1.71	1.66	1.60	1.54	1.51	1.48	1.44	1.40	1.35	1.29
120	2.75	2.35	2.13	1.99	1.90	1.82	1.77	1.72	1.68	1.65	1.60	1.55	1.48	1.45	1.41	1.37	1.32	1.26	1.19
∞	2.71	2.30	2.08	1.94	1.85	1.77	1.72	1.67	1.63	1.60	1.55	1.49	1.42	1.38	1.34	1.30	1.24	1.17	1.00

IV Percentage Points of the *F* Distribution (*Continued*)

$F_{0.05,\nu_1,\nu_2}$

ν_2 \ ν_1	1	2	3	4	5	6	7	8	9	10	12	15	20	24	30	40	60	120	∞
	161.4	199.5	215.7	224.6	230.2	234.0	236.8	238.9	240.5	241.9	243.9	245.9	248.0	249.1	250.1	251.1	252.2	253.3	254.3
2	18.51	19.00	19.16	19.25	19.30	19.33	19.35	19.37	19.38	19.40	19.41	19.43	19.45	19.45	19.46	19.47	19.48	19.49	19.50
3	10.13	9.55	9.28	9.12	9.01	8.94	8.89	8.85	8.81	8.79	8.74	8.70	8.66	8.64	8.62	8.59	8.57	8.55	8.53
4	7.71	6.94	6.59	6.39	6.26	6.16	6.09	6.04	6.00	5.96	5.91	5.86	5.80	5.77	5.75	5.72	5.69	5.66	5.63
5	6.61	5.79	5.41	5.19	5.05	4.95	4.88	4.82	4.77	4.74	4.68	4.62	4.56	4.53	4.50	4.46	4.43	4.40	4.36
6	5.99	5.14	4.76	4.53	4.39	4.28	4.21	4.15	4.10	4.06	4.00	3.94	3.87	3.84	3.81	3.77	3.74	3.70	3.67
7	5.59	4.74	4.35	4.12	3.97	3.87	3.79	3.73	3.68	3.64	3.57	3.51	3.44	3.41	3.38	3.34	3.30	3.27	3.23
8	5.32	4.46	4.07	3.84	3.69	3.58	3.50	3.44	3.39	3.35	3.28	3.22	3.15	3.12	3.08	3.04	3.01	2.97	2.93
9	5.12	4.26	3.86	3.63	3.48	3.37	3.29	3.23	3.18	3.14	3.07	3.01	2.94	2.90	2.86	2.83	2.79	2.75	2.71
10	4.96	4.10	3.71	3.48	3.33	3.22	3.14	3.07	3.02	2.98	2.91	2.85	2.77	2.74	2.70	2.66	2.62	2.58	2.54
11	4.84	3.98	3.59	3.36	3.20	3.09	3.01	2.95	2.90	2.85	2.79	2.72	2.65	2.61	2.57	2.53	2.49	2.45	2.40
12	4.75	3.89	3.49	3.26	3.11	3.00	2.91	2.85	2.80	2.75	2.69	2.62	2.54	2.51	2.47	2.43	2.38	2.34	2.30
13	4.67	3.81	3.41	3.18	3.03	2.92	2.83	2.77	2.71	2.67	2.60	2.53	2.46	2.42	2.38	2.34	2.30	2.25	2.21
14	4.60	3.74	3.34	3.11	2.96	2.85	2.76	2.70	2.65	2.60	2.53	2.46	2.39	2.35	2.31	2.27	2.22	2.18	2.13
15	4.54	3.68	3.29	3.06	2.90	2.79	2.71	2.64	2.59	2.54	2.48	2.40	2.33	2.29	2.25	2.20	2.16	2.11	2.07
16	4.49	3.63	3.24	3.01	2.85	2.74	2.66	2.59	2.54	2.49	2.42	2.35	2.28	2.24	2.19	2.15	2.11	2.06	2.01
17	4.45	3.59	3.20	2.96	2.81	2.70	2.61	2.55	2.49	2.45	2.38	2.31	2.23	2.19	2.15	2.10	2.06	2.01	1.96
18	4.41	3.55	3.16	2.93	2.77	2.66	2.58	2.51	2.46	2.41	2.34	2.27	2.19	2.15	2.11	2.06	2.02	1.97	1.92
19	4.38	3.52	3.13	2.90	2.74	2.63	2.54	2.48	2.42	2.38	2.31	2.23	2.16	2.11	2.07	2.03	1.98	1.93	1.88
20	4.35	3.49	3.10	2.87	2.71	2.60	2.51	2.45	2.39	2.35	2.28	2.20	2.12	2.08	2.04	1.99	1.95	1.90	1.84
21	4.32	3.47	3.07	2.84	2.68	2.57	2.49	2.42	2.37	2.32	2.25	2.18	2.10	2.05	2.01	1.96	1.92	1.87	1.81
22	4.30	3.44	3.05	2.82	2.66	2.55	2.46	2.40	2.34	2.30	2.23	2.15	2.07	2.03	1.98	1.94	1.89	1.84	1.78
23	4.28	3.42	3.03	2.80	2.64	2.53	2.44	2.37	2.32	2.27	2.20	2.13	2.05	2.01	1.96	1.91	1.86	1.81	1.76
24	4.26	3.40	3.01	2.78	2.62	2.51	2.42	2.36	2.30	2.25	2.18	2.11	2.03	1.98	1.94	1.89	1.84	1.79	1.73
25	4.24	3.39	2.99	2.76	2.60	2.49	2.40	2.34	2.28	2.24	2.16	2.09	2.01	1.96	1.92	1.87	1.82	1.77	1.71
26	4.23	3.37	2.98	2.74	2.59	2.47	2.39	2.32	2.27	2.22	2.15	2.07	1.99	1.95	1.90	1.85	1.80	1.75	1.69
27	4.21	3.35	2.96	2.73	2.57	2.46	2.37	2.31	2.25	2.20	2.13	2.06	1.97	1.93	1.88	1.84	1.79	1.73	1.67
28	4.20	3.34	2.95	2.71	2.56	2.45	2.36	2.29	2.24	2.19	2.12	2.04	1.96	1.91	1.87	1.82	1.77	1.71	1.65
29	4.18	3.33	2.93	2.70	2.55	2.43	2.35	2.28	2.22	2.18	2.10	2.03	1.94	1.90	1.85	1.81	1.75	1.70	1.64
30	4.17	3.32	2.92	2.69	2.53	2.42	2.33	2.27	2.21	2.16	2.09	2.01	1.93	1.89	1.84	1.79	1.74	1.68	1.62
40	4.08	3.23	2.84	2.61	2.45	2.34	2.25	2.18	2.12	2.08	2.00	1.92	1.84	1.79	1.74	1.69	1.64	1.58	1.51
60	4.00	3.15	2.76	2.53	2.37	2.25	2.17	2.10	2.04	1.99	1.92	1.84	1.75	1.70	1.65	1.59	1.53	1.47	1.39
120	3.92	3.07	2.68	2.45	2.29	2.17	2.09	2.02	1.96	1.91	1.83	1.75	1.66	1.61	1.55	1.55	1.43	1.35	1.25
∞	3.84	3.00	2.60	2.37	2.21	2.10	2.01	1.94	1.88	1.83	1.75	1.67	1.57	1.52	1.46	1.39	1.32	1.22	1.00

Degrees of Freedom for the Numerator (ν_1); Degrees of Freedom for the Denominator (ν_2)

IV Percentage Points of the F Distribution (*Continued*)

$F_{0.025,\nu_1,\nu_2}$

Degrees of Freedom for the Numerator (ν_1); Degrees of Freedom for the Denominator (ν_2)

ν_2 \ ν_1	1	2	3	4	5	6	7	8	9	10	12	15	20	24	30	40	60	120	∞
1	647.8	799.5	864.2	899.6	921.8	937.1	948.2	956.7	963.3	968.6	976.7	984.9	993.1	997.2	1001	1006	1010	1014	1018
2	38.51	39.00	39.17	39.25	39.30	39.33	39.36	39.37	39.39	39.40	39.41	39.43	39.45	39.46	39.46	39.47	39.48	39.49	39.50
3	17.44	16.04	15.44	15.10	14.88	14.73	14.62	14.54	14.47	14.42	14.34	14.25	14.17	14.12	14.08	14.04	13.99	13.95	13.90
4	12.22	10.65	9.98	9.60	9.36	9.20	9.07	8.98	8.90	8.84	8.75	8.66	8.56	8.51	8.46	8.41	8.36	8.31	8.26
5	10.01	8.43	7.76	7.39	7.15	6.98	6.85	6.76	6.68	6.62	6.52	6.43	6.33	6.28	6.23	6.18	6.12	6.07	6.02
6	8.81	7.26	6.60	6.23	5.99	5.82	5.70	5.60	5.52	5.46	5.37	5.27	5.17	5.12	5.07	5.01	4.96	4.90	4.85
7	8.07	6.54	5.89	5.52	5.29	5.12	4.99	4.90	4.82	4.76	4.67	4.57	4.47	4.42	4.36	4.31	4.25	4.20	4.14
8	7.57	6.06	5.42	5.05	4.82	4.65	4.53	4.43	4.36	4.30	4.20	4.10	4.00	3.95	3.89	3.84	3.78	3.73	3.67
9	7.21	5.71	5.08	4.72	4.48	4.32	4.20	4.10	4.03	3.96	3.87	3.77	3.67	3.61	3.56	3.51	3.45	3.39	3.33
10	6.94	5.46	4.83	4.47	4.24	4.07	3.95	3.85	3.78	3.72	3.62	3.52	3.42	3.37	3.31	3.26	3.20	3.14	3.08
11	6.72	5.26	4.63	4.28	4.04	3.88	3.76	3.66	3.59	3.53	3.43	3.33	3.23	3.17	3.12	3.06	3.00	2.94	2.88
12	6.55	5.10	4.47	4.12	3.89	3.73	3.61	3.51	3.44	3.37	3.28	3.18	3.07	3.02	2.96	2.91	2.85	2.79	2.72
13	6.41	4.97	4.35	4.00	3.77	3.60	3.48	3.39	3.31	3.25	3.15	3.05	2.95	2.89	2.84	2.78	2.72	2.66	2.60
14	6.30	4.86	4.24	3.89	3.66	3.50	3.38	3.29	3.21	3.15	3.05	2.95	2.84	2.79	2.73	2.67	2.61	2.55	2.49
15	6.20	4.77	4.15	3.80	3.58	3.41	3.29	3.20	3.12	3.06	2.96	2.86	2.76	2.70	2.64	2.59	2.52	2.46	2.40
16	6.12	4.69	4.08	3.73	3.50	3.34	3.22	3.12	3.05	2.99	2.89	2.79	2.68	2.63	2.57	2.51	2.45	2.38	2.32
17	6.04	4.62	4.01	3.66	3.44	3.28	3.16	3.06	2.98	2.92	2.82	2.72	2.62	2.56	2.50	2.44	2.38	2.32	2.25
18	5.98	4.56	3.95	3.61	3.38	3.22	3.10	3.01	2.93	2.87	2.77	2.67	2.56	2.50	2.44	2.38	2.32	2.26	2.19
19	5.92	4.51	3.90	3.56	3.33	3.17	3.05	2.96	2.88	2.82	2.72	2.62	2.51	2.45	2.39	2.33	2.27	2.20	2.13
20	5.87	4.46	3.86	3.51	3.29	3.13	3.01	2.91	2.84	2.77	2.68	2.57	2.46	2.41	2.35	2.29	2.22	2.16	2.09
21	5.83	4.42	3.82	3.48	3.25	3.09	2.97	2.87	2.80	2.73	2.64	2.53	2.42	2.37	2.31	2.25	2.18	2.11	2.04
22	5.79	4.38	3.78	3.44	3.22	3.05	2.93	2.84	2.76	2.70	2.60	2.50	2.39	2.33	2.27	2.21	2.14	2.08	2.00
23	5.75	4.35	3.75	3.41	3.18	3.02	2.90	2.81	2.73	2.67	2.57	2.47	2.36	2.30	2.24	2.18	2.11	2.04	1.97
24	5.72	4.32	3.72	3.38	3.15	2.99	2.87	2.78	2.70	2.64	2.54	2.44	2.33	2.27	2.21	2.15	2.08	2.01	1.94
25	5.69	4.29	3.69	3.35	3.13	2.97	2.85	2.75	2.68	2.61	2.51	2.41	2.30	2.24	2.18	2.12	2.05	1.98	1.91
26	5.66	4.27	3.67	3.33	3.10	2.94	2.82	2.73	2.65	2.59	2.49	2.39	2.28	2.22	2.16	2.09	2.03	1.95	1.88
27	5.63	4.24	3.65	3.31	3.08	2.92	2.80	2.71	2.63	2.57	2.47	2.36	2.25	2.19	2.13	2.07	2.00	1.93	1.85
28	5.61	4.22	3.63	3.29	3.06	2.90	2.78	2.69	2.61	2.55	2.45	2.34	2.23	2.17	2.11	2.05	1.98	1.91	1.83
29	5.59	4.20	1.61	3.27	3.04	2.88	2.76	2.67	2.59	2.53	2.43	2.32	2.21	2.15	2.09	2.03	1.96	1.89	1.81
30	5.57	4.18	3.59	3.25	3.03	2.87	2.75	2.65	2.57	2.51	2.41	2.31	2.20	2.14	2.07	2.01	1.94	1.87	1.79
40	5.42	4.05	3.46	3.13	2.90	2.74	2.62	2.53	2.45	2.39	2.29	2.18	2.07	2.01	1.94	1.88	1.80	1.72	1.64
60	5.29	3.93	3.34	3.01	2.79	2.63	2.51	2.41	2.33	2.27	2.17	2.06	1.94	1.88	1.82	1.74	1.67	1.58	1.48
120	5.15	3.80	3.23	2.89	2.67	2.52	2.39	2.30	2.22	2.16	2.05	1.94	1.82	1.76	1.69	1.61	1.53	1.43	1.31
∞	5.02	3.69	3.12	2.79	2.57	2.41	2.29	2.19	2.11	2.05	1.94	1.83	1.71	1.64	1.57	1.48	1.39	1.27	1.00

IV Percentage Points of the *F* Distribution (*Continued*)

$F_{0.01,\nu_1,\nu_2}$

Degrees of Freedom for the Numerator (ν_1); Degrees of Freedom for the Denominator (ν_2)

ν_2 \ ν_1	1	2	3	4	5	6	7	8	9	10	12	15	20	24	30	40	60	120	∞
1	4052	4999.5	5403	5625	5764	5859	5928	5982	6022	6056	6106	6157	6209	6235	6261	6287	6313	6339	6366
2	98.50	99.00	99.17	**99.25**	99.30	99.33	99.36	99.37	99.39	99.40	99.42	99.43	99.45	99.46	99.47	99.47	99.48	99.49	99.50
3	34.12	30.82	29.46	**28.71**	28.24	27.91	27.67	27.49	27.35	27.23	27.05	26.87	26.69	26.00	26.50	26.41	26.32	26.22	26.13
4	21.20	18.00	16.69	**15.98**	15.52	15.21	14.98	**14.80**	14.66	14.55	14.37	14.20	14.02	13.93	13.84	13.75	13.65	13.56	13.46
5	16.26	13.27	12.06	11.39	10.97	10.67	10.46	10.29	10.16	10.05	9.89	9.72	9.55	9.47	9.38	9.29	9.20	9.11	9.02
6	13.75	10.92	9.78	9.15	8.75	8.47	8.26	8.10	7.98	7.87	7.72	7.56	7.40	7.31	7.23	7.14	7.06	6.97	6.88
7	12.25	9.55	8.45	7.85	7.46	7.19	6.99	6.84	6.72	6.62	6.47	6.31	6.16	6.07	5.99	5.91	5.82	5.74	5.65
8	11.26	8.65	7.59	7.01	6.63	6.37	6.18	6.03	5.91	5.81	5.67	5.52	5.36	5.28	5.20	5.12	5.03	4.95	4.86
9	10.56	8.02	6.99	6.42	6.06	5.80	5.61	5.47	5.35	5.26	5.11	4.96	4.81	4.73	4.65	4.57	4.48	4.40	4.31
10	10.04	7.56	6.55	5.99	5.64	5.39	5.20	5.06	4.94	4.85	4.71	4.56	4.41	4.33	4.25	4.17	4.08	4.00	3.91
11	9.65	7.21	6.22	5.67	5.32	5.07	4.89	4.74	4.63	4.54	4.40	4.25	4.10	4.02	3.94	3.86	3.78	3.69	3.60
12	9.33	6.93	5.95	5.41	5.06	4.82	4.64	4.50	4.39	4.30	4.16	4.01	3.86	3.78	3.70	3.62	3.54	3.45	3.36
13	9.07	6.70	5.74	5.21	4.86	4.62	4.44	4.30	4.19	4.10	3.96	3.82	3.66	3.59	3.51	3.43	3.34	3.25	3.17
14	8.86	6.51	5.56	5.04	4.69	4.46	4.28	4.14	4.03	3.94	3.80	3.66	3.51	3.43	3.35	3.27	3.18	3.09	3.00
15	8.68	6.36	5.42	4.89	4.56	4.32	4.14	4.00	3.89	3.80	3.67	3.52	3.37	3.29	3.21	3.13	3.05	2.96	2.87
16	8.53	6.23	5.29	4.77	4.44	4.20	4.03	**3.89**	3.78	3.69	3.55	3.41	3.26	3.18	3.10	3.02	2.93	2.84	2.75
17	8.40	6.11	5.18	4.67	4.34	4.10	3.93	**3.79**	3.68	3.59	3.46	3.31	3.16	3.08	3.00	2.92	2.83	2.75	2.65
18	8.29	6.01	5.09	4.58	4.25	4.01	3.84	**3.71**	3.60	3.51	3.37	3.23	3.08	3.00	2.92	2.84	2.75	2.66	2.57
19	8.18	5.93	5.01	4.50	4.17	3.94	3.77	**3.63**	3.52	3.43	3.30	3.15	3.00	2.92	2.84	2.76	2.67	2.58	2.49
20	8.10	5.85	4.94	4.43	4.10	3.87	3.70	3.56	3.46	3.37	3.23	3.09	2.94	2.86	2.78	2.69	2.61	2.52	2.42
21	8.02	5.78	4.87	4.37	4.04	3.81	3.64	3.51	3.40	3.31	3.17	3.03	2.88	2.80	2.72	2.64	2.55	2.46	2.36
22	7.95	5.72	4.82	4.31	3.99	3.76	3.59	3.45	3.35	3.26	3.12	2.98	2.83	2.75	2.67	2.58	2.50	2.40	2.31
23	7.88	5.66	4.76	4.26	3.94	3.71	3.54	3.41	3.30	3.21	3.07	2.93	2.78	2.70	2.62	2.54	2.45	2.35	2.26
24	7.82	5.61	4.72	4.22	3.90	3.67	3.50	3.36	3.26	3.17	3.03	2.89	2.74	2.66	2.58	2.49	2.40	2.31	2.21
25	7.77	5.57	4.68	4.18	3.85	3.63	3.46	3.32	3.22	3.13	2.99	2.85	2.70	2.62	2.54	2.45	2.36	2.27	2.17
26	7.72	5.53	4.64	4.14	3.82	3.59	3.42	3.29	3.18	3.09	2.96	2.81	2.66	2.58	2.50	2.42	2.33	2.23	2.13
27	7.68	5.49	4.60	4.11	3.78	3.56	3.39	3.26	3.15	3.06	2.93	2.78	2.63	2.55	2.47	2.38	2.29	2.20	2.10
28	7.64	5.45	4.57	4.07	3.75	3.53	3.36	3.23	3.12	3.03	2.90	2.75	2.60	2.52	2.44	2.34	2.26	2.17	2.06
29	7.60	5.42	4.54	4.04	3.73	3.50	3.33	3.20	3.09	3.00	2.87	2.73	2.57	2.49	2.41	2.33	2.23	2.14	2.03
30	7.56	5.39	4.51	4.02	3.70	3.47	3.30	3.17	3.07	2.98	2.84	2.70	2.55	2.47	2.39	2.30	2.21	2.11	2.01
40	7.31	5.18	4.31	3.83	3.51	3.29	3.12	2.99	2.89	2.80	2.66	2.52	2.37	2.29	2.20	2.11	2.02	1.92	1.80
60	7.08	4.98	4.13	3.65	3.34	3.12	2.95	2.82	2.72	2.63	2.50	2.35	2.20	2.12	2.03	1.94	1.84	1.73	1.60
120	6.85	4.79	3.95	3.48	3.17	2.96	2.79	2.66	2.56	2.47	2.34	2.19	2.03	1.95	1.86	1.76	1.66	1.53	1.38
∞	6.63	4.61	3.78	3.32	3.02	2.80	2.64	2.51	2.41	2.32	2.18	2.04	1.88	1.79	1.70	1.59	1.47	1.32	1.00

V Operating Characteristic Curves for the Fixed Effects Model Analysis of Variance[a]

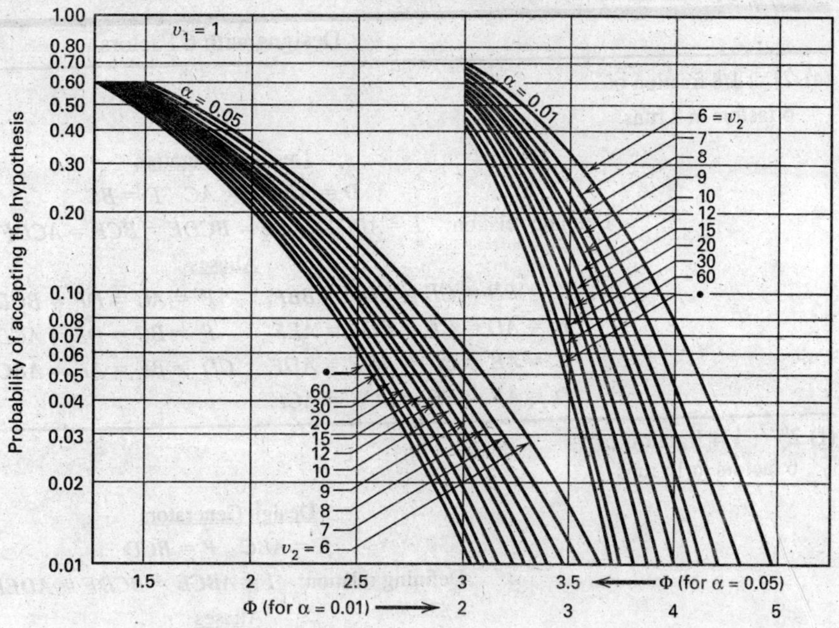

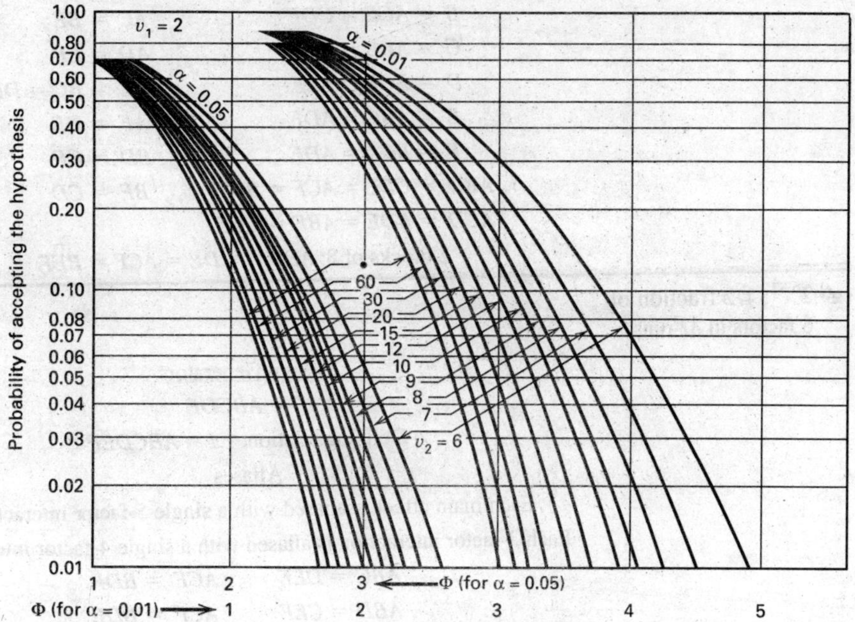

v_1 = Numerator degrees of freedom, v_2 = Denominator degrees of freedom

[a]Adapted from Biometrika. E.S. Pearson and H.O. Hartley. *Biometrika Tables for Statisticians*, Vol. 2 (1972) by permission of Oxford University Press.

V Alias Relationships for 2^{k-p} Fractional Factorial Designs with $k \leq 15$ and $n \leq 64$ (*Continued*)

Designs with 6 Factors

(e) 2^{6-3}; 1/8 fraction of 6 factors in 8 runs — **Resolutio**

Design Generators

$D = AB \quad E = AC \quad F = BC$

Defining relation: $I = ABD = ACE = BCDE = BCF = ACDF = ABEF = DEF$

Aliases

$A = BD = CE = CDF = BEF$

$B = AD = CF = CDE = AEF$

$C = AE = BF = BDE = ADF$

$D = AB = EF = BCE = ACF$

$E = AC = DF = BCD = ABF$

$F = BC = DE = ACD = ABE$

$CD = BE = AF = ABC = ADE = BDF = CEF$

(f) 2^{6-2}; 1/4 fraction of 6 factors in 16 runs — **Resolutior**

Design Generators

$E = ABC \quad F = BCD$

Defining relation: $I = ABCE = BCDF = ADEF$

Aliases

$A = BCE = DEF$

$B = ACE = CDF$

$C = ABE = BDF$

$D = BCF = AEF$

$E = ABC = ADF$

$F = BCD = ADE$

$ABD = CDE = ACF = BEF$

$ACD = BDE = ABF = CEF$

$AB = CE$

$AC = BE$

$AD = EF$

$AE = BC = DF$

$AF = DE$

$BD = CF$

$BF = CD$

2 blocks of 8: $ABD = CDE = ACF = BEF$

(g) 2^{6-1}; 1/2 fraction of 6 factors in 32 runs — **Resolution** '

Design Generators

$F = ABCDE$

Defining relation: $I = ABCDEF$

Aliases

Each main effect is aliased with a single 5-factor interaction.

Each 2-factor interaction is aliased with a single 4-factor interaction.

$ABC = DEF$

$ABD = CEF$

$ABE = CDF$

$ABF = CDE$

$ACD = BEF$

$ACE = BDF$

$ACF = BDE$

$ADE = BCF$

$ADF = BCE$

$AEF = BCD$

2 blocks of 16: $ABC = DEF$

4 blocks of 8: $AB = CDEF$

$ACD = BEF$

$AEF = BCD$

V Operating Characteristic Curves for the Fixed Effects Model Analysis of Variance (*Continued*)

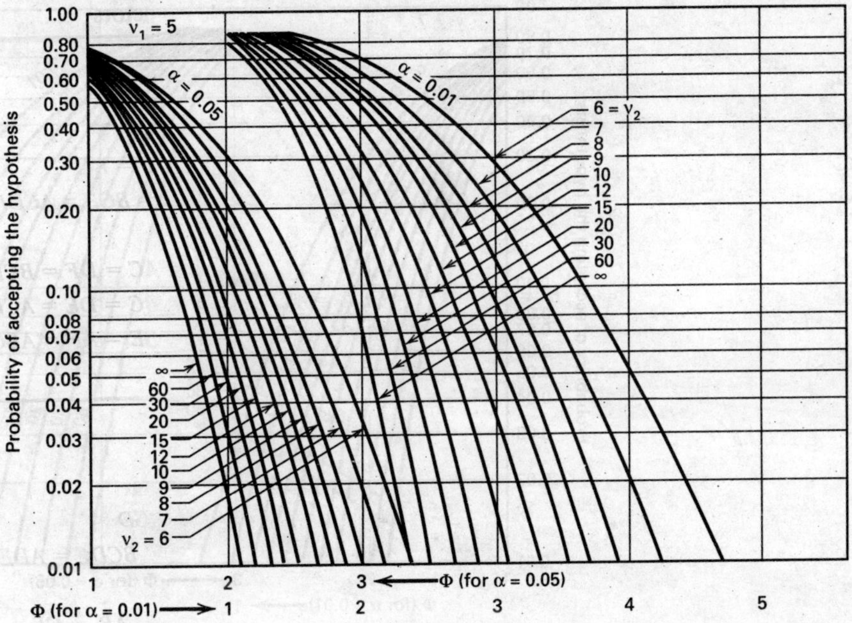

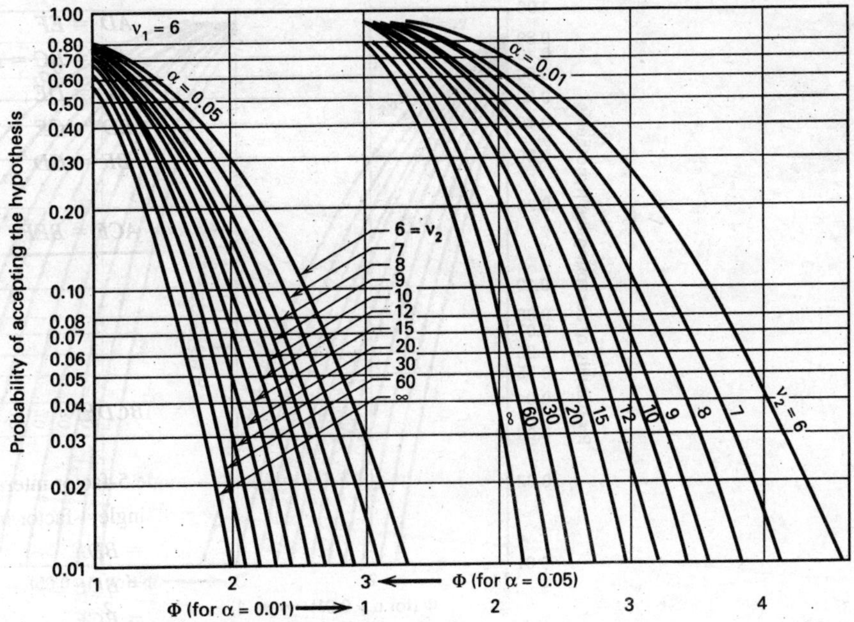

V Operating Characteristic Curves for the Fixed Effects Model Analysis of Variance (*Continued*)

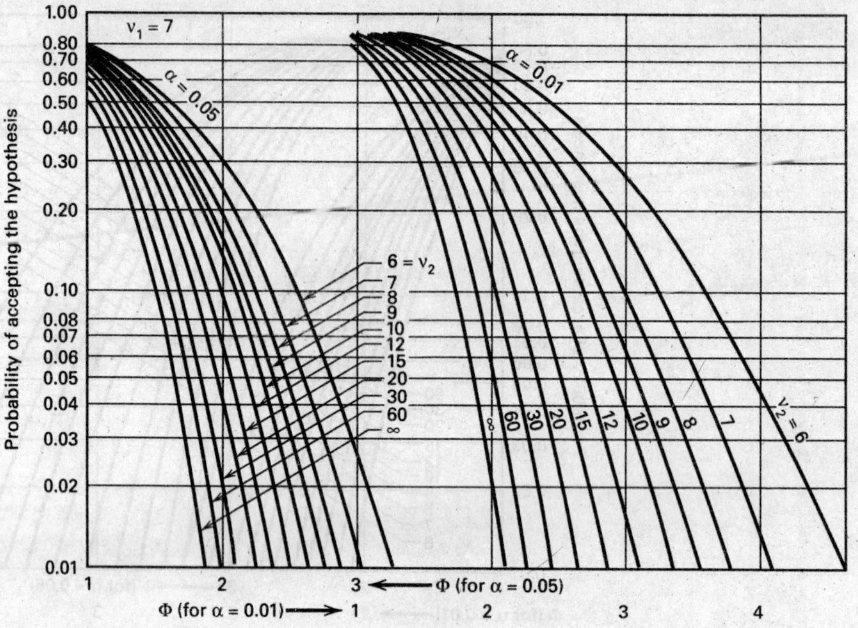

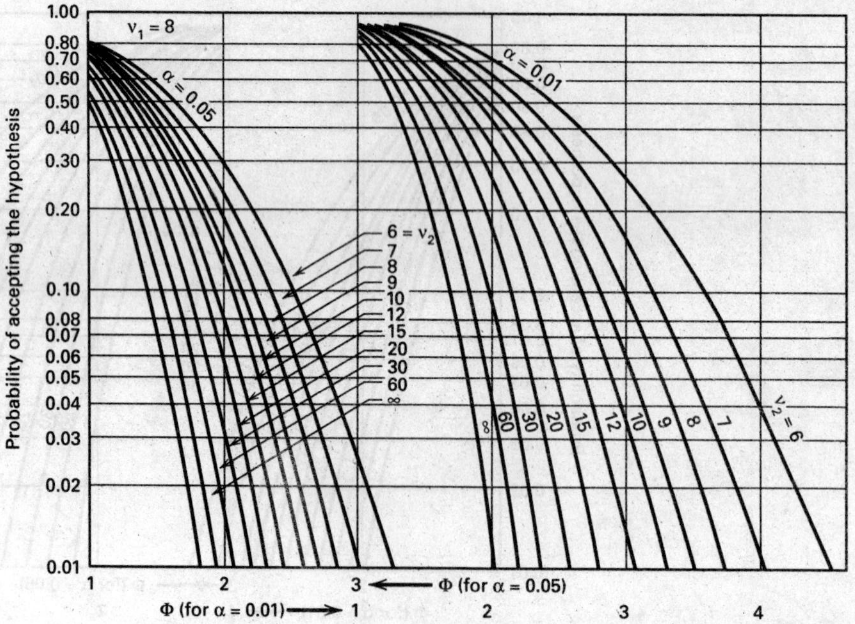

VI Percentage Points of the Studentized Range Statistic[a]

$q_{0.01}(p, f)$

f	p: 2	3	4	5	6	7	8	9	10	11	12	13	14	15	16	17	18	19	20
1	90	135	164	186	202	216	227	237	246	253	260	266	272	272	282	286	290	294	298
2	14.0	19.0	22.3	24.7	26.6	28.2	29.5	30.7	31.7	32.6	33.4	31.4	34.8	35.4	36.0	36.5	37.0	37.5	37.9
3	8.26	10.6	12.2	13.3	14.2	15.0	15.6	16.2	16.7	17.1	17.5	17.9	18.2	18.5	18.8	19.1	19.3	19.5	19.8
4	6.51	8.12	9.17	9.96	10.6	11.1	11.5	11.9	12.3	12.6	12.8	13.1	13.3	13.5	13.7	13.9	14.1	14.2	14.4
5	5.70	6.97	7.80	8.42	8.91	9.32	9.67	9.97	10.24	10.48	10.70	10.89	11.08	11.24	11.40	11.55	11.68	11.81	11.93
6	5.24	6.33	7.03	7.56	7.97	8.32	8.61	8.87	9.10	9.30	9.49	9.65	9.81	9.95	10.08	10.21	10.32	10.43	10.54
7	4.95	5.92	6.54	7.01	7.37	7.68	7.94	8.17	8.37	8.55	8.71	8.86	9.00	9.12	9.24	9.35	9.46	9.55	9.65
8	4.74	5.63	6.20	6.63	6.96	7.24	7.47	7.68	7.87	8.03	8.18	8.31	8.44	8.55	8.66	8.76	8.85	8.94	9.03
9	4.60	5.43	5.96	6.35	6.66	6.91	7.13	7.32	7.49	7.65	7.78	7.91	8.03	8.13	8.23	8.32	8.41	8.49	8.57
10	4.48	5.27	5.77	6.14	6.43	6.67	6.87	7.05	7.21	7.36	7.48	7.60	7.71	7.81	7.91	7.99	8.07	8.15	8.22
11	4.39	5.14	5.62	5.97	6.25	6.48	6.67	6.84	6.99	7.13	7.25	7.36	7.46	7.56	7.65	7.73	7.81	7.88	7.95
12	4.32	5.04	5.50	5.84	6.10	6.32	6.51	6.67	6.81	6.94	7.06	7.17	7.26	7.36	7.44	7.52	7.59	7.66	7.73
13	4.26	4.96	5.40	5.73	5.98	6.19	6.37	6.53	6.67	6.79	6.90	7.01	7.10	7.19	7.27	7.34	7.42	7.48	7.55
14	4.21	4.89	5.32	5.63	5.88	6.08	6.26	6.41	6.54	6.66	6.77	6.87	6.96	7.05	7.12	7.20	7.27	7.33	7.39
15	4.17	4.83	5.25	5.56	5.80	5.99	6.16	6.31	6.44	6.55	6.66	6.76	6.84	6.93	7.00	7.07	7.14	7.20	7.26
16	4.13	4.78	5.19	5.49	5.72	5.92	6.08	6.22	6.35	6.46	6.56	6.66	6.74	6.82	6.90	6.97	7.03	7.09	7.15
17	4.10	4.74	5.14	5.43	5.66	5.85	6.01	6.15	6.27	6.38	6.48	6.57	6.66	6.73	6.80	6.87	6.94	7.00	7.05
18	4.07	4.70	5.09	5.38	5.60	5.79	5.94	6.08	6.20	6.31	6.41	6.50	6.58	6.65	6.72	6.79	6.85	6.91	6.96
19	4.05	4.67	5.05	5.33	5.55	5.73	5.89	6.02	6.14	6.25	6.34	6.43	6.51	6.58	6.65	6.72	6.78	6.84	6.89
20	4.02	4.64	5.02	5.29	5.51	5.69	5.84	5.97	6.09	6.19	6.29	6.37	6.45	6.52	6.59	6.65	6.71	6.76	6.82
24	3.96	4.54	4.91	5.17	5.37	5.54	5.69	5.81	5.92	6.02	6.11	6.19	6.26	6.33	6.39	6.45	6.51	6.56	6.61
30	3.89	4.45	4.80	5.05	5.24	5.40	5.54	5.65	5.76	5.85	5.93	6.01	6.08	6.14	6.20	6.26	6.31	6.36	6.41
40	3.82	4.37	4.70	4.93	5.11	5.27	5.39	5.50	5.60	5.69	5.77	5.84	5.90	5.96	6.02	6.07	6.12	6.17	6.21
60	3.76	4.28	4.60	4.82	4.99	5.13	5.25	5.36	5.45	5.53	5.60	5.67	5.73	5.79	5.84	5.89	5.93	5.98	6.02
120	3.70	4.20	4.50	4.71	4.87	5.01	5.12	5.21	5.30	5.38	5.44	5.51	5.56	5.61	5.66	5.71	5.75	5.79	5.83
∞	3.64	4.12	4.40	4.60	4.76	4.88	4.99	5.08	5.16	5.23	5.29	5.35	5.40	5.45	5.49	5.54	5.57	5.61	5.65

f = Degrees of freedom.

[a]From J. M. May, "Extended and Corrected Tables of the Upper Percentage Points of the Studentized Range," *Biometrika*, Vol. 39, pp. 192–193, 1952. Reproduced by permission of the trustees of *Biometrika*.

VI Percentage Points of the Studentized Range Statistic (*Continued*)

$q_{0.05}(p, f)$

f	p = 2	3	4	5	6	7	8	9	10	11	12	13	14	15	16	17	18	19	20
1	18.1	26.7	32.8	37.2	40.5	43.1	45.4	47.3	49.1	50.6	51.9	53.2	54.3	55.4	56.3	57.2	58.0	58.8	59.6
2	6.09	8.28	9.80	10.89	11.73	12.43	13.03	13.54	13.99	14.39	14.75	15.08	15.38	15.65	15.91	16.14	16.36	16.57	16.77
3	4.50	5.88	6.83	7.51	8.04	8.47	8.85	9.18	9.46	9.72	9.95	10.16	10.35	10.52	10.69	10.84	10.98	11.12	11.24
4	3.93	5.00	5.76	6.31	6.73	7.06	7.35	7.60	7.83	8.03	8.21	8.37	8.52	8.67	8.80	8.92	9.03	9.14	9.24
5	3.64	4.60	5.22	5.67	6.03	6.33	6.58	6.80	6.99	7.17	7.32	7.47	7.60	7.72	7.83	7.93	8.03	8.12	8.21
6	3.46	4.34	4.90	5.31	5.63	5.89	6.12	6.32	6.49	6.65	6.79	6.92	7.04	7.14	7.24	7.34	7.43	7.51	7.59
7	3.34	4.16	4.68	5.06	5.35	5.59	5.80	5.99	6.15	6.29	6.42	6.54	6.65	6.75	6.84	6.93	7.01	7.08	7.16
8	3.26	4.04	4.53	4.89	5.17	5.40	5.60	5.77	5.92	6.05	6.18	6.29	6.39	6.48	6.57	6.65	6.73	6.80	6.87
9	3.20	3.95	4.42	4.76	5.02	5.24	5.43	5.60	5.74	5.87	5.98	6.09	6.19	6.28	6.36	6.44	6.51	6.58	6.65
10	3.15	3.88	4.33	4.66	4.91	5.12	5.30	5.46	5.60	5.72	5.83	5.93	6.03	6.12	6.20	6.27	6.34	6.41	6.47
11	3.11	3.82	4.26	4.58	4.82	5.03	5.20	5.35	5.49	5.61	5.71	5.81	5.90	5.98	6.06	6.14	6.20	6.27	6.33
12	3.08	3.77	4.20	4.51	4.75	4.95	5.12	5.27	5.40	5.51	5.61	5.71	5.80	5.88	5.95	6.02	6.09	6.15	6.21
13	3.06	3.73	4.15	4.46	4.69	4.88	5.05	5.19	5.32	5.43	5.53	5.63	5.71	5.79	5.86	5.93	6.00	6.06	6.11
14	3.03	3.70	4.11	4.41	4.64	4.83	4.99	5.13	5.25	5.36	5.46	5.56	5.64	5.72	5.79	5.86	5.92	5.98	6.03
15	3.01	3.67	4.08	4.37	4.59	4.78	4.94	5.08	5.20	5.31	5.40	5.49	5.57	5.65	5.72	5.79	5.85	5.91	5.96
16	3.00	3.65	4.05	4.34	4.56	4.74	4.90	5.03	5.15	5.26	5.35	5.44	5.52	5.59	5.66	5.73	5.79	5.84	5.90
17	2.98	3.62	4.02	4.31	4.52	4.70	4.86	4.99	5.11	5.21	5.31	5.39	5.47	5.55	5.61	5.68	5.74	5.79	5.84
18	2.97	3.61	4.00	4.28	4.49	4.67	4.83	4.96	5.07	5.17	5.27	5.35	5.43	5.50	5.57	5.63	5.69	5.74	5.79
19	2.96	3.59	3.98	4.26	4.47	4.64	4.79	4.92	5.04	5.14	5.23	5.32	5.39	5.46	5.53	5.59	5.65	5.70	5.75
20	2.95	3.58	3.96	4.24	4.45	4.62	4.77	4.90	5.01	5.11	5.20	5.28	5.36	5.43	5.50	5.56	5.61	5.66	5.71
24	2.92	3.53	3.90	4.17	4.37	4.54	4.68	4.81	4.92	5.01	5.10	5.18	5.25	5.32	5.38	5.44	5.50	5.55	5.59
30	2.89	3.48	3.84	4.11	4.30	4.46	4.60	4.72	4.83	4.92	5.00	5.08	5.15	5.21	5.27	5.33	5.38	5.43	5.48
40	2.86	3.44	3.79	4.04	4.23	4.39	4.52	4.63	4.74	4.82	4.90	4.98	5.05	5.11	5.17	5.22	5.27	5.32	5.36
60	2.83	3.40	3.74	3.98	4.16	4.31	4.44	4.55	4.65	4.73	4.81	4.88	4.94	5.00	5.06	5.11	5.15	5.20	5.24
120	2.80	3.36	3.69	3.92	4.10	4.24	4.36	4.47	4.56	4.64	4.71	4.78	4.84	4.90	4.95	5.00	5.04	5.09	5.13
∞	2.77	3.32	3.63	3.86	4.03	4.17	4.29	4.39	4.47	4.55	4.62	4.68	4.74	4.80	4.84	4.88	4.93	4.97	5.01

V Alias Relationships for 2^{k-p} Fractional Factorial Designs with $k \le 15$ and $n \le 64$ (*Continued*)

Designs with 10 Factors

(r) 2^{10-6}; 1/64 fraction of 10 factors in 16 runs — **Resolution III**

Design Generators

$E = ABC \quad F = BCD \quad G = ACD \quad H = ABD \quad J = ABCD \quad K = AB$

Defining relation: $I = ABCE = BCDF = ADEF = ACDG = BDEG = ABFG = CEFG = ABDH$
$= CDEH = ACFH = BEFH = BCGH = AEGH = DFGH = ABCDEFGH = ABCDJ$
$= DEJ = AFJ = BCEFJ = BGJ = ACEGJ = CDFGI = ABDEFGJ = CHJ$
$= ABEHJ = BDFHJ = ACDEFHJ = ADGHJ = BCDEGHJ = ABCFGHJ = EFGHJ = ABK$
$= CEK = ACDFK = BDEFK = BCDGK = ADEGK = FGK = ABCEFGK = DHK$
$= ABCDEHK = BCFHK = AEFHK = ACGHK = BEGHK = ABDFGHK = CDEFGHK = CDJK$
$= ABDEJK = BFJK = ACEFJK = AGJK = BCEGJK = ABCDFGJK = DEFGJK = ABCHJK$
$= EHJK = ADFHJK = BCDEFHJK = BDGHJK = ACDEGHJK = CFGHJK = ABEFGHJK$

Aliases

$A = FJ = BK$ $\quad J = DE = AF = BG = CH$
$B = GJ = AK$ $\quad K = AB = CE = FG = DH$
$C = HJ = EK$ $\quad AC = BE = DG = FH$
$D = EJ = HK$ $\quad AD = EF = CG = BH$
$E = DJ = CK$ $\quad AE = BC = DF = GH$
$F = AJ = GK$ $\quad AG = CD = BF = EH = JK$
$G = BJ = FK$ $\quad AH = BD = CF = EG$
$H = CJ = DK$

2 blocks of 8: $AG = CD = BF = EH = JK$

(s) 2^{10-5}; 1/32 fraction of 10 factors in 32 runs — **Resolution IV**

Design Generators

$F = ABCD \quad G = ABCE \quad H = ABDE \quad J = ACDE \quad K = BCDE$

Defining relation: $I = ABCDF = ABCEG = DEFG = ABDEH = CEFH = CDGH = ABFGH = ACDEJ$
$= BEFJ = BDGJ = ACFGJ = BCHJ = ADFHJ = AEGHJ = BCDEFGHJ = BCDEK$
$= AEFK = ADGK = BCFGK = ACHK = BDFHK = BEGHK = ACDEFGHK = ABJK$
$= CDFJK = CEGJK = ABDEFGJK = DEHJK = ABCEFHJK = ABCDGHJK = FGHJK$

Aliases

$A = EFK = DGK = CHK = BJK$ $\quad AH = BDE = BFG = DFJ = EGJ = CK$
$B = EFJ = DGJ = CHJ = AJK$ $\quad AJ = CDE = CFG = DFH = EGH = BK$
$C = EFH = DGH = BHJ = AHK$ $\quad AK = EF = DG = CH = BJ$
$D = EFG = CGH = BGJ = AGK$ $\quad BC = ADF = AEG = HJ = DEK = FGK$
$E = DFG = CFH = BFJ = AFK$ $\quad BD = ACF = AEH = GJ = CEK = FHK$
$F = DEG = CEH = BEJ = AEK$ $\quad BE = ACG = ADH = FJ = CDK = GHK$
$G = DEF = CDH = BDJ = ADK$ $\quad BF = ACD = AGH = EJ = CGK = DHK$
$H = CEF = CDG = BCJ = ACK$ $\quad BG = ACE = AFH = DJ = CFK = EHK$
$J = BEF = BDG = BCH = ABK$ $\quad BH = ADE = AFG = CJ = DFK = EGK$
$K = AEF = ADG = ACH = ABJ$ $\quad CD = ABF = GH = AEJ = BEK = FJK$
$AB = CDF = CEG = DEH = FGH = JK$ $\quad CE = ABG = FH = ADJ = BDK = GJK$
$AC = BDF = BEG = DEJ = FGJ = HK$ $\quad CF = ABD = EH = AGJ = BGK = DJK$
$AD = BCF = BEH = CEJ = FHJ = GK$ $\quad CG = ABE = DH = AFJ = BFK = EJK$
$AE = BCG = BDH = CDJ = GHJ = FK$ $\quad DE = FG = ABH = ACJ = BCK = HJK$
$AF = BCD = BGH = CGJ = DHJ = EK$ $\quad DF = ABC = EG = AHJ = BHK = CJK$
$AG = BCE = BFH = CFJ = EHJ = DK$

2 blocks of 16: $AK = EF = DG = CH = BJ$

4 blocks of 8: $AK = EF = DG = CH = BJ$

$AJ = CDE = CFG = DFH = EGH = BK$

$AB = CDF = CEG = DEH = FGH = JK$

VII Critical Values for Dunnett's Test for Comparing Treatments with a Control (*Continued*)

$d_{0.05}(a-1, f)$

One-Sided Comparisons

f	$a-1$ = Number of Treatment Means (Excluding Control) 1	2	3	4	5	6	7	8	9
5	2.02	2.44	2.68	2.85	2.98	3.08	3.16	3.24	3.30
6	1.94	2.34	2.56	2.71	2.83	2.92	3.00	3.07	3.12
7	1.89	2.27	2.48	2.62	2.73	2.82	2.89	2.95	3.01
8	1.86	2.22	2.42	2.55	2.66	2.74	2.81	2.87	2.92
9	1.83	2.18	2.37	2.50	2.60	2.68	2.75	2.81	2.86
10	1.81	2.15	2.34	2.47	2.56	2.64	2.70	2.76	2.81
11	1.80	2.13	2.31	2.44	2.53	2.60	2.67	2.72	2.77
12	1.78	2.11	2.29	2.41	2.50	2.58	2.64	2.69	2.74
13	1.77	2.09	2.27	2.39	2.48	2.55	2.61	2.66	2.71
14	1.76	2.08	2.25	2.37	2.46	2.53	2.59	2.64	2.69
15	1.75	2.07	2.24	2.36	2.44	2.51	2.57	2.62	2.67
16	1.75	2.06	2.23	2.34	2.43	2.50	2.56	2.61	2.65
17	1.74	2.05	2.22	2.33	2.42	2.49	2.54	2.59	2.64
18	1.73	2.04	2.21	2.32	2.41	2.48	2.53	2.58	2.62
19	1.73	2.03	2.20	2.31	2.40	2.47	2.52	2.57	2.61
20	1.72	2.03	2.19	2.30	2.39	2.46	2.51	2.56	2.60
24	1.71	2.01	2.17	2.28	2.36	2.43	2.48	2.53	2.57
30	1.70	1.99	2.15	2.25	2.33	2.40	2.45	2.50	2.54
40	1.68	1.97	2.13	2.23	2.31	2.37	2.42	2.47	2.51
60	1.67	1.95	2.10	2.21	2.28	2.35	2.39	2.44	2.48
120	1.66	1.93	2.08	2.18	2.26	2.32	2.37	2.41	2.45
∞	1.64	1.92	2.06	2.16	2.23	2.29	2.34	2.38	2.42

$d_{0.01}(a-1, f)$

One-Sided Comparisons

f	$a-1$ = Number of Treatment Means (Excluding Control) 1	2	3	4	5	6	7	8	9
5	3.37	3.90	4.21	4.43	4.60	4.73	4.85	4.94	5.03
6	3.14	3.61	3.88	4.07	4.21	4.33	4.43	4.51	4.59
7	3.00	3.42	3.66	3.83	3.96	4.07	4.15	4.23	4.30
8	2.90	3.29	3.51	3.67	3.79	3.88	3.96	4.03	4.09
9	2.82	3.19	3.40	3.55	3.66	3.75	3.82	3.89	3.94
10	2.76	3.11	3.31	3.45	3.56	3.64	3.71	3.78	3.83
11	2.72	3.06	3.25	3.38	3.48	3.56	3.63	3.69	3.74
12	2.68	3.01	3.19	3.32	3.42	3.50	3.56	3.62	3.67
13	2.65	2.97	3.15	3.27	3.37	3.44	3.51	3.56	3.61
14	2.62	2.94	3.11	3.23	3.32	3.40	3.46	3.51	3.56
15	2.60	2.91	3.08	3.20	3.29	3.36	3.42	3.47	3.52
16	2.58	2.88	3.05	3.17	3.26	3.33	3.39	3.44	3.48
17	2.57	2.86	3.03	3.14	3.23	3.30	3.36	3.41	3.45
18	2.55	2.84	3.01	3.12	3.21	3.27	3.33	3.38	3.42
19	2.54	2.83	2.99	3.10	3.18	3.25	3.31	3.36	3.40
20	2.53	2.81	2.97	3.08	3.17	3.23	3.29	3.34	3.38
24	2.49	2.77	2.92	3.03	3.11	3.17	3.22	3.27	3.31
30	2.46	2.72	2.87	2.97	3.05	3.11	3.16	3.21	3.24
40	2.42	2.68	2.82	2.92	2.99	3.05	3.10	3.14	3.18
60	2.39	2.64	2.78	2.87	2.94	3.00	3.04	3.08	3.12
120	2.36	2.60	2.73	2.82	2.89	2.94	2.99	3.03	3.06
∞	2.33	2.56	2.68	2.77	2.84	2.89	2.93	2.97	3.00

VIII Coefficients of Orthogonal Polynomials[a]

	n = 3		n = 4			n = 5				n = 6					n = 7					
X_j	P_1	P_2	P_1	P_2	P_3	P_1	P_2	P_3	P_4	P_1	P_2	P_3	P_4	P_5	P_1	P_2	P_3	P_4	P_5	P_6
1	−1	1	−3	1	−1	−2	2	−1	1	−5	5	−5	1	−1	−3	5	−1	3	−1	1
2	0	−2	−1	−1	3	−1	−1	2	−4	−3	−1	7	−3	5	−2	0	1	−7	4	−6
3	1	1	1	−1	−3	0	−2	0	6	−1	−4	4	2	−10	−1	−3	1	1	−5	15
4			3	1	1	1	−1	−2	−4	1	−4	−4	2	10	0	−4	0	6	0	−20
5						2	2	1	1	3	−1	−7	−3	−5	1	−3	−1	1	5	15
6										5	5	5	1	1	2	0	−1	−7	−4	−6
7															3	5	1	3	1	1
$\sum_{j=1}^{n}\{P_i(X_j)\}^2$	2	6	20	4	20	10	14	10	70	70	84	180	28	252	28	84	6	154	84	924
λ	1	3	2	1	$\frac{10}{3}$	1	1	$\frac{5}{6}$	$\frac{35}{12}$	2	$\frac{3}{2}$	$\frac{5}{3}$	$\frac{7}{12}$	$\frac{21}{10}$	1	1	$\frac{1}{6}$	$\frac{7}{12}$	$\frac{7}{20}$	$\frac{77}{60}$

	n = 8						n = 9						n = 10					
X_j	P_1	P_2	P_3	P_4	P_5	P_6	P_1	P_2	P_3	P_4	P_5	P_6	P_1	P_2	P_3	P_4	P_5	P_6
1	−7	7	−7	7	−7	1	−4	28	−14	14	−4	4	−9	6	−42	18	−6	3
2	−5	1	5	−13	23	−5	−3	7	7	−21	11	−17	−7	2	14	−22	14	−11
3	−3	−3	7	−3	−17	9	−2	−8	13	−11	−4	22	−5	−1	35	−17	−1	10
4	−1	−5	3	9	−15	−5	−1	−17	9	9	−9	1	−3	−3	31	3	−11	6
5	1	−5	−3	9	15	−5	0	−20	0	18	0	−20	−1	−4	12	18	−6	−8
6	3	−3	−7	−3	17	9	1	−17	−9	9	9	1	1	−4	−12	18	6	−8
7	5	1	−5	−13	−23	−5	2	−8	−13	−11	4	22	3	−3	−31	3	11	6
8	7	7	7	7	7	1	3	7	−7	−21	−11	−17	5	−1	−35	−17	1	10
9							4	28	14	14	4	4	7	2	−14	−22	−14	−11
10													9	6	42	18	6	3
$\sum_{j=1}^{n}\{P_i(X_j)\}^2$	168	168	264	616	2184	264	60	2772	990	2002	468	1980	330	132	8580	2860	780	660
λ	2	1	$\frac{2}{3}$	$\frac{7}{12}$	$\frac{7}{10}$	$\frac{11}{60}$	1	3	$\frac{5}{6}$	$\frac{7}{12}$	$\frac{3}{20}$	$\frac{11}{60}$	2	$\frac{1}{2}$	$\frac{5}{3}$	$\frac{5}{12}$	$\frac{1}{10}$	$\frac{11}{240}$

[a]Adapted from Biometrika. E.S. Pearson and H.O. Hartley. *Biometrika Tables for Statisticians*, Vol. 1, 3rd edition (1966) by permission of Oxford University Press.

IX Alias Relationships for 2^{k-p} Fractional Factorial Designs with $k \leq 15$ and $n \leq 64$

Designs with 3 Factors

(a) 2^{3-1}; 1/2 fraction of 3 factors in 4 runs — **Resolution III**

Design Generators

$C = AB$

Defining relation: $I = ABC$

Aliases

$A = BC$
$B = AC$
$C = AB$

Designs with 4 Factors

(b) 2^{4-1}; 1/2 fraction of 4 factors in 8 runs — **Resolution IV**

Design Generators

$D = ABC$

Defining relation: $I = ABCD$

Aliases

$A = BCD$
$B = ACD$
$C = ABD$
$D = ABC$
$AB = CD$
$AC = BD$
$AD = BC$

Designs with 5 Factors

(c) 2^{5-2}; 1/4 fraction of 5 factors in 8 runs — **Resolution III**

Design Generators

$D = AB \quad E = AC$

Defining relation: $I = ABD = ACE = BCDE$

Aliases

$A = BD = CE$
$B = AD = CDE$
$C = AE = BDE$
$D = AB = BCE$
$E = AC = BCD$
$BC = DE = ACD = ABE$
$CD = BE = ABC = ADE$

(d) 2^{5-1}; 1/2 fraction of 5 factors in 16 runs — **Resolution V**

Design Generators

$E = ABCD$

Defining relation: $I = ABCDE$

Aliases

Each main effect is aliased with a single 4-factor interaction.

$AB = CDE$	$BD = ACE$
$AC = BDE$	$BE = ACD$
$AD = BCE$	$CD = ABE$
$AE = BCD$	$CE = ABD$
$BC = ADE$	$DE = ABC$

2 blocks of 8: $AB = CDE$

V Alias Relationships for 2^{k-p} Fractional Factorial Designs with $k \le 15$ and $n \le 64$ (*Continued*)

Designs with 12 Factors

; 1/256 fraction **Resolution III**

factors in 16 runs

Design Generators

$E = ABC \quad F = ABD \quad G = ACD \quad H = BCD$

$J = ABCD \quad K = AB \quad L = AC \quad M = AD$

Aliases

$A = HJ = BK = CL = DM$

$B = GJ = AK = EL = FM$

$C = FJ = EK = AL = GM$

$D = EJ = FK = GL = AM$

$E = DJ = CK = BL = HM$

$F = CJ = DK = HL = BM$

$G = BJ = HK = DL = CM$

$H = AJ = GK = FL = EM$

$J = DE = CF = BG = AH$

$K = AB = CE = DF = GH$

$L = AC = BE = DG = FH$

$M = AD = BF = CG = EH$

$AE = BC = FG = DH = KL = JM$

$AF = BD = EG = CH = JL = KM$

$AG = EF = CD = BH = JK = LM$

2 blocks of 8: $AE = BC = FG = DH = KL = JM$

Designs with 13 Factors

; 1/512 fraction of **Resolution III**

ctors in 16 runs

Design Generators

$E = ABC \quad F = ABD \quad G = ACD \quad H = BCD$

$J = ABCD \quad K = AB \quad L = AC \quad M = AD \quad N = BC$

Aliases

$A = HJ = BK = CL = DM = EN$

$B = GJ = AK = EL = FM = CN$

$C = FJ = EK = AL = GM = BN$

$D = EJ = FK = GL = AM = HN$

$E = DJ = CK = BL = HM = AN$

$F = CJ = DK = HL = BM = GN$

$G = BJ = HK = DL = CM = FN$

$H = AJ = GK = FL = EM = DN$

$J = DE = CF = BG = AH = MN$

$K = AB = CE = DF = GH = LN$

$L = AC = BE = DG = FH = KN$

$M = AD = BF = CG = EH = JN$

$N = BC = AE = FG = DH = KL = JM$

$AF = BD = EG = CH = JL = KM$

$AG = EF = CD = BH = JK = LM$

2 blocks of 8: $AF = BD = EG = CH = JL = KM$

V Alias Relationships for 2^{k-p} Fractional Factorial Designs with $k \le 15$ and $n \le 64$ (*Continued*)

Designs with 14 Factors

(y) 2^{14-10}; 1/1024 fraction of 14 factors in 16 runs

Resolution III

Design Generators

$E = ABC \quad F = ABD \quad G = ACD \quad H = BCD \quad J = ABCD$

$K = AB \quad L = AC \quad M = AD \quad N = BC \quad O = BD$

Aliases

$A = HJ = BK = CL = DM = EN = FO$
$B = GJ = AK = EL = FM = CN = DO$
$C = FJ = EK = AL = GM = BN = HO$
$D = EJ = FK = GL = AM = HN = BO$
$E = DJ = CK = BL = HM = AN = GO$
$F = CJ = DK = HL = BM = GN = AO$
$G = BJ = HK = DL = CM = FN = EO$
$H = AJ = GK = FL = EM = DN = CO$
$J = DE = CF = BG = AH = MN = LO$
$K = AB = CE = DF = GH = LN = MO$
$L = AC = BE = DG = FH = KN = JO$
$M = AD = BF = CG = EH = JN = KO$
$N = BC = AE = FG = DH = KL = JM$
$O = BD = AF = EG = CH = JL = KM$
$AG = EF = CD = BH = JK = LM = NO$
2 blocks of 8: $AG = EF = CD = BH = JK = LM = NO$

Designs with 15 Factors

(z) 2^{15-11}; 1/2048 fraction of 5 factors in 16 runs

Resolution III

Design Generators

$E = ABC \quad F = ABD \quad G = ACD \quad H = BCD \quad J = ABCD$

$K = AB \quad L = AC \quad M = AD \quad N = BC \quad O = BD \quad P = CD$

Aliases

$A = HJ = BK = CL = DM = EN = FO = GP$
$B = GJ = AK = EL = FM = CN = DO = HP$
$C = FJ = EK = AL = GM = BN = HO = DP$
$D = EJ = FK = GL = AM = HN = BO = CP$
$E = DJ = CK = BL = HM = AN = GO = FP$
$F = CJ = DK = HL = BM = GN = AO = EP$
$G = BJ = HK = DL = CM = FN = EO = AP$
$H = AJ = GK = FL = EM = DN = CO = BP$
$J = DE = CF = BG = AH = MN = LO = KP$
$K = AB = CE = DF = GH = LN = MO = JP$
$L = AC = BE = DG = FH = KN = JO = MP$
$M = AD = BF = CG = EH = JN = KO = LP$
$N = BC = AE = FG = DH = KL = JM = OP$
$O = BD = AF = EG = CH = JL = KM = NP$
$P = CD = EF = AG = BH = JK = LM = NO$

Bibliography

Abraham, B., H. Chipman, and K. Vijayan (1999). "Some Risks in the Construction and Analysis of Supersaturated Designs." *Technometrics*, Vol. 41, pp. 135–141.

Addelman, S. (1961). "Irregular Fractions of the 2^n Factorial Experiments." *Technometrics*, Vol. 3, pp. 479–496.

Addelman, S. (1962). "Orthogonal Main Effect-Plans for Asymmetric Factorial Experiments." *Technometrics*, Vol. 4, pp. 21–46.

Addelman, S. (1963). "Techniques for Constructing Fractional Replicate Plans." *Journal of the American Statistical Association*, Vol. 58, pp. 45–71.

Almimi, A. A., M. Kulahci, and D. C. Montgomery (2009). "Checking the Adequacy of Fit of Models from Split-Plot Designs." *Journal of Quality Technology,* Vol. 41, pp. 272–284.

Almimi, A. A., M. Kulahci, and D. C. Montgomery (2008a). "Follow-Up Designs to Resolve confounding in Split-Plot Experiments." *Journal of Quality Technology,* Vol. 40, pp. 154–166.

Almimi, A. A., M. Kulahci, and D. C. Montgomery (2008b). "Estimation of Missing Observations in Two-level Split-Plot Designs." *Quality and Reliability Engineering International,* Vol. 24, pp. 127–152.

Anderson, V. L., and R. A. McLean (1974). *Design of Experiments: A Realistic Approach.* Dekker, New York.

Andere-Rendon, J., D. C. Montgomery, and D. A. Rollier (1997). "Design of Mixture Experiments Using Bayesian D-Optimality." *Journal of Quality Technology*, Vol. 29, pp. 451–463.

Anderson-Cook, C. M., C. M. Borror, and D. C. Montgomery (2009). "Response Surface Design Evaluation and Comparison" (with discussion). *Journal of Statistical Planning and Inference*, Vol. 139, pp. 629–674.

Anscombe, F. J. (1960). "Rejection of Outliers." *Technometrics*, Vol. 2, pp. 123–147.

Anscombe, F. J., and J. W. Tukey (1963). "The Examination and Analysis of Residuals." *Technometrics*, Vol. 5, pp. 141–160.

Bainbridge, T. R. (1965). "Staggered, Nested Designs for Estimating Variance Components." *Industrial Quality Control*, Vol. 22, pp. 12–20.

Bancroft, T. A. (1968). *Topics in Intermediate Statistical Methods.* Iowa State University Press, Ames, Iowa.

Barlett, M. S. (1947). "The Use of Transformations." *Biometrics*, Vol. 3, pp. 39–52.

Barnett, V., and T. Lewis (1994). *Outliers in Statistical Data.* 3rd edition. Wiley, New York.

Barton, R. R (1994), "Metamodels: A State of the Art Review", *Winter Simulation Conference*, pp. 237–277.

Barton, R. R. (1992), "Metamodels for Simulation Input-Output Relations", *Winter Simulation Conference*, pp. 289–299.

Bennett, C. A., and N. L. Franklin (1954). *Statistical Analysis in Chemistry and the Chemical Industry*. Wiley, New York.

Bingham, D., and R. R. Sitter (1999). "Minimum Aberration Two-Level Fractional Factorial Split-Plot Designs." *Technometrics*, Vol. 41, pp. 62–70.

Bisgaard, S. (1998–1999). "Conditional Inference Chart for Small Unreplicated Two-Level Factorial Experiments." *Quality Engineering*, Vol. 11, pp. 267–271.

Bisgaard, S. (2000). "The Design and Analysis of $2^{k-p} \times 2^{q-r}$ Split Plot Experiments." *Journal of Quality Technology*, Vol. 32, pp. 39–56.

Bisgaard, S. and H. T. Fuller, (1994–95), "Analysis of Factorial Experiments with Defects or Defectives as the Response", *Quality Engineering*, Vol. 7, pp. 429–443.

Booth, K. H., and D. R. Cox, (1962). "Some Systematic Supersaturated Designs." *Technometrics*, Vol. 4, pp. 489–495.

Bowker, A. H., and G. J. Lieberman (1972). *Engineering Statistics*. 2nd edition. Prentice Hall, Englewood Cliffs, NJ.

Box, G. E. P. (1954a). "Some Theorems on Quadratic Forms Applied in the Study of Analysis of Variance Problems: I. Effect of Inequality of Variance in the One-Way Classification." *Annals of Mathematical Statistics*, Vol. 25, pp. 290–302.

Box, G. E. P. (1954b). "Some Theorems on Quadratic Forms Applied in the Study of Analysis of Variance Problems: II. Effect of Inequality of Variance and of Correlation of Errors in the Two-Way Classification." *Annals of Mathematical Statistics*, Vol. 25, pp. 484–498.

Box, G. E. P. (1957). "Evolutionary Operation: A Method for Increasing Industrial Productivity." *Applied Statistics*, Vol. 6, pp. 81–101.

Box, G. E. P. (1988). "Signal-to-Noise Ratios, Performance Criteria, and Transformation." *Technometrics*, Vol. 30, pp. 1–40.

Box, G. E. P. (1992–1993). "Sequential Experimentation and Sequential Assembly of Designs." *Quality Engineering*, Vol. 5, No. 2, pp. 321–330.

Box, G. E. P. (1999). "Statistics as a Catalyst to Learning by Scientific Method Part II—A Discussion" (with Discussion). *Journal of Quality Technology*, Vol. 31, pp. 16–29.

Box, G. E. P., and D. W. Behnken (1960). "Some New Three Level Designs for the Study of Quantitative Variables." *Technometrics*, Vol. 2, pp. 455–476.

Box, G. E. P., S. Bisgaard, and C. A. Fung (1988). "An Explanation and Critique of Taguchi's Contributions to Quality Engineering." *Quality and Reliability Engineering International*, Vol. 4, pp. 123–131.

Box, G. E. P., and D. R. Cox (1964). "An Analysis of Transformations." *Journal of the Royal Statistical Society* B, Vol. 26, pp. 211–243.

Box, G. E. P., and N. R. Draper (1969). *Evolutionary Operation*. Wiley, New York.

Box, G. E. P., and N. R. Draper (2007). Response Surfaces, Mixtures, and Ridge Analysis, Wiley, New York.

Box, G. E. P., and J. S. Hunter (1957). "Multifactor Experimental Designs for Exploring Response Surfaces." *Annals of Mathematical Statistics*, Vol. 28, pp. 195–242.

Box, G. E. P., and J. S. Hunter (1961a). "The 2^{k-p} Fractional Factorial Designs, Part I." *Technometrics*, Vol. 3, pp. 311–352.

Box, G. E. P., and J. S. Hunter (1961b). "The 2^{k-p} Fractional Factorial Designs, Part II." *Technometrics*, Vol. 3, pp. 449–458.

Box, G. E. P., W. G. Hunter, and J. S. Hunter (2005). *Statistics for Experimenters*. 2nd Edition Wiley, New York.

Box, G. E. P., and P. Y. T. Liu (1999). "Statistics as a Catalyst to Learning by Scientific Method Part I—An Example." *Journal of Quality Technology*, Vol. 31, pp. 1–15.

Box, G. E. P., and R. D. Meyer (1986). "An Analysis of Unreplicated Fractional Factorials." *Technometrics*, Vol. 28, pp. 11–18.

Box, G. E. P., and K. G. Wilson (1951). "On the Experimental Attainment of Optimum Conditions." *Journal of the Royal Statistical Society*, B, Vol. 13, pp. 1–45.

Box, J. F. (1978). *R. A. Fisher: The Life of a Scientist*. Wiley, New York.

Burdick, R. K., C. M. Borror, and D. C. Montgomery. (2003). "A Review of Methods for Measurement Systems Capability Analysis." *Journal of Quality Technology*, Vol. 35, No. 4, pp. 342–354.

Burdick, R. K., and F. A. Graybill (1992). *Confidence Intervals on Variance Components*. Dekker, New York.

Bursztyn, D., and D. Steinberg (2006). "Comparison of Designs for Computer Experiments." *Journal of Statistical Planning and Inference*, Vol. 136, pp. 1103–1119.

Byrne, D. M., and S. Taguchi (1987). "The Taguchi Approach to Parameter Design." *Quality Progress*, December, pp. 19–26.

Capehart, S. R., A. Keha, M. Kulahci, and D. C. Montgomery (2011). "Designing Fractional Factorial Split-Plot Experiments using Integer Programming." *International Journal of Experimental Design and Process Optimisation*, Vol. 2, pp. 34–47.

Carmer, S. G., and M. R. Swanson (1973). "Evaluation of Ten Pairwise Multiple Comparison Procedures by Monte Carlo Methods." *Journal of the American Statistical Association*, Vol. 68, No. 314, pp. 66–74.

Chaloner, K., and K. Larntz (1989). "Optimal Bayesian Designs Applied to Logistic Regression Experiments." *Journal of Statistical Planning and Inference*, Vol. 21, pp. 191–208.

Cochran, W. G. (1947). "Some Consequences when the Assumptions for the Analysis of Variance Are Not Satisfied." *Biometrics*, Vol. 3, pp. 22–38.

Cochran, W. G. (1957). "Analysis of Covariance: Its Nature and Uses." *Biometrics*, Vol. 13, No. 3, pp. 261–281.

Cochran, W. G., and G. M. Cox (1957). *Experimental Designs*. 2nd edition. Wiley, New York.

Coleman, D. E., and D. C. Montgomery (1993). "A Systematic Approach to Planning for a Designed Industrial Experiment" (with Discussion). *Technometrics*, Vol. 35, pp. 1–27.

Connor, W. S., and M. Zelen (1959). *Fractional Factorial Experimental Designs for Factors at Three Levels*. Applied Mathematics Series, National Bureau of Standards, Washington, DC No. 54.

Conover, W. J. (1980). *Practical Nonparametric Statistics*. 2nd edition. Wiley, New York.

Conover, W. J., and R. L. Iman (1976). "On Some Alternative Procedures Using Ranks for the Analysis of Experimental Designs." *Communications in Statistics*, Vol. A5, pp. 1349–1368.

Conover, W. J., and R. L. Iman (1981). "Rank Transformations as a Bridge Between Parametric and Nonparametric Statistics" (with Discussion). *The American Statistician*, Vol. 35, pp. 124–133.

Conover, W. J., M. E. Johnson, and M. M. Johnson (1981). "A Comparative Study of Tests for Homogeneity of Variances, with Applications to the Outer Continental Shelf Bidding Data." *Technometrics*, Vol. 23, pp. 351–361.

Cook, D. R. (1977). "Detection of Influential Observations in Linear Regression." *Technometrics*, Vol. 19, pp. 15–18.

Cook, D. R. (1979). "Influential Observations in Linear Regression." *Journal of the American Statistical Association*, Vol. 74, pp. 169–174.

Cornell, J. A. (2002). *Experiments with Mixtures: Designs, Models, and the Analysis of Mixture Data*. 3rd edition. Wiley, New York.

Cornfield, J., and J. W. Tukey (1956). "Average Value of Mean Squares in Factorials," *Annals of Mathematical Statistics*, Vol. 27, pp. 907–949.

Daniel, C. (1959). "Use of Half-Normal Plots in Interpreting Factorial Two Level Experiments." *Technometrics*, Vol. 1, pp. 311–342.

Daniel, C. (1976). *Applications of Statistics to Industrial Experimentation*. Wiley, New York.

Davies, O. L. (1956). *Design and Analysis of Industrial Experiments*. 2nd edition. Hafner Publishing Company, New York.

Derringer, G., and R. Suich (1980). "Simultaneous Optimization of Several Response Variables." *Journal of Quality Technology*, Vol. 12, pp. 214–219.

Dolby, J. L. (1963). "A Quick Method for Choosing a Transformation." *Technometrics*, Vol. 5, pp. 317–326.

Draper, N. R., and W. G. Hunter (1969). "Transformations: Some Examples Revisited." *Technometrics*, Vol. 11, pp. 23–40.

DuMouchel, W., and B. Jones (1994). "A Simple Bayesian Modification of D-Optimal Designs to Reduce Dependence on an Assumed Model." *Technometrics*, Vol. 36, pp. 37–47.

Duncan, A. J. (1986). *Quality Control and Industrial Statistics*. 5th edition. Richard D. Irwin, Homewood, IL.

Duncan, D. B. (1955). "Multiple Range and Multiple *F* Tests." *Biometrics*, Vol. 11, pp. 1–42.

Dunnett, C. W. (1964). "New Tables for Multiple Comparisons with a Control." *Biometrics*, Vol. 20, pp. 482–491.

Edwards, D. J., and R. W. Mee (2008). "Empirically Determined *P*-Values for Lenth *t*-Statistics." *Journal of Quality Technology*, Vol. 40, No. 4, pp. 368–380.

Eisenhart, C. (1947). "The Assumptions Underlying the Analysis of Variance." *Biometrics*, Vol. 3, pp. 1–21.

Fang, K. T. (1980), "The Uniform Design: Application of Number-Theoretic Methods in Experimental Design," *Acta Math. Appl. Sinica*., Vol. 3, pp. 363–372.

Fang, K. T., R. Li, and A. Sudjianto (2006), *Design and Modeling for Computer Experiments*, Taylor & Francis Group, Boca Raton, FL.

Fisher, R. A. (1958). *Statistical Methods for Research Workers*. 13th edition. Oliver and Boyd, Edinburgh.

Fisher, R. A. (1966). *The Design of Experiments*. 8th edition. Hafner Publishing Company, New York.

Fisher, R. A., and F. Yates (1953). *Statistical Tables for Biological, Agricultural, and Medical Research*. 4th edition. Oliver and Boyd, Edinburgh.

Fries, A., and W. G. Hunter (1980). "Minimum Aberration 2^{k-p} Designs." *Technometrics*, Vol. 22, pp. 601–608.

Gaylor, D. W., and T. D. Hartwell (1969). "Expected Mean Squares for Nested Classifications." *Biometrics*, Vol. 25, pp. 427–430.

Gaylor, D. W., and F. N. Hopper (1969). "Estimating the Degrees of Freedom for Linear Combinations of Mean Squares by Satterthwaite's Formula." *Technometrics*, Vol. 11, No. 4, pp. 699–706.

Giovannitti-Jensen, A., and R. H. Myers (1989). "Graphical Assessment of the Prediction Capability of Response Surface Designs." *Technometrics*, Vol. 31, pp. 159–171.

Gotwalt, C. M., B. A. Jones, and D. M. Steinberg (2009). "Fast Computation of Designs Robust to Parameter Uncertainty for Nonlinear Settings." *Technometrics,* Vol. 51, pp. 86–95.

Graybill, F. A. (1961). *An Introduction to Linear Statistical Models*. Vol. 1. McGraw-Hill, New York.

Graybill, F. A., and C. M. Wang (1980). "Confidence Intervals on Nonnegative Linear Combinations of Variances." *Journal of the American Statistical Association*, Vol. 75, pp. 869–873.

Graybill, F. A., and D. L. Weeks (1959). "Combining Interblock and Intrablock Information in Balanced Incomplete Blocks." *Annals of Mathematical Statistics*, Vol. 30, pp. 799–805.

Hall, M. Jr. (1961). *Hadamard Matrix of Order 16*, Jet Propulsion Laboratory Research Summary, 1, pp. 21–26.

Hamada, M., and N. Balakrishnan (1998). "Analyzing Unreplicated Factorial Experiments: A Review with Some New Proposals" (with Discussion). *Statistica Sinica*, Vol. 8, pp. 1–41.

Hamada, M., and C. F. J. Wu (1992). "Analysis of Designed Experiments with Complex Aliasing." *Journal of Quality Technology*, Vol. 24, No. 3, pp. 130–137.

Hill, W. G., and W. G. Hunter (1966). "A Review of Response Surface Methodology: A Literature Survey." *Technometrics*, Vol. 8, pp. 571–590.

Hines, W. W., D. C. Montgomery, D. M. Goldsman, and C. M. Borror (2003). *Probability and Statistics in Engineering*. 4th edition. Wiley, New York.

Hocking, R. R. (1973). "A Discussion of the Two-Way Mixed Model." *The American Statistician*, Vol. 27, No. 4, pp. 148–152.

Hocking, R. R., O. P. Hackney, and F. M. Speed (1978). "The Analysis of Linear Models with Unbalanced Data." in *Contributions to Survey Sampling and Applied Statistics*, H. A. David (ed.). Academic Press, New York.

Hocking, R. R., and F. M. Speed (1975). "A Full Rank Analysis of Some Linear Model Problems." *Journal of the American Statistical Association*, Vol. 70, pp. 706–712.

Holcomb, D. R. and W. M. Carlyle, (2002), "Some Notes on the Construction and Evaluation of Supersaturated Designs", *Quality and Reliability Engineering International*, Vol. 18, pp. 299–304.

Holcomb, D. R., D. C. Montgomery, and W. M. Carlyle (2003). "Analysis of Supersaturated Designs." *Journal of Quality Technology*, Vol. 35, No. 1, pp. 13–27.

Hsu, J. (1996), *Multiple Comparisons: Theory and Methods*, Chapman and Hall/CRC, Boca Raton, FL.

Huang, P., D. Chen, and J. Voelkel (1999). "Minimum Aberration Two-Level Split-Plot Designs." *Technometrics*, Vol. 41, pp. 314–326.

Hunter, J. S. (1985). "Statistical Design Applied to Product Design." *Journal of Quality Technology*, Vol. 17, pp. 210–221.

Hunter, J. S. (1989). "Let's All Beware the Latin Square." *Quality Engineering*, Vol. 1, pp. 453–465.

John, J. A., and P. Prescott (1975). "Critical Values of a Test to Detect Outliers in Factorial Experiments." *Applied Statistics*, Vol. 24, pp. 56–59.

John, P. W. M. (1961). "The Three-Quarter Replicates of 2^4 and 2^5 Designs." *Biometrics*, Vol. 17, pp. 319–321.

John, P. W. M. (1962). "Three-Quarter Replicates of 2^n "Designs." *Biometrics*, Vol. 18, pp. 171–184.

John, P. W. M. (1964). "Blocking a $3(2^{n-k})$ Design." *Technometrics*, Vol. 6, pp. 371–376.

John, P. W. M. (1971). *Statistical Design and Analysis of Experiments*. Macmillan, New York.

Johnson, M. E., L. M. Moore, and D. Ylvisaker (1990), "Minimax and Maxmin Distance Design", *Journal of Statistical Planning and Inference, Vol.* 26, pp 131–148.

Johnson, R. T., and D. C. Montgomery (2009). "Choice of Second-Order Response Surface Designs for Logistic and Poisson Regression Models." *International Journal of Experimental Design and Process Optimization*, Vol. 1, pp. 2–23.

Johnson, R. T., and D. C. Montgomery (2010). "Designing Experiments for Nonlinear Models—An Introduction." *Quality and Reliability Engineering International*, Vol. 26, pp. 431–441.

Johnson, L., and S. Burrows (2011). "For Starbucks, It's in the Bag." *Quality* Progress, March, pp. 18–23.

Jones, B., and R. T. Johnson (2009). "Design and Analysis for the Gaussian Process Model." *Quality and Reliability Engineering International*, Vol. 25, pp. 515–524.

Jones, B., and D. C. Montgomery (2010). "Alternatives to Resolution IV Screening Designs in 16 Runs." *International Journal of Experimental Design and Process Optimisation*, Vol. 1, No. 4, pp. 285–295.

Jones, B., and C. J. Nachtsheim (2009). "Split-Plot Designs: What, Why, and How." *Journal of Quality Technology*, Vol. 41, pp. 340–361.

Jones, B., and C. J. Nachtsheim (2011a). "Efficient Designs with Minimal Aliasing." *Technometrics*, Vol. 53, pp. 62–71.

Jones, B., and C. J. Nachtsheim (2011b). "A Class of Three-Level Designs for Definitive Screening in the Presence of Second-Order Effects." *Journal of Quality Technology*, Vol. 43, pp. 1–15.

Jones, B., Lin, D. K. J., and C. J. Nachtsheim (2008). "Bayesian *D*-optimal Supersaturated Designs." *Journal of Statistical Planning and Inference*, Vol. 138, pp. 86–92.

Kackar, R. N. (1985). "Off-Line Quality Control, Parameter Design, and the Taguchi Method." *Journal of Quality Technology*, Vol. 17, pp. 176–188.

Kempthorne, O. (1952). *The Design and Analysis of Experiments*. Wiley, New York.

Keuls, M. (1952). "The Use of the Studentized Range in Connection with an Analysis of Variance." *Euphytica*, Vol. 1, pp. 112–122.

Khuri, A. I., and J. A. Cornell (1996). *Response Surfaces: Designs and Analyses.* 2nd edition. Dekker, New York.

Kiefer, J. (1961). "Optimum Designs in Regression Problems II." *Annals of Mathematical Statistics,* Vol. 32, pp. 298–325.

Kiefer, J., and J. Wolfowitz (1959). "Optimum Designs in Regression Problems." *Annals of Mathematical Statistics*, Vol. 30, pp. 271–294.

Kruskal, W. H., and W. A. Wallis (1952). "Use of Ranks on One Criterion Variance Analysis." *Journal of the American Statistical Association*, Vol. 47, pp. 583–621 (corrections appear in Vol. 48, pp. 907–911).

Larntz, K., and P. Whitcomb (1998). "Use of Replication in Almost Unreplicated Factorials." Presented at the Fall Technical Conference, Corning, NY.

Lenth, R. V. (1989). "Quick and Easy Analysis of Unreplicated Factorials." *Technometrics*, Vol. 31, pp. 469–473.

Leon, R. V., A. C. Shoemaker, and R. N. Kackar (1987). "Performance Measures Independent of Adjustment." *Technometrics*, Vol. 29, pp. 253–265.

Levene, H. (1960). "Robust Tests for Equality of Variance." in *Contributions to Probability and Statistics*, Z. Olkin (ed.). Stanford University Press, Palo Alto, CA, pp. 278–292.

Li, W. W., and C. F. J. Wu (1997). "Columnwise-Pairwise Algorithms with Applications to the Construction of Supersaturated Designs." *Technometrics*, Vol. 39, pp. 171–179.

Li, X., Sudarsanam, N., and D. D. Frey (2006). "Regularities in Data from Factorial Experiments." *Complexity*, Vol. 11, No. 5, pp. 33–45.

Lin, D. K. J. (1993). "A New Class of Supersaturated Designs." *Technometrics*, Vol. 35, pp. 28–31.

Lin, D. K. J. (1995). "Generating Systematic Supersaturated Designs." *Technometrics*, Vol. 37, pp. 213–223.

Lin, D. K. J. (2000). "Recent Developments in Supersaturated Designs." in *Statistical Process Monitoring and Optimization*, S. H. Park and G. G. Vining (eds). Marcel Dekker, New York, pp. 305–319.

Loughin, T. M. (1998). "Calibration of the Lenth Test for Unreplicated Factorial Designs." *Journal of Quality Technology*, Vol. 30, pp. 171–175.

Loughin, T. M., and W. Noble (1997). "A Permutation Test for Effects in an Unreplicated Factorial Design." *Technometrics*, Vol. 39, pp. 180–190.

Macharia, H., and P. Goos (2010). "D-Optimal and D-Efficient Equivalent Estimation Second-Order Split-Plot Designs." *Journal of Quality Technology*, Vol. 42, pp. 358–372.

Margolin, B. H. (1967). "Systematic Methods of Analyzing 2^n3^m Factorial Experiments with Applications." *Technometrics*, Vol. 9, pp. 245–260.

Margolin, B. H. (1969). "Results on Factorial Designs of Resolution IV for the 2^n and 2^n3^m Series." *Technometrics*, Vol. 11, pp. 431–444.

McCullagh, P., and J. A. Nelder (1989). *Generalized Linear Models*. 2nd edition. Chapman and Hall, New York.

McKay, N. D., W. J. Conover, and R. J. Beckman (1979), "A Comparison of Three Methods for Selecting Values of Input Variables in the Analysis of Output from a Computer Code", *Technometrics*, Vol. 21, pp 239–245.

Mee, R. W. (2009). *A Comprehensive Guide to Factorial Two-Level Experimentation*. Springer, New York.

Mee, R. W. (2004). "Efficient Two-Level Designs for Estimating All Main Effects and Two-Factor Interactions." *Journal of Quality Technology*, Vol. 36, pp. 400–412.

Mee, R. W., and Peralta, M. (2000). "Semifolding 2^{k-p} Designs." *Technometrics*, Vol. 42, No. 2, pp. 122–143.

Meyer, R. K., and C. J. Nachtsheim (1995). "The Coordinate-Exchange Algorithm for Constructing Exact Optimal Experimental Designs." *Technometrics*, Vol. 37, pp. 60–69.

Miller, R. G. (1991). *Simultaneous Statistical Inference*. Springer–Verlag, New York.

Miller, R. G., Jr (1977). "Developments in Multiple Comparisons, 1966–1976." *Journal of the American Statistical Association*, Vol. 72, pp. 779–788.

Milliken, G. A., and D. E. Johnson (1984). *Analysis of Messy Data, Vol. 1*. Van Nostrand Reinhold, New York.

Mitchell, T. J. (1974). "An Algorithm for the Construction of *D*-Optimal Experimental Designs." *Technometrics*, Vol. 16, pp. 203–210.

Mitchell, T. J. and M. D. Morris (1992), "The Spatial Correlation Approach to Response Surface Estimation", *Winter Simulation Conference*, pp. 565–571.

Montgomery, D. C. (2009). *Introduction to Statistical Quality Control*. 6th edition. Wiley, New York.

Montgomery, D. C. (1999). "Experimental Design for Product and Process Design and Development." *Journal of the Royal Statistical Society* D, Vol. 48, pp. 159–177.

Montgomery, D. C., C. M. Borror, and J. D. Stanley (1997–1998). "Some Cautions in the Use of Plackett– Burman Designs." *Quality Engineering*, Vol. 10, pp. 371– 381.

Montgomery, D. C., E. A. Peck, and G. G. Vining (2006). *Introduction to Linear Regression Analysis*. 4th edition. Wiley, New York.

Montgomery, D. C., and G. C. Runger (1993a). "Gauge Capability Analysis and Designed Experiments, Part I: Basic Methods." *Quality Engineering*, Vol. 6, pp. 115–135.

Montgomery, D. C., and G. C. Runger (1993b). "Gauge Capability Analysis and Designed Experiments, Part II: Experimental Design Models and Variance Component Estimation." *Quality Engineering*, Vol. 6, pp. 289–305.

Montgomery, D. C., and G. C. Runger (1996). "Foldovers of 2^{k-p} Resolution IV Experimental Designs." *Journal of Quality Technology*, Vol. 28, pp. 446–450.

Montgomery, D. C., and G. C. Runger (2011). *Applied Statistics and Probability for Engineers*. 5th edition. Wiley, New York.

Myers, R. H. (1990). *Classical and Modern Regression with Applications*. 2nd edition. PNS-Kent, Boston.

Myers, R. H., D. C. Montgomery and C. M. Anderson-Cook (2009). *Response Surface Methodology: Process and Product Optimization Using Designed Experiments*. 3rd edition. Wiley, New York.

Myers, R. H., and D. C. Montgomery (1997). "A Tutorial on Generalized Linear Models." *Journal of Quality Technology*, Vol. 29, pp. 274–291.

Myers, R. H., D. C. Montgomery, G. G. Vining and T. J. Robinson (2010). *Generalized Linear Models with Applications in Engineering and the Sciences*. 2nd edition Wiley, New York.

Myers, R. H., D. C. Montgomery, G. G. Vining, C. M. Borror, and S. M. Kowalski (2004). "Response Surface Methodology: A Retrospective and Literature Survey." *Journal of Quality Technology*, Vol. 36, pp. 53–77.

Nair, V. N., (editor) (1992). "Taguchi's Parameter Design: A Panel Discussion." *Technometrics*, Vol. 34, pp. 127–161.

Nelder, J. A., and R. W. M. Wedderburn (1972). "Generalized Linear Models." *Journal of the Royal Statistical Society* A, Vol. 135, pp. 370–384.

Nelson, L. S. (1995a). "Using Nested Designs I: Estimation of Standard Deviations." *Journal of Quality Technology*, Vol. 27, No. 2, pp. 169–171.

Nelson, L. S. (1995b). "Using Nested Designs II: Confidence Limits for Standard Deviations." *Journal of Quality Technology*, Vol. 27, No. 3, pp. 265–267.

Nelson, L. S. (1983). "Variance Estimation Using Staggered, Nested Designs." *Journal of Quality Technology*, Vol. 15, pp. 195–198.

Nelson, P. R. (1989). "Multiple Comparison of Means Using Simultaneous Confidence Intervals." *Journal of Quality Technology*, Vol. 21, No. 4, pp. 232–241.

Newman, D. (1939). "The Distribution of the Range in Samples from a Normal Population, Expressed in Terms of an Independent Estimate of Standard Deviation." *Biometrika*, Vol. 31, pp. 20–30.

O'Neill, R., and G. B. Wetherill (1971). "The Present State of Multiple Comparison Methods." *Journal of the Royal Statistical Society* B, Vol. 33, pp. 218–241.

Ostle, B. (1963). *Statistics in Research*. 2nd edition. Iowa State Press, Ames, Iowa.

Pearson, E. S., and H. O. Hartley (1966). *Biometrika Tables for Statisticians*, Vol. 1. 3rd edition. Cambridge University Press, Cambridge.

Pearson, E. S., and H. O. Hartley (1972). *Biometrika Tables for Statisticians*, Vol. 2. Cambridge University Press, Cambridge.

Pignatiello, J. J., Jr. and J. S. Ramberg (1992). "Top Ten Triumphs and Tragedies of Genichi Taguchi." *Quality Engineering*, Vol. 4, pp. 211–225.

Plackett, R. L., and J. P. Burman (1946). "The Design of Optimum Multifactorial Experiments." *Biometrika*, Vol. 33, pp. 305–325.

Quenouille, M. H. (1953). *The Design and Analysis of Experiments*. Charles Griffin and Company, London.

Sacks, J., W. J. Welch, T. J. Mithchell, and H. P. Wynn (1989). "Design and Analysis of Computer Experiments," *Statistical Science* 4(4), pp. 409–423.

Santner, T. J., B. J. Williams, and W. I. Notz (2003). *The Design and Analysis of Computer Experiments*, Springer Series in Statistics Springer–Verlag, New York.

Satterthwaite, F. E. (1946). "An Approximate Distribution of Estimates of Variance Components." *Biometrics Bull.*, Vol. 2, pp. 110–112.

Satterthwaite, F. E. (1959). "Random Balance Experimentation" (with discussion). *Technometrics*, Vol. 1, pp. 111–137.

Scheffé, H. (1953). "A Method for Judging All Contrasts in the Analysis of Variance." *Biometrika*, Vol. 40, pp. 87–104.

Scheffé, H. (1956a). "A 'Mixed Model' for the Analysis of Variance." *Annals of Mathematical Statistics*, Vol. 27, pp. 23–36.

Scheffé, H. (1956b). "Alternative Models for the Analysis of Variance." *Annals of Mathematical Statistics*, Vol. 27, pp. 251–271.

Scheffé, H. (1959). *The Analysis of Variance*. Wiley, New York.

Searle, S. R. (1971a). *Linear Models*. Wiley, New York.

Searle, S. R. (1971b). "Topics in Variance Component Estimation." *Biometrics*, Vol. 27, pp. 1–76.

Searle, S. R. (1987). *Linear Models for Unbalanced Data*. Wiley, New York.

Searle, S. R., G. Casella, and G. E. McCulloch (1992). *Variance Components*. Wiley, New York.

Searle, S. R., and R. F. Fawcett (1970). "Expected Mean Squares in Variance Component Models Having Finite Populations," *Biometrics*, Vol. 26, pp. 243–254.

Searle, S. R., F. M. Speed, and H. V. Henderson (1981). "Some Computational and Model Equivalences in Analyses of Variance of Unequal-Subclass-Numbers Data." *The American Statistician*, Vol. 35, pp. 16–33.

Simpson, T. and J. Peplinski (1997), "On the Use of Statistics in Design and the Implications for Deterministic Computer Simulation", *ASME Design Engineering Technical Conference*, pp. 1–12.

Smith, H. F. (1957). "Interpretations of Adjusted Treatment Means and Regressions in Analysis of Covariance." *Biometrics*, Vol. 13, No. 3, pp. 282–308.

Smith, J. R., and J. M. Beverly (1981). "The Use and Analysis of Staggered Nested Factorial Designs." *Journal of Quality Technology*, Vol. 13, pp. 166–173.

Speed, F. M., and R. R. Hocking (1976). "The Use of the *R* ( )-Notation with Unbalanced Data." *The American Statistician*, Vol. 30, pp. 30–33.

Speed, F. M., R. R. Hocking, and O. P. Hackney (1978). "Methods of Analysis of Linear Models with Unbalanced Data." *Journal of the American Statistical Association*, Vol. 73, pp. 105–112.

Stefansky, W. (1972). "Rejecting Outliers in Factorial Designs." *Technometrics*, Vol. 14, pp. 469–479.

Taguchi, G. (1987). *System of Experimental Design: Engineering Methods to Optimize Quality and Minimize Cost*, UNIPUB, White Plains, NY.

Taguchi, G., (1991). *Introduction to Quality Engineering*. Asian Productivity Organization, UNIPUB, White Plains, NY.

Taguchi, G., and Y. Wu (1980). *Introduction to Off-Line Quality Control*. Central Japan Quality Control Association, Nagoya, Japan.

Ting, N., R. K. Burdick, F. A. Graybill, S. Jeyaratnam, and T.-F. C. Lu (1990). "Confidence Intervals on Linear Combinations of Variance Components That Are Unrestricted in Sign." *Journal of Statistical Computation and Simulation*, Vol. 35, pp. 135–143.

Tukey, J. W. (1949a). "One Degree of Freedom for Non-Additivity." *Biometrics*, Vol. 5, pp. 232–242.

Tukey, J. W. (1949b). "Comparing Individual Means in the Analysis of Variance." *Biometrics*, Vol. 5, pp. 99–114.

Tukey, J. W. (1951). "Quick and Dirty Methods in Statistics, Part II. Simple Analysis for Standard Designs." *Proceedings of the Fifth Annual Convention, American Society for Quality Control*, pp. 189–197.

Tukey, J. W. (1953). "The Problem of Multiple Comparisons." Unpublished Notes, Princeton University.

Vining, G. G., S. M. Kowalski, and D. C. Montgomery (2005). "Response Surface Designs Within a Split-Plot Structure." *Journal of Quality Technology*, Vol. 37, pp. 115–129.

Vining, G. G., and S. M. Kowalski (2008). "Exact Inference for Response Surface Designs Within a Split-Plot Structure." *Journal of Quality Technology*, Vol. 40, pp. 394–406.

Winer, B. J. (1971). *Statistical Principles in Experimental Design*. 2nd edition. McGraw-Hill, New York.

Wu, C. F. J. (1993). "Construction of Supersaturated Designs Through Partially Aliased Interactions." *Biometrika*, Vol. 80, pp. 661–669.

Xiao, L., D.K.J. Lin, and F. Bai (2012). "Constructing Difinitive Screening Designs Using Conference Matrices." *Journal of Quality Technology*, Vol. 44, pp. 2–8.

Yates, F. (1934). "The Analysis of Multiple Classifications with Unequal Numbers in the Different Classes." *Journal of the American Statistical Association*, Vol. 29, pp. 52–66.

Yates, F. (1937). *Design and Analysis of Factorial Experiments*. Tech. Comm. No. 35, Imperial Bureau of Soil Sciences, London.

Yates, F. (1940). "The Recovery of Interblock Information in Balanced Incomplete Block Designs." *Annals of Eugenics*, Vol. 10, pp. 317–325.

Ye, K., and M. Hamada (2000). "Critical Values of the Lenth Method for Unreplicated Factorial Designs." *Journal of Quality Technology*, Vol. 32, pp. 57–66.

Index

Davies, O. L. (1956). *Design and Analysis of Industrial Experiments*. 2nd edition. Hafner Publishing Company, New York.

Derringer, G., and R. Suich (1980). "Simultaneous Optimization of Several Response Variables." *Journal of Quality Technology*, Vol. 12, pp. 214–219.

Dolby, J. L. (1963). "A Quick Method for Choosing a Transformation." *Technometrics*, Vol. 5, pp. 317–326.

Draper, N. R., and W. G. Hunter (1969). "Transformations: Some Examples Revisited." *Technometrics*, Vol. 11, pp. 23–40.

DuMouchel, W., and B. Jones (1994). "A Simple Bayesian Modification of D-Optimal Designs to Reduce Dependence on an Assumed Model." *Technometrics*, Vol. 36, pp. 37–47.

Duncan, A. J. (1986). *Quality Control and Industrial Statistics*. 5th edition. Richard D. Irwin, Homewood, IL.

Duncan, D. B. (1955). "Multiple Range and Multiple *F* Tests." *Biometrics*, Vol. 11, pp. 1–42.

Dunnett, C. W. (1964). "New Tables for Multiple Comparisons with a Control." *Biometrics*, Vol. 20, pp. 482–491.

Edwards, D. J., and R. W. Mee (2008). "Empirically Determined *P*-Values for Lenth *t*-Statistics." *Journal of Quality Technology*, Vol. 40, No. 4, pp. 368–380.

Eisenhart, C. (1947). "The Assumptions Underlying the Analysis of Variance." *Biometrics*, Vol. 3, pp. 1–21.

Fang, K. T. (1980), "The Uniform Design: Application of Number-Theoretic Methods in Experimental Design," *Acta Math. Appl. Sinica*., Vol. 3, pp. 363–372.

Fang, K. T., R. Li, and A. Sudjianto (2006), *Design and Modeling for Computer Experiments*, Taylor & Francis Group, Boca Raton, FL.

Fisher, R. A. (1958). *Statistical Methods for Research Workers*. 13th edition. Oliver and Boyd, Edinburgh.

Fisher, R. A. (1966). *The Design of Experiments*. 8th edition. Hafner Publishing Company, New York.

Fisher, R. A., and F. Yates (1953). *Statistical Tables for Biological, Agricultural, and Medical Research*. 4th edition. Oliver and Boyd, Edinburgh.

Fries, A., and W. G. Hunter (1980). "Minimum Aberration 2^{k-p} Designs." *Technometrics*, Vol. 22, pp. 601–608.

Gaylor, D. W., and T. D. Hartwell (1969). "Expected Mean Squares for Nested Classifications." *Biometrics*, Vol. 25, pp. 427–430.

Gaylor, D. W., and F. N. Hopper (1969). "Estimating the Degrees of Freedom for Linear Combinations of Mean Squares by Satterthwaite's Formula." *Technometrics*, Vol. 11, No. 4, pp. 699–706.

Giovannitti-Jensen, A., and R. H. Myers (1989). "Graphical Assessment of the Prediction Capability of Response Surface Designs." *Technometrics*, Vol. 31, pp. 159–171.

Gotwalt, C. M., B. A. Jones, and D. M. Steinberg (2009). "Fast Computation of Designs Robust to Parameter Uncertainty for Nonlinear Settings." *Technometrics*, Vol. 51, pp. 86–95.

Graybill, F. A. (1961). *An Introduction to Linear Statistical Models*. Vol. 1. McGraw-Hill, New York.

Graybill, F. A., and C. M. Wang (1980). "Confidence Intervals on Nonnegative Linear Combinations of Variances." *Journal of the American Statistical Association*, Vol. 75, pp. 869–873.

Graybill, F. A., and D. L. Weeks (1959). "Combining Interblock and Intrablock Information in Balanced Incomplete Blocks." *Annals of Mathematical Statistics*, Vol. 30, pp. 799–805.

Hall, M. Jr. (1961). *Hadamard Matrix of Order 16*, Jet Propulsion Laboratory Research Summary, 1, pp. 21–26.

Hamada, M., and N. Balakrishnan (1998). "Analyzing Unreplicated Factorial Experiments: A Review with Some New Proposals" (with Discussion). *Statistica Sinica*, Vol. 8, pp. 1–41.

Hamada, M., and C. F. J. Wu (1992). "Analysis of Designed Experiments with Complex Aliasing." *Journal of Quality Technology*, Vol. 24, No. 3, pp. 130–137.

Hill, W. G., and W. G. Hunter (1966). "A Review of Response Surface Methodology: A Literature Survey." *Technometrics*, Vol. 8, pp. 571–590.

Hines, W. W., D. C. Montgomery, D. M. Goldsman, and C. M. Borror (2003). *Probability and Statistics in Engineering*. 4th edition. Wiley, New York.

Hocking, R. R. (1973). "A Discussion of the Two-Way Mixed Model." *The American Statistician*, Vol. 27, No. 4, pp. 148–152.

Hocking, R. R., O. P. Hackney, and F. M. Speed (1978). "The Analysis of Linear Models with Unbalanced Data." in *Contributions to Survey Sampling and Applied Statistics*, H. A. David (ed.). Academic Press, New York.

Hocking, R. R., and F. M. Speed (1975). "A Full Rank Analysis of Some Linear Model Problems." *Journal of the American Statistical Association*, Vol. 70, pp. 706–712.

Holcomb, D. R. and W. M. Carlyle, (2002), "Some Notes on the Construction and Evaluation of Supersaturated Designs", *Quality and Reliability Engineering International*, Vol. 18, pp. 299–304.

Holcomb, D. R., D. C. Montgomery, and W. M. Carlyle (2003). "Analysis of Supersaturated Designs." *Journal of Quality Technology*, Vol. 35, No. 1, pp. 13–27.

Hsu, J. (1996), *Multiple Comparisons: Theory and Methods*, Chapman and Hall/CRC, Boca Raton, FL.

Huang, P., D. Chen, and J. Voelkel (1999). "Minimum Aberration Two-Level Split-Plot Designs." *Technometrics*, Vol. 41, pp. 314–326.

Hunter, J. S. (1985). "Statistical Design Applied to Product Design." *Journal of Quality Technology*, Vol. 17, pp. 210–221.

Hunter, J. S. (1989). "Let's All Beware the Latin Square." *Quality Engineering*, Vol. 1, pp. 453–465.

John, J. A., and P. Prescott (1975). "Critical Values of a Test to Detect Outliers in Factorial Experiments." *Applied Statistics*, Vol. 24, pp. 56–59.

John, P. W. M. (1961). "The Three-Quarter Replicates of 2^4 and 2^5 Designs." *Biometrics*, Vol. 17, pp. 319–321.

John, P. W. M. (1962). "Three-Quarter Replicates of 2^n "Designs." *Biometrics*, Vol. 18, pp. 171–184.

John, P. W. M. (1964). "Blocking a $3(2^{n-k})$ Design." *Technometrics*, Vol. 6, pp. 371–376.

John, P. W. M. (1971). *Statistical Design and Analysis of Experiments*. Macmillan, New York.

Johnson, M. E., L. M. Moore, and D. Ylvisaker (1990), "Minimax and Maxmin Distance Design", *Journal of Statistical Planning and Inference, Vol.* 26, pp 131–148.

Johnson, R. T., and D. C. Montgomery (2009). "Choice of Second-Order Response Surface Designs for Logistic and Poisson Regression Models." *International Journal of Experimental Design and Process Optimization*, Vol. 1, pp. 2–23.

Johnson, R. T., and D. C. Montgomery (2010). "Designing Experiments for Nonlinear Models—An Introduction." *Quality and Reliability Engineering International*, Vol. 26, pp. 431–441.

Johnson, L., and S. Burrows (2011). "For Starbucks, It's in the Bag." *Quality* Progress, March, pp. 18–23.

Jones, B., and R. T. Johnson (2009). "Design and Analysis for the Gaussian Process Model." *Quality and Reliability Engineering International*, Vol. 25, pp. 515–524.

Jones, B., and D. C. Montgomery (2010). "Alternatives to Resolution IV Screening Designs in 16 Runs." *International Journal of Experimental Design and Process Optimisation*, Vol. 1, No. 4, pp. 285–295.

Jones, B., and C. J. Nachtsheim (2009). "Split-Plot Designs: What, Why, and How." *Journal of Quality Technology*, Vol. 41, pp. 340–361.

Jones, B., and C. J. Nachtsheim (2011a). "Efficient Designs with Minimal Aliasing." *Technometrics*, Vol. 53, pp. 62–71.

Jones, B., and C. J. Nachtsheim (2011b). "A Class of Three-Level Designs for Definitive Screening in the Presence of Second-Order Effects." *Journal of Quality Technology*, Vol. 43, pp. 1–15.

Jones, B., Lin, D. K. J., and C. J. Nachtsheim (2008). "Bayesian *D*-optimal Supersaturated Designs." *Journal of Statistical Planning and Inference*, Vol. 138, pp. 86–92.

Kackar, R. N. (1985). "Off-Line Quality Control, Parameter Design, and the Taguchi Method." *Journal of Quality Technology*, Vol. 17, pp. 176–188.

Kempthorne, O. (1952). *The Design and Analysis of Experiments*. Wiley, New York.

Keuls, M. (1952). "The Use of the Studentized Range in Connection with an Analysis of Variance." *Euphytica*, Vol. 1, pp. 112–122.

Khuri, A. I., and J. A. Cornell (1996). *Response Surfaces: Designs and Analyses*. 2nd edition. Dekker, New York.

Kiefer, J. (1961). "Optimum Designs in Regression Problems II." *Annals of Mathematical Statistics*, Vol. 32, pp. 298–325.

Kiefer, J., and J. Wolfowitz (1959). "Optimum Designs in Regression Problems." *Annals of Mathematical Statistics*, Vol. 30, pp. 271–294.

Kruskal, W. H., and W. A. Wallis (1952). "Use of Ranks on One Criterion Variance Analysis." *Journal of the American Statistical Association*, Vol. 47, pp. 583–621 (corrections appear in Vol. 48, pp. 907–911).

Larntz, K., and P. Whitcomb (1998). "Use of Replication in Almost Unreplicated Factorials." Presented at the Fall Technical Conference, Corning, NY.

Lenth, R. V. (1989). "Quick and Easy Analysis of Unreplicated Factorials." *Technometrics*, Vol. 31, pp. 469–473.

Leon, R. V., A. C. Shoemaker, and R. N. Kackar (1987). "Performance Measures Independent of Adjustment." *Technometrics*, Vol. 29, pp. 253–265.

Levene, H. (1960). "Robust Tests for Equality of Variance." in *Contributions to Probability and Statistics*, Z. Olkin (ed.). Stanford University Press, Palo Alto, CA, pp. 278–292.

Li, W. W., and C. F. J. Wu (1997). "Columnwise-Pairwise Algorithms with Applications to the Construction of Supersaturated Designs." *Technometrics*, Vol. 39, pp. 171–179.

Li, X., Sudarsanam, N., and D. D. Frey (2006). "Regularities in Data from Factorial Experiments." *Complexity*, Vol. 11, No. 5, pp. 33–45.

Lin, D. K. J. (1993). "A New Class of Supersaturated Designs." *Technometrics*, Vol. 35, pp. 28–31.

Lin, D. K. J. (1995). "Generating Systematic Supersaturated Designs." *Technometrics*, Vol. 37, pp. 213–223.

Lin, D. K. J. (2000). "Recent Developments in Supersaturated Designs." in *Statistical Process Monitoring and Optimization*, S. H. Park and G. G. Vining (eds). Marcel Dekker, New York, pp. 305–319.

Loughin, T. M. (1998). "Calibration of the Lenth Test for Unreplicated Factorial Designs." *Journal of Quality Technology*, Vol. 30, pp. 171–175.

Loughin, T. M., and W. Noble (1997). "A Permutation Test for Effects in an Unreplicated Factorial Design." *Technometrics*, Vol. 39, pp. 180–190.

Macharia, H., and P. Goos (2010). "D-Optimal and D-Efficient Equivalent Estimation Second-Order Split-Plot Designs." *Journal of Quality Technology*, Vol. 42, pp. 358–372.

Margolin, B. H. (1967). "Systematic Methods of Analyzing 2^n3^m Factorial Experiments with Applications." *Technometrics*, Vol. 9, pp. 245–260.

Margolin, B. H. (1969). "Results on Factorial Designs of Resolution IV for the 2^n and 2^n3^m Series." *Technometrics*, Vol. 11, pp. 431–444.

McCullagh, P., and J. A. Nelder (1989). *Generalized Linear Models*. 2nd edition. Chapman and Hall, New York.

McKay, N. D., W. J. Conover, and R. J. Beckman (1979), "A Comparison of Three Methods for Selecting Values of Input Variables in the Analysis of Output from a Computer Code", *Technometrics*, Vol. 21, pp 239–245.

Mee, R. W. (2009). *A Comprehensive Guide to Factorial Two-Level Experimentation*. Springer, New York.

Mee, R. W. (2004). "Efficient Two-Level Designs for Estimating All Main Effects and Two-Factor Interactions." *Journal of Quality Technology*, Vol. 36, pp. 400–412.

Mee, R. W., and Peralta, M. (2000). "Semifolding 2^{k-p} Designs." *Technometrics*, Vol. 42, No. 2, pp. 122–143.

Meyer, R. K., and C. J. Nachtsheim (1995). "The Coordinate-Exchange Algorithm for Constructing Exact Optimal Experimental Designs." *Technometrics*, Vol. 37, pp. 60–69.

Miller, R. G. (1991). *Simultaneous Statistical Inference*. Springer–Verlag, New York.

Miller, R. G., Jr (1977). "Developments in Multiple Comparisons, 1966–1976." *Journal of the American Statistical Association*, Vol. 72, pp. 779–788.

Milliken, G. A., and D. E. Johnson (1984). *Analysis of Messy Data, Vol. 1*. Van Nostrand Reinhold, New York.

Mitchell, T. J. (1974). "An Algorithm for the Construction of *D*-Optimal Experimental Designs." *Technometrics*, Vol. 16, pp. 203–210.

Mitchell, T. J. and M. D. Morris (1992), "The Spatial Correlation Approach to Response Surface Estimation", *Winter Simulation Conference*, pp. 565–571.

Montgomery, D. C. (2009). *Introduction to Statistical Quality Control*. 6th edition. Wiley, New York.

Montgomery, D. C. (1999). "Experimental Design for Product and Process Design and Development." *Journal of the Royal Statistical Society* D, Vol. 48, pp. 159–177.

Montgomery, D. C., C. M. Borror, and J. D. Stanley (1997–1998). "Some Cautions in the Use of Plackett– Burman Designs." *Quality Engineering*, Vol. 10, pp. 371– 381.

Montgomery, D. C., E. A. Peck, and G. G. Vining (2006). *Introduction to Linear Regression Analysis*. 4th edition. Wiley, New York.

Montgomery, D. C., and G. C. Runger (1993a). "Gauge Capability Analysis and Designed Experiments, Part I: Basic Methods." *Quality Engineering*, Vol. 6, pp. 115–135.

Montgomery, D. C., and G. C. Runger (1993b). "Gauge Capability Analysis and Designed Experiments, Part II: Experimental Design Models and Variance Component Estimation." *Quality Engineering*, Vol. 6, pp. 289–305.

Montgomery, D. C., and G. C. Runger (1996). "Foldovers of 2^{k-p} Resolution IV Experimental Designs." *Journal of Quality Technology*, Vol. 28, pp. 446–450.

Montgomery, D. C., and G. C. Runger (2011). *Applied Statistics and Probability for Engineers*. 5th edition. Wiley, New York.

Myers, R. H. (1990). *Classical and Modern Regression with Applications*. 2nd edition. PNS-Kent, Boston.

Myers, R. H., D. C. Montgomery and C. M. Anderson-Cook (2009). *Response Surface Methodology: Process and Product Optimization Using Designed Experiments*. 3rd edition. Wiley, New York.

Myers, R. H., and D. C. Montgomery (1997). "A Tutorial on Generalized Linear Models." *Journal of Quality Technology*, Vol. 29, pp. 274–291.

Myers, R. H., D. C. Montgomery, G. G. Vining and T. J. Robinson (2010). *Generalized Linear Models with Applications in Engineering and the Sciences*. 2nd edition Wiley, New York.

Myers, R. H., D. C. Montgomery, G. G. Vining, C. M. Borror, and S. M. Kowalski (2004). "Response Surface Methodology: A Retrospective and Literature Survey." *Journal of Quality Technology*, Vol. 36, pp. 53–77.

Nair, V. N., (editor) (1992). "Taguchi's Parameter Design: A Panel Discussion." *Technometrics*, Vol. 34, pp. 127–161.

Nelder, J. A., and R. W. M. Wedderburn (1972). "Generalized Linear Models." *Journal of the Royal Statistical Society* A, Vol. 135, pp. 370–384.

Nelson, L. S. (1995a). "Using Nested Designs I: Estimation of Standard Deviations." *Journal of Quality Technology*, Vol. 27, No. 2, pp. 169–171.

Nelson, L. S. (1995b). "Using Nested Designs II: Confidence Limits for Standard Deviations." *Journal of Quality Technology*, Vol. 27, No. 3, pp. 265–267.

Nelson, L. S. (1983). "Variance Estimation Using Staggered, Nested Designs." *Journal of Quality Technology*, Vol. 15, pp. 195–198.

Nelson, P. R. (1989). "Multiple Comparison of Means Using Simultaneous Confidence Intervals." *Journal of Quality Technology*, Vol. 21, No. 4, pp. 232–241.

Newman, D. (1939). "The Distribution of the Range in Samples from a Normal Population, Expressed in Terms of an Independent Estimate of Standard Deviation." *Biometrika*, Vol. 31, pp. 20–30.

O'Neill, R., and G. B. Wetherill (1971). "The Present State of Multiple Comparison Methods." *Journal of the Royal Statistical Society* B, Vol. 33, pp. 218–241.

Ostle, B. (1963). *Statistics in Research*. 2nd edition. Iowa State Press, Ames, Iowa.

Pearson, E. S., and H. O. Hartley (1966). *Biometrika Tables for Statisticians*, Vol. 1. 3rd edition. Cambridge University Press, Cambridge.

Pearson, E. S., and H. O. Hartley (1972). *Biometrika Tables for Statisticians*, Vol. 2. Cambridge University Press, Cambridge.

Pignatiello, J. J., Jr. and J. S. Ramberg (1992). "Top Ten Triumphs and Tragedies of Genichi Taguchi." *Quality Engineering*, Vol. 4, pp. 211–225.

Plackett, R. L., and J. P. Burman (1946). "The Design of Optimum Multifactorial Experiments." *Biometrika*, Vol. 33, pp. 305–325.

Quenouille, M. H. (1953). *The Design and Analysis of Experiments*. Charles Griffin and Company, London.

Sacks, J., W. J. Welch, T. J. Mitchell, and H. P. Wynn (1989). "Design and Analysis of Computer Experiments." *Statistical Science* **4**(4), pp. 409–423.

Santner, T. J., B. J. Williams, and W. I. Notz (2003). *The Design and Analysis of Computer Experiments*, Springer Series in Statistics Springer–Verlag, New York.

Satterthwaite, F. E. (1946). "An Approximate Distribution of Estimates of Variance Components." *Biometrics Bull.*, Vol. 2, pp. 110–112.

Satterthwaite, F. E. (1959). "Random Balance Experimentation" (with discussion). *Technometrics*, Vol. 1, pp. 111–137.

Scheffé, H. (1953). "A Method for Judging All Contrasts in the Analysis of Variance." *Biometrika*, Vol. 40, pp. 87–104.

Scheffé, H. (1956a). "A 'Mixed Model' for the Analysis of Variance." *Annals of Mathematical Statistics*, Vol. 27, pp. 23–36.

Scheffé, H. (1956b). "Alternative Models for the Analysis of Variance." *Annals of Mathematical Statistics*, Vol. 27, pp. 251–271.

Scheffé, H. (1959). *The Analysis of Variance*. Wiley, New York.

Searle, S. R. (1971a). *Linear Models*. Wiley, New York.

Searle, S. R. (1971b). "Topics in Variance Component Estimation." *Biometrics*, Vol. 27, pp. 1–76.

Searle, S. R. (1987). *Linear Models for Unbalanced Data*. Wiley, New York.

Searle, S. R., G. Casella, and G. E. McCulloch (1992). *Variance Components*. Wiley, New York.

Searle, S. R., and R. F. Fawcett (1970). "Expected Mean Squares in Variance Component Models Having Finite Populations," *Biometrics*, Vol. 26, pp. 243–254.

Searle, S. R., F. M. Speed, and H. V. Henderson (1981). "Some Computational and Model Equivalences in Analyses of Variance of Unequal-Subclass-Numbers Data." *The American Statistician*, Vol. 35, pp. 16–33.

Simpson, T. and J. Peplinski (1997), "On the Use of Statistics in Design and the Implications for Deterministic Computer Simulation", *ASME Design Engineering Technical Conference*, pp. 1–12.

Smith, H. F. (1957). "Interpretations of Adjusted Treatment Means and Regressions in Analysis of Covariance." *Biometrics*, Vol. 13, No. 3, pp. 282–308.

Smith, J. R., and J. M. Beverly (1981). "The Use and Analysis of Staggered Nested Factorial Designs." *Journal of Quality Technology*, Vol. 13, pp. 166–173.

Speed, F. M., and R. R. Hocking (1976). "The Use of the *R* ( )-Notation with Unbalanced Data." *The American Statistician*, Vol. 30, pp. 30–33.

Speed, F. M., R. R. Hocking, and O. P. Hackney (1978). "Methods of Analysis of Linear Models with Unbalanced Data." *Journal of the American Statistical Association*, Vol. 73, pp. 105–112.

Stefansky, W. (1972). "Rejecting Outliers in Factorial Designs." *Technometrics*, Vol. 14, pp. 469–479.

Taguchi, G. (1987). *System of Experimental Design: Engineering Methods to Optimize Quality and Minimize Cost*, UNIPUB, White Plains, NY.

Taguchi, G., (1991). *Introduction to Quality Engineering*. Asian Productivity Organization, UNIPUB, White Plains, NY.

Taguchi, G., and Y. Wu (1980). *Introduction to Off-Line Quality Control*. Central Japan Quality Control Association, Nagoya, Japan.

Ting, N., R. K. Burdick, F. A. Graybill, S. Jeyaratnam, and T.-F. C. Lu (1990). "Confidence Intervals on Linear Combinations of Variance Components That Are Unrestricted in Sign." *Journal of Statistical Computation and Simulation*, Vol. 35, pp. 135–143.

Tukey, J. W. (1949a). "One Degree of Freedom for Non-Additivity." *Biometrics*, Vol. 5, pp. 232–242.

Tukey, J. W. (1949b). "Comparing Individual Means in the Analysis of Variance." *Biometrics*, Vol. 5, pp. 99–114.

Tukey, J. W. (1951). "Quick and Dirty Methods in Statistics, Part II. Simple Analysis for Standard Designs." *Proceedings of the Fifth Annual Convention, American Society for Quality Control*, pp. 189–197.

Tukey, J. W. (1953). "The Problem of Multiple Comparisons." Unpublished Notes, Princeton University.

Vining, G. G., S. M. Kowalski, and D. C. Montgomery (2005). "Response Surface Designs Within a Split-Plot Structure." *Journal of Quality Technology*, Vol. 37, pp. 115–129.

Vining, G. G., and S. M. Kowalski (2008). "Exact Inference for Response Surface Designs Within a Split-Plot Structure." *Journal of Quality Technology*, Vol. 40, pp. 394–406.

Winer, B. J. (1971). *Statistical Principles in Experimental Design*. 2nd edition. McGraw-Hill, New York.

Wu, C. F. J. (1993). "Construction of Supersaturated Designs Through Partially Aliased Interactions." *Biometrika*, Vol. 80, pp. 661–669.

Xiao, L., D.K.J. Lin, and F. Bai (2012). "Constructing Difinitive Screening Designs Using Conference Matrices." *Journal of Quality Technology*, Vol. 44, pp. 2–8.

Yates, F. (1934). "The Analysis of Multiple Classifications with Unequal Numbers in the Different Classes." *Journal of the American Statistical Association*, Vol. 29, pp. 52–66.

Yates, F. (1937). *Design and Analysis of Factorial Experiments*. Tech. Comm. No. 35, Imperial Bureau of Soil Sciences, London.

Yates, F. (1940). "The Recovery of Interblock Information in Balanced Incomplete Block Designs." *Annals of Eugenics*, Vol. 10, pp. 317–325.

Ye, K., and M. Hamada (2000). "Critical Values of the Lenth Method for Unreplicated Factorial Designs." *Journal of Quality Technology*, Vol. 32, pp. 57–66.

Index